Lecture Notes in Applied Mathematics and Mechanics

Volume 2

About this Series

The Lecture Notes in Applied Mathematics and Mechanics LAMM are intended for an interdisciplinary readership in the fields of applied mathematics and mechanics. This series is published under the auspices of the International Association of Applied Mathematics and Mechanics (IAAMM; German GAMM).

Topics of interest include for example focus areas of the IAAMM such as: foundations of mechanics, thermodynamics, material theory and modeling, multibody dynamics, structural mechanics, solid mechanics, biomechanics, damage, fracture, multiscale modeling and homogenization, fluid mechanics, gas dynamics, laminar flows and transition, turbulence and reactive flows, interface flows, acoustics, waves, applied analysis, mathematical modeling, calculus of variations, variational principles applied operator theory, evolutionary equations, applied stochastics, systems with uncertainty, dynamical systems, control theory, optimization, applied and numerical linear algebra, analysis and numerics of ordinary and partial differential equations.

Each contribution to the series is intended to be accessible to researchers in mathematics and mechanics and is written in English. The aim of the series is to provide introductory texts for modern developments in applied mathematics and mechanics contributing to cross-fertilization. The Lecture Notes are aimed at researchers as well as advanced masters and PhD students in both mechanics and mathematics. Contributions to the series are self-contained and focused on a few central themes. The goal of each contribution is the communication of modern ideas and principles rather than on completeness or detailed proofs. Like lecture notes from a course, a well-chosen example is preferable to an abstract framework that cannot be comprehended without deeper involvement. The typical length of each contribution is between 100 and 300 pages. If the lecture notes represent the proceedings of a summer school with several contributors, a unified, consistent presentation and style are required (e.g., common notation). In exceptional cases, doctoral theses may be accepted, if they fulfill the above-mentioned criteria.

Potential contributors should contact the appropriate editor with a title, table of contents, and a sample chapter. Full manuscripts accepted by the editors will then be peer-reviewed.

More information about this series at http://www.springer.com/series/11915

Paul Steinmann

Geometrical Foundations of Continuum Mechanics

An Application to First- and Second-Order Elasticity and Elasto-Plasticity

Springer

Paul Steinmann
University of Erlangen-Nuremberg
Erlangen
Germany

ISSN 2197-6724 ISSN 2197-6732 (electronic)
Lecture Notes in Applied Mathematics and Mechanics
ISBN 978-3-662-46459-5 ISBN 978-3-662-46460-1 (eBook)
DOI 10.1007/978-3-662-46460-1

Library of Congress Control Number: 2015934930

Springer Heidelberg New York Dordrecht London

Printed on acid-free paper

Springer-Verlag GmbH Berlin Heidelberg is part of Springer Science+Business Media
(www.springer.com)

$$2\left[\mathbf{A} + \mathbf{E}^5\right]$$

Preface

The kinematics of geometrically nonlinear continuum mechanics is deeply rooted in differential geometry. An appreciation thereof is thus particularly illuminating. This is especially true for some generalized models of continuum mechanics, for example, second-order (gradient-type) elasticity and elasto-plasticity. The overarching purpose of this book is thus to illustrate the roots of the geometrically nonlinear kinematics of (generalized) continuum mechanics in differential geometry.

To this end, firstly several relevant concepts from differential geometry, such as connection, parallel transport, torsion, curvature, and metric for holonomic and anholonomic coordinate transformations are reiterated. The notation and the selection of these topics are of course mainly motivated by their relation to the geometrically nonlinear kinematics of continuum mechanics. Then, secondly, the kinematics of geometrically nonlinear continuum mechanics are considered. Together with the discussion on the integrability conditions for the distortions and double-distortions, the concepts of dislocation, disclination and point-defect density tensors are introduced. After touching on the possible interpretations of nonlinear first- and second-order elasticity using concepts from differential geometry, a detailed discussion of the kinematics of (multiplicative) first- and second-order elasto-plasticity is given. The discussion naturally culminates in a comprehensive set of different types of dislocation, disclination and point-defect density tensors. Potentially, these can be used to model densities of geometrically necessary defects and the accompanying hardening in crystalline materials. Continuum elasto-plasticity formulations of this kind fall into the class of generalized (gradient-type) continuum models.

Upon pursuing my habilitation thesis at the University of Hannover some 20 years ago I came across the exciting works of, e.g., Kondo, Bilby, Kröner, Anthony, de Wit, and associates ([91, 92, 17, 18, 94, 98, 95, 6, 7, 8, 202]). These deal with the intimate relation of differential geometry and the continuum theory of defects including dislocations, disclinations and point-defects. Since the dislocation density tensor is a suited vehicle to introduce some sort of physically motivated gradients of the plastic variables into a formulation of elasto-plasticity, I was keen to develop a corresponding theory that later became part of my habilitation thesis and also resulted in [181]. Afterwards I continued with my group at the University of Kaiserslautern for some time along these lines, see [128, 116, 115]. However soon in the new millennium the topic became somewhat dormant for me. In 2009 Holm Altenbach and Victor Eremeyev asked me to contribute six lectures to a CISM course to be held in 2011 in Udine, Italy. This course was dedicated to *Generalized Continua – from the Theory to Engineering Applications*; a theme that I was actively working on with my group at the University of Erlangen-Nuremberg. However, since all possible lecture topics that I could think of at that time had already been proposed by the other lectures of that course, I came back to the challenging relation between differential geometry and generalized continuum mechanics. During the ensuing intense preparation of the CISM course I first gave an advanced lecture on differential geometry at the University of Erlangen-Nuremberg in spring 2011. Subsequently, upon writing the corresponding contribution to the CISM lecture notes in autumn 2011, see [186], I felt that the topic would need more room to be developed comprehensively. In particular I could not squeeze into [186] all the material that I had prepared for the lectures. Thus I got the naïve idea to add just a few chapters - a bit of extra material here and there - and to extend [186] quickly into a little booklet. That was in winter 2011, but since thinks go as they go, the manuscript grew and grew, and time flew by without mercy. By now, some three years later, the material has matured sufficiently so that further editorial iterations do not result in noticeable progress anymore. I have to admit though, that this is clearly not yet a perfect book, indeed I strongly fear that it needs further editing and polishing (so if you find a typo, an omission, an inconsistency in notation, or even an error, please send me a corresponding note; I will be happy to take your input into account. At this point I especially want to thank the anonymous reviewers and the editors for their valuable comments to help improving the manuscript). But I certainly do hope that the present status already sheds a fresh and clarifying light on the *Geometrical Foundations of Continuum Mechanics*, and that it will be especially helpful concerning *Application to First- and Second-Order Elasticity and Elasto-Plasticity.*

Cape Town, November 2014 *Paul Steinmann*

Acknowledgements

I greatly acknowledge the endless patience and tolerance of my family during the many evenings and weekends that it took me to obsessively write this book. I am aware that I will never be able to make up for this!

Contents

Notation in Part I

$\cdot$	Single Contraction
$:$	Double Contraction
$\therefore$	Triple Contraction
grad	Gradient
div	Divergence
curl	Curl
$[\![\{\bullet\}]\!]$	Jump of $\{\bullet\}$
δ_{ab}	Kronecker Symbol
$\boldsymbol{e}_a$	Cartesian Base Vectors
$\mathbf{1} = \delta_{ab}\boldsymbol{e}_a \otimes \boldsymbol{e}_b$	Second-Order Unit Tensor
$\boldsymbol{\mathsf{i}}^{\text{sym}} = \frac{1}{2}[\delta_{ac}\delta_{bd} + \delta_{ad}\delta_{bc}]\, \boldsymbol{e}_a \otimes \boldsymbol{e}_b \otimes \boldsymbol{e}_c \otimes \boldsymbol{e}_d$	Symmetric Fourth-Order Unit Tensor
$\boldsymbol{\mathsf{i}}^{\text{skw}} = \frac{1}{2}[\delta_{ac}\delta_{bd} - \delta_{ad}\delta_{bc}]\, \boldsymbol{e}_a \otimes \boldsymbol{e}_b \otimes \boldsymbol{e}_c \otimes \boldsymbol{e}_d$	Skew Symmetric Fourth-Order Unit Tensor
e_{abc}	Levi-Civita/Permutation Symbol
$\boldsymbol{e} = e_{abc}\boldsymbol{e}_a \otimes \boldsymbol{e}_b \otimes \boldsymbol{e}_c$	Levi-Civita/Permutation Tensor
$\boldsymbol{u} = u_a\boldsymbol{e}_a$	Displacement Field
$\boldsymbol{v} = v_a\boldsymbol{e}_a$	Velocity Field
$\boldsymbol{\epsilon} = \epsilon_{ab}\ \boldsymbol{e}_a \otimes \boldsymbol{e}_b$	Symmetric Strain
$\boldsymbol{\omega} = \omega_{ab}\ \boldsymbol{e}_a \otimes \boldsymbol{e}_b$	Skew-Symmetric Rotation
$\boldsymbol{\kappa} = \kappa_{ab}\ \boldsymbol{e}_a \otimes \boldsymbol{e}_b$	Nye Lattice Curvature
$\boldsymbol{\beta} = \beta_{ab}\ \boldsymbol{e}_a \otimes \boldsymbol{e}_b$	(Single) Distortion
$\boldsymbol{\gamma} = \gamma_{abc}\boldsymbol{e}_a \otimes \boldsymbol{e}_b \otimes \boldsymbol{e}_c$	Double-Distortion
$\boldsymbol{\lambda} = \lambda_{abc}\boldsymbol{e}_a \otimes \boldsymbol{e}_b \otimes \boldsymbol{e}_c$	Relative Double-Distortion

$\boldsymbol{\sigma} = \sigma_{ab}\ \boldsymbol{e}_a \otimes \boldsymbol{e}_b$	Cauchy (Single) Stress
$\boldsymbol{\sigma} = \sigma_{abc} \boldsymbol{e}_a \otimes \boldsymbol{e}_b \otimes \boldsymbol{e}_c$	Double-Stress
$\boldsymbol{n} = n_a \boldsymbol{e}_a$	Outwards Pointing Surface Normal
$\boldsymbol{b} = b_a \boldsymbol{e}_a$	Body Force Density
$\boldsymbol{t} = t_a \boldsymbol{e}_a$	Prescribed (Single) Traction
$\boldsymbol{t}' = t'_a \boldsymbol{e}_a$	Prescribed Double-Traction
$\mathbb{b}$	Burgers Vector Density
$\mathbb{f}$	Frank Tensor Density
$\boldsymbol{\alpha} = \alpha_{ab}\ \boldsymbol{e}_a \otimes \boldsymbol{e}_b$	Dislocation Density
${}^{\mathrm{I}}\boldsymbol{\alpha}$	Primary Dislocation Density
${}^{\mathrm{II}}\boldsymbol{\alpha}$	Secondary Dislocation Density
$\boldsymbol{\vartheta} = \vartheta_{abc} \boldsymbol{e}_a \otimes \boldsymbol{e}_b \otimes \boldsymbol{e}_c$	Disclination Density
$\boldsymbol{\theta} = \theta_{ab}\ \boldsymbol{e}_a \otimes \boldsymbol{e}_b$	
$\boldsymbol{\pi} = \pi_{abc} \boldsymbol{e}_a \otimes \boldsymbol{e}_b \otimes \boldsymbol{e}_c$	Point-Defect Density
Ψ	Free Energy Density
$\mathcal{D}$	Dissipation Power Density
$\mathcal{R}$	Non-Locality Residual
G	Shear Modulus
$\ell_\bullet$	Characteristic Lengths
$\boldsymbol{p}_\varsigma = \boldsymbol{s}_\varsigma \otimes \boldsymbol{m}_\varsigma$	Schmid (Projection) Tensor
σ_ς	Schmid Stress
$\sigma^\star_\varsigma$	Generalized Schmid Stress
ϕ_ς	Yield Surface
γ_ς	Plastic Multiplier/Slip Rate
φ_ς	Equivalent Stress
σ_y	Initial Yield Stress
H	(Isotropic) Drag Stress
κ	(Isotropic) Hardening Variable

Notation in Part II

$\mathcal{M}$	Manifold
$\mathcal{P}$	Point in Manifold
$T_{\mathcal{P}}\mathcal{M}$	Tangent Space to Manifold
$T^*_{\mathcal{P}}\mathcal{M}$	Cotangent Space to Manifold
I, J, K, L, i, j, k, l	General Indices
$\mathcal{X}^I, x^i$	General Coordinates
$\mathcal{V}^I, v^i$	Coefficients of a Vector
$\mathcal{V}_I, v_i$	Coefficients of a Covector
$\mathcal{Y}^I, y^i$	General Coordinate Transformations
$\mathcal{F}^i{}_J, f^I{}_j$	Jacobian of General Coordinate Transformations
$,$	Partial Derivative
$.$	Anholonomic Partial Derivative
$\vert$	Covariant Derivative
$\mathrm{d}\{\bullet\}$	Differential of $\{\bullet\}$
$\mathrm{D}\{\bullet\}$	Covariant Differential of $\{\bullet\}$
$\mathrm{p}\{\bullet\}$	Parallel Transport of $\{\bullet\}$
$\delta_{IJ}, \delta^{IJ}, \delta_{ij}, \delta^{ij}$	Kronecker Symbol

$\mathcal{L}^I{}_{JK}, \mathcal{L}_{IJK}, l^i{}_{jk}, l_{ijk}$	Linear (Affine) Connection
$\mathcal{T}^I{}_{JK}, \mathcal{T}_{IJK}, t^i{}_{jk}, t_{ijk}$	Torsion
$\mathcal{T}_I, t_i$	Torsion Covector
$\mathcal{A}^I{}_{JK}, \mathcal{A}_{IJK}, a^i{}_{jk}, a_{ijk}$	Anholonomic Object
$\mathcal{D}^i{}_{JK}, \mathcal{D}_{iJK}, d^I{}_{jk}, d_{Ijk}$	Dislocation Density
$\mathcal{R}^I{}_{JKL}, \mathcal{R}_{IJKL}, r^i{}_{jkl}, r_{ijkl}$	Curvature
$\mathcal{R}^I{}_J, \mathcal{R}_{IJ}, r^i{}_j, r_{ij}$	First Ricci Tensor
$\mathcal{V}^I{}_J, \mathcal{V}_{IJ}, v^i{}_j, v_{ij}$	Second Ricci Tensor
$\mathcal{R}, r$	Ricci Scalar
$\mathcal{M}_{IJ}, \mathcal{M}^{IJ}, m_{ij}, m^{ij}$	Metric
$\mathcal{M}^I{}_{JK}, \mathcal{M}_{IJK}, m^i{}_{jk}, m_{ijk}$	Riemann Connection
	Levi-Civita Connection
$\mathcal{K}^I{}_{JK}, \mathcal{K}_{IJK}, k^i{}_{jk}, k_{ijk}$	Contortion
$\mathcal{G}^{IJ}, g^{ij}$	Einstein Tensor
$\mathcal{M}^I{}_{JKL}, \mathcal{M}_{IJKL}, m^i{}_{jkl}, m_{ijkl}$	Riemann Curvature
$\mathcal{Q}_{IJK}, q_{ijk}$	Non-Metricity
$\mathcal{Q}_I, q_i$	Non-Metricity Covector
$\mathcal{O}_{IJK}, o_{ijk}$	Distention

$\boldsymbol{g}_i = \boldsymbol{\partial}_{\chi^i}$		Coordinate Basis
$\boldsymbol{g}^i = \mathbf{d}\chi^i$		Dual Basis
$\boldsymbol{v}^\sharp$		Vector
$\boldsymbol{v}^\flat$		Covector
$\mathbf{d}\vartheta$		Gradient of ϑ
$\partial_{\boldsymbol{v}^\sharp}(\vartheta)$		Directional Derivative of ϑ
$\nabla\boldsymbol{v}^\sharp$		Covariant Derivative of $\boldsymbol{v}^\sharp$
$[\boldsymbol{v}^\sharp, \boldsymbol{w}^\sharp]$		Commutator of $\boldsymbol{v}^\sharp$ and $\boldsymbol{w}^\sharp$
$\boldsymbol{t}^\natural$		Torsion
$\boldsymbol{a}^\natural$		Anholonomic Object
$\boldsymbol{r}^\natural$		Curvature
$\mathsf{g}_i = \boldsymbol{\partial}_{\chi^i}$	Frame, Simple	Vector
$\mathsf{v} = v^i\mathsf{g}_i$	General	Vector
$\mathsf{g}^i = \mathbf{d}\chi^i$	Coframe, Simple	(Scalar-Valued) One-Form
$\mathsf{v} = v_i\mathsf{g}^i$	General	(Scalar-Valued) One-Form
$\wedge$		Exterior Product
$\lrcorner$		Interior Product
$\mathbf{d}$		Exterior Derivative
$\mathbf{D}$		Covariant Exterior Derivative
$\mathsf{l}^i{}_j, \mathsf{l}_{ij}$		Connection One-Form
$\mathsf{t}^i, \mathsf{t}_i$		Torsion Two-Form
$\mathsf{a}^i, \mathsf{a}_i$		Anholonomity Two-Form
$\mathsf{r}^i{}_j, \mathsf{r}_{ij}$		Curvature Two-Form
m_{ij}		Riemann One-Form
k_{ij}		Contortion One-Form
b_{ij}		Bend-Twist One-Form

$\mathcal{E}^{IJKL}, \mathcal{E}_{IJKL}$	Permutation Tensor in Four-Space
$^\star\mathcal{T}^{IJK}$	Dual Torsion
$^\star\mathcal{K}^{IJK}$	Dual Contortion
$^\star\mathcal{R}^{IJ}{}_{KL}{}^\star = \mathcal{K}^{IJ}{}_{KL}$	Double-Dual Curvature
$\mathcal{E}^{IJK}, \mathcal{E}_{IJK}$	Permutation Tensor in Three-Space
$^\star\mathcal{T}^{IJ}$	Dual Torsion
$^\star\mathcal{K}^{IJ}$	Dual Contortion
$^\star\mathcal{A}^{IJ}$	Dual Anholonomic Object
$^\star\mathcal{D}^{iJ}$	Dual Dislocation Density
$^\star\mathcal{R}^{I}{}_{J}{}^\star = \mathcal{K}^{I}{}_{J}$	Double-Dual Curvature
$\mathcal{G}^I, \mathcal{G}_I$	Axial Vector of Einstein Tensor
$\mathcal{E}^{IJ}, \mathcal{E}_{IJ}$	Permutation Tensor in Two-Space
$^\star\mathcal{T}^{I}$	Dual Torsion
$^\star\mathcal{K}^{I}$	Dual Contortion
$^\star\mathcal{R}^\star = \mathcal{K}$	Gaussian/Double-Dual Curvature

Notation in Part III

$\mathscr{P}$	Physical Point
$\mathscr{B}$	Physical Body
$B_t \subset \mathbb{R}^3, \mathcal{B}_t \subset \mathbb{E}^3$	Spatial Configuration
$B_0 \subset \mathbb{R}^3, \mathcal{B}_0 \subset \mathbb{E}^3$	Material Configuration
I, J, K, L, i, j, k, l	General Indices
A, B, C, D, a, b, c, d	Cartesian Indices
$\mathcal{x}^i, \mathcal{X}^I$	General Coordinates in Euclidean Space
x^a, X^A	Cartesian Coordinates in Euclidean Space
$\mathfrak{X}^{\mathfrak{i}}$	Convected Coordinates in Euclidean Space
$\hat{\mathcal{y}}^i, \hat{\mathcal{Y}}^I$	General $\leftarrow$ Cartesian Coordinates
$\mathcal{j}^i{}_b, \mathcal{J}^I{}_B$	Jacobian of General $\leftarrow$ Cartesian Coordinates
$\boldsymbol{g}_i, \boldsymbol{g}^i, \boldsymbol{G}_I, \boldsymbol{G}^I$	General Base Vectors in Euclidean Space
$\mathcal{g}_{ij}, \mathcal{g}^{ij}, \mathcal{G}_{IJ}, \mathcal{G}^{IJ}$	General Metric in Euclidean Space
$\mathcal{n}^i{}_{jk}, \mathcal{n}_{ijk}, \mathcal{N}^I{}_{JK}, \mathcal{N}_{IJK}$	Christoffel Symbols in Euclidean Space
$\hat{y}^a, \hat{Y}^A$	General $\rightarrow$ Cartesian Coordinates
$j^a{}_j, J^A{}_J$	Jacobian of General $\rightarrow$ Cartesian Coordinates
$\boldsymbol{e}_a, \boldsymbol{e}^a, \boldsymbol{E}_A, \boldsymbol{E}^A$	Cartesian Base Vectors in Euclidean Space
$\delta_{ab}, \delta^{ab}, \delta_{AB}, \delta^{AB}$	Cartesian Metric in Euclidean Space
	Kronecker Symbol in Euclidean Space
$e^{abc}, e_{abc}, E^{ABC}, E_{ABC}$	Permutation Symbol in Euclidean Space
$\boldsymbol{E}, \boldsymbol{e}$	Permutation Tensor in Euclidean Space
$\hat{X}^A, \hat{x}^a$	Convected $\rightarrow$ Cartesian Coordinates
$\hat{\mathfrak{X}}^{\mathfrak{i}}$	Convected $\leftarrow$ Cartesian Coordinates
$\boldsymbol{g}_{\mathfrak{i}}, \boldsymbol{g}^{\mathfrak{i}}, \boldsymbol{G}_{\mathfrak{i}}, \boldsymbol{G}^{\mathfrak{i}}$	Convected Base Vectors in Euclidean Space

$,$	Partial Derivative in Euclidean Space
$.$	Anholonomic Partial Derivative in Euclidean Space
$;$	Covariant Derivative in Euclidean Space
$:$	Total Covariant Derivative in Euclidean Space
Grad, grad	Gradient in Euclidean Space
Div, div	Divergence in Euclidean Space
Curl, curl	Curl in Euclidean Space
Inc, inc	Incompatibility in Euclidean Space
$\mathscr{y}^i, \mathscr{Y}^I$	Map of General ↔ General Coordinates
$\mathscr{F}^i{}_J, \mathscr{f}^I{}_j$	(Single) Distortion in General Coordinates
$\mathscr{G}^i{}_{JK}, \mathscr{g}^I{}_{jk}$	Double-Distortion in General Coordinates
$\mathscr{H}^i{}_{JKL}, \mathscr{h}^I{}_{jkl}$	Triple-Distortion in General Coordinates
$\mathsf{y}^i, \mathsf{Y}^I$	Map of General ↔ General Coordinates
$\mathsf{F}^i{}_J, \mathsf{f}^I{}_j$	(Single) Distortion in General Coordinates
$\mathsf{G}^i{}_{JK}, \mathsf{g}^I{}_{jk}$	Double-Distortion in General Coordinates
$\mathsf{H}^i{}_{JKL}, \mathsf{h}^I{}_{jkl}$	Triple-Distortion in General Coordinates
y^A, Y^a	Map of Cartesian ↔ Cartesian Coordinates
$F^a{}_B, f^A{}_b$	(Single) Distortion in Cartesian Coordinates
$G^a{}_{BC}, g^A{}_{bc}$	Double-Distortion in Cartesian Coordinates
$H^a{}_{BCD}, h^A{}_{bcd}$	Triple-Distortion in Cartesian Coordinates
$\boldsymbol{x}, \boldsymbol{X}$	Position Vector in Euclidean Space
$\boldsymbol{y}, \boldsymbol{Y}$	Deformation Map in Euclidean Space
$\boldsymbol{F}, \boldsymbol{f}$	(Single) Distortion in Euclidean Space
$\boldsymbol{G}, \boldsymbol{g}$	Double-Distortion in Euclidean Space
$\boldsymbol{H}, \boldsymbol{h}$	Triple-Distortion in Euclidean Space

${}^{\#}\mathcal{L}^{I}{}_{JK}$, ${}^{\#}\ell^{i}{}_{jk}$	Non-Integrable Part of Connection in General Coordinates
${}^{\#}\mathcal{G}^{i}{}_{JK}$, ${}^{\#}\mathscr{g}^{I}{}_{jk}$	Relative Double-Distortion in General Coordinates
$\mathcal{K}^{i}{}_{JKL}$, $\mathcal{k}^{I}{}_{jkl}$	Disclination Density in General Coordinates
${}^{\#}L^{A}{}_{BC}$, ${}^{\#}l^{a}{}_{bc}$	Non-Integrable Part of Connection in Cartesian Coordinates
${}^{\#}G^{a}{}_{BC}$, ${}^{\#}g^{A}{}_{bc}$	Relative Double-Distortion in Cartesian Coordinates
$K^{a}{}_{BDC}$, $k^{A}{}_{bdc}$	Disclination Density in Cartesian Coordinates
${}^{\mathrm{I}}\boldsymbol{D}$, ${}^{\mathrm{I}}\boldsymbol{d}$	Primary Dislocation Density in Euclidean Space
${}^{\mathrm{II}}\boldsymbol{D}$, ${}^{\mathrm{II}}\boldsymbol{d}$	Secondary Dislocation Density in Euclidean Space
$\boldsymbol{K}, \boldsymbol{k}$	Disclination Density in Euclidean Space
$\boldsymbol{Q}, \boldsymbol{q}$	Point-Defect Density in Euclidean Space
$\boldsymbol{C}, \boldsymbol{c}$	Cauchy-Green and Finger Strain in Euclidean Space
$\boldsymbol{\Gamma}, \boldsymbol{\gamma}$	Connection in Euclidean Space
$\boldsymbol{D}$, $\boldsymbol{d}$	Piola-Type Dislocation Density in Euclidean Space
$\boldsymbol{A}$, $\boldsymbol{a}$	Cauchy-Type Dislocation Density in Euclidean Space
${}^{\mathrm{II}}\boldsymbol{A}$, ${}^{\mathrm{II}}\boldsymbol{a}$	Secondary Cauchy-Type Dislocation Density in Euclidean Space
$\boldsymbol{H}_{\tau}$, $\boldsymbol{h}_{\tau}$	Incompatibility Density in Euclidean Space
${}^{\mathrm{I}}\boldsymbol{H}_{\tau}$, ${}^{\mathrm{I}}\boldsymbol{h}_{\tau}$	Primary Incompatibility Density in Euclidean Space
${}^{\mathrm{II}}\boldsymbol{H}$, ${}^{\mathrm{II}}\boldsymbol{h}$	Secondary Incompatibility Density in Euclidean Space
$\boldsymbol{K}, \boldsymbol{k}$	Dual Contortion in Euclidean Space

Christian Morgenstern [1871 to 1914] was a German poet and author, who was born in Munich and who died of tuberculosis in Tirol. He was traveling most of his life through Germany and its southern neighboring countries in a quest for the improvement of his health conditions. As an inheritance from his mother his health was constantly threatened by tuberculosis. He was a *true representative of Anthroposophy* and follower of Rudolf Steiner. During his lifetime his work was not much acknowledged, however many of his poems (that were in large parts only published posthumously) became famous later and enjoy immense popularity till today. The following chapters are preceded by a personal selection of some of Christian Morgenstern's fantastic nonsense poems. These masterpieces of wordplay and humorous poetry are celebrated also for containing lots of subtle truth and metaphysics.

Part I
Prologue

...introduces generalized plasticity formulations within geometrically linear first- and second-order continuum mechanics. Thereby the need to define defect density tensors based on geometrical integrability arguments and their use as internal variables in generalized plasticity is motivated.

Chapter 1
Motivation: Linear Crystal Plasticity

DER MEILENSTEIN

Tief im dunklen Walde steht er
und auf ihm mit schwarzer Farbe,
daß des Wandrers Geist nicht darbe:
Dreiundzwanzig Kilometer

Seltsam ist und schier zum Lachen,
daß es diesen Text nicht gibt,
wenn es keinem Blick beliebt,
ihn durch sich zu Text zu machen.

Und noch weiter vorgestellt:
Was wohl ist er – ungesehen?
Ein uns völlig fremd Geschehen.
Erst das Auge schafft die Welt.

Christian Morgenstern

Abstract. Mono-crystals display various defects that may potentially act as obstacles to further evolution of inelastic deformations, i.e. to further plastic flow. These are translational defects in terms of (primary and secondary) dislocations, rotational defects in terms of disclinations, and (dilatational) point-defects in terms of lattice vacancies or interstitial atoms. Formulations of generalized crystal plasticity incorporate the densities of these defects in order to capture the hardening of the material. In particular the inclusion of the defect densities other than the dislocation density requires to root the formulation in a second-order continuum description.

1.1 Introduction

In crystal plasticity the amount of accumulated defects such as dislocations (translational defects), disclinations (rotational defects) and point-defects (dilatational defects) in an otherwise perfect crystalline lattice are responsible for the state of hardening that the crystalline material displays. A particularly wonderful image of an edge dislocation in a platinum-nickel alloy is displayed in Fig. 1.1.

As the paradigmatic example the total amount of arrested dislocations is decomposed into *statistically stored dislocations* (SSD) and *geometrically necessary dislocations* (GND). The former are then assumed responsible for isotropic hardening. The latter are necessary to support the plastic part of the deformation and form an (additional) obstacle to further dislocation flow. Geometrically necessary dislocations may be subdivided further into

P. Steinmann, *Geometrical Foundations of Continuum Mechanics*,
Lecture Notes in Applied Mathematics and Mechanics 2, DOI: 10.1007/978-3-662-46460-1_1

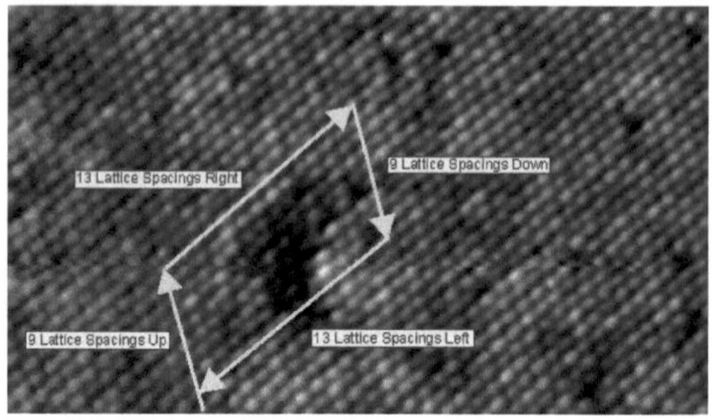

Fig. 1.1 Scanning tunneling microcopy (STM) of a platinum-nickel alloy displaying an atomically resolved dislocation. The total width of the image is 10 nm. The highlighted path denotes a Burgers circuit, the resulting Burgers gap is clearly visible. (Original STM image published in [166, 165], usage of raw image with kind permission of the authors)

dislocations responsible for a macroscopically stress free curvature of the crystal lattice, and dislocations responsible for macroscopic residual stresses, both after the removal of external loads. The illuminating relation between the stress free curvature of the crystal lattice and the part of the dislocation density that is geometrically necessary to support this curvature was established by [152]. Both contributions to the geometrically necessary dislocations, i.e. those resulting in a stress free curvature of the crystal lattice and those resulting in residual stresses, constitute additional contributions to the hardening of the crystalline material. Thus, geometrically necessary dislocations obviously have to be taken into account when modeling of plasticity.

However, in general ***all*** kinds of *geometrically necessary defects* have to be taken into account to describe the hardening behavior of crystalline materials more realistically and thus more accurately.

A consideration of the continuum version of geometrically necessary dislocations, i.e. the *dislocation density tensor*, in a thermodynamically consistent modeling framework inevitably results in a form of gradient crystal plasticity, see [181], [128]. The dislocation density tensor, however, is intimately related to one of the key concepts in non-Riemann differential geometry, i.e. the third-order torsion tensor as introduced by Cartan [22]. For anholonomic coordinates, as in the case of crystal plasticity, the Cartan torsion coincides moreover with the so-called anholonomic object of differential geometry. The important relation between the continuum description of dislocation density and a non-Riemann geometry was discovered by [91] and [17, 18]. Prior to this, differential geometry was instrumental in the development of the Theory of General Relativity/Gravitation, see [142]. Important contributions to the elaboration of differential geometry in this context have been made by [167, 168].

Kröner [94] proposed a geometrically linear continuum theory of residual stresses based on the concept of dislocation densities. Motivated by insights into differential geometry, the corresponding extension to the geometrically nonlinear case was developed by [98] and [95]. It turned out that the interplay between continuum mechanics and differential geometry is extremely helpful: firstly, rather involved relations of the geometrically nonlinear kinematics of continuum mechanics such as the connection between the dislocation density and the St. Venant compatibility conditions for the strains could be clarified; and secondly, generalized continuum formulations that consider more general defects besides dislocations, such as disclinations, point-defects (vacancies and/or interstitial atoms in the crystal lattice), and the distribution of quasi-dislocations caused by, e.g., inhomogeneous temperature distributions, electric or magnetic fields, and the like, are motivated by the existence of other, more involved, types of differential geometries, see e.g. the contributions by [6, 7, 8, 205]. A comprehensive account of the geometrically linearized version of the continuum theory of general defects in crystal lattices is found in [202]. Further interesting contributions to the continuum theory of dislocations/defects are e.g. by [92], [151], [201], [96], [55], [51] and more recently by [41, 42, 45, 43, 44].

After the prolific developments in the 1950's to 1970's the topic became somewhat dormant, but since the 1990's there has been a renewed interest. This had to do with, on the one hand, the intense research on possibilities to overcome the pathological dependencies on the discretization that computational solutions, mainly based on the finite element method, displayed for the simulation of inelastic materials with a softening response. The incorporation (in one way or another) of gradients of the inelastic variables into the modeling, see e.g. [5], has a regularizing effect that results in discretization-independent simulations, see e.g. [116]. On the other hand, the continuing trend towards miniaturization made clear that the inelastic response of a material especially is length scale (size) dependent. Again, size dependence can be included into the modeling by incorporating gradients of the inelastic variables. For an overview of a variety of possibilities to arrive at a generalized model of plasticity see, e.g., [82]; as an example, e.g. [61] and [66] strongly advocate micromorphic continuum formulations.

However, purely phenomenological approaches for generalized models of continuum mechanics are somewhat unsatisfying if a clear link to the underlying physics is lacking. The plasticity of crystalline materials is a notable exception, as the mechanisms of plasticity are well understood to depend on the concepts of dislocations and dislocation flow. The flow of dislocations causes the plastic deformation process while the accumulating arrest of single dislocations represents an obstacle that has to be overcome if ongoing flow of dislocations is to occur. To better capture the underlying physics of crystalline material was the main motivation for the proposal in [181] to include the dislocation density tensor as an additional argument in the free energy density. As a consequence a gradient-type crystal plasticity formulation emerges. Subsequently, many more or less related formulations considering specific versions of dislocation and disclination density tensors were pursued, among them the important contributions by, e.g., [109, 111, 110], [206], [4, 3],

[101, 102, 103, 104, 106, 107, 108], [28], [71], [192], [14], [160], [39, 40, 37, 38], [65], [203, 205] (and many more). Other aspects such as the gauge theory of dislocations as treated, e.g., by [105] or nonsingular stress and strain fields of dislocations and disclinations embedded in nonlinear and gradient elasticity, see [50], [107], are exciting topics of current research activities.

It is the aim of the subsequent exposition to extend previous modeling approaches to the hardening in generalized crystal plasticity (such as [181, 128]), that solely incorporate the dislocation density tensor, to more complex continuum models that additionally take into account more general defects such as disclination and point-defect densities, see also [187].

1.2 First-Order Continuum

As a motivation of the subsequent investigations and following the exposition in [128] a geometrically linear first-order continuum shall be considered. Thereby, the internal and external working are assumed as

$$\mathcal{W}^{\text{int}} = \int_{\mathcal{B}} \boldsymbol{\sigma} : \operatorname{grad} \boldsymbol{v} \, \mathrm{d}v \quad \text{and} \quad \mathcal{W}^{\text{ext}} = \int_{\mathcal{B}} \boldsymbol{b} \cdot \boldsymbol{v} \, \mathrm{d}v + \int_{\partial\mathcal{B}} \boldsymbol{t} \cdot \boldsymbol{v} \, \mathrm{d}a. \tag{1.1}$$

The terminology *first-order continuum* derives from the presence of the first gradient of the velocity field $\boldsymbol{v}$ in the definition of the internal working. Here $\mathcal{B} \subset \mathbb{E}^3$ denotes the configuration of the body $\mathscr{B}$, i.e. the set of all position vectors $\boldsymbol{x} \in \mathbb{E}^3$ that are occupied by the body $\mathscr{B}$ in Euclidean space $\mathbb{E}^3$. Furthermore $\boldsymbol{\beta} := \operatorname{grad} \boldsymbol{u}$ defines the *distortion*, $\boldsymbol{\sigma}$ denotes the (Cauchy) stress, $\boldsymbol{b}$ are distributed body forces (per unit volume) and $\boldsymbol{t}$ represents the prescribed traction (per unit area), $\boldsymbol{n}$ is the outwards-pointing normal to the boundary. Requiring the internal working to equal the external working for all velocity fields $\boldsymbol{v}$ results in the corresponding equilibrium conditions in the bulk and at the boundary

$$-\operatorname{div}\boldsymbol{\sigma} = \boldsymbol{b} \quad \text{in} \quad \mathcal{B} \quad \text{and} \quad \boldsymbol{\sigma} \cdot \boldsymbol{n} = \boldsymbol{t} \quad \text{on} \quad \partial\mathcal{B}. \tag{1.2}$$

Proof. Integrating the internal working by parts renders

$$\int_{\mathcal{B}} \boldsymbol{\sigma} : \operatorname{grad} \boldsymbol{v} \, \mathrm{d}v = \int_{\mathcal{B}} \big[\operatorname{div}(\boldsymbol{v} \cdot \boldsymbol{\sigma}) - \boldsymbol{v} \cdot \operatorname{div}\boldsymbol{\sigma} \big] \, \mathrm{d}v.$$

Applying next the Gauss (bulk divergence) theorem then results in

$$\int_{\mathcal{B}} \boldsymbol{\sigma} : \operatorname{grad} \boldsymbol{v} \, \mathrm{d}v = \int_{\partial\mathcal{B}} \boldsymbol{v} \cdot \boldsymbol{\sigma} \cdot \boldsymbol{n} \, \mathrm{d}a - \int_{\mathcal{B}} \boldsymbol{v} \cdot \operatorname{div}\boldsymbol{\sigma} \, \mathrm{d}v.$$

Equating the internal and external working then renders the result. □

1.2.1 Integrability Condition for the Distortion

Integrating the differential displacement $\mathrm{d}\boldsymbol{u}$ along a closed circuit $\mathcal{C}$ (encircling the area $\mathcal{A}$) in the spirit of a *Burgers circuit*, see Fig. 1.2, determines a possible incompatibility of the (possibly non-continuous, multivalued) displacement in terms of the resulting *displacement closure gap*

$$[\![\boldsymbol{u}]\!] := \oint_{\mathcal{C}} \mathrm{d}\boldsymbol{u}. \tag{1.3}$$

Next, expressing the differential displacement $\mathrm{d}\boldsymbol{u}$ in terms of the *distortion* (that in general may be unrelated to the displacement)

$$\mathrm{d}\boldsymbol{u} := \boldsymbol{\beta} \cdot \mathrm{d}\boldsymbol{x} \tag{1.4}$$

allows to re-express the displacement closure gap as the *circular integral* of the distortion

$$[\![\boldsymbol{u}]\!] = \oint_{\mathcal{C}} \boldsymbol{\beta} \cdot \mathrm{d}\boldsymbol{x}. \tag{1.5}$$

Then the *Stokes theorem* reformulates the circular integral into an area integral over the area $\mathcal{A}$ encircled by the circuit $\mathcal{C}$, thereby involving the $\operatorname{curl} \boldsymbol{\beta}$ of the distortion

$$[\![\boldsymbol{u}]\!] = \int_{\mathcal{A}} \operatorname{curl} \boldsymbol{\beta} \cdot \mathrm{d}\boldsymbol{a}. \tag{1.6}$$

Here $\mathrm{d}\boldsymbol{a} = \boldsymbol{n}\,\mathrm{d}a$ denotes the vectorial area element to $\mathcal{A}$ consisting of the unit normal $\boldsymbol{n}$ and the scalar area element $\mathrm{d}a$. As a conclusion, Eq. 1.6 allows to compute the resulting displacement closure gap for a *given* distortion $\boldsymbol{\beta}$.

Remark 1.1. In Cartesian coordinates with the orthonormal base vectors $\boldsymbol{e}_a$ (no distinction between co- and contravariant objects necessary) the $\operatorname{curl} \boldsymbol{\beta}$ of the distortion obeys the following coordinate representation

$$\operatorname{curl} \boldsymbol{\beta} := -\operatorname{grad} \boldsymbol{\beta} : \boldsymbol{e} = -\beta_{ab,c} e_{bcd} \boldsymbol{e}_a \otimes \boldsymbol{e}_d. \tag{1.7}$$

Here, e_{bcd} denotes the (Levi-Civita) permutation symbol fully skew symmetric in the indices b, c, d. Due to the skew-symmetry of the permutation symbol, $\operatorname{curl} \boldsymbol{\beta}$ is essentially equivalent to the (right) skew-symmetric part of $\operatorname{grad} \boldsymbol{\beta}$,

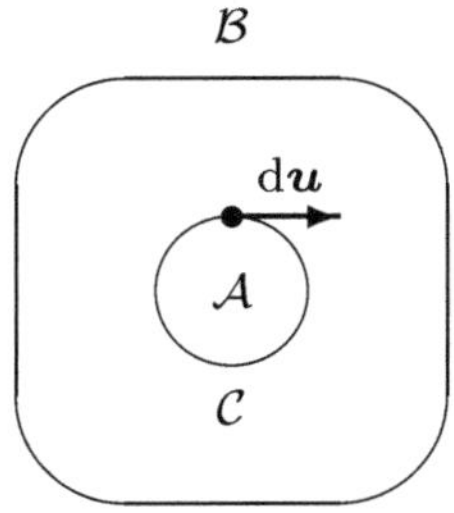

Fig. 1.2 Integrating the differential displacement $\mathrm{d}\boldsymbol{u}$ along a closed circuit $\mathcal{C}$ encircling the area $\mathcal{A}$ in the configuration $\mathcal{B}$ determines the resulting *displacement closure gap* $[\![\boldsymbol{u}]\!]$.

i.e. to $\beta_{a[b,c]}$. (Note the symbolic notation for third-order tensors, such as $\boldsymbol{e}$ in terms of *large* bold fonts.) □

Discussion 1.1. The arguments leading to the expression for the displacement closure gap in Eq. 1.6 may be reversed in order to render the *point-wise* integrability condition for a *given* distortion $\boldsymbol{\beta}$ to be integrable to a continuous, single-valued vector field $\boldsymbol{u}^*$.

Thereby the reasoning is as follows:

Requiring that the displacement closure gap in Eq. 1.6 vanishes for *arbitrary* circuits $\mathcal{C}$ and thus encircled areas $\mathcal{A}$ within the configuration $\mathcal{B}$, the point-wise integrability condition for the distortion $\boldsymbol{\beta}$ reads

$$\operatorname{curl}\boldsymbol{\beta} \doteq \mathbf{0} \quad \rightsquigarrow \quad \boldsymbol{\beta} = \operatorname{grad}\boldsymbol{u}^*. \tag{1.8}$$

In other words, an integrable distortion that results in a vanishing displacement closure gap for arbitrary circuits $\mathcal{C}$ within the configuration $\mathcal{B}$ is expressible as the gradient of a vector field $\boldsymbol{u}^*$. Note, however, that based on the compatibility condition in Eq. 1.8 the field $\boldsymbol{u}^*$ is only determined from a given $\boldsymbol{\beta}$ up to an arbitrary constant term. □

1.2.2 Dislocation Density Tensor

In the sequel the deviation of the distortion from being integrable shall be considered as dislocation density. Thereby the integrability condition for the distortion is the continuum kinematics' counterpart of the materials science notion of the *Burgers circuit* and the resulting *Burgers vector* in a crystalline material, whereby the total Burgers vector corresponds to the resultant displacement closure gap in a (finite) Burgers circuit, refer to Fig. 1.3. Thus the displacement closure gap may be re-interpreted as the integral over its *displacement closure gap density* per unit area (that may also be denoted as *Burgers vector density*)

$$[\![\boldsymbol{u}]\!] \doteq \int_{\mathcal{A}} \mathbb{b}\, da, \tag{1.9}$$

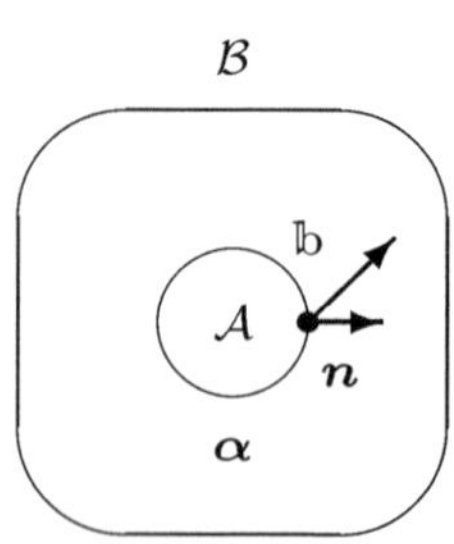

Fig. 1.3 Integrating the displacement closure gap (Burgers vector) density $\mathbb{b}$ over the area $\mathcal{A}$ determines the total amount of dislocations piercing through $\mathcal{A}$. This quantity is here denoted the total displacement closure gap (Burgers vector).

whereby the Burgers vector density $\mathbb{b}$ is computed from the second-order *dislocation density tensor* $\boldsymbol{\alpha}$ in the spirit of a Cauchy theorem

$$\mathbb{b} := \boldsymbol{\alpha} \cdot \boldsymbol{n}. \tag{1.10}$$

Interestingly, the relation of the Burgers vector density and the dislocation density tensor in terms of a Cauchy theorem compares the dislocation density tensor *formally* with the stress tensor. The dislocation density tensor is generally non-symmetric as follows from its definition that arises from the comparison with Eq. 1.6

$$\boldsymbol{\alpha} := \operatorname{curl} \boldsymbol{\beta} = -\beta_{ab,c} e_{bcd} \boldsymbol{e}_a \otimes \boldsymbol{e}_d. \tag{1.11}$$

Note that a vanishing dislocation density tensor $\boldsymbol{\alpha} = \boldsymbol{0}$ agrees with the previously stated integrability condition for the distortion $\beta_{a[b,c]} \doteq 0$. As a conclusion by identifying the closure gap with the integral of its density the definition of the dislocation density tensor in terms of the distortion is obtained. In the differential geometry setting discussed in the subsequent parts of this treatise the dislocation density tensor will be identified with the so-called *anholonomic object*.

Remark 1.2. Observe that the dislocation density tensor satisfies the following differential identity

$$\operatorname{div} \boldsymbol{\alpha} = \boldsymbol{0}. \tag{1.12}$$

The above holds trivially true due to $\alpha_{ad,d} = -\beta_{ab,cd} e_{bcd} \equiv 0$. □

Remark 1.3. In Cartesian coordinates the coefficients α_{ab} of the dislocation density tensor decompose into the density of *screw dislocations* corresponding to translational/first kind *Volterra dislocations* (of type three) as given by the diagonal terms $\alpha_{ab}, a = b$

$$\begin{bmatrix} \bullet & \circ & \circ \\ \circ & \bullet & \circ \\ \circ & \circ & \bullet \end{bmatrix},$$

and the density of *edge dislocations* corresponding to translational/first kind *Volterra dislocations* (of type one and two) as given by the off-diagonal terms $\alpha_{ab}, a \neq b$

$$\begin{bmatrix} \circ & \bullet & \bullet \\ \bullet & \circ & \bullet \\ \bullet & \bullet & \circ \end{bmatrix}.$$

For edge dislocations the Burgers vector is perpendicular to the *dislocation line* whereas for screw dislocation the Burgers vector is parallel to the dislocation line, see also the discussion by [94]. Thus the columns of α_{ab} denote the coefficients of the Burgers vector $\mathbb{b}_b = \alpha_{ab} \boldsymbol{e}_a$ penetrating a *unit* surface area with orientation given by its surface normal $\boldsymbol{e}_b$. Here $\boldsymbol{e}_b$ may be thought of as being tangent to the dislocation lines penetrating the surface, thus the normal part of $\mathbb{b}_b$, i.e. $\boldsymbol{e}_a \cdot \mathbb{b}_b, a = b$ denotes the corresponding screw dislocations whereas the two tangent parts of $\mathbb{b}_b$, i.e. $\boldsymbol{e}_a \cdot \mathbb{b}_b, a \neq b$ represent the corresponding edge dislocations. Finally, for a *single* dislocation the dislocation density tensor may be regarded a *distribution*

$$\boldsymbol{\alpha} = \delta(\boldsymbol{x})[\![\boldsymbol{u}]\!] \otimes \mathbb{t}$$

in terms of the Dirac-delta $\delta = \delta(\boldsymbol{x})$ distribution and the unit tangent $\mathbb{t}$ to the corresponding dislocation line. □

Remark 1.4. Volterra dislocations may be categorized into translational and rotational, or, likewise, into Volterra dislocations of first and second kind. Thereby translational/first kind and rotational/second kind Volterra dislocations taken together are sub-categorized into six types, see Tab. 1.1: Volterra dislocations of type one to three are translational/first kind defects and characterize edge and screw dislocations, respectively, whereas Volterra dislocations of type four to six are rotational/second kind defects and characterize twist and wedge disclinations, respectively. These defects may easily be visualized by imaging a thick-walled tube with an axial cut. Then six conceptually different possibilities to produce a gap between the so created cut-surfaces may be considered: firstly translational gaps may be obtained by relative translation of the two cut-surfaces either perpendicular or parallel to the tube's axis; likewise rotational gaps may be obtained by relative rotation of the two cut-surfaces either perpendicular or parallel to the tube's axis. □

Table 1.1 Categorization of the six types of Volterra dislocations into translational/first kind and rotational/second kind

Volterra Dislocations	Translational/First Kind	Rotational/Second Kind
Type 1	Edge Dislocations	
Type 2	Edge Dislocations	
Type 3	Screw Dislocations	
Type 4		Twist Disclinations
Type 5		Twist Disclinations
Type 6		Wedge Disclinations

1.2.3 Kinematics of Crystal Plasticity

In continuum *crystal plasticity* the total distortion is additively decomposed into plastic and elastic contributions

$$\operatorname{grad} \boldsymbol{u} = \widetilde{\boldsymbol{\beta}} + \overline{\boldsymbol{\beta}}. \tag{1.13}$$

For a motivation of this decomposition recall the relation between the geometrically linear strain tensor and the symmetric part of the distortion

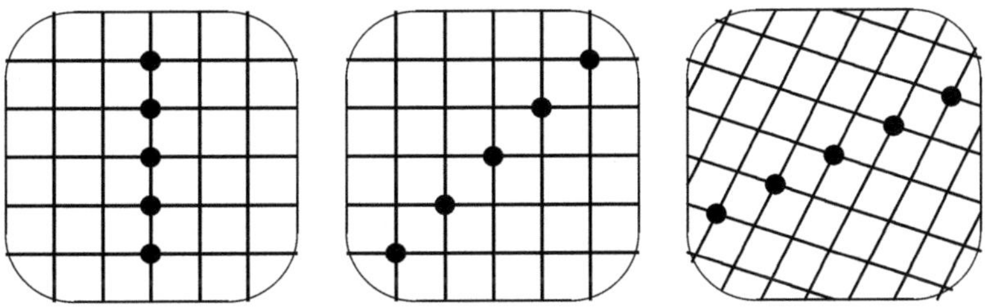

Fig. 1.4 Decomposition of the distortion $\boldsymbol{\beta}$ in an idealized crystalline lattice into plastic and elastic contributions $\widetilde{\boldsymbol{\beta}}$ and $\overline{\boldsymbol{\beta}}$, respectively. Due to the dislocation flow along the slip directions the plastic distortion $\widetilde{\boldsymbol{\beta}}$ leaves the crystalline lattice unaffected, the elastic distortion $\overline{\boldsymbol{\beta}}$, however, distorts the lattice and is thus responsible for the generation of stress.

$\boldsymbol{\epsilon} := \boldsymbol{\beta}^{\mathrm{sym}}$. However, in contrast to *phenomenological elasto-plasticity* that relies on the additive decomposition of the symmetric strain $\boldsymbol{\epsilon}$ into plastic and elastic contributions, the kinematics of crystal plasticity dictate a flow rule for the evolution of the plastic distortion $\widetilde{\boldsymbol{\beta}}$ thus suggesting the decomposition in Eq. 1.13. Accordingly the plastic and elastic dislocation density tensors $\widetilde{\boldsymbol{\alpha}}$ and $\overline{\boldsymbol{\alpha}}$, respectively, follow as

$$\widetilde{\boldsymbol{\alpha}} := \operatorname{curl} \widetilde{\boldsymbol{\beta}} \quad \text{and} \quad \overline{\boldsymbol{\alpha}} := \operatorname{curl} \overline{\boldsymbol{\beta}}. \tag{1.14}$$

Based on the plastic and elastic dislocation density tensors the condition for the integrability of the total distortion reads as

$$\widetilde{\boldsymbol{\alpha}} + \overline{\boldsymbol{\alpha}} = \mathbf{0}. \tag{1.15}$$

As a consequence of their definitions in Eq. 1.14 in terms of the curl operator the so-called *continuity equation* for the plastic and elastic dislocation density tensors follows trivially as

$$\operatorname{div} \widetilde{\boldsymbol{\alpha}} = \mathbf{0} \quad \text{and} \quad \operatorname{div} \overline{\boldsymbol{\alpha}} = \mathbf{0}. \tag{1.16}$$

Observe the formal similarity of the continuity equation for the dislocation density tensors with the (homogeneous) mechanical equilibrium equation. In continuum dislocation theory, however, the interpretation of the continuity equation is the impossibility for dislocation lines to end within the volume, i.e. the dislocation density is source free, compare to the arguments in [94].

Nye Lattice Curvature. The elastic distortion $\overline{\boldsymbol{\beta}}$ may be additively decomposed into its symmetric and skew symmetric parts, i.e. into the elastic strain $\overline{\boldsymbol{\epsilon}}$ and the elastic rotation $\overline{\boldsymbol{\omega}}$, respectively

$$\overline{\boldsymbol{\beta}} = \overline{\boldsymbol{\epsilon}} + \overline{\boldsymbol{\omega}}. \tag{1.17}$$

The Nye lattice curvature tensor $\overline{\boldsymbol{\kappa}}$ is then given by the gradient of the axial vector $\overline{\boldsymbol{\omega}}^\times$ of the elastic rotation

$$\overline{\boldsymbol{\kappa}} := \text{grad}\, \overline{\boldsymbol{\omega}}^\times = \bar{\omega}^\times_{a,d} \boldsymbol{e}_a \otimes \boldsymbol{e}_d. \tag{1.18}$$

Here the axial vector $\overline{\boldsymbol{\omega}}^\times$ of the elastic rotation $\overline{\boldsymbol{\omega}}$ is computed in terms of the (Levi-Civita) permutation tensor

$$\overline{\boldsymbol{\omega}}^\times := -\frac{1}{2}\boldsymbol{e} : \overline{\boldsymbol{\omega}} = -\frac{1}{2} e_{abc}\bar{\omega}_{bc}\boldsymbol{e}_a \quad \rightsquigarrow \quad \overline{\boldsymbol{\kappa}} = -\frac{1}{2} e_{abc}\bar{\omega}_{bc,d}\boldsymbol{e}_a \otimes \boldsymbol{e}_d. \tag{1.19}$$

For a macroscopically stress-free configuration the elastic distortion consists exclusively in the elastic rotation $\overline{\boldsymbol{\beta}} = \overline{\boldsymbol{\omega}}$. Accordingly, the corresponding elastic dislocation density computes as

$$\overline{\boldsymbol{\alpha}} = -\bar{\omega}_{ab,c} e_{bcd} \boldsymbol{e}_a \otimes \boldsymbol{e}_d. \tag{1.20}$$

The (spin) tensor $\overline{\boldsymbol{\omega}}$ associated to the axial vector $\overline{\boldsymbol{\omega}}^\times$ is again given in terms of the (Levi-Civita) permutation tensor as

$$\overline{\boldsymbol{\omega}} := -\boldsymbol{e} \cdot \overline{\boldsymbol{\omega}}^\times = -e_{abe}\bar{\omega}^\times_e \boldsymbol{e}_a \otimes \boldsymbol{e}_b \quad \rightsquigarrow \quad \overline{\boldsymbol{\alpha}} = e_{abe}\bar{\omega}^\times_{e,c} e_{bcd}\boldsymbol{e}_a \otimes \boldsymbol{e}_d. \tag{1.21}$$

Thus, with $e_{eab}e_{bcd} = \delta_{ec}\delta_{ad} - \delta_{ed}\delta_{ac}$ the famous finding by [152] relating the stress-free lattice curvature to the elastic dislocation density follows as

$$\overline{\boldsymbol{\alpha}} = [\bar{\omega}^\times_{c,c}\delta_{ad} - \bar{\omega}^\times_{d,a}]\boldsymbol{e}_a \otimes \boldsymbol{e}_d = [\overline{\boldsymbol{\kappa}}^t : \boldsymbol{1}]\boldsymbol{1} - \overline{\boldsymbol{\kappa}}^t. \tag{1.22}$$

Here $\boldsymbol{1} = \delta_{ab}\boldsymbol{e}_a \otimes \boldsymbol{e}_b$ denotes the second-order unit tensor, whereby the Cartesian coefficients δ_{ab} coincide with the coefficients of the Kronecker symbol. Vice versa, with $\overline{\boldsymbol{\alpha}} : \boldsymbol{1} = 2\overline{\boldsymbol{\kappa}} : \boldsymbol{1}$ and $\boldsymbol{1} : \boldsymbol{1} = 3$ the Nye lattice curvature is finally given in terms of the dislocation density by

$$\overline{\boldsymbol{\kappa}} = \frac{1}{2}[\overline{\boldsymbol{\alpha}}^t : \boldsymbol{1}]\boldsymbol{1} - \overline{\boldsymbol{\alpha}}^t. \tag{1.23}$$

As a conclusion it may be stated that a stress-free Nye lattice curvature necessitates a distributed (elastic) dislocation density. Due to the integrability condition in Eq. 1.15 the elastic dislocation density tensor may likewise be substituted by the negative plastic dislocation density tensor.

2d Example: Idealized Crystal. As a simple two-dimensional example motivated from the discussion in [202] the idealized crystal depicted in Fig. 1.5 is considered. It is endowed with two *slip systems* characterized by the *slip directions* $\boldsymbol{s}_\varsigma$ and the *slip plane normals* $\boldsymbol{m}_\varsigma$, defined as unit vectors

$$\boldsymbol{s}_{1/2} = \frac{1}{\sqrt{2}}\begin{bmatrix} 1 \\ \pm 1 \end{bmatrix} \quad \text{and} \quad \boldsymbol{m}_{1/2} = \frac{1}{\sqrt{2}}\begin{bmatrix} \mp 1 \\ 1 \end{bmatrix}. \tag{1.24}$$

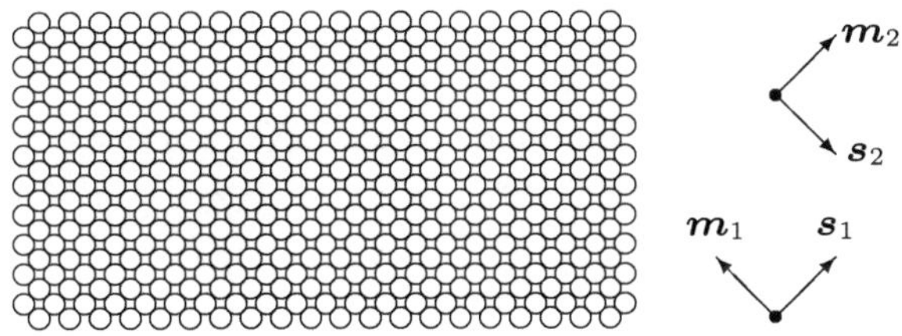

Fig. 1.5 Idealized crystal with two slip systems that allow plastic glide along the diagonals.

together with the corresponding *Schmid (projection) tensors* $\boldsymbol{p}_\varsigma := \boldsymbol{s}_\varsigma \otimes \boldsymbol{m}_\varsigma$

$$\boldsymbol{p}_{1/2} = \frac{1}{2}\begin{bmatrix} \mp 1 & 1 \\ -1 & \pm 1 \end{bmatrix}. \tag{1.25}$$

For the two-dimensional setting the only non-vanishing coefficients of the dislocation density tensors defined by $\alpha_{ad} = -\beta_{ab,c}e_{bcd}$ are obviously

$$\alpha_{13} = \beta_{12,1} - \beta_{11,2} \quad \text{and} \quad \alpha_{23} = \beta_{22,1} - \beta_{21,2}. \tag{1.26}$$

Correspondingly the only non-vanishing coefficients of the elastic Nye lattice curvature defined by $\bar{\kappa}_{ad} = -\frac{1}{2}e_{abc}\bar{\omega}_{bc,d}$ are

$$2\bar{\kappa}_{31} = -\bar{\omega}_{12,1} + \bar{\omega}_{21,1} \quad \text{and} \quad 2\bar{\kappa}_{32} = -\bar{\omega}_{12,2} + \bar{\omega}_{21,2}. \tag{1.27}$$

The displacement field of a globally compatible bending deformation is given in terms of the two parameters c_1 and c_2

$$\boldsymbol{u} = -c_1 x_1 x_2\, \boldsymbol{e}_1 + \frac{1}{2}[c_1 x_1^2 + c_2 x_2^2]\, \boldsymbol{e}_2 \tag{1.28}$$

with the coefficients of the corresponding total distortion computed as

$$[\beta_{ab}] = \begin{bmatrix} -c_1 x_2 & -c_1 x_1 \\ c_1 x_1 & c_2 x_2 \end{bmatrix} \quad \text{with} \quad \alpha_{13} = \alpha_{23} \equiv 0. \tag{1.29}$$

Then for a purely **elastic** distortion with $\overline{\boldsymbol{\beta}} = \boldsymbol{\beta}$ as depicted in Fig. 1.6 the total and the elastic dislocation tensors vanish, whereas the (elastic) lattice curvature is constant

$$\boldsymbol{\alpha} = \overline{\boldsymbol{\alpha}} = \boldsymbol{0} \quad \text{with} \quad \bar{\kappa}_{31} = c_1. \tag{1.30}$$

Vice versa for a purely **plastic** distortion with $\widetilde{\boldsymbol{\beta}} = \boldsymbol{\beta}$ as depicted in Fig. 1.7 the total and the plastic dislocation tensors vanish, likewise the (elastic) lattice curvature is zero

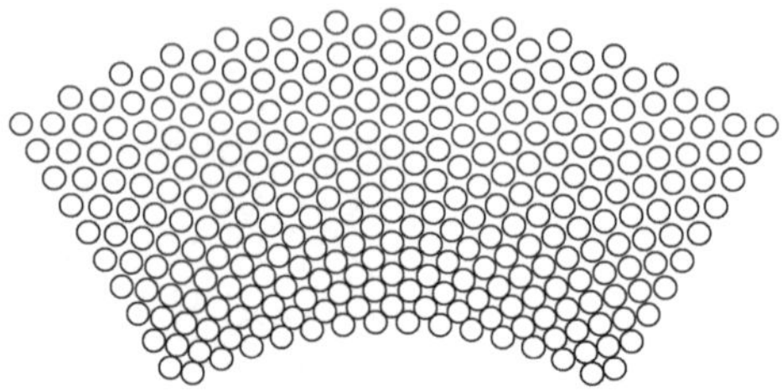

Fig. 1.6 Purely elastic distortion of an idealized crystal. The total and the elastic dislocation tensors vanish, the (elastic) lattice curvature is constant.

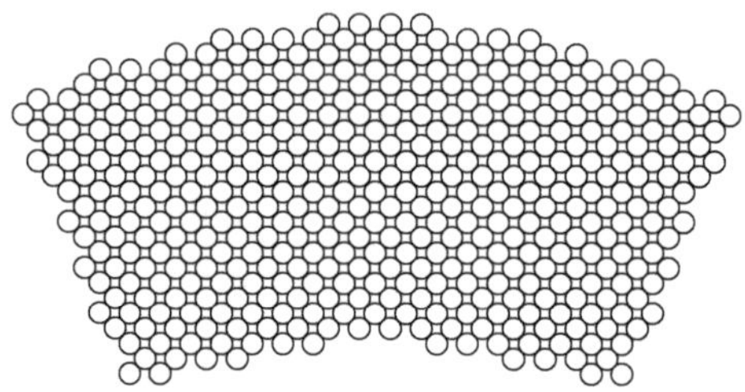

Fig. 1.7 Purely plastic distortion of an idealized crystal. The total and the plastic dislocation tensors vanish, the (elastic) lattice curvature is zero.

$$\boldsymbol{\alpha} = \widetilde{\boldsymbol{\alpha}} = \mathbf{0} \quad \text{with} \quad \bar{\kappa}_{31} = 0. \tag{1.31}$$

An incompressible plastic distortion is obtained by setting $c_2 = c_1$, thus the plastic distortion is realized by the following combination of Schmid tensors

$$\widetilde{\boldsymbol{\beta}} = c_1 x_2 [\boldsymbol{p}_1 - \boldsymbol{p}_2] - c_1 x_1 [\boldsymbol{p}_1 + \boldsymbol{p}_2]. \tag{1.32}$$

Finally, for a possible macroscopically **stress-free** distortion as depicted in Fig. 1.8 with the elastic and plastic contributions

$$\overline{\boldsymbol{\beta}} = \begin{bmatrix} 0 & -c_1 x_1 \\ c_1 x_1 & 0 \end{bmatrix} \quad \text{and} \quad \widetilde{\boldsymbol{\beta}} = \begin{bmatrix} -c_1 x_2 & 0 \\ 0 & c_2 x_2 \end{bmatrix} \tag{1.33}$$

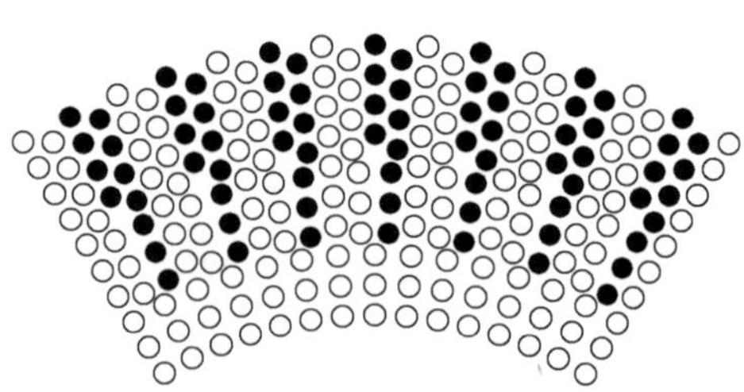

Fig. 1.8 Macroscopically stress-free distortion of an idealized crystal. The total and the plastic dislocation tensors together with the (elastic) lattice curvature are constant. Dislocations are formed by ending lattice lines that are here marked by full circles.

the plastic and elastic dislocation densities together with the (elastic) lattice curvature take constant values

$$\tilde{\alpha}_{13} = -\bar{\alpha}_{13} = c_1 \quad \text{with} \quad \bar{\kappa}_{31} = c_1. \tag{1.34}$$

An incompressible plastic distortion is obtained by setting $c_2 = c_1$, thus the plastic distortion is realized by the following combination of Schmid tensors

$$\widetilde{\boldsymbol{\beta}} = c_1 x_2 [\boldsymbol{p}_1 - \boldsymbol{p}_2]. \tag{1.35}$$

Note finally that although the distortion is macroscopically stress-free, the dislocation cores at the end of the lattice lines in Fig. 1.8 (consisting of the five closest atoms) experience locally severe lattice distortions.

1.2.4 Generalized Crystal Plasticity

The key idea to formulate a physically motivated theory of *generalized crystal plasticity in a first-order continuum* is to observe that the *geometrically necessary dislocation density* contributes to the hardening of the material. Thus the incorporation of the plastic dislocation density into the free energy density emerges as a natural concept

$$\Psi = \Psi^{\mathrm{mac}}(\boldsymbol{\epsilon} - \widetilde{\boldsymbol{\epsilon}}) + \Psi^{\mathrm{mic}}(\kappa) + \Psi^{\mathrm{dis}}(\widetilde{\boldsymbol{\alpha}}). \tag{1.36}$$

Here the macroscopic part Ψ^{mac} of the free energy density depends on the elastic strain (symmetric distortion) as given by $\boldsymbol{\epsilon} - \widetilde{\boldsymbol{\epsilon}}$, the microscopic part Ψ^{mic} captures the hardening due to *statistically stored dislocations* in terms of a single scalar hardening variable κ (for simplicity), and the dislocation part Ψ^{dis} captures hardening due to geometrically necessary dislocations. Then, according to [53] the Clausius-Duhem (dissipation power density) inequality is formulated including a so-called non-locality residual $\mathcal{R}$

$$\mathcal{D} = \boldsymbol{\sigma} : \partial_t \boldsymbol{\epsilon} - \partial_t \Psi + \mathcal{R} \geq 0. \tag{1.37}$$

A standard Coleman-Noll exploitation then renders firstly the symmetric macroscopic (Cauchy) stress

$$\boldsymbol{\sigma} := \frac{\partial \Psi}{\partial \boldsymbol{\epsilon}}. \tag{1.38}$$

Secondly, the *thermo-dynamically driving forces* are obtained as conjugated to the plastic variables. These are the *plastic stress* (that agrees here identically with the equilibrium stress) and the so-called *drag stress*

$$\widetilde{\boldsymbol{\sigma}} := -\frac{\partial \Psi}{\partial \widetilde{\boldsymbol{\epsilon}}} \equiv \boldsymbol{\sigma} \quad \text{and} \quad H := \frac{\partial \Psi}{\partial \kappa}. \tag{1.39}$$

Finally, the dislocation density is conjugated to the *dislocation stress* as

$$\boldsymbol{\sigma}^{\alpha} := \frac{\partial \Psi}{\partial \widetilde{\boldsymbol{\alpha}}}. \tag{1.40}$$

Thus the *reduced* dissipation power density reads

$$\mathcal{D} = \boldsymbol{\sigma} : \partial_t \widetilde{\boldsymbol{\beta}} - H \partial_t \kappa - \boldsymbol{\sigma}^{\alpha} : \partial_t \widetilde{\boldsymbol{\alpha}} + \mathcal{R} \geq 0. \tag{1.41}$$

Expressing the rate of the plastic dislocation density in terms of the plastic distortion and applying the product rule of differentiation then results in the abbreviated bi-linear expression

$$\mathcal{D} = \boldsymbol{\sigma}^{\star} : \partial_t \widetilde{\boldsymbol{\beta}} - H \partial_t \kappa \geq 0. \tag{1.42}$$

Here, the *generalized* driving stress is defined as

$$\boldsymbol{\sigma}^{\star} := \boldsymbol{\sigma} - \operatorname{curl} \boldsymbol{\sigma}^{\alpha}. \tag{1.43}$$

Note that the dislocation-induced stress acts as a kind of *back-stress*. Moreover the non-locality residual has here been defined in the format

$$\mathcal{R} := - \operatorname{div} \left(\partial_t \widetilde{\boldsymbol{\beta}} : [\boldsymbol{\sigma}^{\alpha} \cdot \boldsymbol{e}] \right). \tag{1.44}$$

Proof. The term involving the dislocation stress may be reformulated as

$$-\boldsymbol{\sigma}^{\alpha} : \partial_t \widetilde{\boldsymbol{\alpha}} \equiv -\boldsymbol{\sigma}^{\alpha} : \operatorname{curl} \partial_t \widetilde{\boldsymbol{\beta}} = - \operatorname{curl} \boldsymbol{\sigma}^{\alpha} : \partial_t \widetilde{\boldsymbol{\beta}} + \operatorname{div} \left(\partial_t \widetilde{\boldsymbol{\beta}} : [\boldsymbol{\sigma}^{\alpha} \cdot \boldsymbol{e}] \right).$$

Likewise, in index notation this expression reads for Cartesian coordinates

$$\sigma^{\alpha}_{ad} [\partial_t \widetilde{\beta}_{ab,c} e_{bcd}] = [\sigma^{\alpha}_{ad,c} e_{dcb}] \partial_t \widetilde{\beta}_{ab} + \left[\partial_t \widetilde{\beta}_{ab} [\sigma^{\alpha}_{ad} e_{dbc}] \right]_{,c}.$$

Clearly, the non-locality residual has to cancel the divergence term on the right-hand-side. □

Then considering the so-called *insulation condition*, see [158, 159], identifies suited boundary conditions for the plastic distortion

$$\int_{\mathcal{B}^{\mathrm{p}}} \mathcal{R} \, \mathrm{d}v = \int_{\partial \mathcal{B}^{\mathrm{p}}} \left[\partial_t \widetilde{\boldsymbol{\beta}} : [\boldsymbol{\sigma}^{\alpha} \cdot \widehat{\boldsymbol{n}}] \right] \mathrm{d}a \doteq 0. \tag{1.45}$$

Here $\widehat{\boldsymbol{n}} = -\boldsymbol{e} \cdot \boldsymbol{n}$ with $\boldsymbol{n}$ the outwards pointing normal to the elastic-plastic boundary. The assumption of homogeneous Neumann-type boundary conditions for $\boldsymbol{\sigma}^{\alpha} \cdot \widehat{\boldsymbol{n}}$ is a valid choice for enforcing the insulation condition identically, for a more detailed discussion please refer to [158, 159]. Then a multi-surface yield condition in terms of the generalized driving stress $\boldsymbol{\sigma}^{\star}$ and the drag stress H may be introduced for each slip system ς as

$$\phi_{\varsigma}(\boldsymbol{\sigma}^{\star}, H) = \varphi_{\varsigma}(\boldsymbol{\sigma}^{\star}) - [\sigma_y + H] \leq 0. \tag{1.46}$$

The postulate of maximum dissipation results in a minimization problem with inequality constraints that are enforced by Lagrange or rather plastic multipliers γ_ς, i.e.

$$\mathcal{L} := -\mathcal{D} + \sum_\varsigma \gamma_\varsigma \phi_\varsigma \to \min . \tag{1.47}$$

Consequently the associated evolution rules for the plastic distortion and the scalar hardening variable follow as

$$\partial_t \widetilde{\boldsymbol{\beta}} = \sum_\varsigma \gamma_\varsigma \frac{\partial \varphi_\varsigma}{\partial \boldsymbol{\sigma}^\star} \quad \text{and} \quad \partial_t \kappa = \sum_\varsigma \gamma_\varsigma . \tag{1.48}$$

The simplest possibility to express the contribution of the plastic dislocation density to the free energy density is a quadratic dependence

$$\Psi^{\text{dis}} := \frac{1}{2} G \ell_\alpha^2 \widetilde{\boldsymbol{\alpha}} : \widetilde{\boldsymbol{\alpha}}. \tag{1.49}$$

Here G denotes the shear modulus and ℓ_α is a characteristic length that has to be introduced for dimensional consistency. Then the dislocation-stress is computed as

$$\boldsymbol{\sigma}^\alpha = G \ell_\alpha^2 \operatorname{curl} \widetilde{\boldsymbol{\beta}}. \tag{1.50}$$

Moreover, the generalized driving stress then follows simply by applying the curl operator to the plastic distortion twice (this is however *not* the *incompatibility operator* inc as introduced by [94], that rather follows by applying the *transpose* of the curl operator twice, i.e. $\operatorname{inc}(\{\bullet\}) := \operatorname{curl}^t \operatorname{curl}^t(\{\bullet\})$)

$$\boldsymbol{\sigma}^\star = \boldsymbol{\sigma} - G \ell_\alpha^2 \operatorname{curl}(\operatorname{curl} \widetilde{\boldsymbol{\beta}}). \tag{1.51}$$

The simplest φ_ς coincides with a *generalized Schmid stress* $\sigma_\varsigma^\star$ and is formulated in terms of the Schmid tensor $\boldsymbol{s}_\varsigma \otimes \boldsymbol{m}_\varsigma$ for each slip system ς

$$\varphi_\varsigma \equiv \sigma_\varsigma^\star = \boldsymbol{\sigma}^\star : [\boldsymbol{s}_\varsigma \otimes \boldsymbol{m}_\varsigma]. \tag{1.52}$$

Thus the generalized Schmid stress is motivated as a driving force for plastic evolution on slip system ς with

$$\sigma_\varsigma^\star = \sigma_\varsigma - G \ell_\alpha^2 \boldsymbol{s}_\varsigma \cdot \operatorname{curl}(\operatorname{curl} \widetilde{\boldsymbol{\beta}}) \cdot \boldsymbol{m}_\varsigma. \tag{1.53}$$

Here $\sigma_\varsigma = \boldsymbol{s}_\varsigma \cdot \boldsymbol{\sigma} \cdot \boldsymbol{m}_\varsigma$ denotes the common Schmid stress, i.e. the ordinary resolved shear stress at slip system ς. Interestingly, in this formulation the evolution law for the plastic distortion or rather the flow rule is unaffected by the generalization to incorporate the dependence on the dislocation density

$$\partial_t \widetilde{\boldsymbol{\beta}} = \sum_\varsigma \gamma_\varsigma \boldsymbol{s}_\varsigma \otimes \boldsymbol{m}_\varsigma . \tag{1.54}$$

It shall be noted that a characteristic length ℓ_α has been incorporated into the formulation due to the consideration of the plastic dislocation density. Moreover, second derivatives of the plastic distortion contribute to the generalized Schmid stress that serves as the driving force for inelastic evolution. These are the typical ingredients of a generalized plasticity formulation that has here been obtained based on the physically motivated concept of geometrically necessary dislocations.

1.3 Second-Order Continuum

As a further motivation a generalized, i.e. second-order continuum formulation shall be considered that incorporates in addition to the common stress $\boldsymbol{\sigma}$ an additional double-stress $\boldsymbol{\sigma}$ in the sense of [140]. Thereby the internal and external working are assumed as

$$\begin{aligned} \mathcal{W}^{\text{int}} &= \int_{\mathcal{B}} \left[\boldsymbol{\sigma} : \operatorname{grad} \boldsymbol{v} + \boldsymbol{\sigma} \,\therefore\, \operatorname{grad}(\operatorname{grad} \boldsymbol{v})\right] \mathrm{d}v, \\ \mathcal{W}^{\text{ext}} &= \int_{\mathcal{B}} \boldsymbol{b} \cdot \boldsymbol{v} \,\mathrm{d}v + \int_{\partial\mathcal{B}} \left[\boldsymbol{t} \cdot \boldsymbol{v} + \boldsymbol{t}' \cdot \boldsymbol{v}'\right] \mathrm{d}a. \end{aligned} \tag{1.55}$$

Here $\boldsymbol{\beta} := \operatorname{grad} \boldsymbol{u}$ and $\boldsymbol{\gamma} := \operatorname{grad}(\operatorname{grad} \boldsymbol{u})$ denote the distortion and the *double-distortion*, respectively, $\boldsymbol{b}$ are distributed body forces (per unit volume) and $\boldsymbol{t}$ and $\boldsymbol{t}'$ represent the prescribed traction and double-traction, respectively, (per unit area) and $\boldsymbol{n}$ is the outwards-pointing normal to the boundary. Moreover, $\boldsymbol{v}' := \operatorname{grad} \boldsymbol{v} \cdot \boldsymbol{n}$ abbreviates the normal gradient of the velocity field $\boldsymbol{v}$. (Note again the symbolic notation for third-order tensors, such as $\boldsymbol{\sigma}$ or $\boldsymbol{\gamma}$ in terms of *large* bold fonts.) The terminology *second-order continuum* derives from the presence of the second gradient of the velocity field $\boldsymbol{v}$ in the definition of the internal working. Requiring the internal working to equal the external working for all velocity fields $\boldsymbol{v}$ results in the corresponding equilibrium conditions in the bulk and at the boundary

$$\begin{aligned} -\operatorname{div}(\boldsymbol{\sigma} - \operatorname{div}\boldsymbol{\sigma}) &= \boldsymbol{b} \quad \text{in} \quad \mathcal{B}, \\ [\boldsymbol{\sigma} - \operatorname{div}\boldsymbol{\sigma}] \cdot \boldsymbol{n} - \varsigma(\boldsymbol{\sigma} \cdot \boldsymbol{n}) &= \boldsymbol{t} \quad \text{on} \quad \partial\mathcal{B}, \\ [\boldsymbol{\sigma} \cdot \boldsymbol{n}] \cdot \boldsymbol{n} &= \boldsymbol{t}' \quad \text{on} \quad \partial\mathcal{B}. \end{aligned} \tag{1.56}$$

In the above, $\varsigma(\boldsymbol{\sigma} \cdot \boldsymbol{n}) := \widehat{\operatorname{div}}(\widehat{[\boldsymbol{\sigma} \cdot \boldsymbol{n}]}) = -[\widehat{\operatorname{div}}\, \boldsymbol{n}][\boldsymbol{\sigma} \cdot \boldsymbol{n}] \cdot \boldsymbol{n} + \widehat{\operatorname{div}}(\boldsymbol{\sigma} \cdot \boldsymbol{n})$ takes into account the contribution of the double-stress $\boldsymbol{\sigma}$ to the traction due to the curvature of the boundary. Here the *surface divergence* $\widehat{\operatorname{div}}(\bullet) = \operatorname{grad}(\bullet) : \widehat{\mathbf{1}}$ is defined in terms of the *surface projection operator* $\widehat{\mathbf{1}} = \mathbf{1} - \boldsymbol{n} \otimes \boldsymbol{n}$, with $\mathbf{1}$ denoting the second-order unit tensor and $\widehat{[\boldsymbol{\sigma} \cdot \boldsymbol{n}]} := [\boldsymbol{\sigma} \cdot \boldsymbol{n}] : \widehat{\mathbf{1}}$.

Proof. Integrating the second term in the internal working by parts twice renders firstly

$$\int_{\mathcal{B}} \boldsymbol{\sigma} \,\vdots\, \mathrm{grad}(\mathrm{grad}\,\boldsymbol{v})\,\mathrm{d}v = \int_{\mathcal{B}} \left[\mathrm{div}(\mathrm{grad}\,\boldsymbol{v} : \boldsymbol{\sigma}) - \mathrm{grad}\,\boldsymbol{v} : \mathrm{div}\boldsymbol{\sigma}\right] \mathrm{d}v$$
$$= \int_{\mathcal{B}} \left[\mathrm{div}(\mathrm{grad}\,\boldsymbol{v} : \boldsymbol{\sigma} - \boldsymbol{v} \cdot \mathrm{div}\boldsymbol{\sigma}) + \boldsymbol{v} \cdot \mathrm{div}(\mathrm{div}\boldsymbol{\sigma})\right] \mathrm{d}v.$$

Applying next the Gauss (bulk divergence) theorem results in

$$\int_{\mathcal{B}} \boldsymbol{\sigma} \,\vdots\, \mathrm{grad}(\mathrm{grad}\,\boldsymbol{v})\,\mathrm{d}v = \int_{\partial\mathcal{B}} \left[\mathrm{grad}\,\boldsymbol{v} : \boldsymbol{\sigma} - \boldsymbol{v} \cdot \mathrm{div}\boldsymbol{\sigma}\right] \cdot \boldsymbol{n}\,\mathrm{d}a + \int_{\mathcal{B}} \boldsymbol{v} \cdot \mathrm{div}(\mathrm{div}\boldsymbol{\sigma})\,\mathrm{d}v.$$

For smooth and closed surfaces the first surface term on the right-hand-side is finally re-expressed in terms of the surface divergence theorem

$$\int_{\partial\mathcal{B}} \mathrm{grad}\,\boldsymbol{v} : [\boldsymbol{\sigma} \cdot \boldsymbol{n}]\,\mathrm{d}a = \int_{\partial\mathcal{B}} \left[[\mathrm{grad}\,\boldsymbol{v} \cdot \boldsymbol{n}] \cdot [\boldsymbol{\sigma} \cdot \boldsymbol{n}] \cdot \boldsymbol{n} - \boldsymbol{v} \cdot \varsigma(\boldsymbol{\sigma} \cdot \boldsymbol{n})\right] \mathrm{d}a.$$

For non-smooth and/or non-closed surfaces the surface divergence theorem will result in additional contributions to the boundary conditions. □

1.3.1 Integrability Condition for the Double-Distortion

Integrating the *differential distortion* $\mathrm{d}\boldsymbol{\beta}$ along a closed circuit $\mathcal{C}$ (encircling the area $\mathcal{A}$), see Fig. 1.9, determines a possible incompatibility of the (possibly non-continuous, multi-valued) distortion in terms of the resulting *distortion closure gap*

$$[\![\boldsymbol{\beta}]\!] := \oint_{\mathcal{C}} \mathrm{d}\boldsymbol{\beta}. \tag{1.57}$$

Next, expressing the differential distortion $\mathrm{d}\boldsymbol{\beta}$ in terms of the *double-distortion* (that in general may be unrelated to the distortion)

$$\mathrm{d}\boldsymbol{\beta} := \boldsymbol{\gamma} \cdot \mathrm{d}\boldsymbol{x} \tag{1.58}$$

allows to re-express the distortion closure gap as the *circular integral* of the double-distortion

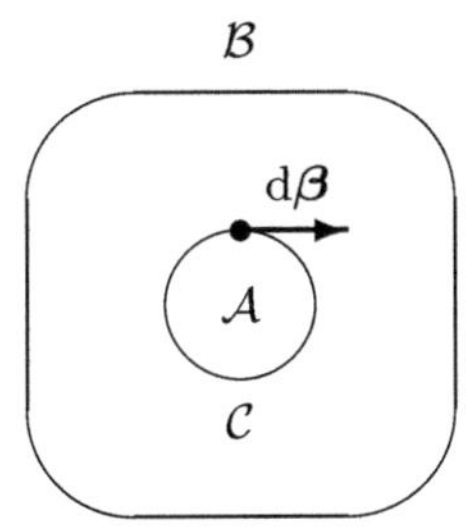

Fig. 1.9 Integrating the differential distortion $\mathrm{d}\boldsymbol{\beta}$ along a closed circuit $\mathcal{C}$ encircling the area $\mathcal{A}$ in the configuration $\mathcal{B}$ determines the resulting *distortion closure gap* $[\![\boldsymbol{\beta}]\!]$.

$$[\![\boldsymbol{\beta}]\!] = \oint_{\mathcal{C}} \boldsymbol{\gamma} \cdot \mathrm{d}\boldsymbol{x}. \tag{1.59}$$

Then the *Stokes theorem* reformulates the circular integral by involving the $\operatorname{curl} \boldsymbol{\gamma}$ of the double-distortion

$$[\![\boldsymbol{\beta}]\!] = \int_{\mathcal{A}} \operatorname{curl} \boldsymbol{\gamma} \cdot \mathrm{d}\boldsymbol{a}. \tag{1.60}$$

As a conclusion, Eq. 1.60 allows to compute the resulting distortion closure gap for a *given* double-distortion $\boldsymbol{\gamma}$.

Remark 1.5. In Cartesian coordinates with the orthonormal base vectors $\boldsymbol{e}_a$ the $\operatorname{curl} \boldsymbol{\gamma}$ of the double-distortion obeys the following coordinate representation

$$\operatorname{curl} \boldsymbol{\gamma} := - \operatorname{grad} \boldsymbol{\gamma} : \boldsymbol{e} = -\gamma_{abc,d} e_{cde} \boldsymbol{e}_a \otimes \boldsymbol{e}_b \otimes \boldsymbol{e}_e. \tag{1.61}$$

Due to the skew-symmetry of the permutation symbol, $\operatorname{curl} \boldsymbol{\gamma}$ is essentially equivalent to the (right) skew-symmetric part of $\operatorname{grad} \boldsymbol{\gamma}$, i.e. to $\gamma_{ab[c,d]}$. □

Discussion 1.2. The arguments leading to the expression for the distortion closure gap in Eq. 1.60 may be reversed in order to render the *point-wise* integrability condition for a *given* double-distortion $\boldsymbol{\gamma}$ to be integrable to a continuous, single-valued (second-order) tensor field $\boldsymbol{\beta}^*$.

Thereby the reasoning is as follows:

Requiring that the distortion closure gap in Eq. 1.60 vanishes for *arbitrary* circuits $\mathcal{C}$ and thus encircled areas $\mathcal{A}$ within the configuration $\mathcal{B}$, the point-wise integrability condition for the double-distortion $\boldsymbol{\gamma}$ reads

$$\operatorname{curl} \boldsymbol{\gamma} \doteq \boldsymbol{0} \quad \rightsquigarrow \quad \boldsymbol{\gamma} = \operatorname{grad} \boldsymbol{\beta}^*. \tag{1.62}$$

In other words, an integrable double-distortion that results in a vanishing distortion closure gap for arbitrary circuits $\mathcal{C}$ within the configuration $\mathcal{B}$ is expressible as the gradient of a tensor field $\boldsymbol{\beta}^*$. Note, however, that based on the compatibility condition in Eq. 1.62 the field $\boldsymbol{\beta}^*$ is only determined from a given $\boldsymbol{\gamma}$ up to an arbitrary constant term. □

Discussion 1.3. With Eq. 1.62 satisfied, as an additional requirement of integrability and based on the discussion in the previous section, the condition for the tensor field $\boldsymbol{\beta}^*$ being integrable to a continuous, single-valued vector field $\boldsymbol{u}^{**}$ is likewise expressible in terms of the double-distortion

$$\boldsymbol{\gamma} : \boldsymbol{e} \overset{1.62}{=} \operatorname{grad} \boldsymbol{\beta}^* : \boldsymbol{e} \doteq \boldsymbol{0} \quad \rightsquigarrow \quad \boldsymbol{\beta}^* = \operatorname{grad} \boldsymbol{u}^{**}. \tag{1.63}$$

Note, however, that based on the compatibility condition in Eqs. 1.63 the field $\boldsymbol{u}^{**}$ is only determined from a given $\boldsymbol{\beta}^*$ up to an arbitrary constant term. □

Discussion 1.4. The condition $\operatorname{curl}\boldsymbol{\gamma} = \mathbf{0}$ in Eqs. 1.62 does not necessarily provide integrability of $\boldsymbol{\gamma}$ to a *given*, i.e. *particular* distortion $\boldsymbol{\beta}$, since every $\boldsymbol{\lambda}$ in $\boldsymbol{\gamma} = \operatorname{grad}\boldsymbol{\beta} + \boldsymbol{\lambda}$ with $\operatorname{curl}\boldsymbol{\lambda} = \mathbf{0}$ will also satisfy $\operatorname{curl}\boldsymbol{\gamma} = \mathbf{0}$. Thus the additional constraint

$$\boldsymbol{\lambda} := \boldsymbol{\gamma} - \operatorname{grad}\boldsymbol{\beta} \doteq \mathbf{0} \tag{1.64}$$

has to be considered for the *relative* double-distortion if the integrability of $\boldsymbol{\gamma}$ into the *given* distortion $\boldsymbol{\beta}$ shall be enforced. The constraint $\boldsymbol{\lambda} \doteq \mathbf{0}$ may furthermore be decomposed into its (left) symmetric and skew-symmetric contribution $\lambda_{(ab)c} := \gamma_{(ab)c} - \epsilon_{ab,c} \doteq 0$ and $\lambda_{[ab]c} := \gamma_{[ab]c} - \omega_{ab,c} \doteq 0$, thus involving the symmetric strain and the skew-symmetric (continuum) rotation, respectively. Alternatively, it may be shown that the constraint may also be satisfied by simultaneously satisfying

$$\lambda_{(ab)c} \doteq 0 \quad \text{and} \quad \lambda_{a[bc]} \doteq 0. \tag{1.65}$$

The above skew-symmetries and symmetries can only be satisfied simultaneously by a vanishing $\lambda_{abc} \equiv 0$ since successive index permutation results in

$$\lambda_{abc} = -\lambda_{bac} = -\lambda_{bca} = \lambda_{cba} = \lambda_{cab} = -\lambda_{acb} = -\lambda_{abc}. \tag{1.66}$$

Finally, the constraint $\lambda_{a[bc]} \doteq 0$ is obviously enforced by requiring $\beta_{a[b,c]} \doteq 0$ in addition to $\gamma_{a[bc]} \doteq 0$, i.e.

$$\boldsymbol{\gamma} : \boldsymbol{e} \doteq \mathbf{0} \quad \text{and} \quad \operatorname{curl}\boldsymbol{\beta} \doteq \mathbf{0} \quad \rightsquigarrow \quad \boldsymbol{\lambda} : \boldsymbol{e} \doteq \mathbf{0}. \tag{1.67}$$

Thereby, the condition $\beta_{a[b,c]} \doteq 0$ has been substituted equivalently by $\operatorname{curl}\boldsymbol{\beta} \doteq \mathbf{0}$. □

1.3.2 Defect Density Tensors

In the sequel the deviations of the distortion and the double-distortion from being integrable shall be considered as various types of defect densities.

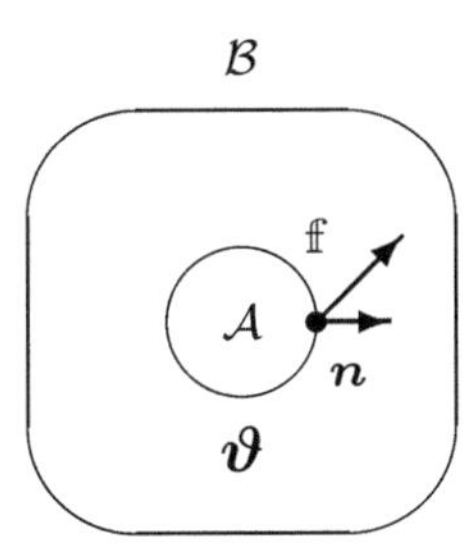

Fig. 1.10 Integrating the distortion closure gap (Frank tensor) density f over the area $\mathcal{A}$ determines the total amount of disclinations piercing through $\mathcal{A}$. This quantity is here denoted the total distortion closure gap (Frank tensor).

Disclination Density Tensor. The distortion closure gap is a generalization of the continuum kinematics' counterpart of the materials science notion of the total *Frank vector* in a crystalline material, whereby the total Frank vector corresponds to the resultant orientation gap of tangents to a (finite) Burgers circuit. These kind of rotational defects have been observed already early in a variety of materials such as liquid crystals, superconductors of type two, and polymers, see [63, 198, 114] and the review [163]. The distortion closure gap may thus be re-interpreted as the integral over its *distortion closure gap density* per unit area (that may also be denoted as *Frank tensor density*), see Fig. 1.10

$$[\![\boldsymbol{\beta}]\!] \doteq \int_{\mathcal{A}} \mathbb{f}\, da, \tag{1.68}$$

whereby the Frank tensor density $\mathbb{f}$ is computed from the (generalized) third-order *disclination density tensor* $\boldsymbol{\vartheta}$ in the spirit of a Cauchy theorem

$$\mathbb{f} := \boldsymbol{\vartheta} \cdot \boldsymbol{n}. \tag{1.69}$$

Thereby, the (generalized) disclination density tensor follows from the comparison with Eq. 1.60 as

$$\boldsymbol{\vartheta} := \operatorname{curl} \boldsymbol{\gamma} = -\gamma_{abc,d} e_{cde} \boldsymbol{e}_a \otimes \boldsymbol{e}_b \otimes \boldsymbol{e}_e. \tag{1.70}$$

Thus identifying the distortion closure gap with the integral of its density renders the definition of the (generalized) disclination density tensor in terms of the double-distortion. In the differential geometry setting discussed in the subsequent parts of this treatise the disclination density tensor will be identified (modulo further terms that arise due to a geometrically nonlinear description) with the so-called *curvature tensor*.

Remark 1.6. Observe that the disclination density tensor satisfies the following differential identity

$$\operatorname{div} \boldsymbol{\vartheta} = \boldsymbol{0}. \tag{1.71}$$

The above holds trivially true due to $\vartheta_{abe,e} = -\gamma_{abc,de} e_{cde} \equiv 0$. □

Remark 1.7. In most of the literature (see, e.g., [7]) the distortion $\boldsymbol{\beta} = \boldsymbol{\epsilon} + \boldsymbol{\omega}$ is decomposed into its symmetric and its skew-symmetric contribution, i.e. into the strain $\boldsymbol{\epsilon} := \boldsymbol{\beta}^{\text{sym}}$ and the (continuum) rotation $\boldsymbol{\omega} := \boldsymbol{\beta}^{\text{skw}}$, respectively. Then for the definition of the double-distortion only the skew-symmetric part of the distortion is considered, i.e. $\boldsymbol{\gamma} := \operatorname{grad} \boldsymbol{\omega}$. Since $\boldsymbol{\omega} = -\boldsymbol{e} \cdot \boldsymbol{\omega}^{\times}$ is skew-symmetric it may be substituted by its axial vector $\boldsymbol{\omega}^{\times}$, likewise the so defined double-distortion $\boldsymbol{\gamma} = -\boldsymbol{e} \cdot \boldsymbol{\gamma}^{\times}$ is (left) skew-symmetric and may thus be substituted by the continuum curvature $\boldsymbol{\kappa} := \boldsymbol{\gamma}^{\times} = \operatorname{grad} \boldsymbol{\omega}^{\times}$. Based on these prerequisites the secondary dislocation density (or rather the torsion tensor, see below) follows as ${}^{\text{II}}\boldsymbol{\alpha} := \operatorname{curl} \boldsymbol{\epsilon} - \boldsymbol{\gamma} : \boldsymbol{e} = \operatorname{curl} \boldsymbol{\epsilon} + [\boldsymbol{\kappa}^t : \boldsymbol{1}]\boldsymbol{1} - \boldsymbol{\kappa}^t$, whereas the (left) skew-symmetric disclination density is introduced as $\boldsymbol{\vartheta} := \operatorname{curl} \boldsymbol{\gamma}$ with $\boldsymbol{\vartheta} = -\boldsymbol{e} \cdot \boldsymbol{\vartheta}^{\times}$. The latter may likewise be substituted by the disclination density (second-order) tensor $\boldsymbol{\theta} := \boldsymbol{\vartheta}^\times = \operatorname{curl}\boldsymbol{\gamma}^\times = \operatorname{curl}\boldsymbol{\kappa}$. These translational and rotational defect densities satisfy jointly the following differential identities

$$\operatorname{div}{}^{\mathrm{II}}\boldsymbol{\alpha} - \boldsymbol{\vartheta} : \mathbf{1} = \operatorname{div}{}^{\mathrm{II}}\boldsymbol{\alpha} - \boldsymbol{e} : \boldsymbol{\theta} = \mathbf{0} \quad \text{and} \quad \operatorname{div}\boldsymbol{\vartheta} = -\boldsymbol{e} \cdot \operatorname{div}\boldsymbol{\theta} = \mathbf{0}.$$

Obviously, for $\boldsymbol{\vartheta} = \mathbf{0}$ the double-distortion $\boldsymbol{\gamma} = \operatorname{grad}\boldsymbol{\omega}^*$ (and thus the continuum curvature $\boldsymbol{\kappa}$) follows as the gradient of a (continuum) rotation $\boldsymbol{\omega}^*$. Note that for given $\boldsymbol{\omega}$ an arbitrary $\boldsymbol{\omega}^* - \boldsymbol{\omega}$ remains undetermined. Thus if the integrability of the double-distortion (and thus the continuum curvature) into a particular (continuum) rotation shall be enforced, the additional constraint $\boldsymbol{\lambda} = \boldsymbol{\gamma} - \operatorname{grad}\boldsymbol{\omega} = \mathbf{0}$ has to be considered. If in addition ${}^{\mathrm{II}}\boldsymbol{\alpha} = \mathbf{0}$ holds with ${}^{\mathrm{II}}\boldsymbol{\alpha} = \operatorname{curl}(\boldsymbol{\epsilon} + \boldsymbol{\omega}^*) = \operatorname{curl}\boldsymbol{\beta}^*$ due to $\boldsymbol{\vartheta} = \mathbf{0}$, the tensor field $\boldsymbol{\beta}^* = \operatorname{grad}\boldsymbol{u}^{**}$ follows as the gradient of a vector field $\boldsymbol{u}^{**}$. □

Remark 1.8. In Cartesian coordinates the coefficients θ_{ab} of the disclination density tensor decompose into the density of *wedge disclinations* corresponding to rotational/second kind *Volterra dislocations* (of type six) as given by the diagonal terms $\theta_{ab}, a = b$

$$\begin{bmatrix} \bullet & \circ & \circ \\ \circ & \bullet & \circ \\ \circ & \circ & \bullet \end{bmatrix},$$

and the density of *twist disclinations* corresponding to rotational/second kind *Volterra dislocations* (of type four and five) as given by the off-diagonal terms $\theta_{ab}, a \neq b$

$$\begin{bmatrix} \circ & \bullet & \bullet \\ \bullet & \circ & \bullet \\ \bullet & \bullet & \circ \end{bmatrix}.$$

For twist dislocations the Frank vector is perpendicular to the *disclination line* whereas for wedge disclinations the Frank vector is parallel to the disclination line, see also the extensive discussion in [97]. Thus the columns of θ_{ab} denote the coefficients of the Frank vector $\mathbb{f}_b = \theta_{ab}\boldsymbol{e}_a$ penetrating a *unit* surface area with orientation given by its surface normal $\boldsymbol{e}_b$. Here $\boldsymbol{e}_b$ may be thought of as being tangent to the disclination lines penetrating the surface, thus the normal part of $\mathbb{f}_b$, i.e. $\boldsymbol{e}_a \cdot \mathbb{f}_b, a = b$ denotes the corresponding wedge disclinations whereas the two tangent parts of $\mathbb{f}_b$, i.e. $\boldsymbol{e}_a \cdot \mathbb{f}_b, a \neq b$ represent the corresponding twist disclinations. Finally, for a *single* disclination the disclination density tensor may be regarded a *distribution*

$$\boldsymbol{\theta} = \delta(\boldsymbol{x})[\![\boldsymbol{\omega}^\times]\!] \otimes \mathbb{t}$$

in terms of the Dirac-delta $\delta = \delta(\boldsymbol{x})$ distribution and the unit tangent $\mathbb{t}$ to the corresponding disclination line. □

Primary Dislocation Density Tensor. The *primary dislocation density tensor* is defined as before in terms of the distortion

$$^{\mathrm{I}}\boldsymbol{\alpha} := \operatorname{curl}\boldsymbol{\beta} = -\beta_{ab,c} e_{bcd} \boldsymbol{e}_a \otimes \boldsymbol{e}_d. \tag{1.72}$$

Note that a vanishing primary dislocation density tensor $^{\mathrm{I}}\boldsymbol{\alpha} = \mathbf{0}$ agrees with the previously stated constraint $\beta_{a[b,c]} \doteq 0$. In the differential geometry setting discussed in the subsequent parts of this treatise the primary dislocation density tensor will be identified with the so-called *anholonomic object.*

Point-Defect Density Tensor. Furthermore, the additional integrability condition $\lambda_{(ab)c} \doteq 0$ motivates the introduction of the *point-defect density tensor* in the format

$$\boldsymbol{\pi} := \mathbf{i}^{\mathrm{sym}} : \boldsymbol{\lambda} = \lambda_{(ab)c} \boldsymbol{e}_a \otimes \boldsymbol{e}_b \otimes \boldsymbol{e}_c. \tag{1.73}$$

Here $\mathbf{i}^{\mathrm{sym}} = 1/2[\delta_{ac}\delta_{bd} + \delta_{ad}\delta_{bc}]\boldsymbol{e}_a \otimes \boldsymbol{e}_b \otimes \boldsymbol{e}_c \otimes \boldsymbol{e}_d$ denotes the fourth-order symmetric unit tensor. Typical examples for point-defects are lattice vacancies, interstitial atoms and the-like. In the differential geometry setting discussed in the subsequent parts of this treatise the point-defect density tensor will be identified with the so-called *non-metricity tensor.*

Secondary Dislocation Density Tensor. Finally, the remaining integrability condition $\boldsymbol{\gamma} : \boldsymbol{e} \doteq \mathbf{0}$ motivates the introduction of the *secondary dislocation density tensor* in the format

$$^{\mathrm{II}}\boldsymbol{\alpha} := -\boldsymbol{\gamma} : \boldsymbol{e} = -\gamma_{abc} e_{bcd} \boldsymbol{e}_a \otimes \boldsymbol{e}_d. \tag{1.74}$$

Note that a vanishing secondary dislocation density tensor $^{\mathrm{II}}\boldsymbol{\alpha} = \mathbf{0}$ in addition to a vanishing primary dislocation density tensor $^{\mathrm{I}}\boldsymbol{\alpha} = \mathbf{0}$ agrees with the previously stated constraint $\lambda_{a[bc]} \doteq 0$. In the differential geometry setting discussed in the subsequent parts of this treatise the secondary dislocation density tensor will be identified with the so-called *torsion tensor.*

Remark 1.9. Observe that the secondary dislocation density tensor and the disclination density tensor satisfy jointly the following differential identities

$$\operatorname{div}{}^{\mathrm{II}}\boldsymbol{\alpha} - \boldsymbol{\vartheta} : \mathbf{1} = \mathbf{0} \quad \text{and} \quad \operatorname{div}\boldsymbol{\vartheta} = \mathbf{0}. \tag{1.75}$$

The above holds trivially true due to $^{\mathrm{II}}\alpha_{ad,d} = -\gamma_{abc,d} e_{bcd} = \vartheta_{abb}$. □

1.3.3 Kinematics of Crystal Plasticity

For crystal plasticity in a second-order continuum the total distortion and the total double-distortion are likewise additively decomposed into plastic and elastic contributions

$$\operatorname{grad}\boldsymbol{u} = \widetilde{\boldsymbol{\beta}} + \overline{\boldsymbol{\beta}} \quad \text{and} \quad \operatorname{grad}\boldsymbol{\beta} = \widetilde{\boldsymbol{\gamma}} + \overline{\boldsymbol{\gamma}}. \tag{1.76}$$

Accordingly the primary plastic and elastic dislocation density tensors ${}^{\mathrm{I}}\widetilde{\boldsymbol{\alpha}}$, ${}^{\mathrm{I}}\overline{\boldsymbol{\alpha}}$, respectively, and the plastic and elastic disclination density tensors $\widetilde{\boldsymbol{\vartheta}}$, $\overline{\boldsymbol{\vartheta}}$, respectively, follow as

$$ {}^{\mathrm{I}}\widetilde{\boldsymbol{\alpha}} = \operatorname{curl}\widetilde{\boldsymbol{\beta}}, \quad {}^{\mathrm{I}}\overline{\boldsymbol{\alpha}} = \operatorname{curl}\overline{\boldsymbol{\beta}} \quad \text{and} \quad \widetilde{\boldsymbol{\vartheta}} = \operatorname{curl}\widetilde{\boldsymbol{\gamma}}, \quad \overline{\boldsymbol{\vartheta}} = \operatorname{curl}\overline{\boldsymbol{\gamma}}. \tag{1.77} $$

Likewise, the plastic and elastic point-defect density tensors $\widetilde{\boldsymbol{\pi}}$ and $\overline{\boldsymbol{\pi}}$, respectively, and the secondary plastic and elastic dislocation density tensors ${}^{\mathrm{II}}\widetilde{\boldsymbol{\alpha}}$ and ${}^{\mathrm{II}}\overline{\boldsymbol{\alpha}}$, respectively, read as

$$ \widetilde{\boldsymbol{\pi}} = \mathsf{i}^{\mathrm{sym}} : \widetilde{\boldsymbol{\lambda}}, \quad \overline{\boldsymbol{\pi}} = \mathsf{i}^{\mathrm{sym}} : \overline{\boldsymbol{\lambda}} \quad \text{and} \quad {}^{\mathrm{II}}\widetilde{\boldsymbol{\alpha}} = -\widetilde{\boldsymbol{\gamma}} : \boldsymbol{e}, \quad {}^{\mathrm{II}}\overline{\boldsymbol{\alpha}} = -\overline{\boldsymbol{\gamma}} : \boldsymbol{e}. \tag{1.78} $$

Based on the primary and secondary plastic and elastic dislocation, disclination and point-defect density tensors the conditions for the integrability of the total distortion and double-distortion read as

$$ {}^{\mathrm{I}}\widetilde{\boldsymbol{\alpha}} + {}^{\mathrm{I}}\overline{\boldsymbol{\alpha}} = \mathbf{0}, \quad \widetilde{\boldsymbol{\vartheta}} + \overline{\boldsymbol{\vartheta}} = \mathbf{0}, \quad \widetilde{\boldsymbol{\pi}} + \overline{\boldsymbol{\pi}} = \mathbf{0} \quad \text{and} \quad {}^{\mathrm{II}}\widetilde{\boldsymbol{\alpha}} + {}^{\mathrm{II}}\overline{\boldsymbol{\alpha}} = \mathbf{0}. \tag{1.79} $$

Finally, as a consequence of their definitions, e.g. in terms of the curl operator, the so-called *continuity equations* for the secondary plastic and elastic dislocation density tensors and the plastic and elastic disclination density tensors follow trivially as

$$ \begin{aligned} \operatorname{div} {}^{\mathrm{II}}\widetilde{\boldsymbol{\alpha}} - \widetilde{\boldsymbol{\vartheta}} : \mathbf{1} = \mathbf{0} \quad &\text{and} \quad \operatorname{div}\widetilde{\boldsymbol{\vartheta}} = \mathbf{0}, \\ \operatorname{div} {}^{\mathrm{II}}\overline{\boldsymbol{\alpha}} - \overline{\boldsymbol{\vartheta}} : \mathbf{1} = \mathbf{0} \quad &\text{and} \quad \operatorname{div}\overline{\boldsymbol{\vartheta}} = \mathbf{0}. \end{aligned} $$

A detailed discussion of these continuity equations may be found, e.g., in [7].

1.3.4 Generalized Crystal Plasticity

The key idea to formulate a physically motivated theory of *generalized crystal plasticity in a second-order continuum* is to observe that all *geometrically necessary defect densities*, i.e. the primary and secondary dislocation, disclination and point-defect densities, contribute to the hardening of the material. Thus the incorporation of the plastic defect densities into the free energy density emerges as a natural concept

$$ \Psi = \Psi^{\mathrm{mac}}(\boldsymbol{\epsilon} - \widetilde{\boldsymbol{\epsilon}}, \boldsymbol{\gamma} - \widetilde{\boldsymbol{\gamma}}) + \Psi^{\mathrm{mic}}(\kappa) + \Psi^{\mathrm{def}}({}^{\mathrm{I}}\widetilde{\boldsymbol{\alpha}}, {}^{\mathrm{II}}\widetilde{\boldsymbol{\alpha}}, \widetilde{\boldsymbol{\vartheta}}, \widetilde{\boldsymbol{\pi}}). \tag{1.80} $$

Here the macroscopic part Ψ^{mac} of the free energy density depends on the elastic strain (symmetric distortion) and double-distortion as given by $\boldsymbol{\epsilon} - \widetilde{\boldsymbol{\epsilon}}$ and $\boldsymbol{\gamma} - \widetilde{\boldsymbol{\gamma}}$, respectively, the microscopic part Ψ^{mic} captures again the hardening due to *statistically stored defects* in terms of a single scalar hardening variable κ (for simplicity), and the defect part Ψ^{def} now captures the hardening due to all geometrically necessary defects (primary and secondary dislocations, disclinations, point-defects). Then, according to [53, 158] the Clausius-Duhem (dissipation power density) inequality is formulated including the non-locality residual $\mathcal{R}$

$$\mathcal{D} = \boldsymbol{\sigma} : \partial_t \boldsymbol{\epsilon} + \boldsymbol{\sigma} \cdot\!: \partial_t \boldsymbol{\gamma} - \partial_t \Psi + \mathcal{R} \geq 0. \tag{1.81}$$

The standard Coleman-Noll exploitation then renders firstly the stress and double-stress as

$$\boldsymbol{\sigma} := \frac{\partial \Psi}{\partial \boldsymbol{\epsilon}} \quad \text{and} \quad \boldsymbol{\sigma} := \frac{\partial \Psi}{\partial \boldsymbol{\gamma}}. \tag{1.82}$$

Observe that the (Cauchy) stress $\boldsymbol{\sigma}$ is conjugated to the strain and is thus symmetric. Secondly, the *thermo-dynamically driving forces* are obtained as conjugated to the plastic variables. These are the *plastic stress* and *plastic double-stress* (that agree here identically with the equilibrium stress and double stress)

$$\tilde{\boldsymbol{\sigma}} := -\frac{\partial \Psi}{\partial \tilde{\boldsymbol{\epsilon}}} \equiv \boldsymbol{\sigma} \quad \text{and} \quad \tilde{\boldsymbol{\sigma}} := -\frac{\partial \Psi}{\partial \tilde{\boldsymbol{\gamma}}} \equiv \boldsymbol{\sigma} \tag{1.83}$$

together with the so-called *drag stress*

$$H := \frac{\partial \Psi}{\partial \kappa}. \tag{1.84}$$

Moreover, the primary and secondary dislocation densities are conjugated to the *primary* and *secondary dislocation stress*, respectively, as

$${}^{\mathrm{I}}\boldsymbol{\sigma}^{\alpha} := \frac{\partial \Psi}{\partial\, {}^{\mathrm{I}}\tilde{\boldsymbol{\alpha}}} \quad \text{and} \quad {}^{\mathrm{II}}\boldsymbol{\sigma}^{\alpha} := \frac{\partial \Psi}{\partial\, {}^{\mathrm{II}}\tilde{\boldsymbol{\alpha}}}. \tag{1.85}$$

Finally the disclination and point-defect densities are conjugated to the *disclination double-stress* and the (left) symmetric *point-defect double-stress*, respectively, as

$$\boldsymbol{\sigma}^{\vartheta} := \frac{\partial \Psi}{\partial \tilde{\boldsymbol{\vartheta}}} \quad \text{and} \quad \boldsymbol{\sigma}^{\pi} := \frac{\partial \Psi}{\partial \tilde{\boldsymbol{\pi}}}. \tag{1.86}$$

Thus the *reduced* dissipation power density reads

$$\begin{aligned} \mathcal{D} = {} & \boldsymbol{\sigma} : \partial_t \tilde{\boldsymbol{\beta}} + \boldsymbol{\sigma} \cdot\!: \partial_t \tilde{\boldsymbol{\gamma}} - H \partial_t \kappa \\ & - {}^{\mathrm{I}}\boldsymbol{\sigma}^{\alpha} : \partial_t\, {}^{\mathrm{I}}\tilde{\boldsymbol{\alpha}} - {}^{\mathrm{II}}\boldsymbol{\sigma}^{\alpha} : \partial_t\, {}^{\mathrm{II}}\tilde{\boldsymbol{\alpha}} - \boldsymbol{\sigma}^{\vartheta} \cdot\!: \partial_t \tilde{\boldsymbol{\vartheta}} - \boldsymbol{\sigma}^{\pi} \cdot\!: \partial_t \tilde{\boldsymbol{\pi}} + \mathcal{R} \geq 0. \end{aligned} \tag{1.87}$$

Expressing the rates of the plastic defect densities in terms of the plastic distortion and double-distortion and applying the product rule of differentiation then results in the abbreviated bi-linear expression

$$\mathcal{D} = \boldsymbol{\sigma}^{\star} : \partial_t \tilde{\boldsymbol{\beta}} + \boldsymbol{\sigma}^{\star} \cdot\!: \partial_t \tilde{\boldsymbol{\gamma}} - H \partial_t \kappa \geq 0. \tag{1.88}$$

Here, the *generalized* driving stress and double-stress are defined as

$$\boldsymbol{\sigma}^\star := [\boldsymbol{\sigma} - \operatorname{div}\boldsymbol{\sigma}^\pi] - \operatorname{curl}{}^{\mathrm{I}}\boldsymbol{\sigma}^\alpha, \qquad (1.89)$$
$$\boldsymbol{\sigma}^\star := [\boldsymbol{\sigma} - \boldsymbol{\sigma}^\pi + {}^{\mathrm{II}}\boldsymbol{\sigma}^\alpha \cdot \boldsymbol{e}] - \operatorname{curl}\boldsymbol{\sigma}^\vartheta.$$

Note that the defect-induced stresses act collectively as a kind of *back-stress.* Moreover the non-locality residual is defined in the format

$$\mathcal{R} := -\operatorname{div}\left(\partial_t\widetilde{\boldsymbol{\beta}} : [{}^{\mathrm{I}}\boldsymbol{\sigma}^\alpha \cdot \boldsymbol{e} + \boldsymbol{\sigma}^\pi] + \partial_t\widetilde{\boldsymbol{\gamma}} \,\cdot\!\!:\, [\boldsymbol{\sigma}^\vartheta \cdot \boldsymbol{e}]\right). \qquad (1.90)$$

Proof. The term involving the disclination double-stress expands into

$$-\boldsymbol{\sigma}^\vartheta \,\cdot\!\!:\, \partial_t\widetilde{\boldsymbol{\vartheta}} \equiv -\boldsymbol{\sigma}^\vartheta \,\cdot\!\!:\, \operatorname{curl}\partial_t\widetilde{\boldsymbol{\gamma}} = -\operatorname{curl}\boldsymbol{\sigma}^\vartheta \,\cdot\!\!:\, \partial_t\widetilde{\boldsymbol{\gamma}} + \operatorname{div}\left(\partial_t\widetilde{\boldsymbol{\gamma}} \,\cdot\!\!:\, [\boldsymbol{\sigma}^\vartheta \cdot \boldsymbol{e}]\right).$$

Likewise, in index notation this expression reads for Cartesian coordinates

$$\sigma^\vartheta_{abe}[\partial_t\widetilde{\gamma}_{abc,d}e_{cde}] = [\sigma^\vartheta_{abe,d}e_{edc}]\partial_t\widetilde{\gamma}_{abc} + \left[\partial_t\widetilde{\gamma}_{abc}[\sigma^\vartheta_{abe}e_{ecd}]\right]_{,d}.$$

The term involving the point-defect double stress is re-expressed as

$$-\boldsymbol{\sigma}^\pi \,\cdot\!\!:\, \partial_t\widetilde{\boldsymbol{\pi}} \equiv -\boldsymbol{\sigma}^\pi \,\cdot\!\!:\, \left[\mathbf{i}^{\mathrm{sym}} : [\partial_t\widetilde{\boldsymbol{\gamma}} - \operatorname{grad}\partial_t\widetilde{\boldsymbol{\beta}}]\right]$$
$$= -\boldsymbol{\sigma}^\pi \,\cdot\!\!:\, \partial_t\widetilde{\boldsymbol{\gamma}} - \partial_t\widetilde{\boldsymbol{\beta}} : \operatorname{div}\boldsymbol{\sigma}^\pi + \operatorname{div}\left(\partial_t\widetilde{\boldsymbol{\beta}} : \boldsymbol{\sigma}^\pi\right).$$

In index notation this expression reads for Cartesian coordinates

$$-\sigma^\pi_{(ab)c}[\partial_t\widetilde{\gamma}_{abc} - \partial_t\beta_{ab,c}] = -\sigma^\pi_{(ab)c}\partial_t\widetilde{\gamma}_{abc} - \partial_t\widetilde{\beta}_{ab}\sigma^\pi_{(ab)c,c} + \left[\partial_t\widetilde{\beta}_{ab}\sigma^\pi_{(ab)c}\right]_{,c}.$$

Finally, the term involving the secondary dislocation stress re-writes simply as

$$-{}^{\mathrm{II}}\boldsymbol{\sigma}^\alpha : \partial_t\widetilde{\boldsymbol{\alpha}} \equiv {}^{\mathrm{II}}\boldsymbol{\sigma}^\alpha : [\partial_t\widetilde{\boldsymbol{\gamma}} : \boldsymbol{e}] = [{}^{\mathrm{II}}\boldsymbol{\sigma}^\alpha \cdot \boldsymbol{e}] \,\cdot\!\!:\, \partial_t\widetilde{\boldsymbol{\gamma}}.$$

The index notation in Cartesian coordinates follows as

$${}^{\mathrm{II}}\sigma^\alpha_{ad}[\partial_t\widetilde{\gamma}_{abc}e_{bcd}] = [{}^{\mathrm{II}}\sigma^\alpha_{ad}e_{dbc}]\partial_t\widetilde{\gamma}_{abc}.$$

Obviously, the non-locality residual has to cancel all the divergence terms on the right-hand-sides. □

Then considering the so-called *insulation condition*, see [158, 159], identifies suited boundary conditions for the plastic distortion and double-distortion

$$\int_{\mathcal{B}^{\mathrm{p}}} \mathcal{R}\,\mathrm{d}v = \int_{\partial\mathcal{B}^{\mathrm{p}}} \left[\partial_t\widetilde{\boldsymbol{\beta}} : [{}^{\mathrm{I}}\boldsymbol{\sigma}^\alpha \cdot \widehat{\boldsymbol{n}} - \boldsymbol{\sigma}^\pi \cdot \boldsymbol{n}] + \partial_t\widetilde{\boldsymbol{\gamma}} \,\cdot\!\!:\, [\boldsymbol{\sigma}^\vartheta \cdot \widehat{\boldsymbol{n}}]\right] \mathrm{d}a \doteq 0. \qquad (1.91)$$

Here $\widehat{\boldsymbol{n}} := -\boldsymbol{e} \cdot \boldsymbol{n}$ with $\boldsymbol{n}$ the outwards pointing normal to the elastic-plastic boundary. The assumption of homogeneous Neumann-type boundary conditions for ${}^{\mathrm{I}}\boldsymbol{\sigma}^\alpha \cdot \widehat{\boldsymbol{n}}$, $\boldsymbol{\sigma}^\pi \cdot \boldsymbol{n}$ and $\boldsymbol{\sigma}^\vartheta \cdot \widehat{\boldsymbol{n}}$ is a valid choice for enforcing the insulation condition identically, for a more detailed discussion please refer to [158, 159]. Then a multi-surface yield condition in terms of the generalized driving stress

$\boldsymbol{\sigma}^\star$ and double-stress $\boldsymbol{\sigma}^\star$ together with the drag stress H may be introduced for each slip system ς as

$$\phi_\varsigma(\boldsymbol{\sigma}^\star, \boldsymbol{\sigma}^\star, H) = \varphi_\varsigma(\boldsymbol{\sigma}^\star, \boldsymbol{\sigma}^\star) - [\sigma_y + H] \le 0. \tag{1.92}$$

The postulate of maximum dissipation results in a minimization problem with inequality constraints that are enforced by Lagrange or rather plastic multipliers γ_ς, i.e. $\mathcal{L} := -\mathcal{D} + \sum_\varsigma \gamma_\varsigma \phi_\varsigma \to \min$. Consequently the associated evolution rules for the plastic distortion and plastic double-distortion together with the scalar hardening variable follow as

$$\partial_t \widetilde{\boldsymbol{\beta}} = \sum_\varsigma \gamma_\varsigma \frac{\partial \varphi_\varsigma}{\partial \boldsymbol{\sigma}^\star}, \quad \partial_t \widetilde{\boldsymbol{\gamma}} = \sum_\varsigma \gamma_\varsigma \frac{\partial \varphi_\varsigma}{\partial \boldsymbol{\sigma}^\star} \quad \text{and} \quad \partial_t \kappa = \sum_\varsigma \gamma_\varsigma. \tag{1.93}$$

The simplest possibility to express the contribution of the plastic defect density to the free energy density is a quadratic dependence

$$\begin{aligned} \Psi^{\text{def}} :=& \frac{1}{2} G\, {}^{\text{I}}\ell_\alpha^2\, {}^{\text{I}}\widetilde{\boldsymbol{\alpha}} : {}^{\text{I}}\widetilde{\boldsymbol{\alpha}} + \frac{1}{2} G\, {}^{\text{II}}\ell_\alpha^2\, {}^{\text{II}}\widetilde{\boldsymbol{\alpha}} : {}^{\text{II}}\widetilde{\boldsymbol{\alpha}} \\ &+ \frac{1}{2} G \ell_\vartheta^4 \widetilde{\boldsymbol{\vartheta}} \cdot\!: \widetilde{\boldsymbol{\vartheta}} + \frac{1}{2} K \ell_\pi^2 \widetilde{\boldsymbol{\pi}} \cdot\!: \widetilde{\boldsymbol{\pi}}. \end{aligned} \tag{1.94}$$

Here G and K denote the shear and bulk modulus, respectively, and the $\ell_\bullet$'s (including the ${}^{\text{I}}\ell_\alpha$ and ${}^{\text{II}}\ell_\alpha$) are (in general different) characteristic lengths that have to be introduced for dimensional consistency. Then the defect-induced stresses are computed as

$$\begin{aligned} {}^{\text{I}}\boldsymbol{\sigma}^\alpha &= G\, {}^{\text{I}}\ell_\alpha^2 \operatorname{curl} \widetilde{\boldsymbol{\beta}}, \\ -\, {}^{\text{II}}\boldsymbol{\sigma}^\alpha &= G\, {}^{\text{II}}\ell_\alpha^2 \widetilde{\boldsymbol{\gamma}} : \boldsymbol{e}, \\ \boldsymbol{\sigma}^\vartheta &= G \ell_\vartheta^4 \operatorname{curl} \widetilde{\boldsymbol{\gamma}}, \\ \boldsymbol{\sigma}^\pi &= K \ell_\pi^2 \mathsf{i}^{\text{sym}} : [\widetilde{\boldsymbol{\gamma}} - \operatorname{grad} \widetilde{\boldsymbol{\beta}}]. \end{aligned} \tag{1.95}$$

Moreover, the generalized driving stress and double-stress then follow partly by applying the curl operator to the plastic distortion and double-distortion, respectively, twice (note again that this is *not* the same as the *incompatibility operator* inc, that denotes twice the application of the *transposed* curl operator, see [94])

$$\begin{aligned} \boldsymbol{\sigma}^\star &= \big[\boldsymbol{\sigma} - K \ell_\pi^2 \mathsf{i}^{\text{sym}} : [\operatorname{div} \widetilde{\boldsymbol{\gamma}} - \Delta \widetilde{\boldsymbol{\beta}}]\big] - G\, {}^{\text{I}}\ell_\alpha^2 \operatorname{curl}(\operatorname{curl} \widetilde{\boldsymbol{\beta}}) \\ \boldsymbol{\sigma}^\star &= \big[\boldsymbol{\sigma} - K \ell_\pi^2 \mathsf{i}^{\text{sym}} : [\widetilde{\boldsymbol{\gamma}} - \operatorname{grad} \widetilde{\boldsymbol{\beta}}] - G\, {}^{\text{II}}\ell_\alpha^2 \widetilde{\boldsymbol{\gamma}} : \mathsf{i}^{\text{skw}}\big] - G \ell_\vartheta^4 \operatorname{curl}(\operatorname{curl} \widetilde{\boldsymbol{\gamma}}). \end{aligned} \tag{1.96}$$

Here $\mathsf{i}^{\text{skw}} = 1/2[\delta_{ac}\delta_{bd} - \delta_{ad}\delta_{bc}]\boldsymbol{e}_a \otimes \boldsymbol{e}_b \otimes \boldsymbol{e}_c \otimes \boldsymbol{e}_d$ denotes the fourth-order skew-symmetric unit tensor. The simplest φ_ς coincides with a *generalized Schmid stress* $\sigma_\varsigma^\star$ and is formulated in terms of the Schmid tensors $\boldsymbol{s}_\varsigma \otimes \boldsymbol{m}_\varsigma$ and (as just one possible example) $\boldsymbol{s}_\varsigma \otimes \boldsymbol{n}_\varsigma \otimes \boldsymbol{m}_\varsigma$ for each slip system ς

$$\varphi_\varsigma \equiv \sigma_\varsigma^\star = \boldsymbol{\sigma}^\star : [\boldsymbol{s}_\varsigma \otimes \boldsymbol{m}_\varsigma] + \ell_\varsigma^{-1} \boldsymbol{\sigma}^\star \mathbin{\vdots} [\boldsymbol{s}_\varsigma \otimes \boldsymbol{n}_\varsigma \otimes \boldsymbol{m}_\varsigma]. \tag{1.97}$$

The vectors $\boldsymbol{s}_\varsigma$, $\boldsymbol{m}_\varsigma$, and $\boldsymbol{n}_\varsigma$ denote the slip direction, the slip plane normal, and the slip plane bi-normal of slip system ς, respectively, ℓ_ς is another characteristic length that has to be introduced again for dimensional consistency. Thus the generalized Schmid stress is motivated as a driving force for plastic evolution on slip system ς with

$$\begin{aligned} \sigma_\varsigma^\star = {} & \sigma_\varsigma \\ & - K\ell_\pi^2 [\boldsymbol{s}_\varsigma \otimes \boldsymbol{m}_\varsigma]^{\text{sym}} : [\operatorname{div} \widetilde{\boldsymbol{\gamma}} - \Delta \widetilde{\boldsymbol{\beta}}] \\ & - G^{\text{I}} \ell_\alpha^2 \boldsymbol{s}_\varsigma \cdot \operatorname{curl}(\operatorname{curl} \widetilde{\boldsymbol{\beta}}) \cdot \boldsymbol{m}_\varsigma \\ & + \sigma_\varsigma \\ & - K\ell_\pi^2 \ell_\varsigma^{-1} [\boldsymbol{s}_\varsigma \otimes \boldsymbol{n}_\varsigma]^{\text{sym}} : [\widetilde{\boldsymbol{\gamma}} - \operatorname{grad} \widetilde{\boldsymbol{\beta}}] \cdot \boldsymbol{m}_\varsigma \\ & - G^{\text{II}} \ell_\alpha^2 \ell_\varsigma^{-1} \boldsymbol{s}_\varsigma \cdot \widetilde{\boldsymbol{\gamma}} : [\boldsymbol{n}_\varsigma \otimes \boldsymbol{m}_\varsigma]^{\text{skw}} \\ & - G\ell_\vartheta^4 \ell_\varsigma^{-1} \boldsymbol{s}_\varsigma \cdot \operatorname{curl}(\operatorname{curl} \widetilde{\boldsymbol{\gamma}}) : [\boldsymbol{n}_\varsigma \otimes \boldsymbol{m}_\varsigma]. \end{aligned} \tag{1.98}$$

Here $\sigma_\varsigma := \boldsymbol{s}_\varsigma \cdot \boldsymbol{\sigma} \cdot \boldsymbol{m}_\varsigma$ denotes the common Schmid stress, i.e. the ordinary resolved shear stress at slip system ς, likewise $\sigma_\varsigma := \ell_\varsigma^{-1} \boldsymbol{s}_\varsigma \cdot \boldsymbol{\sigma} : [\boldsymbol{n}_\varsigma \otimes \boldsymbol{m}_\varsigma]$ is the contribution (as yet one example of various possibilities) to the resolved shear stress at slip system ς due to the double-stress. Interestingly, in this formulation the evolution law for the plastic distortion and double-distortion or rather the flow rule is unaffected by the generalization to incorporate the dependence on the defect densities

$$\partial_t \widetilde{\boldsymbol{\beta}} = \sum_\varsigma \gamma_\varsigma \boldsymbol{s}_\varsigma \otimes \boldsymbol{m}_\varsigma \quad \text{and} \quad \partial_t \widetilde{\boldsymbol{\gamma}} = \sum_\varsigma \gamma_\varsigma \ell_\varsigma^{-1} \boldsymbol{s}_\varsigma \otimes \boldsymbol{n}_\varsigma \otimes \boldsymbol{m}_\varsigma. \tag{1.99}$$

It shall be noted that the characteristic lengths $\ell_\bullet$ have been incorporated into the formulation due to the consideration of the plastic defect densities. Moreover, second derivatives of the plastic distortion and double-distortion contribute to the generalized Schmid stress that serves as the driving force for inelastic evolution. These are the typical ingredients of a generalized plasticity formulation that has here been obtained based on the physically motivated concept of geometrically necessary defect densities.

Part II
Differential Geometry

...gives a comprehensive and self-contained exposition of the relevant basics of differential geometry. Thereby most of the extensive presentation relies on heavy but easy-to-follow index notation, however the relation to the more modern but also more abstract coordinate-free invariant formulation and exterior calculus is likewise highlighted.

Chapter 2
Preliminaries

DIE WISSENSCHAFT

So beschließen beide denn
nach so manchem Doch und Wenn,

sich mit ihren Theorien
vor die Wissenschaft zu knien.

Doch die Wissenschaft, man weiß es,
achtet nicht des Laienfleißes.

Hier auch schürzt sie nur den Mund,
murmelt von 'Phantasmen' und

beugt sich wieder dann auf ihre
wichtigen Spezialpapiere.

"Komm," spricht Palmström, "Kamerad, -
alles Feinste bleibt - privat!"

Christian Morgenstern

Abstract. The treatment of geometry goes back to the ancient Greek, thereby the findings documented in the 13 volumes of Euclid's Elements defined the state of affairs for some 2000 years. The advent of differential geometry is associated with the Habilitation lecture of Riemann in 1854. Its further development enabled and cumulated in the formulation of Einstein's Theory of General Relativity/Gravitation some sixty years later in 1915. However the necessity for a non-Euclidean geometry may be motivated already from simply considering the failure of some of the corner stones in Euclidean geometry, for example the parallel axiom, on a two-dimensional curved manifold such as a sphere. Differential manifolds may be classified in terms of the two fundamental quantities connection and metric that in turn give rise to the three most essential tensors in differential geometry describing torsion, curvature, and non-metricity.

2.1 History of Differential Geometry

In the following short bio-sketches of some of the most important protagonists in the history of differential geometry are given. Thereby, necessarily, neither the selection of these protagonists nor the representation of their achievements and the account of their curriculum vitae is exhaustive. All information assembled below is public domain and easily accessible. An excellent source for more extended and detailed expositions, also for many more scientist from mathematics and physics, is the comprehensive web page *http://www-history.mcs.st-andrews.ac.uk* from where also all the portraits are taken from.

P. Steinmann, *Geometrical Foundations of Continuum Mechanics*,
Lecture Notes in Applied Mathematics and Mechanics 2, DOI: 10.1007/978-3-662-46460-1_2

Pythagoras [about 570 to about 495] was a Greek philosopher, mathematician and also religious leader from the Aegean island of Samos. His dedication to numbers is often quoted as *Number is the ruler of forms and ideas, and the cause of gods and daemons* or shorter *Number rules the universe.* The famous Pythagoras theorem relating the length of the sides in a right-angled triangle as $a^2+b^2=c^2$, which most probably was already known to the Babylonians a millennium earlier but might have been proven first by Pythagoras, can be considered as one of the earliest statements of a geometrical relation valid in a flat two-dimensional space.

Euclid [about 360 to about 280] was a Greek mathematician from Alexandria at the time Ptolemy I was reigning Egypt. His book *The Elements* strongly influenced mathematics, especially geometry and how it was taught for some two millennia. There Euclid presented the mathematical knowledge available at his time in a logically coherent, and still valid formal approach, especially towards geometry (the Euclidean geometry), based on definitions, postulates, propositions and proofs. In particular his fifth postulate: *That, if a straight line falling on two straight lines makes the interior angles on the same side less than two right angles, the two straight lines, if produced indefinitely, meet on that side on which the angles are less than two right angles* led to an intense debate on its axiomatic, i.e. unprovable character. It became clear only during the 19th century that a geometry, a so-called non-Euclidean geometry, can be constructed without the fifth axiom, thus underpinning indeed its axiomatic character.

René Descartes [1596 to 1650] was an influencial French philosopher, natural scientist and mathematician, who was born in La Haye en Touraine and who died in Stockholm. As a philosopher his dictum *cogito ergo sum* introducing the consciousness as a philosophical topic became famous. His theory explaining gravitation by the help of aether vortexes was later disproved by Newtonian mechanics. In mathematics he is mostly known as the father of analytical or rather Cartesian geometry. Thereby the application of algebra to geometry intimately relies on the introduction of a coordinate system and is instrumental to analyze the interplay between geometrical objects and their representation by algebraic equations. An orthonormal coordinates system in Euclidean space is therefore denoted as a Cartesian coordinate system in his honor.

Nikolai Ivanovich Lobachevsky [1792 to 1856] was a Russian mathematician, who was born in Nizhny Novgorod and who died in Kazan. Upon working on a geometry that does not require Euclid's fifth postulate he developed a non-Euclidean geometry that is nowadays called hyperbolic or Lobachevskian geometry. For this type of geometry more than one line parallel to a reference line can be extended through a given point (that is not contained in the reference line). A striking consequence is that the sum of angles in a triangle adds up to less than 180 degrees. He reported on his ideas as early as 1826 and was the first to publish on this revolutionary type of geometry in 1829. In appreciation of his insights he was later sometimes regarded as the *Copernicus of Geometry.*

János Bolyai [1802 to 1860] was a Hungarian mathematician, who was born in Klausenburg (now Cluj-Napoca in Romania) and who died in Neumarkt am Mieresch (now Târgu-Mureş in Romania). He was obsessed with Euclid's fifth postulate to an extent that his father warned him *For God's sake, I beseech you, give it up.* However, eventually he managed to devise consistent geometries that do not rely on the fifth postulate. Thus he could respond to his father that *Out of nothing I have created a strange new universe*, indeed the universe of non-Euclidean geometry. In 1832 his system of non-Euclidean geometry was published in an appendix to a book on mathematics (the Tentamen) by his father. After reading this appendix Carl Friedrich Gauss commented in a letter *I regard this young geometer Bolyai as a genius of the first order.*

Carl Friedrich Gauss [1777 to 1855] was a German mathematician and physical scientist, who was born in Braunschweig and who died in Göttingen. His contributions were extremely influential in areas as widespread as analysis, differential geometry, number theory, statistics, astronomy, electrostatics, geodesy, geophysics and optics. His practical occupation with the geodesic survey of the Kingdom of Hanover challenged his interest in differential geometry. As a paradigmatic example the notion of the Gaussian curvature

of a surface is an instrumental insight. In particular Gauss' famous *Theorema Egregium* (the remarkable theorem) stating that the intrinsic curvature of a surface can be determined solely by measurements on the surface without knowledge of the embedding space is a cornerstone of the differential geometry of surfaces. He also claimed in a letter to the father of János Bolyai that he developed non-Euclidean geometry earlier (which is nowadays believed to be true), however he never published it for fear of intense controversies. Due to his significant achievements to mathematics and the sciences (that are boldly documented by countless theorems that carry his name) he is referred to as *Princeps mathematicorum.*

Bernhard Riemann [1826 to 1866] was a German mathematician, who was born in Breselenz and who died in Selasca at Lago Maggiore, Italy. After having studied mathematics in Göttingen and Berlin the advisor for his doctoral thesis on complex analysis was Carl Friedrich Gauss. It was also Gauss who proposed to him to pursue a Habilitation in Göttingen. The regulations for the completion of the Habilitation process required to give an inauguration lecture. According to the regulations Bernhard Riemann had to propose three possible topics for this lecture from which Gauss chose the lecture on geometry which was the one Riemann had prepared the least. After going through the painful process of developing a completely novel topic that later proved to be revolutionary, Riemann delivered his Habilitation lecture *On the hypotheses that lie at the foundations of geometry* in 1854. In his lecture he laid out the foundations for what is now called *Riemannian Geometry* by extending the results of Gauss on the differential geometry of surfaces to spaces of arbitrary dimensions. There he introduced the concepts of the metric and curvature tensors on a manifold as key ingredients of differential geometry. Apparently, however, the lecture was too far ahead of its time and the comprehension of its audience. It is reported that *only Gauss was able to appreciate the depth of his thoughts.* It took some sixty years until Riemann's ideas were fully understood when they constituted the mathematical backbone for the development of the Theory of General Relativity/Gravitation by Albert Einstein. Besides his lasting contributions to differential geometry Bernhard Riemann was also extremely influential in many other areas such as analysis and complex analysis, analytic number theory, and mathematical physics. He is considered one of the most important mathematicians, whereby especially his works on geometry anticipated and enabled the advent of Einstein's geometrodynamics in the

20th century. Bernhard Riemann died in 1866 at the Lago Maggiore from tuberculosis at the age of only 39.

Elwin Bruno Christoffel [1829 to 1900] was a German mathematician and physicist, who was born in Monschau (formerly Montjoie) and who died in Strasbourg. After studying in Berlin, partly under the guidance of Peter Gustav Dirichlet [1805-1859], he became a professor at the Polytechnic School in Zurich and later at the University of Strasbourg. Besides further remarkable mathematical achievements he contributed important concepts to the by then developing tensor calculus, in particular he may be considered the founder of the concept of *covariant differentiation*, [33]. The connection that renders the covariant derivative of the metric in a Riemannian manifold zero is nowadays denoted the Levi-Civita connection, however its coefficients are the *Christoffel symbols*, see [179].

Gregorio Ricci-Curbastro [1853 to 1925] was an Italian mathematician, who was born in Lugo di Romagna and who died in Bologna. His fame is mainly based on his eminent contributions to the *absolute differential calculus*, the so-called Ricci calculus, as the foundation of tensor calculus. Thereby, the concept of covariant differentiation as introduced by Elwin Bruno Christoffel was instrumental for his insights. His student was Tullio Levi-Civita with whom he published his most influential single memoir *Méthodes de calcul différentiel absolu et leurs applications*, [161], that constitutes a detailed description of tensor calculus. The contribution circles around geometrical objects in Riemannian manifolds that are independent of any particular coordinate system. It is remarkable that this was his only publication that he authored under his abbreviated name Ricci, however due to its impact *tensor calculus* and *Ricci calculus* soon became synonyms. He introduced in particular the Ricci curvature tensor that should later play a decisive role in Einstein's Theory of General Relativity/Gravitation.

Luigi Bianchi [1856 to 1928] was an Italian mathematician, who was born in Parma and who died in Pisa. He became a Professor and colleague of Gregorio Ricci-Curbastro at the Scuola Normale Superiore at Pisa in 1881. There he dealt mainly with problems of differential geometry that later found application to the Theory of General Relativity/Gravitation. As an example

he (re-)discovered in 1902 the so-called (algebraic and differential) Bianchi identities for the Riemann curvature tensor. These, as reported by Tullio Levi-Civita, however seem to have been discovered already 22 years earlier by Gregorio Ricci-Curbastro who had apparently forgotten his discovery by then. In particular the differential Bianchi identity plays an important role in the Theory of General Relativity/Gravitation by expressing the conservation of the energy-momentum tensor in curved space-time.

Tullio Levi-Civita [1873 to 1941] was an Italian mathematician, who was born in Padua and who died in Rome. He was the student and coworker of Gregorio Ricci-Curbastro with whom he contributed important developments to the absolute differential calculus. As two examples, the permutation tensor, fully skew symmetric in all index permutations and the symmetric and metric connection in a Riemannian manifold carry his name. His formulation of tensors calculus was in particular instrumental for Albert Einstein to develop the Theory of General Relativity/Gravitation. It is reported that Tullio Levi-Civita even corrected some mathematical errors in Einstein's work in their extended correspondence. Einstein responded in one of his letters *I admire the elegance of your method of computation; it must be nice to ride through these fields upon the horse of true mathematics while the like of us have to make our way laboriously on foot.* As a convinced scientific internationalist he was one of the founders of the *International Congresses of Applied Mechanics* after World War I. The end of his life was overshadowed by the upcoming fascism in Italy that led to his demission from his professorship in Rome due to his Jewish background.

Élie Joseph Cartan [1869 to 1951] was a French mathematician, who was born in Dolomieu, Rhône-Alpes, and who died in Paris. He was born as a son of a blacksmith in a rather poor family. However, due to his exceptional talent he was able to achieve a higher education and rise through the French system of that times to a professorship at the Sorbonne for Analysis, Rational Mechanics, and Geometry. Along his way he developed into one of the most original and influential mathematicians in the 20th century. Besides

contributing important works to Lie groups and differential equations he is the inventor of skew symmetric *differential forms* and the *exterior calculus* that proved instrumental, e.g. for the modern approach to *differential geometry*. Out of his four children his son Henri Cartan [1904-2008] became also an important mathematician and one of the founders of the Bourbaki group, that helped to promote Élie Cartan's findings and methods to their present dominating status. The torsion of a connected manifold, an important notion in differential geometry, is often denoted the *Cartan torsion* in his honor.

Albert Einstein [1879 to 1955] was a German physicist, who was born in Ulm, and who died in Princeton, USA. Even though his most famous finding is perhaps the *Matter-Energy Equivalence* as imprinted in the emblematic relation $E = mc^2$, in 1921 he received the Nobel prize in physics for his revolutionary work on the *Photoelectric Effect* that was instrumental to early developments in quantum theory. These two paradigm changing publications appeared together with two further milestone contributions on *Special Relativity* and on *Brownian Motion* in the journal 'Annalen der Physik' in 1905, his so-called *annus mirabilis.* After some eight years of research between 1907 and 1915 he published his *Theory of General Relativity/Gravitation* in a paper on *The Field Equations of Gravitation* at the end of the year 1915. The Theory of General Relativity/Gravitation relies heavily on concepts from differential geometry, thus it is sometimes also called the Theory of Geometrodynamics. Thereby the gravitational attraction between masses results from the curvature of the space-time geometry, that is in turn due to the presence of matter and energy ('space-time tells mass how to move and mass tells space-time how to curve', see [36]). These insights are engraved in the famous Einstein Field Equations that relate the Einstein tensor to the Energy-Momentum tensor. The Einstein tensor is expressed in terms of the Ricci tensor and the Ricci scalar that in turn are related to the Riemann curvature tensor of space-time geometry. The Riemann curvature tensor finally depends on the metric tensor of this space-time geometry. Thus the Einstein Field Equations are indeed a geometrized formulation that relates geometrical concepts to the matter-energy content of space-time. Furthermore the geodesics in the space-time geometry as the analogy in a curved manifold to a straight line in a flat manifold characterize the equations of free motion for moving matter, e.g. a test particle. With the Nazis coming to power in 1933 in Germany Einstein decided, also due to his Jewish background, not to return from a visit to the USA where he continued to work until his death in Princeton in 1955. In his later days Einstein was heavily involved in political, pacifist and philosophical discussions.

All portraits taken from:
http://www-history.mcs.st-andrews.ac.uk/PictDisplay/*name*.html

2.2 Necessity of Differential Geometry

Flat Euclidean geometry is typically characterized by the following elementary observations:

- Two parallel straight lines do only meet at infinity.
- The sum of angles in a triangle is 180°.
- The ratio of the circumference of a circle to its radius is 2π.

These observations are essentially outlined in the Elements of Euclid [58]. However, as a counter example, in a two-dimensional curved geometry defined on a sphere of radius R none of these observations hold true anymore, refer, e.g., to [59]:

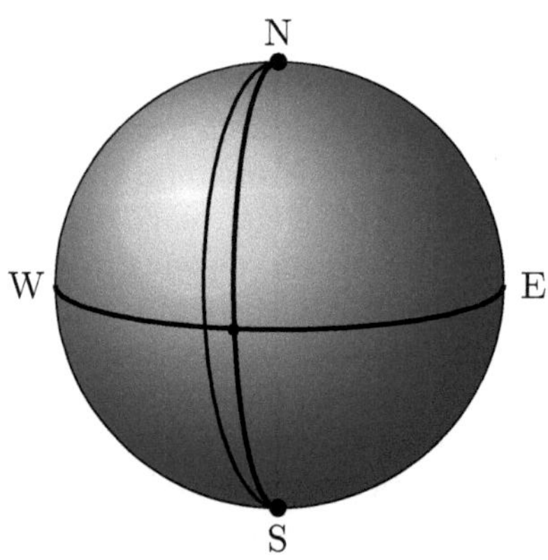

Parallel straight lines do meet after a finite distance on a sphere: Two meridians infinitesimal apart when crossing the equator may be considered parallel straight lines in the infinitesimal vicinity of the equator, i.e. in the tangent plane to the equator. However, due to the curvature they meet after a finite distance $\pi R/2$, i.e. exactly after a quarter of the sphere's circumference, at the north pole and at the south pole.

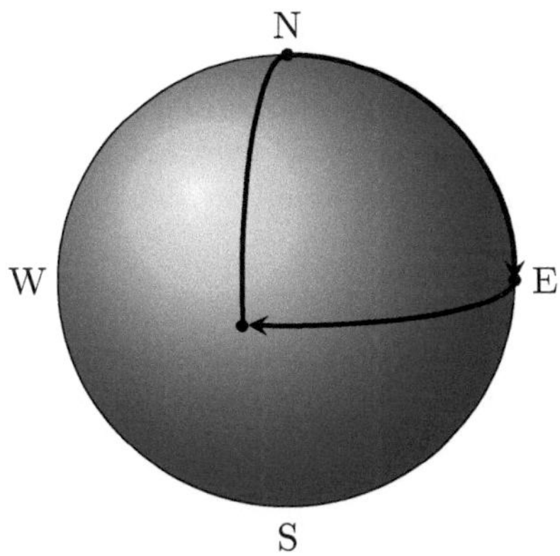

The sum of angles in a triangle is greater than $180°$ ***on a sphere:*** Starting from the equator, a triangle that is laid out by i) moving on the zero meridian northbound towards the north pole, ii) there turning orthogonally to the right and then moving southbound along the $270°$ meridian back to the equator, iii) there turning again orthogonally to the right and proceeding to the starting point consists of three right angles, i.e. the sum of angles in that triangle is $270°$. Only infinitesimal small triangles, i.e. triangles in the tangent plane to the equator, do possess an angle sum of $180°$.

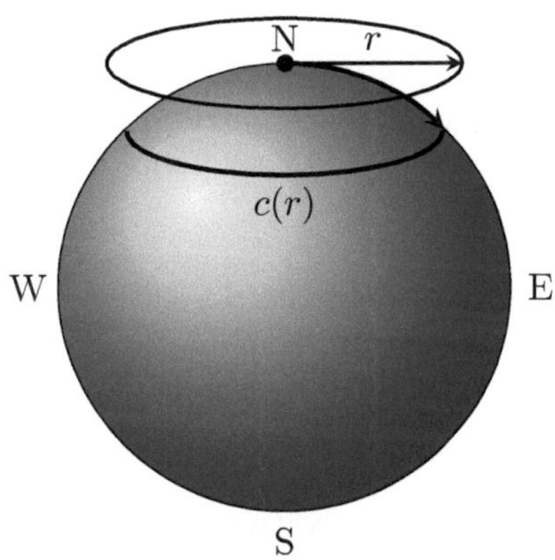

The ratio of the circumference of a circle to its radius is less than 2π ***on a sphere:*** By moving southbound a constant distance r from the north pole on all possible meridians of the sphere lays out a circle with a circumference

$$c(r) := 2\pi R \sin(r/R),$$

thus the ratio $\rho(r) := c(r)/r$ of the circumference $c(r)$ of the circle to its apparent radius r is

$$\rho(r) = 2\pi R/r \sin(r/R).$$

Only for infinitesimal apparent radius $r \to 0$, i.e. for a circle in the tangent plane to the north pole, the ratio $\rho(0) = 2\pi$ coincides with the 2π in an Euclidean geometry, for $r = \pi R/2$, i.e. a quarter of the sphere's circumference, $\rho(\pi R/2) = 4$ takes its minimum value. The difference

$$e(r) := r - c(r)/[2\pi]$$

is denoted the excess radius in a curved geometry, e.g. on a sphere

$$e(r) = r[1 - R/r \sin(r/R)]$$

with $e(0) = 0$ and $e(\pi R/2) = R[\pi/2 - 1] > 0$.

It were observations of this kind that inspired Carl Friedrich Gauss, Nicolai Ivanovich Lobaschevsky, János Bolyai, and their successors, spearheaded by Bernhard Riemann, to develop non-Euclidean geometries. Eventually these developments enabled Albert Einstein to formulate his Theory of General Relativity/Gravitation.

2.3 Classification of Differential Geometry

Differential geometry deals with geometric objects on differential manifolds. Thereby (affine) manifolds may be classified in terms of two fundamental notions from differential geometry that will be discussed in detail in the following chapters, these are:

- ***Connection:*** *(or rather the affine connection)*
  The connection defines covariant differentiation and parallel transport on a manifold. It allows comparison of two vectors (covectors) that are infinitesimal apart.
- ***Metric:***
  The metric defines the inner product on a manifold. It allows to measure the length of and the angle between vectors (covectors).

Based on the *metric* and/or the *connection* three important tensorial quantities in differential geometry result. These are:

- ***Torsion:***
  The torsion is defined as the (right) skew symmetric contribution to the connection. It measures whether parallel transported coordinate differentials do form a parallelogram.
- ***Curvature:***
  The curvature is defined as the skew symmetric partial derivative of the connection corrected by skew symmetric quadratic combinations of the connection. It measures the closure gap after parallel transport of a vector (covector) along infinitesimal circuits.
- ***Non-Metricity:***
  The non-metricity is defined as the covariant derivative of the metric. It measures the length change of and the angle change between vectors (covectors) upon parallel transport.

These issues and their precise meaning will be discussed in the subsequent chapters in great detail. The connection and the metric together with the torsion, the curvature, and the non-metricity then allow for the classification of (affine) manifolds (spaces/geometries[1]) as outlined in Tab. 2.1:

- ***Connected Manifold:***
  Manifold equipped with a connection.
- ***Metric Manifold:***
  Connected manifold additionally equipped with a metric.

Then the following attributes may be defined:

- ***Symmetric Manifold:***
  Connected or *metric* manifold with zero *torsion.*

[1] Here *manifold*, *space*, and *geometry* are used synonymously. In Part III distinction will be made between general manifolds to describe distributed defects in continuum bodies and the embedding of these continuum bodies in Euclidean space.

- ***Flat Manifold:*** *(manifold with teleparallelism)*
  Connected or *metric* manifold with zero *curvature.*
- ***Metrically Connected Manifold:*** *(metrically compatible manifold)*
  Metric manifold with zero *non-metricity* (possesses *metric (compatible) connection*).

Typical examples are the following manifolds (spaces/geometries) that are partly denoted by the names of researchers that have either introduced and/or popularized the corresponding manifolds:

- ***Euclid:***
  Metric manifold: symmetric, flat, and metrically connected.
- ***Riemann:***
  Metric manifold: symmetric, non-flat, and metrically connected.
- ***Cartan:*** *(Weitzenböck)*
  Metric manifold: non-symmetric, flat, and metrically connected.
- ***Riemann-Cartan:*** *(Einstein-Cartan)*
  Metric manifold: non-symmetric, non-flat, and metrically connected.
- ***Weyl:***
  Metric manifold: symmetric, non-flat, and non-metrically connected.
- ***General Metric:***
  Metric manifold: non-symmetric, non-flat, and non-metrically connected.
- ***General Connected:***
  Connected manifold: non-symmetric, and non-flat.

It is interesting to note that many of these manifolds (spaces/geometries) have corresponding counterparts in the kinematics of various (generalized) continuum theories:

- ***Euclid:***
  Elasticity.
- ***Riemann:***
  Continuum-disclination-based elastoplasticity.
- ***Cartan:***
  Continuum-dislocation-based elastoplasticity.
- ***Riemann-Cartan:***
  Continuum-disclination/continuum-dislocation-based elastoplasticity.
- ***General Metric:***
  Continuum-defect-based elastoplasticity.
  (Here general (point) defects such as the distribution of quasi-dislocations caused, e.g., by inhomogeneous temperature distributions, electric or magnetic fields, vacancies, interstitial atoms and the like are essentially characterized by the non-metricity, see [8] and more recently [37]).

Finally the notion of a manifold is intimately related to its description in terms of *general coordinates* and corresponding *coordinate transformations* between the underlying coordinate systems. Thereby *holonomic* coordinate transformations are discriminated from *anholonomic* coordinate transformations. Then anholonomic coordinate transformations are characterized in terms of the

- ***Anholonomic Object:***
  The anholonomic object captures the non-integrability of coordinate transformations (or rather the transformation of *coordinate differentials*). For the case of anholonomic coordinates it contributes correction terms to the expressions for the torsion and the curvature (however the non-metricity is unaffected).

Comprehensive monographs that are essentially dedicated to differential geometry and its applications are, e.g., due to Schouten [167, 168], Sternberg [190], H. Cartan [26], Spivak [179], Bloom [19], É. Cartan [25], Schutz [171], Abraham et al. [1], Flanders [60], Lovelock and Rund [117], Kreyszig [93], Marsden and Hughes [122], Lee [113], Misner et al. [142], Hehl [78], Frankel [64], Ciarlet [35], O'Neill [154], and Epstein [54], among many others.

Table 2.1 Possible classification of (affine) manifolds (spaces/geometries) based on two fundamental objects from differential geometry, i.e. the (affine) *connection* and the *metric*. Manifolds endowed with a metric and/or a connection may furthermore be characterized by the three important tensorial quantities resulting from the metric and/or the connection, i.e. the *torsion*, the *curvature* and the *non-metricity* tensor. The torsion tensor discriminates *non-symmetric* from *symmetric* manifolds, the curvature tensor distinguishes *non-flat (curved)* from *flat (non-curved)* manifolds, and the non-metricity tensor qualifies a manifold as being either *non-metrically* or *metrically connected*. Some of these combinations are denoted by the names of researchers that have either introduced and/or popularized the corresponding manifolds.

		Torsion	Curvature	Non-Metricity	Manifold
(Affine) Connection		yes	yes		*General Connected*
		no	yes		
		yes	no		
		no	no		
	Metric	yes	yes	yes	*General Metric*
		no	yes	yes	*Weyl*
		yes	no	yes	
		no	no	yes	
		yes	yes	no	*Riemann-Cartan*
		no	yes	no	*Riemann*
		yes	no	no	*Cartan*
		no	no	no	*Euclid*

Chapter 3
Geometry on Connected Manifolds

ER

Er kratzte sinnend sich den Hinterkopf
mit seinem Kleinenfingernagel, den er
so lange nicht beschnitten hatte, bis
derselbe rings um unsre Erdensphäre
gewachsen war und ihm am Ende jener
den längst inzwischen kahl gewordenen Schopf
hinreichte (Ziel zugleich und Hindernis) -
ob es nicht kürzer auch gegangen wäre.

Christian Morgenstern

Abstract. The differential geometry on manifolds is considered in an abstract setting without resorting to the concept of a metric. To this end, vectors and covectors are distinguished according to their transformation behavior upon changes of coordinates. Since partial derivatives of vectors and covectors do not transform like tensors, the concept of the covariant derivative of tensors, obeying proper tensor transformation behavior, is motivated. This is achieved by introducing the connection, a third-order non-tensorial object, as one of the most important objects of differential geometry. The covariant derivative also serves to identify what is considered as the parallel transport of tensors. Moreover, the (right) skew symmetric contribution to the connection is denoted the torsion, a third-order tensor that discriminates symmetric from non-symmetric manifolds. Along these lines also the anholonomic object is introduced as a third-order tensor that is related to the concept of the dislocation density in the sequel. Finally, the fourth-order curvature tensor is derived from considering the parallel transport of vectors and covectors along infinitesimal circuits in the manifold. Various aspects of the curvature tensor, in particular the so-called Bianchi identities and the Ricci tensors, are carefully discussed. For the sake of transparency the exposition follows mainly an index notation, however, in order to relate to more modern representations, the main concepts are also given in a coordinate-free invariant formulation and in terms of elements of exterior calculus.

3.1 Manifolds

Central to the following discussions is the notion of a manifold. Thereby, the key idea of a manifold is to allow for general coordinate systems and corresponding transformations between these coordinate systems, see e.g. the discussion in [122]. Correspondingly and more formal is the following

P. Steinmann, *Geometrical Foundations of Continuum Mechanics*,
Lecture Notes in Applied Mathematics and Mechanics 2, DOI: 10.1007/978-3-662-46460-1_3

Definition 3.1. A smooth n_{dm}-dimensional manifold is a set $\mathcal{M}$ such that for each point $\mathcal{P} \in \mathcal{M}$ there is a subset $\mathcal{U}$ of $\mathcal{M}$ containing $\mathcal{P}$, and a one-to-one mapping called chart (coordinate system) $\{\chi^I\}$ from $\mathcal{U}$ onto an open set in $\mathbb{R}^{n_{dm}}$. Multiple charts may be needed to cover the manifold. Coordinate transformations $\{\chi^I\} \rightarrow \{\chi^i\}$ (on a region of $\mathcal{M}$) are infinitely differentiable, i.e. C^∞. A collection of charts covering $\mathcal{M}$ is called an atlas. □

Example 3.1. As a simple example for a manifold consider either a circle or a sphere that can only be covered by at least two charts. Thus the corresponding atlas also consists of at least two charts. For a 1-sphere $\mathbb{S}^1$ in $\mathbb{R}^2$, i.e. $\mathbb{S}^1 = \{\mathbf{r} | \mathbf{r} \in \mathbb{R}^2, \|\mathbf{r}\| = 1\}$, the atlas with the minimum number of charts consists for example in the stereographic projections from the south pole $\mathbb{S}^1 \setminus S \rightarrow \mathbb{R}^1$ and the north pole $\mathbb{S}^1 \setminus N \rightarrow \mathbb{R}^1$ onto tangents (the real line with coordinates χ in the open interval $\chi \in (-\infty, +\infty)$) to the respective opposite pole.

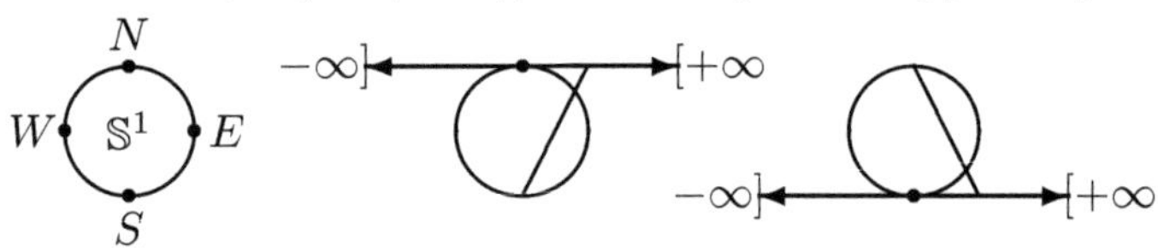

Another possible atlas consists in the four orthogonal projections from the northern, western, southern, and eastern semi-1-spheres onto corresponding tangents (the real line with coordinates χ in the open interval $\chi \in (-1, +1)$).

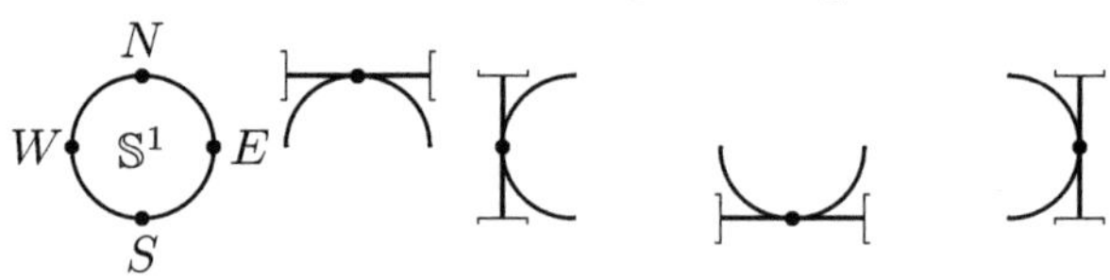

□

Abstracting from of our usual idea of a (Euclidean) space a manifold may also be considered as a *generalized space*. Thus less formal is the alternative

Definition 3.2. A system that is assigned to n_{dm} variables $\chi^1, \chi^2, \cdots, \chi^{n_{dm}}$ is a point $\mathcal{P}$ of an n_{dm}-dimensional manifold $\mathcal{M}$. The n_{dm} variables, i.e. numbers $\chi^1, \chi^2, \cdots, \chi^{n_{dm}}$ are the coordinates of the point $\mathcal{P}$. The set of all points $\mathcal{P}$ then defines the manifold $\mathcal{M}$. □

To illuminate this viewpoint consider as specific examples: (i) a mechanical system with n_{dm} generalized coordinates $\chi^1, \chi^2, \cdots, \chi^{n_{dm}}$, (ii) the set of ellipsoids with $n_{dm} = 3$ half-axes χ^1, χ^2, χ^3, or (iii) as the most basic case simply the ordinary n_{dm}-dimensional Euclidean space.

As a final, rather technical, aspect it is noted that a manifold $\mathcal{M}$ has to satisfy the axioms of a *Hausdorff space*.

Remark 3.1. A *topological space* is given in the

Definition 3.3. A topological space is a set $\mathcal{X}$ together with $\mathcal{T}$ (a collection of subsets of $\mathcal{X}$) satisfying the following axioms:

T1: The empty set $\emptyset$ and $\mathcal{X}$ are in $\mathcal{T}$
T2: The union of any collection of sets in $\mathcal{T}$ is also in $\mathcal{T}$
T3: The intersection of any pair of sets in $\mathcal{T}$ is also in $\mathcal{T}$

For a *trivial topology* the collection $\mathcal{T}$ consists of only the two subsets of $\mathcal{X}$ required by the axioms. □

As an example the set $\mathcal{X} = \{1, 2, 3, 4\}$ and the collection $\mathcal{T} = \{\emptyset, \{1, 2, 3, 4\}\}$ constitute a trivial topology. A non-trivial topology is for example given by the set $\mathcal{X} = \{1, 2, 3, 4\}$ and the collection $\mathcal{T} = \{\emptyset, \{2\}, \{1, 2\}, \{2, 3\}, \{1, 2, 3\}, \{1, 2, 3, 4\}\}$.

Then a *Hausdorff space* is given in the

Definition 3.4. A Hausdorff space is a topological space in which distinct points have disjoint neighborhoods, thus

H1: $\mathcal{X}$ is a Hausdorff space iff any two distinct points of $\mathcal{X}$ can be separated by neighborhoods

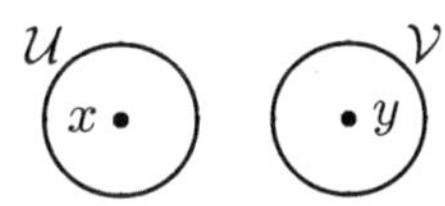

Recall that points x and y in a topological space $\mathcal{X}$ can be separated by neighborhoods if there exists a neighborhood $\mathcal{U}$ of x and a neighborhood $\mathcal{V}$ of y such that $\mathcal{U}$ and $\mathcal{V}$ are disjoint ($\mathcal{U} \cap \mathcal{V} = \emptyset$). For a more detailed exposition of these fundamentals please refer, e.g., to [2]. □

3.1.1 Differentials

Let a chart (coordinate system) consist of n_{dm} coordinates

$$\mathcal{X}^1, \mathcal{X}^2, \cdots, \mathcal{X}^{n_{dm}} =: \{\mathcal{X}^I\}. \tag{3.1}$$

Then a coordinate transformation from the n_{dm} coordinates $\{\mathcal{X}^I\}$ to a new n_{dm}-dimensional set of coordinates $\{\chi^i\}$ is given by the (one-to-one) mapping

$$\chi^i = y^i(\{\mathcal{X}^J\}) \quad \text{with} \quad \mathcal{X}^J = \mathcal{Y}^J(\{\chi^i\}). \tag{3.2}$$

Consequently the chain rule allows to work out the transformation behavior of coordinate differentials simply as

$$\mathrm{d}\chi^i = \frac{\partial y^i}{\partial \mathcal{X}^J} \mathrm{d}\mathcal{X}^J =: \mathcal{F}^i{}_J \, \mathrm{d}\mathcal{X}^J \quad \text{and} \quad \mathrm{d}\mathcal{X}^J = \frac{\partial \mathcal{Y}^J}{\partial \chi^i} \mathrm{d}\chi^i =: f^J{}_i \, \mathrm{d}\chi^i. \tag{3.3}$$

Remark 3.2. Please note that it is on purpose that the notation for coordinate mappings and their Jacobians resembles notation typically used in the kinematics of continuum mechanics, see section 6.2. Thus to unify terminology, coordinates $\mathcal{X}^I$ and χ^i will also be addressed as material and spatial coordinates, respectively. □

3.1.2 Gradients

Consider next a (scalar-valued) field, i.e. a function of n_{dm} coordinates, that depends on either of the n_{dm}-dimensional coordinate systems

$$\vartheta = \Theta(\{\mathcal{X}^J\}) = \theta(\{\chi^i\}) \circ y^i(\{\mathcal{X}^J\}). \tag{3.4}$$

Here $\circ$ denotes composition of mappings. Then the total differential involves the gradient of the field with respect to the coordinates

$$\mathrm{d}\vartheta = \frac{\partial \Theta}{\partial \mathcal{X}^J}\,\mathrm{d}\mathcal{X}^J = \frac{\partial \theta}{\partial \chi^i}\,\mathrm{d}\chi^i. \tag{3.5}$$

Thus by either using the chain rule or by incorporating the coordinate differentials as derived in Eq. 3.3 the transformation of gradients follows as

$$\frac{\partial \Theta}{\partial \mathcal{X}^J} = \frac{\partial \theta}{\partial \chi^i}\frac{\partial y^i}{\partial \mathcal{X}^J} = \frac{\partial \theta}{\partial \chi^i}\mathcal{F}^i{}_J \quad \text{and} \quad \frac{\partial \theta}{\partial \chi^i} = \frac{\partial \Theta}{\partial \mathcal{X}^J}\frac{\partial \mathcal{Y}^J}{\partial \chi^i} = \frac{\partial \Theta}{\partial \mathcal{X}^J}f^J{}_i. \tag{3.6}$$

In conclusion it shall be recognized carefully that differentials and gradients obey different transformation behavior upon a change of coordinates.

3.1.3 Co- and Contravariant Transformations

We may next attach n_{dm}-dimensional tuple $\mathcal{V}^J(\{\mathcal{X}^K\})$ and $\mathcal{V}_J(\{\mathcal{X}^K\})$ to each point $\mathcal{P}$ of $\mathcal{M}$. Then $\mathcal{V}^J$ are denoted the contravariant coefficients (of a vector) while $\mathcal{V}_J$ are the covariant coefficients (of a covector), both evaluated at point $\mathcal{P}$ with coordinates $\{\mathcal{X}^K\}$. Obviously these have to be distinguished by their transformation behavior upon a change of coordinates:

Contravariant coefficients (of a vector) transform like differentials

$$v^i = \mathcal{F}^i{}_J \mathcal{V}^J \quad \text{and} \quad \mathcal{V}^J = f^J{}_i v^i, \tag{3.7}$$

whereas covariant coefficients (of a covector) transform like gradients

$$\mathcal{V}_J = v_i \mathcal{F}^i{}_J \quad \text{and} \quad v_i = \mathcal{V}_J f^J{}_i. \tag{3.8}$$

Note that contravariant and covariant coefficients obey different, i.e. in a sense inverse transformation behavior upon a change of coordinates.

3.1.4 Tensors

Sloppily speaking tensors are objects with multiple indices that respect the following

Definition 3.5. Coefficients of tensors change in a 'proper way' with coordinate transformations. □

As an example the previously introduced vectors and covectors may be regarded as first-order tensors with transformation properties

$$u^i = \mathcal{F}^i{}_J \mathcal{U}^J \quad \text{and} \quad u_i = f^J{}_i \mathcal{U}_J. \tag{3.9}$$

Consequently four different types of (simple) second-order tensors may be constructed from dyadic products of first-order tensors and may be distinguished by their transformation behavior

$$\begin{aligned} t^{ij} &:= u^i v^j = \mathcal{F}^i{}_K \mathcal{U}^K \mathcal{V}^L \mathcal{F}^j{}_L =: \mathcal{F}^i{}_K \mathcal{T}^{KL} \mathcal{F}^j{}_L, \\ t_{ij} &:= u_i v_j = f^K{}_i \mathcal{U}_K \mathcal{V}_L f^L{}_j =: f^K{}_i \mathcal{T}_{KL} f^L{}_j, \\ t^i{}_j &:= u^i v_j = \mathcal{F}^i{}_K \mathcal{U}^K \mathcal{V}_L f^L{}_j =: \mathcal{F}^i{}_K \mathcal{T}^K{}_L f^L{}_j, \\ t_i{}^j &:= u_i v^j = f^K{}_i \mathcal{U}_K \mathcal{V}^L \mathcal{F}^j{}_L =: f^K{}_i \mathcal{T}_K{}^L \mathcal{F}^j{}_L. \end{aligned} \tag{3.10}$$

Clearly these transformations do also hold for general second-order tensors that are constructed from a sum of simple second-order tensors. The extension to higher-order tensors follows the same pattern and is thus straightforward.

Remark 3.3. Fully contravariant, fully covariant, contra-covariant, and co-contravariant, the latter two collectively being referred to as mixed-variant, types of second-order tensors may be distinguished. The extension of this terminology to higher-order tensors gets however a bit cumbersome for the detailed denomination of mixed-variant tensors. Thus a tensor with m contravariant and n covariant indices is alternatively denoted a (m, n)-tensor. □

3.1.5 Affine Tangent Space

In general no vectors (nor covectors) are defined in a manifold $\mathcal{M}$. However a n_{dm}-dimensional vector space (the tangent space $T_{\mathcal{P}}\mathcal{M}$), satisfying the axioms of an affine vector space, may be attached to each point $\mathcal{P}$ of an n_{dm}-dimensional manifold $\mathcal{M}$. It thus follows from the (sloppy)

Definition 3.6. The tangent space $T_{\mathcal{P}}\mathcal{M}$ consists of all vectors $\mathcal{V}^I$ emanating from $\mathcal{P}$. □

Moreover at each point $\mathcal{P}$ a (covariant) coordinate basis of the affine tangent space denoted by $\boldsymbol{\partial}_{X^I}$ with $I = 1 \cdots n_{dm}$ may be introduced.

Remark 3.4. Likewise a n_{dm}-dimensional affine vector space (the cotangent space $T^*_{\mathcal{P}}\mathcal{M}$) may be attached to each point $\mathcal{P}$ of an n_{dm}-dimensional manifold $\mathcal{M}$. It follows from the (sloppy)

Definition 3.7. The cotangent space $T^*_{\mathcal{P}}\mathcal{M}$ consists of all covectors (one-forms) $\mathcal{V}_I$ emanating from $\mathcal{P}$. □

Finally at each point $\mathcal{P}$ a (contravariant) dual basis of the affine cotangent space denoted by $\mathbf{d}\mathcal{X}^I$ with $I = 1 \cdots n_{dm}$ may be introduced. □

As rather elementary (however already specialized) examples consider 1-dimensional curves and 2-dimensional surfaces embedded into the Euclidean ambient space: Then for a parameter curve $\chi^I = \chi^i(t)$ the 1-dimensional tangent space, i.e. a straight line, follows from the assignment $\mathrm{d}\chi^i \leftrightarrow \mathrm{d}\boldsymbol{x} = \mathrm{d}\chi^i \boldsymbol{\partial}_{\chi^i} = \mathrm{d}\chi^i \boldsymbol{g}_i$ with $\mathrm{d}\chi^i = \partial_t \chi^i \,\mathrm{d}t$. Likewise the 2-dimensional tangent space of the surface is given by its tangent plane spanned by $\boldsymbol{g}_1$ and $\boldsymbol{g}_2$. However, in general a manifold and its tangent space do not necessitate the concept of an embedding Euclidean space.

Remark 3.5. An *affine vector space* is given in the

Definition 3.8. A set $\mathcal{A}$ of elements (vectors) $\boldsymbol{a}, \boldsymbol{b}, \boldsymbol{c}, \cdots$ that satisfy the following axioms defines an affine vector space, its elements are then denoted as affine vectors:

A1: Addition of vectors:
- a) $\boldsymbol{a} + \boldsymbol{b} = \boldsymbol{b} + \boldsymbol{a}$ commutativity
- b) $\boldsymbol{a} + [\boldsymbol{b} + \boldsymbol{c}] = [\boldsymbol{a} + \boldsymbol{b}] + \boldsymbol{c} = \boldsymbol{a} + \boldsymbol{b} + \boldsymbol{c}$ associativity
- c) $\boldsymbol{a} + \boldsymbol{0} = \boldsymbol{a}$ null element
- d) $\boldsymbol{a} + [-\boldsymbol{a}] = \boldsymbol{0}$ negative element

A2: Multiplication of vectors with scalars:
- a) $1\boldsymbol{a} = \boldsymbol{a}$
- b) $\alpha(\beta\boldsymbol{a}) = (\alpha\beta)\boldsymbol{a} = \alpha\beta\boldsymbol{a}$ associativity
- c) $[\alpha + \beta]\boldsymbol{a} = \alpha\boldsymbol{a} + \beta\boldsymbol{a}$ distributivity for scalar addition
- d) $\alpha[\boldsymbol{a} + \boldsymbol{b}] = \alpha\boldsymbol{a} + \beta\boldsymbol{a}$ distributivity for vector addition

Thus in an affine vector space addition of vectors and multiplication of vectors with scalars are defined. □

As an example the columns and rows of a matrix satisfy the axioms of an affine vector space. □

3.2 Connection

3.2.1 Partial Derivatives

Based on the transformation rule for contravariant first-order tensors and the chain rule the partial derivatives (PD) of vectors with respect to the coordinates are computed as

$$\begin{aligned} v^i &= \mathcal{F}^i{}_J \mathcal{V}^J \rightarrow v^i{}_{,k} = \mathcal{F}^i{}_J \mathcal{V}^J{}_{,L} f^L{}_k + \underline{\mathcal{F}^i{}_{M,L} \mathcal{V}^M f^L{}_k}, \\ \mathcal{V}^I &= f^I{}_j v^j \rightarrow \mathcal{V}^I{}_{,K} = f^I{}_j v^j{}_{,l} \mathcal{F}^l{}_K + \underline{f^I{}_{m,l} v^m \mathcal{F}^l{}_K}. \end{aligned} \tag{3.11}$$

Likewise, based on the transformation rule for covariant first-order tensors and the chain rule the partial derivatives of covectors with respect to the coordinates follow as

$$\begin{aligned} v_i &= f^J{}_i \mathcal{V}_J \rightarrow v_{i,k} = f^J{}_i \mathcal{V}_{J,L} f^L{}_k + \underline{\mathcal{V}_M f^M{}_{i,k}}, \\ \mathcal{V}_I &= \mathcal{F}^j{}_I v_j \rightarrow \mathcal{V}_{I,K} = \mathcal{F}^j{}_I v_{j,l} \mathcal{F}^l{}_K + \underline{v_m \mathcal{F}^m{}_{I,K}}. \end{aligned} \tag{3.12}$$

It is obvious from the discussion in the preceding section and the representation in Eq. 3.10 that the underlined terms conflict with the transformation rules for second-order tensors. As a result it may be stated that the partial derivative of a vector or a covector does not result in a second-order tensor.

3.2.2 Covariant Derivatives

Thus the challenge is to find a correction to the partial derivative of a vector or a covector so as to reinstall the transformation behavior of second-order tensors. As a result an alternative derivative with respect to the coordinates (indicated by a vertical bar $|$) is sought for vectors that transforms as

$$v^i{}_{|k} \doteq \mathcal{F}^i{}_J \mathcal{V}^J{}_{|L} f^L{}_k \quad \text{and} \quad \mathcal{V}^I{}_{|K} \doteq f^I{}_j v^j{}_{|l} \mathcal{F}^l{}_K. \tag{3.13}$$

Likewise a corresponding derivative for covectors is sought with the following transformation behavior

$$v_{i|k} \doteq f^J{}_i \mathcal{V}_{J|L} f^L{}_k \quad \text{and} \quad \mathcal{V}_{I|K} \doteq \mathcal{F}^j{}_I v_{j|l} \mathcal{F}^l{}_K. \tag{3.14}$$

If such derivatives may be found the resulting operation shall be called *covariant derivative* (CD). A suited Ansatz to solve the above problem is to introduce third-order objects $\mathcal{L}^I{}_{KL}$ and $\ell^i{}_{jk}$, the so-called linear (or affine) *connection*. Then the connection allows to reinstall the transformation behavior of the covariant derivatives of vectors and covectors provided the connection satisfies the following non-tensorial transformation properties

$$\begin{aligned} \mathcal{F}^i{}_{M,L} f^L{}_k &= \mathcal{F}^i{}_J \mathcal{L}^J{}_{ML} f^L{}_k - \ell^i{}_{nk} \mathcal{F}^n{}_M, \\ f^I{}_{m,l} \mathcal{F}^l{}_K &= f^I{}_j \ell^j{}_{ml} \mathcal{F}^l{}_K - \mathcal{L}^I{}_{NK} f^N{}_m. \end{aligned} \tag{3.15}$$

By rearrangement these non-tensorial transformation properties of the connection may also be stated equivalently as

$$
\begin{aligned}
f^M{}_{i,k} &= f^M{}_n \ell^n{}_{ik} - \mathcal{L}^M{}_{JL} f^J{}_i f^L{}_k, \\
\mathcal{F}^m{}_{I,K} &= \mathcal{F}^m{}_N \mathcal{L}^N{}_{IK} - \ell^m{}_{jl} \mathcal{F}^j{}_I \mathcal{F}^l{}_K.
\end{aligned}
\tag{3.16}
$$

By combining the transformation behavior of the connection in Eq. 3.15 with that of the partial derivative of a vector in Eq. 3.11 the covariant derivative of a vector is eventually given by

$$
v^i{}_{|j} = v^i{}_{,j} + \ell^i{}_{mj} v^m \quad \text{and} \quad \mathcal{V}^I{}_{|J} = \mathcal{V}^I{}_{,J} + \mathcal{L}^I{}_{MJ} \mathcal{V}^M. \tag{3.17}
$$

Please observe that the position for the running index m or M, respectively, and thus the precise arrangement of indices in all later expressions that involve the connection varies in the literature, however once defined as in the above it only matters to consequently stick to this convention in the sequel. (Note that here the index and sign conventions advocated, e.g., by [142] are adopted.) Likewise the covariant derivative of a covector follows from inserting the transformation in Eq. 3.16 into Eq. 3.12 to render

$$
v_{i|j} = v_{i,j} - v_m \ell^m{}_{ij} \quad \text{and} \quad \mathcal{V}_{I|J} = \mathcal{V}_{I,J} - \mathcal{V}_M \mathcal{L}^M{}_{IJ}. \tag{3.18}
$$

Then the covariant derivatives of the four types of (simple) second-order tensors follow from the product rule applied to their dyadic representation

$$
\begin{aligned}
\mathcal{T}^{IJ}{}_{|K} &= \mathcal{T}^{IJ}{}_{,K} + \mathcal{L}^I{}_{MK} \mathcal{T}^{MJ} + \mathcal{L}^J{}_{MK} \mathcal{T}^{IM}, \\
\mathcal{T}^I{}_{J|K} &= \mathcal{T}^I{}_{J,K} + \mathcal{L}^I{}_{MK} \mathcal{T}^M{}_J - \mathcal{L}^M{}_{JK} \mathcal{T}^I{}_M, \\
\mathcal{T}_{IJ|K} &= \mathcal{T}_{IJ,K} - \mathcal{L}^M{}_{IK} \mathcal{T}_{MJ} - \mathcal{L}^M{}_{JK} \mathcal{T}_{IM}, \\
\mathcal{T}_I{}^J{}_{|K} &= \mathcal{T}_I{}^J{}_{,K} - \mathcal{L}^M{}_{IK} \mathcal{T}_M{}^J + \mathcal{L}^J{}_{MK} \mathcal{T}_I{}^M.
\end{aligned}
\tag{3.19}
$$

Again these expressions do also hold for general second-order tensors that are constructed from a sum of simple second-order tensors. Recall that covariant differentiation allows tensorial transformation upon coordinate changes, thus

$$
\begin{aligned}
t^{ij}{}_{|k} &= \mathcal{F}^i{}_I \mathcal{T}^{IJ}{}_{|K} \mathcal{F}^j{}_J f^K{}_k, \\
t^i{}_{j|k} &= \mathcal{F}^i{}_I \mathcal{T}^I{}_{J|K} f^J{}_j f^K{}_k, \\
t_{ij|k} &= f^I{}_i \mathcal{T}_{IJ|K} f^J{}_j f^K{}_k, \\
t_i{}^j{}_{|k} &= f^I{}_i \mathcal{T}_I{}^J{}_{|K} \mathcal{F}^j{}_J f^K{}_k.
\end{aligned}
\tag{3.20}
$$

The covariant derivatives of higher-order tensors (and objects) follow likewise, e.g. for third-order objects as occurring in the sequel one finds

$$
\begin{aligned}
\mathcal{T}^{IJ}{}_{L|K} &= \mathcal{T}^{IJ}{}_{L,K} + \mathcal{L}^I{}_{MK} \mathcal{T}^{MJ}{}_L + \mathcal{L}^J{}_{MK} \mathcal{T}^{IM}{}_L - \mathcal{L}^M{}_{LK} \mathcal{T}^{IJ}{}_M, \\
\mathcal{T}^I{}_{JL|K} &= \mathcal{T}^I{}_{JL,K} + \mathcal{L}^I{}_{MK} \mathcal{T}^M{}_{JL} - \mathcal{L}^M{}_{JK} \mathcal{T}^I{}_{ML} - \mathcal{L}^M{}_{LK} \mathcal{T}^I{}_{JM}, \\
\mathcal{T}_{IJL|K} &= \mathcal{T}_{IJL,K} - \mathcal{L}^M{}_{IK} \mathcal{T}_{MJL} - \mathcal{L}^M{}_{JK} \mathcal{T}_{IML} - \mathcal{L}^M{}_{LK} \mathcal{T}_{IJM}.
\end{aligned}
\tag{3.21}
$$

Again, it is worth noting that coordinate changes result in tensorial transformation, e.g.

$$
\begin{aligned}
t^i{}_{jl|k} &= \mathcal{F}^i{}_I \mathcal{T}^I{}_{JL|K} f^J{}_j f^L{}_l f^K{}_k, \\
t_{ijl|k} &= f^I{}_i \mathcal{T}_{IJL|K} f^J{}_j f^L{}_l f^K{}_k.
\end{aligned}
\tag{3.22}
$$

Based on its definition the covariant derivative (CD) obeys a number of important rules, for example:

D1: The CD of scalars coincides with the PD of scalars,
D2: The CD obeys the distribution rule,
D3: The CD obeys the Leibniz (product) rule.

Proof. The first and second rule are obvious, the proof of the last rule is based on the application of the partial derivative to the contraction of a vector and a covector into a scalar (whereby opposite but otherwise identical indices follow the Einstein summation rule)

$$[\mathcal{V}^I \mathcal{V}_I]_{,J} = \mathcal{V}^I{}_{,J} \mathcal{V}_I + \mathcal{V}^I \mathcal{V}_{I,J}. \tag{3.23}$$

Since based on the first rule the partial and the covariant derivatives of scalars coincide it also holds that

$$[\mathcal{V}^I \mathcal{V}_I]_{|J} = [\mathcal{V}^I{}_{,J} + \mathcal{L}^I{}_{MJ} \mathcal{V}^M] \mathcal{V}_I + \mathcal{V}^I [\mathcal{V}_{I,J} - \mathcal{V}_M \mathcal{L}^M{}_{IJ}] \doteq [\mathcal{V}^I \mathcal{V}_I]_{,J}. \tag{3.24}$$

Comparing the two results in Eqs. 3.23 and 3.24 and noting that $\mathcal{L}^I{}_{MJ} \mathcal{V}^M \mathcal{V}_I \equiv \mathcal{V}^I \mathcal{V}_M \mathcal{L}^M{}_{IJ}$ concludes the proof. □

Finally, the covariant derivatives of, e.g., second-order tensors with respect to two different connections, say $\mathcal{L}^I{}_{JK}$ and $\mathcal{Y}^I{}_{JK}$ with $\mathcal{X}^I{}_{JK} = \mathcal{L}^I{}_{JK} - \mathcal{Y}^I{}_{JK}$, follow obviously as

$$
\begin{aligned}
\mathcal{T}^{IJ}{}_{|K} &= \mathcal{T}^{IJ}{}_{\|K} + \mathcal{X}^I{}_{MK} \mathcal{T}^{MJ} + \mathcal{X}^J{}_{MK} \mathcal{T}^{IM}, \\
\mathcal{T}^I{}_{J|K} &= \mathcal{T}^I{}_{J\|K} + \mathcal{X}^I{}_{MK} \mathcal{T}^M{}_J - \mathcal{X}^M{}_{JK} \mathcal{T}^I{}_M, \\
\mathcal{T}_{IJ|K} &= \mathcal{T}_{IJ\|K} - \mathcal{X}^M{}_{IK} \mathcal{T}_{MJ} - \mathcal{X}^M{}_{JK} \mathcal{T}_{IM}, \\
\mathcal{T}_I{}^J{}_{|K} &= \mathcal{T}_I{}^J{}_{\|K} - \mathcal{X}^M{}_{IK} \mathcal{T}_M{}^J + \mathcal{X}^J{}_{MK} \mathcal{T}_I{}^M.
\end{aligned}
\tag{3.25}
$$

Here covariant derivatives with respect to $\mathcal{Y}^I{}_{JK}$ have been denoted by $\|$.

3.2.3 Parallel Transport of Vectors and Covectors

It shall be observed that in general tangent spaces $T_{\mathcal{P}}\mathcal{M}$ and cotangent spaces $T^*_{\mathcal{P}}\mathcal{M}$ at different points $\mathcal{P}$ of a manifold $\mathcal{M}$ are not connected. However, if a covariant derivative of vectors $\mathcal{V}^I{}_{|K} = \mathcal{V}^I{}_{,K} + \mathcal{L}^I{}_{JK} \mathcal{V}^J$ and covectors $\mathcal{V}_{J|K} = \mathcal{V}_{J,K} - \mathcal{V}_I \mathcal{L}^I{}_{JK}$ is introduced based on a linear connection $\mathcal{L}^I{}_{JK}$, the notion of *parallel transport* may be defined. Then the bundle $T\mathcal{M}$ of tangent spaces

$T_{\mathcal{P}}\mathcal{M}$ and the bundle $T^*\mathcal{M}$ of cotangent spaces $T^*_{\mathcal{P}}\mathcal{M}$ constitute affinely connected spaces.

Thereby the motivation for the notion of parallel transport $\mathrm{p}\mathcal{V}^I$ of a vector $\mathcal{V}^I$ is as follows[1]: The comparison of two vectors $\mathcal{V}^I(\{\mathcal{X}^J\})$ and $\mathcal{V}^I(\{\mathcal{X}^J + \mathrm{d}\mathcal{X}^J\})$ in two different (infinitesimal close) tangent spaces attached to $\{\mathcal{X}^J\}$ and $\{\mathcal{X}^J + \mathrm{d}\mathcal{X}^J\}$, respectively, necessitates first a parallel (back) transport of $\mathcal{V}^I(\{\mathcal{X}^J + \mathrm{d}\mathcal{X}^J\})$ to $\{\mathcal{X}^J\}$. Thereby this parallel transport is assumed proportional to $\mathcal{V}^K$ and $\mathrm{d}\mathcal{X}^J$, i.e. $\mathrm{p}\mathcal{V}^I := -\mathcal{L}^I{}_{KJ}\mathcal{V}^K\,\mathrm{d}\mathcal{X}^J$, the minus sign (and the sequence of indices) being convention. The argument holds likewise for covectors. From these considerations we may derive for vectors the

Definition 3.9. The transport along a parameter curve $\mathcal{X}^J(t)$ of a vector $\mathcal{V}^I$ that is attached to a manifold is called parallel if the covariant derivative (or rather the *covariant differential* $\mathrm{D}\mathcal{V}^I$) of $\mathcal{V}^I$ vanishes

$$\mathcal{V}^I{}_{|J} = \mathcal{V}^I{}_{,J} + \mathcal{L}^I{}_{KJ}\mathcal{V}^K = 0 \quad \text{with} \quad \mathrm{D}\mathcal{V}^I := \mathcal{V}^I{}_{|J}\,\mathrm{d}\mathcal{X}^J = 0. \tag{3.26}$$

Thus for a parallel transport the change of the vector in the direction of the parameter curve (directional derivative) follows as

$$\mathrm{d}\mathcal{V}^I := \mathcal{V}^I{}_{,J}\,\mathrm{d}\mathcal{X}^J \equiv -\mathcal{L}^I{}_{KJ}\mathcal{V}^K\,\mathrm{d}\mathcal{X}^J =: \mathrm{p}\mathcal{V}^I. \tag{3.27}$$

Note carefully, however, that $\mathrm{d}\mathcal{V}^I$ and $\mathrm{p}\mathcal{V}^I$ do only coincide as in the above in the case of parallel transport. □

Alternatively, for covectors we have the

Definition 3.10. The transport along a parameter curve $\mathcal{X}^K(t)$ of a covector $\mathcal{V}_J$ that is attached to a manifold is called parallel if the covariant derivative (or rather the covariant differential) of $\mathcal{V}_J$ vanishes

$$\mathcal{V}_{J|K} = \mathcal{V}_{J,K} - \mathcal{V}_I\mathcal{L}^I{}_{JK} = 0 \quad \text{with} \quad \mathrm{D}\mathcal{V}_J := \mathcal{V}_{J|K}\,\mathrm{d}\mathcal{X}^K = 0. \tag{3.28}$$

Thus for parallel transport the change of the covector in the direction of the parameter curve (directional derivative) follows as

$$\mathrm{d}\mathcal{V}_J := \mathcal{V}_{J,K}\,\mathrm{d}\mathcal{X}^K \equiv \mathcal{V}_I\mathcal{L}^I{}_{JK}\,\mathrm{d}\mathcal{X}^K =: \mathrm{p}\mathcal{V}_J. \tag{3.29}$$

Note again that $\mathrm{d}\mathcal{V}_J$ and $\mathrm{p}\mathcal{V}_J$ do only coincide as in the above in the case of parallel transport. □

As a conclusion it may be stated that for a covariant derivative the change of a vector $\mathcal{V}^I$ due to its partial derivative with respect to the coordinates, i.e.

$$\mathrm{d}\mathcal{V}^I := \mathcal{V}^I{}_{,J}\,\mathrm{d}\mathcal{X}^J = \mathcal{V}^I(\{\mathcal{X}^J + \mathrm{d}\mathcal{X}^J\}) - \mathcal{V}^I(\{\mathcal{X}^J\})$$

---

[1] The notation $\mathrm{p}\mathcal{V}^I$ for the parallel transport of $\mathcal{V}^I$ is motivated by simply rotating the common notation for a differential d upside down to obtain p.

is corrected by the contribution of the parallel transport

$$\mathrm{p}\mathcal{V}^I := -\mathcal{L}^I{}_{KJ}\mathcal{V}^K \,\mathrm{d}\mathcal{X}^J$$

to render the covariant differential

$$\mathrm{D}\mathcal{V}^I := \mathrm{d}\mathcal{V}^I - \mathrm{p}\mathcal{V}^I = \left[\mathcal{V}^I{}_{,J} + \mathcal{L}^I{}_{KJ}\mathcal{V}^K\right] \mathrm{d}\mathcal{X}^J.$$

It shall be noted that the same arguments hold likewise for covectors, i.e. for a covariant derivative the change of a covector $\mathcal{V}_J$ due to its partial derivative with respect to the coordinates, i.e.

$$\mathrm{d}\mathcal{V}_J := \mathcal{V}_{J,K}\,\mathrm{d}\mathcal{X}^K = \mathcal{V}_J(\{\mathcal{X}^K + \mathrm{d}\mathcal{X}^K\}) - \mathcal{V}_J(\{\mathcal{X}^K\})$$

is corrected by the contribution of the parallel transport

$$\mathrm{p}\mathcal{V}_J := \mathcal{V}_I\mathcal{L}^I{}_{JK}\,\mathrm{d}\mathcal{X}^K$$

to render eventually the covariant differential

$$\mathrm{D}\mathcal{V}_J := \mathrm{d}\mathcal{V}_J - \mathrm{p}\mathcal{V}_J = \left[\mathcal{V}_{J,K} - \mathcal{V}_I\mathcal{L}^I{}_{JK}\right] \mathrm{d}\mathcal{X}^K.$$

Based on its definition parallel transport (PT) obeys a number of important rules, for example:

P1: The PT of scalars is zero,
P2: The PT obeys the distribution rule,
P3: The PT obeys the Leibniz (product) rule.

Proof. The first and second rule are obvious, the proof of the last rule is based on its application to the parallel transport of the contraction of a vector and a covector into scalar

$$\mathrm{p}[\mathcal{V}^I\mathcal{V}_I] = \mathrm{p}\mathcal{V}^I\mathcal{V}_I + \mathcal{V}^I\,\mathrm{p}\mathcal{V}_I.$$

Since based on the first rule the parallel transport of scalars vanishes it also holds that

$$\mathrm{p}[\mathcal{V}^I\mathcal{V}_I] = -\mathcal{L}^I{}_{MK}\mathcal{V}^M\,\mathrm{d}\mathcal{X}^K\mathcal{V}_I + \mathcal{V}^I\mathcal{V}_M\mathcal{L}^M{}_{IK}\,\mathrm{d}\mathcal{X}^K \doteq 0.$$

Noting that $\mathcal{L}^I{}_{MK}\mathcal{V}^M\mathcal{V}_I = \mathcal{V}^I\mathcal{V}_M\mathcal{L}^M{}_{IK}$ concludes the proof. □

Example 3.2. Consider the example of the two-dimensional non-flat, i.e. *curved manifold* represented by the *surface of a sphere* in Fig. 3.1. For $[0,\pi] \ni \chi^1$ the *polar angle*, and $(-\pi,\pi] \ni \chi^2$ the *azimuthal angle* the coefficients of the linear (affine) connection on the surface of a sphere read

$$\ell^1_{11} = 0, \quad \ell^1_{12} = \ell^1_{21} = 0, \quad \ell^1_{22} = -\sin\chi^1\cos\chi^1,$$

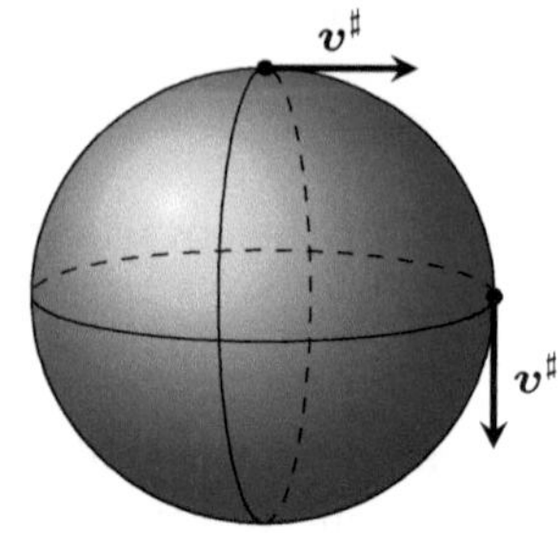

Fig. 3.1 Parallel transport of a vector along a meridian on the two-dimensional surface of a sphere as an example of a non-flat, i.e. curved manifold

$$\Gamma^2_{11} = 0, \quad \Gamma^2_{12} = \Gamma^2_{21} = \cot \chi^1, \quad \Gamma^2_{22} = 0.$$

Then starting at the north-pole with $\{\chi^i\} = \{0, 0\}$ the parallel transport of a vector v^i with $\{v^i(0, 0)\} = \{1, 0\}$ along the geodesic given by a meridian $\chi^1 \in [0, \pi/2]$ and $\chi^2 =$ constant (thus the increment of the coordinates on the meridian reads $\{\,\mathrm{d}\chi^i\} = \{\,\mathrm{d}\chi^1, 0\}$) renders

$$\mathrm{p}v^1 = 0 = \mathrm{d}v^1 \qquad \mathrm{p}v^2 = 0 = \mathrm{d}v^2.$$

As a conclusion the parallel transport of the vector $\{v^i(0, 0)\} = \{1, 0\}$ along a meridian leaves its coefficients v^i unchanged, however seen from the perspective of the embedding space the vector suffers a change in orientation, see the representation in Fig. 3.1. □

Example 3.3. Consider the example of the *flat manifold* represented by the two-dimensional *Euclidean space* parameterized in terms of polar coordinates in Fig. 3.2. For $0 \leq \chi^1$ the *radial distance*, and $[0, 2\pi) \ni \chi^2$ the *polar angle* the coefficients of the linear (affine) connection that coincide in Euclidean space with the Christoffel symbols read

$$\Gamma^1_{11} = 0, \quad \Gamma^1_{12} = \Gamma^1_{21} = 0, \quad \Gamma^1_{22} = -\chi^1,$$

$$\Gamma^2_{11} = 0, \quad \Gamma^2_{12} = \Gamma^2_{21} = [\chi^1]^{-1}, \quad \Gamma^2_{22} = 0.$$

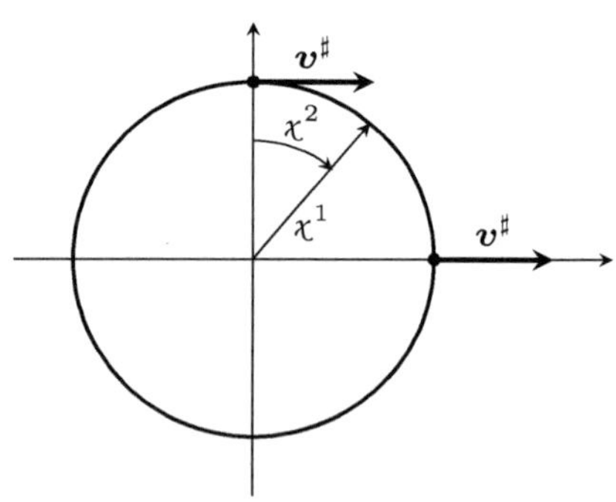

Fig. 3.2 Parallel transport of a vector in the two-dimensional plane parameterized in terms of polar coordinates as an example of a flat, i.e. non-curved manifold

Then starting at the north-pole with $\{\chi^i\} = \{r, 0\}$ the parallel transport of the vector v^i with $\{v^i(r,0)\} = \{0,1\}$ along the quarter-circle given by $\chi^1 = r = \text{constant}$, and $\chi^2 \in [0, \pi/2]$ (thus the increment of the coordinates on the circle reads $\{\,\mathrm{d}\chi^i\} = \{0, \,\mathrm{d}\chi^2\}$) renders

$$\mathrm{p}v^1 = rv^2 \,\mathrm{d}\chi^2 = \mathrm{d}v^1 \qquad \mathrm{p}v^2 = -[r]^{-1}v^1 \,\mathrm{d}\chi^2 = \mathrm{d}v^2.$$

This system may be rewritten as

$$\frac{\mathrm{d}^2 v^1}{\mathrm{d}\chi^2 \,\mathrm{d}\chi^2} = r\frac{\mathrm{d}v^2}{\mathrm{d}\chi^2} = -v^1 \qquad \frac{\mathrm{d}^2 v^2}{\mathrm{d}\chi^2 \,\mathrm{d}\chi^2} = [r]^{-1}\frac{\mathrm{d}v^1}{\mathrm{d}\chi^2} = -v^2$$

and has the solution in terms of the integration constants $c^1 = 0$, $c^2 = r$, $s^1 = r$, and $s^2 = 0$

$$v^1 = c^1 \cos\chi^2 + s^1 \sin\chi^2 = \sin\chi^2 \qquad v^2 = c^2 \cos\chi^2 + s^2 \sin\chi^2 = \cos\chi^2.$$

Thus upon parallel transport from $\chi^2 = 0$ to $\chi^2 = \pi/2$ the coefficients of v^i with respect to the polar coordinate system are changed from $\{v^i(r,0)\} = \{0,1\}$ to $\{v^i(r,\pi/2)\} = \{1,0\}$, however seen from the perspective of a Cartesian coordinate system the vector does not suffer any change in orientation, see the representation in Fig. 3.2. □

3.2.4 Geodesics as Straightest Curves

Geodesics are an important concept in differential geometry, they are parameter curves $\chi^K = \chi^K(t)$ (in terms of the *affine parameter* t in the sense of [168, 142]) that are as 'straight' as possible. Geodesics as the straightest curves in a manifold $\mathcal{M}$ follow from the

Definition 3.11. Tangent vectors $\dot{\chi}^I = \mathrm{d}\chi^I/\,\mathrm{d}t$ at every point $\chi^K(t)$ of a geodesic remain tangent upon a parallel transport along the geodesic. This statement is equivalent to the requirement that the covariant differential of $\dot{\chi}^I$ along a geodesic vanishes. Thus geodesics satisfy

$$\mathrm{D}\dot{\chi}^I = \dot{\chi}^I{}_{,K} \,\mathrm{d}\chi^K + \mathcal{L}^I{}_{JK}\dot{\chi}^J \,\mathrm{d}\chi^K = 0. \tag{3.30}$$

Noting that $\ddot{\chi}^I \,\mathrm{d}t = \dot{\chi}^I{}_{,K} \,\mathrm{d}\chi^K$ and dividing by $\mathrm{d}t$ renders finally the differential equation defining a geodesic

$$\mathrm{D}\dot{\chi}^I/\,\mathrm{d}t = \ddot{\chi}^I + \mathcal{L}^I{}_{JK}\dot{\chi}^J\dot{\chi}^K = 0. \tag{3.31}$$

Geodesics are also denoted (parameter) curves with the property of *auto-parallelism* for their tangent vectors. Note that geodesics are obviously independent of the skew symmetry in the indices J, K (i.e. the torsion that is introduced in the next subsection). □

3.2.5 Scalar Product and 'Angle' with Geodesics

3.2.5.1 Scalar Product

It may easily been proven that in a manifold (space/geometry) the scalar product $\mathcal{S} := \mathcal{V}^I \mathcal{W}_I$ of a vector $\mathcal{V}^I$ and a covector $\mathcal{W}_I$ is preserved upon an infinitesimal parallel transport along a parameter curve $\mathcal{X}^K(t)$ since firstly

$$\mathrm{p}\mathcal{S} = \mathrm{p}\mathcal{V}^I \mathcal{W}_I + \mathcal{V}^I \,\mathrm{p}\mathcal{W}_I. \tag{3.32}$$

Using secondly the definition of parallel transport renders the result

$$\mathrm{p}\mathcal{S} = -\mathcal{L}^I{}_{MK} \mathcal{V}^M \,\mathrm{d}\mathcal{X}^K \mathcal{W}_I + \mathcal{V}^I \mathcal{W}_M \mathcal{L}^M{}_{IK} \,\mathrm{d}\mathcal{X}^K \equiv 0. \tag{3.33}$$

Note that this result holds without the presence of a metric, to be introduced only later, and is of course in concord with the vanishing parallel transport of a scalar as established in rule **P1**.

3.2.5.2 'Angle' with Geodesics

Based on the above result an interpretation of parallel transport in a manifold (space/geometry) is possible: A covector $\mathcal{V}_I$ that is transported parallel along a geodesic $\mathcal{X}^I(t)$, parameterized by the arc-length t, preserves its scalar product with the tangent vector $\dot{\mathcal{X}}^I$ to the geodesic, i.e.

$$\frac{\mathrm{d}}{\mathrm{d}t}\left[\mathcal{V}_I \dot{\mathcal{X}}^I\right] = 0. \tag{3.34}$$

The scalar product of $\mathcal{V}_I$ and $\dot{\mathcal{X}}^I$ is here sloppily associated with the 'angle' between the covector and the tangent vector, even though their 'length' is not defined yet without the presence of a metric, to be introduced only later.

Proof. Performing firstly the differentiation on the left side of Eq. 3.34 with respect to the curve parameter t renders

$$\mathcal{V}_{I,K} \dot{\mathcal{X}}^I \dot{\mathcal{X}}^K + \mathcal{V}_I \ddot{\mathcal{X}}^I.$$

Secondly, upon subtracting and adding $\mathcal{V}_M \mathcal{L}^M{}_{IK} \dot{\mathcal{X}}^I \dot{\mathcal{X}}^K$ and reordering of the various terms the above result reads as

$$\left[\mathcal{V}_{I,K} - \mathcal{V}_M \mathcal{L}^M{}_{IK}\right] \dot{\mathcal{X}}^I \dot{\mathcal{X}}^K + \mathcal{V}_I \left[\ddot{\mathcal{X}}^I + \mathcal{L}^I{}_{JK} \dot{\mathcal{X}}^J \dot{\mathcal{X}}^K\right]$$

The terms in the first bracket define the covariant derivative of $\mathcal{V}_I$, however the covariant differential vanishes for the assumed parallel transport

$$\mathrm{D}\mathcal{V}_I = \left[\mathcal{V}_{I,K} - \mathcal{V}_M \mathcal{L}^M{}_{IK}\right] \dot{\mathcal{X}}^K \,\mathrm{d}t \equiv 0.$$

The terms in the second bracket are the defining equations for a geodesic and thus, according to the hypothesis, also vanish

$$\ddot{\mathcal{X}}^I + \mathcal{L}^I{}_{JK}\dot{\mathcal{X}}^J\dot{\mathcal{X}}^K \equiv 0.$$

Thus the left side of Eq. 3.34 combines to zero which concludes the proof. □

3.2.6 Integrable Connection

A special case occurs whenever the covariant and the partial derivative coincide for a particular coordinate system, which only happens for a (possibly anholonomic) Cartesian coordinate system with parallel transport of Euclidean type i.e. in a *flat manifold*. Then the connection $L^A{}_{BC}$ (or $l^a{}_{bc}$, respectively,) vanishes identically in the Cartesian-type coordinates $\{X^A\}$ (or $\{x^a\}$, respectively,) and the corresponding connection in the transformed coordinates $\{\chi^i\}$ (or $\{\mathcal{X}^I\}$, respectively,) consequently reads

$$\ell^i_{jk} = F^i{}_A f^A{}_{j,k} \quad (\text{or} \quad \mathcal{L}^I{}_{JK} = f^I{}_a F^a{}_{J,K}, \text{ respectively}). \tag{3.35}$$

Vice versa, connections of the format as in the above allow a coordinate transformation to a (possibly anholonomic) coordinate system with parallel transport of Euclidean type. Connections of this type are denoted as *integrable connections*, the reason for this terminology becoming clear only after the concept of *curvature* has been introduced.

As an example for the importance of this particular type of connection consider the covariant derivative with respect to χ^i in terms of an integrable connection ℓ^i_{jk}. After going through the derivations it turns out that it simply denotes the (mixed-variant) push-forward of the partial derivative with respect to (possibly anholonomic) Cartesian-type coordinates X^A

$$v^i{}_{|j} = F^i{}_A V^A{}_{,B} f^B{}_j. \tag{3.36}$$

Proof. The covariant derivative of v^i expands in terms of the connection as

$$v^i{}_{|j} = v^i{}_{,j} + \ell^i_{mj} v^m = v^i{}_{,j} + F^i{}_A f^A{}_{m,j} v^m.$$

Multiplying the above with $f^A{}_i$ renders

$$f^A{}_i v^i{}_{,j} + f^A{}_{m,j} v^m = [f^A{}_m v^m]_{,j} = V^A{}_{,B} f^B{}_j.$$

Multiplying finally the above with $F^i{}_A$ then proves the result. □

If the $\{X^A\}$ (or $\{x^a\}$, respectively,) are holonomic Cartesian coordinates in flat Euclidean space, i.e. $F^i{}_A \equiv \chi^i{}_{,A}$ (or $f^I{}_a \equiv \mathcal{X}^I{}_{,a}$, respectively,) integrable connections as defined in the above coincide with the corresponding *Christoffel symbols* associated with holonomic curvilinear coordinates $\{\chi^i\}$ (or $\{\mathcal{X}^I\}$, respectively). In the sequel, for better distinction, Christoffel symbols shall be denoted as

$$n^i{}_{jk} := F^i{}_A f^A{}_{j,k} \quad (\text{or} \quad \mathcal{N}^I{}_{JK} := f^I{}_a F^a{}_{J,K}, \text{ respectively}). \tag{3.37}$$

It is remarked that the Christoffel symbols are occasionally also denoted as Levi-Civita connection, see the comment by [179] and also terminology in [122, 37].

Addendum 3.1. Pseudo-Integrable Connection

A *pseudo-integrable connection* takes almost the same format as an integrable connection, however the Jacobian of the coordinate transformation $f^I{}_m$ and the *linear tangent map* $\mathcal{F}^{\star m}{}_J$ are independent from each other

$$\mathcal{L}^I{}_{JK} = f^I{}_m \mathcal{F}^{\star m}{}_{J,K}.$$

The independent linear tangent map $\mathcal{F}^{\star m}{}_J$ may nevertheless be related to the Jacobian of the coordinate transformation $\mathcal{F}^m{}_J$ as

$$\mathcal{F}^{\star m}{}_J = \mathcal{F}^m{}_J + \mathcal{H}^{\star m}{}_J.$$

As a consequence a pseudo-integrable connection may alternatively be expressed as the sum of an integrable and a non-integrable contribution

$$\mathcal{L}^I{}_{JK} = f^I{}_m \mathcal{F}^m{}_{J,K} + f^I{}_m \mathcal{H}^{\star m}{}_{J,K} = f^I{}_m \mathcal{F}^m{}_{J,K} + \mathcal{L}^{\star I}{}_{JK}.$$

Of course, any connection may be decomposed into an integrable and a non-integrable contribution $\mathcal{L}^{\star I}{}_{JK}$, however the non-integrable contribution to a pseudo-integrable connection derives from a tensor field $\mathcal{H}^{\star m}{}_J$ in the particular format $\mathcal{L}^{\star I}{}_{JK} = f^I{}_m \mathcal{H}^{\star m}{}_{J,K}$.

3.2.7 Cartan Transport of Affine Tangent Space

Another type of transport may be defined by considering the points of an affine tangent space rather than its vectors. Recall that an affine tangent space $T_{\mathcal{P}}\mathcal{M}$ may be attached to each point $\mathcal{P}$ (with coordinates $\{\mathcal{X}^K\}$) of a manifold $\mathcal{M}$. The vector from the origin of $T_{\mathcal{P}}\mathcal{M}$ at $\mathcal{P}$ to any other point in $T_{\mathcal{P}}\mathcal{M}$ is denoted the *radius vector* and has coordinates $\mathcal{Z}^I$. Then the transport of $T_{\mathcal{P}}\mathcal{M}$ along a parameter curve $\mathcal{X}^K(t)$ into a neighboring point $\mathcal{P}' \in \mathcal{M}$ at $\{\mathcal{X}^K + \mathrm{d}\mathcal{X}^K\}$ is called a *Cartan transport* if:

a) the tangent space $T_{\mathcal{P}}\mathcal{M}$ is transported along $\mathrm{d}\mathcal{X}^K$ such that it coincides momentarily with the affine tangent space $T_{\mathcal{P}'}\mathcal{M}$ at $\mathcal{P}'$,
b) and the radius vector $\mathcal{Z}^I$ is firstly transported parallel into $T_{\mathcal{P}'}\mathcal{M}$,
c) and is secondly elongated by the (negative) increment in the coordinates $\{\mathcal{X}^I\}$, i.e. by $-\,\mathrm{d}\mathcal{X}^I$.

Thus the Cartan transported radius vector reads

$$\mathcal{Z}^I(\{\mathcal{X}^K + \mathrm{d}\mathcal{X}^K\}) = \mathcal{Z}^I(\{\mathcal{X}^K\}) + \mathrm{p}\mathcal{Z}^I(\mathcal{X}^K) - \mathrm{d}\mathcal{X}^I. \tag{3.38}$$

As a consequence the change of the radius vector in the direction of the parameter curve (directional derivative) $\mathrm{d}\mathcal{Z}^I := \mathcal{Z}^I(\{\mathcal{X}^K + \mathrm{d}\mathcal{X}^K\}) - \mathcal{Z}^I(\{\mathcal{X}^K\})$ follows as

$$\mathrm{d}\mathcal{Z}^I = \mathrm{p}\mathcal{Z}^I - \mathrm{d}\mathcal{X}^I = -\mathcal{L}^I{}_{JK}\mathcal{Z}^J\,\mathrm{d}\mathcal{X}^K - \mathrm{d}\mathcal{X}^I. \tag{3.39}$$

Remarkably, this definition may be reformulated alternatively as a differential equation for the evolution of the radius vector upon Cartan transport, see [199], as

$$\mathrm{D}\mathcal{Z}^I + \mathrm{d}\mathcal{X}^I = 0. \tag{3.40}$$

The Cartan transport in a two-dimensional manifold is conceptually depicted in Fig. 3.3. Alternatively, Cartan transport of a tangent space may be thought of rolling the tangent space along the (curved) manifold from $\mathcal{P}$ to $\mathcal{P}'$ without sliding, see [79].

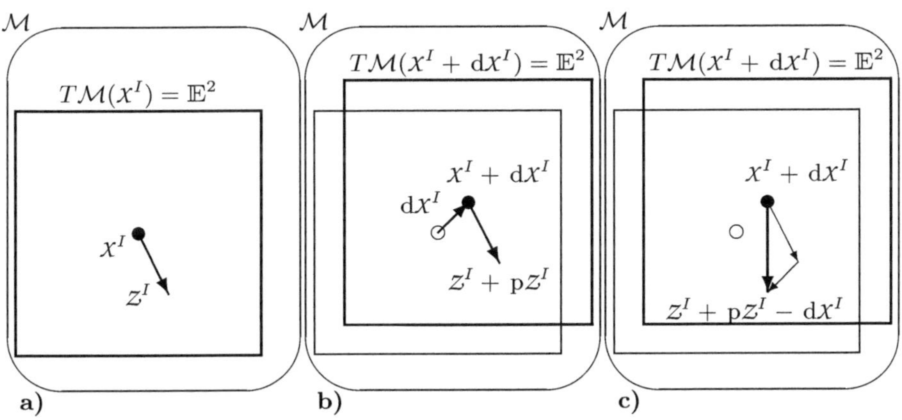

Fig. 3.3 Cartan transport in a two-dimensional manifold $\mathcal{M}$ (depicted by the oval): a) Two-dimensional (Euclidean) tangent space $T\mathcal{M}$ (depicted by the square) at $\{\mathcal{X}^I\}$ containing the radius vector $\mathcal{Z}^I$, b) Cartan transport of $T\mathcal{M}$ along $\mathrm{d}\mathcal{X}^I$ to coincide momentarily with tangent space at $\{\mathcal{X}^I + \mathrm{d}\mathcal{X}^I\}$ together with parallel transport of $\mathcal{Z}^I$ into $\mathcal{Z}^I + \mathrm{p}\mathcal{Z}^I$, c) radius vector $\mathcal{Z}^I + \mathrm{p}\mathcal{Z}^I - \mathrm{d}\mathcal{X}^I$ at $\{\mathcal{X}^I + \mathrm{d}\mathcal{X}^I\}$

3.3 Torsion

3.3.1 Transformation of Connection

As a motivation for the introduction of the *torsion* remember that the linear (or affine) connections $\mathcal{L}^I{}_{KL}$ and $\ell^i{}_{jk}$ do not transform like a tensor, but according to Eqs. 3.15 and 3.16 transform rather like

$$\begin{aligned} \mathcal{l}^i_{\ jk} &= \mathcal{F}^i_{\ I}\ \mathcal{L}^I_{\ JK}\ f^J_{\ j}\ f^K_{\ k} + \mathcal{F}^i_{\ I}\ f^I_{\ j,k}, \\ \mathcal{L}^I_{\ JK} &= f^I_{\ i}\ \ \mathcal{l}^i_{\ jk}\ \ \mathcal{F}^j_{\ J}\ \mathcal{F}^k_{\ K} + f^I_{\ i}\ \mathcal{F}^i_{\ J,K}. \end{aligned} \tag{3.41}$$

Observe that it is the second term in each line that conflicts with a tensorial transformation behaviour. By resorting to the following easy to proof relations for the partial derivatives of the tangent maps

$$\mathcal{F}^i_{\ I} f^I_{\ j,k} = -\mathcal{F}^i_{\ J,K} f^J_{\ j} f^K_{\ k} \quad \text{and} \quad f^I_{\ i}\mathcal{F}^i_{\ J,K} = -f^I_{\ j,k}\mathcal{F}^j_{\ J}\mathcal{F}^k_{\ K} \tag{3.42}$$

it is useful in the sequel to express the transformation of the connections also alternatively as

$$\begin{aligned} \mathcal{l}^i_{\ jk} &= \mathcal{F}^i_{\ I}\ \mathcal{L}^I_{\ JK}\ f^J_{\ j}\ f^K_{\ k} - \mathcal{F}^i_{\ J,K}\ f^J_{\ j}\ f^K_{\ k}, \\ \mathcal{L}^I_{\ JK} &= f^I_{\ i}\ \ \mathcal{l}^i_{\ jk}\ \ \mathcal{F}^j_{\ J}\ \mathcal{F}^k_{\ K} - f^I_{\ j,k}\ \ \mathcal{F}^j_{\ J}\ \mathcal{F}^k_{\ K}. \end{aligned} \tag{3.43}$$

It shall be observed carefully that these transformations of the connections are valid for *holonomic* as well as *anholonomic* coordinate transformations. Here holonomic and anholonomic refers to the *integrability* and *non-integrability* of the tangent map $\mathcal{F}^i_{\ I}$ (or likewise $f^I_{\ i}$) into a map $x^i = y^i(X^I)$ (or likewise $X^I = \mathcal{Y}^I(x^i)$).

3.3.2 Holonomic Transformation

It is obvious from the previous discussion that the connections $\mathcal{L}^I_{\ KL}$ and $\mathcal{l}^i_{\ jk}$ do not transform like third-order tensors. Under a holonomic change of coordinates, however, due to the symmetry of the second partial derivatives contained in the second terms of Eq. 3.43 its (right) skew symmetric contribution does

$$\begin{aligned} \mathcal{F}^i_{\ [J,K]} &= 0 \to\ \ \mathcal{l}^i_{\ [jk]} &&= \mathcal{F}^i_{\ I}\ \mathcal{L}^I_{\ [JK]}\ f^J_{\ j}\ f^K_{\ k}, \\ f^I_{\ [j,k]} &= 0 \to \mathcal{L}^I_{\ [JK]} &&= f^I_{\ i}\ \ \mathcal{l}^i_{\ [jk]}\ \ \mathcal{F}^j_{\ J}\ \mathcal{F}^k_{\ K}. \end{aligned} \tag{3.44}$$

Recall that skew symmetry in an index pair is denoted by square brackets, i.e. for example $\mathcal{F}^i_{\ [J,K]}$ denotes $\mathcal{F}^i_{\ [J,K]} := [\mathcal{F}^i_{\ J,K} - \mathcal{F}^i_{\ K,J}]/2$.
Now as a new object, *twice* the skew symmetric part of the connection is called the (Cartan) *torsion* or rather the *torsion tensor*

$$\mathcal{T}^I_{\ JK} := 2\mathcal{L}^I_{\ [JK]} \quad \text{and} \quad t^i_{\ jk} := 2\mathcal{l}^i_{\ [jk]}. \tag{3.45}$$

Remark 3.6. Note carefully that the torsion tensor is here introduced as *twice* the (right) skew symmetric connection, which is in contrast to a fair amount of literature [167, 168, 200, 6, 7, 8, 202, 37]. It turns out, however, that the scaling by the factor 2 results in a nice consistency with the representation of the *curvature* tensor, to be introduced later. □

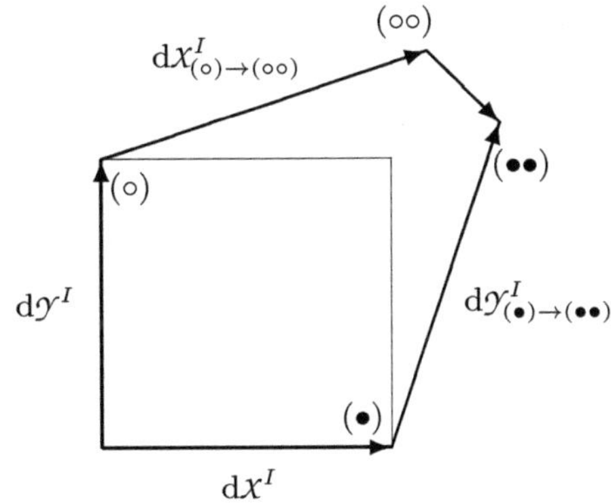

Fig. 3.4 In a manifold with torsion parallel transport of coordinate differentials along each other results in a pentagon

The consequence of the existence of torsion in a manifold can be highlighted by considering the situation sketched in Fig. 3.4, compare also to [168]. The parallel transport of two coordinate differentials $\mathrm{d}\mathcal{X}^I$ and $\mathrm{d}\mathcal{Y}^I$ along each other results in a pentagon formed by $\mathrm{d}\mathcal{X}^I$ and $\mathrm{d}\mathcal{Y}^I$ together with the parallel transported coordinate differentials

$$\mathrm{d}\mathcal{X}^I_{(\circ)\to(\circ\circ)} = \mathrm{d}\mathcal{X}^I - \mathcal{L}^I{}_{JK}\,\mathrm{d}\mathcal{X}^J\,\mathrm{d}\mathcal{Y}^K \tag{3.46}$$

and

$$\mathrm{d}\mathcal{Y}^I_{(\bullet)\to(\bullet\bullet)} = \mathrm{d}\mathcal{Y}^I - \mathcal{L}^I{}_{JK}\,\mathrm{d}\mathcal{Y}^J\,\mathrm{d}\mathcal{X}^K. \tag{3.47}$$

From the situation sketched in Fig. 3.4 it is thus clear that torsion measures the closure gap

$$\begin{aligned}\mathrm{d}\mathcal{X}^I + \mathrm{d}\mathcal{Y}^I_{(\bullet)\to(\bullet\bullet)} - \mathrm{d}\mathcal{Y}^I - \mathrm{d}\mathcal{X}^I_{(\circ)\to(\circ\circ)} = \\ \mathcal{L}^I{}_{JK}\,\mathrm{d}\mathcal{X}^J\,\mathrm{d}\mathcal{Y}^K - \mathcal{L}^I{}_{JK}\,\mathrm{d}\mathcal{Y}^J\,\mathrm{d}\mathcal{X}^K = \mathcal{T}^I{}_{JK}\,\mathrm{d}\mathcal{X}^J\,\mathrm{d}\mathcal{Y}^K.\end{aligned} \tag{3.48}$$

As a result infinitesimal parallelograms constructed from parallel transported coordinate differentials do only exist in manifolds (spaces/geometries) with vanishing torsion.

3.3.3 Anholonomic Transformation

Recall that the connections $\mathcal{L}^I{}_{KL}$ and ℓ^i_{jk} do not transform like third-order tensors. Under an anholonomic change of coordinates its (right) skew symmetric contribution thus transforms as

$$\begin{aligned}\mathcal{F}^i{}_{[J,K]} \neq 0 \to 2\ \ \ell^i_{[jk]} + \ a^i{}_{jk} &= \mathcal{F}^i_I\ \mathcal{T}^I{}_{JK}\ f^J_j\ f^K{}_k, \\ f^I{}_{[j,k]} \neq 0 \to 2\mathcal{L}^I{}_{[JK]} + \mathcal{A}^I{}_{JK} &= f^I_i\ \ t^i{}_{jk}\ \ \mathcal{F}^j{}_J\ \mathcal{F}^k{}_K,\end{aligned} \tag{3.49}$$

whereby the torsion in the holonomic coordinates follows the standard definition

$$\mathcal{T}^I{}_{JK} = 2\mathcal{L}^I{}_{[JK]} \quad \text{and} \quad t^i{}_{jk} = 2\ell^i{}_{[jk]}. \tag{3.50}$$

The additional contribution appearing in the transformation due to the lack of integrability is called the *anholonomic object*:

$$a^i{}_{jk} := 2\mathcal{F}^i{}_{[J,K]}f^J{}_j f^K{}_k \quad \text{and} \quad \mathcal{A}^I{}_{JK} := 2f^I{}_{[j,k]}\mathcal{F}^j{}_J\mathcal{F}^k{}_K. \tag{3.51}$$

The factor 2 in the above expressions shall be noted carefully. Moreover, observe that in the above expressions either the (spatial) coordinates χ^i in the first row of Eq. 3.49 or the (material) coordinates X^I in the second row of Eq. 3.49 are anholonomic.

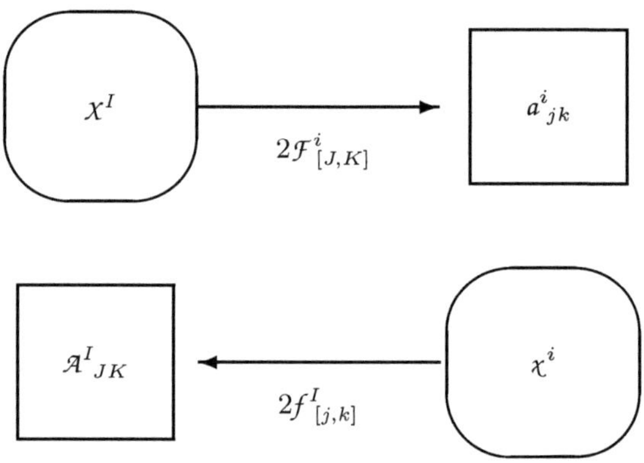

Fig. 3.5 The anholonomic object characterizes the non-integrability of the tangent map, i.e. the transformation of coordinate differentials. In the top figure χ^i are anholonomic while X^I are holonomic; in the bottom figure the situation is reversed, i.e. X^I are anholonomic and χ^i are holonomic.

Based on the anholonomic object and the representation in Eq. 3.49 the torsion in a space that is equipped with anholonomic coordinates follows from the

Definition 3.12. The torsion in an anholonomic space with either anholonomic (spatial) coordinate χ^i or anholonomic (material) coordinates X^I, respectively, is given as

$$t^i{}_{jk} := 2\ell^i{}_{[jk]} + a^i{}_{jk} \quad \text{and} \quad \mathcal{T}^I{}_{JK} := 2\mathcal{L}^I{}_{[JK]} + \mathcal{A}^I{}_{JK}. \tag{3.52}$$

The situation is highlighted in Fig. 3.5. □

It will be shown in the sequel, that the anholonomic objects may be associated with *dislocation density tensors*, see also [38]. Thereby, quite like in the definition of the various stress measures in nonlinear continuum mechanics, Piola-type anholonomic objects corresponding to two-point description dislocation density tensors together with Cauchy-type anholonomic objects follow from the

Definition 3.13. The Piola-type anholonomic object corresponding to the two-point description dislocation density tensor is given by

$$\mathcal{D}^{i}{}_{JK} := 2\mathcal{F}^{i}{}_{[J,K]} \quad \text{and} \quad d^{I}{}_{jk} := 2f^{I}{}_{[j,k]}. \tag{3.53}$$

Consequently the anholonomic object introduced previously is of Cauchy-type and results from a convection (push-forward/pull-back) by the corresponding tangent map

$$a^{i}{}_{jk} := \mathcal{D}^{i}{}_{JK} f^{J}{}_{j} f^{K}{}_{k} \quad \text{and} \quad \mathcal{A}^{I}{}_{JK} := d^{I}{}_{jk} \mathcal{F}^{j}{}_{J} \mathcal{F}^{k}{}_{K}. \tag{3.54}$$

The situation is highlighted in Fig. 3.6. □

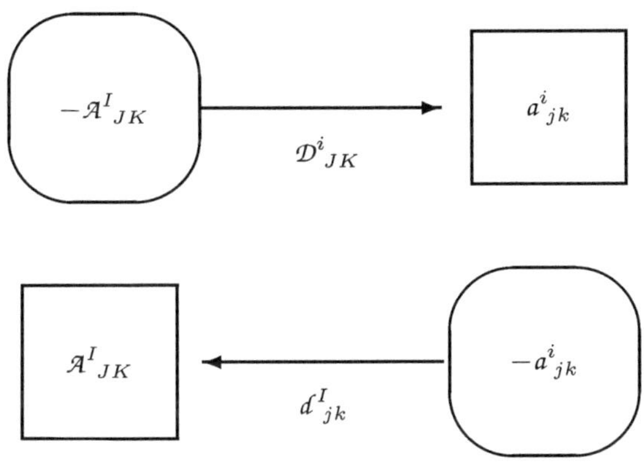

Fig. 3.6 The anholonomic object characterizes the non-integrability of the tangent map, i.e. the transformation of coordinate differentials. In the top figure x^{i} are anholonomic while X^{I} are holonomic; in the bottom figure the situation is reversed, i.e. X^{I} are anholonomic and x^{i} are holonomic.

Finally, Piola-Kirchhoff-type anholonomic objects may be regarded either as the pull-back/push-forward of the Piola-type or the Cauchy-type anholonomic objects, respectively, due to the

Definition 3.14. The Piola-Kirchhoff-type anholonomic object follows from the convection (pull-back/push-forward) of the Piola-type anholonomic object by the corresponding tangent map

$$-\mathcal{A}^{I}{}_{JK} = f^{I}{}_{i} \mathcal{D}^{i}{}_{JK} \quad \text{and} \quad -a^{i}{}_{jk} = \mathcal{F}^{i}{}_{I} d^{I}{}_{jk}. \tag{3.55}$$

It coincides with the definition of the previously introduced Cauchy-type anholonomic object if the following *anholonomic partial derivatives* are involved

$$f^{I}{}_{i} \mathcal{F}^{i}{}_{[J,K]} =: -f^{I}{}_{[j.k]} \mathcal{F}^{j}{}_{J} \mathcal{F}^{k}{}_{K} \quad \text{and} \quad \mathcal{F}^{i}{}_{I} f^{I}{}_{[j,k]} =: -\mathcal{F}^{i}{}_{[J.K]} f^{J}{}_{j} f^{K}{}_{k}. \tag{3.56}$$

Here the anholonomic partial derivatives are denoted by a dot instead of by a comma and are defined as $f^{I}{}_{j.k} := f^{I}{}_{j,K} f^{K}{}_{k}$ and $\mathcal{F}^{i}{}_{J.K} = \mathcal{F}^{i}{}_{J,k} \mathcal{F}^{k}{}_{K}$, respectively. Recall that in the above expressions either the (spatial) coordinates

χ^i or the (material) coordinates X^I are anholonomic. The situation is highlighted in Fig. 3.6. □

Remark 3.7. From the above definitions it is clear that the terminology *Cauchy-type* and *Piola-Kirchhoff-type* is used interchangeably if instead of the tangent map $\mathcal{F}^i_{\ I} : \mathrm{d}X^I \mapsto \mathrm{d}\chi^i$ the (inverse) tangent map $f^I_{\ i} : \mathrm{d}\chi^i \mapsto \mathrm{d}X^I$ is considered. □

3.3.4 Skew Symmetric Second Covariant Derivative

For holonomic coordinates the second covariant derivative of a (scalar-valued) field $\vartheta = \Theta(\{X^I\})$ is computed as the covariant derivative of the (covariant) first-order tensor (covector) represented by $\Theta_{|J} = \Theta_{,J}$, see the definition in Eq. 3.18.2

$$\Theta_{|JK} = \Theta_{,JK} - \Theta_{|I}\mathcal{L}^I_{\ JK}. \tag{3.57}$$

Taking into account the symmetry of the second partial derivatives renders the skew symmetric contribution to the second covariant derivative of a scalar-valued field in terms of the torsion

$$2\Theta_{|[JK]} = -2\Theta_{|I}\mathcal{L}^I_{\ [JK]} = -\Theta_{|I}\mathcal{T}^I_{\ JK}. \tag{3.58}$$

In the case of anholonomic (material) coordinates X^I the second partial derivative of a scalar-valued field computes as

$$\Theta_{,JK} = [\theta_{,j}\mathcal{F}^j_{\ J}]_{,k}\mathcal{F}^k_{\ K} = \theta_{,jk}\mathcal{F}^j_{\ J}\mathcal{F}^k_{\ K} - \Theta_{,I}f^I_{\ j,k}\mathcal{F}^j_{\ J}\mathcal{F}^k_{\ K}. \tag{3.59}$$

Thus the skew symmetric contribution to the second covariant derivative of a scalar-valued field reads as

$$2\Theta_{|[JK]} = -\Theta_{|I}[2\mathcal{L}^I_{\ [JK]} + \mathcal{A}^I_{\ JK}] = -\Theta_{|I}\mathcal{T}^I_{\ JK}. \tag{3.60}$$

The last equality follows from the definition of the torsion for anholonomic coordinates. The skew symmetric second covariant derivative of a (scalar-valued) field $\vartheta = \theta(\{\chi^i\})$ follows likewise as

$$2\theta_{|[jk]} = -\theta_{|i}t^i_{\ jk}. \tag{3.61}$$

The above relations holds obviously also for the case of anholonomic (spatial) χ^i with $t^i_{\ jk} = 2\ell^i_{\ [jk]} + a^i_{\ jk}$. In summary the second covariant derivative of a (scalar-valued) field is only symmetric in a symmetric manifold with vanishing torsion.

3.3.5 Torsion Covector, Semi-symmetric Connection

It proves convenient in the sequel to introduce in addition the *torsion covector* by the

Definition 3.15. The torsion covector follows as the trace (here the contraction of the two outer indices) of the torsion tensor

$$\mathcal{T}_J := \delta_I{}^K \mathcal{T}^I{}_{JK} = \mathcal{T}^I{}_{JI} = -\mathcal{T}^I{}_{IJ}. \tag{3.62}$$

The definition of the torsion covector in, e.g., [167] differs from Eq. 3.62 by a scalar factor composed from the dimension n_{dm} of the manifold. □

As a special case the notion of a *semi-symmetric connection* is introduced by the

Definition 3.16. For a semi-symmetric connection the torsion tensor follows in terms of the torsion covector as

$$\mathcal{T}^I{}_{JK} := \frac{2}{1-n_{dm}} \delta^I{}_{[J} \mathcal{T}_{K]} = \frac{1}{1-n_{dm}} [\delta^I{}_J \mathcal{T}_K - \delta^I{}_K \mathcal{T}_J]. \tag{3.63}$$

Obviously contracting the so defined $\mathcal{T}^I{}_{JK}$ by $\delta_I{}^K$ reproduces the torsion covector $\mathcal{T}_J$ from Eq. 3.62. □

Remark 3.8. For a semi-symmetric connection the closure gap for the pentagon resulting from two coordinate differentials and their parallel transport along each other computes as

$$\mathcal{T}^I{}_{JK}\, \mathrm{d}\mathcal{X}^J\, \mathrm{d}\mathcal{Y}^K = \frac{1}{1-n_{dm}} [\,\mathrm{d}\mathcal{X}^I\, \mathrm{d}\mathcal{Y}^M - \mathrm{d}\mathcal{Y}^I\, \mathrm{d}\mathcal{X}^M] \mathcal{T}_M.$$

Consequently the closure gap vanishes if $\mathrm{d}\mathcal{X}^M \mathcal{T}_M = 0$ and $\mathrm{d}\mathcal{Y}^M \mathcal{T}_M = 0$ holds simultaneously, i.e. if the coordinate differentials $\mathrm{d}\mathcal{X}^M$ and $\mathrm{d}\mathcal{Y}^M$ are from the $n_{dm} - 1$ dimensional subspace orthogonal to the torsion covector $\mathcal{T}_M$. Thus the connection is symmetric with infinitesimal parallelograms made up from parallel transported coordinate differentials exclusively for this subspace. □

3.3.6 Torsion of an Integrable Connection

For a connection $L^A{}_{BC} \equiv 0$ (or $l^a{}_{bc} \equiv 0$, respectively,) related to the (possibly anholonomic) Cartesian-type coordinates $\{X^A\}$ (or $\{x^a\}$, respectively,) and the corresponding integrable connection $\ell^i_{jk} = F^i{}_A f^A{}_{j,k}$ (or $\mathcal{L}^I{}_{JK} = f^I{}_a F^a{}_{J,K}$), respectively) in the transformed (holonomic) coordinates $\{\mathcal{x}^i\}$ (or $\{\mathcal{X}^I\}$, respectively,) the corresponding torsion in the (possibly anholonomic) Cartesian-type coordinates coincides with the anholonomic object

$$T^A{}_{BC} = 2L^A{}_{[BC]} + A^A{}_{BC} \equiv A^A{}_{BC} \quad (\text{or} \quad t^a{}_{bc} = 2l^a{}_{[bc]} + a^a{}_{bc} \equiv a^a{}_{bc}), \tag{3.64}$$

whereby the anholonomic object is defined as

$$A^A{}_{BC} := 2f^A{}_{[j,k]} F^j{}_B F^k{}_C \quad (\text{or} \quad a^a{}_{bc} := 2F^a{}_{[J,K]} f^J{}_b f^K{}_c). \tag{3.65}$$

Accordingly, in the transformed (holonomic) coordinates the tensorially transformed torsion follows as

$$t^i{}_{jk} = 2\ell^i{}_{[jk]} \equiv 2F^i{}_A f^A{}_{[j,k]} \quad (\text{or} \quad \mathcal{T}^I{}_{JK} = 2\mathcal{L}^I{}_{[JK]} \equiv 2f^I{}_a F^a{}_{[J,K]}). \tag{3.66}$$

Here the properly defined transformed anholonomic object coincides (modulo the sign) with the torsion

$$a^i{}_{jk} := -2F^i{}_A f^A{}_{[j,k]} \equiv -t^i{}_{jk} \quad (\text{or} \quad \mathcal{A}^I{}_{JK} := -2f^I{}_a F^a{}_{[J,K]} \equiv -\mathcal{T}^I{}_{JK}). \tag{3.67}$$

Note the minus sign in the definition of the transformed anholonomic object. Flat but anholonomic manifolds (spaces/geometries) play an important role in the kinematics of crystalline materials as considered in Part III of this treatise. There the anholonomic objects are related to dislocation density tensors.

3.4 Curvature

The notion of *curvature* or rather the *curvature tensor* is central to the differential geometry of manifolds. Formally the curvature tensor is introduced by the following

Definition 3.17. Based on the linear connection a fourth-order object, the curvature tensor, is defined as:

$$\mathcal{R}^I{}_{JKL} := \mathcal{L}^I{}_{JL,K} - \mathcal{L}^I{}_{JK,L} + \mathcal{L}^I{}_{MK}\mathcal{L}^M{}_{JL} - \mathcal{L}^I{}_{ML}\mathcal{L}^M{}_{JK}. \tag{3.68}$$

A mnemonic trick or rather symbolic instruction of how to compute the coefficients of the curvature tensor is as follows, see [36]

$$\mathcal{R}^I{}_{JKL} := -\begin{vmatrix} \mathcal{L}^I{}_{JK} & \mathcal{L}^I{}_{JL} \\ \partial_{X^K} & \partial_{X^L} \end{vmatrix} - \begin{vmatrix} \mathcal{L}^M{}_{JK} & \mathcal{L}^M{}_{JL} \\ \mathcal{L}^I{}_{MK} & \mathcal{L}^I{}_{ML} \end{vmatrix}. \tag{3.69}$$

The tensorial transformation properties of the curvature tensor will be demonstrated in more detail later. □

From its definition the curvature tensor obeys the following skew symmetries

$$\mathcal{R}^I{}_{JKL} = 2\mathcal{L}^I{}_{J[L,K]} + 2\mathcal{L}^I{}_{M[K}\mathcal{L}^M{}_{JL]} = \mathcal{R}^I{}_{J[KL]}. \tag{3.70}$$

Remark 3.9. Note carefully that here, in contrast to most of the literature on differential geometry, the notation for skew symmetrization of the two indices in the term quadratic in the connection is used in the following format

$$2\mathcal{L}^I{}_{M[K}\mathcal{L}^M{}_{JL]} := \mathcal{L}^I{}_{MK}\mathcal{L}^M{}_{JL} - \mathcal{L}^I{}_{ML}\mathcal{L}^M{}_{JK}. \tag{3.71}$$

This somewhat less heavy notation is here preferred over the traditional $\mathcal{L}^I{}_{M[K}\mathcal{L}^M{}_{|J|L]}$. □

Expressing the partial derivative of the connection in terms of its covariant derivative, i.e.

$$2\mathcal{L}^I{}_{J[L,K]} = 2\mathcal{L}^I{}_{J[L|K]} - 2\mathcal{L}^I{}_{M[K}\mathcal{L}^M{}_{JL]} + 2\mathcal{L}^M{}_{J[K}\mathcal{L}^I{}_{ML]} + \mathcal{T}^M{}_{LK}\mathcal{L}^I{}_{JM}$$

allows eventually to express the curvature tensor alternatively as

$$\mathcal{R}^I{}_{JKL} = 2\mathcal{L}^I{}_{J[L|K]} - 2\mathcal{L}^I{}_{M[K}\mathcal{L}^M{}_{JL]} - \mathcal{L}^I{}_{JM}\mathcal{T}^M{}_{KL}. \tag{3.72}$$

Observe that the representation of the curvature tensor in terms of the covariant derivative of the connection involves also the previously introduced torsion.

Addendum 3.2. Identities for the Curvature Tensor

From a straightforward but tedious analysis the following identities for the curvature tensor may be proven (see, e.g. [167], however with pre-factors 2 and 4 due to a different scaling of the definition for the torsion):

C1: $\mathcal{R}^I{}_{J(KL)} = 0$
C2: $\mathcal{R}^I{}_{[\![JKL]\!]} = -\mathcal{T}^I{}_{[\![JK|L]\!]} + \mathcal{T}^I{}_{M[\![J}\mathcal{T}^M{}_{KL]\!]}$
C3: $\mathcal{R}^I{}_{J[\![KL|M]\!]} = \mathcal{R}^I{}_{JN[\![K}\mathcal{T}^N{}_{LM]\!]}$

Here, symmetry in an index pair is denoted by round brackets, i.e. for example $\mathcal{R}^I{}_{J(KL)} := [\mathcal{R}^I{}_{JKL} + \mathcal{R}^I{}_{JLK}]/2$. Moreover $[\![JKL]\!]$ denotes the full skew symmetry (all possible swaps of two indices with alternating signs) in three indices, i.e. $3![\![JKL]\!] := JKL - KJL + KLJ - JLK + LJK - LKJ$. Note that the above identities hold without the presence of a metric that will only be introduced later (and that will result in slightly modified identities). Due to **C1** (due to the skew symmetry of the curvature tensor in the indices K, L it holds for the index combination $3![\![JKL]\!] := 2[JKL + LJK + KLJ]$) they may alternatively be stated as

c2: $\mathcal{R}^I{}_{JKL} + \mathcal{R}^I{}_{LJK} + \mathcal{R}^I{}_{KLJ} = -\left[\mathcal{T}^I{}_{JK|L} + \mathcal{T}^I{}_{LJ|K} + \mathcal{T}^I{}_{KL|J}\right] + \left[\mathcal{T}^I{}_{MJ}\mathcal{T}^M{}_{KL} + \mathcal{T}^I{}_{ML}\mathcal{T}^M{}_{JK} + \mathcal{T}^I{}_{MK}\mathcal{T}^M{}_{LJ}\right]$

c3: $\mathcal{R}^I{}_{JKL|M} + \mathcal{R}^I{}_{JMK|L} + \mathcal{R}^I{}_{JLM|K} = \left[\mathcal{R}^I{}_{JNK}\mathcal{T}^N{}_{LM} + \mathcal{R}^I{}_{JNM}\mathcal{T}^N{}_{KL} + \mathcal{R}^I{}_{JNL}\mathcal{T}^N{}_{MK}\right]$

C1 simply holds due to the very definition of the curvature tensor. **C2** (**c2**) and **C3** (**c3**) may be verified explicitly by carefully writing down the lengthy expressions. Performing this tiring task is left to the reader. **C2** (**c2**) denotes the *first (algebraic) Bianchi identity*, correspondingly **C3** (**c3**) is called the *second (differential) Bianchi identity*. Observe that the right-hand sides of the Bianchi identities vanish identically for a symmetric manifold (space/geometry) without torsion.

Less formal and more operational is the alternative

Definition 3.18. The curvature tensor determines the change of a vector $\mathcal{V}^I$ for a parallel transport along infinitesimal circuits (closed curves) as

$$\Delta\mathcal{V}^I = \mathcal{R}^I{}_{JKL}\mathcal{V}^J \,\mathrm{d}\mathcal{X}^K \,\mathrm{d}\mathcal{Y}^L. \tag{3.73}$$

Thus, upon transporting a vector $\mathcal{V}^I$ parallel along infinitesimal circuits (closed curves) if suffers a change $\Delta\mathcal{V}^I$ that depends on the curvature tensor, i.e. on the curvature of the manifold.

Moreover, the curvature tensor also determines the skew symmetric contribution to the second covariant derivatives of a vector

$$2\mathcal{V}^I{}_{|[KL]} = -\mathcal{R}^I{}_{JKL}\mathcal{V}^J - \mathcal{V}^I{}_{|M}\mathcal{T}^M{}_{KL}. \tag{3.74}$$

Observe that the skew second covariant derivative of a vector involves in particular the torsion. □

Similar statements for the curvature tensor hold in terms of covectors, they follow from the

Definition 3.19. The curvature tensor determines the change of a covector $\mathcal{V}_J$ for a parallel transport along infinitesimal circuits (closed curves) as

$$\Delta\mathcal{V}_J = -\mathcal{V}_I\mathcal{R}^I{}_{JKL} \,\mathrm{d}\mathcal{X}^K \,\mathrm{d}\mathcal{Y}^L. \tag{3.75}$$

Thus, upon transporting a covector $\mathcal{V}_J$ parallel along infinitesimal circuits (closed curves) if suffers a change $\Delta\mathcal{V}_J$ that depends on the curvature tensor, i.e. on the curvature of the manifold.

In addition, the curvature tensor also determines the skew symmetric contribution to the second covariant derivatives of a covector

$$2\mathcal{V}_{J|[KL]} = \mathcal{V}_I\mathcal{R}^I{}_{JKL} - \mathcal{V}_{J|M}\mathcal{T}^M{}_{KL}. \tag{3.76}$$

Note that the skew second covariant derivative of a covector involves in particular the torsion. □

In the sequel both, the change of vectors $\mathcal{V}^I$ and covectors $\mathcal{V}_I$ for a *parallel transport along infinitesimal circuits (closed curves)* and the *skew symmetric second covariant derivatives* of vectors $\mathcal{V}^I$ and covectors $\mathcal{V}_I$, relations that are also denoted the *Ricci identities*, shall be investigated.

3.4.1 Parallel Transport along Infinitesimal Circuits

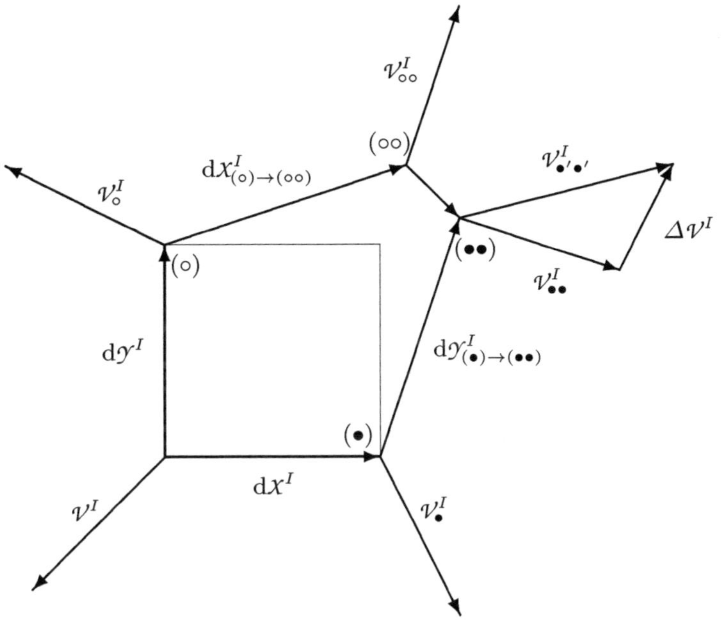

Fig. 3.7 Parallel transport of a vector $\mathcal{V}^I$ along an infinitesimal circuit (closed curve). Due to the curvature $\mathcal{V}^I$ suffers a change $\Delta\mathcal{V}^I$. The closed curve consists of the coordinate differentials $\mathrm{d}\mathcal{X}^I$ and $\mathrm{d}\mathcal{Y}^I$ together with their parallel transport along each other $\mathrm{d}\mathcal{Y}^I_{(\bullet)\to(\bullet\bullet)}$ and $\mathrm{d}\mathcal{X}^I_{(\circ)\to(\circ\circ)}$ and the resulting closure gap, compare Fig. 3.4.

Parallel Transport of Vectors along Infinitesimal Circuits. By referring to Fig. 3.7 the proof of $\Delta\mathcal{V}^I = \mathcal{R}^I{}_{JKL}\mathcal{V}^J\,\mathrm{d}\mathcal{X}^K\,\mathrm{d}\mathcal{Y}^L$ may be sketched in nine steps:

1. Parallel transport of $\mathcal{V}^I$ to $(\bullet)$ and connection $\mathcal{L}^I{}_{JK\bullet}$ at $(\bullet)$,
2. Parallel transport of $\mathcal{V}^I_\bullet$ to $(\bullet\bullet)$,
3. Retain terms up to quadratic order,
4. Parallel transport of $\mathcal{V}^I$ to $(\circ)$ and connection $\mathcal{L}^I{}_{JK\circ}$ at $(\circ)$,
5. Parallel transport of $\mathcal{V}^I_\circ$ to $(\circ\circ)$,
6. Retain terms up to quadratic order,
7. Parallel transport from $(\circ\circ)$ to $(\bullet\bullet)$,
   retain terms up to quadratic order,
8. Subtract,
9. Express Result in terms of the curvature tensor.

Proof. The above steps shall now be outlined in more detail:

1. Parallel transport of $\mathcal{V}^I$ to $(\bullet)$ and connection $\mathcal{L}^I{}_{JK\bullet}$ at $(\bullet)$:

$$\mathcal{V}^I_\bullet = \mathcal{V}^I - \mathcal{L}^I{}_{JK}\mathcal{V}^J \,\mathrm{d}\mathcal{X}^K \quad \text{and} \quad \mathcal{L}^I{}_{JK\bullet} = \mathcal{L}^I{}_{JK} + \mathcal{L}^I{}_{JK,L}\,\mathrm{d}\mathcal{X}^L.$$

2. Parallel transport of $\mathcal{V}^I_\bullet$ to $(\bullet\bullet)$:

$$\mathcal{V}^I_{\bullet\bullet} = \mathcal{V}^I_\bullet - \mathcal{L}^I{}_{JK\bullet}\mathcal{V}^J_\bullet \left[\mathrm{d}\mathcal{Y}^K - \mathcal{L}^K{}_{OP}\,\mathrm{d}\mathcal{Y}^O\,\mathrm{d}\mathcal{X}^P\right] =$$

$$\left[\mathcal{V}^I - \mathcal{L}^I{}_{JK}\mathcal{V}^J\,\mathrm{d}\mathcal{X}^K\right] - \left[\mathcal{L}^I{}_{JK} + \mathcal{L}^I{}_{JK,L}\,\mathrm{d}\mathcal{X}^L\right] \times$$

$$\left[\mathcal{V}^J - \mathcal{L}^J{}_{MN}\mathcal{V}^M\,\mathrm{d}\mathcal{X}^N\right] \times \left[\mathrm{d}\mathcal{Y}^K - \mathcal{L}^K{}_{OP}\,\mathrm{d}\mathcal{Y}^O\,\mathrm{d}\mathcal{X}^P\right].$$

3. Retain terms up to quadratic order in the coordinate differentials:

$$\mathcal{V}^I_{\bullet\bullet} = \mathcal{V}^I - \mathcal{L}^I{}_{JK}\mathcal{V}^J\,\mathrm{d}\mathcal{X}^K - \mathcal{L}^I{}_{JK}\mathcal{V}^J\,\mathrm{d}\mathcal{Y}^K - \mathcal{L}^I{}_{JK,L}\,\mathrm{d}\mathcal{X}^L\mathcal{V}^J\,\mathrm{d}\mathcal{Y}^K$$

$$+\mathcal{L}^I{}_{JK}\mathcal{L}^J{}_{MN}\mathcal{V}^M\,\mathrm{d}\mathcal{X}^N\,\mathrm{d}\mathcal{Y}^K + \mathcal{L}^I{}_{JK}\mathcal{L}^K{}_{OP}\mathcal{V}^J\,\mathrm{d}\mathcal{Y}^O\,\mathrm{d}\mathcal{X}^P.$$

4. Parallel transport of $\mathcal{V}^I$ to $(\circ)$ and connection $\mathcal{L}^I{}_{JK\circ}$ at $(\circ)$:

$$\mathcal{V}^I_\circ = \mathcal{V}^I - \mathcal{L}^I{}_{JK}\mathcal{V}^J\,\mathrm{d}\mathcal{Y}^K \quad \text{and} \quad \mathcal{L}^I{}_{JK\circ} = \mathcal{L}^I{}_{JK} + \mathcal{L}^I{}_{JK,L}\,\mathrm{d}\mathcal{Y}^L.$$

5. Parallel transport of $\mathcal{V}^I_\circ$ to $(\circ\circ)$:

$$\mathcal{V}^I_{\circ\circ} = \mathcal{V}^I_\circ - \mathcal{L}^I{}_{JK\circ}\mathcal{V}^J_\circ\left[\mathrm{d}\mathcal{X}^K - \mathcal{L}^K{}_{OP}\,\mathrm{d}\mathcal{X}^O\,\mathrm{d}\mathcal{Y}^P\right] =$$

$$\left[\mathcal{V}^I - \mathcal{L}^I{}_{JK}\mathcal{V}^J\,\mathrm{d}\mathcal{Y}^K\right] - \left[\mathcal{L}^I{}_{JK} + \mathcal{L}^I{}_{JK,L}\,\mathrm{d}\mathcal{Y}^L\right] \times$$

$$\left[\mathcal{V}^J - \mathcal{L}^J{}_{MN}\mathcal{V}^M\,\mathrm{d}\mathcal{Y}^N\right] \times \left[\mathrm{d}\mathcal{X}^K - \mathcal{L}^K{}_{OP}\,\mathrm{d}\mathcal{X}^O\,\mathrm{d}\mathcal{Y}^P\right].$$

6. Retain terms up to quadratic order in the coordinate differentials:

$$\mathcal{V}^I_{\circ\circ} = \mathcal{V}^I - \mathcal{L}^I{}_{JK}\mathcal{V}^J\,\mathrm{d}\mathcal{Y}^K - \mathcal{L}^I{}_{JK}\mathcal{V}^J\,\mathrm{d}\mathcal{X}^K - \mathcal{L}^I{}_{JK,L}\,\mathrm{d}\mathcal{Y}^L\mathcal{V}^J\,\mathrm{d}\mathcal{X}^K$$

$$+\mathcal{L}^I{}_{JK}\mathcal{L}^J{}_{MN}\mathcal{V}^M\,\mathrm{d}\mathcal{Y}^N\,\mathrm{d}\mathcal{X}^K + \mathcal{L}^I{}_{JK}\mathcal{L}^K{}_{OP}\mathcal{V}^J\,\mathrm{d}\mathcal{X}^O\,\mathrm{d}\mathcal{Y}^P.$$

7. Subtract:

$$\mathcal{V}^I_{\circ\circ} - \mathcal{V}^I_{\bullet\bullet} = -2\mathcal{L}^I{}_{J[K,L]}\,\mathrm{d}\mathcal{Y}^L\mathcal{V}^J\,\mathrm{d}\mathcal{X}^K$$

$$+2\mathcal{L}^I{}_{J[K}\mathcal{L}^J{}_{MN]}\mathcal{V}^M\,\mathrm{d}\mathcal{Y}^N\,\mathrm{d}\mathcal{X}^K + 2\mathcal{L}^I{}_{JK}\mathcal{L}^K{}_{[OP]}\mathcal{V}^J\,\mathrm{d}\mathcal{X}^O\,\mathrm{d}\mathcal{Y}^P.$$

8. Parallel transport from $(\circ\circ)$ to $(\bullet\bullet)$, retain terms up to quadratic order in the coordinate differentials:

$$\mathcal{V}^I_{\bullet'\bullet'} = \mathcal{V}^I_{\circ\circ} - \mathcal{L}^I{}_{JK}\mathcal{V}^J\left[2\mathcal{L}^K{}_{[OP]}\,\mathrm{d}\mathcal{X}^O\,\mathrm{d}\mathcal{Y}^P\right].$$

9. Subtract:

$$\mathcal{V}^I_{\bullet'\bullet'} - \mathcal{V}^I_{\bullet\bullet} = -2\mathcal{L}^I{}_{J[K,L]}\,\mathrm{d}\mathcal{Y}^L\mathcal{V}^J\,\mathrm{d}\mathcal{X}^K + 2\mathcal{L}^I{}_{J[K}\mathcal{L}^J{}_{MN]}\mathcal{V}^M\,\mathrm{d}\mathcal{Y}^N\,\mathrm{d}\mathcal{X}^K.$$

Thus, in summary the change of the vector $\mathcal{V}^I$ may be expressed in terms of the curvature tensor as defined in Eq. 3.70

$$\Delta\mathcal{V}^I := \mathcal{V}^I_{\bullet'\bullet'} - \mathcal{V}^I_{\bullet\bullet} =: \mathcal{R}^I{}_{JKL}\,\mathcal{V}^J\,\mathrm{d}\mathcal{X}^K\,\mathrm{d}\mathcal{Y}^L.$$

Clearly, from the above derivation and in accordance with the definition in Eq. 3.70 the curvature tensor is eventually recognized as

$$\frac{1}{2}\mathcal{R}^I{}_{JKL} = \mathcal{L}^I{}_{J[L,K]} + \mathcal{L}^I{}_{M[K}\mathcal{L}^M{}_{JL]}.$$

This concludes the proof. □

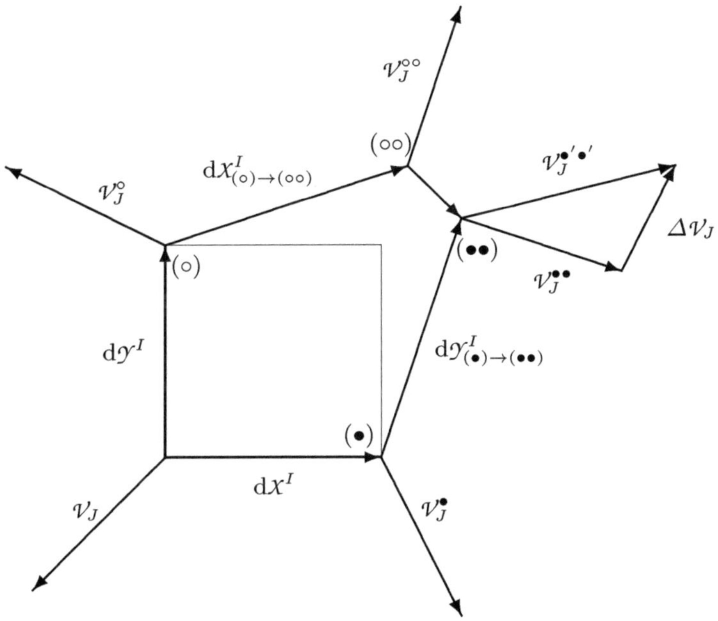

Fig. 3.8 Parallel transport of a covector $\mathcal{V}_J$ along an infinitesimal circuit (closed curve). Due to the curvature $\mathcal{V}_J$ suffers a change $\Delta\mathcal{V}_J$. The closed curve consists of the coordinate differentials $\mathrm{d}\mathcal{X}^I$ and $\mathrm{d}\mathcal{Y}^I$ together with their parallel transport along each other $\mathrm{d}\mathcal{Y}^I_{(\bullet)\to(\bullet\bullet)}$ and $\mathrm{d}\mathcal{X}^I_{(\circ)\to(\circ\circ)}$ and the resulting closure gap, compare Fig. 3.4.

Parallel Transport of Covectors along Infinitesimal Circuits. By referring to Fig. 3.8 the proof of $\Delta\mathcal{V}_J = -\mathcal{V}_I\mathcal{R}^I{}_{JKL}\,\mathrm{d}\mathcal{X}^K\,\mathrm{d}\mathcal{Y}^L$ follows corresponding steps as in the above.

Proof. These steps shall now be outlined in more detail:

1. Parallel transport of $\mathcal{V}_J$ to $(\bullet)$ and connection $\mathcal{L}^I{}_{JK\bullet}$ at $(\bullet)$:

$$\mathcal{V}_J^{\bullet} = \mathcal{V}_J + \mathcal{V}_I \mathcal{L}^I{}_{JK}\, \mathrm{d}\mathcal{X}^K \quad \text{and} \quad \mathcal{L}^I{}_{JK\bullet} = \mathcal{L}^I{}_{JK} + \mathcal{L}^I{}_{JK,L}\, \mathrm{d}\mathcal{X}^L .$$

2. Parallel transport of $\mathcal{V}_J^{\bullet}$ to $(\bullet\bullet)$:

$$\mathcal{V}_J^{\bullet\bullet} = \mathcal{V}_J^{\bullet} + \mathcal{V}_I^{\bullet} \mathcal{L}^I{}_{JK\bullet} \left[\mathrm{d}\mathcal{Y}^K - \mathcal{L}^K{}_{OP}\, \mathrm{d}\mathcal{Y}^O\, \mathrm{d}\mathcal{X}^P \right] =$$

$$\left[\mathcal{V}_J + \mathcal{V}_I \mathcal{L}^I{}_{JK}\, \mathrm{d}\mathcal{X}^K \right] + \left[\mathcal{V}_I + \mathcal{V}_M \mathcal{L}^M{}_{IN}\, \mathrm{d}\mathcal{X}^N \right] \times$$
$$\left[\mathcal{L}^I{}_{JK} + \mathcal{L}^I{}_{JK,L}\, \mathrm{d}\mathcal{X}^L \right] \times \left[\mathrm{d}\mathcal{Y}^K - \mathcal{L}^K{}_{OP}\, \mathrm{d}\mathcal{Y}^O\, \mathrm{d}\mathcal{X}^P \right] .$$

3. Retain terms up to quadratic order in the coordinate differentials:

$$\mathcal{V}_J^{\bullet\bullet} = \mathcal{V}_J + \mathcal{V}_I \mathcal{L}^I{}_{JK}\, \mathrm{d}\mathcal{X}^K + \mathcal{V}_I \mathcal{L}^I{}_{JK}\, \mathrm{d}\mathcal{Y}^K + \mathcal{V}_I \mathcal{L}^I{}_{JK,L}\, \mathrm{d}\mathcal{X}^L\, \mathrm{d}\mathcal{Y}^K$$

$$+ \mathcal{V}_M \mathcal{L}^M{}_{IN} \mathcal{L}^I{}_{JK}\, \mathrm{d}\mathcal{X}^N\, \mathrm{d}\mathcal{Y}^K - \mathcal{V}_I \mathcal{L}^I{}_{JK} \mathcal{L}^K{}_{OP}\, \mathrm{d}\mathcal{Y}^O\, \mathrm{d}\mathcal{X}^P .$$

4. Parallel transport of $\mathcal{V}_J$ to $(\circ)$ and connection $\mathcal{L}^I{}_{JK\circ}$ at $(\circ)$:

$$\mathcal{V}_J^{\circ} = \mathcal{V}_J + \mathcal{V}_I \mathcal{L}^I{}_{JK}\, \mathrm{d}\mathcal{Y}^K \quad \text{and} \quad \mathcal{L}^I{}_{JK\circ} = \mathcal{L}^I{}_{JK} + \mathcal{L}^I{}_{JK,L}\, \mathrm{d}\mathcal{Y}^L .$$

5. Parallel transport of $\mathcal{V}_J^{\circ}$ to $(\circ\circ)$:

$$\mathcal{V}_J^{\circ\circ} = \mathcal{V}_J^{\circ} + \mathcal{V}_I^{\circ} \mathcal{L}^I{}_{JK\circ} \left[\mathrm{d}\mathcal{X}^K - \mathcal{L}^K{}_{OP}\, \mathrm{d}\mathcal{X}^O\, \mathrm{d}\mathcal{Y}^P \right] =$$

$$\left[\mathcal{V}_J + \mathcal{V}_I \mathcal{L}^I{}_{JK}\, \mathrm{d}\mathcal{Y}^K \right] + \left[\mathcal{V}_I + \mathcal{V}_M \mathcal{L}^M{}_{IN}\, \mathrm{d}\mathcal{Y}^N \right] \times$$
$$\left[\mathcal{L}^I{}_{JK} + \mathcal{L}^I{}_{JK,L}\, \mathrm{d}\mathcal{Y}^L \right] \times \left[\mathrm{d}\mathcal{X}^K - \mathcal{L}^K{}_{OP}\, \mathrm{d}\mathcal{X}^O\, \mathrm{d}\mathcal{Y}^P \right] .$$

6. Retain terms up to quadratic order in the coordinate differentials:

$$\mathcal{V}_J^{\circ\circ} = \mathcal{V}_J + \mathcal{V}_I \mathcal{L}^I{}_{JK}\, \mathrm{d}\mathcal{Y}^K + \mathcal{V}_I \mathcal{L}^I{}_{JK}\, \mathrm{d}\mathcal{X}^K + \mathcal{V}_I \mathcal{L}^I{}_{JK,L}\, \mathrm{d}\mathcal{Y}^L\, \mathrm{d}\mathcal{X}^K$$

$$+ \mathcal{V}_M \mathcal{L}^M{}_{IN} \mathcal{L}^I{}_{JK}\, \mathrm{d}\mathcal{Y}^N\, \mathrm{d}\mathcal{X}^K - \mathcal{V}_I \mathcal{L}^I{}_{JK} \mathcal{L}^K{}_{OP}\, \mathrm{d}\mathcal{X}^O\, \mathrm{d}\mathcal{Y}^P .$$

7. Subtract:

$$\mathcal{V}_J^{\circ\circ} - \mathcal{V}_J^{\bullet\bullet} = -2 \mathcal{V}_I \mathcal{L}^I{}_{J[L,K]}\, \mathrm{d}\mathcal{Y}^L\, \mathrm{d}\mathcal{X}^K$$
$$-2 \mathcal{V}_M \mathcal{L}^M{}_{I[K} \mathcal{L}^I{}_{JN]}\, \mathrm{d}\mathcal{Y}^N\, \mathrm{d}\mathcal{X}^K - 2 \mathcal{V}_I \mathcal{L}^I{}_{JK} \mathcal{L}^K{}_{[OP]}\, \mathrm{d}\mathcal{X}^O\, \mathrm{d}\mathcal{Y}^P .$$

8. Parallel transport from $(\circ\circ)$ to $(\bullet\bullet)$, retain terms up to quadratic order in the coordinate differentials:

$$\mathcal{V}_J^{\bullet'\bullet'} = \mathcal{V}_J^{\circ\circ} + \mathcal{V}_I \mathcal{L}^I{}_{JK} \left[2 \mathcal{L}^K{}_{[OP]}\, \mathrm{d}\mathcal{X}^O\, \mathrm{d}\mathcal{Y}^P \right] .$$

9. Subtract:

$$\mathcal{V}_J^{\bullet'\bullet'} - \mathcal{V}_J^{\bullet\bullet} = -2 \mathcal{V}_I \mathcal{L}^I{}_{J[L,K]}\, \mathrm{d}\mathcal{Y}^L\, \mathrm{d}\mathcal{X}^K - 2 \mathcal{V}_M \mathcal{L}^M{}_{I[K} \mathcal{L}^I{}_{JN]}\, \mathrm{d}\mathcal{Y}^N\, \mathrm{d}\mathcal{X}^K$$

Thus, in summary the change of the covector $\mathcal{V}_J$ may be expressed in terms of the curvature tensor as defined in Eq. 3.70

$$\Delta \mathcal{V}_J := \mathcal{V}_J^{\bullet'\bullet'} - \mathcal{V}_J^{\bullet\bullet} =: -\mathcal{V}_I \mathcal{R}^I{}_{JKL}\, \mathrm{d}\mathcal{X}^K\, \mathrm{d}\mathcal{Y}^L.$$

Clearly, from the above derivation and in accordance with the definition in Eq. 3.70 the curvature tensor is again recognized as

$$\frac{1}{2}\mathcal{R}^I{}_{JKL} = \mathcal{L}^I{}_{J[L,K]} + \mathcal{L}^I{}_{M[K}\mathcal{L}^M{}_{JL]}.$$

This concludes the proof. □

3.4.2 Skew Symmetric Second Covariant Derivatives

Skew Symmetric Second Covariant Derivative of Vectors. The second covariant derivative of a vector is computed as the covariant derivative of the (mixed-variant) second-order tensor represented by $\mathcal{V}^I{}_{|J}$, see the definition in Eq. 3.19.2

$$\mathcal{V}^I{}_{|JK} = \mathcal{V}^I{}_{|J,K} + \mathcal{L}^I{}_{MK}\mathcal{V}^M{}_{|J} - \mathcal{L}^M{}_{JK}\mathcal{V}^I{}_{|M}. \tag{3.77}$$

Involving next the covariant derivative of a vector in Eq. 3.17 inflates the above expression to

$$\begin{aligned}\mathcal{V}^I{}_{|JK} = \mathcal{V}^I{}_{,JK} &+ \mathcal{L}^I{}_{MJ}\mathcal{V}^M{}_{,K} + \mathcal{L}^I{}_{MJ,K}\mathcal{V}^M \\ &+ \mathcal{L}^I{}_{MK}\mathcal{V}^M{}_{,J} + \mathcal{L}^I{}_{MK}\mathcal{L}^M{}_{NJ}\mathcal{V}^N \\ &- \mathcal{L}^M{}_{JK}\mathcal{V}^I{}_{|M}.\end{aligned} \tag{3.78}$$

Likewise, changing the sequence of the indices JK renders the corresponding result

$$\begin{aligned}\mathcal{V}^I{}_{|KJ} = \mathcal{V}^I{}_{,KJ} &+ \mathcal{L}^I{}_{MK}\mathcal{V}^M{}_{,J} + \mathcal{L}^I{}_{MK,J}\mathcal{V}^M \\ &+ \mathcal{L}^I{}_{MJ}\mathcal{V}^M{}_{,K} + \mathcal{L}^I{}_{MJ}\mathcal{L}^M{}_{NK}\mathcal{V}^N \\ &- \mathcal{L}^M{}_{KJ}\mathcal{V}^I{}_{|M}.\end{aligned} \tag{3.79}$$

Finally, subtracting the two results in Eqs. 3.78 and 3.79 and taking into account the symmetry of the second partial derivatives renders the skew symmetric contribution to the second covariant derivative of a vector in terms of the curvature and the torsion

$$\mathcal{V}^I{}_{|[JK]} = \underbrace{\left[\mathcal{L}^I{}_{N[J,K]} + \mathcal{L}^I{}_{M[K}\mathcal{L}^M{}_{NJ]}\right]}_{-\mathcal{R}^I{}_{NJK}/2}\mathcal{V}^N - \underbrace{\mathcal{L}^M{}_{[JK]}}_{\mathcal{T}^M{}_{JK}/2}\mathcal{V}^I{}_{|M}. \tag{3.80}$$

Skew Symmetric Second Covariant Derivative of Covectors. A similar result may be derived for the second covariant derivative of a covector. Here the second covariant derivative of a covector is computed as the covariant derivative of the (covariant) second-order tensor represented by $\mathcal{V}_{I|J}$, see the definition in Eq. 3.19.3

$$\mathcal{V}_{I|JK} = \mathcal{V}_{I|J,K} - \mathcal{L}^{M}{}_{IK}\mathcal{V}_{M|J} - \mathcal{L}^{M}{}_{JK}\mathcal{V}_{I|M}. \tag{3.81}$$

Incorporating the covariant derivative of a covector in Eq. 3.18 expands the above expression to

$$\begin{aligned}\mathcal{V}_{I|JK} = \mathcal{V}_{I,JK} &- \mathcal{L}^{M}{}_{IJ}\mathcal{V}_{M,K} - \mathcal{L}^{M}{}_{IJ,K}\mathcal{V}_{M} \\ &- \mathcal{L}^{M}{}_{IK}\mathcal{V}_{M,J} + \mathcal{L}^{M}{}_{IK}\mathcal{L}^{N}{}_{MJ}\mathcal{V}_{N} \\ &- \mathcal{L}^{M}{}_{JK}\mathcal{V}_{I|M}.\end{aligned} \tag{3.82}$$

Changing next the sequence of the indices JK renders the corresponding result

$$\begin{aligned}\mathcal{V}_{I|KJ} = \mathcal{V}_{I,KJ} &- \mathcal{L}^{M}{}_{IK}\mathcal{V}_{M,J} - \mathcal{L}^{M}{}_{IK,J}\mathcal{V}_{M} \\ &- \mathcal{L}^{M}{}_{IJ}\mathcal{V}_{M,K} + \mathcal{L}^{M}{}_{IJ}\mathcal{L}^{N}{}_{MK}\mathcal{V}_{N} \\ &- \mathcal{L}^{M}{}_{KJ}\mathcal{V}_{I|M}.\end{aligned} \tag{3.83}$$

Subtracting finally the two results in Eqs. 3.82 and 3.83 and taking into account the symmetry of the second partial derivatives renders the skew symmetric contribution to the second covariant derivative of a covector in terms of the curvature and the torsion

$$\mathcal{V}_{I|[JK]} = \underbrace{\left[\mathcal{L}^{N}{}_{I[K,J]} + \mathcal{L}^{N}{}_{M[J}\mathcal{L}^{M}{}_{IK]}\right]}_{\mathcal{R}^{N}{}_{IJK}/2}\mathcal{V}_{N} - \underbrace{\mathcal{L}^{M}{}_{[JK]}}_{\mathcal{T}^{M}{}_{JK}/2}\mathcal{V}_{I|M}. \tag{3.84}$$

Skew Symmetric Second Covariant Derivative of Tensors. The skew symmetric contribution to the second covariant derivatives satisfies the Leibniz (product) rule, see [167], Chapter III.4. Thus the skew second covariant derivatives of the four types of (simple) second-order tensors follow from the product rule applied to their dyadic representation

$$\begin{aligned}2\mathcal{T}^{IJ}{}_{|[KL]} &= -\,\mathcal{R}^{I}{}_{NKL}\mathcal{T}^{NJ} - \mathcal{R}^{J}{}_{NKL}\mathcal{T}^{IN} - \mathcal{T}^{IJ}{}_{|M}\mathcal{T}^{M}{}_{KL},\\ 2\mathcal{T}^{I}{}_{J|[KL]} &= -\,\mathcal{R}^{I}{}_{NKL}\mathcal{T}^{N}{}_{J} + \mathcal{R}^{N}{}_{JKL}\mathcal{T}^{I}{}_{N} - \mathcal{T}^{I}{}_{J|M}\mathcal{T}^{M}{}_{KL},\\ 2\mathcal{T}_{IJ|[KL]} &= \mathcal{R}^{N}{}_{IKL}\mathcal{T}_{NJ} + \mathcal{R}^{N}{}_{JKL}\mathcal{T}_{IN} - \mathcal{T}_{IJ|M}\mathcal{T}^{M}{}_{KL},\\ 2\mathcal{T}_{I}{}^{J}{}_{|[KL]} &= \mathcal{R}^{N}{}_{IKL}\mathcal{T}_{N}{}^{J} - \mathcal{R}^{J}{}_{NKL}\mathcal{T}_{I}{}^{N} - \mathcal{T}_{I}{}^{J}{}_{|M}\mathcal{T}^{M}{}_{KL}.\end{aligned} \tag{3.85}$$

Note that these expressions do also hold for general second-order tensors that are constructed from a sum of simple second-order tensors. The covariant derivatives of higher-order tensors (and objects) follow accordingly.

Addendum 3.3. Skew Symmetric Second Covariant Derivatives

In summary it may be stated that the skew symmetric second covariant derivative of tensors of arbitrary order involves the curvature and the torsion, the only exception being tensors of zeroth-order:

0st-Order $\quad 2\Theta_{|[IJ]} = -\Theta_{|M} \mathcal{T}^M{}_{IJ}$

1st-Order
$$2\mathcal{V}^I{}_{|[JK]} = -\mathcal{R}^I{}_{MJK}\mathcal{V}^M - \mathcal{V}^I{}_{|M}\mathcal{T}^M{}_{JK}$$
$$2\mathcal{V}_{I|[JK]} = \mathcal{R}^N{}_{IJK}\mathcal{V}_M - \mathcal{V}_{I|M}\mathcal{T}^M{}_{JK}$$

2st-Order
$$2\mathcal{T}^{IJ}{}_{|[KL]} = -\mathcal{R}^I{}_{MKL}\mathcal{T}^{MJ} - \mathcal{R}^J{}_{MKL}\mathcal{T}^{IM} - \mathcal{T}^{IJ}{}_{|M}\mathcal{T}^M{}_{KL}$$
$$2\mathcal{T}^I{}_{J|[KL]} = -\mathcal{R}^I{}_{MKL}\mathcal{T}^M{}_J + \mathcal{R}^M{}_{JKL}\mathcal{T}^I{}_M - \mathcal{T}^I{}_{J|M}\mathcal{T}^M{}_{KL}$$
$$2\mathcal{T}_{IJ|[KL]} = \mathcal{R}^M{}_{IKL}\mathcal{T}_{MJ} + \mathcal{R}^M{}_{JKL}\mathcal{T}_{IM} - \mathcal{T}_{IJ|M}\mathcal{T}^M{}_{KL}$$
$$2\mathcal{T}_I{}^J{}_{|[KL]} = \mathcal{R}^M{}_{IKL}\mathcal{T}_M{}^J - \mathcal{R}^J{}_{MKL}\mathcal{T}_I{}^M - \mathcal{T}_I{}^J{}_{|M}\mathcal{T}^M{}_{KL}$$

n^{th}-Order $\quad \vdots$

3.4.3 Cartan Transport along Infinitesimal Circuits

The Cartan transport of radius vectors $\mathcal{Z}^I$ along an infinitesimal circuit results in

$$\Delta\mathcal{Z}^I = \left[\mathcal{R}^I{}_{JKL}\mathcal{Z}^J - \mathcal{T}^I{}_{KL}\right] \mathrm{d}\mathcal{X}^K \,\mathrm{d}\mathcal{Y}^L. \tag{3.86}$$

Observe the additional contribution of the torsion as compared to the result for the parallel transport of vectors along infinitesimal circuits. Note in particular that for an initially vanishing radius vector $\mathcal{Z}^I \equiv 0$ the Cartan transport along an infinitesimal circuit renders an alternative interpretation of the torsion tensor

$$\Delta\mathcal{Z}^I = -\mathcal{T}^I{}_{KL}\,\mathrm{d}\mathcal{X}^K \,\mathrm{d}\mathcal{Y}^L. \tag{3.87}$$

Proof. The proof of the above results for the Cartan transport follows essentially the steps 1 to 9 pursued earlier for the parallel transport of vectors along an infinitesimal circuit. Nevertheless, for the sake of concreteness and in order to highlight the relevant modifications these steps shall be outlined in detail:

1. Cartan transport of $\mathcal{Z}^I$ to $(\bullet)$ and connection $\mathcal{L}^I{}_{JK\bullet}$ at $(\bullet)$:

$$\mathcal{Z}^I_\bullet = \mathcal{Z}^I - \mathcal{L}^I{}_{JK}\mathcal{Z}^J\,\mathrm{d}\mathcal{X}^K - \mathrm{d}\mathcal{X}^I \quad \text{and} \quad \mathcal{L}^I{}_{JK\bullet} = \mathcal{L}^I{}_{JK} + \mathcal{L}^I{}_{JK,L}\,\mathrm{d}\mathcal{X}^L.$$

2. Cartan transport of $\mathcal{Z}^I_\bullet$ to $(\bullet\bullet)$:

$$\mathcal{Z}^I_{\bullet\bullet} = \mathcal{Z}^I_\bullet - \mathcal{L}^I{}_{JK\bullet}\mathcal{Z}^J_\bullet\left[\mathrm{d}\mathcal{Y}^K - \mathcal{L}^K{}_{OP}\,\mathrm{d}\mathcal{Y}^O\,\mathrm{d}\mathcal{X}^P\right] - \left[\mathrm{d}\mathcal{Y}^I - \mathcal{L}^I{}_{JK}\,\mathrm{d}\mathcal{Y}^J\,\mathrm{d}\mathcal{X}^K\right] =$$

$$\left[\mathcal{Z}^I - \mathcal{L}^I{}_{JK}\mathcal{Z}^J\,\mathrm{d}\mathcal{X}^K - \mathrm{d}\mathcal{X}^I\right] - \left[\mathcal{L}^I{}_{JK} + \mathcal{L}^I{}_{JK,L}\,\mathrm{d}\mathcal{X}^L\right]\times$$
$$\left[\mathcal{Z}^J - \mathcal{L}^J{}_{MN}\mathcal{Z}^M\,\mathrm{d}\mathcal{X}^N - \mathrm{d}\mathcal{X}^J\right]\times\left[\mathrm{d}\mathcal{Y}^K - \mathcal{L}^K{}_{OP}\,\mathrm{d}\mathcal{Y}^O\,\mathrm{d}\mathcal{X}^P\right]$$
$$-\left[\mathrm{d}\mathcal{Y}^I - \mathcal{L}^I{}_{JK}\,\mathrm{d}\mathcal{Y}^J\,\mathrm{d}\mathcal{X}^K\right].$$

3. Retain terms up to quadratic order in the coordinate differentials:

$$\mathcal{Z}^I_{\bullet\bullet} = \mathcal{Z}^I - \mathcal{L}^I{}_{JK}\mathcal{Z}^J[\,\mathrm{d}\mathcal{X}^K + \mathrm{d}\mathcal{Y}^K] - [\,\mathrm{d}\mathcal{X}^I + \mathrm{d}\mathcal{Y}^I] + 2\mathcal{L}^I{}_{(JK)}\,\mathrm{d}\mathcal{X}^J\,\mathrm{d}\mathcal{Y}^K$$

$$-\mathcal{L}^I{}_{JK,L}\,\mathrm{d}\mathcal{X}^L\mathcal{Z}^J\,\mathrm{d}\mathcal{Y}^K + \mathcal{L}^I{}_{JK}\mathcal{L}^J{}_{MN}\mathcal{Z}^M\,\mathrm{d}\mathcal{X}^N\,\mathrm{d}\mathcal{Y}^K + \mathcal{L}^I{}_{JK}\mathcal{L}^K{}_{OP}\mathcal{Z}^J\,\mathrm{d}\mathcal{Y}^O\,\mathrm{d}\mathcal{X}^P.$$

4. Cartan transport of $\mathcal{Z}^I$ to $(\circ)$ and connection $\mathcal{L}^I{}_{JK\circ}$ at $(\circ)$:

$$\mathcal{Z}^I_\circ = \mathcal{Z}^I - \mathcal{L}^I{}_{JK}\mathcal{Z}^J\,\mathrm{d}\mathcal{Y}^K - \mathrm{d}\mathcal{Y}^I \quad\text{and}\quad \mathcal{L}^I{}_{JK\circ} = \mathcal{L}^I{}_{JK} + \mathcal{L}^I{}_{JK,L}\,\mathrm{d}\mathcal{Y}^L.$$

5. Cartan transport of $\mathcal{Z}^I_\circ$ to $(\circ\circ)$:

$$\mathcal{Z}^I_{\circ\circ} = \mathcal{Z}^I_\circ - \mathcal{L}^I{}_{JK\circ}\mathcal{Z}^J_\circ\left[\mathrm{d}\mathcal{X}^K - \mathcal{L}^K{}_{OP}\,\mathrm{d}\mathcal{X}^O\,\mathrm{d}\mathcal{Y}^P\right] - \left[\mathrm{d}\mathcal{X}^I - \mathcal{L}^I{}_{JK}\,\mathrm{d}\mathcal{X}^J\,\mathrm{d}\mathcal{Y}^K\right] =$$

$$\left[\mathcal{Z}^I - \mathcal{L}^I{}_{JK}\mathcal{Z}^J\,\mathrm{d}\mathcal{Y}^K - \mathrm{d}\mathcal{Y}^I\right] - \left[\mathcal{L}^I{}_{JK} + \mathcal{L}^I{}_{JK,L}\,\mathrm{d}\mathcal{Y}^L\right]\times$$
$$\left[\mathcal{Z}^J - \mathcal{L}^J{}_{MN}\mathcal{Z}^M\,\mathrm{d}\mathcal{Y}^N - \mathrm{d}\mathcal{Y}^J\right]\times\left[\mathrm{d}\mathcal{X}^K - \mathcal{L}^K{}_{OP}\,\mathrm{d}\mathcal{X}^O\,\mathrm{d}\mathcal{Y}^P\right]$$
$$-\left[\mathrm{d}\mathcal{X}^I - \mathcal{L}^I{}_{JK}\,\mathrm{d}\mathcal{X}^J\,\mathrm{d}\mathcal{Y}^K\right].$$

6. Retain terms up to quadratic order in the coordinate differentials:

$$\mathcal{Z}^I_{\circ\circ} = \mathcal{Z}^I - \mathcal{L}^I{}_{JK}\mathcal{Z}^J[\,\mathrm{d}\mathcal{Y}^K + \mathrm{d}\mathcal{X}^K] - [\,\mathrm{d}\mathcal{Y}^I + \mathrm{d}\mathcal{X}^I] + 2\mathcal{L}^I{}_{(JK)}\,\mathrm{d}\mathcal{Y}^J\,\mathrm{d}\mathcal{X}^K$$

$$-\mathcal{L}^I{}_{JK,L}\,\mathrm{d}\mathcal{Y}^L\mathcal{Z}^J\,\mathrm{d}\mathcal{X}^K + \mathcal{L}^I{}_{JK}\mathcal{L}^J{}_{MN}\mathcal{Z}^M\,\mathrm{d}\mathcal{Y}^N\,\mathrm{d}\mathcal{X}^K + \mathcal{L}^I{}_{JK}\mathcal{L}^K{}_{OP}\mathcal{Z}^J\,\mathrm{d}\mathcal{X}^O\,\mathrm{d}\mathcal{Y}^P.$$

7. Subtract:

$$\mathcal{Z}^I_{\circ\circ} - \mathcal{Z}^I_{\bullet\bullet} = -2\mathcal{L}^I{}_{J[K,L]}\,\mathrm{d}\mathcal{Y}^L\mathcal{Z}^J\,\mathrm{d}\mathcal{X}^K$$
$$+2\mathcal{L}^I{}_{J[K}\mathcal{L}^J{}_{MN]}\mathcal{Z}^M\,\mathrm{d}\mathcal{Y}^N\,\mathrm{d}\mathcal{X}^K + 2\mathcal{L}^I{}_{JK}\mathcal{L}^K{}_{[OP]}\mathcal{Z}^J\,\mathrm{d}\mathcal{X}^O\,\mathrm{d}\mathcal{Y}^P.$$

8. Cartan transport from $(\circ\circ)$ to $(\bullet\bullet)$, retain terms up to quadratic order in the coordinate differentials:

$$\mathcal{Z}^I_{\bullet'\bullet'} = \mathcal{Z}^I_{\circ\circ} - \mathcal{L}^I{}_{JK}\mathcal{Z}^J\left[2\mathcal{L}^K{}_{[OP]}\,\mathrm{d}\mathcal{X}^O\,\mathrm{d}\mathcal{Y}^P\right] - 2\mathcal{L}^I{}_{[JK]}\,\mathrm{d}\mathcal{X}^J\,\mathrm{d}\mathcal{Y}^K.$$

9. Subtract:

$$\mathcal{Z}^I_{\bullet'\bullet'} - \mathcal{Z}^I_{\bullet\bullet} =$$
$$-2\mathcal{L}^I{}_{J[K,L]}\,\mathrm{d}\mathcal{Y}^L\mathcal{Z}^J\,\mathrm{d}\mathcal{X}^K + 2\mathcal{L}^I{}_{J[K}\mathcal{L}^J{}_{MN]}\mathcal{Z}^M\,\mathrm{d}\mathcal{Y}^N\,\mathrm{d}\mathcal{X}^K - 2\mathcal{L}^I{}_{[JK]}\,\mathrm{d}\mathcal{X}^J\,\mathrm{d}\mathcal{Y}^K.$$

Thus, in summary the change of the radius vector $\mathcal{Z}^I$ upon Cartan transport along a closed circuit may be expressed in terms of the curvature tensor and the torsion tensor as defined in Eqs. 3.70 and 3.45

$$\Delta\mathcal{Z}^I := \mathcal{Z}^I_{\bullet'\bullet'} - \mathcal{Z}^I_{\bullet\bullet} =: \left[\mathcal{R}^I{}_{JKL}\mathcal{Z}^J - \mathcal{T}^I{}_{KL}\right] \mathrm{d}\mathcal{X}^K \,\mathrm{d}\mathcal{Y}^L .$$

This concludes the proof. □

3.4.4 Geodesic Deviation

The deviation of neighboring geodesics or, more precise, the relative acceleration of the separation vector $\mathcal{W}^I$ between two infinitesimal close geodesics, i.e. between the parameter curves $\mathcal{X}^I(t)$ (the so-called *fiducial geodesic*) and $\mathcal{X}^I(t)+\mathcal{W}^I(t)$, is given by the *equation of geodesic deviation* (a generalization of the classical *Jacobi equation* for the distance between two geodesics on a two-dimensional manifold, see below). The equation of geodesic deviation relates the *second covariant parameter derivative* of the separation vector $\mathcal{W}^I$ with respect to the affine parameter t to the curvature tensor $\mathcal{R}^I{}_{JKL}$ (and, for a non-symmetric manifold (space/geometry), to the *first covariant parameter derivative* of a term involving the torsion)

$$\frac{\mathrm{D}^2}{\mathrm{d}t^2}\mathcal{W}^I = -\mathcal{R}^I{}_{JKL}\dot{\mathcal{X}}^J\mathcal{W}^K\dot{\mathcal{X}}^L - \frac{\mathrm{D}}{\mathrm{d}t}\left[\mathcal{T}^I{}_{JK}\dot{\mathcal{X}}^J\mathcal{W}^K\right]. \tag{3.88}$$

The equation of geodesic deviation plays an important role in the theory of gravitation since it allows to determine the curvature tensor of space-time by observing the free fall of nearby test particles in the gravitational field.

Proof. The first and second covariant parameter derivatives of the separation vector $\mathcal{V}^I$ read

$$\frac{\mathrm{D}}{\mathrm{d}t}\mathcal{W}^I = \mathcal{W}^I{}_{|J}\dot{\mathcal{X}}^J \quad\text{and}\quad \frac{\mathrm{D}^2}{\mathrm{d}t^2}\mathcal{W}^I = \mathcal{W}^I{}_{|JL}\dot{\mathcal{X}}^J\dot{\mathcal{X}}^L + \mathcal{W}^I{}_{|J}\frac{\mathrm{D}}{\mathrm{d}t}\dot{\mathcal{X}}^J = \mathcal{W}^I{}_{|JL}\dot{\mathcal{X}}^J\dot{\mathcal{X}}^L$$

The term $\mathrm{D}/\,\mathrm{d}t\,\dot{\mathcal{X}}^J = 0$ vanishes due to the definition of the parameter curve $\mathcal{X}^J(t)$ being geodesic, see Eq. 3.31. Next the term containing the second covariant derivative $\mathcal{W}^I{}_{|JL}$ is expanded in analogy to Eq. 3.78 to render

$$\begin{aligned}\frac{\mathrm{D}^2}{\mathrm{d}t^2}\mathcal{W}^I &= \mathcal{W}^I{}_{,JL}\dot{\mathcal{X}}^J\dot{\mathcal{X}}^L + 2\mathcal{L}^I{}_{KJ}\dot{\mathcal{X}}^J\dot{\mathcal{W}}^K \\ &\quad + \left[\mathcal{L}^I{}_{KJ,L} + \mathcal{L}^I{}_{NL}\mathcal{L}^N{}_{KJ}\right]\dot{\mathcal{X}}^J\mathcal{W}^K\dot{\mathcal{X}}^L - \mathcal{W}^I{}_{|N}\mathcal{L}^N{}_{JL}\dot{\mathcal{X}}^J\dot{\mathcal{X}}^L .\end{aligned}$$

Due to the second partial parameter derivative $\ddot{\mathcal{W}}^I = \mathcal{W}^I{}_{,JL}\dot{\mathcal{X}}^J\dot{\mathcal{X}}^L + \mathcal{W}^I{}_{,N}\ddot{\mathcal{X}}^N$, the definition of $\mathcal{X}^N$ being geodesic such that $\ddot{\mathcal{X}}^N = -\mathcal{L}^N{}_{JL}\dot{\mathcal{X}}^J\dot{\mathcal{X}}^L$, and due to the definition of the covariant derivative $\mathcal{W}^I{}_{|N} = \mathcal{W}^I{}_{,N} + \mathcal{L}^I{}_{KN}\mathcal{W}^K$ the first and the last terms on the right hand side cancel partly to render

$$\frac{\mathrm{D}^2}{\mathrm{d}t^2}\mathcal{W}^I = \ddot{\mathcal{W}}^I + 2\mathcal{L}^I{}_{KJ}\dot{\mathcal{X}}^J\dot{\mathcal{W}}^K + \left[\mathcal{L}^I{}_{KJ,L} + \mathcal{L}^I{}_{NL}\mathcal{L}^N{}_{KJ} - \mathcal{L}^I{}_{KN}\mathcal{L}^N{}_{JL}\right]\dot{\mathcal{X}}^J\mathcal{W}^K\dot{\mathcal{X}}^L.$$

Next the two infinitesimal close geodesics (the fiducial geodesics and its neighboring geodesic), i.e. the parameter curves $\mathcal{X}^I(t)$ and $\mathcal{X}^I(t) + \mathcal{W}^I(t)$, satisfy their respective defining equations

$$\frac{\mathrm{D}}{\mathrm{d}t}\mathcal{X}^I = \ddot{\mathcal{X}}^I + \mathcal{L}^I{}_{JL}\dot{\mathcal{X}}^J\dot{\mathcal{X}}^L = 0 \quad \text{and}$$

$$\frac{\mathrm{D}}{\mathrm{d}t}[\mathcal{X}^I + \mathcal{W}^I] = [\ddot{\mathcal{X}}^I + \ddot{\mathcal{W}}^I] + [\mathcal{L}^I{}_{JL} + \mathcal{L}^I{}_{JL,K}\mathcal{W}^K][\dot{\mathcal{X}}^J + \dot{\mathcal{W}}^J][\dot{\mathcal{X}}^L + \dot{\mathcal{W}}^L] = 0.$$

Expanding the second equation in the above up to terms linear in $\mathcal{W}^K, \dot{\mathcal{W}}^J$, $\dot{\mathcal{W}}^L$ and subtracting the first equation renders eventually the basic version of the geodesic deviation equation

$$\ddot{\mathcal{W}}^I = -2\mathcal{L}^I{}_{(KJ)}\dot{\mathcal{X}}^J\dot{\mathcal{W}}^K - \mathcal{L}^I{}_{JL,K}\dot{\mathcal{X}}^J\mathcal{W}^K\dot{\mathcal{X}}^L.$$

Combining the last with the second last expression and introducing the torsion tensor $\mathcal{T}^I{}_{JK} := 2\mathcal{L}^I{}_{[JK]}$ results in

$$\frac{\mathrm{D}^2}{\mathrm{d}t^2}\mathcal{W}^I = -2\left[\mathcal{L}^I{}_{J[L,K]} + \mathcal{L}^I{}_{N[K}\mathcal{L}^N{}_{JL]}\right]\dot{\mathcal{X}}^J\mathcal{W}^K\dot{\mathcal{X}}^L - \mathcal{T}^I{}_{JK}\dot{\mathcal{X}}^J\dot{\mathcal{W}}^K - \left[\mathcal{T}^I{}_{JK,L} + \mathcal{L}^I{}_{NL}\mathcal{T}^N{}_{JK} - \mathcal{T}^I{}_{NK}\mathcal{L}^N{}_{JL}\right]\dot{\mathcal{X}}^J\mathcal{W}^K\dot{\mathcal{X}}^L.$$

The first term on the right hand side may be re-expressed in terms of the curvature tensor $\mathcal{R}^I{}_{JKL}$, the four remaining terms condense into the covariant parameter derivative of $\mathcal{T}^I{}_{JK}\dot{\mathcal{X}}^J\mathcal{W}^K$ since

(i) $$\mathcal{T}^I{}_{JK,L}\dot{\mathcal{X}}^L = \dot{\mathcal{T}}^I{}_{JK},$$

(ii) $$\mathcal{T}^I{}_{JK}\dot{\mathcal{W}}^K + \dot{\mathcal{T}}^I{}_{JK}\mathcal{W}^K = [\mathcal{T}^I{}_{JK}\mathcal{W}^K]_{,L}\dot{\mathcal{X}}^L,$$

(iii) $$[\mathcal{T}^I{}_{JK}\mathcal{W}^K]_{|L} = [\mathcal{T}^I{}_{JK}\mathcal{W}^K]_{,L} + \mathcal{L}^I{}_{NL}[\mathcal{T}^N{}_{JK}\mathcal{W}^K] - \mathcal{L}^N{}_{JL}[\mathcal{T}^I{}_{NK}\mathcal{W}^K],$$

(iv) $$\frac{\mathrm{D}}{\mathrm{d}t}[\mathcal{T}^I{}_{JK}\mathcal{W}^K] = [\mathcal{T}^I{}_{JK}\mathcal{W}^K]_{|L}\dot{\mathcal{X}}^L,$$

(v) $$\frac{\mathrm{D}}{\mathrm{d}t}[\mathcal{T}^I{}_{JK}\mathcal{W}^K]\dot{\mathcal{X}}^J = \frac{\mathrm{D}}{\mathrm{d}t}[\mathcal{T}^I{}_{JK}\dot{\mathcal{X}}^J\mathcal{W}^K].$$

Thus the format for the equation of geodesic deviation as claimed in the above is established. □

Remark 3.10. The equation of geodesic deviation may alternatively be expressed in terms of the *Jacobi curvature tensor*, see [142]

$$\frac{\mathrm{D}^2}{\mathrm{d}t^2}\mathcal{W}^I = -\mathcal{J}^I{}_{KJL}\mathcal{W}^K\dot{\mathcal{X}}^J\dot{\mathcal{X}}^L - \frac{\mathrm{D}}{\mathrm{d}t}\left[\mathcal{T}^I{}_{JK}\dot{\mathcal{X}}^J\mathcal{W}^K\right].$$

The Jacobi curvature tensor is related to the curvature tensor by symmetrizing its second and third index

$$\mathcal{J}^I{}_{KJL} = \mathcal{J}^I{}_{K(JL)} := \mathcal{R}^I{}_{(JKL)}.$$

The Jacobi curvature and the curvature tensor are shown to be equivalent in terms of their information content. The Jacobi curvature tensor satisfies a set of identities similar to the algebraic and differential Bianchi identities for the curvature tensor, whereby symmetries instead of skew symmetries in the indices play the relevant role for these alternative identities. □

Addendum 3.4. Jacobi Equation

On a two-dimensional surface two nearby geodesics, the fiducial geodesic and a test geodesic, that are parallel at the arc-length parameter t_0 of the fiducial geodesic are originally separated by the geodesic deviation $\mathcal{W}_0$ (orthogonal distance of the test geodesic from the fiducial geodesic). For a curved surface the two geodesics are in general no longer parallel at any other arc-length parameter t. Then the geodesic deviation $\mathcal{W}(t)$ follows the *Jacobi equation* (of geodesic deviation) describing the acceleration of separation

$$\ddot{\mathcal{W}} + \mathcal{K}\mathcal{W} = 0.$$

Here $\mathcal{K}$ is the *Gaussian* curvature of the two-dimensional surface, see Chapter 5, the dot denotes the derivative with respect to the arc-length parameter t. As an example the geodesic deviation on a sphere with radius R reads $\mathcal{W} = \mathcal{W}_0 \cos(t/R)$, thus $\ddot{\mathcal{W}} = -\mathcal{W}/R^2$ (i.e. the Gaussian curvature is $\mathcal{K} = 1/R^2$).

3.4.5 Ricci Tensors

By contraction of either the first and the third or the first and the second index of the fourth-order mixed-variant curvature tensor $\mathcal{R}^I{}_{JKL}$ second-order, covariant tensors, the (first) Ricci tensor $\mathcal{R}_{JL}$ and the (less common) second Ricci tensor $\mathcal{V}_{KL}$, respectively, may be introduced. They follow from the

Definition 3.20. Based on the connection second-order objects, the (first) and second Ricci tensors, respectively, are defined as

$$\mathcal{R}_{JL} := \mathcal{L}^I{}_{JL,I} - \mathcal{L}^I{}_{JI,L} + \mathcal{L}^I{}_{MI}\mathcal{L}^M{}_{JL} - \mathcal{L}^I{}_{ML}\mathcal{L}^M{}_{JI} \tag{3.89}$$

and

$$\mathcal{V}_{KL} := \mathcal{L}^I{}_{IL,K} - \mathcal{L}^I{}_{IK,L} \tag{3.90}$$

Here the terms $\mathcal{L}^I{}_{MK}\mathcal{L}^M{}_{IL} - \mathcal{L}^I{}_{ML}\mathcal{L}^M{}_{IK} = 0$ have been equated to zero. □

By condensing the skew symmetries in the indices I, L and K, L, respectively, the (first) and second Ricci tensors read alternatively

$$\mathcal{R}_{JL} = 2\mathcal{L}^{I}{}_{J[L,I]} + 2\mathcal{L}^{I}{}_{M[I}\mathcal{L}^{M}{}_{JL]} \quad \text{and} \quad \mathcal{V}_{KL} = 2\mathcal{L}^{I}{}_{I[L,K]} \tag{3.91}$$

As a consequence the relation of the second-order (first) and second Ricci tensors to the fourth-order curvature tensor by contraction of either the first and the third or the first and the second index, respectively, is obvious

$$\mathcal{R}_{JL} := \mathcal{R}^{I}{}_{JIL} = \delta_{I}{}^{K}\mathcal{R}^{I}{}_{JKL} \quad \text{and} \quad \mathcal{V}_{KL} := \mathcal{R}^{I}{}_{IKL} = \delta_{I}{}^{J}\mathcal{R}^{I}{}_{JKL}. \tag{3.92}$$

Contractions of contra-and covariant indices are effected by the Kronecker symbol $\delta_I{}^K$ or $\delta_I{}^J$, respectively (recall that a metric is not yet available).

Addendum 3.5. Identities for the Curvature Tensor

As a consequence of the identities **C1**, **C2** (**c2**), **C3** (**c3**) for the curvature tensor, by contracting indices I, K the following identities hold for the *(first) Ricci tensor*

R1: $\mathcal{R}_{JL} = -\mathcal{R}^{I}{}_{JLI}$
R2: $2\mathcal{R}_{[JL]} - \mathcal{V}_{JL} = \mathcal{T}^{I}{}_{JL|I} - 2\mathcal{T}_{[J|L]} - \mathcal{T}_{M}\mathcal{T}^{M}{}_{JL}$
R3: $2\mathcal{R}_{J[L|M]} + \mathcal{R}^{I}{}_{JLM|I} = -\mathcal{R}_{JN}\mathcal{T}^{N}{}_{LM} + 2\mathcal{R}^{I}{}_{JN[L}\mathcal{T}^{N}{}_{M]I}$

R2 and **R3** may also be called the *contracted Bianchi identities* as valid without the presence of a metric, see [142].

Moreover, by contracting indices I, J of the curvature tensor in the identities **C1**, **C2** (**c2**), **C3** (**c3**), additional identities hold for the *second Ricci tensor*

V1: $\mathcal{V}_{KL} = \mathcal{R}^{I}{}_{IKL} = -\mathcal{V}_{LK}$
V2: $\mathcal{V}_{KL} - 2\mathcal{R}_{[KL]} = -\mathcal{T}^{I}{}_{KL|I} + 2\mathcal{T}_{[K|L]} + \mathcal{T}_{M}\mathcal{T}^{M}{}_{KL}$
V3: $\mathcal{V}_{[\![KL|M]\!]} = \mathcal{V}_{N[\![K}\mathcal{T}^{N}{}_{LM]\!]}$

Note that identities **V2** and **R2** trivially coincide with a change of sign.

3.4.6 Additive Connection

Sometimes it proves convenient to decompose the connection additively. Then for an *additive connection* in the format

$$\mathcal{L}^{I}{}_{JL} := \mathcal{X}^{I}{}_{JL} + \mathcal{Y}^{I}{}_{JL} \tag{3.93}$$

the fourth-order curvature tensor follows in terms of the covariant derivative with respect to the full connection $\mathcal{L}^{I}{}_{JL}$ as

$$\mathcal{R}^{I}{}_{JKL} = \mathcal{X}^{I}{}_{JKL} + 2\mathcal{Y}^{I}{}_{J[L|K]} - 2\mathcal{Y}^{I}{}_{M[K}\mathcal{Y}^{M}{}_{JL]} - \mathcal{Y}^{I}{}_{JM}\mathcal{T}^{M}{}_{KL}. \tag{3.94}$$

Here, $\mathcal{X}^I{}_{JKL}$ denotes the curvature tensor that is based on the connection $\mathcal{X}^I{}_{JL}$ in the format of Eq. 3.70.

Proof. First the term of the curvature tensor quadratic in the connection $\mathcal{L}^{\bullet}{}_{\bullet\bullet}$ is expanded into the four terms combining the connections $\mathcal{X}^{\bullet}{}_{\bullet\bullet}$ and $\mathcal{Y}^{\bullet}{}_{\bullet\bullet}$

$$\begin{aligned} 2\mathcal{L}^I{}_{M[K}\mathcal{L}^M{}_{JL]} &= 2\mathcal{X}^I{}_{M[K}\mathcal{X}^M{}_{JL]} \\ &+ 2\mathcal{X}^I{}_{M[K}\mathcal{Y}^M{}_{JL]} \\ &+ 2\mathcal{Y}^I{}_{M[K}\mathcal{X}^M{}_{JL]} \\ &+ 2\mathcal{Y}^I{}_{M[K}\mathcal{Y}^M{}_{JL]}. \end{aligned}$$

Next the skew symmetrizing brackets for the indices K, L in the last three terms are inflated to render

$$\begin{aligned} 2\mathcal{L}^I{}_{M[K}\mathcal{L}^M{}_{JL]} &= 2\mathcal{X}^I{}_{M[K}\mathcal{X}^M{}_{JL]} \\ &+ \underline{\mathcal{X}^I{}_{MK}\mathcal{Y}^M{}_{JL}} - \mathcal{X}^I{}_{ML}\mathcal{Y}^M{}_{JK} \\ &+ \mathcal{Y}^I{}_{MK}\mathcal{X}^M{}_{JL} - \underline{\mathcal{Y}^I{}_{ML}\mathcal{X}^M{}_{JK}} \\ &+ \underline{\mathcal{Y}^I{}_{MK}\mathcal{Y}^M{}_{JL}} - \underline{\mathcal{Y}^I{}_{ML}\mathcal{Y}^M{}_{JK}}. \end{aligned}$$

Then the underlined terms in the above may be regrouped pairwise to form terms including the connection $\mathcal{L}^{\bullet}{}_{\bullet\bullet}$ which are again underlined in the below

$$\begin{aligned} 2\mathcal{L}^I{}_{M[K}\mathcal{L}^M{}_{JL]} &= 2\mathcal{X}^I{}_{M[K}\mathcal{X}^M{}_{JL]} \\ &+ \underline{\mathcal{L}^I{}_{MK}\mathcal{Y}^M{}_{JL}} \\ &- \underline{\mathcal{Y}^I{}_{ML}\mathcal{L}^M{}_{JK}} \\ &- \mathcal{X}^I{}_{ML}\mathcal{Y}^M{}_{JK} \\ &+ \mathcal{Y}^I{}_{MK}\mathcal{X}^M{}_{JL}. \end{aligned}$$

The underlined terms in the above are further expanded in terms of a skew symmetrization in the indices K, L and the corresponding 'correction' terms underlined in the second and third line in the below

$$\begin{aligned} 2\mathcal{L}^I{}_{M[K}\mathcal{L}^M{}_{JL]} &= 2\mathcal{X}^I{}_{M[K}\mathcal{X}^M{}_{JL]} \\ &+ 2\mathcal{L}^I{}_{M[K}\mathcal{Y}^M{}_{JL]} + \underline{\mathcal{L}^I{}_{ML}\mathcal{Y}^M{}_{JK}} \\ &- 2\mathcal{Y}^I{}_{M[L}\mathcal{L}^M{}_{JK]} - \underline{\mathcal{Y}^I{}_{MK}\mathcal{L}^M{}_{JL}} \\ &- \underline{\mathcal{X}^I{}_{ML}\mathcal{Y}^M{}_{JK}} \\ &+ \underline{\mathcal{Y}^I{}_{MK}\mathcal{X}^M{}_{JL}}. \end{aligned}$$

The underlined 'correction' terms in the second and third line are then combined with the underlined terms in the last two lines in the above to finally produce the two underlined terms below

$$
\begin{aligned}
2\mathcal{L}^I{}_{M[K}\mathcal{L}^M{}_{JL]} = & \; 2\mathcal{X}^I{}_{M[K}\mathcal{X}^M{}_{JL]} \\
& + 2\mathcal{L}^I{}_{M[K}\mathcal{Y}^M{}_{JL]} \\
& - 2\mathcal{Y}^I{}_{M[L}\mathcal{L}^M{}_{JK]} \\
& + \underline{\mathcal{Y}^I{}_{ML}\mathcal{Y}^M{}_{JK}} \\
& - \underline{\mathcal{Y}^I{}_{MK}\mathcal{Y}^M{}_{JL}}.
\end{aligned}
$$

The two underlined terms in the above are recognized as $2\mathcal{Y}^I{}_{M[L}\mathcal{Y}^M{}_{JK]}$. Next the partial derivative of the total connection $\mathcal{L}^\bullet{}_{\bullet\bullet}$ is combined with the above expression to render

$$
\begin{aligned}
2\mathcal{L}^I{}_{J[L,K]} + 2\mathcal{L}^I{}_{M[K}\mathcal{L}^M{}_{JL]} = & \; 2\mathcal{X}^I{}_{J[L,K]} + 2\mathcal{X}^I{}_{M[K}\mathcal{X}^M{}_{JL]} \\
& + \underline{2\mathcal{Y}^I{}_{J[L,K]}} \\
& + \underline{2\mathcal{L}^I{}_{M[K}\mathcal{Y}^M{}_{JL]}} \\
& - \underline{2\mathcal{Y}^I{}_{M[L}\mathcal{L}^M{}_{JK]}} \\
& - \underline{2\mathcal{Y}^I{}_{JM}\mathcal{L}^M{}_{[LK]}} \\
& + 2\mathcal{Y}^I{}_{JM}\mathcal{L}^M{}_{[LK]} \\
& + 2\mathcal{Y}^I{}_{M[L}\mathcal{Y}^M{}_{JK]}.
\end{aligned}
$$

Here, the fifth and sixth line that add up to zero are additionally included. It may then be observed that the underlined terms in the above define the covariant derivative of $\mathcal{Y}^\bullet{}_{\bullet\bullet}$, i.e. the underlined term in the below, the sixth line may be re-expressed in terms of the torsion. Thus, eventually the result is obtained as

$$
\begin{aligned}
2\mathcal{L}^I{}_{J[L,K]} + 2\mathcal{L}^I{}_{M[K}\mathcal{L}^M{}_{JL]} = & \; \mathcal{X}^I{}_{JKL} + \underline{2\mathcal{Y}^I{}_{J[L|K]}} \\
& - \mathcal{Y}^I{}_{JM}\mathcal{T}^M{}_{[KL]} - 2\mathcal{Y}^I{}_{M[K}\mathcal{Y}^M{}_{JL]}
\end{aligned}
$$

Again the skew symmetries in the indices K, L have been exploited in the last step. □

Alternatively, we may also write directly

$$
\mathcal{R}^I{}_{JKL} = \mathcal{X}^I{}_{JKL} + \mathcal{Y}^I{}_{JKL} + 2\mathcal{X}^I{}_{M[K}\mathcal{Y}^M{}_{JL]} + 2\mathcal{Y}^I{}_{M[K}\mathcal{X}^M{}_{JL]}. \tag{3.95}
$$

Obviously, the latter result follows trivially from the format of the curvature tensor in Eq. 3.70.

3.4.7 Transformation of the Curvature Tensor

Due to the tensor property of the curvature the following convection or rather pull-back ($\mathcal{Y}$)/push-forward (y) relations hold in the case of holonomic coordinate transformations:

Curvature($\mathcal{Y}$(connection))=$\mathcal{Y}$(curvature(connection))

curvature(y(Connection))=y(Curvature(Connection))

As an example the (tensorial) pull-back of the spatial curvature expressed in terms of the spatial connection equals the material curvature expressed in terms of the (non-tensorial) pull-back of the spatial connection. The corresponding relation holds if spatial and material objects are exchanged. However, in the case of anholonomic coordinate transformations extra contributions in terms of the anholonomic object arise.

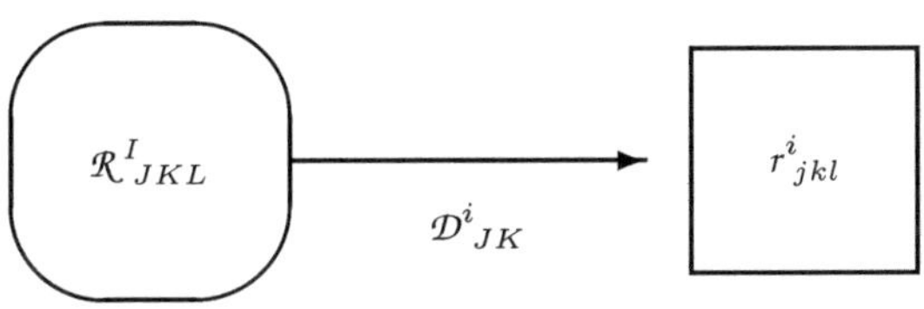

Fig. 3.9 Transformation of the curvature tensor for the case of holonomic $\mathcal{X}^I$ and anholonomic χ^i

The tensorial transformation of the curvature tensor upon changing the coordinate system between holonomic coordinates $\mathcal{X}^I$ and anholonomic coordinates χ^i is stated as

$$\mathcal{R}^I{}_{JKL} = f^I{}_i r^i{}_{jkl} \mathcal{F}^j{}_J \mathcal{F}^k{}_K \mathcal{F}^l{}_L. \tag{3.96}$$

Thereby, for holonomic $\mathcal{X}^I$ and anholonomic χ^i the curvature tensor $\mathcal{R}^I{}_{JKL}$ follows the standard definition in Eqs. 3.68 and 3.70 whereas the curvature tensor $r^i{}_{jkl}$ involves extra contributions (underlined in the sequel) in terms of the connection and the anholonomic object

$$r^i{}_{jkl} = 2\ell^i{}_{j[l.k]} + 2\ell^i{}_{m[k}\ell^m{}_{jl]} + \underline{\ell^i{}_{jm} a^m{}_{lk}}. \tag{3.97}$$

The situation is depicted in Fig. 3.9.

Remark 3.11. Alternatively, by resorting to the relation between the partial and the covariant derivative of the connection in analogy to Eq. 3.21.1

$$2\ell^i{}_{j[l.k]} = 2\ell^i{}_{j[l|k]} \underline{-2\ell^i{}_{m[k}\ell^m{}_{jl]}} - 2\ell^i{}_{m[k}\ell^m{}_{jl]} + 2\ell^i{}_{jm}\ell^m{}_{[lk]}$$

and noting that the underlined term cancels out when inserting into Eq. 3.97, the curvature tensor for anholonomic χ^i expressed in terms of the covariant derivative reads

$$r^i{}_{jkl} = 2\ell^i{}_{j[l|k]} - 2\ell^i{}_{m[k}\ell^m{}_{jl]} + \ell^i{}_{jm}t^m{}_{lk}. \tag{3.98}$$

Here, according to Eq. 3.52.1 $t^m{}_{lk} := a^m{}_{lk} + 2\ell^m{}_{[lk]}$ denotes the torsion in the anholonomic space with coordinates χ^i. □

Likewise, the tensorial transformation of the curvature tensor upon changing the coordinate system between holonomic coordinates χ^i and anholonomic coordinates $\mathcal{X}^I$ is stated as

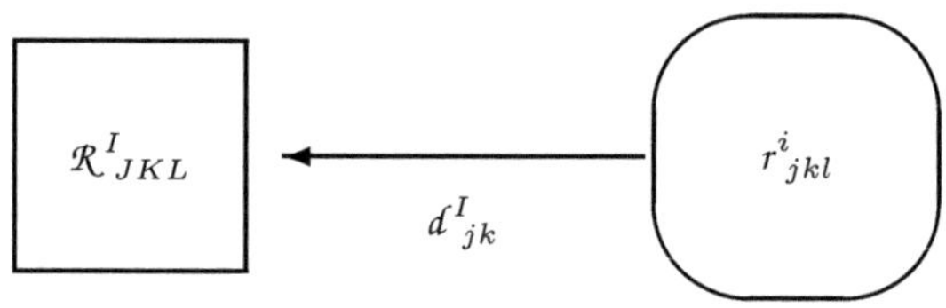

Fig. 3.10 Transformation of the curvature tensor for the case of holonomic χ^i and anholonomic $\mathcal{X}^I$

$$r^i{}_{jkl} = \mathcal{F}^i{}_I \mathcal{R}^I{}_{JKL} f^J{}_j f^K{}_k f^L{}_l. \tag{3.99}$$

Then, for holonomic χ^i and anholonomic $\mathcal{X}^I$ the curvature tensor $r^i{}_{jkl}$ follows the standard definition corresponding to Eqs. 3.68 and 3.70 whereas the curvature tensor $\mathcal{R}^I{}_{JKL}$ involves extra contributions (underlined in the sequel) in terms of the connection and the anholonomic object

$$\mathcal{R}^I{}_{JKL} = 2\mathcal{L}^I{}_{J[L.K]} + 2\mathcal{L}^I{}_{M[K}\mathcal{L}^M{}_{JL]} + \underline{\mathcal{L}^I{}_{JM}\mathcal{A}^M{}_{LK}}. \tag{3.100}$$

The situation is depicted in Fig. 3.10.

Remark 3.12. Alternatively, by resorting to the relation between the partial and the covariant derivative of the connection according to Eq. 3.21.1

$$2\mathcal{L}^I{}_{J[L.K]} = 2\mathcal{L}^I{}_{J[L|K]} \underline{- 2\mathcal{L}^I{}_{M[K}\mathcal{L}^M{}_{JL]}} - 2\mathcal{L}^I{}_{M[K}\mathcal{L}^M{}_{JL]} + 2\mathcal{L}^I{}_{JM}\mathcal{L}^M{}_{[LK]}$$

and noting that the underlined term cancels out when inserting into Eq. 3.100, the curvature tensor for anholonomic $\mathcal{X}^I$ expressed in terms of the covariant derivative reads

$$\mathcal{R}^I{}_{JKL} = 2\mathcal{L}^I{}_{J[L|K]} - 2\mathcal{L}^I{}_{M[K}\mathcal{L}^M{}_{JL]} + \mathcal{L}^I{}_{JM}\mathcal{T}^M{}_{LK}. \tag{3.101}$$

Here, according to Eq. 3.52.2 $\mathcal{T}^M{}_{LK} := \mathcal{A}^M{}_{LK} + 2\mathcal{L}^M{}_{[LK]}$ denotes the torsion in the anholonomic space with coordinates $\mathcal{X}^I$. □

Proof. To proof the above assertions in Eqs. 3.97 and 3.100 the transformation of the connection according to Eq. 3.41 has to be inserted into the standard definition of the curvature tensor in Eq. 3.68 or 3.70.

As an example the case of holonomic $\mathcal{X}^I$ and anholonomic $\mathcal{x}^i$ shall be considered in detail. To start with, the transformation of the connection reads as

$$\mathcal{L}^I{}_{JL} = \mathcal{f}^I{}_i \mathcal{l}^i{}_{jl} \mathcal{F}^j{}_J \mathcal{F}^l{}_L + \mathcal{f}^I{}_i \mathcal{F}^i{}_{J,L}.$$

Computing the partial derivative of the connection as needed in the definition of the curvature renders the lengthy expression

$$\begin{aligned}
\mathcal{L}^I{}_{JL,K} &= \mathcal{f}^I{}_{i,K} \mathcal{l}^i{}_{jl} \mathcal{F}^j{}_J \mathcal{F}^l{}_L \\
&+ \mathcal{f}^I{}_i \mathcal{l}^i{}_{jl.k} \mathcal{F}^j{}_J \mathcal{F}^k{}_K \mathcal{F}^l{}_L \\
&+ \mathcal{f}^I{}_i \mathcal{l}^i{}_{jl} \mathcal{F}^j{}_{J,K} \mathcal{F}^l{}_L \\
&+ \mathcal{f}^I{}_i \mathcal{l}^i{}_{jl} \mathcal{F}^j{}_J \mathcal{F}^l{}_{L,K} \\
&+ \mathcal{f}^I{}_{i,K} \mathcal{F}^i{}_{J,L} \\
&+ \mathcal{f}^I{}_i \mathcal{F}^i{}_{J,LK}.
\end{aligned}$$

Unfortunately, upon skew symmetrization in L and K only one single term drops out

$$\begin{aligned}
\mathcal{L}^I{}_{J[L,K]} &= \mathcal{f}^I{}_{i,[K} \mathcal{l}^i{}_{jl} \mathcal{F}^j{}_J \mathcal{F}^l{}_{L]} \\
&+ \mathcal{f}^I{}_i \underline{\mathcal{l}^i{}_{j[l.k]}} \mathcal{F}^j{}_J \mathcal{F}^k{}_K \mathcal{F}^l{}_L \\
&+ \mathcal{f}^I{}_i \mathcal{l}^i{}_{jl} \mathcal{F}^j{}_{J,[K} \mathcal{F}^l{}_{L]} \\
&+ \mathcal{f}^I{}_i \mathcal{l}^i{}_{jl} \mathcal{F}^j{}_J \mathcal{F}^l{}_{[L,K]} \\
&+ \mathcal{f}^I{}_{i,[K} \mathcal{F}^i{}_{J,L]}.
\end{aligned}$$

Observe that the underlined term is already part of the sought for curvature tensor $\mathcal{r}^i{}_{jkl}$. Next the term of the curvature quadratic in the connection shall be computed. To this end the transformation of the connection is recalled once again with the right set of indices

$$\begin{aligned}
\mathcal{L}^I{}_{MK} &= \mathcal{f}^I{}_i \mathcal{l}^i{}_{mk} \mathcal{F}^m{}_M \mathcal{F}^k{}_K + \mathcal{f}^I{}_i \mathcal{F}^i{}_{M,K}, \\
\mathcal{L}^M{}_{JL} &= \mathcal{f}^M{}_m \mathcal{l}^m{}_{jl} \mathcal{F}^j{}_J \mathcal{F}^l{}_L + \mathcal{f}^M{}_m \mathcal{F}^m{}_{J,L}.
\end{aligned}$$

Then multiplication of the two representations of the connection in the above results in the multi-term expression

$$\mathcal{L}^I{}_{MK}\mathcal{L}^M{}_{JL} = f^I{}_i c^i{}_{mk}\mathcal{F}^m{}_M\mathcal{F}^k{}_K f^M{}_n c^n{}_{jl}\mathcal{F}^j{}_J\mathcal{F}^l{}_L$$
$$+ f^I{}_i c^i{}_{mk}\mathcal{F}^m{}_M\mathcal{F}^k{}_K f^M{}_n\mathcal{F}^n{}_{J,L}$$
$$+ f^I{}_i\mathcal{F}^i{}_{M,K} f^M{}_m c^m{}_{jl}\mathcal{F}^j{}_J\mathcal{F}^l{}_L$$
$$+ f^I{}_i\mathcal{F}^i{}_{M,K} f^M{}_m\mathcal{F}^m{}_{J,L}.$$

Here many terms may be simplified by taking out multiplications of the tangent map by its inverse and by substituting partial derivatives of the tangent map by those of its inverse in the spirit of Eq. 3.56, thus

$$\mathcal{L}^I{}_{MK}\mathcal{L}^M{}_{JL} = f^I{}_i \underline{c^i{}_{mk} c^m{}_{jl}}\mathcal{F}^j{}_J\mathcal{F}^k{}_K\mathcal{F}^l{}_L$$
$$+ f^I{}_i c^i{}_{jl}\mathcal{F}^j{}_{J,L}\mathcal{F}^l{}_K$$
$$- f^I{}_{i,K} c^i{}_{jl}\mathcal{F}^j{}_J\mathcal{F}^l{}_L$$
$$- f^I{}_{i,K}\mathcal{F}^i{}_{J,L}.$$

Observe that the underlined term will be another part of the sought for curvature tensor $r^i{}_{jkl}$. Upon skew symmetrization in L and K no further term drops out

$$\mathcal{L}^I{}_{M[K}\mathcal{L}^M{}_{JL]} = f^I{}_i \underline{c^i{}_{m[k} c^m{}_{jl]}}\mathcal{F}^j{}_J\mathcal{F}^k{}_K\mathcal{F}^l{}_L$$
$$- f^I{}_i c^i{}_{jl}\mathcal{F}^j{}_{J,[K}\mathcal{F}^l{}_{L]}$$
$$- f^I{}_{i,[K} c^i{}_{jl}\mathcal{F}^j{}_J\mathcal{F}^l{}_{L]}$$
$$- f^I{}_{i,[K}\mathcal{F}^i{}_{J,L]}.$$

However, if we combine the above results so as to produce the curvature tensor $\mathcal{R}^I{}_{JKL}$ many terms drop out and the resulting expression reads as

$$\frac{1}{2}\mathcal{R}^I{}_{JKL} := \mathcal{L}^I{}_{J[L,K]} + \mathcal{L}^I{}_{M[K}\mathcal{L}^M{}_{JL]}$$
$$= f^I{}_i\left[c^i{}_{j[l.k]} + c^i{}_{m[k} c^m{}_{jl]} + c^i{}_{jm}\mathcal{F}^m{}_{[N,M]} f^N{}_l f^M{}_k\right]\mathcal{F}^j{}_J\mathcal{F}^k{}_K\mathcal{F}^l{}_L.$$

Inserting finally the definition of the anholonomic object

$$a^m{}_{lk} := 2\mathcal{F}^m{}_{[N,M]} f^N{}_l f^M{}_k$$

into the above result concludes the proof. □

3.4.8 Curvature of an Integrable Connection

The particular case of an *integrable connection*, i.e. the connection derives from a two-point tensor $F^a{}_J$ with inverse $f^I{}_a$ as $\mathcal{L}^I{}_{JK} = f^I{}_a F^a{}_{J,K}$, results in a vanishing curvature tensor

$$\mathcal{R}^I{}_{JKL} \equiv 0. \tag{3.102}$$

An integrable connection arises in a *flat manifold* by a transformation from (possibly anholonomic) Cartesian-type coordinates x^a to (holonomic) curvilinear coordinates X^J with $\mathrm{d}x^a = F^a{}_J\,\mathrm{d}X^J$ and $l^a{}_{bc} = 0$.

Proof. Derive the second partial derivative of $F^a{}_J$

$$F^a{}_{J,LK} = \left[F^a{}_M \mathcal{L}^M{}_{JL}\right]_{,K} = F^a{}_M\left[\mathcal{L}^M{}_{JL,K} + f^M{}_b F^b{}_{N,K}\mathcal{L}^N{}_{JL}\right].$$

Then due to the symmetry of the second partial derivatives the result follows

$$0 = 2 f^I{}_a F^a{}_{J,[LK]} = 2\mathcal{L}^I{}_{J[L,K]} + 2\mathcal{L}^I{}_{N[K}\mathcal{L}^N{}_{JL]} =: \mathcal{R}^I{}_{JKL}.$$

Integrable connections are related to the property of *teleparallelism* and find intensive applications in the continuum theory of crystal plasticity. □

Addendum 3.6. Curvature of a Pseudo-Integrable Connection

The curvature tensor of a *pseudo-integrable connection* may be computed easily based on the additive decomposition into its integrable and its non-integrable part. Since the curvature tensor of the integrable part vanishes identically, the curvature tensor of a pseudo-integrable connection reads

$$\mathcal{R}^I{}_{JKL} = 2\mathcal{L}^{\star I}{}_{J[L|K]} - 2\mathcal{L}^{\star I}{}_{M[K}\mathcal{L}^{\star M}{}_{JL]} - \mathcal{L}^{\star I}{}_{JM}\mathcal{T}^M{}_{KL}.$$

The skew symmetric covariant derivative of $\mathcal{L}^{\star I}{}_{JL}$ may be expressed in terms of the partial derivative as

$$2\mathcal{L}^{\star I}{}_{J[L|K]} = 2\mathcal{L}^{\star I}{}_{J[L,K]} + 2\mathcal{L}^I{}_{M[K}\mathcal{L}^{\star M}{}_{JL]} + 2\mathcal{L}^{\star I}{}_{M[K}\mathcal{L}^M{}_{JL]} + \mathcal{L}^{\star I}{}_{JM}\mathcal{T}^M{}_{KL},$$

thus the curvature tensor may be re-written alternatively as

$$\mathcal{R}^I{}_{JKL} = 2\mathcal{L}^{\star I}{}_{J[L,K]} + 2[\mathcal{L}^I{}_{M[K} - \mathcal{L}^{\star I}{}_{M[K}]\mathcal{L}^{\star M}{}_{JL]} + 2\mathcal{L}^{\star I}{}_{M[K}\mathcal{L}^M{}_{JL]}.$$

Computing the skew symmetric part of the partial derivative of $\mathcal{L}^{\star I}{}_{JL} = f^I{}_m \mathcal{H}^{\star m}{}_{J,L}$ as

$$2\mathcal{L}^{\star I}{}_{J[L,K]} = 2 f^I{}_m \mathcal{H}^{\star m}{}_{J,[LK]} - 2[\mathcal{L}^I{}_{M[K} - \mathcal{L}^{\star I}{}_{M[K}]\mathcal{L}^{\star M}{}_{JL]}$$

and observing that the skew part of second partial derivatives vanishes $\mathcal{H}^{\star m}{}_{J,[LK]} = 0$ reduces the curvature tensor of a pseudo-integrable connection eventually to

$$\mathcal{R}^I{}_{JKL} = 2\mathcal{L}^{\star I}{}_{M[K}\mathcal{L}^M{}_{JL]}.$$

Clearly for a vanishing non-integrable contribution $\mathcal{L}^{\star I}{}_{MK} = 0$ the corresponding curvature tensor also vanishes.

Supplement 3.1. Invariant Formulation

The representation of vectors and covectors (and tensors of any order, see below) in terms of their coefficients depends on the chosen coordinate system (or rather the coordinate chart). Manipulations of these objects in terms of a corresponding index notation follow an easy and straightforward logic, however sometimes at the expense of cumbersome and lengthy expressions (in [179] this representation is denoted in a picturesque way *the debauch of indices*). The introduction of a coordinate basis and a dual basis allows to resort to an *invariant formulation* that is valid independently of the chosen coordinate system and that allows a more compact and elegant notation. In general physical laws ought to be valid independently of the chosen coordinate system, thus they have to have a representation in terms of an invariant formulation.

Coordinate Basis and Dual Basis in Operational Notation:

The *coordinate basis* in a manifold $\mathcal{M}$ corresponding to the coordinate system $\{\chi^i\}$ is denoted by $\boldsymbol{\partial}_{\chi^i}$, whereas the *dual basis* is denoted by $\mathbf{d}\chi^i$, see [122] (Note the use of boldface $\boldsymbol{\partial}$ and $\mathbf{d}$ in order to establish a notational similarity to the more familiar notion of *base vectors* in an Euclidean space, however, this is absolutely not mandatory but rather a matter of taste). Due to their differential operator type definition the coordinate basis and the dual basis transform upon a coordinate change as

$$\boldsymbol{\partial}_{X^I} = \boldsymbol{\partial}_{\chi^j} \frac{\partial \chi^j}{\partial X^I} = \boldsymbol{\partial}_{\chi^j} \mathcal{F}^j{}_I \quad \text{and} \quad \mathbf{d}X^I = \frac{\partial X^I}{\partial \chi^j} \mathbf{d}\chi^j = f^I{}_j \mathbf{d}\chi^j .$$

Observe that the coordinate basis $\boldsymbol{\partial}_{\chi^i}$ transforms in analogy to the covariant coefficients of a covector whereas the dual basis $\mathbf{d}\chi^i$ transforms in analogy to the contravariant coefficients of a vector (thus a combination of contravariant coefficients v^i with the inversely transforming coordinate basis $\boldsymbol{\partial}_{\chi^i}$ and likewise the combination of covariant coefficients v_i with the inversely transforming dual basis $\mathbf{d}\chi^i$, respectively, renders corresponding results that are independent of the coordinate system, i.e. that remain invariant under a coordinate change; this property can be taken as the definition of a tensorial quantity).

Vectors and Covectors:

In an arbitrary manifold (in contrast to an Euclidean space) there are no (co- and contravariant) base vectors available, thus vectors and covectors have to be defined *operationally* (i.e. as the directional derivative operator and the gradient operator, respectively). The coordinate representations of a vector, say $\boldsymbol{v}^\sharp$, and of a covector, say $\boldsymbol{v}^\flat$, (which is also denoted a scalar-valued *one-form*, i.e. a linear form that maps vectors into the set of real numbers $\mathbb{R}$, see the more extensive introduction of the terminology and notation below), respectively, follow correspondingly in operational form as

$$\boldsymbol{v}^\sharp = v^i\boldsymbol{\partial}_{\chi^i} \quad \text{and} \quad \boldsymbol{v}^\flat = v_i\mathbf{d}\chi^i \quad \text{with} \quad v^i = \mathbf{d}\chi^i(\boldsymbol{v}^\sharp) \quad \text{and} \quad v_i = \boldsymbol{v}^\flat(\boldsymbol{\partial}_{\chi^i}).$$

Here, $\sharp$ and $\flat$ have merely been introduced to distinguish vectors from covectors in symbolic notation (traditionally covectors or rather one-forms are distinguished by using Greek letters), round brackets () denote *application to*. Thereby, obviously, the following *duality relation* holds

$$\mathbf{d}\chi^i(\boldsymbol{\partial}_{\chi^j}) = \frac{\partial \chi^i}{\partial \chi^j} = \delta^i{}_j.$$

Differential or Gradient:
The *differential* or rather *gradient* $\mathbf{d}\vartheta$ of a (scalar-valued) function $\vartheta = \theta(\{\chi^j\})$ is defined as

$$\mathbf{d}\vartheta = \frac{\partial\theta}{\partial\chi^j}\mathbf{d}\chi^j.$$

It allows to express the *directional derivative* $\partial_{\boldsymbol{v}^\sharp}(\vartheta)$ of the function ϑ in the direction of the vector $\boldsymbol{v}^\sharp$, i.e.

$$\partial_{\boldsymbol{v}^\sharp}(\vartheta) := v^i\frac{\partial\theta}{\partial\chi^i} = \boldsymbol{v}^\sharp(\vartheta)$$

either in terms of the *directional derivative operator* $\partial_{\boldsymbol{v}^\sharp} := v^i\boldsymbol{\partial}_{\chi^i}$ along the vector $\boldsymbol{v}^\sharp$, or, equivalently, as the application of the differential or rather gradient $\mathbf{d}\vartheta$ of the function ϑ to the vector $\boldsymbol{v}^\sharp$, i.e.

$$\boldsymbol{v}^\sharp(\vartheta) \equiv \mathbf{d}\vartheta(\boldsymbol{v}^\sharp).$$

Thus it holds in particular for the coordinate basis $\boldsymbol{\partial}_{\chi^m} = \delta^i{}_m\boldsymbol{\partial}_{\chi^i}$ as the paradigm of a vector that

$$\boldsymbol{\partial}_{\chi^m}(\vartheta) \equiv \mathbf{d}\vartheta(\boldsymbol{\partial}_{\chi^m}) = \delta^i{}_m\frac{\partial\theta}{\partial\chi^i} = \frac{\partial\theta}{\partial\chi^m}.$$

Discussion:
Note that the coordinate representation of a vector $\boldsymbol{v}^\sharp = v^i\boldsymbol{\partial}_{\chi^i}$ thus takes the interpretation as the directional derivative operator $\partial_{\boldsymbol{v}^\sharp}$, likewise the contravariant coefficients of a vector $v^i = \mathbf{d}\chi^i(\boldsymbol{v}^\sharp) = \boldsymbol{v}^\sharp(\chi^i)$ take the interpretation as the directional derivative of the coordinate functions χ^i along $\boldsymbol{v}^\sharp$. Along the same lines, the coordinate representation of a covector $\boldsymbol{v}^\flat = v_i\mathbf{d}\chi^i$ takes the interpretation as the differential or rather the gradient of a (scalar-valued) function ϑ with $v_i := \partial\theta/\partial\chi^i$.

Coordinate Basis and Dual Basis in Abbreviated Notation:
In the sequel the notation for the coordinate basis and dual basis will be significantly alleviated by resorting to the abbreviations

$$\boldsymbol{g}_i := \boldsymbol{\partial}_{\chi^i} \quad \text{and} \quad \boldsymbol{g}^i := \mathbf{d}\chi^i \quad \text{and} \quad \boldsymbol{g}_I := \boldsymbol{\partial}_{\mathcal{X}^I} \quad \text{and} \quad \boldsymbol{g}^I := \mathbf{d}\mathcal{X}^I,$$

however without sacrificing their operational definition. Consequently, the transformation behavior of the coordinate basis and dual basis reads in terms of these abbreviations

$$\boldsymbol{g}_I = \boldsymbol{g}_i \mathcal{F}^i_{\ I} \quad \text{and} \quad \boldsymbol{g}^I = f^I_{\ i} \boldsymbol{g}^i.$$

Note that it is only in an Euclidean space parameterized by curvilinear coordinates $\{\chi^i\}$ that the coordinate basis and dual basis $\boldsymbol{g}_i = \boldsymbol{\partial}_{\chi^i}$ and $\boldsymbol{g}^i = \mathbf{d}\chi^i$ coincide with the co- and contravariant *base vectors* (*tangent vectors* to *coordinate lines* and *normal vectors* to *coordinate surfaces*) that in turn may be related to the orthonormal Cartesian base vectors $\boldsymbol{e}_a$ and $\boldsymbol{e}^a$, respectively, see section 6.2.

Mapping between Tangent Spaces to Different Manifolds:
Suppose that the coordinates in two *different* manifolds (of the same dimension), say $\mathcal{S}$ (the spatial manifold) and $\mathcal{M}$ (the material manifold), are related by coordinate maps as in Eq. 3.2 and consequently coefficients of vectors follow transformation rules as in Eqs. 3.3 and 3.7. Then vectors $\boldsymbol{v}^\sharp = v^i \boldsymbol{g}_i$ and $\boldsymbol{V}^\sharp = \mathcal{V}^I \boldsymbol{G}_I$ from the tangent spaces to $\mathcal{S}$ and $\mathcal{M}$ are related by the (linear) tangent map $\boldsymbol{F}^\natural$ (a *two-point* $(1,1)$-tensor) as

$$\boldsymbol{v}^\sharp = \boldsymbol{F}^\natural(\boldsymbol{V}^\sharp) = \mathcal{F}^i_{\ I} \mathcal{V}^I \boldsymbol{g}_i = v^i \boldsymbol{g}_i = \mathcal{V}^I \boldsymbol{g}_I \quad \text{with} \quad \boldsymbol{F}^\natural := \mathcal{F}^i_{\ I} \boldsymbol{g}_i \otimes \boldsymbol{G}^I = \boldsymbol{g}_I \otimes \boldsymbol{G}^I.$$

Note carefully that $\boldsymbol{v}^\sharp$ and $\boldsymbol{V}^\sharp$ are indeed different vectors (that allow, however, separately different coordinate representations) from tangent spaces to different manifolds $\mathcal{S}$ and $\mathcal{M}$. Likewise, covectors $\boldsymbol{v}^\flat = v_i \boldsymbol{g}^i$ and $\boldsymbol{V}^\flat = \mathcal{V}_I \boldsymbol{G}^I$ from the cotangent spaces to $\mathcal{S}$ and $\mathcal{M}$ are related by the (linear) tangent map $\boldsymbol{f}^\natural$ (a two-point $(1,1)$-tensor) as

$$\boldsymbol{v}^\flat = \boldsymbol{V}^\flat(\boldsymbol{f}^\natural) = \mathcal{V}_I f^I_{\ i} \boldsymbol{g}^i = v_i \boldsymbol{g}^i = \mathcal{V}_I \boldsymbol{g}^I \quad \text{with} \quad \boldsymbol{f}^\natural := f^I_{\ i} \boldsymbol{G}_I \otimes \boldsymbol{g}^i = \boldsymbol{G}_i \otimes \boldsymbol{g}^i.$$

Mappings between the tangent and cotangent spaces to different manifolds by $\boldsymbol{F}^\natural$ and/or $\boldsymbol{f}^\natural$ are denoted as *pull-back* and *push forward*, respectively, and will be of utmost importance in section 6.2 on the kinematics of finitely deforming continuum bodies.

Linear (Koszul) Connection:
Following more extensive and formal expositions of differential geometry as, e.g., by [179] a linear (Koszul) connection on a smooth manifold $\mathcal{M}$ is a mapping ∇ that associates to each pair of C^∞ vector fields $\boldsymbol{v}^\sharp = v^i \boldsymbol{g}_i = v^j \boldsymbol{g}_j$ and $\boldsymbol{w}^\sharp = w^k \boldsymbol{g}_k$ on $\mathcal{M}$ a third C^∞ vector field $\nabla_{\boldsymbol{w}^\sharp} \boldsymbol{v}^\sharp$ called *covariant derivative* of $\boldsymbol{v}^\sharp$ along $\boldsymbol{w}^\sharp$ with

$$\nabla_{\boldsymbol{w}^\sharp} \boldsymbol{v}^\sharp = [v^i_{\ ,k} + \Gamma^i_{jk} v^j] w^k \boldsymbol{g}_i = v^i_{\ |k} w^k \boldsymbol{g}_i.$$

A linear (Koszul) connection satisfies obviously:

K1: $\nabla_{\boldsymbol{w}^\sharp}\boldsymbol{v}^\sharp$ is linear in $\boldsymbol{w}^\sharp$, i.e. $\nabla_{[\boldsymbol{w}_1^\sharp+\boldsymbol{w}_2^\sharp]}\boldsymbol{v}^\sharp = \nabla_{\boldsymbol{w}_1^\sharp}\boldsymbol{v}^\sharp + \nabla_{\boldsymbol{w}_2^\sharp}\boldsymbol{v}^\sharp$,

K2: $\nabla_{\boldsymbol{w}^\sharp}\boldsymbol{v}^\sharp$ is linear in $\boldsymbol{v}^\sharp$, i.e. $\nabla_{\boldsymbol{w}^\sharp}[\boldsymbol{v}_1^\sharp + \boldsymbol{v}_2^\sharp] = \nabla_{\boldsymbol{w}^\sharp}\boldsymbol{v}_1^\sharp + \nabla_{\boldsymbol{w}^\sharp}\boldsymbol{v}_2^\sharp$,

K3: $\nabla_{[\vartheta\boldsymbol{w}^\sharp]}\boldsymbol{v}^\sharp = \vartheta\nabla_{\boldsymbol{w}^\sharp}\boldsymbol{v}^\sharp$ for differentiable, scalar-valued functions $\vartheta = \theta(\{\chi^i\})$,

K4: $\nabla_{\boldsymbol{w}^\sharp}[\vartheta\boldsymbol{v}^\sharp] = \boldsymbol{w}^\sharp(\vartheta)\boldsymbol{v}^\sharp + \vartheta\nabla_{\boldsymbol{w}^\sharp}\boldsymbol{v}^\sharp$ for differentiable, scalar-valued functions $\vartheta = \theta(\{\chi^i\})$.

Examples:

As an example it is interesting to note that due to the above definition it holds in a smooth manifold $\mathcal{M}$ in particular for the coordinate basis $\boldsymbol{g}_m = \delta^i{}_m\boldsymbol{g}_i = \delta^j{}_m\boldsymbol{g}_j$ and $\boldsymbol{g}_n = \delta^k{}_n\boldsymbol{g}_k$ that

$$\nabla_{\boldsymbol{g}_n}\boldsymbol{g}_m = [\delta^i{}_{m,k} + \Gamma^i_{jk}\delta^j{}_m]\delta^k{}_n\boldsymbol{g}_i = \Gamma^i_{mn}\boldsymbol{g}_i.$$

Note the different sequence of indices m, n on the left and right hand side of the above equation. Thus, as a consequence of **K3**, the covariant derivative of the coordinate basis $\boldsymbol{g}_m$ along the vector $\boldsymbol{w}^\sharp$ follows as

$$\nabla_{\boldsymbol{w}^\sharp}\boldsymbol{g}_m = \Gamma^i_{mn}w^n\boldsymbol{g}_i.$$

Observe that **K4** is the analogy to the product (Leibniz) rule of differentiation, its application to the coordinate representation $\boldsymbol{v}^\sharp = v^i\boldsymbol{g}_i$ of a vector thus renders

$$\nabla_{\boldsymbol{w}^\sharp}[v^i\boldsymbol{g}_i] = \partial_{\boldsymbol{w}^\sharp}(v^i)\boldsymbol{g}_i + v^j\nabla_{\boldsymbol{w}^\sharp}\boldsymbol{g}_j = [v^i{}_{,k} + \Gamma^i_{jk}v^j]w^k\boldsymbol{g}_i.$$

Furthermore, based on **K3**, the application of the *covariant derivative operator* $\nabla_{\boldsymbol{w}^\sharp}$ along $\boldsymbol{w}^\sharp = w^k\boldsymbol{g}_k$ (in terms of the *covariant coordinate basis* $\boldsymbol{\nabla}_{\chi^k}$), i.e.

$$\nabla_{\boldsymbol{w}^\sharp} = w^k\nabla_{\boldsymbol{g}_k} =: w^k\boldsymbol{\nabla}_{\chi^k},$$

to a vector $\boldsymbol{v}^\sharp$ results in the covariant derivative of $\boldsymbol{v}^\sharp$ along $\boldsymbol{w}^\sharp$ (see also [179].2, Chapter 6)

$$\nabla_{\boldsymbol{w}^\sharp}(\boldsymbol{v}^\sharp) = w^k v^i{}_{|k}\boldsymbol{g}_i = \nabla_{\boldsymbol{w}^\sharp}\boldsymbol{v}^\sharp \quad \text{with} \quad \boldsymbol{\nabla}_{\chi^k}(\boldsymbol{v}^\sharp) := v^i{}_{|k}\boldsymbol{g}_i.$$

Covariant Derivative:

Finally the *covariant derivative* or rather *gradient* of a vector field $\boldsymbol{v}^\sharp$ is a second-order tensor field $\nabla\boldsymbol{v}^\sharp$ with

$$\nabla\boldsymbol{v}^\sharp = v^i{}_{|k}\boldsymbol{g}_i \otimes \boldsymbol{g}^k \quad \text{and} \quad \nabla\boldsymbol{v}^\sharp(\boldsymbol{w}^\sharp) = v^i{}_{|k}w^k\boldsymbol{g}_i = \nabla_{\boldsymbol{w}^\sharp}\boldsymbol{v}^\sharp.$$

Note also the relation between the covariant derivative of $\boldsymbol{v}^\sharp$ applied to the coordinate basis $\boldsymbol{g}_k$ and the application of the covariant derivative operator along $\boldsymbol{g}_k$ to the vector field $\boldsymbol{v}^\sharp$, i.e. $\nabla\boldsymbol{v}^\sharp(\boldsymbol{g}_k) = \boldsymbol{\nabla}_{\chi^k}(\boldsymbol{v}^\sharp)$. The covariant derivative satisfies:

D1*: $\nabla\vartheta = \mathbf{d}\vartheta$ for a differentiable, scalar-valued function $\vartheta = \theta(\{\chi^i\})$,
D2*: $\nabla[\boldsymbol{v}^\sharp + \boldsymbol{w}^\sharp] = \nabla\boldsymbol{v}^\sharp + \nabla\boldsymbol{w}^\sharp$,
D3*: $\nabla[\boldsymbol{v}^\sharp \otimes \boldsymbol{w}^\sharp] = \nabla\boldsymbol{v}^\sharp \otimes \boldsymbol{w}^\sharp + \boldsymbol{v}^\sharp \otimes \nabla\boldsymbol{w}^\sharp$.

Coordinate Transformation of the Linear (Koszul) Connection:
Under a coordinate transformation the coordinates of vectors and the corresponding coordinate basis transform as

$$\mathcal{V}^I = f^I{}_i v^i, \quad (w^k = \mathcal{F}^k{}_K \mathcal{W}^K) \quad \text{and} \quad \boldsymbol{g}_i = f^I{}_i \boldsymbol{g}_I \quad \text{with} \quad v^i \boldsymbol{g}_i = \mathcal{V}^I \boldsymbol{g}_I.$$

It is then straightforward to show that the covariant derivative of $\boldsymbol{v}^\sharp$ along $\boldsymbol{w}^\sharp$ may be expressed in the identical format in both coordinate systems

$$[v^i{}_{,k} + \ell^i_{jk} v^j] w^k \boldsymbol{g}_i \equiv [\mathcal{V}^I{}_{,K} + \mathcal{L}^I{}_{JK} \mathcal{V}^J] \mathcal{W}^K \boldsymbol{g}_I$$

provided the coefficients of the linear connection transform (in a non-tensorial fashion) as

$$\ell^i_{jk} = \mathcal{F}^i{}_I \mathcal{L}^I{}_{JK} f^J{}_j f^K{}_k + \mathcal{F}^i{}_M f^M{}_{j,k} = \mathcal{F}^i{}_I \mathcal{L}^I{}_{JK} f^J{}_j f^K{}_k - \mathcal{F}^i{}_{J,K} f^J{}_j f^K{}_k.$$

It shall be noted, that (as long as the partial derivatives are related by $\mathcal{V}^I{}_{,K} = \mathcal{V}^I{}_{,k} \mathcal{F}^k{}_K$) the transformation property for the linear connection holds regardless whether the (first-order distortions) $\mathcal{F}^i{}_I$ are compatible with a coordinate map $\chi^i = y^i(\mathcal{X}^I)$ between holonomic coordinate systems $\mathcal{X}^I$ and χ^i or if they are postulated independently. For a proof of the above statement for the covariant derivative consider the partial derivative

$$\mathcal{V}^I{}_{,K} = f^I{}_i [v^i{}_{,k} + \mathcal{F}^i{}_M f^M{}_{j,k} v^j] \mathcal{F}^k{}_K.$$

Multiplication of the above by $\mathcal{W}^K$, i.e.

$$\mathcal{V}^I{}_{,K} \mathcal{W}^K = [v^i{}_{,k} + \mathcal{F}^i{}_M f^M{}_{j,k} v^j] w^k f^I{}_i$$

and by also multiplying $\mathcal{L}^I{}_{JK} \mathcal{V}^J$ by $\mathcal{W}^K$, i.e.

$$[\mathcal{L}^I{}_{JK} \mathcal{V}^J] \mathcal{W}^K = [\mathcal{F}^i{}_M \mathcal{L}^M_{JK} f^J{}_j f^K{}_k v^j] w^k f^I{}_i$$

renders finally the transformation result for the covariant derivative stated in the above.

Commutator:
For a holonomic coordinate system the definitions in the above allow to introduce the *commutator* or rather the *Lie bracket* of two vector fields as the vector field

$$[\boldsymbol{v}^\sharp, \boldsymbol{w}^\sharp] := [\partial_{\boldsymbol{v}^\sharp}, \partial_{\boldsymbol{w}^\sharp}] := \partial_{\boldsymbol{v}^\sharp}\partial_{\boldsymbol{w}^\sharp} - \partial_{\boldsymbol{w}^\sharp}\partial_{\boldsymbol{v}^\sharp} = \left[v^j \frac{\partial w^i}{\partial \chi^j} - w^j \frac{\partial v^i}{\partial \chi^j} \right] \boldsymbol{g}_i.$$

As an example, the application of the commutator to a (scalar-valued) function $\vartheta = \theta(\{\chi^i\})$ renders

$$[\boldsymbol{v}^\sharp, \boldsymbol{w}^\sharp](\vartheta) = \partial_{\boldsymbol{v}^\sharp}(\partial_{\boldsymbol{w}^\sharp}(\vartheta)) - \partial_{\boldsymbol{w}^\sharp}(\partial_{\boldsymbol{v}^\sharp}(\vartheta)) = \left[v^j \frac{\partial w^i}{\partial \chi^j} - w^j \frac{\partial v^i}{\partial \chi^j}\right] \frac{\partial \theta}{\partial \chi^i}.$$

It is easy to verify that the Lie bracket satisfies the following algebraic identities

L1: $[\boldsymbol{v}^\sharp, \boldsymbol{v}^\sharp] = \boldsymbol{0}$
L2: $[\boldsymbol{v}^\sharp, \boldsymbol{w}^\sharp] + [\boldsymbol{w}^\sharp, \boldsymbol{v}^\sharp] = \boldsymbol{0}$
L3: $[\boldsymbol{u}^\sharp + \boldsymbol{v}^\sharp, \boldsymbol{w}^\sharp] = [\boldsymbol{u}^\sharp, \boldsymbol{w}^\sharp] + [\boldsymbol{v}^\sharp, \boldsymbol{w}^\sharp]$
L4: $\left[[\boldsymbol{u}^\sharp, \boldsymbol{v}^\sharp], \boldsymbol{w}^\sharp\right] + \left[[\boldsymbol{v}^\sharp, \boldsymbol{w}^\sharp], \boldsymbol{u}^\sharp\right] + \left[[\boldsymbol{w}^\sharp, \boldsymbol{u}^\sharp], \boldsymbol{v}^\sharp\right] = \boldsymbol{0}$ (Jacobi identity)

Coordinate Transformation of Commutator:
Recall that the coordinates of vectors and the corresponding coordinate basis transform as

$$w^i = \mathcal{F}^i{}_J \mathcal{W}^J, \quad (\mathcal{V}^K = f^K{}_k v^k) \quad \text{and} \quad \boldsymbol{g}_i = f^J{}_i \boldsymbol{g}_J \quad \text{with} \quad v^i \boldsymbol{g}_i = \mathcal{V}^J \boldsymbol{g}_J.$$

Note that in general these relation need not be based on the $\mathcal{F}^i{}_J$ being compatible with a map $\chi^i = y^i(\{\mathcal{X}^J\})$ between holonomic coordinate systems $\mathcal{X}^J$ and χ^i but are assumed true also for arbitrary, i.e. in particular incompatible $\mathcal{F}^i{}_J$. Here we shall assume the general case of anholonomic χ^i whereas the $\mathcal{X}^J$ define a holonomic coordinate system, thus in general $\mathcal{F}^i{}_{[J,K]} \neq 0$. Therefore in general the partial derivative with respect to the χ^k has to be understood as the partial derivative with respect to the holonomic $\mathcal{X}^K$ followed by a multiplication with $f^K{}_k$, i.e. $w^i{}_{,k} := w^i{}_{,K} f^K{}_k$. Then the partial derivatives contained in the commutator of $\boldsymbol{v}^\sharp$ and $\boldsymbol{w}^\sharp$ transform as

$$w^i{}_{,k} = [\mathcal{F}^i{}_{J,K} \mathcal{W}^J + \mathcal{F}^i{}_J \mathcal{W}^J{}_{,K}] f^K{}_k = \mathcal{F}^i{}_{J,K} f^J{}_j f^K{}_k w^j + f^K{}_k \mathcal{W}^J{}_{,K} \mathcal{F}^i{}_J.$$

Multiplication with v^k and rearranging the various terms results in

$$[v^k w^i{}_{,k} - \mathcal{F}^i{}_{J,K} f^J{}_j f^K{}_k w^j v^k] f^J{}_i = \mathcal{V}^K \mathcal{W}^J{}_{,K}$$

Next, with the definition of the *anholonomic object* (that vanishes identically in the holonomic case)

$$a^i{}_{jk} := 2\mathcal{F}^i{}_{[J,K]} f^J{}_j f^K{}_k$$

it then follows that the commutator of $\boldsymbol{v}^\sharp$ and $\boldsymbol{w}^\sharp$ may be expressed in the holonomic and the anholonomic coordinate systems, respectively, as

$$[\boldsymbol{v}^\sharp, \boldsymbol{w}^\sharp] = \left[\mathcal{V}^K \frac{\partial \mathcal{W}^J}{\partial \mathcal{X}^K} - \mathcal{W}^K \frac{\partial \mathcal{V}^J}{\partial \mathcal{X}^K}\right] \boldsymbol{g}_J = \left[v^k \frac{\partial w^i}{\partial \chi^k} - w^k \frac{\partial v^i}{\partial \chi^k}\right] \boldsymbol{g}_i - a^i{}_{jk} w^j v^k \boldsymbol{g}_i.$$

The commutator or rather Lie bracket will be of importance for the definition of the torsion and the curvature tensors in the sequel. Inserting in particular the anholonomic coordinate basis $\boldsymbol{g}_m = \delta^i{}_m f^I{}_i \boldsymbol{g}_I = \delta^k{}_m f^K{}_k \boldsymbol{g}_K$ and $\boldsymbol{g}_n = \delta^i{}_n f^I{}_i \boldsymbol{g}_I = \delta^j{}_n f^J{}_j \boldsymbol{g}_J = \delta^k{}_n f^K{}_k \boldsymbol{g}_K$ into the definition of the commutator and noticing, e.g., $\delta^i{}_{m,k} = 0$ renders the remarkable result

$$[\boldsymbol{g}_m, \boldsymbol{g}_n] = -a^i{}_{nm} \boldsymbol{g}_i = a^i{}_{mn} \boldsymbol{g}_i \quad \text{with} \quad a^k{}_{mn} = [\boldsymbol{g}_m, \boldsymbol{g}_n](\chi^k).$$

Torsion:
Operationally the torsion tensor (a $(1,2)$-tensor) may be defined for a holonomic coordinate system by its operation on two vector fields $\boldsymbol{v}^\sharp$ and $\boldsymbol{w}^\sharp$ as follows

$$\boldsymbol{t}^\natural(\boldsymbol{v}^\sharp, \boldsymbol{w}^\sharp) := \nabla_{\boldsymbol{v}^\sharp} \boldsymbol{w}^\sharp - \nabla_{\boldsymbol{w}^\sharp} \boldsymbol{v}^\sharp - [\boldsymbol{v}^\sharp, \boldsymbol{w}^\sharp] = 2\Gamma^i_{[jk]} w^j v^k \boldsymbol{g}_i.$$

Note that the sequence of the vectors in $\boldsymbol{t}^\natural(\boldsymbol{v}^\sharp, \boldsymbol{w}^\sharp)$ refers to first applying $\boldsymbol{t}^\natural$ to $\boldsymbol{v}^\sharp$ to obtain a $(1,1)$-tensor $\boldsymbol{t}^\natural(\boldsymbol{v}^\sharp)$, and secondly applying $\boldsymbol{t}^\natural(\boldsymbol{v}^\sharp)$ to $\boldsymbol{w}^\sharp$ to obtain the vector ($(1,0)$-tensor) $\boldsymbol{t}^\natural(\boldsymbol{v}^\sharp, \boldsymbol{w}^\sharp)$. In coordinate representation the above definition spells out as

$$[[w^i{}_{,k} + \Gamma^i_{jk} w^j] v^k - [v^i{}_{,k} + \Gamma^i_{jk} v^j] w^k - [v^k w^i{}_{,k} - w^k v^i{}_{,k}]] \boldsymbol{g}_i = 2\Gamma^i_{[jk]} w^j v^k \boldsymbol{g}_i.$$

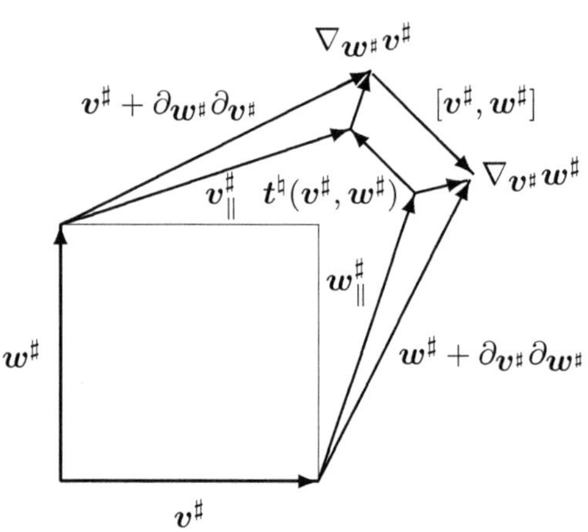

Geometrical interpretation of torsion, commutator, parallel transport, and covariant derivative (motivated by a similar representation in [79]). Here $\boldsymbol{v}^\sharp_\parallel := [v^i - \Gamma^i_{jk} v^j w^k] \boldsymbol{g}_i$ and $\boldsymbol{w}^\sharp_\parallel := [w^i - \Gamma^i_{jk} w^j v^k] \boldsymbol{g}_i$ denote the vectors $\boldsymbol{v}^\sharp$ and $\boldsymbol{w}^\sharp$ after parallel transport along $\boldsymbol{w}^\sharp$ and $\boldsymbol{v}^\sharp$, respectively. Likewise, $\partial_{\boldsymbol{w}^\sharp} \partial_{\boldsymbol{v}^\sharp} = w^k v^i{}_{,k} \boldsymbol{g}_i$ and $\partial_{\boldsymbol{v}^\sharp} \partial_{\boldsymbol{w}^\sharp} = v^k w^i{}_{,k} \boldsymbol{g}_i$, thus the covariant derivatives read $\nabla_{\boldsymbol{w}^\sharp} \boldsymbol{v}^\sharp = [v^i{}_{,k} + \Gamma^i_{jk} v^j] w^k \boldsymbol{g}_i$ and $\nabla_{\boldsymbol{v}^\sharp} \boldsymbol{w}^\sharp = [w^i{}_{,k} + \Gamma^i_{jk} w^j] v^k \boldsymbol{g}_i$, respectively. Note the sequence of indices for $\boldsymbol{t}^\natural(\boldsymbol{v}^\sharp, \boldsymbol{w}^\sharp) = 2\Gamma^i_{[jk]} w^j v^k \boldsymbol{g}_i$, thus possessing the inverse direction as compared to the representation in Fig. 3.4.

Coordinate Transformation of Torsion:
The non-tensorial transformation behavior of the (linear) connection under a change of coordinates results in

$$[\mathscr{l}^i_{jk} w^j v^k + \mathscr{F}^i_{J,K} f^J_j f^K_k w^j v^k] f^I_i = \mathscr{L}^I_{JK} \mathscr{W}^J \mathscr{V}^K.$$

Combined with the previously obtained result for the partial derivative

$$[w^i_{,k} v^k - \mathscr{F}^i_{J,K} f^J_j f^K_k w^j v^k] f^I_i = \mathscr{W}^I_{,K} \mathscr{V}^K$$

the covariant derivatives contained in the torsion applied to the vector fields $\boldsymbol{v}^\sharp$ and $\boldsymbol{w}^\sharp$ transform under a change of coordinates as

$$[w^i_{,k} + \mathscr{l}^i_{jk} w^j] v^k \boldsymbol{g}_i = [\mathscr{W}^I_{,K} + \mathscr{L}^I_{JK} \mathscr{W}^J] \mathscr{V}^K \boldsymbol{g}_I.$$

This is of course no surprise, since the covariant derivative of, e.g., $\boldsymbol{v}^\sharp$ along $\boldsymbol{w}^\sharp$ has earlier been introduced so as to transform tensorially. Thus, for the general case of anholonomic χ^i and holonomic $\mathscr{X}^J$ and due to the result for the transformation of the commutator the torsion tensor as following from its operational definition expands into

$$\boldsymbol{t}^\natural(\boldsymbol{v}^\sharp, \boldsymbol{w}^\sharp) := \nabla_{\boldsymbol{v}^\sharp} \boldsymbol{w}^\sharp - \nabla_{\boldsymbol{w}^\sharp} \boldsymbol{v}^\sharp - [\boldsymbol{v}^\sharp, \boldsymbol{w}^\sharp] = \left[2\mathscr{l}^i_{[jk]} + a^i_{jk}\right] w^j v^k \boldsymbol{g}_i.$$

Inserting in particular the anholonomic coordinate basis $\boldsymbol{g}_m$ and $\boldsymbol{g}_n$ into the definition of the torsion renders

$$\boldsymbol{t}^\natural(\boldsymbol{g}_m, \boldsymbol{g}_n) := \nabla_{\boldsymbol{g}_m} \boldsymbol{g}_n - \nabla_{\boldsymbol{g}_n} \boldsymbol{g}_m - [\boldsymbol{g}_m, \boldsymbol{g}_n] = \left[2\mathscr{l}^i_{[nm]} + a^i_{nm}\right] \boldsymbol{g}_i.$$

Finally, with $t^i_{jk} := 2\mathscr{l}^i_{[jk]} + a^i_{jk}$ the torsion tensor reads in coordinate representation

$$\boldsymbol{t}^\natural = t^i_{jk} \boldsymbol{g}_i \otimes \boldsymbol{g}^j \otimes \boldsymbol{g}^k \quad \text{with} \quad \boldsymbol{t}^\natural(\boldsymbol{g}_k, \boldsymbol{g}_j) = \boldsymbol{t}^\natural : [\boldsymbol{g}_j \otimes \boldsymbol{g}_k] = t^i_{jk} \boldsymbol{g}_i.$$

Mapping between Tangent Spaces to Different Manifolds:
For two different manifolds $\mathcal{S}$ and $\mathcal{M}$ with tangent spaces related by the tangent maps $\boldsymbol{F}^\natural : T\mathcal{M} \to \mathcal{T}\mathcal{S}$ and $\boldsymbol{f}^\natural : T\mathcal{S} \to \mathcal{T}\mathcal{M}$, respectively, with $\boldsymbol{v}^\sharp = \boldsymbol{F}^\natural(\boldsymbol{V}^\sharp)$, and $\boldsymbol{w}^\sharp = \boldsymbol{F}^\natural(\boldsymbol{W}^\sharp)$, the corresponding torsion tensors are related by

$$\boldsymbol{T}^\natural(\boldsymbol{V}^\sharp, \boldsymbol{W}^\sharp) = \boldsymbol{f}^\natural(\boldsymbol{t}^\natural(\boldsymbol{v}^\sharp, \boldsymbol{w}^\sharp)).$$

Thus in a coordinate representation the following pull-back/push-forward relation for the torsion tensors holds, finally, as

$$\boldsymbol{T}^\natural = \mathscr{T}^I_{JK} \boldsymbol{G}_I \otimes \boldsymbol{G}^J \otimes \boldsymbol{G}^K = f^I_i t^i_{jk} \mathscr{F}^j_J \mathscr{F}^k_K \boldsymbol{G}_I \otimes \boldsymbol{G}^J \otimes \boldsymbol{G}^K.$$

Thereby, with e.g. $\boldsymbol{W}^\sharp = \mathscr{W}^J \boldsymbol{G}_J = f^J_j w^j \boldsymbol{G}_J = w^j \boldsymbol{G}_j$, it also holds that

$$\boldsymbol{f}^\natural(\boldsymbol{t}^\natural(\boldsymbol{v}^\sharp, \boldsymbol{w}^\sharp)) = [t^i_{jk} \boldsymbol{G}_i \otimes \boldsymbol{G}^j \otimes \boldsymbol{G}^k] : \left[\boldsymbol{W}^\sharp \otimes \boldsymbol{V}^\sharp\right].$$

Accordingly, the anholonomic objects of $\mathcal{S}$ and $\mathcal{M}$ are related by

$$\boldsymbol{A}^{\natural}(\boldsymbol{V}^{\sharp},\boldsymbol{W}^{\sharp}) = -\boldsymbol{f}^{\natural}\big(\boldsymbol{a}^{\natural}(\boldsymbol{v}^{\sharp},\boldsymbol{w}^{\sharp})\big) = -[a^{i}_{\ jk}\boldsymbol{G}_i \otimes \boldsymbol{G}^j \otimes \boldsymbol{G}^k] : \big[\boldsymbol{W}^{\sharp} \otimes \boldsymbol{V}^{\sharp}\big].$$

Here the minus sign has been incorporated by convention.

Curvature:
In symbolic notation the $(1,3)$ curvature tensor is defined operationally as

$$\begin{aligned}
\boldsymbol{r}^{\natural}(\boldsymbol{u}^{\sharp},\boldsymbol{v}^{\sharp},\boldsymbol{w}^{\sharp}) &:= -\mathfrak{r}_{[\boldsymbol{u}^{\sharp},\boldsymbol{v}^{\sharp}]}(\boldsymbol{w}^{\sharp}) \\
&:= -\big[[\nabla_{\boldsymbol{u}^{\sharp}},\nabla_{\boldsymbol{v}^{\sharp}}] - \nabla_{[\boldsymbol{u}^{\sharp},\boldsymbol{v}^{\sharp}]}\big]\boldsymbol{w}^{\sharp} \\
&:= -\big[\nabla_{\boldsymbol{u}^{\sharp}}\nabla_{\boldsymbol{v}^{\sharp}} - \nabla_{\boldsymbol{v}^{\sharp}}\nabla_{\boldsymbol{u}^{\sharp}} - \nabla_{[\boldsymbol{u}^{\sharp},\boldsymbol{v}^{\sharp}]}\big]\boldsymbol{w}^{\sharp}.
\end{aligned}$$

Note that the sequence of vector arguments in $\boldsymbol{r}^{\natural}(\boldsymbol{u}^{\sharp},\boldsymbol{v}^{\sharp},\boldsymbol{w}^{\sharp})$ refers to the *sequence* of application of $\boldsymbol{r}^{\natural}()$ to the individual vectors, thus in coordinates the coefficients of $\boldsymbol{u}^{\sharp},\boldsymbol{v}^{\sharp},\boldsymbol{w}^{\sharp}$ contract with the fourth, third and second index of $\boldsymbol{r}^{\natural}$, respectively (that is why there is an additional minus sign in front of the *curvature operator* $\mathfrak{r}_{[\boldsymbol{u}^{\sharp},\boldsymbol{v}^{\sharp}]} := [\nabla_{\boldsymbol{u}^{\sharp}},\nabla_{\boldsymbol{v}^{\sharp}}] - \nabla_{[\boldsymbol{u}^{\sharp},\boldsymbol{v}^{\sharp}]}$ as compared to most text books, e.g. [142], that indicate by the sequence of vector arguments in $\boldsymbol{r}^{\natural}()$ the *slots* of $\boldsymbol{r}^{\natural}$, thus the current *sequence-notation* would translate into the *slot-notation* as $\boldsymbol{r}^{\natural}(\boldsymbol{w}^{\sharp},\boldsymbol{v}^{\sharp},\boldsymbol{u}^{\sharp}) = -\mathfrak{r}_{[\boldsymbol{u}^{\sharp},\boldsymbol{v}^{\sharp}]}(\boldsymbol{w}^{\sharp}) = \mathfrak{r}_{[\boldsymbol{v}^{\sharp},\boldsymbol{u}^{\sharp}]}(\boldsymbol{w}^{\sharp}) = \big[[\nabla_{\boldsymbol{v}^{\sharp}},\nabla_{\boldsymbol{u}^{\sharp}}] - \nabla_{[\boldsymbol{v}^{\sharp},\boldsymbol{u}^{\sharp}]}\big]\boldsymbol{w}^{\sharp}$, which however breaks a bit notational symmetry between the left and right hand side of the curvature definition). In coordinate representation the individual contributions to the curvature tensor spell out as

$$\begin{aligned}
\nabla_{\boldsymbol{u}^{\sharp}}\nabla_{\boldsymbol{v}^{\sharp}}\boldsymbol{w}^{\sharp}(\boldsymbol{g}^i) &= \underline{[\ell^{i}_{\ jk,l} + \ell^{i}_{\ ml}\ell^{m}_{\ jk}]w^j v^k u^l} \\
&\quad + \underline{\underline{[w^{i}_{\ ,kl} + 2\ell^{i}_{\ m(k}w^{m}_{\ ,l)}]v^k u^l}} \\
&\quad + \underline{\underline{\underline{[\ell^{i}_{\ jm}v^{m}_{\ ,l}]w^j u^l + [w^{i}_{\ ,m}v^{m}_{\ ,l}]u^l}}}, \\
\nabla_{\boldsymbol{v}^{\sharp}}\nabla_{\boldsymbol{u}^{\sharp}}\boldsymbol{w}^{\sharp}(\boldsymbol{g}^i) &= \underline{[\ell^{i}_{\ jl,k} + \ell^{i}_{\ mk}\ell^{m}_{\ jl}]w^j v^k u^l} \\
&\quad + \underline{\underline{[w^{i}_{\ ,lk} + 2\ell^{i}_{\ m(l}w^{m}_{\ ,k)}]v^k u^l}} \\
&\quad + \underline{\underline{\underline{[\ell^{i}_{\ jm}u^{m}_{\ ,k}]w^j v^k + [w^{i}_{\ ,m}u^{m}_{\ ,k}]v^k}}}.
\end{aligned}$$

The single underlined terms combine into the coefficients $r^{i}_{\ jkl} := 2\ell^{i}_{\ j[l,k]} + 2\ell^{i}_{\ m[k}\ell^{m}_{\ jl]}$ of the curvature tensor

$$\boldsymbol{r}^{\natural} := r^{i}_{\ jkl}\boldsymbol{g}_i \otimes \boldsymbol{g}^j \otimes \boldsymbol{g}^k \otimes \boldsymbol{g}^l,$$

the double underline terms cancel due to symmetry in k, l, and the triple underlined terms cancel with

$$\nabla_{[\boldsymbol{u}^{\sharp},\boldsymbol{v}^{\sharp}]}\boldsymbol{w}^{\sharp}(\boldsymbol{g}^i) = [\ell^{i}_{\ jm}v^{m}_{\ ,l}]w^j u^l - [\ell^{i}_{\ jm}u^{m}_{\ ,k}]w^j v^k + [w^{i}_{\ ,m}v^{m}_{\ ,l}]u^l - [w^{i}_{\ ,m}u^{m}_{\ ,k}]v^k.$$

Coordinate Transformation of Curvature:
Under a coordinate transformation from holonomic coordinates $\mathcal{X}^I$ to anholonomic coordinates χ^i the expressions for $\nabla_{\boldsymbol{u}^\sharp}\nabla_{\boldsymbol{v}^\sharp}\boldsymbol{w}^\sharp(\boldsymbol{g}^i)$ and $\nabla_{\boldsymbol{v}^\sharp}\nabla_{\boldsymbol{u}^\sharp}\boldsymbol{w}^\sharp(\boldsymbol{g}^i)$ remain formally unchanged, however the second partial derivatives

$$w^i{}_{,kl} = w^i{}_{,KL} f^K{}_k f^L{}_l + w^i{}_{,m} \mathcal{F}^m{}_K f^K{}_{k,l}$$
$$w^i{}_{,lk} = w^i{}_{,LK} f^L{}_l f^K{}_k + w^i{}_{,m} \mathcal{F}^m{}_L f^L{}_{l,k}$$

loose their symmetry, i.e. their skew symmetric contribution combines into

$$2w^i{}_{,[kl]} = \underline{w^i{}_{,m} a^m{}_{lk}}.$$

Recall next the expression for the commutator under a coordinate transformation from holonomic coordinates $\mathcal{X}^I$ to anholonomic coordinates χ^i

$$[\boldsymbol{u}^\sharp, \boldsymbol{v}^\sharp] = [u^l v^m{}_{,l} - v^k u^m{}_{,k} + a^m{}_{lk} u^l v^k]\boldsymbol{g}_m.$$

Then, compared to the holonomic case, the expression for $\nabla_{[\boldsymbol{u}^\sharp,\boldsymbol{v}^\sharp]}\boldsymbol{w}^\sharp(\boldsymbol{g}^i)$ is enlarged by two extra terms

$$\nabla_{[\boldsymbol{u}^\sharp,\boldsymbol{v}^\sharp]}\boldsymbol{w}^\sharp(\boldsymbol{g}^i) = \cdots + [\underline{w^i{}_{,m} a^m{}_{lk}} + \Gamma^i{}_{jm} a^m{}_{lk} w^j] v^k u^l.$$

Combining all terms and observing that the underlined terms cancel each other renders finally the coefficients of the curvature tensor in anholonomic coordinates as

$$r^i{}_{jkl} := 2\Gamma^i{}_{j[l,k]} + 2\Gamma^i{}_{m[k}\Gamma^m{}_{jl]} + \Gamma^i{}_{jm} a^m{}_{lk}.$$

Mapping between Tangent Spaces to Different Manifolds:
For two different manifolds $\mathcal{S}$ and $\mathcal{M}$ with tangent spaces related by the tangent maps $\boldsymbol{F}^\natural : T\mathcal{M} \to T\mathcal{S}$ and $\boldsymbol{f}^\natural : T\mathcal{S} \to T\mathcal{M}$, respectively, and with $\boldsymbol{u}^\sharp = \boldsymbol{F}^\natural(\boldsymbol{U}^\sharp)$, $\boldsymbol{v}^\sharp = \boldsymbol{F}^\natural(\boldsymbol{V}^\sharp)$, and $\boldsymbol{w}^\sharp = \boldsymbol{F}^\natural(\boldsymbol{W}^\sharp)$, the corresponding curvature tensors are related by

$$\boldsymbol{R}^\natural(\boldsymbol{U}^\sharp, \boldsymbol{V}^\sharp, \boldsymbol{W}^\sharp) = \boldsymbol{f}^\natural(\boldsymbol{r}^\natural(\boldsymbol{u}^\sharp, \boldsymbol{v}^\sharp, \boldsymbol{w}^\sharp)).$$

Thus in a coordinate representation the following pull-back/push-forward relation for the curvature tensors holds

$$\boldsymbol{R}^\natural = \mathcal{R}^I{}_{JKL}\boldsymbol{G}_I \otimes \boldsymbol{G}^J \otimes \boldsymbol{G}^K \otimes \boldsymbol{G}^L = f^I{}_i r^i{}_{jkl} \mathcal{F}^j{}_J \mathcal{F}^k{}_K \mathcal{F}^l{}_L \boldsymbol{G}_I \otimes \boldsymbol{G}^J \otimes \boldsymbol{G}^K \otimes \boldsymbol{G}^L.$$

Thereby, with e.g. $\boldsymbol{W}^\sharp = \mathcal{W}^J\boldsymbol{G}_J = f^J{}_j w^j \boldsymbol{G}_J = w^j \boldsymbol{G}_j$, it also holds that

$$\boldsymbol{f}^\natural(\boldsymbol{r}^\natural(\boldsymbol{u}^\sharp, \boldsymbol{v}^\sharp, \boldsymbol{w}^\sharp)) = [r^i{}_{jkl}\boldsymbol{G}_i \otimes \boldsymbol{G}^j \otimes \boldsymbol{G}^k \otimes \boldsymbol{G}^l] : [\boldsymbol{W}^\sharp \otimes \boldsymbol{V}^\sharp \otimes \boldsymbol{U}^\sharp].$$

Supplement 3.2. Exterior Calculus

Exterior calculus is a powerful tool that allows to condense the expressions and manipulations as occurring, e.g., in differential geometry to the extreme, i.e. expressions and manipulations that look quite cumbersome in index notation reduce to concise and elegant statements in exterior calculus. The following representation builds heavily on and mixes items from the classical textbooks by [117], [142], and [122]. Additional helpful information may be found, e.g., in [60], [179], [26], [20] and of course in the original contributions by [24].

Exterior Product of Forms:

Based on the simple (vector-valued) one-forms $\mathbf{g}^i := \mathbf{d}\chi^i$ (the dual basis in a manifold $\mathcal{M}$) general (scalar-valued) one-forms are expanded as $\mathsf{v} = v_i \mathbf{g}^i$ (Note that the *bold sans serif* font is used here to explicitly denote forms). Scalar-valued refers to the fact that there is no other index except the single covariant index that relates to the simple one-form $\mathbf{g}^i$. Then the *exterior product* (wedge product) of two general (scalar-valued) one-forms v and w is defined as

$$\mathsf{v} \wedge \mathsf{w} := \mathsf{v} \otimes \mathsf{w} - \mathsf{w} \otimes \mathsf{v} = 2! v_{[i} w_{j]} \mathbf{g}^i \otimes \mathbf{g}^j.$$

Consequently, with $2! v_{[i} w_{j]} := v_i w_j - w_i v_j$ the following skew symmetry holds upon swapping the sequence of the two one-forms involved

$$\begin{aligned} \mathsf{v} \wedge \mathsf{w} &= +v_i w_j \mathbf{g}^i \wedge \mathbf{g}^j \\ = -\mathsf{w} \wedge \mathsf{v} &= -w_i v_j \mathbf{g}^i \wedge \mathbf{g}^j = v_{[i} w_{j]} \mathbf{g}^i \wedge \mathbf{g}^j. \end{aligned}$$

The exterior product of two general (scalar-valued) one-forms is also the definition of a *simple (scalar-valued) two-form*, likewise *simple (scalar-valued) three-forms* are defined as

$$\begin{aligned} \mathsf{u} \wedge \mathsf{v} \wedge \mathsf{w} := \quad & \mathsf{u} \otimes \mathsf{v} \otimes \mathsf{w} - \mathsf{v} \otimes \mathsf{u} \otimes \mathsf{w} + \mathsf{v} \otimes \mathsf{w} \otimes \mathsf{u} \\ & - \mathsf{u} \otimes \mathsf{w} \otimes \mathsf{v} + \mathsf{w} \otimes \mathsf{u} \otimes \mathsf{v} - \mathsf{w} \otimes \mathsf{v} \otimes \mathsf{u} \\ = \quad & 3! u_{[\![i} v_j w_{k]\!]} \mathbf{g}^i \otimes \mathbf{g}^j \otimes \mathbf{g}^k. \end{aligned}$$

With $3! u_{[\![i} v_j w_{k]\!]} := u_i v_j w_k - v_i u_j w_k + v_i w_j u_k - u_i w_j v_k + w_i u_j v_k - w_i v_j u_k$ a simple (scalar-valued) three-form is fully skew symmetric, i.e. any possible swap of two of the constituting simple (scalar-valued) one-forms alternates the sign

$$\begin{aligned} \mathsf{u} \wedge \mathsf{v} \wedge \mathsf{w} &= +u_i v_j w_k \mathbf{g}^i \wedge \mathbf{g}^j \wedge \mathbf{g}^k \\ = -\mathsf{v} \wedge \mathsf{u} \wedge \mathsf{w} &= -v_i u_j w_k \mathbf{g}^i \wedge \mathbf{g}^j \wedge \mathbf{g}^k \\ = +\mathsf{v} \wedge \mathsf{w} \wedge \mathsf{u} &= +v_i w_j u_k \mathbf{g}^i \wedge \mathbf{g}^j \wedge \mathbf{g}^k \\ = -\mathsf{u} \wedge \mathsf{w} \wedge \mathsf{v} &= -u_i w_j v_k \mathbf{g}^i \wedge \mathbf{g}^j \wedge \mathbf{g}^k \\ = +\mathsf{w} \wedge \mathsf{u} \wedge \mathsf{v} &= +w_i u_j v_k \mathbf{g}^i \wedge \mathbf{g}^j \wedge \mathbf{g}^k \\ = -\mathsf{w} \wedge \mathsf{v} \wedge \mathsf{u} &= -w_i v_j u_k \mathbf{g}^i \wedge \mathbf{g}^j \wedge \mathbf{g}^k = u_{[\![i} v_j w_{k]\!]} \mathbf{g}^i \wedge \mathbf{g}^j \wedge \mathbf{g}^k. \end{aligned}$$

This construction rule may then be extended in analogy to simple (scalar-valued) p-forms with $1 \leq p \leq n_{dm}$. *General* (scalar-valued) p-forms consequently follow as *fully skew symmetric* pth-order covariant tensors

$$\mathbf{s} = s_{k_1 k_2 \cdots k_p} \mathbf{g}^{k_1} \otimes \mathbf{g}^{k_2} \otimes \cdots \otimes \mathbf{g}^{k_p} = \frac{1}{p!} s_{k_1 k_2 \cdots k_p} \mathbf{g}^{k_1} \wedge \mathbf{g}^{k_2} \wedge \cdots \wedge \mathbf{g}^{k_p}.$$

The exterior product of arbitrary (scalar-valued) p- and q-forms is both associative and distributive, however for (scalar-valued) p-forms $\mathbf{s}$ and (scalar-valued) q-forms $\mathbf{t}$ the exterior product follows a *modified commutation law*

$$\mathbf{s} \wedge \mathbf{t} = [-1]^{pq} \mathbf{t} \wedge \mathbf{s}.$$

Clearly, the exterior product of two general (scalar-valued) one-forms in the above is nothing but the special case of the modified commutation law with $p = q = 1$.

Tensor-Valued Forms:
It is worth mentioning that more general p-forms may appear with coefficients of a $(m, n+p)$-tensor, skew symmetric in the p covariant indices, given by

$$\begin{aligned}\mathbf{s} = \;& s^{i_1 \cdots i_m}{}_{j_1 \cdots j_n k_1 \cdots k_p} \\ & \mathbf{g}_{i_1} \otimes \cdots \otimes \mathbf{g}_{i_m} \otimes \mathbf{g}^{j_1} \otimes \cdots \otimes \mathbf{g}^{j_n} \otimes \mathbf{g}^{k_1} \otimes \cdots \otimes \mathbf{g}^{k_p} \\ = \;& \frac{1}{p!} s^{i_1 \cdots i_m}{}_{j_1 \cdots j_n k_1 \cdots k_p} \\ & \mathbf{g}_{i_1} \otimes \cdots \otimes \mathbf{g}_{i_m} \otimes \mathbf{g}^{j_1} \otimes \cdots \otimes \mathbf{g}^{j_n} \otimes \mathbf{g}^{k_1} \wedge \cdots \wedge \mathbf{g}^{k_p}.\end{aligned}$$

These forms may be denoted as (m, n)-*tensor-valued p-forms*, as examples consider a scalar and a vector field as $(0, 0)$-tensor-valued (scalar-valued) and $(1, 0)$-tensor-valued (vector-valued) zero-forms. Likewise, the p-forms discussed in the preceding section may be denoted more precisely as $(0, 0)$-tensor-valued (scalar-valued) p-forms. An alternative notation that focuses entirely on the form part to a (m, n)-tensor-valued p-forms reads

$$\begin{aligned}\mathbf{s}^{i_1 \cdots i_m}{}_{j_1 \cdots j_n} = \;& s^{i_1 \cdots i_m}{}_{j_1 \cdots j_n k_1 \cdots k_p} \mathbf{g}^{k_1} \otimes \cdots \otimes \mathbf{g}^{k_p} \\ = \;& \frac{1}{p!} s^{i_1 \cdots i_m}{}_{j_1 \cdots j_n k_1 \cdots k_p} \mathbf{g}^{k_1} \wedge \cdots \wedge \mathbf{g}^{k_p}.\end{aligned}$$

Condensing indices into multi-indices $\boldsymbol{i} := \{i_1 \cdots i_m\}$ and $\boldsymbol{j} := \{j_1 \cdots j_n\}$ allows to re-write a (m, n)-tensor-valued p-form in an abbreviated fashion

$$\mathbf{s}^{\boldsymbol{i}}{}_{\boldsymbol{j}} = \frac{1}{p!} s^{\boldsymbol{i}}{}_{\boldsymbol{j} k_1 \cdots k_p} \mathbf{g}^{k_1} \wedge \cdots \wedge \mathbf{g}^{k_p}.$$

Note that (m,n)-tensor-valued *zero-forms* are simply denoted by the *coefficients* $s^{i}{}_{j} = s^{i_1,\cdots,i_m}{}_{j_1,\cdots,j_n}$ of the (m,n)-tensor. As before the exterior product of arbitrary tensor-valued p- and q-forms is both associative and distributive, likewise for (m,n)-tensor-valued p-forms $\mathbf{s}^{i}{}_{j}$ and (r,s)-tensor-valued q-forms $\mathbf{t}^{k}{}_{l}$ the exterior product follows again the *modified commutation law*

$$\mathbf{s}^{i}{}_{j} \wedge \mathbf{t}^{k}{}_{l} = [-1]^{pq} \mathbf{t}^{k}{}_{l} \wedge \mathbf{s}^{i}{}_{j}.$$

Tensor-valued forms play an important role in a concise representation of differential geometry, see below.

Interior Product of Vectors with Forms:
Even in the absence of a metric the *interior product* of a vector $\mathbf{z} = z^{i}\mathbf{g}_{i}$ with (scalar-valued) p-forms may be computed, whereby $\rfloor$ denotes the interior product operator. The interior product of a vector $\mathbf{z}$ with a (scalar-valued) zero-form $\vartheta = \theta(\{\chi^{i}\})$, i.e. a function, and a (scalar-valued) one-form $\mathbf{u}$, i.e. a covector $\mathbf{u} = u_i \mathbf{g}^{i}$, are defined as

$$\mathbf{z} \rfloor \vartheta := 0 \quad \text{and} \quad \mathbf{z} \rfloor \mathbf{u} = [z^{j}\mathbf{g}_{j}] \rfloor [u_i \mathbf{g}^{i}] := \frac{1}{0!} z^{j} u_{j} \quad \text{with} \quad \mathbf{g}_{j} \rfloor \mathbf{g}^{i} := \delta_{j}{}^{i}.$$

Moreover, the interior product of a vector $\mathbf{z}$ with a (scalar-valued) two-form $\mathbf{v}$, i.e. a skew symmetric second-order covariant tensor $\mathbf{v} = v_{ij}\mathbf{g}^{i} \otimes \mathbf{g}^{j}$ with $v_{ij} \equiv v_{[ij]}$, renders

$$\mathbf{z} \rfloor \mathbf{v} = [z^{k}\mathbf{g}_{k}] \rfloor \frac{1}{2!} [v_{ij}\mathbf{g}^{i} \wedge \mathbf{g}^{j}] = [z^{k}\mathbf{g}_{k}] \rfloor [v_{ij}\mathbf{g}^{i} \otimes \mathbf{g}^{j}] = \frac{1}{1!} z^{k} v_{kj} \mathbf{g}^{j}.$$

Likewise, the interior product of a vector $\mathbf{z}$ with a (scalar-valued) three-form $\mathbf{w}$, i.e. a fully skew symmetric third-order covariant tensor $\mathbf{w} = w_{ijk}\mathbf{g}^{i} \otimes \mathbf{g}^{j} \otimes \mathbf{g}^{k}$ with $w_{ijk} \equiv w_{[\![ijk]\!]}$, renders

$$\mathbf{z} \rfloor \mathbf{w} = [z^{l}\mathbf{g}_{l}] \rfloor \frac{1}{3!} [w_{ijk}\mathbf{g}^{i} \wedge \mathbf{g}^{j} \wedge \mathbf{g}^{k}] = [z^{l}\mathbf{g}_{l}] \rfloor [w_{ijk}\mathbf{g}^{i} \otimes \mathbf{g}^{j} \otimes \mathbf{g}^{k}] = \frac{1}{2!} z^{l} w_{ljk} \mathbf{g}^{j} \wedge \mathbf{g}^{k}.$$

Finally, the interior product of a vector $\mathbf{z}$ with a (scalar-valued) p-form $\mathbf{s}$, i.e. a fully skew symmetric pth-order covariant tensor $\mathbf{s} = s_{m_1 m_2 \cdots m_p} \mathbf{g}^{m_1} \otimes \mathbf{g}^{m_2} \otimes \cdots \otimes \mathbf{g}^{m_p} = 1/p!\, s_{m_1 m_2 \cdots m_p} \mathbf{g}^{m_1} \wedge \mathbf{g}^{m_2} \wedge \cdots \wedge \mathbf{g}^{m_p}$, renders

$$\mathbf{z} \rfloor \mathbf{s} = \frac{1}{[p-1]!} z^{n} s_{n m_2 \cdots m_p} \mathbf{g}^{m_2} \wedge \cdots \wedge \mathbf{g}^{m_p}.$$

The interior product of vectors with (scalar-valued) p-forms is both associative and distributive, however the interior product of vectors with the exterior product of (scalar-valued) p-forms $\mathbf{s}$ and (scalar-valued) q-forms $\mathbf{t}$ computes as

$$\mathbf{z} \rfloor [\mathbf{s} \wedge \mathbf{t}] = [\mathbf{z} \rfloor \mathbf{s}] \wedge \mathbf{t} + [-1]^{p} \mathbf{s} \wedge [\mathbf{z} \rfloor \mathbf{t}].$$

Multiple Interior Product of Vectors with Forms:
Applying the interior product of vectors $\mathbf{z}_r = z_r^{n_r}\mathbf{g}_{n_r}, r = 1\cdots q \leq p$ subsequently, whereby $\mathbf{s}$ is a (scalar-valued) p-form, renders

$$\mathbf{z}_q\rfloor\cdots\mathbf{z}_2\rfloor\mathbf{z}_1\rfloor\mathbf{s} = \frac{1}{[p-q]!}z_q^{n_q}\cdots z_2^{n_2}z_1^{n_1}s_{n_1n_2\cdots n_qm_{q+1}\cdots m_p}\mathbf{g}^{m_{q+1}}\wedge\cdots\wedge\mathbf{g}^{m_p}.$$

An alternative useful notation for the *multiple interior product* is

$$\mathbf{z}_q\rfloor\cdots\mathbf{z}_2\rfloor\mathbf{z}_1\rfloor\mathbf{s} = [\mathbf{z}_q\otimes\cdots\mathbf{z}_2\otimes\mathbf{z}_1]\rfloor^{(q)}\mathbf{s}.$$

Due to the full skew symmetry of $\mathbf{s}$ every swap in the sequence of vectors $\mathbf{z}_r$ changes the sign of the result. As an example for $q = 2$ it holds that

$$\mathbf{z}_2\rfloor\mathbf{z}_1\rfloor\mathbf{s} = [\mathbf{z}_2\otimes\mathbf{z}_1]\rfloor^{(2)}\mathbf{s} = -[\mathbf{z}_1\otimes\mathbf{z}_2]\rfloor^{(2)}\mathbf{s} = -\mathbf{z}_1\rfloor\mathbf{z}_2\rfloor\mathbf{s}.$$

Interior Product of Coordinate Bases with Forms:
Computing the interior product of the coordinate basis $\mathbf{g}_n$ with a (scalar-valued) p-form $\mathbf{s}$ results in

$$\mathbf{g}_n\rfloor\mathbf{s} = \frac{1}{[p-1]!}s_{nm_2\cdots m_p}\mathbf{g}^{m_2}\wedge\cdots\wedge\mathbf{g}^{m_p}.$$

Consequently, the multiple interior product of coordinate bases $\mathbf{g}_{n_r}$ with $\mathbf{s}$ renders

$$\mathbf{g}_{n_q}\rfloor\cdots\mathbf{g}_{n_2}\rfloor\mathbf{g}_{n_1}\rfloor\mathbf{s} = \frac{1}{[p-q]!}s_{n_1n_2\cdots n_qm_{q+1}\cdots m_p}\mathbf{g}^{m_{q+1}}\wedge\cdots\wedge\mathbf{g}^{m_p}.$$

Thus setting in particular $q = p$ extracts the covariant coefficients of the fully skew symmetric pth-order covariant tensor $\mathbf{s}$

$$\mathbf{g}_{n_p}\rfloor\cdots\mathbf{g}_{n_2}\rfloor\mathbf{g}_{n_1}\rfloor\mathbf{s} = s_{n_1n_2\cdots n_p}.$$

Here the alternative notation for the multiple interior product reads

$$\mathbf{g}_{n_q}\rfloor\cdots\mathbf{g}_{n_2}\rfloor\mathbf{g}_{n_1}\rfloor\mathbf{s} = [\mathbf{g}_{n_q}\otimes\cdots\mathbf{g}_{n_2}\otimes\mathbf{g}_{n_1}]\rfloor^{(q)}\mathbf{s}.$$

As a consequence the multiple interior product of a qth-order contravariant tensor $\mathbf{y} = y^{n_q\cdots n_2n_1}\mathbf{g}_{n_q}\otimes\cdots\mathbf{g}_{n_2}\otimes\mathbf{g}_{n_1}$ with $\mathbf{s}$ is expressed as

$$\mathbf{y}\rfloor^{(q)}\mathbf{s} = \frac{1}{[p-q]!}y^{n_q\cdots n_2n_1}s_{n_1n_2\cdots n_qm_{q+1}\cdots m_p}\mathbf{g}^{m_{q+1}}\wedge\cdots\wedge\mathbf{g}^{m_p}$$

Setting in particular $q = p$ results in the full contraction of the contravariant coefficients of $\mathbf{y}$ and the covariant coefficients of $\mathbf{s}$

$$\mathbf{y}\rfloor^{(p)}\mathbf{s} = y^{n_p\cdots n_2n_1}s_{n_1n_2\cdots n_p}.$$

Finally, for a p-vector $\mathbf{y}$, i.e. $\mathbf{y}$ is a fully skew symmetric pth-order contravariant tensor the multiple interior product $\mathbf{y}\rfloor^{(p)}\mathbf{s}$ simplifies to $p!\, y^{n_p \cdots n_2 n_1} s_{n_1 n_2 \cdots n_p}$ with $n_1 < n_2 < \cdots < n_p$.

Exterior Derivative of Forms:
In the case of a (scalar-valued) one-form $\mathbf{v}$ its *exterior derivative* $\mathbf{d}$ renders a (scalar-valued) two-form denoted by $\mathbf{dv}$ with definition

$$\mathbf{dv} := \frac{\partial v_i}{\partial \chi^j} \mathbf{g}^j \wedge \mathbf{g}^i = -\frac{\partial v_i}{\partial \chi^j} \mathbf{g}^i \wedge \mathbf{g}^j.$$

Note thus the relation of the coefficients of $\mathbf{g}^j \wedge \mathbf{g}^i$ with the curl of $\mathbf{v}$

$$\mathbf{dv} = \frac{1}{2}\left[\frac{\partial v_i}{\partial \chi^j} - \frac{\partial v_j}{\partial \chi^i}\right] \mathbf{g}^j \wedge \mathbf{g}^i.$$

Likewise in the case of a (scalar-valued) two-form $\mathbf{t}$ the *exterior derivative* $\mathbf{d}$ renders a (scalar-valued) three-form denoted by $\mathbf{dt}$ with definition

$$\mathbf{dt} := \frac{\partial t_{ij}}{\partial \chi^k} \mathbf{g}^k \wedge \mathbf{g}^i \wedge \mathbf{g}^j = \frac{\partial t_{ij}}{\partial \chi^k} \mathbf{g}^i \wedge \mathbf{g}^j \wedge \mathbf{g}^k.$$

The exterior derivative $\mathbf{d}$ may then be extended in analogy to general (scalar-valued) p-forms $\mathbf{t}$ with $1 \le p \le n_{dm}$ to render a (scalar-valued) $[p+1]$-form defined as

$$\begin{aligned} \mathbf{dt} &:= \frac{\partial t_{i_1 i_2 \cdots i_p}}{\partial \chi^j} \mathbf{g}^j \wedge \mathbf{g}^{i_1} \wedge \mathbf{g}^{i_2} \wedge \cdots \wedge \mathbf{g}^{i_p}. \\ &:= [-1]^p \frac{\partial t_{i_1 i_2 \cdots i_p}}{\partial \chi^j} \mathbf{g}^{i_1} \wedge \mathbf{g}^{i_2} \wedge \cdots \wedge \mathbf{g}^{i_p} \wedge \mathbf{g}^j. \end{aligned}$$

Note that, as a matter of consistency, the exterior derivative $\mathbf{d}\vartheta$ of a function $\vartheta = \theta(\{\chi^i\})$, i.e. a scalar-valued zero-form, coincides with the previously defined differential of the function, i.e.

$$\mathbf{d}\vartheta = \frac{\partial \theta}{\partial \chi^i} \mathbf{g}^i.$$

Moreover, due to the symmetry of the second partial derivatives, the double application of the exterior derivative to any (scalar-valued) p-form with coefficients of class C^2 renders

$$\mathbf{ddt} = \mathbf{0} \quad \text{with} \quad \mathbf{ddt} = -\frac{\partial^2 t_{i_1 i_2 \cdots i_p}}{\partial \chi^j \partial \chi^k} \mathbf{g}^{i_1} \wedge \mathbf{g}^{i_2} \wedge \cdots \wedge \mathbf{g}^{i_p} \wedge \mathbf{g}^j \wedge \mathbf{g}^k.$$

Differential p-forms are denoted *closed* if $\mathbf{dt} = \mathbf{0}$ and *exact* if there exists a $[p-1]$-form $\mathbf{s}$ such that $\mathbf{t} = \mathbf{ds}$. These results are related to the *Poincaré Lemma* stating that every exact p-form is closed, the converse only being true locally. The exterior derivative of the exterior product of (scalar-valued) p-forms $\mathbf{s}$ and q-forms $\mathbf{t}$ renders

$$\mathbf{d}(\mathbf{s} \wedge \mathbf{t}) = \mathbf{d}\mathbf{s} \wedge \mathbf{t} + [-1]^p \mathbf{s} \wedge \mathbf{d}\mathbf{t}.$$

In analogy the exterior derivative of the exterior product of (m,n)-tensor-valued p-forms $\mathbf{s}^i{}_j$ and (r,s)-tensor-valued q-forms $\mathbf{t}^k{}_l$ reads

$$\mathbf{d}(\mathbf{s}^i{}_j \wedge \mathbf{t}^k{}_l) = \mathbf{d}\mathbf{s}^i{}_j \wedge \mathbf{t}^k{}_l + [-1]^p \mathbf{s}^i{}_j \wedge \mathbf{d}\mathbf{t}^k{}_l.$$

Covariant Exterior Derivative of Forms:
Due to the skew-symmetry of the exterior product the coefficients of the exterior derivative of a scalar-valued one-form transform tensorially upon a change of holonomic coordinates

$$\mathbf{d}\mathbf{v} = -\frac{\partial v_i}{\partial \chi^j} \mathbf{g}^i \wedge \mathbf{g}^j = -\left[f^I{}_i \frac{\partial \mathcal{V}_I}{\partial \mathcal{X}^J} f^J{}_j + \mathcal{V}_M \frac{\partial f^M{}_i}{\partial \chi^j} \right] \mathbf{g}^i \wedge \mathbf{g}^j = -\frac{\partial \mathcal{V}_I}{\partial \mathcal{X}^J} \mathbf{g}^I \wedge \mathbf{g}^J.$$

This property carries over to the exterior derivative of scalar-valued p-forms, however the exterior derivative of arbitrary tensor-valued forms is generally not tensorial. Thus in order to construct a (tensorial) *covariant exterior derivative* of forms the connection ℓ^i_{jk} and the concept of covariant differentiation have to be invoked. To this end first the following (non-tensorial) one-form, the so-called *connection one-form* is introduced based on the coefficients of the connection

$$\mathbf{l}^i{}_j := \ell^i_{jk} \mathbf{g}^k.$$

Then the covariant exterior derivative of tensorially transforming vector- and covector-valued p-forms renders tensorially transforming vector- and covector-valued $[p+1]$-forms, respectively, as

$$\mathbf{D}\mathbf{t}^i = \mathbf{d}\mathbf{t}^i + \mathbf{l}^i{}_j \wedge \mathbf{t}^j, \quad \text{and} \quad \mathbf{D}\mathbf{t}_j = \mathbf{d}\mathbf{t}_j - \mathbf{l}^i{}_j \wedge \mathbf{t}_i.$$

It is interesting to note that the covariant exterior derivative of vector- and covector-valued zero-forms, i.e. of vector and covector fields v^i and v_j, respectively, coincides in particular with the covariant derivative of those fields

$$\mathbf{D}v^i = \mathbf{d}v^i + \mathbf{l}^i{}_j v^j = [v^i{}_{,k} + \ell^i_{jk} v^j] \mathbf{g}^k = v^i{}_{|k} \mathbf{g}^k,$$
$$\mathbf{D}v_j = \mathbf{d}v_j - \mathbf{l}^i{}_j v_i = [v_{j,k} - \ell^i_{jk} v_i] \mathbf{g}^k = v_{j|k} \mathbf{g}^k.$$

Moreover, the covariant exterior derivative of tensorially transforming second-order tensor-valued p-forms renders tensorially transforming second-order tensor-valued $[p+1]$-forms

$$
\begin{aligned}
\mathbf{Dt}^{ij} &= \mathbf{dt}^{ij} + \mathbf{l}^{i}{}_{m} \wedge \mathbf{t}^{mj} + \mathbf{l}^{j}{}_{m} \wedge \mathbf{t}^{im}, \\
\mathbf{Dt}^{i}{}_{j} &= \mathbf{dt}^{i}{}_{j} + \mathbf{l}^{i}{}_{m} \wedge \mathbf{t}^{m}{}_{j} - \mathbf{l}^{m}{}_{j} \wedge \mathbf{t}^{i}{}_{m}, \\
\mathbf{Dt}_{ij} &= \mathbf{dt}_{ij} - \mathbf{l}^{m}{}_{i} \wedge \mathbf{t}_{mj} - \mathbf{l}^{m}{}_{j} \wedge \mathbf{t}_{im}, \\
\mathbf{Dt}_{i}{}^{j} &= \mathbf{dt}_{i}{}^{j} - \mathbf{l}^{m}{}_{i} \wedge \mathbf{t}_{m}{}^{j} + \mathbf{l}^{j}{}_{m} \wedge \mathbf{t}_{i}{}^{m}.
\end{aligned}
$$

Likewise, the corresponding results hold for third-order tensor-valued p-forms

$$
\begin{aligned}
\mathbf{Dt}^{i}{}_{jl} &= \mathbf{dt}^{i}{}_{jl} + \mathbf{l}^{i}{}_{m} \wedge \mathbf{t}^{m}{}_{jl} - \mathbf{l}^{m}{}_{j} \wedge \mathbf{t}^{i}{}_{ml} - \mathbf{l}^{m}{}_{l} \wedge \mathbf{t}^{i}{}_{jm} \\
\mathbf{Dt}_{ijl} &= \mathbf{dt}_{ijl} - \mathbf{l}^{m}{}_{i} \wedge \mathbf{t}_{mjl} - \mathbf{l}^{m}{}_{j} \wedge \mathbf{t}_{iml} - \mathbf{l}^{m}{}_{l} \wedge \mathbf{t}_{ijm}.
\end{aligned}
$$

Finally the covariant exterior derivative of the exterior product of (m, n)-tensor-valued p-forms $\mathbf{s}^{i}{}_{j}$ and (r, s)-tensor-valued q-forms $\mathbf{t}^{k}{}_{l}$ reads

$$
\mathbf{D}(\mathbf{s}^{i}{}_{j} \wedge \mathbf{t}^{k}{}_{l}) = \mathbf{Ds}^{i}{}_{j} \wedge \mathbf{t}^{k}{}_{l} + [-1]^{p} \mathbf{s}^{i}{}_{j} \wedge \mathbf{Dt}^{k}{}_{l}.
$$

Sub-Manifolds:
A parametric representation of a p-dimensional *sub-manifold* $\widehat{\mathcal{M}}$ embedded in the n_{dm}-dimensional manifold $\mathcal{M}$ is given by

$$
\chi^{i} = \chi^{i}(\hat{\chi}^{k}) \quad \text{with} \quad i = 1, 2, \cdots, n_{dm} \quad \text{and} \quad k = 1, 2, \cdots, p \leq n_{dm}.
$$

Here χ^{i} are the n_{dm} coordinates of the manifold $\mathcal{M}$ and $\hat{\chi}^{k}$ denote the p coordinates of the sub-manifold $\widehat{\mathcal{M}}$. Accordingly the p one-forms $\hat{\mathbf{g}}^{k} := \mathbf{d}\hat{\chi}^{k}$ in the tangent space $T_{\mathcal{P}}\widehat{\mathcal{M}}$ determine the n_{dm} one-forms $\mathbf{g}^{i} := \mathbf{d}\chi^{i}$ in the tangent space $T_{\mathcal{P}}\mathcal{M}$ as

$$
\mathbf{g}^{i} = \chi^{i}{}_{,k} \hat{\mathbf{g}}^{k}.
$$

Thus $T_{\mathcal{P}}\widehat{\mathcal{M}}$ may be considered a linear *sub-space* of $T_{\mathcal{P}}\mathcal{M}$ spanned by the p linearly independent vectors $\chi^{i}{}_{,1}, \chi^{i}{}_{,2}, \cdots, \chi^{i}{}_{,p}$.

Integration of Forms:
The Stokes' theorem is concerned with the integration of p-forms $\mathbf{t}$ defined on the n_{dm}-dimensional manifold $\mathcal{M}$ over measurable subsets $\widehat{\mathcal{U}} \subset \widehat{\mathcal{M}}$ of p-dimensional sub-manifolds $\widehat{\mathcal{M}}$ with $p \leq n_{dm}$ that are embedded in $\mathcal{M}$

$$
\int_{\widehat{\mathcal{U}}} \mathbf{t} = \int_{\widehat{\mathcal{U}}} t_{i_1 \cdots i_p}(\{\chi^{j}\}) \mathbf{g}^{i_1} \wedge \cdots \wedge \mathbf{g}^{i_p},
$$

thus the corresponding integrals have to be defined first. To this end, the parametric representation of $\widehat{\mathcal{U}}$ is invoked to re-express the p-form in the integrand

$$
\int_{\widehat{\mathcal{U}}} \mathbf{t} = \int_{\widehat{\mathcal{U}}} t_{i_1 \cdots i_p}(\{\chi^{j}(\hat{\chi}^{l})\}) \chi^{i_1}{}_{,k_1} \cdots \chi^{i_p}{}_{,k_p} \hat{\mathbf{g}}^{k_1} \wedge \cdots \wedge \hat{\mathbf{g}}^{k_p}.
$$

With $\hat{t}_{k_1\cdots k_p} := t_{i_1\cdots i_p}\chi^{i_1}{}_{,k_1}\cdots\chi^{i_p}{}_{,k_p}$ the p-form $\mathbf{t}$ on $\mathcal{M}$ becomes a p-form $\hat{\mathbf{t}}$ on $\widehat{\mathcal{M}}$, thus

$$\int_{\widehat{\mathcal{U}}}\hat{\mathbf{t}} = \int_{\widehat{\mathcal{U}}}\hat{t}_{k_1\cdots k_p}(\{\hat{\chi}^l\})\hat{\mathbf{g}}^{k_1}\wedge\cdots\wedge\hat{\mathbf{g}}^{k_p}.$$

Next, it is instrumental to note that the p-form $\hat{\mathbf{t}}$ on the p-dimensional (sub-) manifold $\widehat{\mathcal{M}}$ allows the following alternative (unique) representation

$$\hat{\mathbf{t}} = \hat{t}_{k_1\cdots k_p}(\{\hat{\chi}^l\})\hat{\mathbf{g}}^{k_1}\wedge\cdots\wedge\hat{\mathbf{g}}^{k_p} = \hat{\theta}(\{\hat{\chi}^l\})\hat{\mathbf{g}}^{k=1}\wedge\cdots\wedge\hat{\mathbf{g}}^{k=p}.$$

Here, $\hat{\theta}(\{\hat{\chi}^l\})$ collects the corresponding terms of $\hat{t}_{k_1\cdots k_p}(\{\hat{\chi}^l\})$. Then, with $\hat{\mathbf{g}}^i = \mathbf{d}\hat{\chi}^i$ the integral of the p-form $\hat{\mathbf{t}}$ over a measurable subset $\widehat{\mathcal{U}}\subset\widehat{\mathcal{M}}$ is finally defined as

$$\int_{\widehat{\mathcal{U}}}\hat{\mathbf{t}} = \int_{\widehat{\mathcal{U}}}\hat{\theta}(\{\hat{\chi}^l\})\hat{\mathbf{g}}^{k=1}\wedge\cdots\wedge\hat{\mathbf{g}}^{k=p} := \int_{\widehat{\mathcal{U}}}\hat{\theta}(\{\hat{\chi}^l\})\,\mathrm{d}\hat{\chi}^{k=1}\cdots\mathrm{d}\hat{\chi}^{k=p}.$$

Here the right-hand side denotes the elementary definition of p-fold integration.

As an example, for $p=1$ and with $\hat{\theta}=\hat{t}=t_i\chi^{i\prime}$ and $\mathrm{d}\chi^i := \chi^{i\prime}\,\mathrm{d}\hat{\chi}$ integration of a one-form over a one-dimensional $\widehat{\mathcal{U}}$ renders

$$\int_{\widehat{\mathcal{U}}}\mathbf{t} = \int_{\widehat{\mathcal{U}}}t_i\mathbf{g}^i = \int_{\widehat{\mathcal{U}}}t_i\chi^{i\prime}\hat{\mathbf{g}} = \int_{\widehat{\mathcal{U}}}t_i\,\mathrm{d}\chi^i.$$

Likewise, for $p=2$ and with $\hat{\theta}=\hat{t}_{[12]}=t_{[ij]}\chi^i{}_{,1}\chi^j{}_{,2}$ and $\mathrm{d}_1\chi^i := \chi^i{}_{,1}\,\mathrm{d}\hat{\chi}^1$ and $\mathrm{d}_2\chi^j := \chi^j{}_{,2}\,\mathrm{d}\hat{\chi}^2$ integration of a two-form over a two-dimensional $\widehat{\mathcal{U}}$ results in

$$\int_{\widehat{\mathcal{U}}}\mathbf{t} = \int_{\widehat{\mathcal{U}}}\frac{1}{2}t_{ij}\mathbf{g}^i\wedge\mathbf{g}^j = \int_{\widehat{\mathcal{U}}}t_{[ij]}\chi^i{}_{,1}\chi^j{}_{,2}\hat{\mathbf{g}}^1\wedge\hat{\mathbf{g}}^2 = \int_{\widehat{\mathcal{U}}}t_{[ij]}\,\mathrm{d}_1\chi^i\,\mathrm{d}_2\chi^j.$$

Finally, it is interesting to note that a p-form $\hat{\mathbf{g}}^K$ constructed from the vector-valued one-forms $\hat{\mathbf{g}}^K = \hat{f}^K{}_k\hat{\mathbf{g}}^k$ with $K,k=1,\cdots,p$ allows the representation

$$\hat{\mathbf{g}}^{K=1}\wedge\cdots\wedge\hat{\mathbf{g}}^{K=p} = \det[\hat{f}^K{}_k]\hat{\mathbf{g}}^{k=1}\wedge\cdots\wedge\hat{\mathbf{g}}^{k=p}.$$

This relation corresponds to the transformation rule for the *volume-p-form* in a p-dimensional (sub-)manifold $\widehat{\mathcal{M}}$, see [122]. Then, the integral of a scalar function $\hat{\vartheta} = \hat{\Theta}(\{\hat{X}^L\}) = \hat{\theta}(\{\hat{\chi}^l\})$ over a measurable subset $\widehat{\mathcal{U}}\subset\widehat{\mathcal{M}}$ expands into

$$\int_{\widehat{\mathcal{U}}}\hat{\Theta}(\{\hat{X}^L\})\hat{\mathbf{g}}^{K=1}\wedge\cdots\wedge\hat{\mathbf{g}}^{K=p} = \int_{\widehat{\mathcal{U}}}\hat{\theta}(\{\hat{\chi}^l\})\det[\hat{f}^K{}_k]\hat{\mathbf{g}}^{k=1}\wedge\cdots\wedge\hat{\mathbf{g}}^{k=p}.$$

With $\hat{\mathbf{g}}^i = \mathbf{d}\hat{\chi}^i$ the last integral may then be computed based on the above recipe for integration on manifolds as

$$\int_{\widehat{\mathcal{U}}} \hat{\theta}(\{\hat{\chi}^l\}) \det[\hat{f}^K{}_k] \hat{\mathbf{g}}^{k=1} \wedge \cdots \wedge \hat{\mathbf{g}}^{k=p} := \int_{\widehat{\mathcal{U}}} \hat{\theta}(\{\hat{\chi}^l\}) \det[\hat{f}^K{}_k] \, \mathrm{d}\hat{\chi}^{k=1} \cdots \mathrm{d}\hat{\chi}^{k=p}.$$

The integration of differential forms on manifolds is the prerequisite for the derivation of the *Stokes theorem*, see e.g. [179, 60, 117].

Generalized Stokes' Theorem:
For a p-form $\mathbf{t}$ defined on the n_{dm}-dimensional manifold $\mathcal{M}$ with $[p+1] \leq n_{dm}$ and for a subset $\widehat{\mathcal{U}}$ of the $[p+1]$-dimensional sub-manifold $\widehat{\mathcal{M}}$ with positively oriented p-dimensional $\partial\widehat{\mathcal{U}}$ the *generalized Stokes' theorem* reads

$$\int_{\widehat{\mathcal{U}}} \mathbf{dt} = \int_{\partial\widehat{\mathcal{U}}} \mathbf{t}.$$

Thus the integral of the p-form $\mathbf{t}$ over the p-dimensional boundary $\partial\widehat{\mathcal{U}}$ equals the integral of the $[p+1]$-form $\mathbf{dt}$ over its the $[p+1]$-dimensional interior $\widehat{\mathcal{U}}$!

(In the above *positively oriented* refers to a coordinate chart $\{\hat{\chi}^i\}$ on $\widehat{\mathcal{U}}$ with $\partial\widehat{\mathcal{U}}$ given by $\hat{\chi}^1 = 0$ and $\widehat{\mathcal{U}}$ being represented by $\hat{\chi}^1 < 0$.) A more explicit version reads

$$\int_{\widehat{\mathcal{U}}} [-1]^p \frac{\partial t_{i_1 \cdots i_p}}{\partial \chi^j} \mathbf{g}^{i_1} \wedge \cdots \wedge \mathbf{g}^{i_p} \wedge \mathbf{g}^j = \int_{\partial\widehat{\mathcal{U}}} t_{i_1 \cdots i_p} \mathbf{g}^{i_1} \wedge \cdots \wedge \mathbf{g}^{i_p}.$$

The generalized Stokes' theorem contains as special cases the common integral theorems, e.g. due to Gauss, Green, Stokes, and elementary calculus.

As an example, for a one-dimensional curve $\widehat{\mathcal{U}}$ with curve parameter $\hat{\chi}$, $\partial\widehat{\mathcal{U}}$ its endpoints at $\hat{\chi}_0$ and $\hat{\chi}_1$, and the scalar-valued zero-form (function) $\vartheta = \hat{\theta}(\hat{\chi})$ it holds that

$$\int_{\widehat{\mathcal{U}}} \mathbf{d}\hat{\theta} := \int_{\hat{\chi}_0}^{\hat{\chi}_1} \hat{\theta}' \, \mathrm{d}\hat{\chi} \quad \text{and} \quad \int_{\partial\widehat{\mathcal{U}}} \hat{\theta} := \hat{\theta}(\hat{\chi}_1) - \hat{\theta}(\hat{\chi}_0).$$

Likewise, for a two-dimensional surface $\widehat{\mathcal{U}}$ embedded in three-dimensional Euclidean space, $\partial\widehat{\mathcal{U}}$ its closed boundary curve, and the (scalar-valued) one-form $\mathbf{v}$ (i.e. the covector $\boldsymbol{v}^\flat$) the subsequent identifications may be made

$$\int_{\widehat{\mathcal{U}}} \mathbf{dv} := \int_{\widehat{\mathcal{U}}} \operatorname{curl} \boldsymbol{v}^\flat \cdot \mathrm{d}\boldsymbol{a}^\flat \quad \text{and} \quad \int_{\partial\widehat{\mathcal{U}}} \mathbf{v} := \int_{\partial\widehat{\mathcal{U}}} \boldsymbol{v}^\flat \cdot \mathrm{d}\boldsymbol{l}^\sharp.$$

Here, $\mathrm{d}\boldsymbol{a}^\flat = \boldsymbol{n}^\flat \, \mathrm{d}a$ and $\mathrm{d}\boldsymbol{l}^\sharp$ denote the covectorial area element of the surface and the vectorial line element of its boundary curve, respectively, and $\operatorname{curl} \boldsymbol{v}^\flat$ renders the vectorial curl of $\boldsymbol{v}^\flat$, for a detailed derivation see Part III of this treatise. Finally, for a n_{dm}-dimensional Riemannian manifold $\mathcal{M}$, for a definition see below, its positively oriented $[n_{dm}-1]$-dimensional boundary manifold $\partial\mathcal{M}$, and the $[n_{dm}-1]$-form $[\boldsymbol{v}^\sharp \cdot \boldsymbol{n}^\flat]\mathbf{d}a$ it follows that (see [122])

$$\int_{\mathcal{U}} \mathbf{d}([\boldsymbol{v}^\sharp \cdot \boldsymbol{n}^\flat]\mathbf{d}a) := \int_{\mathcal{U}} \operatorname{div}\boldsymbol{v}^\sharp \, \mathrm{d}v \quad \text{and} \quad \int_{\partial\mathcal{U}} [\boldsymbol{v}^\sharp \cdot \boldsymbol{n}^\flat]\mathbf{d}a := \int_{\partial\mathcal{U}} \boldsymbol{v}^\sharp \cdot \mathrm{d}\boldsymbol{a}^\flat .$$

Here, $\mathbf{d}a$ denotes the area $[n_{dm}-1]$-form on the $[n_{dm}-1]$-dimensional manifold $\partial\mathcal{M}$, and $\mathrm{d}v$ and $\mathrm{d}a$ are the ordinary (scalar-valued) volume and area element in $\mathcal{M}$ and $\partial\mathcal{M}$, respectively.

Application to Hamiltonian Equations:
For a dynamical system with n_{dm} degrees of freedom consider the following (scalar-valued) one-form in the *state space* with $2n_{dm}+1$ dimensions

$$\mathsf{w} = w_i \mathbf{g}^i = p_\alpha \mathbf{d}q^\alpha - \mathcal{H}\mathbf{d}t \quad \text{with} \quad [w_i] = \big[[0_\alpha], p_1, \cdots, p_{n_{dm}}, -\mathcal{H}\big],$$

whereby $\{\chi^1, \cdots, \chi^{n_{dm}}\} = \{p_1, \cdots, p_{n_{dm}}\}$ denote the *generalized momenta*, $\chi^{n_{dm}+1}, \cdots, \chi^{2n_{dm}} = \{q^1, \cdots, q^{n_{dm}}\}$ denote the *generalized coordinates*, and $\chi^{2n_{dm}+1} = t$ denotes the *time coordinate*. Moreover, Latin indices $i = 1, \cdots, 2n_{dm}+1$ number variables in the state space, accordingly Greek indices $\alpha = 1, \cdots, n_{dm}$ number variables in either the *configuration space* or the *cotangent (momentum) space*. The *autonomous Hamiltonian* (no explicit dependence on time) of the system is given by $\mathcal{H} = \mathcal{H}(q^\alpha, p_\beta)$. The one-form w is occasionally also denoted the *Poincaré-Cartan* integral invariant of a Hamiltonian system, see [10], and is the differential form analogy of the *Lagrangian* $\mathcal{L}(q^\alpha, p_\beta)\,\mathrm{d}t = p_\alpha\,\mathrm{d}q^\alpha - \mathcal{H}(q^\alpha, p_\beta)\,\mathrm{d}t$ (whereby the velocities have been exchanged in favor of the momenta). Recall that Hamilton's canonical equations in *phase space* read

$$\dot{p}_\alpha = -\partial_{q^\alpha}\mathcal{H} \quad \text{and} \quad \dot{q}^\alpha = \partial_{p_\alpha}\mathcal{H}.$$

The exterior derivative $\mathbf{d}\mathsf{w}$ renders the (closed and exact) fundamental two-form

$$\mathbf{d}\mathsf{w} = -w_{i,j}\mathbf{g}^i \wedge \mathbf{g}^j = \mathbf{d}p_\alpha \wedge \mathbf{d}q^\alpha - \mathbf{d}\mathcal{H} \wedge \mathbf{d}t := \frac{1}{2} s_{ij}\mathbf{g}^i \wedge \mathbf{g}^j$$

with the (skew symmetric) coefficients $s_{ij} \equiv s_{[ij]}$ of the above two-form, see [10]

$$[s_{ij}] := \begin{bmatrix} [0_{\alpha\beta}] & -[\delta_{\alpha\beta}] & [\partial_{p_\alpha}\mathcal{H}] \\ [\delta_{\alpha\beta}] & [0_{\alpha\beta}] & [\partial_{q^\alpha}\mathcal{H}] \\ -[\partial_{p_\alpha}\mathcal{H}] & -[\partial_{q^\alpha}\mathcal{H}] & 0 \end{bmatrix}.$$

Define next the Hamiltonian *vector field* (extended by the time dimension), a one-vector, as

$$\mathsf{v} = v^i \mathbf{g}_i := \dot{\chi}^i \mathbf{g}_i = \dot{p}_\alpha \boldsymbol{\partial}_{p_\alpha} + \dot{q}^\alpha \boldsymbol{\partial}_{q^\alpha} + \boldsymbol{\partial}_t \quad \text{with} \quad [v_i] := \big[[\dot{p}_\alpha], [\dot{q}^\alpha], 1\big].$$

Then it is straightforward to demonstrate that the following one-form vanishes identically if Hamilton's canonical equations hold

$$u_i \mathbf{g}^i := \mathbf{dw}(\mathbf{v}) = \mathbf{0} \quad \text{with} \quad [u_i] := \begin{bmatrix} -[\dot{q}^\alpha - \partial_{p_\alpha} \mathcal{H}] \\ [\dot{p}_\alpha + \partial_{q^\alpha} \mathcal{H}] \\ -\partial_{p_\alpha} \mathcal{H} \dot{p}_\alpha - \partial_{q^\alpha} \mathcal{H} \dot{q}^\alpha \end{bmatrix}.$$

Of course the statement $\mathbf{dw}(\mathbf{v}) = \mathbf{0}$ is nothing but the differential form analogy to *Hamilton's principle of least action.* The geometric setting of Hamiltonian equations allows deep and illuminating analyses of dynamical systems.

Application to Maxwell Equations:
Consider the following (scalar-valued) two-forms in (flat) four-space, whereby x^1, x^2, x^3 denote the *space coordinates* and x^4 denotes the *time coordinate*

$$\mathbf{f} = \frac{1}{2} f_{ij} \mathbf{g}^i \wedge \mathbf{g}^j \quad \text{and} \quad \mathbf{m} = \frac{1}{2} m_{ij} \mathbf{g}^i \wedge \mathbf{g}^j.$$

(These are denoted the *Faraday* and the *Maxwell* 'machine', respectively, in the slang of [142].) If the (skew symmetric) coefficients of these two-forms are identified with the contributions to the electro-magnetic fields, i.e. the *electric field* $\boldsymbol{E}^\flat$, the *magnetic induction* $\boldsymbol{B}^\flat$, the *magnetic field* $\boldsymbol{H}^\flat$, and the *dielectric displacement* $\boldsymbol{D}^\flat$ as

$$[f_{ij}] := \begin{bmatrix} 0 & +B_z & -B_y & +E_x \\ -B_z & 0 & +B_x & +E_y \\ +B_y & -B_x & 0 & +E_z \\ -E_x & -E_y & -E_z & 0 \end{bmatrix} \quad \text{and} \quad [m_{ij}] := \begin{bmatrix} 0 & +D_z & -D_y & -H_x \\ -D_z & 0 & +D_x & -H_y \\ +D_y & -D_x & 0 & -H_z \\ +H_x & +H_y & +H_z & 0 \end{bmatrix},$$

and if the coefficients of the three-form $\mathbf{j}$ are identified with the *electric current density* $\boldsymbol{J}^\flat$, and the *(free) charge density* ρ as

$$\mathbf{j} := -\rho \mathbf{g}^x \wedge \mathbf{g}^y \wedge \mathbf{g}^z + J_x \mathbf{g}^y \wedge \mathbf{g}^z \wedge \mathbf{g}^t + J_y \mathbf{g}^z \wedge \mathbf{g}^x \wedge \mathbf{g}^t + J_z \mathbf{g}^x \wedge \mathbf{g}^y \wedge \mathbf{g}^t$$

the *Maxwell equations* may be put into an extremely concise format in terms of the exterior derivatives of $\mathbf{f}$ and $\mathbf{m}$ (see, e.g. [60])

$$\mathbf{df} = \mathbf{0} \quad \text{and} \quad \mathbf{dm} + \mathbf{j} = \mathbf{0}.$$

Based on $\mathbf{df} = \frac{1}{2} f_{ij,k} \mathbf{g}^i \wedge \mathbf{g}^j \wedge \mathbf{g}^k$ the first of these equations comprises the *magnetic Gauss* and the *Faraday* law, respectively, i.e.

$$\begin{aligned} B_{x,x} + B_{y,y} + B_{z,z} &= 0 \quad \rightsquigarrow \quad \operatorname{div} \boldsymbol{B}^\flat = 0, \\ B_{x,t} - E_{y,z} + E_{z,y} &= 0 \\ B_{y,t} - E_{z,x} + E_{x,z} &= 0 \quad \rightsquigarrow \quad \dot{\boldsymbol{B}}^\flat + \operatorname{curl} \boldsymbol{E}^\flat = \mathbf{0}, \\ B_{z,t} - E_{x,y} + E_{y,x} &= 0 \end{aligned}$$

accordingly, based on $\mathbf{dm} = \frac{1}{2}m_{ij,k}\mathbf{g}^i \wedge \mathbf{g}^j \wedge \mathbf{g}^k$, the second of the above equations comprises the *electric Gauss* and the *Ampere* law, respectively, i.e.

$$\begin{aligned} D_{x,x} + D_{y,y} + D_{z,z} - \rho &= 0 \quad \rightsquigarrow \quad \mathrm{div}\boldsymbol{D}^\flat = \rho, \\ D_{x,t} + H_{y,z} - H_{z,y} + J_x &= 0 \\ D_{y,t} + H_{z,x} - H_{x,z} + J_y &= 0 \quad \rightsquigarrow -\dot{\boldsymbol{D}}^\flat + \mathrm{curl}\,\boldsymbol{H}^\flat = \boldsymbol{J}^\flat. \\ D_{z,t} + H_{x,y} - H_{y,x} + J_z &= 0 \end{aligned}$$

Here, the Maxwell equations have been expressed in *natural* Lorentz-Heaviside units omitting all ε_0, μ_0 (or $c^2 = 1/[\varepsilon_0\mu_0]$), and 4π. Since $\mathbf{df} = \mathbf{0}$ it is concluded that $\mathbf{f}$ is *closed* and (under certain technical conditions) also *exact*, thus it may be expressed as $\mathbf{f} = \mathbf{da}$ in terms of a one-form $\mathbf{a}$ that serves as an *electro-magnetic four-potential*

$$\mathbf{a} = a_i\mathbf{g}^i \quad \text{with} \quad [a_i] := [A_x, A_y, A_z, -U] \quad \rightsquigarrow \quad \mathbf{da} = -a_{i,j}\mathbf{g}^i \wedge \mathbf{g}^j$$

Here $\boldsymbol{A}^\flat$ and U denote the common *vector* and *scalar (electric) potential*, respectively, of classical electromagnetism. Thus the magnetic induction $\boldsymbol{B}^\flat$ and the electric field $\boldsymbol{E}^\flat$ follow from $\mathbf{f} = \mathbf{da}$ with the electro-magnetic four-potential $\mathbf{a}$ as

$$\begin{aligned} B_x &= A_{z,y} - A_{y,z}, & & E_x = -U_{,x} - A_{x,t}, \\ B_y &= A_{x,z} - A_{z,x}, & \text{and} \quad & E_y = -U_{,y} - A_{y,t}, \\ B_z &= A_{y,x} - A_{x,y}, & & E_z = -U_{,z} - A_{z,t}. \end{aligned}$$

Likewise, from $\mathbf{dm} = -\mathbf{j}$ it is additionally concluded that $\mathbf{dj} = \mathbf{0}$ which expands into

$$\rho_{,t} + J_{x,x} + J_{y,y} + J_{z,z} = 0 \quad \rightsquigarrow \quad \dot{\rho} + \mathrm{div}\boldsymbol{J}^\flat = 0.$$

This is the *charge conservation equation* that is here a direct consequence of the Maxwell equations. Finally it is interesting to note that upon defining the *four-momentum* $\mathbf{p} = p_i\mathbf{g}^i$, a one-form, and the *four-velocity* $\mathbf{v} = v^i\mathbf{g}_i$, a one-vector, with

$$[p_i] := [p_x, p_y, p_z, -W] \quad \text{and} \quad [v^i] := [v^x, v^y, v^z, 1],$$

a generalization of the classical *Lorentz* force acting on a test particle with charge q moving in a electromagnetic field reads

$$\dot{\mathbf{p}} = q\mathbf{f}(\mathbf{v}) \quad \rightsquigarrow \quad \begin{aligned} \dot{p}_x &= q[E_x + v^y B_z - v^z B_y] \\ \dot{p}_y &= q[E_y + v^z B_x - v^x B_z] \quad \rightsquigarrow \dot{\boldsymbol{p}} = q[\boldsymbol{E}^\flat + \boldsymbol{v}^\sharp \times \boldsymbol{B}^\flat], \\ \dot{p}_z &= q[E_z + v^x B_y - v^y B_x] \\ \dot{W} &= q[v^x E_x + v^y E_y + v^z E_z] \rightsquigarrow \dot{W} = q\boldsymbol{v}^\sharp \cdot \boldsymbol{E}^\flat. \end{aligned}$$

Here $W = qU$ denotes the (negative) work performed by moving the test particle against the electromagnetic field from infinity to its present position. Note that the magnetic contribution to the Lorentz force does not perform work.

Torsion and Curvature Forms:
The *torsion two-form* $\mathbf{t}^i$ is defined as the exterior product of the (dual basis) one-form $\mathbf{g}^j$ and the connection one-form $\mathbf{l}^i{}_j$ or rather the two-form based on the right skew symmetric torsion tensor, respectively

$$\mathbf{t}^i := -\mathbf{l}^i{}_j \wedge \mathbf{g}^j = \ell^i_{jk}\mathbf{g}^j \wedge \mathbf{g}^k = \frac{1}{2}t^i{}_{jk}\mathbf{g}^j \wedge \mathbf{g}^k.$$

The above denotes the first Cartan *equation of structure*. It is interesting to note that the torsion-two form is also the result of applying the covariant exterior derivative to the (dual basis) one-form $\mathbf{g}^i$

$$\mathbf{D}\mathbf{g}^i = \mathbf{d}\mathbf{g}^i + \mathbf{l}^i{}_j \wedge \mathbf{g}^j = \mathbf{l}^i{}_j \wedge \mathbf{g}^j = -\mathbf{t}^i.$$

Here $\mathbf{d}\mathbf{g}^i = \mathbf{d}\mathbf{d}\chi^i = \mathbf{0}$ has been invoked, moreover it is remarkable that $\mathbf{D}\mathbf{g}^i = \mathbf{0}$ holds only in a symmetric manifold. An alternative expression for the $(1,0)$-tensor-valued torsion two-form $\mathbf{t}$ reads

$$\mathbf{t} = \mathbf{t}^i\mathbf{g}_i = \frac{1}{2}t^i{}_{jk}\mathbf{g}_i \otimes \mathbf{g}^j \wedge \mathbf{g}^k = t^i{}_{jk}\mathbf{g}_i \otimes \mathbf{g}^j \otimes \mathbf{g}^k.$$

Moreover the corresponding *torsion one-form* is defined as the interior product of the (coordinate basis) vector $\mathbf{g}_i$ and the torsion two-form $\mathbf{t}^i$ or rather the one-form based on the torsion covector, respectively

$$\mathbf{g}_i \rfloor \mathbf{t}^i = t^i{}_{ik}\mathbf{g}^k = -t_k\mathbf{g}^k.$$

The torsion two-form for the special case of a semi-symmetric connection (a *trator* in the terminology of e.g. [105]) expands in terms of a given torsion one-form $t_k\mathbf{g}^k$ as

$$\mathbf{t}^i := \frac{1}{1-n_{dm}}\mathbf{g}^i \wedge [t_k\mathbf{g}^k] = \frac{1}{1-n_{dm}}\delta^i{}_j t_k\mathbf{g}^j \wedge \mathbf{g}^k = \frac{2}{1-n_{dm}}\delta^i{}_{[j}t_{k]}\mathbf{g}^j \otimes \mathbf{g}^k.$$

Obviously, computing the interior product of the coordinate basis with the so defined torsion two-form reproduces the given torsion one-form

$$\mathbf{g}_i \rfloor \mathbf{t}^i = \frac{2}{1-n_{dm}}\delta^i{}_{[i}t_{k]}\mathbf{g}^k = -t_k\mathbf{g}^k.$$

Next, computing the covariant exterior derivative of the torsion two-form renders

$$\mathbf{D}\mathbf{t}^i = \mathbf{d}\mathbf{t}^i + \mathbf{l}^i{}_m \wedge \mathbf{t}^m = -\left[\mathbf{d}\mathbf{l}^i{}_j + \mathbf{l}^i{}_m \wedge \mathbf{l}^m{}_j\right] \wedge \mathbf{g}^j.$$

On the one hand the definition of the *curvature two-form* $\mathbf{r}^i{}_j$ renders the second Cartan *equation of structure* as

$$\mathbf{r}^i{}_j := \mathbf{dl}^i{}_j + \mathbf{l}^i{}_m \wedge \mathbf{l}^m{}_j = \left[\ell^i{}_{jl,k} + \ell^i{}_{mk}\ell^m{}_{jl}\right] \mathbf{g}^k \wedge \mathbf{g}^l = \frac{1}{2} r^i{}_{jkl} \mathbf{g}^k \wedge \mathbf{g}^l.$$

An alternative expression for the $(1,1)$-tensor-valued curvature two-form $\mathbf{r}$ reads

$$\mathbf{r} = \mathbf{r}^i{}_j \mathbf{g}_i \otimes \mathbf{g}^j = \frac{1}{2} r^i{}_{jkl} \mathbf{g}_i \otimes \mathbf{g}^j \otimes \mathbf{g}^k \wedge \mathbf{g}^l = r^i{}_{jkl} \mathbf{g}_i \otimes \mathbf{g}^j \otimes \mathbf{g}^k \otimes \mathbf{g}^l.$$

Moreover the *curvature one-form* $\mathbf{r}_j$ follows as the interior product of the (coordinate basis) vector $\mathbf{g}_i$ and the curvature two-form $\mathbf{r}^i{}_j$ or rather the one-form based on the (first) Ricci tensor, respectively

$$\mathbf{r}_j := \mathbf{g}_i \rfloor \mathbf{r}^i{}_j = r^i{}_{jil} \mathbf{g}^l = r_{jl} \mathbf{g}^l.$$

Note that the two-form based on the second Ricci tensor simply reads as $\mathbf{r}^i{}_i$. On the other hand the covariant exterior derivative of the torsion two-form may be expanded directly in terms of the torsion tensor

$$\mathbf{Dt}^i = \frac{1}{2} \left[t^i{}_{jk,l} + \ell^i{}_{mj} t^m{}_{kl}\right] \mathbf{g}^j \wedge \mathbf{g}^k \wedge \mathbf{g}^l = \frac{1}{2} \left[t^i{}_{jk|l} - t^i{}_{mj} t^m{}_{kl}\right] \mathbf{g}^j \wedge \mathbf{g}^k \wedge \mathbf{g}^l.$$

As a result the *first (algebraic) Bianchi identity* expressed in terms of forms takes an extremely condensed format

$$\mathbf{Dt}^i = -\mathbf{r}^i{}_j \wedge \mathbf{g}^j.$$

Next, the covariant exterior derivative of the curvature two-form follows as

$$\mathbf{Dr}^i{}_j = \mathbf{dr}^i{}_j + \mathbf{l}^i{}_m \wedge \mathbf{r}^m{}_j - \mathbf{l}^m{}_j \wedge \mathbf{r}^i{}_m.$$

On the one hand the covariant exterior derivative of the curvature two-form may be expanded directly in terms of the curvature tensor

$$\begin{aligned}\mathbf{Dr}^i{}_j &= \frac{1}{2} \left[r^i{}_{jkl,m} + \ell^i{}_{nk} r^n{}_{jlm} - \ell^n{}_{jk} r^i{}_{nlm}\right] \mathbf{g}^k \wedge \mathbf{g}^l \wedge \mathbf{g}^m \\ &= \frac{1}{2} \left[r^i{}_{jkl|m} - r^i{}_{jnk} t^n{}_{lm}\right] \mathbf{g}^k \wedge \mathbf{g}^l \wedge \mathbf{g}^m.\end{aligned}$$

On the other hand, the exterior derivative of the curvature two-form reads

$$\mathbf{dr}^i{}_j = \mathbf{dl}^i{}_m \wedge \mathbf{l}^m{}_j - \mathbf{l}^i{}_m \wedge \mathbf{dl}^m{}_j = \mathbf{r}^i{}_m \wedge \mathbf{l}^m{}_j - \mathbf{l}^i{}_m \wedge \mathbf{r}^m{}_j.$$

Here the terms $\mathbf{l}^i{}_n \wedge \mathbf{l}^n{}_m \wedge \mathbf{l}^m{}_j - \mathbf{l}^i{}_m \wedge \mathbf{l}^m{}_n \wedge \mathbf{l}^n{}_j$ canceled each other to zero. Thus, with $\mathbf{r}^i{}_m \wedge \mathbf{l}^m{}_j = \mathbf{l}^m{}_j \wedge \mathbf{r}^i{}_m$, also the *second (differential) Bianchi identity* expressed in terms of forms takes the extremely condensed format

$$\mathbf{Dr}^{i}{}_{j} = \mathbf{0}.$$

Recall finally that the covariant exterior derivative of vector- and covector-valued zero-forms, i.e. of vector and covector fields v^i and v_j, respectively, render the one-forms

$$\mathbf{D}v^i = v^i{}_{|k}\mathbf{g}^k, \quad \text{and} \quad \mathbf{D}v_j = v_{j|k}\mathbf{g}^k.$$

Then on the one hand the second covariant exterior derivative of vector- and covector-valued zero-forms follows by the product rule as the two-forms

$$\mathbf{DD}v^i = \mathbf{D}v^i{}_{|k} \wedge \mathbf{g}^k + v^i{}_{|m}\mathbf{Dg}^m, \quad \text{and} \quad \mathbf{DD}v_j = \mathbf{D}v_{j|k} \wedge \mathbf{g}^k + v_{j|m}\mathbf{Dg}^m$$

and results in the expressions

$$\mathbf{DD}v^i = -v^i{}_{|kl}\mathbf{g}^k \wedge \mathbf{g}^l - v^i{}_{|m}\mathbf{t}^m, \quad \text{and} \quad \mathbf{DD}v_j = -v_{j|kl}\mathbf{g}^k \wedge \mathbf{g}^l - v_{j|m}\mathbf{t}^m.$$

On the other hand $\mathbf{D}v^i$ and $\mathbf{D}v_j$ are (1,0)- and (0,1)-tensor-valued one-forms, thus their covariant exterior derivative reads as

$$\mathbf{DD}v^i = \mathbf{dD}v^i + \mathbf{l}^i{}_m \wedge \mathbf{D}v^m = \mathbf{d}[\mathbf{d}v^i + \mathbf{l}^i{}_j v^j] + \mathbf{l}^i{}_m \wedge [\mathbf{d}v^m + \mathbf{l}^m{}_j v^j] = +\mathbf{r}^i{}_j v^j,$$
$$\mathbf{DD}v_j = \mathbf{dD}v_j - \mathbf{l}^m{}_j \wedge \mathbf{D}v_m = \mathbf{d}[\mathbf{d}v_j - \mathbf{l}^i{}_j v_i] - \mathbf{l}^m{}_j \wedge [\mathbf{d}v_m - \mathbf{l}^i{}_m v_i] = -\mathbf{r}^i{}_j v_i\,.$$

Combining the two expressions for $\mathbf{DD}v^i$ and $\mathbf{DD}v_j$, respectively, renders finally the *Ricci identities* for the second covariant derivatives of vector and covector fields

$$\left[2v^i{}_{|kl} + r^i{}_{jkl}v^j + v^i{}_{|m}t^m{}_{kl}\right]\mathbf{g}^k \wedge \mathbf{g}^l = \mathbf{0},$$
$$\left[2v_{j|kl} - r^i{}_{jkl}v_i + v_{j|m}t^m{}_{kl}\right]\mathbf{g}^k \wedge \mathbf{g}^l = \mathbf{0}.$$

Cartan's Repère Mobile (Moving Frame):
In [23] Cartan developed a formulation of differential geometry in terms of differential forms and the exterior calculus that is based on the notion of a moving frame (and an associated coframe), the so called *repère mobile* (repère=frame [of reference]), that is in general independent of the coordinate basis (and the associated dual basis) of a manifold and that proved to be very powerful in the development and representation of, e.g., the Theory of General Relativity/Gravitation. The use of the repère mobile also allows for the consideration of anholonomic moving frames without any conceptual difficulties. Often the particular case of an *orthonormal moving frame* is considered which is no restriction since an orthonormal moving frame may always be constructed from an arbitrary moving frame by the help of the Gram-Schmidt orthonormalization, [179]. To establish the link to the differential forms representation previously established based on the coordinate basis (and the associated dual basis) of a manifold in terms of holonomic

coordinates $\mathcal{X}^I$ the expressions for the connection one-form together with the anholonomity, the torsion, and the curvature two-forms and the corresponding Bianchi identities shall now be revisited in terms of an arbitrary moving frame (and coframe).

Frame and Coframe:
At each point $\mathcal{P}$ of a manifold $\mathcal{M}$ the corresponding tangent space $T_{\mathcal{P}}\mathcal{M}$ may be spanned by an *arbitrary frame* $\mathbf{g}_i$ that is expressible in terms of the *natural frame* $\mathbf{g}_I = \boldsymbol{\partial}_{\mathcal{X}^I}$ as

$$\mathbf{g}_i = f^I{}_i \mathbf{g}_I.$$

The frame field $\mathbf{g}_i(\{\mathcal{X}^I\})$, i.e. the frame at every $\mathcal{P}$ of $\mathcal{M}$ is then called a *moving frame*. Note that a moving frame need not be holonomic, i.e. the commutator $[\mathbf{g}_i, \mathbf{g}_j]$ need not necessarily vanish. Moreover, in the cotangent space $T^*_{\mathcal{P}}\mathcal{M}$ the *arbitrary coframe* $\mathbf{g}^i$ is expressed in terms of the *natural coframe* $\mathbf{g}^I = \mathbf{d}\mathcal{X}^I$ as

$$\mathbf{g}^i = \mathcal{F}^i{}_I \mathbf{g}^I.$$

The coframe field $\mathbf{g}^i(\{\mathcal{X}^I\})$, i.e. the coframe at every $\mathcal{P}$ of $\mathcal{M}$ is then called a *moving coframe*. The moving frame and coframe are dual with respect to their *interior product* $\rfloor$

$$\mathbf{g}_i \rfloor \mathbf{g}^j = f^K{}_i \mathcal{F}^j{}_K = \delta^j{}_i.$$

The differential geometry of $\mathcal{M}$ may then be characterized completely by studying how the frame and coframe are moving.

Connection One-Form:
The *connection one-form* $\mathbf{l}^i{}_j$ and the connection one-form $\mathbf{L}^I{}_J$ in the holonomic coordinates $\mathcal{X}^I$ are defined as

$$\mathbf{l}^i{}_j := \ell^i{}_{jk} \mathbf{g}^k \quad \text{and} \quad \mathbf{L}^I{}_J := \mathcal{L}^I{}_{JK} \mathbf{g}^K.$$

Thereby, the coefficients of $\mathbf{l}^i{}_j$ in the coframe basis $\mathbf{g}^k$ follow from the non-tensorial transformation rule for the connection as

$$\mathbf{l}^i{}_j = [\mathcal{F}^i{}_I \mathcal{L}^I{}_{JK} - \mathcal{F}^i{}_{J,K}] f^J{}_j f^K{}_k \mathbf{g}^k.$$

As a consequence the connection one-forms $\mathbf{l}^i{}_j$ and $\mathbf{L}^I{}_J$ are related by

$$\mathbf{l}^i{}_j = [\mathcal{F}^i{}_I \mathcal{L}^I{}_{JK} - \mathcal{F}^i{}_{J,K}] f^J{}_j \mathbf{g}^K = [\mathcal{F}^i{}_I \mathbf{L}^I{}_J - \mathbf{d}\mathcal{F}^i{}_J] f^J{}_j.$$

Here the one-form $\mathbf{d}\mathcal{F}^i{}_J$ is the differential of the $\mathcal{F}^i{}_J(\{\mathcal{X}^K\})$ considered as a function of the holonomic coordinates $\mathcal{X}^K$

$$\mathbf{d}\mathcal{F}^i{}_J = \mathcal{F}^i{}_{J,K} \mathbf{g}^K.$$

The above representation for $\mathsf{l}^i{}_j$ in terms of $\mathsf{L}^I{}_J$ and $\mathbf{d}\mathcal{F}^i{}_J$ is particularly useful in the sequel.

Anholonomity Two-Form:
The *anholonomity two-form* $\mathbf{a}^i$ is introduced as the (negative) exterior derivative of the coframe

$$\mathbf{a}^i = \frac{1}{2} a^i{}_{jk} \mathbf{g}^j \wedge \mathbf{g}^k := -\mathbf{d}\mathbf{g}^i.$$

Thereby, the coefficients of $\mathbf{a}^i$ in the coframe basis $\mathbf{g}^j \wedge \mathbf{g}^k$ follow directly as

$$\mathbf{a}^i = -\mathcal{F}^i{}_{J,K} \mathbf{g}^K \wedge \mathbf{g}^J = \mathcal{F}^i{}_{[J,K]} \mathbf{g}^J \wedge \mathbf{g}^K = \mathcal{F}^i{}_{[J,K]} f^J{}_j f^K{}_k \mathbf{g}^j \wedge \mathbf{g}^k.$$

Thus the anholonomity two-form follows alternatively as the exterior product

$$\mathbf{a}^i = -\mathbf{d}\mathcal{F}^i{}_J \wedge \mathbf{g}^J.$$

The alternative representation for $\mathbf{a}^i$ in terms of $\mathbf{d}\mathcal{F}^i{}_J$ and $\mathbf{g}^J$ is particularly useful in the sequel.

Torsion Two-Form:
The *torsion two-form* $\mathbf{t}^i$ is defined as the (negative) covariant exterior derivative of the coframe

$$\mathbf{t}^i = \frac{1}{2} t^i{}_{jk} \mathbf{g}^j \wedge \mathbf{g}^k := -\mathbf{D}\mathbf{g}^i = -\mathbf{d}\mathbf{g}^i - \mathsf{l}^i{}_j \wedge \mathbf{g}^j = \mathbf{a}^i - \mathsf{l}^i{}_j \wedge \mathbf{g}^j.$$

This is the first Cartan equation of structure. Thereby, the coefficients of $\mathbf{t}^i$ in the coframe basis $\mathbf{g}^j \wedge \mathbf{g}^k$ read as

$$\mathbf{t}^i = \mathcal{F}^i{}_I \mathcal{L}^I{}_{[JK]} \mathbf{g}^J \wedge \mathbf{g}^K = \mathcal{F}^i{}_I \mathcal{L}^I{}_{[JK]} f^J{}_j f^K{}_k \mathbf{g}^j \wedge \mathbf{g}^k.$$

With $\mathbf{d}\mathbf{g}^I = \mathbf{d}\mathbf{d}\mathcal{X}^I \equiv \mathbf{0}$ the torsion two-form T^I in the holonomic coordinates $\mathcal{X}^I$ is defined as

$$\mathsf{T}^I = \frac{1}{2} \mathcal{T}^I{}_{JK} \mathbf{g}^J \wedge \mathbf{g}^K := -\mathsf{L}^I{}_J \wedge \mathbf{g}^J = \mathcal{L}^I{}_{[JK]} \mathbf{g}^J \wedge \mathbf{g}^K.$$

As a consequence the torsion two-forms $\mathbf{t}^i$ and T^I are related by

$$\mathbf{t}^i = \mathbf{a}^i - \left[\mathcal{F}^i{}_I \mathsf{L}^I{}_J - \mathbf{d}\mathcal{F}^i{}_J\right] \wedge \mathbf{g}^J = \mathcal{F}^i{}_I \mathsf{T}^I.$$

Based on this representation of $\mathbf{t}^i$ its coefficients in the coframe basis $\mathbf{g}^j \wedge \mathbf{g}^k$ as given in the above are easily deduced.

Curvature Two-Form:
The *curvature two-form* $\mathbf{r}^i{}_j$ is defined in terms of the connection one-form as

$$\mathbf{r}^i{}_j = \frac{1}{2} r^i{}_{jkl} \mathbf{g}^k \wedge \mathbf{g}^l := \mathbf{d}\mathbf{l}^i{}_j + \mathbf{l}^i{}_m \wedge \mathbf{l}^m{}_j.$$

This is the second Cartan equation of structure. Thereby, with the exterior derivative $\mathbf{d}\mathbf{l}^i{}_j = \ell^i_{jl,k} \mathbf{g}^k \wedge \mathbf{g}^l - \ell^i_{jm} \mathbf{a}^m$ the coefficients of $\mathbf{r}^i{}_j$ in the coframe basis $\mathbf{g}^k \wedge \mathbf{g}^l$ read as

$$\mathbf{r}^i{}_j = \left[\ell^i_{jl,k} + \ell^i_{mk} \ell^m_{jl} - \frac{1}{2} \ell^i_{jm} a^m{}_{kl} \right] \mathbf{g}^k \wedge \mathbf{g}^l.$$

Note that the one-form $\mathbf{l}^i{}_j$ may be considered the exterior product of the one-form $\left[\mathcal{F}^i{}_I \mathbf{L}^I{}_J - \mathbf{d}\mathcal{F}^i{}_J \right]$ and the zero-form $f^J{}_j$, thus the exterior derivative of $\mathbf{l}^i{}_j$ follows as

$$\mathbf{d}\mathbf{l}^i{}_j = \mathbf{d}\left[\mathcal{F}^i{}_I \mathbf{L}^I{}_J - \mathbf{d}\mathcal{F}^i{}_J \right] f^J{}_j + [-1]^1 \left[\mathcal{F}^i{}_I \mathbf{L}^I{}_J - \mathbf{d}\mathcal{F}^i{}_J \right] \wedge \mathbf{d} f^J{}_j.$$

With the following set of intermediate results

$$\mathbf{d}\left[\mathcal{F}^i{}_I \mathbf{L}^I{}_J \right] = \mathbf{d}\mathcal{F}^i{}_I \wedge \mathbf{L}^I{}_J + [-1]^0 \mathcal{F}^i{}_I \mathbf{d}\mathbf{L}^I{}_J, \quad \mathbf{d}\mathbf{d}\mathcal{F}^i{}_J = \mathbf{0}, \quad \mathbf{d} f^J{}_j = -f^J{}_m \mathbf{d}\mathcal{F}^m{}_M f^M{}_j$$

the exterior derivative of $\mathbf{l}^i{}_j$ is then expressed as

$$\mathbf{d}\mathbf{l}^i{}_j = \left[\mathbf{d}\mathcal{F}^i{}_I \wedge \mathbf{L}^I{}_J + \mathcal{F}^i{}_I \underline{\mathbf{d}\mathbf{L}^I{}_J} \right] f^J{}_j + \left[\mathcal{F}^i{}_I \mathbf{L}^I{}_J - \mathbf{d}\mathcal{F}^i{}_J \right] \wedge \left[f^J{}_m \mathbf{d}\mathcal{F}^m{}_M f^M{}_j \right].$$

Moreover the term quadratic in the connection one-form expands into

$$\begin{aligned} \mathbf{l}^i{}_m \wedge \mathbf{l}^m{}_j = {} & \mathcal{F}^i{}_I \underline{\mathbf{L}^I{}_M \wedge \mathbf{L}^M{}_J} f^J{}_j \\ & - \mathbf{d}\mathcal{F}^i{}_I \wedge \mathbf{L}^I{}_J f^J{}_j - \mathcal{F}^i{}_I \mathbf{L}^I{}_J f^J{}_m \wedge \mathbf{d}\mathcal{F}^m{}_M f^M{}_j + \mathbf{d}\mathcal{F}^i{}_J f^J{}_m \wedge \mathbf{d}\mathcal{F}^m{}_M f^M{}_j. \end{aligned}$$

By adding the last two expressions all but the underlined terms cancel. Finally, the curvature two-form $\mathbf{R}^I{}_J$ in the holonomic coordinates X^I is defined as

$$\mathbf{R}^I{}_J = \frac{1}{2} \mathcal{R}^I{}_{JKL} \mathbf{g}^K \wedge \mathbf{g}^L := \mathbf{d}\mathbf{L}^I{}_J + \mathbf{L}^I{}_M \wedge \mathbf{L}^M{}_J = \left[\mathcal{L}^I{}_{JL,K} + \mathcal{L}^I{}_{MK} \mathcal{L}^M{}_{JL} \right] \mathbf{g}^K \wedge \mathbf{g}^L.$$

As a consequence the curvature two-forms $\mathbf{r}^i{}_j$ and $\mathbf{R}^I{}_J$ are related by

$$\mathbf{r}^i{}_j = \mathcal{F}^i{}_I \left[\mathbf{d}\mathbf{L}^I{}_J + \mathbf{L}^I{}_M \wedge \mathbf{L}^M{}_J \right] f^J{}_j = \mathcal{F}^i{}_I \mathbf{R}^I{}_J f^J{}_j.$$

Clearly this is simply the statement of tensorial transformation behavior for the curvature tensor.

Bianchi Identities:
The first (algebraic) and second (differential) Bianchi identities read in the holonomic coordinates X^I

$$\mathbf{D}\mathbf{T}^I = -\mathbf{R}^I{}_J \wedge \mathbf{g}^J \quad \text{and} \quad \mathbf{D}\mathbf{R}^I{}_J = \mathbf{0}.$$

Then, firstly, with $\mathbf{a}^i = -\mathbf{d}\mathbf{g}^i$ and thus $\mathbf{d}\mathbf{a}^i = -\mathbf{d}\mathbf{d}\mathbf{g}^i = \mathbf{0}$ the covariant exterior derivative of $\mathbf{t}^i = \mathbf{a}^i - \mathbf{l}^i{}_j \wedge \mathbf{g}^j$ renders

$$\mathbf{D}\mathbf{t}^i = \mathbf{d}\mathbf{t}^i + \mathbf{l}^i{}_j \wedge \mathbf{t}^j = -\mathbf{d}\mathbf{l}^i{}_j \wedge \mathbf{g}^j - [-1]^1 \mathbf{l}^i{}_j \wedge \mathbf{d}\mathbf{g}^j + \mathbf{l}^i{}_j \wedge \mathbf{a}^j - \mathbf{l}^i{}_m \wedge \mathbf{l}^m{}_j \wedge \mathbf{g}^j = -\mathbf{r}^i{}_j \wedge \mathbf{g}^j.$$

Secondly, with $\mathbf{r}^i{}_j = \mathbf{d}\mathbf{l}^i{}_j + \mathbf{l}^i{}_m \wedge \mathbf{l}^m{}_j$, $\mathbf{d}\mathbf{d}\mathbf{l}^i{}_j = \mathbf{0}$ and thus $\mathbf{d}\mathbf{r}^i{}_j = \mathbf{r}^i{}_m \wedge \mathbf{l}^m{}_j - \mathbf{l}^i{}_m \wedge \mathbf{r}^m{}_j$ the covariant exterior derivative of $\mathbf{r}^i{}_j$ renders

$$\mathbf{D}\mathbf{r}^i{}_j = \mathbf{d}\mathbf{r}^i{}_j + \mathbf{l}^i{}_m \wedge \mathbf{r}^m{}_j - \mathbf{r}^i{}_m \wedge \mathbf{l}^m{}_j = \mathbf{0}.$$

Thus, in summary, the first (algebraic) and second (differential) Bianchi identities may be stated alternatively as

$$\mathbf{D}\mathbf{t}^i = -\mathbf{r}^i{}_j \wedge \mathbf{g}^j \quad \text{and} \quad \mathbf{D}\mathbf{r}^i{}_j = \mathbf{0}.$$

This result is remarkable insofar as the structure of the Bianchi identities (and also of their ingredients, i.e. the torsion and curvature two-forms) when expressed in terms of differential forms is independent of the frame and coframe being natural or arbitrary. This seems a good motivation for the introduction of the repère mobile, i.e. of an arbitrary frame and coframe.

Chapter 4
Geometry on Metric Manifolds

DIE ZWEI PARALLELEN

Es gingen zwei Parallelen
ins Endlose hinaus,
zwei kerzengerade Seelen
und aus solidem Haus.

Sie wollten sich nicht schneiden
bis an ihr seliges Grab:
Das war nun einmal der beiden
geheimer Stolz und Stab.

Doch als sie zehn Lichtjahre
gewandert neben sich hin,
da wards dem einsamen Paare
nicht irdisch mehr zu Sinn.

War'n sie noch Parallelen?
Sie wußtens selber nicht,-
sie flossen nur wie zwei Seelen
zusammen durch ewiges Licht.

Das ewige Licht durchdrang sie,
da wurden sie eins in ihm;
die Ewigkeit verschlang sie
als wie zwei Seraphim.

Christian Morgenstern

Abstract. The *metric* is an important tensorial object that introduces more structure into a (differential) manifold. The metric coefficients allow for example to determine the length of parameter curves in the manifold and make it possible to relate corresponding co- and contravariant objects defined on the manifold. Thus the *inner product* of either co- or contravariant quantities and in particular the angle between two vectors or (covectors) may be computed. Adopting the Ricci postulate of vanishing covariant derivative of the metric coefficients results in a decomposition of the fully covariant connection into its Riemann part, that depends exclusively on the metric, and the contortion, that depends exclusively on the torsion. Moreover, in this case the length of vectors and their angle with respect to geodesics are preserved upon parallel transport. Based on the metric coefficients the fully covariant curvature tensor, displaying left and right skew symmetry, is introduced and the corresponding modifications of the Bianchi identities are highlighted. Likewise the metric allows to introduce the mixed-variant Ricci tensor together with corresponding identities. Finally the metric enables to compute the Ricci scalar from the previous curvature tensors. The particular case of a symmetric,

P. Steinmann, *Geometrical Foundations of Continuum Mechanics*,
Lecture Notes in Applied Mathematics and Mechanics 2, DOI: 10.1007/978-3-662-46460-1_4

metrically connected manifold represents a Riemann geometry with associated Riemann curvature tensor as fundamental for example in the Einstein Theory of General Relativity/Gravitation. A generalization is obtained by allowing for metric manifolds with non-vanishing covariant derivative of its metric coefficients. The presence of the non-metricity is then subsequently reflected by extra terms in the decomposition of the connection and in the explicit representation of the curvature tensor, the Ricci tensor, the Ricci scalar, and the corresponding Bianchi identities for the various curvature quantities.

4.1 Metric

The main characteristic of a metric manifold (space/geometry), i.e. a manifold equipped with a metric, is given in the

Definition 4.1. If a n_{dm}-dimensional (differentiable) manifold $\mathcal{M}$ is equipped with a symmetric field of metric coefficients $\mathcal{M}_{IJ}(\mathcal{X}^1, \mathcal{X}^2, \cdots, \mathcal{X}^{n_{dm}})$ such that the arc-length of a parameter curve $\mathcal{X}^I = \mathcal{X}^I(t)$ between parameter values t_a and t_b is given by

$$S(t_b) - S(t_a) = \int_{t_a}^{t_b} \sqrt{\dot{\mathcal{X}}^I \mathcal{M}_{IJ} \dot{\mathcal{X}}^J}\,\mathrm{d}t \tag{4.1}$$

the manifold $\mathcal{M}$ is metric. Moreover, its tangent space $T_{\mathcal{P}}\mathcal{M}$ at $\mathcal{P}$ is an Euclidean vector space. □

Thereby the metric coefficients $\mathcal{M}_{IJ}$ shall obey the following properties:

- $\mathcal{M}_{IJ}$ is *symmetric*, i.e.

$$\mathcal{M}_{IJ} = \mathcal{M}_{JI} = \mathcal{M}_{(IJ)} \quad \text{with} \quad \mathcal{M}_{[IJ]} = 0$$

- $\mathcal{M}_{IJ}$ is *positive definite*, i.e.

$$\mathcal{V}^I \mathcal{M}_{IJ} \mathcal{V}^J > 0 \quad \forall \{\mathcal{V}^K\} \neq \{0\}$$

- $\mathcal{M}_{IJ}$ transforms as second-order tensor, i.e.

$$m_{kl} = f^I{}_k \mathcal{M}_{IJ} f^J{}_l$$

The first property is obvious since any skew symmetric contributions would not contribute to a quadratic form as needed for the determination of the length. The second property is specific to the later application to (three-dimensional) continuum mechanics, relativity and general relativity allow also for indefinite metrics, see, e.g., [142]. Finally the proof of the third property is straightforward from the transformation behaviour of the coordinate differentials, see Eq. 3.3:

$$\mathrm{d}S^2 = \mathrm{d}\mathcal{X}^I \mathcal{M}_{IJ}\, \mathrm{d}\mathcal{X}^J = \mathrm{d}\chi^k f^I{}_k \mathcal{M}_{IJ} f^J{}_l\, \mathrm{d}\chi^l = \mathrm{d}\chi^k m_{kl}\, \mathrm{d}\chi^l. \qquad (4.2)$$

In addition to the covariant metric coefficients $\mathcal{M}_{IJ}$ it is useful to also introduce contravariant metric coefficients $\mathcal{M}^{IJ}$. The matrix arrangement of the contravariant metric coefficients $[\mathcal{M}^{IJ}]$ follows from the (algebraic) inversion of the corresponding matrix arrangement of the covariant metric coefficients $[\mathcal{M}_{IJ}]$. The contravariant metric coefficients $\mathcal{M}^{IJ}$ may be used to *raise* covariant indices, e.g. $\mathcal{V}^I = \mathcal{M}^{IJ}\mathcal{V}_J$, likewise the covariant metric coefficients $\mathcal{M}_{IJ}$ are used to *lower* contravariant indices, e.g. $\mathcal{V}_I = \mathcal{M}_{IJ}\mathcal{V}^J$, thus making the distinction between vectors and covectors somewhat superfluous in a metric manifold (space/geometry).

Remark 4.1. In a manifold *without a metric* it is only possible to compute the *scalar product* of vectors and covectors, i.e. to evaluate pairings $\mathcal{V}^I\mathcal{V}_I$ of elements from the tangent and the cotangent space to the manifold. In a manifold *with a metric*, however, it is also possible to evaluate pairings $\mathcal{V}^I\mathcal{M}_{IJ}\mathcal{V}^J = \mathcal{V}_I\mathcal{M}^{IJ}\mathcal{V}_J$ of elements that are either both from the tangent or both from the cotangent space to the manifold, respectively, in terms of of the *inner product*. For the computation of the inner product a metric is inevitably necessary. □

Remark 4.2. An *Euclidean vector space* is given in the

Definition 4.2. A set $\mathcal{E}$ of elements (vectors) $\boldsymbol{a}, \boldsymbol{b}, \boldsymbol{c}, \cdots$ that satisfy in addition to axioms A1 and A2 of an affine vector space the following axiom defines an Euclidean vector space, its elements are then denoted as Euclidean vectors:

E1: Scalar product of vectors:
a) $\boldsymbol{a}\cdot\boldsymbol{b} = \boldsymbol{b}\cdot\boldsymbol{a}$ commutativity
b) $[\alpha\boldsymbol{a}]\cdot\boldsymbol{b} = \boldsymbol{a}\cdot[\alpha\boldsymbol{b}]$ associativity for scalar multiplication
c) $\boldsymbol{a}\cdot[\boldsymbol{b}+\boldsymbol{c}] = \boldsymbol{a}\cdot\boldsymbol{b}+\boldsymbol{a}\cdot\boldsymbol{c}$ distributivity
d) $\boldsymbol{a}\cdot\boldsymbol{b} = 0 \quad \forall\boldsymbol{b} \quad \to \quad \boldsymbol{a} = \boldsymbol{0}$ null element

Thus, besides addition of vectors and multiplication of vectors with scalars, in particular the scalar product is additionally defined in an Euclidean vector space. □

Observe that in an Euclidean space it is always possible to find Cartesian coordinates X^A such that the metric coincides with the Kronecker symbol/Cartesian metric, i.e. $M_{AB} \equiv \delta_{AB}$ and $M^{AB} \equiv \delta^{AB}$, respectively. In this situation no distinction between covariant and contravariant indices is necessary at all. □

4.2 Metric Connection

4.2.1 Ricci Postulate

As immediate consequence the introduction of a metric allows to formulate the

Postulate 4.1 (Ricci Postulate). For a *metric connection* the covariant derivative of the *covariant* metric coefficients vanishes, i.e.

$$\mathcal{M}_{IJ|K} = \mathcal{M}_{IJ,K} - \mathcal{L}^{M}{}_{IK}\mathcal{M}_{MJ} - \mathcal{L}^{M}{}_{JK}\mathcal{M}_{IM} = 0. \tag{4.3}$$

From resolving the above equality the partial derivative of the covariant metric may thus alternatively be stated in terms of the (left) symmetric part of the *fully covariant connection*

$$\mathcal{M}_{IJ,K} = \mathcal{L}_{JIK} + \mathcal{L}_{IJK} = 2\mathcal{L}_{(IJ)K}. \tag{4.4}$$

Note that the covariant metric has been used to lower the contravariant indices of the connection. The above representation will be useful in the sequel. A metric manifold that is additionally equipped with a metric connection is denoted a *metrically connected manifold (space/geometry)* in the sequel. □

Due to the Ricci postulate the multiplication of the covariant derivative of a vector by the covariant metric renders the definition of the covariant derivative of a covector

$$\begin{aligned}
\mathcal{M}_{IJ}\mathcal{V}^{I}{}_{|K} &= \mathcal{M}_{IJ}\left[\mathcal{V}^{I}{}_{,K} + \mathcal{L}^{I}{}_{MK}\mathcal{V}^{M}\right] \\
&= \left[\mathcal{M}_{IJ}\mathcal{V}^{I}\right]_{,K} - \mathcal{V}^{I}\mathcal{M}_{IJ,K} + \mathcal{L}_{JMK}\mathcal{V}^{M} \\
&= \mathcal{V}_{J,K} - \mathcal{V}^{I}\left[\mathcal{L}_{JIK} + \mathcal{L}_{IJK}\right] + \mathcal{L}_{JMK}\mathcal{V}^{M} \\
&= \mathcal{V}_{J,K} - \mathcal{V}^{I}\mathcal{L}_{IJK} \\
&= \mathcal{V}_{J,K} - \mathcal{V}_{I}\mathcal{L}^{I}{}_{JK} \\
&= \mathcal{V}_{J|K}.
\end{aligned} \tag{4.5}$$

This is a demonstration that covariant differentiation and raising or lowering indices by the metric do commute for a metric connection.
As a consequence of the Ricci postulate a metric connection is additively decomposed into the *Riemann part of the connection* and the *contortion*:

$$\underbrace{\mathcal{L}_{IJK}}_{\text{Metric Connection}} = \underbrace{\mathcal{M}_{IJK}}_{\text{Riemann}} + \underbrace{\mathcal{K}_{IJK}}_{\text{Contortion}}. \tag{4.6}$$

Proof. The proof follows directly by stating the Ricci postulate three times upon cyclic permutation of the indices and subsequent addition of the resulting expressions to render:

$$
\begin{array}{rl}
\mathcal{M}_{IJ|K} = & \mathcal{M}_{IJ,K} - \mathcal{L}_{JIK} - \mathcal{L}_{IJK} = 0 \\
-\mathcal{M}_{JK|I} = & -\mathcal{M}_{JK,I} + \mathcal{L}_{KJI} + \mathcal{L}_{JKI} = 0 \\
\mathcal{M}_{KI|J} = & \mathcal{M}_{KI,J} - \mathcal{L}_{IKJ} - \mathcal{L}_{KIJ} = 0 \\
\hline\hline
\end{array}
$$

$$
\mathcal{M}_{IJ,K} - \mathcal{M}_{JK,I} + \mathcal{M}_{KI,J} + 2\mathcal{L}_{J[KI]} - 2\mathcal{L}_{K[IJ]} - 2\mathcal{L}_{I(JK)} = 0
$$

Then by dividing by 2 and by adding and subtracting the terms

$$
\mathcal{L}_{IJK} = \mathcal{L}_{I(JK)} + \mathcal{L}_{I[JK]}
$$

the result follows immediately as

$$
\mathcal{L}_{IJK} = \frac{1}{2}\left[\mathcal{M}_{IJ,K} - \mathcal{M}_{JK,I} + \mathcal{M}_{KI,J}\right] + \mathcal{L}_{I[JK]} + \mathcal{L}_{J[KI]} - \mathcal{L}_{K[IJ]}.
$$

This concludes the proof. □

Remark 4.3. Note that the Ricci postulate is commonly stated in terms of the covariant derivative of the covariant metric. However, as a consequence of the Ricci postulate (and as demonstrated in Eq. 4.5), raising and lowering of indices by the metric commute with covariant differentiation. This agrees with the (less common) alternative, contravariant formulation of the

Postulate 4.2 (Ricci Postulate). For a *metric connection* the covariant derivative of the *contravariant* metric coefficients vanishes, i.e.

$$
\mathcal{M}^{IJ}|_K = \mathcal{M}^{IJ}{}_{,K} + \mathcal{L}^{I}{}_{MK}\mathcal{M}^{MJ} + \mathcal{L}^{J}{}_{MK}\mathcal{M}^{IM} = 0.
$$

The partial derivative of the contravariant metric may thus alternatively be stated in terms of the (left) symmetric part of the *mixed-variant connection* that follows from raising the second index

$$
\mathcal{M}^{IJ}{}_{,K} = -\mathcal{L}^{IJ}{}_{K} - \mathcal{L}^{JI}{}_{K} = -2\mathcal{L}^{(IJ)}{}_{K}.
$$

This is an alternative relation valid for the connection in a metrically connected manifold (space/geometry). □

Accordingly the additive decomposition of the *fully contravariant* metric connection into its Riemann part and the contortion now reads

$$
\underbrace{\mathcal{L}^{IJK}}_{\text{Metric Connection}} = \underbrace{\mathcal{M}^{IJK}}_{\text{Riemann}} + \underbrace{\mathcal{K}^{IJK}}_{\text{Contortion}}.
$$

Proof. The proof follows as before by stating the (contravariant) Ricci postulate (multiplied by the contravariant metric) three times upon cyclic permutation of the indices and subsequent addition

$$\begin{aligned} \mathcal{M}^{IJ}|_N \mathcal{M}^{NK} &= \mathcal{M}^{IJ}{}_{,N}\mathcal{M}^{NK} + \mathcal{L}^{IJK} + \mathcal{L}^{JIK} = 0 \\ -\mathcal{M}^{JK}|_L \mathcal{M}^{LI} &= -\mathcal{M}^{JK}{}_{,L}\mathcal{M}^{LI} - \mathcal{L}^{JKI} - \mathcal{L}^{KJI} = 0 \\ \mathcal{M}^{KI}|_M \mathcal{M}^{MJ} &= \mathcal{M}^{KI}{}_{,M}\mathcal{M}^{MJ} + \mathcal{L}^{KIJ} + \mathcal{L}^{IKJ} = 0 \end{aligned}$$

$$\mathcal{M}^{IJ}{}_{,N}\mathcal{M}^{NK} - \mathcal{M}^{JK}{}_{,L}\mathcal{M}^{LI} + \mathcal{M}^{KI}{}_{,M}\mathcal{M}^{MJ} - 2\mathcal{L}^{J[KI]} + 2\mathcal{L}^{K[IJ]} + 2\mathcal{L}^{I(JK)} = 0$$

Then by dividing by 2, subtracting and adding the terms

$$\mathcal{L}^{IJK} = \mathcal{L}^{I(JK)} + \mathcal{L}^{I[JK]}$$

and by incorporating the relation between the fully covariant Riemann connection $2\mathcal{M}_{LMN} = \mathcal{M}_{LM,N} - \mathcal{M}_{MN,L} + \mathcal{M}_{NL,M}$ and its fully contravariant counterpart

$$\mathcal{M}^{IJK} = \mathcal{M}^{IL}\mathcal{M}_{LMN}\mathcal{M}^{MJ}\mathcal{M}^{NK} = -\frac{1}{2}\left[\mathcal{M}^{IJ}{}_{,N}\mathcal{M}^{NK} - \mathcal{M}^{JK}{}_{,L}\mathcal{M}^{LI} + \mathcal{M}^{KI}{}_{,M}\mathcal{M}^{MJ}\right]$$

the result follows as

$$\mathcal{L}^{IJK} = \mathcal{M}^{IL}\left[\mathcal{M}_{LMN} + \mathcal{K}_{LMN}\right]\mathcal{M}^{MJ}\mathcal{M}^{NK}.$$

Here the fully contravariant contortion is given in terms of the fully contravariant connection as

$$\mathcal{K}^{IJK} = \mathcal{M}^{IL}\mathcal{K}_{LMN}\mathcal{M}^{MJ}\mathcal{M}^{NK} = \mathcal{L}^{I[JK]} + \mathcal{L}^{J[KI]} - \mathcal{L}^{K[IJ]}.$$

This concludes the proof. □

The result for the contravariant Ricci postulate is in concord with $[\delta^I{}_J]_{|K} = 0$ and thus $[\mathcal{M}^{IM}\mathcal{M}_{MJ}]_{|K} = 0$ and consequently $\mathcal{M}^{IJ}{}_{|K} = -\mathcal{M}^{IM}\mathcal{M}_{MN|K}\mathcal{M}^{NJ}$ together with the subsequent insertion of the result for $\mathcal{M}_{MN|K}$. □

4.2.1.1 Riemann Connection

The Riemann part of the connection (sometimes also denoted the Levi-Civita connection, whereby the terminology Christoffel connection would historically be more appropriate, see the discussion in [179]) or in short simply the *Riemann connection* follows from the

Definition 4.3. The Riemann connection is exclusively computed from the metric coefficients

$$\mathcal{M}_{IJK} := \frac{1}{2}\left[\mathcal{M}_{IJ,K} - \mathcal{M}_{JK,I} + \mathcal{M}_{KI,J}\right]. \tag{4.7}$$

As will be demonstrated below the Riemann connection is a metric connection. □

It reduces to the Christoffel symbols in Euclidean space and obeys the transformation behavior of the connection (that is obtained by multiplying the transformation of the mixed-variant connection in Eq. 3.41 by the covariant metric $\mathcal{M}_{IJ}$), i.e.

$$\begin{aligned} m_{ijk} &= f^I_{\ i}\left[\mathcal{M}_{IJK} f^J_{\ j} f^K_{\ k} + \mathcal{M}_{IJ} f^J_{\ j,k}\right], \\ \mathcal{M}_{IJK} &= \mathcal{F}^i_{\ I}\left[m_{ijk} \mathcal{F}^j_{\ J} \mathcal{F}^k_{\ K} + m_{ij} \mathcal{F}^j_{\ J,K}\right], \end{aligned} \tag{4.8}$$

thus it is not a third-order tensor. It may be shown by inspection that, based on the symmetry of the metric coefficients, in particular its (right) skew symmetric contribution vanishes

$$\mathcal{M}_{I[JK]} = 0. \tag{4.9}$$

Remark 4.4. The complete set of symmetric and skew symmetric contributions in the various index pairings of the Riemann connection follows as

$$\begin{aligned} \text{Index Pair 23:} \quad & \mathcal{M}_{I(JK)} = \frac{1}{2}\left[\mathcal{M}_{IJ,K} - \mathcal{M}_{JK,I} + \mathcal{M}_{KI,J}\right] \\ & \mathcal{M}_{I[JK]} = 0 \\ \text{Index Pair 31:} \quad & \mathcal{M}_{(IJK)} = \frac{1}{2}\mathcal{M}_{KI,J} \\ & \mathcal{M}_{[IJK]} = \frac{1}{2}\left[\mathcal{M}_{IJ,K} - \mathcal{M}_{JK,I}\right] \\ \text{Index Pair 12:} \quad & \mathcal{M}_{(IJ)K} = \frac{1}{2}\mathcal{M}_{IJ,K} \\ & \mathcal{M}_{[IJ]K} = \frac{1}{2}\left[-\mathcal{M}_{JK,I} + \mathcal{M}_{KI,J}\right] \end{aligned}$$

□

4.2.1.2 Contortion

Furthermore (for holonomic coordinates) the contortion is given in the

Definition 4.4. The contortion depends (linearly) on the metric coefficients and in particular on the torsion tensor

$$\mathcal{K}_{IJK} := \frac{1}{2}\left[\mathcal{T}_{IJK} + \mathcal{T}_{JKI} - \mathcal{T}_{KIJ}\right]. \tag{4.10}$$

The contortion inherits from the torsion the transformation properties of a tensor, thus it is also denoted the *contortion tensor.* □

It may be shown by inspection, based on the symmetry of the metric coefficients and the skew symmetry of the torsion tensor, that in particular the (left) symmetric contribution to the contortion vanishes

$$\mathcal{K}_{(IJ)K} = 0. \tag{4.11}$$

Recall that symmetry in an index pair is denoted by round brackets, i.e. here $\mathcal{K}_{(IJ)K} := [\mathcal{K}_{IJK} + \mathcal{K}_{JIK}]/2$. Finally the conjecture in the above that the Riemann connection is also a metric connection is obviously proven by

$$\mathcal{M}_{IJ,K} \doteq 2\mathcal{L}_{(IJ)K} = 2\mathcal{M}_{(IJ)K} + 2\mathcal{K}_{(IJ)K} \equiv 2\mathcal{M}_{(IJ)K}. \tag{4.12}$$

Remark 4.5. The complete set of symmetric and skew symmetric contributions in the various index pairings of the contortion follows as

$$\begin{aligned}
\text{Index Pair 23:}\quad & \mathcal{K}_{I(JK)} = \frac{1}{2}[\mathcal{T}_{JKI} - \mathcal{T}_{KIJ}] \\
& \mathcal{K}_{I[JK]} = \frac{1}{2}\,\mathcal{T}_{IJK} \\
\text{Index Pair 31:}\quad & \mathcal{K}_{(IJK)} = \frac{1}{2}[\mathcal{T}_{IJK} - \mathcal{T}_{KIJ}] \\
& \mathcal{K}_{[IJK]} = \frac{1}{2}\,\mathcal{T}_{JKI} \\
\text{Index Pair 12:}\quad & \mathcal{K}_{(IJ)K} = 0 \\
& \mathcal{K}_{[IJ]K} = \frac{1}{2}[\mathcal{T}_{IJK} + \mathcal{T}_{JKI} - \mathcal{T}_{KIJ}]
\end{aligned}$$

□

Remark 4.6. Observe that as a consequence of the decomposition of the connection into Riemann connection and contortion tensor together with the (right) symmetry of the Riemann connection

$$\mathcal{L}_{IJK} = \mathcal{M}_{IJK} + \mathcal{K}_{IJK} \quad \text{and} \quad \mathcal{M}_{I[JK]} = 0 \tag{4.13}$$

the *(right) skew symmetric contribution* of the contortion tensor equals the torsion tensor

$$\mathcal{T}_{IJK} := 2\mathcal{L}_{I[JK]} = 2\mathcal{K}_{I[JK]}. \tag{4.14}$$

The *(left) skew symmetry* $\mathcal{K}_{(IJ)K} = 0$ of the contortion is in agreement with its relation to the torsion $\mathcal{T}_{IJK} = 2\mathcal{K}_{I[JK]}$ since the defining equation for the contortion reads

$$\mathcal{K}_{IJK} := \frac{1}{2}\left[\mathcal{T}_{I[JK]} + \mathcal{T}_{J[KI]} - \mathcal{T}_{K[IJ]}\right] \equiv \mathcal{K}_{I[JK]} + \mathcal{K}_{J[KI]} - \mathcal{K}_{K[IJ]}.$$

By writing out the terms and rearrangement, this constraint may be reformulated as

$$\mathcal{K}_{(IK)J} - \mathcal{K}_{(KJ)I} + \mathcal{K}_{(JI)K} = 0.$$

Thus it is clearly satisfied by the (left) skew symmetry of the contortion. □

Remark 4.7. It is also interesting to remark that due to its left skew symmetry the contortion tensor does neither contribute to the projection of the parallel transport of a vector

$$\mathcal{V}_I \, \mathrm{p}\mathcal{V}^I = -\mathcal{V}^L \mathcal{V}^K \mathcal{L}_{LKJ} \, \mathrm{d}\mathcal{X}^J = -\mathcal{V}^L \mathcal{V}^K \mathcal{M}_{LKJ} \, \mathrm{d}\mathcal{X}^J \tag{4.15}$$

nor to the projection of the parallel transport of a covector

$$\mathcal{V}^J \, \mathrm{p}\mathcal{V}_J = \mathcal{V}^J \mathcal{V}^N \mathcal{L}_{NJK} \, \mathrm{d}\mathcal{X}^K = \mathcal{V}^J \mathcal{V}^N \mathcal{M}_{NJK} \, \mathrm{d}\mathcal{X}^K. \tag{4.16}$$

Combining Eqs. 4.15 and 4.16 renders the vanishing parallel transport of a scalar. □

Remark 4.8. Due to its (right) symmetry the Riemann connection may alternatively be expressed as

$$\mathcal{M}_{IJK} = \frac{1}{2} \left[\mathcal{M}_{IK,J} - \mathcal{M}_{KJ,I} + \mathcal{M}_{JI,K} \right] =: \mathcal{M}_{\{IJ,K\}}.$$

Likewise, due to its (left) skew symmetry the contortion takes an analogous format

$$\mathcal{K}_{IJK} = -\frac{1}{2} \left[\mathcal{T}_{IKJ} - \mathcal{T}_{KJI} + \mathcal{T}_{JIK} \right] =: -\mathcal{T}_{\{IJK\}}.$$

Obviously, based on the abbreviation

$$\mathcal{Z}_{\{IJK\}} := \frac{1}{2} \left[\mathcal{Z}_{IKJ} - \mathcal{Z}_{KJI} + \mathcal{Z}_{JIK} \right]$$

for an anti-cyclic permutation of the indices I, J, K (combined with an alternating sign for the three terms) of any expression $\mathcal{Z}_{IJK}$ the decomposition of the connection into Riemann connection and contortion may be written alternatively as

$$\mathcal{L}_{IJK} = \mathcal{M}_{\{IJ,K\}} - \mathcal{T}_{\{IJK\}},$$

see e.g. [167] (however with a different sequence of the indices). □

4.2.2 Inner Product and Angle with Geodesics

4.2.2.1 Inner Product

It may easily been proven that in a metrically connected manifold (space/geometry) the inner product $\mathcal{S} := \mathcal{V}^I \mathcal{M}_{IJ} \mathcal{W}^J$ of two vectors $\mathcal{V}^I$ and $\mathcal{W}^J$ is preserved upon an infinitesimal parallel transport along a parameter curve $\mathcal{X}^K(t)$ since firstly

$$\mathrm{p}\mathcal{S} = \mathcal{M}_{IJ,K} \mathcal{V}^I \mathcal{W}^J \, \mathrm{d}\mathcal{X}^K + \mathcal{M}_{IJ} \, \mathrm{p}\mathcal{V}^I \mathcal{W}^J + \mathcal{M}_{IJ} \mathcal{V}^I \, \mathrm{p}\mathcal{W}^J. \tag{4.17}$$

Using next the definition of parallel transport then renders

$$\mathrm{p}\mathcal{S} = \left[\mathcal{M}_{IJ,K} - \mathcal{L}^M{}_{IK}\mathcal{M}_{MJ} - \mathcal{L}^M{}_{JK}\mathcal{M}_{IM}\right]\mathcal{V}^I\mathcal{W}^J\,\mathrm{d}\mathcal{X}^K. \tag{4.18}$$

Including finally the definition of a metric connection thus results in

$$\mathrm{p}\mathcal{S} = \left[\mathcal{M}_{IJ,K} - 2\mathcal{L}_{(IJ)K}\right]\mathcal{V}^I\mathcal{W}^J\,\mathrm{d}\mathcal{X}^K \equiv 0. \tag{4.19}$$

Note that due to $\mathcal{L}_{(IJ)K} = \mathcal{M}_{(IJ)K}$ the contortion does not contribute to this result. Finally, as a consequence it is obvious that in a metrically connected manifold (space/geometry) in particular the (quadratic) length $\mathcal{V}^2 := \mathcal{V}^I\mathcal{M}_{IJ}\mathcal{V}^J$ of a vector $\mathcal{V}^I$ is preserved upon an infinitesimal parallel transport along a parameter curve $\mathcal{X}^K(t)$. Observe again that the contortion does not contribute, obviously this result is in agreement with Eqs. 4.15 and 4.16.

4.2.2.2 Angle with Geodesics

Based on the above result a picturesque and intuitive description of parallel transport is possible in a metrically connected manifold (space/geometry): A vector $\mathcal{V}^I$ that is transported parallel in a metrically manifold along a geodesic $\mathcal{X}^K(t)$, parameterized by the arc-length t, preserves its angle with the tangent vector $\dot{\mathcal{X}}^J$ to the geodesic (a unit vector due to the parametrization in the arc-length), i.e.

$$\frac{\mathrm{d}}{\mathrm{d}t}\left[\mathcal{V}^I\mathcal{M}_{IJ}\dot{\mathcal{X}}^J\right] = 0. \tag{4.20}$$

An illustration of this property is given in Fig. 4.1 that discusses parallel transport along big circles (geodesics) on a globe (sphere).

Proof. Performing firstly the differentiation on the left side of Eq. 4.20 with respect to the curve parameter t renders

$$\left[\mathcal{M}_{IJ,K}\mathcal{V}^I + \mathcal{M}_{IJ}\mathcal{V}^I{}_{,K}\right]\dot{\mathcal{X}}^J\dot{\mathcal{X}}^K + \mathcal{M}_{IJ}\mathcal{V}^I\ddot{\mathcal{X}}^J.$$

Next, the partial derivative of the metric may be re-expressed as a consequence of the connection being metric, i.e.

$$\mathcal{M}_{IJ,K} = \mathcal{M}_{MJ}\mathcal{L}^M{}_{IK} + \mathcal{M}_{IM}\mathcal{L}^M{}_{JK}.$$

Insertion of this relation into the expanded version of the left side of Eq. 4.20 then results in

$$\left[\left[\mathcal{M}_{MJ}\mathcal{L}^M{}_{IK} + \mathcal{M}_{IM}\mathcal{L}^M{}_{JK}\right]\mathcal{V}^I + \mathcal{M}_{IJ}\mathcal{V}^I{}_{,K}\right]\dot{\mathcal{X}}^J\dot{\mathcal{X}}^K + \mathcal{M}_{IJ}\mathcal{V}^I\ddot{\mathcal{X}}^J.$$

Finally, upon reordering of the various terms this result reads as

$$\mathcal{M}_{IJ}\mathcal{V}^I\left[\mathcal{L}^J{}_{KL}\dot{\mathcal{X}}^K\dot{\mathcal{X}}^L + \ddot{\mathcal{X}}^J\right] + \mathcal{M}_{IJ}\left[\mathcal{L}^I{}_{LK}\mathcal{V}^L + \mathcal{V}^I{}_{,K}\right]\dot{\mathcal{X}}^J\dot{\mathcal{X}}^K.$$

The terms in the first bracket are the defining equations for a geodesic and thus, according to the hypothesis, vanish

$$\mathcal{L}^J{}_{KL}\dot{\mathcal{X}}^K\dot{\mathcal{X}}^L + \ddot{\mathcal{X}}^J \equiv 0.$$

The terms in the second bracket define the covariant derivative of $\mathcal{V}^I$, however the covariant differential vanishes for the assumed parallel transport

$$\mathrm{D}\mathcal{V}^I = \left[\mathcal{V}^I{}_{,K} + \mathcal{L}^I{}_{LK}\mathcal{V}^L\right]\dot{\mathcal{X}}^K\,\mathrm{d}t \equiv 0.$$

Thus the left side of Eq. 4.20 combines to zero which concludes the proof. □

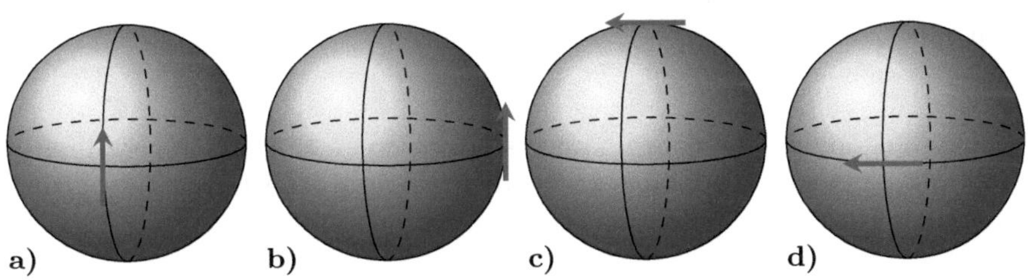

Fig. 4.1 Parallel transport on a globe (sphere): a) a vector is initially positioned at the intersection of the equator with the zeroth meridian and is oriented pointing north; b) parallel transport along the equator into its intersection with the 90th meridian, the vector is still oriented pointing north; c) parallel transport along the 90th meridian into the north pole, the vector is now oriented pointing west; d) parallel transport along the zeroth meridian into its intersection with the equator, the vector is oriented pointing west. Due to the curvature of the globe (sphere) the vector has changed its orientation by +90 degrees after having traveled by parallel transport first east-bound, then north-bound, and finally south-bound back to the starting position in a closed (finite) curve.

4.2.3 Transformation of the Connection

4.2.3.1 Riemann Connection

Due to the non-tensorial transformation property of the Riemann connection (based on the metric) the following convection or rather pull-back ($\mathcal{Y}$)/push-forward (y) relations hold in the case of holonomic coordinate transformations:

Connection($\mathcal{Y}$(metric))=$\mathcal{Y}$(connection(metric))

connection(y(Metric))=y(Connection(Metric))

As an example the (non-tensorial) pull-back of the spatial Riemann connection expressed in terms of the spatial metric equals the material Riemann connection in terms of the (tensorial) pull-back of the spatial metric. The corresponding relations hold if spatial and material objects are exchanged. However in the case of anholonomic coordinate transformations extra contributions in terms of the anholonomic object arise.

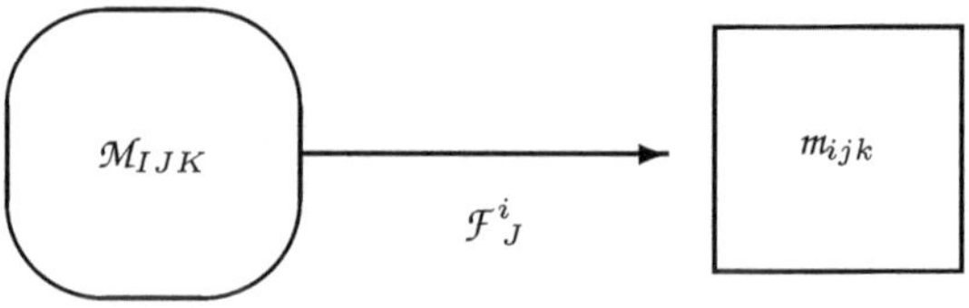

Fig. 4.2 Transformation of the Riemann connection for the case of holonomic X^I and anholonomic χ^i.

Upon changing the coordinate system between holonomic coordinates X^I and anholonomic coordinates χ^i the non-tensorial transformation of the fully covariant Riemann connection in Eq. 4.8 is re-stated

$$\mathcal{M}_{IJK} = \mathcal{F}^i{}_I \left[m_{ijk} \mathcal{F}^j{}_J \mathcal{F}^k{}_K + m_{ij} \mathcal{F}^j{}_{J,K} \right] . \tag{4.21}$$

Thereby, for holonomic X^I and anholonomic χ^i the Riemann connection $\mathcal{M}_{IJK}$ follows the standard definition in Eq. 4.7 whereas the Riemann connection m_{ijk} involves extra contributions in terms of the anholonomic object

$$m_{ijk} := m_{\{ij.k\}} + a_{\{ijk\}} . \tag{4.22}$$

The situation is depicted in Fig. 4.2.

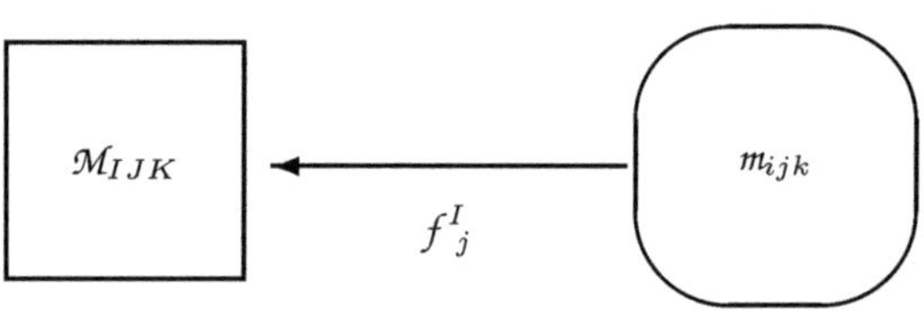

Fig. 4.3 Transformation of the Riemann connection for the case of holonomic χ^i and anholonomic X^I.

Vice versa, upon changing the coordinate system between holonomic coordinates χ^i and anholonomic coordinates X^I the non-tensorial transformation of the fully covariant Riemann connection in Eq. 4.8 is re-stated

$$m_{ijk} = f^I{}_i \left[\mathcal{M}_{IJK} f^J{}_j f^K{}_k + \mathcal{M}_{IJ} f^J{}_{j,k} \right] . \tag{4.23}$$

Thereby, for holonomic χ^i and anholonomic X^I the Riemann connection m_{ijk} follows the standard definition in Eq. 4.7 whereas the Riemann connection $\mathcal{M}_{IJK}$ involves extra contributions in terms of the anholonomic object

$$\mathcal{M}_{IJK} := \mathcal{M}_{\{IJ.K\}} + \mathcal{A}_{\{IJK\}}. \tag{4.24}$$

The situation is depicted in Fig. 4.3.

Proof. As an example the case of holonomic X^I and anholonomic χ^i shall be considered in detail.

Then to proof the assertion in Eq. 4.22 (and likewise in Eq. 4.24) the transformation of the metric has to be inserted into the definition of the Riemann connection in Eq. 4.7. The transformation of the metric reads as

$$\mathcal{M}_{IJ} = \mathcal{F}^i{}_I m_{ij} \mathcal{F}^j{}_J.$$

Computing the partial derivatives of the metric as needed in the definition of the Riemann connection and subsequent addition of the resulting expressions upon exploitation the symmetry of the metric renders

$$\begin{array}{rl}
\mathcal{M}_{IJ,K} = & \mathcal{F}^i{}_{I,K} m_{ij} \mathcal{F}^j{}_J + \mathcal{F}^i{}_I m_{ij.k} \mathcal{F}^j{}_J \mathcal{F}^k{}_K + \mathcal{F}^i{}_I m_{ij} \mathcal{F}^j{}_{J,K} \\
-\mathcal{M}_{JK,I} = & - \mathcal{F}^j{}_{J,I} m_{jk} \mathcal{F}^k{}_K - \mathcal{F}^j{}_J m_{jk.i} \mathcal{F}^k{}_K \mathcal{F}^i{}_I - \mathcal{F}^j{}_J m_{jk} \mathcal{F}^k{}_{K,I} \\
\mathcal{M}_{KI,J} = & \mathcal{F}^k{}_{K,J} m_{ki} \mathcal{F}^i{}_I + \mathcal{F}^k{}_K m_{ki.j} \mathcal{F}^i{}_I \mathcal{F}^j{}_J + \mathcal{F}^k{}_K m_{ki} \mathcal{F}^i{}_{I,J} \\
\hline\hline
\end{array}$$

$$\mathcal{M}_{\{IJ,K\}} = \mathcal{F}^i{}_I m_{\{ij.k\}} \mathcal{F}^j{}_J \mathcal{F}^k{}_K + \mathcal{F}^i{}_I m_{ij} \mathcal{F}^j{}_{(J,K)} - \mathcal{F}^j{}_J m_{jk} \mathcal{F}^k{}_{[K,I]} + \mathcal{F}^k{}_K m_{ki} \mathcal{F}^i{}_{[I,J]}$$

Noting that $\mathcal{M}_{IJK} \equiv \mathcal{M}_{\{IJ,K\}}$ (due to the holonomic X^I), together with adding and subtracting $\mathcal{F}^i{}_I m_{ij} \mathcal{F}^j{}_{[J,K]}$ leads eventually to the result

$$\mathcal{M}_{IJK} = \mathcal{F}^i{}_I \left[m_{ijk} \mathcal{F}^j{}_J \mathcal{F}^k{}_K + m_{ij} \mathcal{F}^j{}_{J,K} \right] \quad \text{with} \quad m_{ijk} := m_{\{ij.k\}} + a_{\{ijk\}}$$

and the combination of fully covariant spatial anholonomic objects

$$a_{\{ijk\}} := -m_{ij} \mathcal{F}^j{}_{[J,K]} f^J{}_j f^K{}_k - m_{jk} \mathcal{F}^k{}_{[K,I]} f^K{}_k f^I{}_i + m_{ki} \mathcal{F}^i{}_{[I,J]} f^I{}_i f^J{}_j.$$

Note the relation $2a_{\{ijk\}} = a_{ikj} - a_{kji} + a_{jik}$ whereby the right skew symmetry of the anholonomic object has been exploited. □

Remark 4.9. The complete set of symmetric and skew symmetric contributions in the various index pairings of, e.g., $\mathcal{A}_{\{IJK\}}$ follows as

$$\begin{aligned}
\text{Index Pair 23:} \quad & \mathcal{A}_{\{I(JK)\}} = -\frac{1}{2}[\mathcal{A}_{JKI} - \mathcal{A}_{KIJ}] \\
& \mathcal{A}_{\{I[JK]\}} = -\frac{1}{2}\,\mathcal{A}_{IJK} \\
\text{Index Pair 31:} \quad & \mathcal{A}_{\{(IJK)\}} = -\frac{1}{2}[\mathcal{A}_{IJK} - \mathcal{A}_{KIJ}] \\
& \mathcal{A}_{\{[IJK]\}} = -\frac{1}{2}\,\mathcal{A}_{JKI} \\
\text{Index Pair 12:} \quad & \mathcal{A}_{\{(IJ)K\}} = 0 \\
& \mathcal{A}_{\{[IJ]K\}} = -\frac{1}{2}[\mathcal{A}_{IJK} + \mathcal{A}_{JKI} - \mathcal{A}_{KIJ}]
\end{aligned}$$

□

Remark 4.10. On the one hand, incorporating the non-tensorial transformation of Riemann connections, the Ricci postulate $\mathcal{M}_{IJ,K} = 2\mathcal{M}_{(IJ)K} \equiv 2\mathcal{M}_{\{(IJ),K\}}$ reads

$$\mathcal{M}_{IJ,K} = 2\mathcal{F}^i{}_I m_{(ij)k} \mathcal{F}^j{}_J \mathcal{F}^k{}_K + 2\mathcal{F}^i{}_{(I} m_{ij} \mathcal{F}^j{}_{J),K}.$$

On the other hand, the partial derivative of the metric $\mathcal{M}_{IJ} = \mathcal{F}^i{}_I m_{ij} \mathcal{F}^j{}_J$ expands directly as

$$\mathcal{M}_{IJ,K} = \mathcal{F}^i{}_I m_{ij.k} \mathcal{F}^j{}_J \mathcal{F}^k{}_K + 2\mathcal{F}^i{}_{(I} m_{ij} \mathcal{F}^j{}_{J),K}.$$

Thus by comparing coefficients it can be concluded that the Ricci postulate $m_{ij.k} = 2m_{(ij)k} \equiv 2m_{\{(ij).k\}}$ (where $a_{\{(ij)k\}} = 0$) holds unchanged in holonomic as well as in anholonomic coordinates ($\mathcal{X}^I$ and x^i, respectively). □

4.2.3.2 Contortion

Due to the tensorial transformation property of the contortion (based on the torsion) the following convection or rather pull-back ($\mathcal{Y}$)/push-forward (y) relations hold in the case of holonomic coordinate transformations:

Contortion($\mathcal{Y}$(torsion))=$\mathcal{Y}$(contortion(torsion))

contortion(y(Torsion))=y(Contortion(Torsion))

As an example the (tensorial) pull-back of the spatial contortion expressed in terms of the spatial torsion equals the material contortion in terms of the (tensorial) pull-back of the spatial torsion. The corresponding relations hold if spatial and material objects are exchanged. However in the case of anholonomic coordinate transformations extra contributions in terms of the anholonomic object arise.

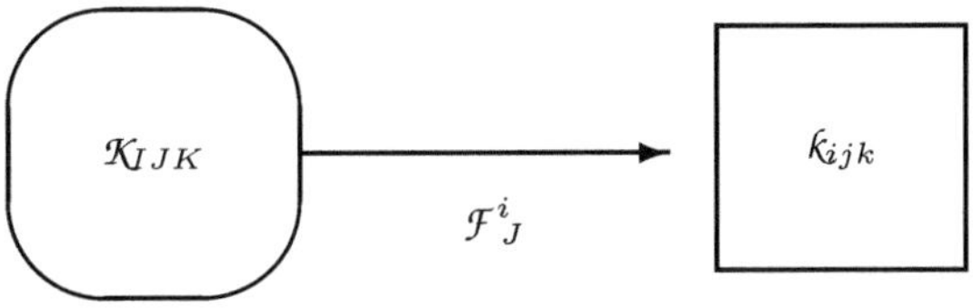

Fig. 4.4 Transformation of the contortion for the case of holonomic $\mathcal{X}^I$ and anholonomic χ^i.

Upon changing the coordinate system between holonomic coordinates $\mathcal{X}^I$ and anholonomic coordinates χ^i the tensorial transformation of the fully covariant contortion follows as

$$\mathcal{K}_{IJK} = \mathcal{F}^i{}_I \mathcal{k}_{ijk} \mathcal{F}^j{}_J \mathcal{F}^k{}_K. \tag{4.25}$$

Then, for holonomic $\mathcal{X}^I$ and anholonomic χ^i the contortion $\mathcal{K}_{IJK}$ follows the standard definition in Eq. 4.10 whereas the contortion $\mathcal{k}_{ijk}$ involves extra contributions in terms of the anholonomic object

$$\mathcal{k}_{ijk} := -2\ell_{\{i[jk]\}} - a_{\{ijk\}} = -t_{\{ijk\}}. \tag{4.26}$$

The situation is depicted in Fig. 4.4.

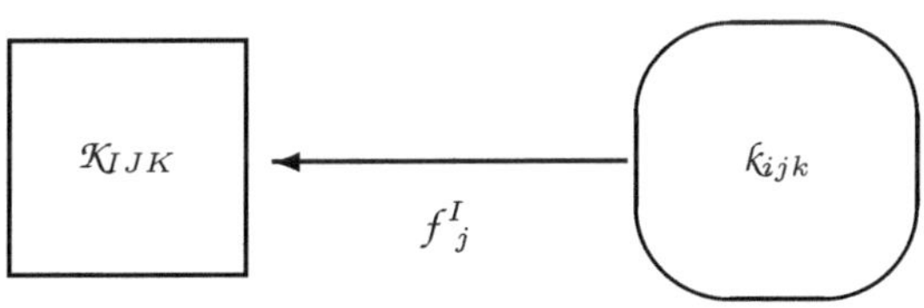

Fig. 4.5 Transformation of the contortion for the case of holonomic χ^i and anholonomic $\mathcal{X}^I$.

Vice versa, upon changing the coordinate system between holonomic coordinates χ^i and anholonomic coordinates $\mathcal{X}^I$ the tensorial transformation of the fully covariant contortion follows as

$$\mathcal{k}_{ijk} = f^I{}_i \mathcal{K}_{IJK} f^J{}_j f^K{}_k. \tag{4.27}$$

Then, for holonomic χ^i and anholonomic $\mathcal{X}^I$ the contortion $\mathcal{k}_{ijk}$ follows the standard definition in Eq. 4.10 whereas the contortion $\mathcal{K}_{IJK}$ involves extra contributions in terms of the anholonomic object

$$\mathcal{K}_{IJK} := -2\mathcal{L}_{\{I[JK]\}} - \mathcal{A}_{\{IJK\}} = -\mathcal{T}_{\{IJK\}}. \tag{4.28}$$

The situation is depicted in Fig. 4.5. Observe that the definitions of the torsion in Eqs. 4.26 and 4.28 involve the corresponding anholonomic objects.

Proof. As an example the case of holonomic $\mathcal{X}^I$ and anholonomic χ^i shall be considered in detail.

Then to proof the above assertion in Eq. 4.26 (and likewise in Eq. 4.28) the transformation of the torsion according to Eq. 3.49 has to be inserted into the definition of the contortion in Eq. 4.10. Then the transformation of the fully covariant torsion reads as

$$\mathcal{T}_{IJK} := 2\mathcal{L}_{I[JK]} = \mathcal{F}^i{}_I t_{ijk} \mathcal{F}^j{}_J \mathcal{F}^k{}_K \quad \text{with} \quad t_{ijk} := 2l_{i[jk]} + a_{ijk}.$$

Consequently, the transformation of the fully covariant contortion follows

$$\mathcal{K}_{IJK} := -2\mathcal{L}_{\{I[JK]\}} = \mathcal{F}^i{}_I k_{ijk} \mathcal{F}^j{}_J \mathcal{F}^k{}_K \quad \text{with} \quad k_{ijk} := -2l_{\{i[jk]\}} - a_{\{ijk\}}.$$

□

Remark 4.11. Adding finally the transformations of the Riemann connection and the contortion, e.g. for holonomic coordinates $\mathcal{X}^I$ and anholonomic coordinates χ^i, cancels the contribution of the corresponding anholonomic object $a_{\{ijk\}}$ and thus results in the common expression also for the transformed (metric) connection in terms of its right skew symmetric contribution $l_{i[jk]}$ as

$$l_{ijk} = m_{ijk} + k_{ijk} = m_{\{ij.k\}} - 2l_{\{i[jk]\}} = m_{\{ij.k\}} + a_{\{ijk\}} - t_{\{ijk\}}.$$

For holonomic coordinates χ^i and anholonomic coordinates $\mathcal{X}^I$ the corresponding result reads

$$\mathcal{L}_{IJK} = \mathcal{M}_{IJK} + \mathcal{K}_{IJK} = \mathcal{M}_{\{IJ.K\}} - 2\mathcal{L}_{\{I[JK]\}} = \mathcal{M}_{\{IJ.K\}} + \mathcal{A}_{\{IJK\}} - \mathcal{T}_{\{IJK\}}.$$

The representation on the outmost right sides follow from explicitly involving the corresponding anholonomic object and the torsion $t_{\{ijk\}}$ or $\mathcal{T}_{\{IJK\}}$, respectively, for the case of anholonomic coordinates. □

Remark 4.12. Since the transformed contortion is (left) skew symmetric with $k_{(ij)k} = 0$ it can be concluded that the Ricci postulate

$$m_{ij.k} = 2l_{(ij)k} \equiv 2m_{\{(ij).k\}}$$

(where $a_{\{(ij)k\}} = 0$ and $t_{\{(ij)k\}} = 0$) holds unchanged in holonomic as well as in anholonomic coordinates ($\mathcal{X}^I$ and χ^i, respectively). □

4.2.4 Integrable Connection

For a manifold that is equipped with the metric $\mathcal{M}_{IJ} = F^a{}_I \delta_{ab} F^b{}_J$, i.e. the *pull-back* of the Cartesian-type metric δ_{ab} (the Kronecker symbol) by the two-point tensor $F^a{}_I$ (that possesses the inverse $f^I{}_a$), and the *integrable connection* $\mathcal{L}^I{}_{JK} = f^I{}_a F^a{}_{J,K}$ a number of interesting results follow:

Lemma 4.1. *The integrable connection $\mathcal{L}^I{}_{JK}$ is a metric connection with respect to the metric $\mathcal{M}_{IJ}$*

$$\mathcal{M}_{IJ,K} = \mathcal{L}_{JIK} + \mathcal{L}_{IJK} = 2\mathcal{L}_{(IJ)K}.$$

The fully covariant versions of the integrable connection are obtained by lowering the first index by the metric coefficients. □

Proof. The fully covariant versions of the integrable connection expand into

$$\mathcal{L}_{JIK} = F^b{}_J \delta_{ba} F^a{}_{I,K} \quad \text{and} \quad \mathcal{L}_{IJK} = F^a{}_I \delta_{ab} F^b{}_{J,K}.$$

The partial derivative of the metric coefficients follows from the product rule

$$\mathcal{M}_{IJ,K} = F^a{}_{I,K} \delta_{ab} F^b{}_J + F^a{}_I \delta_{ab} F^b{}_{J,K}.$$

Observing finally the symmetry of the Kronecker symbol/Cartesian metric $\delta_{ab} = \delta_{ba}$ and comparing expressions concludes the proof. □

Lemma 4.2. *The Riemann connection corresponding to the integrable connection $\mathcal{L}^I{}_{JK}$ and the metric $\mathcal{M}_{IJ}$ reads as*

$$\mathcal{M}_{IJK} = F^a{}_I \delta_{ab} F^b{}_{(J,K)} - F^a{}_J \delta_{ab} F^b{}_{[K,I]} + F^a{}_K \delta_{ab} F^b{}_{[I,J]}.$$

Consequently a (right) symmetric integrable connection is a Riemann connection with vanishing contortion. □

Lemma 4.3. *The contortion corresponding to the integrable connection $\mathcal{L}^I{}_{JK}$ and the metric $\mathcal{M}_{IJ}$ reads as*

$$\mathcal{K}_{IJK} = F^a{}_I \delta_{ab} F^b{}_{[J,K]} + F^a{}_J \delta_{ab} F^b{}_{[K,I]} - F^a{}_K \delta_{ab} F^b{}_{[I,J]}.$$

The addition of the Riemann connection and the contortion corresponding to the integrable connection renders indeed $\mathcal{L}_{IJK}$. □

Proof. The proof follows by stating the partial derivative of the metric three times upon cyclic permutation of the indices

$$\begin{aligned} \mathcal{M}_{IJ,K} &= + \; F^a{}_J \; \delta_{ab} \; F^b{}_{I,K} + \; F^a{}_I \; \delta_{ab} \; F^b{}_{J,K} \\ -\mathcal{M}_{JK,I} &= - \; F^a{}_K \; \delta_{ab} \; F^b{}_{J,I} - \; F^a{}_J \; \delta_{ab} \; F^b{}_{K,I} \\ \mathcal{M}_{KI,J} &= + \; F^a{}_I \; \delta_{ab} \; F^b{}_{K,J} + F^a{}_K \; \delta_{ab} \; F^b{}_{I,J} \end{aligned}$$

Addition of the resulting expressions renders the Riemann connection as

$$\mathcal{M}_{IJK} = F^a{}_I \delta_{ab} F^b{}_{(J,K)} - F^a{}_J \delta_{ab} F^b{}_{[K,I]} + F^a{}_K \delta_{ab} F^b{}_{[I,J]}.$$

The contortion then follows either directly from its definition in terms of the torsion or from the difference between the integrable connection and its Riemann part. □

Lemma 4.4. *For holonomic coordinates X^I and anholonomic coordinates x^a the Riemann connection m_{abc} and the contortion k_{abc} corresponding to an integrable connection $\mathcal{L}^I{}_{JK} = f^I{}_a F^a{}_{J,K}$ read*

$$m_{abc} = a_{\{abc\}} \quad \text{and} \quad k_{abc} = -a_{\{abc\}} \quad \rightsquigarrow \quad l_{abc} = 0.$$

These representations fit to the Ricci postulate $\delta_{ab.c} = 2l_{(ab)c} = 0$ (and thus $\delta_{\{ab.c\}} = 0$) and to the expressions for $\mathcal{M}_{IJK}$ and $\mathcal{K}_{IJK}$ in the above. □

4.3 Curvature Based on a Metric Connection

The presence of a metric together with the connection being metric has considerable impact on the possible representations of the curvature.

4.3.1 Curvature Tensor

The *fully covariant version of the curvature tensor* follows from the

Definition 4.5. The fully covariant curvature tensor is obtained by lowering the first index by the covariant metric coefficients

$$\mathcal{R}_{IJKL} := \mathcal{M}_{IM} \mathcal{R}^M{}_{JKL}. \tag{4.29}$$

Thus from the definition of the mixed-variant curvature tensor it follows:

$$\mathcal{R}_{IJKL} = 2\mathcal{M}_{IM} \mathcal{L}^M{}_{J[L,K]} + 2\mathcal{L}_{IM[K} \mathcal{L}^M{}_{JL]}. \tag{4.30}$$

As a result of the subsequent step by step proof the fully covariant curvature tensor is eventually obtained explicitly as

$$\mathcal{R}_{IJKL} = 2\mathcal{L}_{IJ[L,K]} - 2\mathcal{L}_{MI[K} \mathcal{L}^M{}_{JL]}. \tag{4.31}$$

Observe the different sequence of indices in the term quadratic in the connection and its difference in sign for the mixed- and the covariant curvature tensors, respectively. □

Proof. The proof of the explicit representation for the covariant curvature follows in four steps:

Step 1: Application of the product rule to the first term in Eq. 4.30 and index lowering

$$\mathcal{M}_{IM} \mathcal{L}^M{}_{J[L,K]} = \mathcal{L}_{IJ[L,K]} - \mathcal{M}_{IM,[K} \mathcal{L}^M{}_{JL]}.$$

Step 2: Exploitation of the connection being a metric connection

$$\mathcal{M}_{IM,K} = \mathcal{L}_{MIK} + \mathcal{L}_{IMK}.$$

Step 3: Insert result of step 2 into the second term of the result in step 1

$$\mathcal{M}_{IM,[K}\mathcal{L}^M{}_{JL]} = \left[\mathcal{L}_{MI[K} + \mathcal{L}_{IM[K]}\right]\mathcal{L}^M{}_{JL]}.$$

Step 4: Insert the result of steps 1 to 3 into the definition in Eq. 4.30

$$\mathcal{M}_{IM}\mathcal{L}^M{}_{J[L,K]} + \mathcal{L}_{IM[K}\mathcal{L}^M{}_{JL]} = \mathcal{L}_{IJ[L,K]} - \mathcal{L}_{MI[K}\mathcal{L}^M{}_{JL]}.$$

This concludes the proof. □

It shall be noted further that the fully covariant curvature tensor obeys the following left and right (minor) skew symmetries

$$\mathcal{R}_{IJKL} = \mathcal{R}_{IJ[KL]} = \mathcal{R}_{[IJ][KL]}. \tag{4.32}$$

Proof. The skew symmetry in the index pair KL is a direct result from the definition of the curvature tensor in Eq. 4.31. The assertion of the skew symmetry in the index pair IJ can be proven easily in two steps:

Step 1: Exploit the connection being metric together with the symmetry of second partial derivatives

$$\mathcal{M}_{IJ,KL} = \mathcal{L}_{JIK,L} + \mathcal{L}_{IJK,L} = \mathcal{L}_{JIL,K} + \mathcal{L}_{IJL,K},$$

$$\rightarrow \qquad \mathcal{L}_{IJ[L,K]} = -\mathcal{L}_{JI[L,K]}.$$

Step 2: Rearranging the skew symmetry in the indices of the quadratic term

$$2\mathcal{L}_{MI[K}\mathcal{L}^M{}_{JL]} = \mathcal{L}_{MIK}\mathcal{L}^M{}_{JL} - \mathcal{L}_{MIL}\mathcal{L}^M{}_{JK} = -2\mathcal{L}_{MJ[K}\mathcal{L}^M{}_{IL]}.$$

This concludes the proof. □

Remark 4.13. Alternatively this result may be obtained directly by considering the condition for a metric connection in terms of the Ricci postulate

$$\mathcal{M}_{IJ|K} = 0 \tag{4.33}$$

and computing the skew symmetric part of the second covariant derivative according to Eq. 3.85.3

$$\mathcal{M}_{IJ|[KL]} = \frac{1}{2}\mathcal{M}_{NJ}\mathcal{R}^N{}_{IKL} + \frac{1}{2}\mathcal{M}_{IN}\mathcal{R}^N{}_{JKL} = \mathcal{R}_{(IJ)KL} = 0. \tag{4.34}$$

Thus the vanishing skew symmetric part of the second covariant derivative results in the above stated left (minor) skew symmetry of the covariant curvature tensor. □

Remark 4.14. Again as another alternative the Ricci postulate reads for the contravariant metric

$$\mathcal{M}^{IJ}{}_{|K} = 0.$$

Thus computing the skew symmetric part of the second covariant derivative according to Eq. 3.85.1 results alternatively in

$$\mathcal{M}^{IJ}{}_{|[KL]} = -\frac{1}{2}\mathcal{M}^{NJ}\mathcal{R}^{I}{}_{NKL} - \frac{1}{2}\mathcal{M}^{IN}\mathcal{R}^{J}{}_{NKL} = -\mathcal{R}^{(IJ)}{}_{KL} = 0.$$

□

Addendum 4.1. Identities for the Curvature Tensor

Lowering the first index in **C1** to **C3** by the covariant metric renders the (slightly modified) identities:

C1$^\flat$: $\mathcal{R}_{IJ(KL)} = 0$ and $\mathcal{R}_{(IJ)KL} = 0$
C2$^\flat$: $\mathcal{R}_{I[\![JKL]\!]} = -\mathcal{T}_{I[\![JK|L]\!]} + \mathcal{T}_{IM[\![J}\mathcal{T}^{M}{}_{KL]\!]}$
C3$^\flat$: $\mathcal{R}_{IJ[\![KL|M]\!]} = \mathcal{R}_{IJN[\![K}\mathcal{T}^{N}{}_{LM]\!]}$

Due to **C1** they may alternatively be expanded into

c2$^\flat$: $\mathcal{R}_{IJKL} + \mathcal{R}_{ILJK} + \mathcal{R}_{IKLJ} = -\left[\mathcal{T}_{IJK|L} + \mathcal{T}_{ILJ|K} + \mathcal{T}_{IKL|J}\right] + \left[\mathcal{T}_{IMJ}\mathcal{T}^{M}{}_{KL} + \mathcal{T}_{IML}\mathcal{T}^{M}{}_{JK} + \mathcal{T}_{IMK}\mathcal{T}^{M}{}_{LJ}\right]$

c3$^\flat$: $\mathcal{R}_{IJKL|M} + \mathcal{R}_{IJMK|L} + \mathcal{R}_{IJLM|K} = \left[\mathcal{R}_{IJNK}\mathcal{T}^{N}{}_{LM} + \mathcal{R}_{IJNM}\mathcal{T}^{N}{}_{KL} + \mathcal{R}_{IJNL}\mathcal{T}^{N}{}_{MK}\right]$

C2$^\flat$ (**c2**$^\flat$) and **C3**$^\flat$ (**c3**$^\flat$) denote the fully covariant versions of the *first (algebraic) Bianchi identity* and the *second (differential) Bianchi identity*, respectively.

4.3.2 Ricci Tensors

Due to the existence of the metric the fully covariant Ricci tensors as introduced in Eqs. 3.89 to 3.92 may be modified into *mixed-variant Ricci tensors* based on the

Definition 4.6. The mixed-variant Ricci tensors are obtained by raising the first index by the contravariant metric coefficients

$$\mathcal{R}^{J}{}_{L} := \mathcal{M}^{JN}\mathcal{R}_{NL} \quad \text{and} \quad \mathcal{V}^{K}{}_{L} := \mathcal{M}^{KN}\mathcal{V}_{NL}. \tag{4.35}$$

Thus from the definition of the covariant Ricci tensors it follows:

$$\mathcal{R}^{J}{}_{L} = 2\mathcal{M}^{JN}\mathcal{L}^{M}{}_{N[L,M]} - 2\mathcal{L}^{NJ}{}_{[M}\mathcal{L}^{M}{}_{NL]} \quad \text{and} \quad \mathcal{V}^{K}{}_{L} = 0. \tag{4.36}$$

As a result of the subsequent step by step proof the mixed-variant (first) Ricci tensor is eventually obtained explicitly as

$$\mathcal{R}^{J}{}_{L} = 2\mathcal{L}^{MJ}{}_{[L,M]} + 2\mathcal{L}^{JN}{}_{[M}\mathcal{L}^{M}{}_{NL]}. \tag{4.37}$$

Observe the different sequence of indices in the term quadratic in the connection and its difference in sign for the co- and the mixed-variant (first) Ricci tensor, respectively. □

Remark 4.15. In a three-dimensional Riemannian manifold the curvature tensor is expressed solely in terms of the (first) Ricci tensor (and its trace, viz. the Ricci scalar, see below). □

Proof. The proof of the explicit representation for the mixed-variant (first) Ricci tensor follows in four steps:

Step 1: Application of the product rule to the first term in Eq. 4.36 and index raising

$$\mathcal{M}^{JN}\mathcal{L}^{M}{}_{N[L,M]} = \mathcal{L}^{MJ}{}_{[L,M]} - \mathcal{M}^{JN}{}_{,[M}\mathcal{L}^{M}{}_{NL]}.$$

Step 2: Exploitation of the connection being a metric connection

$$\mathcal{M}^{JN}{}_{,M} = -\mathcal{M}^{JK}\mathcal{M}^{NL}\mathcal{M}_{KL,M} = -\mathcal{L}^{JN}{}_{M} - \mathcal{L}^{NJ}{}_{M}.$$

Step 3: Insert result of step 2 into the second term of the result in step 1

$$-\mathcal{M}^{JN}{}_{,[M}\mathcal{L}^{M}{}_{NL]} = \left[\mathcal{L}^{JN}{}_{[M} + \mathcal{L}^{NJ}{}_{[M}\right]\mathcal{L}^{M}{}_{NL]}.$$

Step 4: Insert the result of steps 1 to 3 into the definition in Eq. 4.36

$$\mathcal{M}^{JN}\mathcal{L}^{M}{}_{N[L,M]} - \mathcal{L}^{NJ}{}_{[M}\mathcal{L}^{M}{}_{NL]} = \mathcal{L}^{MJ}{}_{[L,M]} + \mathcal{L}^{JN}{}_{[M}\mathcal{L}^{M}{}_{NL]}.$$

This concludes the proof. □

From

$$\delta_M{}^K\mathcal{R}^M{}_{JKL} = \delta_M{}^K\mathcal{M}^{MI}\mathcal{R}_{IJKL} = \mathcal{M}^{IK}\mathcal{R}_{IJKL}$$

and

$$\delta_M{}^J\mathcal{R}^M{}_{JKL} = \delta_M{}^J\mathcal{M}^{MI}\mathcal{R}_{IJKL} = \mathcal{M}^{IJ}\mathcal{R}_{IJKL}$$

the relation of the second-order covariant (first) and second Ricci tensors to the fully covariant fourth-order curvature tensor follows by contraction with the contravariant metric

$$\mathcal{R}_{JL} = \mathcal{M}^{IK}\mathcal{R}_{IJKL} \quad \text{and} \quad \mathcal{V}_{KL} = \mathcal{M}^{IJ}\mathcal{R}_{IJKL} \equiv 0. \tag{4.38}$$

The last equality holds due to the skew symmetry of $\mathcal{R}_{IJKL}$ in the indices I, J as a result from the metric connection.

Addendum 4.2. Identities for the Curvature Tensor

As a consequence of the identities **R1** to **R3**, and the resulting skew symmetries of the fully covariant curvature tensor the following identities hold for the *(first) Ricci tensor* in the presence of a metric

R1$^\flat$: $\mathcal{R}_{JL} = -\mathcal{M}^{IK}\mathcal{R}_{IJLK} = \mathcal{M}^{IK}\mathcal{R}_{JILK} = -\mathcal{M}^{IK}\mathcal{R}_{JIKL}$
R2$^\flat$: $2\mathcal{R}_{[JL]} = \mathcal{T}^{I}{}_{JL|I} - 2\mathcal{T}_{[J|L]} - \mathcal{T}_M\mathcal{T}^{M}{}_{JL}$
R3$^\flat$: $2\mathcal{R}_{J[L|M]} + \mathcal{R}^{I}{}_{JLM|I} = -\mathcal{R}_{JN}\mathcal{T}^{N}{}_{LM} + 2\mathcal{R}^{I}{}_{JN[L}\mathcal{T}^{N}{}_{M]I}$

In **R2**$^\flat$ the skew symmetries in $\mathcal{R}_{IJKL}$ as manifested in **C1**$^\flat$ have been exploited, i.e. $\mathcal{V}_{JL} = 0$, see also **V1**$^\flat$. Note further that **R3**$^\flat$ coincides trivially with **R3**. **R2**$^\flat$ and **R3**$^\flat$ are the *contracted Bianchi identities* in the presence of a metric.

Due to **C1**$^\flat$, i.e. the skew symmetry of the curvature tensor in indices I, J, only one essential identity remains for the *second Ricci tensor*

V1$^\flat$: $\mathcal{V}_{KL} = \mathcal{M}^{IJ}\mathcal{R}_{IJLK} = 0$
V2$^\flat$: $-2\mathcal{R}_{[KL]} = -\mathcal{T}^{I}{}_{KL|I} + 2\mathcal{T}_{[K|L]} + \mathcal{T}_M\mathcal{T}^{M}{}_{KL}$

Note that **V2**$^\flat$ coincides trivially with the negative of **R2**$^\flat$. Due to **V1**$^\flat$ a possible **V3**$^\flat$ is a trivial identity ($0 \equiv 0$).

A further contraction of **R1**$^\flat$ to **R3**$^\flat$ with $\mathcal{M}^{JL}$ renders finally the additional identities

R1*: $\mathcal{R} = \mathcal{M}^{IK}\mathcal{M}^{JL}\mathcal{R}_{IJKL} = -\mathcal{M}^{IK}\mathcal{M}^{JL}\mathcal{R}_{IJLK}$
R3*: $\left[\mathcal{R}^{J}{}_{J}\delta^{L}{}_{M} - 2\mathcal{R}^{L}{}_{M}\right]_{|L} = -2\mathcal{R}^{L}{}_{N}\mathcal{T}^{N}{}_{LM} - \mathcal{R}_{ILNM}\mathcal{T}^{NLI}$

Here, $\mathcal{R}$ denotes the *Ricci scalar*, furthermore due to the skew symmetry of **R2**$^\flat$ in indices J, L a possible **R2*** is a trivial identity ($0 \equiv 0$), finally **R3*** may be denoted the *double-contracted second (differential) Bianchi identity*, compare [142].

4.3.3 Ricci Scalar

A further contraction of indices renders finally the *Ricci scalar*, it follows from the

Definition 4.7. Based on the connection and the metric a zeroth-order object, the Ricci scalar, is defined as:

$$\mathcal{R} := \mathcal{L}^{ML}{}_{L,M} - \mathcal{L}^{ML}{}_{M,L} + \mathcal{L}^{LN}{}_{M}\mathcal{L}^{M}{}_{NL} - \mathcal{L}^{LN}{}_{L}\mathcal{L}^{M}{}_{NM}. \tag{4.39}$$

In a two-dimensional Riemannian manifold the curvature tensor is expressed solely in terms of the Ricci scalar and the fourth-order skew symmetric unit tensor in terms of the metric, i.e. $\mathcal{R}_{IJKL} = \frac{1}{2}\mathcal{R}\,[\mathcal{M}_{IK}\mathcal{M}_{JL} - \mathcal{M}_{IL}\mathcal{M}_{JK}]$. Thereby the Ricci scalar relates in particular to the Gaussian curvature $\mathcal{R} = 2\mathcal{K}$ of a smooth surface, see Chapter 5. □

By condensing the skew symmetries in the indices L, M the Ricci scalar reads alternatively

$$\mathcal{R} = 2\mathcal{L}^{ML}{}_{[L,M]} + 2\mathcal{L}^{LN}{}_{[M}\mathcal{L}^{M}{}_{NL]}. \tag{4.40}$$

As a consequence the relation of the Ricci scalar to the second-order mixed-variant Ricci tensor by contraction of the indices is obvious

$$\mathcal{R} = \mathcal{R}^{J}{}_{J} = \delta_{J}{}^{L}\mathcal{R}^{J}{}_{L}. \tag{4.41}$$

Likewise, from $\delta_{I}{}^{L}\mathcal{R}^{I}{}_{L} = \delta_{I}{}^{L}\mathcal{M}^{IJ}\mathcal{R}_{JL} = \mathcal{M}^{JL}\mathcal{R}_{JL}$ the relation of the Ricci scalar to the second-order covariant Ricci tensor follows from contraction with the contravariant metric

$$\mathcal{R} = \mathcal{M}^{JL}\mathcal{R}_{JL}. \tag{4.42}$$

Finally, from $\mathcal{M}^{JL}\mathcal{R}_{JL} = \mathcal{M}^{IK}\mathcal{M}^{JL}\mathcal{R}_{IJKL}$ the Ricci scalar may be related to the fourth-order curvature tensor

$$\mathcal{R} = \mathcal{R}^{KL}{}_{KL} = \mathcal{M}^{IK}\mathcal{M}^{JL}\mathcal{R}_{IJKL}. \tag{4.43}$$

4.3.4 Einstein Tensor

The double-contracted second (differential) Bianchi identity **R3*** motivates to introduce the covariant Einstein tensor[1]

$$\mathcal{G}_{IJ} := \frac{1}{2}\mathcal{R}\,\mathcal{M}_{JI} - \mathcal{R}_{JI}. \tag{4.44}$$

Due to the double-contracted second (differential) Bianchi identity, the Einstein tensor is divergence, i.e. source free with $\mathcal{G}^{IJ}{}_{|J} = 0$ in the absence of torsion, see [142].

Addendum 4.3. Energy-Momentum Tensor

In the Theory of General Relativity/Gravitation the Einstein tensor as defined in the above, however with a negative sign, enters the so-called *Einstein field equations* that relates the Einstein tensor to 8π times the *energy-momentum tensor*. The Einstein field equations may be derived from a variational principle

[1] Due to its prominent role in the Theory of General Relativity/*Gravitation* the Einstein tensor is commonly denoted by $\mathcal{G}_{IJ}$. This, however, shall not be confused with the notation used in later chapters for the double-distortion of continuum kinematics $\mathcal{G}^{i}{}_{KL}$ and related quantities.

of least action, see [81]. Thereby, the action is defined in terms of a Lagrangian density $\mathcal{L} := \sqrt{-\det[\mathcal{M}_{IJ}]}\mathscr{L}$ that is decomposed into a contribution $\mathcal{L}_{\mathrm{g}} \propto \mathcal{R}$ from the (space-time) geometry, as sufficient for empty space, and a contribution $\mathcal{L}_{\mathrm{m}}$ from matter and/or fields, e.g. due to mass and/or electromagnetism. The corresponding action integral A is then considered to be a functional $A = A(\mathcal{M}^{IJ})$ of the metric tensor. The action is extremized if the corresponding Einstein field equations hold. Then the variational derivative of $\mathcal{L}_{\mathrm{g}}$ results in a multiple of the Einstein tensor $\mathcal{G}_{IJ}$, whereas the variational derivative of $\mathcal{L}_{\mathrm{m}}$ defines the *energy-momentum tensor*

$$\mathcal{T}_{IJ} := \mathscr{L}_{\mathrm{m}}\mathcal{M}_{IJ} - 2\frac{\delta\mathscr{L}_{\mathrm{m}}}{\delta\mathcal{M}^{IJ}}.$$

Examples for the energy-momentum tensor for various scenarios (with $\mathcal{E}, \mathcal{P}, \mathcal{V}_I$, and $\mathcal{F}_{IJ}$ here denoting the mass-energy density, the isotropic pressure, the four-velocity, and the electromagnetic field tensor, respectively) are listed in the following:

Particle Dust: $\mathcal{T}_{IJ} = \mathcal{E}\,\mathcal{V}_I\mathcal{V}_J$

Perfect Fluid: $\mathcal{T}_{IJ} = [\mathcal{E} + \mathcal{P}]\mathcal{V}_I\mathcal{V}_J + \mathcal{P}\,\mathcal{M}_{IJ}$

Electromagnetic Field: $\mathcal{T}_{IJ} = \frac{1}{4\pi}\left[\mathcal{F}_{IK}\mathcal{M}^{KL}\mathcal{F}_{JL} - \frac{1}{4}\mathcal{M}_{IJ}\left[\mathcal{F}_{KL}\mathcal{F}^{KL}\right]\right]$

Note that due to the Einstein field equations and due to the equivalence of the curvature tensor and the Einstein tensor in metrically connected four-space the curvature tensor of space-time vanishes identically in regions with vanishing energy-momentum-tensor.

4.3.5 Transformation of the Curvature Tensor

The tensorial transformation of the fully covariant version of the curvature tensor upon changing the coordinate system between coordinates $\mathcal{X}^I$ and coordinates χ^i is stated as

$$\mathcal{R}_{IJKL} = \mathcal{F}^i{}_I r_{ijkl}\mathcal{F}^j{}_J\mathcal{F}^k{}_K\mathcal{F}^l{}_L. \tag{4.45}$$

Thereby, for holonomic $\mathcal{X}^I$ and anholonomic χ^i the fully covariant version of the curvature tensor $\mathcal{R}_{IJKL}$ follows the standard definition in Eq. 4.31 whereas the curvature tensor r_{ijkl} involves extra contributions (underlined in the sequel) in terms of the connection and the anholonomic object

$$r_{ijkl} = 2\ell_{ij[l.k]} - 2\ell_{mi[k}\ell^m{}_{jl]} + \underline{\ell_{ijm}a^m{}_{lk}}. \tag{4.46}$$

The tensorial transformation of the fully covariant version of the curvature tensor upon changing the coordinate system between holonomic coordinates χ^i and anholonomic coordinates $\mathcal{X}^I$ and the resulting curvature tensor

$$\mathcal{R}_{IJKL} = 2\mathcal{L}_{IJ[L.K]} - 2\mathcal{L}_{MI[K}\mathcal{L}^{M}{}_{JL]} + \underline{\mathcal{L}_{IJM}\mathcal{A}^{M}{}_{LK}}. \tag{4.47}$$

involving extra contributions in terms of the connection $\mathcal{L}_{IJM}$ and the anholonomic object $\mathcal{A}^{M}{}_{LK}$ follows in analogy.

Proof. The case of holonomic $\mathcal{X}^I$ and anholonomic χ^i as resulting in Eq. 4.46 is considered in sequel. First the transformation of the connection reads as

$$\mathcal{L}_{IJL} = \mathcal{F}^{i}{}_{I}l_{ijl}\mathcal{F}^{j}{}_{J}\mathcal{F}^{l}{}_{L} + \mathcal{F}^{i}{}_{I}m_{ij}\mathcal{F}^{j}{}_{J,L}.$$

Computing the partial derivative of the connection as needed in the definition of the curvature renders the lengthy expression

$$\begin{aligned}
\mathcal{L}_{IJL,K} = & \mathcal{F}^{i}{}_{I,K}l_{ijl}\mathcal{F}^{j}{}_{J}\mathcal{F}^{l}{}_{L} \\
& + \mathcal{F}^{i}{}_{I}l_{ijl.k}\mathcal{F}^{j}{}_{J}\mathcal{F}^{k}{}_{K}\mathcal{F}^{l}{}_{L} \\
& + \mathcal{F}^{i}{}_{I}l_{ijl}\mathcal{F}^{j}{}_{J,K}\mathcal{F}^{l}{}_{L} \\
& + \mathcal{F}^{i}{}_{I}l_{ijl}\mathcal{F}^{j}{}_{J}\mathcal{F}^{l}{}_{L,K} \\
& + \mathcal{F}^{i}{}_{I,K}m_{ij}\mathcal{F}^{j}{}_{J,L} \\
& + \mathcal{F}^{i}{}_{I}m_{ij.k}\mathcal{F}^{j}{}_{J,L}\mathcal{F}^{k}{}_{K} \\
& + \mathcal{F}^{i}{}_{I}m_{ij}\mathcal{F}^{j}{}_{J,LK}.
\end{aligned}$$

Upon insertion of the Ricci postulate $m_{ij.k} = l_{ijk} + l_{jik}$ and skew symmetrization in L and K two terms drop out

$$\begin{aligned}
\mathcal{L}_{IJ[L,K]} = & \mathcal{F}^{i}{}_{I,[K}l_{ijl}\mathcal{F}^{j}{}_{J}\mathcal{F}^{l}{}_{L]} \\
& + \mathcal{F}^{i}{}_{I}\underline{l_{ij[l.k]}}\mathcal{F}^{j}{}_{J}\mathcal{F}^{k}{}_{K}\mathcal{F}^{l}{}_{L} \\
& + \mathcal{F}^{i}{}_{I}l_{ijl}\mathcal{F}^{j}{}_{J}\mathcal{F}^{l}{}_{[L,K]} \\
& + \mathcal{F}^{i}{}_{I,[K}m_{ij}\mathcal{F}^{j}{}_{J,L]} \\
& + \mathcal{F}^{i}{}_{I}l_{jik}\mathcal{F}^{j}{}_{J,[L}\mathcal{F}^{k}{}_{K]}.
\end{aligned}$$

Observe that the underlined term is already part of the sought for curvature tensor r_{ijkl}. Next the term of the curvature quadratic in the connection shall be computed. To this end the transformation of the connection is recalled once again with the right set of indices

$$\mathcal{L}_{MIK} = \mathcal{F}^{m}{}_{M}\left[l_{mik}\mathcal{F}^{i}{}_{I}\mathcal{F}^{k}{}_{K} + m_{mi}\mathcal{F}^{i}{}_{I,K}\right], \; \mathcal{L}^{M}{}_{JL} = f^{M}{}_{m}\left[l^{m}{}_{jl}\mathcal{F}^{j}{}_{J}\mathcal{F}^{l}{}_{L} + \mathcal{F}^{m}{}_{J,L}\right].$$

Then multiplication of the two representations of the connection in the above and skew symmetrization in L and K results in the multi-term expression

$$\begin{aligned}\mathcal{L}_{MI[K}\mathcal{L}^{M}{}_{JL]} &= \mathcal{F}^{i}{}_{I}\underline{\ell_{mi[k}\ell^{m}{}_{jl]}}\mathcal{F}^{j}{}_{J}\mathcal{F}^{k}{}_{K}\mathcal{F}^{l}{}_{L}\\ &+ \mathcal{F}^{i}{}_{I}\ell_{jik}\mathcal{F}^{j}{}_{J,[L}\mathcal{F}^{k}{}_{K]}\\ &+ \mathcal{F}^{i}{}_{I,[K}\ell_{ijl}\mathcal{F}^{j}{}_{J}\mathcal{F}^{l}{}_{L]}\\ &+ \mathcal{F}^{i}{}_{I,[K}m_{ij}\mathcal{F}^{j}{}_{J,L]}.\end{aligned}$$

Observe that the underlined term will be another part of the sought for curvature tensor r_{ijkl}. If we finally combine the above results so as to produce the curvature tensor $\mathcal{R}_{IJKL}$ many terms drop out and the resulting expression reads as

$$\frac{1}{2}\mathcal{R}_{IJKL} := \mathcal{L}_{IJ[L,K]} - \mathcal{L}_{MI[K}\mathcal{L}^{M}{}_{JL]}$$

$$= \mathcal{F}^{i}{}_{I}\left[\ell_{ij[l.k]} - \ell_{mi[k}\ell^{m}{}_{jl]} + \ell_{ijm}\mathcal{F}^{m}{}_{[N,M]}f^{N}{}_{l}f^{M}{}_{k}\right]\mathcal{F}^{j}{}_{J}\mathcal{F}^{k}{}_{K}\mathcal{F}^{l}{}_{L}.$$

Inserting finally the definition of the anholonomic object $a^{m}{}_{lk}$ into the above result concludes the proof. Clearly this result agrees with

$$r_{ijkl} = m_{in}r^{n}{}_{jkl} = m_{in}\left[2\ell^{n}{}_{j[l.k]} + 2\ell^{n}{}_{m[k}\ell^{m}{}_{jl]} + \ell^{n}{}_{jm}a^{m}{}_{lk}\right].$$

Thus the transformation behavior of the covariant curvature tensor fits to the previously established transformation behavior of the contra/co/co/co-(mixed) variant curvature tensor (which is of course no surprise). □

Remark 4.16. Alternatively, the proof for, e.g., the case in Eq. 4.47 is based on the representation of $\mathcal{R}^{I}{}_{JKL}$ in terms of the covariant derivative in Eq. 3.101. Pre-multiplying Eq. 3.101 with the covariant metric renders firstly

$$\mathcal{R}_{IJKL} = 2\mathcal{L}_{IJ[L|K]} - \underline{2\mathcal{L}_{IM[K}\mathcal{L}^{M}{}_{JL]}} + \underline{\underline{\underline{\mathcal{L}_{IJM}\mathcal{T}^{M}{}_{LK}}}}.$$

Expanding next the covariant derivative $\mathcal{L}_{IJ[L|K]}$ according to Eq. 3.21.3 results in

$$2\mathcal{L}_{IJ[L|K]} = \underline{\underline{\underline{2\mathcal{L}_{IJ[L.K]} - 2\mathcal{L}_{MI[K}\mathcal{L}^{M}{}_{JL]}}}} + \underline{2\mathcal{L}_{IM[K}\mathcal{L}^{M}{}_{JL]}} - \underline{\underline{2\mathcal{L}_{IJM}\mathcal{L}^{M}{}_{[LK]}}}.$$

Then the single underlined terms cancel, the double underlined terms combine into $\mathcal{L}_{IJM}\mathcal{A}^{M}{}_{LK}$, so that together with the triple underlined terms the result is proven. □

Note that due to the tensorial transformation behavior the fully covariant curvature tensor as expressed in anholonomic coordinates inherits the left and right (minor) skew symmetries displayed by the fully covariant curvature tensor as expressed in holonomic coordinates.

Proof. The case of holonomic $\mathcal{X}^I$ and anholonomic χ^i as resulting in Eq. 4.46 is considered as the example. First, the first and second partial derivatives of the metric m_{ij} result in

$$m_{ij.k} := m_{ij,K} f^K{}_k \quad \rightsquigarrow \quad m_{ij.kl} = m_{ij,KL} f^K{}_k f^L{}_l + m_{ij.m} \mathcal{F}^m{}_K f^K{}_{k.l}.$$

Then based on the definition of the anholonomic object $a^m{}_{kl} = -2\mathcal{F}^m{}_K f^K{}_{[k.l]}$ and the Ricci postulate $m_{ij.m} = 2\ell_{(ij)m}$ the skew second derivatives of the metric m_{ij} expand as

$$m_{ij.[kl]} = -\ell_{(ij)m} a^m{}_{kl}.$$

Thus in anholonomic coordinates second partial derivatives are not symmetric! Indeed, for exchanging partial derivatives k and l it holds that

$$m_{ij.kl} = m_{ij.lk} + 2m_{ij.[kl]} \quad \rightsquigarrow \quad 2\ell_{(ij)k.l} = 2\ell_{(ij)l.k} - 2\ell_{(ij)m} a^m{}_{kl}$$

whereby the Ricci postulate has been involved again. Finally this relation may be rearranged to render eventually

$$2\ell_{ij[k.l]} + \ell_{ijm} a^m{}_{kl} = -2\ell_{ji[k.l]} - \ell_{jim} a^m{}_{kl}.$$

Clearly, the expression on the right-hand-side is skew symmetric in i and j. Thus together with the (left) skew symmetry of the quadratic term $2\ell_{mi[k}\ell^m{}_{jl]}$ the (left) skew symmetry of r_{ijkl} is proven. The (right) skew symmetry in k and l is obvious. □

4.4 Riemann Geometry

Definition 4.8. A Riemann geometry is a possibly non-flat, i.e. curved, but symmetric, i.e. torsion-free, metric manifold (space/geometry) that is equipped with a (unique) metric connection (the Riemann connection, sometimes also denoted the Levi-Civita connection, see [179]). □

The consequences of this definition shall be explored in the sequel.

4.4.1 Ricci Lemma

For a Riemann connection the previously stated *Ricci postulate* obtains the stronger status of a *lemma*. The *Ricci lemma* establishes that the covariant derivative in terms of the Riemann connection of, e.g., the covariant metric coefficients vanishes identically, i.e.

$$\mathcal{M}_{IJ|K} = \mathcal{M}_{IJ,K} - \mathcal{M}_{JIK} - \mathcal{M}_{IJK} \equiv 0. \tag{4.48}$$

Proof. The proof follows at once by expanding the Riemann connection in terms of the metric coefficients in the Ricci lemma

$$\begin{aligned}\mathcal{M}_{IJ|K} = \mathcal{M}_{IJ,K} &- \frac{1}{2}\big[\mathcal{M}_{JI,K} - \mathcal{M}_{IK,J} + \mathcal{M}_{KJ,I}\big] \\ &- \frac{1}{2}\big[\mathcal{M}_{IJ,K} - \mathcal{M}_{JK,I} + \mathcal{M}_{KI,J}\big] \equiv 0.\end{aligned}$$

Observing that due to the symmetry of the metric coefficients all terms on the right hand side cancel out concludes the proof. □

Alternatively, the Ricci lemma is stated in terms of the partial derivative of the metric coefficients

$$\mathcal{M}_{IJ,K} = \mathcal{M}_{JIK} + \mathcal{M}_{IJK} = 2\mathcal{M}_{(IJ)K}. \tag{4.49}$$

4.4.2 Fundamental Theorem of Riemann Geometry

For a Riemann manifold the fundamental theorem of Riemann geometry holds, see [122]. It is given in the

Theorem 4.1. *On a Riemann manifold with metric $\mathcal{M}_{IJ}$ the only connection $\mathcal{L}_{IJK}$ which is (right) symmetric, i.e. $\mathcal{L}_{IJK} = \mathcal{L}_{I(JK)}$ (torsion free), and that preserves inner products of vectors upon parallel transport, i.e. $\mathrm{p}\mathcal{S} = 0$, is uniquely given by the Riemann connection $\mathcal{M}_{IJK} = \mathcal{M}_{\{IJ,K\}}$.* □

Proof. Recall the previously established result for the parallel transport of the inner product $\mathcal{S} = \mathcal{V}^I \mathcal{M}_{IJ} \mathcal{W}^J$ of two vectors $\mathcal{V}^I$ and $\mathcal{W}^J$

$$\mathrm{p}\mathcal{S} = \big[\mathcal{M}_{IJ,K} - 2\mathcal{L}_{(IJ)K}\big]\mathcal{V}^I \mathcal{W}^J \,\mathrm{d}\mathcal{X}^K.$$

Setting next $\mathrm{p}\mathcal{S} = 0$ in order to preserve $\mathcal{S}$ results in the requirement for a zero covariant derivative of the metric coefficients

$$\mathcal{M}_{IJ|K} := \mathcal{M}_{IJ,K} - 2\mathcal{L}_{(IJ)K} = \mathcal{M}_{IJ,K} - \mathcal{L}_{JIK} - \mathcal{L}_{IJK} = 0.$$

Proceeding further as in the proof of the Ricci postulate and incorporating finally the required (right) symmetry of the connection proves eventually the result. □

4.4.3 Geodesics: Extremal Curves

On a Riemann manifold(space/geometry) geodesics may also be given an important variational definition in addition to their general definition as the straightest curves in a manifold as introduced previously. This follows from the

Definition 4.9. Geodesics are parameter curves $\mathcal{X}^K = \mathcal{X}^K(t)$ that extremize the distance between two points $\mathcal{P}$ in a Riemann manifold $\mathcal{M}$. They follow from the variational problem

$$I(\{\mathcal{X}^K\}) = \int_{t_1}^{t_2} \sqrt{\dot{\mathcal{X}}^I \mathcal{M}_{IJ} \dot{\mathcal{X}}^J}\, \mathrm{d}t \to \text{stat.} \tag{4.50}$$

Recall that a Riemann manifold is equipped with a (positive definite) Riemann metric $\mathcal{M}_{IJ}$ and a corresponding Riemann connection $\mathcal{M}^I{}_{JK}$. □

Proof. Identifying the curve parameter t with the arclength, the Euler-Lagrange equations corresponding to the variational problem in Eq. 4.50 follow as

$$2\frac{\mathrm{d}}{\mathrm{d}t}\left[\dot{\mathcal{X}}^I \mathcal{M}_{IK}\right] - \mathcal{M}_{IJ,K}\dot{\mathcal{X}}^I \dot{\mathcal{X}}^J = 2\mathcal{M}_{KI}\ddot{\mathcal{X}}^I + \left[2\mathcal{M}_{(IK,J)} - \mathcal{M}_{JI,K}\right]\dot{\mathcal{X}}^I \dot{\mathcal{X}}^J = 0.$$

After rearrangement and based on $2\mathcal{M}_{(IK,J)} = \mathcal{M}_{IK,J} + \mathcal{M}_{JK,I}$ and the symmetry of the metric $\mathcal{M}_{JK}$ together with the definition of the Riemann connection $\mathcal{M}_{KIJ} = \mathcal{M}_{\{KI,J\}}$ and $\mathcal{M}^L{}_{IJ} = \mathcal{M}^{LK}\mathcal{M}_{KIJ}$ the above Euler-Lagrange equations read eventually as

$$\ddot{\mathcal{X}}^L + \mathcal{M}^L{}_{IJ}\dot{\mathcal{X}}^I \dot{\mathcal{X}}^J = 0.$$

These are of course nothing but the ordinary differential equations governing the geodesics in a Riemann geometry. □

Example 4.1. As an example from analytical dynamics the equations of (*free*) motion of a mechanical system with n_{dm} generalized coordinates $\mathcal{X}^1, \mathcal{X}^2, \cdots, \mathcal{X}^{n_{dm}}$ shall be considered. Without external and internal (interaction) forces the equations of motion for a system of mass points, i.e. the equations of *free* motion, trivially degenerate to the conservation of the generalized momenta for each separate mass point. The coordinates $\mathcal{X}^I$ may describe, e.g., the placement $\boldsymbol{r}^1, \boldsymbol{r}^2, \cdots, \boldsymbol{r}^{\nu_{mp}}$ of ν_{mp} mass points $\alpha = 1, 2, \cdots, \nu_{mp}$ in ordinary three-dimensional Euclidean space

$$\boldsymbol{r}^\alpha = \boldsymbol{r}^\alpha(\{\mathcal{X}^K\}).$$

Then the kinetic energy T of this system of mass points is given in terms of the ν_{mp} masses m_α and the corresponding velocities $\dot{\boldsymbol{r}}^\alpha = [\partial \boldsymbol{r}^\alpha / \partial \mathcal{X}^I]\dot{\mathcal{X}}^I$ as

$$T = \frac{1}{2}\sum_{\alpha=1}^{\nu_{mp}} m_\alpha \dot{\boldsymbol{r}}^\alpha \cdot \dot{\boldsymbol{r}}^\alpha = \frac{1}{2}\dot{\mathcal{X}}^I \mathcal{M}_{IJ} \dot{\mathcal{X}}^J$$

whereby the summation convention is implied for the indices I, J. The mechanical problem is thus formulated in terms of a n_{dm}-dimensional Riemann manifold with the coordinates $\{\mathcal{X}^I\}$ and the corresponding metric

$$\mathcal{M}_{IJ}(\{X^K\}) := \sum_{\alpha=1}^{\nu_{mp}} m_\alpha \frac{\partial \boldsymbol{r}^\alpha}{\partial X^I} \cdot \frac{\partial \boldsymbol{r}^\alpha}{\partial X^J}.$$

The equations of (free) motion then follow from the Hamilton principle as

$$\frac{\mathrm{d}}{\mathrm{d}t}\frac{\partial T}{\partial \dot{X}^K} - \frac{\partial T}{\partial X^K} = \frac{\mathrm{d}}{\mathrm{d}t}\left[\dot{X}^I \mathcal{M}_{IK}\right] - \frac{1}{2}\dot{X}^I \mathcal{M}_{IJ,K}\dot{X}^J = 0.$$

and may eventually be rearranged following the steps as in the above proof into

$$\ddot{X}^I + \mathcal{M}^I{}_{JK}\dot{X}^J\dot{X}^K = 0 \quad \rightsquigarrow \quad \mathcal{M}_{IJ}\ddot{X}^J + \mathcal{M}_{IJK}\dot{X}^J\dot{X}^K = 0. \tag{4.51}$$

Clearly, this is nothing but the definition of the geodesic in a n_{dm}-dimensional Riemann manifold equipped with the metric $\mathcal{M}_{IJ}$ and the corresponding Riemann connection $\mathcal{M}^I{}_{JK}$. It is remarkable that the equations of motion for, e.g., a single mass point in general relativity are also expressed as the defining equation for the geodesics in a four-dimensional (one time and three space coordinates) curved, i.e., Riemann manifold, however with a metric determined by the gravitational field, for more details refer for example to [142]. □

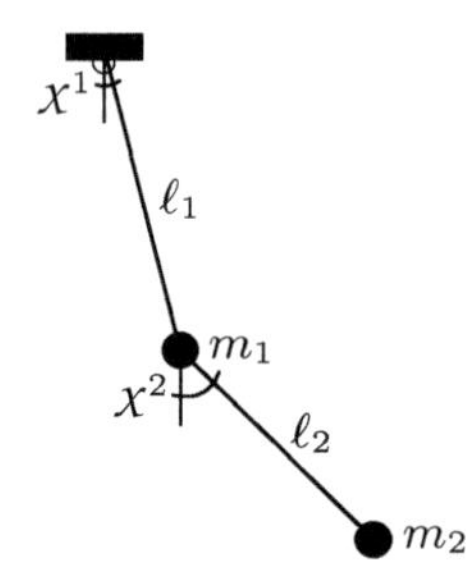

Fig. 4.6 Double pendulum. The two pendulum bars of length ℓ_1 and ℓ_2 are rigid and massless, the two mass point possess mass m_1 and m_2, the hinges are frictionless, the $n_{dm} = 2$ generalized coordinates X^1 and X^2 measure the angle of the pendulum bars with the vertical.

Example 4.2. Consider as an elementary example a double pendulum in Fig. 4.6 with kinetic energy expressed in terms of $n_{dm} = 2$ generalized coordinates X^1 and X^2 denoting the angles between the vertical and the rigid pendulum bars of length ℓ_1 and ℓ_2, respectively

$$T = \frac{1}{2}[m_1 + m_2][\ell_1\dot{X}^1]^2 + m_2\ell_1\ell_2\dot{X}^1\dot{X}^2\cos(X^1 - X^2) + \frac{1}{2}m_2[\ell_2\dot{X}^2]^2.$$

Here (consistent with the representation of the kinetic energy) the placement of the two mass points in the two-dimensional Euclidean space reads

$$\boldsymbol{r}^1 = \ell_1 \begin{bmatrix} \sin \mathcal{X}^1 \\ -\cos \mathcal{X}^1 \end{bmatrix} \quad \text{and} \quad \boldsymbol{r}^2 = \boldsymbol{r}^1 + \ell_2 \begin{bmatrix} \sin \mathcal{X}^2 \\ -\cos \mathcal{X}^2 \end{bmatrix}.$$

The configuration dependent metric $\mathcal{M}_{IJ}$ thus follows as

$$[\mathcal{M}_{IJ}] = \begin{bmatrix} [m_1 + m_2][\ell_1]^2 & m_2 \ell_1 \ell_2 \cos(\mathcal{X}^1 - \mathcal{X}^2) \\ m_2 \ell_1 \ell_2 \cos(\mathcal{X}^1 - \mathcal{X}^2) & m_2 [\ell_2]^2 \end{bmatrix} \quad \text{with} \quad T = \frac{1}{2} \dot{\mathcal{X}}^I \mathcal{M}_{IJ} \dot{\mathcal{X}}^J.$$

The Riemann connection $\mathcal{M}_{IJK}$ corresponding to the metric $\mathcal{M}_{IJ}$ consists of only two non-zero terms

$$\mathcal{M}_{122} = m_2 \ell_1 \ell_2 \sin(\mathcal{X}^1 - \mathcal{X}^2) \quad \text{and} \quad \mathcal{M}_{211} = -m_2 \ell_1 \ell_2 \sin(\mathcal{X}^1 - \mathcal{X}^2).$$

Consequently the nonlinear equations of motion as resulting from the geodesic in Eq. 4.51.2 read

$$\begin{aligned} \mathcal{M}_{11} \ddot{\mathcal{X}}^1 + \mathcal{M}_{12} \ddot{\mathcal{X}}^2 + \mathcal{M}_{122} \dot{\mathcal{X}}^2 \dot{\mathcal{X}}^2 &= 0, \\ \mathcal{M}_{21} \ddot{\mathcal{X}}^1 + \mathcal{M}_{22} \ddot{\mathcal{X}}^2 + \mathcal{M}_{211} \dot{\mathcal{X}}^1 \dot{\mathcal{X}}^1 &= 0. \end{aligned}$$

Inserting the specific coefficients of the metric and the connection finally results in the well-established result

$$\begin{aligned} [m_1 + m_2][\ell_1]^2 \ddot{\mathcal{X}}^1 + m_2 \ell_1 \ell_2 \cos(\mathcal{X}^1 - \mathcal{X}^2) \ddot{\mathcal{X}}^2 + m_2 \ell_1 \ell_2 \sin(\mathcal{X}^1 - \mathcal{X}^2) \dot{\mathcal{X}}^2 \dot{\mathcal{X}}^2 &= 0, \\ m_2 \, [\ell_2]^2 \ddot{\mathcal{X}}^2 + m_2 \ell_1 \ell_2 \cos(\mathcal{X}^1 - \mathcal{X}^2) \ddot{\mathcal{X}}^1 - m_2 \ell_1 \ell_2 \sin(\mathcal{X}^1 - \mathcal{X}^2) \dot{\mathcal{X}}^1 \dot{\mathcal{X}}^1 &= 0. \end{aligned}$$

These are the coupled set of ordinary differential equations describing the free motion of a double pendulum. □

4.4.4 Riemann Curvature Tensor

In a Riemann manifold (space/geometry), since there is no torsion, the fully covariant curvature tensor or rather the Riemann curvature tensor is computed from the Riemann connection solely. The Riemann connection in turn depends only on the metric, see Eq. 4.7. Thus the Riemann curvature tensor allows a representation entirely in terms of the metric and its partial derivatives

$$\begin{aligned} \mathcal{M}_{IJKL} = & \frac{1}{2} [\mathcal{M}_{IL,JK} + \mathcal{M}_{JK,IL} - \mathcal{M}_{IK,JL} - \mathcal{M}_{JL,IK}] \\ & - \left[\mathcal{M}_{M(I,K)} - \frac{1}{2} \mathcal{M}_{IK,M} \right] \mathcal{M}^{MN} \left[\mathcal{M}_{N(J,L)} - \frac{1}{2} \mathcal{M}_{JL,N} \right] \\ & + \left[\mathcal{M}_{M(I,L)} - \frac{1}{2} \mathcal{M}_{IL,M} \right] \mathcal{M}^{MN} \left[\mathcal{M}_{N(J,K)} - \frac{1}{2} \mathcal{M}_{JK,N} \right]. \end{aligned} \tag{4.52}$$

Besides the (minor) skew symmetries discussed in the above it turns out that it obeys the following additional (major) symmetry for exchanging index pairs

$$\mathcal{M}_{[IJ][KL]} = \mathcal{M}_{[KL][IJ]}. \tag{4.53}$$

Thus, as an immediate result, the corresponding Ricci and Einstein tensors associated to the Riemann curvature tensor are always symmetric. As a further consequence from the first (algebraic) Ricci identity and the above symmetry and skew symmetry properties the number of independent entries in the Riemann curvature tensor is reduced to

$$\frac{n_{dm}^2[n_{dm}^2 - 1]}{12}. \tag{4.54}$$

For $n_{dm} = 1, 2, 3, 4$ the number of independent entries computes as $0, 1, 6,$ and 20, respectively.

Proof. The proof of the (major) symmetry assertion follows directly in two steps:

Step 1: Exploit the structure of the Riemann connection, and take into account the symmetry of the metric and that of the second partial derivatives

$$4\mathcal{M}_{IJ[K,L]} = -\mathcal{M}_{JK,IL} + \mathcal{M}_{KI,JL} + \mathcal{M}_{JL,IK} - \mathcal{M}_{LI,JK} = 4\mathcal{M}_{KL[I,J]},$$

$$4\mathcal{M}_{KL[I,J]} = -\mathcal{M}_{LI,KJ} + \mathcal{M}_{IK,LJ} + \mathcal{M}_{LJ,KI} - \mathcal{M}_{JK,LI} = 4\mathcal{M}_{IJ[K,L]}.$$

Step 2: Take the vanishing torsion of the Riemann connection into account

$$2\mathcal{M}_{MI[K}\mathcal{M}^{M}{}_{JL]} = \mathcal{M}_{M(IK)}\mathcal{M}^{M}{}_{(JL)} - \mathcal{M}_{M(IL)}\mathcal{M}^{M}{}_{(JK)} = 2\mathcal{M}_{MK[I}\mathcal{M}^{M}{}_{LJ]},$$

$$2\mathcal{M}_{MK[I}\mathcal{M}^{M}{}_{LJ]} = \mathcal{M}_{M(KI)}\mathcal{M}^{M}{}_{(LJ)} - \mathcal{M}_{M(KJ)}\mathcal{M}^{M}{}_{(LI)} = 2\mathcal{M}_{MI[K}\mathcal{M}^{M}{}_{JL]}.$$

This concludes the proof. □

Addendum 4.4. Identities for the Curvature Tensor

The fully covariant Riemann curvature tensor satisfies the following identities:

M1$^\flat$: $\mathcal{M}_{IJ(KL)} = 0$ and $\mathcal{M}_{(IJ)KL} = 0$
M2$^\flat$: $\mathcal{M}_{I[\![JKL]\!]} = 0$
M3$^\flat$: $\mathcal{M}_{IJ[\![KL|M]\!]} = 0$
M4$^\flat$: $\mathcal{M}_{[IJ][KL]} = \mathcal{M}_{[KL][IJ]}$

Due to **M1$^\flat$** and **M4$^\flat$** identity **M2$^\flat$** is alternatively stated as

M2$^{\flat\flat}$: $\mathcal{M}_{[\![IJKL]\!]} = 0$

Here $[\![IJKL]\!]$ denotes the full skew symmetry (all possible swaps of two indices with alternating signs) in four indices, i.e. 4! index combinations (due to the skew symmetries of the curvature tensor in the indices I, J and K, L together with the symmetry in the index pairs $[IJ], [KL]$ it holds that $4![\![IJKL]\!] := 8[IJKL + ILJK + IKLJ]$). The above identities may thus also be expanded into

$$\mathbf{m2}^\flat: \quad \mathcal{M}_{IJKL} + \mathcal{M}_{ILJK} + \mathcal{M}_{IKLJ} = 0$$
$$\mathbf{m3}^\flat: \quad \mathcal{M}_{IJKL|M} + \mathcal{M}_{IJMK|L} + \mathcal{M}_{IJLM|K} = 0$$

$\mathbf{M2}^\flat$ ($\mathbf{m2}^\flat$) and $\mathbf{M3}^\flat$ ($\mathbf{m3}^\flat$) denote the fully covariant versions of the *first (algebraic) Bianchi identity* and the *second (differential) Bianchi identity*, respectively, for the Riemann curvature tensor.

4.4.5 Transformation of the Curvature Tensor

The tensorial transformation of the Riemann curvature tensor upon changing the coordinate system between coordinates $\mathcal{X}^I$ and coordinates χ^i is stated as

$$\mathcal{M}_{IJKL} = \mathcal{F}^i{}_I \mathcal{m}_{ijkl} \mathcal{F}^j{}_J \mathcal{F}^k{}_K \mathcal{F}^l{}_L. \tag{4.55}$$

Thereby, for holonomic $\mathcal{X}^I$ and anholonomic χ^i the Riemann curvature tensor $\mathcal{M}_{IJKL}$ follows the standard definition in Eq. 4.31 (that specializes for a Riemann connection to Eq. 4.52) whereas the curvature tensor $\mathcal{m}_{ijkl}$ involves extra contributions (underlined in the sequel) in terms of the Riemann connection and the anholonomic object

$$\mathcal{m}_{ijkl} = 2\mathcal{m}_{ij[l.k]} - 2\mathcal{m}_{mi[k}\mathcal{m}^m{}_{jl]} + \underline{\mathcal{m}_{ijm}a^m{}_{lk}}. \tag{4.56}$$

The tensorial transformation of the Riemann curvature tensor upon changing the coordinate system between holonomic coordinates χ^i and anholonomic coordinates $\mathcal{X}^I$ and the resulting curvature tensor

$$\mathcal{M}_{IJKL} = 2\mathcal{M}_{IJ[L.K]} - 2\mathcal{M}_{MI[K}\mathcal{M}^M{}_{JL]} + \underline{\mathcal{M}_{IJM}\mathcal{A}^M{}_{LK}}. \tag{4.57}$$

involving extra contributions in terms of the Riemann connection $\mathcal{M}_{IJM}$ and the anholonomic object $\mathcal{A}^M{}_{LK}$ follows in analogy.

It is noted that the major and minor symmetries of the Riemann curvature tensor as displayed in holonomic coordinates are clearly preserved by the tensorial transformation to its counterpart in anholonomic coordinates!

4.5 Non-Metric Connection

A further generalization is obtained by relaxing the Ricci postulate (requiring the covariant derivative of the metric to vanish identically) in order to allow for *non-metric connections*. Then all results stated in the preceding sections have to be enhanced by the presence of the *non-metricity tensor*. Instead of performing the tedious task to repeat all the lengthy derivations following the detailed steps in all their splendor as in the above, this section simply states the main starting point, i.e. what may be called the *non-metric Ricci postulate*, and its resulting impact on the representation of the connection and the curvature.

4.5.1 Non-Metric Ricci Postulate

As an extension of the Ricci Postulate the presence of a metric in a manifold allows to formulate the

Postulate 4.3 (Non-Metric Ricci Postulate). For a *non-metric connection* the covariant derivative of the *covariant* metric coefficients corrected by the (left) symmetric *non-metricity tensor* $\mathcal{Q}_{IJK} \equiv \mathcal{Q}_{(IJ)K}$ vanishes

$$\mathcal{M}_{IJ|K} + \mathcal{Q}_{IJK} = \mathcal{M}_{IJ,K} - \mathcal{L}^{M}{}_{IK}\mathcal{M}_{MJ} - \mathcal{L}^{M}{}_{JK}\mathcal{M}_{IM} + \mathcal{Q}_{IJK} = 0. \tag{4.58}$$

Here the sign convention and notation for the non-metricity tensor is adopted from [167]. From resolving the above equality the partial derivative of the covariant metric may then be stated in terms of the (left) symmetric part of the fully covariant connection corrected by the (left) symmetric non-metricity tensor

$$\mathcal{M}_{IJ,K} = \mathcal{L}_{JIK} + \mathcal{L}_{IJK} - \mathcal{Q}_{IJK} = 2\mathcal{L}_{(IJ)K} - \mathcal{Q}_{(IJ)K}. \tag{4.59}$$

The underlying metric manifold that is equipped with a non-metric connection is then denoted a *non-metrically connected manifold (space/geometry).* □

As a consequence of the non-metric Ricci postulate the covariant derivatives of vectors and covectors are related by the metric *and* the non-metricity tensor

$$\begin{aligned}
\mathcal{M}_{IJ}\mathcal{V}^{I}{}_{|K} &= \mathcal{M}_{IJ}\left[\mathcal{V}^{I}{}_{,K} + \mathcal{L}^{I}{}_{MK}\mathcal{V}^{M}\right] \\
&= \left[\mathcal{M}_{IJ}\mathcal{V}^{I}\right]_{,K} - \mathcal{V}^{I}\mathcal{M}_{IJ,K} + \mathcal{L}_{JMK}\mathcal{V}^{M} \\
&= \mathcal{V}_{J,K} - \mathcal{V}^{I}\left[\mathcal{L}_{JIK} + \mathcal{L}_{IJK} - \mathcal{Q}_{IJK}\right] + \mathcal{L}_{JMK}\mathcal{V}^{M} \\
&= \mathcal{V}_{J,K} - \mathcal{V}^{I}\mathcal{L}_{IJK} + \mathcal{V}^{I}\mathcal{Q}_{IJK} \\
&= \mathcal{V}_{J,K} - \mathcal{V}_{I}\mathcal{L}^{I}{}_{JK} + \mathcal{V}_{I}\mathcal{Q}^{I}{}_{JK} \\
&= \mathcal{V}_{J|K} + \mathcal{V}_{I}\mathcal{Q}^{I}{}_{JK}.
\end{aligned} \tag{4.60}$$

Thus covariant differentiation and raising or lowering of indices by the metric do not simply commute anymore for the general case of a non-metric connection.

As a further consequence of the non-metric Ricci postulate a metric connection is additively decomposed into the Riemann connection, the *distention* (the non-metric part of the connection), and the contortion

$$\underbrace{\mathcal{L}_{IJK}}_{\text{Non-Metric Connection}} = \underbrace{\mathcal{M}_{\{IJ,K\}}}_{\text{Riemann}} + \underbrace{\mathcal{Q}_{\{IJK\}}}_{\text{Distention}} + \underbrace{[-\mathcal{T}_{\{IJK\}}]}_{\text{Contortion}} . \tag{4.61}$$

Proof. The proof follows readily by stating the non-metric Ricci postulate three times upon cyclic permutation of the indices and subsequent addition of the resulting expressions to render:

$$\begin{aligned} \mathcal{M}_{IJ|K} &= \mathcal{M}_{IJ,K} - \mathcal{L}_{JIK} - \mathcal{L}_{IJK} =: -\mathcal{Q}_{IJK} \\ -\mathcal{M}_{JK|I} &= -\mathcal{M}_{JK,I} + \mathcal{L}_{KJI} + \mathcal{L}_{JKI} =: \mathcal{Q}_{JKI} \\ \mathcal{M}_{KI|J} &= \mathcal{M}_{KI,J} - \mathcal{L}_{IKJ} - \mathcal{L}_{KIJ} =: -\mathcal{Q}_{KIJ} \end{aligned}$$

$$\mathcal{M}_{IJ,K} - \mathcal{M}_{JK,I} + \mathcal{M}_{KI,J} + \mathcal{Q}_{IJK} - \mathcal{Q}_{JKI} + \mathcal{Q}_{KIJ} + 2\mathcal{L}_{J[KI]} - 2\mathcal{L}_{K[IJ]} - 2\mathcal{L}_{I(JK)} = 0$$

Then by dividing by 2 and by adding and subtracting the terms

$$\mathcal{L}_{IJK} = \mathcal{L}_{I(JK)} + \mathcal{L}_{I[JK]}$$

and employing the $\{IJK\}$ abbreviation the result follows immediately as

$$\mathcal{L}_{IJK} = \mathcal{M}_{\{IJ,K\}} + \mathcal{Q}_{\{IJK\}} + \mathcal{L}_{I[JK]} + \mathcal{L}_{J[KI]} - \mathcal{L}_{K[IJ]}.$$

Identifying finally $2\mathcal{L}_{I[JK]} =: \mathcal{T}_{IJK}$ with the torsion concludes the proof. □

Postulate 4.4 (Non-Metric Ricci Postulate). For a *non-metric connection* the covariant derivative of the contravariant metric coefficients corrected by the (left) symmetric *mixed-variant* non-metricity tensor $\mathcal{Q}^{IJ}{}_K = \mathcal{Q}^{(IJ)}{}_K$ vanishes

$$\mathcal{M}^{IJ}|_K - \mathcal{Q}^{IJ}{}_K = \mathcal{M}^{IJ}{}_{,K} + \mathcal{L}^{JI}{}_K + \mathcal{L}^{IJ}{}_K - \mathcal{Q}^{IJ}{}_K = 0.$$

The partial derivative of the contravariant metric may thus alternatively be stated in terms of the (left) symmetric part of the mixed-variant connection corrected by the (left) symmetric mixed-variant non-metricity tensor

$$\mathcal{M}^{IJ}{}_{,K} = -\mathcal{L}^{JI}{}_K - \mathcal{L}^{IJ}{}_K + \mathcal{Q}^{IJ}{}_K = -2\mathcal{L}^{(IJ)}{}_K + \mathcal{Q}^{IJ}{}_K.$$

This is an alternative relation valid for a metric, non-metrically connected manifold (space/geometry). □

Proof. The covariant derivative of the Kronecker symbol $\delta^I{}_K = \mathcal{M}^{IJ}\mathcal{M}_{JK}$ vanishes and thus renders upon applying the Leibniz rule

$$\mathcal{M}^{IJ}{}_{|K} = -\mathcal{M}^{IM}\mathcal{M}_{MN|K}\mathcal{M}^{NJ}.$$

Inserting the result for $\mathcal{M}_{MN|K}$ and observing that

$$-\mathcal{M}^{IM}[\mathcal{M}_{MN,K}\mathcal{M}^{NJ}] = -\mathcal{M}^{IM}[\delta_M{}^J{}_{,K} - \mathcal{M}_{MN}\mathcal{M}^{NJ}{}_{,K}] = \delta^I{}_N\mathcal{M}^{NJ}{}_{,K} = \mathcal{M}^{IJ}{}_{,K}$$

concludes the proof. □

4.5.1.1 Non-Metric Part of the Connection

The non-metric part of the connection, that shall be called the *distention* in the sequel, follows from the

Definition 4.10. The distention is exclusively computed from the non-metricity tensor:

$$\mathcal{O}_{IJK} := \mathcal{Q}_{\{IJK\}} = \frac{1}{2}\left[\mathcal{Q}_{IJK} - \mathcal{Q}_{JKI} + \mathcal{Q}_{KIJ}\right]. \tag{4.62}$$

□

Due to the definition of the non-metricity tensor as a covariant derivative of the metric the distention obeys the transformation behavior of a third-order tensor

$$o_{ijk} = f^I{}_i \mathcal{O}_{IJK} f^J{}_j f^K{}_k, \quad \text{and} \quad \mathcal{O}_{IJK} = \mathcal{F}^i{}_I o_{ijk} \mathcal{F}^j{}_J \mathcal{F}^k{}_K. \tag{4.63}$$

It may be shown by inspection that, based on the (left) symmetry of the non-metricity tensor, in particular the (right) skew symmetric contribution of the distention vanishes

$$\mathcal{O}_{I[JK]} = 0. \tag{4.64}$$

Thus the distention does in particular not contribute to the torsion $\mathcal{T}_{IJK} := 2\mathcal{L}_{I[JK]} = 2\mathcal{K}_{I[JK]}$ which therefore still results from the (right) skew symmetric contribution of the contortion only.

Remark 4.17. The complete set of symmetric and skew symmetric contributions in the various index pairings of the distention follows as:

$$\begin{aligned}
\text{Index Pair 23:} \quad & \mathcal{O}_{I(JK)} = \frac{1}{2}\left[\mathcal{Q}_{IJK} - \mathcal{Q}_{JKI} + \mathcal{Q}_{KIJ}\right] \\
& \mathcal{O}_{I[JK]} = 0 \\
\text{Index Pair 31:} \quad & \mathcal{O}_{(IJK)} = \frac{1}{2}\mathcal{Q}_{KIJ} \\
& \mathcal{O}_{[IJK]} = \frac{1}{2}\left[\mathcal{Q}_{IJK} - \mathcal{Q}_{JKI}\right] \\
\text{Index Pair 12:} \quad & \mathcal{O}_{(IJ)K} = \frac{1}{2}\mathcal{Q}_{IJK} \\
& \mathcal{O}_{[IJ]K} = \frac{1}{2}\left[-\mathcal{Q}_{JKI} + \mathcal{Q}_{KIJ}\right]
\end{aligned}$$

□

4.5.2 Inner Product and Angle with Geodesics

4.5.2.1 Inner Product

In a non-metrically connected manifold (space/geometry) the inner product $\mathcal{S} := \mathcal{V}^I \mathcal{M}_{IJ} \mathcal{W}^J$ of two vectors $\mathcal{V}^I$ and $\mathcal{W}^J$ is *not* preserved upon an infinitesimal parallel transport along a parameter curve $\mathcal{X}^K(t)$. Instead, due to the definition of a non-metric connection, $\mathrm{p}\mathcal{S}$ is related to the non-metricity tensor as

$$\mathrm{p}\mathcal{S} = \left[\mathcal{M}_{IJ,K} - 2\mathcal{L}_{(IJ)K}\right] \mathcal{V}^I \mathcal{W}^J \,\mathrm{d}\mathcal{X}^K \equiv -\mathcal{Q}_{IJK} \mathcal{V}^I \mathcal{W}^J \,\mathrm{d}\mathcal{X}^K. \tag{4.65}$$

Moreover it is also obvious that in a non-metrically connected manifold (space/geometry) in particular the (quadratic) length $\mathcal{V}^2 := \mathcal{V}^I \mathcal{M}_{IJ} \mathcal{V}^J$ of a vector $\mathcal{V}^I$ is not preserved upon an infinitesimal parallel transport along a parameter curve $\mathcal{X}^K(t)$.

$$\mathrm{p}\mathcal{V}^2 = -\mathcal{Q}_{IJK} \mathcal{V}^I \mathcal{V}^J \,\mathrm{d}\mathcal{X}^K. \tag{4.66}$$

Thus the non-metricity tensor captures the change in the (quadratic) length of a vector upon its parallel transport in a non-metrically connected manifold (space/geometry).

4.5.2.2 Angle with Geodesics

A vector $\mathcal{V}^I$ that is transported parallel in a non-metrically connected manifold (space/geometry) along a geodesic $\mathcal{X}^K(t)$, parameterized by the arc-length t, changes its angle with the tangent vector $\dot{\mathcal{X}}^J$ to the geodesic in terms of the non-metricity tensor as

$$\frac{\mathrm{d}}{\mathrm{d}t}\left[\mathcal{V}^I \mathcal{M}_{IJ} \dot{\mathcal{X}}^J\right] = -\mathcal{Q}_{IJK} \mathcal{V}^I \dot{\mathcal{X}}^J \dot{\mathcal{X}}^K. \tag{4.67}$$

Proof. Inserting the partial derivative of the metric for a non-metric connection into the left side of Eq. 4.67 and taking into account the definitions for a geodesic and the parallel transport of a vector concludes the proof. □

4.5.3 Non-Metricity Covector, Semi-metric Connection

For completeness the *non-metricity covector* is introduced by the

Definition 4.11. The non-metricity covector follows as the trace of the non-metricity tensor

$$\mathcal{Q}_K := \mathcal{M}_{IJ} \mathcal{Q}^{IJ}{}_K = \mathcal{Q}^I{}_{IK}. \tag{4.68}$$

The definition of the non-metricity covector in, e.g., [167] differs from Eq. 4.68 by a scalar factor composed from the dimension n_{dm} of the manifold. □

As a special case the notion of a *semi-metric connection* is introduced by the

Definition 4.12. For a semi-metric connection the non-metricity tensor follows in terms of the metric and the non-metricity covector as

$$\mathcal{Q}^{IJ}{}_K := \frac{1}{n_{dm}} \mathcal{M}^{IJ} \mathcal{Q}_K. \tag{4.69}$$

Obviously contracting the so defined $\mathcal{Q}^{IJ}{}_K$ by $\mathcal{M}_{IJ}$ reproduces the non-metricity covector $\mathcal{Q}_K$ from Eq. 4.68. □

Remark 4.18. The connection of a *Weyl geometry* is torsion free, i.e. symmetric, and semi-metric (with real metric), thus the (quadratic) length change of a vector upon parallel transport computes as

$$\mathrm{p}[\mathcal{V}^I \mathcal{M}_{IJ} \mathcal{V}^J] = -\frac{1}{n_{dm}} [\mathcal{V}^I \mathcal{M}_{IJ} \mathcal{V}^J] \mathcal{Q}_K \, \mathrm{d}\mathcal{X}^K.$$

Thus in a Weyl geometry the (quadratic) length of a vector is not preserved upon parallel transport, however, interestingly, the ratio of the (quadratic) length $\mathcal{V} = \mathcal{V}^I \mathcal{M}_{IJ} \mathcal{V}^J$ and $\mathcal{W} = \mathcal{W}^I \mathcal{M}_{IJ} \mathcal{W}^J$ of two vectors $\mathcal{V}^I$ and $\mathcal{W}^I$ at the same point is preserved

$$\mathrm{p}\frac{\mathcal{V}}{\mathcal{W}} = \frac{\mathrm{p}\mathcal{V}}{\mathcal{W}} - \frac{\mathcal{V}}{\mathcal{W}^2}\,\mathrm{p}\mathcal{W} = -\frac{1}{n_{dm}}[\frac{\mathcal{V}}{\mathcal{W}} - \frac{\mathcal{V}}{\mathcal{W}}] \mathcal{Q}_K \, \mathrm{d}\mathcal{X}^K = 0.$$

An equivalent result holds for the change of the angle of a vector upon parallel transport along a geodesic in a Weyl geometry

$$\frac{\mathrm{d}}{\mathrm{d}t}\left[\mathcal{V}^I \mathcal{M}_{IJ} \dot{\mathcal{X}}^J\right] = -\frac{1}{n_{dm}}\left[\mathcal{V}^I \mathcal{M}_{IJ} \dot{\mathcal{X}}^J\right] \mathcal{Q}_K \dot{\mathcal{X}}^K. \tag{4.70}$$

It is straightforward to show that the ratio of the angle of two different vectors with a geodesic is preserved. The relation between the continuum theory of point-defects and a Weyl geometry has been exploited in [204]. □

4.5.4 Transformation of the Non-Metricity Tensor

The tensorial transformation of the fully covariant non-metricity tensor $\mathcal{Q}_{IJK}$ upon changing the coordinate system between coordinates $\mathcal{X}^I$ and coordinates χ^i reads as

$$\mathcal{Q}_{IJK} = \mathcal{F}^i{}_I \, q_{ijk} \, \mathcal{F}^j{}_J \mathcal{F}^k{}_K. \tag{4.71}$$

Thereby the non-metricity tensors expressed in coordinate systems $\mathcal{X}^I$ and χ^i follow as

$$\mathcal{Q}_{IJK} = \mathcal{M}_{IJ,K} - 2\mathcal{L}_{(IJ)K} \quad \text{and} \quad q_{ijk} = m_{ij,k} - 2l_{(ij)k}. \tag{4.72}$$

Remarkably, the tensorial transformation of the non-metricity tensor is completely unaffected whether the coordinate systems $\mathcal{X}^I$ or χ^i are holonomic or anholonomic as long as the appropriate holonomic or anholonomic partial derivative is utilized.

Proof. On the one hand the partial derivative of the tensorial transformation of the metric $\mathcal{M}_{IJ} = \mathcal{F}^i{}_I m_{ij} \mathcal{F}^j{}_J$ expands into

$$\mathcal{M}_{IJ,K} = \mathcal{F}^i{}_I m_{ij,k} \mathcal{F}^j{}_J \mathcal{F}^k{}_K + 2\mathcal{F}^i{}_{(I,K} m_{ij} \mathcal{F}^j{}_{J)}.$$

On the other hand twice the left symmetric part of the connection transforms (non-tensorially) as

$$2\mathcal{L}_{(IJ)K} = 2\mathcal{F}^i{}_I l_{(ij)k} \mathcal{F}^j{}_J \mathcal{F}^k{}_K + 2\mathcal{F}^i{}_{(I} m_{ij} \mathcal{F}^j{}_{J),K}.$$

Considering that due to symmetry $\mathcal{F}^i{}_{(I,K} m_{ij} \mathcal{F}^j{}_{J)} = \mathcal{F}^i{}_{(I} m_{ij} \mathcal{F}^j{}_{J),K}$ holds and subtracting the two results in the above completes the proof. Note that the integrability or non-integrability of $\mathcal{F}^i{}_I$ and thus the issue of holonomic or anholonomic coordinate systems is of no relevance in this argument. □

4.6 Curvature Based on a Non-Metric Connection

The presence of a metric together with the connection being non-metric has considerable impact on the possible representations of the curvature.

4.6.1 Curvature Tensor

The *fully covariant (0,4)-version of the curvature tensor* follows from the

Definition 4.13. The fully covariant (0,4)-curvature tensor is obtained by lowering the first index of the underlying $\mathcal{R}^I{}_{JKL}$. Using the representation of the partial derivative of the metric for a non-metric connection renders eventually

$$\mathcal{R}_{IJKL} = 2\mathcal{L}_{IJ[L,K]} - 2\mathcal{L}_{MI[K}\mathcal{L}^M{}_{JL]} + 2\mathcal{Q}_{MI[K}\mathcal{L}^M{}_{JL]}. \tag{4.73}$$

Observe the occurrence of a term involving the non-metricity tensor. □

For a non-metric connection the fully covariant (0,4)-curvature tensor obeys the following symmetry properties

$$\mathcal{R}_{IJ(KL)} = 0 \quad \text{and} \quad \mathcal{R}_{(IJ)KL} = -\mathcal{Q}_{IJ[K|L]} - \frac{1}{2}\mathcal{Q}_{IJM}\mathcal{T}^M{}_{KL}. \tag{4.74}$$

The symmetry in the indices I, J on the right hand side is a direct consequence from the (left) symmetry of the non-metricity tensor.

Proof. The result follows by considering the condition for a non-metric connection in terms of the non-metric Ricci postulate $\mathcal{M}_{IJ|K} = -\mathcal{Q}_{IJK}$ and computing the skew symmetric part of the second covariant derivative of $\mathcal{M}_{IJ}$ according to Eq. 3.85.3

$$\mathcal{M}_{IJ|[KL]} = \mathcal{R}_{(IJ)KL} + \frac{1}{2}\mathcal{Q}_{IJM}\mathcal{T}^{M}{}_{KL} = -\mathcal{Q}_{IJ[K|L]}.$$

□

Expanding the covariant derivative $\mathcal{Q}_{IJ[K|L]}$ according to Eq. 3.21 into the partial derivative $\mathcal{Q}_{IJ[K,L]}$ and three further terms containing the connection renders the alternative representation for the (left) symmetric contribution to the fully covariant (0,4)-curvature tensor

$$\mathcal{R}_{(IJ)KL} = -\mathcal{Q}_{IJ[K,L]} + \mathcal{Q}_{MI[K}\mathcal{L}^{M}{}_{JL]} + \mathcal{Q}_{MJ[K}\mathcal{L}^{M}{}_{IL]}. \tag{4.75}$$

As a consequence the (left) skew symmetric contribution to the fully covariant (0,4)-curvature tensor may be computed from

$$\mathcal{R}_{[IJ]KL} = \mathcal{R}_{IJKL} - \mathcal{R}_{(IJ)KL} = -\mathcal{R}_{[JI]KL} \tag{4.76}$$

and results in the final representation

$$\mathcal{R}_{[IJ]KL} = 2\mathcal{L}_{IJ[L,K]} - 2\mathcal{L}_{MI[K}\mathcal{L}^{M}{}_{JL]} - \mathcal{Q}_{IJ[L,K]} + \mathcal{Q}_{MI[K}\mathcal{L}^{M}{}_{JL]} - \mathcal{Q}_{MJ[K}\mathcal{L}^{M}{}_{IL]}.$$

The skew symmetry of the above expression in I, J then follows straightforwardly from the

Proof. Consider the condition for a non-metric connection in terms of the non-metric Ricci postulate $\mathcal{M}_{IJ,L} = \mathcal{L}_{JIL} + \mathcal{L}_{IJL} - \mathcal{Q}_{IJL}$ and equate the skew symmetric part of the second partial derivative of $\mathcal{M}_{IJ}$ to zero

$$\mathcal{M}_{IJ,[LK]} = 0 \quad \rightsquigarrow \quad 2\mathcal{L}_{IJ[L,K]} = -2\mathcal{L}_{JI[L,K]} + 2\mathcal{Q}_{JI[L,K]}.$$

Furthermore, from raising and lowering the index M it holds that

$$-2\mathcal{L}_{MI[K}\mathcal{L}^{M}{}_{JL]} = 2\mathcal{L}_{MJ[K}\mathcal{L}^{M}{}_{IL]}.$$

Then by inserting the two above relations into the representation for $\mathcal{R}_{[IJ]KL}$ renders eventually

$$\mathcal{R}_{[IJ]KL} = -2\mathcal{L}_{JI[L,K]} + 2\mathcal{L}_{MJ[K}\mathcal{L}^{M}{}_{IL]} + \mathcal{Q}_{JI[L,K]} - \mathcal{Q}_{MJ[K}\mathcal{L}^{M}{}_{IL]} + \mathcal{Q}_{MI[K}\mathcal{L}^{M}{}_{JL]}.$$

Observing finally that the right hand side corresponds to $-\mathcal{R}_{[JI]KL}$ concludes the proof. □

Alternatively, the *mixed-variant (2,2)-version of the curvature tensor* follows from the

Definition 4.14. The mixed-variant (2,2)-curvature tensor is obtained by raising the second index of the underlying $\mathcal{R}^I{}_{JKL}$. Using the representation of the partial derivative of the metric for a non-metric connection renders eventually

$$\mathcal{R}^{IJ}{}_{KL} = 2\mathcal{L}^{IJ}{}_{[L,K]} - 2\mathcal{L}^I{}_{M[K}\mathcal{L}^{JM}{}_{L]} + 2\mathcal{L}^I{}_{M[K}\mathcal{Q}^{MJ}{}_{L]}.$$

Observe the occurrence of a term involving the non-metricity tensor. □

For a non-metric connection the mixed-variant (2,2)-version of the curvature tensor displays the following symmetries

$$\mathcal{R}^{IJ}{}_{(KL)} = 0 \quad \text{and} \quad \mathcal{R}^{(IJ)}{}_{KL} = -\mathcal{Q}^{IJ}{}_{[K|L]} - \frac{1}{2}\mathcal{Q}^{IJ}{}_M\mathcal{T}^M{}_{KL}.$$

The symmetry in the indices I, J on the right hand side is an immediate consequence from the (left) symmetry of the non-metricity tensor.

Proof. On the one hand this result is consistent with $\mathcal{R}^{(IJ)}{}_{KL} = \mathcal{M}^{IM}\mathcal{M}^{JN}\mathcal{R}_{(MN)KL}$ since

$$\begin{aligned}
\mathcal{M}^{IM}\mathcal{M}^{JN}\mathcal{Q}_{MN[K|L]} &= \mathcal{Q}^{IJ}{}_{[K|L]} - \mathcal{Q}_M{}^J{}_{[K}\mathcal{M}^{IM}{}_{|L]} - \mathcal{Q}^I{}_{N[K}\mathcal{M}^{NJ}{}_{|L]} \\
&= \mathcal{Q}^{IJ}{}_{[K|L]} - \mathcal{Q}_M{}^J{}_{[K}\mathcal{Q}^{IM}{}_{L]} - \mathcal{Q}^I{}_{N[K}\mathcal{Q}^{NJ}{}_{L]} \\
&= \mathcal{Q}^{IJ}{}_{[K|L]} + \mathcal{Q}^{MJ}{}_{[L}\mathcal{Q}^I{}_{MK]} - \mathcal{Q}^I{}_{N[K}\mathcal{Q}^{NJ}{}_{L]} \equiv \mathcal{Q}^{IJ}{}_{[K|L]}.
\end{aligned}$$

On the other hand, upon considering $\mathcal{M}^{IJ}{}_{|K} = \mathcal{Q}^{IJ}{}_K$, the above result is likewise consistent with computing the skew symmetric part of the second covariant derivative of $\mathcal{M}^{IJ}$ according to Eq. 3.85.1, i.e.

$$\mathcal{M}^{IJ}{}_{|[KL]} = -\mathcal{R}^{(IJ)}{}_{KL} - \frac{1}{2}\mathcal{Q}^{IJ}{}_M\mathcal{T}^M{}_{KL} = \mathcal{Q}^{IJ}{}_{[K|L]}.$$

□

Expanding the covariant derivative $\mathcal{Q}^{IJ}{}_{[K|L]}$ according to Eq. 3.21 into the partial derivative $\mathcal{Q}^{IJ}{}_{[K,L]}$ and three further terms containing the connection renders the alternative representation for the (left) symmetric contribution to the mixed-variant (2,2)-curvature tensor

$$\mathcal{R}^{(IJ)}{}_{KL} = -\mathcal{Q}^{IJ}{}_{[K,L]} + \mathcal{L}^I{}_{M[K}\mathcal{Q}^{JM}{}_{L]} + \mathcal{L}^J{}_{M[K}\mathcal{Q}^{IM}{}_{L]}.$$

As a consequence the (left) skew symmetric contribution to the mixed-variant (2,2)-curvature tensor may be computed from

$$\mathcal{R}^{[IJ]}{}_{KL} = \mathcal{R}^{IJ}{}_{KL} - \mathcal{R}^{(IJ)}{}_{KL} = -\mathcal{R}^{[JI]}{}_{KL} \tag{4.77}$$

and results in the final representation

$$\mathcal{R}^{[IJ]}{}_{KL} = 2\mathcal{L}^{IJ}{}_{[L,K]} - 2\mathcal{L}^{I}{}_{M[K}\mathcal{L}^{JM}{}_{L]} - \mathcal{Q}^{IJ}{}_{[L,K]} + \mathcal{L}^{I}{}_{M[K}\mathcal{Q}^{JM}{}_{L]} - \mathcal{L}^{J}{}_{M[K}\mathcal{Q}^{IM}{}_{L]}.$$

The skew symmetry of the above expression in I, J then follows straightforwardly from the

Proof. Consider the condition for a non-metric connection in terms of the non-metric Ricci postulate $\mathcal{M}^{IJ}{}_{,L} = -\mathcal{L}^{JI}{}_{L} - \mathcal{L}^{IJ}{}_{L} + \mathcal{Q}^{IJ}{}_{L}$ and equate the skew symmetric part of the second partial derivative of $\mathcal{M}^{IJ}$ to zero

$$\mathcal{M}^{IJ}{}_{,[LK]} = 0 \quad \rightsquigarrow \quad 2\mathcal{L}^{IJ}{}_{[L,K]} = -2\mathcal{L}^{JI}{}_{[L,K]} + 2\mathcal{Q}^{JI}{}_{[L,K]}.$$

Furthermore, from raising and lowering the index M it holds that

$$-2\mathcal{L}^{I}{}_{M[K}\mathcal{L}^{JM}{}_{L]} = 2\mathcal{L}^{J}{}_{M[K}\mathcal{L}^{IM}{}_{L]}.$$

Then by inserting the two above relations into the representation for $\mathcal{R}^{[IJ]}{}_{KL}$ renders eventually

$$\mathcal{R}^{[IJ]}{}_{KL} = -2\mathcal{L}^{JI}{}_{[L,K]} + 2\mathcal{L}^{J}{}_{M[K}\mathcal{L}^{IM}{}_{L]} + \mathcal{Q}^{JI}{}_{[L,K]} - \mathcal{L}^{J}{}_{M[K}\mathcal{Q}^{IM}{}_{L]} + \mathcal{L}^{I}{}_{M[K}\mathcal{Q}^{JM}{}_{L]}.$$

Observing finally that the right hand side corresponds to $-\mathcal{R}^{[JI]}{}_{KL}$ concludes the proof. □

Addendum 4.5. Identities for the Curvature Tensor

Lowering the first index in **C1** to **C3** by the covariant metric in the presence of non-metricity renders the identities:

C1$^{\flat\flat}$: $\mathcal{R}_{IJ(KL)} = 0$ and $\mathcal{R}_{(IJ)KL} = -\mathcal{Q}_{IJ[K|L]} - \frac{1}{2}\mathcal{Q}_{IJM}\mathcal{T}^{M}{}_{KL}$
C2$^{\flat\flat}$: $\mathcal{R}_{I[\![JKL]\!]} = -\mathcal{T}_{I[\![JK|L]\!]} + \mathcal{T}_{IM[\![J}\mathcal{T}^{M}{}_{KL]\!]} - \mathcal{Q}_{IM[\![J}\mathcal{T}^{M}{}_{KL]\!]}$
C3$^{\flat\flat}$: $\mathcal{R}_{IJ[\![KL|M]\!]} = \mathcal{R}_{IJN[\![K}\mathcal{T}^{N}{}_{LM]\!]} - \mathcal{M}_{IN}\mathcal{R}_{OJ[\![KL}\mathcal{Q}^{NO}{}_{M]\!]}$

Due to **C1** they may alternatively be expanded into

c2$^{\flat\flat}$:
$$\begin{aligned}\mathcal{R}_{IJKL} + \mathcal{R}_{ILJK} + \mathcal{R}_{IKLJ} = & -\left[\mathcal{T}_{IJK|L} + \mathcal{T}_{ILJ|K} + \mathcal{T}_{IKL|J}\right] \\ & + \left[\mathcal{T}_{IMJ}\mathcal{T}^{M}{}_{KL} + \mathcal{T}_{IML}\mathcal{T}^{M}{}_{JK} + \mathcal{T}_{IMK}\mathcal{T}^{M}{}_{LJ}\right] \\ & - \left[\mathcal{Q}_{IMJ}\mathcal{T}^{M}{}_{KL} + \mathcal{Q}_{IML}\mathcal{T}^{M}{}_{JK} + \mathcal{Q}_{IMK}\mathcal{T}^{M}{}_{LJ}\right]\end{aligned}$$

c3$^{\flat\flat}$:
$$\begin{aligned}\mathcal{R}_{IJKL|M} + \mathcal{R}_{IJMK|L} + \mathcal{R}_{IJLM|K} = & \\ & \left[\mathcal{R}_{IJNK}\mathcal{T}^{N}{}_{LM} + \mathcal{R}_{IJNM}\mathcal{T}^{N}{}_{KL} + \mathcal{R}_{IJNL}\mathcal{T}^{N}{}_{MK}\right] \\ & -\mathcal{M}_{IN}\left[\mathcal{R}_{OJKL}\mathcal{Q}^{NO}{}_{M} + \mathcal{R}_{OJMK}\mathcal{Q}^{NO}{}_{L} + \mathcal{R}_{OJLM}\mathcal{Q}^{NO}{}_{K}\right]\end{aligned}$$

C2$^{\flat\flat}$ (**c2**$^{\flat\flat}$) and **C3**$^{\flat\flat}$ (**c3**$^{\flat\flat}$) denote the fully covariant versions of the *first (algebraic) Bianchi identity* and the *second (differential) Bianchi identity* in the presence of non-metricity, respectively.

4.6.2 Ricci Tensors

For a manifold with non-metric connection the Ricci tensors follow from the

Definition 4.15. The fully covariant first and second Ricci tensors are related to the fully covariant (0,4)-curvature tensor by

$$\mathcal{R}_{JL} = \mathcal{M}^{IK}\mathcal{R}_{IJKL} \quad \text{and} \quad \mathcal{V}_{KL} = \mathcal{M}^{IJ}\mathcal{R}_{IJKL} = \mathcal{M}^{IJ}\mathcal{R}_{(IJ)KL}. \tag{4.78}$$

By exploiting the format of the fully covariant curvature tensor in Eq. 4.73 (that involves the non-metricity tensor) and by observing the rules for partial derivatives of the metric in a non-metrically connected manifold (space/geometry) the first and second Ricci tensor eventually result in a format that is unchanged as compared to Eq. 3.91

$$\mathcal{R}_{JL} = 2\mathcal{L}^K{}_{J[L,K]} + 2\mathcal{L}^K{}_{I[K}\mathcal{L}^I{}_{JL]} \quad \text{and} \quad \mathcal{V}_{KL} = 2\mathcal{L}^I{}_{I[L,K]}. \tag{4.79}$$

Moreover, the mixed-variant first Ricci tensor is computed from the mixed-variant (2,2)-curvature tensor as

$$\mathcal{R}^J{}_L = \delta_I{}^K\mathcal{R}^{IJ}{}_{KL} = 2\mathcal{L}^{KJ}{}_{[L,K]} - 2\mathcal{L}^K{}_{M[K}\mathcal{L}^{JM}{}_{L]} + 2\mathcal{L}^K{}_{M[K}\mathcal{Q}^{MJ}{}_{L]}. \tag{4.80}$$

Alternatively, the mixed-variant Ricci tensors are related to the fully covariant Ricci tensors by $\mathcal{R}^J{}_L = \mathcal{M}^{JN}\mathcal{R}_{NL}$ and $\mathcal{V}^K{}_L = \mathcal{M}^{KN}\mathcal{V}_{NL}$. □

Proof. Based on the fully covariant curvature tensor in Eq. 4.73 the first Ricci tensor writes as

$$\mathcal{R}_{JL} = 2\mathcal{M}^{IK}\mathcal{L}_{IJ[L,K]} - 2\mathcal{L}_M{}^K{}_{[K}\mathcal{L}^M{}_{JL]} + 2\mathcal{Q}_M{}^K{}_{[K}\mathcal{L}^M{}_{JL]}.$$

Expanding $2\mathcal{M}^{IK}\mathcal{L}_{IJ[L,K]}$ by the help of $\mathcal{M}^{IJ}{}_{,K} = \mathcal{Q}^{IJ}{}_K - 2\mathcal{L}^{(IJ)}{}_K$ into

$$2\mathcal{L}^K{}_{J[L,K]} - 2\mathcal{L}_{IJ[L}\mathcal{M}^{IK}{}_{,K]} = 2\mathcal{L}^K{}_{J[L,K]} + 2\left[\mathcal{L}^K{}_{I[K} + \mathcal{L}_I{}^K{}_{[K} - \mathcal{Q}_I{}^K{}_{[K]}\right]\mathcal{L}^I{}_{JL]}$$

reduces the fully covariant first Ricci tensor to its format in Eq. 4.79. □

It is interesting to note that the fully covariant second Ricci tensor is alternatively expressed entirely in terms of the non-metricity covector

$$\mathcal{V}_{KL} = \mathcal{Q}_{[L,K]}. \tag{4.81}$$

Proof. Based on the (left) symmetric contribution of the fully covariant curvature tensor in Eq. 4.74, the second Ricci tensor results in

$$\mathcal{V}_{KL} = -\mathcal{M}^{IJ}\mathcal{Q}_{IJ[K|L]} - \frac{1}{2}\mathcal{Q}^I{}_{IM}\mathcal{T}^M{}_{KL}.$$

Expanding $\mathcal{M}^{IJ}\mathcal{Q}_{IJ[K|L]}$ by the help of $\mathcal{M}^{IJ}{}_{|L} = \mathcal{Q}^{IJ}{}_L$ into

$$\mathcal{Q}^I{}_{I[K|L]} - \mathcal{Q}_{IJ[K}\mathcal{M}^{IJ}{}_{|L]} = \mathcal{Q}^I{}_{I[K|L]} - \mathcal{Q}_{IJ[K}\mathcal{Q}^{IJ}{}_{L]} \equiv \mathcal{Q}^I{}_{I[K|L]}$$

renders the fully covariant second Ricci tensor finally in terms of the non-metricity covector and the torsion tensor

$$\mathcal{V}_{KL} = -\mathcal{Q}_{[K|L]} - \frac{1}{2}\mathcal{Q}_M\mathcal{T}^M{}_{KL} = -\mathcal{Q}_{[K,L]}.$$

In the last step the definition of the covariant derivative of a covector was exploited. It is interesting to note that this result is in agreement with Eq. 4.79.2 due to the decomposition of the connection into its Riemann part, the distention and the contortion together with their symmetry properties

$$2\mathcal{L}^I{}_{I[L,K]} = 2\big[\mathcal{M}^{IJ}[\mathcal{M}_{IJ[L} + \mathcal{O}_{IJ[L} + \mathcal{K}_{IJ[L]}]\big]_{,K]} = \mathcal{Q}_{[L,K]}. \tag{4.82}$$

The term involving the Riemann connection $2\mathcal{M}_{(IJ)L} = \mathcal{M}_{IJ,L}$ vanishes due to the symmetry of second partial derivatives $\mathcal{M}_{IJ,[LK]} = 0$ (and $\mathcal{M}^{IJ}{}_{,[K}\mathcal{M}_{IJ,L]} = 0$), the term involving the contortion $2\mathcal{K}_{(IJ)L} = 0$ vanishes due to its left symmetry, and the term involving the distention $2\mathcal{O}_{(IJ)L} = \mathcal{Q}_{IJL}$ reduces to the non-metricity tensor, respectively to the non-metricity covector $\mathcal{M}^{IJ}\mathcal{Q}_{IJL} = \mathcal{Q}_L$. □

Remark 4.19. Based on the (left) symmetric contribution to the mixed-variant (2,2)-curvature tensor the corresponding mixed-variant first Ricci tensor ${}^{()}\mathcal{R}^J{}_L$ reads as

$${}^{()}\mathcal{R}^J{}_L := \mathcal{R}^{(KJ)}{}_{KL} = -\mathcal{Q}^{JK}{}_{[K|L]} - \frac{1}{2}\mathcal{Q}^{JK}{}_M\mathcal{T}^M{}_{KL}. \tag{4.83}$$

Accordingly, the mixed-variant first Ricci tensor corresponding to the (left) skew symmetric contribution to the mixed-variant (2,2)-curvature tensor follows as ${}^{[]}\mathcal{R}^J{}_L := \mathcal{R}^J{}_L - {}^{()}\mathcal{R}^J{}_L$. □

Addendum 4.6. Identities for the Curvature Tensor

As a consequence of the identities $\mathbf{C1}^{\flat\flat}$, $\mathbf{C2}^{\flat\flat}$ ($\mathbf{c2}^{\flat\flat}$), $\mathbf{C3}^{\flat\flat}$ ($\mathbf{c3}^{\flat\flat}$) for the fully covariant curvature tensor, by contracting with $\mathcal{M}^{IK}$ the following identities hold for the *(first) Ricci tensor*

$\mathbf{R1}^{\flat\flat}$: $\mathcal{R}_{JL} = \mathcal{M}^{IK}\mathcal{R}_{IJKL} = -\mathcal{M}^{IK}\mathcal{R}_{IJLK}$
$\mathbf{R2}^{\flat\flat}$: $2\mathcal{R}_{[JL]} - \mathcal{V}_{JL} = \mathcal{T}^I{}_{JL|I} - 2\mathcal{T}_{[J|L]} - \mathcal{T}_M\mathcal{T}^M{}_{JL}$
$\mathbf{R3}^{\flat\flat}$: $2\mathcal{R}_{J[L|M]} + \mathcal{R}^I{}_{JLM|I} = -\mathcal{R}_{JN}\mathcal{T}^N{}_{LM} + 2\mathcal{R}^I{}_{JN[L}\mathcal{T}^N{}_{M]I}$

$\mathbf{R2}^{\flat\flat}$ and $\mathbf{R3}^{\flat\flat}$ are identical to the *contracted Bianchi identities* **R2** and **R3**.

Moreover, by contracting the fully covariant curvature tensor in the identities $\mathbf{C1}^{\flat\flat}$, $\mathbf{C2}^{\flat\flat}$ ($\mathbf{c2}^{\flat\flat}$), $\mathbf{C3}^{\flat\flat}$ ($\mathbf{c3}^{\flat\flat}$) with $\mathcal{M}^{IJ}$, additional identities hold for the *second Ricci tensor*

$$\begin{aligned}
&\mathbf{V1}^{\flat\flat}: && \mathcal{V}_{KL} = \mathcal{M}^{IJ}\mathcal{R}_{IJKL} = -\mathcal{M}^{IJ}\mathcal{R}_{IJLK} = -\mathcal{V}_{LK} \\
&\mathbf{V2}^{\flat\flat}: && \mathcal{V}_{KL} - 2\mathcal{R}_{[KL]} = -\mathcal{T}^{I}{}_{KL|I} + 2\mathcal{T}_{[K|L]} + \mathcal{T}_{M}\mathcal{T}^{M}{}_{KL} \\
&\mathbf{V3}^{\flat\flat}: && \mathcal{V}_{[\![KL|M]\!]} = \mathcal{V}_{N[\![K}\mathcal{T}^{N}{}_{LM]\!]}
\end{aligned}$$

$\mathbf{V2}^{\flat\flat}$ and $\mathbf{V3}^{\flat\flat}$ are identical to **V2** and **V3**, moreover identities $\mathbf{V2}^{\flat\flat}$ and $\mathbf{R2}^{\flat\flat}$ trivially coincide with a change of sign.

A further contraction of $\mathbf{R1}^{\flat\flat}$ to $\mathbf{R3}^{\flat\flat}$ with $\mathcal{M}^{JL}$ renders the additional identities

$$\begin{aligned}
&\mathbf{R1}^{\star\star}: && \mathcal{R} = \mathcal{M}^{IK}\mathcal{M}^{JL}\mathcal{R}_{IJKL} = -\mathcal{M}^{IK}\mathcal{M}^{JL}\mathcal{R}_{IJLK} \\
&\mathbf{R3}^{\star\star}: && \left[{}^{\Box}\mathcal{R}^{J}{}_{J}\delta^{L}{}_{M} - 2\,{}^{\Box}\mathcal{R}^{L}{}_{M}\right]_{|L} = -2\,{}^{\Box}\mathcal{R}^{L}{}_{N}\mathcal{T}^{N}{}_{LM} - \mathcal{R}_{[IL]NM}\mathcal{T}^{NLI} \\
& && \qquad +2\;\mathcal{R}^{L}{}_{[N}\mathcal{Q}^{N}{}_{LM]} + \mathcal{R}_{IL\,NM}\mathcal{Q}^{NLI}
\end{aligned}$$

Recall that ${}^{\Box}\mathcal{R}^{L}{}_{M} := \mathcal{R}^{[NL]}{}_{NM} = \mathcal{R}^{L}{}_{M} + \mathcal{Q}^{LN}{}_{[N|M]} + \frac{1}{2}\mathcal{Q}^{LK}{}_{N}\mathcal{T}^{N}{}_{KM}$ defines the (first) *Ricci tensor* based on the (left) skew symmetric contribution to the curvature tensor. Due to the skew symmetry of $\mathbf{R2}^{\flat\flat}$ in indices J, L a possible $\mathbf{R2}^{\star\star}$ is a trivial identity ($0 \equiv 0$), furthermore $\mathbf{R3}^{\star\star}$ is the *double-contracted second (differential) Bianchi identity* in the presence of non-metricity. Note that the first line of $\mathbf{R3}^{\star\star}$ is exclusively expressed in terms of the (left) skew symmetric contribution to the curvature tensor.

Note finally, that due to their skew symmetry in indices K, L a contraction of $\mathbf{V1}^{\flat\flat}$ to $\mathbf{V3}^{\flat\flat}$ with $\mathcal{M}^{KL}$ renders only trivial identities ($0 \equiv 0$).

4.6.3 Ricci Scalar

For a manifold with non-metric connection the Ricci scalar follows from the

Definition 4.16. The Ricci scalar is related to the fully covariant (0,4)-curvature tensor and to the (first) Ricci tensor by

$$\mathcal{R} = \mathcal{M}^{IK}\mathcal{M}^{JL}\mathcal{R}_{IJKL} = \mathcal{M}^{JL}\mathcal{R}_{JL} = \mathcal{R}^{L}{}_{L}. \tag{4.84}$$

By observing the format of the (first) Ricci tensor in either Eq. 3.91 or Eqs. 4.79 and 4.80 the Ricci scalar takes the explicit format

$$\mathcal{R} = 2\mathcal{L}^{ML}{}_{[L,M]} + 2\mathcal{L}^{LN}{}_{[M}\mathcal{L}^{M}{}_{NL]} - 2\mathcal{Q}^{LN}{}_{[M}\mathcal{L}^{M}{}_{NL]}. \tag{4.85}$$

Observe the additional term as compared to Eq. 4.40 which is due to the non-metricity. □

Remark 4.20. Note that the Ricci scalar ${}^{()}\mathcal{R}$ corresponding to the mixed-variant first Ricci tensor ${}^{()}\mathcal{R}^{J}{}_{L}$ based on the (left) symmetric contribution to the mixed-variant (2,2)-curvature tensor vanishes identically

$$ {}^{()}\mathcal{R} = {}^{()}\mathcal{R}^{L}{}_{L} = -\mathcal{Q}^{KL}{}_{[K|L]} - \frac{1}{2}\mathcal{Q}^{KL}{}_{M}\mathcal{T}^{M}{}_{KL} \equiv 0. $$

This identify follows immediately from the symmetries and skew symmetries in the indices K, L. As a consequence the Ricci scalar may likewise be computed solely from the mixed-variant first Ricci tensor ${}^{[]}\mathcal{R}^{J}{}_{L}$ based on the (left) skew symmetric contribution to the mixed-variant (2,2)-curvature tensor as $\mathcal{R} = {}^{[]}\mathcal{R} = {}^{[]}\mathcal{R}^{L}{}_{L}$. □

Addendum 4.7. Metrified Manifold

For a non-metrically connected manifold with metric $\mathcal{M}_{IJ}$ and non-metric connection $\mathcal{L}_{IJK}$ an associated *metrified* manifold may be introduced that is endowed with the identical metric but with an associated *metrified* connection

$$ \acute{\mathcal{L}}_{IJK} := \mathcal{L}_{IJK} - \frac{1}{2}\mathcal{Q}_{IJK} \quad \text{with} \quad \mathcal{M}_{IJ,K} = 2\acute{\mathcal{L}}_{(IJ)K}. $$

The metrified connection induces the corresponding metrified covariant derivative, covariant differential, and parallel transport, e.g. of a covector as

$$ \mathcal{V}_{J/K} := \mathcal{V}_{J,K} - \mathcal{V}_{I}\acute{\mathcal{L}}^{I}{}_{JK}, \quad \acute{\mathrm{D}}\mathcal{V}_{J} := \mathcal{V}_{J/K}\,\mathrm{d}\mathcal{X}^{K} \quad \text{and} \quad \acute{\mathrm{p}}\mathcal{V}_{J} := \mathcal{V}_{I}\acute{\mathcal{L}}^{I}{}_{JK}\,\mathrm{d}\mathcal{X}^{K}. $$

Consequently the inner product of two vectors together with the angle between a vector and the tangent to a geodesic in the metrified manifold are preserved, if both are evaluated upon parallel transport in terms of the metrified connection

$$ \acute{\mathrm{p}}[\mathcal{V}^{I}\mathcal{M}_{IJ}\mathcal{W}^{J}] = 0 \quad \text{and} \quad \frac{\mathrm{d}}{\mathrm{d}t}[\mathcal{V}^{I}\mathcal{M}_{IJ}\dot{\mathcal{X}}^{J}] = 0. $$

Based on the Ricci postulate of vanishing covariant derivative of the metric $\mathcal{M}_{IJ/K} = 0$ in terms of the metrified connection $\acute{\mathcal{L}}_{IJK}$, the latter allows decomposition $\acute{\mathcal{L}}_{IJK} = \acute{\mathcal{M}}_{IJK} + \acute{\mathcal{K}}_{IJK}$ into its Riemann part $\acute{\mathcal{M}}_{IJK}$, that is identical to the Riemann part $\acute{\mathcal{M}}_{IJK} \equiv \mathcal{M}_{IJK}$ of the original non-metric connection, and the metrified contortion

$$ \acute{\mathcal{K}}_{IJK} := -\acute{\mathcal{T}}_{\{IJK\}} = \mathcal{K}_{IJK} + \mathcal{O}_{IJK} - \frac{1}{2}\mathcal{Q}_{IJK} \quad \text{with} \quad \acute{\mathcal{T}}_{IJK} := \mathcal{T}_{IJK} - \mathcal{Q}_{I[JK]}. $$

Finally, based on the representation for additive connections in Eq. 3.94 the curvature tensor of the metrified connection may be related to the curvature tensor of the original non-metric connection

$$ \acute{\mathcal{R}}_{IJKL} = \mathcal{R}_{IJKL} - \mathcal{Q}_{IJ[L/K]} - \frac{1}{2}\mathcal{Q}_{MI[K}\mathcal{Q}^{M}{}_{JL]} + \frac{1}{2}\mathcal{Q}_{IJM}\acute{\mathcal{T}}^{M}{}_{KL}. $$

Here the fully covariant metrified curvature tensor is obtained by multiplying the mixed-variant metrified curvature tensor as obtained from Eq. 3.94 with the covariant metric and exploiting that the metrified covariant derivative of the metric $\mathcal{M}_{IJ/K} = 0$ vanishes.

Supplement 4.1. Exterior Calculus

Riemann and Contortion Forms:
The Ricci postulate of a metric (compatible) connection may be expressed in terms of the (0, 2)-tensor-valued connection one-form as

$$\frac{1}{2}\mathbf{d}m_{ij} = \frac{1}{2}m_{ij,k}\mathbf{g}^k = \mathbf{l}_{(ij)}.$$

As a consequence the *Riemann one-form* $\mathbf{m}_{ij}$ allows representation as

$$\mathbf{m}_{ij} = m_{ijk}\mathbf{g}^k := \frac{1}{2}\mathbf{d}m_{ij} - \mathbf{g}_{[i}\rfloor[\mathbf{d}m_{j]k} \wedge \mathbf{g}^k].$$

Likewise, the *contortion one-form* $\mathbf{k}_{ij}$ may be expressed in terms of the torsion two-form $\mathbf{t}_i$ as

$$\mathbf{k}_{ij} = -k_{ijk}\mathbf{g}^k := \mathbf{g}_{[i}\rfloor\mathbf{t}_{j]} - \frac{1}{2}\mathbf{g}_i\rfloor\mathbf{g}_j\rfloor[\mathbf{t}_k \wedge \mathbf{g}^k].$$

Vice versa, the torsion two-form $\mathbf{t}_i$ is finally computed from the contortion one-form $\mathbf{k}_{ij}$ as

$$\mathbf{t}_i = \mathbf{k}_{ij} \wedge \mathbf{g}^j \quad \text{with} \quad \mathbf{l}_{ij} = \mathbf{m}_{ij} - \mathbf{k}_{ij} \quad \text{and} \quad \mathbf{m}_{ij} \wedge \mathbf{g}^j = \mathbf{0}.$$

Bend-Twist Form:
Based on the representation for $\mathbf{l}_{(ij)}$ in terms of $\mathbf{d}m_{ij}$ an alternative representation of the connection one-form $\mathbf{l}_{ij}$ reads

$$\mathbf{l}_{ij} = \frac{1}{2}\mathbf{d}m_{ij} - \mathbf{b}_{ij} \quad \text{with} \quad \mathbf{b}_{ij} := -\mathbf{l}_{[ij]} = \mathbf{k}_{ij} + \mathbf{g}_{[i}\rfloor[\mathbf{d}m_{j]k} \wedge \mathbf{g}^k].$$

Here $\mathbf{b}_{ij} = b_{ijk}\mathbf{g}^k$ denotes the so-called *bend-twist one-form*, see e.g. [202].

Curvature Form:
Based on the Ricci postulate the exterior derivative of the (0, 2)-tensor-valued connection one-form $\mathbf{l}_{ij} = m_{in}\mathbf{l}^n{}_j$ expands as

$$\mathbf{d}\mathbf{l}_{ij} = \mathbf{l}_{im} \wedge \mathbf{l}^m{}_j + \mathbf{l}_{mi} \wedge \mathbf{l}^m{}_j + m_{in}\mathbf{d}\mathbf{l}^n{}_j.$$

Accordingly the (0, 2)-tensor-valued curvature two-form $\mathbf{r}_{ij} = m_{in}\mathbf{r}^n{}_j$ reads

$$\mathbf{r}_{ij} = \mathbf{d}\mathbf{l}_{ij} - \mathbf{l}_{mi} \wedge \mathbf{l}^m{}_j = \left[l_{ijl,k} - l_{mik}l^m{}_{jl}\right]\mathbf{g}^k \wedge \mathbf{g}^l = \frac{1}{2}r_{ijkl}\mathbf{g}^k \wedge \mathbf{g}^l.$$

Observe the different sequence of indices in the term quadratic in the connection one-form and its difference in sign for the (1, 1)- and the (0, 2)-tensor-valued curvature two-forms, respectively.

Cartan's Spiral Staircase:
Cartan's Spiral (Helical) Staircase [22] is a particular example of a three-dimensional metric manifold with torsion (and curvature), i.e. a Riemann-Cartan manifold, that is embedded in three-dimensional Euclidean space.

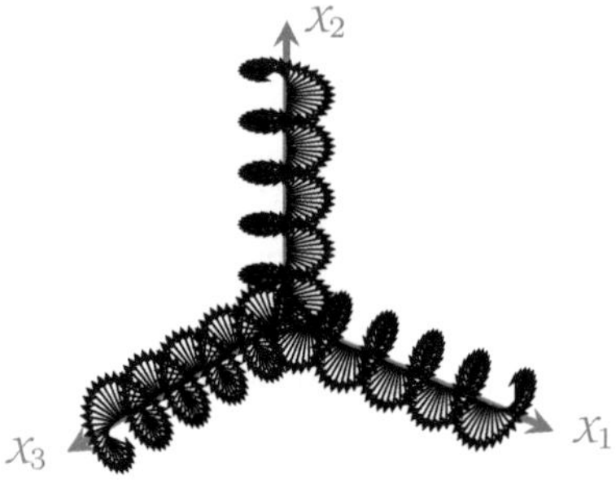

A comprehensive treatise along with various applications of Cartan's construction as proposed in [22] is provided by [79] and [105]. A visualization of the spiral staircase is given by rigid orthonormal triads that move helically along the Cartesian coordinate lines of the underlying Euclidean space, whereby the winding sense around the coordinate lines is positive and the winding 'speed' is constant (the present visualization is motivated by a similar representation in [79].

These helically winding triads then define the *trivial* frame and coframe that in this case coincide with the natural frame and coframe, i.e.

$$\mathbf{g}_i = \delta^I{}_i \mathbf{g}_I \quad \text{and} \quad \mathbf{g}^i = \delta^i{}_I \mathbf{g}^I .$$

Here the Kronecker symbols $\delta^I{}_i$ and $\delta^i{}_I$ simply exchange (upper and lower case) indices. Since the triads are rigid, the corresponding (constant) metric is Euclidean with

$$m_{ij} = \delta_{ij} \quad \text{and} \quad \mathcal{M}_{IJ} = \delta_{IJ} .$$

Again, δ_{ij} and δ_{IJ} simply denote Kronecker symbols. Parallel transport of vectors and covectors is now defined as preserving coefficients with respect to the rotated triads, i.e. vectors and covectors rigidly rotate with the triads, consequently the connection and its corresponding one-form read as

$$\ell_{ijk} = m_{in} \ell^n{}_{jk} = w\, e_{ijk} =: w_{ijk} \quad \text{and} \quad \mathbf{l}_{ij} = m_{in} \mathbf{l}^n{}_j = w_{ijk} \mathbf{g}^k =: \mathbf{w}_{ij} .$$

Here w denotes the constant winding 'speed' (angle per unit length) and e_{ijk} is the fully skew-symmetric permutation (Levi-Civita) symbol (Recall that here $m = \det[m_{ij}] = 1$). As a consequence of the fully skew-symmetric connection the parallel transport of a vector v^j along a particular $\mathbf{g}^k$ (that is aligned, for example, with the Cartesian coordinate line $\mathcal{X}^K$) results in an infinitesimal rotation with angle $w|\mathbf{g}^k|$ around the unit-direction $\mathbf{g}^k/|\mathbf{g}^k|$, whereby the winding sense is positive

$$\mathbf{d}v_i = m_{in} \mathbf{d}v^n = -w_{ijk} v^j \mathbf{g}^k = -\mathbf{w}_{ij} v^j .$$

It is remarked that due to the full skew-symmetry of the connection geodesic lines and coordinate lines coincide [79]. Based on the connection the torsion and its corresponding two-form read

$$t_{ijk} = m_{in} t^n{}_{jk} = 2w_{ijk} \quad \text{and} \quad \mathbf{t}_i = m_{in}\mathbf{t}^n = -\mathbf{w}_{ij} \wedge \mathbf{g}^j = w_{ijk}\mathbf{g}^j \wedge \mathbf{g}^k =: 2\mathbf{w}_i.$$

Finally the curvature and its corresponding two-form read

$$r_{ijkl} = m_{in} r^n{}_{jkl} = 2w_{im[k} w^m{}_{jl]} = w^2[\delta_{il}\delta_{kj} - \delta_{ik}\delta_{lj}]$$

and

$$\mathbf{r}_{ij} = m_{in}\mathbf{r}^n{}_j = \mathbf{w}_{im} \wedge \mathbf{w}^m{}_j = w^2\frac{1}{2}[\delta_{il}\delta_{kj} - \delta_{ik}\delta_{lj}]\mathbf{g}^k \wedge \mathbf{g}^l = -w^2\mathbf{g}_i \wedge \mathbf{g}_j.$$

The corresponding Ricci tensor and scalar read $r_{jl} = -2w^2\delta_{jl}$ and $r = -6w^2$, respectively. An alternative representation is obtained by considering the standard basis for zero-, one-, two- and three-forms in three dimensions

$$1, \quad \mathbf{g}^i, \quad \mathbf{g}^i \wedge \mathbf{g}^j, \quad \mathbf{g}^i \wedge \mathbf{g}^j \wedge \mathbf{g}^k,$$

along with the *Hodge dual basis* for three-, two-, one- and zero-forms

$$\begin{aligned}
{}^\star(1) &:= \frac{1}{3!} e_{ijk}\mathbf{g}^i \wedge \mathbf{g}^j \wedge \mathbf{g}^k,\\
{}^\star(\mathbf{g}_i) &:= \frac{1}{2!} e_{ijk}\mathbf{g}^j \wedge \mathbf{g}^k,\\
{}^\star(\mathbf{g}_i \wedge \mathbf{g}_j) &:= \frac{1}{1!} e_{ijk}\mathbf{g}^k,\\
{}^\star(\mathbf{g}_i \wedge \mathbf{g}_j \wedge \mathbf{g}_k) &:= \frac{1}{0!} e_{ijk}.
\end{aligned}$$

Here * denotes the Hodge star operator which applied twice renders

$${}^{\star\star}(1) = 1, \quad {}^{\star\star}(\mathbf{g}_i) = \mathbf{g}_i, \quad {}^{\star\star}(\mathbf{g}_i \wedge \mathbf{g}_j) = \mathbf{g}_i \wedge \mathbf{g}_j, \quad {}^{\star\star}(\mathbf{g}_i \wedge \mathbf{g}_j \wedge \mathbf{g}_k) = \mathbf{g}_i \wedge \mathbf{g}_j \wedge \mathbf{g}_k,$$

moreover the Euclidean metric with $m = 1$ has already been incorporated in the scaling of the Levi-Civita/permutation symbol. Then the above results take the following alternative representation

$$\mathbf{l}_{ij} = w^\star(\mathbf{g}_i \wedge \mathbf{g}_j) \quad \text{and} \quad \mathbf{t}_i = 2w^\star(\mathbf{g}_i) = \mathbf{k}_{ij} \wedge \mathbf{g}^j \quad \text{with} \quad \mathbf{k}_{ij} = -w^\star(\mathbf{g}_i \wedge \mathbf{g}_j).$$

In the above the contortion one-form $\mathbf{k}_{ij}$ has finally been introduced for completeness. By comparing the connection and the contortion it is revealed that the Riemann part of the connection $\mathbf{m}_{ij} = \mathbf{l}_{ij} + \mathbf{k}_{ij}$ (i.e. the Levi-Civita connection based on the metric) vanishes identically for the Cartan Spiral Staircase manifold.

Chapter 5
Representations in Four-, Three-, Two-Space

DAS PERLHUHN

Das Perlhuhn zählt: Eins, zwei, drei, vier ...
Was zählt es wohl, das gute Tier,
dort unter den dunklen Erlen?

Es zählt, von Wissensdrang gejückt,
(die es sowohl wie uns entzückt:)
die Anzahl seiner Perlen.

Christian Morgenstern

Abstract. The main results for metric and metrically connected manifolds of arbitrary dimension as presented in the previous chapter shall now be specialized and detailed to the cases of four-, three- and two-dimensional manifolds. (The case of one-dimensional manifolds, i.e. curves is not of relevance here, since the curvature tensor, that measures solely the internal geometry of a manifold, degenerates to zero in one dimension.) The above manifolds will be denoted as four-, three- and two-space, respectively, in the sequel. Four-, three- and two-spaces have most relevant applications such as gravitation, continuum physics (mechanics/electromagnetism) and surfaces. It is interesting to observe the similarities of the different cases that are only fully appreciated by such a deductive approach. For these concrete cases it proves convenient to exploit skew symmetries in the torsion tensor, the contortion tensor and the curvature tensor by introducing the dual torsion tensor, the dual contortion tensor, and the double-dual curvature tensor along with the corresponding double-dual Ricci tensor and scalar. Moreover, the relation between the dual torsion and contortion tensors as well as between the double-dual curvature tensor and the Einstein tensor will be highlighted. Finally, the Bianchi identities will be re-examined in terms of the dual quantities.

5.1 Representation in Four-Space

As the most prominent application of differential geometry, the theory of gravitation is formulated in terms of four-dimensional time-space coordinates. Thus it proves convenient to introduce the *fourth-order* Levi-Civita tensor $\mathcal{E}^{IJKL}$ or $\mathcal{E}_{IJKL}$, i.e. the fourth-order permutation tensor fully skew symmetric (all possible swaps of two indices with alternating signs) in the four indices I, J, K, L.

P. Steinmann, *Geometrical Foundations of Continuum Mechanics*,
Lecture Notes in Applied Mathematics and Mechanics 2, DOI: 10.1007/978-3-662-46460-1_5

Addendum 5.1. Fourth-Order Levi-Civita/Permutation Tensor

Swapping two indices of the fourth-order Levi-Civita or rather permutation tensor $\mathcal{E}^{IJKL}$ changes the sign, the absolute value of the non-zero coefficients is given by ${}^{-1}\sqrt{\mathcal{M}}$ with $\mathcal{M} = \det[\mathcal{M}_{IJ}] > 0$ the determinant of the matrix arrangement of the covariant metric coefficients $\mathcal{M}_{IJ}$. Here, for simplicity, a positive definite metric $\mathcal{M}_{IJ}$ is assumed, which is in contrast to the indefinite metric used in the theory of gravitation, see for example [36].

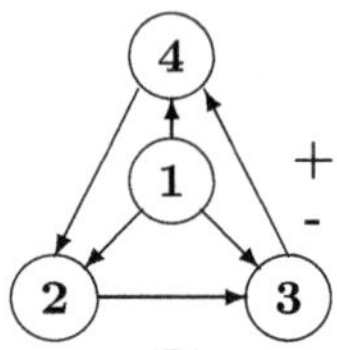

\+ 1423 1234 1342
\- 1243 1432 1324

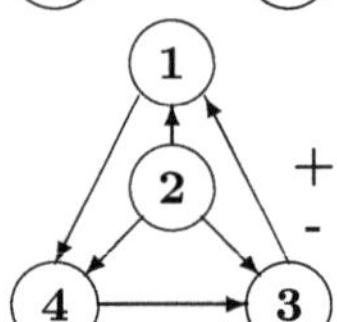

\+ 2143 2431 2314
\- 2413 2134 2341

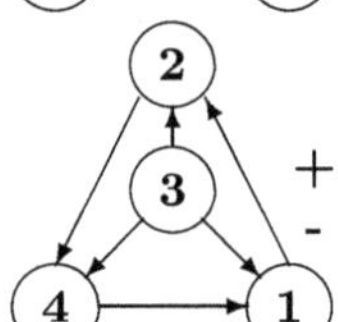

\+ 3241 3412 3124
\- 3421 3214 3142

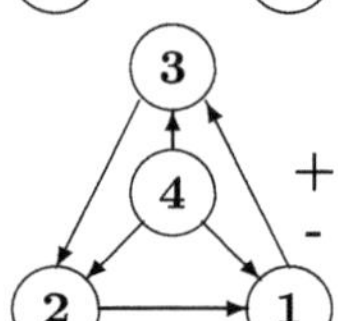

\+ 4321 4213 4132
\- 4231 4312 4123

The sketch in the above is a graphical representation of the non-zero coefficients of the fourth-order Levi-Civita or rather permutation tensor $\mathcal{E}^{IJKL}$: The indices might be thought as being arranged in the vertices of a tetragon. Following the arrows for the sequence $IJKL$ results in a positive sign, changing the orientation for the outer circuit in the JKL renders a negative sign.

Thereby, the fourth-order co- and contravariant permutation tensors satisfy the following relations in terms of δ-tensors:

(i) upon a *dyadic combination*

$$\begin{aligned} \mathcal{E}^{IJKL}\mathcal{E}_{MNOP} &=: 0!\,\delta^{IJKL}{}_{MNOP}, \\ \mathcal{E}_{IJKL}\mathcal{E}^{MNOP} &=: 0!\,\delta_{IJKL}{}^{MNOP}, \end{aligned} \tag{5.1}$$

(ii) upon a *single contraction*

$$\mathfrak{E}^{OIJK}\mathfrak{E}_{OLMN} = 0!\,\delta^{OIJK}{}_{OLMN} =: 1!\,\delta^{IJK}{}_{LMN},$$
$$\mathfrak{E}_{OIJK}\mathfrak{E}^{OLMN} = 0!\,\delta_{OIJK}{}^{OLMN} =: 1!\,\delta_{IJK}{}^{LMN}, \tag{5.2}$$

(iii) upon a *double contraction*

$$\mathfrak{E}^{MNIJ}\mathfrak{E}_{MNKL} = 1!\,\delta^{OIJ}{}_{OKL} =: 2!\,\delta^{IJ}{}_{KL},$$
$$\mathfrak{E}_{MNIJ}\mathfrak{E}^{MNKL} = 1!\,\delta_{OIJ}{}^{OKL} =: 2!\,\delta_{IJ}{}^{KL}, \tag{5.3}$$

(iv) upon a *triple contraction*

$$\mathfrak{E}^{MNOI}\mathfrak{E}_{MNOJ} = 2!\,\delta^{OI}{}_{OJ} = 3!\,\delta^{I}{}_{J},$$
$$\mathfrak{E}_{MNOI}\mathfrak{E}^{MNOJ} = 2!\,\delta_{OI}{}^{OJ} = 3!\,\delta_{I}{}^{J}, \tag{5.4}$$

(v) upon a *quadruple contraction*

$$\mathfrak{E}^{IJKL}\mathfrak{E}_{IJKL} = 3!\,\delta^{O}{}_{O} = 4! \quad \text{and} \quad \mathfrak{E}_{IJKL}\mathfrak{E}^{IJKL} = 3!\,\delta_{O}{}^{O} = 4!. \tag{5.5}$$

These relations will frequently be exploited in the sequel.

Addendum 5.2. δ-Tensors

The δ-tensors are related to the common second-order Kronecker-δ by:

$$\begin{aligned}
&\delta^{IJKL}{}_{MNOP} := \\
&+ \delta^I{}_M\delta^J{}_N\delta^K{}_O\delta^L{}_P - \delta^I{}_N\delta^J{}_M\delta^K{}_O\delta^L{}_P + \delta^I{}_N\delta^J{}_O\delta^K{}_M\delta^L{}_P - \delta^I{}_N\delta^J{}_O\delta^K{}_P\delta^L{}_M \\
&+ \delta^I{}_M\delta^J{}_O\delta^K{}_P\delta^L{}_N - \delta^I{}_O\delta^J{}_M\delta^K{}_P\delta^L{}_N + \delta^I{}_O\delta^J{}_P\delta^K{}_M\delta^L{}_N - \delta^I{}_O\delta^J{}_P\delta^K{}_N\delta^L{}_M \\
&+ \delta^I{}_M\delta^J{}_P\delta^K{}_N\delta^L{}_O - \delta^I{}_P\delta^J{}_M\delta^K{}_N\delta^L{}_O + \delta^I{}_P\delta^J{}_N\delta^K{}_M\delta^L{}_O - \delta^I{}_P\delta^J{}_N\delta^K{}_O\delta^L{}_M \\
&- \delta^I{}_M\delta^J{}_N\delta^K{}_P\delta^L{}_O + \delta^I{}_N\delta^J{}_M\delta^K{}_P\delta^L{}_O - \delta^I{}_O\delta^J{}_N\delta^K{}_M\delta^L{}_P + \delta^I{}_O\delta^J{}_N\delta^K{}_P\delta^L{}_M \\
&- \delta^I{}_M\delta^J{}_O\delta^K{}_N\delta^L{}_P + \delta^I{}_O\delta^J{}_M\delta^K{}_N\delta^L{}_P - \delta^I{}_P\delta^J{}_O\delta^K{}_M\delta^L{}_N + \delta^I{}_P\delta^J{}_O\delta^K{}_N\delta^L{}_M \\
&- \delta^I{}_M\delta^J{}_P\delta^K{}_O\delta^L{}_N + \delta^I{}_P\delta^J{}_M\delta^K{}_O\delta^L{}_N - \delta^I{}_N\delta^J{}_P\delta^K{}_M\delta^L{}_O + \delta^I{}_N\delta^J{}_P\delta^K{}_O\delta^L{}_M
\end{aligned}$$

$$\begin{aligned}
&\delta^{IJK}{}_{LMN} := \\
&+ \delta^I{}_L\delta^J{}_M\delta^K{}_N - \delta^I{}_M\delta^J{}_L\delta^K{}_N \\
&+ \delta^I{}_M\delta^J{}_N\delta^K{}_L - \delta^I{}_L\delta^J{}_N\delta^K{}_M \\
&+ \delta^I{}_N\delta^J{}_L\delta^K{}_M - \delta^I{}_N\delta^J{}_M\delta^K{}_L
\end{aligned}$$

$$\begin{aligned}
&\delta^{IJ}{}_{KL} := \\
&+ \delta^I{}_K\delta^J{}_L - \delta^I{}_L\delta^J{}_K
\end{aligned}$$

Observe the systematic construction of the δ-tensors by keeping the sequence of the contravariant indices and fully permuting the covariant indices.

5.1.1 Torsion and Contortion

In four-space, due to its (right) skew symmetry the third-order torsion tensor

$$\mathcal{T}^I{}_{MN} = \mathcal{T}^I{}_{[MN]}$$

may equivalently be represented by the third-order *dual torsion tensor*, the inverse relation following immediately from the properties of the fourth-order permutation symbol

$$^\star\mathcal{T}^{IJK} := -\frac{1}{2}\mathcal{T}^I{}_{OP}\mathcal{E}^{OPJK} \quad \text{and} \quad \mathcal{T}^I{}_{MN} = -\frac{1}{2}{}^\star\mathcal{T}^{IJK}\mathcal{E}_{JKMN}. \tag{5.6}$$

Accordingly the *torsion covector* has the representation

$$\mathcal{T}_I = -\frac{1}{2}\mathcal{E}_{IKLM}{}^\star\mathcal{T}^{KLM} = \frac{1}{4}\mathcal{E}_{IKLM}\mathcal{T}^K{}_{OP}\mathcal{E}^{OPLM} = \mathcal{T}^K{}_{IK}. \tag{5.7}$$

Moreover, in metrically connected four-space due to its (left) skew symmetry the third-order contortion tensor

$$\mathcal{K}_{IJ}{}^K = \mathcal{K}_{[IJ]}{}^K$$

may equivalently be represented by the third-order *dual contortion tensor*, again the inverse relation following from the properties of the fourth-order permutation symbol

$$^\star\mathcal{K}^{MNK} := -\frac{1}{2}\mathcal{E}^{MNOP}\mathcal{K}_{OP}{}^K \quad \text{and} \quad \mathcal{K}_{IJ}{}^K = -\frac{1}{2}\mathcal{E}_{IJMN}{}^\star\mathcal{K}^{MNK}. \tag{5.8}$$

Recall that the (right) skew symmetric part of the fully covariant contortion tensor coincides with half of the fully covariant torsion tensor, i.e.

$$\mathcal{K}_{I[JK]} = \frac{1}{2}\mathcal{T}_{IJK}.$$

Thus, as a consequence, in metrically connected four-space the third-order (mixed-variant) dual torsion and contortion tensors are related by

$$^\star\mathcal{T}_I{}^{JK} = 2\delta_I{}^{[J\star}\mathcal{K}^{MK]}{}_M - {}^\star\mathcal{K}^{JK}{}_I. \tag{5.9}$$

Proof. Inserting the relation between the torsion and the contortion into the definition of the dual torsion firstly results in

$$^\star\mathcal{T}_I{}^{JK} = -\mathcal{K}_{IOP}\mathcal{E}^{OPJK} = \frac{1}{2}\mathcal{E}_{IOMN}{}^\star\mathcal{K}^{MN}{}_P\mathcal{E}^{OPJK} = -\frac{1}{2}\delta_{IMN}{}^{PJK\star}\mathcal{K}^{MN}{}_P.$$

A further more detailed evaluation in terms of the δ-tensors then results in

$$\begin{aligned}-2^\star\mathcal{T}_I{}^{JK} = \delta_{IMN}{}^{PJK\star}\mathcal{K}^{MN}{}_P = [&\delta_I{}^P\delta_M{}^J\delta_N{}^K - \delta_I{}^J\delta_M{}^P\delta_N{}^K\\ &+ \delta_I{}^J\delta_M{}^K\delta_N{}^P - \delta_I{}^P\delta_M{}^K\delta_N{}^J\\ &+ \delta_I{}^K\delta_M{}^P\delta_N{}^J - \delta_I{}^K\delta_M{}^J\delta_N{}^P]^\star\mathcal{K}^{MN}{}_P\\ &= 2^\star\mathcal{K}^{JK}{}_I - 2\delta_I{}^{J\star}\mathcal{K}^{MK}{}_M + 2\delta_I{}^{K\star}\mathcal{K}^{MJ}{}_M.\end{aligned}$$

Observe that ${}^\star\mathcal{T}_M{}^{MK} = 2^\star\mathcal{K}^{MK}{}_M$ follows from contracting the first two indices in the above expression for the dual torsion. □

Finally, from the previous result the inverse relation between the dual contortion and the torsion tensors reads

$$^\star\mathcal{K}^{JK}{}_I = \delta_I{}^{[J\star}\mathcal{T}_M{}^{MK]} - {}^\star\mathcal{T}_I{}^{JK}. \tag{5.10}$$

These are Nye-type relations between the dual torsion and contortion as valid in four-space.

5.1.2 Curvature

The fourth-order (co-contravariant) curvature tensor based on a metric connection obeys the (minor) skew symmetries

$$\mathcal{R}_{IJ}{}^{KL} = \mathcal{R}_{IJ}{}^{[KL]} = \mathcal{R}_{[IJ]}{}^{[KL]}.$$

Thus it may likewise be represented in four-space by the fourth-order (contra-covariant) *double-dual curvature tensor*

$$^\star\mathcal{R}^{MN}{}_{OP}{}^\star := \frac{1}{4}\mathcal{E}^{MNIJ}\mathcal{R}_{IJ}{}^{KL}\mathcal{E}_{KLOP} =: \mathcal{K}^{MN}{}_{OP}. \tag{5.11}$$

Vice versa the fourth-order curvature tensor may be extracted from the fourth-order double-dual curvature tensor as

$$\mathcal{R}_{IJ}{}^{KL} = \frac{1}{4}\mathcal{E}_{IJMN}\mathcal{K}^{MN}{}_{OP}\mathcal{E}^{OPKL}. \tag{5.12}$$

Next the (co-contravariant) second-order *Ricci tensor* is computed from contracting the indices I, K of the curvature tensor, thus its relation to the double-dual curvature tensor follows from

$$\mathcal{R}_J{}^L := \mathcal{R}_{IJ}{}^{IL} = \frac{1}{4}\mathcal{E}_{MNIJ}\mathcal{E}^{OPIL}\mathcal{K}^{MN}{}_{OP} = \frac{1}{4}\delta_{JMN}{}^{OPL}\mathcal{K}^{MN}{}_{OP}. \tag{5.13}$$

The detailed relation of the Ricci tensor to the double-dual curvature tensor is then revealed as

$$
\begin{aligned}
4\mathcal{R}_J{}^L = \delta_{JMN}{}^{OPL}\mathcal{K}^{MN}{}_{OP} &= [\delta_J{}^O\delta_M{}^P\delta_N{}^L - \delta_J{}^P\delta_M{}^O\delta_N{}^L \\
&+ \delta_J{}^P\delta_M{}^L\delta_N{}^O - \delta_J{}^O\delta_M{}^L\delta_N{}^P \\
&+ \delta_J{}^L\delta_M{}^O\delta_N{}^P - \delta_J{}^L\delta_M{}^P\delta_N{}^O]\mathcal{K}^{MN}{}_{OP} \\
&= 2\mathcal{K}^{MN}{}_{MN}\delta_J{}^L - 4\mathcal{K}^{ML}{}_{MJ}.
\end{aligned}
$$

A further contraction of the indices J, L renders finally the *Ricci scalar* expressed in terms of the double-dual curvature tensor

$$
\mathcal{R} := \mathcal{R}_J{}^J = \mathcal{K}^{MN}{}_{MN}. \tag{5.14}
$$

Vice versa the (contra-covariant) second-order *double-dual Ricci tensor* is computed from contracting the indices M, O of the double-dual curvature tensor, thus its relation to the curvature tensor follows from

$$
\mathcal{K}^N{}_P := \mathcal{K}^{MN}{}_{MP} = \frac{1}{4}\mathcal{E}^{MNIJ}\mathcal{E}_{KLMP}\mathcal{R}_{IJ}{}^{KL} = \frac{1}{4}\delta^{NIJ}{}_{KLP}\mathcal{R}_{IJ}{}^{KL}. \tag{5.15}
$$

The detailed relation of the double-dual Ricci tensor to the curvature tensor is then revealed as

$$
\begin{aligned}
4\mathcal{K}^N{}_P = \delta^{NIJ}{}_{KLP}\mathcal{R}_{IJ}{}^{KL} &= [\delta^N{}_K\delta^I{}_L\delta^J{}_P - \delta^N{}_L\delta^I{}_K\delta^J{}_P \\
&+ \delta^N{}_L\delta^I{}_P\delta^J{}_K - \delta^N{}_K\delta^I{}_P\delta^J{}_L \\
&+ \delta^N{}_P\delta^I{}_K\delta^J{}_L - \delta^N{}_P\delta^I{}_L\delta^J{}_K]\mathcal{R}_{IJ}{}^{KL} \\
&= 2\mathcal{R}_{IJ}{}^{IJ}\delta^N{}_P - 4\mathcal{R}_{IP}{}^{IN}.
\end{aligned}
$$

A further contraction of the indices N, P renders finally the *double-dual Ricci scalar* expressed in terms of the curvature tensor

$$
\mathcal{K} := \mathcal{K}^N{}_N = \mathcal{R}_{IJ}{}^{IJ}. \tag{5.16}
$$

Summarizing the following relations between the Ricci and double-dual Ricci tensors and scalars hold in metrically connected four-space

$$
\mathcal{R}_J{}^L = \frac{1}{2}\mathcal{K}\,\delta^L{}_J - \mathcal{K}^L{}_J \quad \Leftrightarrow \quad \mathcal{K}^N{}_P = \frac{1}{2}\mathcal{R}\,\delta_P{}^N - \mathcal{R}_P{}^N \quad \text{with} \quad \mathcal{R} \equiv \mathcal{K}. \tag{5.17}
$$

Remark 5.1. The relation between the Ricci tensor (and scalar) and the double-dual Ricci tensor in Eq. 5.17.2 identifies the latter as the *Einstein tensor* as introduced in Eq. 4.44, i.e.

$$
\mathcal{G}^N{}_P \equiv \mathcal{K}^N{}_P = \frac{1}{2}\mathcal{R}\,\delta_P{}^N - \mathcal{R}_P{}^N. \tag{5.18}
$$

The relation between the Einstein and the Ricci tensor is a particularly nice example for the usefulness of having introduced the double-dual curvature tensor. □

5.1.3 Bianchi Identities

First Bianchi Identity. Projecting the fully contravariant double-dual curvature tensor by the fourth-order permutation tensor renders

$$\mathcal{E}_{IJKL}\mathcal{K}^{JKLO} = \mathcal{R}_{IJKL}\mathcal{E}^{JKLO}. \tag{5.19}$$

Proof. The result follows directly from expressing the double-dual curvature in terms of the curvature

$$\begin{aligned}\mathcal{E}_{IMNJ}\mathcal{K}^{MNJO} &= \frac{1}{4}\mathcal{E}_{IMNJ}\mathcal{E}^{MNRS}\mathcal{R}_{RSKL}\mathcal{E}^{KLJO} = \frac{1}{2}\delta_{IJ}{}^{RS}\mathcal{R}_{RSKL}\mathcal{E}^{JKLO}\\ &= \mathcal{R}_{IJKL}\mathcal{E}^{JKLO}.\end{aligned}$$

Finally the 'silent' indices M, N, J are exchanged by J, K, L on the left hand side. □

Due to the skew symmetries of the fourth-order Levi-Civita permutation tensor the right hand side of Eq. 5.19 may likewise be expressed by the left hand side of identity $\mathbf{C2}^\flat$

$$\mathcal{R}_{IJKL}\mathcal{E}^{JKLO} = \mathcal{R}_{I[\![JKL]\!]}\mathcal{E}^{JKLO}.$$

Thus, finally, the right hand side of Eq. 5.19 may be expressed in terms of the identity $\mathbf{C2}^\flat$, the resulting representation is thereby

$$\mathcal{E}_{IJKL}\mathcal{K}^{JKLO} = 2^\star\mathcal{T}_I{}^{MO}{}_{|M} - 2\mathcal{T}_{IMN}{}^\star\mathcal{T}^{MNO}. \tag{5.20}$$

Proof. Due to the full skew symmetry of the permutation tensor identity $\mathbf{C2}^\flat$ is equivalent to

$$\mathcal{R}_{IJKL}\mathcal{E}^{JKLO} = -\mathcal{T}_{IJK|L}\mathcal{E}^{JKLO} + \mathcal{T}_{IMJ}\mathcal{T}^M{}_{KL}\mathcal{E}^{JKLO}$$

Inserting finally the relation between the torsion and the dual torsion in terms of the permutation tensor concludes the proof. □

In summary Eq. 5.20 is an alternative representation of the *first (algebraic) Bianchi identity* in four-space stated in terms of the double-dual curvature tensor rather than in terms of the curvature tensor.

Second Bianchi Identity. Taking the covariant divergence of the fully contravariant double-dual curvature tensor renders

$$4\mathcal{K}^{MNOP}{}_{|P} = \mathcal{E}^{MNIJ}\mathcal{R}_{IJKL|P}\mathcal{E}^{KLOP}. \tag{5.21}$$

Due to the skew symmetries of the Levi-Civita permutation tensor the right hand side of Eq. 5.21 may likewise be expressed by the left hand side of identity $\mathbf{C3}^\flat$

$$\mathcal{E}^{MNIJ}\mathcal{R}_{IJKL|P}\mathcal{E}^{KLOP} = \mathcal{E}^{MNIJ}\mathcal{R}_{IJ[\![KL|P]\!]}\mathcal{E}^{KLOP}.$$

Thus, finally, the right hand side of Eq. 5.21 may be expressed in terms of the identity $\mathbf{C3}^\flat$, thereby the resulting representation follows eventually as

$$\mathcal{K}^{MNOP}{}_{|P} = \frac{1}{2}\mathcal{K}^{MNIJ}\mathcal{E}_{IJKL}{}^{\star}\mathcal{T}^{KLO}. \tag{5.22}$$

Proof. Due to the full skew symmetry of the permutation tensor identity $\mathbf{C3}^\flat$ is equivalent to

$$\mathcal{R}_{IJKL|P}\mathcal{E}^{KLOP} = -\mathcal{R}_{IJNK}\mathcal{T}^{N}{}_{LP}\mathcal{E}^{LPKO} = 2\mathcal{R}_{IJKL}{}^{\star}\mathcal{T}^{KLO}.$$

Projecting the result by $\mathcal{E}^{MNIJ}$ and expressing the curvature tensor in terms of the double-dual curvature tensor finally concludes the proof. □

In summary Eq. 5.22 is an alternative representation of the *second (differential) Bianchi identity* in four-space stated in terms of the double-dual curvature tensor rather than in terms of the curvature tensor.

5.2 Representation in Three-Space

In later sections the previous concepts of manifolds and differential geometry shall be applied to the geometrically nonlinear kinematics of continuum mechanics, thus all considerations may be restricted to three-dimensional (metric) spaces. To this end it proves convenient to introduce the *third-order Levi-Civita* tensor $\mathcal{E}^{IJK}$ or $\mathcal{E}_{IJK}$, i.e. the third-order permutation tensor fully skew symmetric (all possible swaps of two indices with alternating signs) in the three indices I, J, K.

Addendum 5.3. Swaping two indices of the third-order Levi-Civita or rather permutation tensor $\mathcal{E}^{IJK}$ changes the sign, the absolute value of the non-zero coefficients is given by $\sqrt[-1]{\mathcal{M}}$ with $\mathcal{M} = \det[\mathcal{M}_{IJ}] > 0$ the determinant of the matrix arrangement of the covariant metric coefficients $\mathcal{M}_{IJ}$.

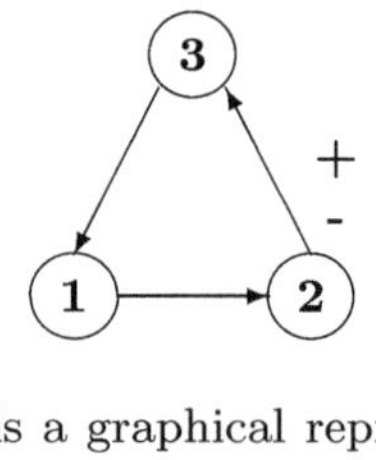

+	123	231	312
-	213	132	321

The sketch in the above is a graphical representation of the non-zero coefficients of the third-order Levi-Civita or rather permutation tensor $\mathcal{E}^{IJK}$: The indices might be thought as being arranged in the vertices of a triangle. Following the arrows for the sequence IJK results in a positive sign, changing the orientation for the circuit in the IJK renders a negative sign.

Thereby, the third-order co- and contravariant permutation tensors satisfy the following relations in terms of δ-tensors:

(i) upon a *dyadic combination*

$$\mathcal{E}^{IJK}\mathcal{E}_{LMN} =: 0!\,\delta^{IJK}{}_{LMN} \quad \text{and} \quad \mathcal{E}_{IJK}\mathcal{E}^{LMN} =: 0!\,\delta_{IJK}{}^{LMN}, \tag{5.23}$$

(ii) upon a *single contraction*

$$\begin{aligned} \mathcal{E}^{MIJ}\mathcal{E}_{MKL} &= 0!\,\delta^{MIJ}{}_{MKL} =: 1!\,\delta^{IJ}{}_{KL}, \\ \mathcal{E}_{MIJ}\mathcal{E}^{MKL} &= 0!\,\delta_{MIJ}{}^{MKL} =: 1!\,\delta_{IJ}{}^{KL}, \end{aligned} \tag{5.24}$$

(iii) upon a *double contraction*

$$\begin{aligned} \mathcal{E}^{MNI}\mathcal{E}_{MNJ} &= 1!\,\delta^{MI}{}_{MJ} = 2!\,\delta^{I}{}_{J}, \\ \mathcal{E}_{MNI}\mathcal{E}^{MNJ} &= 1!\,\delta_{MI}{}^{MJ} = 2!\,\delta_{I}{}^{J}, \end{aligned} \tag{5.25}$$

(iv) upon a *triple contraction*

$$\mathcal{E}^{IJK}\mathcal{E}_{IJK} = 2!\,\delta^{M}{}_{M} = 3! \quad \text{and} \quad \mathcal{E}_{IJK}\mathcal{E}^{IJK} = 2!\,\delta_{M}{}^{M} = 3!. \tag{5.26}$$

These relations will frequently be exploited in the sequel.

5.2.1 Torsion and Contortion

In three-space, due to its (right) skew symmetry the third-order torsion tensor may equivalently be represented by the second-order *dual torsion tensor*, the inverse relation following immediately from the properties of the third-order permutation symbol

$${}^{\star}\mathcal{T}^{IJ} := -\frac{1}{2}\mathcal{T}^{I}{}_{OP}\mathcal{E}^{OPJ} \quad \text{and} \quad \mathcal{T}^{I}{}_{MN} = -{}^{\star}\mathcal{T}^{IJ}\mathcal{E}_{JMN}. \tag{5.27}$$

Accordingly the *torsion covector* has the representation

$$\mathcal{T}_{I} = -\mathcal{E}_{IKL}{}^{\star}\mathcal{T}^{KL} = \frac{1}{2}\mathcal{E}_{IKL}\mathcal{T}^{K}{}_{MN}\mathcal{E}^{MNL} = \mathcal{T}^{K}{}_{IK}. \tag{5.28}$$

Remark 5.2. In passing it is interesting to note that in three-space the torsion covector may be interpreted as (twice) the *axial vector* of the skew symmetric part of the second-order (contravariant) dual torsion tensor. As a consequence it holds that

$$\mathcal{T}_{I}{}^{\star}\mathcal{T}^{[IJ]} = 0 \quad \text{thus} \quad \mathcal{T}_{I}{}^{\star}\mathcal{T}^{IJ} = \mathcal{T}_{I}{}^{\star}\mathcal{T}^{JI}. \tag{5.29}$$

In conclusion $\mathcal{T}_{I}$ is in the *null-space* of its associated *spin tensor* ${}^{\star}\mathcal{T}^{[IJ]}$. □

Moreover, in metrically connected three-space due to its (left) skew symmetry the third-order contortion tensor may equivalently be represented by the second-order *dual contortion tensor*, again the inverse relation following from the properties of the third-order permutation symbol

$$^{\star}\mathcal{K}^{MN} := -\frac{1}{2}\mathcal{E}^{MOP}\mathcal{K}_{OP}{}^{N} \quad \text{and} \quad \mathcal{K}_{IJ}{}^{N} = -\mathcal{E}_{IJM}{}^{\star}\mathcal{K}^{MN}. \tag{5.30}$$

As a consequence of the relation between the (right) skew symmetric part of the fully covariant contortion tensor and the fully covariant torsion tensor, the second-order (mixed-variant) dual torsion and contortion tensors in metrically connected three-space are related by

$$^{\star}\mathcal{T}_{I}{}^{J} = {}^{\star}\mathcal{K}^{M}{}_{M}\delta_{I}{}^{J} - {}^{\star}\mathcal{K}^{J}{}_{I}. \tag{5.31}$$

Proof. Inserting the relation between the torsion and the contortion into the definition of the dual torsion firstly results in

$$^{\star}\mathcal{T}_{I}{}^{J} = -\mathcal{K}_{IOP}\mathcal{E}^{OPJ} = \mathcal{E}_{IOM}{}^{\star}\mathcal{K}^{M}{}_{P}\mathcal{E}^{OPJ} = -\delta_{IM}{}^{PJ\star}\mathcal{K}^{M}{}_{P}.$$

A further more detailed evaluation in terms of the δ-tensors then results in

$$-{}^{\star}\mathcal{T}_{I}{}^{J} = \delta_{IM}{}^{PJ\star}\mathcal{K}^{M}{}_{P} = [\delta_{I}{}^{P}\delta_{M}{}^{J} - \delta_{I}{}^{J}\delta_{M}{}^{P}]^{\star}\mathcal{K}^{M}{}_{P} = {}^{\star}\mathcal{K}^{J}{}_{I} - \delta_{I}{}^{J\star}\mathcal{K}^{M}{}_{M}.$$

Observe that ${}^{\star}\mathcal{T}_{M}{}^{M} = 2{}^{\star}\mathcal{K}^{M}{}_{M}$ follows from contracting the first two indices in the above expression for the dual torsion. □

Finally, from the previous result the inverse relation between the dual contortion and torsion tensors reads

$$^{\star}\mathcal{K}^{J}{}_{I} = \frac{1}{2}{}^{\star}\mathcal{T}_{M}{}^{M}\delta_{I}{}^{J} - {}^{\star}\mathcal{T}_{I}{}^{J}. \tag{5.32}$$

It turns out that these are the analogues in differential geometry to the relations in Eqs. 1.22 and 1.23 between the stress free curvature of a crystalline lattice and the density of dislocations geometrically necessary to support this irreversible deformation as discovered by [152].

5.2.1.1 Anholonomic Objects

Part III will focus on the application of differential geometry to continuum mechanics in three-space. There, the anholonomic objects take the interpretation as dislocation densities. In this context it is observed that, due to their (right) skew symmetry, the (Cauchy- and Piola-type) anholonomic objects

$$a^{i}{}_{op} = \mathcal{D}^{i}{}_{OP}f^{O}{}_{o}f^{P}{}_{p} \quad \text{and} \quad \mathcal{A}^{I}{}_{OP} = d^{I}{}_{op}\mathcal{F}^{o}{}_{O}\mathcal{F}^{p}{}_{P} \tag{5.33}$$

may equivalently be represented in three-dimensional space by second-order (dual) anholonomic objects of Piola-type

$$^\star d^{Ij} := -\frac{1}{2} d^I{}_{op} e^{opj} \quad \text{and} \quad ^\star \mathcal{D}^{iJ} := -\frac{1}{2} \mathcal{D}^i{}_{OP} \mathcal{E}^{OPJ}, \tag{5.34}$$

and second-order (dual) anholonomic objects of Cauchy-type

$$^\star a^{ij} := -\frac{1}{2} a^i{}_{op} e^{opj} \quad \text{and} \quad ^\star \mathcal{A}^{IJ} := -\frac{1}{2} \mathcal{A}^I{}_{OP} \mathcal{E}^{OPJ}. \tag{5.35}$$

These anholonomic objects will be considered as dislocation density tensors in continuum mechanics. The inverse relations read

$$d^I{}_{mn} = -^\star d^{Ij} e_{jmn} \quad \text{and} \quad \mathcal{D}^i{}_{MN} = -^\star \mathcal{D}^{iJ} \mathcal{E}_{JMN}, \tag{5.36}$$

and

$$a^i{}_{mn} = -^\star a^{ij} e_{jmn} \quad \text{and} \quad \mathcal{A}^I{}_{MN} = -^\star \mathcal{A}^{IJ} \mathcal{E}_{JMN}. \tag{5.37}$$

The second-order (dual) anholonomic objects are thus related by

$$^\star a^{ij} = -\frac{1}{2} \mathcal{D}^i{}_{OP} f^O{}_o f^P{}_p e^{opj} = -j \left[\frac{1}{2} \mathcal{D}^i{}_{OP} \mathcal{E}^{OPJ} \right] \mathcal{F}^j{}_J = j\, ^\star \mathcal{D}^{iJ} \mathcal{F}^j{}_J, \tag{5.38}$$

and

$$^\star \mathcal{A}^{IJ} = -\frac{1}{2} d^I{}_{op} \mathcal{F}^o{}_O \mathcal{F}^p{}_P \mathcal{E}^{OPJ} = -\jmath \left[\frac{1}{2} d^I{}_{op} e^{opj} \right] f^J{}_j = \jmath\, ^\star d^{Ij} f^J{}_j. \tag{5.39}$$

Here the transformation behaviour of the third-order permutation symbol upon a change of coordinates has been involved

$$f^O{}_o f^P{}_p e^{opj} = j\, \mathcal{E}^{OPJ} \mathcal{F}^j{}_J \quad \text{and} \quad \mathcal{F}^o{}_O \mathcal{F}^p{}_P \mathcal{E}^{OPJ} = \jmath\, e^{opj} f^J{}_j, \tag{5.40}$$

whereby $j := \det[f^I{}_i]$ and $\jmath := \det[\mathcal{F}^i{}_I]$ denote the Jacobian determinants of the matrix arrangement of the tangent maps $f^I{}_i$ and $\mathcal{F}^i{}_I$, respectively. Note that the relations between the second-order Cauchy- and Piola-type anholonomic objects in Eq. 5.38 correspond to Piola transformations well-established in geometrically nonlinear continuum mechanics.

5.2.2 Curvature

The fourth-order (co-contravariant) curvature tensor based on a metric connection obeys (minor) skew symmetries. Thus it may likewise be represented in three-space by the second-order (contra-covariant) *double-dual curvature tensor*

$$^\star \mathcal{R}^M{}_N{}^\star := \frac{1}{4} \mathcal{E}^{MIJ} \mathcal{R}_{IJ}{}^{KL} \mathcal{E}_{KLN} =: \mathcal{K}^M{}_N. \tag{5.41}$$

Vice versa the fourth-order curvature tensor may be extracted from the second-order double-dual curvature tensor as

$$\mathcal{R}_{IJ}{}^{KL} = \mathcal{E}_{IJM}\mathcal{K}^{M}{}_{N}\mathcal{E}^{NKL}. \tag{5.42}$$

Next the (co-contravariant) second-order *Ricci tensor* is computed from contracting the indices I, K of the curvature tensor, thus its relation to the double-dual curvature tensor follows from

$$\mathcal{R}_{J}{}^{L} := \mathcal{R}_{IJ}{}^{IL} = \mathcal{E}_{IJM}\mathcal{E}^{NIL}\mathcal{K}^{M}{}_{N} = \delta_{JM}{}^{LN}\mathcal{K}^{M}{}_{N}. \tag{5.43}$$

The detailed relation of the Ricci tensor to the double-dual curvature tensor is then revealed as

$$\mathcal{R}_{J}{}^{L} = \delta_{JM}{}^{LN}\mathcal{K}^{M}{}_{N} = [\delta_{J}{}^{L}\delta_{M}{}^{N} - \delta_{J}{}^{N}\delta_{M}{}^{L}]\mathcal{K}^{M}{}_{N} = \mathcal{K}^{M}{}_{M}\delta_{J}{}^{L} - \mathcal{K}^{L}{}_{J}.$$

A further contraction of the indices J, L renders finally the *Ricci scalar* expressed in terms of the double-dual curvature tensor

$$\mathcal{R} := \mathcal{R}_{J}{}^{J} = 2\mathcal{K}^{M}{}_{M}. \tag{5.44}$$

Vice versa, the second-order (contra-covariant) double-dual curvature tensor is related to the curvature tensor as

$$\mathcal{K}^{M}{}_{N} := \frac{1}{4}\mathcal{E}^{MIJ}\mathcal{E}_{KLN}\mathcal{R}_{IJ}{}^{KL} = \frac{1}{4}\delta^{MIJ}{}_{KLN}\mathcal{R}_{IJ}{}^{KL}. \tag{5.45}$$

The detailed relation of the double-dual curvature tensor to the curvature tensor is then revealed as

$$\begin{aligned} 4\mathcal{K}^{M}{}_{N} = \delta^{MIJ}{}_{KLN}\mathcal{R}_{IJ}{}^{KL} &= [\delta^{M}{}_{K}\delta^{I}{}_{L}\delta^{J}{}_{N} - \delta^{M}{}_{L}\delta^{I}{}_{K}\delta^{J}{}_{N} \\ &+ \delta^{M}{}_{L}\delta^{I}{}_{N}\delta^{J}{}_{K} - \delta^{M}{}_{K}\delta^{I}{}_{N}\delta^{J}{}_{L} \\ &+ \delta^{M}{}_{N}\delta^{I}{}_{K}\delta^{J}{}_{L} - \delta^{M}{}_{N}\delta^{I}{}_{L}\delta^{J}{}_{K}]\mathcal{R}_{IJ}{}^{KL} \\ &= 2\mathcal{R}_{IJ}{}^{IJ}\delta^{M}{}_{N} - 4\mathcal{R}_{IN}{}^{IM}. \end{aligned}$$

A further contraction of the indices M, N renders finally the *double-dual Ricci scalar* expressed in terms of the curvature tensor

$$\mathcal{K} := \mathcal{K}^{M}{}_{M} = \frac{1}{2}\mathcal{R}_{IJ}{}^{IJ}. \tag{5.46}$$

Summarizing the following relations between the Ricci tensor (and scalar) and the double-dual curvature tensor hold in metrically connected three-space

$$\mathcal{R}_{J}{}^{L} = \mathcal{K}\delta^{L}{}_{J} - \mathcal{K}^{L}{}_{J} \quad \Leftrightarrow \quad \mathcal{K}^{M}{}_{N} = \frac{1}{2}\mathcal{R}\delta_{N}{}^{M} - \mathcal{R}_{N}{}^{M} \quad \text{with} \quad \frac{1}{2}\mathcal{R} \equiv \mathcal{K}. \tag{5.47}$$

Remark 5.3. The relation between the Ricci tensor (and scalar) and the double-dual curvature tensor in Eq. 5.47.2 identifies the latter as the *Einstein tensor* as introduced in Eq. 4.44, i.e.

$$\mathcal{G}^M{}_N \equiv \mathcal{K}^M{}_N = \frac{1}{2}\mathcal{R}\,\delta_N{}^M - \mathcal{R}_N{}^M . \tag{5.48}$$

Observe the missing factor $\frac{1}{2}$ in the inverse relation in Eq. 5.47.1 that vice versa expresses the Ricci tensor in terms of the Einstein tensor. □

Remark 5.4. As a consequence of Eqs. 5.42 and 5.47.2, in three-space the curvature tensor may be represented in terms of the Ricci tensor and the Ricci scalar as

$$\mathcal{R}_{IJ}{}^{KL} = \frac{1}{2}\mathcal{R}\,\left[\delta_I{}^K\delta_J{}^L - \delta_I{}^L\delta_J{}^K\right] - \mathcal{E}_{IJM}\mathcal{R}_N{}^M\mathcal{E}^{NKL} .$$

Since the Ricci scalar derives from the Ricci tensor the curvature tensor in metrically connected and torsion free three-space is given in terms of only six independent coefficients. □

5.2.3 Bianchi Identities

First Bianchi Identity. Projecting the fully contravariant double-dual curvature tensor by the third-order permutation tensor renders

$$\mathcal{E}_{IJK}\mathcal{K}^{JK} = -\frac{1}{2}\mathcal{R}_{IJKL}\mathcal{E}^{JKL} . \tag{5.49}$$

Proof. The result follows directly from expressing the double-dual curvature in terms of the curvature

$$\begin{aligned}\mathcal{E}_{IMJ}\mathcal{K}^{MJ} &= \frac{1}{4}\mathcal{E}_{IMJ}\mathcal{E}^{MNO}\mathcal{R}_{NOKL}\mathcal{E}^{KLJ} = -\frac{1}{4}\delta_{IJ}{}^{NO}\mathcal{R}_{NOKL}\mathcal{E}^{JKL}\\ &= -\frac{1}{2}\mathcal{R}_{IJKL}\mathcal{E}^{JKL} .\end{aligned}$$

Finally the indices M, J are exchanged by J, K on the left hand side. □

Due to the skew symmetries of the fourth-order Levi-Civita permutation tensor the right hand side of Eq. 5.49 may likewise be expressed by the left hand side of identity $\mathbf{C2}^\flat$

$$\mathcal{R}_{IJKL}\mathcal{E}^{JKL} = \mathcal{R}_{I[\![JKL]\!]}\mathcal{E}^{JKL} .$$

Thus, finally, the right hand side of Eq. 5.49 may be expressed in terms of the identity $\mathbf{C2}^\flat$, the resulting representation is thereby

$$\mathcal{E}_{IJK}\mathcal{K}^{JK} = -{}^\star\mathcal{T}_I{}^M{}_{|M} + \mathcal{T}_{IMN}{}^\star\mathcal{T}^{MN} . \tag{5.50}$$

Proof. Due to the full skew symmetry of the permutation tensor identity $\mathbf{C2}^\flat$ is equivalent to

$$\mathcal{R}_{IJKL}\mathcal{E}^{JKL} = -\mathcal{T}_{IJK|L}\mathcal{E}^{JKL} + \mathcal{T}_{IMJ}\mathcal{T}^M{}_{KL}\mathcal{E}^{JKL}$$

Inserting finally the relation between the torsion and the dual torsion in terms of the permutation tensor concludes the proof. □

In summary Eq. 5.50 is an alternative representation of the contracted *first (algebraic) Bianchi identity* in three-space stated in terms of the double-dual curvature tensor rather than in terms of the curvature tensor. Finally it shall be observed that in three space the double-dual curvature tensor $\mathcal{K}^{JK}$ coincides with the Einstein tensor $\mathcal{G}^{JK}$. Note that in his Eq. 98 [6] rearranges Eq. 5.50 alternatively as one of a set of partial differential equations for the dual torsion and the Einstein tensor

$$^{\star}\mathcal{T}_I{}^M{}_{|M} + \mathcal{G}^{MN}\mathcal{E}_{MNI} + {}^{\star}\mathcal{T}_I{}^L\mathcal{E}_{LMN}{}^{\star}\mathcal{T}^{MN} = 0.$$

Remark 5.5. It is interesting to note that in three-space the co- and the contravariant version for the *axial vector* of the Einstein tensor follows as

$$\mathcal{G}_I := -\frac{1}{2}\mathcal{E}_{IJK}\mathcal{G}^{JK} \quad \text{and} \quad \mathcal{G}^I := -\frac{1}{2}\mathcal{E}^{IJK}\mathcal{G}_{JK}.$$

Thus the first (algebraic) Bianchi identity in Eq. 5.50 may alternatively be expressed as

$$2\mathcal{G}_I = {}^{\star}\mathcal{T}_I{}^M{}_{|M} - \mathcal{T}_{IMN}{}^{\star}\mathcal{T}^{MN} \quad \text{and} \quad 2\mathcal{G}^I = {}^{\star}\mathcal{T}^{IM}{}_{|M} - \mathcal{T}^I{}_{MN}{}^{\star}\mathcal{T}^{MN}.$$

Therefore the Einstein tensor is in particular symmetric for vanishing torsion, since its axial vector then vanishes identically, i.e.

$$\mathcal{G}_I \equiv 0 \quad \text{and} \quad \mathcal{G}^I \equiv 0. \tag{5.51}$$

However, the above is of course also generally true for a *flat geometry* with vanishing curvature tensor, since then $\mathcal{R}_{IJKL} = 0$ identically. □

Remark 5.6. Due to Eq. 4.44 the skew symmetric parts or rather the axial vectors of the Einstein tensor and the Ricci tensor coincide modulo the sign

$$\mathcal{G}^I = -\frac{1}{2}\mathcal{G}_{[LJ]}\mathcal{E}^{LJI} = \frac{1}{2}\mathcal{R}_{[JL]}\mathcal{E}^{LJI}.$$

Thus contracting the right hand side of $\mathbf{R2}^\flat$ with $\mathcal{E}^{LJI}$ (and dividing by 4) renders

$$\frac{1}{2}{}^{\star}\mathcal{T}^{MI}{}_{|M} + \frac{1}{2}\mathcal{T}_{J|M}\mathcal{E}^{JMI} - \frac{1}{2}\mathcal{T}_K{}^{\star}\mathcal{T}^{KI} = \frac{1}{2}{}^{\star}\mathcal{T}^{IM}{}_{|M} - \frac{1}{2}\mathcal{T}^I{}_{MN}{}^{\star}\mathcal{T}^{MN}.$$

With $\mathcal{T}_J = -\mathcal{E}_{JKL}{}^{\star}\mathcal{T}^{[KL]}$ the second term on the left hand side allows the representation $\mathcal{T}_{J|M}\mathcal{E}^{JMI} = -2{}^{\star}\mathcal{T}^{[MI]}{}_{|M}$. Moreover, since $\mathcal{T}_K$ is in the null-space of ${}^{\star}\mathcal{T}^{[KI]}$, the third term on the left hand side may be rewritten as $\mathcal{T}_K{}^{\star}\mathcal{T}^{KI} = \mathcal{T}_K{}^{\star}\mathcal{T}^{IK} = -\mathcal{E}_{KMN}{}^{\star}\mathcal{T}^{MN}{}^{\star}\mathcal{T}^{IK} = \mathcal{T}^I{}_{MN}{}^{\star}\mathcal{T}^{MN}$. Combining these intermediate relations renders the result on the right hand side.

Thus, identities $\mathbf{C2}^\flat$ and $\mathbf{R2}^\flat$ are consistent with the representation of the contracted first (algebraic) Bianchi identity in Eq. 5.50. □

Proof. Alternatively, a direct proof of the assertion in Eq. 5.50 follows in four steps:

Step 1: Write down the fully covariant and the mixed-variant curvature tensors as being defined before

$$\begin{aligned}\mathcal{R}_{IJKL} &= 2\mathcal{L}_{IJ[L,K]}\underline{-2\mathcal{L}_{MI[K}\mathcal{L}^{M}{}_{JL]}},\\ \mathcal{R}^{I}{}_{JKL} &= 2\mathcal{L}^{I}{}_{J[L,K]}\underline{+2\mathcal{L}^{I}{}_{M[K}\mathcal{L}^{M}{}_{JL]}}.\end{aligned}$$

Step 2: Relate the partial derivative of the connection to its covariant derivative

$$\begin{aligned}2\mathcal{L}_{IJ[L,K]} &= 2\mathcal{L}_{IJ[L|K]}\underline{+2\mathcal{L}_{MI[K}\mathcal{L}^{M}{}_{JL]}} - 2\mathcal{L}_{IM[K}\mathcal{L}^{M}{}_{JL]} - 2\mathcal{L}_{IJM}\mathcal{L}^{M}{}_{[KL]},\\ 2\mathcal{L}^{I}{}_{J[L,K]} &= 2\mathcal{L}^{I}{}_{J[L|K]}\underline{-2\mathcal{L}^{I}{}_{M[K}\mathcal{L}^{M}{}_{JL]}} - 2\mathcal{L}^{I}{}_{M[K}\mathcal{L}^{M}{}_{JL]} - 2\mathcal{L}^{I}{}_{JM}\mathcal{L}^{M}{}_{[KL]}.\end{aligned}$$

Step 3: Combine the results in step 1 and step 2 and note that the underlined terms cancel out to produce an alternative representation of the curvature tensors in terms of the covariant rather than the partial derivative

$$\begin{aligned}\mathcal{R}_{IJKL} &= 2\mathcal{L}_{IJ[L|K]} - 2\mathcal{L}_{IM[K}\mathcal{L}^{M}{}_{JL]} - \mathcal{L}_{IJM}\mathcal{T}^{M}{}_{KL},\\ \mathcal{R}^{I}{}_{JKL} &= 2\mathcal{L}^{I}{}_{J[L|K]} - 2\mathcal{L}^{I}{}_{M[K}\mathcal{L}^{M}{}_{JL]} - \mathcal{L}^{I}{}_{JM}\mathcal{T}^{M}{}_{KL}.\end{aligned}$$

Step 4: Project the last result with $\mathcal{E}^{JKL}$

$$\begin{aligned}\mathcal{R}_{IJKL}\mathcal{E}^{JKL} &= \left[-2\mathcal{L}_{I[LJ]|K} + 2\mathcal{L}_{IMJ}\mathcal{L}^{M}{}_{[KL]} - \mathcal{L}_{IJM}\mathcal{T}^{M}{}_{KL}\right]\mathcal{E}^{JKL},\\ \mathcal{R}^{I}{}_{JKL}\mathcal{E}^{JKL} &= \left[-2\mathcal{L}^{I}{}_{[LJ]|K} + 2\mathcal{L}^{I}{}_{MJ}\mathcal{L}^{M}{}_{[KL]} - \mathcal{L}^{I}{}_{JM}\mathcal{T}^{M}{}_{KL}\right]\mathcal{E}^{JKL}.\end{aligned}$$

and exploiting $\mathcal{E}^{JKL}{}_{|M} = 0$ (being true since $\mathcal{E}^{JKL}$ is scaled by $\sqrt{\det[\mathcal{M}_{IJ}]}$) renders finally

$$\begin{aligned}-\frac{1}{2}\mathcal{R}_{IJKL}\mathcal{E}^{JKL} &= -{}^{\star}\mathcal{T}_{I}{}^{K}{}_{|K} + \mathcal{T}_{IMJ}{}^{\star}\mathcal{T}^{MJ},\\ -\frac{1}{2}\mathcal{R}^{I}{}_{JKL}\mathcal{E}^{JKL} &= -{}^{\star}\mathcal{T}^{IK}{}_{|K} + \mathcal{T}^{I}{}_{MJ}{}^{\star}\mathcal{T}^{MJ}.\end{aligned}$$

Here the skew symmetries in the various index pairings of J, K, L have been re-assigned. □

Second Bianchi Identity. Taking the covariant divergence of the fully contravariant double-dual curvature tensor renders

$$4\mathcal{K}^{MN}{}_{|N} = \mathcal{E}^{MIJ}\mathcal{R}_{IJKL|N}\mathcal{E}^{KLN}. \tag{5.52}$$

Due to the skew symmetries of the Levi-Civita permutation tensor the right hand side of Eq. 5.52 may likewise be expressed by the left hand side of identity $\mathbf{C3}^\flat$

$$\mathcal{E}^{MIJ}\mathcal{R}_{IJKL|N}\mathcal{E}^{KLN} = \mathcal{E}^{MIJ}\mathcal{R}_{IJ[\![KL|N]\!]}\mathcal{E}^{KLN}.$$

Thus, finally, the right hand side of Eq. 5.52 may be expressed in terms of the identity $\mathbf{C3}^\flat$, thereby the resulting representation follows eventually as

$$\mathcal{K}^{MN}{}_{|N} = -\mathcal{K}^{MN}\mathcal{E}_{NOP}{}^{\star}\mathcal{T}^{OP}. \tag{5.53}$$

Proof. Due to the full skew symmetry of the permutation tensor identity $\mathbf{C3}^\flat$ is equivalent to

$$\mathcal{R}_{IJKL|N}\mathcal{E}^{KLN} = \mathcal{R}_{IJMK}\mathcal{T}^{M}{}_{LN}\mathcal{E}^{LNK} = -2\mathcal{R}_{IJKL}{}^{\star}\mathcal{T}^{KL}.$$

Projecting the result by $\mathcal{E}^{MIJ}$ and expressing the curvature tensor in terms of the double-dual curvature tensor finally concludes the proof. □

In summary Eq. 5.53 is an alternative representation of the double-contracted *second (differential) Bianchi identity* in three-space stated in terms of the double-dual curvature tensor rather than in terms of the curvature tensor. Recall that in three space the double-dual curvature tensor $\mathcal{K}^{JK}$ coincides with the Einstein tensor $\mathcal{G}^{JK}$. Then Eq. 5.53 agrees with Eq. 99 as stated by [6] as one of a set of differential equations for the Einstein tensor and the dual torsion tensor

$$\mathcal{G}^{MN}{}_{|N} + \mathcal{G}^{MN}\mathcal{E}_{NOP}{}^{\star}\mathcal{T}^{OP} = 0.$$

Finally an alternative arrangement of Eq. 5.53 in terms of the torsion covector reads

$$\mathcal{G}^{MN}{}_{|N} = \mathcal{G}^{MN}\mathcal{T}_{N}.$$

Remark 5.7. The contravariant version of the left hand side of identity $\mathbf{R3}^\star$ may be expressed in terms of the fully contravariant Einstein tensor

$$[\mathcal{R}\,\mathcal{M}^{NM} - 2\mathcal{R}^{NM}]_{|N} = 2\mathcal{G}^{MN}{}_{|N}$$

and thus coincides with twice the left hand side of Eq. 5.53. Vice versa twice the right hand side of Eq. 5.53 may be rearranged as

$$\begin{aligned}
-2\mathcal{G}^{MN}\mathcal{E}_{NOP}{}^{\star}\mathcal{T}^{OP} &= \frac{1}{2}\delta^{MIJ}{}_{OPL}\mathcal{R}_{IJ}{}^{KL}\mathcal{T}_{K}{}^{OP} \\
&= \frac{1}{2}[\delta^{M}{}_{O}\delta^{I}{}_{P}\delta^{J}{}_{L} - \delta^{M}{}_{P}\delta^{I}{}_{O}\delta^{J}{}_{L} \\
&+ \quad \delta^{M}{}_{P}\delta^{I}{}_{L}\delta^{J}{}_{O} - \delta^{M}{}_{O}\delta^{I}{}_{L}\delta^{J}{}_{P} \\
&+ \quad \delta^{M}{}_{L}\delta^{I}{}_{O}\delta^{J}{}_{P} - \delta^{M}{}_{L}\delta^{I}{}_{P}\delta^{J}{}_{O}]\mathcal{R}_{IJ}{}^{KL}\mathcal{T}_{K}{}^{OP} \\
&= -2\mathcal{R}_{NO}\mathcal{T}^{ONM} - \mathcal{R}_{PON}{}^{M}\mathcal{T}^{NOP}
\end{aligned}$$

and therefore agrees with the contravariant version of the right hand side of identity $\mathbf{R3}^{\star}$.

Thus, identities $\mathbf{C3}^{\flat}$ and $\mathbf{R3}^{\star}$ are consistent with the representation of the double-contracted second (differential) Bianchi identity in Eq. 5.53. □

Proof. The direct proof of the result in Eq. 5.53 follows in 7 steps:

Step 1: Expand the covariant derivative of the curvature tensor

$$\mathcal{R}^{I}{}_{JKL|N} = \mathcal{R}^{I}{}_{JKL,N} + \mathcal{L}^{I}{}_{ON}\mathcal{R}^{O}{}_{JKL} - \mathcal{L}^{O}{}_{JN}\mathcal{R}^{I}{}_{OKL} - 2\mathcal{L}^{O}{}_{[KN}\mathcal{R}^{I}{}_{JOL]}$$

Step 2: Expanded the first term on the rhs of Step 1 and project with $\mathcal{E}^{KLN}$

$$\mathcal{R}^{I}{}_{JKL,N} = 2\left[\mathcal{L}^{I}{}_{JL,KN} + \mathcal{L}^{I}{}_{OK,N}\mathcal{L}^{O}{}_{JL} + \mathcal{L}^{I}{}_{OK}\mathcal{L}^{O}{}_{JL,N}\right]\mathcal{E}^{KLN}$$

resulting in

$$\mathcal{R}^{I}{}_{JKL,N} = 2\left[\underline{\mathcal{L}^{O}{}_{JL}\mathcal{L}^{I}{}_{OK,N}} + \underline{\underline{\mathcal{L}^{I}{}_{OK}\mathcal{L}^{O}{}_{JL,N}}}\right]\mathcal{E}^{KLN}.$$

Step 3: Expand the second term on the rhs of Step 1 and project with $\mathcal{E}^{KLN}$

$$+\mathcal{L}^{I}{}_{ON}\mathcal{R}^{O}{}_{JKL}\mathcal{E}^{KLN} = 2\mathcal{L}^{I}{}_{ON}\left[\mathcal{L}^{O}{}_{JL,K} + \mathcal{L}^{O}{}_{PK}\mathcal{L}^{P}{}_{JL}\right]\mathcal{E}^{KLN}$$

resulting in

$$+\mathcal{L}^{I}{}_{ON}\mathcal{R}^{O}{}_{JKL}\mathcal{E}^{KLN} = -2\left[\underline{\underline{\mathcal{L}^{I}{}_{OK}\mathcal{L}^{O}{}_{JL,N}}} + \underline{\underline{\underline{\mathcal{L}^{I}{}_{ON}\mathcal{L}^{O}{}_{PL}\mathcal{L}^{P}{}_{JK}}}}\right]\mathcal{E}^{KLN}$$

Step 4: Expand the third term on the rhs of Step 1 and project with $\mathcal{E}^{KLN}$

$$-\mathcal{L}^{O}{}_{JN}\mathcal{R}^{I}{}_{OKL}\mathcal{E}^{KLN} = -2\mathcal{L}^{O}{}_{JN}\left[\mathcal{L}^{I}{}_{OL,K} + \mathcal{L}^{I}{}_{PK}\mathcal{L}^{P}{}_{OL}\right]\mathcal{E}^{KLN}$$

resulting in

$$-\mathcal{L}^{O}{}_{JN}\mathcal{R}^{I}{}_{OKL}\mathcal{E}^{KLN} = -2\left[\underline{\mathcal{L}^{O}{}_{JL}\mathcal{L}^{I}{}_{OK,N}} - \underline{\underline{\underline{\mathcal{L}^{I}{}_{ON}\mathcal{L}^{O}{}_{PL}\mathcal{L}^{P}{}_{JK}}}}\right]\mathcal{E}^{KLN}$$

Step 5: Project the fourth term on the rhs of Step 1 with $\mathcal{E}^{KLN}$

$$-2\mathcal{L}^O{}_{[KN}\mathcal{R}^I{}_{JOL]}\mathcal{E}^{KLN} = 2\mathcal{L}^O{}_{KN}\mathcal{E}^{KNL}\mathcal{R}^I{}_{JOL} = -2\mathcal{R}^I{}_{JOL}{}^\star\mathcal{T}^{OL}.$$

Step 6: Sum the results in Steps 2 to 5 noting in particular that the sum of Steps 2 to 4 add to zero

$$\mathcal{R}^I{}_{JKL|N}\mathcal{E}^{KLN} = -2\mathcal{R}^I{}_{JOL}{}^\star\mathcal{T}^{OL}.$$

Step 7: Lower the first index by the covariant metric and project with $\mathcal{E}^{MIJ}$

$$\mathcal{E}^{MIJ}\mathcal{R}_{IJKL|N}\mathcal{E}^{KLN} = \mathcal{E}^{MIJ}\mathcal{R}_{IJOL}\mathcal{E}^{OLN}\mathcal{T}_N.$$

Dividing finally by four renders eventually

$$\mathcal{G}^{MN}{}_{|N} = \mathcal{G}^{MN}\mathcal{T}_N.$$

This concludes the proof. □

5.3 Representation in Two-Space

Differential geometry has initially been developed to describe and analyze the special case of two-dimensional manifolds, i.e. (curved) *surfaces*. This section is intended to relate the traditional terminology from the differential geometry of surfaces to the more general concepts of differential geometry on manifolds. Recall that the presentation in the forgoing chapters is independent of the dimension of the manifold, thus they may in particular be restricted to two-dimensional (metric) spaces. Here it proves convenient to introduce the *second-order Levi-Civita* tensor $\mathcal{E}^{IJ}$ or $\mathcal{E}_{IJ}$, i.e. the second-order permutation tensor fully skew symmetric in the two indices I, J.

Addendum 5.4. Swaping the indices of the second-order Levi-Civita or rather permutation tensor $\mathcal{E}^{IJ}$ changes the sign, the absolute value of the non-zero coefficients is given by $\sqrt{\mathcal{M}}^{-1}$ with $\mathcal{M} = \det[\mathcal{M}_{IJ}] > 0$ the determinant of the matrix arrangement of the covariant metric coefficients $\mathcal{M}_{IJ}$.

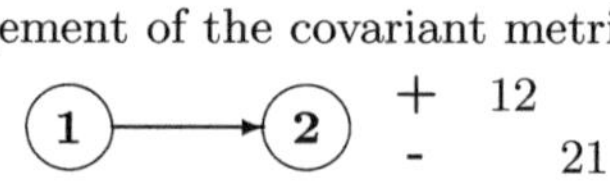

The sketch in the above is a graphical representation of the non-zero coefficients of the second-order Levi-Civita or rather permutation tensor $\mathcal{E}^{IJ}$: The indices might be thought as being arranged in the vertices of a line. Following the arrow for the sequence IJ results in a positive sign, the opposite sequence of the IJ renders a negative sign.

Thereby, the second-order co- and contravariant permutation tensors satisfy the following relations in terms of δ-tensors:

(i) upon a *dyadic combination*

$$\mathcal{E}^{IJ}\mathcal{E}_{KL} = 0!\,\delta^{IJ}{}_{KL} \quad \text{and} \quad \mathcal{E}_{IJ}\mathcal{E}^{KL} = 0!\,\delta_{IJ}{}^{KL}, \tag{5.54}$$

(ii) upon a *single contraction*

$$\begin{aligned} \mathcal{E}^{MI}\mathcal{E}_{MJ} &= 0!\,\delta^{MI}{}_{MJ} = 1!\,\delta^{I}{}_{J}, \\ \mathcal{E}_{MI}\mathcal{E}^{MJ} &= 0!\,\delta_{MI}{}^{MJ} = 1!\,\delta_{I}{}^{J}, \end{aligned} \tag{5.55}$$

(iii) upon a *double contraction*

$$\mathcal{E}^{IJ}\mathcal{E}_{IJ} = 1!\,\delta^{M}{}_{M} = 2! \quad \text{and} \quad \mathcal{E}_{IJ}\mathcal{E}^{IJ} = 1!\,\delta_{M}{}^{M} = 2!. \tag{5.56}$$

These relations will frequently be exploited in the sequel.

5.3.1 Torsion and Contortion

Following similar arguments as in the previous sections, the third-order torsion tensor in two-space may equivalently be represented by the first-order *dual torsion tensor*, the inverse relation then follows immediately from the properties of the second-order permutation symbol

$$^{\star}\mathcal{T}^{I} := -\frac{1}{2}\mathcal{T}^{I}{}_{OP}\mathcal{E}^{OP} \quad \text{and} \quad \mathcal{T}^{I}{}_{MN} = -{}^{\star}\mathcal{T}^{I}\mathcal{E}_{MN}. \tag{5.57}$$

Accordingly the *torsion covector* has the representation

$$\mathcal{T}_{I} = -\mathcal{E}_{IJ}{}^{\star}\mathcal{T}^{J} = \frac{1}{2}\mathcal{E}_{IJ}\mathcal{T}^{J}{}_{MN}\mathcal{E}^{MN} = \mathcal{T}^{J}{}_{IJ}. \tag{5.58}$$

Remark 5.8. In passing it is interesting to note that the torsion covector and the dual torsion tensor are orthogonal in two-space

$$\mathcal{T}_{I}{}^{\star}\mathcal{T}^{I} = -\mathcal{E}_{IJ}{}^{\star}\mathcal{T}^{J\star}\mathcal{T}^{I} = 0. \tag{5.59}$$

Likewise the torsion covector and the dual torsion tensor possess identical length. □

Moreover, in metrically connected two-space due to its (left) skew symmetry the third-order contortion tensor may equivalently be represented by the first-order *dual contortion tensor*, again the inverse relation results from the properties of the second-order permutation symbol

$$^\star\mathcal{K}^N := -\frac{1}{2}\mathcal{E}^{OP}\mathcal{K}_{OP}{}^N \quad \text{and} \quad \mathcal{K}_{IJ}{}^N = -\mathcal{E}_{IJ}{}^\star\mathcal{K}^N. \tag{5.60}$$

As a consequence of the relation between the (right) skew symmetric part of the fully covariant contortion tensor and the fully covariant torsion tensor, the first-order (covariant) dual torsion and contortion tensors in metrically connected two-space are related by

$$^\star\mathcal{T}_I = -{}^\star\mathcal{K}_I \quad \text{and} \quad {}^\star\mathcal{K}_I = -{}^\star\mathcal{T}_I. \tag{5.61}$$

Proof. Inserting the relation between the torsion and the contortion into the definition of the dual torsion results in

$$^\star\mathcal{T}_I = -\mathcal{K}_{IOP}\mathcal{E}^{OP} = \mathcal{E}_{IO}{}^\star\mathcal{K}_P\mathcal{E}^{OP} = -\delta_I{}^{P\star}\mathcal{K}_P = -{}^\star\mathcal{K}_I.$$

Thus in two-space the distinction between dual torsion and dual contortion and thus between torsion and contortion disappears (modulo the sign). □

5.3.2 Curvature

Due to its (minor) skew symmetries the fourth-order (co-contravariant) curvature tensor based on a metric connection may likewise be represented in two-dimensional space by the *double-dual curvature* (scalar)

$$^\star\mathcal{R}^\star := \frac{1}{4}\mathcal{E}^{IJ}\mathcal{R}_{IJ}{}^{KL}\mathcal{E}_{KL} =: \mathcal{K}. \tag{5.62}$$

Vice versa the fourth-order curvature tensor may be extracted from the double-dual curvature as

$$\mathcal{R}_{IJ}{}^{KL} = \mathcal{K}\,\mathcal{E}_{IJ}\mathcal{E}^{KL}. \tag{5.63}$$

Next the (co-contravariant) second-order *Ricci tensor* is computed from contracting the indices I, K of the curvature tensor, thus its relation to the double-dual curvature follows from

$$\mathcal{R}_J{}^L := \mathcal{R}_{IJ}{}^{IL} = \mathcal{K}\,\mathcal{E}_{IJ}\mathcal{E}^{IL} = \mathcal{K}\,\delta_J{}^L. \tag{5.64}$$

A further contraction of the indices J, L renders finally the *Ricci scalar* expressed in terms of the double-dual curvature

$$\mathcal{R} := \mathcal{R}_J{}^J = 2\mathcal{K}. \tag{5.65}$$

Vice versa the double-dual curvature is related to the curvature tensor by

$$\mathcal{K} := \frac{1}{4}\mathcal{E}^{IJ}\mathcal{E}_{KL}\mathcal{R}_{IJ}{}^{KL} = \frac{1}{4}\delta^{IJ}{}_{KL}\mathcal{R}_{IJ}{}^{KL}. \tag{5.66}$$

The detailed relation of the double-dual curvature to the curvature tensor is then revealed as

$$4\mathcal{K} = \delta^{IJ}{}_{KL}\mathcal{R}_{IJ}{}^{KL} = [\delta^{I}{}_{K}\delta^{J}{}_{L} - \delta^{I}{}_{L}\delta^{J}{}_{K}]\mathcal{R}_{IJ}{}^{KL} = 2\mathcal{R}_{IJ}{}^{IJ}.$$

Summarizing the following relations between the Ricci tensor (and scalar) and the double-dual curvature hold in metrically connected two-space

$$\mathcal{R}_{J}{}^{L} = \mathcal{K}\delta^{L}{}_{J} \quad \Leftrightarrow \quad \mathcal{K} = \frac{1}{2}\mathcal{R}. \tag{5.67}$$

Remark 5.9. Based on Eq. 5.67 it turns out that the *Einstein tensor* vanishes identically in two-space

$$\mathcal{G}^{J}{}_{L} = \frac{1}{2}\mathcal{R}\,\delta_{L}{}^{J} - \mathcal{R}_{L}{}^{J} \equiv 0.$$

The Einstein tensor is an example for tensorial quantities in differential geometry that do only show up for higher dimensions $n_{dm} > 2$. □

Remark 5.10. As a consequence of Eqs. 5.63 and 5.67.2, in two-space the curvature tensor may be represented exclusively in terms of the Ricci scalar as

$$\mathcal{R}_{IJ}{}^{KL} = \frac{1}{2}\mathcal{R}\left[\delta_{I}{}^{K}\delta_{J}{}^{L} - \delta_{I}{}^{L}\delta_{J}{}^{K}\right].$$

Thus the curvature tensor in metrically connected and torsion free two-space is given in terms of only one independent coefficient. Thereby, the only non-vanishing coefficient of the fully covariant $\mathcal{R}_{IJKL}$ (and of $\mathcal{R}_{JL}$) is

$$\mathcal{R}_{1212} = -\mathcal{R}_{2112} = \mathcal{R}_{2121} = -\mathcal{R}_{1221} \equiv \frac{1}{2}\mathcal{R}\,\mathcal{M}$$

with the coordinate representation

$$\mathcal{R}_{1212} = \mathcal{L}_{122,1} - \mathcal{L}_{121,2} - \mathcal{L}_{111}\mathcal{L}^{1}{}_{22} + \mathcal{L}_{112}\mathcal{L}^{1}{}_{21} - \mathcal{L}_{211}\mathcal{L}^{2}{}_{22} + \mathcal{L}_{212}\mathcal{L}^{2}{}_{21}.$$

and the determinant $\mathcal{M} := \det[\mathcal{M}_{IJ}] = \mathcal{M}_{11}\mathcal{M}_{22} - \mathcal{M}_{12}\mathcal{M}_{21}$ of the covariant metric coefficients. □

5.3.3 Bianchi Identities

In two-space the curvature tensor and the Ricci tensor take a particular simple format in terms of the Ricci scalar, i.e. the Ricci scalar multiplied by the fourth-order skew symmetric unit tensor and the Ricci scalar multiplied by half of the second-order unit tensor, respectively. Moreover the corresponding Einstein tensor vanishes identically. As a consequence the identities $\mathbf{R2}^{\flat}$, $\mathbf{R3}^{\flat}$

and **R3*** degenerate to trivial identities ($0 \equiv 0$). The proof is straightforward and is thus left to the reader.

5.3.4 Gaussian Curvature and Mean Curvature

Essential quantities characterizing the geometry of a surface, i.e. a two-dimensional Riemannian manifold, are given in the

Definition 5.1. The principal values of the *second fundamental form* $\mathcal{B}_{IJ}$ (the second order surface curvature tensor) of a surface correspond to the surface's *principal curvatures* κ_1, κ_2 (the curvature of the minimal and maximal circles that may be fitted to arbitrary plane 'normal cuts' of the smooth surface in the sense of the Meusnier theorem). The associated principal directions of $\mathcal{B}_{IJ}$ define the *principal curvature coordinates.* The *Gaussian curvature* $\mathcal{K} := \kappa_1\kappa_2$ of a surface follows as the determinant $\det[\mathcal{B}_I{}^J] = \det[\mathcal{B}_{IK}]\det[\mathcal{M}^{KJ}] =: \mathcal{B}/\mathcal{M}$ of its second fundamental form $\mathcal{B}^I{}_J$. Twice the *mean curvature* $2\mathcal{H} = \kappa_1 + \kappa_2$ of a surface follows as the trace $\mathcal{B}_{IJ}\mathcal{M}^{IJ}$ of its second fundamental form $\mathcal{B}_{IJ}$. □

Any smooth surface may locally (i.e. at $\bar{\mathcal{X}}^1 = \bar{\mathcal{X}}^2 = 0$) be approximated by a quadratic surface that is expressed in terms of (local) principal curvature coordinates $\bar{\mathcal{X}}^1, \bar{\mathcal{X}}^2$ as

$$\bar{\mathcal{X}}^3 = \frac{\kappa_1}{2}[\bar{\mathcal{X}}^1]^2 + \frac{\kappa_2}{2}[\bar{\mathcal{X}}^2]^2. \tag{5.68}$$

It may then be shown that the Gaussian curvature $\mathcal{K}$ and the Ricci curvature scalar $\mathcal{R}$ of the smooth surface are related by

$$\mathcal{K} := \kappa_1\kappa_2 \equiv \frac{\bar{\mathcal{R}}_{1212}}{\bar{\mathcal{M}}}|_{\bar{\mathcal{X}}^1=\bar{\mathcal{X}}^2=0} =: \frac{1}{2}\mathcal{R}. \tag{5.69}$$

Proof. Based on the parametrization of the quadratic approximation surface in Eq. 5.68 the differential of $\bar{\mathcal{X}}^3$ reads

$$\mathrm{d}\bar{\mathcal{X}}^3 = \kappa_1\bar{\mathcal{X}}^1\,\mathrm{d}\bar{\mathcal{X}}^1 + \kappa_2\bar{\mathcal{X}}^2\,\mathrm{d}\bar{\mathcal{X}}^2.$$

Consequently the square of $\mathrm{d}\bar{\mathcal{X}}^3$ is expressed as

$$[\,\mathrm{d}\bar{\mathcal{X}}^3]^2 = [\kappa_1\bar{\mathcal{X}}^1]^2[\,\mathrm{d}\bar{\mathcal{X}}^1]^2 + [\kappa_2\bar{\mathcal{X}}^2]^2[\,\mathrm{d}\bar{\mathcal{X}}^2]^2 + 2\kappa_1\kappa_2\bar{\mathcal{X}}^1\bar{\mathcal{X}}^2\,\mathrm{d}\bar{\mathcal{X}}^1\,\mathrm{d}\bar{\mathcal{X}}^2.$$

Thus the square of the arclength increment $\mathrm{d}\bar{\mathcal{S}}$ on the quadratic approximation surface follows as

$$\mathrm{d}\bar{\mathcal{S}}^2 = [1 + [\kappa_1\bar{\mathcal{X}}^1]^2][\,\mathrm{d}\bar{\mathcal{X}}^1]^2 + [1 + [\kappa_2\bar{\mathcal{X}}^2]^2][\,\mathrm{d}\bar{\mathcal{X}}^2]^2 + 2\kappa_1\kappa_2\bar{\mathcal{X}}^1\bar{\mathcal{X}}^2\,\mathrm{d}\bar{\mathcal{X}}^1\,\mathrm{d}\bar{\mathcal{X}}^2.$$

From the representation $\mathrm{d}\bar{\mathcal{S}}^2 = \mathrm{d}\bar{\mathcal{X}}^I\,\bar{\mathcal{M}}_{IJ}\,\mathrm{d}\bar{\mathcal{X}}^J$ the covariant metric coefficients $\bar{\mathcal{M}}_{IJ}$ may be determined as

$$[\bar{\mathcal{M}}_{IJ}] = \begin{bmatrix} 1+[\kappa_1\bar{X}^1]^2 & \kappa_1\kappa_2\bar{X}^1\bar{X}^2 \\ \kappa_1\kappa_2\bar{X}^1\bar{X}^2 & 1+[\kappa_2\bar{X}^2]^2 \end{bmatrix} \quad \text{with} \quad [\bar{\mathcal{M}}_{IJ}]|_{\bar{X}^1=\bar{X}^2=0} = \begin{bmatrix} 1 & 0 \\ 0 & 1 \end{bmatrix}.$$

Accordingly the determinant $\bar{\mathcal{M}}$ of the metric coefficients is found as

$$\bar{\mathcal{M}} = 1 + [\kappa_1\bar{X}^1]^2 + [\kappa_2\bar{X}^2]^2 \quad \text{with} \quad \bar{\mathcal{M}}|_{\bar{X}^1=\bar{X}^2=0} = 1.$$

Next the partial derivatives of the metric coefficients are computed by

$$[\bar{\mathcal{M}}_{IJ,1}] = \begin{bmatrix} 2[\kappa_1]^2\bar{X}^1 & \kappa_1\kappa_2\bar{X}^2 \\ \kappa_1\kappa_2\bar{X}^2 & 0 \end{bmatrix} \quad \text{and} \quad [\bar{\mathcal{M}}_{IJ,2}] = \begin{bmatrix} 0 & \kappa_1\kappa_2\bar{X}^1 \\ \kappa_1\kappa_2\bar{X}^1 & 2[\kappa_2]^2\bar{X}^2 \end{bmatrix}.$$

Based on these preliminaries the coefficients of the fully covariant Riemann connection follow

$$[\bar{\mathcal{M}}_{IJ1}] = \begin{bmatrix} [\kappa_1]^2\bar{X}^1 & 0 \\ \kappa_1\kappa_2\bar{X}^2 & 0 \end{bmatrix} \quad \text{with} \quad [\bar{\mathcal{M}}_{IJ1}]|_{\bar{X}^1=\bar{X}^2=0} = \begin{bmatrix} 0 & 0 \\ 0 & 0 \end{bmatrix}.$$

$$[\bar{\mathcal{M}}_{IJ2}] = \begin{bmatrix} 0 & \kappa_1\kappa_2\bar{X}^1 \\ 0 & [\kappa_2]^2\bar{X}^2 \end{bmatrix} \quad \text{with} \quad [\bar{\mathcal{M}}_{IJ2}]|_{\bar{X}^1=\bar{X}^2=0} = \begin{bmatrix} 0 & 0 \\ 0 & 0 \end{bmatrix}.$$

Obviously the mixed-variant Riemann connection thus reads at the origin

$$[\bar{\mathcal{M}}^I{}_{J1}]|_{\bar{X}^1=\bar{X}^2=0} = \begin{bmatrix} 0 & 0 \\ 0 & 0 \end{bmatrix} \quad \text{and} \quad [\bar{\mathcal{M}}^I{}_{J2}]|_{\bar{X}^1=\bar{X}^2=0} = \begin{bmatrix} 0 & 0 \\ 0 & 0 \end{bmatrix}.$$

Next the partial derivatives of the fully covariant Riemann connection are computed as

$$[\bar{\mathcal{M}}_{IJ1,1}] = \begin{bmatrix} [\kappa_1]^2 & 0 \\ 0 & 0 \end{bmatrix} \quad \text{and} \quad [\bar{\mathcal{M}}_{IJ1,2}] = \begin{bmatrix} 0 & 0 \\ \kappa_1\kappa_2 & 0 \end{bmatrix},$$

$$[\bar{\mathcal{M}}_{IJ2,1}] = \begin{bmatrix} 0 & \kappa_1\kappa_2 \\ 0 & 0 \end{bmatrix} \quad \text{and} \quad [\bar{\mathcal{M}}_{IJ2,2}] = \begin{bmatrix} 0 & 0 \\ 0 & [\kappa_2]^2 \end{bmatrix}.$$

Putting everything together renders finally the only non-vanishing coefficient of the curvature tensor related to the quadratic approximation surface

$$\bar{\mathcal{R}}_{1212}|_{\bar{X}^1=\bar{X}^2=0} = [\bar{\mathcal{M}}_{122,1} - \bar{\mathcal{M}}_{121,2}]|_{\bar{X}^1=\bar{X}^2=0} = \kappa_1\kappa_2.$$

As a consequence the Ricci scalar relates to the Gaussian curvature $\mathcal{K}$ by

$$\mathcal{R} = 2\kappa_1\kappa_2 =: 2\mathcal{K}.$$

The relation between the Ricci scalar and the Gaussian curvature lies at the heart of the *theorema egregium* by Gauss. □

Example 5.1. Geometry on a Sphere
In $\mathbb{E}^3$ the surface of a unit sphere, see Fig. 5.1, is given by the Cartesian coordinates r^a of the radial vector expressed in terms of spherical coordinates

$\mathcal{X}^1$ and $\mathcal{X}^2$ denoting the angle spanned by the vertical axis and the radial vector, and the angle in the equatorial plane, respectively

$$[r^a(\mathcal{X}^I)] = [\sin(\mathcal{X}^1)\cos(\mathcal{X}^2), \sin(\mathcal{X}^1)\sin(\mathcal{X}^2), \cos(\mathcal{X}^1)].$$

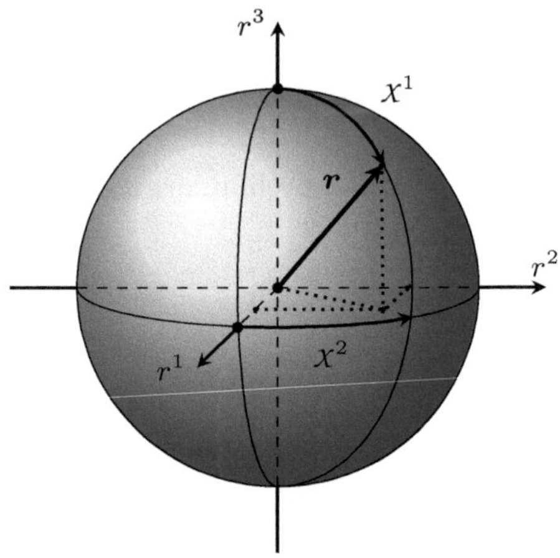

Fig. 5.1 Unit sphere $S(1)$ with $|\boldsymbol{r}| = 1$ in Euclidean space parameterized in terms of spherical coordinates $\mathcal{X}^1$ and $\mathcal{X}^2$. Thereby $\mathcal{X}^1$ denotes the angle spanned by the vertical r^3-axis and the radial vector $\boldsymbol{r}$ whereas $\mathcal{X}^2$ denotes the angle spanned by the r^1-axis and the projection of the radial vector $\boldsymbol{r}$ to the equatorial plane.

Thus the increment of the radial vector is computed as

$$[\,\mathrm{d}r^a] = \begin{bmatrix} \cos(\mathcal{X}^1)\cos(\mathcal{X}^2)\,\mathrm{d}\mathcal{X}^1 - \sin(\mathcal{X}^1)\sin(\mathcal{X}^2)\,\mathrm{d}\mathcal{X}^2 \\ \cos(\mathcal{X}^1)\sin(\mathcal{X}^2)\,\mathrm{d}\mathcal{X}^1 + \sin(\mathcal{X}^1)\cos(\mathcal{X}^2)\,\mathrm{d}\mathcal{X}^2 \\ -\sin(\mathcal{X}^1)\,\mathrm{d}\mathcal{X}^1 \end{bmatrix}.$$

The square of $\mathrm{d}r^a$, i.e. $\mathrm{d}r^a\,\delta_{ab}\,\mathrm{d}r^b$ reads

$$\begin{aligned}
&[\cos(\mathcal{X}^1)\cos(\mathcal{X}^2)\,\mathrm{d}\mathcal{X}^1]^2 - 2\cos(\mathcal{X}^1)\cos(\mathcal{X}^2)\sin(\mathcal{X}^1)\sin(\mathcal{X}^2)\,\mathrm{d}\mathcal{X}^1\,\mathrm{d}\mathcal{X}^2 + \\
&[\sin(\mathcal{X}^1)\sin(\mathcal{X}^2)\,\mathrm{d}\mathcal{X}^2]^2 + \\
&[\cos(\mathcal{X}^1)\sin(\mathcal{X}^2)\,\mathrm{d}\mathcal{X}^1]^2 + 2\cos(\mathcal{X}^1)\cos(\mathcal{X}^2)\sin(\mathcal{X}^1)\sin(\mathcal{X}^2)\,\mathrm{d}\mathcal{X}^1\,\mathrm{d}\mathcal{X}^2 + \\
&[\sin(\mathcal{X}^1)\cos(\mathcal{X}^2)\,\mathrm{d}\mathcal{X}^2]^2 + \\
&[\sin(\mathcal{X}^1)\,\mathrm{d}\mathcal{X}^1]^2
\end{aligned}$$

After straightforward simplifications, the square of the arclength increment on a unit sphere thus follows eventually as

$$\mathrm{d}\mathcal{S}^2 = \mathrm{d}r^a\,\delta_{ab}\,\mathrm{d}r^b = [\,\mathrm{d}\mathcal{X}^1]^2 + \sin^2(\mathcal{X}^1)[\,\mathrm{d}\mathcal{X}^2]^2.$$

Therefore the co- and contravariant metric coefficients follow as

$$[\mathcal{M}_{IJ}] = \begin{bmatrix} 1 & 0 \\ 0 & \sin^2(\mathcal{X}^1) \end{bmatrix} \quad \text{and} \quad [\mathcal{M}^{IJ}] = \begin{bmatrix} 1 & 0 \\ 0 & \sin^{-2}(\mathcal{X}^1) \end{bmatrix}.$$

Consequently the determinant $\mathcal{M}$ of the covariant metric coefficients reads

$$\mathcal{M} = \sin^2(\mathcal{X}^1).$$

Next, with the abbreviation $a := \sin(\mathcal{X}^1)\cos(\mathcal{X}^1)$, the partial derivatives of the covariant metric coefficients are computed

$$[\mathcal{M}_{IJ,1}] = \begin{bmatrix} 0 & 0 \\ 0 & 2a \end{bmatrix} \quad \text{and} \quad [\mathcal{M}_{IJ,2}] = \begin{bmatrix} 0 & 0 \\ 0 & 0 \end{bmatrix}.$$

Thus the fully covariant Riemann connection follows as

$$[\mathcal{M}_{IJ1}] = \begin{bmatrix} 0 & 0 \\ 0 & a \end{bmatrix} \quad \text{and} \quad [\mathcal{M}_{IJ2}] = \begin{bmatrix} 0 & -a \\ a & 0 \end{bmatrix}.$$

Consequently, with the abbreviation $b := \tan^{-1}(\mathcal{X}^1)$, the mixed-variant Riemann connection reads

$$[\mathcal{M}^I{}_{J1}] = \begin{bmatrix} 0 & 0 \\ 0 & b \end{bmatrix} \quad \text{and} \quad [\mathcal{M}^I{}_{J2}] = \begin{bmatrix} 0 & -a \\ b & 0 \end{bmatrix}.$$

Then, with the abbreviation $c := \cos^2(\mathcal{X}^1) - \sin^2(\mathcal{X}^1)$, the partial derivatives of the fully covariant Riemann connection are computed as

$$[\mathcal{M}_{IJ1,1}] = \begin{bmatrix} 0 & 0 \\ 0 & c \end{bmatrix}, [\mathcal{M}_{IJ1,2}] = [\mathcal{M}_{IJ2,2}] = \begin{bmatrix} 0 & 0 \\ 0 & 0 \end{bmatrix}, [\mathcal{M}_{IJ2,1}] = \begin{bmatrix} 0 & -c \\ c & 0 \end{bmatrix}.$$

Putting all ingredients together, the only non-vanishing coefficient of the curvature tensors follows as

$$\mathcal{R}_{1212} = -c - 0 - 0 + 0 - 0 + ab = \sin^2(\mathcal{X}^1) \equiv \mathcal{M}.$$

Consequently the Gaussian curvature is computed as the inverse of the (unit) radius

$$\frac{1}{2}\mathcal{R} = \mathcal{K} = 1.$$

The solutions to the two equations describing the geodesics on a sphere

$$\ddot{\mathcal{X}}^1 - a\dot{\mathcal{X}}^2\dot{\mathcal{X}}^2 = 0 \quad \text{and} \quad \ddot{\mathcal{X}}^2 + 2b\dot{\mathcal{X}}^1\dot{\mathcal{X}}^2 = 0$$

are given by, e.g., $\mathcal{X}^1 = \pi/2$ =constant and $\mathcal{X}^2$ =linear in the affine parameter, thus describing an equatorial big circle or by $\mathcal{X}^2$ =constant and $\mathcal{X}^1$ =linear in the affine parameter, thus describing a meridian big circle. □

Example 5.2. Geometry on a Cylinder
In $\mathbb{E}^3$ the surface of a unit cylinder, see Fig. 5.2, is given by the Cartesian coordinates r^a of the radial vector expressed in terms of cylindrical coordinates $\mathcal{X}^1$ and $\mathcal{X}^2$ denoting the circumferential angle and the axial height, respectively

$$[r^a(\mathcal{X}^I)] = [\cos(\mathcal{X}^1), \sin(\mathcal{X}^1), \mathcal{X}^2].$$

Thus the increment of the radius vector is computed as

$$[\,\mathrm{d}r^a] = \left[-\sin(\mathcal{X}^1)\,\mathrm{d}\mathcal{X}^1, \cos(\mathcal{X}^1)\,\mathrm{d}\mathcal{X}^1,\, \mathrm{d}\mathcal{X}^2\right].$$

The square of $\mathrm{d}r^a$, i.e. $\mathrm{d}r^a \delta_{ab}\,\mathrm{d}r^b$ then reads

$$[\sin(\mathcal{X}^1)\,\mathrm{d}\mathcal{X}^1]^2 + [\cos(\mathcal{X}^1)\,\mathrm{d}\mathcal{X}^1]^2 + [\,\mathrm{d}\mathcal{X}^2]^2.$$

After straightforward simplification, the square of the arclength increment on a unit cylinder thus follows eventually as

$$\mathrm{d}\mathcal{S}^2 = \mathrm{d}r^a \delta_{ab}\,\mathrm{d}r^b = [\,\mathrm{d}\mathcal{X}^1]^2 + [\,\mathrm{d}\mathcal{X}^2]^2.$$

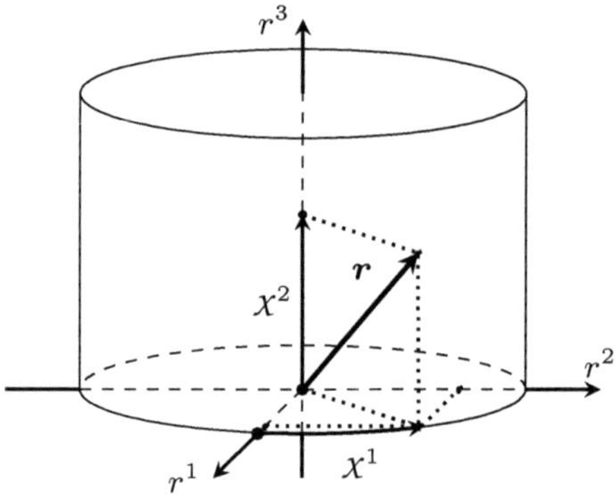

Fig. 5.2 Unit cylinder with $|\boldsymbol{r}| = 1$ in Euclidean space parameterized in terms of cylindrical coordinates $\mathcal{X}^1$ and $\mathcal{X}^2$. Thereby $\mathcal{X}^1$ denotes the circumferential angle spanned by the r^1-axis and the projection of $\boldsymbol{r}$ to the basis plane whereas $\mathcal{X}^2$ denotes the axial height.

Thus the constant metric coefficients follow trivially as

$$[\mathcal{M}_{IJ}] = \begin{bmatrix} 1 & 0 \\ 0 & 1 \end{bmatrix} \quad \text{and} \quad [\mathcal{M}^{IJ}] = \begin{bmatrix} 1 & 0 \\ 0 & 1 \end{bmatrix},$$

and render the corresponding constant determinant of the metric coefficients

$$\mathcal{M} = 1.$$

Consequently, the Gaussian curvature turns out to vanish

$$\frac{1}{2}\mathcal{R} = \mathcal{K} = 0$$

demonstrating that the curvature tensor or rather the Ricci scalar, respectively, measures only the internal curvature. □

Supplement 5.1. Geometry on a Surface

Surface theory has important applications in the modeling of shells and membranes both in engineering as well as in the sciences. Moreover, the investigation of the geometry on surfaces serves as a blueprint for the differential geometry on manifolds.

First Fundamental Form and Metric:

In $\mathbb{E}^3$ a two-dimensional (smooth) surface $\mathcal{S}$ is given by its position vectors $\boldsymbol{X}$ parameterized in terms of two surface coordinates χ^I with $I = 1, 2$ as

$$\boldsymbol{X} = \boldsymbol{X}(\chi^I).$$

The corresponding tangent vectors $\boldsymbol{A}_I \in T\mathcal{S}$ to the surface coordinate lines χ^I, i.e. the covariant (natural) surface basis vectors are given by

$$\boldsymbol{A}_I = \partial_{\chi^I} \boldsymbol{X} =: \boldsymbol{X}_{,I}.$$

The associated contravariant (dual) surface basis vectors $\boldsymbol{A}^I$ are defined by the Kronecker property $\delta^I{}_J = \boldsymbol{A}^I \cdot \boldsymbol{A}_J$ and are explicitly related to the covariant surface basis vectors $\boldsymbol{A}_I$ by the co- and contravariant surface metric coefficients $\mathcal{A}_{IJ}$ (coefficients of the first fundamental form for the surface) and $\mathcal{A}^{IJ}$, respectively, as

$$\boldsymbol{A}_I = \mathcal{A}_{IJ} \boldsymbol{A}^J \quad \text{with} \quad \mathcal{A}_{IJ} = \boldsymbol{A}_I \cdot \boldsymbol{A}_J = [\mathcal{A}^{IJ}]^{-1}$$

$$\boldsymbol{A}^I = \mathcal{A}^{IJ} \boldsymbol{A}_J \quad \text{with} \quad \mathcal{A}^{IJ} = \boldsymbol{A}^I \cdot \boldsymbol{A}^J = [\mathcal{A}_{IJ}]^{-1}.$$

Contra- and covariant base vectors $\boldsymbol{A}^3$ and $\boldsymbol{A}_3$, normal to $T\mathcal{S}$, are defined by

$$\boldsymbol{A}^3 := \boldsymbol{A}_1 \times \boldsymbol{A}_2 \quad \text{and} \quad \boldsymbol{A}_3 := [\mathcal{A}^{33}]^{-1} \boldsymbol{A}^3 \quad \text{so that} \quad \boldsymbol{A}^3 \cdot \boldsymbol{A}_3 = 1.$$

Thereby, the corresponding contra- and covariant metric coefficients $\mathcal{A}^{33}$ and $\mathcal{A}_{33}$ follow

$$\mathcal{A}^{33} = |\boldsymbol{A}_1 \times \boldsymbol{A}_2|^2 = \det[\mathcal{A}_{IJ}] = [\det[\mathcal{A}^{IJ}]]^{-1} = [\mathcal{A}_{33}]^{-1}.$$

Accordingly, the surface normal $\boldsymbol{N}$ is computed as

$$\boldsymbol{N} = [\mathcal{A}_{33}]^{1/2} \boldsymbol{A}^3 = [\mathcal{A}^{33}]^{1/2} \boldsymbol{A}_3.$$

Finally the surface metric tensor in invariant representation is given by

$$\boldsymbol{A} = \mathcal{A}_{IJ} \boldsymbol{A}^I \otimes \boldsymbol{A}^J \quad \text{and} \quad \boldsymbol{A}^{-1} = \mathcal{A}^{IJ} \boldsymbol{A}_I \otimes \boldsymbol{A}_J.$$

Second Fundamental Form and Curvature:

The (symmetric) covariant coefficients of the second-order surface curvature tensor (coefficients of the second fundamental form for the surface) then follow as

$$\mathcal{B}_{IJ} = \boldsymbol{A}_{I,J} \cdot \boldsymbol{N} = -\boldsymbol{A}_I \cdot \boldsymbol{N}_{,J} = \mathcal{B}_{JI} \quad \text{with} \quad \boldsymbol{B} = \mathcal{B}_{IJ} \boldsymbol{A}^I \otimes \boldsymbol{A}^J = -\boldsymbol{N}_{,J} \otimes \boldsymbol{A}^J.$$

Next, the Gauss and Weingarten formulae rendering the derivatives of the base vectors read

$$\boldsymbol{A}_{J,L} = \boldsymbol{A}_I \mathcal{A}^I{}_{JL} + \boldsymbol{N} \mathcal{B}_{JL} \quad \text{and} \quad \boldsymbol{N}_{,L} = -\boldsymbol{A}_I \mathcal{B}^I{}_L = -\boldsymbol{A}^I \mathcal{B}_{IL}.$$

Here $\mathcal{A}^I{}_{JK} := \boldsymbol{A}^I \cdot \boldsymbol{A}_{J,K}$ denote the (right symmetric) surface Christoffel symbols. Then the integrability conditions for $\boldsymbol{A}_{J,L}$ and $\boldsymbol{N}_{,L}$, i.e. the condition for the Gauss and Weingarten formulae to have solutions so that for given $\mathcal{A}_{IJ}$ and $\mathcal{B}_{IJ}$ there exists a surface such that $\mathcal{A}_{IJ}$ and $\mathcal{B}_{IJ}$ are indeed the coefficients of the corresponding fundamental forms, follow from $\boldsymbol{A}_{J,[LK]} \doteq \boldsymbol{0}$ and expand into

$$\boldsymbol{A}_I \left[\mathcal{A}^I{}_{J[L,K]} + \mathcal{A}^I{}_{M[K} \mathcal{A}^M{}_{JL]} - \mathcal{B}^I{}_{[K} \mathcal{B}_{JL]} \right] + \boldsymbol{N} \left[\mathcal{B}_{J[L,K]} - \mathcal{B}_{M[L} \mathcal{A}^M{}_{JK]} \right] \doteq \boldsymbol{0}.$$

Since $\boldsymbol{A}_I$ and $\boldsymbol{N}$ are linearly independent, the terms in the brackets have to vanish separately, i.e.

$$\mathcal{A}^I{}_{JKL} \doteq \mathcal{B}^I{}_K \mathcal{B}_{JL} - \mathcal{B}^I{}_L \mathcal{B}_{JK} \quad \text{and} \quad \mathcal{B}_{J[L|K]} \doteq 0.$$

The first condition expresses the fourth-order mixed-variant Riemann curvature tensor of the surface

$$\mathcal{A}^I{}_{JKL} := 2\mathcal{A}^I{}_{J[L,K]} + 2\mathcal{A}^I{}_{M[K} \mathcal{A}^M{}_{JL]}$$

in terms of the coefficients of the second fundamental form (resulting eventually in the *theorema egregium* of Gauss) whereas the second condition represents the Mainardi-Codazzi formula expressed in terms of the covariant derivative on the surface

$$\mathcal{B}_{J[L|K]} = \mathcal{B}_{J[L,K]} - \mathcal{B}_{M[L} \mathcal{A}^M{}_{JK]} - \mathcal{B}_{JM} \mathcal{A}^M{}_{[LK]} \equiv \mathcal{B}_{J[L,K]} - \mathcal{B}_{M[L} \mathcal{A}^M{}_{JK]}.$$

Here the (right) skew symmetry of the surface Christoffel symbols $\mathcal{A}^M{}_{[LK]} = 0$ has been incorporated. The only non-vanishing coefficient $\mathcal{A}_{1212}$ of the fully covariant Riemann curvature tensor of the surface $\mathcal{A}_{IJKL}$ then follows from

$$\mathcal{A}_{IJKL} = \mathcal{B}_{IK} \mathcal{B}_{JL} - \mathcal{B}_{IL} \mathcal{B}_{JK} \quad \rightsquigarrow \quad \mathcal{A}_{1212} = \mathcal{B}_{11} \mathcal{B}_{22} - \mathcal{B}_{12} \mathcal{B}_{21} =: \mathcal{B}$$

and is related to the Gaussian curvature of the surface by

$$\mathcal{K} = \kappa_1 \kappa_2 = \frac{\mathcal{B}}{\mathcal{A}} \equiv \frac{\mathcal{A}_{1212}}{\mathcal{A}}.$$

This is the *theorema egregium* of Gauss: since $\mathcal{A}_{1212}$ and $\mathcal{A} := \det[\mathcal{A}_{IJ}]$ are expressed exclusively in terms of the covariant metric coefficients and their partial derivatives, the Gaussian curvature of a surface may be determined solely by measurements on the surface. Thus no knowledge of the embedding space for the surface is necessary to determine its *internal geometry*. Obviously, the internal curvature of a surface as measured by $\mathcal{K}$ vanishes if at least one of the principal curvatures is zero, i.e. if the surface is developable on a plane. If only one principal curvatures is zero the surface possesses merely *external curvature* and thus is isometric to a flat two-dimensional Euclidean space, i.e. a plane. As an example for a surface with external curvature and vanishing internal curvature consider, e.g., the case of a cylindrical surface.

Curves on a Surface and Principal Curvature of the Surface:
In general a one-dimensional (smooth) parameter curve $\mathcal{C}$ on a (smooth) surface $\mathcal{S}$ is given by its position vectors $\boldsymbol{X}$ parameterized in terms of the arc length t

$$\boldsymbol{X} = \boldsymbol{X}(t) = \boldsymbol{X}(x^I(t)).$$

The corresponding *tangent* vector $\dot{\boldsymbol{X}} \in T\mathcal{C}$ to the curve, together with the *principal normal* vector $\boldsymbol{P}$, orthogonal to $T\mathcal{C}$ and pointing towards the center of the *curvature circle*, are defined by

$$\dot{\boldsymbol{X}} = \boldsymbol{A}_I \dot{x}^I \quad \text{and} \quad \kappa \boldsymbol{P} = \ddot{\boldsymbol{X}} = \boldsymbol{A}_{I,J} \dot{x}^I \dot{x}^J + \boldsymbol{A}_I \ddot{x}^I \quad \text{with} \quad \kappa = |\ddot{\boldsymbol{X}}|.$$

Here κ denotes the scalar valued curvature of the curve, i.e. the inverse of the *curvature radius* that is defined as the radius of the (curvature) circle that may locally be inscribed to the curve. The relation between $\ddot{\boldsymbol{X}}$ and $\boldsymbol{P}$ in terms of κ is one of the three Frénet-Serret formulae for the derivative of the curve basis vectors (the tangent, the principal normal, and the binormal). Due to the parametrization of the curve in its arc length t the tangent vector $\dot{\boldsymbol{X}}$ has unit length and the square of the curve's line element is computed as

$$\mathrm{d}\boldsymbol{X} \cdot \mathrm{d}\boldsymbol{X} = \dot{\boldsymbol{X}} \cdot \dot{\boldsymbol{X}}\, \mathrm{d}t^2 = \mathrm{d}t^2 = \boldsymbol{A}_I \cdot \boldsymbol{A}_J \, \mathrm{d}x^I \, \mathrm{d}x^J = \mathcal{A}_{IJ} \, \mathrm{d}x^I \, \mathrm{d}x^J.$$

Then the angle between the surface normal $\boldsymbol{N}$ and the principal normal $\boldsymbol{P}$ follows from their scalar product

$$\kappa \boldsymbol{N} \cdot \boldsymbol{P} = \boldsymbol{N} \cdot \boldsymbol{A}_{I,J} \dot{x}^I \dot{x}^J = \mathcal{B}_{IJ} \frac{\mathrm{d}x^I}{\mathrm{d}t} \frac{\mathrm{d}x^J}{\mathrm{d}t} = \frac{\mathcal{B}_{IJ} \, \mathrm{d}x^I \, \mathrm{d}x^J}{\mathcal{A}_{IJ} \, \mathrm{d}x^I \, \mathrm{d}x^J}.$$

The plane intersection curve of a surface with a plane passing through the normal $\boldsymbol{N}$ to the surface at a point $\mathcal{P}$ of the surface, i.e. $\boldsymbol{P} \| \boldsymbol{N}$, is in particular called a *normal section* of the surface. The tangent to the normal section evaluated at the same point $\mathcal{P}$ of the surface is denoted $\boldsymbol{M}$. The curvature κ_N of the normal section, i.e. the *normal curvature* then solely depends on the orientation of the tangent $\boldsymbol{M}$ at point $\mathcal{P}$ and may be expressed as the inverse of the *normal curvature radius*, i.e. the radius of the (normal curvature) circle that may locally be inscribed to the normal section. It then follows from the above that the normal curvature follows in terms of the first and second fundamental form as

$$\kappa_N := \kappa \boldsymbol{N} \cdot \boldsymbol{P} = \frac{\mathcal{B}_{IJ}\,\mathrm{d}X^I\,\mathrm{d}X^J}{\mathcal{A}_{IJ}\,\mathrm{d}X^I\,\mathrm{d}X^J} \quad \text{with} \quad \boldsymbol{P} \| \boldsymbol{N}.$$

The generalization of this result to the curvature of intersection curves (plane sections) of the surface with arbitrary planes passing through the tangent $\dot{\boldsymbol{X}}$ to the normal section at the point $\mathcal{P}$ of the surface is the Meusnier theorem:

The circles of curvature of all plane sections of the surface with common tangent at the point $\mathcal{P}$ lie on a sphere, i.e. the curvature sphere with radius coinciding with the curvature radius of the normal section, [93].

Thus the normal section possesses the maximal curvature radius of all plane sections with common tangent $\dot{\boldsymbol{X}}$ at point $\mathcal{P}$ of the surface. Moreover varying the orientation of the tangent $\dot{\boldsymbol{X}}$ at point $\mathcal{P}$ identifies the principal curvatures $\kappa_1 := \min \kappa_N$ and $\kappa_2 := \max \kappa_N$ and the associated directions of the principal curvature coordinates. The principal curvatures and the associated principal directions follow jointly from the general eigenvalue problem

$$[\boldsymbol{B} - \kappa_{1,2}\boldsymbol{A}] \cdot \dot{\boldsymbol{X}} = \boldsymbol{0}.$$

Thus the Gaussian curvature and twice the mean curvature follow as $\mathcal{K} = \det \boldsymbol{B} / \det \boldsymbol{A} = \kappa_1 \kappa_2$ and $2\mathcal{H} = \boldsymbol{B} : \boldsymbol{A}^{-1} = \kappa_1 + \kappa_2$, respectively.

Part III
Nonlinear Continuum Mechanics

...applies the previously discussed concepts of differential geometry to geometrically nonlinear first- and second-order continuum mechanics. Thereby the aim is to investigate the underlying geometric structure of first- and second-order elastoplasticity and to derive suited defect density measures. These defect density measures could then be used in the modeling of the hardening due to geometrically necessary defects such as dislocations and disclinations in crystalline materials.

Chapter 6
Continuum Kinematics

DAS GRAB DES HUNDES

Gestern war ich in dem Tal,
wo der Hund begraben liegt.
Trat erst durch ein Felsportal
und dann, wo nach links es biegt.

Vorwärts drang ich ungestört
noch um ein Erkleckliches -
Ist auch niemand da, der hört?
Denn nun tat ich Schreckliches:

Hob den Stein, auf welchem steht,
welchem steht: Hier liegt der Hund -
hob den Stein auf, hob ihn - und -
sah - oh, die ihr da seid, geht!

Sah - sah die Idee des Hunds,
sah den Hund, den Hund an sich.
Reichen wir die Hände uns;
dies ist wirklich fürchterlich.

Wie sie aussah, die Idee?
Bitte, bändigt euren Mund.
Denn ich kann nicht sagen meh,
als daß sie aussah wie ein - Hund.

Christian Morgenstern

Abstract. A continuum body is always embedded in three-dimensional Euclidean space, however its kinematics may be described in either rectilinear (Cartesian) or curvilinear coordinates. Expressed in curvilinear coordinates the differential geometry of flat Euclidean space is captured by the Christoffel symbols that take the role of a symmetric and integrable metric connection with associated zero curvature tensor. The position of a physical point together with the distortion, the double-distortion, and the triple-distortion are essential quantities to describe the continuum kinematics. Thus their representation is carefully elaborated. The kinematics of a continuum body are further characterized by an embedded non-Euclidean manifold. The embedded manifold is represented by a connection that additively decomposes into an integrable and a non-integrable contribution. The integrability conditions for the distortion and the double-distortion prove to be governed by the anholonomic object, the torsion, the curvature, and the non-metricity of the embedded manifold. To describe the deviation from integrability four defect density tensors, i.e. the primary and the secondary dislocation density tensors, the disclination density tensor, and the point-defect density tensor are

P. Steinmann, *Geometrical Foundations of Continuum Mechanics*,
Lecture Notes in Applied Mathematics and Mechanics 2, DOI: 10.1007/978-3-662-46460-1_6

introduced. Various types of continua with defects may be classified based on these defect density tensors.

The chapter also contains a comprehensive account on tensor calculus in Euclidean space in an extended supplement (which also introduces the symbolic notation used extensively throughout Part III).

6.1 Coordinates in Euclidean Space

The kinematics of (nonlinear) continuum mechanics and their relation to the previously introduced concepts of coordinates and their transformations within differential geometry are sketched in Fig. 6.1. Thereby this representation is in particular inspired by the exposition in [77], Chapter 1. Further details may also be found, e.g., in [122]. Thereby it is important to note firstly that although a continuum body is always embedded in symmetric, flat and metrically-connected Euclidean space, both rectilinear (Cartesian) and curvilinear coordinates are a possibility. Moreover, in his excellent overview article [202] distinguishes spaces from geometries: *..., it is a characteristic feature of defect theory that these geometries always exist in Euclidean space. Hence, in this paper we are not dealing with curved space, but with curved geometry*

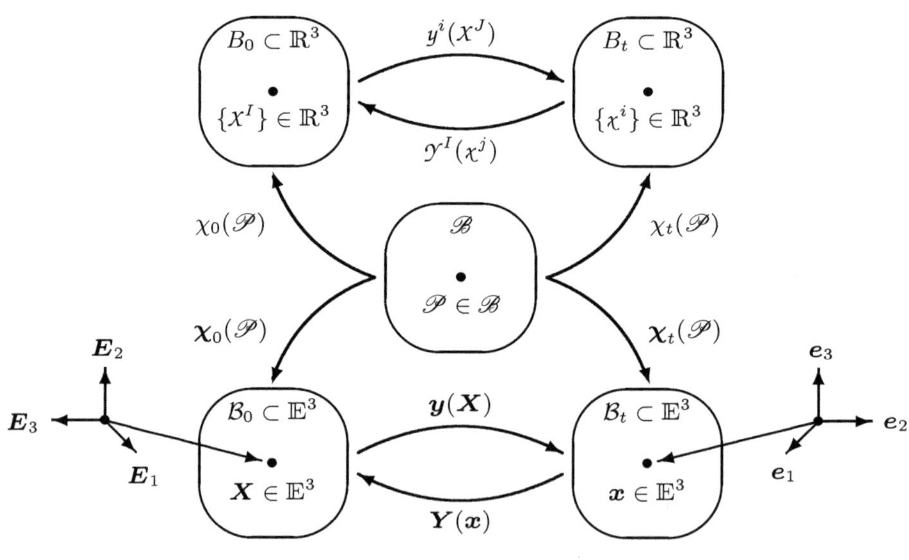

Fig. 6.1 Kinematics of geometrically nonlinear continuum mechanics. *Physical body* $\mathscr{B}$ consisting of *physical points* $\mathscr{P}$, their assignment to material and spatial *coordinates* in $\mathbb{R}^3$ and *position vectors* in $\mathbb{E}^3$, and its *placement* into material and spatial *configurations*. The material and spatial coordinates are related by an invertible and continuously differentiable *coordinate transformation*, whereas the material and spatial position vectors are related by an invertible and continuously differentiable *point transformation.*

in flat space. Thus the Euclidean embedding of the continuum body does not preclude to work with non-Euclidean geometries in order to describe, e.g. distributed defects.

6.1.1 Spatial and Material Coordinates

To set the stage, a *physical body* $\mathscr{B}$ consists in a set of *physical points* $\mathscr{P}$ (think of individual atoms or molecules)

$$\mathscr{B} = \{\mathscr{P}\}. \tag{6.1}$$

Then at time t the *assignment* of the physical points $\mathscr{P}$ to *spatial coordinates*, i.e. triplets $\{x^i(t)\} \in \mathbb{R}^3$ may be stated by the help of the map χ_t as

$$\{\mathscr{P}, t\} \mapsto \{x^1(t), x^2(t), x^3(t)\} = \chi(\mathscr{P}, t) = \chi_t(\mathscr{P}). \tag{6.2}$$

Accordingly, the *placement* of the whole physical body $\mathscr{B}$ into the *spatial configuration* $B_t \subset \mathbb{R}^3$ reads at time t as

$$\{\mathscr{B}, t\} \mapsto B_t = \chi(\mathscr{B}, t) = \chi_t(\mathscr{B}). \tag{6.3}$$

Alternatively, by the map χ_0 at time $t = t_0 = 0$, the physical points $\mathscr{P}$ may be assigned to *material coordinates*, i.e. triplets $\{X^I = x^i(0)\} \in \mathbb{R}^3$

$$\{\mathscr{P}, 0\} \mapsto \{X^1, X^2, X^3\} = \chi(\mathscr{P}, 0) = \chi_0(\mathscr{P}). \tag{6.4}$$

Thus, the placement of the physical body $\mathscr{B}$ into the *material configuration* $B_0 \subset \mathbb{R}^3$ follows at time $t = t_0 = 0$ as

$$\{\mathscr{B}, 0\} \mapsto B_0 = \chi(\mathscr{B}, 0) = \chi_0(\mathscr{B}). \tag{6.5}$$

The relation between the material and the spatial coordinates is then given in terms of an invertible and continuously differentiable *coordinate transformation*

$$\{X^J\} \mapsto x^i = y^i(X^J) \quad \text{and} \quad \{x^i\} \mapsto X^I = \mathcal{Y}^I(x^j). \tag{6.6}$$

It shall be noted that within Euclidean space these coordinate transformations may be regarded as an example of Eq. 3.2.

6.1.2 Spatial and Material Position Vectors

Recall that for the present purposes a continuum body is always embedded into the ordinary three-dimensional ambient space that is mathematically described by an Euclidean space $\mathbb{E}^3$. Then, at time t, the physical points $\mathscr{P}$ may be associated by the map $\boldsymbol{\chi}_t$ to *spatial position vectors* $\boldsymbol{x}(t) \in \mathbb{E}^3$

$$\{\mathscr{P}, t\} \mapsto \boldsymbol{x}(t) = \boldsymbol{\chi}(\mathscr{P}, t) = \boldsymbol{\chi}_t(\mathscr{P}). \tag{6.7}$$

Accordingly, the placement of the whole physical body $\mathscr{B}$ into the spatial configuration $\mathcal{B}_t \subset \mathbb{E}^3$ reads at time t as

$$\{\mathscr{B}, t\} \mapsto \mathcal{B}_t = \boldsymbol{\chi}(\mathscr{B}, t) = \boldsymbol{\chi}_t(\mathscr{B}). \tag{6.8}$$

Likewise, by the map $\boldsymbol{\chi}_0$ at time $t = t_0 = 0$, the physical points $\mathscr{P}$ may be assigned to *material position vectors* $\boldsymbol{X} = \boldsymbol{x}(0) \in \mathbb{E}^3$

$$\{\mathscr{P}, t\} \mapsto \boldsymbol{X} = \boldsymbol{\chi}(\mathscr{P}, 0) = \boldsymbol{\chi}_0(\mathscr{P}). \tag{6.9}$$

Thus the placement of the physical body $\mathscr{B}$ into material configuration $\mathcal{B}_0 \subset \mathbb{E}^3$ reads at time $t_0 = t = 0$

$$\{\mathscr{B}, 0\} \mapsto \mathcal{B}_0 = \boldsymbol{\chi}(\mathscr{B}, 0) = \boldsymbol{\chi}_0(\mathscr{B}). \tag{6.10}$$

Finally, by composition, the material and spatial position vectors are related by an invertible and continuously differentiable *point transformation*

$$\boldsymbol{X} \mapsto \boldsymbol{x} = \boldsymbol{y}(\boldsymbol{X}) \quad \text{and} \quad \boldsymbol{x} \mapsto \boldsymbol{X} = \boldsymbol{Y}(\boldsymbol{x}). \tag{6.11}$$

The point transformation is also frequently denoted in a descriptive fashion as the *deformation map*.

6.1.3 Rectilinear versus Curvilinear Coordinates

In the embedding ambient (symmetric, flat, and metrically connected) Euclidean space *Cartesian coordinates* X^A and x^a and corresponding orthonormal base vectors $\boldsymbol{E}_A$ and $\boldsymbol{e}_a$ may be used, thus the material and spatial position vectors may be written as

$$\boldsymbol{X} = X^A \boldsymbol{E}_A = \boldsymbol{X}(X^A) \quad \text{and} \quad \boldsymbol{x} = x^a \boldsymbol{e}_a = \boldsymbol{x}(x^a), \tag{6.12}$$

respectively. Note that indices A, B, C, D or a, b, c, d are here reserved for Cartesian coordinates and Cartesian base vectors, respectively. The curvilinear coordinates and accompanying base vectors are equipped with indices I, J, K, L or i, j, k, l, respectively. Then the relation between the *rectilinear* (Cartesian) coordinates X^A and x^a and the *curvilinear* coordinates $\mathcal{X}^J$ and χ^j as depicted in Fig. 6.2 is stated as

$$X^A = \hat{Y}^A(\mathcal{X}^J) \quad \text{with} \quad \mathcal{X}^I = \hat{\mathcal{Y}}^I(X^B) \quad \text{and}$$
$$x^a = \hat{y}^a(\chi^j) \quad \text{with} \quad \chi^i = \hat{y}^i(x^b). \tag{6.13}$$

The corresponding *Jacobians* of these coordinate mappings are denoted by

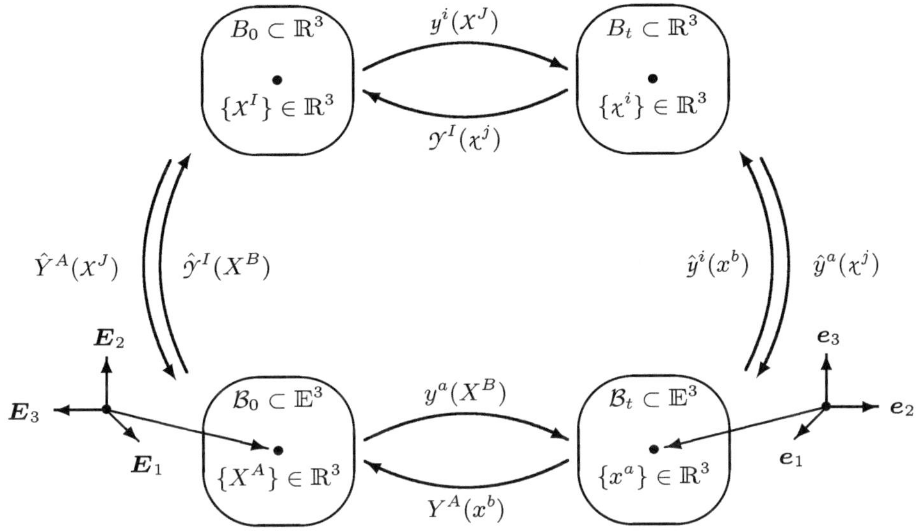

Fig. 6.2 Rectilinear (Cartesian) versus curvilinear coordinates. Assignment of *rectilinear* (Cartesian) coordinates X^A and x^a to *curvilinear* coordinates $\mathcal{X}^J$ and χ^j such that the material and spatial *position vectors* follow as $\boldsymbol{X} = X^A \boldsymbol{E}_A = \hat{Y}^A(\mathcal{X}^J)\boldsymbol{E}_A$ and $\boldsymbol{x} = x^a \boldsymbol{e}_a = \hat{y}^a(\chi^j)\boldsymbol{e}_a$, respectively.

$$J^A{}_J = \hat{Y}^A{}_{,J} \quad \text{with} \quad \jmath^I{}_B = \hat{\mathcal{Y}}^I{}_{,B} \quad \text{and}$$

$$j^a{}_j = \hat{y}^a{}_{,j} \quad \text{with} \quad j^i{}_b = \hat{y}^i{}_{,b}. \qquad (6.14)$$

Due to the above relations between the rectilinear and the curvilinear coordinates the material and spatial position vectors

$$\boldsymbol{X} = \hat{Y}^A(\mathcal{X}^J)\boldsymbol{E}_A = \boldsymbol{X}(\mathcal{X}^J) \quad \text{and} \quad \boldsymbol{x} = \hat{y}^a(\chi^j)\boldsymbol{e}_a = \boldsymbol{x}(\chi^j) \qquad (6.15)$$

may finally be expressed as functions of the curvilinear coordinates.

6.1.4 Spatial and Material Euclidean Connection

Next, material and spatial *tangent vectors* $\boldsymbol{G}_I := \boldsymbol{X}_{,I} = X^A{}_{,I}\boldsymbol{E}_A$ and $\boldsymbol{g}_i := \boldsymbol{x}_{,i} = x^a{}_{,i}\boldsymbol{e}_a$ to material and spatial *coordinate lines* $\boldsymbol{X} = \boldsymbol{X}(\mathcal{X}^I)$ (with $\mathcal{X}^J = \text{const}$ and $\mathcal{X}^K = \text{const}$) and $\boldsymbol{x} = \boldsymbol{x}(\chi^i)$ (with $\chi^j = \text{const}$ and $\chi^k = \text{const}$) together with material and spatial *normal vectors* $\boldsymbol{G}^J := \text{Grad}\,\mathcal{X}^J = \mathcal{X}^J{}_{,B}\boldsymbol{E}^B$ and $\boldsymbol{g}^j := \text{grad}\,\chi^j = \chi^j{}_{,b}\boldsymbol{e}^b$ to material and spatial *coordinate surfaces* $\mathcal{X}^J(\boldsymbol{X}) = \text{const}$ and $\chi^j(\boldsymbol{x}) = \text{const}$ may be defined in Euclidean space if material and spatial curvilinear coordinates $\{\mathcal{X}^K\}$ and $\{\chi^k\}$ are utilized, thus

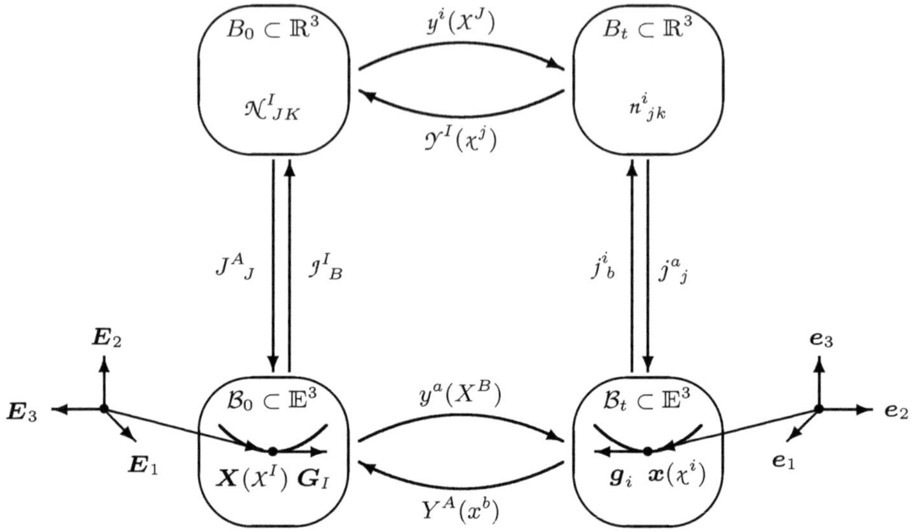

Fig. 6.3 Christoffel symbols, Jacobians of coordinate mappings, and covariant base vectors. The covariant base vectors follow as *tangent vectors* to coordinate lines, the corresponding contravariant base vectors are the *normal vectors* to coordinate surfaces. The Christoffel symbols take the interpretation as the Euclidean connection.

$$
\begin{aligned}
\boldsymbol{G}_I &= J^A{}_I \boldsymbol{E}_A, & \boldsymbol{g}_i &= \jmath^a{}_i \boldsymbol{e}_a, \\
\boldsymbol{G}^J &= \jmath^J{}_B \boldsymbol{E}^B, & \boldsymbol{g}^j &= j^j{}_b \boldsymbol{e}^b.
\end{aligned} \tag{6.16}
$$

Moreover, material and spatial *Christoffel symbols* associated with the curvilinear coordinates in the material and spatial configuration are introduced based on the Jacobians of the coordinate mappings in the above as

$$
\mathcal{N}^I{}_{JK} := \jmath^I{}_A J^A{}_{J,K} \quad \text{and} \quad n^i{}_{jk} := j^i{}_a \jmath^a{}_{j,k}. \tag{6.17}
$$

Thereby it is easy to verify that the Christoffel symbols allow essentially to express the partial derivatives of the material and spatial base vectors as

$$
\begin{aligned}
\boldsymbol{G}_{J,K} &= \mathcal{N}^I{}_{JK} \boldsymbol{G}_I, & \boldsymbol{g}_{j,k} &= n^i{}_{jk} \boldsymbol{g}_i, \\
\boldsymbol{G}^I{}_{,K} &= -\mathcal{N}^I{}_{JK} \boldsymbol{G}^J, & \boldsymbol{g}^i{}_{,k} &= -n^i{}_{jk} \boldsymbol{g}^j.
\end{aligned} \tag{6.18}
$$

Note that in an Euclidean space the Christoffel symbols take the role of the connection, that shall henceforth be denoted the material and spatial *Euclidean connection*. Due to the possibility to use Cartesian coordinates in Euclidean space, the Euclidean connection is by definition an integrable connection. Next covariant differentiation of material and spatial vectors and covectors in

Euclidean space based on the Euclidean connection, i.e. *Euclidean covariant differentiation* (denoted by a semicolon ;), is defined as

$$\mathcal{V}^I{}_{;K} = \mathcal{V}^I{}_{,K} + \mathcal{N}^I{}_{JK}\mathcal{V}^J, \qquad v^i{}_{;k} = v^i{}_{,k} + n^i{}_{jk}v^j \tag{6.19}$$
$$\mathcal{V}_{J;K} = \mathcal{V}_{J,K} - \mathcal{V}_I\mathcal{N}^I{}_{JK}, \qquad v_{j;k} = v_{j,k} - v_i n^i{}_{jk}.$$

Consequently, the material gradient of, e.g., a vector field $\boldsymbol{V}(\boldsymbol{X})$ and the spatial gradient of, e.g., a vector field $\boldsymbol{v}(\boldsymbol{x})$, respectively, read

$$\begin{aligned} \operatorname{Grad}\boldsymbol{V} &= \mathcal{V}^I{}_{;J}\ \boldsymbol{G}_I \otimes \boldsymbol{G}^J & \operatorname{grad}\boldsymbol{v} &= v^i{}_{;j}\ \boldsymbol{g}_i \otimes \boldsymbol{g}^j \\ &= J^A{}_I\mathcal{V}^I{}_{;J}\jmath^J{}_B\ \boldsymbol{E}_A \otimes \boldsymbol{E}^B & &= j^a{}_i v^i{}_{;j}\jmath^j{}_b\ \boldsymbol{e}_a \otimes \boldsymbol{e}^b \\ &= \mathcal{V}_{I;J}\ \boldsymbol{G}^I \otimes \boldsymbol{G}^J & &= v_{i;j}\ \boldsymbol{g}^i \otimes \boldsymbol{g}^j \\ &= \jmath^I{}_A\mathcal{V}_{I;J}\jmath^J{}_B\ \boldsymbol{E}^A \otimes \boldsymbol{E}^B & &= \jmath^i{}_a v_{i;j}\jmath^j{}_b\ \boldsymbol{e}^a \otimes \boldsymbol{e}^b. \end{aligned} \tag{6.20}$$

In summary the material and spatial Euclidean covariant derivatives in terms of the curvilinear coordinates may be related to the ordinary partial derivative in terms of the rectilinear (Cartesian) coordinates by

$$V^A{}_{,B} = J^A{}_I\mathcal{V}^I{}_{;J}\jmath^J{}_B, \qquad v^a{}_{,b} = j^a{}_i v^i{}_{;j}\jmath^j{}_b, \tag{6.21}$$
$$V_{A,B} = \jmath^I{}_A\mathcal{V}_{I;J}\jmath^J{}_B, \qquad v_{a,b} = \jmath^i{}_a v_{i;j}\jmath^j{}_b.$$

Finally material and spatial co- and contravariant metric coefficients (that serve among other purposes to lower and raise indices) are defined as

$$G_{IJ} = \boldsymbol{G}_I \cdot \boldsymbol{G}_J = J^A{}_I\delta_{AB}J^B{}_J \qquad g_{ij} = \boldsymbol{g}_i \cdot \boldsymbol{g}_j = j^a{}_i\delta_{ab}j^b{}_j \tag{6.22}$$
$$G^{IJ} = \boldsymbol{G}^I \cdot \boldsymbol{G}^J = \jmath^I{}_A\delta^{AB}\jmath^J{}_B \qquad g^{ij} = \boldsymbol{g}^i \cdot \boldsymbol{g}^j = \jmath^i{}_a\delta^{ab}\jmath^j{}_b.$$

Then the Euclidean covariant derivatives of the co- and contravariant metric coefficients vanish identically

$$G_{IJ;K} = G_{IJ,K} - 2\mathcal{N}_{(IJ)K} \equiv 0 \qquad g_{ij;k} = g_{ij,k} - 2n_{(ij)k} \equiv 0 \tag{6.23}$$
$$G^{IJ}{}_{;K} = G^{IJ}{}_{,K} + 2\mathcal{N}^{(IJ)}{}_K \equiv 0 \qquad g^{ij}{}_{;k} = g^{ij}{}_{,k} + 2n^{(ij)}{}_k \equiv 0$$

with the fully co- and contravariant Christoffel symbols that expand into

$$\mathcal{N}_{IJK} = J^A{}_I\delta_{AB}J^B{}_{J,K} \qquad n_{ijk} = j^a{}_i\delta_{ab}j^b{}_{j,k} \tag{6.24}$$
$$\mathcal{N}^{IJ}{}_K = -\jmath^I{}_{A,K}\delta^{AB}\jmath^J{}_B \qquad n^{ij}{}_k = -\jmath^i{}_{a,k}\delta^{ab}\jmath^j{}_b.$$

Thus, as a consequence, the material and spatial Euclidean connections are of course metric connections.

6.1.5 Spatial and Material Euclidean Curvature

The material and spatial curvature tensor corresponding to the Euclidean connection shall be denoted the *Euclidean curvature* in the sequel. Clearly the Euclidean curvature vanishes identically

$$0 \equiv 2\mathcal{N}^I{}_{J[L,K]} + 2\mathcal{N}^I{}_{M[K}\mathcal{N}^M{}_{JL]} \quad \text{and} \quad 0 \equiv 2n^i{}_{j[l,k]} + 2n^i{}_{m[k}n^m{}_{jl]}. \tag{6.25}$$

The Euclidean curvature tensor being zero is of course in full agreement with the corresponding Euclidean connection being integrable.

Proof. Considering first the partial derivatives of the Euclidean connection or rather the Christoffel symbols

$$\mathcal{N}^I{}_{JK} = \mathcal{J}^I{}_A J^A{}_{J,K} \quad \text{and} \quad n^i{}_{jk} = \mathrm{j}^i{}_a j^a{}_{j,k}$$

results in

$$\mathcal{N}^I{}_{JK,L} = \mathcal{J}^I{}_A J^A{}_{J,KL} + \underline{\mathcal{J}^I{}_{A,B} J^B{}_L} J^A{}_{J,K} \quad \text{and} \quad n^i{}_{jk,l} = \mathrm{j}^i{}_a j^a{}_{j,kl} + \underline{\mathrm{j}^i{}_{a,b} j^b{}_l} j^a{}_{j,k}.$$

Next by the product and chain rules it holds for the underlined terms in the second summands

$$\underline{\mathcal{J}^I{}_{A,B} J^B{}_L} = -\mathcal{J}^I{}_C J^C{}_{M,L} \mathcal{J}^M{}_A \quad \text{and} \quad \underline{\mathrm{j}^i{}_{a,b} j^b{}_l} = -\mathrm{j}^i{}_c j^c{}_{m,l} \mathrm{j}^m{}_a$$

Taking again into account the expression for the Christoffel symbols

$$\mathcal{N}^I{}_{ML} = \mathcal{J}^I{}_C J^C{}_{M,L} \quad \text{and} \quad n^i{}_{ml} = \mathrm{j}^i{}_c j^c{}_{m,l}$$

and inserting these representations into the above partial derivatives of the Christoffel symbols renders

$$\mathcal{N}^I{}_{JK,L} = \mathcal{J}^I{}_A J^A{}_{J,KL} - \mathcal{N}^I{}_{ML} \mathcal{J}^M{}_A J^A{}_{J,K} \quad \text{and} \quad n^i{}_{jk,l} = \mathrm{j}^i{}_a j^a{}_{j,kl} - n^i{}_{ml} \mathrm{j}^m{}_a j^a{}_{j,k}.$$

Observing finally once again the expression for the Christoffel symbols

$$\mathcal{N}^M{}_{JK} = \mathcal{J}^M{}_A J^A{}_{J,K} \quad \text{and} \quad n^m{}_{jk} = \mathrm{j}^m{}_a j^a{}_{j,k}$$

and reordering terms renders eventually

$$\mathcal{J}^I{}_A J^A{}_{J,KL} = \mathcal{N}^I{}_{JK,L} + \mathcal{N}^I{}_{ML}\mathcal{N}^M{}_{JK} \tag{6.26}$$

and

$$\mathrm{j}^i{}_a j^a{}_{j,kl} = n^i{}_{jk,l} + n^i{}_{ml} n^m{}_{jk}. \tag{6.27}$$

The symmetry of the second partial derivatives then concludes the proof. □

6.1.6 Convected Coordinates

A special case occurs for the case of convected (curvilinear) coordinates, see Fig. 6.4, which are defined as

$$\mathfrak{x}^{\mathfrak{i}} := \chi^i \equiv \delta^i{}_I X^I. \tag{6.28}$$

Thus the coordinate transformation $\chi^i = y^i(X^J)$ degenerates to the unit map $\chi^i = \delta^i{}_J X^J$ or likewise $\mathfrak{x}^{\mathfrak{i}} = \delta^{\mathfrak{i}}{}_{\mathfrak{j}}\mathfrak{x}^{\mathfrak{j}}$ with Jacobian $\delta^i{}_J = \delta^{\mathfrak{i}}{}_{\mathfrak{j}}$ given by the Kronecker symbol. Here the fraktur font is used to denote quantities and especially indices related to the convective coordinates. The relation between the rectilinear (Cartesian) coordinates and the convected coordinates follow as

$$X^A = \hat{X}^A(\mathfrak{x}^{\mathfrak{j}}) \quad \text{and} \quad x^a = \hat{x}^a(\mathfrak{x}^{\mathfrak{j}}) \quad \text{with} \quad \mathfrak{x}^{\mathfrak{i}} = \hat{\mathfrak{x}}^{\mathfrak{i}}(X^B) = \hat{\mathfrak{x}}^{\mathfrak{i}}(x^b). \tag{6.29}$$

Then the material and spatial position vectors are parameterized in terms of the convected coordinates

$$\boldsymbol{X} = \hat{X}^A(\mathfrak{x}^{\mathfrak{j}})\boldsymbol{E}_A = \boldsymbol{X}(\mathfrak{x}^{\mathfrak{j}}) \quad \text{and} \quad \boldsymbol{x} = \hat{x}^a(\mathfrak{x}^{\mathfrak{j}})\boldsymbol{e}_a = \boldsymbol{x}(\mathfrak{x}^{\mathfrak{j}}). \tag{6.30}$$

Consequently, the material and spatial co- and contravariant base vectors compute as

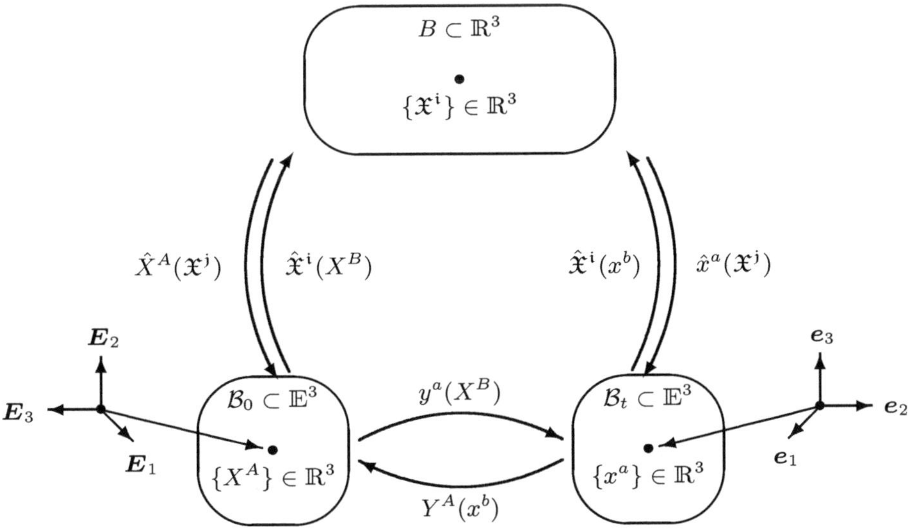

Fig. 6.4 Rectilinear (Cartesian) versus convected (curvilinear) coordinates. Assignment of *rectilinear* (Cartesian) coordinates X^A and x^a to *convected (curvilinear) coordinates* $\mathfrak{x}^{\mathfrak{j}}$ such that the material and spatial *position vectors* follow as $\boldsymbol{X} = X^A\boldsymbol{E}_A = \hat{X}^A(\mathfrak{x}^{\mathfrak{j}})\boldsymbol{E}_A$ and $\boldsymbol{x} = x^a\boldsymbol{e}_a = \hat{x}^a(\mathfrak{x}^{\mathfrak{j}})\boldsymbol{e}_a$, respectively.

$$\boldsymbol{G}_{\mathfrak{i}} = \boldsymbol{X}_{,\mathfrak{i}} \quad \boldsymbol{G}^{\mathfrak{j}} = \operatorname{Grad} \mathfrak{X}^{\mathfrak{j}} \quad \text{and} \quad \boldsymbol{g}_{\mathfrak{i}} = \boldsymbol{x}_{,\mathfrak{i}} \quad \boldsymbol{g}^{\mathfrak{j}} = \operatorname{grad} \mathfrak{X}^{\mathfrak{j}}. \tag{6.31}$$

The use of convected coordinates results in particularly easy pull-back/push-forward relations, since only the base vectors of tensor fields have to be exchanged while their coefficients remain unchanged.

6.2 Position and Distortions

Besides the point transformation $\boldsymbol{x} = \boldsymbol{y}(\boldsymbol{X})$ between material and spatial position vectors, and likewise the coordinate transformation $\chi^i = y^i(\mathcal{X}^J)$ between material and spatial coordinates, the *distortion* is an important kinematical concept in continuum mechanics. In the following the sequence of compatible (single-,) double-, and triple-distortions shall be represented in symbolic notation as well as in rectilinear (Cartesian) and curvilinear coefficients, respectively. Thereby, careful distinction is made between the ordinary curvilinear and *tensorially transforming* curvilinear coordinate representations. Moreover, the relation between these two representations is elaborated in much detail. Tensorially transforming coefficients are indicated notation-wise by sans serif font. For notational consistency the point transformation is also represented in terms of the coefficients of a properly transforming vector field $\mathsf{x}^i = \mathsf{y}^i(\mathcal{X}^J)$, i.e. the *position*, rather than in terms of the coordinate transformation. An overview on the various notations and relations adopted is given in tables 6.1 to 6.3.

Table 6.1 Representation of the position (**P**), the distortion (**D**), the double-distortion (**D-D**), and the triple-distortion (**T-D**) in terms of rectilinear (Cartesian) coefficients and corresponding base vectors

P	$\boldsymbol{x}$		$= x^a$	$\boldsymbol{e}_a$	x^a	$= y^a(X^B)$
D	$\boldsymbol{F}$	$= \operatorname{Grad} \boldsymbol{y}$	$= F^a{}_B$	$\boldsymbol{e}_a \otimes \boldsymbol{E}^B$	$F^a{}_B$	$= y^a{}_{,B}$
D-D	$\boldsymbol{G}$	$= \operatorname{Grad} \boldsymbol{F}$	$= G^a{}_{BC}$	$\boldsymbol{e}_a \otimes \boldsymbol{E}^B \otimes \boldsymbol{E}^C$	$G^a{}_{BC}$	$= F^a{}_{B,C}$
T-D	$\boldsymbol{H}$	$= \operatorname{Grad} \boldsymbol{G}$	$= H^a{}_{BCD}$	$\boldsymbol{e}_a \otimes \boldsymbol{E}^B \otimes \boldsymbol{E}^C \otimes \boldsymbol{E}^D$	$H^a{}_{BCD}$	$= G^a{}_{BC,D}$

Table 6.2 Representation of the position (**P**), the distortion (**D**), the double-distortion (**D-D**), and the triple-distortion (**T-D**) in terms of tensorially transforming curvilinear coefficients and corresponding base vectors

P	$\boldsymbol{x}$		$= \mathsf{x}^i$	$\boldsymbol{g}_i$	x^i	$= \mathsf{y}^i(\mathcal{X}^J)$
D	$\boldsymbol{F}$	$= \operatorname{Grad} \boldsymbol{y}$	$= \mathsf{F}^i{}_J$	$\boldsymbol{g}_i \otimes \boldsymbol{G}^J$	$\mathsf{F}^i{}_J$	$= \mathsf{y}^i{}_{;J}$
D-D	$\boldsymbol{G}$	$= \operatorname{Grad} \boldsymbol{F}$	$= \mathsf{G}^i{}_{JK}$	$\boldsymbol{g}_i \otimes \boldsymbol{G}^J \otimes \boldsymbol{G}^K$	$\mathsf{G}^i{}_{JK}$	$= \mathsf{F}^i{}_{J:K}$
T-D	$\boldsymbol{H}$	$= \operatorname{Grad} \boldsymbol{G}$	$= \mathsf{H}^i{}_{JKL}$	$\boldsymbol{g}_i \otimes \boldsymbol{G}^J \otimes \boldsymbol{G}^K \otimes \boldsymbol{G}^L$	$\mathsf{H}^i{}_{JKL}$	$= \mathsf{G}^i{}_{JK:L}$

Table 6.3 Vectorially transforming curvilinear coefficients of the position (**P**) and definition of the tensorially transforming curvilinear coefficients of the distortion (**D**), the double-distortion (**D-D**), and the triple-distortion (**T-D**) as Euclidean covariant derivatives of the position (**P**), the distortion (**D**), and the double-distortion (**D-D**)

P	$\mathsf{y}^i(\mathcal{X}^J) = y^a \mathcal{j}^i{}_a$
D	$\mathsf{y}^i{}_{;J} = \mathsf{y}^i{}_{,J} + n^i{}_{mj}\, \mathsf{y}^m\, \mathcal{F}^j{}_J$
D-D	$\mathsf{F}^i{}_{J:K} = \mathsf{F}^i{}_{J,K} + n^i{}_{mk}\, \mathsf{F}^m{}_J\, \mathcal{F}^k{}_K - \mathsf{F}^i{}_M\, \mathcal{N}^M{}_{JK}$
T-D	$\mathsf{G}^i{}_{JK:L} = \mathsf{G}^i{}_{JK,L} + n^i{}_{ml}\, \mathsf{G}^m{}_{JK}\, \mathcal{F}^l{}_L - \mathsf{G}^i{}_{MK}\, \mathcal{N}^M{}_{JL} - \mathsf{G}^i{}_{JM}\, \mathcal{N}^M{}_{KL}$

6.2.1 Position

In an Euclidean space the relation between the Cartesian base vectors $\boldsymbol{e}_a$ and $\boldsymbol{E}_A$ and the tangent vectors $\boldsymbol{g}_i$ and $\boldsymbol{G}_I$ in terms of the Jacobians in Eq. 6.14 makes it possible to *formally* introduce a vectorial change of coordinates for the position vector in the following fashion

$$\boldsymbol{x} = x^a \boldsymbol{e}_a =: \mathsf{x}^i \boldsymbol{g}_i \quad \text{and} \quad \boldsymbol{X} = X^A \boldsymbol{E}_A =: \mathsf{X}^I \boldsymbol{G}_I. \tag{6.32}$$

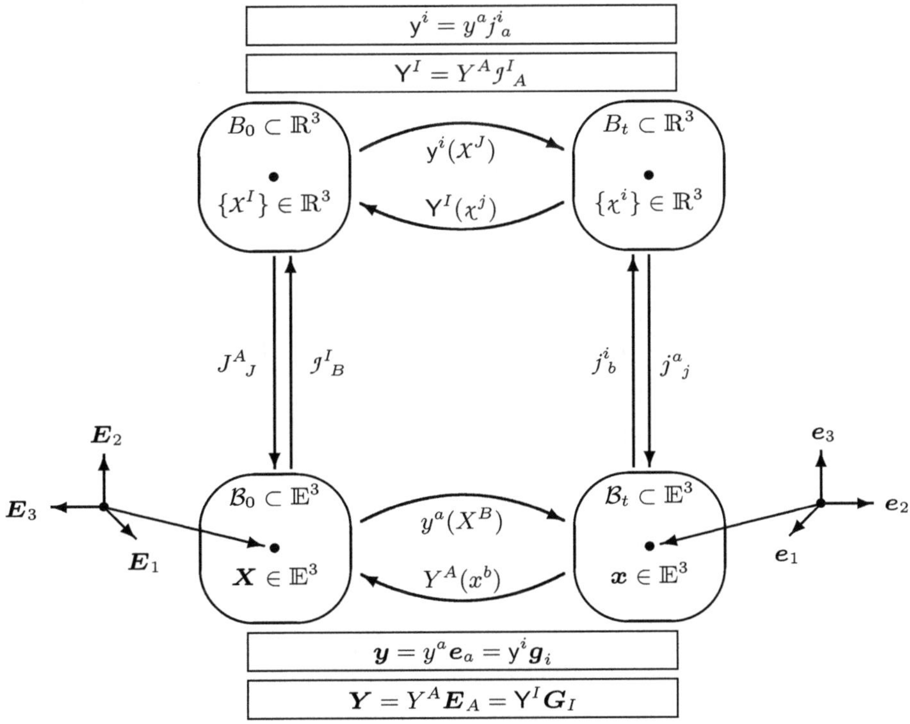

Fig. 6.5 Representation of the position in rectilinear (Cartesian) and vectorially transforming curvilinear coefficients

Here the vectorially transforming *position vector coordinates* follow as

$$\mathsf{x}^i = \mathsf{y}^i(\mathcal{x}^j) = y^a \jmath^i{}_a \quad \text{and} \quad \mathsf{X}^I = \mathsf{Y}^I(\mathcal{X}^J) = Y^A \jmath^I{}_A. \tag{6.33}$$

Then from the product rule and the definition of the Christoffel symbols the partial derivatives of the position vector coordinates with respect to the curvilinear coordinates renders

$$\mathsf{y}^i{}_{,k} = \delta^i{}_k - n^i{}_{jk}\mathsf{y}^j \quad \text{and} \quad \mathsf{Y}^I{}_{,K} = \delta^I{}_K - \mathcal{N}^I{}_{JK}\mathsf{Y}^J. \tag{6.34}$$

The position vector coordinates y^i and y^I with respect to the tangent vectors $\boldsymbol{g}_i$ and $\boldsymbol{G}_I$ will be helpful in the sequel to maintain a consistent notation when considering (single-,) double-, and triple-distortions. In the following the position vector coordinates shall simply be denoted as the *position*.

6.2.2 Distortion

The curvilinear and the rectilinear (Cartesian) coefficients of the compatible *distortion* (or the single-distortion) are introduced as the *Jacobians*, i.e.

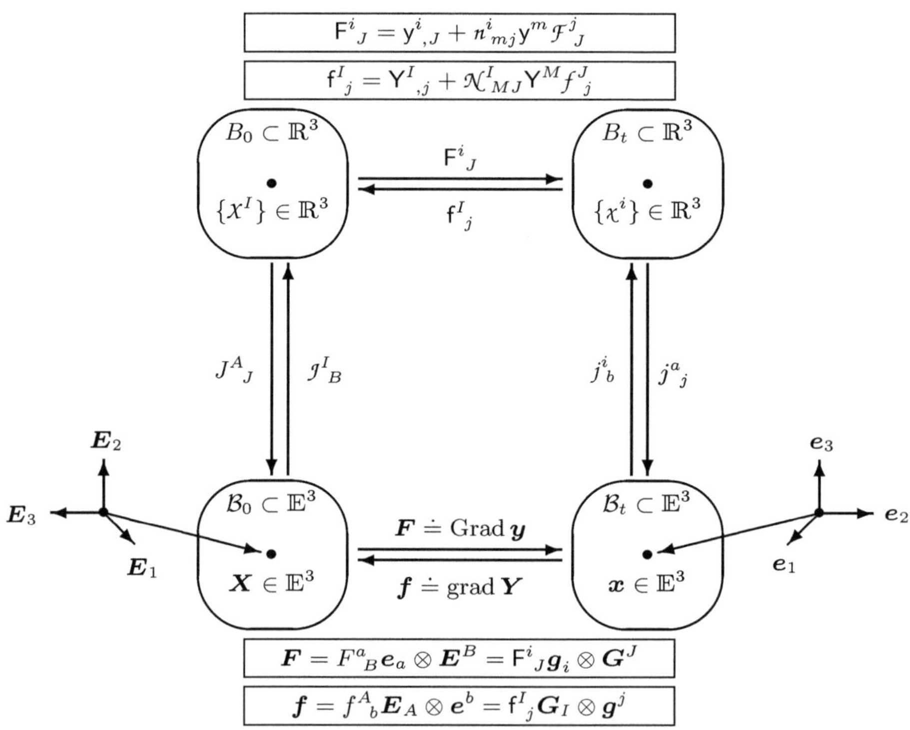

Fig. 6.6 Representation of the distortion in rectilinear (Cartesian) and tensorially transforming curvilinear coefficients

the first *partial derivatives* of the curvilinear and the rectilinear (Cartesian) *coordinate transformations*, respectively

$$\mathcal{F}^i{}_J := y^i{}_{,J} \quad \text{and} \quad f^I{}_j := \mathcal{Y}^I{}_{,j}, \tag{6.35}$$
$$F^a{}_B := y^a{}_{,B} \quad \text{and} \quad f^A{}_b := Y^A{}_{,b}.$$

Moreover, the tensorially transforming (Euclidean) *covariant derivative* of the *position* in terms of the curvilinear coordinates are computed as

$$\mathsf{F}^i{}_J := \mathsf{y}^i{}_{;J} = \mathsf{y}^i{}_{,J} + n^i{}_{mj}\mathsf{y}^m \mathcal{F}^j{}_J, \tag{6.36}$$
$$\mathsf{f}^I{}_j := \mathsf{Y}^I{}_{;j} = \mathsf{Y}^I{}_{,j} + \mathcal{N}^I{}_{MJ}\mathsf{Y}^M f^J{}_j.$$

Observe that the (Euclidean) *covariant derivative* (denoted by a semicolon ;) takes into account, e.g., varying spatial tangent vectors of curvilinear coordinate systems. Finally the corresponding *distortion* (two-point) tensors follow accordingly as the first-order gradients of the *point transformation*

$$\boldsymbol{F} = \operatorname{Grad} \boldsymbol{y} = \underbrace{y^a{}_{,B}}_{F^a{}_B} \boldsymbol{e}_a \otimes \boldsymbol{E}^B = \underbrace{\mathsf{y}^i{}_{;J}}_{\mathsf{F}^i{}_J} \boldsymbol{g}_i \otimes \boldsymbol{G}^J, \tag{6.37}$$
$$\boldsymbol{f} = \operatorname{grad} \boldsymbol{Y} = \underbrace{Y^A{}_{,b}}_{f^A{}_b} \boldsymbol{E}_A \otimes \boldsymbol{e}^b = \underbrace{\mathsf{Y}^I{}_{;j}}_{\mathsf{f}^I{}_j} \boldsymbol{G}_I \otimes \boldsymbol{g}^j.$$

The distortion tensors $\boldsymbol{F}$ and $\boldsymbol{f}$ are also denoted as *deformation gradients*.

Discussion 6.1. Relation to Curvilinear Coordinates
By recalling the partial derivatives $\mathsf{y}^i{}_{,J}$ and $\mathsf{Y}^I{}_{,j}$ in Eq. 6.34, the covariant derivatives of the vectorially transforming position are finally expressed as

$$\mathsf{F}^i{}_J = \mathsf{y}^i{}_{;J} = y^i{}_{,J} = \mathcal{F}^i{}_J \quad \text{and} \quad \mathsf{f}^I{}_j = \mathsf{Y}^I{}_{;j} = \mathcal{Y}^I{}_{,j} = f^I{}_j.$$

The coincidence of the covariant derivative of the position y^i or Y^I, respectively, and the partial derivative of the coordinate transformation y^i or $\mathcal{Y}^I$, respectively, is a remarkable result. It agrees with the discussion in [122] stating that, due to y^i and $\mathcal{Y}^I$ being coordinate maps, no covariant derivative is involved in the definition of the distortion $\mathcal{F}^i{}_J$ and $f^I{}_j$. Note that here the sans serif font is only introduced (admittedly somewhat superfluous) for $\mathsf{F}^i{}_J := \mathcal{F}^i{}_J$ and $\mathsf{f}^I{}_j := f^I{}_j$ in order to allow for notational consistency with the double- and triple-distortions treated in the sequel. □

Discussion 6.2. Relation to Convected Coordinates
In convected coordinates $\mathfrak{X}^{\mathsf{i}} := \chi^i \equiv \delta^i{}_I X^I$ the coordinate transformation and consequently the distortion degenerate to the unit map $\chi^i = \delta^i{}_J X^I$ and $\mathcal{F}^i{}_J \equiv \delta^i{}_J$, respectively. Thus the deformation gradients expressed in convected coordinates expand into

$$\boldsymbol{F} = \boldsymbol{g}_{\mathsf{i}} \otimes \boldsymbol{G}^{\mathsf{i}} \quad \text{and} \quad \boldsymbol{f} = \boldsymbol{G}_{\mathsf{i}} \otimes \boldsymbol{g}^{\mathsf{i}}.$$

As a consequence the distortion tensor is exclusively expressed in terms of the spatial and material base vectors when evaluated in convected coordinates. □

6.2.3 Double-Distortion

The curvilinear and the rectilinear (Cartesian) coefficients of the compatible *double-distortion* are introduced as the *Hessians*, i.e. as the second *partial derivatives* of the curvilinear and the rectilinear (Cartesian) *coordinate transformations*, respectively

$$\mathcal{G}^i{}_{JK} := \mathcal{F}^i{}_{J,K} \quad \text{and} \quad g^I{}_{jk} := f^I{}_{j,k} \qquad (6.38)$$
$$G^a{}_{BC} := F^a{}_{B,C} \quad \text{and} \quad g^A{}_{bc} := f^I{}_{b,c}.$$

Moreover, the tensorially transforming (Euclidean) *total covariant derivatives* of the *distortion* in terms of the curvilinear coordinates are computed as

$$\mathsf{G}^i{}_{JK} := \mathsf{F}^i{}_{J:K} = \mathsf{F}^i{}_{J,K} + n^i{}_{mk}\mathsf{F}^m{}_J\mathcal{F}^k{}_K - \mathsf{F}^i{}_M\mathcal{N}^M{}_{JK}, \qquad (6.39)$$
$$\mathsf{g}^I{}_{jk} := \mathsf{f}^I{}_{j:k} = \mathsf{f}^I{}_{j,k} + \mathcal{N}^I{}_{MK}\mathsf{f}^M{}_j f^K{}_k - \mathsf{f}^I{}_m n^m{}_{jk}.$$

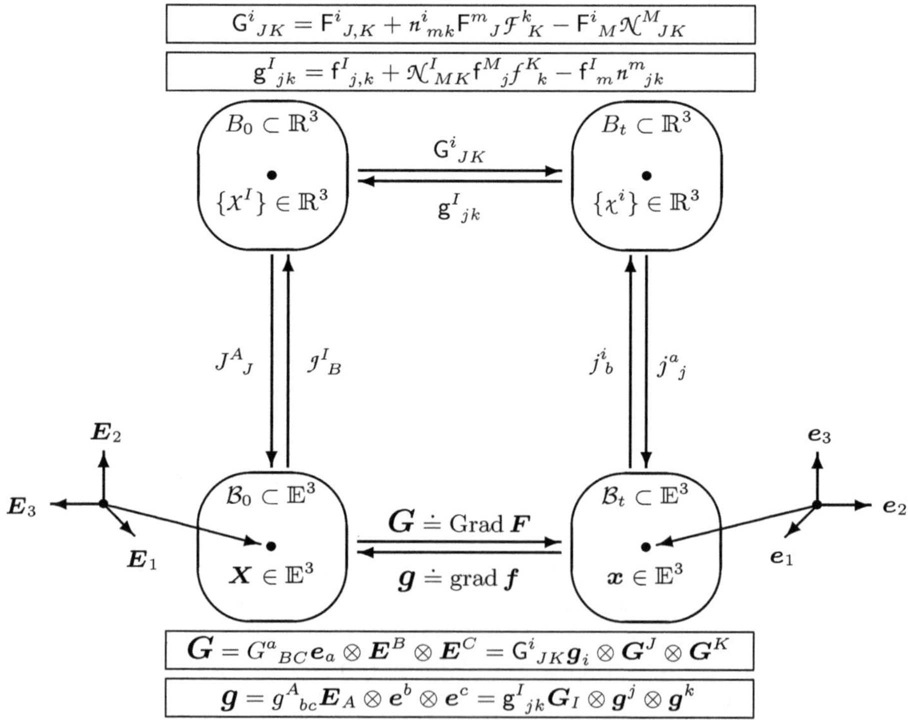

Fig. 6.7 Representation of the double-distortion in rectilinear (Cartesian) and tensorially transforming curvilinear coefficients

Note that the (Euclidean) total covariant derivative (denoted by a double dot :) simultaneously takes into account varying material and spatial tangent and normal vectors of the material and spatial curvilinear coordinate systems, for more details see for example [56]. Finally, the corresponding *double-distortion* (two-point) tensors follow accordingly as the second-order gradients of the point transformation

$$\boldsymbol{G} = \operatorname{Grad} \boldsymbol{F} = \underbrace{F^a{}_{B,C}}_{G^a{}_{BC}} \boldsymbol{e}_a \otimes \boldsymbol{E}^B \otimes \boldsymbol{E}^C = \underbrace{\mathsf{F}^i{}_{J:K}}_{\mathsf{G}^i{}_{JK}} \boldsymbol{g}_i \otimes \boldsymbol{G}^J \otimes \boldsymbol{G}^K, \quad (6.40)$$

$$\boldsymbol{g} = \operatorname{grad} \boldsymbol{f} = \underbrace{f^A{}_{b,c}}_{g^A{}_{bc}} \boldsymbol{E}_A \otimes \boldsymbol{e}^b \otimes \boldsymbol{e}^c = \underbrace{\mathsf{f}^I{}_{j:k}}_{\mathsf{g}^I{}_{jk}} \boldsymbol{G}_I \otimes \boldsymbol{g}^j \otimes \boldsymbol{g}^k.$$

The double-distortion tensors $\boldsymbol{G}$ and $\boldsymbol{g}$ are also denoted as *second (-order) deformation gradients.*

Discussion 6.3. Relation to Curvilinear Coordinates
In order to explicitly consider (for example) the case of a *non-integrable* distortion $\boldsymbol{F}$ its tensorially transforming and its curvilinear coefficients are expanded as

$$\mathsf{F}^i{}_J = \mathsf{y}^i{}_{;J} + {}^{\#}\mathsf{F}^i{}_J \quad \text{and} \quad \mathcal{F}^i{}_J = y^i{}_{,J} + {}^{\#}\mathcal{F}^i{}_J.$$

Here ${}^{\#}\mathsf{F}^i{}_J \equiv {}^{\#}\mathcal{F}^i{}_J$ identifies the *non-integrable* contribution to the distortion (that may be denoted the *relative distortion* whereby the overhead for its notation is only introduced for notational consistency). However, since (for example) $\mathsf{y}^i{}_{;J} \equiv y^i{}_{,J}$ holds, the distortions, being either integrable or non-integrable, generally allow the representation

$$\mathsf{F}^i{}_J \equiv \mathcal{F}^i{}_J \quad \text{and} \quad \mathsf{f}^I{}_j \equiv f^I{}_j.$$

Thus the double-distortions in curvilinear coordinates are alternatively expressed as

$$\begin{aligned}
\mathsf{G}^i{}_{JK} = \mathsf{F}^i{}_{J:K} &= \mathcal{F}^i{}_{J,K} + n^i{}_{mk}\mathcal{F}^m{}_{(J}\mathcal{F}^k{}_{K)} - \mathcal{F}^i{}_M \mathcal{N}^M{}_{(JK)}, \\
&= \mathcal{G}^i{}_{JK} + n^i{}_{mk}\mathcal{F}^m{}_{(J}\mathcal{F}^k{}_{K)} - \mathcal{F}^i{}_M \mathcal{N}^M{}_{(JK)}, \\
\mathsf{g}^I{}_{jk} = \mathsf{f}^I{}_{j:k} &= f^I{}_{j,k} + \mathcal{N}^I{}_{MK} f^M{}_{(j} f^K{}_{k)} - f^I{}_m n^m{}_{(jk)}, \\
&= g^I{}_{jk} + \mathcal{N}^I{}_{MK} f^M{}_{(j} f^K{}_{k)} - f^I{}_m n^m{}_{(jk)}.
\end{aligned}$$

Observe that only the first terms on the right hand sides possess skew symmetric contributions in indices J, K or j, k, respectively. Note that the sans serif font is used here to distinguish the total covariant derivative of the distortion $\mathsf{G}^i{}_{JK} = \mathsf{F}^i{}_{J:K}$ from the corresponding partial derivative $\mathcal{G}^i{}_{JK} = \mathcal{F}^i{}_{J,K}$.
□

Discussion 6.4. Relation to Convected Coordinates
For integrable distortions the second (-order) deformation gradients expressed in convected coordinates expand into

$$\boldsymbol{G} = [n^{\mathfrak{i}}{}_{\mathfrak{j}\mathfrak{k}} - \mathcal{N}^{\mathfrak{i}}_{\mathfrak{j}\mathfrak{k}}]\boldsymbol{g}_{\mathfrak{i}} \otimes \boldsymbol{G}^{\mathfrak{j}} \otimes \boldsymbol{G}^{\mathfrak{k}} \quad \text{and} \quad \boldsymbol{g} = [\mathcal{N}^{\mathfrak{i}}_{\mathfrak{j}\mathfrak{k}} - n^{\mathfrak{i}}{}_{\mathfrak{j}\mathfrak{k}}]\boldsymbol{G}_{\mathfrak{i}} \otimes \boldsymbol{g}^{\mathfrak{j}} \otimes \boldsymbol{g}^{\mathfrak{k}}.$$

Note that the coefficients of the spatial and material second (-order) deformation gradients differ only by the sign when expressed in convected coordinates. □

6.2.4 Triple-Distortion

The curvilinear and the rectilinear (Cartesian) coefficients of the compatible *triple-distortion* are introduced as the third *partial derivatives* of the curvilinear and the rectilinear (Cartesian) *coordinate transformations*, respectively

$$\begin{aligned} \mathcal{H}^i{}_{JKL} &:= \mathcal{G}^i{}_{JK,L} \quad \text{and} \quad \hbar^I{}_{jkl} := \mathscr{g}^I{}_{jk,l} \\ H^a{}_{BCD} &:= G^a{}_{BC,D} \quad \text{and} \quad h^A{}_{bcd} := g^I{}_{bc,d}. \end{aligned} \tag{6.41}$$

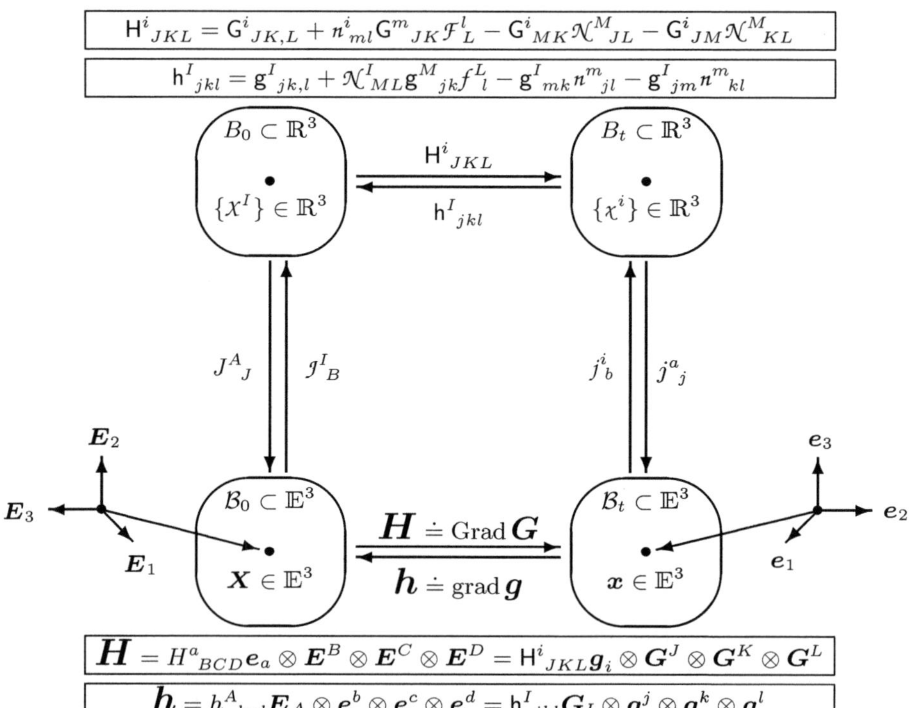

Fig. 6.8 Representation of the triple-distortion in rectilinear (Cartesian) and tensorially transforming curvilinear coefficients

Moreover, the tensorially transforming (Euclidean) *total covariant derivatives* of the *triple-distortion* in terms of the curvilinear coordinates are computed as

$$\begin{aligned}
\mathsf{H}^i{}_{JKL} &:= \mathsf{G}^i{}_{JK:L} \\
&= \mathsf{G}^i{}_{JK,L} + n^i{}_{ml}\mathsf{G}^m{}_{JK}\mathcal{F}^l{}_L - \mathsf{G}^i{}_{MK}\mathcal{N}^M{}_{JL} - \mathsf{G}^i{}_{JM}\mathcal{N}^M{}_{KL}, \\
\mathsf{h}^I{}_{jkl} &:= \mathsf{g}^I{}_{jk:l} \\
&= \mathsf{g}^I{}_{jk,l} + \mathcal{N}^I{}_{ML}\mathsf{g}^M{}_{jk}f^L{}_l - \mathsf{g}^I{}_{mk}n^m{}_{jl} - \mathsf{g}^I{}_{jm}n^m{}_{kl}.
\end{aligned} \tag{6.42}$$

Finally, the corresponding *triple-distortion* (two-point) tensors follow accordingly as the third-order gradients of the point transformation

$$\begin{aligned}
\boldsymbol{H} &= \operatorname{Grad}\boldsymbol{G} = \underbrace{G^a{}_{BC,D}}_{H^a{}_{BCD}}\boldsymbol{e}_a \otimes \boldsymbol{E}^B \otimes \boldsymbol{E}^C \otimes \boldsymbol{E}^D \\
&= \underbrace{\mathsf{G}^i{}_{JK:L}}_{\mathsf{H}^i{}_{JKL}}\boldsymbol{g}_i \otimes \boldsymbol{G}^J \otimes \boldsymbol{G}^K \otimes \boldsymbol{G}^L, \\
\boldsymbol{h} &= \operatorname{grad}\boldsymbol{g} = \underbrace{g^A{}_{bc,d}}_{h^A{}_{bcd}}\boldsymbol{E}_A \otimes \boldsymbol{e}^b \otimes \boldsymbol{e}^c \otimes \boldsymbol{e}^d \\
&= \underbrace{\mathsf{g}^I{}_{jk:l}}_{\mathsf{h}^I{}_{jkl}}\boldsymbol{G}_I \otimes \boldsymbol{g}^j \otimes \boldsymbol{g}^k \otimes \boldsymbol{g}^l.
\end{aligned} \tag{6.43}$$

The triple-distortion tensors $\boldsymbol{H}$ and $\boldsymbol{h}$ are also denoted as *third (-order) deformation gradients.*

Discussion 6.5. Relation to Curvilinear Coordinates
Next, to express in more detail (for example) the partial derivatives $\mathsf{G}^i{}_{JK,L}$ of the double-distortion $\mathsf{G}^i{}_{JK}$ its tensorially transforming coefficients and the curvilinear coefficients $\mathcal{G}^i{}_{JK}$ are expanded as

$$\mathsf{G}^i{}_{JK} = \mathsf{F}^i{}_{J:K} + {}^{\#}\mathsf{G}^i{}_{JK} \quad \text{and} \quad \mathcal{G}^i{}_{JK} = \mathcal{F}^i{}_{J,K} + {}^{\#}\mathcal{G}^i{}_{JK}.$$

Here ${}^{\#}\mathsf{G}^i{}_{JK} \equiv {}^{\#}\mathcal{G}^m{}_{JL}$ identifies the *non-integrable* contribution to the double-distortion that shall be denoted the *relative double-distortion.* The overhead in notation for the relative double-distortion is only introduced for notational consistency. As a result $\mathsf{G}^i{}_{JK}$ allows the representation

$$\mathsf{G}^i{}_{JK} = \mathcal{G}^i{}_{JK} + n^i{}_{jk}\mathcal{F}^j{}_J\mathcal{F}^k{}_K - \mathcal{F}^i{}_I\mathcal{N}^I{}_{JK}.$$

Then $\mathsf{G}^i{}_{JK,L}$ expands into six terms

$$
\begin{aligned}
\mathsf{G}^i{}_{JK,L} &\overset{①}{=} \mathcal{G}^i{}_{JK,L} \\
&\overset{②}{+} n^i{}_{jk,l}\mathcal{F}^j{}_J\mathcal{F}^k{}_K\mathcal{F}^l{}_L \\
&\overset{③}{+} n^i{}_{jk}\mathcal{F}^j{}_{J,L}\mathcal{F}^k{}_K \\
&\overset{④}{+} n^i{}_{jk}\mathcal{F}^j{}_J\mathcal{F}^k{}_{K,L} \\
&\overset{⑤}{-} \mathcal{F}^i{}_{I,L}\mathcal{N}^I{}_{JK} \\
&\overset{⑥}{-} \mathcal{F}^i{}_I\mathcal{N}^I{}_{JK,L}.
\end{aligned}
$$

Setting $\mathcal{G}^i{}_{JK,L} = \mathcal{H}^i{}_{JKL}$ in ①, inserting $n^i{}_{jk,l}$ and $\mathcal{N}^I{}_{JK,L}$ from Eqs. 6.27 and 6.26 into ② and ⑥, and expanding $\mathcal{F}^m{}_{J,L} = \mathsf{F}^m{}_{J:L} - n^m{}_{jl}\mathcal{F}^j{}_J\mathcal{F}^l{}_L + \mathcal{F}^m{}_M\mathcal{N}^M{}_{JL}$ and $\mathcal{F}^i{}_{M,L} = \mathsf{F}^i{}_{M:L} - n^i{}_{ml}\mathcal{F}^m{}_M\mathcal{F}^l{}_L + \mathcal{F}^i{}_N\mathcal{N}^N{}_{ML}$ in ③ and ⑤ renders

$$
\begin{aligned}
\mathsf{G}^i{}_{JK,L} &\overset{①}{=} \mathcal{H}^i{}_{JKL} \\
&\overset{②}{+} \underline{j^i{}_a j^a{}_{j,kl}\mathcal{F}^j{}_J\mathcal{F}^k{}_K\mathcal{F}^l{}_L} - \underline{\underline{\underline{n^i{}_{ml}n^m{}_{jk}\mathcal{F}^j{}_J\mathcal{F}^k{}_K\mathcal{F}^l{}_L}}} \\
&\overset{③}{+} \underline{\underline{n^i{}_{mk}\mathcal{F}^k{}_K\mathsf{F}^m{}_{J:L}}} - \underline{\underline{\underline{n^i{}_{mk}n^m{}_{jl}\mathcal{F}^j{}_J\mathcal{F}^k{}_K\mathcal{F}^l{}_L}}} + \underline{\underline{\underline{\underline{n^i{}_{mk}\mathcal{F}^m{}_M\mathcal{F}^k{}_K\mathcal{N}^M{}_{JL}}}}} \\
&\overset{④}{+} n^i{}_{jk}\mathcal{F}^j{}_J\mathcal{F}^k{}_{K,L} \\
&\overset{⑤}{-} \underline{\underline{\mathsf{F}^i{}_{M:L}\mathcal{N}^M{}_{JK}}} + \underline{\underline{\underline{\underline{n^i{}_{ml}\mathcal{F}^m{}_M\mathcal{F}^l{}_L\mathcal{N}^M{}_{JK}}}}} - \underline{\underline{\underline{\underline{\mathcal{F}^i{}_N\mathcal{N}^N{}_{ML}\mathcal{N}^M{}_{JK}}}}} \\
&\overset{⑥}{-} \underline{\mathcal{F}^i{}_M\mathcal{J}^M_A J^A{}_{J,KL}} + \underline{\underline{\underline{\underline{\mathcal{F}^i{}_N\mathcal{N}^N{}_{ML}\mathcal{N}^M{}_{JK}}}}}.
\end{aligned}
$$

Then the single underlined terms are symmetric in indices K, L per se, the double underlined terms combine with corresponding terms from the total covariant derivative of the double-distortion in Eq. 6.42

$$
n^i{}_{ml}\mathcal{F}^l{}_L\mathsf{G}^m{}_{JK} + \underline{\underline{n^i{}_{mk}\mathcal{F}^k{}_K\mathsf{F}^m{}_{J:L}}} = 2n^i{}_{jk}\mathcal{F}^k{}_{(K}\mathsf{G}^m{}_{JL)} - n^i{}_{mk}\mathcal{F}^k{}_K\,{}^{\#}\mathsf{G}^m{}_{JL}
$$

and

$$
-\mathsf{G}^i{}_{MK}\mathcal{N}^M{}_{JL} - \underline{\underline{\mathsf{F}^i{}_{M:L}\mathcal{N}^M{}_{JK}}} = -2\mathsf{G}^i{}_{M(L}\mathcal{N}^M{}_{JK)} + {}^{\#}\mathsf{G}^i{}_{ML}\mathcal{N}^M{}_{JK}
$$

into expressions symmetric in K, L and contributions due to the relative double-distortion ${}^{\#}\mathsf{G}^m{}_{JL} = \mathsf{G}^m{}_{JL} - \mathsf{F}^m{}_{J:L}$, the triple and quadruple underlined terms combine into expressions symmetric in K, L, respectively, and the quintuple underlined terms cancel each other. As a result, and after reordering of terms, the total covariant derivative of the double-distortions is finally expressed in curvilinear coordinates as

$$
\begin{aligned}
\mathsf{H}^i{}_{JKL} = \mathsf{G}^i{}_{JK:L} = {} & \mathcal{H}^i{}_{JKL} + n^i{}_{jk}\mathcal{F}^j{}_J\mathcal{F}^k{}_{K,L} \\
& - \underline{\underline{n^i{}_{mk}\mathcal{F}^k{}_K\,{}^{\#}\mathsf{G}^m{}_{JL}}} + \underline{\underline{{}^{\#}\mathsf{G}^i{}_{ML}\mathcal{N}^M{}_{JK}}} \\
& + \underline{j^i{}_a j^a{}_{j,kl}\mathcal{F}^j{}_J\mathcal{F}^k{}_{(K}\mathcal{F}^l{}_{L)}} - \underline{\mathcal{F}^i{}_M \mathcal{J}^M{}_A J^A{}_{J,(KL)}} \\
& + \underline{2n^i{}_{mk}\mathcal{F}^k{}_{(K}\mathsf{G}^m{}_{JL)}} - \underline{2\mathsf{G}^i{}_{M(L}\mathcal{N}^M{}_{JK)}} - \mathsf{G}^i{}_{JM}\mathcal{N}^M{}_{(KL)} \\
& - \underline{\underline{\underline{2n^i{}_{ml}n^m{}_{jk}\mathcal{F}^j{}_J\mathcal{F}^k{}_{(K}\mathcal{F}^l{}_{L)}}}} + \underline{\underline{\underline{\underline{2n^i{}_{mk}\mathcal{F}^m{}_M\mathcal{F}^k{}_{(K}\mathcal{N}^M{}_{JL)}}}}} \\
\mathsf{h}^I{}_{jkl} = \mathsf{g}^I{}_{jk:l} = {} & \mathcal{h}^I{}_{jkl} + \mathcal{N}^I{}_{JK}f^J{}_j f^K{}_{k,l} \\
& - \underline{\underline{\mathcal{N}^I{}_{MK}f^K{}_k\,{}^{\#}\mathsf{g}^M{}_{jl}}} + \underline{\underline{{}^{\#}\mathsf{g}^I{}_{ml}n^m{}_{jk}}} \\
& + \underline{\mathcal{J}^I{}_A J^A{}_{J,KL}f^J{}_j f^K{}_{(k}f^L{}_{l)}} - \underline{f^I{}_m j^m{}_a j^a{}_{j,(kl)}} \\
& + \underline{2\mathcal{N}^I{}_{MK}f^K{}_{(k}\mathsf{g}^M{}_{jl)}} - \underline{2\mathsf{g}^I{}_{m(l}n^m{}_{jk)}} - \mathsf{g}^I{}_{jm}n^m{}_{(kl)} \\
& - \underline{\underline{\underline{2\mathcal{N}^I{}_{ML}\mathcal{N}^M{}_{JK}f^J{}_j f^K{}_{(k}f^L{}_{l)}}}} + \underline{\underline{\underline{\underline{2\mathcal{N}^I{}_{MK}f^M{}_m f^K{}_{(k}n^m{}_{jl)}}}}}
\end{aligned}
$$

Observe that only the first four terms on the right hand sides possess skew symmetric contributions in indices K, L or k, l, respectively. However, if for example the additional constraints ${}^{\#}\mathsf{G}^m{}_{JL} \doteq 0$ and $\mathcal{F}^k{}_{[K,L]} \doteq 0$, that shall be discussed later, are satisfied $\mathsf{H}^i{}_{J[KL]}$ and $\mathcal{H}^i{}_{J[KL]}$ coincide. □

Discussion 6.6. Relation to Convected Coordinates
For integrable distortions and double-distortions the third (-order) deformation gradients expressed in convected coordinates expand into

$$
\boldsymbol{H} = [n^{\mathfrak{i}}{}_{\mathfrak{j}\mathfrak{k}} - \mathcal{N}^{\mathfrak{i}}_{\mathfrak{j}\mathfrak{k}}]_{:\mathfrak{l}}\,\boldsymbol{g}_{\mathfrak{i}} \otimes \boldsymbol{G}^{\mathfrak{j}} \otimes \boldsymbol{G}^{\mathfrak{k}} \otimes \boldsymbol{G}^{\mathfrak{l}} \quad \text{and} \quad \boldsymbol{h} = [\mathcal{N}^{\mathfrak{i}}_{\mathfrak{j}\mathfrak{k}} - n^{\mathfrak{i}}{}_{\mathfrak{j}\mathfrak{k}}]_{:\mathfrak{l}}\,\boldsymbol{G}_{\mathfrak{i}} \otimes \boldsymbol{g}^{\mathfrak{j}} \otimes \boldsymbol{g}^{\mathfrak{k}} \otimes \boldsymbol{g}^{\mathfrak{l}}.
$$

Thereby covariant derivatives of the difference in Christoffel symbols compute as

$$
\begin{aligned}
[n^{\mathfrak{i}}{}_{\mathfrak{j}\mathfrak{k}} - \mathcal{N}^{\mathfrak{i}}_{\mathfrak{j}\mathfrak{k}}]_{:\mathfrak{l}} = {} & [n^{\mathfrak{i}}{}_{\mathfrak{j}\mathfrak{k}} - \mathcal{N}^{\mathfrak{i}}_{\mathfrak{j}\mathfrak{k}}]_{,\mathfrak{l}} \\
& + [n^{\mathfrak{m}}{}_{\mathfrak{j}\mathfrak{k}} - \mathcal{N}^{\mathfrak{m}}{}_{\mathfrak{j}\mathfrak{k}}]n^{\mathfrak{i}}{}_{\mathfrak{m}\mathfrak{l}} - [n^{\mathfrak{i}}{}_{\mathfrak{m}\mathfrak{k}} - \mathcal{N}^{\mathfrak{i}}_{\mathfrak{m}\mathfrak{k}}]\mathcal{N}^{\mathfrak{m}}_{\mathfrak{j}\mathfrak{l}} - [n^{\mathfrak{i}}{}_{\mathfrak{j}\mathfrak{m}} - \mathcal{N}^{\mathfrak{i}}_{\mathfrak{j}\mathfrak{m}}]\mathcal{N}^{\mathfrak{m}}{}_{\mathfrak{k}\mathfrak{l}}, \\
[\mathcal{N}^{\mathfrak{i}}_{\mathfrak{j}\mathfrak{k}} - n^{\mathfrak{i}}{}_{\mathfrak{j}\mathfrak{k}}]_{:\mathfrak{l}} = {} & [\mathcal{N}^{\mathfrak{i}}_{\mathfrak{j}\mathfrak{k}} - n^{\mathfrak{i}}{}_{\mathfrak{j}\mathfrak{k}}]_{,\mathfrak{l}} \\
& + [\mathcal{N}^{\mathfrak{m}}{}_{\mathfrak{j}\mathfrak{k}} - n^{\mathfrak{m}}{}_{\mathfrak{j}\mathfrak{k}}]\mathcal{N}^{\mathfrak{i}}_{\mathfrak{m}\mathfrak{l}} - [\mathcal{N}^{\mathfrak{i}}_{\mathfrak{m}\mathfrak{k}} - n^{\mathfrak{i}}{}_{\mathfrak{m}\mathfrak{k}}]n^{\mathfrak{m}}{}_{\mathfrak{j}\mathfrak{l}} - [\mathcal{N}^{\mathfrak{i}}_{\mathfrak{j}\mathfrak{m}} - n^{\mathfrak{i}}{}_{\mathfrak{j}\mathfrak{m}}]n^{\mathfrak{m}}{}_{\mathfrak{k}\mathfrak{l}}.
\end{aligned}
$$

Observe carefully that due to the definition of the covariant derivative the coefficients of the spatial and material third (-order) deformation gradients differ more than only by the sign when expressed in convected coordinates.
□

6.3 Embedded General Metric Manifold

A continuum body is always embedded in three-dimensional Euclidean space (thus either rectilinear (Cartesian) or curvilinear coordinates may be used), nevertheless its kinematics may be described in terms of an embedded general metric manifold (space/geometry). As an example for the embedding of a metric manifold with torsion and curvature in Euclidean space consider the case of the *Cartan Spiral Staircase* as briefly discussed in Part II. A linear connection suitable to characterize the kinematics of a generalized continuum may then be defined either in material rectilinear (Cartesian) coordinates

$$L^A{}_{BC} = \left[f^A{}_a l^a{}_{bc} - f^A{}_{b,c} \right] F^b{}_B F^c{}_C, \tag{6.44}$$

or in spatial rectilinear (Cartesian) coordinates

$$l^a{}_{bc} = \left[F^a{}_A L^A{}_{BC} - F^a{}_{B,C} \right] f^B{}_b f^C{}_c, \tag{6.45}$$

or likewise in material curvilinear coordinates

$$\mathcal{L}^I{}_{JK} = f^I{}_i \left[\ell^i{}_{jk} \mathcal{F}^j{}_J \mathcal{F}^k{}_K + \mathcal{F}^i{}_{J,K} \right] = \mathcal{J}^I{}_A L^A{}_{BC} J^B{}_J J^C{}_K + \mathcal{N}^I{}_{JK}, \tag{6.46}$$

or in spatial curvilinear coordinates

$$\ell^i{}_{jk} = \mathcal{F}^i{}_I \left[\mathcal{L}^I{}_{JK} f^J{}_j f^K{}_k + f^I{}_{j,k} \right] = j^i{}_a l^a{}_{bc} j^b{}_j j^c{}_k + n^i{}_{jk}, \tag{6.47}$$

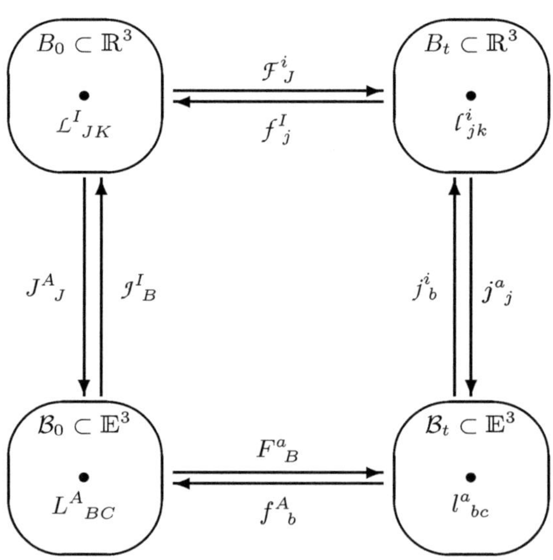

Fig. 6.9 A general metric manifold embedded in Euclidean space is characterized by its linear connection either in material and spatial rectilinear (Cartesian) or likewise in material and spatial curvilinear coordinates

see Fig. 6.8. Thereby, these different representations are related according to the transformation rule for linear connections as discussed in Part II. Note in particular that $\mathcal{N}^I{}_{JK}$ and $n^i{}_{jk}$ denote material and spatial Christoffel symbols that appear whenever curvilinear coordinates are used.

6.3.1 Holonomic Material Configuration

In the case of inelasticity, that shall be treated later, the situation in which the material configuration is endowed with holonomic coordinates whereas the corresponding spatial configuration is characterized by anholonomic coordinates is often encounter (keyword: looking forward from the material configuration to the intermediate configuration). As it turns out this situation may be captured by a material connection that consists of an integrable and a non-integrable part. The consequences for the relevant quantities from differential geometry such as the connection itself, the torsion, the non-metricity, and the curvature shall be examined in the sequel for material and spatial rectilinear (Cartesian) as well as for material and spatial curvilinear coordinates.

6.3.1.1 Connection

Rectilinear Coordinates. In material rectilinear (Cartesian) coordinates a linear connection $L^A{}_{BC}$ is proposed as an additive connection, i.e. as the addition of two terms

$$L^A{}_{BC} := {}^{\#}L^A{}_{BC} + f^A{}_a F^a{}_{B,C}. \tag{6.48}$$

Here the first and second term are denoted the *non-integrable* and the *integrable* part of the connection. It is sometimes convenient to factor out $f^A{}_a$ in the representation for the material connection

$$L^A{}_{BC} = f^A{}_a \big[{}^{\#}G^a{}_{BC} + F^a{}_{B,C} \big] \quad \text{with} \quad {}^{\#}L^A{}_{BC} =: f^A{}_a \, {}^{\#}G^a{}_{BC}. \tag{6.49}$$

Thus an alternative format for the material connection reads

$$L^A{}_{BC} = f^A{}_a G^a{}_{BC} \quad \text{with} \quad G^a{}_{BC} := {}^{\#}G^a{}_{BC} + F^a{}_{B,C}. \tag{6.50}$$

Here the double-distortion $G^a{}_{BC}$ and the relative double-distortion ${}^{\#}G^a{}_{BC}$, i.e. the deviation of $G^a{}_{BC}$ from the (compatible) gradient of the distortion $F^a{}_{B,C}$, have been introduced. Based on its non-tensorial transformation behavior the material connection $L^A{}_{BC}$ transforms into the corresponding connection in spatial rectilinear (Cartesian) coordinates as

$$l^a{}_{bc} = {}^{\#}l^a{}_{bc} \quad \text{with} \quad {}^{\#}l^a{}_{bc} := F^a{}_A \, {}^{\#}L^A{}_{BC} f^B{}_b f^C{}_c. \tag{6.51}$$

Alternatively, the spatial connection has the representation

$$l^a{}_{bc} = \big[G^a{}_{BC} - F^a{}_{B,C}\big] f^B{}_b f^C{}_c = {}^\#G^a{}_{BC} f^B{}_b f^C{}_c. \tag{6.52}$$

Observe that the *non-integrable* part of the connection in spatial rectilinear (Cartesian) coordinates follows by a tensorial transformation from its material counterpart and coincides here identically with the spatial connection itself. It thus vanishes obviously for a material connection that consists of only an integrable part. Recall that a non-vanishing spatial connection indicates *parallel transport* of non-Euclidean type.

Curvilinear Coordinates. A further transformation into spatial curvilinear coordinates then incorporates the corresponding spatial Christoffel symbols

$$\mathscr{l}^i{}_{jk} = {}^\#\mathscr{l}^i{}_{jk} + \mathscr{n}^i{}_{jk} \quad \text{with} \quad {}^\#\mathscr{l}^i{}_{jk} := \mathscr{j}^i{}_a\, {}^\#l^a{}_{bc}\, \mathscr{j}^b{}_j \mathscr{j}^c{}_k. \tag{6.53}$$

Accordingly, the transformation into material curvilinear coordinates renders

$$\mathcal{L}^I{}_{JK} = \mathscr{f}^I{}_i \mathcal{G}^i{}_{JK} + \mathscr{f}^I{}_i \mathscr{n}^i{}_{jk} \mathcal{F}^j{}_J \mathcal{F}^k{}_K \quad \text{with} \quad \mathcal{G}^i{}_{JK} := {}^\#\mathcal{G}^i{}_{JK} + \mathcal{F}^i{}_{J,K}. \tag{6.54}$$

Here the double-distortion $\mathcal{G}^i{}_{JK}$ and the relative double-distortion ${}^\#\mathcal{G}^I{}_{JK}$ in curvilinear coordinates have been introduced. Alternatively, starting from $L^A{}_{BC}$ in material rectilinear coordinates $\mathcal{L}^I{}_{JK}$ is obtained as

$$\mathcal{L}^I{}_{JK} = \mathscr{f}^I{}_i \mathsf{G}^i{}_{JK} + \mathcal{N}^I{}_{JK} \quad \text{with} \quad \mathsf{G}^i{}_{JK} := {}^\#\mathsf{G}^i{}_{JK} + \mathsf{F}^i{}_{J:K}. \tag{6.55}$$

Here $\mathsf{F}^i{}_{J:K} \equiv \mathcal{F}^i{}_{J:K}$ denotes the tensorial transformation of $F^a{}_{B,C}$ in terms of the Euclidean covariant derivative, and ${}^\#\mathsf{G}^i{}_{JK} \equiv {}^\#\mathcal{G}^i{}_{JK}$ is set for notational consistency. In the above the non-integrable part of the material connection is consistently defined as

$${}^\#\mathcal{L}^I{}_{JK} := \mathscr{f}^I{}_i\, {}^\#\mathcal{G}^i{}_{JK} = \mathscr{f}^I{}_i\, {}^\#\mathscr{l}^i{}_{jk} \mathcal{F}^j{}_J \mathcal{F}^k{}_K = \mathcal{J}^I{}_A\, {}^\#L^A{}_{BC} J^B{}_J J^C{}_K. \tag{6.56}$$

The two seemingly different representations for $\mathcal{L}^I{}_{JK}$ do indeed coincide due to the definition of the Euclidean covariant derivative of the distortion $\mathsf{F}^i{}_{J:K} \equiv \mathcal{F}^i{}_{J:K} := \mathcal{F}^i{}_{J,K} + \mathscr{n}^i{}_{jk} \mathcal{F}^j{}_J \mathcal{F}^k{}_K - \mathcal{F}^i{}_I \mathcal{N}^I{}_{JK}$.

6.3.1.2 Torsion

Rectilinear Coordinates. Suppose next that the rectilinear (Cartesian) material coordinates are *holonomic* whereas the corresponding spatial coordinates are *anholonomic*. Then the material and spatial torsion follow as

$$T^A{}_{BC} := 2 f^A{}_a \big[{}^\#G^A{}_{[BC]} + F^a{}_{[B,C]}\big] \quad \text{and} \quad t^a{}_{bc} := 2\, {}^\#l^a{}_{[bc]} + a^a{}_{bc}. \tag{6.57}$$

The second term in the spatial torsion is the contribution from the spatial anholonomic object $a^a{}_{bc} := 2F^a{}_{[B,C]} f^B{}_b f^C{}_c$ (recall that the corresponding material anholonomic object follows from a tensorial pull-back plus a change of sign as $A^A{}_{BC} := -2f^A{}_a F^a{}_{[B,C]}$). Obviously, the material and spatial torsion are related by a tensorial transformation

$$t^a{}_{bc} := F^a{}_A T^A{}_{BC} f^B{}_b f^C{}_c. \tag{6.58}$$

Note that for a vanishing non-integrable part of the connection the torsion agrees (modulo the sign) with the anholonomic object.

Curvilinear Coordinates. Thanks to the right symmetry of the Christoffel symbols, the material torsion in curvilinear coordinates reads

$$\mathcal{T}^I{}_{JK} := 2f^I{}_i\big[{}^\#\mathcal{G}^i{}_{[JK]} + \mathcal{F}^i{}_{[J,K]}\big] \quad \text{and} \quad t^i{}_{jk} = 2\,{}^\#\ell^i{}_{jk} + a^i{}_{jk}. \tag{6.59}$$

Since the torsion transforms tensorially, its representations in rectilinear and curvilinear coordinates do formally coincide.

6.3.1.3 Non-Metricity

Rectilinear Coordinates. Suppose next the following material and spatial metric

$$M_{AB} = F^a{}_A \delta_{ab} F^b{}_B \quad \text{and} \quad m_{ab} = \delta_{ab}. \tag{6.60}$$

The spatial metric i_{ab} indicates *distance determination* of Euclidean type. Then the fully covariant version of the connection reads as

$$L_{ABC} = {}^\#L_{ABC} + F^a{}_A \delta_{ab} F^b{}_{B,C} \quad \text{with} \quad {}^\#L_{ABC} := M_{AM}\,{}^\#L^M{}_{BC}. \tag{6.61}$$

The covariant derivative of the metric with respect to the connection thus results in

$$M_{AB|C} = F^a{}_{A,C}\delta_{ab}F^b{}_B + F^a{}_A\delta_{ab}F^b{}_{B,C} - 2L_{(AB)C} = -2\,{}^\#L_{(AB)C}. \tag{6.62}$$

Consequently, the non-metricity tensor $Q_{ABC} \equiv 2\,{}^\#L_{(AB)C}$ coincides with the left symmetric contribution to the non-integrable part of the connection. Thus the connection $L^A{}_{BC}$ is only a metric connection for a vanishing left symmetric contribution of its non-integrable part ${}^\#L_{(AB)C} \equiv 0$.

6.3.1.4 Curvature

Rectilinear Coordinates. In rectilinear (Cartesian) coordinates the material curvature based on the additive connection $L^A{}_{BC}$ follows in terms of the covariant derivative with respect to $L^A{}_{BC}$ as

$$R^A{}_{BCD} := 2\,{}^\#L^A{}_{B[D|C]} - 2\,{}^\#L^A{}_{M[C}\,{}^\#L^M{}_{BD]} - {}^\#L^A{}_{BM}T^M{}_{CD}. \tag{6.63}$$

Observe that the curvature is exclusively expressed in terms of the non-integrable part of the connection since the curvature tensor based on the integrable part vanishes identically. Likewise, the spatial curvature allows representation in terms of the covariant derivative with respect to ${l^a}_{bc}$ as

$$ {r^a}_{bcd} := 2\,{{}^\#l^a}_{b[d|c]} - 2\,{{}^\#l^a}_{m[c}\,{{}^\#l^m}_{bd]} - {{}^\#l^a}_{bm}{t^m}_{cd}. \tag{6.64}$$

Here the contribution of the spatial anholonomic object ${a^a}_{bc}$ to the torsion and the curvature in anholonomic spatial coordinates have been taken into account. Clearly, the material and spatial curvature are related by a tensorial transformation

$$ {r^a}_{bcd} := {F^a}_A {R^A}_{BCD} {f^B}_b {f^C}_c {f^D}_d. \tag{6.65}$$

Note the formally identical structure for the material and spatial curvature which is due to the tensorial transformation behavior of each of its terms. The material curvature may be re-expressed by expanding the covariant derivative

$$ {{}^\#L^A}_{B[D|C]} = {{}^\#L^A}_{B[D,C]} + {L^A}_{M[C}\,{{}^\#L^M}_{BD]} + {{}^\#L^A}_{M[C}{L^M}_{BD]} + \frac{1}{2}\,{{}^\#L^A}_{BM}{T^M}_{CD}$$

in terms of the partial derivative plus further correction terms to render

$$\begin{aligned} {R^A}_{BCD} &= 2\,{{}^\#L^A}_{B[D,C]} + 2\,{{}^\#L^A}_{M[C}\,{{}^\#L^M}_{BD]} \\ &\quad + 2\big[{L^A}_{M[C} - {{}^\#L^A}_{M[C}\big]\,{{}^\#L^M}_{BD]} + 2\,{{}^\#L^A}_{M[C}\big[{L^M}_{BD]} - {{}^\#L^M}_{BD]}\big] \end{aligned}$$

Alternatively, expressing the partial derivative of ${{}^\#L^A}_{BD} = {f^A}_a\,{{}^\#G^a}_{BD}$ as

$$ {{}^\#L^A}_{B[D,C]} = {f^A}_a\,{{}^\#G^a}_{B[D,C]} - \big[{L^A}_{M[C} - {{}^\#L^A}_{M[C}\big]\,{{}^\#L^M}_{BD]}$$

allows to rewrite the material curvature eventually as

$$ {R^A}_{BCD} = 2{f^A}_a\big[{G^a}_{B[D,C]} + {{}^\#G^a}_{M[C}{L^M}_{BD]}\big]. \tag{6.66}$$

Note that ${{}^\#G^a}_{B[D,C]}$ has been substituted equivalently by ${G^a}_{B[D,C]}$ due to the symmetry of the second partial derivatives ${F^a}_{B,[DC]} \equiv 0$. Finally the fourth-order tensor

$$ {K^a}_{BDC} := {G^a}_{B[D,C]} \equiv {{}^\#G^a}_{B[D,C]} \tag{6.67}$$

is introduced that will later be identified as capturing the (spatial) *disclination density*.

Curvilinear Coordinates. It can be shown that, based on the material connection in curvilinear coordinates as discussed in the above, the corresponding curvature tensor

$$ {\mathcal{R}^I}_{JKL} = {\jmath^I}_A {R^A}_{BCD} {J^B}_J {J^C}_K {J^D}_L \tag{6.68}$$

may consistently be expressed in terms of the double-distortion and the relative double-distortion in curvilinear coordinates

$$\mathcal{R}^I{}_{JKL} = 2\mathsf{f}^I{}_i\big[\mathsf{G}^i{}_{J[L:K]} + {}^\#\mathsf{G}^i{}_{M[K}\mathsf{L}^M{}_{JL]}\big]. \tag{6.69}$$

Here, the abbreviation $\mathsf{L}^M{}_{JL} := \mathsf{f}^M{}_m\mathsf{G}^m{}_{JL} = \mathsf{f}^M{}_m[\mathsf{F}^m{}_{J:L} + {}^\#\mathsf{G}^m{}_{JL}]$ has been invoked for notational consistency. In addition the fourth-order tensors

$$\mathsf{K}^i{}_{JLK} := \mathsf{G}^i{}_{J[L:K]} \quad \text{and} \quad \mathcal{K}^i{}_{JLK} := \mathcal{G}^i{}_{J[L,K]} \equiv {}^\#\mathcal{G}^i{}_{J[L,K]} \tag{6.70}$$

are defined as the *disclination density* tensors in curvilinear coordinates.

Proof. Based on the above representation for the material connection $\mathcal{L}^I{}_{JK} = f^I{}_i n^i{}_{jk}\mathcal{F}^j{}_J\mathcal{F}^k{}_K + f^I{}_i\mathcal{G}^i{}_{JK}$ and the material curvature

$$\mathcal{R}^I{}_{JKL} := 2\mathcal{L}^I{}_{J[L,K]} + 2\mathcal{L}^I{}_{M[K}\mathcal{L}^M{}_{JL]}$$

the proof follows simply from a tedious computation. To this end the first term on the right hand side of the definition for the material curvature in curvilinear coordinates is expanded into

$$\begin{array}{ll|l|l}
\mathcal{L}^I{}_{J[L,K]} = & f^I{}_{i,[K}\, n^i{}_{jl}\,\mathcal{F}^j{}_J\,\mathcal{F}^l{}_{L]} & = & - f^I{}_i\,\mathcal{F}^i{}_{M,[K}\, f^M{}_m\, n^m{}_{jl}\,\mathcal{F}^j{}_J\,\mathcal{F}^l{}_{L]} \quad \textcircled{a}\\
& + f^I{}_i\, n^i{}_{jl,k}\,\mathcal{F}^j{}_J\mathcal{F}^k{}_{[K}\,\mathcal{F}^l{}_{L]} & & + f^I{}_i\, n^i{}_{j[l,k]}\,\mathcal{F}^j{}_J\,\mathcal{F}^k{}_K\,\mathcal{F}^l{}_L \quad \textcircled{b}\\
& + f^I{}_i\, n^i{}_{jl}\,\mathcal{F}^j{}_{J,[K}\,\mathcal{F}^l{}_{L]} & & - f^I{}_i\, n^i{}_{jk}\,\mathcal{F}^k{}_{[K}\,\mathcal{F}^j{}_{J,L]} \quad \textcircled{c}\\
& + f^I{}_i\, n^i{}_{jl}\,\mathcal{F}^j{}_J\,\mathcal{F}^l{}_{[L,K]} & & - f^I{}_i\, n^i{}_{jk}\,\mathcal{F}^j{}_J\,\mathcal{F}^k{}_{[K,L]} \quad \textcircled{d}\\
& + f^I{}_{i,[K}\,\mathcal{G}^i{}_{JL]} & & - f^I{}_i\,\mathcal{F}^i{}_{M,[K}\, f^M{}_m\,\mathcal{G}^m{}_{JL]} \quad \textcircled{e}\\
& + f^I{}_i\,\mathcal{G}^i{}_{J[L,K]} & & + f^I{}_i\,\mathcal{K}^i{}_{JLK} \quad \textcircled{f}
\end{array}$$

Likewise, based on the representations for the material connection in curvilinear coordinates

$$\begin{aligned}
\mathcal{L}^I{}_{MK} &= f^I{}_i n^i{}_{mk}\mathcal{F}^m{}_M\mathcal{F}^k{}_K + f^I{}_i\mathcal{G}^i{}_{MK}\\
\mathcal{L}^M{}_{JL} &= f^M{}_n n^n{}_{jl}\mathcal{F}^j{}_J\mathcal{F}^l{}_L \;\; + f^M{}_n\mathcal{G}^n{}_{JL}
\end{aligned}$$

the second term on the right hand side of the definition for the curvature tensor unfolds into

$$\begin{array}{l|l|l}
\mathcal{L}^I{}_{M[K}\mathcal{L}^M{}_{JL]} = & & \\
f^I{}_i\mathcal{G}^i{}_{M[K}\; f^M{}_n n^n{}_{jl}\mathcal{F}^j{}_J\mathcal{F}^l{}_{L]} & = & f^I{}_i\mathcal{G}^i{}_{M[K}\; f^M{}_m n^m{}_{jl}\mathcal{F}^j{}_J\mathcal{F}^l{}_{L]} \quad \textcircled{a}'\\
+ f^I{}_i n^i{}_{mk}\mathcal{F}^m{}_M\mathcal{F}^k{}_{[K}\; f^M{}_n n^n{}_{jl}\mathcal{F}^j{}_J\mathcal{F}^l{}_{L]} & & + f^I{}_i n^i{}_{m[k}n^m{}_{jl]}\mathcal{F}^j{}_J\;\mathcal{F}^k{}_K\mathcal{F}^l{}_L \quad \textcircled{b}'\\
+ f^I{}_i n^i{}_{mk}\mathcal{F}^m{}_M\mathcal{F}^k{}_{[K}\; f^M{}_n\mathcal{G}^n{}_{JL]} & & + f^I{}_i n^i{}_{jk}\;\mathcal{F}^k{}_{[K}\mathcal{G}^j{}_{JL]} \quad \textcircled{c}'\\
+ f^I{}_i\mathcal{G}^i{}_{M[K}\; f^M{}_n\mathcal{G}^n{}_{JL]} & & + f^I{}_i\mathcal{G}^i{}_{M[K}\; f^M{}_m\mathcal{G}^m{}_{JL]} \quad \textcircled{e}'
\end{array}$$

Note that the terms Ⓑ and Ⓑ′ add up to $n^i{}_{j[l,k]} + n^i{}_{m[k}n^m{}_{jl]} = 0$ thus simply reflecting the flatness of the embedding Euclidean space. Based on the representation for the triple-distortion $\mathsf{H}^i{}_{JLK} = \mathsf{G}^i{}_{JL:K}$ the disclination density $\mathcal{K}^i{}_{JLK}$ in curvilinear coordinates is expressed as

$$\underbrace{\mathcal{K}^i{}_{JLK}}_{\textcircled{f}} = \mathsf{G}^i{}_{J[L:K]} - \underbrace{n^i{}_{jk}\mathcal{F}^k{}_{[K}\,{}^\#\mathcal{G}^j{}_{JL]}}_{\textcircled{c}''} + \underbrace{n^i{}_{jk}\mathcal{F}^j{}_J\mathcal{F}^k{}_{[K,L]}}_{\textcircled{d}''} - \underbrace{{}^\#\mathcal{G}^i{}_{M[K}\mathcal{N}^M{}_{JL]}}_{\textcircled{g}''}.$$

Then terms ⓒ, ⓒ′ and ⓒ″ as well as ⓓ and ⓓ″ cancel each other. The remaining terms ⓐ, ⓐ′ and ⓔ, ⓔ′ together with ⓖ″ combine into

$$f^I{}_i \Big[\mathsf{G}^i{}_{J[L:K]} + {}^{\#}\mathcal{G}^i{}_{M[K} \big[\mathcal{L}^M{}_{JL]} - \mathcal{N}^M{}_{JL]} \big] \Big] =: \mathsf{f}^I{}_i \Big[\mathsf{G}^i{}_{J[L:K]} + {}^{\#}\mathsf{G}^i{}_{M[K} \mathsf{L}^M{}_{JL]} \Big].$$

Here the identifications $f^I{}_i \equiv \mathsf{f}^I{}_i$, ${}^{\#}\mathcal{G}^i{}_{MK} \equiv {}^{\#}\mathsf{G}^i{}_{MK}$ and $\mathcal{L}^M{}_{JL} - \mathcal{N}^M{}_{JL} = \mathsf{f}^M{}_m[\mathsf{F}^m{}_{J:L} + {}^{\#}\mathsf{G}^m{}_{JL}] = \mathsf{f}^M{}_m \mathsf{G}^m{}_{JL} =: \mathsf{L}^m{}_{JL}$ have been made for notational consistency. □

6.3.2 *Holonomic Spatial Configuration*

The situation in which the spatial configuration is endowed with holonomic coordinates whereas the corresponding material configuration is characterized by anholonomic coordinates is also often encounter in the case of inelasticity (keyword: looking backward from the spatial configuration to the intermediate configuration). This situation may be captured by a spatial connection that consists of an integrable and a non-integrable part. The resulting relations are equivalent to those discussed in the previous subsection if the role of the spatial and the material configuration (and thus the notation in terms of upper and lower case letters) is simply reversed. However, for the sake of completeness and an explicit representation the relevant expressions shall be briefly discussed for spatial and material rectilinear (Cartesian) coordinates in the sequel.

6.3.2.1 Connection

Rectilinear Coordinates. In spatial rectilinear (Cartesian) coordinates a linear connection $l^a{}_{bc}$ is proposed as an additive connection, i.e. as the addition of two terms

$$l^a{}_{bc} := {}^{\#}l^a{}_{bc} + F^a{}_A f^A{}_{b,c}. \tag{6.71}$$

Here the first and second term are the *non-integrable* and the *integrable* part of the connection. An alternative format for the spatial connection reads

$$l^a{}_{bc} = F^a{}_A g^A{}_{bc} \quad \text{with} \quad g^A{}_{bc} := {}^{\#}g^A{}_{bc} + f^A{}_{b,c}. \tag{6.72}$$

Here the double-distortion $g^A{}_{bc}$ and the relative double-distortion ${}^{\#}g^A{}_{bc}$ have been introduced. Based on its non-tensorial transformation behavior the spatial connection $l^a{}_{bc}$ transforms into the corresponding connection in material rectilinear (Cartesian) coordinates as

$$L^A{}_{BC} = {}^{\#}L^A{}_{BC} \quad \text{with} \quad {}^{\#}L^A{}_{BC} := f^A{}_a\, {}^{\#}l^a{}_{bc} F^b{}_B F^c{}_C. \tag{6.73}$$

Alternatively, the material connection has the representation

$$L^A{}_{BC} = \big[g^A{}_{bc} - f^A{}_{b,c} \big] F^b{}_B F^c{}_C = {}^{\#}g^A{}_{bc} F^b{}_B F^c{}_C. \tag{6.74}$$

The *non-integrable* part of the connection in material rectilinear (Cartesian) coordinates follows by a tensorial transformation from its spatial counterpart and coincides here identically with the material connection itself. It thus vanishes obviously for a spatial connection that consists of only an integrable part. A non-vanishing material connection indicates *parallel transport* of non-Euclidean type.

6.3.2.2 Torsion

Rectilinear Coordinates. Suppose next that the rectilinear (Cartesian) spatial coordinates are *holonomic* whereas the corresponding material coordinates are *anholonomic*. Then the spatial and material torsion follow as

$$t^a{}_{bc} := 2F^a{}_A\big[\,{}^{\#}g^a{}_{[bc]} + f^A{}_{[b,c]}\big] \quad \text{and} \quad T^A{}_{BC} := 2\,{}^{\#}L^A{}_{[BC]} + A^A{}_{BC}. \tag{6.75}$$

The second term in the material torsion is the contribution from the material anholonomic object $A^A{}_{BC} := 2f^A{}_{[b,c]}F^b{}_BF^c{}_C$ (recall that the corresponding spatial anholonomic object follows from a tensorial push-forward plus a change of sign as $a^a{}_{bc} := -2F^a{}_Af^A{}_{[b,c]}$). Obviously, the spatial and material torsion are related by a tensorial transformation

$$T^A{}_{bc} := f^A{}_a t^a{}_{bc} F^b{}_B F^c{}_C. \tag{6.76}$$

Note that for a vanishing non-integrable part of the connection the torsion agrees (modulo the sign) with the anholonomic object.

6.3.2.3 Non-Metricity

Rectilinear Coordinates. Suppose next the following spatial and material metric

$$m_{ab} = f^A{}_a \delta_{AB} f^B{}_b \quad \text{and} \quad M_{AB} = \delta_{AB}. \tag{6.77}$$

The material metric I_{AB} indicates *distance determination* of Euclidean type. Then the fully covariant version of the connection reads as

$$l_{abc} = {}^{\#}l_{abc} + f^A{}_a \delta_{AB} f^B{}_{b,c} \quad \text{with} \quad {}^{\#}l_{abc} := m_{am}\,{}^{\#}l^m{}_{bc}. \tag{6.78}$$

The covariant derivative of the metric with respect to the connection thus results in

$$m_{ab|c} = f^A{}_{a,c}\delta_{AB}f^B{}_b + f^A{}_a\delta_{AB}f^B{}_{b,c} - 2l_{(ab)c} = -2\,{}^{\#}l_{(ab)c}. \tag{6.79}$$

Consequently, the non-metricity tensor $q_{abc} \equiv 2\,{}^{\#}l_{(ab)c}$ coincides with the left symmetric contribution to the non-integrable part of the connection. Thus the connection $l^a{}_{bc}$ is only a metric connection for a vanishing left symmetric contribution of its non-integrable part ${}^{\#}l_{(ab)c} \equiv 0$.

6.3.2.4 Curvature

Rectilinear Coordinates. In rectilinear (Cartesian) coordinates the spatial curvature based on the additive connection $l^a{}_{bc}$ follows in terms of the covariant derivative with respect to $l^a{}_{bc}$ as

$$r^a{}_{bcd} := 2\,{}^{\#}l^a{}_{b[d|c]} - 2\,{}^{\#}l^a{}_{m[c}\,{}^{\#}l^m{}_{bd]} - {}^{\#}l^a{}_{bm}t^m{}_{cd}. \tag{6.80}$$

Observe that the curvature is exclusively expressed in terms of the non-integrable part of the connection since the curvature tensor based on the integrable part vanishes identically. Likewise, the material curvature allows representation in terms of the covariant derivative with respect to $L^A{}_{BC}$ as

$$R^A{}_{BCD} := 2\,{}^{\#}L^A{}_{B[D|C]} - 2\,{}^{\#}L^A{}_{M[C}\,{}^{\#}L^M{}_{BD]} - {}^{\#}L^A{}_{BM}T^M{}_{CD}. \tag{6.81}$$

Here the contribution of the material anholonomic object $A^A{}_{BC}$ to the torsion and the curvature in anholonomic material coordinates have been taken into account. Clearly, the spatial and material curvature are related by a tensorial transformation

$$R^A{}_{BCD} := f^A{}_a r^a{}_{bcd} F^b{}_B F^c{}_C F^d{}_D. \tag{6.82}$$

Note the formally identical structure for the spatial and material curvature which is due to the tensorial transformation behavior of each of its terms. The spatial curvature may be re-expressed by expanding the covariant derivative

$${}^{\#}l^a{}_{b[d|c]} = {}^{\#}l^a{}_{b[d,c]} + l^a{}_{m[c}\,{}^{\#}l^m{}_{bd]} + {}^{\#}l^a{}_{m[c}l^m{}_{bd]} + \frac{1}{2}\,{}^{\#}l^a{}_{bm}t^m{}_{cd}$$

in terms of the partial derivative plus further correction terms to render

$$\begin{aligned} r^a{}_{bcd} &= 2\,{}^{\#}l^a{}_{b[d,c]} + 2\,{}^{\#}l^a{}_{m[c}\,{}^{\#}l^m{}_{bd]} \\ &+ 2\big[l^a{}_{m[c} - {}^{\#}l^a{}_{m[c]}\big]\,{}^{\#}l^m{}_{bd]} + 2\,{}^{\#}l^a{}_{m[c}\big[l^m{}_{bd]} - {}^{\#}l^m{}_{bd]}\big] \end{aligned}$$

Alternatively, expressing the partial derivative of ${}^{\#}l^a{}_{bd} = F^a{}_A\,{}^{\#}g^A{}_{bd}$ as

$${}^{\#}l^a{}_{b[d,c]} = F^a{}_A\,{}^{\#}g^A{}_{b[d,c]} - \big[l^a{}_{m[c} - {}^{\#}l^a{}_{m[c]}\big]\,{}^{\#}l^m{}_{bd]}$$

allows to rewrite the spatial curvature eventually as

$$r^a{}_{bcd} = 2F^a{}_A\big[g^A{}_{b[d,c]} + {}^{\#}g^A{}_{m[c}l^m{}_{bd]}\big]. \tag{6.83}$$

Note that ${}^{\#}g^A{}_{b[d,c]}$ has been substituted equivalently by $g^A{}_{b[d,c]}$ due to the symmetry of the second partial derivatives $f^A{}_{b,[dc]} \equiv 0$. Finally the fourth-order tensor

$$k^A{}_{bdc} := g^A{}_{b[d,c]} \equiv {}^{\#}g^A{}_{b[d,c]} \tag{6.84}$$

is introduced that will later be identified as capturing the (material) *disclination density*.

6.4 Integrability of Distortion and Double-Distortion

It is of considerable interest to relate the conditions i) for the integrability of the distortion into a compatible vector field (the deformation) and ii) for the integrability of the double-distortion into a compatible tensor field (the distortion) to the previously discussed concepts of differential geometry. To this end it is enlightening to first reconsider the concepts of coordinate, parallel and Cartan transport along finite circuits based on the connection of the embedded general metric manifold. Subsequently the integrability conditions are motivated from considering possible closure gaps of the deformation and the distortion upon either Burgers or Frank circuits. Thereby, the following consideration shall be restricted to the case of holonomic coordinates in the material configuration and possibly anholonomic coordinates in the spatial configuration. Clearly, the reverse case then follows in a straightforward manner.

Addendum 6.1. Stokes' Theorem in Euclidean Space
Based on the general exposition in Part II *Stokes' theorem* reads for a two-dimensional surface $\mathcal{A}_t := \widehat{\mathcal{U}}$ embedded in three-dimensional Euclidean space with $\mathcal{C}_t := \partial\widehat{\mathcal{U}}$ its closed boundary curve, and the (scalar-valued) one-form v (i.e. the covector $\boldsymbol{v}^\flat$)

$$-2\int_{\mathcal{A}_t} v_{[i,j]}\,\mathrm{d}_1\chi^i\,\mathrm{d}_2\chi^j = \oint_{\mathcal{C}_t} v_i\,\mathrm{d}\chi^i.$$

In the line integral on the right-hand-side $\mathrm{d}\chi^i := \mathrm{d}l^i$ is identified as the line element $\mathrm{d}\boldsymbol{l}^\sharp$ in the tangent space to $\mathcal{C}_t$. Due to the skew symmetry in the indices i, j the area integral on the left-hand-side may be rewritten in terms of the permutation symbol

$$-2\int_{\mathcal{A}_t} v_{[i,j]}\,\mathrm{d}_1\chi^i\,\mathrm{d}_2\chi^j = -\int_{\mathcal{A}_t} v_{i,j}\,e^{ijm}\,e_{mkl}\,\mathrm{d}_1\chi^k\,\mathrm{d}_2\chi^l.$$

Here the single contraction of the permutation symbol agrees with twice the fourth-order skew-symmetric unit tensor $e^{ijm}\,e_{mkl} = \delta^i{}_k\delta^j{}_l - \delta^i{}_l\delta^j{}_k$. Next, $\mathrm{d}_1\chi^k =: \mathrm{d}l_1^k$ and $\mathrm{d}_2\chi^l =: \mathrm{d}l_2^l$ are identified as line elements $\mathrm{d}\boldsymbol{l}_1^\sharp$ and $\mathrm{d}\boldsymbol{l}_2^\sharp$ in the tangent space to $\mathcal{A}_t$ thus identifying $e_{mkl}\,\mathrm{d}_1\chi^k\,\mathrm{d}_2\chi^l =: \mathrm{d}a_m$ as the corresponding covectorial area element $\mathrm{d}\boldsymbol{a}^\flat$. (The bi-vector $-e^{ijm}\,\mathrm{d}a_m =: \mathrm{d}a^{ij}$ denotes in addition the skew symmetric second-order contravariant tensorial area element $\mathrm{d}\boldsymbol{a}^\sharp$.) Moreover, $-v_{i,j}\,e^{ijm}$ are the coefficients of the curl of $\boldsymbol{v}^\flat$, i.e. of the vector $\operatorname{curl}\boldsymbol{v}^\flat$.

In summary Stokes' theorem in three-dimensional Euclidean space is expressed in symbolic notation as

$$\int_{\mathcal{A}_t} \operatorname{curl}\boldsymbol{v}^\flat \cdot \mathrm{d}\boldsymbol{a}^\flat = \oint_{\mathcal{C}_t} \boldsymbol{v}^\flat \cdot \mathrm{d}\boldsymbol{l}^\sharp.$$

(On the left-hand-side $\operatorname{curl} \boldsymbol{v}^\flat \cdot \mathrm{d}\boldsymbol{a}^\flat$ may alternatively be expressed in terms of the tensorial area element $\mathrm{d}\boldsymbol{a}^\sharp$ as $\operatorname{grad} \boldsymbol{v}^\flat : \mathrm{d}\boldsymbol{a}^\sharp$).

6.4.1 Coordinate Transport along Finite Circuits

Integrating the spatial coordinate increment $\mathrm{d}x^a$ along a finite circuit $\mathcal{C}_t$ (a process that shall be denoted *coordinate transport*) renders the equivalent to a *Burgers circuit* in terms of the anholonomic object

$$[\![x^a]\!] = \frac{1}{2} \int_{\mathcal{A}_t} a^a{}_{cd} \,\mathrm{d}a^{cd}. \tag{6.85}$$

The above expression highlights the role of the anholonomic object for capturing the resulting spatial *closure gap* for coordinate transport along a (closed) finite circuit. Thus only for holonomic spatial coordinates with zero anholonomic object the resulting closure gap does vanish identically.

Proof. Integrating the coordinate increment $\mathrm{d}x^a = \delta^a{}_c \,\mathrm{d}x^c$ along a finite circuit $\mathcal{C}_t$ renders

$$[\![x^a]\!] := \oint_{\mathcal{C}_t} \mathrm{d}x^a = \oint_{\mathcal{C}_t} \delta^a{}_c \,\mathrm{d}x^c.$$

In the case of anholonomic coordinates x^c it is necessary to first perform a transformation to the holonomic coordinates X^C in terms of the distortion $F^c{}_C$, i.e. $\mathrm{d}x^c = F^c{}_C \,\mathrm{d}X^C$, to re-express the jump of x^a as

$$[\![x^a]\!] = \oint_{\mathcal{C}_0} F^a{}_C \,\mathrm{d}X^C.$$

In a positively oriented manifold Stokes' theorem may then be applied to the circuit integral, thus rendering

$$[\![x^a]\!] = \int_{\mathcal{A}_0} F^a{}_{C,D} \,\mathrm{d}A^{CD}.$$

Taking into account the skew symmetry implied by the tensorial area element $\mathrm{d}A^{CD} = \mathrm{d}A^{[CD]}$ invokes the (right) skew symmetric Cauchy-type dislocation density tensor $a^a{}_{cd}$ via

$$F^a{}_{[C,D]} = \frac{1}{2} a^a{}_{cd} F^c{}_C F^d{}_D.$$

By observing further that $F^c{}_C F^d{}_D \, \mathrm{d}A^{CD} = \mathrm{d}a^{cd}$ the above integral may conveniently be re-written as

$$[\![x^a]\!] = \frac{1}{2} \int_{\mathcal{A}_t} a^a{}_{cd} \, \mathrm{d}a^{cd}.$$

As a conclusion the final result is expressed in terms of the anholonomic object due to spatial anholonomic coordinates. □

6.4.2 Parallel Transport along Finite Circuits

Parallel transport of a spatial vector v^a along a finite circuit $\mathcal{C}_t$ renders the equivalent to a *Frank circuit* in terms of the curvature tensor

$$[\![v^a]\!] = \frac{1}{2} \int_{\mathcal{A}_t} r^a{}_{bcd} v^b \, \mathrm{d}a^{cd}. \tag{6.86}$$

The above expression highlights the role of the curvature tensor for capturing the resulting spatial *orientation gap* for a parallel transport of a vector along a (closed) finite circuit. Thus in a flat manifold (space/geometry) with zero spatial curvature tensor also the resulting orientation gap vanishes identically; a case denoted as teleparallelism.

Proof. Upon parallel transport the increment $\mathrm{d}v^a = v^a{}_{.c} \, \mathrm{d}x^c$ of the vector v^a in a geometry with connection $l^a{}_{bc}$ reads

$$\mathrm{d}v^a = -l^a{}_{bc} v^b \, \mathrm{d}x^c.$$

Thus integrating the increment of the vector along a finite circuit $\mathcal{C}_t$ renders

$$[\![v^a]\!] := \oint_{\mathcal{C}_t} \mathrm{d}v^a = -\oint_{\mathcal{C}_t} l^a{}_{bc} v^b \, \mathrm{d}x^c.$$

In the case of anholonomic coordinates x^c it is necessary to first perform a transformation to the holonomic coordinates X^C in terms of the distortion $F^c{}_C$, i.e. $\mathrm{d}x^c = F^c{}_C \, \mathrm{d}X^C$, to re-express the jump of the vector v^a after parallel transport along a finite circuit $\mathcal{C}_0$ as

$$[\![v^a]\!] = -\oint_{\mathcal{C}_0} l^a{}_{bc} v^b F^c{}_C \, \mathrm{d}X^C.$$

In a positively oriented manifold Stokes' theorem may then be applied to the circuit integral, thus rendering

$$[\![v^a]\!] = -\int_{\mathcal{A}_0} \left[l^a{}_{bc}v^bF^c{}_C\right]_{,D} \mathrm{d}A^{CD}.$$

Taking into account the skew symmetry implied by the tensorial area element $\mathrm{d}A^{CD} = \mathrm{d}A^{[CD]}$ the partial derivative of the integrand is then expanded by the product rule as

$$\left[l^a{}_{bc}v^bF^c{}_{[C}\right]_{,D]} = \frac{1}{2}\left[2l^a{}_{b[c.d]}v^b + 2l^a{}_{e[c}v^e{}_{.d]} + l^a{}_{be}v^ba^e{}_{cd}\right]F^c{}_CF^d{}_D.$$

Here, the (right) skew symmetric Cauchy-type dislocation density tensor $a^e{}_{cd}$ is invoked via

$$F^e{}_{[C,D]} = \frac{1}{2}a^e{}_{cd}F^c{}_CF^d{}_D.$$

By observing further that $F^c{}_CF^d{}_D\,\mathrm{d}A^{CD} = \mathrm{d}a^{cd}$ the above integral may conveniently be re-written as

$$[\![v^a]\!] = -\frac{1}{2}\int_{\mathcal{A}_t} \left[2l^a{}_{b[c.d]}v^b + 2l^a{}_{e[c}v^e{}_{.d]} + l^a{}_{be}v^ba^e{}_{cd}\right] \mathrm{d}a^{cd}.$$

Upon finally expressing $v^e{}_{.d} = -l^e{}_{bd}v^b$ according to a parallel transport the second term in the integrand expands into

$$2l^a{}_{e[c}v^e{}_{.d]} = -2l^a{}_{e[c}l^e{}_{bd]}v^b.$$

Putting everything together and ordering terms thus renders eventually

$$[\![v^a]\!] = \frac{1}{2}\int_{\mathcal{A}_t} \left[2l^a{}_{b[d,c]} + 2l^a{}_{e[c}l^e{}_{bd]} - l^a{}_{be}a^e{}_{cd}\right]v^b\,\mathrm{d}a^{cd}.$$

As a conclusion the final result as stated in the above may be expressed in terms of the spatial curvature tensor as evaluated for anholonomic spatial coordinates. □

6.4.3 Cartan Transport along Finite Circuits

Cartan transport of a spatial radius vector z^a along a finite circuit $\mathcal{C}_t$ renders the equivalent to a generalized *Burgers circuit* in terms of the curvature and the torsion tensor

$$[\![z^a]\!] = \frac{1}{2}\int_{\mathcal{A}_t} \left[r^a{}_{bcd}z^b - t^a{}_{cd}\right] \mathrm{d}a^{cd}. \tag{6.87}$$

The above expression highlights the role of the combination of the curvature tensor and the torsion tensor for capturing the resulting spatial *closure gap* for a Cartan transport along a (closed) finite circuit. Thus in a flat manifold (space/geometry) with zero curvature tensor the resulting closure gap is exclusively due to the torsion tensor; likewise in a symmetric manifold (space/geometry) with zero torsion tensor the resulting closure gap is exclusively due to the curvature tensor (and coincides in this case formally with the resulting orientation gap); only in a flat and symmetric manifold (space/geometry) with zero curvature and torsion tensor the resulting closure gap does vanish identically.

Proof. Upon Cartan transport the increment $\mathrm{d}z^a = z^a{}_{.c}\,\mathrm{d}x^c$ of the radius vector z^a in a geometry with connection $l^a{}_{bc}$ reads

$$\mathrm{d}z^a = -\left[l^a{}_{bc}z^b + \delta^a{}_c\right]\mathrm{d}x^c.$$

Thus integrating the increment of the radius vector along a finite circuit $\mathcal{C}_t$ renders

$$[\![z^a]\!] := \oint_{\mathcal{C}_t} \mathrm{d}z^a = -\oint_{\mathcal{C}_t} \left[l^a{}_{bc}z^b + \delta^a{}_c\right]\mathrm{d}x^c.$$

In the case of anholonomic coordinates x^c it is necessary to first perform a transformation to the holonomic coordinates X^C in terms of the distortion $F^c{}_C$, i.e. $\mathrm{d}x^c = F^c{}_C\,\mathrm{d}X^C$, to re-express the jump of the radius vector z^a after Cartan transport along a finite circuit $\mathcal{C}_0$ as

$$[\![z^a]\!] = -\oint_{\mathcal{C}_0} \left[l^a{}_{bc}z^bF^c{}_C + F^a{}_C\right]\mathrm{d}X^C.$$

In a positively oriented manifold Stokes' theorem may then be applied to the circuit integral, thus rendering

$$[\![z^a]\!] = -\int_{\mathcal{A}_0} \left[l^a{}_{bc}z^bF^c{}_C + F^a{}_C\right]_{,D}\mathrm{d}A^{CD}.$$

Taking into account the skew symmetry implied by the tensorial area element $\mathrm{d}A^{CD} = \mathrm{d}A^{[CD]}$ the partial derivative of the first term in the integrand is then expanded by the product rule as

$$\left[l^a{}_{bc}z^bF^c{}_{[C}\right]_{,D]} = \frac{1}{2}\left[2l^a{}_{b[c.d]}z^b + 2l^a{}_{e[c}z^e{}_{.d]} + l^a{}_{be}z^ba^e{}_{cd}\right]F^c{}_CF^d{}_D.$$

Here, as well as for the partial derivative of the second term in the integrand the (right) skew symmetric Cauchy-type dislocation density tensor $a^a{}_{cd}$ is invoked via

$$F^a{}_{[C,D]} = \frac{1}{2}a^a{}_{cd}F^c{}_CF^d{}_D.$$

By observing further that $F^c{}_CF^d{}_D\,\mathrm{d}A^{CD} = \mathrm{d}a^{cd}$ the above integral may conveniently be re-written as

$$[\![z^a]\!] = -\frac{1}{2}\int\limits_{\mathcal{A}_t} \left[2l^a{}_{b[c.d]}z^b + 2l^a{}_{e[c}z^e{}_{.d]} + l^a{}_{be}z^b a^e{}_{cd} + a^a{}_{cd}\right] \mathrm{d}a^{cd}.$$

Upon finally expressing $z^e{}_{.d} = -[l^e{}_{bd}z^b + \delta^e{}_d]$ according to a Cartan transport the second term in the integrand expands into

$$2l^a{}_{e[c}z^e{}_{.d]} = -2\left[l^a{}_{e[c}l^e{}_{bd]}z^b + l^a{}_{[dc]}\right].$$

Putting everything together and ordering terms thus renders eventually

$$[\![z^a]\!] = \frac{1}{2}\int\limits_{\mathcal{A}_t} \left[\left[2l^a{}_{b[d.c]} + 2l^a{}_{e[c}l^e{}_{bd]} - l^a{}_{be}a^e{}_{cd}\right]z^b - \left[2l^a{}_{[cd]} + a^a{}_{cd}\right]\right] \mathrm{d}a^{cd}.$$

As a conclusion the final result as stated in the above may be expressed in terms of the curvature tensor and the torsion tensor as evaluated for anholonomic spatial coordinates. □

6.4.4 Integrability Condition for the Distortion

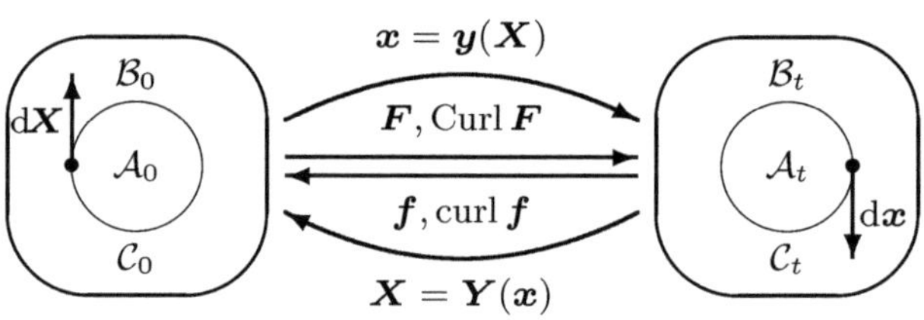

Fig. 6.10 Integrability conditions for the distortions
Right: The spatial closure failure $[\![\boldsymbol{y}]\!]$ in the spatial configuration $\mathcal{B}_t$ results from the line integral of the spatial line element $\mathrm{d}\boldsymbol{x}$ along any arbitrary closed circuit $\mathcal{C}_t$. The area surrounded by $\mathcal{C}_t$ is denoted $\mathcal{A}_t$, the corresponding vectorial area element is $\mathrm{d}\boldsymbol{a} = \boldsymbol{n}\,\mathrm{d}a$.
Left: The material closure failure $[\![\boldsymbol{Y}]\!]$ in the material configuration $\mathcal{B}_0$ results from the line integral of the material line element $\mathrm{d}\boldsymbol{X}$ along any arbitrary closed circuit $\mathcal{C}_0$. The area surrounded by $\mathcal{C}_0$ is denoted $\mathcal{A}_0$, the corresponding vectorial area element is $\mathrm{d}\boldsymbol{A} = \boldsymbol{N}\,\mathrm{d}A$.

Integrating the differential (forward) deformation $\mathrm{d}\boldsymbol{y}$ along a closed circuit $\mathcal{C}_t$ (encircling the area $\mathcal{A}_t$), see Fig. 6.10, determines a possible incompatibility of the (possibly non-continuous, multivalued) deformation in terms of the closure gap $[\![\boldsymbol{y}]\!]$, likewise integrating the differential (backward) deformation $\mathrm{d}\boldsymbol{Y}$ along a closed circuit $\mathcal{C}_0$ (encircling the area $\mathcal{A}_0$), see Fig. 6.10, determines a possible incompatibility of the (possibly non-continuous, multivalued) deformation in terms of the closure gap $[\![\boldsymbol{Y}]\!]$

$$[\![\boldsymbol{y}]\!] := \oint_{\mathcal{C}_t} \mathrm{d}\boldsymbol{y} \quad \text{and} \quad [\![\boldsymbol{Y}]\!] := \oint_{\mathcal{C}_0} \mathrm{d}\boldsymbol{Y}. \tag{6.88}$$

By relating the differential (forward and backward) deformations $\mathrm{d}\boldsymbol{y}$ and $\mathrm{d}\boldsymbol{Y}$ to the (forward and backward) distortions $\boldsymbol{F}$ and $\boldsymbol{f}$ (that are in general unrelated to the deformation)

$$\mathrm{d}\boldsymbol{y} := \boldsymbol{F} \cdot \mathrm{d}\boldsymbol{X} \quad \text{and} \quad \mathrm{d}\boldsymbol{Y} := \boldsymbol{f} \cdot \mathrm{d}\boldsymbol{x} \tag{6.89}$$

the closure gaps are re-expressed with a change of the integration domain as the circular integrals of the distortions as

$$[\![\boldsymbol{y}]\!] = \oint_{\mathcal{C}_0} \boldsymbol{F} \cdot \mathrm{d}\boldsymbol{X} \quad \text{and} \quad [\![\boldsymbol{Y}]\!] = \oint_{\mathcal{C}_t} \boldsymbol{f} \cdot \mathrm{d}\boldsymbol{x}. \tag{6.90}$$

Then, by the help of Stokes' theorem, the closure gaps are reformulated on the areas $\mathcal{A}_0$ and $\mathcal{A}_t$ surrounded by $\mathcal{C}_0$ and $\mathcal{C}_t$, respectively, and equipped with (vectorial) surface elements $\mathrm{d}\boldsymbol{A} := \boldsymbol{N}\,\mathrm{d}A$ and $\mathrm{d}\boldsymbol{a} := \boldsymbol{n}\,\mathrm{d}a$, respectively, as

$$[\![\boldsymbol{y}]\!] = \int_{\mathcal{A}_0} \operatorname{Curl}\boldsymbol{F} \cdot \mathrm{d}\boldsymbol{A} \quad \text{and} \quad [\![\boldsymbol{Y}]\!] = \int_{\mathcal{A}_t} \operatorname{curl}\boldsymbol{f} \cdot \mathrm{d}\boldsymbol{a}. \tag{6.91}$$

In summary, Eqs. 6.91.1 and 6.91.2 allow to determine the resulting (forward and backward) closure gaps for given distortions $\boldsymbol{F}$ and $\boldsymbol{f}$, respectively.

Discussion 6.7. Requiring that the closure gaps in Eqs. 6.91 vanish for arbitrary circuits $\mathcal{C}_0$ or $\mathcal{C}_t$, respectively, the *pointwise* conditions for the distortions $\boldsymbol{F}$ or $\boldsymbol{f}$, respectively, to be integrable to continuous, single-valued vector fields $\boldsymbol{y}^*$ or $\boldsymbol{Y}^*$, respectively, read

$$\boldsymbol{0} \doteq \operatorname{Curl}\boldsymbol{F} \quad \rightsquigarrow \quad \boldsymbol{F} = \operatorname{Grad}\boldsymbol{y}^* \quad \text{and}$$
$$\boldsymbol{0} \doteq \operatorname{curl}\boldsymbol{f} \quad \rightsquigarrow \quad \boldsymbol{f} = \operatorname{grad}\boldsymbol{Y}^*. \tag{6.92}$$

Thus an integrable distortion $\boldsymbol{F}$ that results in a vanishing closure gap $[\![\boldsymbol{y}]\!]$ for arbitrary circuits $\mathcal{C}_0$ within the material configuration $\mathcal{B}_0$ is expressible as the material gradient $\boldsymbol{F} = \operatorname{Grad}\boldsymbol{y}^*$ of a vector field $\boldsymbol{y}^*(\boldsymbol{X})$. Likewise an integrable distortion $\boldsymbol{f}$ that results in a vanishing closure gap $[\![\boldsymbol{Y}]\!]$ for arbitrary circuits $\mathcal{C}_t$ within the spatial configuration $\mathcal{B}_t$ is expressible as the spatial gradient $\boldsymbol{f} = \operatorname{grad}\boldsymbol{Y}^*$ of a vector field $\boldsymbol{Y}^*(\boldsymbol{x})$. Note, however, that based on the corresponding compatibility condition in Eq. 6.92 the field $\boldsymbol{y}^*$ or $\boldsymbol{Y}^*$, respectively, is only determined from a given distortion up to an arbitrary constant term. □

Coordinate Representation. In rectilinear (Cartesian) coordinates the condition for the distortions $F^a{}_B$ and $f^A{}_b$, respectively, to be integrable into vector fields $y^a(X^B)$ and $Y^A(x^b)$, respectively, derive simply from the symmetry of the second partial derivatives, i.e.

$$F^a{}_{[B,C]} \doteq 0 \quad \text{and} \quad f^A{}_{[b,c]} \doteq 0. \tag{6.93}$$

Since in curvilinear coordinates the Euclidean total covariant derivative (indicated by a double dot :) of $\mathsf{F}^i{}_J$ and $\mathsf{f}^I{}_j$, i.e. $\mathsf{F}^i{}_{J:K}$ and $\mathsf{f}^I{}_{j:k}$, may be considered a transformation of the $F^a{}_{B,C}$ or $f^A{}_{b,c}$, respectively, i.e.

$$F^a{}_{B,C} = j^a{}_i \mathsf{F}^i{}_{J:K} J^J{}_B J^K{}_C \quad \text{and} \quad f^A{}_{b,c} = J^A{}_I \mathsf{f}^I{}_{j:k} j^j{}_b j^k{}_c, \tag{6.94}$$

and since this is a similarity transformation in the indices J, K and j, k, respectively, that preserves the symmetry and skew symmetry properties, the integrability conditions for the distortions in curvilinear coordinates may be stated likewise as

$$\mathsf{F}^i{}_{[J:K]} \doteq 0 \quad \text{and} \quad \mathsf{f}^I{}_{[j:k]} \doteq 0. \tag{6.95}$$

The above integrability conditions will be related to the dislocation density tensors in the sequel.

Remark 6.1. Recall that the Euclidean total covariant derivatives of the distortions $\mathsf{F}^i{}_J$ and $\mathsf{f}^I{}_j$, respectively, follow as

$$\mathsf{F}^i{}_{J:K} = \mathcal{F}^i{}_{J,K} + \{\bullet\}^i{}_{(JK)} \quad \text{and} \quad \mathsf{f}^I{}_{j:k} = f^I{}_{j,k} + \{\bullet\}^i{}_{(jk)}.$$

Since, however, all terms on the right-hand-side except the first terms are (right) symmetric, the integrability condition for the distortion expressed in curvilinear coordinates may likewise be stated in terms of the ordinary partial derivative

$$\mathcal{F}^i{}_{[J,K]} \doteq 0 \quad \text{and} \quad f^I{}_{[j,k]} \doteq 0.$$

Thus the integrability condition for the distortion takes a similar format in curvilinear coordinates as compared to the expression in tensorially transforming curvilinear coordinates. □

6.4.5 Integrability Condition for the Double-Distortion

Integrating the differential (forward) distortion $\mathrm{d}\boldsymbol{F}$ along a closed circuit $\mathcal{C}_t$ (encircling the area $\mathcal{A}_t$), see Fig. 6.11, determines a possible incompatibility of the (possibly non-continuous, multivalued) distortion in terms of the distortion closure gap $[\![\boldsymbol{F}]\!]$, likewise integrating the differential (backward) distortion $\mathrm{d}\boldsymbol{f}$ along a closed circuit $\mathcal{C}_0$ (encircling the area $\mathcal{A}_0$), see Fig. 6.11, determines a possible incompatibility of the (possibly non-continuous, multivalued) distortion in terms of the distortion closure gap $[\![\boldsymbol{f}]\!]$

$$[\![\boldsymbol{F}]\!] := \oint_{\mathcal{C}_t} \mathrm{d}\boldsymbol{F} \quad \text{and} \quad [\![\boldsymbol{f}]\!] := \oint_{\mathcal{C}_0} \mathrm{d}\boldsymbol{f}. \tag{6.96}$$

By relating the differential (forward and backward) distortions $\mathrm{d}\boldsymbol{F}$ and $\mathrm{d}\boldsymbol{f}$ to the (forward and backward) double-distortions $\boldsymbol{G}$ and $\boldsymbol{g}$ (that are in general unrelated to the distortion)

$$\mathrm{d}\boldsymbol{F} := \boldsymbol{G} \cdot \mathrm{d}\boldsymbol{X} \quad \text{and} \quad \mathrm{d}\boldsymbol{f} := \boldsymbol{g} \cdot \mathrm{d}\boldsymbol{x} \tag{6.97}$$

the distortion closure gaps are re-expressed with a change of the integration domain as the circular integrals of the double-distortions as

$$[\![\boldsymbol{F}]\!] = \oint_{\mathcal{C}_0} \boldsymbol{G} \cdot \mathrm{d}\boldsymbol{X} \quad \text{and} \quad [\![\boldsymbol{f}]\!] = \oint_{\mathcal{C}_t} \boldsymbol{g} \cdot \mathrm{d}\boldsymbol{x}. \tag{6.98}$$

Then, by the help of Stokes' theorem, the distortion closure gaps are reformulated on the areas $\mathcal{A}_0$ and $\mathcal{A}_t$ surrounded by $\mathcal{C}_0$ and $\mathcal{C}_t$, respectively, and equipped with (vectorial) surface elements $\mathrm{d}\boldsymbol{A} := \boldsymbol{N}\,\mathrm{d}A$ and $\mathrm{d}\boldsymbol{a} := \boldsymbol{n}\,\mathrm{d}a$, respectively, as

$$[\![\boldsymbol{F}]\!] = \int_{\mathcal{A}_0} \mathrm{Curl}\,\boldsymbol{G} \cdot \mathrm{d}\boldsymbol{A} \quad \text{and} \quad [\![\boldsymbol{f}]\!] = \int_{\mathcal{A}_t} \mathrm{curl}\,\boldsymbol{g} \cdot \mathrm{d}\boldsymbol{a}. \tag{6.99}$$

In summary, Eqs. 6.99.1 and 6.99.2 allow to determine the resulting (forward and backward) distortion closure gaps for given double-distortions $\boldsymbol{G}$ and $\boldsymbol{g}$, respectively.

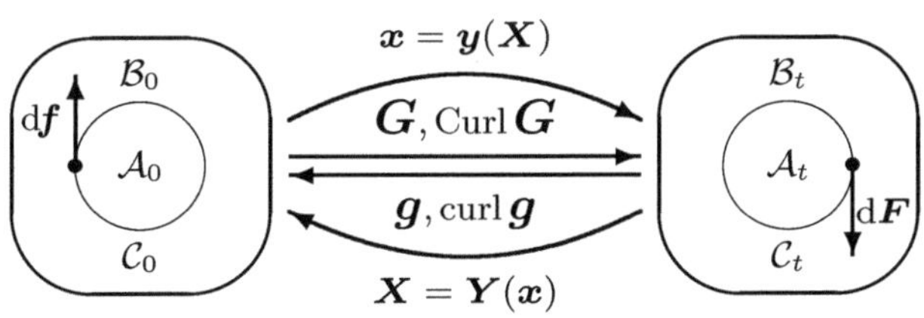

Fig. 6.11 Integrability conditions for the double-distortions
Right: Closure failure $[\![\boldsymbol{F}]\!]$ in the spatial configuration $\mathcal{B}_t$ as resulting from the line integral of the spatial distortion element $\mathrm{d}\boldsymbol{F}$ along any arbitrary closed circuit $\mathcal{C}_t$.
Left: Closure failure $[\![\boldsymbol{f}]\!]$ in the material configuration $\mathcal{B}_0$ as resulting from the line integral of the material distortion element $\mathrm{d}\boldsymbol{f}$ along any arbitrary closed circuit $\mathcal{C}_0$.

Next, on the one hand, either integrating $\mathrm{d}[\boldsymbol{F} \cdot \boldsymbol{X}]$ along a closed circuit $\mathcal{C}_t$ for holonomic $\boldsymbol{X}$ or, likewise, integrating $\mathrm{d}[\boldsymbol{f} \cdot \boldsymbol{x}]$ along a closed circuit $\mathcal{C}_0$ for holonomic $\boldsymbol{x}$, renders

$$\oint_{\mathcal{C}_0} \mathrm{d}[\boldsymbol{F}\cdot\boldsymbol{X}] = [\![\boldsymbol{F}\cdot\boldsymbol{X}]\!] = [\![\boldsymbol{F}]\!]\cdot\boldsymbol{X}_0 \quad \text{or} \quad \oint_{\mathcal{C}_t} \mathrm{d}[\boldsymbol{f}\cdot\boldsymbol{x}] = [\![\boldsymbol{f}\cdot\boldsymbol{x}]\!] = [\![\boldsymbol{f}]\!]\cdot\boldsymbol{x}_0. \tag{6.100}$$

Here $\boldsymbol{X}_0$ and $\boldsymbol{x}_0$ denote the position of the starting points of the circuit integrals. On the other hand, expanding the integrands, and substituting either $\mathrm{d}\boldsymbol{F} = \boldsymbol{G}\cdot\mathrm{d}\boldsymbol{X}$ or $\mathrm{d}\boldsymbol{f} = \boldsymbol{g}\cdot\mathrm{d}\boldsymbol{x}$ leads to

$$[\![\boldsymbol{F}]\!]\cdot\boldsymbol{X}_0 = \oint_{\mathcal{C}_0} \boldsymbol{F}\cdot\mathrm{d}\boldsymbol{X} + \oint_{\mathcal{C}_0} \boldsymbol{G} : [\boldsymbol{X}\otimes\mathrm{d}\boldsymbol{X}] \quad \text{and}$$

$$[\![\boldsymbol{f}]\!]\cdot\boldsymbol{x}_0 = \oint_{\mathcal{C}_t} \boldsymbol{f}\cdot\mathrm{d}\boldsymbol{x} + \oint_{\mathcal{C}_t} \boldsymbol{g} : [\boldsymbol{x}\otimes\mathrm{d}\boldsymbol{x}]. \tag{6.101}$$

Expressing the first terms on the respective right-hand sides either by $[\![\boldsymbol{y}]\!]$ or by $[\![\boldsymbol{Y}]\!]$, and applying Stokes' theorem to the second terms on the respective right-hand sides results eventually in

$$[\![\boldsymbol{F}]\!]\cdot\boldsymbol{X}_0 = [\![\boldsymbol{y}]\!] + \int_{\mathcal{A}_0} \operatorname{Curl}\boldsymbol{G} : [\boldsymbol{X}\otimes\mathrm{d}\boldsymbol{A}] + \int_{\mathcal{A}_0} \boldsymbol{G} : \boldsymbol{E}\cdot\mathrm{d}\boldsymbol{A} \quad \text{and}$$

$$[\![\boldsymbol{f}]\!]\cdot\boldsymbol{x}_0 = [\![\boldsymbol{Y}]\!] + \int_{\mathcal{A}_t} \operatorname{curl}\boldsymbol{g} : [\boldsymbol{x}\otimes\mathrm{d}\boldsymbol{a}] + \int_{\mathcal{A}_t} \boldsymbol{g} : \boldsymbol{e}\cdot\mathrm{d}\boldsymbol{a}. \tag{6.102}$$

Thus a possible non-integrability of the double-distortion results in an incompatibility of both the deformation and the distortion.

Discussion 6.8. Requiring that the distortion closure gaps in Eqs. 6.99 vanish for arbitrary circuits $\mathcal{C}_0$ or $\mathcal{C}_t$, respectively, the *pointwise* conditions for the double-distortions $\boldsymbol{G}$ or $\boldsymbol{g}$, respectively, to be integrable to continuous, single-valued tensor fields $\boldsymbol{F}^*$ or $\boldsymbol{f}^*$, respectively, read

$$\boldsymbol{0} \doteq \operatorname{Curl}\boldsymbol{G} \quad \rightsquigarrow \quad \boldsymbol{G} = \operatorname{Grad}\boldsymbol{F}^* \quad \text{and}$$

$$\boldsymbol{0} \doteq \operatorname{curl}\boldsymbol{g} \quad \rightsquigarrow \quad \boldsymbol{g} = \operatorname{grad}\boldsymbol{f}^*. \tag{6.103}$$

Thus an integrable double-distortion $\boldsymbol{G}$ that results in a vanishing closure gap $[\![\boldsymbol{F}]\!]$ for arbitrary circuits $\mathcal{C}_0$ within the material configuration $\mathcal{B}_0$ is expressible as the material gradient $\boldsymbol{G} = \operatorname{Grad}\boldsymbol{F}^*$ of a tensor field $\boldsymbol{F}^*(\boldsymbol{X})$. Likewise an integrable double-distortion $\boldsymbol{g}$ that results in a vanishing closure gap $[\![\boldsymbol{f}]\!]$ for arbitrary circuits $\mathcal{C}_t$ within the spatial configuration $\mathcal{B}_t$ is expressible as the spatial gradient $\boldsymbol{g} = \operatorname{grad}\boldsymbol{f}^*$ of a tensor field $\boldsymbol{f}^*(\boldsymbol{x})$. Note, however, that based on the corresponding compatibility condition in Eq. 6.103 the field $\boldsymbol{F}^*$ or $\boldsymbol{f}^*$, respectively, is only determined from a given double-distortion up to an arbitrary constant term. □

Discussion 6.9. With Eq. 6.103.1 satisfied, as an additional requirement of integrability and based on the discussion in the previous section, the condition for the tensor field $\boldsymbol{F}^*$ being integrable to a continuous, single-valued vector field $\boldsymbol{y}^{**}$ is likewise expressible in terms of the double-distortion

$$\boldsymbol{G} : \boldsymbol{E} \overset{6.103.1}{=} \operatorname{Grad} \boldsymbol{F}^* : \boldsymbol{E} \doteq \boldsymbol{0} \quad \rightsquigarrow \quad \boldsymbol{F}^* = \operatorname{Grad} \boldsymbol{y}^{**}. \tag{6.104}$$

Note, however, that based on the compatibility condition in Eqs. 6.104 the field $\boldsymbol{y}^{**}$ is only determined from a given $\boldsymbol{F}^*$ up to an arbitrary constant term. Likewise with Eq. 6.103.2 satisfied the condition for the tensor field $\boldsymbol{f}^*$ being integrable to a continuous, single-valued vector field $\boldsymbol{Y}^{**}$ is likewise expressible in terms of the double-distortion

$$\boldsymbol{g} : \boldsymbol{e} \overset{6.103.2}{=} \operatorname{grad} \boldsymbol{f}^* : \boldsymbol{e} \doteq \boldsymbol{0} \quad \rightsquigarrow \quad \boldsymbol{f}^* = \operatorname{grad} \boldsymbol{Y}^{**}. \tag{6.105}$$

Note, however, that based on the compatibility condition in Eqs. 6.105 the field $\boldsymbol{Y}^{**}$ is only determined from a given $\boldsymbol{f}^*$ up to an arbitrary constant term. □

Discussion 6.10. The condition $\operatorname{Curl} \boldsymbol{G} = \boldsymbol{0}$ in Eq. 6.103.1 does not necessarily provide integrability of $\boldsymbol{G}$ to a *given*, i.e. *particular* distortion $\boldsymbol{F}$, since every ${}^{\#}\boldsymbol{G}$ in $\boldsymbol{G} = \operatorname{Grad} \boldsymbol{F} + {}^{\#}\boldsymbol{G}$ with $\operatorname{Curl} {}^{\#}\boldsymbol{G} = \boldsymbol{0}$ will also satisfy $\operatorname{Curl} \boldsymbol{G} = \boldsymbol{0}$. Thus the additional constraint has to be considered for the *relative* double-distortion or likewise the non-integrable part of the connection

$${}^{\#}\boldsymbol{G} := \boldsymbol{G} - \operatorname{Grad} \boldsymbol{F} \doteq \boldsymbol{0} \quad \rightsquigarrow \quad {}^{\#}L_{ABC} = F^a{}_A \delta_{am} {}^{\#}G^m{}_{BC} = 0 \tag{6.106}$$

if the integrability of $\boldsymbol{G}$ into the *given* distortion $\boldsymbol{F}$ shall be enforced. The constraint ${}^{\#}L_{ABC} \doteq 0$ may also be satisfied by simultaneously satisfying

$${}^{\#}L_{(AB)C} \doteq 0 \quad \text{and} \quad {}^{\#}L_{A[BC]} \doteq 0 \quad \rightsquigarrow \quad {}^{\#}L_{ABC} \doteq -{}^{\#}L_{ABC}. \tag{6.107}$$

Recall that the (left) symmetric contribution to the non-integrable part of the connection determines the material non-metricity $Q_{ABC} = 2\,{}^{\#}L_{(AB)C}$. Finally, the constraint ${}^{\#}L_{A[BC]} \doteq 0$ or rather ${}^{\#}G^a{}_{[BC]} \doteq 0$ is obviously enforced by requiring $F^a{}_{[B,C]} \doteq 0$ in addition to $G^a{}_{[BC]} \doteq 0$, thus rendering

$$\boldsymbol{G} : \boldsymbol{E} \doteq \boldsymbol{0} \quad \text{and} \quad \operatorname{Curl} \boldsymbol{F} \doteq \boldsymbol{0} \quad \rightsquigarrow \quad {}^{\#}\boldsymbol{G} : \boldsymbol{E} \doteq \boldsymbol{0}. \tag{6.108}$$

Thereby, the condition $F^a{}_{[B,C]} \doteq 0$ has been substituted by $\operatorname{Curl} \boldsymbol{F} \doteq \boldsymbol{0}$. Likewise the condition $\operatorname{curl} \boldsymbol{g} = \boldsymbol{0}$ in Eq. 6.103.2 does not necessarily provide integrability of $\boldsymbol{g}$ to a *given*, i.e. *particular* distortion $\boldsymbol{f}$, since every ${}^{\#}\boldsymbol{g}$ in $\boldsymbol{g} = \operatorname{grad} \boldsymbol{f} + {}^{\#}\boldsymbol{g}$ with $\operatorname{curl} {}^{\#}\boldsymbol{g} = \boldsymbol{0}$ will also satisfy $\operatorname{curl} \boldsymbol{g} = \boldsymbol{0}$. Thus the additional constraint

$${}^{\#}\boldsymbol{g} := \boldsymbol{g} - \operatorname{grad} \boldsymbol{f} \doteq \boldsymbol{0} \quad \rightsquigarrow \quad {}^{\#}l_{abc} = f^A{}_a \delta_{AM} {}^{\#}g^M{}_{bc} = 0 \tag{6.109}$$

has to be considered for the *relative* double-distortion if the integrability of $\boldsymbol{g}$ into the *given* distortion $\boldsymbol{f}$ shall be enforced. The constraint ${}^{\#}l_{abc} \doteq 0$ may also be satisfied by simultaneously satisfying

$$ {}^{\#}l_{(ab)c} \doteq 0 \quad \text{and} \quad {}^{\#}l_{a[bc]} \doteq 0 \quad \rightsquigarrow \quad {}^{\#}l_{abc} \doteq -{}^{\#}l_{abc}. \tag{6.110} $$

Here the (left) symmetric contribution to the non-integrable part of the connection determines the spatial non-metricity $q_{abc} = 2\,{}^{\#}l_{(ab)c}$. Finally, the constraint ${}^{\#}l_{a[bc]} \doteq 0$ or rather ${}^{\#}g^{A}{}_{[bc]} \doteq 0$ is obviously enforced by requiring $f^{A}{}_{[b,c]} \doteq 0$ in addition to $g^{A}{}_{[bc]} \doteq 0$, thus rendering

$$ \boldsymbol{g} : \boldsymbol{e} \doteq \boldsymbol{0} \quad \text{and} \quad \operatorname{curl} \boldsymbol{f} \doteq \boldsymbol{0} \quad \rightsquigarrow \quad {}^{\#}\boldsymbol{g} : \boldsymbol{e} \doteq \boldsymbol{0}. \tag{6.111} $$

Thereby, the condition $f^{A}{}_{[b,c]} \doteq 0$ has been substituted by $\operatorname{curl} \boldsymbol{f} \doteq \boldsymbol{0}$. □

Coordinate Representation. In rectilinear (Cartesian) coordinates the condition for the double-distortions $G^{a}{}_{BC}$ and $g^{A}{}_{bc}$, respectively, to be integrable into distortions $F^{a}{}_{B}(X^{C})$ and $f^{A}{}_{b}(x^{c})$, respectively, derive simply from the symmetry of the second partial derivatives, i.e.

$$ G^{a}{}_{B[C,D]} \doteq 0 \quad \text{and} \quad g^{A}{}_{b[c,d]} \doteq 0. \tag{6.112} $$

Since in curvilinear coordinates the Euclidean total covariant derivative of $\mathsf{G}^{i}{}_{JK}$ and $\mathsf{g}^{I}{}_{jk}$, i.e. $\mathsf{G}^{i}{}_{JK:L}$ and $\mathsf{g}^{I}{}_{jk:l}$, may be considered a transformation of the $G^{a}{}_{BC,D}$ and $g^{A}{}_{bc,d}$, respectively, i.e.

$$ G^{a}{}_{BC,D} = j^{a}{}_{i} \mathsf{G}^{i}{}_{JK:L} \mathcal{J}^{J}{}_{B} \mathcal{J}^{K}{}_{C} \mathcal{J}^{L}{}_{D} \quad \text{and} \quad g^{A}{}_{bc,d} = J^{A}{}_{I} \mathsf{g}^{I}{}_{jk:l} j^{j}{}_{b} j^{k}{}_{c} j^{l}{}_{d}, \tag{6.113} $$

and since this is a similarity transformation in the indices K, L and k, l, respectively, that preserves the symmetry and skew symmetry properties, the integrability conditions for the double-distortions in curvilinear coordinates may be stated likewise as

$$ \mathsf{G}^{i}{}_{J[K:L]} \doteq 0 \quad \text{and} \quad \mathsf{g}^{I}{}_{j[k:l]} \doteq 0. \tag{6.114} $$

These integrability conditions will be related to the disclination density tensors in the sequel.

Remark 6.2. Recall that the Euclidean total covariant derivatives of the double-distortions $\mathsf{G}^{i}{}_{JK}$ and $\mathsf{g}^{I}{}_{jk}$, respectively, follow as

$$ \mathsf{G}^{i}{}_{JK:L} = \mathcal{G}^{i}{}_{JK,L} + n^{i}{}_{jk} \mathcal{F}^{j}{}_{J} \mathcal{F}^{k}{}_{K,L} - n^{i}{}_{mk} \mathcal{F}^{k}{}_{K} {}^{\#}\mathcal{G}^{m}{}_{JL} + {}^{\#}\mathcal{G}^{i}{}_{ML} \mathcal{N}^{M}{}_{JK} + \{\bullet\}^{i}{}_{J(KL)}, $$

and

$$ \mathsf{g}^{I}{}_{jk:l} = g^{I}{}_{jk,l} + \mathcal{N}^{I}{}_{JK} f^{J}{}_{j} f^{K}{}_{k,l} - \mathcal{N}^{I}{}_{MK} f^{K}{}_{k} {}^{\#}g^{M}{}_{jl} + {}^{\#}g^{I}{}_{ml} n^{m}{}_{jk} + \{\bullet\}^{i}{}_{j(kl)}. $$

Since, however, all terms on the right-hand-side except the first four terms are (right) symmetric, the integrability condition for the double-distortion expressed in curvilinear coordinates may likewise be stated as

$$\mathcal{G}^i{}_{J[K,L]} + n^i{}_{jk}\mathcal{F}^j{}_J\mathcal{F}^k{}_{[K,L]} - n^i{}_{mk}\mathcal{F}^k{}_{[K}\,{}^{\#}\mathcal{G}^m{}_{JL]} + {}^{\#}\mathcal{G}^i{}_{M[L}\mathcal{N}^M{}_{JK]} \doteq 0,$$

and

$$g^I{}_{j[k,l]} + \mathcal{N}^I{}_{JK}f^J{}_jf^K{}_{[k,l]} - \mathcal{N}^I{}_{MK}f^K{}_{[k}\,{}^{\#}g^M{}_{jl]} + {}^{\#}g^I{}_{m[l}n^m{}_{jk]} \doteq 0.$$

Thus the integrability condition for the double-distortion takes a quite involved format in curvilinear coordinates as compared to the expression in tensorially transforming curvilinear coordinates. □

6.4.6 Defect Density Tensors

Based on the above discussion the failure of the distortion and the double-distortion to be integrable shall be considered as different types of defect density tensors. Thereby the various defect density tensors as motivated for the geometrically linear case in Part I, Chapter 1 will here be re-considered for the more challenging geometrically nonlinear case. For an in-depth discussion of the physical background of defect densities in materials science please refer to Part I, Chapter 1 and the literature cited therein.

Primary Dislocation Density Tensors. Failure of the distortions being integrable is equivalent to either spatial or material non-vanishing Burgers vectors and is locally captured by the *primary dislocation density tensors*

$${}^{\mathrm{I}}\boldsymbol{D} := \operatorname{Curl}\boldsymbol{F} \quad \text{and} \quad {}^{\mathrm{I}}\boldsymbol{d} := \operatorname{curl}\boldsymbol{f}. \tag{6.115}$$

Thus the (Piola-type) primary dislocation densities tensors in rectilinear coordinates

$${}^{\mathrm{I}}D^{aD} = -F^a{}_{[B,C]}E^{BCD} \quad \text{and} \quad {}^{\mathrm{I}}d^{Ad} = -f^A{}_{[b,c]}e^{bcd}$$

correspond to the anholonomic objects of the previously discussed connection

$$A^A{}_{BC} = -2f^A{}_aF^a{}_{[B,C]} \quad \text{and} \quad a^a{}_{bc} = -2F^a{}_Af^A{}_{[b,c]}.$$

Remark 6.3. Observe that the primary dislocation density tensors satisfy the following differential identities

$$\operatorname{Div}{}^{\mathrm{I}}\boldsymbol{D} = \boldsymbol{0} \quad \text{and} \quad \operatorname{div}{}^{\mathrm{I}}\boldsymbol{d} = \boldsymbol{0}. \tag{6.116}$$

The above holds trivially true due to ${}^{\mathrm{I}}D^{aD}{}_{,D} = -F^a{}_{B,CD}E^{BCD} \equiv 0$ and ${}^{\mathrm{I}}d^{Ad}{}_{,d} = -f^A{}_{b,cd}e^{bcd} \equiv 0$, respectively. □

Disclination Density Tensors. Failure of the double-distortions being integrable is equivalent to either spatial or material non-vanishing Frank vectors and is locally captured by the *disclination density tensors*

$$\boldsymbol{K} := \operatorname{Curl} \boldsymbol{G} \quad \text{and} \quad \boldsymbol{k} := \operatorname{curl} \boldsymbol{g}. \tag{6.117}$$

The (Piola-type) disclination density tensors in rectilinear coordinates

$$K^a{}_B{}^E = -G^a{}_{B[C,D]} E^{CDE} \quad \text{and} \quad k^A{}_b{}^e = -g^A{}_{b[c,d]} e^{cde}$$

correspond (modulo the sign) to the disclination densities of the previously discussed connection

$$K^a{}_{BCD} = G^a{}_{B[C,D]} \quad \text{and} \quad k^A{}_{abc} = g^A{}_{a[b,c]}.$$

Remark 6.4. Observe that the disclination density tensors satisfy the following differential identities

$$\operatorname{Div} \boldsymbol{K} = \boldsymbol{0} \quad \text{and} \quad \operatorname{div} \boldsymbol{k} = \boldsymbol{0}. \tag{6.118}$$

The above holds trivially true due to $K^a{}_B{}^E{}_{,E} = -G^a{}_{BC,DE} E^{CDE} \equiv 0$ and $k^A{}_b{}^e{}_{,e} = -g^A{}_{bc,de} e^{cde} \equiv 0$, respectively. □

Point-Defect Density Tensors. Failure of the metric being compatible with the connection is captured by the *point-defect density tensors*

$$\boldsymbol{Q} := 2\, {}^{\#}L_{(AB)C} \boldsymbol{E}^A \otimes \boldsymbol{E}^B \otimes \boldsymbol{E}^C \quad \text{and} \quad \boldsymbol{q} := 2\, {}^{\#}l_{(ab)c} \boldsymbol{e}^a \otimes \boldsymbol{e}^b \otimes \boldsymbol{e}^c. \tag{6.119}$$

The point-defect density tensors $\boldsymbol{Q}$ and $\boldsymbol{q}$ coincide with the non-metricities Q_{ABC} and q_{abc} defined as the deviation of the non-integrable part of the connection from being (left) skew symmetric.

Secondary Dislocation Density Tensors. Failure of the double-distortions being (right) symmetric is captured in terms of the *secondary dislocation density tensors*

$${}^{\mathrm{II}}\boldsymbol{D} := -\boldsymbol{G} : \boldsymbol{E} \quad \text{and} \quad {}^{\mathrm{II}}\boldsymbol{d} := -\boldsymbol{g} : \boldsymbol{e}. \tag{6.120}$$

The (Piola-type) two-point secondary dislocation density tensors in rectilinear coordinates

$${}^{\mathrm{II}}D^{aD} = -G^a{}_{[BC]} E^{BCD} \quad \text{and} \quad {}^{\mathrm{II}}d^{Ad} = -g^A{}_{[bc]} e^{bcd}$$

correspond (modulo the sign) to the torsion of the previously discussed connection

$$T^A{}_{BC} = 2 f^A{}_a G^a{}_{[BC]} \quad \text{and} \quad t^a{}_{bc} = 2 F^a{}_A g^A{}_{[bc]}.$$

Remark 6.5. Observe that the secondary dislocation density tensors and the disclination density tensors satisfy jointly the following differential identities

$$\operatorname{Div}{}^{\mathrm{II}}\boldsymbol{D} = \boldsymbol{K} : \boldsymbol{1} \quad \text{and} \quad \operatorname{Div}\boldsymbol{K} = \boldsymbol{0} \quad \text{and}$$
$$\operatorname{div}{}^{\mathrm{II}}\boldsymbol{d} = \boldsymbol{k} : \boldsymbol{1} \quad \text{and} \quad \operatorname{div}\boldsymbol{k} = \boldsymbol{0}. \quad (6.121)$$

The above holds trivially true due to ${}^{\mathrm{II}}D^{aD}{}_{,D} = -G^{a}{}_{BC,D}E^{BCD} \equiv K^{a}{}_{B}{}^{B}$ and ${}^{\mathrm{II}}d^{Ad}{}_{,d} = -g^{A}{}_{bc,d}e^{bcd} \equiv k^{A}{}_{b}{}^{b}$, respectively. □

Remark 6.6. In terms of the disclination density tensors and the secondary dislocation density tensors the resulting Burgers vectors due to the non-integrability of the double-distortion read as

$$[\![\boldsymbol{y}]\!] = [\![\boldsymbol{F}]\!] \cdot \boldsymbol{X}_0 - \int_{\mathcal{A}_0} \boldsymbol{K} : [\boldsymbol{X} \otimes \mathrm{d}\boldsymbol{A}] + \int_{\mathcal{A}_0} {}^{\mathrm{II}}\boldsymbol{D} \cdot \mathrm{d}\boldsymbol{A} \quad \text{and}$$
$$[\![\boldsymbol{Y}]\!] = [\![\boldsymbol{f}]\!] \cdot \boldsymbol{x}_0 - \int_{\mathcal{A}_t} \boldsymbol{k} : [\boldsymbol{x} \otimes \mathrm{d}\boldsymbol{a}] + \int_{\mathcal{A}_t} {}^{\mathrm{II}}\boldsymbol{d} \cdot \mathrm{d}\boldsymbol{a}. \quad (6.122)$$

These relations are in formal agreement with similar statements in [7] (Eq. 26) for the geometrically linear case of a micropolar continuum. □

Table 6.4 Classification based on defect density tensors for the cases of holonomic or anholonomic coordinates and integrable or non-integrable connections. For holonomic coordinates and an integrable connection all defect density tensors vanish identically. For anholonomic coordinates and an integrable connection only the primary and secondary dislocation density tensors are different from zero and, moreover, do coincide. For anholonomic coordinates and a non-integrable connection all defect density tensors are different from zero.

		${}^{\mathrm{I}}\boldsymbol{D}$	${}^{\mathrm{II}}\boldsymbol{D}$	$\boldsymbol{K}$	$\boldsymbol{Q}$
Holonomic Coordinates $^{\#}\boldsymbol{F} = \boldsymbol{0}$	**Integrable Connection** $^{\#}\boldsymbol{G} = \boldsymbol{0}$	0	0	0	0
	Non-Integrable Connection $^{\#}\boldsymbol{G} \neq \boldsymbol{0}$	0	$-{}^{\#}\boldsymbol{G} : \boldsymbol{E}$	$\operatorname{Curl}{}^{\#}\boldsymbol{G}$	$\operatorname{sym}{}^{\#}\boldsymbol{L}^{\flat}$
Anholonomic Coordinates $^{\#}\boldsymbol{F} \neq \boldsymbol{0}$	**Integrable Connection** $^{\#}\boldsymbol{G} = \boldsymbol{0}$	$\operatorname{Curl}{}^{\#}\boldsymbol{F}$	$\operatorname{Curl}{}^{\#}\boldsymbol{F}$	0	0
	Non-Integrable Connection $^{\#}\boldsymbol{G} \neq \boldsymbol{0}$	$\operatorname{Curl}{}^{\#}\boldsymbol{F}$	$\operatorname{Curl}{}^{\#}\boldsymbol{F} - {}^{\#}\boldsymbol{G} : \boldsymbol{E}$	$\operatorname{Curl}{}^{\#}\boldsymbol{G}$	$\operatorname{sym}{}^{\#}\boldsymbol{L}^{\flat}$

6.4.7 Discussion of Continua with Defects

The key quantities characterizing a metric manifold (space, geometry) are the metric and the connection. Within continuum mechanics the (material) metric may always be associated with the right Cauchy-Green-type strain tensor expressed in terms of the distortion, i.e. $\boldsymbol{C} := \boldsymbol{F}^t \cdot \boldsymbol{F}$ with gradient $\operatorname{Grad} \boldsymbol{C} = 2\mathsf{I}^{\text{sym}} : [\boldsymbol{F}^t \cdot \operatorname{Grad} \boldsymbol{F}]$. Motivated by the irrotational $\oplus$ solenoidal (Helmholtz) decomposition of second-order tensors the distortion is regarded as the summation of a holonomic and an anholonomic contribution, i.e. $\boldsymbol{F} = \operatorname{Grad} \boldsymbol{\varphi} + {}^{\#}\boldsymbol{F}$. Moreover, the connection is directly related to the double-distortion that, again motivated by the irrotational $\oplus$ solenoidal (Helmholtz) decomposition of third-order tensors, is in turn decomposed into an integrable and a non-integrable contribution, i.e. $\boldsymbol{G} = \operatorname{Grad} \boldsymbol{F} + {}^{\#}\boldsymbol{G}$. Then the fully covariant connection $\boldsymbol{L}^{\flat} := \boldsymbol{F}^t \cdot \boldsymbol{G}$ expands as $\boldsymbol{L}^{\flat} = \boldsymbol{F}^t \cdot \operatorname{Grad} \boldsymbol{F} + {}^{\#}\boldsymbol{L}^{\flat}$ with non-integrable contribution ${}^{\#}\boldsymbol{L}^{\flat} := \boldsymbol{F}^t \cdot {}^{\#}\boldsymbol{G}$. Therefore it follows for the left symmetric $2\mathsf{I}^{\text{sym}} : \boldsymbol{L}^{\flat} = \operatorname{Grad} \boldsymbol{C} + 2\mathsf{I}^{\text{sym}} : {}^{\#}\boldsymbol{L}^{\flat}$. Summarizing, the key kinematic quantities of continuum mechanics that reflect the differential geometry of an embedded general metric manifold are

$$\boldsymbol{F} = \operatorname{Grad} \boldsymbol{\varphi} + {}^{\#}\boldsymbol{F} \quad \text{and} \quad \boldsymbol{G} = \operatorname{Grad} \operatorname{Grad} \boldsymbol{\varphi} + \operatorname{Grad} {}^{\#}\boldsymbol{F} + {}^{\#}\boldsymbol{G}. \tag{6.123}$$

Consequently, the defect density tensors introduced in the above are expressed in terms of the anholonomic contribution to the distortion ${}^{\#}\boldsymbol{F}$ and the non-integrable contribution to the double-distortion ${}^{\#}\boldsymbol{G}$ as

$$\begin{aligned} {}^{\text{I}}\boldsymbol{D} &= \operatorname{Curl} {}^{\#}\boldsymbol{F}, \\ {}^{\text{II}}\boldsymbol{D} &= \operatorname{Curl} {}^{\#}\boldsymbol{F} - {}^{\#}\boldsymbol{G} : \boldsymbol{E}, \\ \boldsymbol{K} &= \operatorname{Curl} {}^{\#}\boldsymbol{G}, \\ \boldsymbol{Q} &= 2\mathsf{I}^{\text{sym}} : {}^{\#}\boldsymbol{L}^{\flat} =: {}^{\text{sym}}\, {}^{\#}\boldsymbol{L}^{\flat}. \end{aligned} \tag{6.124}$$

These representations allow for a classification of continua with defects in terms of the anholonomic ${}^{\#}\boldsymbol{F}$ and the non-integrable ${}^{\#}\boldsymbol{G}$ as displayed in Tab. 6.4. It is interesting to observe that for the case of anholonomic coordinates and an integrable connection the anholonomic object and the torsion and thus ${}^{\text{I}}\boldsymbol{D}$ and ${}^{\text{II}}\boldsymbol{D}$ coincide. This case is at the heart of the classical discussion on dislocation densities in a metric manifold (space, geometry) with teleparallelism, e.g. [17, 18, 94, 98, 95].

In the case of non-integrable connections further sub-classification is possible based on the representation of the non-integrable contribution to the double-distortion ${}^{\#}\boldsymbol{G}$.

Irrotational $\oplus$ Solenoidal Decomposition of ${}^{\#}\boldsymbol{G}$: On the one hand motivated by its irrotational $\oplus$ solenoidal (Helmholtz) decomposition ${}^{\#}\boldsymbol{G}$

may be regarded as the summation of a gradient and a non-gradient contribution

$$ {}^{\#}\boldsymbol{G} = \operatorname{Grad} {}^{\#\#}\boldsymbol{F} + {}^{\#\#}\boldsymbol{G}. \tag{6.125} $$

The irrotational $\oplus$ solenoidal (Helmholtz) decomposition of the non-integrable contribution ${}^{\#}\boldsymbol{G}$ to the double-distortion allows the sub-classification into pseudo-integrable and non-pseudo-integrable connections as displayed in Tab. 6.5. Thereby the disclination density tensor $\boldsymbol{K}$ vanishes for a pseudo-integrable connection, however the non-metricity tensor $\boldsymbol{Q}$ is in general different from zero.

Table 6.5 Sub-classification of defect density tensors for the cases of holonomic or anholonomic coordinates and non-integrable connections that are either pseudo-integrable or non-pseudo-integrable. For pseudo-integrable connections the disclination density tensor $\boldsymbol{K}$ vanishes, however the non-metricity tensor $\boldsymbol{Q} = 2\mathsf{I}^{\text{sym}} : {}^{\#}\boldsymbol{L}^{\flat}$ is in general different from zero.

			${}^{\mathrm{I}}\boldsymbol{D}$	${}^{\mathrm{II}}\boldsymbol{D}$	$\boldsymbol{K}$
Holonomic Coordinates	**Non-Integrable Connection**	*Pseudo* ${}^{\#\#}\boldsymbol{G} = \boldsymbol{0}$	$\boldsymbol{0}$	$\operatorname{Curl} {}^{\#\#}\boldsymbol{F}$	$\boldsymbol{0}$
${}^{\#}\boldsymbol{F} = \boldsymbol{0}$	${}^{\#}\boldsymbol{G} \neq \boldsymbol{0}$	*Non-Pseudo* ${}^{\#\#}\boldsymbol{G} \neq \boldsymbol{0}$	$\boldsymbol{0}$	$\operatorname{Curl} {}^{\#\#}\boldsymbol{F} - {}^{\#\#}\boldsymbol{G} : \boldsymbol{E}$	$\operatorname{Curl} {}^{\#\#}\boldsymbol{G}$
Anholonomic Coordinates	**Non-Integrable Connection**	*Pseudo* ${}^{\#\#}\boldsymbol{G} = \boldsymbol{0}$	$\operatorname{Curl} {}^{\#}\boldsymbol{F}$	$\operatorname{Curl} {}^{\#}\boldsymbol{F} + \operatorname{Curl} {}^{\#\#}\boldsymbol{F}$	$\boldsymbol{0}$
${}^{\#}\boldsymbol{F} \neq \boldsymbol{0}$	${}^{\#}\boldsymbol{G} \neq \boldsymbol{0}$	*Non-Pseudo* ${}^{\#\#}\boldsymbol{G} \neq \boldsymbol{0}$	$\operatorname{Curl} {}^{\#}\boldsymbol{F}$	$\operatorname{Curl} {}^{\#}\boldsymbol{F} + \operatorname{Curl} {}^{\#\#}\boldsymbol{F} - {}^{\#\#}\boldsymbol{G} : \boldsymbol{E}$	$\operatorname{Curl} {}^{\#\#}\boldsymbol{G}$

Skew Symmetric $\oplus$ Symmetric Decomposition of ${}^{\#}\boldsymbol{G}$: On the other hand a decomposition of the fully covariant ${}^{\#}\boldsymbol{L}^{\flat} = \boldsymbol{F}^t \cdot {}^{\#}\boldsymbol{G}$ into left skew symmetric and left symmetric contributions is a valid option

$$ {}^{\#}\boldsymbol{L}^{\flat} = \boldsymbol{F}^t \cdot {}^{\#}\boldsymbol{G} = \mathsf{I}^{\text{skw}} : {}^{\#}\boldsymbol{L}^{\flat} + \mathsf{I}^{\text{sym}} : {}^{\#}\boldsymbol{L}^{\flat}. \tag{6.126} $$

The skew symmetric $\oplus$ symmetric decomposition of the non-integrable contribution ${}^{\#}\boldsymbol{G}$ to the double-distortion or rather of the non-integrable contribution ${}^{\#}\boldsymbol{L}^{\flat}$ to the fully covariant connection allows the sub-classification into metric and non-metric connections as displayed in Tab. 6.6. Thereby the non-metricity tensor $\boldsymbol{Q}$ vanishes for a metric connection, however the disclination density tensor $\boldsymbol{K}$ is in general different from zero. Note that non-metricity indicates a modified counting of crystallographic steps in a crystal due to various point defects, see [96].

Table 6.6 Sub-classification of defect density tensors for the cases of holonomic or anholonomic coordinates and non-integrable connections that are either metric or non-metric. For metric connections the non-metricity tensor $\boldsymbol{Q}$ vanishes, however the disclination density tensor $\boldsymbol{K} = \text{Curl}\ {}^\#\boldsymbol{G}$ is in general different from zero.

			${}^{\text{I}}\boldsymbol{D}$	${}^{\text{II}}\boldsymbol{D}$	$\boldsymbol{Q}$
Holonomic Coordinates ${}^\#\boldsymbol{F} = 0$	**Non-Integrable Connection** ${}^\#\boldsymbol{G} \neq 0$	*Metric* $\text{sym}\ {}^\#\boldsymbol{L}^\flat = 0$	0	$-{}^\#\boldsymbol{G} : \boldsymbol{E}$	0
		Non-Metric $\text{sym}\ {}^\#\boldsymbol{L}^\flat \neq 0$	0	$-{}^\#\boldsymbol{G} : \boldsymbol{E}$	$\text{sym}\ {}^\#\boldsymbol{L}^\flat$
Anholonomic Coordinates ${}^\#\boldsymbol{F} \neq 0$	**Non-Integrable Connection** ${}^\#\boldsymbol{G} \neq 0$	*Metric* $\text{sym}\ {}^\#\boldsymbol{L}^\flat = 0$	$\text{Curl}\ {}^\#\boldsymbol{F}$	$\text{Curl}\ {}^\#\boldsymbol{F} - {}^\#\boldsymbol{G} : \boldsymbol{E}$	0
		Non-Metric $\text{sym}\ {}^\#\boldsymbol{L}^\flat \neq 0$	$\text{Curl}\ {}^\#\boldsymbol{F}$	$\text{Curl}\ {}^\#\boldsymbol{F} - {}^\#\boldsymbol{G} : \boldsymbol{E}$	$\text{sym}\ {}^\#\boldsymbol{L}^\flat$

Example 6.1. A typical Ansatz for the non-integrable contribution to the fully covariant connection ${}^\#\boldsymbol{L}^\flat$ is formulated in terms of a proper orthogonal tensor field $\boldsymbol{R}$, i.e. a (micro) rotation, with left skew symmetric $\boldsymbol{R}^t \cdot \text{Grad}\,\boldsymbol{R}$, and a vector field $\boldsymbol{\Delta}$, i.e. a (micro) dilatation, so that

$$ {}^\#\boldsymbol{L}^\flat = \boldsymbol{R}^t \cdot \text{Grad}\,\boldsymbol{R} + \frac{1}{2}\boldsymbol{C} \otimes \boldsymbol{\Delta}. $$

Observe that the left symmetric contribution $\frac{1}{2}\boldsymbol{C}\otimes\boldsymbol{\Delta}$ takes the interpretation as a semi-metric connection with $\frac{3}{2}\boldsymbol{\Delta}$ the non-metricity covector. Then the left symmetric contribution to ${}^\#\boldsymbol{L}^\flat$ reads as

$$ 2\mathsf{I}^{\text{sym}} : {}^\#\boldsymbol{L}^\flat = \boldsymbol{C} \otimes \boldsymbol{\Delta}, $$

whereas the right skew symmetric contribution to ${}^\#\boldsymbol{L}^\flat$ expands into

$$ -{}^\#\boldsymbol{L}^\flat : \boldsymbol{E} = \boldsymbol{R}^t \cdot \text{Curl}\,\boldsymbol{R} - \frac{1}{2}\boldsymbol{C} \rtimes \boldsymbol{\Delta}. $$

Thus in this case $\boldsymbol{Q} = \boldsymbol{C} \otimes \boldsymbol{\Delta}$ and $\boldsymbol{F}^t \cdot [{}^{\text{II}}\boldsymbol{D} - {}^{\text{I}}\boldsymbol{D}] = \boldsymbol{R}^t \cdot \text{Curl}\,\boldsymbol{R} - \frac{1}{2}\boldsymbol{C} \rtimes \boldsymbol{\Delta}$, respectively. Connections of this type, with $\boldsymbol{R}$ and $\boldsymbol{\Delta}$ related to micromorphic degrees of freedom, have been advocated, e.g. by [37]. □

Supplement 6.1. Tensor Calculus in Euclidean Space

A n_{dm}-dimensional Euclidean space $\mathbb{E}^{n_{dm}}$ allows for rectilinear (Cartesian) as well as for curvilinear coordinates, moreover no distinction is necessary between vectors and covectors since $\mathbb{E}^{n_{dm}}$ coincides with both its tangent as well as its cotangent space. In an Euclidean (vector) space the multiplication of a scalar with a vector $\boldsymbol{v} = \alpha\boldsymbol{u}$, the addition of vectors $\boldsymbol{w} = \boldsymbol{u} + \boldsymbol{v}$, and the scalar product of vectors $\boldsymbol{u} \cdot \boldsymbol{v}$ are axiomatically defined. Tensor calculus as instrumental for the formulation of continuum mechanics can likewise be represented in symbolic (coordinate free) and coordinate notation. Comprehensive expositions of tensor calculus in Euclidean space are found, e.g., in [174, 85, 164, 90]. Here only some of the most relevant aspects of tensor calculus in *three-dimensional* Euclidean space shall be outlined briefly.

Affine Space and Affine Vector Space:

A *point space* $\mathbb{P}$ consists of a set of elements called *points* that are denoted by $\mathbf{x}, \mathbf{y}, \mathbf{z}, \cdots \in \mathbb{P}$. Points may for example be thought of as locations in a city like e.g. the Railway Station, the main building of the University, or the Chair of Applied Mechanics in Erlangen. (In the sequel a continuous distribution of points shall be assumed). The arrow from point $\mathbf{x}$ to point $\mathbf{y}$, i.e. the difference between $\mathbf{y}$ and $\mathbf{x}$, defines a unique *vector* $\boldsymbol{v} := \overrightarrow{\mathbf{xy}}$ that is an element from the *vector space* $\mathbb{V}$. Then the point $\mathbf{x}$ may be translated by the vector $\boldsymbol{v}$ into the new point $\mathbf{y}$; an operation that is denoted $\mathbf{y} = \mathbf{x} \oplus \boldsymbol{v}$. Thereby the operation $\oplus$ indicates the translation of a point by a vector that shall be denoted the addition of a vector to a point in the sequel; note that addition of two points makes no sense. The triplet $(\mathbb{P}, \mathbb{V}, \oplus)$ is denoted an *affine (point) space* if the operation $\oplus$ satisfies the following axioms:

a) $\forall(\mathbf{x}, \mathbf{y}) \rightsquigarrow \mathbf{y} = \mathbf{x} \oplus \boldsymbol{v}$ with unique $\boldsymbol{v} = \overrightarrow{\mathbf{xy}}$ uniqueness
b) $\mathbf{x} = \mathbf{x} \oplus \mathbf{0}$ identity
c) $[\mathbf{x} \oplus \boldsymbol{u}] \oplus \boldsymbol{v} = \mathbf{x} \oplus [\boldsymbol{u} + \boldsymbol{v}]$ associativity

Introducing furthermore a particular but arbitrary point $\mathbf{o} \in \mathbb{P}$ as the origin allows to identify each point $\mathbf{x}, \mathbf{y}, \mathbf{z}, \cdots \in \mathbb{P}$ with the corresponding position vector $\boldsymbol{x}, \boldsymbol{y}, \boldsymbol{z}, \cdots \in \mathbb{V}$, respectively, so that $\mathbf{x} = \mathbf{o} \oplus \boldsymbol{x}, \mathbf{y} = \mathbf{o} \oplus \boldsymbol{y}, \mathbf{z} = \mathbf{o} \oplus \boldsymbol{z}, \cdots$ holds. The points $\mathbf{x}, \mathbf{y}, \mathbf{z}, \cdots \in \mathbb{P}$ may thus equivalently be substituted by the position vectors $\boldsymbol{x}, \boldsymbol{y}, \boldsymbol{z}, \cdots \in \mathbb{V}$; the resulting triplet $(\mathbb{V}, \mathbb{V}, +)$ then forms an *affine vector space*. Note that here the operation $\oplus$ coincides simply with the ordinary vector addition. In summary, in contrast to the situation in a linear vector space, in an affine vector space no point of the underlying affine (point) space is distinguished, thus any point may serve likewise as the origin.

Base Vectors, Vectors and Metric Coefficients:

Three-dimensional Euclidean space $\mathbb{E}^3$ allows for sets of three linearly independent vectors $\boldsymbol{g}_i \neq \mathbf{0}$ such that $\alpha^i \boldsymbol{g}_i = \mathbf{0}$ holds only for all $\alpha^i = 0$. Here and in the following the Einstein summation convention for repeated indices is implied. Then every additional vector $\boldsymbol{v} \in \mathbb{E}^3$ is linearly dependent, i.e. $\lambda \boldsymbol{v} + \alpha^i \boldsymbol{g}_i = \mathbf{0}$ with $\lambda \neq 0$ and at least one $\alpha^i \neq 0$ and may be represented by $\boldsymbol{v} = v^i \boldsymbol{g}_i$. Here v^i are the *contravariant coefficients* of $\boldsymbol{v}$ with respect to the *covariant base vectors* $\boldsymbol{g}_i$ (natural basis).

Based on the orthonormality $\boldsymbol{g}_i \cdot \boldsymbol{g}^j = \delta_i^{\ j}$ with $\delta_i^{\ j}$ the Kronecker symbol, i.e $[\delta_i^{\ j}]$ is the diagonal unit matrix $\lceil 1, 1, 1 \rfloor$, the corresponding *contravariant base vectors* $\boldsymbol{g}^j$ (dual basis) are introduced, thus a vector also allows representation $\boldsymbol{v} = v_j \boldsymbol{g}^j$ with v_j its *covariant coefficients.* The co- and contravariant coefficients of $\boldsymbol{v}$ are extracted by $v_i = \boldsymbol{v} \cdot \boldsymbol{g}_i$ and $v^j = \boldsymbol{v} \cdot \boldsymbol{g}^j$, respectively.

Moreover, the co- and contravariant base vectors allow to compute the corresponding co- and contravariant *metric coefficients* $g_{ij} = \boldsymbol{g}_i \cdot \boldsymbol{g}_j$ and $g^{ij} = \boldsymbol{g}^i \cdot \boldsymbol{g}^j$, respectively. Then the matrix arrangement $[g_{ij}]$ is inverse to the matrix arrangement $[g^{ij}]$, for linearly independent $\boldsymbol{g}_i$ it holds that $g := \det[g_{ij}] \neq 0$. The metric coefficients may be used to *raise* and *lower* indices of the coefficients of a vector $v^i = g^{ij} v_j$ or $v_i = g_{ij} v^j$, and likewise of the base vectors $\boldsymbol{g}^i = g^{ij} \boldsymbol{g}_j$ or $\boldsymbol{g}_i = g_{ij} \boldsymbol{g}^j$.

Finally a Cartesian base system is in particular characterized by a set of orthonormal base vectors $\boldsymbol{e}_a = \boldsymbol{e}^a$ with metric coefficients $e_{ab} = e^{ab}$ coinciding with the Kronecker symbol $\delta_{ab} = \delta^{ab}$.

Products of Vectors:

The *scalar product* of vectors $\boldsymbol{u}$ and $\boldsymbol{v}$ is denoted as $p := \boldsymbol{u} \cdot \boldsymbol{v} \in \mathbb{R}$. It satisfies e.g. the following rules:

P1.1: *Scalar product of vectors:*

a)	$[\boldsymbol{u} + \boldsymbol{v}] \cdot \boldsymbol{w} = \boldsymbol{u} \cdot \boldsymbol{w} + \boldsymbol{v} \cdot \boldsymbol{w}$	distributivity	(vector addition)
	$\boldsymbol{w} \cdot [\boldsymbol{u} + \boldsymbol{v}] = \boldsymbol{w} \cdot \boldsymbol{u} + \boldsymbol{w} \cdot \boldsymbol{v}$		
b)	$\boldsymbol{u} \cdot [\alpha \boldsymbol{v}] = \alpha [\boldsymbol{u} \cdot \boldsymbol{v}] = \alpha \boldsymbol{u} \cdot \boldsymbol{v}$	associativity	(scalar multiplication)
	$[\alpha \boldsymbol{u}] \cdot \boldsymbol{v} = \alpha [\boldsymbol{u} \cdot \boldsymbol{v}] = \alpha \boldsymbol{u} \cdot \boldsymbol{v}$		
c)	$\boldsymbol{u} \cdot \boldsymbol{v} = \boldsymbol{v} \cdot \boldsymbol{u}$	commutativity	(vector multiplication)
d)	$\boldsymbol{u} \cdot \boldsymbol{v} = 0 \quad \rightarrow \boldsymbol{u} \perp \boldsymbol{v}$	orthogonality	(vector multiplication)

Then the *norm of a vector* $|\boldsymbol{v}|^2 = \boldsymbol{v} \cdot \boldsymbol{v}$ may be computed as its length, and the *Schwarz* and the *triangle inequality* $|\boldsymbol{u} \cdot \boldsymbol{v}| \leq |\boldsymbol{u}| \, |\boldsymbol{v}|$ and $|\boldsymbol{u} + \boldsymbol{v}| \leq |\boldsymbol{u}| + |\boldsymbol{v}|$, respectively, hold for this norm. Moreover the *angle* φ enclosed by $\boldsymbol{u}$ and $\boldsymbol{v}$ may be defined via $\boldsymbol{u} \cdot \boldsymbol{v} := |\boldsymbol{u}| \, |\boldsymbol{v}| \cos \varphi$. The scalar product of vectors has coordinate representation $\boldsymbol{u} \cdot \boldsymbol{v} = u^i v_i = u_i v^i$. Thus the scalar product of vectors follows as the summation of the products of the corresponding coefficients.

The *vector product* of vectors $\boldsymbol{u}$ and $\boldsymbol{v}$ is denoted as $\boldsymbol{p} := \boldsymbol{u} \times \boldsymbol{v} \in \mathbb{E}^3$ and is defined so that $\boldsymbol{p}$ is orthogonal to the plane spanned by $\boldsymbol{u}$ and $\boldsymbol{v}$ in a right-handed sense and has magnitude $|\boldsymbol{p}| := |\boldsymbol{u}| \, |\boldsymbol{v}| \sin \varphi$. It satisfies e.g. the following rules:

P1.2: *Vector product of vectors:*

a) $[\boldsymbol{u}+\boldsymbol{v}]\times\boldsymbol{w}=\boldsymbol{u}\times\boldsymbol{w}+\boldsymbol{v}\times\boldsymbol{w}$ $\boldsymbol{w}\times[\boldsymbol{u}+\boldsymbol{v}]=\boldsymbol{w}\times\boldsymbol{u}+\boldsymbol{w}\times\boldsymbol{v}$	distributivity	(vector addition)
b) $\boldsymbol{u}\times[\alpha\boldsymbol{v}]=\alpha[\boldsymbol{u}\times\boldsymbol{v}]=\alpha\boldsymbol{u}\times\boldsymbol{v}$ $[\alpha\boldsymbol{u}]\times\boldsymbol{v}=\alpha[\boldsymbol{u}\times\boldsymbol{v}]=\alpha\boldsymbol{u}\times\boldsymbol{v}$	associativity	(scalar multiplication)
c) $\boldsymbol{u}\times\boldsymbol{v}=-\boldsymbol{v}\times\boldsymbol{u}$	anti-commutativity	(vector multiplication)
d) $\boldsymbol{u}\times\boldsymbol{v}=\boldsymbol{0}\quad\rightarrow\boldsymbol{u}\parallel\boldsymbol{v}$	coaxiality	(vector multiplication)

It is important to observe that the vector product is *not commutative*! The vector product of vectors has coordinate representation $\boldsymbol{u}\times\boldsymbol{v}=u^iv^j\varepsilon_{ijk}\boldsymbol{g}^k=u_iv_j\varepsilon^{ijk}\boldsymbol{g}_k$. Here $\varepsilon_{ijk}=\sqrt{g}e_{ijk}$ and $\varepsilon^{ijk}=\sqrt[-2]{g}e^{ijk}$ denote the co- and contravariant Levi-Civita permutation symbols, see below.

The *scalar triple product* of vectors $\boldsymbol{u}$, $\boldsymbol{v}$ and $\boldsymbol{w}$ is denoted as $p:=[\boldsymbol{uvw}]\in\mathbb{R}$ and expands into $p=[\boldsymbol{u}\times\boldsymbol{v}]\cdot\boldsymbol{w}$. It satisfies the following rule:

P1.3: *Scalar triple product of vectors:*

$$[\boldsymbol{uvw}]=\quad[\boldsymbol{wuv}]=\quad[\boldsymbol{vwu}]\qquad\text{permutativity (even)}$$
$$=-[\boldsymbol{vuw}]=-[\boldsymbol{uwv}]=-[\boldsymbol{wvu}]\ \text{anti-permutativity (odd)}$$

The scalar triple product of vectors has coordinate representation $[\boldsymbol{uvw}]=u^iv^jw^k\varepsilon_{ijk}=u_iv_jw_k\varepsilon^{ijk}$ in terms of the co- and contravariant Levi-Civita permutation symbols. The scalar triple product of the co- and contravariant base vectors renders in particular $[\boldsymbol{g}_1\boldsymbol{g}_2\boldsymbol{g}_3]=\sqrt{g}$ and $[\boldsymbol{g}^1\boldsymbol{g}^2\boldsymbol{g}^3]=\sqrt[-2]{g}$, respectively. Moreover, the co- and contravariant Levi-Civita symbols allow representation as $\varepsilon_{ijk}=[\boldsymbol{g}_i\boldsymbol{g}_j\boldsymbol{g}_k]$ and $\varepsilon^{ijk}=[\boldsymbol{g}^i\boldsymbol{g}^j\boldsymbol{g}^k]$. Thus a compact representation for the vector product of base vectors reads $\boldsymbol{g}^i\times\boldsymbol{g}^j=\varepsilon^{ijk}\boldsymbol{g}_k$ and $\boldsymbol{g}_i\times\boldsymbol{g}_j=\varepsilon_{ijk}\boldsymbol{g}^k$.

The *vector triple product* of vectors $\boldsymbol{u}$, $\boldsymbol{v}$ and $\boldsymbol{w}$ is denoted as $\boldsymbol{p}:=[\boldsymbol{u}\times\boldsymbol{v}]\times\boldsymbol{w}=\boldsymbol{w}\times[\boldsymbol{v}\times\boldsymbol{u}]\in\mathbb{E}^3$ and expands into $\boldsymbol{p}=[\boldsymbol{u}\cdot\boldsymbol{w}]\boldsymbol{v}-[\boldsymbol{v}\cdot\boldsymbol{w}]\boldsymbol{u}$. It satisfies the following rule:

P1.4: *Vector triple product of vectors:*

$$[\boldsymbol{u}\times\boldsymbol{v}]\times\boldsymbol{w}+[\boldsymbol{w}\times\boldsymbol{u}]\times\boldsymbol{v}+[\boldsymbol{v}\times\boldsymbol{w}]\times\boldsymbol{u}=\boldsymbol{0}\ \text{Lagrange identity}$$
$$\boldsymbol{w}\times[\boldsymbol{v}\times\boldsymbol{u}]+\boldsymbol{v}\times[\boldsymbol{u}\times\boldsymbol{w}]+\boldsymbol{u}\times[\boldsymbol{w}\times\boldsymbol{v}]=\boldsymbol{0}$$

The vector triple product of vectors has coordinate representation $[\boldsymbol{u}\times\boldsymbol{v}]\times\boldsymbol{w}=u^iv^jw_k\varepsilon_{ijm}\varepsilon^{mkl}\boldsymbol{g}_l$ in terms of the the co- and contravariant Levi-Civita permutation symbols. Here the single contraction of the permutation symbol renders $\varepsilon_{ijm}\varepsilon^{mkl}=\delta_i^{\ k}\delta_j^{\ l}-\delta_i^{\ l}\delta_j^{\ k}$ thus the coordinate representation takes the expression $[\boldsymbol{u}\times\boldsymbol{v}]\times\boldsymbol{w}=[[u^kw_k]v^l-[v^kw_k]u^l]\boldsymbol{g}_l$.

Second-Order Tensors:

For vectors $\boldsymbol{u},\boldsymbol{v}\in\mathbb{E}^3$ a second-order tensor $\boldsymbol{T}$ may (for example) be defined either as a linear map of vectors into vectors $\boldsymbol{v}=\boldsymbol{T}\cdot\boldsymbol{u}$, a dyadic product of vectors $\boldsymbol{T}=\boldsymbol{u}\otimes\boldsymbol{v}$ (a simple tensor) or a multilinear (bilinear) form of vectors $\alpha_T(\boldsymbol{u},\boldsymbol{v})=\boldsymbol{u}\cdot\boldsymbol{T}\cdot\boldsymbol{v}\in\mathbb{R}$. Thereby a second-order tensor satisfies the following equivalent axioms:

T2.1: *Linear map of vectors into vectors:*

a)	$\boldsymbol{T}\cdot[\boldsymbol{u}+\boldsymbol{v}]=\boldsymbol{T}\cdot\boldsymbol{u}+\boldsymbol{T}\cdot\boldsymbol{v}$	distributivity	(vector addition)
b)	$[\boldsymbol{S}+\boldsymbol{T}]\cdot\boldsymbol{u}=\boldsymbol{S}\cdot\boldsymbol{u}+\boldsymbol{T}\cdot\boldsymbol{u}$	distributivity	(tensor addition)
c)	$\boldsymbol{T}\cdot[\alpha\boldsymbol{u}]=\alpha[\boldsymbol{T}\cdot\boldsymbol{u}]=\alpha\boldsymbol{T}\cdot\boldsymbol{u}$	associativity	(scalar multiplication)
d)	$[\alpha\boldsymbol{T}]\cdot\boldsymbol{u}=\alpha[\boldsymbol{T}\cdot\boldsymbol{u}]=\alpha\boldsymbol{T}\cdot\boldsymbol{u}$	associativity	(scalar multiplication)
e)	$\alpha\boldsymbol{T}=\boldsymbol{T}\alpha$	commutativity	(scalar multiplication)
f)	$\boldsymbol{0}\cdot\boldsymbol{u}=\boldsymbol{0}$	null element	(null tensor)
g)	$\boldsymbol{1}\cdot\boldsymbol{u}=\boldsymbol{u}$	identity	(identity tensor)

T2.2: *Dyadic product of vectors:*

a)	$[\boldsymbol{u}\otimes\boldsymbol{v}]\cdot\boldsymbol{w}=[\boldsymbol{v}\cdot\boldsymbol{w}]\boldsymbol{u}$	contraction	
b)	$\boldsymbol{u}\otimes[\boldsymbol{v}+\boldsymbol{w}]=\boldsymbol{u}\otimes\boldsymbol{v}+\boldsymbol{u}\otimes\boldsymbol{w}$	distributivity	(vector addition)
c)	$[\alpha\boldsymbol{u}]\otimes\boldsymbol{v}=\boldsymbol{u}\otimes[\alpha\boldsymbol{v}]=\alpha[\boldsymbol{u}\otimes\boldsymbol{v}]$	associativity	(scalar multiplication)

T2.3: *Multilinear form of vectors:*

a)	$\alpha_T(\boldsymbol{u},\boldsymbol{v}+\boldsymbol{w})=\alpha_T(\boldsymbol{u},\boldsymbol{v})+\alpha_T(\boldsymbol{u},\boldsymbol{w})$	distributivity	(vector addition)
b)	$\alpha_{T+S}(\boldsymbol{u},\boldsymbol{v})=\alpha_T(\boldsymbol{u},\boldsymbol{v})+\alpha_S(\boldsymbol{u},\boldsymbol{v})$	distributivity	(tensor addition)
c)	$\alpha_T(\boldsymbol{u},\beta\boldsymbol{v})=\beta\alpha_T(\boldsymbol{u},\boldsymbol{v})$	associativity	(scalar multiplication)
	$\alpha_T(\beta\boldsymbol{u},\boldsymbol{v})=\beta\alpha_T(\boldsymbol{u},\boldsymbol{v})$		
d)	$\alpha_{\beta T}(\boldsymbol{u},\boldsymbol{v})=\beta\alpha_T(\boldsymbol{u},\boldsymbol{v})$	associativity	(scalar multiplication)
e)	$\alpha_{\beta T}(\boldsymbol{u},\boldsymbol{v})=\alpha_{T\beta}(\boldsymbol{u},\boldsymbol{v})$	commutativity	(scalar multiplication)
f)	$\alpha_0(\boldsymbol{u},\boldsymbol{v})=0$	null element	(null tensor)
g)	$\alpha_1(\boldsymbol{u},\boldsymbol{v})=\boldsymbol{u}\cdot\boldsymbol{v}$	identity	(identity tensor)

The nine-dimensional tensorial product space $\mathbb{E}^3\otimes\mathbb{E}^3$ allows for sets of nine linearly independent dyadic products $\boldsymbol{g}_i\otimes\boldsymbol{g}_j\neq\boldsymbol{0}$ in terms of the linearly independent base vectors $\boldsymbol{g}_i$ such that $\alpha^{ij}\boldsymbol{g}_i\otimes\boldsymbol{g}_j=\boldsymbol{0}$ holds only for all $\alpha^{ij}=0$. Then every additional second-order tensor $\boldsymbol{T}\in\mathbb{E}^3\otimes\mathbb{E}^3$ is linearly dependent, i.e. $\lambda\boldsymbol{T}+\alpha^{ij}\boldsymbol{g}_i\otimes\boldsymbol{g}_j=\boldsymbol{0}$ with $\lambda\neq 0$ and at least one $\alpha^{ij}\neq 0$ and may be represented by $\boldsymbol{T}=T^{ij}\boldsymbol{g}_i\otimes\boldsymbol{g}_j$. Here T^{ij} are the contravariant coefficients of $\boldsymbol{T}$ with respect to the covariant base dyads $\boldsymbol{g}_i\otimes\boldsymbol{g}_j$.

Based on the corresponding contra- and mixedvariant base dyads $\boldsymbol{g}^i\otimes\boldsymbol{g}^j$, $\boldsymbol{g}_i\otimes\boldsymbol{g}^j$, and $\boldsymbol{g}^i\otimes\boldsymbol{g}_j$ a second-order tensor also allows representation $\boldsymbol{T}=T_{ij}\boldsymbol{g}^i\otimes\boldsymbol{g}^j$, $\boldsymbol{T}=T^i{}_j\boldsymbol{g}_i\otimes\boldsymbol{g}^j$, and $\boldsymbol{T}=T_i{}^j\boldsymbol{g}^i\otimes\boldsymbol{g}_j$ with T_{ij}, $T^i{}_j$, and $T_i{}^j$ its co- and mixedvariant coefficients. The co-, contra- and mixedvariant coefficients of $\boldsymbol{T}$ are extracted by $T_{ij}=\boldsymbol{g}_i\cdot\boldsymbol{T}\cdot\boldsymbol{g}_j$, $T^{ij}=\boldsymbol{g}^i\cdot\boldsymbol{T}\cdot\boldsymbol{g}^j$, $T_i{}^j=\boldsymbol{g}_i\cdot\boldsymbol{T}\cdot\boldsymbol{g}^j$, and $T^i{}_j=\boldsymbol{g}^i\cdot\boldsymbol{T}\cdot\boldsymbol{g}_j$, respectively. The coefficients of $\boldsymbol{T}$ may be arranged in a quadratic matrix $[T_{ij}]$, $[T_{ij}]$, $[T_i{}^j]$, or $[T^i{}_j]$, $\boldsymbol{T}$ has full rank for a second-order tensor constructed from at least three linearly independent dyads.

The metric coefficients may be used to raise and lower indices of the coefficients of a second-order tensor $T^{ij}=T^i{}_l g^{lj}=g^{ik}T_k{}^j=g^{ik}T_{kl}g^{lj}$, $T_{ij}=T_i{}^l g_{lj}=g_{ik}T^k{}_j=g_{ik}T^{kl}g_{lj}$, $T^i{}_j=T^{il}g_{lj}=g^{ik}T_{kj}=g^{ik}T_k{}^l g_{lj}$, or $T_i{}^j=T_{il}g^{lj}=g_{ik}T^{kj}=g_{ik}T^k{}_l g^{lj}$, respectively.

The linear map of vectors into vectors has coordinate representation $\boldsymbol{T}\cdot\boldsymbol{u}=T^{ij}u_j\boldsymbol{g}_i=T_{ij}u^j\boldsymbol{g}^i=T^i{}_ju^j\boldsymbol{g}_i=T_i{}^ju_j\boldsymbol{g}^i$. Moreover, the dyadic product of vectors has coordinate representation $\boldsymbol{u} \otimes \boldsymbol{v} = u^i v^j \boldsymbol{g}_i \otimes \boldsymbol{g}_j = u_i v_j \boldsymbol{g}^i \otimes \boldsymbol{g}^j = u^i v_j \boldsymbol{g}_i \otimes \boldsymbol{g}^j = u_i v^j \boldsymbol{g}^i \otimes \boldsymbol{g}_j$. Likewise, the multilinear form of vectors has coordinate representation $\alpha_T(\boldsymbol{u}, \boldsymbol{v}) = u_i T^{ij} v_j = u^j T_{ij} v^j = u_i T^i{}_j v^j = u^i T_i{}^j v_j$.

Products of Second-Order Tensors:
The *scalar product* of general second-order tensors $\boldsymbol{S}$ and $\boldsymbol{T}$ is denoted as $p := \boldsymbol{S} : \boldsymbol{T} \in \mathbb{R}$. It satisfies e.g. the following rules:

P2.1: *Scalar product of second-order tensors:*

a) $[\boldsymbol{R} + \boldsymbol{S}] : \boldsymbol{T} = \boldsymbol{R} : \boldsymbol{T} + \boldsymbol{S} : \boldsymbol{T}$ distributivity (tensor addition)
$\boldsymbol{T} : [\boldsymbol{R} + \boldsymbol{S}] = \boldsymbol{T} : \boldsymbol{R} + \boldsymbol{T} : \boldsymbol{S}$

b) $\boldsymbol{S} : [\alpha \boldsymbol{T}] = \alpha [\boldsymbol{S} : \boldsymbol{T}] = \alpha \boldsymbol{S} : \boldsymbol{T}$ associativity (scalar multiplication)
$[\alpha \boldsymbol{S}] : \boldsymbol{T} = \alpha [\boldsymbol{S} : \boldsymbol{T}] = \alpha \boldsymbol{S} : \boldsymbol{T}$

c) $\boldsymbol{S} : \boldsymbol{T} = \boldsymbol{T} : \boldsymbol{S}$ commutativity (tensor multiplication)

d) $\boldsymbol{S} : \boldsymbol{T} = 0 \quad \rightarrow \boldsymbol{S} \perp \boldsymbol{T}$ orthogonality (tensor multiplication)

Then the *norm of a second-order tensor* $|\boldsymbol{T}|^2 = \boldsymbol{T} : \boldsymbol{T}$ may be computed as its length, and the Schwarz and the triangle inequality $|\boldsymbol{S} : \boldsymbol{T}| \leq |\boldsymbol{S}|\, |\boldsymbol{T}|$ and $|\boldsymbol{S} + \boldsymbol{T}| \leq |\boldsymbol{S}| + |\boldsymbol{T}|$, respectively, hold for this norm. Moreover the angle φ enclosed by $\boldsymbol{S}$ and $\boldsymbol{T}$ may be defined via $\boldsymbol{S} : \boldsymbol{T} := |\boldsymbol{S}|\, |\boldsymbol{T}| \cos \varphi$. For the special case that $\boldsymbol{T} = \boldsymbol{u} \otimes \boldsymbol{v}$ is a simple second-order tensor the scalar product renders $p = \boldsymbol{S} : [\boldsymbol{u} \otimes \boldsymbol{v}] = \boldsymbol{u} \cdot \boldsymbol{S} \cdot \boldsymbol{v}$, if in addition also $\boldsymbol{S} = \boldsymbol{a} \otimes \boldsymbol{b}$ is a simple second-order tensor the scalar product results in $p = [\boldsymbol{a} \otimes \boldsymbol{b}] : [\boldsymbol{u} \otimes \boldsymbol{v}] = [\boldsymbol{u} \cdot \boldsymbol{a}][\boldsymbol{b} \cdot \boldsymbol{v}]$, thus e.g. for for the covariant base dyads $[\boldsymbol{g}_i \otimes \boldsymbol{g}_j] : [\boldsymbol{g}_k \otimes \boldsymbol{g}_l] = g_{ik} g_{jl}$ holds in particular. Consequently, the scalar product of second-order tensors has coordinate representation $\boldsymbol{S} : \boldsymbol{T} = S^{ij} T_{ij} = S_{ij} T^{ij} = S^i{}_j T_i{}^j = S_i{}^j T^i{}_j$. Thus the scalar product of second-order tensors follows as the summation of the products of the corresponding coefficients.

The *(left) vector cross product* of general second-order tensors $\boldsymbol{S}$ and $\boldsymbol{T}$ is denoted as $\boldsymbol{p} := \boldsymbol{S} \ltimes \boldsymbol{T} \in \mathbb{E}^3$ and is defined so that the scalar product $\boldsymbol{p} \cdot \boldsymbol{w}$ with $\boldsymbol{w} = \boldsymbol{u} \times \boldsymbol{v}$ results in $[\boldsymbol{S}^t \cdot \boldsymbol{u}] \cdot [\boldsymbol{T}^t \cdot \boldsymbol{v}] - [\boldsymbol{S}^t \cdot \boldsymbol{v}] \cdot [\boldsymbol{T}^t \cdot \boldsymbol{u}]$. It satisfies e.g. the following rules:

P2.2: *(Left) Vector cross product of second-order tensors:*

a) $[\boldsymbol{R} + \boldsymbol{S}] \ltimes \boldsymbol{T} = \boldsymbol{R} \ltimes \boldsymbol{T} + \boldsymbol{S} \ltimes \boldsymbol{T}$ distributivity (tensor addition)
$\boldsymbol{T} \ltimes [\boldsymbol{R} + \boldsymbol{S}] = \boldsymbol{T} \ltimes \boldsymbol{R} + \boldsymbol{T} \ltimes \boldsymbol{S}$

b) $\boldsymbol{S} \ltimes [\alpha \boldsymbol{T}] = \alpha [\boldsymbol{S} \ltimes \boldsymbol{T}] = \alpha \boldsymbol{S} \ltimes \boldsymbol{T}$ associativity (scalar multiplication)
$[\alpha \boldsymbol{S}] \ltimes \boldsymbol{T} = \alpha [\boldsymbol{S} \ltimes \boldsymbol{T}] = \alpha \boldsymbol{S} \ltimes \boldsymbol{T}$

c) $\boldsymbol{S} \ltimes \boldsymbol{T} = -\boldsymbol{T} \ltimes \boldsymbol{S}$ anti-commutativity (tensor multiplication)

It is important to observe that the (left) vector cross product is *not commutative*! For the special case that $\boldsymbol{T} = \boldsymbol{u} \otimes \boldsymbol{v}$ is a simple second-order tensor the (left) vector cross product renders $\boldsymbol{p} = \boldsymbol{S} \ltimes [\boldsymbol{u} \otimes \boldsymbol{v}] = -\operatorname{spn} \boldsymbol{u} \cdot [\boldsymbol{S} \cdot \boldsymbol{v}]$ (here $\operatorname{spn} \boldsymbol{u}$ associates a skew symmetric tensor with an axial vector $\boldsymbol{u}$, see below), if in addition also $\boldsymbol{S} = \boldsymbol{a} \otimes \boldsymbol{b}$ is a simple second-order tensor the (left) vector cross product results in $\boldsymbol{p} = [\boldsymbol{a} \otimes \boldsymbol{b}] \ltimes [\boldsymbol{u} \otimes \boldsymbol{v}] = [\boldsymbol{a} \times \boldsymbol{u}][\boldsymbol{b} \cdot \boldsymbol{v}]$, thus e.g. for the covariant base dyads $[\boldsymbol{g}_i \otimes \boldsymbol{g}_l] \ltimes [\boldsymbol{g}_j \otimes \boldsymbol{g}_m] = g_{lm} \varepsilon_{ijk} \boldsymbol{g}^k$ holds in particular. Consequently, the (left) vector cross product of second-order tensors has coordinate representation $\boldsymbol{S} \ltimes \boldsymbol{T} = S^{il}T^{jm}g_{lm}\varepsilon_{ijk}\boldsymbol{g}^k = S_{il}T_{jm}g^{lm}\varepsilon^{ijk}\boldsymbol{g}_k$ in terms of the the co- and contravariant Levi-Civita permutation symbols.

The *(right) vector cross product* of general second-order tensors $\boldsymbol{S}$ and $\boldsymbol{T}$ is denoted as $\boldsymbol{p} := \boldsymbol{S} \rtimes \boldsymbol{T} \in \mathbb{E}^3$ and is defined so that the scalar product $\boldsymbol{p} \cdot \boldsymbol{w}$ with $\boldsymbol{w} = \boldsymbol{u} \times \boldsymbol{v}$ results in $[\boldsymbol{S} \cdot \boldsymbol{u}] \cdot [\boldsymbol{T} \cdot \boldsymbol{v}] - [\boldsymbol{S} \cdot \boldsymbol{v}] \cdot [\boldsymbol{T} \cdot \boldsymbol{u}]$. It satisfies e.g. the following rules:

***P2.3:** (Right) Vector cross product of second-order tensors:*

a) $[\boldsymbol{R} + \boldsymbol{S}] \rtimes \boldsymbol{T} = \boldsymbol{R} \rtimes \boldsymbol{T} + \boldsymbol{S} \rtimes \boldsymbol{T}$ distributivity (tensor addition)
$\boldsymbol{T} \rtimes [\boldsymbol{R} + \boldsymbol{S}] = \boldsymbol{T} \rtimes \boldsymbol{R} + \boldsymbol{T} \rtimes \boldsymbol{S}$

b) $\boldsymbol{S} \rtimes [\alpha \boldsymbol{T}] = \alpha[\boldsymbol{S} \rtimes \boldsymbol{T}] = \alpha \boldsymbol{S} \rtimes \boldsymbol{T}$ associativity (scalar multiplication)
$[\alpha \boldsymbol{S}] \rtimes \boldsymbol{T} = \alpha[\boldsymbol{S} \rtimes \boldsymbol{T}] = \alpha \boldsymbol{S} \rtimes \boldsymbol{T}$

c) $\boldsymbol{S} \rtimes \boldsymbol{T} = -\boldsymbol{T} \rtimes \boldsymbol{S}$ anti-commutativity (tensor multiplication)

It is important to observe that the (right) vector cross product is *not commutative*! For the special case that $\boldsymbol{T} = \boldsymbol{u} \otimes \boldsymbol{v}$ is a simple second-order tensor the (right) vector cross product renders $\boldsymbol{p} = \boldsymbol{S} \rtimes [\boldsymbol{u} \otimes \boldsymbol{v}] = [\boldsymbol{u} \cdot \boldsymbol{S}] \cdot \operatorname{spn} \boldsymbol{v}$, if in addition also $\boldsymbol{S} = \boldsymbol{a} \otimes \boldsymbol{b}$ is a simple second-order tensor the (right) vector cross product results in $\boldsymbol{p} = [\boldsymbol{a} \otimes \boldsymbol{b}] \rtimes [\boldsymbol{u} \otimes \boldsymbol{v}] = [\boldsymbol{a} \cdot \boldsymbol{u}][\boldsymbol{b} \times \boldsymbol{v}]$, thus e.g. for the covariant base dyads $[\boldsymbol{g}_i \otimes \boldsymbol{g}_l] \rtimes [\boldsymbol{g}_j \otimes \boldsymbol{g}_m] = g_{ij}\varepsilon_{lmn}\boldsymbol{g}^n$ holds in particular. Consequently, the (right) vector cross product of second-order tensors has coordinate representation $\boldsymbol{S} \rtimes \boldsymbol{T} = S^{il}T^{jm}g_{ij}\varepsilon_{lmn}\boldsymbol{g}^n = S_{il}T_{jm}g^{ij}\varepsilon^{lmn}\boldsymbol{g}_n$ in terms of the the co- and contravariant Levi-Civita permutation symbols.

The *tensor cross product* of general second-order tensors $\boldsymbol{S}$ and $\boldsymbol{T}$ is denoted as $\boldsymbol{P} := \boldsymbol{S} \times\!\!\times \boldsymbol{T} \in \mathbb{E}^3 \otimes \mathbb{E}^3$ and is defined so that the linear map $\boldsymbol{P} \cdot \boldsymbol{w}$ with $\boldsymbol{w} = \boldsymbol{u} \times \boldsymbol{v}$ results in $[\boldsymbol{S} \cdot \boldsymbol{u}] \times [\boldsymbol{T} \cdot \boldsymbol{v}] - [\boldsymbol{S} \cdot \boldsymbol{v}] \times [\boldsymbol{T} \cdot \boldsymbol{u}]$. It satisfies e.g. the following rules:

***P2.4:** Tensor cross product of second-order tensors:*

a) $[\boldsymbol{R} + \boldsymbol{S}] \times\!\!\times \boldsymbol{T} = \boldsymbol{R} \times\!\!\times \boldsymbol{T} + \boldsymbol{S} \times\!\!\times \boldsymbol{T}$ distributivity (tensor addition)
$\boldsymbol{T} \times\!\!\times [\boldsymbol{R} + \boldsymbol{S}] = \boldsymbol{T} \times\!\!\times \boldsymbol{R} + \boldsymbol{T} \times\!\!\times \boldsymbol{S}$

b) $\boldsymbol{S} \times\!\!\times [\alpha \boldsymbol{T}] = \alpha[\boldsymbol{S} \times\!\!\times \boldsymbol{T}] = \alpha \boldsymbol{S} \times\!\!\times \boldsymbol{T}$ associativity (scalar multiplication)
$[\alpha \boldsymbol{S}] \times\!\!\times \boldsymbol{T} = \alpha[\boldsymbol{S} \times\!\!\times \boldsymbol{T}] = \alpha \boldsymbol{S} \times\!\!\times \boldsymbol{T}$

c) $\boldsymbol{S} \times\!\!\times \boldsymbol{T} = \boldsymbol{T} \times\!\!\times \boldsymbol{S}$ commutativity (tensor multiplication)

d) $[\boldsymbol{S} \times\!\!\times \boldsymbol{T}]^t = \boldsymbol{T}^t \times\!\!\times \boldsymbol{S}^t$ transposition

For the special case that $\boldsymbol{T} = \boldsymbol{u} \otimes \boldsymbol{v}$ is a simple second-order tensor the tensor cross product renders $\boldsymbol{P} = \boldsymbol{S} \times\!\!\times [\boldsymbol{u} \otimes \boldsymbol{v}] = -\operatorname{spn} \boldsymbol{u} \cdot \boldsymbol{S} \cdot \operatorname{spn} \boldsymbol{v}$, if in addition also $\boldsymbol{S} = \boldsymbol{a} \otimes \boldsymbol{b}$ is a simple second-order tensor the tensor cross product results in $\boldsymbol{P} = [\boldsymbol{a} \otimes \boldsymbol{b}] \times\!\!\times [\boldsymbol{u} \otimes \boldsymbol{v}] = [\boldsymbol{a} \times \boldsymbol{u}] \otimes [\boldsymbol{b} \times \boldsymbol{v}]$, thus e.g. for the covariant base dyads $[\boldsymbol{g}_i \otimes \boldsymbol{g}_l] \times\!\!\times [\boldsymbol{g}_j \otimes \boldsymbol{g}_m] = \varepsilon_{ijk}\varepsilon_{lmn}\boldsymbol{g}^k \otimes \boldsymbol{g}^n$ holds in particular. Consequently, the tensor cross product of second-order tensors has coordinate representation $\boldsymbol{S} \times\!\!\times \boldsymbol{T} = S^{il}T^{jm}\varepsilon_{ijk}\varepsilon_{lmn}\boldsymbol{g}^k \otimes \boldsymbol{g}^n = S_{il}T_{jm}\varepsilon^{ijk}\varepsilon^{lmn}\boldsymbol{g}_k \otimes \boldsymbol{g}_n$ in terms of the the co- and contravariant Levi-Civita permutation symbols. Finally, for $\boldsymbol{S} = \boldsymbol{T}$ the definition in the above renders $[\boldsymbol{T} \times\!\!\times \boldsymbol{T}] \cdot [\boldsymbol{u} \times \boldsymbol{v}] = 2[\boldsymbol{T} \cdot \boldsymbol{u}] \times [\boldsymbol{T} \cdot \boldsymbol{v}]$.

The *scalar triple product* of general second-order tensors $\boldsymbol{R}$, $\boldsymbol{S}$ and $\boldsymbol{T}$ is denoted as $p := [\boldsymbol{RST}] \in \mathbb{R}$ and expands into $p = [\boldsymbol{R} \times\!\!\times \boldsymbol{S}] : \boldsymbol{T}$. It satisfies the following rule:

P2.5: *Scalar triple product of second-order tensors:*

$[\boldsymbol{RST}] = [\boldsymbol{TRS}] = [\boldsymbol{STR}]$ permutativity (even)

$= [\boldsymbol{SRT}] = [\boldsymbol{RTS}] = [\boldsymbol{TSR}]$ permutativity (odd)

For e.g. the covariant base dyads $\big[[\boldsymbol{g}_i \otimes \boldsymbol{g}_l] \times\!\!\times [\boldsymbol{g}_j \otimes \boldsymbol{g}_m]\big] : [\boldsymbol{g}_k \otimes \boldsymbol{g}_n] = \varepsilon_{ijk}\varepsilon_{lmn}$ holds in particular. Consequently, the scalar triple product of general second-order tensors has coordinate representation $[\boldsymbol{RST}] = R^{il}S^{jm}T^{kn}\varepsilon_{ijk}\varepsilon_{lmn} = R_{il}S_{jm}T_{kn}\varepsilon^{ijk}\varepsilon^{lmn}$ in terms of the the co- and contravariant Levi-Civita permutation symbols. Finally, by setting $\boldsymbol{R} = \boldsymbol{S} = \boldsymbol{T}$ it follows from the identity $3! = \varepsilon^{lmn}\varepsilon_{lmn}$ and $\varepsilon^{lmn}[\boldsymbol{TTT}] = 3!T^{il}T^{jm}T^{kn}\varepsilon_{ijk}$ that $[\boldsymbol{uvw}][\boldsymbol{TTT}] = 3!\big[[\boldsymbol{T}\cdot\boldsymbol{u}][\boldsymbol{T}\cdot\boldsymbol{v}][\boldsymbol{T}\cdot\boldsymbol{w}]\big]$ holds.

The *(left) tensor cross product* of vectors $\boldsymbol{v}$ and general second-order tensors $\boldsymbol{T}$ is denoted as $\boldsymbol{P} := \boldsymbol{v} \ltimes \boldsymbol{T} \in \mathbb{E}^3 \otimes \mathbb{E}^3$ and is defined so that the linear maps $\boldsymbol{u}\cdot\boldsymbol{P}$ and $\boldsymbol{P}\cdot\boldsymbol{w}$ result in $[\boldsymbol{u}\times\boldsymbol{v}]\cdot\boldsymbol{T}$ and $\boldsymbol{v}\times[\boldsymbol{T}\cdot\boldsymbol{w}]$. It satisfies e.g. the following rules:

P2.6: *(Left) Tensor cross product of vectors and second-order tensors:*

a)	$[\boldsymbol{u}+\boldsymbol{v}] \ltimes \boldsymbol{T} = \boldsymbol{u} \ltimes \boldsymbol{T} + \boldsymbol{v} \ltimes \boldsymbol{T}$	distributivity	(vector addition)
	$\boldsymbol{v} \ltimes [\boldsymbol{S}+\boldsymbol{T}] = \boldsymbol{v} \ltimes \boldsymbol{S} + \boldsymbol{v} \ltimes \boldsymbol{T}$	distributivity	(tensor addition)
b)	$\boldsymbol{v} \ltimes [\alpha\boldsymbol{T}] = \alpha[\boldsymbol{v} \ltimes \boldsymbol{T}] = \alpha\boldsymbol{v} \ltimes \boldsymbol{T}$	associativity	(scalar multiplication)
	$[\alpha\boldsymbol{v}] \ltimes \boldsymbol{T} = \alpha[\boldsymbol{v} \ltimes \boldsymbol{T}] = \alpha\boldsymbol{v} \ltimes \boldsymbol{T}$		
c)	$-[\boldsymbol{v} \ltimes \boldsymbol{T}]^t = \boldsymbol{T} \ltimes \boldsymbol{v}$	transposition	
d)	$-[\boldsymbol{v} \ltimes \boldsymbol{1}]^t = \boldsymbol{1} \ltimes \boldsymbol{v} = \boldsymbol{v} \ltimes \boldsymbol{1}$	skew symmetry	

For the special case that $\boldsymbol{T} = \boldsymbol{a}\otimes\boldsymbol{b}$ is a simple second-order tensor the (left) tensor cross product renders $\boldsymbol{P} = \boldsymbol{v} \ltimes [\boldsymbol{a}\otimes\boldsymbol{b}] = [\boldsymbol{v}\times\boldsymbol{a}]\otimes\boldsymbol{b}$, thus e.g. for the covariant base dyads $\boldsymbol{g}_i \ltimes [\boldsymbol{g}_j \otimes \boldsymbol{g}_l] = \varepsilon_{ijk}\boldsymbol{g}^k \otimes \boldsymbol{g}_l$ holds in particular. Consequently, the tensor cross product of vectors and second-order tensors has coordinate representation $\boldsymbol{v} \ltimes \boldsymbol{T} = v^iT^{jl}\varepsilon_{ijk}\boldsymbol{g}^k \otimes \boldsymbol{g}_l = v_iT_{jl}\varepsilon^{ijk}\boldsymbol{g}_k \otimes \boldsymbol{g}^l$ in terms of the the co- and contravariant Levi-Civita permutation symbols.

The *(right) tensor cross product* of general second-order tensors $\boldsymbol{T}$ and vectors $\boldsymbol{v}$ is denoted as $\boldsymbol{P} := \boldsymbol{T} \rtimes \boldsymbol{v} \in \mathbb{E}^3 \otimes \mathbb{E}^3$ and is defined so that the linear maps $\boldsymbol{u}\cdot\boldsymbol{P}$ and $\boldsymbol{P}\cdot\boldsymbol{w}$ result in $[\boldsymbol{u}\cdot\boldsymbol{T}]\times\boldsymbol{v}$ and $\boldsymbol{T}\cdot[\boldsymbol{v}\times\boldsymbol{w}]$. It satisfies e.g. the following rules:

P2.7: *(Right) Tensor cross product of second-order tensors and vectors:*

a)	$\boldsymbol{T} \rtimes [\boldsymbol{u}+\boldsymbol{v}] = \boldsymbol{T} \rtimes \boldsymbol{u} + \boldsymbol{T} \rtimes \boldsymbol{v}$	distributivity	(vector addition)
	$[\boldsymbol{S}+\boldsymbol{T}] \rtimes \boldsymbol{v} = \boldsymbol{S} \rtimes \boldsymbol{v} + \boldsymbol{T} \rtimes \boldsymbol{v}$	distributivity	(tensor addition)
b)	$[\alpha\boldsymbol{T}] \rtimes \boldsymbol{v} = \alpha[\boldsymbol{T} \rtimes \boldsymbol{v}] = \alpha\boldsymbol{T} \rtimes \boldsymbol{v}$	associativity	(scalar multiplication)
	$\boldsymbol{T} \rtimes [\alpha\boldsymbol{v}] = \alpha[\boldsymbol{T} \rtimes \boldsymbol{v}] = \alpha\boldsymbol{T} \rtimes \boldsymbol{v}$		
c)	$-[\boldsymbol{T} \rtimes \boldsymbol{v}]^t = \boldsymbol{v} \rtimes \boldsymbol{T}$	transposition	
d)	$-[\boldsymbol{1} \rtimes \boldsymbol{v}]^t = \boldsymbol{v} \rtimes \boldsymbol{1} = \boldsymbol{1} \rtimes \boldsymbol{v}$	skew symmetry	

For the special case that $\boldsymbol{T} = \boldsymbol{a}\otimes\boldsymbol{b}$ is a simple second-order tensor the (right) tensor cross product renders $\boldsymbol{P} = [\boldsymbol{a}\otimes\boldsymbol{b}] \rtimes \boldsymbol{v} = \boldsymbol{a}\otimes[\boldsymbol{b}\times\boldsymbol{v}]$, thus e.g. for the covariant base dyads $[\boldsymbol{g}_i \otimes \boldsymbol{g}_j] \rtimes \boldsymbol{g}_k = \varepsilon_{jkl}\boldsymbol{g}_i \otimes \boldsymbol{g}^l$ holds in particular. Consequently, the tensor cross product of second-order tensors and vectors has coordinate representation $\boldsymbol{T} \rtimes \boldsymbol{v} = T^{ij}v^k\varepsilon_{jkl}\boldsymbol{g}_i \otimes \boldsymbol{g}^l = T_{ij}v_k\varepsilon^{jkl}\boldsymbol{g}^i \otimes \boldsymbol{g}_l$ in terms of the the co- and contravariant Levi-Civita permutation symbols.

The *tensor dot product* of general second-order tensors $\boldsymbol{S}$ and $\boldsymbol{T}$ is denoted as $\boldsymbol{P} := \boldsymbol{S} \cdot \boldsymbol{T} \in \mathbb{E}^3 \otimes \mathbb{E}^3$. It satisfies e.g. the following rules:

***P2.8:** Tensor dot product of second-order tensors:*

a)	$[\boldsymbol{R}+\boldsymbol{S}]\cdot\boldsymbol{T} = \boldsymbol{R}\cdot\boldsymbol{T}+\boldsymbol{S}\cdot\boldsymbol{T}$	distributivity	(tensor addition)
	$\boldsymbol{T}\cdot[\boldsymbol{R}+\boldsymbol{S}] = \boldsymbol{T}\cdot\boldsymbol{R}+\boldsymbol{T}\cdot\boldsymbol{S}$		
b)	$[\boldsymbol{R}\cdot\boldsymbol{S}]\cdot\boldsymbol{T} = \boldsymbol{R}\cdot[\boldsymbol{S}\cdot\boldsymbol{T}] = \boldsymbol{R}\cdot\ \boldsymbol{S}\cdot\boldsymbol{T}$	associativity	(tensor multiplication)
c)	$[\boldsymbol{S}\cdot\boldsymbol{T}]\cdot\boldsymbol{u} = \boldsymbol{S}\cdot[\boldsymbol{T}\cdot\boldsymbol{u}] = \boldsymbol{S}\cdot\ \boldsymbol{T}\cdot\boldsymbol{u}$	associativity	(vector multiplication)
d)	$\boldsymbol{S}\cdot[\alpha\boldsymbol{T}] = \alpha[\boldsymbol{S}\cdot\boldsymbol{T}] = \alpha\,\boldsymbol{S}\cdot\boldsymbol{T}$	associativity	(scalar multiplication)
	$[\alpha\boldsymbol{S}]\cdot\boldsymbol{T} = \alpha[\boldsymbol{S}\cdot\boldsymbol{T}] = \alpha\,\boldsymbol{S}\cdot\boldsymbol{T}$		
e)	$\boldsymbol{S}\cdot\boldsymbol{T} \neq \boldsymbol{T}\cdot\boldsymbol{S}$	non-commutativity	(tensor multiplication)
f)	$\boldsymbol{0}\cdot\boldsymbol{T} = \boldsymbol{T}\cdot\boldsymbol{0} = \boldsymbol{0}$	null element	(null tensor)
g)	$\boldsymbol{1}\cdot\boldsymbol{T} = \boldsymbol{T}\cdot\boldsymbol{1} = \boldsymbol{T}$	identity	(identity tensor)

It is important to observe that the tensor dot product is *not commutative*! For the special case that $\boldsymbol{T} = \boldsymbol{u} \otimes \boldsymbol{v}$ is a simple second-order tensor the tensor dot product renders $\boldsymbol{P} = \boldsymbol{S} \cdot [\boldsymbol{u} \otimes \boldsymbol{v}] = [\boldsymbol{S} \cdot \boldsymbol{u}] \otimes \boldsymbol{v}$, if in addition also $\boldsymbol{S} = \boldsymbol{a} \otimes \boldsymbol{b}$ is a simple second-order tensor the tensor dot product results in $\boldsymbol{P} = [\boldsymbol{a} \otimes \boldsymbol{b}] \cdot [\boldsymbol{u} \otimes \boldsymbol{v}] = [\boldsymbol{b} \cdot \boldsymbol{u}]\boldsymbol{a} \otimes \boldsymbol{v}$, thus e.g. for the covariant base dyads $[\boldsymbol{g}_i \otimes \boldsymbol{g}_j] \cdot [\boldsymbol{g}_k \otimes \boldsymbol{g}_l] = g_{jk}\boldsymbol{g}_i \otimes \boldsymbol{g}_l$ holds. Consequently, the tensor dot product of second-order tensors has coordinate representation $\boldsymbol{S} \cdot \boldsymbol{T} = S^{ij}T_{jl}\boldsymbol{g}_i \otimes \boldsymbol{g}^l = S_{ij}T^{jl}\boldsymbol{g}^i \otimes \boldsymbol{g}_l = S^i{}_jT^j{}_l\boldsymbol{g}_i \otimes \boldsymbol{g}^l = S_i{}^jT_j{}^l\boldsymbol{g}^i \otimes \boldsymbol{g}_l$. Thus the tensor dot product of second-order tensors follows as the product of the matrix arrangement of the corresponding coefficients and the dyadic product of the outer base vectors.

Special Second-Order Tensors and Algebraic Operators:

The second-order *identity tensor* is denoted as $\boldsymbol{1}$ and is defined so that $\boldsymbol{u} =: \boldsymbol{1}\cdot\boldsymbol{u}$. It expands in a Cartesian base system in terms of the Kronecker symbol as $\delta_{ab} = \delta^{ab} = \delta_a{}^b = \delta^a{}_b$ with $[\delta_{ab}] = [\delta^{ab}] = [\delta_a{}^b] = [\delta^a{}_b]$ the diagonal unit matrix $\lceil 1, 1, 1 \rfloor$ as $\boldsymbol{1} = \delta_{ab}\boldsymbol{e}^a \otimes \boldsymbol{e}^b = \delta^{ab}\boldsymbol{e}_a \otimes \boldsymbol{e}_b = \delta_a{}^b\boldsymbol{e}^a \otimes \boldsymbol{e}_b = \delta^a{}_b\boldsymbol{e}_a \otimes \boldsymbol{e}^b$. In co-, contra-, and mixedvariant base dyads the identity tensor is expressed in terms of the metric coefficients and thus coincides identically with the *metric tensors* $\boldsymbol{j} := g_{ij}\boldsymbol{g}^i \otimes \boldsymbol{g}^j$, $\boldsymbol{i} := g^{ij}\boldsymbol{g}_i \otimes \boldsymbol{g}_j$, and $\boldsymbol{1} := \delta_i{}^j\boldsymbol{g}^i \otimes \boldsymbol{g}_j = \delta^i{}_j\boldsymbol{g}_i \otimes \boldsymbol{g}^j$ whereby $\delta_i{}^j = \delta^i{}_j$ denotes again the Kronecker symbol with $[\delta_i{}^j] = [\delta^i{}_j]$ the diagonal unit matrix $\lceil 1, 1, 1 \rfloor$. The second-order identity tensor is an *isotropic second-order tensor* since the metric coefficients are invariant under rigid rotations of the base system (by an orthogonal tensor $\boldsymbol{R}$ so that $g_{ij} =$

$\boldsymbol{g}_i \cdot \boldsymbol{g}_j = \boldsymbol{g}_i \cdot \boldsymbol{R}^t \cdot \boldsymbol{R} \cdot \boldsymbol{g}_j$). Indeed scalar multiples of $\mathbf{1}$ are the only isotropic second-order tensors.

The *transpose* of a second-order tensor $\boldsymbol{T}$ is denoted as $\boldsymbol{T}^t$ and is defined so that $\boldsymbol{u} \cdot \boldsymbol{T} \cdot \boldsymbol{v} =: \boldsymbol{v} \cdot \boldsymbol{T}^t \cdot \boldsymbol{u}$. It satisfies e.g. the following rules:

S2.1: *Transpose of second-order tensors:*
a) $[\boldsymbol{S} + \boldsymbol{T}]^t = \boldsymbol{S}^t + \boldsymbol{T}^t$
b) $[\alpha \boldsymbol{T}]^t = \alpha \boldsymbol{T}^t$
c) $[\boldsymbol{S} \cdot \boldsymbol{T}]^t = \boldsymbol{T}^t \cdot \boldsymbol{S}^t$
d) $[\boldsymbol{u} \otimes \boldsymbol{v}]^t = \boldsymbol{v} \otimes \boldsymbol{u}$

The transpose of second-order tensors has coordinate representation $\boldsymbol{T}^t = T_{ji}\boldsymbol{g}^i \otimes \boldsymbol{g}^j = T^{ji}\boldsymbol{g}_i \otimes \boldsymbol{g}_j = T^j{}_i\boldsymbol{g}^i \otimes \boldsymbol{g}_j = T_j{}^i\boldsymbol{g}_i \otimes \boldsymbol{g}^j$. A tensor is called *symmetric* if it equals its transpose $\boldsymbol{T}^{\text{sym}} := \boldsymbol{T} \equiv \boldsymbol{T}^t$ and *skew symmetric* if it equals its negative transpose $\boldsymbol{T}^{\text{skw}} := \boldsymbol{T} \equiv -\boldsymbol{T}^t$.

The *symmetric* part of a second-order tensor is denoted $\boldsymbol{T}^{\text{sym}}$ and is computed from $2\boldsymbol{T}^{\text{sym}} := \boldsymbol{T} + \boldsymbol{T}^t$. Then $\alpha_{\boldsymbol{T}^{\text{sym}}}(\boldsymbol{w}, \boldsymbol{w}) = \boldsymbol{w} \cdot \boldsymbol{T}^{\text{sym}} \cdot \boldsymbol{w} = \boldsymbol{w} \cdot \boldsymbol{T} \cdot \boldsymbol{w} = \alpha_{\boldsymbol{T}}(\boldsymbol{w}, \boldsymbol{w})$ holds for all $\boldsymbol{w} \neq \mathbf{0}$. The *skew symmetric* part of a second-order tensor is denoted $\boldsymbol{T}^{\text{skw}}$ and is computed from $2\boldsymbol{T}^{\text{skw}} := \boldsymbol{T} - \boldsymbol{T}^t$. Then $\alpha_{\boldsymbol{T}^{\text{skw}}}(\boldsymbol{w}, \boldsymbol{w}) = \boldsymbol{w} \cdot \boldsymbol{T}^{\text{skw}} \cdot \boldsymbol{w} = 0$ holds for all $\boldsymbol{w} \neq \mathbf{0}$. Finally, every second-order tensor $\boldsymbol{T}$ allows unique *additive decomposition* $\boldsymbol{T} = \boldsymbol{T}^{\text{sym}} + \boldsymbol{T}^{\text{skw}}$ into its symmetric and skew symmetric contributions.

A second-order tensor $\boldsymbol{T}$ is classified as either positive definite, positive semidefinite, indefinite, negative semidefinite, or negative definite, respectively, if its *quadratic form* $\alpha := \alpha_{\boldsymbol{T}}(\boldsymbol{w}, \boldsymbol{w}) = \boldsymbol{w} \cdot \boldsymbol{T} \cdot \boldsymbol{w}$ renders $\alpha > 0$, $\alpha \geq 0$, $\alpha \gtreqless 0$, $\alpha \leq 0$, or $\alpha < 0$, respectively, for all $\boldsymbol{w} \neq \mathbf{0}$.

The *determinant* of a second-order tensor $\boldsymbol{T}$ is denoted as $\det \boldsymbol{T}$ and is defined from $\det \boldsymbol{T}\,[\boldsymbol{uvw}] := \big[[\boldsymbol{T} \cdot \boldsymbol{u}][\boldsymbol{T} \cdot \boldsymbol{v}][\boldsymbol{T} \cdot \boldsymbol{w}]\big]$. It satisfies e.g. the following rules:

S2.2: *Determinant of second-order tensors:*
a) $\det(\boldsymbol{S} \cdot \boldsymbol{T}) = \det \boldsymbol{S} \det \boldsymbol{T}$
b) $\det \boldsymbol{T}^{-1} = [\det \boldsymbol{T}]^{-1}$
c) $\det \boldsymbol{T}^t = \det \boldsymbol{T}$
d) $\det(\alpha \boldsymbol{T}) = \alpha^3 \det \boldsymbol{T}$

The determinant of second-order tensors has coordinate representation $3! \det \boldsymbol{T} = \varepsilon_{ijk}\varepsilon_{lmn}T^{il}T^{jm}T^{kn} = \varepsilon^{ijk}\varepsilon^{lmn}T_{il}T_{jm}T_{kn}$. For a rank 1 modification $\boldsymbol{T} = \boldsymbol{S} + \alpha \boldsymbol{u} \otimes \boldsymbol{v}$ with $\det \boldsymbol{S} \neq 0$ it holds that $\det \boldsymbol{T} = \det \boldsymbol{S}[1 + \alpha \boldsymbol{v} \cdot \boldsymbol{S}^{-1} \cdot \boldsymbol{u}]$.

The *cofactor* of a second-order tensor $\boldsymbol{T}$ is denoted as $\operatorname{cof} \boldsymbol{T}$ and is defined from $\operatorname{cof} \boldsymbol{T} \cdot [\boldsymbol{u} \times \boldsymbol{v}] := [\boldsymbol{T} \cdot \boldsymbol{u}] \times [\boldsymbol{T} \cdot \boldsymbol{v}]$. It satisfies e.g. the following rules:

S2.3: *Cofactor of second-order tensors:*
a) $\det(\operatorname{cof} \boldsymbol{T}) = [\det \boldsymbol{T}]^2$
b) $\boldsymbol{T} : \operatorname{cof} \boldsymbol{T} = 3 \det \boldsymbol{T}$
c) $\boldsymbol{T}^t \cdot \operatorname{cof} \boldsymbol{T} = \det \boldsymbol{T}\, \mathbf{1}$
d) $[\operatorname{cof} \boldsymbol{T}]^t = \operatorname{cof} \boldsymbol{T}^t$
e) $\operatorname{cof}(\boldsymbol{S} \cdot \boldsymbol{T}) = \operatorname{cof} \boldsymbol{T} \cdot \operatorname{cof} \boldsymbol{S}$

For invertible $\boldsymbol{T}$, see below, it holds that $\operatorname{cof}\boldsymbol{T} = \det\boldsymbol{T}\boldsymbol{T}^{-t}$. The cofactor of second-order tensors has coordinate representation $2!\,[\operatorname{cof}\boldsymbol{T}]_{il} = \varepsilon_{ijk}\varepsilon_{lmn}T^{jm}T^{kn}$ or $2!\,[\operatorname{cof}\boldsymbol{T}]^{il} = \varepsilon^{ijk}\varepsilon^{lmn}T_{jm}T_{kn}$, respectively. For a rank 1 modification $\boldsymbol{T} = \boldsymbol{S} + \alpha\boldsymbol{u}\otimes\boldsymbol{v}$ it holds that $\operatorname{cof}\boldsymbol{T} = \operatorname{cof}\boldsymbol{S} - \alpha\operatorname{spn}\boldsymbol{u}\cdot\boldsymbol{S}\cdot\operatorname{spn}\boldsymbol{v}$.

The *inverse* of a second-order tensor $\boldsymbol{T}$ is denoted as $\boldsymbol{T}^{-1}$ and is defined so that $\boldsymbol{u} = \boldsymbol{T}\cdot\boldsymbol{v}$ leads to $\boldsymbol{v} = \boldsymbol{T}^{-1}\cdot\boldsymbol{u}$. It satisfies e.g. the following rules:

S2.4: *Inverse of second-order tensors:*

a) $\boldsymbol{T}\cdot\boldsymbol{T}^{-1} = \boldsymbol{T}^{-1}\cdot\boldsymbol{T} = \boldsymbol{1}$

b) $[\boldsymbol{T}^{-1}]^{-1} = \boldsymbol{T}$

c) $[\alpha\boldsymbol{T}]^{-1} = \alpha^{-1}\boldsymbol{T}^{-1}$

d) $[\boldsymbol{T}^{-1}]^t = [\boldsymbol{T}^t]^{-1} = \boldsymbol{T}^{-t}$

e) $[\boldsymbol{S}\cdot\boldsymbol{T}]^{-1} = \boldsymbol{T}^{-1}\cdot\boldsymbol{S}^{-1}$

The necessary and sufficient condition for the inverse to exist is a non-vanishing determinant $\det\boldsymbol{T} \neq 0$. The inverse of second-order tensors has coordinate representation $2!\det\boldsymbol{T}\,[\boldsymbol{T}^{-t}]_{il} = \varepsilon_{ijk}\varepsilon_{lmn}T^{jm}T^{kn}$ or $2!\det\boldsymbol{T}\,[\boldsymbol{T}^{-t}]^{il} = \varepsilon^{ijk}\varepsilon^{lmn}T_{jm}T_{kn}$, respectively. For a rank 1 modification $\boldsymbol{T} = \boldsymbol{S} + \alpha\boldsymbol{u}\otimes\boldsymbol{v}$ with $\det\boldsymbol{S} \neq 0$ it holds that $\det\boldsymbol{T}[\boldsymbol{T}^{-1} - \boldsymbol{S}^{-1}] = -\alpha\det\boldsymbol{S}[\boldsymbol{S}^{-1}\cdot\boldsymbol{u}]\otimes[\boldsymbol{v}\cdot\boldsymbol{S}^{-1}]$.

The *trace* of a second-order tensor is denoted $\operatorname{trc}\boldsymbol{T}$ and is defined as $\operatorname{trc}\boldsymbol{T} := \boldsymbol{T} : \boldsymbol{1}$. It satisfies e.g. the following rules:

S2.5: *Trace of second-order tensors:*

a) $\operatorname{trc}(\boldsymbol{u}\otimes\boldsymbol{v}) = \boldsymbol{u}\cdot\boldsymbol{v}$

b) $\operatorname{trc}\ \boldsymbol{T}^t = \operatorname{trc}\ \boldsymbol{T}$

c) $\operatorname{trc}(\boldsymbol{S}\cdot\boldsymbol{T}) = \operatorname{trc}(\boldsymbol{T}\cdot\boldsymbol{S})$

d) $\operatorname{trc}(\boldsymbol{S}\cdot\boldsymbol{T}^t) = \boldsymbol{S} : \boldsymbol{T}$

e) $\operatorname{trc}\ \boldsymbol{1} = 3$

Since $\varepsilon_{ijk}\delta_{mn} = \varepsilon_{ijm}\delta_{nk} + \varepsilon_{kim}\delta_{nj} + \varepsilon_{jkm}\delta_{ni}$ holds, see e.g. [164], an interesting alternative expression for the trace of $\boldsymbol{T}$ reads $[\boldsymbol{u}\boldsymbol{v}\boldsymbol{w}]\operatorname{trc}\boldsymbol{T} = [\boldsymbol{u}\boldsymbol{v}[\boldsymbol{A}\cdot\boldsymbol{w}]] + [\boldsymbol{w}\boldsymbol{u}[\boldsymbol{A}\cdot\boldsymbol{v}]] + [\boldsymbol{v}\boldsymbol{w}[\boldsymbol{A}\cdot\boldsymbol{u}]]$.

The *spherical* part of a second-order tensor is denoted as $\operatorname{sph}\boldsymbol{T}$ and is defined as $3\operatorname{sph}\boldsymbol{T} := \operatorname{trc}\boldsymbol{T}\boldsymbol{1}$. Then $\operatorname{trc}(\operatorname{sph}\boldsymbol{T}) = \operatorname{trc}\boldsymbol{T}$ holds. The *deviatoric* part of a second-order tensor is denoted as $\operatorname{dev}\boldsymbol{T}$ and is defined as $\operatorname{dev}\boldsymbol{T} := \boldsymbol{T} - \operatorname{sph}\boldsymbol{T}$. Then $\operatorname{trc}(\operatorname{dev}\boldsymbol{T}) = 0$ holds. Finally, every second-order tensor $\boldsymbol{T}$ allows unique *additive decomposition* $\boldsymbol{T} = \operatorname{sph}\boldsymbol{T} + \operatorname{dev}\boldsymbol{T}$ into its spherical and deviatoric contributions.

A *orthogonal* tensor $\boldsymbol{R}$ satisfies $\boldsymbol{R}^t = \boldsymbol{R}^{-1}$ with $\det\boldsymbol{R} = \pm 1$, whereby $\det\boldsymbol{R} = +1$ identifies in particular a *special orthogonal* tensor. Then every invertible second-order tensor $\boldsymbol{T}$ with $\det\boldsymbol{T} \neq 0$ allows unique *multiplicative (polar) decomposition* $\boldsymbol{T} = \boldsymbol{R}\cdot\boldsymbol{U} = \boldsymbol{V}\cdot\boldsymbol{R}$ into a symmetric positive definite tensor $\boldsymbol{U} = \boldsymbol{U}^{\text{sym}}$ or $\boldsymbol{V} = \boldsymbol{V}^{\text{sym}}$, respectively, with $\det\boldsymbol{U} = \det\boldsymbol{V} = \det\boldsymbol{T} \neq 0$ and a special orthogonal tensor $\boldsymbol{R}$. The proof follows from i) the identity $\boldsymbol{T} = \boldsymbol{T}^{-t}\cdot\boldsymbol{T}^t\cdot\boldsymbol{T}$, ii) the definitions $\boldsymbol{U}\cdot\boldsymbol{U} := \boldsymbol{T}^t\cdot\boldsymbol{T}$ and $\boldsymbol{R} := \boldsymbol{T}^{-t}\cdot\boldsymbol{U}$, iii) the symmetry property $[\boldsymbol{U}\cdot\boldsymbol{U}]^t = \boldsymbol{T}^t\cdot\boldsymbol{T} = \boldsymbol{U}\cdot\boldsymbol{U}$, iv) the positive definiteness $\boldsymbol{w}\cdot\boldsymbol{U}\cdot\boldsymbol{U}\cdot\boldsymbol{w} = |\boldsymbol{T}\cdot\boldsymbol{w}|^2 > 0$ for all $\boldsymbol{w} \neq \boldsymbol{0}$ with $\det\boldsymbol{T} \neq 0$, and finally v) the orthogonality $\boldsymbol{R}\cdot\boldsymbol{R}^t = \boldsymbol{T}^{-t}\cdot\boldsymbol{U}\cdot\boldsymbol{U}\cdot\boldsymbol{T}^{-1} = \boldsymbol{T}^{-t}\cdot\boldsymbol{T}^t\cdot\boldsymbol{T}\cdot\boldsymbol{T}^{-1} = \boldsymbol{1}$.

The *volumetric* part of a second-order tensor is denoted as $\operatorname{vol}\boldsymbol{T}$ and is defined as $\operatorname{vol}\boldsymbol{T} := \sqrt[3]{\det\boldsymbol{T}}\boldsymbol{1}$. Then $\det(\operatorname{vol}\boldsymbol{T}) = \det\boldsymbol{T}$ holds. The *isochoric* part of a second-order tensor is denoted as $\operatorname{iso}\boldsymbol{T}$ and is defined as $\operatorname{iso}\boldsymbol{T} := \sqrt[-3]{\det\boldsymbol{T}}\boldsymbol{T}$. Then $\det(\operatorname{iso}\boldsymbol{T}) = 1$ holds. Finally, every second-order tensor $\boldsymbol{T}$ allows unique *multiplicative decomposition* $\boldsymbol{T} = \operatorname{vol}\boldsymbol{T}\cdot\operatorname{iso}\boldsymbol{T}$ into its volumetric and isochoric contributions.

In *Voigt matrix notation* a second-order tensor $\boldsymbol{T} \in \mathbb{E}^3 \otimes \mathbb{E}^3$ is equivalently represented by a *column matrix* $\underline{t} \in \mathbb{E}^9$. This is achieved by the assignment of the nine Cartesian coefficients of $\boldsymbol{T}$ to the nine entries of $\underline{t}$ by the map $\underline{t} = v_*(\boldsymbol{T})$ so that $[t_\mathsf{a}] = [T_{11}, T_{22}, T_{33}, T_{23}, T_{31}, T_{12}, T_{32}, T_{13}, T_{21}]^\mathsf{T}$, likewise the inverse map is denoted by $\boldsymbol{T} = v^*(\underline{t})$. The assignments $t_\mathsf{a} = \mathsf{b}_{\mathsf{a}bc}T_{bc}$ and $T_{bc} = t_\mathsf{a}\mathsf{b}_{\mathsf{a}bc}$ are expressed in terms of the Boolean operator $\mathsf{b}_{\mathsf{a}bc}$; obviously $\mathsf{b}_{\mathsf{a}bc}$ has $9 \times 3 \times 3$ entries that are either 1 or 0

		T_{11}	T_{22}	T_{33}	T_{23}	T_{31}	T_{12}	T_{32}	T_{13}	T_{21}
		$\mathsf{b}_{\mathsf{a}11}$	$\mathsf{b}_{\mathsf{a}22}$	$\mathsf{b}_{\mathsf{a}33}$	$\mathsf{b}_{\mathsf{a}23}$	$\mathsf{b}_{\mathsf{a}31}$	$\mathsf{b}_{\mathsf{a}12}$	$\mathsf{b}_{\mathsf{a}32}$	$\mathsf{b}_{\mathsf{a}13}$	$\mathsf{b}_{\mathsf{a}21}$
t_1	b_{1bc}	1	0	0	0	0	0	0	0	0
t_2	b_{2bc}	0	1	0	0	0	0	0	0	0
t_3	b_{3bc}	0	0	1	0	0	0	0	0	0
t_4	b_{4bc}	0	0	0	1	0	0	0	0	0
t_5	b_{5bc}	0	0	0	0	1	0	0	0	0
t_6	b_{6bc}	0	0	0	0	0	1	0	0	0
t_7	b_{7bc}	0	0	0	0	0	0	1	0	0
t_8	b_{8bc}	0	0	0	0	0	0	0	1	0
t_9	b_{9bc}	0	0	0	0	0	0	0	0	1

Thereby (with $2\langle x\rangle = x + |x|$ the MaCaulay brackets and $\operatorname{mod}(x, y)$ the modulo function) the entries for $\mathsf{a} = 1, 2, 3$ and $b, c = 1, 2, 3$ read $\mathsf{b}_{\mathsf{a}bc} = \langle e_{\mathsf{a}\mathfrak{b}\mathfrak{c}}\rangle$ with $\mathfrak{b} := 1 + \operatorname{mod}(b, 3) = 2, 3, 1$ and $\mathfrak{c} := 1 + \operatorname{mod}([c + 1], 3) = 3, 1, 2$; the entries for $\mathsf{a} = 4, 5, 6$ and $b, c = 1, 2, 3$ read $\mathsf{b}_{\mathsf{a}bc} = \langle +e_{\mathfrak{a}bc}\rangle$ with $\mathfrak{a} := \mathsf{a} - 3 = 1, 2, 3$; and the entries for $\mathsf{a} = 7, 8, 9$ and $b, c = 1, 2, 3$ read $\mathsf{b}_{\mathsf{a}bc} = \langle -e_{\mathfrak{a}bc}\rangle$ with $\mathfrak{a} := \mathsf{a} - 6 = 1, 2, 3$. Note that $\mathsf{b}_{\mathsf{a}cd}\mathsf{b}_{\mathsf{b}cd} \to \underline{1} \in \mathbb{E}^9 \otimes \mathbb{E}^9$, with $\underline{1}$ the 9×9 unit matrix.

A symmetric second-order tensor $\boldsymbol{T} = \boldsymbol{T}^{\mathrm{sym}}$ may be represented by a column matrix $\underline{t} \in \mathbb{E}^6$ with $[t_\mathsf{a}] = [T_{11}, T_{22}, T_{33}, \sqrt{2}T_{23}, \sqrt{2}T_{31}, \sqrt{2}T_{12}]$ so that the norm $\boldsymbol{T}^{\mathrm{sym}} : \boldsymbol{T}^{\mathrm{sym}} = \underline{t}^\mathsf{T}\underline{t}$. The assignments $t_\mathsf{a} = \mathsf{b}^{\mathrm{sym}}_{\mathsf{a}bc}T_{bc}$ and $T_{bc} = t_\mathsf{a}\mathsf{b}^{\mathrm{sym}}_{\mathsf{a}bc}$ are expressed in terms of the operator $\mathsf{b}^{\mathrm{sym}}_{\mathsf{a}bc}$; obviously $\mathsf{b}^{\mathrm{sym}}_{\mathsf{a}bc}$ has $6 \times 3 \times 3$ entries

		T_{11}	T_{22}	T_{33}	T_{23}	T_{31}	T_{12}	T_{32}	T_{13}	T_{21}
		$\mathsf{b}^{\mathrm{sym}}_{\mathsf{a}11}$	$\mathsf{b}^{\mathrm{sym}}_{\mathsf{a}22}$	$\mathsf{b}^{\mathrm{sym}}_{\mathsf{a}33}$	$\mathsf{b}^{\mathrm{sym}}_{\mathsf{a}23}$	$\mathsf{b}^{\mathrm{sym}}_{\mathsf{a}31}$	$\mathsf{b}^{\mathrm{sym}}_{\mathsf{a}12}$	$\mathsf{b}^{\mathrm{sym}}_{\mathsf{a}32}$	$\mathsf{b}^{\mathrm{sym}}_{\mathsf{a}13}$	$\mathsf{b}^{\mathrm{sym}}_{\mathsf{a}21}$
t_1	$\mathsf{b}^{\mathrm{sym}}_{1bc}$	1	0	0	0	0	0	0	0	0
t_2	$\mathsf{b}^{\mathrm{sym}}_{2bc}$	0	1	0	0	0	0	0	0	0
t_3	$\mathsf{b}^{\mathrm{sym}}_{3bc}$	0	0	1	0	0	0	0	0	0
t_4	$\mathsf{b}^{\mathrm{sym}}_{4bc}$	0	0	0	$\sqrt[-2]{2}$	0	0	$\sqrt[-2]{2}$	0	0
t_5	$\mathsf{b}^{\mathrm{sym}}_{5bc}$	0	0	0	0	$\sqrt[-2]{2}$	0	0	$\sqrt[-2]{2}$	0
t_6	$\mathsf{b}^{\mathrm{sym}}_{6bc}$	0	0	0	0	0	$\sqrt[-2]{2}$	0	0	$\sqrt[-2]{2}$

Thereby the entries for $\mathsf{a} = 1, 2, 3$ and $b, c = 1, 2, 3$ read $\mathsf{b}^{\text{sym}}_{\mathsf{a}bc} = \mathsf{b}_{\mathsf{a}bc}$; and the entries for $\mathsf{a} = 4, 5, 6$ and $b, c = 1, 2, 3$ read $\sqrt{2}\mathsf{b}^{\text{sym}}_{\mathsf{a}bc} = |e_{\mathfrak{a}bc}|$ with $\mathfrak{a} := \mathsf{a} - 3 = 1, 2, 3$. As an example, the symmetric identity tensor $\mathbf{1}$ maps into $[\delta_{\mathsf{a}}] = [1, 1, 1, 0, 0, 0]$. Note finally that $\mathsf{b}^{\text{sym}}_{\mathsf{a}cd}\mathsf{b}^{\text{sym}}_{\mathsf{b}cd} \rightarrow \underline{1} \in \mathbb{E}^6 \otimes \mathbb{E}^6$, with $\underline{1}$ the 6×6 unit matrix.

A skew symmetric second-order tensor $\boldsymbol{T} = \boldsymbol{T}^{\text{skw}}$ may be represented by a column matrix $\underline{t} \in \mathbb{E}^3$ with $[t_{\mathsf{a}}] = [\sqrt{2}T_{32}, \sqrt{2}T_{13}, \sqrt{2}T_{21}]$ so that the norm $\boldsymbol{T}^{\text{skw}} : \boldsymbol{T}^{\text{skw}} = \underline{t}^{\mathsf{T}}\underline{t}$. The assignments $t_{\mathsf{a}} = \mathsf{b}^{\text{skw}}_{\mathsf{a}bc}T_{bc}$ and $T_{bc} = t_{\mathsf{a}}\mathsf{b}^{\text{skw}}_{\mathsf{a}bc}$ are expressed in terms of the operator $\mathsf{b}^{\text{skw}}_{\mathsf{a}bc}$; obviously $\mathsf{b}^{\text{skw}}_{\mathsf{a}bc}$ has $3 \times 3 \times 3$ entries

		T_{11}	T_{22}	T_{33}	T_{23}	T_{31}	T_{12}	T_{32}	T_{13}	T_{21}
		$\mathsf{b}^{\text{skw}}_{\mathsf{a}11}$	$\mathsf{b}^{\text{skw}}_{\mathsf{a}22}$	$\mathsf{b}^{\text{skw}}_{\mathsf{a}33}$	$\mathsf{b}^{\text{skw}}_{\mathsf{a}23}$	$\mathsf{b}^{\text{skw}}_{\mathsf{a}31}$	$\mathsf{b}^{\text{skw}}_{\mathsf{a}12}$	$\mathsf{b}^{\text{skw}}_{\mathsf{a}32}$	$\mathsf{b}^{\text{skw}}_{\mathsf{a}13}$	$\mathsf{b}^{\text{skw}}_{\mathsf{a}21}$
t_1	$\mathsf{b}^{\text{skw}}_{1bc}$	0	0	0	$-\sqrt[-2]{2}$	0	0	$\sqrt[-2]{2}$	0	0
t_2	$\mathsf{b}^{\text{skw}}_{2bc}$	0	0	0	0	$-\sqrt[-2]{2}$	0	0	$\sqrt[-2]{2}$	0
t_3	$\mathsf{b}^{\text{skw}}_{3bc}$	0	0	0	0	0	$-\sqrt[-2]{2}$	0	0	$\sqrt[-2]{2}$

Thereby the entries for $\mathsf{a} = 1, 2, 3$ and $b, c = 1, 2, 3$ read $\sqrt{2}\mathsf{b}^{\text{skw}}_{\mathsf{a}bc} = -e_{\mathsf{a}bc}$. As an example, the skew symmetric symplectic tensor $\mathbb{J} := \text{spn}(\boldsymbol{e}_1 + \boldsymbol{e}_2 + \boldsymbol{e}_3)$ maps into $[j_{\mathsf{a}}] = [\sqrt{2}, \sqrt{2}, \sqrt{2}]$. Note finally that $\mathsf{b}^{\text{skw}}_{\mathsf{a}cd}\mathsf{b}^{\text{skw}}_{\mathsf{b}cd} \rightarrow \underline{1} \in \mathbb{E}^3 \otimes \mathbb{E}^3$, with $\underline{1}$ the 3×3 unit matrix.

Spectral Properties of Second-Order Tensors:
The left and right *special eigenvalue problems* for a second-order tensor $\boldsymbol{T}$ read $\boldsymbol{m} \cdot \boldsymbol{T} = \boldsymbol{m}\lambda$ and $\boldsymbol{T} \cdot \boldsymbol{n} = \lambda\boldsymbol{n}$ whereby λ denotes the *eigenvalue* and $\boldsymbol{m}$ and $\boldsymbol{n}$ are the corresponding left and right *eigenvectors*, respectively. Then the unique condition for *non-trivial solutions* to both $\boldsymbol{m} \cdot [\boldsymbol{T} - \mathbf{1}\lambda] = \mathbf{0}$ and $[\boldsymbol{T} - \lambda\mathbf{1}] \cdot \boldsymbol{n} = \mathbf{0}$ is $\det(\boldsymbol{T} - \lambda\mathbf{1}) = 0$ resulting in the *characteristic equation* $p(\lambda) := \det \boldsymbol{T} + \operatorname{cof} \boldsymbol{T} : [-\lambda\mathbf{1}] + \boldsymbol{T} : \operatorname{cof}(-\lambda\mathbf{1}) + \det(-\lambda\mathbf{1}) \doteq 0$. The characteristic equation expands into the third-order polynomial $p(\lambda) = I_3 - I_2\lambda + I_1\lambda^2 - \lambda^3$ in terms of the three *(principal) invariants* $I_3 := \det \boldsymbol{T}$, $I_2 := \operatorname{cof} \boldsymbol{T} : \mathbf{1}$, and $I_1 := \boldsymbol{T} : \mathbf{1}$ and admits solutions λ^a, $a = 1, 2, 3$, with corresponding left and right eigenvectors $\boldsymbol{m}_a$ and $\boldsymbol{n}_a$. In the sequel is assumed that all eigenvalues λ^a are distinct, the case of multiple eigenvalues necessitates a special treatment that may be by-passed in practice by perturbing equal eigenvalues. Since for $\lambda^a \neq \lambda^b$ it holds that $\lambda^a\boldsymbol{m}_a \cdot \boldsymbol{n}_b = [\boldsymbol{m}_a \cdot \boldsymbol{T}] \cdot \boldsymbol{n}_b = \boldsymbol{m}_a \cdot [\boldsymbol{T} \cdot \boldsymbol{n}_b] = \lambda^b\boldsymbol{m}_a \cdot \boldsymbol{n}_b$ it is concluded that $\boldsymbol{m}_a \cdot \boldsymbol{n}_b = 0$ for $a \neq b$, thus the left and right eigenvectors may be orthonormalized with $\boldsymbol{m}_a \cdot \boldsymbol{n}_b = \delta_{ab}$. As a consequence the left and right eigenvectors may be considered *reciprocal base vectors* in the sense of contra- and covariant base vectors as discussed before.

Since the coefficients and thus the invariants of $\boldsymbol{T}$ are *real* at least one of the three eigenvalues λ^a and thus the corresponding left and right eigenvector $\boldsymbol{m}_a$ and $\boldsymbol{n}_a$ is real. The two remaining eigenvalues λ^b and λ^c and thus the corresponding left and right eigenvectors $\boldsymbol{m}_b$, $\boldsymbol{m}_c$ and $\boldsymbol{n}_b$, $\boldsymbol{n}_c$ are either *conjugate complex* or also real. A *symmetric* $\boldsymbol{T} = \boldsymbol{T}^t$ admits only real eigenvalues λ^a and corresponding eigenvectors $\boldsymbol{m}_a \equiv \boldsymbol{n}_a$. Due to the symmetry of $\boldsymbol{T}$ the proof follows directly from contracting $\boldsymbol{T} \cdot [\boldsymbol{n}^{\Re} \pm i\boldsymbol{n}^{\Im}] = [\lambda^{\Re} \pm i\lambda^{\Im}][\boldsymbol{n}^{\Re} \pm i\boldsymbol{n}^{\Im}]$

with $[\boldsymbol{n}^{\Re} \mp i\boldsymbol{n}^{\Im}]$ to render $\boldsymbol{n}^{\Re} \cdot \boldsymbol{T} \cdot \boldsymbol{n}^{\Re} + \boldsymbol{n}^{\Im} \cdot \boldsymbol{T} \cdot \boldsymbol{n}^{\Im} = [\lambda^{\Re} \pm i\lambda^{\Im}][\boldsymbol{n}^{\Re} \cdot \boldsymbol{n}^{\Re} + \boldsymbol{n}^{\Im} \cdot \boldsymbol{n}^{\Im}]$, it is thus concluded that $\lambda^{\Im} = 0$ and therefore also $\boldsymbol{n}^{\Im} = \boldsymbol{0}$. A *skew symmetric* $\boldsymbol{T} = -\boldsymbol{T}^t$ with axial vector $\boldsymbol{t}$ (so that $\boldsymbol{T} \cdot \boldsymbol{u} = \boldsymbol{t} \times \boldsymbol{u}$ and $\boldsymbol{T} \cdot \boldsymbol{t} = \boldsymbol{0}$) possesses one zero eigenvalue with corresponding eigenvector $\boldsymbol{t}$ and the two conjugate complex eigenvalues $\pm i|\boldsymbol{t}|$. A general *non-symmetric* $\boldsymbol{T}$ may possess either only real eigenvalues or one real and two conjugate complex eigenvalues.

Next the left and right *modal tensors* $\boldsymbol{Q}_m := \boldsymbol{m}_a \otimes \boldsymbol{e}^a$ and $\boldsymbol{Q}_n := \boldsymbol{n}_b \otimes \boldsymbol{e}^b$ are introduced with the obvious relation $\boldsymbol{Q}_m^t \cdot \boldsymbol{Q}_n = \boldsymbol{Q}_n^t \cdot \boldsymbol{Q}_m = \boldsymbol{e}^a \otimes \boldsymbol{e}^a = \boldsymbol{1}$ and, since the identity tensor also allows the alternative representation $\boldsymbol{1} = \boldsymbol{n}_a \otimes \boldsymbol{m}_a$ (compare to the above representation of the identity tensor in mixedvariant base vectors), in particular with $\boldsymbol{Q}_n \cdot \boldsymbol{Q}_m^t = \boldsymbol{n}_a \otimes \boldsymbol{m}_a = \boldsymbol{1}$. Then the *Cayley-Hamilton theorem* states that every second-order tensor satisfies its own characteristic equation $\boldsymbol{P}(\boldsymbol{T}) := I_3\boldsymbol{1} - I_2\boldsymbol{T} + I_1\boldsymbol{T} \cdot \boldsymbol{T} - \boldsymbol{T} \cdot \boldsymbol{T} \cdot \boldsymbol{T} = \boldsymbol{0}$. The proof follows in a straightforward manner from either $\boldsymbol{m}p(\lambda) = \boldsymbol{0}$ or $p(\lambda)\boldsymbol{n} = \boldsymbol{0}$ and inserting $\boldsymbol{m}\lambda = \boldsymbol{m} \cdot \boldsymbol{T}$ or $\lambda\boldsymbol{n} = \boldsymbol{T} \cdot \boldsymbol{n}$ recursively to obtain either $\boldsymbol{m} \cdot \boldsymbol{P}(\boldsymbol{T}) = \boldsymbol{0}$ or $\boldsymbol{P}(\boldsymbol{T}) \cdot \boldsymbol{n} = \boldsymbol{0}$, respectively. By extending these relation to either $\boldsymbol{Q}_m^t \cdot \boldsymbol{P}(\boldsymbol{T}) = \boldsymbol{0}$ or $\boldsymbol{P}(\boldsymbol{T}) \cdot \boldsymbol{Q}_n = \boldsymbol{0}$ and pre- or post-multiplying with either $\boldsymbol{Q}_n$ or $\boldsymbol{Q}_m^t$, respectively, renders the result. The Cayley-Hamilton theorem is very useful to compute, e.g., any integer power of $\boldsymbol{T}$ recursively from $\boldsymbol{P}(\boldsymbol{T}) \cdot \boldsymbol{T}$, the determinant of $\boldsymbol{T}$ from $\boldsymbol{P}(\boldsymbol{T}) : \boldsymbol{1}$ or the cofactor and, if it exists, the inverse of $\boldsymbol{T}$ from $\boldsymbol{P}(\boldsymbol{T}) \cdot \boldsymbol{T}^{-1}$.

The *spectral representation* $\boldsymbol{T} = \sum_a \lambda^a \boldsymbol{n}_a \otimes \boldsymbol{m}_a$ follows from the representation of the identity tensor in terms of the left and right eigenvectors since $\boldsymbol{T} = \boldsymbol{1} \cdot \boldsymbol{T} = \sum_a \boldsymbol{n}_a \otimes [\boldsymbol{m}_a \cdot \boldsymbol{T}]$ or $\boldsymbol{T} = \boldsymbol{T} \cdot \boldsymbol{1} = \sum_a [\boldsymbol{T} \cdot \boldsymbol{n}_a] \otimes \boldsymbol{m}_a$, respectively. Then for the case or real eigenvalues an *isotropic tensor function* $f(\boldsymbol{T})$ allows the representation $f(\boldsymbol{T}) = \sum_a f(\lambda^a)\boldsymbol{n}_a \otimes \boldsymbol{m}_a$ with $f(\lambda)$ a converging series expansion of a function. Typical applications are the exponent $\exp(\boldsymbol{T})$, the logarithm $\ln(\boldsymbol{T})$, and the sine or cosine $\sin(\boldsymbol{T})$ or $\cos(\boldsymbol{T})$ of a second-order tensor. Finally, for distinct eigenvalues the *Serrin formula* allows a closed form expression for the eigendyads $\boldsymbol{n}_a \otimes \boldsymbol{m}_a = D_a^{-1} \prod_{b=1\backslash a}^{3} [\boldsymbol{T} - \lambda^b \boldsymbol{1}]$ with $D_a = \prod_{b=1\backslash a}^{3} [\lambda^a - \lambda^b]$. The proof follows in a straightforward manner from noting that $\boldsymbol{T} - \lambda^a \boldsymbol{1} = \sum_{b=1\backslash a}^{3} [\lambda^b - \lambda^a] \boldsymbol{n}_b \otimes \boldsymbol{m}_b$ and then computing $\prod_{b=1\backslash a}^{3} [\boldsymbol{T} - \lambda^b \boldsymbol{1}]$ resulting in the above relation. The Serrin formula is very useful when computing the derivative of the eigendyads of $\boldsymbol{T}$ with respect to $\boldsymbol{T}$.

For invertible $\boldsymbol{T}$ the *multiplicative (polar) decomposition* into the symmetric positive definite $\boldsymbol{U} = \sum_a \lambda^a \boldsymbol{n}_a^U \otimes \boldsymbol{n}_a^U$ and $\boldsymbol{V} = \sum_a \lambda^a \boldsymbol{n}_a^V \otimes \boldsymbol{n}_a^V$ and the special orthogonal $\boldsymbol{R} = \sum_a \boldsymbol{n}_a^V \otimes \boldsymbol{n}_a^U$ allows the representation $\boldsymbol{T} = \sum_a \lambda^a \boldsymbol{n}_a^V \otimes \boldsymbol{n}_a^U$. The characteristic equations and thus the eigenvalues λ^a of $\boldsymbol{U}$ and $\boldsymbol{V}$ coincide, the orthonormalized eigentriad $\boldsymbol{n}_a^U$ of $\boldsymbol{U}$ is rigidly rotated into the orthonormalized eigentriad $\boldsymbol{n}_a^V$ of $\boldsymbol{V}$ by the rotation tensor $\boldsymbol{R}$.

Representations in Terms of the Third-Order Permutation Tensor: The third-order Levi-Civita *permutation tensor* is expressed alternatively as $\boldsymbol{e} := \varepsilon_{ijk}\boldsymbol{g}^i \otimes \boldsymbol{g}^j \otimes \boldsymbol{g}^k$, and $\boldsymbol{e} := \varepsilon^{ijk}\boldsymbol{g}_i \otimes \boldsymbol{g}_j \otimes \boldsymbol{g}_k$, or $\boldsymbol{e} := e_{abc}\boldsymbol{e}^a \otimes \boldsymbol{e}^b \otimes \boldsymbol{e}^c$. Here the *fully skew symmetric* Levi-Civita *permutation symbol* computes alternatively from $\varepsilon_{ijk} := [\boldsymbol{g}_i\boldsymbol{g}_j\boldsymbol{g}_k]$, and $\varepsilon^{ijk} := [\boldsymbol{g}^i\boldsymbol{g}^j\boldsymbol{g}^k]$, or $e_{abc} := [\boldsymbol{e}_a\boldsymbol{e}_b\boldsymbol{e}_c]$. Thus it holds in particular that $e_{abc} = \pm 1$ for even/odd permutations of abc, and $e_{abc} = 0$ else. Then $\varepsilon_{ijk} = \sqrt{g}e_{ijk}$ and $\varepsilon^{ijk} = \sqrt[-2]{g}e^{ijk}$ with $e_{ijk} \equiv e_{abc}\delta^a{}_i\delta^b{}_j\delta^c{}_k$ and $e^{ijk} \equiv e^{abc}\delta^i{}_a\delta^j{}_b\delta^k{}_c$. The permutation-tensor is the only *isotropic third-order tensor.* Single, double, and triple contractions of the permutation tensor with itself render $\boldsymbol{e}\cdot\boldsymbol{e} = 2!\mathsf{I}^{\text{skw}}$, $\boldsymbol{e} : \boldsymbol{e} = 2!\mathbf{1}$, and $\boldsymbol{e} \cdot\!: \boldsymbol{e} = 3!$ with I^{skw} the fourth-order *skew symmetric unit tensor*, likewise for the permutation symbol $\varepsilon^{ijm}\varepsilon_{mkl} = \delta^i{}_k\delta^j{}_l - \delta^i{}_l\delta^j{}_k$, $\varepsilon^{imn}\varepsilon_{mnl} = 2!\delta^i{}_l$, and $\varepsilon^{mno}\varepsilon_{mno} = 3!$ hold.

The permutation tensor allows to express the *vector product* and the *scalar triple product* of vectors $\boldsymbol{u},\boldsymbol{v},\boldsymbol{w}$ as $\boldsymbol{u}\times\boldsymbol{v} = \boldsymbol{e} : [\boldsymbol{u}\otimes\boldsymbol{v}] = [\boldsymbol{u}\otimes\boldsymbol{v}] : \boldsymbol{e}$ and $[\boldsymbol{uvw}] = \boldsymbol{u}\cdot\boldsymbol{e} : [\boldsymbol{v}\otimes\boldsymbol{w}] = [\boldsymbol{u}\otimes\boldsymbol{v}] : \boldsymbol{e}\cdot\boldsymbol{w}$, respectively. Likewise, based on the properties of the permutation tensor the *triple vector product* of vectors $\boldsymbol{u},\boldsymbol{v},\boldsymbol{w}$ expands into $[\boldsymbol{u}\times\boldsymbol{v}]\times\boldsymbol{w} = -2[\boldsymbol{u}\otimes\boldsymbol{v}] : \mathsf{I}^{\text{skw}}\cdot\boldsymbol{w} = -2\boldsymbol{w}\cdot\mathsf{I}^{\text{skw}} : [\boldsymbol{v}\otimes\boldsymbol{u}] = \boldsymbol{w}\times[\boldsymbol{v}\times\boldsymbol{u}]$. Moreover, the (left and right) vector product and the tensor cross product of second-order tensors allow representations $\boldsymbol{S}\ltimes\boldsymbol{T} = [\boldsymbol{S}\cdot\boldsymbol{T}^t] : \boldsymbol{e} = \boldsymbol{e} : [\boldsymbol{S}\boxtimes\boldsymbol{T}] : \mathbf{1}$, $\boldsymbol{S}\rtimes\boldsymbol{T} = [\boldsymbol{S}^t\cdot\boldsymbol{T}] : \boldsymbol{e} = \mathbf{1} : [\boldsymbol{S}\boxtimes\boldsymbol{T}] : \boldsymbol{e}$, and $\boldsymbol{S}\divideontimes\boldsymbol{T} = [\boldsymbol{e}\cdot\boldsymbol{S}] : [\boldsymbol{T}\cdot\boldsymbol{e}] = \boldsymbol{e} : [\boldsymbol{S}\boxtimes\boldsymbol{T}] : \boldsymbol{e}$, respectively. Finally, the (left and right) tensor cross product of vectors and second-order tensors are expressed as $\boldsymbol{v}\ltimes\boldsymbol{T} = \boldsymbol{e} : [\boldsymbol{v}\otimes\boldsymbol{T}]$, and $\boldsymbol{T}\rtimes\boldsymbol{v} = [\boldsymbol{T}\otimes\boldsymbol{v}] : \boldsymbol{e}$, respectively.

The *cofactor* and the *determinant* of a second-order tensor $\boldsymbol{T}$ are expressed in symbolic notation as $2!\operatorname{cof}\boldsymbol{T} = \boldsymbol{T}\divideontimes\boldsymbol{T}$, and $3!\det\boldsymbol{T} = [\boldsymbol{T}\divideontimes\boldsymbol{T}] : \boldsymbol{T} = [\boldsymbol{TTT}]$, respectively (obviously $3! = [\mathbf{1}\divideontimes\mathbf{1}] : \mathbf{1} = [\mathbf{111}]$). The *inverse* of $\boldsymbol{T}$, if it exists, may then be computed from $\boldsymbol{T}^{-1} = \operatorname{cof}\boldsymbol{T}^t/\det\boldsymbol{T}^t$. Finally the *trace* of a second-order tensor $\boldsymbol{T}$ and its cofactor $\operatorname{cof}\boldsymbol{T}$ follow as $2!\boldsymbol{T} : \mathbf{1} = [\boldsymbol{T}\divideontimes\mathbf{1}] : \mathbf{1} = [\boldsymbol{T}\mathbf{11}]$ and $2!\operatorname{cof}\boldsymbol{T} : \mathbf{1} = [\boldsymbol{T}\divideontimes\boldsymbol{T}] : \mathbf{1} = [\boldsymbol{TT}\mathbf{1}]$, respectively.

A *skew symmetric* second-order tensor $\boldsymbol{W}\in\text{so}(3)$ and its corresponding *axial vector* $\boldsymbol{w}\in\mathbb{E}^3$ are related by $2\boldsymbol{w} = 2\operatorname{axl}\boldsymbol{W} := -\boldsymbol{e} : \boldsymbol{W} = \mathbf{1}\ltimes\boldsymbol{W} = \boldsymbol{W}\rtimes\mathbf{1}$ and $\boldsymbol{W} = \operatorname{spn}\boldsymbol{w} := -\boldsymbol{e}\cdot\boldsymbol{w} = \mathbf{1}\ltimes\boldsymbol{w} = \boldsymbol{w}\rtimes\mathbf{1}$. Then for example $\boldsymbol{u}\cdot\boldsymbol{W} = \boldsymbol{u}\times\boldsymbol{w}$, $\boldsymbol{W}\cdot\boldsymbol{v} = \boldsymbol{w}\times\boldsymbol{v}$, $\boldsymbol{w}\cdot\boldsymbol{W} = \boldsymbol{W}\cdot\boldsymbol{w} = \mathbf{0}$, $\boldsymbol{W}^2 := \boldsymbol{W}\cdot\boldsymbol{W} = \boldsymbol{w}\otimes\boldsymbol{w} - |\boldsymbol{w}|^2\mathbf{1}$, and $\boldsymbol{W}^3 := \boldsymbol{W}\cdot\boldsymbol{W}\cdot\boldsymbol{W} = -|\boldsymbol{w}|^2\boldsymbol{W}$ hold. Thus arbitrary integer powers of $\boldsymbol{W}$ allow representation $\boldsymbol{W}^{2m} = -[-1]^m|\boldsymbol{w}|^{2m}[\boldsymbol{d}\otimes\boldsymbol{d} - \mathbf{1}]$ for $m\geq 1$ and $\boldsymbol{W}^{2m+1} = [-1]^m|\boldsymbol{w}|^{2m+1}\operatorname{spn}\boldsymbol{d}$ for $m\geq 0$ with $\boldsymbol{w} = |\boldsymbol{w}|\boldsymbol{d}$ and the normalized direction vector $\boldsymbol{d}\in\mathbb{S}^2$. Due to $\boldsymbol{W} : \mathbf{1} = 0$ the first invariant of $\boldsymbol{W}$ vanishes $I_1(\boldsymbol{W}) = 0$, moreover since $\operatorname{cof}\boldsymbol{W} = 2[\mathsf{I}^{\text{skw}}\cdot\boldsymbol{w}] : [\boldsymbol{w}\cdot\mathsf{I}^{\text{skw}}] = \boldsymbol{w}\otimes\boldsymbol{w}$ the second invariant of $\boldsymbol{W}$ computes as $I_2(\boldsymbol{W}) = |\boldsymbol{w}|^2$, and finally since $\det\boldsymbol{W} \doteq \det\boldsymbol{W}^t \equiv -\det\boldsymbol{W}$ the third invariant of $\boldsymbol{W}$ also vanishes $I_3(\boldsymbol{W}) = 0$. Consequently the characteristic equation for $\boldsymbol{W}$ reads $[|\boldsymbol{w}|^2 + \lambda^2]\lambda = 0$ with solution $\lambda = \{0, \pm i|\boldsymbol{w}|\}$. The eigenvectors are given by $\sqrt{2}\boldsymbol{m}_{2/1} = \sqrt{2}\boldsymbol{n}_{1/2} = -\boldsymbol{d}_1 \pm i\boldsymbol{d}_2$ and $\boldsymbol{m}_3 = \boldsymbol{n}_3 = \boldsymbol{d} =: \boldsymbol{d}_3$ with $\boldsymbol{d}_1$ and $\boldsymbol{d}_2$ arbitrary orthonormalized vectors in the plane perpendicular to $\boldsymbol{d}$ so that

$\boldsymbol{d}_1, \boldsymbol{d}_2, \boldsymbol{d}_3$ form a right-handed triad. Then $\sqrt{2}\boldsymbol{W} \cdot \boldsymbol{n}_{1/2} = -|\boldsymbol{w}|\boldsymbol{d}_3 \times \boldsymbol{d}_1 \pm i|\boldsymbol{w}|\boldsymbol{d}_3 \times \boldsymbol{d}_2 = -|\boldsymbol{w}|\boldsymbol{d}_2 \mp i|\boldsymbol{w}|\boldsymbol{d}_1 = \pm i|\boldsymbol{w}|\sqrt{2}\boldsymbol{n}_{1/2}$.

The *exterior product* of vectors $\boldsymbol{u}$ and $\boldsymbol{v}$ renders a skew symmetric tensor and is defined as $\boldsymbol{u} \wedge \boldsymbol{v} := \boldsymbol{u} \otimes \boldsymbol{v} - \boldsymbol{v} \otimes \boldsymbol{u} = \boldsymbol{W}$ with coordinate representation $\boldsymbol{u} \wedge \boldsymbol{v} = [u^i v^j - v^i u^j]\boldsymbol{g}_i \otimes \boldsymbol{g}_j = [u_i v_j - v_i u_j]\boldsymbol{g}^i \otimes \boldsymbol{g}^j$. Then the corresponding axial vector follows as $\boldsymbol{w} = \operatorname{axl}(\boldsymbol{u} \wedge \boldsymbol{v}) = \boldsymbol{v} \times \boldsymbol{u}$ so that $\boldsymbol{W} = \operatorname{spn}(\boldsymbol{v} \times \boldsymbol{u}) = \boldsymbol{u} \wedge \boldsymbol{v}$.

A *special orthogonal* second-order tensor $\boldsymbol{R} \in \mathrm{SO}(3)$ defines a finite *rotation* with magnitude $w = |\boldsymbol{w}| \in [0, 2\pi)$ and rotation sense and axis given by a normalized direction vector $\boldsymbol{d}$ with $\boldsymbol{w} := w\boldsymbol{d}$. A finite *rotation tensor* may then be computed as the tensor exponential $\boldsymbol{R} = \exp(\boldsymbol{W})$ of the skew symmetric tensor $\boldsymbol{W} = \operatorname{spn} \boldsymbol{w}$ and allows the *Euler-Rodriguez* representation $\boldsymbol{R}(w, \boldsymbol{d}) = \cos w \boldsymbol{1} + [1 - \cos w]\boldsymbol{d} \otimes \boldsymbol{d} + \sin w \operatorname{spn} \boldsymbol{d}$ with $\boldsymbol{R} \cdot \boldsymbol{d} = \boldsymbol{d}$. The proof follows from the series expansion of the tensor exponent $\exp(\boldsymbol{W}) = \sum_{k=0}^{\infty} \boldsymbol{W}^k / k! = \boldsymbol{1} + \sum_{m=0}^{\infty} \boldsymbol{W}^{2m+1}/[2m+1]! + \sum_{m=1}^{\infty} \boldsymbol{W}^{2m}/[2m]!$ that due to the above representations expands further into $\exp(\boldsymbol{W}) = \boldsymbol{1} + \sum_{m=0}^{\infty}[-1]^m w^{2m+1}/[2m+1]! \operatorname{spn} \boldsymbol{d} - \sum_{m=1}^{\infty}[-1]^m w^{2m}/[2m]![\boldsymbol{d} \otimes \boldsymbol{d} - \boldsymbol{1}]$. Identifying next $\sin w := \sum_{m=0}^{\infty}[-1]^m w^{2m+1}/[2m+1]!$ and $\cos w - 1 := \sum_{m=1}^{\infty}[-1]^m w^{2m}/[2m]!$ renders the result. The first invariant of $\boldsymbol{R}$ computes as $I_1(\boldsymbol{R}) = 2\cos w + 1$, moreover since $\operatorname{cof} \boldsymbol{R} = \boldsymbol{R}$ also the second invariant of $\boldsymbol{R}$ computes as $I_2(\boldsymbol{R}) = 2\cos w + 1$, and finally since $\det \boldsymbol{R} = 1$ the third invariant of $\boldsymbol{R}$ computes as $I_3(\boldsymbol{R}) = 1$. Consequently the characteristic equation for $\boldsymbol{R}$ reads $[1 - \lambda][1 - 2\cos w\lambda + \lambda^2] = 0$ with solution $\lambda = \{1, \cos w \pm i \sin w\}$. Note that the eigenvalues of $\boldsymbol{R}$ may also be expressed in terms of the eigenvalues of $\boldsymbol{W}$ as $\lambda = \{\exp(0), \exp(\pm iw)\}$. The eigenvectors are given by $\sqrt{2}\boldsymbol{m}_{2/1} = \sqrt{2}\boldsymbol{n}_{1/2} = -\boldsymbol{d}_1 \pm i\boldsymbol{d}_2$ and $\boldsymbol{m}_3 = \boldsymbol{n}_3 = \boldsymbol{d} =: \boldsymbol{d}_3$ with $\boldsymbol{d}_1$ and $\boldsymbol{d}_2$ arbitrary orthonormalized vectors in the plane perpendicular to $\boldsymbol{d}$ so that $\boldsymbol{d}_1, \boldsymbol{d}_2, \boldsymbol{d}_3$ form a right-handed triad. Then $\sqrt{2}\boldsymbol{R} \cdot \boldsymbol{n}_{1/2} = -[\cos w \pm i \sin w]\boldsymbol{d}_1 - [\sin w \mp i \cos w]\boldsymbol{d}_2 = [\cos w \pm i \sin w]\sqrt{2}\boldsymbol{n}_{1/2}$.

Fourth-Order Tensors:

For second-order tensors $\boldsymbol{S}, \boldsymbol{T} \in \mathbb{E}^3 \otimes \mathbb{E}^3$ a fourth-order tensor D may (for example) be defined either as a linear map of second-order tensors into second-order tensors $\boldsymbol{T} = \mathsf{D} : \boldsymbol{S}$, a dyadic product of second-order tensors $\mathsf{D} = \boldsymbol{S} \otimes \boldsymbol{T}$ (a simple fourth-order tensor), $\mathsf{D} = \boldsymbol{S} \boxtimes \boldsymbol{T}$ or $\mathsf{D} = \boldsymbol{S} \boxdot \boldsymbol{T}$, respectively, or a multilinear form of vectors (bilinear form of simple second-order tensors) $\alpha_{\mathsf{D}}(\boldsymbol{S}, \boldsymbol{T}) = \boldsymbol{S} : \mathsf{D} : \boldsymbol{T} \in \mathbb{R}$. Thereby a fourth-order tensor satisfies the following equivalent axioms:

T4.1: Linear map of second-order tensors into second-order tensors:

a)	$\mathsf{D} : [\boldsymbol{S} + \boldsymbol{T}] = \mathsf{D} : \boldsymbol{S} + \mathsf{D} : \boldsymbol{T}$	distributivity	(tensor addition)
b)	$[\mathsf{C} + \mathsf{D}] : \boldsymbol{T} = \mathsf{C} : \boldsymbol{T} + \mathsf{D} : \boldsymbol{T}$	distributivity	(tensor addition)
c)	$\mathsf{D} : [\alpha \boldsymbol{T}] = \alpha[\mathsf{D} : \boldsymbol{T}] = \alpha \mathsf{D} : \boldsymbol{T}$	associativity	(scalar multiplication)
d)	$[\alpha \mathsf{D}] : \boldsymbol{T} = \alpha[\mathsf{D} : \boldsymbol{T}] = \alpha \mathsf{D} : \boldsymbol{T}$	associativity	(scalar multiplication)
e)	$\alpha \mathsf{D} = \mathsf{D} \alpha$	commutativity	(scalar multiplication)
f)	$\mathsf{O} : \boldsymbol{T} = \boldsymbol{0}$	null element	(null tensor)
g)	$\mathsf{I} : \boldsymbol{T} = \boldsymbol{T}$	identity	(identity tensor)

T4.2: Dyadic product of second-order tensors:

a)	$[\boldsymbol{R}\otimes\boldsymbol{S}]:\boldsymbol{T}=[\boldsymbol{S}:\boldsymbol{T}]\boldsymbol{R}$	contraction
b)	$\boldsymbol{R}\otimes[\boldsymbol{S}+\boldsymbol{T}]=\boldsymbol{R}\otimes\boldsymbol{S}+\boldsymbol{R}\otimes\boldsymbol{T}$	distributivity (tensor addition)
c)	$[\alpha\boldsymbol{S}]\otimes\boldsymbol{T}=\boldsymbol{S}\otimes[\alpha\boldsymbol{T}]=\alpha[\boldsymbol{S}\otimes\boldsymbol{T}]$	associativity (scalar multiplication)
d)	$[\boldsymbol{S}\boxtimes\boldsymbol{T}]:\boldsymbol{R}=\boldsymbol{S}\cdot\boldsymbol{R}\cdot\boldsymbol{T}^t$	right multiplication
e)	$[\boldsymbol{S}\boxdot\boldsymbol{T}]:\boldsymbol{R}=\boldsymbol{S}\cdot\boldsymbol{R}^t\cdot\boldsymbol{T}^t$	
f)	$\boldsymbol{R}:[\boldsymbol{S}\boxtimes\boldsymbol{T}]=\boldsymbol{S}^t\cdot\boldsymbol{R}\cdot\boldsymbol{T}$	left multiplication
g)	$\boldsymbol{R}:[\boldsymbol{S}\boxdot\boldsymbol{T}]=\boldsymbol{S}^t\cdot\boldsymbol{R}^t\cdot\boldsymbol{T}$	

T4.3: Multilinear form of vectors (simple second-order tensors):

a)	$\alpha_{\mathsf{D}}(\boldsymbol{R},\boldsymbol{S}+\boldsymbol{T})=\alpha_{\mathsf{D}}(\boldsymbol{R},\boldsymbol{S})+\alpha_{\mathsf{D}}(\boldsymbol{R},\boldsymbol{T})$	distributivity (tensor addition)
b)	$\alpha_{\mathsf{D}+\mathsf{C}}(\boldsymbol{S},\boldsymbol{T})=\alpha_{\mathsf{D}}(\boldsymbol{S},\boldsymbol{T})+\alpha_{\mathsf{C}}(\boldsymbol{S},\boldsymbol{T})$	distributivity (tensor addition)
c)	$\alpha_{\mathsf{D}}(\boldsymbol{S},\beta\boldsymbol{T})=\beta\alpha_{\mathsf{D}}(\boldsymbol{S},\boldsymbol{T})$	associativity (scalar multiplication)
	$\alpha_{\mathsf{D}}(\beta\boldsymbol{S},\boldsymbol{T})=\beta\alpha_{\mathsf{D}}(\boldsymbol{S},\boldsymbol{T})$	
d)	$\alpha_{\beta\mathsf{D}}(\boldsymbol{S},\boldsymbol{T})=\beta\alpha_{\mathsf{D}}(\boldsymbol{S},\boldsymbol{T})$	associativity (scalar multiplication)
e)	$\alpha_{\beta\mathsf{D}}(\boldsymbol{S},\boldsymbol{T})=\alpha_{\mathsf{D}\beta}(\boldsymbol{S},\boldsymbol{T})$	commutativity(scalar multiplication)
f)	$\alpha_{\mathsf{O}}(\boldsymbol{S},\boldsymbol{T})=0$	null element (null tensor)
g)	$\alpha_{\mathsf{I}}(\boldsymbol{S},\boldsymbol{T})=\boldsymbol{S}:\boldsymbol{T}$	identity (identity tensor)

The 81-dimensional tensorial product space $\mathbb{E}^3\otimes\mathbb{E}^3\otimes\mathbb{E}^3\otimes\mathbb{E}^3$ allows for sets of 81 linearly independent dyadic products $\boldsymbol{g}_i\otimes\boldsymbol{g}_j\otimes\boldsymbol{g}_k\otimes\boldsymbol{g}_l\neq\boldsymbol{0}$ in terms of the linearly independent base vectors $\boldsymbol{g}_i$ such that $\alpha^{ijkl}\boldsymbol{g}_i\otimes\boldsymbol{g}_j\otimes\boldsymbol{g}_k\otimes\boldsymbol{g}_l=\boldsymbol{0}$ holds only for all $\alpha^{ijkl}=0$. Then every additional fourth-order tensor $\mathsf{D}\in\mathbb{E}^3\otimes\mathbb{E}^3\otimes\mathbb{E}^3\otimes\mathbb{E}^3$ is linearly dependent, i.e. $\lambda\mathsf{D}+\alpha^{ijkl}\boldsymbol{g}_i\otimes\boldsymbol{g}_j\otimes\boldsymbol{g}_k\otimes\boldsymbol{g}_l=\boldsymbol{0}$ with $\lambda\neq0$ and at least one $\alpha^{ijkl}\neq0$ and may be represented by $\mathsf{D}=\mathsf{D}^{ijkl}\boldsymbol{g}_i\otimes\boldsymbol{g}_j\otimes\boldsymbol{g}_k\otimes\boldsymbol{g}_l$. Here D^{ijkl} are the contravariant coefficients of D with respect to the covariant base dyads $\boldsymbol{g}_i\otimes\boldsymbol{g}_j\otimes\boldsymbol{g}_k\otimes\boldsymbol{g}_l$.

Based on the corresponding contravariant base dyads $\boldsymbol{g}^i\otimes\boldsymbol{g}^j\otimes\boldsymbol{g}^k\otimes\boldsymbol{g}^l$ a fourth-order tensor allows for example also representation $\mathsf{D}=\mathsf{D}_{ijkl}\boldsymbol{g}^i\otimes\boldsymbol{g}^j\otimes\boldsymbol{g}^k\otimes\boldsymbol{g}^l$ with D_{ijkl} its covariant coefficients.

The covariant coefficients of D are extracted by $\mathsf{D}_{ijkl}=[\boldsymbol{g}_i\otimes\boldsymbol{g}_j]:\mathsf{D}:[\boldsymbol{g}_k\otimes\boldsymbol{g}_l]$, likewise various mixedvariant representations are possible. The metric coefficients may be used to raise and lower indices of the coefficients of a fourth-order tensor, e.g. $\mathsf{D}^{ijkl}=g^{im}g^{jn}\mathsf{D}_{mnop}g^{ok}g^{pl}$.

The linear map of second-order tensors into second-order tensors has coordinate representation $\mathsf{D}:\boldsymbol{S}=\mathsf{D}^{ijkl}S_{kl}\boldsymbol{g}_i\otimes\boldsymbol{g}_j=\mathsf{D}_{ijkl}S^{kl}\boldsymbol{g}^i\otimes\boldsymbol{g}^j$. Moreover, the dyadic products of second-order tensors have coordinate representation $\boldsymbol{S}\otimes\boldsymbol{T}=S^{ij}T^{kl}\boldsymbol{g}_i\otimes\boldsymbol{g}_j\otimes\boldsymbol{g}_k\otimes\boldsymbol{g}_l=S_{ij}T_{kl}\boldsymbol{g}^i\otimes\boldsymbol{g}^j\otimes\boldsymbol{g}^k\otimes\boldsymbol{g}^l$, $\boldsymbol{S}\boxtimes\boldsymbol{T}=S^{ik}T^{jl}\boldsymbol{g}_i\otimes\boldsymbol{g}_j\otimes\boldsymbol{g}_k\otimes\boldsymbol{g}_l=S_{ik}T_{jl}\boldsymbol{g}^i\otimes\boldsymbol{g}^j\otimes\boldsymbol{g}^k\otimes\boldsymbol{g}^l$, and $\boldsymbol{S}\boxdot\boldsymbol{T}=S^{il}T^{jk}\boldsymbol{g}_i\otimes\boldsymbol{g}_j\otimes\boldsymbol{g}_k\otimes\boldsymbol{g}_l=S_{il}T_{jk}\boldsymbol{g}^i\otimes\boldsymbol{g}^j\otimes\boldsymbol{g}^k\otimes\boldsymbol{g}^l$, respectively. Likewise, the multilinear form of vectors in terms of simple second-order tensors has coordinate representation $\alpha_{\mathsf{D}}(\boldsymbol{S},\boldsymbol{T})=S_{ij}\mathsf{D}^{ijkl}T_{kl}=S^{ij}\mathsf{D}_{ijkl}T^{kl}$.

Products of Fourth-Order Tensors:
The *scalar product* of general fourth-order tensors C and D is denoted as $p := \mathsf{C} :: \mathsf{D} \in \mathbb{R}$. It satisfies e.g. the following rules:

P4.1: *Scalar product of fourth-order tensors:*

a)	$[\mathsf{B}+\mathsf{C}] :: \mathsf{D} = \mathsf{B} :: \mathsf{D} + \mathsf{C} :: \mathsf{D}$	distributivity	(tensor addition)
	$\mathsf{D} :: [\mathsf{B}+\mathsf{C}] = \mathsf{D} :: \mathsf{B} + \mathsf{D} :: \mathsf{C}$		
b)	$\mathsf{C} :: [\alpha\mathsf{D}] = \alpha[\mathsf{C} :: \mathsf{D}] = \alpha\mathsf{C} :: \mathsf{D}$	associativity	(scalar multiplication)
	$[\alpha\mathsf{C}] :: \mathsf{D} = \alpha[\mathsf{C} :: \mathsf{D}] = \alpha\mathsf{C} :: \mathsf{D}$		
c)	$\mathsf{C} :: \mathsf{D} = \mathsf{D} :: \mathsf{C}$	commutativity	(tensor multiplication)
d)	$\mathsf{C} :: \mathsf{D} = 0 \quad \to \mathsf{C} \perp \mathsf{D}$	orthogonality	(tensor multiplication)

Then the *norm of a fourth-order tensor* $|\mathsf{D}|^2 = \mathsf{D} :: \mathsf{D}$ may be computed as its length, and the Schwarz and the triangle inequality $|\mathsf{C} :: \mathsf{D}| \le |\mathsf{C}|\,|\mathsf{D}|$ and $|\mathsf{C}+\mathsf{D}| \le |\mathsf{C}| + |\mathsf{D}|$, respectively, hold for this norm. Moreover the angle φ enclosed by C and D may be defined via $\mathsf{C} :: \mathsf{D} := |\mathsf{C}|\,|\mathsf{D}|\cos\varphi$. For the special case that $\mathsf{D} = \boldsymbol{S}\otimes\boldsymbol{T}$ is a simple fourth-order tensor the scalar product renders $p = \mathsf{C} :: [\boldsymbol{S}\otimes\boldsymbol{T}] = \boldsymbol{S} : \mathsf{C} : \boldsymbol{T}$, if in addition also $\mathsf{C} = \boldsymbol{U}\otimes\boldsymbol{V}$ is a simple fourth-order tensor the scalar product results in $p = [\boldsymbol{U}\otimes\boldsymbol{V}] :: [\boldsymbol{S}\otimes\boldsymbol{T}] = [\boldsymbol{S} : \boldsymbol{U}][\boldsymbol{V} : \boldsymbol{T}]$. Finally the scalar product of fourth-order tensors has coordinate representation $\mathsf{C} :: \mathsf{D} = \mathsf{C}^{ijkl}\mathsf{D}_{ijkl} = \mathsf{C}_{ijkl}\mathsf{D}^{ijkl}$. Thus the scalar product of fourth-order tensors follows as the summation of the products of the corresponding coefficients.

The *tensor dot product* of general fourth-order tensors C and D is denoted as $\mathsf{P} := \mathsf{C} : \mathsf{D} \in \mathbb{E}^3\otimes\mathbb{E}^3\otimes\mathbb{E}^3\otimes\mathbb{E}^3$. It satisfies e.g. the following rules:

P4.2: *Tensor dot product of fourth-order tensors:*

a)	$[\mathsf{B}+\mathsf{C}] : \mathsf{D} = \mathsf{B} : \mathsf{D} + \mathsf{C} : \mathsf{D}$	distributivity	(tensor addition)
	$\mathsf{D} : [\mathsf{B}+\mathsf{C}] = \mathsf{D} : \mathsf{B} + \mathsf{D} : \mathsf{C}$		
b)	$[\mathsf{B} : \mathsf{C}] : \mathsf{D} = \mathsf{B} : [\mathsf{C} : \mathsf{D}] = \mathsf{B} : \mathsf{C} : \mathsf{D}$	associativity	(tensor multiplication)
c)	$[\mathsf{B} : \mathsf{C}] : \boldsymbol{S} = \mathsf{B} : [\mathsf{C} : \mathsf{S}] = \mathsf{B} : \mathsf{C} : \mathsf{S}$	associativity	(tensor multiplication)
d)	$\mathsf{C} : [\alpha\mathsf{D}] = \alpha[\mathsf{C} : \mathsf{D}] = \alpha\,\mathsf{C} : \mathsf{D}$	associativity	(scalar multiplication)
	$[\alpha\mathsf{C}] : \mathsf{D} = \alpha[\mathsf{C} : \mathsf{D}] = \alpha\,\mathsf{C} : \mathsf{D}$		
e)	$\mathsf{C} : \mathsf{D} \ne \mathsf{D} : \mathsf{C}$	non-commutativity	(tensor multiplication)
f)	$\mathsf{O} : \mathsf{D} = \mathsf{D} : \mathsf{O} = \mathsf{O}$	null element	(null tensor)
g)	$\mathsf{I} : \mathsf{D} = \mathsf{D} : \mathsf{I} = \mathsf{D}$	identity	(identity tensor)

It is important to observe that the tensor dot product is *not commutative*! For the special case that $\mathsf{D} = \boldsymbol{S}\otimes\boldsymbol{T}$ is a simple fourth-order tensor the tensor dot product renders $\mathsf{P} = \mathsf{C} : [\boldsymbol{S}\otimes\boldsymbol{T}] = [\mathsf{C} : \boldsymbol{S}]\otimes\boldsymbol{T}$, if in addition also $\mathsf{C} = \boldsymbol{U}\otimes\boldsymbol{V}$ is a simple fourth-order tensor the tensor dot product results in $\mathsf{P} = [\boldsymbol{U}\otimes\boldsymbol{V}] : [\boldsymbol{S}\otimes\boldsymbol{T}] = [\boldsymbol{V} : \boldsymbol{S}]\boldsymbol{U}\otimes\boldsymbol{T}$. Finally the tensor dot product of fourth-order tensors has coordinate representation $\mathsf{C} : \mathsf{D} = \mathsf{C}^{ijkl}\mathsf{D}_{klmn}\boldsymbol{g}_i\otimes\boldsymbol{g}_j\otimes\boldsymbol{g}^m\otimes\boldsymbol{g}^n = \mathsf{C}_{ijkl}\mathsf{D}^{klmn}\boldsymbol{g}^i\otimes\boldsymbol{g}^j\otimes\boldsymbol{g}_m\otimes\boldsymbol{g}_n$.

Various *tensor cross products* of general fourth-order tensors $\mathbf{C}$ and $\mathbf{D}$ may be defined in analogy to the case of second-order tensors. However the results of this tedious task are neither particularly enlightening nor needed in the sequel.

Special Fourth-Order Tensors and Algebraic Operators:
The fourth-order *dyadic identity tensors* are denoted as $\mathbf{I}^{\boxtimes} := \mathbf{1} \boxtimes \mathbf{1} =: \mathbf{I}$, $\mathbf{I}^{\boxdot} := \mathbf{1} \boxdot \mathbf{1} =: \mathbf{I}^t$, and $\mathbf{I}^{\otimes} := \mathbf{1} \otimes \mathbf{1}$ and are defined so that $\mathbf{I}^{\boxtimes} : \boldsymbol{T} = \boldsymbol{T}$, and $\mathbf{I}^{\boxdot} : \boldsymbol{T} = \boldsymbol{T}^t$. They expand in a Cartesian base system in terms of the Kronecker symbol as $\mathbf{I}^{\boxtimes} = \delta_{ac}\delta_{bd}\boldsymbol{e}^a \otimes \boldsymbol{e}^b \otimes \boldsymbol{e}^c \otimes \boldsymbol{e}^d$, $\mathbf{I}^{\boxdot} = \delta_{ad}\delta_{bc}\boldsymbol{e}^a \otimes \boldsymbol{e}^b \otimes \boldsymbol{e}^c \otimes \boldsymbol{e}^d$, and $\mathbf{I}^{\otimes} = \delta_{ab}\delta_{cd}\boldsymbol{e}^a \otimes \boldsymbol{e}^b \otimes \boldsymbol{e}^c \otimes \boldsymbol{e}^d$. In e.g. contravariant base dyads the dyadic identity tensors are expressed in terms of the metric coefficients $\mathbf{I}^{\boxtimes} = g_{ik}g_{jl}\boldsymbol{g}^i \otimes \boldsymbol{g}^j \otimes \boldsymbol{g}^k \otimes \boldsymbol{g}^l$, $\mathbf{I}^{\boxdot} = g_{il}g_{jk}\boldsymbol{g}^i \otimes \boldsymbol{g}^j \otimes \boldsymbol{g}^k \otimes \boldsymbol{g}^l$, and $\mathbf{I}^{\otimes} = g_{ij}g_{kl}\boldsymbol{g}^i \otimes \boldsymbol{g}^j \otimes \boldsymbol{g}^k \otimes \boldsymbol{g}^l$. Scalar multiples of fourth-order dyadic identity tensors and combinations thereof are the only *isotropic fourth-order tensors.*

The fourth-order *symmetric* and *skew symmetric identity tensors* are denoted as $2\mathbf{I}^{\text{sym}} := \mathbf{I}^{\boxtimes} + \mathbf{I}^{\boxdot} = \mathbf{1}\boxtimes\mathbf{1} + \mathbf{1}\boxdot\mathbf{1}$, and $2\mathbf{I}^{\text{skw}} := \mathbf{I}^{\boxtimes} - \mathbf{I}^{\boxdot} = \mathbf{1}\boxtimes\mathbf{1} - \mathbf{1}\boxdot\mathbf{1}$, and are defined so that $\mathbf{I}^{\text{sym}} : \boldsymbol{T} = \boldsymbol{T}^{\text{sym}}$, and $\mathbf{I}^{\text{skw}} : \boldsymbol{T} = \boldsymbol{T}^{\text{skw}}$. All fourth-order identity tensors are major symmetric, $\mathbf{I}^{\boxdot}$ is the (right) minor transpose of $\mathbf{I}^{\boxtimes}$, the symmetric and skew symmetric identity tensors $\mathbf{I}^{\text{sym}}$ and $\mathbf{I}^{\text{skw}}$ possess in addition minor (right) symmetry and skew symmetry, respectively, see below.

The *major transpose* of a fourth-order tensor $\mathbf{D}$ is denoted as $\mathbf{D}^T$ and is defined so that $\boldsymbol{S} : \mathbf{D} : \boldsymbol{T} =: \boldsymbol{T} : \mathbf{D}^T : \boldsymbol{S}$. It satisfies e.g. the following rules:

***S4.1:** Major transpose of fourth-order tensors:*
a) $[\mathbf{C} + \mathbf{D}]^T = \mathbf{C}^T + \mathbf{D}^T$
b) $[\alpha\mathbf{D}]^T = \alpha\mathbf{D}^T$
c) $[\mathbf{C} : \mathbf{D}]^T = \mathbf{D}^T : \mathbf{C}^T$
d) $[\boldsymbol{S} \otimes \boldsymbol{T}]^T = \boldsymbol{T} \otimes \boldsymbol{S}$

The major transpose of fourth-order tensors has coordinate representation $\mathbf{D}^T = \mathsf{D}_{klij}\boldsymbol{g}^i \otimes \boldsymbol{g}^j \otimes \boldsymbol{g}^k \otimes \boldsymbol{g}^l = \mathsf{D}^{klij}\boldsymbol{g}_i \otimes \boldsymbol{g}_j \otimes \boldsymbol{g}_k \otimes \boldsymbol{g}_l$. A fourth-order tensor is called *major symmetric* if it equals its major transpose $\mathbf{D}^{\text{SYM}} := \mathbf{D} \equiv \mathbf{D}^T$ and *major skew symmetric* if it equals its negative major transpose $\mathbf{D}^{\text{SKW}} := \mathbf{D} \equiv -\mathbf{D}^T$.

The *minor (right) transpose* of a fourth-order tensor $\mathbf{D}$ is denoted as $\mathbf{D}^t$ and is defined so that $\boldsymbol{S} : \mathbf{D} : \boldsymbol{T} =: \boldsymbol{S} : \mathbf{D}^t : \boldsymbol{T}^t$. It satisfies e.g. the following rules:

***S4.2:** Minor (right) transpose of fourth-order tensors:*
a) $[\mathbf{C} + \mathbf{D}]^t = \mathbf{C}^t + \mathbf{D}^t$
b) $[\alpha\mathbf{D}]^t = \alpha\mathbf{D}^t$
c) $[\boldsymbol{S} \otimes \boldsymbol{T}]^t = \boldsymbol{S} \otimes \boldsymbol{T}^t$

The minor (right) transpose of fourth-order tensors has coordinate representation $\mathbf{D}^t = \mathsf{D}_{ijlk}\boldsymbol{g}^i \otimes \boldsymbol{g}^j \otimes \boldsymbol{g}^k \otimes \boldsymbol{g}^l = \mathsf{D}^{ijlk}\boldsymbol{g}_i \otimes \boldsymbol{g}_j \otimes \boldsymbol{g}_k \otimes \boldsymbol{g}_l$. A fourth-order tensor is called *minor (right) symmetric* if it equals its minor (right) transpose $\mathbf{D}^{\mathrm{sym}} := \mathbf{D} \equiv \mathbf{D}^t$ and *minor (right) skew symmetric* if it equals its negative minor (right) transpose $\mathbf{D}^{\mathrm{skw}} := \mathbf{D} \equiv -\mathbf{D}^t$. *Minor (left) transposition* and *minor (left) symmetry and skew symmetry* are defined in analogy.

The *major symmetric* part of a fourth-order tensor is denoted $\mathbf{D}^{\mathrm{SYM}}$ and is computed from $2\mathbf{D}^{\mathrm{SYM}} := \mathbf{D} + \mathbf{D}^T$. Then $\alpha_{\mathbf{D}^{\mathrm{SYM}}}(\boldsymbol{T},\boldsymbol{T}) = \boldsymbol{T} : \mathbf{D}^{\mathrm{SYM}} : \boldsymbol{T} = \boldsymbol{T} : \mathbf{D} : \boldsymbol{T} = \alpha_{\mathbf{D}}(\boldsymbol{T},\boldsymbol{T})$ holds for all $\boldsymbol{T} \neq \mathbf{0}$. The *major skew symmetric* part of a fourth-order tensor is denoted $\mathbf{D}^{\mathrm{SKW}}$ and is computed from $2\mathbf{D}^{\mathrm{SKW}} := \mathbf{D} - \mathbf{D}^T$. Then $\alpha_{\mathbf{D}^{\mathrm{SKW}}}(\boldsymbol{T},\boldsymbol{T}) = \boldsymbol{T} : \mathbf{D}^{\mathrm{SKW}} : \boldsymbol{T} = 0$ holds for all $\boldsymbol{T} \neq \mathbf{0}$. Finally, every fourth-order tensor $\mathbf{D}$ allows unique *additive decomposition* $\mathbf{D} = \mathbf{D}^{\mathrm{SYM}} + \mathbf{D}^{\mathrm{SKW}}$ into its major symmetric and major skew symmetric contributions.

The *minor (right) symmetric* part of a fourth-order tensor is denoted $\mathbf{D}^{\mathrm{sym}}$ and is computed from $2\mathbf{D}^{\mathrm{sym}} := \mathbf{D} + \mathbf{D}^t$. The *minor (right) skew symmetric* part of a fourth-order tensor is denoted $\mathbf{D}^{\mathrm{skw}}$ and is computed from $2\mathbf{D}^{\mathrm{skw}} := \mathbf{D} - \mathbf{D}^t$. Finally, every fourth-order tensor $\mathbf{D}$ allows unique *additive decomposition* $\mathbf{D} = \mathbf{D}^{\mathrm{sym}} + \mathbf{D}^{\mathrm{skw}}$ into its minor symmetric and minor skew symmetric contributions. The *minor (left) symmetric and skew symmetric* parts of a fourth-order tensor are defined in analogy.

A fourth-order tensor $\mathbf{D}$ is classified as either positive definite, positive semidefinite, indefinite, negative semidefinite, or negative definite, respectively, if its *quadratic form* $\alpha := \alpha_{\mathbf{D}}(\boldsymbol{T},\boldsymbol{T}) = \boldsymbol{T} : \mathbf{D} : \boldsymbol{T}$ renders $\alpha > 0$, $\alpha \geq 0$, $\alpha \gtreqless 0$, $\alpha \leq 0$, or $\alpha < 0$, respectively, for all $\boldsymbol{T} \neq \mathbf{0}$.

In *Voigt matrix notation* a fourth-order tensor $\mathbf{D} \in \mathbb{E}^3 \otimes \mathbb{E}^3 \otimes \mathbb{E}^3 \otimes \mathbb{E}^3$ is equivalently represented by a *matrix* $\underline{D} \in \mathbb{E}^9 \otimes \mathbb{E}^9$. This is achieved by the assignment of the 81 Cartesian coefficients of $\mathbf{D}$ to the 81 entries of $\underline{D}$ by the map $\underline{D} = v_*(\mathbf{D})$ so that $[D_{\mathsf{ef}}] =$

$$\begin{bmatrix} \mathsf{D}_{1111} & \mathsf{D}_{1122} & \mathsf{D}_{1133} & \mathsf{D}_{1123} & \mathsf{D}_{1131} & \mathsf{D}_{1112} & \mathsf{D}_{1132} & \mathsf{D}_{1113} & \mathsf{D}_{1121} \\ \mathsf{D}_{2211} & \mathsf{D}_{2222} & \mathsf{D}_{2233} & \mathsf{D}_{2223} & \mathsf{D}_{2231} & \mathsf{D}_{2212} & \mathsf{D}_{2232} & \mathsf{D}_{2213} & \mathsf{D}_{2221} \\ \mathsf{D}_{3311} & \mathsf{D}_{3322} & \mathsf{D}_{3333} & \mathsf{D}_{3323} & \mathsf{D}_{3331} & \mathsf{D}_{3312} & \mathsf{D}_{3332} & \mathsf{D}_{3313} & \mathsf{D}_{3321} \\ \mathsf{D}_{2311} & \mathsf{D}_{2322} & \mathsf{D}_{2333} & \mathsf{D}_{2323} & \mathsf{D}_{2331} & \mathsf{D}_{2312} & \mathsf{D}_{2332} & \mathsf{D}_{2313} & \mathsf{D}_{2321} \\ \mathsf{D}_{3111} & \mathsf{D}_{3122} & \mathsf{D}_{3133} & \mathsf{D}_{3123} & \mathsf{D}_{3131} & \mathsf{D}_{3112} & \mathsf{D}_{3132} & \mathsf{D}_{3113} & \mathsf{D}_{3121} \\ \mathsf{D}_{1211} & \mathsf{D}_{1222} & \mathsf{D}_{1233} & \mathsf{D}_{1223} & \mathsf{D}_{1231} & \mathsf{D}_{1212} & \mathsf{D}_{1232} & \mathsf{D}_{1213} & \mathsf{D}_{1221} \\ \mathsf{D}_{3211} & \mathsf{D}_{3222} & \mathsf{D}_{3233} & \mathsf{D}_{3223} & \mathsf{D}_{3231} & \mathsf{D}_{3212} & \mathsf{D}_{3232} & \mathsf{D}_{3213} & \mathsf{D}_{3221} \\ \mathsf{D}_{1311} & \mathsf{D}_{1322} & \mathsf{D}_{1333} & \mathsf{D}_{1323} & \mathsf{D}_{1331} & \mathsf{D}_{1312} & \mathsf{D}_{1332} & \mathsf{D}_{1313} & \mathsf{D}_{1321} \\ \mathsf{D}_{2111} & \mathsf{D}_{2122} & \mathsf{D}_{2133} & \mathsf{D}_{2123} & \mathsf{D}_{2131} & \mathsf{D}_{2112} & \mathsf{D}_{2132} & \mathsf{D}_{2113} & \mathsf{D}_{2121} \end{bmatrix}$$

Vice versa the inverse map is denoted by $\mathbf{D} = v^*(\underline{D})$. The assignments $D_{\mathsf{ef}} = \mathsf{b}_{\mathsf{e}ab}\mathsf{D}_{abcd}\mathsf{b}_{\mathsf{f}cd}$ and $\mathsf{D}_{abcd} = \mathsf{b}_{\mathsf{e}ab}D_{\mathsf{ef}}\mathsf{b}_{\mathsf{f}cd}$ are expressed in terms of the previously introduced Boolean operator $\mathsf{b}_{\mathsf{a}bc}$. As an example, the identity tensor $\mathbf{I} = \mathbf{I}^{\boxtimes}$ maps into the 9×9 unit matrix $\underline{1} \in \mathbb{E}^9 \otimes \mathbb{E}^9$, with the inverse map $\mathsf{b}_{\mathsf{e}ab}\mathsf{b}_{\mathsf{e}cd} \to \mathbf{I}$.

A minor (left and right) symmetric fourth-order tensor $\mathsf{D} = {}^{\text{sym}}\mathsf{D}^{\text{sym}}$ may be represented by a matrix $\underline{D} \in \mathbb{E}^6 \otimes \mathbb{E}^6$ with $D_{\mathsf{ef}} = \mathsf{b}^{\text{sym}}_{\mathsf{e}ab}\mathsf{D}_{abcd}\mathsf{b}^{\text{sym}}_{\mathsf{f}cd}$, i.e. $[D_{\mathsf{ef}}] =$

$$\begin{bmatrix} \mathsf{D}_{11\,11} & \mathsf{D}_{11\,22} & \mathsf{D}_{11\,33} & \sqrt{2}\mathsf{D}_{11\,(23)} & \sqrt{2}\mathsf{D}_{11\,(31)} & \sqrt{2}\mathsf{D}_{11\,(12)} \\ \mathsf{D}_{22\,11} & \mathsf{D}_{22\,22} & \mathsf{D}_{22\,33} & \sqrt{2}\mathsf{D}_{22\,(23)} & \sqrt{2}\mathsf{D}_{22\,(31)} & \sqrt{2}\mathsf{D}_{22\,(12)} \\ \mathsf{D}_{33\,11} & \mathsf{D}_{33\,22} & \mathsf{D}_{33\,33} & \sqrt{2}\mathsf{D}_{33\,(23)} & \sqrt{2}\mathsf{D}_{33\,(31)} & \sqrt{2}\mathsf{D}_{33\,(12)} \\ \sqrt{2}\mathsf{D}_{(23)11} & \sqrt{2}\mathsf{D}_{(23)22} & \sqrt{2}\mathsf{D}_{(23)33} & 2\mathsf{D}_{(23)(23)} & 2\mathsf{D}_{(23)(31)} & 2\mathsf{D}_{(23)(12)} \\ \sqrt{2}\mathsf{D}_{(31)11} & \sqrt{2}\mathsf{D}_{(31)22} & \sqrt{2}\mathsf{D}_{(31)33} & 2\mathsf{D}_{(31)(23)} & 2\mathsf{D}_{(31)(31)} & 2\mathsf{D}_{(31)(12)} \\ \sqrt{2}\mathsf{D}_{(12)11} & \sqrt{2}\mathsf{D}_{(12)22} & \sqrt{2}\mathsf{D}_{(12)33} & 2\mathsf{D}_{(12)(23)} & 2\mathsf{D}_{(12)(31)} & 2\mathsf{D}_{(12)(12)} \end{bmatrix}$$

so that $\boldsymbol{T}^{\text{sym}} : \mathsf{D} : \boldsymbol{T}^{\text{sym}} = \underline{t}^{\mathsf{T}}\underline{D}\underline{t}$. Vice versa $\mathsf{D}_{abcd} = \mathsf{b}^{\text{sym}}_{\mathsf{e}ab} D_{\mathsf{ef}} \mathsf{b}^{\text{sym}}_{\mathsf{f}cd}$ holds. As an example, the identity tensor I^{sym} maps into $\underline{1} \in \mathbb{E}^6 \otimes \mathbb{E}^6$, with the inverse map $\mathsf{b}^{\text{sym}}_{\mathsf{e}ab}\mathsf{b}^{\text{sym}}_{\mathsf{e}cd} \to \mathsf{I}^{\text{sym}}$.

A minor (left and right) skew symmetric fourth-order tensor $\mathsf{D} = {}^{\text{skw}}\mathsf{D}^{\text{skw}}$ may be represented by a matrix $\underline{D} \in \mathbb{E}^3 \otimes \mathbb{E}^3$ with $D_{\mathsf{ef}} = \mathsf{b}^{\text{skw}}_{\mathsf{e}ab}\mathsf{D}_{abcd}\mathsf{b}^{\text{skw}}_{\mathsf{f}cd}$; i.e.

$$[D_{\mathsf{ef}}] = \begin{bmatrix} 2\mathsf{D}_{[32][32]} & 2\mathsf{D}_{[32][13]} & 2\mathsf{D}_{[32][21]} \\ 2\mathsf{D}_{[13][32]} & 2\mathsf{D}_{[13][13]} & 2\mathsf{D}_{[13][21]} \\ 2\mathsf{D}_{[21][32]} & 2\mathsf{D}_{[21][13]} & 2\mathsf{D}_{[21][21]} \end{bmatrix}$$

so that $\boldsymbol{T}^{\text{skw}} : \mathsf{D} : \boldsymbol{T}^{\text{skw}} = \underline{t}^{\mathsf{T}}\underline{D}\underline{t}$. Vice versa $\mathsf{D}_{abcd} = \mathsf{b}^{\text{skw}}_{\mathsf{e}ab} D_{\mathsf{ef}} \mathsf{b}^{\text{skw}}_{\mathsf{f}cd}$ holds. As an example, the identity tensor I^{skw} maps into $\underline{1} \in \mathbb{E}^3 \otimes \mathbb{E}^3$, with the inverse map $\mathsf{b}^{\text{skw}}_{\mathsf{e}ab}\mathsf{b}^{\text{skw}}_{\mathsf{e}cd} \to \mathsf{I}^{\text{skw}}$.

The *determinant* of a fourth-order tensor D is denoted as $\det \mathsf{D}$ and is defined as $\det \mathsf{D} := \det \underline{D}$. It satisfies e.g. the following rules:

***S4.3:** Determinant of fourth-order tensors:*

a) $\det(\mathsf{C} : \mathsf{D}) := \det(\underline{C}\underline{D}) = \det \underline{C} \det \underline{D} =: \det \mathsf{C} \det \mathsf{D}$
b) $\det \mathsf{D}^{-1} := \det \underline{D}^{-1} = [\det \underline{D}]^{-1} =: [\det \mathsf{D}]^{-1}$
c) $\det \mathsf{D}^{T} =: \det \underline{D}^{\mathsf{T}} = \det \underline{D} =: \det \mathsf{D}$
d) $\det(\alpha\mathsf{D}) =: \det(\alpha\underline{D}) = \alpha^9 \det \underline{D} =: \alpha^9 \det \mathsf{D}$

The determinant of a nine-dimensional matrix $\underline{D}$ has coordinate representation $9! \det \underline{D} = e^{\mathsf{a}_1\mathsf{a}_2\cdots\mathsf{a}_9} e^{\mathsf{b}_1\mathsf{b}_2\cdots\mathsf{b}_9} D_{\mathsf{a}_1\mathsf{b}_1} D_{\mathsf{a}_2\mathsf{b}_2} \cdots D_{\mathsf{a}_9\mathsf{b}_9}$. Here $e^{\mathsf{a}_1\mathsf{a}_2\cdots\mathsf{a}_9}$ denotes the fully skew symmetric *ninth-order permutation symbol* with $\mathsf{a}_\mathsf{i} = 1, 2, \cdots 9$ for all $\mathsf{i} = 1, 2, \cdots 9$. It is ± 1 for all even/odd permutations of $\mathsf{a}_1\mathsf{a}_2\cdots\mathsf{a}_9 = 12\cdots 9$ and zero else.

The *cofactor* of a fourth-order tensor D is denoted as $\operatorname{cof} \mathsf{D}$ and is defined as $\operatorname{cof} \mathsf{D} := v^*(\operatorname{cof} \underline{D})$. It satisfies e.g. the following rules:

***S4.4:** Cofactor of fourth-order tensors:*

a) $\det(\operatorname{cof} \mathsf{D}) := \det \operatorname{cof} \underline{D} = [\det \underline{D}]^8 =: [\det \mathsf{D}]^8$
b) $\mathsf{D} :: \operatorname{cof} \mathsf{D} := |\underline{D} \operatorname{cof} \underline{D}| = 9 \det \underline{D} =: 9 \det \mathsf{D}$
c) $\mathsf{D}^{T} : \operatorname{cof} \mathsf{D} := v^*(\underline{D}^{\mathsf{T}} \operatorname{cof} \underline{D}) = \det \underline{D}\, v^*(\underline{1}) =: \det \mathsf{D}\, \mathsf{I}$
d) $[\operatorname{cof} \mathsf{D}]^{T} := v^*([\operatorname{cof} \underline{D}]^{\mathsf{T}}) = v^*(\operatorname{cof} \underline{D}^{\mathsf{T}}) =: \operatorname{cof} \mathsf{D}^{T}$
e) $\operatorname{cof}(\mathsf{C} : \mathsf{D}) := v^*(\operatorname{cof}(\underline{C}\underline{D})) = v^*(\operatorname{cof} \underline{D} \operatorname{cof} \underline{C}) =: \operatorname{cof} \mathsf{D} : \operatorname{cof} \mathsf{C}$

For invertible $\mathbf{D} = \upsilon^*(\underline{D})$, see below, it holds that $\operatorname{cof}\mathbf{D} = \det\mathbf{D}\mathbf{D}^{-T} = \upsilon^*(\det\underline{D}\underline{D}^{\mathsf{T}})$. The cofactor of a nine-dimensional matrix $\underline{D}$ has coordinate representation $8![\operatorname{cof}\underline{D}]^{\mathsf{a}_1\mathsf{b}_1} = e^{\mathsf{a}_1\mathsf{a}_2\cdots\mathsf{a}_9}e^{\mathsf{b}_1\mathsf{b}_2\cdots\mathsf{b}_9}D_{\mathsf{a}_2\mathsf{b}_2}\cdots D_{\mathsf{a}_9\mathsf{b}_9}$.

The *inverse* of a fourth-order tensor $\mathbf{D}$ is denoted as $\mathbf{D}^{-1}$ and is defined as $\mathbf{D}^{-1} := \upsilon^*(\underline{D}^{-1})$. It satisfies e.g. the following rules:

S4.5: *Inverse of fourth-order tensors:*

a) $\mathbf{D} : \mathbf{D}^{-1} := \upsilon^*(\underline{D}\underline{D}^{-1}) = \upsilon^*(\underline{D}^{-1}\underline{D}) =: \mathbf{D}^{-1} : \mathbf{D} = \mathbf{I}$

b) $[\mathbf{D}^{-1}]^{-1} := \upsilon^*([\underline{D}^{-1}]^{-1}) = \upsilon^*(\underline{D}) =: \mathbf{D}$

c) $[\alpha\mathbf{D}]^{-1} := \upsilon^*([\alpha\underline{D}]^{-1}) = \upsilon^*(\alpha^{-1}\underline{D}^{-1}) =: \alpha^{-1}\mathbf{D}^{-1}$

d) $[\mathbf{D}^{-1}]^{T} := \upsilon^*([\underline{D}^{-1}]^{\mathsf{T}}) = \upsilon^*(\underline{D}^{-\mathsf{T}}) =: \mathbf{D}^{-T}$

$[\mathbf{D}^{T}]^{-1} := \upsilon^*([\underline{D}^{\mathsf{T}}]^{-1}) = \upsilon^*(\underline{D}^{-\mathsf{T}}) =: \mathbf{D}^{-T}$

e) $[\mathbf{C} : \mathbf{D}]^{-1} := \upsilon^*([\underline{C}\underline{D}]^{-1}) = \upsilon^*(\underline{D}^{-1}\underline{C}^{-1}) =: \mathbf{D}^{-1} : \mathbf{C}^{-1}$

The necessary and sufficient condition for the inverse to exist is a non-vanishing determinant $\det\mathbf{D} = \det\underline{D} \neq 0$. The inverse of a nine-dimensional matrix $\underline{D}$ has coordinate representation $8!\det\underline{D}\,[\underline{D}^{-\mathsf{T}}]^{\mathsf{a}_1\mathsf{b}_1} = e^{\mathsf{a}_1\mathsf{a}_2\cdots\mathsf{a}_9}e^{\mathsf{b}_1\mathsf{b}_2\cdots\mathsf{b}_9}$ $D_{\mathsf{a}_2\mathsf{b}_2}\cdots D_{\mathsf{a}_9\mathsf{b}_9}$.

The determinant, cofactor, and inverse of minor (left and right) either symmetric or skew symmetric fourth-order tensors $\mathbf{D}$ are defined in analogy. In particular a fourth-order tensor $\mathbf{D}$ that is *fully minor (left and right) skew symmetric* allows representation in terms of its *axial tensor* $2\boldsymbol{D} := \boldsymbol{e} : \mathbf{D} : \boldsymbol{e}$ as $2\mathbf{D} = \boldsymbol{e}\cdot\boldsymbol{D}\cdot\boldsymbol{e}$. Then its *fully minor (left and right) skew symmetric inverse* computes as $2\mathbf{D}^{-1} = \boldsymbol{e}\cdot\boldsymbol{D}^{-1}\cdot\boldsymbol{e}$ so that $\mathbf{D}^{-1} : \mathbf{D} = \mathbf{I}^{\text{skw}}$. (Note the factor 2 as opposed to the conventional factor 4 in the definition of the Einstein tensor.)

The *dyadic traces* of a fourth-order tensor are denoted $\operatorname{trc}^{\otimes}\mathbf{D}$, $\operatorname{trc}^{\boxtimes}\mathbf{D}$, and $\operatorname{trc}^{\boxdot}\mathbf{D}$ and are defined as $\operatorname{trc}^{\otimes}\mathbf{D} := \mathbf{D} :: \mathbf{I}^{\otimes} = \mathsf{D}_{aadd}$, $\operatorname{trc}^{\boxtimes}\mathbf{D} := \mathbf{D} :: \mathbf{I}^{\boxtimes} = \mathsf{D}_{abab}$, and $\operatorname{trc}^{\boxdot}\mathbf{D} := \mathbf{D} :: \mathbf{I}^{\boxdot} = \mathsf{D}_{abba}$, respectively. Likewise, the *symmetric* and *skew symmetric traces* of a fourth-order tensor are denoted $\operatorname{trc}^{\text{sym}}\mathbf{D}$, and $\operatorname{trc}^{\text{skw}}\mathbf{D}$ and are defined as $\operatorname{trc}^{\text{sym}}\mathbf{D} := \mathbf{D} :: \mathbf{I}^{\text{sym}} = \mathsf{D}_{(ab)(ab)}$, and $\operatorname{trc}^{\text{skw}}\mathbf{D} := \mathbf{D} :: \mathbf{I}^{\text{skw}} = \mathsf{D}_{[ab][ab]}$, respectively.

The fourth-order *spherical projection tensor* is denoted as $\mathbf{P}_{\text{sph}}$ and is computed from $3\mathbf{P}_{\text{sph}} := \mathbf{I}^{\otimes}$. Then $\mathbf{P}_{\text{sph}} : \boldsymbol{T} = \operatorname{sph}\boldsymbol{T}$, $\mathbf{P}_{\text{sph}} : \operatorname{dev}\boldsymbol{T} = \mathbf{0}$, $\mathbf{P}_{\text{sph}} : \mathbf{P}_{\text{sph}} = \mathbf{P}_{\text{sph}}$, and $\mathbf{P}_{\text{sph}} : \mathbf{P}_{\text{dev}} = \mathbf{O}$ hold. The fourth-order *deviatoric projection tensor* is denoted as $\mathbf{P}_{\text{dev}}$ and is computed from $\mathbf{P}_{\text{dev}} := \mathbf{I}^{\boxtimes} - \mathbf{P}_{\text{sph}}$. Then $\mathbf{P}_{\text{dev}} : \boldsymbol{T} = \operatorname{dev}\boldsymbol{T}$, $\mathbf{P}_{\text{dev}} : \operatorname{sph}\boldsymbol{T} = \mathbf{0}$, $\mathbf{P}_{\text{dev}} : \mathbf{P}_{\text{dev}} = \mathbf{P}_{\text{dev}}$, and $\mathbf{P}_{\text{dev}} : \mathbf{P}_{\text{sph}} = \mathbf{O}$ hold. Finally, the fourth-order *symmetric deviatoric projection tensor* may be introduced as $\mathbf{P}^{\text{sym}}_{\text{dev}} := \mathbf{I}^{\text{sym}} - \mathbf{P}_{\text{sph}}$ with $\mathbf{P}^{\text{sym}}_{\text{dev}} : \boldsymbol{T} = \operatorname{dev}\boldsymbol{T}^{\text{sym}}$, $\mathbf{P}^{\text{sym}}_{\text{dev}} : \operatorname{sph}\boldsymbol{T} = \boldsymbol{O}$, $\mathbf{P}^{\text{sym}}_{\text{dev}} : \mathbf{P}^{\text{sym}}_{\text{dev}} = \mathbf{P}^{\text{sym}}_{\text{dev}}$, and $\mathbf{P}^{\text{sym}}_{\text{dev}} : \mathbf{P}_{\text{sph}} = \mathbf{O}$.

Spectral Properties of Fourth-Order Tensors:

The left and right *special eigenvalue problems* for a fourth-order tensor $\mathbf{D}$ read $\boldsymbol{M} : \mathbf{D} = \boldsymbol{M}\lambda$ and $\mathbf{D} : \boldsymbol{N} = \lambda\boldsymbol{N}$ whereby λ denotes the *eigenvalue* and

$\boldsymbol{M}$ and $\boldsymbol{N}$ are the corresponding left and right second-order *eigentensors*, respectively. In terms of the Voigt matrix notation these read equivalently as $\underline{m}^{\mathsf{T}}\underline{D} = \underline{m}^{\mathsf{T}}\lambda$ and $\underline{D}\underline{n} = \lambda\underline{n}$. Then the unique condition for *non-trivial solutions* to both $\underline{m}^{\mathsf{T}}[\underline{D} - \underline{1}\lambda] = \underline{0}$ and $[\underline{D} - \lambda\underline{1}]\underline{n} = \underline{0}$ is $\det(\underline{D} - \lambda\underline{1}) = 0$ resulting in the *characteristic equation* that expands into the ninth-order polynomial $p(\lambda) = I_9 - I_8\lambda + I_7\lambda^2 \cdots + I_1\lambda^8 - \lambda^9$ in terms of the nine *(principal) invariants*

$$
\begin{aligned}
9!0!I_9 &:= e^{\mathsf{a}_1\mathsf{a}_2\cdots\mathsf{a}_8\mathsf{a}_9}\; e^{\mathsf{b}_1\mathsf{b}_2\cdots\mathsf{b}_9}\; D_{\mathsf{a}_1\mathsf{b}_1}\; D_{\mathsf{a}_2\mathsf{b}_2} \cdots D_{\mathsf{a}_8\mathsf{b}_8}\; D_{\mathsf{a}_9\mathsf{b}_9}\\
8!1!I_8 &:= e^{\mathsf{a}_1\mathsf{a}_2\cdots\mathsf{a}_8\mathsf{a}_9}\; e^{\mathsf{b}_1\mathsf{b}_2\cdots\mathsf{b}_9}\; \delta_{\mathsf{a}_1\mathsf{b}_1}\; D_{\mathsf{a}_2\mathsf{b}_2} \cdots D_{\mathsf{a}_8\mathsf{b}_8}\; D_{\mathsf{a}_9\mathsf{b}_9}\\
7!2!I_7 &:= e^{\mathsf{a}_1\mathsf{a}_2\cdots\mathsf{a}_8\mathsf{a}_9}\; e^{\mathsf{b}_1\mathsf{b}_2\cdots\mathsf{b}_9}\; \delta_{\mathsf{a}_1\mathsf{b}_1}\; \delta_{\mathsf{a}_2\mathsf{b}_2} \cdots D_{\mathsf{a}_8\mathsf{b}_8}\; D_{\mathsf{a}_9\mathsf{b}_9}\\
&\;\vdots\\
1!8!I_1 &:= e^{\mathsf{a}_1\mathsf{a}_2\cdots\mathsf{a}_8\mathsf{a}_9}\; e^{\mathsf{b}_1\mathsf{b}_2\cdots\mathsf{b}_9}\; \delta_{\mathsf{a}_1\mathsf{b}_1}\; \delta_{\mathsf{a}_2\mathsf{b}_2} \cdots \delta_{\mathsf{a}_8\mathsf{b}_8}\; D_{\mathsf{a}_9\mathsf{b}_9}
\end{aligned}
$$

It admits the solutions λ^{a} with corresponding left and right eigenvectors $\underline{m}_{\mathsf{a}}$ and $\underline{n}_{\mathsf{a}}$, $\mathsf{a} = 1, \cdots, 9$ which shall be assumed to be distinct in the sequel. Since for $\lambda^{\mathsf{a}} \neq \lambda^{\mathsf{b}}$ it holds that $\lambda^{\mathsf{a}}\underline{m}_{\mathsf{a}}^{\mathsf{T}}\underline{n}_{\mathsf{b}} = [\underline{m}_{\mathsf{a}}^{\mathsf{T}}\underline{D}]\underline{n}_{\mathsf{b}} = \underline{m}_{\mathsf{a}}^{\mathsf{T}}[\underline{D}\underline{n}_{\mathsf{b}}] = \lambda^{\mathsf{b}}\underline{m}_{\mathsf{a}}^{\mathsf{T}}\underline{n}_{\mathsf{b}}$ it is concluded that $\underline{m}_{\mathsf{a}}^{\mathsf{T}}\underline{n}_{\mathsf{b}} = 0$ for $\mathsf{a} \neq \mathsf{b}$, thus the left and right eigenvectors may be orthonormalized with $\underline{m}_{\mathsf{a}}^{\mathsf{T}}\underline{n}_{\mathsf{b}} = \delta_{\mathsf{ab}}$. As a consequence the left and right eigenvectors may be considered *reciprocal base vectors*

A general *non-symmetric* $\underline{D}$ may possess *real* and/or *complex* eigenvalues and associated left and right eigenvectors. In the sequel only the case of a *symmetric* $\underline{D} = \underline{D}^{\mathsf{T}}$ corresponding to a major symmetric $\mathbf{D} = \mathbf{D}^{T}$ shall be considered. A symmetric $\underline{D}$ admits only real eigenvalues λ^{a} and corresponding eigenvectors $\underline{m}_{\mathsf{a}} \equiv \underline{n}_{\mathsf{a}}$. Then the *spectral representation* $\underline{D} = \sum_{\mathsf{a}} \lambda^{\mathsf{a}}\underline{n}_{\mathsf{a}}\underline{m}_{\mathsf{a}}^{\mathsf{T}}$ follows from the representation of the unit matrix in terms of the left and right eigenvectors since $\underline{D} = \underline{1}\underline{D} = \sum_{\mathsf{a}} \underline{n}_{\mathsf{a}}[\underline{m}_{\mathsf{a}}^{\mathsf{T}}\underline{D}]$ or $\underline{D} = \underline{D}\underline{1} = \sum_{\mathsf{a}}[\underline{D}\underline{n}_{\mathsf{a}}]\underline{m}_{\mathsf{a}}^{\mathsf{T}}$, respectively. These results may finally be mapped back to the *major symmetric* fourth-order tensor $\mathbf{D} = \mathbf{D}^{T}$ as $\mathbf{D} = \sum_{a} \lambda^{a}\boldsymbol{N}_{a} \otimes \boldsymbol{M}_{a}$. The spectral representation of fourth-order tensors is particularly useful when analyzing fourth-order anisotropic elasticity tensors in terms of the so-called *Kelvin modes*, refer e.g. to the extensive discussions in [46, 127, 191, 47, 11, 30, 119].

Scalar-, Vector-, and Tensor-Functions:

Functions may be considered as instructions of how to map elements from their definition domain into their image space. Depending on the definition domain scalar-, vector-, and tensor-functions and likewise depending on the image space scalar-, vector-, and tensor-valued functions shall be distinguished.

Value	Argument	Function	Domain	Image
Scalar	Scalar	$f(x)$:	$\mathbb{R}$	$\rightarrow \mathbb{R}$
Vector	Scalar	$\boldsymbol{f}(x)$:	$\mathbb{R}$	$\rightarrow \mathbb{E}^3$
Tensor	Scalar	$\boldsymbol{F}(x)$:	$\mathbb{R}$	$\rightarrow \mathbb{E}^3 \otimes \mathbb{E}^3$
Scalar	Vector	$f(\boldsymbol{x})$:	$\mathbb{E}^3$	$\rightarrow \mathbb{R}$
Vector	Vector	$\boldsymbol{f}(\boldsymbol{x})$:	$\mathbb{E}^3$	$\rightarrow \mathbb{E}^3$
Tensor	Vector	$\boldsymbol{F}(\boldsymbol{x})$:	$\mathbb{E}^3$	$\rightarrow \mathbb{E}^3 \otimes \mathbb{E}^3$
Scalar	Tensor	$f(\boldsymbol{X})$:	$\mathbb{E}^3 \otimes \mathbb{E}^3$	$\rightarrow \mathbb{R}$
Vector	Tensor	$\boldsymbol{f}(\boldsymbol{X})$:	$\mathbb{E}^3 \otimes \mathbb{E}^3$	$\rightarrow \mathbb{E}^3$
Tensor	Tensor	$\boldsymbol{F}(\boldsymbol{X})$:	$\mathbb{E}^3 \otimes \mathbb{E}^3$	$\rightarrow \mathbb{E}^3 \otimes \mathbb{E}^3$

As typical examples, scalar-, vector-, and tensor-valued scalar-functions may capture a dependence on time, correspondingly scalar-, vector-, and tensor-valued vector-functions may describe a dependence on the position vector, and scalar-, vector-, and tensor-valued tensor-functions may denote a dependence on the strain tensor.

Linearization, Directional Derivative, and Differential:
The linear part or rather the *linearization* of an ordinary function $f(x)$ evaluated at $x = x_0$ is written as $\operatorname{lin} f(\Delta x; x_0) := f|_0 + \operatorname{dir} f(\Delta x; x_0)$ with $f|_0 := f(x_0)$ and $\Delta x := x - x_0$. Here $\operatorname{dir} f(\Delta x; x_0) := \mathrm{d}_\alpha[f(x_0 + \alpha\Delta x)]_{\alpha=0}$ denotes the *directional derivative* that is given in terms of the derivative with respect to the argument $\partial_x f|_0$ as $\operatorname{dir} f(\Delta x; x_0) =: \partial_x f|_0 \Delta x$. Likewise the (total) *differential* of $f(x)$ may be defined as $\mathrm{d}f := \operatorname{dir} f(\mathrm{d}x; x)$. As an example consider the function $f(x) = x^2$, then $f(x_0 + \alpha\Delta x) = x_0^2 + 2x_0\alpha\Delta x + \alpha^2[\Delta x]^2$ and thus $\mathrm{d}_\alpha[f(x_0+\alpha\Delta x)]_{\alpha=0} = [2x_0\Delta x + 2\alpha[\Delta x]^2]_{\alpha=0} = 2x_0\Delta x = \partial_x f|_0 \Delta x$. These concepts may be extended to arbitrary types of functions as the ones discussed in the above.

$$
\begin{aligned}
\operatorname{dir} f(\Delta x; x_0) &:= \mathrm{d}_\alpha[\, f(x_0 + \alpha\Delta x)]_{\alpha=0} &&=: \partial_x && f|_0 \Delta x \\
\operatorname{dir} \boldsymbol{f}(\Delta x; x_0) &:= \mathrm{d}_\alpha[\, \boldsymbol{f}(x_0 + \alpha\Delta x)]_{\alpha=0} &&=: \partial_x && \boldsymbol{f}|_0 \Delta x \\
\operatorname{dir} \boldsymbol{F}(\Delta x; x_0) &:= \mathrm{d}_\alpha[\boldsymbol{F}(x_0 + \alpha\Delta x)]_{\alpha=0} &&=: \partial_x && \boldsymbol{F}|_0 \Delta x \\
\hline
\operatorname{dir} f(\Delta \boldsymbol{x}; \boldsymbol{x}_0) &:= \mathrm{d}_\alpha[\, f(\boldsymbol{x}_0 + \alpha\Delta \boldsymbol{x})]_{\alpha=0} &&=: \partial_{\boldsymbol{x}} && f|_0 \cdot \Delta \boldsymbol{x} \\
\operatorname{dir} \boldsymbol{f}(\Delta \boldsymbol{x}; \boldsymbol{x}_0) &:= \mathrm{d}_\alpha[\, \boldsymbol{f}(\boldsymbol{x}_0 + \alpha\Delta \boldsymbol{x})]_{\alpha=0} &&=: \partial_{\boldsymbol{x}} && \boldsymbol{f}|_0 \cdot \Delta \boldsymbol{x} \\
\operatorname{dir} \boldsymbol{F}(\Delta \boldsymbol{x}; \boldsymbol{x}_0) &:= \mathrm{d}_\alpha[\boldsymbol{F}(\boldsymbol{x}_0 + \alpha\Delta \boldsymbol{x})]_{\alpha=0} &&=: \partial_{\boldsymbol{x}} && \boldsymbol{F}|_0 \cdot \Delta \boldsymbol{x} \\
\hline
\operatorname{dir} f(\Delta \boldsymbol{X}; \boldsymbol{X}_0) &:= \mathrm{d}_\alpha[\, f(\boldsymbol{X}_0 + \alpha\Delta \boldsymbol{X})]_{\alpha=0} &&=: \partial_{\boldsymbol{X}} && f|_0 : \Delta \boldsymbol{X} \\
\operatorname{dir} \boldsymbol{f}(\Delta \boldsymbol{X}; \boldsymbol{X}_0) &:= \mathrm{d}_\alpha[\, \boldsymbol{f}(\boldsymbol{X}_0 + \alpha\Delta \boldsymbol{X})]_{\alpha=0} &&=: \partial_{\boldsymbol{X}} && \boldsymbol{f}|_0 : \Delta \boldsymbol{X} \\
\operatorname{dir} \boldsymbol{F}(\Delta \boldsymbol{X}; \boldsymbol{X}_0) &:= \mathrm{d}_\alpha[\boldsymbol{F}(\boldsymbol{X}_0 + \alpha\Delta \boldsymbol{X})]_{\alpha=0} &&=: \partial_{\boldsymbol{X}} && \boldsymbol{F}|_0 : \Delta \boldsymbol{X}
\end{aligned}
$$

Note that here functions may depend generically on either scalar-, vector- or tensor-valued arguments. Thereby the dependence of functions (or rather the coefficients of functions) on a scalar-, vector- or tensor-valued argument x, $\boldsymbol{x}$ or $\boldsymbol{X}$ is via its coefficients x, $x^j = \boldsymbol{x} \cdot \boldsymbol{g}^j$ or $X^{jk} = \boldsymbol{g}^j \cdot \boldsymbol{X} \cdot \boldsymbol{g}^k$, respectively. Thus the (partial) derivative of scalar-, vector-, and tensor-functions with respect to their argument follows from the chain rule, whereby $\partial_x x = 1$, $\partial_{\boldsymbol{x}} x^j = \boldsymbol{g}^j$, and $\partial_{\boldsymbol{X}} X^{jk} = \boldsymbol{g}^j \otimes \boldsymbol{g}^k$, respectively. As a consequence, for coordinate representations of functions, the (partial) derivative with respect to their argument expands in terms of the *covariant derivative* indicated here generically by $\wr$ (wave) in order to capture a possible dependence of the base vectors $\boldsymbol{g}_j$ and $\boldsymbol{g}^i$ on the coefficients x, x^k or X^{kl}, respectively

$f(x) = f$	$\partial_x f = f_{\wr}$
$\boldsymbol{f}(x) = f_i \; \boldsymbol{g}^i$	$\partial_x \boldsymbol{f} = f_{i\wr} \; \boldsymbol{g}^i$
$= f^i \; \boldsymbol{g}_i$	$= f^i_{\;\wr} \; \boldsymbol{g}_i$
$\boldsymbol{F}(x) = f_{ij} \; \boldsymbol{g}^i \otimes \boldsymbol{g}^j$	$\partial_x \boldsymbol{F} = f_{ij\wr} \; \boldsymbol{g}^i \otimes \boldsymbol{g}^j$
$= f^{ij} \; \boldsymbol{g}_i \otimes \boldsymbol{g}_j$	$= f^{ij}_{\;\;\wr} \; \boldsymbol{g}_i \otimes \boldsymbol{g}_j$
$\vdots$	$\vdots$
$f(\boldsymbol{x}) = f$	$\partial_{\boldsymbol{x}} f = f_{\wr i} \; \boldsymbol{g}^i$
$\boldsymbol{f}(\boldsymbol{x}) = f_i \; \boldsymbol{g}^i$	$\partial_{\boldsymbol{x}} \boldsymbol{f} = f_{i\wr j} \; \boldsymbol{g}^i \otimes \boldsymbol{g}^j$
$= f^i \; \boldsymbol{g}_i$	$= f^i_{\;\wr j} \; \boldsymbol{g}_i \otimes \boldsymbol{g}^j$
$\boldsymbol{F}(\boldsymbol{x}) = f_{ij} \; \boldsymbol{g}^i \otimes \boldsymbol{g}^j$	$\partial_{\boldsymbol{x}} \boldsymbol{F} = f_{ij\wr k} \; \boldsymbol{g}^i \otimes \boldsymbol{g}^j \otimes \boldsymbol{g}^k$
$= f^{ij} \; \boldsymbol{g}_i \otimes \boldsymbol{g}_j$	$= f^{ij}_{\;\;\wr k} \; \boldsymbol{g}_i \otimes \boldsymbol{g}_j \otimes \boldsymbol{g}^k$
$\vdots$	$\vdots$
$f(\boldsymbol{X}) = f$	$\partial_{\boldsymbol{X}} f = f_{\wr\overline{ij}} \; \boldsymbol{g}^i \otimes \boldsymbol{g}^j$
$\boldsymbol{f}(\boldsymbol{X}) = f_i \; \boldsymbol{g}^i$	$\partial_{\boldsymbol{X}} \boldsymbol{f} = f_{i\wr\overline{jk}} \; \boldsymbol{g}^i \otimes \boldsymbol{g}^j \otimes \boldsymbol{g}^k$
$= f^i \; \boldsymbol{g}_i$	$= f^i_{\;\wr\overline{jk}} \; \boldsymbol{g}_i \otimes \boldsymbol{g}^j \otimes \boldsymbol{g}^k$
$\boldsymbol{F}(\boldsymbol{X}) = f_{ij} \; \boldsymbol{g}^i \otimes \boldsymbol{g}^j$	$\partial_{\boldsymbol{X}} \boldsymbol{F} = f_{ij\wr\overline{kl}} \; \boldsymbol{g}^i \otimes \boldsymbol{g}^j \otimes \boldsymbol{g}^k \otimes \boldsymbol{g}^l$
$= f^{ij} \; \boldsymbol{g}_i \otimes \boldsymbol{g}_j$	$= f^{ij}_{\;\;\wr\overline{kl}} \; \boldsymbol{g}_i \otimes \boldsymbol{g}_j \otimes \boldsymbol{g}^k \otimes \boldsymbol{g}^l$
$\vdots$	$\vdots$

However for scalar-valued functions the covariant and the partial derivative coincide, i.e. for $f(x)$, $f(\boldsymbol{x})$, and $f(\boldsymbol{X})$ it holds that $f_{\wr} := \partial_x f =: f'$, $f_{\wr i} := \partial_{x^i} f =: f_{,i}$, and $f_{\wr ij} := \partial_{X^{ij}} f =: f_{,\overline{ij}}$, respectively. Here the notation ${}_{,\overline{ij}}$ refers to the partial derivative with respect to the tensor-valued argument X_{ij}. The (partial) derivative of the base vectors $\boldsymbol{g}_j$ and $\boldsymbol{g}^i$ with respect to the coefficients x, x^k or X^{kl}, respectively, is conveniently expressed in terms of the base vectors $\boldsymbol{g}_i$ and $\boldsymbol{g}^j$ themselves as

$\partial_x \; \boldsymbol{g}_j = \boldsymbol{g}_j' =: \boldsymbol{g}_i n^i_{\;j}$	$\partial_x \; \boldsymbol{g}^i = \boldsymbol{g}^{i\prime} =: -\boldsymbol{g}^j n^i_{\;j}$
$\partial_{x^k} \; \boldsymbol{g}_j = \boldsymbol{g}_{j,k} =: \boldsymbol{g}_i n^i_{\;jk}$	$\partial_{x^k} \; \boldsymbol{g}^i = \boldsymbol{g}^i_{\;,k} =: -\boldsymbol{g}^j n^i_{\;jk}$
$\partial_{X^{kl}} \; \boldsymbol{g}_j = \boldsymbol{g}_{j,\overline{kl}} =: \boldsymbol{g}_i n^i_{\;j\overline{kl}}$	$\partial_{X^{kl}} \; \boldsymbol{g}^i = \boldsymbol{g}^i_{\;,\overline{kl}} =: -\boldsymbol{g}^j n^i_{\;j\overline{kl}}$

Here the two-, three-, and fourth-order objects $n^i{}_j$, $n^i{}_{jk}$, and $n^i{}_{j\overline{kl}}$, respectively, are Christoffel-type symbols (for derivatives with respect to the coefficients x, x^k or X^{kl}, respectively) and compute from

$n^i{}_j = \boldsymbol{g}^i \cdot \boldsymbol{g}_j{}' = -\boldsymbol{g}^{i\prime} \cdot \boldsymbol{g}_j$	$g_{ij}{}' = 2n_{(ij)}$	$g^{ij\prime} = -2n^{(ij)}$
$n^i{}_{jk} = \boldsymbol{g}^i \cdot \boldsymbol{g}_{j,k} = -\boldsymbol{g}^i{}_{,k} \cdot \boldsymbol{g}_j$	$g_{ij,k} = 2n_{(ij)k}$	$g^{ij}{}_{,k} = -2n^{(ij)}{}_k$
$n^i{}_{j\overline{kl}} = \boldsymbol{g}^i \cdot \boldsymbol{g}_{j,\overline{kl}} = -\boldsymbol{g}^i{}_{,\overline{kl}} \cdot \boldsymbol{g}_j$	$g_{ij,\overline{kl}} = 2n_{(ij)\overline{kl}}$	$g^{ij}{}_{,\overline{kl}} = -2n^{(ij)}{}_{\overline{kl}}$

It follows immediately that the Christoffel-type symbols also serve to express the (partial) derivative of the co- and contravariant metric coefficients. Note that traditionally it is only the $n^i{}_{jk}$ that are denoted as Christoffel symbols. Then the covariant derivative of scalar-, vector-, and tensor-valued scalar-, vector-, and tensor-functions, respectively, is detailed as

$f(x)$	$f_{\wr} = f'$
$\boldsymbol{f}(x)$	$f_{i\wr} = f_i{}' - f_m\, n^m{}_i$
	$f^i{}_{\wr} = f^{i\prime} + f^m\, n^i{}_m$
$\boldsymbol{F}(x)$	$f_{ij\wr} = f_{ij}{}' - f_{mj}\, n^m{}_i - f_{im}\, n^m{}_j$
	$f^{ij}{}_{\wr} = f^{ij\prime} + f^{mj}\, n^i{}_m + f^{im}\, n^j{}_m$
	$\vdots$
$f(\boldsymbol{x})$	$f_{\wr i} = f_{,i}$
$\boldsymbol{f}(\boldsymbol{x})$	$f_{i\wr j} = f_{i,j} - f_m\, n^m{}_{ij}$
	$f^i{}_{\wr j} = f^i{}_{,j} + f^m\, n^i{}_{mj}$
$\boldsymbol{F}(\boldsymbol{x})$	$f_{ij\wr k} = f_{ij,k} - f_{mj}\, n^m{}_{ik} - f_{im}\, n^m{}_{jk}$
	$f^{ij}{}_{\wr k} = f^{ij}{}_{,k} + f_{mj}\, n^i{}_{mk} + f_{im}\, n^j{}_{mk}$
	$\vdots$
$f(\boldsymbol{X})$	$f_{\wr\overline{ij}} = f_{,\overline{ij}}$
$\boldsymbol{f}(\boldsymbol{X})$	$f_{i\wr\overline{jk}} = f_{i,\overline{jk}} - f_m\, n^m{}_{i\overline{jk}}$
	$f^i{}_{\wr\overline{jk}} = f^i{}_{,\overline{jk}} + f^m\, n^i{}_{m\overline{jk}}$
$\boldsymbol{F}(\boldsymbol{X})$	$f_{ij\wr\overline{kl}} = f_{ij,\overline{kl}} - f_{mj}\, n^m{}_{i\overline{kl}} - f_{im}\, n^m{}_{j\overline{kl}}$
	$f^{ij}{}_{\wr\overline{kl}} = f^{ij}{}_{,\overline{kl}} + f^{mj}\, n^i{}_{m\overline{kl}} + f^{im}\, n^j{}_{m\overline{kl}}$
	$\vdots$

The dependence of the base vectors on a scalar-valued argument x occurs for example for curves embedded in $\mathbb{E}^3$. If the $\boldsymbol{g}_j$ are associated with the tangent, the normal, and the binormal vector of the accompanying base triad, and x denotes the arc-length of the curve the matrix arrangement $[n^i{}_j]$ is a skew symmetric matrix in terms of the curve's curvature $n^2{}_1 = -n^1{}_2$ and torsion $n^3{}_2 = -n^2{}_3$ in line with the Frénet-Serret formulae. The dependence of the base vectors on a vector-valued argument $\boldsymbol{x}$ occurs if $\boldsymbol{x}$ denotes the position vector in $\mathbb{E}^3$. In this particular case the *Euclidean covariant derivative* is denoted by ${}_{\wr} = {}_{;}$ (semicolon). The admittedly unconventional dependence of the base vectors on a tensor-valued argument $\boldsymbol{X}$ is only included for completeness. It is finally noted that the (partial) derivative of products of functions with respect to their argument follows from the usual product rule.

Differential Operators (Gradient, Divergence, Curl):
For the case of scalar-, vector-, and tensor-valued vector-functions with the vector-valued argument $\boldsymbol{x}$ denoting the position vector in $\mathbb{E}^3$ important differential operators may be introduced:

The *gradient* of a scalar-, vector-, and tensor-valued vector-function renders a vector-, tensor-, and (third-order) tensor-valued vector-function, respectively (i.e. the gradient raises the tensorial order)

$$
\begin{array}{l|l}
f(\boldsymbol{x}) & \operatorname{grad} f(\boldsymbol{x}) := \partial_{\boldsymbol{x}} f = f_{;i} \; \boldsymbol{g}^i \\
\hline
\boldsymbol{f}(\boldsymbol{x}) & \operatorname{grad} \boldsymbol{f}(\boldsymbol{x}) := \partial_{\boldsymbol{x}} \boldsymbol{f} = f_{i;j} \; \boldsymbol{g}^i \otimes \boldsymbol{g}^j \\
 & \qquad\qquad\qquad\qquad = f^i{}_{;j} \; \boldsymbol{g}_i \otimes \boldsymbol{g}^j \\
\hline
\boldsymbol{F}(\boldsymbol{x}) & \operatorname{grad} \boldsymbol{F}(\boldsymbol{x}) := \partial_{\boldsymbol{x}} \boldsymbol{F} = f_{ij;k} \; \boldsymbol{g}^i \otimes \boldsymbol{g}^j \otimes \boldsymbol{g}^k \\
 & \qquad\qquad\qquad\qquad = f^{ij}{}_{;k} \; \boldsymbol{g}_i \otimes \boldsymbol{g}_j \otimes \boldsymbol{g}^k \\
 & \qquad\qquad \vdots \qquad\qquad \vdots
\end{array}
$$

The *divergence* of a vector-, and tensor-valued vector-function renders a scalar-, and vector-valued vector-function, respectively (i.e. the divergence lowers the tensorial order)

$$
\begin{array}{l|l}
\boldsymbol{f}(\boldsymbol{x}) & \operatorname{div} \boldsymbol{f}(\boldsymbol{x}) := \operatorname{grad} \boldsymbol{f} : \mathbf{1} = f_{i;i} \\
 & \qquad\qquad\qquad\qquad = f^i{}_{;i} \\
\hline
\boldsymbol{F}(\boldsymbol{x}) & \operatorname{div} \boldsymbol{F}(\boldsymbol{x}) := \operatorname{grad} \boldsymbol{F} : \mathbf{1} = f_{ij;j} \; \boldsymbol{g}^i \\
 & \qquad\qquad\qquad\qquad = f^{ij}{}_{;j} \; \boldsymbol{g}_i \\
 & \qquad\qquad \vdots \qquad\qquad \vdots
\end{array}
$$

The *curl* of a vector-, and tensor-valued vector-function renders a vector-, and tensor-valued vector-function, respectively (i.e. the curl preserves the tensorial order)

$$
\begin{array}{l|l}
\boldsymbol{f}(\boldsymbol{x}) & \operatorname{curl} \boldsymbol{f}(\boldsymbol{x}) := -\operatorname{grad} \boldsymbol{f} : \boldsymbol{e} = -f_{i;j} \; \varepsilon^{ijk} \; \boldsymbol{g}_k \\
 & \qquad\qquad\qquad\qquad = -f^i{}_{;j} \; \varepsilon_i{}^{jk} \; \boldsymbol{g}_k \\
\hline
\boldsymbol{F}(\boldsymbol{x}) & \operatorname{curl} \boldsymbol{F}(\boldsymbol{x}) := -\operatorname{grad} \boldsymbol{F} : \boldsymbol{e} = -f_{ij;k} \; \varepsilon^{jkl} \; \boldsymbol{g}^i \otimes \boldsymbol{g}_l \\
 & \qquad\qquad\qquad\qquad = -f^{ij}{}_{;k} \; \varepsilon_j{}^{kl} \; \boldsymbol{g}_i \otimes \boldsymbol{g}_l \\
 & \qquad\qquad \vdots \qquad\qquad \vdots
\end{array}
$$

The *Laplace operator* applied to a scalar-, vector-, and tensor-valued vector-function renders a scalar-, vector-, and tensor-valued vector-function, respectively (i.e. application of the Laplace operator preserves the tensorial order)

$f(\boldsymbol{x})$	$\operatorname{lapl}\ f(\boldsymbol{x}) := \operatorname{div}\operatorname{grad} f \ = f_{;ii}$
$\boldsymbol{f}(\boldsymbol{x})$	$\operatorname{lapl}\ \boldsymbol{f}(\boldsymbol{x}) := \operatorname{div}\operatorname{grad} \boldsymbol{f} \ = f_{i;jj}\ \boldsymbol{g}^i$ $= f^i{}_{;jj}\ \boldsymbol{g}_i$
$\boldsymbol{F}(\boldsymbol{x})$	$\operatorname{lapl}\ \boldsymbol{F}(\boldsymbol{x}) := \operatorname{div}\operatorname{grad} \boldsymbol{F} = f_{ij;kk}\ \boldsymbol{g}^i \otimes \boldsymbol{g}^j$ $= f^{ij}{}_{;kk}\ \boldsymbol{g}_i \otimes \boldsymbol{g}_j$
	$\vdots \qquad \vdots$

Integral Theorems:
Circuit integrals of vector-, and tensor-valued vector-functions along closed circuits $\mathcal{C}$ that are equipped with line elements $\mathrm{d}\boldsymbol{x}$ are transformed into surface integrals over the encircled area $\mathcal{A}$ that is equipped with the oriented area element $\mathrm{d}\boldsymbol{a} = \boldsymbol{n}\,\mathrm{d}a$ by the *Stokes* theorem

$$\oint_{\mathcal{C}} \boldsymbol{f} \cdot \mathrm{d}\boldsymbol{x} = \int_{\mathcal{A}} \operatorname{curl} \boldsymbol{f} \cdot \mathrm{d}\boldsymbol{a} \quad \text{and} \quad \oint_{\mathcal{C}} \boldsymbol{F} \cdot \mathrm{d}\boldsymbol{x} = \int_{\mathcal{A}} \operatorname{curl} \boldsymbol{F} \cdot \mathrm{d}\boldsymbol{a}.$$

Scalar-valued vector-functions may be treated by setting $\boldsymbol{F} = f\boldsymbol{1}$ with $\boldsymbol{F} \cdot \mathrm{d}\boldsymbol{x} = f\,\mathrm{d}\boldsymbol{x}$ and noting that $\operatorname{curl}(f\boldsymbol{1}) = \boldsymbol{e}\cdot\operatorname{grad} f$ thus rendering $\operatorname{curl}(f\boldsymbol{1})\cdot \mathrm{d}\boldsymbol{a} = \mathrm{d}\boldsymbol{a} \times \operatorname{grad} f$. Likewise expressions such as $\boldsymbol{T} := \boldsymbol{f} \ltimes \boldsymbol{F} = \boldsymbol{e} : [\boldsymbol{f} \otimes \boldsymbol{F}]$ with $\boldsymbol{T} \cdot \mathrm{d}\boldsymbol{x} = \boldsymbol{f} \times [\boldsymbol{F} \cdot \mathrm{d}\boldsymbol{x}]$ may be handled with $\operatorname{curl} \boldsymbol{T} = \boldsymbol{e} : [\operatorname{grad} \boldsymbol{f} \boxtimes \boldsymbol{F}] : \boldsymbol{e} + \boldsymbol{f} \ltimes \operatorname{curl} \boldsymbol{F}$ whereby $\boldsymbol{e} : [\operatorname{grad} \boldsymbol{f} \boxtimes \boldsymbol{F}] : \boldsymbol{e} = \operatorname{grad} \boldsymbol{f} \times\!\!\!\times \boldsymbol{F}$. For the special case $\boldsymbol{T} = \boldsymbol{f} \ltimes \boldsymbol{1}$, i.e. $\boldsymbol{T}\cdot \mathrm{d}\boldsymbol{x} = \boldsymbol{f} \times \mathrm{d}\boldsymbol{x}$ this results with $\operatorname{curl} \boldsymbol{T} = \operatorname{div}\boldsymbol{f}\boldsymbol{1} - \operatorname{grad}^t \boldsymbol{f}$ in $\operatorname{curl} \boldsymbol{T} \cdot \mathrm{d}\boldsymbol{a} = \operatorname{div}\boldsymbol{f}\,\mathrm{d}\boldsymbol{a} - \mathrm{d}\boldsymbol{a} \cdot \operatorname{grad} \boldsymbol{f}$, thus

$$\oint_{\mathcal{C}} f\,\mathrm{d}\boldsymbol{x} = \int_{\mathcal{A}} \mathrm{d}\boldsymbol{a} \times \operatorname{grad} f \quad \text{and} \quad \oint_{\mathcal{C}} \boldsymbol{f} \times \mathrm{d}\boldsymbol{x} = \int_{\mathcal{A}} [\operatorname{div}\boldsymbol{f}\,\mathrm{d}\boldsymbol{a} - \mathrm{d}\boldsymbol{a} \cdot \operatorname{grad} \boldsymbol{f}].$$

Surface integrals of vector-, and tensor-valued vector-functions over areas $\mathcal{A}$ that are equipped with oriented area elements $\mathrm{d}\boldsymbol{a} = \boldsymbol{n}\,\mathrm{d}a$ are transformed into volume integrals over the contained volume $\mathcal{V}$ that is equipped with volume element $\mathrm{d}v$ by the *Gauss* theorem

$$\int_{\mathcal{A}} \boldsymbol{f} \cdot \mathrm{d}\boldsymbol{a} = \int_{\mathcal{V}} \operatorname{div}\boldsymbol{f}\,\mathrm{d}v \quad \text{and} \quad \int_{\mathcal{A}} \boldsymbol{F} \cdot \mathrm{d}\boldsymbol{a} = \int_{\mathcal{V}} \operatorname{div}\boldsymbol{F}\,\mathrm{d}v.$$

Scalar-valued vector-functions may be treated by setting $\boldsymbol{F} = f\boldsymbol{1}$ with $\boldsymbol{F} \cdot \mathrm{d}\boldsymbol{a} = f\,\mathrm{d}\boldsymbol{a}$ and noting that $\operatorname{div}(f\boldsymbol{1}) = \boldsymbol{1} \cdot \operatorname{grad} f = \operatorname{grad} f$. Setting in particular f =const with $\operatorname{grad} f = 0$ renders the *area theorem* for closed surfaces. Likewise expressions such as $\boldsymbol{T} := \boldsymbol{f} \ltimes \boldsymbol{F} = \boldsymbol{e} : [\boldsymbol{f} \otimes \boldsymbol{F}]$ with $\boldsymbol{T} \cdot \mathrm{d}\boldsymbol{a} = \boldsymbol{f} \times [\boldsymbol{F} \cdot \mathrm{d}\boldsymbol{a}]$ may be handled with $\operatorname{div}\boldsymbol{T} = \boldsymbol{e} : [\operatorname{grad} \boldsymbol{f} \boxtimes \boldsymbol{F}] : \boldsymbol{1} + \boldsymbol{f} \times \operatorname{div}\boldsymbol{F}$ whereby $\boldsymbol{e} : [\operatorname{grad} \boldsymbol{f} \boxtimes \boldsymbol{F}] : \boldsymbol{1} = \operatorname{grad} \boldsymbol{f} \ltimes \boldsymbol{F} = \boldsymbol{e} : [\operatorname{grad} \boldsymbol{f} \cdot \boldsymbol{F}^t]$. For the special case $\boldsymbol{T} = \boldsymbol{f} \ltimes \boldsymbol{1}$, i.e. $\boldsymbol{T} \cdot \mathrm{d}\boldsymbol{a} = \boldsymbol{f} \times \mathrm{d}\boldsymbol{a}$ this results in $\operatorname{div}\boldsymbol{T} = -\operatorname{curl} \boldsymbol{f}$ thus

$$\int_{\mathcal{A}} f \, \mathrm{d}\boldsymbol{a} = \int_{\mathcal{V}} \operatorname{grad} f \, \mathrm{d}v \quad \text{and} \quad \int_{\mathcal{A}} \boldsymbol{f} \times \mathrm{d}\boldsymbol{a} = -\int_{\mathcal{V}} \operatorname{curl} \boldsymbol{f} \, \mathrm{d}v.$$

Addendum 6.2. Skew Symmetric ⊕ Symmetric Decomposition
The additive decomposition of second-order tensors into their skew symmetric and symmetric contributions is classical. Its extension to third-order tensors is of relevance for the discussion of the integrability of the distortion and the double-distortion.

Second-Order Tensors:
An arbitrary second-order tensor $\boldsymbol{\phi}$ allows *unique* decomposition into a skew symmetric and a symmetric contribution

$$\boldsymbol{\phi} = \boldsymbol{\phi}^{\text{skw}} + \boldsymbol{\phi}^{\text{sym}} =: -\boldsymbol{\varphi} \cdot \boldsymbol{e} + \boldsymbol{\vartheta} \equiv -\boldsymbol{e} \cdot \boldsymbol{\varphi} + \boldsymbol{\vartheta}.$$

Here $\boldsymbol{\phi}^{\text{skw}} := -\boldsymbol{\varphi} \cdot \boldsymbol{e} \equiv -\boldsymbol{e} \cdot \boldsymbol{\varphi}$ is the skew symmetric, and $\boldsymbol{\phi}^{\text{sym}} := \boldsymbol{\vartheta}$ is the symmetric contribution to $\boldsymbol{\phi}$ with

$$\boldsymbol{\phi} : \mathbf{i}^{\text{skw}} = \mathbf{i}^{\text{skw}} : \boldsymbol{\phi} =: \boldsymbol{\phi}^{\text{skw}} \quad \text{and} \quad \boldsymbol{\phi} : \mathbf{i}^{\text{sym}} = \mathbf{i}^{\text{sym}} : \boldsymbol{\phi} =: \boldsymbol{\phi}^{\text{sym}}.$$

The axial vector $\boldsymbol{\varphi}$ of the skew symmetric contribution to $\boldsymbol{\phi}$ is extracted as

$$-\boldsymbol{\phi} : \boldsymbol{e} = -[-\boldsymbol{\varphi} \cdot \boldsymbol{e}] : \boldsymbol{e} = 2\boldsymbol{\varphi} = -\boldsymbol{e} : [-\boldsymbol{e} \cdot \boldsymbol{\varphi}] = -\boldsymbol{e} : \boldsymbol{\phi}.$$

Vice versa the skew symmetric contribution to $\boldsymbol{\phi}$ is computed from its axial vector by $\boldsymbol{\phi}^{\text{skw}} = -\boldsymbol{\varphi} \cdot \boldsymbol{e} \equiv -\boldsymbol{e} \cdot \boldsymbol{\varphi}$.

Third-Order Tensors:
On the one hand an arbitrary third-order tensor $\boldsymbol{\Phi}$ allows representation as

$$\boldsymbol{\Phi} = \Phi^{ABC} \boldsymbol{E}_A \otimes \boldsymbol{E}_B \otimes \boldsymbol{E}_C = \boldsymbol{E}_A \otimes \boldsymbol{\phi}_{\text{r}}^A.$$

Then the three tensors $\boldsymbol{\phi}_{\text{r}}^A := \Phi^{ABC} \boldsymbol{E}_B \otimes \boldsymbol{E}_C$ with $A = 1, 2, 3$ are expressed based on their unique skew symmetric and symmetric contributions as

$$\boldsymbol{\phi}_{\text{r}}^A = -\boldsymbol{\varphi}_{\text{r}}^A \cdot \boldsymbol{e} + \boldsymbol{\vartheta}_{\text{r}}^A,$$

thus rendering the corresponding decomposition of the third-order tensor $\boldsymbol{\Phi}$ into its *right* skew symmetric and *right* symmetric contributions as

$$\boldsymbol{\Phi} = \boldsymbol{E}_A \otimes [-\boldsymbol{\varphi}_{\text{r}}^A \cdot \boldsymbol{e} + \boldsymbol{\vartheta}_{\text{r}}^A] = -[\boldsymbol{E}_A \otimes \boldsymbol{\varphi}_{\text{r}}^A] \cdot \boldsymbol{e} + [\boldsymbol{E}_A \otimes \boldsymbol{\vartheta}_{\text{r}}^A] =: -\boldsymbol{\phi}_{\text{r}} \cdot \boldsymbol{e} + \boldsymbol{\Theta}_{\text{r}}.$$

The next step is to note that the second-order tensor $\boldsymbol{\phi}_{\text{r}} := \boldsymbol{E}_A \otimes \boldsymbol{\varphi}_{\text{r}}^A$ allows in turn corresponding unique decomposition into its skew symmetric and symmetric contributions, thus resulting in

$$\boldsymbol{\Phi} = -[-\boldsymbol{\varphi}_{\text{r}} \cdot \boldsymbol{e}] \cdot \boldsymbol{e} - \boldsymbol{\vartheta}_{\text{r}} \cdot \boldsymbol{e} + \boldsymbol{\Theta}_{\text{r}}.$$

Here the third-order tensor $\boldsymbol{\Theta}_{\text{r}} := \boldsymbol{E}_A \otimes \boldsymbol{\vartheta}_{\text{r}}^A$ is *right* symmetric. For the special case that the three symmetric tensors $\boldsymbol{\vartheta}_{\text{r}}^A$ allow representation as scalar multiples $\boldsymbol{\vartheta}_{\text{r}}^A = \theta_{\text{r}}^A \boldsymbol{\vartheta}_{\text{r}}'$ of a common symmetric tensor $\boldsymbol{\vartheta}_{\text{r}}'$ the right symmetric $\boldsymbol{\Theta}_{\mathrm{r}}$ expands into the dyadic product $\boldsymbol{\Theta}_{\mathrm{r}} = \boldsymbol{E}_A \otimes \theta_{\mathrm{r}}^A \boldsymbol{\vartheta}'_{\mathrm{r}} = \boldsymbol{\theta}_{\mathrm{r}} \otimes \boldsymbol{\vartheta}'_{\mathrm{r}}$ with $\boldsymbol{\theta}_{\mathrm{r}} := \theta_{\mathrm{r}}^A \boldsymbol{E}_A$. Finally, the axial vector $\boldsymbol{\varphi}_{\mathrm{r}}$, the symmetric tensor $\boldsymbol{\vartheta}_{\mathrm{r}}$, and the right symmetric $\boldsymbol{\Theta}_{\mathrm{r}}$ are extracted from $\boldsymbol{\Phi}$ by

$$4\boldsymbol{\varphi}_{\mathrm{r}} = -[-\boldsymbol{\Phi} : \boldsymbol{e}] : \boldsymbol{e} \quad \text{and} \quad 2\boldsymbol{\vartheta}_{\mathrm{r}} = [-\boldsymbol{\Phi} : \boldsymbol{e}] : \mathbf{i}^{\mathrm{sym}} \quad \text{and} \quad \boldsymbol{\Theta}_{\mathrm{r}} = \boldsymbol{\Phi} : \mathbf{i}^{\mathrm{sym}}.$$

On the other hand an arbitrary third-order tensor $\boldsymbol{\Phi}$ allows representation as

$$\boldsymbol{\Phi} = \Phi^{ABC} \boldsymbol{E}_A \otimes \boldsymbol{E}_B \otimes \boldsymbol{E}_C = \boldsymbol{\phi}_{\mathrm{l}}^C \otimes \boldsymbol{E}_C.$$

Then the three tensors $\boldsymbol{\phi}_{\mathrm{l}}^C := \Phi^{ABC} \boldsymbol{E}_A \otimes \boldsymbol{E}_B$ with $C = 1, 2, 3$ are expressed based on their unique skew symmetric and symmetric contributions as

$$\boldsymbol{\phi}_{\mathrm{l}}^C = -\boldsymbol{e} \cdot \boldsymbol{\varphi}_{\mathrm{l}}^C + \boldsymbol{\vartheta}_{\mathrm{l}}^C,$$

thus rendering the corresponding decomposition of the third-order tensor $\boldsymbol{\Phi}$ into its *left* skew symmetric and *left* symmetric contributions as

$$\boldsymbol{\Phi} = [-\boldsymbol{e} \cdot \boldsymbol{\varphi}_{\mathrm{l}}^C + \boldsymbol{\vartheta}_{\mathrm{l}}^C] \otimes \boldsymbol{E}_C = -\boldsymbol{e} \cdot [\boldsymbol{\varphi}_{\mathrm{l}}^C \otimes \boldsymbol{E}_C] + [\boldsymbol{\vartheta}_{\mathrm{l}}^C \otimes \boldsymbol{E}_C] =: -\boldsymbol{e} \cdot \boldsymbol{\phi}_{\mathrm{l}} + \boldsymbol{\Theta}_{\mathrm{l}}.$$

The next step is to note that the second-order tensor $\boldsymbol{\phi}_{\mathrm{l}} := \boldsymbol{\varphi}_{\mathrm{l}}^C \otimes \boldsymbol{E}_C$ allows in turn corresponding unique decomposition into its skew symmetric and symmetric contributions, thus resulting in

$$\boldsymbol{\Phi} = -\boldsymbol{e} \cdot [-\boldsymbol{e} \cdot \boldsymbol{\varphi}_{\mathrm{l}}] - \boldsymbol{e} \cdot \boldsymbol{\vartheta}_{\mathrm{l}} + \boldsymbol{\Theta}_{\mathrm{l}}.$$

Here the third-order tensor $\boldsymbol{\Theta}_{\mathrm{l}} := \boldsymbol{\vartheta}_{\mathrm{l}}^C \otimes \boldsymbol{E}_C$ is *left* symmetric. For the special case that the three symmetric tensors $\boldsymbol{\vartheta}_{\mathrm{l}}^C$ allow representation as scalar multiples $\boldsymbol{\vartheta}_{\mathrm{l}}^C = \boldsymbol{\vartheta}'_{\mathrm{l}} \theta_{\mathrm{l}}^C$ of a common symmetric tensor $\boldsymbol{\vartheta}'_{\mathrm{l}}$ the left symmetric $\boldsymbol{\Theta}_{\mathrm{l}}$ expands into the dyadic product $\boldsymbol{\Theta}_{\mathrm{l}} = \boldsymbol{\vartheta}'_{\mathrm{l}} \theta_{\mathrm{l}}^C \otimes \boldsymbol{E}_C = \boldsymbol{\vartheta}'_{\mathrm{l}} \otimes \boldsymbol{\theta}_{\mathrm{l}}$ with $\boldsymbol{\theta}_{\mathrm{l}} := \theta_{\mathrm{l}}^C \boldsymbol{E}_C$. Finally, the axial vector $\boldsymbol{\varphi}_{\mathrm{l}}$, the symmetric tensor $\boldsymbol{\vartheta}_{\mathrm{l}}$, and the left symmetric $\boldsymbol{\Theta}_{\mathrm{l}}$ are extracted from $\boldsymbol{\Phi}$ by

$$4\boldsymbol{\varphi}_{\mathrm{l}} = -\boldsymbol{e} : [-\boldsymbol{e} : \boldsymbol{\Phi}] \quad \text{and} \quad 2\boldsymbol{\vartheta}_{\mathrm{l}} = \mathbf{i}^{\mathrm{sym}} : [-\boldsymbol{e} : \boldsymbol{\Phi}] \quad \text{and} \quad \boldsymbol{\Theta}_{\mathrm{l}} = \mathbf{i}^{\mathrm{sym}} : \boldsymbol{\Phi}.$$

It shall be noted carefully that the decomposition of a third-order tensor $\boldsymbol{\Phi}$ into either right or left skew symmetric and symmetric contributions involves different, say right and left $\boldsymbol{\varphi}_{\mathrm{r,l}}$, $\boldsymbol{\vartheta}_{\mathrm{r,l}}$, and $\boldsymbol{\Theta}_{\mathrm{r,l}}$. Thereby the coupled relations between the right and left quantities may be revealed by straightforward but tedious manipulations, typical intermediate steps are thereby

$$\boldsymbol{\Theta}_{\mathrm{r}} = \boldsymbol{\Phi} : \mathbf{i}^{\mathrm{sym}} \quad \text{and} \quad 2\boldsymbol{\phi}_{\mathrm{r}} = -\boldsymbol{\Phi} : \boldsymbol{e} = \mathbf{1} \mathbin{\times\!\!\!\times} \boldsymbol{\phi}_{\mathrm{l}} - \boldsymbol{\Theta}_{\mathrm{l}} : \boldsymbol{e}.$$

Here $\mathbf{1} \mathbin{\times\!\!\!\times} \boldsymbol{\phi}_{\mathrm{l}}$ expands into $[\boldsymbol{\phi}_{\mathrm{l}} : \mathbf{1}]\mathbf{1} - \boldsymbol{\phi}_{\mathrm{l}}^t$, moreover the special case $\boldsymbol{\Theta}_{\mathrm{l}} = \boldsymbol{\vartheta}'_{\mathrm{l}} \otimes \boldsymbol{\theta}_{\mathrm{l}}$ allows to rewrite $\boldsymbol{\Theta}_{\mathrm{l}} : \boldsymbol{e} = \boldsymbol{\vartheta}'_{\mathrm{l}} \rtimes \boldsymbol{\theta}_{\mathrm{l}}$. Finally, for completeness it is remarked that it is also possible to decompose Φ^{ABC} into $\Phi^{ABC} = \Phi^{[ABC]} + \Phi^{(ABC)}$.

Addendum 6.3. Irrotational $\oplus$ Solenoidal Decomposition
The Helmholtz decomposition of vector fields into irrotational (curl free) and solenoidal (divergence free) parts is classical and proves instrumental for example in electro-magnetism. The extension of the Helmholtz decomposition to tensor fields of arbitrary order has been advocated in [49]. In summary, the Helmholtz decomposition of arbitrary tensor fields of order n follows from a recursive use of the classical Helmholtz theorem and results in a beautiful representation based on repeated application of gradient and curl operators to scalar, vector, and tensor fields up to order n. Due to their relevance in the discussion of the integrability of the distortion and the double-distortion, here only the cases of vector fields together with second- and third-order tensor fields are briefly re-iterated.

Vector Fields:
Under certain technical conditions on the boundedness and regularity of the solution domain $\mathcal{V} \in \mathbb{E}^3$ and the continuity and smoothness of a vector field $\boldsymbol{\varphi} = \boldsymbol{\varphi}(\boldsymbol{X})$ the classical *Helmholtz theorem* for the decomposition of C^1 vector fields, [80], reads

$$\boldsymbol{\varphi} = \operatorname{Grad} \varphi(\boldsymbol{X}) + \operatorname{Curl} \boldsymbol{\theta}(\boldsymbol{X}).$$

Here $\operatorname{Grad} \varphi$ is the irrotational (curl free), and $\operatorname{Curl} \boldsymbol{\theta}$ is the solenoidal (divergence free) contribution to $\boldsymbol{\varphi}$ with

$$\operatorname{Div} \boldsymbol{\varphi} \equiv \operatorname{Div} \operatorname{Grad} \varphi(\boldsymbol{X}) \quad \text{and} \quad \operatorname{Curl} \boldsymbol{\varphi} \equiv \operatorname{Curl} \operatorname{Curl} \boldsymbol{\theta}(\boldsymbol{X}).$$

The proof of the Helmholtz theorem follows from a differential identity for an auxiliary C^2 vector field $\boldsymbol{\alpha} = \boldsymbol{\alpha}(\boldsymbol{X})$ on $\mathcal{V}$ that in turn is the solution to a Poisson equation with right-hand-side given by the vector field $\boldsymbol{\varphi} = \boldsymbol{\varphi}(\boldsymbol{X})$, i.e.

$$\operatorname{Div} \operatorname{Grad} \boldsymbol{\alpha} \equiv \operatorname{Grad} \operatorname{Div} \boldsymbol{\alpha}(\boldsymbol{X}) - \operatorname{Curl} \operatorname{Curl} \boldsymbol{\alpha}(\boldsymbol{X}) \doteq \boldsymbol{\varphi}(\boldsymbol{X}).$$

Obviously, for this setting the scalar and vector potentials φ and $\boldsymbol{\theta}$ are related to the auxiliary vector field $\boldsymbol{\alpha}$ by

$$\varphi := \operatorname{Div} \boldsymbol{\alpha}(\boldsymbol{X}) \quad \text{and} \quad \boldsymbol{\theta} := -\operatorname{Curl} \boldsymbol{\alpha}(\boldsymbol{X}).$$

Since under the previously mentioned technical conditions (in particular $\boldsymbol{\varphi}(\boldsymbol{X}) \to 0$ faster than $|\boldsymbol{X}|^{-2} \to 0$ for the below integral to converge) the solution to the above Poisson equation is given by

$$\boldsymbol{\alpha}(\boldsymbol{X}) = -\frac{1}{4\pi} \int_{\mathcal{V}} \frac{\boldsymbol{\varphi}(\boldsymbol{X}')}{|\boldsymbol{X} - \boldsymbol{X}'|} \, \mathrm{d}V',$$

the scalar and the vector potentials φ and $\boldsymbol{\theta}$ for the vector field $\boldsymbol{\varphi}$ may be expressed in terms of its scalar and vector sources $\operatorname{Div} \boldsymbol{\varphi}$ and $\operatorname{Curl} \boldsymbol{\varphi}$ as

$$\varphi(\boldsymbol{X}) = -\frac{1}{4\pi} \int_{\mathcal{V}} \frac{\operatorname{Div}' \boldsymbol{\varphi}(\boldsymbol{X}')}{|\boldsymbol{X} - \boldsymbol{X}'|} \, \mathrm{d}V' \quad \text{and} \quad \boldsymbol{\theta}(\boldsymbol{X}) = \frac{1}{4\pi} \int_{\mathcal{V}} \frac{\operatorname{Curl}' \boldsymbol{\varphi}(\boldsymbol{X}')}{|\boldsymbol{X} - \boldsymbol{X}'|} \, \mathrm{d}V'.$$

The above result relies on $\operatorname{Grad} = -\operatorname{Grad}'$ and integration by parts whereby it is exploited that by assumption $\boldsymbol{\varphi}(\boldsymbol{X}) \to \boldsymbol{0}$ faster than $|\boldsymbol{X}|^{-2} \to 0$. In order to eliminate the over-determinacy in representing three coefficients of $\boldsymbol{\varphi}$ in terms of one scalar potential φ and three coefficients of a vector potential $\boldsymbol{\theta}$ an additional *Gauge condition* is usually postulated for the vector potential

$$\operatorname{Div}\boldsymbol{\theta}(\boldsymbol{X}) = 0.$$

Finally, it is noted that the Helmholtz decomposition of vector fields into irrotational and solenoidal contributions is *unique* only up to the addition and subtraction of the gradient of a harmonic scalar field $\varphi_{\mathrm{h}} = \varphi_{\mathrm{h}}(\boldsymbol{X})$ with $\operatorname{Div}\operatorname{Grad}\varphi_{\mathrm{h}} \equiv 0$. Obviously, for $\boldsymbol{\varphi} = \operatorname{Grad}(\varphi + \varphi_{\mathrm{h}}) + [\operatorname{Curl}\boldsymbol{\theta} - \operatorname{Grad}\varphi_{\mathrm{h}}]$ it holds that $\operatorname{Div}\operatorname{Grad}(\varphi + \varphi_{\mathrm{h}}) = \operatorname{Div}\operatorname{Grad}\varphi$ while $\operatorname{Curl}\operatorname{Grad}(\varphi + \varphi_{\mathrm{h}}) = \boldsymbol{0}$, and $\operatorname{Curl}(\operatorname{Curl}\boldsymbol{a} - \operatorname{Grad}\varphi_{\mathrm{h}}) = \operatorname{Curl}\operatorname{Curl}\boldsymbol{\theta}$ while $\operatorname{Div}(\operatorname{Curl}\boldsymbol{\theta} - \operatorname{Grad}\varphi_{\mathrm{h}}) = 0$, respectively.

Second-Order Tensor Fields:
An arbitrary second-order (two-point) tensor field $\boldsymbol{\phi} = \boldsymbol{\phi}(\boldsymbol{X})$ allows representation $\boldsymbol{\phi} = \phi^{aB}(\boldsymbol{X})\boldsymbol{e}_a \otimes \boldsymbol{E}_B = \boldsymbol{e}_a \otimes \boldsymbol{\varphi}^a(\boldsymbol{X})$. Then the three vector fields $\boldsymbol{\varphi}^a := \phi^{aB}(\boldsymbol{X})\boldsymbol{E}_B$ with $a = 1, 2, 3$ are expressed based on the vectorial Helmholtz decomposition as $\boldsymbol{\varphi}^a = \operatorname{Grad}\varphi^a + \operatorname{Curl}\boldsymbol{\theta}^a$, thus rendering the corresponding Helmholtz decomposition of the (second-order) tensor field $\boldsymbol{\phi} = \boldsymbol{\phi}(\boldsymbol{X})$, see [49], as

$$\begin{aligned}\boldsymbol{\phi} = \boldsymbol{e}_a \otimes [\operatorname{Grad}\varphi^a + \operatorname{Curl}\boldsymbol{\theta}^a] &= \operatorname{Grad}(\boldsymbol{e}_a\varphi^a) + \operatorname{Curl}(\boldsymbol{e}_a \otimes \boldsymbol{\theta}^a)\\ &= \operatorname{Grad}\boldsymbol{\varphi}(\boldsymbol{X}) + \operatorname{Curl}\boldsymbol{\vartheta}'(\boldsymbol{X}).\end{aligned}$$

The final step is to note that the vector potential $\boldsymbol{\varphi} := \boldsymbol{e}_a\varphi^a$ and the tensor potential $\boldsymbol{\vartheta}' := \boldsymbol{e}_a \otimes \boldsymbol{\theta}^a$ allow in turn corresponding Helmholtz decompositions, thus resulting in

$$\boldsymbol{\phi} = \operatorname{Grad}\operatorname{Grad}\varphi(\boldsymbol{X}) + \operatorname{Grad}\operatorname{Curl}\boldsymbol{\theta}(\boldsymbol{X}) + \operatorname{Curl}\operatorname{Curl}\boldsymbol{\vartheta}(\boldsymbol{X}).$$

To eliminate the over-determinacy in representing nine coefficients of $\boldsymbol{\phi}$ in terms of one scalar potential φ, three coefficients of a vector potential $\boldsymbol{\theta}$, and nine coefficients of a tensor potential $\boldsymbol{\vartheta}$ additional (three + one) *Gauge conditions* have been proposed in [49] as

$$\operatorname{Div}\boldsymbol{\vartheta}(\boldsymbol{X}) = \boldsymbol{0} \quad \text{and} \quad \operatorname{Div}\operatorname{Div}\boldsymbol{\vartheta}(\boldsymbol{X}) = 0.$$

It seems that the last Gauge condition could alternatively be substituted by $\operatorname{Div}\boldsymbol{\theta} = 0$ in order to introduce a sufficient number of constraints.

Third-Order Tensor Fields:
An arbitrary third-order (two-point) tensor field $\boldsymbol{\Phi} = \boldsymbol{\Phi}(\boldsymbol{X})$ allows representation $\boldsymbol{\Phi} = \Phi^{aBC}(\boldsymbol{X})\boldsymbol{e}_a \otimes \boldsymbol{E}_B \otimes \boldsymbol{E}_C = \boldsymbol{e}_a \otimes \boldsymbol{\phi}^a(\boldsymbol{X})$. Then the three tensor fields $\boldsymbol{\phi}^a := \Phi^{aBC}(\boldsymbol{X})\boldsymbol{E}_B \otimes \boldsymbol{E}_C$ with $a = 1, 2, 3$ are expressed based on the tensorial Helmholtz decomposition as $\boldsymbol{\phi}^a = \operatorname{Grad}\boldsymbol{\varphi}^a + \operatorname{Curl}\boldsymbol{\vartheta}'^a$, thus rendering the corresponding Helmholtz decomposition of the third-order tensor field $\boldsymbol{\Phi} = \boldsymbol{\Phi}(\boldsymbol{X})$, see [49], as

$$\begin{aligned}\boldsymbol{\Phi} = \boldsymbol{e}_a \otimes [\operatorname{Grad}\boldsymbol{\varphi}^a + \operatorname{Curl}\boldsymbol{\vartheta}'^a] &= \operatorname{Grad}(\boldsymbol{e}_a \otimes \boldsymbol{\varphi}^a) + \operatorname{Curl}(\boldsymbol{e}_a \otimes \boldsymbol{\vartheta}'^a)\\ &= \operatorname{Grad}\boldsymbol{\phi}(\boldsymbol{X}) + \operatorname{Curl}\boldsymbol{\Theta}''(\boldsymbol{X}).\end{aligned}$$

The next step is to note that the (second-order) tensor potential $\boldsymbol{\phi} := \boldsymbol{e}_a \otimes \boldsymbol{\varphi}^a$ and the third-order tensor potential $\boldsymbol{\Theta}'' := \boldsymbol{e}_a \otimes \boldsymbol{\vartheta}'^a$ allow in turn corresponding Helmholtz decompositions, thus resulting in

$$\boldsymbol{\Phi} = \operatorname{Grad}\operatorname{Grad}\boldsymbol{\varphi}(\boldsymbol{X}) + \operatorname{Grad}\operatorname{Curl}\boldsymbol{\vartheta}'(\boldsymbol{X}) + \operatorname{Curl}\operatorname{Curl}\boldsymbol{\Theta}'(\boldsymbol{X}).$$

The final step is to note again that the vector potential $\boldsymbol{\varphi}$, the (second-order) tensor potential $\boldsymbol{\vartheta}'$, and the third-order tensor potential $\boldsymbol{\Theta}'$ also allow corresponding Helmholtz decompositions, rendering the final representation

$$\begin{aligned}\boldsymbol{\Phi} = {} & \operatorname{Grad}\operatorname{Grad}\operatorname{Grad}\varphi(\boldsymbol{X}) + \operatorname{Grad}\operatorname{Grad}\operatorname{Curl}\boldsymbol{\theta}(\boldsymbol{X}) \\ & + \operatorname{Grad}\operatorname{Curl}\operatorname{Curl}\boldsymbol{\vartheta}(\boldsymbol{X}) + \operatorname{Curl}\operatorname{Curl}\operatorname{Curl}\boldsymbol{\Theta}(\boldsymbol{X}).\end{aligned}$$

To eliminate the over-determinacy in representing 27 coefficients of $\boldsymbol{\Phi}$ in terms of one scalar potential φ, three coefficients of a vector potential $\boldsymbol{\theta}$, nine coefficients of a (second-order) tensor potential $\boldsymbol{\vartheta}$, and 27 coefficients of a third-order tensor potential $\boldsymbol{\Theta}$ additional (nine + three + one) *Gauge conditions* have been proposed in [49] as

$$\operatorname{Div}\boldsymbol{\Theta}(\boldsymbol{X}) = \boldsymbol{0} \quad \text{and} \quad \operatorname{Div}\operatorname{Div}\boldsymbol{\Theta}(\boldsymbol{X}) = \boldsymbol{0} \quad \text{and} \quad \operatorname{Div}\operatorname{Div}\operatorname{Div}\boldsymbol{\Theta}(\boldsymbol{X}) = 0.$$

It seems that the last two Gauge conditions could alternatively be substituted by $\operatorname{Div}\boldsymbol{\vartheta} = \boldsymbol{0}$ and $\operatorname{Div}\boldsymbol{\theta} = 0$ in order to introduce a sufficient number of constraints.

Addendum 6.4. Material and Spatial Permutation Tensors

The permutation tensor expressed in curvilinear coordinates might easily create some confusion. Thus the various representations and transformations of the permutation tensor shall be outlined carefully in the sequel.

Material Permutation Tensor:

In the tangent space to the material configuration the material version of the permutation tensor has the following representation in terms of rectilinear and curvilinear coordinates, respectively

$$\boldsymbol{E} := E^{ABC}\boldsymbol{E}_A \otimes \boldsymbol{E}_B \otimes \boldsymbol{E}_C = \mathsf{E}^{IJK}\boldsymbol{G}_I \otimes \boldsymbol{G}_J \otimes \boldsymbol{G}_K.$$

For even and odd permutations of ABC and IJK the material permutations symbols E^{ABC} and E^{IJK}, respectively, take values

$$E^{ABC} = \begin{cases} +1 & \text{even}, \\ -1 & \text{odd}, \\ 0 & \text{else}, \end{cases} \quad \text{and} \quad \mathsf{E}^{IJK} = \begin{cases} +\sqrt[-2]{G} & \text{even}, \\ -\sqrt[-2]{G} & \text{odd}, \\ 0 & \text{else}, \end{cases}$$

whereby $G = \det[G_{IJ}]$ denotes the determinant of the matrix arrangement of the metric coefficients $G_{IJ} := \boldsymbol{G}_I \cdot \boldsymbol{G}_J$ related to the curvilinear coordinates in the material configuration. Thereby the material permutation symbol E^{IJK}

in curvilinear coordinates is expressed in terms of the permutation symbol $\mathcal{E}^{IJK}$ in the material curvilinear parameter space by

$$\mathsf{E}^{IJK} = \sqrt[-2]{G}\mathcal{E}^{IJK} \quad \text{with} \quad \mathcal{E}^{IJK} = \begin{cases} +1 & \text{even,} \\ -1 & \text{odd,} \\ 0 & \text{else.} \end{cases}$$

Moreover, due to the tensorial transformation properties of the base vectors it also holds that

$$E^{ABC} = \mathsf{E}^{IJK} J^{A}{}_{I} J^{B}{}_{J} J^{C}{}_{K} = \sqrt[-2]{G}\mathcal{E}^{IJK} J^{A}{}_{I} J^{B}{}_{J} J^{C}{}_{K}.$$

Note that since the covariant base vectors follow as $\boldsymbol{G}_I = J^{A}{}_{I}\boldsymbol{E}_A$, the metric coefficients may also be expressed as $G_{IJ} = J^{A}{}_{I}\delta_{AB}J^{B}{}_{J}$ and thus it holds that $G = \det^2[J^{A}{}_{I}]$.

Spatial Permutation Tensor:
In the tangent space to the spatial configuration the spatial version of the permutation tensor has the following representation in terms of rectilinear and curvilinear coordinates, respectively

$$\boldsymbol{e} := e^{abc}\boldsymbol{e}_a \otimes \boldsymbol{e}_b \otimes \boldsymbol{e}_c = \mathsf{e}^{ijk}\boldsymbol{g}_i \otimes \boldsymbol{g}_j \otimes \boldsymbol{g}_k$$

For even and odd permutations of abc and ijk the spatial permutations symbols e^{abc} and e^{ijk}, respectively, take values

$$e^{abc} = \begin{cases} +1 & \text{even,} \\ -1 & \text{odd,} \\ 0 & \text{else,} \end{cases} \quad \text{and} \quad \mathsf{e}^{ijk} = \begin{cases} +\sqrt[-2]{g} & \text{even,} \\ -\sqrt[-2]{g} & \text{odd,} \\ 0 & \text{else,} \end{cases}$$

whereby $g = \det[g_{ij}]$ denotes the determinant of the matrix arrangement of the metric coefficients $g_{ij} := \boldsymbol{g}_i \cdot \boldsymbol{g}_j$ related to the curvilinear coordinates in the spatial configuration. Thereby the spatial permutation symbol e^{ijk} in curvilinear coordinates is expressed in terms of the permutation symbol $\mathcal{e}^{ijk}$ in the spatial curvilinear parameter space by

$$\mathsf{e}^{ijk} = \sqrt[-2]{g}\mathcal{e}^{ijk} \quad \text{with} \quad \mathcal{e}^{ijk} = \begin{cases} +1 & \text{even,} \\ -1 & \text{odd,} \\ 0 & \text{else.} \end{cases}$$

Moreover, due to the tensorial transformation properties of the base vectors it also holds that

$$e^{abc} = \mathsf{e}^{ijk} j^{\cdot a}{}_{i} j^{\cdot b}{}_{j} j^{\cdot c}{}_{k} = \sqrt[-2]{g}\mathcal{e}^{ijk} j^{\cdot a}{}_{i} j^{\cdot b}{}_{j} j^{\cdot c}{}_{k}.$$

Note that since the covariant base vectors follow as $\boldsymbol{g}_i = j^{a}{}_{i}\boldsymbol{e}_a$, the metric coefficients may also be expressed as $g_{ij} = j^{a}{}_{i}\delta_{ab}j^{b}{}_{j}$ and thus it holds that $g = \det^2[j^{a}{}_{i}]$.

Spatial versus Material Permutation Tensor:
Finally the spatial and material permutation tensors are convected as

$$\boldsymbol{e} = j[\boldsymbol{F} \boxtimes \boldsymbol{F}] : \boldsymbol{E} \cdot \boldsymbol{F}^t \quad \text{and} \quad \boldsymbol{E} = J[\boldsymbol{f} \boxtimes \boldsymbol{f}] : \boldsymbol{e} \cdot \boldsymbol{f}^t,$$

whereby $J := \det \boldsymbol{F}$ and $j := \det \boldsymbol{f}$ denote the determinants of the distortions. These expand into

$$J = \sqrt[2]{g/G}\det[\mathcal{F}^i_{\ I}] \quad \text{and} \quad j = \sqrt[2]{G/g}\det[f^I_{\ i}]$$

with $\mathcal{J} := \det[\mathcal{F}^i_{\ I}]$ and $j := \det[f^I_{\ i}]$, respectively. Moreover $\boxtimes$ denotes a non-standard dyadic product that is highlighted in the corresponding index notation

$$e^{abc} = jE^{ABC}F^a_{\ A}F^b_{\ B}F^c_{\ C} \quad \text{and} \quad \mathsf{e}^{ijk} = j\mathsf{E}^{IJK}\mathsf{F}^i_{\ I}\mathsf{F}^j_{\ J}\mathsf{F}^k_{\ K}.$$

As a consequence, the permutation symbols in the spatial and material curvilinear parameter spaces are related by

$$e^{ijk} = j\,\mathcal{E}^{IJK}\mathcal{F}^i_{\ I}\mathcal{F}^j_{\ J}\mathcal{F}^k_{\ K}.$$

The above transformation behavior of the permutation symbols upon coordinate changes underpins that the permutation tensor is an isotropic third-order tensor (indeed the only one).

Addendum 6.5. Material and Spatial Curl Operators

Material Curl Operators:
The material *curl operator* with respect to the material coordinates as applied to the deformation gradient $\boldsymbol{F}$ allows the following coordinate representation

$$\operatorname{Curl} \boldsymbol{F} := D^{aC}\boldsymbol{e}_a \otimes \boldsymbol{E}_C = \mathsf{D}^{iK}\boldsymbol{g}_i \otimes \boldsymbol{G}_K,$$

whereby the coefficients of $\operatorname{Curl} \boldsymbol{F}$ are given in terms of the permutation symbols expressed in material rectilinear and curvilinear coordinates, respectively

$$D^{aC} := -F^a_{\ A,B}E^{ABC} \quad \text{and} \quad \mathsf{D}^{iK} := -\mathsf{F}^i_{\ I:J}\mathsf{E}^{IJK} \equiv -\sqrt[-2]{G}\mathcal{F}^i_{\ I,J}\mathcal{E}^{IJK}.$$

Note that expressed in curvilinear coordinates the Euclidean total covariant derivative may be substituted by the ordinary partial derivative due to the (right) symmetry of the spatial and material Christoffel symbols.

Spatial Curl Operators:
Likewise, the spatial curl operator with respect to the spatial coordinates as applied to the deformation gradient $\boldsymbol{f}$ allows the following representation

$$\operatorname{curl} \boldsymbol{f} := d^{Ac}\boldsymbol{E}_A \otimes \boldsymbol{e}_c = \mathsf{d}^{Ik}\boldsymbol{G}_I \otimes \boldsymbol{g}_k,$$

whereby here the coefficients of $\operatorname{curl} \boldsymbol{f}$ are given in terms of the permutation symbols expressed in spatial rectilinear and curvilinear coordinates, respectively

$$d^{Ac} := -f^A{}_{a,b}e^{abc} \quad \text{and} \quad \mathsf{d}^{Ik} := -\mathsf{f}^I{}_{i:j}\mathsf{e}^{ijk} \equiv -\sqrt[-2]{g}f^I{}_{i,j}e^{ijk}.$$

Note again that expressed in curvilinear coordinates the Euclidean total covariant derivative may be substituted by the ordinary partial derivative due to the (right) symmetry of the spatial and material Christoffel symbols.

Spatial versus Material Curl Operators:
Finally, it may be shown by the chain rule applied to, e.g. $\boldsymbol{f} \cdot \boldsymbol{F} = \mathbf{1}$ that the spatial and material curl operators are convected as

$$\operatorname{Curl} \boldsymbol{F} = -J\boldsymbol{F} \cdot \operatorname{curl} \boldsymbol{f} \cdot \boldsymbol{f}^t \quad \text{and} \quad \operatorname{curl} \boldsymbol{f} = -j\boldsymbol{f} \cdot \operatorname{Curl} \boldsymbol{F} \cdot \boldsymbol{F}^t.$$

These are the key relations that allow to determine the relation between, e.g. the spatial and material dislocation density tensors.

Chapter 7
Elasticity

GEBURTSAKT DER PHILOSOPHIE

Erschrocken staunt der Heide Schaf mich an,
als sähs in mir den ersten Menschenmann.
Sein Blick steckt an; wir stehen wie im Schlaf;
mir ist, ich säh zum ersten Mal ein Schaf.

Christian Morgenstern

Abstract. The previous concepts of differential geometry, in particular aspects related to compatibility, shall be applied to the kinematics of first- and second-order (nonlinear) elasticity. Thereby it shall be noted that both the material and spatial configurations of first- and second-order elasticity ought to be compatible, see Fig. 7.1. Then, two cases may be considered: firstly the material configuration is assumed compatible and the conditions on the deformation for the spatial configuration to also be compatible are sought; secondly the situation is reversed: the spatial configuration is assumed compatible and the conditions on the (inverse) deformation for the material configuration to be compatible are sought. Thereby distinction can be made between the straightforward conditions for the distortion (and double-distortion) being integrable and the more involved conditions for the metric (and double-metric) being integrable. The latter leads to nonlinear (and extended) versions of the famous St-Venant compatibility conditions.

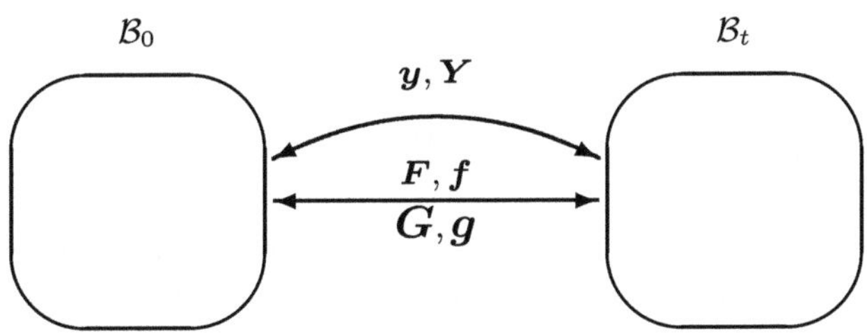

Fig. 7.1 The material and spatial configurations of first- and second-order elasticity are compatible. Thus conditions for both the direct and the inverse deformation to be compatible may be established.

Comprehensive accounts on first- and second-order elasticity in Euclidean space are provided for the sake of reference at the end of the chapter in two extended supplements.

P. Steinmann, *Geometrical Foundations of Continuum Mechanics*,
Lecture Notes in Applied Mathematics and Mechanics 2, DOI: 10.1007/978-3-662-46460-1_7

7.1 First-Order Continuum

The kinematics of elasticity in a first-order continuum are essentially characterized by the distortion. The importance of the distortion is particularly obvious from the observation that it represents the sole argument of the *elastically stored energy density function* as the key ingredient of a generic *master constitutive relation*. Thus for a given distortion the conditions for its integrability into a compatible deformation map are of utmost interest. These conditions have been discussed already, nevertheless without explicitly referring to elasticity, in the previous chapter and thus need not to be derived in detail again. However, due to the requirements of *objectivity* and *material symmetry* it is rather the corresponding *(strain) metric*, being quadratic in the distortion, that enters the elastically stored energy density function as its argument. Therefore, as an alternative to the conditions for the distortion being integrable, the rather involved conditions for the (strain) metric being integrable into a compatible deformation map need to be established in addition.

7.1.1 Integrable Distortion

For first-order elasticity integrability of the distortion renders the primary conditions for the spatial and the material configuration being compatible.

Looking Forward: Compatibility of Spatial Configuration. For a spatial configuration of first-order elasticity to be compatible, i.e. the integration of the spatial line element $\mathrm{d}\boldsymbol{y} = \boldsymbol{F} \cdot \mathrm{d}\boldsymbol{X}$ along arbitrary closed material circuits $\mathcal{C}_0$ does never result in a spatial closure gap $[\![\boldsymbol{y}]\!] \neq \mathbf{0}$, the distortion $\boldsymbol{F}$ ought to be integrable. For *simply-connected* bodies this condition is expressed as

$$\mathrm{Curl}\,\boldsymbol{F} \doteq \mathbf{0}. \tag{7.1}$$

Recall that this condition takes equivalent expressions $F^a{}_{[B,C]} \doteq 0$ or $\mathsf{F}^i{}_{[J:K]} = \mathcal{F}^i{}_{[J,K]} \doteq 0$ in rectilinear (Cartesian) or curvilinear coordinates, respectively.

Looking Backward: Compatibility of Material Configuration. Likewise, for a material configuration of first-order elasticity to be compatible, i.e. the integration of the material line element $\mathrm{d}\boldsymbol{Y} = \boldsymbol{f} \cdot \mathrm{d}\boldsymbol{x}$ along arbitrary closed spatial circuits $\mathcal{C}_t$ does never result in a material closure gap $[\![\boldsymbol{Y}]\!] \neq \mathbf{0}$, the distortion $\boldsymbol{f}$ ought to be integrable. For *simply-connected* bodies this condition is expressed as

$$\mathrm{curl}\,\boldsymbol{f} \doteq \mathbf{0}. \tag{7.2}$$

Recall that this condition takes equivalent expressions $f^A{}_{[b,c]} \doteq 0$ or $\mathsf{f}^I{}_{[j:k]} = f^I{}_{[j,k]} \doteq 0$ in rectilinear (Cartesian) or curvilinear coordinates, respectively.

Discussion 7.1. Topologically a simply-connected body is defined so that arbitrary closed curves fully contained in the body can be continuously contracted to a point situated within the body (as a counterexample consider a torus in three dimensions). As a demonstration of the limitation of the integrability condition for the distortion to simply-connected bodies consider the following distortion expressed in terms of material rectilinear (Cartesian) coordinates

$$[F^a{}_B] = \begin{bmatrix} 1 & 0 & 0 \\ 0 & 1 & 0 \\ \dfrac{X^2}{[X^1]^2 + [X^2]^2} & \dfrac{-X^1}{[X^1]^2 + [X^2]^2} & 1 \end{bmatrix} \quad \rightsquigarrow \quad F^a{}_{[B,C]} \equiv 0.$$

Then the corresponding deformation is obtained by integration as

$$y^1(\{X^I\}) = X^1, \qquad y^2(\{X^I\}) = X^2 \quad \text{and} \quad y^3(\{X^I\}) = X^3 - \arctan\frac{X^2}{X^1}.$$

Note that in each plane perpendicular to the X^3-axis the angle $\arctan(X^2/X^1)$ between the X^1-axis and the radial line connecting $(0,0)$ with (X^1, X^2) is constant for each point on that line. Thus the deformation displays a singularity at every $(0,0,X^3)$, accordingly the X^3-axis has to be excluded in the following examples. Next, for $0 < R_\text{i} < R_\text{e}, L > 0, \alpha < 2\pi$ consider:

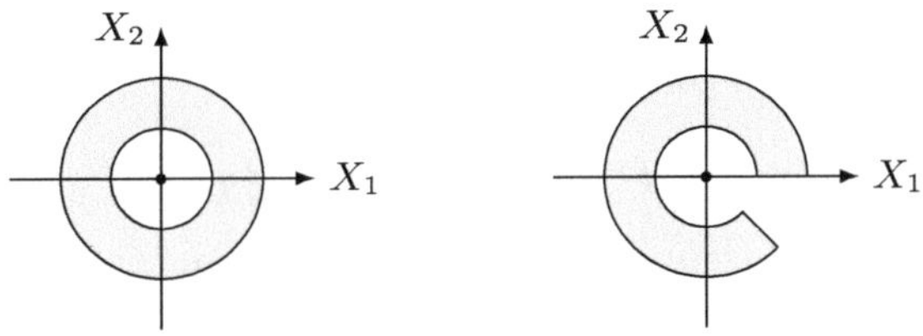

Fig. 7.2 Left: non-simply-connected body. Right: simply-connected body.

i) a non-simply-connected body defined by

$$\mathcal{B}_0 := \{R_\text{i}^2 < [X^1]^2 + [X^2]^2 < R_\text{e}^2\} \times (-L, +L),$$

ii) a simply-connected body defined by

$$\mathcal{B}_0 := \{X^1 = R\cos\theta, X^2 = R\sin\theta : R_\text{i} < R < R_\text{e}, 0 < \theta < \alpha\} \times (-L, +L).$$

The deformation on the non-simply-connected body i) displays a jump by 2π in X^3-direction along $X^2 = 0$, $X^1 > 0$ and is thus not compatible, despite $\operatorname{Curl}\boldsymbol{F}$ being zero. Thus the compatibility condition $\operatorname{Curl}\boldsymbol{F} = \boldsymbol{0}$ is not sufficient for a non-simply-connected body. Vice-versa, for α infinitesimal close to 2π the deformation on the simply-connected body ii) represents roughly the warping of a longitudinally slitted thick-walled tube due to torsion, a deformation that is compatible. □

7.1.2 Integrable Metric

For a number of reasons, see the supplement on first-order elasticity below, it is often not primarily the distortion but rather the (strain) metric, being quadratic in the distortion, that enters the constitutive relations of first-order elasticity. Thus the conditions on the (strain) metric being integrable into a compatible deformation map need to be considered. The derivation of these conditions are a bit more involved and result in the so-called *St-Venant compatibility conditions.*

Looking Forward: Compatibility of Spatial Configuration. The compatibility condition for the spatial configuration in terms of the material (strain) metric results from the following reasoning: For a spatial configuration of elasticity to be compatible distance determination and parallel transport ought to be of Euclidean type, see Fig. 7.3, thus Cartesian coordinates may be selected with the spatial metric coefficients coinciding with the Kronecker symbol and a vanishing spatial connection

$$m_{ab} = \delta_{ab} \quad \text{and} \quad l^a{}_{bc} = 0. \tag{7.3}$$

The convection of the spatial metric and the spatial connection by the deformation map then renders the *Cauchy-Green tensor* and an *integrable connection* in the material configuration

$$\mathcal{M}_{IJ} \equiv C_{IJ} := F^a{}_I \delta_{ab} F^b{}_J \quad \text{and} \quad \mathcal{L}^I{}_{JK} = f^I{}_a F^a{}_{J,K}. \tag{7.4}$$

Furthermore, it may easily be shown that the fully covariant material connection $\mathcal{L}_{IJK} = C_{IM}\mathcal{L}^M{}_{JK}$ is metric with respect to C_{IJ}, i.e.

$$C_{IJ,K} = F^a{}_{I,K}\delta_{ab}F^b{}_J + F^a{}_I\delta_{ab}F^b{}_{J,K} = 2\mathcal{L}_{(IJ)K}. \tag{7.5}$$

Consequently, and due to the integrability of $F^a{}_J$ with $F^a{}_{[J,K]} = 0$ that results in a torsionless $\mathcal{L}^I{}_{JK} = \mathcal{L}^I{}_{(JK)}$ (compare Eq. 4.9) the material connection $\mathcal{L}_{IJK}$ coincides exclusively with the material Riemann connection

$$\mathcal{M}_{IJK}(C) := \frac{1}{2}[C_{IJ,K} - C_{JK,I} + C_{KI,J}] \tag{7.6}$$

based on the metric C_{IJ} while the corresponding material contortion vanishes identically:

$$\begin{array}{rl} C_{IJ,K} = & + \, F^a{}_J \, \delta_{ab} \, F^b{}_{I,K} + F^a{}_I \, \delta_{ab} \, F^b{}_{J,K} \\ -C_{JK,I} = & - \, F^a{}_K \, \delta_{ab} \, F^b{}_{J,I} - F^a{}_J \, \delta_{ab} \, F^b{}_{K,I} \\ C_{KI,J} = & + \, F^a{}_I \, \delta_{ab} \, F^b{}_{K,J} + F^a{}_K \, \delta_{ab} \, F^b{}_{I,J} \\ \hline\hline \end{array}$$

$$\rightarrow \quad \mathcal{M}_{IJK}(\mathcal{C}) = F^a{}_I \delta_{ab} F^b{}_{(J,K)} \quad \text{and} \quad \mathcal{K}_{IJK} \equiv 0. \tag{7.7}$$

Next, since the material connection is integrable, its associated material Riemann curvature tensor vanishes identically

$$\mathcal{M}_{IJKL}(\mathcal{C}) := 2\mathcal{M}_{IJ[L,K]} - 2\mathcal{M}_{MI[K}\mathcal{M}^M{}_{JL]} \equiv 0. \tag{7.8}$$

This finding coincides of course with the postulated Euclidean character of the spatial configuration

$$0 \equiv m_{abcd}(i) = m_{abcd}(y(\mathcal{C})) = y(\mathcal{M}_{IJKL}(\mathcal{C})). \tag{7.9}$$

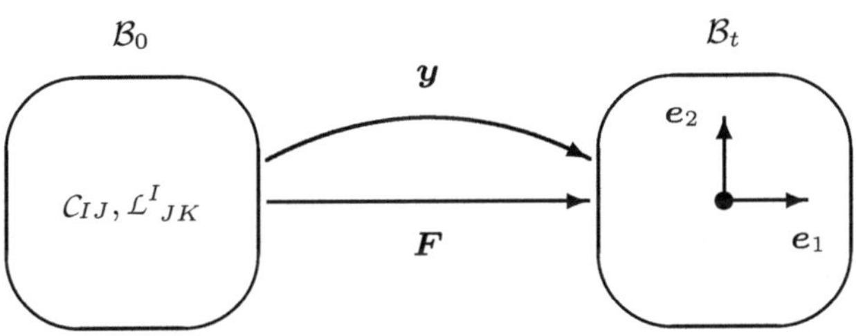

Fig. 7.3 In the spatial configuration of elasticity distance determination and parallel transport are of Euclidean type. The convection of the metric and the connection renders the Cauchy-Green tensor and an integrable connection in the material configuration.

The previous reasoning may reversed as follows, see Fig. 7.4: for the case that the material Riemann curvature tensor $\mathcal{M}_{IJKL}$ based on the Riemann connection $\mathcal{M}_{IJK}$ vanishes it may be concluded that the Riemann connection coincides with an integrable connection $\mathcal{M}_{IJK} \equiv F^a{}_I \delta_{ab} F^b{}_{J,K} = \mathcal{L}_{IJK}$. Thereby the Riemann connection $\mathcal{M}_{IJK}$ and thus also the integrable connection $\mathcal{L}_{IJK}$ is metric with respect to the material metric $\mathcal{M}_{IJ} = F^a{}_I \delta_{ab} F^b{}_J$. However, since every metric connection allows representation as $\mathcal{L}_{IJK} = \mathcal{M}_{IJK} + \mathcal{K}_{IJK}$ it may be concluded that the corresponding contortion $\mathcal{K}_{IJK}$ and thus the torsion $\mathcal{T}_{IJK} = 2\mathcal{K}_{I[JK]}$ vanishes identically. Therefore, also in agreement with the right symmetry of the Riemann connection $\mathcal{M}_{IJK} = \mathcal{M}_{I(JK)}$ the integrable connection is also right symmetric $\mathcal{L}_{IJK} = F^a{}_I \delta_{ab} F^b{}_{(J,K)}$. Finally $F^b{}_{J,K} = F^b{}_{(J,K)}$ implies the integrability of the distortion $F^b{}_J = y^b{}_{,J}$ into a compatible deformation map $y^b = y^b(X^J)$. Thus in the material configuration a vanishing Riemann curvature tensor based on a Riemann connection

Vanishing Riemann Curvature

$$\mathcal{M}_{IJKL} = 2\mathcal{M}_{IJ[L,K]} - 2\mathcal{M}_{MI[K}\mathcal{M}^{M}{}_{JL]} \equiv 0$$

$$\Downarrow$$

Integrable Metric Connection

$$\mathcal{M}_{IJK} = \mathcal{M}_{I(JK)} \equiv F^{a}{}_{I} i_{ab} F^{b}{}_{J,K}$$

$$\Downarrow$$

Vanishing Torsion

$$F^{b}{}_{J,K} = F^{b}{}_{(J,K)}$$

$$\Downarrow$$

Integrable Distortion

$$F^{b}{}_{J} = y^{b}{}_{,J}$$

Fig. 7.4 Sketch of the reasoning to deduce the relation between a vanishing material Riemann curvature and the integrability of the material metric into a compatible deformation map

assures integrability of the material metric into a compatible deformation map, i.e. into a compatible spatial configuration.

In summary the kinematics of elasticity may be considered a flat Riemann geometry over the material configuration $\mathcal{B}_0$. A flat Riemann geometry is of course an Euclidean geometry. Concluding, flatness assures integrability, i.e. the Cauchy-Green tensor may be derived from a vector valued deformation map. This is the nonlinear version of the *St. Venant compatibility conditions* expressed in terms of the material metric.

Remark 7.1. For geometrically linear elasticity in Cartesian coordinates the St.-Venant compatibility conditions for the symmetric strain tensor are expressed in terms of the second-order symmetric incompatibility density tensor

$$\boldsymbol{\eta} = \eta_{mn}\boldsymbol{e}_m \otimes \boldsymbol{e}_n \doteq \boldsymbol{0} \quad \text{with} \quad \eta_{mn} := e_{mab}\epsilon_{ac,bd}e_{cdn}.$$

The incompatibility density tensor may be expressed in terms of the Kröner incompatibility operator, see [94], as

$$\boldsymbol{\eta} := \operatorname{inc}\boldsymbol{\epsilon} \quad \text{with} \quad \operatorname{inc}\boldsymbol{\epsilon} := \operatorname{curl}^t \operatorname{curl}^t \boldsymbol{\epsilon} \quad \text{and} \quad \operatorname{curl}^t \boldsymbol{\epsilon} := -\epsilon_{ac,d}e_{cdn}\boldsymbol{e}_n \otimes \boldsymbol{e}_a.$$

Thus the additional solenoidal constraint holds for the incompatibility density tensor

$$\operatorname{div}\boldsymbol{\eta} = \boldsymbol{0}.$$

The six St. Venant compatibility conditions in the coefficients of $\boldsymbol{\eta}$ may be expanded as

$$\eta_{11} = \epsilon_{22,33} + \epsilon_{33,22} - 2\epsilon_{23,32},$$
$$\eta_{22} = \epsilon_{33,11} + \epsilon_{11,33} - 2\epsilon_{31,13},$$
$$\eta_{33} = \epsilon_{11,22} + \epsilon_{22,11} - 2\epsilon_{12,21},$$
$$\eta_{12} = \epsilon_{13,32} + \epsilon_{23,31} - \epsilon_{33,12} - \epsilon_{12,33},$$
$$\eta_{23} = \epsilon_{21,13} + \epsilon_{31,12} - \epsilon_{11,23} - \epsilon_{23,11},$$
$$\eta_{31} = \epsilon_{32,21} + \epsilon_{12,23} - \epsilon_{22,31} - \epsilon_{31,22}.$$

In summary the six strain-displacement relations in $\boldsymbol{\epsilon} = \text{grad}^{\text{sym}}\, \boldsymbol{u}$ may be integrated into the three displacements in $\boldsymbol{u}$ subject to the six compatibility conditions in $\boldsymbol{\eta} = \text{inc}\, \boldsymbol{\epsilon} \doteq \mathbf{0}$ that itself are constrained by the three solenoidal conditions in $\text{div}\boldsymbol{\eta} = \mathbf{0}$. Thus in the case of compatibility there are enough equations (6-[6-3]=3) to determine the three unknown displacement coefficients from a given strain field. □

Looking Backward: Compatibility of Material Configuration. Next the situation is reversed, thus the compatibility condition for the material configuration in terms of the spatial (strain) metric results from the following reasoning: For the material configuration of elasticity to be compatible distance determination and parallel transport ought to be of Euclidean type, see Fig. 7.5, thus Cartesian coordinates may be selected with the material metric coefficients agreeing with the Kronecker-symbol and a vanishing material connection

$$M_{AB} = \delta_{AB} \quad \text{and} \quad L^{A}{}_{BC} = 0. \tag{7.10}$$

Then the convection of the material metric and the material connection by the (inverse) deformation map renders the *Finger tensor* and an integrable connection in the spatial configuration

$$m_{ij} \equiv c_{ij} := f^{A}{}_{i}\delta_{AB}f^{B}{}_{j} \quad \text{and} \quad \ell^{i}{}_{jk} = F^{i}{}_{A}f^{A}{}_{j,k}. \tag{7.11}$$

Moreover, it may be shown easily that the fully covariant spatial connection $\ell_{ijk} = c_{im}\ell^{m}{}_{jk}$ is metric with respect to c_{ij}, i.e.

$$c_{ij,k} = f^{A}{}_{i,k}\delta_{AB}f^{B}{}_{j} + f^{A}{}_{i}I_{AB}f^{B}{}_{j,k} = 2\ell_{(ij)k}. \tag{7.12}$$

As a consequence, and due to the integrability of $f^{A}{}_{i}$ with $f^{A}{}_{[i,j]} = 0$ that results in a torsionless $\ell^{i}{}_{jk} = \ell^{i}{}_{(jk)}$ (refer again to Eq. 4.9), the spatial connection ℓ_{ijk} coincides with the Riemann connection

$$m_{ijk}(c) := \frac{1}{2}[c_{ij,k} - c_{jk,i} + c_{ki,j}] \tag{7.13}$$

based on the metric c_{ij} while the corresponding spatial contortion vanishes identically:

$$
\begin{array}{rl}
\mathfrak{c}_{ij,k} = & + f^A{}_j\ \delta_{AB}\ f^B{}_{i,k} + f^A{}_i\ \delta_{AB}\ f^B{}_{j,k} \\
-\mathfrak{c}_{jk,i} = & - f^A{}_k\ \delta_{AB}\ f^B{}_{j,i} - f^A{}_j\ \delta_{AB}\ f^B{}_{k,i} \\
\mathfrak{c}_{ki,j} = & + f^A{}_i\ \delta_{AB}\ f^B{}_{k,j} + f^A{}_k\ \delta_{AB}\ f^B{}_{i,j} \\
\hline\hline
\end{array}
$$

$$\rightarrow \quad \mathfrak{m}_{ijk}(\mathfrak{c}) = f^A{}_i \delta_{AB} f^B{}_{(j,k)} \quad \text{and} \quad \mathfrak{k}_{ijk} \equiv 0. \tag{7.14}$$

Moreover, since the spatial connection is integrable, its associated spatial Riemann curvature tensor vanishes identically

$$\mathfrak{m}_{ijkl}(\mathfrak{c}) := 2\mathfrak{m}_{ij[l,k]} - 2\mathfrak{m}_{mi[k}\mathfrak{m}^m{}_{jl]} \equiv 0. \tag{7.15}$$

Clearly, this finding coincides with the Euclidean character of the material configuration

$$0 \equiv M_{ABCD}(I) = M_{ABCD}(Y(\mathfrak{c})) = Y(\mathfrak{m}_{ijkl}(\mathfrak{c})). \tag{7.16}$$

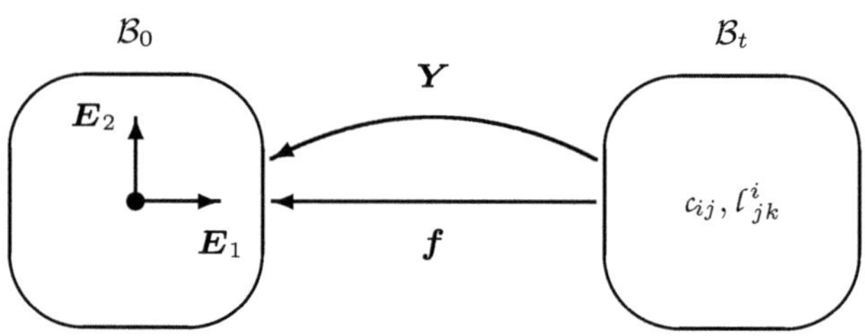

Fig. 7.5 In the material configuration of elasticity distance determination and parallel transport are of Euclidean type. The convection of the metric and the connection renders the Finger tensor and an integrable connection in the spatial configuration.

The previous reasoning may reversed as follows, see Fig. 7.6: for the case that the spatial Riemann curvature tensor $\mathfrak{m}_{ijkl}$ based on the Riemann connection $\mathfrak{m}_{ijk}$ vanishes it may be concluded that the Riemann connection coincides with an integrable connection $\mathfrak{m}_{ijk} \equiv f^A{}_i \delta_{AB} f^B{}_{j,k} = \mathfrak{l}_{ijk}$. Thereby the Riemann connection $\mathfrak{m}_{ijk}$ and thus also the integrable connection $\mathfrak{l}_{ijk}$ is metric with respect to the spatial metric $\mathfrak{m}_{ij} = f^A{}_i \delta_{AB} f^B{}_j$. However, since every metric connection allows representation as $\mathfrak{l}_{ijk} = \mathfrak{m}_{ijk} + \mathfrak{k}_{ijk}$ it may be concluded that the corresponding contortion $\mathfrak{k}_{ijk}$ and thus the torsion $\mathfrak{t}_{ijk} = 2\mathfrak{k}_{i[jk]}$ vanishes identically. Therefore, also in agreement with the right symmetry of the Riemann connection $\mathfrak{m}_{ijk} = \mathfrak{m}_{i(jk)}$ the integrable connection is also right symmetric $\mathfrak{l}_{ijk} = f^A{}_i \delta_{AB} f^B{}_{(j,k)}$. Finally $f^B{}_{j,k} = f^B{}_{(j,k)}$ implies the integrability of the (inverse) distortion $f^B{}_j = Y^B{}_{,j}$ into a compatible (inverse) deformation map $Y^B = Y^B(\mathfrak{x}^j)$. Thus in the spatial configuration a vanishing Riemann curvature tensor based on a Riemann connection assures integrability of the

Vanishing Riemann Curvature

$$m_{ijkl} = 2m_{ij[l,k]} - 2m_{mi[k}m^{m}{}_{jl]} \equiv 0$$

$\Downarrow$

Integrable Metric Connection

$$m_{ijk} = m_{i(jk)} \equiv f^{A}{}_{i} I_{AB} f^{B}{}_{j,k}$$

$\Downarrow$

Vanishing Torsion

$$f^{B}{}_{j,k} = f^{B}{}_{(j,k)}$$

$\Downarrow$

Integrable (Inverse) Distortion

$$f^{B}{}_{j} = Y^{B}{}_{,j}$$

Fig. 7.6 Sketch of the reasoning to deduce the relation between a vanishing spatial Riemann curvature and the integrability of the spatial metric into a compatible (inverse) deformation map

spatial metric into a compatible (inverse) deformation map, i.e a compatible material configuration.

In summary the kinematics of elasticity may alternatively be considered a flat Riemann geometry (i.e. an Euclidean geometry) over the spatial configuration $\mathcal{B}_t$. To conclude flatness assures integrability, i.e. the Finger tensor may be derived from a vector valued (inverse) deformation map. This is again the nonlinear version of the St. Venant compatibility conditions.

Remark 7.2. As outlined in [181] a curved membrane defines a two dimensional Riemannian geometry with a non-vanishing curvature tensor that is here represented by the Gauss curvature. The initially unstrained membrane is then deformed into a two dimensional Euclidean space, i.e. into a plane. Let this plane define the spatial configuration $\mathcal{B}_t$. Obviously the applied strains $\boldsymbol{c}$ can only relax into a *incompatible*, i.e. disconnected configuration $\mathcal{B}_0$ within the planar Euclidean geometry. Thus the configuration $\mathcal{B}_t$ that is strained in the *flat* Euclidean geometry can only relax into a *compatible* configuration within a *curved* Riemannian geometry. □

7.2 Second-Order Continuum

The kinematics of elasticity in a second-order continuum are essentially characterized by the distortion and the double-distortion. The importance of the distortion and the double-distortion is especially clear from the fact that they represent the arguments of the elastically stored energy density function as the key ingredients of a generic master constitutive relation. Thus for a given distortion and a given double-distortion the conditions for their integrability into a compatible deformation map are of utmost interest. Again, these conditions have been discussed already, nevertheless without explicitly referring to second-order elasticity, in the previous chapter and thus need not be re-derived in detail. However, due to the requirements of *objectivity* and *material symmetry* it is rather the corresponding *(strain) metric*, being quadratic in the distortion, and the *connection* that enter the elastically stored energy density function as its argument. Therefore, as an alternative to the conditions for the distortion and the double-distortion being integrable, the more involved conditions for the (strain) metric and the connection being integrable into a compatible deformation map need to be established in addition.

7.2.1 Integrable Distortion and Double-Distortion

For second-order continuum elasticity integrability of the distortion *and* the double-distortion renders the primary conditions for the spatial and the material configuration being compatible.

Looking Forward: Compatibility of Spatial Configuration. For a spatial configuration of second-order elasticity to be compatible, i.e. i) the integration of the spatial line element $\mathrm{d}\boldsymbol{y}$ along arbitrary closed spatial circuits $\mathcal{C}_t$ does never result in a spatial closure gap $[\![\boldsymbol{y}]\!] \neq \boldsymbol{0}$, ii) the integration of the (forward) differential distortion along arbitrary closed spatial circuits $\mathcal{C}_t$ does never result in a distortion closure gap $[\![\boldsymbol{F}]\!] \neq \boldsymbol{0}$, and iii) the double-distortion $\boldsymbol{G}$ is compatible with the distortion $\boldsymbol{F}$; i) the distortion $\boldsymbol{F}$ ought to be integrable, ii) the double-distortion $\boldsymbol{G}$ ought to be integrable, and iii.a) the double-distortion $\boldsymbol{G}$ ought to be right symmetric together with iii.b) the material non-metricity ought to vanish.
For simply-connected bodies conditions i) and ii) are expressed as

$$\operatorname{Curl}\boldsymbol{F} \doteq \boldsymbol{0} \quad \text{and} \quad \operatorname{Curl}\boldsymbol{G} \doteq \boldsymbol{0}. \tag{7.17}$$

Recall that these conditions take equivalent expressions $F^a{}_{[B,C]} \doteq 0$ and $G^a{}_{B[C,D]} \doteq 0$ or $\mathsf{F}^i{}_{[J:K]} \doteq 0$ and $\mathsf{G}^i{}_{J[K:L]} \doteq 0$ in rectilinear (Cartesian) or curvilinear coordinates, respectively.
Moreover conditions iii.a) and iii.b) are stated as

$$\boldsymbol{G} : \boldsymbol{E} \doteq \boldsymbol{0} \quad \text{and} \quad \mathsf{I}^{\mathrm{sym}} : \big[\boldsymbol{F}^t \cdot [\boldsymbol{G} - \operatorname{Grad}\boldsymbol{F}]\big] \doteq \boldsymbol{0}. \tag{7.18}$$

Note that these conditions take equivalent expressions $G^a{}_{[BC]} \doteq 0$ and $F^a{}_{(A}\delta_{ab}[G^b{}_{B)C} - F^b{}_{B),C}] \doteq 0$ or $\mathsf{G}^i{}_{[JK]} \doteq 0$ and $\mathsf{F}^i{}_{(I}\delta_{ij}[\mathsf{G}^j{}_{J)K} - \mathsf{F}^j{}_{J):K}] \doteq 0$ in rectilinear (Cartesian) or curvilinear coordinates, respectively.

Looking Backward: Compatibility of Material Configuration. Likewise, for a material configuration of second-order elasticity to be compatible,

i.e. i) the integration of the material line element $\mathrm{d}\boldsymbol{Y}$ along arbitrary closed material circuits $\mathcal{C}_0$ does never result in a material closure gap $[\![\boldsymbol{Y}]\!] \neq \mathbf{0}$, ii) the integration of the (backward) differential distortion along arbitrary closed material circuits $\mathcal{C}_0$ does never result in a distortion closure gap $[\![\boldsymbol{f}]\!] \neq \mathbf{0}$, and iii) the double-distortion $\boldsymbol{g}$ is compatible with the distortion $\boldsymbol{f}$; i) the distortion $\boldsymbol{f}$ ought to be integrable, ii) the double-distortion $\boldsymbol{g}$ ought to be integrable, and iii.a) the double-distortion $\boldsymbol{g}$ ought to be right symmetric together with iii.b) the spatial non-metricity ought to vanish.
For simply-connected bodies conditions i) and ii) are expressed as

$$\operatorname{curl} \boldsymbol{f} \doteq \mathbf{0} \quad \text{and} \quad \operatorname{curl} \boldsymbol{g} \doteq \mathbf{0}. \tag{7.19}$$

Recall that these conditions take equivalent expressions $f^A{}_{[b,c]} \doteq 0$ and $g^A{}_{b[c,d]} \doteq 0$ or $\mathsf{f}^I{}_{[j:k]} \doteq 0$ and $\mathsf{g}^I{}_{j[k:l]} \doteq 0$ in rectilinear (Cartesian) or curvilinear coordinates, respectively.
Moreover conditions iii.a) and iii.b) are stated as

$$\boldsymbol{g} : \boldsymbol{e} \doteq \mathbf{0} \quad \text{and} \quad \mathsf{i}^{\text{sym}} : \left[\boldsymbol{f}^t \cdot [\boldsymbol{g} - \operatorname{grad} \boldsymbol{f}]\right] \doteq \mathbf{0}. \tag{7.20}$$

Note that these conditions take equivalent expressions $g^A{}_{[bc]} \doteq 0$ and $f^A{}_{(a}\delta_{AB}[g^B{}_{b)c} - f^B{}_{b),c}] \doteq 0$ or $\mathsf{g}^I{}_{[jk]} \doteq 0$ and $\mathsf{f}^I{}_{(i}\delta_{IJ}[\mathsf{g}^J{}_{j)k} - \mathsf{f}^J{}_{j):k}] \doteq 0$ in rectilinear (Cartesian) or curvilinear coordinates, respectively.

7.2.2 Integrable Metric and Connection

For a number of reasons, see the supplement on second-order elasticity below, it is often not primarily the distortion and the double-distortion but rather the (strain) metric, being quadratic in the distortion, and the (double-strain) double-metric, viz. the connection that enter the constitutive relations of second-order elasticity. Thus the conditions on the (strain) metric and the (double-strain) double-metric, viz. the connection being integrable into a compatible deformation map need to be considered. The derivation of these conditions are a bit more involved and result in an extended version of the classical St-Venant compatibility conditions.

Looking Forward: Compatibility of Spatial Configuration. The condition i) for the material metric $\mathcal{M}_{IJ} \equiv \mathcal{C}_{IJ}$ to be compatible with a deformation map ${}^{\mathcal{M}}y^a = {}^{\mathcal{M}}y^a(X^J)$ has already been discussed in terms of the corresponding material Riemann curvature tensor $\mathcal{M}_{IJKL}$, see the sketch of the reasoning in Fig. 7.4. It thus remains to state the conditions for ii) the material connection $\mathcal{L}^I{}_{JK}$ to be integrable into a distortion $F^a{}_J$, iii) this distortion to be integrable into a deformation map ${}^{\mathcal{L}}y^a = {}^{\mathcal{L}}y^a(X^J)$, and iv) the material metric $\mathcal{M}_{IJ}$ to be compatible with the material connection $\mathcal{L}^I{}_{JK}$. Thereby condition iv) ensures that the deformation map ${}^{\mathcal{M}}y^a$ compatible with the material metric and the deformation map ${}^{\mathcal{L}}y^a$ compatible with the

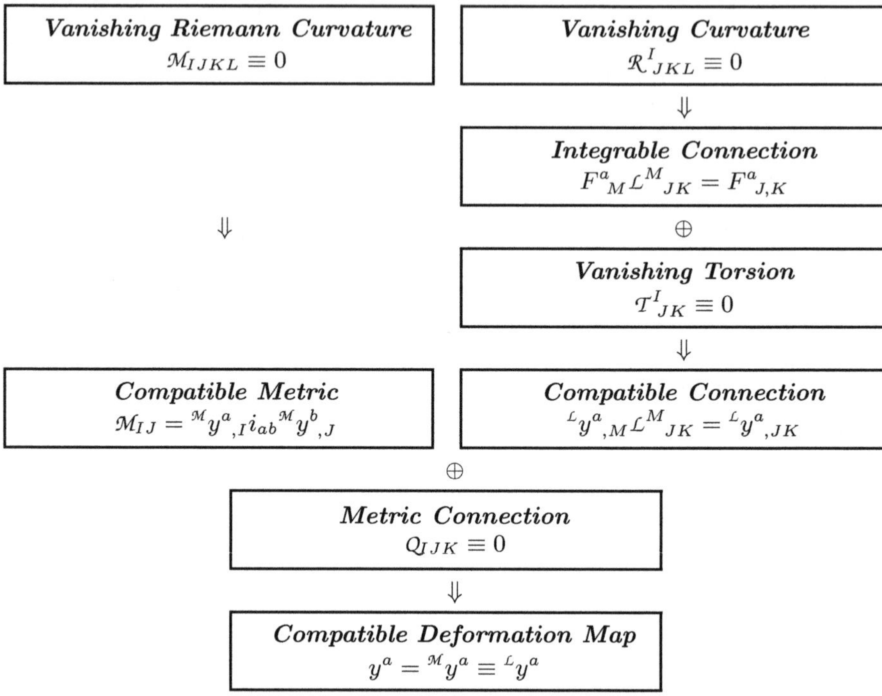

Fig. 7.7 Sketch of the reasoning to deduce the relation between vanishing i) material Riemann curvature, ii) material curvature, iii) material torsion, and iv) material non-metricity and the integrability of the material metric and the material connection into a compatible deformation map

material connection do indeed coincide.

Condition i): The condition for the material metric $\mathcal{M}_{IJ} \equiv C_{IJ}$ to be compatible with a deformation map ${}^{\mathcal{M}}y^a = {}^{\mathcal{M}}y^a(X^J)$ reads in terms of the corresponding Riemann curvature tensor $\mathcal{M}_{IJKL}$ expressed by the Riemann connection $\mathcal{M}_{IJK}$ associated with the material metric

$$\mathcal{M}_{IJKL} = 2\mathcal{M}_{IJ[L,K]} - 2\mathcal{M}_{MI[K}\mathcal{M}^M{}_{JL]} \doteq 0. \tag{7.21}$$

Condition ii): The condition for the material connection $\mathcal{L}^I{}_{JK}$ to be integrable into a distortion $F^a{}_J$ so that $\mathcal{L}^I{}_{JK} = f^I{}_a F^a{}_{J,K}$ is stated in terms of the material curvature tensor $\mathcal{R}^I{}_{JKL}$ associated with the material connection

$$\mathcal{R}^I{}_{JKL} = 2\mathcal{L}^I{}_{J[L,K]} + 2\mathcal{L}^I{}_{M[K}\mathcal{L}^M{}_{JL]} \doteq 0. \tag{7.22}$$

Condition iii): The condition for the distortion $F^a{}_J$ as consistent with condition ii) to be integrable into a deformation map ${}^{\mathcal{L}}y^a = {}^{\mathcal{L}}y^a(X^J)$ is stated in terms of the material torsion tensor $\mathcal{T}^I{}_{JK}$ associated with the material connection

$$\mathcal{T}^I{}_{JK} = 2\mathcal{L}^I{}_{[JK]} \overset{\text{ii)}}{=} 2f^I{}_a F^a{}_{[J,K]} \doteq 0. \tag{7.23}$$

Condition iv): The condition for the material metric $\mathcal{M}_{IJ}$ to be compatible with the material connection $\mathcal{L}^I{}_{JK}$ is stated in terms of the material non-metricity tensor $\mathcal{Q}_{IJK}$ associated with the material connection and the material metric

$$\mathcal{Q}_{IJK} = 2\mathcal{M}_{(IM}\mathcal{L}^M{}_{J)K} - \mathcal{M}_{IJ,K} \overset{\text{ii)}}{=} 2\mathcal{M}_{(IM}f^M{}_a F^a{}_{J),K} - \mathcal{M}_{IJ,K} \doteq 0. \tag{7.24}$$

A sketch of the above reasoning is displayed in Fig. 7.7.

Looking Backward: Compatibility of Material Configuration. The condition i) for the spatial metric $m_{ij} \equiv c_{ij}$ to be compatible with an inverse deformation map ${}^m Y^A = {}^m Y^A(\chi^j)$ has already been discussed in terms of the corresponding spatial Riemann curvature tensor m_{ijkl}, see the sketch of the reasoning in Fig. 7.6. It thus remains to state the conditions for ii) the spatial connection $\ell^i{}_{jk}$ to be integrable into an inverse distortion $f^A{}_j$, iii) this distortion to be integrable into an inverse deformation map ${}^\ell Y^A = {}^\ell Y^A(\chi^j)$, and iv) the spatial metric m_{ij} to be compatible with the spatial connection $\ell^i{}_{jk}$. Thereby condition iv) ensures that the inverse deformation map ${}^m Y^A$ compatible with the spatial metric and the inverse deformation map ${}^\ell Y^A$ compatible with the spatial connection do indeed coincide.

Condition i): The condition for the spatial metric $m_{ij} \equiv c_{ij}$ to be compatible with an inverse deformation map ${}^m Y^A = {}^m Y^A(\chi^j)$ reads in terms of the corresponding Riemann curvature tensor m_{ijkl} expressed by the Riemann connection m_{ijk} associated with the material metric

$$m_{ijkl} = 2m_{ij[l,k]} - 2m_{mi[k}m^m{}_{jl]} \doteq 0. \tag{7.25}$$

Condition ii): The condition for the spatial connection $\ell^i{}_{jk}$ to be integrable into an inverse distortion $f^A{}_j$ so that $\ell^i{}_{jk} = F^i{}_A f^A{}_{j,k}$ is stated in terms of the spatial curvature tensor $r^i{}_{jkl}$ associated with the spatial connection

$$r^i{}_{jkl} = 2\ell^i{}_{j[l,k]} + 2\ell^i{}_{m[k}\ell^m{}_{jl]} \doteq 0. \tag{7.26}$$

Condition iii): The condition for the inverse distortion $f^A{}_j$ as consistent with condition ii) to be integrable into an inverse deformation map ${}^\ell Y^A = {}^\ell Y^A(\chi^j)$ is stated in terms of the spatial torsion tensor $t^i{}_{jk}$ associated with the spatial connection

$$t^i{}_{jk} = 2\ell^i{}_{[jk]} \overset{\text{ii)}}{=} 2F^i{}_A f^A{}_{[j,k]} \doteq 0. \tag{7.27}$$

Condition iv): The condition for the spatial metric m_{ij} to be compatible with the spatial connection ℓ^i_{jk} is stated in terms of the spatial non-metricity tensor q_{ijk} associated with the spatial connection and the spatial metric

$$q_{ijk} = 2m_{(im}\ell^m_{\ j)k} - m_{ij,k} \overset{\text{ii)}}{=} 2m_{(im}F^m_{\ A}f^A_{\ j),k} - m_{ij,k} \doteq 0. \tag{7.28}$$

A sketch of the above reasoning is displayed in Fig. 7.8.

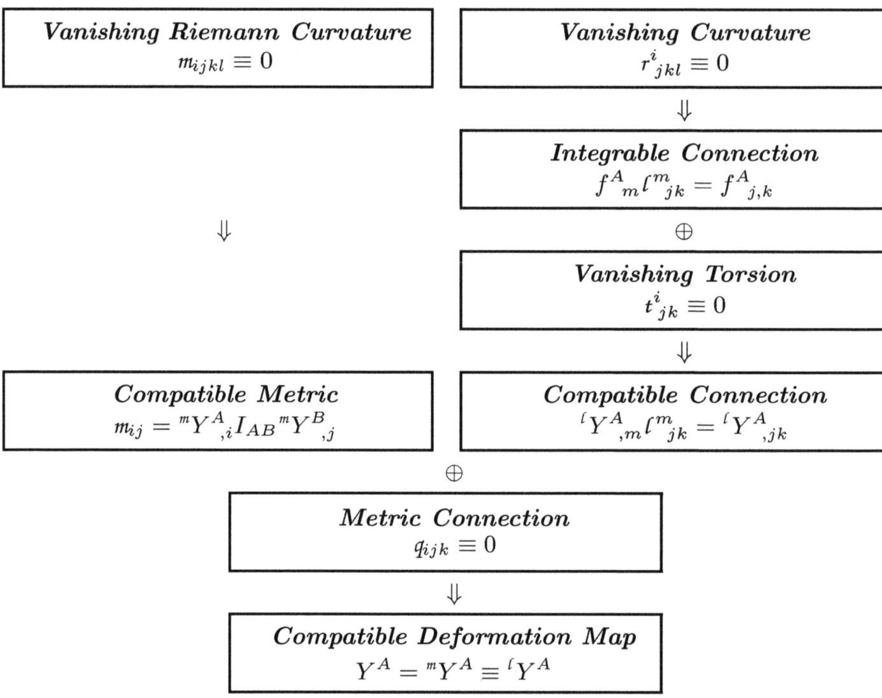

Fig. 7.8 Sketch of the reasoning to deduce the relation between vanishing i) spatial Riemann curvature, ii) spatial curvature, iii) spatial torsion, and iv) spatial non-metricity and the integrability of the spatial metric and the spatial connection into a compatible (inverse) deformation map

Supplement 7.1. First-Order Elasticity in Euclidean Space
Elasticity is perhaps the best studied and analyzed subject of first-order continuum mechanics. Excellent treatises devoted to various aspects of the formulation and mathematical analysis of nonlinear elasticity are e.g. the monographs [67, 153, 122, 34, 9, 83]. General classical accounts on nonlinear continuum mechanics are due to [120, 68, 200, 29, 178], more recent treatises on continuum mechanics modeling are e.g. represented by [173, 77, 72, 194, 148]. In the sequel a brief account, that is by no way comprehensive, on the essential ingredients of nonlinear continuum elasticity in Euclidean space shall be given. These are grouped into i) the kinematics of finite deformations, ii) the mechanical and thermodynamical balance laws, and iii) the constitutive relations. Thereby, many aspects of the formulation are generic and are thus also valid for inelastic continuum mechanics.

Deformation Gradient:
The first gradient of the nonlinear deformation map $\boldsymbol{x} = \boldsymbol{y}(\boldsymbol{X})$ with respect to the material coordinates $\boldsymbol{X}$ is denoted the *deformation gradient* $\boldsymbol{F} := \operatorname{Grad} \boldsymbol{y}$. It is a second-order non-symmetric two-point (spatial/material) tensor and may be computed from the directional derivative $\mathrm{d}\boldsymbol{x} := \mathrm{d}_\epsilon \boldsymbol{y}(\boldsymbol{X} + \epsilon\, \mathrm{d}\boldsymbol{X})|_{\epsilon=0}$. Thus, since $\boldsymbol{F}$ maps elements $\mathrm{d}\boldsymbol{X}$ of the material tangent space $T\mathcal{B}_0$ linearly into elements $\mathrm{d}\boldsymbol{x}$ of the spatial tangent space $T\mathcal{B}_t$ it is also denoted the *linear tangent map*. The terminology *distortion* includes also the case of incompatible $\boldsymbol{F}$, i.e. the generic case that $\boldsymbol{F}$ does not derive as the gradient of a vector field $\boldsymbol{y}(\boldsymbol{X})$. The determinant (Jacobian) $J := \det \boldsymbol{F}$, the cofactor $\operatorname{cof} \boldsymbol{F} := J\boldsymbol{F}^{-t}$, and the deformation gradient $\boldsymbol{F}$ map material volume elements $\mathrm{d}V$, area elements $\mathrm{d}\boldsymbol{A}$ (Nanson formula), and line elements $\mathrm{d}\boldsymbol{X}$ into their corresponding spatial counterparts

$$\mathrm{d}v = J\,\mathrm{d}V \quad \text{and} \quad \mathrm{d}\boldsymbol{a} = \operatorname{cof} \boldsymbol{F} \cdot \mathrm{d}\boldsymbol{A} \quad \text{and} \quad \mathrm{d}\boldsymbol{x} = \boldsymbol{F} \cdot \mathrm{d}\boldsymbol{X}.$$

Obviously, in order for the inverse deformation map $\boldsymbol{X} = \boldsymbol{Y}(\boldsymbol{x})$ to exist and in order to avoid self penetration the Jacobian J has to be strictly positive, i.e. $J \in (0, \infty)$. For the ease of notation the inverse of the deformation gradient shall be denoted as $\boldsymbol{f} := \boldsymbol{F}^{-1}$ in the sequel. Observe that the gradients of a convected level set function $S(\boldsymbol{X}) = s(\boldsymbol{x}) \circ \boldsymbol{y}(\boldsymbol{X})$ with respect to material and spatial coordinates are related by covariant transformation as $\mathbf{d}S = \boldsymbol{F}^t \cdot \mathbf{d}s$ with gradients defined as $\mathbf{d}S := \operatorname{Grad} S$ and $\mathbf{d}s := \operatorname{grad} s$ (likewise material and spatial line elements are related by contravariant transformation). Finally the polar decomposition theorem allows representation of the deformation gradient as

$$\boldsymbol{F} = \boldsymbol{R} \cdot \boldsymbol{U} = \boldsymbol{v} \cdot \boldsymbol{R}.$$

Here the continuum rotation $\boldsymbol{R}$ is a proper orthogonal tensor with $\boldsymbol{R} = \boldsymbol{R}^{-t}$ and $\det \boldsymbol{R} = +1$, and the right and left stretch tensors $\boldsymbol{U}$ and $\boldsymbol{v}$ are symmetric and positive definite with $\boldsymbol{U} = \boldsymbol{U}^t > 0$ and $\boldsymbol{v} = \boldsymbol{v}^t > 0$ (here $\boldsymbol{U} > 0$ and $\boldsymbol{v} > 0$ is a sloppy notation to indicate that all eigenvalues of $\boldsymbol{U}$ and $\boldsymbol{v}$ are positive).

Covariant Strain Measures:
In order to introduce *covariant* strain measures the (quadratic) length of the spatial line element $\mathrm{d}\boldsymbol{x}$ (a vector) is computed from

$$|\mathrm{d}\boldsymbol{x}|^2 = \mathrm{d}\boldsymbol{x} \cdot \boldsymbol{j} \cdot \mathrm{d}\boldsymbol{x} = \mathrm{d}\boldsymbol{X} \cdot \boldsymbol{F}^t \cdot \boldsymbol{F} \cdot \mathrm{d}\boldsymbol{X} =: \mathrm{d}\boldsymbol{X} \cdot \boldsymbol{C} \cdot \mathrm{d}\boldsymbol{X} > 0 \quad \forall \mathrm{d}\boldsymbol{X} \neq \boldsymbol{0}.$$

Here the spatial description covariant unit tensor $\boldsymbol{j}$ and the material description, symmetric and positive definite covariant *right Cauchy-Green* strain tensor $\boldsymbol{C} := \boldsymbol{F}^t \cdot \boldsymbol{F}$ have been introduced. Likewise, the (quadratic) length of the material line element $\mathrm{d}\boldsymbol{X}$ follows as

$$|\mathrm{d}\boldsymbol{X}|^2 = \mathrm{d}\boldsymbol{X} \cdot \boldsymbol{J} \cdot \mathrm{d}\boldsymbol{X} = \mathrm{d}\boldsymbol{x} \cdot \boldsymbol{f}^t \cdot \boldsymbol{f} \cdot \mathrm{d}\boldsymbol{x} =: \mathrm{d}\boldsymbol{x} \cdot \boldsymbol{c} \cdot \mathrm{d}\boldsymbol{x} > 0 \quad \forall \mathrm{d}\boldsymbol{x} \neq \boldsymbol{0}.$$

Here the material description covariant unit tensor $\boldsymbol{J}$ and the spatial description, symmetric and positive definite covariant *Finger* strain tensor $\boldsymbol{c} := \boldsymbol{f}^t \cdot \boldsymbol{f}$ have been introduced. On the one hand the comparison of the (quadratic) lengths of the spatial and material line elements may be thus expressed as

$$|\mathrm{d}\boldsymbol{x}|^2 - |\mathrm{d}\boldsymbol{X}|^2 = \mathrm{d}\boldsymbol{X} \cdot [\boldsymbol{C} - \boldsymbol{J}] \cdot \mathrm{d}\boldsymbol{X} =: 2\,\mathrm{d}\boldsymbol{X} \cdot \boldsymbol{E} \cdot \mathrm{d}\boldsymbol{X}.$$

Here the material description, symmetric and indefinite covariant *Green-Lagrange* strain tensor $2\boldsymbol{E} := \boldsymbol{C} - \boldsymbol{J}$ has been introduced. On the other hand the comparison of the (quadratic) lengths of the spatial and material line elements may be stated as

$$|\mathrm{d}\boldsymbol{x}|^2 - |\mathrm{d}\boldsymbol{X}|^2 = \mathrm{d}\boldsymbol{x} \cdot [\boldsymbol{j} - \boldsymbol{c}] \cdot \mathrm{d}\boldsymbol{x} =: 2\,\mathrm{d}\boldsymbol{x} \cdot \boldsymbol{e} \cdot \mathrm{d}\boldsymbol{x}.$$

Here the spatial description, symmetric and indefinite covariant *Almansi* strain tensor $2\boldsymbol{e} := \boldsymbol{j} - \boldsymbol{c}$ has been introduced. Finally the above material and spatial description covariant strain measures and unit tensors are related by

$$\boldsymbol{E} = \boldsymbol{F}^t \cdot \boldsymbol{e} \cdot \boldsymbol{F} \quad \text{and} \quad \boldsymbol{C} = \boldsymbol{F}^t \cdot \boldsymbol{j} \cdot \boldsymbol{F} \quad \text{and} \quad \boldsymbol{J} = \boldsymbol{F}^t \cdot \boldsymbol{c} \cdot \boldsymbol{F}.$$

The covariant transformation behavior qualifies the above strain measures and unit tensors as covariant tensors.

Contravariant Strain Measures:
In order to introduce *contravariant* strain measures the (quadratic) length of the material level set normal $\mathbf{d}S$ (a one-form) is computed from

$$|\mathbf{d}S|^2 = \mathbf{d}S \cdot \boldsymbol{I} \cdot \mathbf{d}S = \mathbf{d}s \cdot \boldsymbol{F} \cdot \boldsymbol{F}^t \cdot \mathbf{d}s =: \mathbf{d}s \cdot \boldsymbol{b} \cdot \mathbf{d}s > 0 \quad \forall \mathbf{d}s \neq \boldsymbol{0}.$$

Here the material description contravariant unit tensor $\boldsymbol{I}$ and the spatial description, symmetric and positive definite contravariant *left Cauchy-Green* strain tensor $\boldsymbol{b} := \boldsymbol{F} \cdot \boldsymbol{F}^t$ have been introduced. Likewise, the (quadratic) length of the spatial level set normal $\mathbf{d}s$ follows as

$$|\mathrm{d}\boldsymbol{s}|^2 = \mathrm{d}\boldsymbol{s} \cdot \boldsymbol{i} \cdot \mathrm{d}\boldsymbol{s} = \mathrm{d}\boldsymbol{S} \cdot \boldsymbol{f} \cdot \boldsymbol{f}^t \cdot \mathrm{d}\boldsymbol{S} =: \mathrm{d}\boldsymbol{S} \cdot \boldsymbol{B} \cdot \mathrm{d}\boldsymbol{S} > 0 \quad \forall \mathrm{d}\boldsymbol{S} \neq \boldsymbol{0}.$$

Here the spatial description contravariant unit tensor $\boldsymbol{i}$ and the material description, symmetric and positive definite contravariant *Piola* strain tensor $\boldsymbol{B} := \boldsymbol{f} \cdot \boldsymbol{f}^t$ have been introduced. On the one hand the comparison of the (quadratic) lengths of the material and spatial level set normals may thus be expressed as

$$|\mathrm{d}\boldsymbol{S}|^2 - |\mathrm{d}\boldsymbol{s}|^2 = \mathrm{d}\boldsymbol{s} \cdot [\boldsymbol{b} - \boldsymbol{i}] \cdot \mathrm{d}\boldsymbol{s} =: 2\mathrm{d}\boldsymbol{s} \cdot \boldsymbol{k} \cdot \mathrm{d}\boldsymbol{s}.$$

Here the spatial description, symmetric and indefinite contravariant *Karni-Reiner* strain tensor $2\boldsymbol{k} := \boldsymbol{b} - \boldsymbol{i}$ has been introduced. On the other hand the comparison of the (quadratic) lengths of the material and spatial level set normals may be stated as

$$|\mathrm{d}\boldsymbol{S}|^2 - |\mathrm{d}\boldsymbol{s}|^2 = \mathrm{d}\boldsymbol{S} \cdot [\boldsymbol{I} - \boldsymbol{B}] \cdot \mathrm{d}\boldsymbol{S} =: 2\mathrm{d}\boldsymbol{S} \cdot \boldsymbol{K} \cdot \mathrm{d}\boldsymbol{S}.$$

Here the material description, symmetric and indefinite covariant *Karni-Reiner* strain tensor $2\boldsymbol{K} := \boldsymbol{I} - \boldsymbol{B}$ has been introduced. Finally the above spatial and material description contravariant strain measures and unit tensors are related by

$$\boldsymbol{k} = \boldsymbol{F} \cdot \boldsymbol{K} \cdot \boldsymbol{F}^t \quad \text{and} \quad \boldsymbol{b} = \boldsymbol{F} \cdot \boldsymbol{I} \cdot \boldsymbol{F}^t \quad \text{and} \quad \boldsymbol{i} = \boldsymbol{F} \cdot \boldsymbol{B} \cdot \boldsymbol{F}^t.$$

The contravariant transformation behavior qualifies the above strain measures and unit tensors as contravariant tensors.

Spectral Representation of Strain Measures:

On the one hand the symmetric and positive definite right and left Cauchy-Green strain tensors $\boldsymbol{C}$ and $\boldsymbol{b}$ allow spectral representation in terms of the real and positive *principal stretches* λ_a and the corresponding Lagrangian and Eulerian dyads $\boldsymbol{N}_a \otimes \boldsymbol{N}_a$ and $\boldsymbol{n}_a \otimes \boldsymbol{n}_a$, respectively, of the *principal directions*

$$\boldsymbol{C} = \sum_{a=1}^{3} [\lambda_a]^2 \boldsymbol{N}_a \otimes \boldsymbol{N}_a \quad \text{and} \quad \boldsymbol{b} = \sum_{a=1}^{3} [\lambda_a]^2 \boldsymbol{n}_a \otimes \boldsymbol{n}_a.$$

The Lagrangian and Eulerian principal directions $\boldsymbol{N}_a$ and $\boldsymbol{n}_a$ (the eigenvectors) satisfy the orthonormality conditions

$$\boldsymbol{N}_a \cdot \boldsymbol{N}_b = \delta_{ab} \quad \text{and} \quad \boldsymbol{n}_a \cdot \boldsymbol{n}_b = \delta_{ab} \quad \text{with} \quad a, b = 1, 2, 3.$$

Consequently the contra- and covariant, material and spatial unit tensors $\boldsymbol{I}, \boldsymbol{J}$ and $\boldsymbol{i}, \boldsymbol{j}$, respectively, allow spectral representation

$$\boldsymbol{I} = \boldsymbol{J} = \sum_{a=1}^{3} \boldsymbol{N}_a \otimes \boldsymbol{N}_a \quad \text{and} \quad \boldsymbol{i} = \boldsymbol{j} = \sum_{a=1}^{3} \boldsymbol{n}_a \otimes \boldsymbol{n}_a.$$

The square of the principal stretches $[\lambda_a]^2$ (the eigenvalues) follow from the characteristic equation

$$[\lambda_a]^6 - I_1[\lambda_a]^4 + I_2[\lambda_a]^2 - I_3 = 0$$

in terms of the three *principal invariants*

$$I_1 = \boldsymbol{C} : \boldsymbol{I} = \boldsymbol{j} : \boldsymbol{b} \quad \text{and} \quad I_2 = \operatorname{cof} \boldsymbol{C} : \boldsymbol{J} = \boldsymbol{i} : \operatorname{cof} \boldsymbol{b} \quad \text{and} \quad I_3 = \det \boldsymbol{C} = \det \boldsymbol{b}.$$

Due to the spectral representation the principal invariants are alternatively expressible in terms of the *spectral invariants* $[\lambda_a]^2$ as

$$\begin{aligned} I_1 &= [\lambda_1]^2 + [\lambda_2]^2 + [\lambda_3]^2 \\ I_2 &= [\lambda_1]^2[\lambda_2]^2 + [\lambda_2]^2[\lambda_3]^2 + [\lambda_3]^2[\lambda_1]^2 \\ I_3 &= [\lambda_1]^2[\lambda_2]^2[\lambda_3]^2. \end{aligned}$$

Since the right and left Cauchy-Green strain tensors possess identical principal invariants I_a they also share the same eigenvalues $[\lambda_a]^2$, however the corresponding eigenvectors satisfying $\boldsymbol{C}\cdot\boldsymbol{N}_a = [\lambda_a]^2\boldsymbol{N}_a$ and $\boldsymbol{b}\cdot\boldsymbol{n}_a = [\lambda_a]^2\boldsymbol{n}_a$, respectively, differ.

Spectral Representation of Inverse Strain Measures:
On the other hand the symmetric and positive definite Piola and Finger strain tensors $\boldsymbol{B} \equiv \boldsymbol{C}^{-1}$ and $\boldsymbol{c} \equiv \boldsymbol{b}^{-1}$, that may be considered the *inverses* of the right and left Cauchy-Green strain tensors, allow spectral representation in terms of the real and positive *principal inverse stretches* Λ_a and the corresponding Lagrangian and Eulerian dyads $\boldsymbol{N}_a \otimes \boldsymbol{N}_a$ and $\boldsymbol{n}_a \otimes \boldsymbol{n}_a$, respectively

$$\boldsymbol{B} = \sum_{a=1}^{3} [\Lambda_a]^2 \boldsymbol{N}_a \otimes \boldsymbol{N}_a \quad \text{and} \quad \boldsymbol{c} = \sum_{a=1}^{3} [\Lambda_a]^2 \boldsymbol{n}_a \otimes \boldsymbol{n}_a.$$

The square of the principal inverse stretches $[\Lambda_a]^2$ follow from the characteristic equation

$$[\Lambda_a]^6 - i_1[\Lambda_a]^4 + i_2[\Lambda_a]^2 - i_3 = 0$$

in terms of the three *principal 'inverse' invariants*

$$i_1 = \boldsymbol{B} : \boldsymbol{J} = \boldsymbol{i} : \boldsymbol{c} \quad \text{and} \quad i_2 = \operatorname{cof} \boldsymbol{B} : \boldsymbol{I} = \boldsymbol{j} : \operatorname{cof} \boldsymbol{c} \quad \text{and} \quad i_3 = \det \boldsymbol{B} = \det \boldsymbol{c}.$$

Due to the spectral representation the principal 'inverse' invariants are alternatively expressible in terms of the *spectral inverse invariants* $[\Lambda_a]^2$ as

$$i_1 = [\Lambda_1]^2 + [\Lambda_2]^2 + [\Lambda_3]^2$$
$$i_2 = [\Lambda_1]^2[\Lambda_2]^2 + [\Lambda_2]^2[\Lambda_3]^2 + [\Lambda_3]^2[\Lambda_1]^2$$
$$i_3 = [\Lambda_1]^2[\Lambda_2]^2[\Lambda_3]^2.$$

Since $\Lambda_a = [\lambda_a]^{-1}$ it is easy to verify that the principal 'inverse' invariants i_a and the principal invariants I_a are related by

$$i_1 = \frac{I_2}{I_3} \quad \text{and} \quad i_2 = \frac{I_1}{I_3} \quad \text{and} \quad i_3 = \frac{1}{I_3}.$$

Since the Piola and Finger strain tensors possess identical principal invariants i_a they also share the same eigenvalues $[\Lambda_a]^2$, however the corresponding eigenvectors satisfying $\boldsymbol{B} \cdot \boldsymbol{N}_a = [\Lambda_a]^2 \boldsymbol{N}_a$ and $\boldsymbol{c} \cdot \boldsymbol{n}_a = [\Lambda_a]^2 \boldsymbol{n}_a$, respectively, differ.

Family of Generalized Material Strain Measures:
On the one hand the spectral representation of the material strain measures $\boldsymbol{C}$ and $\boldsymbol{B}$ allow the introduction of a family of generalized material strain measures, the so-called material family of Seth strain measures, see [172]

$$\boldsymbol{E}_{(m)} := \frac{1}{m} \sum_{a=1}^{3} [[\lambda_a]^m - 1] \boldsymbol{N}_a \otimes \boldsymbol{N}_a = \frac{1}{m} [\boldsymbol{C}^{\frac{m}{2}} - \boldsymbol{J}],$$
$$\boldsymbol{K}_{(m)} := \frac{1}{m} \sum_{a=1}^{3} [1 - [\Lambda_a]^m] \boldsymbol{N}_a \otimes \boldsymbol{N}_a = \frac{1}{m} [\boldsymbol{I} - \boldsymbol{B}^{\frac{m}{2}}].$$

The above definitions hold for exponents $m \geq 1$; observe that the Green-Lagrange strain $\boldsymbol{E}$ and the material Karni-Reiner strain $\boldsymbol{K}$ are obtained for $m = 2$, i.e. $\boldsymbol{E} = \boldsymbol{E}_{(2)}$ and $\boldsymbol{K} = \boldsymbol{K}_{(2)}$, respectively. For $m = 1$ the strain measure $\boldsymbol{E}_{(1)} = \boldsymbol{U} - \boldsymbol{J}$ is denoted the *Biot* strain. Since $\ln \lambda = \lim_{m \to 0} \frac{1}{m}[\lambda^m - 1]$ and likewise $\ln \Lambda = \lim_{m \to 0} \frac{1}{m}[\Lambda^m - 1]$ the case $m = 0$ renders

$$\boldsymbol{E}_{(0)} := \sum_{a=1}^{3} \ln \lambda_a \boldsymbol{N}_a \otimes \boldsymbol{N}_a \equiv - \sum_{a=1}^{3} \ln \Lambda_a \boldsymbol{N}_a \otimes \boldsymbol{N}_a := \boldsymbol{K}_{(0)}.$$

The strain measure $\boldsymbol{E}_{(0)} = \frac{1}{2} \ln \boldsymbol{C} = -\frac{1}{2} \ln \boldsymbol{B} = \boldsymbol{K}_{(0)}$ is commonly denoted the material *Hencky* strain.

Family of Generalized Spatial Strain Measures:
On the other hand the spectral representation of the spatial strain measures $\boldsymbol{b}$ and $\boldsymbol{c}$ allow the introduction of a family of generalized spatial strain measures, the so-called spatial family of Seth strain measures, see [172]

$$\boldsymbol{k}_{(m)} := \frac{1}{m}\sum_{a=1}^{3}\left[[\lambda_a]^m - 1\right]\boldsymbol{n}_a \otimes \boldsymbol{n}_a = \frac{1}{m}\left[\boldsymbol{b}^{\frac{m}{2}} - \boldsymbol{i}\right],$$

$$\boldsymbol{e}_{(m)} := \frac{1}{m}\sum_{a=1}^{3}\left[1 - [\Lambda_a]^m\right]\boldsymbol{n}_a \otimes \boldsymbol{n}_a = \frac{1}{m}\left[\boldsymbol{j} - \boldsymbol{c}^{\frac{m}{2}}\right].$$

The above definitions hold for exponents $m \geq 1$; observe that the spatial Karni-Reiner strain $\boldsymbol{k}$ and the Almansi strain $\boldsymbol{e}$ are obtained for $m = 2$, i.e. $\boldsymbol{k} = \boldsymbol{k}_{(2)}$ and $\boldsymbol{e} = \boldsymbol{e}_{(2)}$, respectively. With the same limits for $m \to 0$ the case $m = 0$ renders

$$\boldsymbol{k}_{(0)} := \sum_{a=1}^{3}\ln\lambda_a \boldsymbol{n}_a \otimes \boldsymbol{n}_a \equiv -\sum_{a=1}^{3}\ln\Lambda_a \boldsymbol{n}_a \otimes \boldsymbol{n}_a := \boldsymbol{e}_{(0)}.$$

The strain measure $\boldsymbol{k}_{(0)} = \frac{1}{2}\ln\boldsymbol{b} = -\frac{1}{2}\ln\boldsymbol{c} = \boldsymbol{e}_{(0)}$ is commonly denoted the spatial *Hencky* strain.

Spectral Representation and Polar Decomposition:
Due to the polar decomposition of the deformation gradient $\boldsymbol{F} = \boldsymbol{R}\cdot\boldsymbol{U} = \boldsymbol{v}\cdot\boldsymbol{R}$ into the proper orthogonal *continuum rotation tensor* $\boldsymbol{R}$ and the symmetric, positive definite right and left *stretch tensors* $\boldsymbol{U}$ and $\boldsymbol{v}$, respectively, the right and left Cauchy-Green strain tensors allow representation

$$\boldsymbol{C} = \boldsymbol{U}^2 \quad \text{and} \quad \boldsymbol{b} = \boldsymbol{v}^2.$$

Thus the spectral representation of the right and left stretch tensor follows as

$$\boldsymbol{U} = \sum_{a=1}^{3}\lambda_a \boldsymbol{N}_a \otimes \boldsymbol{N}_a \quad \text{and} \quad \boldsymbol{v} = \sum_{a=1}^{3}\lambda_a \boldsymbol{n}_a \otimes \boldsymbol{n}_a.$$

Obviously the determinant of the deformation gradient reads as $J = \det\boldsymbol{F} = \det\boldsymbol{U} = \det\boldsymbol{v} = \lambda_1\lambda_2\lambda_3$. Furthermore, since the three Lagrangian principal directions $\boldsymbol{N}_a$ and the three Eulerian principal directions $\boldsymbol{n}_a$ represent orthonormal triads with different orientation in three space, they are related through a proper orthogonal transformation by the continuum rotation tensor

$$\boldsymbol{n}_a = \boldsymbol{R}\cdot\boldsymbol{N}_a \quad \text{with} \quad \boldsymbol{R} = \sum_{a=1}^{3}\boldsymbol{n}_a \otimes \boldsymbol{N}_a \quad \text{and} \quad \det\boldsymbol{R} = 1.$$

Thus as a consequence of the polar decomposition theorem the deformation gradient eventually expands into quantities from the spectral representation

$$\boldsymbol{F} = \sum_{a=1}^{3}\lambda_a \boldsymbol{n}_a \otimes \boldsymbol{N}_a \quad \text{and} \quad \boldsymbol{f} = \sum_{a=1}^{3}\Lambda_a \boldsymbol{N}_a \otimes \boldsymbol{n}_a.$$

Based on the representation of the continuum rotation tensor the following relations between material and spatial strain measures and unit tensors hold

$$\boldsymbol{b} = \boldsymbol{R}\cdot\boldsymbol{C}\cdot\boldsymbol{R}^t \quad \text{and} \quad \boldsymbol{c} = \boldsymbol{R}\cdot\boldsymbol{B}\cdot\boldsymbol{R}^t \quad \text{and} \quad \boldsymbol{i} = \boldsymbol{R}\cdot\boldsymbol{J}\cdot\boldsymbol{R}^t \quad \text{and} \quad \boldsymbol{j} = \boldsymbol{R}\cdot\boldsymbol{I}\cdot\boldsymbol{R}^t.$$

Likewise the generalized material and spatial strain measures of the Seth family are related by

$$\boldsymbol{k}_{(m)} = \boldsymbol{R}\cdot\boldsymbol{E}_{(m)}\cdot\boldsymbol{R}^t \quad \text{and} \quad \boldsymbol{e}_{(m)} = \boldsymbol{R}\cdot\boldsymbol{K}_{(m)}\cdot\boldsymbol{R}^t.$$

Finally, by selecting the material line element as $\mathrm{d}\boldsymbol{X} = \boldsymbol{N}_a$, the material area element as $\mathrm{d}\boldsymbol{A} = \boldsymbol{N}_a \times \boldsymbol{N}_b = \boldsymbol{N}_c$, and the material volume element as $\mathrm{d}V = [\boldsymbol{N}_a \times \boldsymbol{N}_b]\cdot\boldsymbol{N}_c = 1$ with cyclic permutations of $a, b, c = 1, 2, 3$ the following relations may be established

$$\begin{aligned} \lambda_a \boldsymbol{n}_a &= \boldsymbol{F}\cdot\boldsymbol{N}_a, \\ \lambda_a\lambda_b[\boldsymbol{n}_a \times \boldsymbol{n}_b] &= \operatorname{cof}\boldsymbol{F}\cdot[\boldsymbol{N}_a \times \boldsymbol{N}_b], \\ \lambda_a\lambda_b\lambda_c[\boldsymbol{n}_a \times \boldsymbol{n}_b]\cdot\boldsymbol{n}_c &= J[\boldsymbol{N}_a \times \boldsymbol{N}_b]\cdot\boldsymbol{N}_c. \end{aligned}$$

In summary the polar decomposition may be interpreted as describing the affine mapping of a unit sphere (with half axes $\boldsymbol{N}_a$) into an ellipsoid with half axes $\lambda_a\boldsymbol{n}_a$. Thereby either the unit sphere is firstly stretched along the Lagrangian principal directions $\boldsymbol{N}_a$ into an ellipsoid and subsequently the ellipsoid with half axes $\lambda_a\boldsymbol{N}_a$ is rotated rigidly into the Eulerian principal directions $\boldsymbol{n}_a$, or the sphere is firstly rotated rigidly from the Lagrangian principal directions $\boldsymbol{N}_a$ into the Eulerian principal directions $\boldsymbol{n}_a$ and subsequently the sphere is stretched along the Eulerian principal directions into an ellipsoid with half axes $\lambda_a\boldsymbol{n}_a$.

Relation between Strain Measures:
The various co- and contravariant, spatial and material strain measures obey various interesting relations among each other. These are either given by inversion of co- or contravariant strain measures, covariant pull-back or push-forward, contravariant pull-back or push-forward, or by backward of forward rotation. The application of these operations to a generic strain measure $\{\bullet\}$ shall symbolically be denoted as

$$\begin{aligned} &\circlearrowright\!: \ \{\bullet\}^{-1} &&\text{and} &&\circlearrowleft\!: \ \{\bullet\}^{-1} \\ &\swarrow\!: \boldsymbol{F}^t\cdot\{\bullet\}\cdot\boldsymbol{F} &&\text{and} &&\searrow\!: \boldsymbol{f}^t\cdot\{\bullet\}\cdot\boldsymbol{f} \\ &\nwarrow\!: \boldsymbol{f}\cdot\{\bullet\}\cdot\boldsymbol{f}^t &&\text{and} &&\nearrow\!: \boldsymbol{F}\cdot\{\bullet\}\cdot\boldsymbol{F}^t \\ &\curvearrowleft\!: \boldsymbol{R}^t\cdot\{\bullet\}\cdot\boldsymbol{R} &&\text{and} &&\curvearrowright\!: \boldsymbol{R}\cdot\{\bullet\}\cdot\boldsymbol{R}^t. \end{aligned}$$

An overview on the intricate relations between the elementary co- and contravariant strain measures and unit tensors $\boldsymbol{C}$, $\boldsymbol{B}$, $\boldsymbol{J}$, $\boldsymbol{I}$,and $\boldsymbol{c}$, $\boldsymbol{b}$, $\boldsymbol{j}$, $\boldsymbol{i}$ is given in the following table:

	$\boldsymbol{C}$	$\boldsymbol{B}$	$\boldsymbol{J}$	$\boldsymbol{I}$	$\boldsymbol{c}$	$\boldsymbol{b}$	$\boldsymbol{j}$	$\boldsymbol{i}$
$\boldsymbol{C} =$	$\bullet$	$\circlearrowright$	—	—	$\circlearrowright \oplus \curvearrowleft$	$\curvearrowleft$	$\swarrow$	$\circlearrowright \oplus \swarrow$
$\boldsymbol{B} =$	$\circlearrowleft$	$\bullet$	—	—	$\curvearrowleft$	$\circlearrowright \oplus \curvearrowleft$	$\circlearrowleft \oplus \nwarrow$	$\nwarrow$
$\boldsymbol{J} =$	—	—	$\bullet$	$\circlearrowright$	$\swarrow$	$\nwarrow$	$\circlearrowleft \oplus \curvearrowleft$	$\curvearrowleft$
$\boldsymbol{I} =$	—	—	$\circlearrowleft$	$\bullet$	$\swarrow$	$\nwarrow$	$\curvearrowleft$	$\circlearrowright \oplus \curvearrowleft$
$\boldsymbol{c} =$	$\circlearrowleft \oplus \curvearrowright$	$\curvearrowright$	$\searrow$	$\circlearrowright \oplus \searrow$	$\bullet$	$\circlearrowright$	—	—
$\boldsymbol{b} =$	$\curvearrowright$	$\circlearrowright \oplus \curvearrowright$	$\circlearrowleft \oplus \nearrow$	$\nearrow$	$\circlearrowleft$	$\bullet$	—	—
$\boldsymbol{j} =$	$\searrow$	$\nearrow$	$\circlearrowleft \oplus \curvearrowright$	$\curvearrowright$	—	—	$\bullet$	$\circlearrowright$
$\boldsymbol{i} =$	$\searrow$	$\nearrow$	$\curvearrowright$	$\circlearrowright \oplus \curvearrowright$	—	—	$\circlearrowleft$	$\bullet$

Clearly, inversion $\circlearrowleft$ or $\circlearrowright$ corresponds to inverting the eigenvalues and preserving the eigendyads, whereas rotation $\curvearrowleft$ or $\curvearrowright$ corresponds to preserving the eigenvalues and exchanging the eigendyads. Likewise, in curvilinear coordinates pull-back $\swarrow$ or $\nwarrow$, and push-forward $\searrow$ or $\nearrow$ correspond to preserving the curvilinear coefficients and exchanging the corresponding base vectors.

Surface Jacobian:
The material and spatial vector valued area elements $\mathrm{d}\boldsymbol{A}$ and $\mathrm{d}\boldsymbol{a}$ expand in terms of the corresponding scalar valued area elements $\mathrm{d}A$ and $\mathrm{d}a$, and the outwards pointing material and spatial surface unit normals $\boldsymbol{N}$ and $\boldsymbol{n}$, respectively, as

$$\mathrm{d}\boldsymbol{A} := \boldsymbol{N}\,\mathrm{d}A \quad \text{and} \quad \mathrm{d}\boldsymbol{a} := \boldsymbol{n}\,\mathrm{d}a.$$

Then, based on the Nanson formula $\mathrm{d}\boldsymbol{a} = \operatorname{cof}\boldsymbol{F} \cdot \mathrm{d}\boldsymbol{A}$ the *surface Jacobian* $\widehat{J}$ relates the material and spatial scalar valued area elements as

$$\mathrm{d}a = \widehat{J}\,\mathrm{d}A \quad \text{with} \quad \widehat{J} := J\Lambda_N \quad \text{and} \quad \Lambda_N^2 := \boldsymbol{N} \cdot \boldsymbol{B} \cdot \boldsymbol{N} = |\boldsymbol{N} \cdot \boldsymbol{f}|^2.$$

Here Λ_N takes the interpretation as the inverse stretch of a fibre aligned in the material configuration with the material line element $\boldsymbol{N}$. It is thus inverse to the stretch λ_n of a fibre aligned in the spatial configuration with the spatial line element $\boldsymbol{n}$ whereby $\lambda_n^2 := \boldsymbol{n} \cdot \boldsymbol{b} \cdot \boldsymbol{n} = |\boldsymbol{n} \cdot \boldsymbol{F}|^2$.

Velocity Gradients:
The *material time derivative* of generic fields $\{\bullet\}$ that are parameterized in $\boldsymbol{X}$ and t, i.e. the time derivative at fixed material coordinates $\boldsymbol{X}$ is denoted by

$$\mathrm{D}_t\{\bullet\} := \partial_t\{\bullet\}|_{\boldsymbol{X}}.$$

It is often convenient to use the common superposed dot notation instead, i.e.

$$\mathrm{D}_t\{\bullet\} \equiv \dot{\overline{\{\bullet\}}}.$$

Based on the velocity field $\boldsymbol{v} = \boldsymbol{v}(\boldsymbol{X}, t) := \dot{\boldsymbol{y}}(\boldsymbol{X}, t)$ and $\boldsymbol{v}(\boldsymbol{x}, t) = \boldsymbol{v}(\boldsymbol{X}, t) \circ \boldsymbol{Y}(\boldsymbol{x}, t)$ the contra/co- (mixed) variant two-point and spatial *velocity gradients* are then introduced as

$$\boldsymbol{\Lambda} := \operatorname{Grad} \boldsymbol{v} = \dot{\boldsymbol{F}} \quad \text{and} \quad \boldsymbol{l} := \operatorname{grad} \boldsymbol{v} = \dot{\boldsymbol{F}} \cdot \boldsymbol{f}.$$

The two-point (spatial/material) and spatial velocity gradients map either the material or the spatial line element into the material time derivative of the spatial line element

$$\dot{\overline{\mathrm{d}\boldsymbol{x}}} = \boldsymbol{\Lambda} \cdot \mathrm{d}\boldsymbol{X} = \boldsymbol{l} \cdot \mathrm{d}\boldsymbol{x}.$$

Furthermore, since $\boldsymbol{l}$ is a contra/co- (mixed) variant spatial second-order tensor field, the following one- and two-sided pull-backs are possible

$$\boldsymbol{\Lambda} = \boldsymbol{l} \cdot \boldsymbol{F} \quad \text{and} \quad \boldsymbol{\lambda} := \boldsymbol{f} \cdot \boldsymbol{l} \quad \text{and} \quad \boldsymbol{L} := \boldsymbol{f} \cdot \boldsymbol{l} \cdot \boldsymbol{F}.$$

Observe that the two-point (material/spatial) $\boldsymbol{\lambda}$ and the material $\boldsymbol{L}$ do not represent any gradient of the velocity field, nevertheless for simplicity they shall also be addressed as 'velocity gradients'. The relation between the various velocity gradients is depicted in the following table:

	$\boldsymbol{l} = \operatorname{grad} \boldsymbol{v}$	$\boldsymbol{\Lambda} = \operatorname{Grad} \boldsymbol{v}$	$\boldsymbol{L}$	$\boldsymbol{\lambda}$
$\boldsymbol{l}$	$\bullet$	$\boldsymbol{\Lambda} \cdot \boldsymbol{f}$	$\boldsymbol{F} \cdot \boldsymbol{L} \cdot \boldsymbol{f}$	$\boldsymbol{F} \cdot \boldsymbol{\lambda}$
$\boldsymbol{\Lambda}$	$\boldsymbol{l} \cdot \boldsymbol{F}$	$\bullet$	$\boldsymbol{F} \cdot \boldsymbol{L}$	$\boldsymbol{F} \cdot \boldsymbol{\lambda} \cdot \boldsymbol{F}$
$\boldsymbol{L}$	$\boldsymbol{f} \cdot \boldsymbol{l} \cdot \boldsymbol{F}$	$\boldsymbol{f} \cdot \boldsymbol{\Lambda}$	$\bullet$	$\boldsymbol{\lambda} \cdot \boldsymbol{F}$
$\boldsymbol{\lambda}$	$\boldsymbol{f} \cdot \boldsymbol{l}$	$\boldsymbol{f} \cdot \boldsymbol{\Lambda} \cdot \boldsymbol{f}$	$\boldsymbol{L} \cdot \boldsymbol{f}$	$\bullet$

It is interesting to note that $\boldsymbol{\lambda}$ agrees with the (negative) material time derivative of the inverse deformation gradient, i.e. $\boldsymbol{\lambda} = -\,\mathrm{D}_t \boldsymbol{f}$.

Spatial Rate-of-Deformation and Spin Tensors:
On the one hand the fully co- and contravariant versions of the *spatial velocity gradient* allow decomposition into symmetric and skew symmetric contributions

$$\boldsymbol{j} \cdot \boldsymbol{l} = [\boldsymbol{j} \cdot \boldsymbol{l}]^{\mathrm{sym}} + [\boldsymbol{j} \cdot \boldsymbol{l}]^{\mathrm{skw}} \quad \text{and} \quad \boldsymbol{l} \cdot \boldsymbol{i} = [\boldsymbol{l} \cdot \boldsymbol{i}]^{\mathrm{sym}} + [\boldsymbol{l} \cdot \boldsymbol{i}]^{\mathrm{skw}}.$$

Here the multiplications with the spatial unit tensors $\boldsymbol{j}$ and $\boldsymbol{i}$ are basically a notational reminders that either the co- or the contravariant version of the mixed-variant $\boldsymbol{l}$ is considered. Then the symmetric contribution is denoted the *spatial rate-of-deformation tensor* $\boldsymbol{d}$ whereas the skew symmetric contribution is denoted the *spatial spin tensor* $\boldsymbol{w}$, i.e.

$$\boldsymbol{d} := [\boldsymbol{j} \cdot \boldsymbol{l}]^{\mathrm{sym}} = [\boldsymbol{l} \cdot \boldsymbol{i}]^{\mathrm{sym}} \quad \text{and} \quad \boldsymbol{w} := [\boldsymbol{j} \cdot \boldsymbol{l}]^{\mathrm{skw}} = [\boldsymbol{l} \cdot \boldsymbol{i}]^{\mathrm{skw}}.$$

Observe that here no special notation is adopted to distinguish the co- and contravariant versions of $\boldsymbol{d}$ and $\boldsymbol{w}$. Then the covariant pull-back of $\boldsymbol{j} \cdot \boldsymbol{l}$ involves the covariant $\boldsymbol{C}$ as a metric and renders

$$\boldsymbol{C} \cdot \boldsymbol{L} = [\boldsymbol{C} \cdot \boldsymbol{L}]^{\text{sym}} + [\boldsymbol{C} \cdot \boldsymbol{L}]^{\text{skw}} = \boldsymbol{F}^t \cdot [\boldsymbol{d} + \boldsymbol{w}] \cdot \boldsymbol{F} = \boldsymbol{F}^t \cdot [\boldsymbol{j} \cdot \boldsymbol{l}] \cdot \boldsymbol{F}.$$

Likewise the contravariant pull-back of $\boldsymbol{l} \cdot \boldsymbol{i}$ involves the contravariant $\boldsymbol{B}$ as a metric and results in

$$\boldsymbol{L} \cdot \boldsymbol{B} = [\boldsymbol{L} \cdot \boldsymbol{B}]^{\text{sym}} + [\boldsymbol{L} \cdot \boldsymbol{B}]^{\text{skw}} = \boldsymbol{f} \cdot [\boldsymbol{d} + \boldsymbol{w}] \cdot \boldsymbol{f}^t = \boldsymbol{f} \cdot [\boldsymbol{l} \cdot \boldsymbol{i}] \cdot \boldsymbol{f}^t.$$

In summary the decomposition of the spatial velocity gradient into symmetric and skew symmetric contributions renders spatial rate-of-deformation and spin tensors, respectively, together with corresponding pull-back quantities. Thereby, the formal similarities and relations between the various tensors is intriguing.

Material Rate-of-Deformation and Spin Tensors:
On the other hand the fully co- and contravariant versions of the *material velocity gradient* allow decomposition into symmetric and skew symmetric contributions

$$\boldsymbol{J} \cdot \boldsymbol{L} = [\boldsymbol{J} \cdot \boldsymbol{L}]^{\text{sym}} + [\boldsymbol{J} \cdot \boldsymbol{L}]^{\text{skw}} \quad \text{and} \quad \boldsymbol{L} \cdot \boldsymbol{I} = [\boldsymbol{L} \cdot \boldsymbol{I}]^{\text{sym}} + [\boldsymbol{L} \cdot \boldsymbol{I}]^{\text{skw}}.$$

Here the multiplications with the material unit tensors $\boldsymbol{J}$ and $\boldsymbol{I}$ are again essentially a notational reminders that either the co- or the contravariant version of the mixed-variant $\boldsymbol{L}$ is considered. Then the symmetric contribution is denoted the *material rate-of-deformation tensor* $\boldsymbol{D}$ whereas the skew symmetric contribution is denoted the *material spin tensor* $\boldsymbol{W}$, i.e.

$$\boldsymbol{D} := [\boldsymbol{J} \cdot \boldsymbol{L}]^{\text{sym}} = [\boldsymbol{L} \cdot \boldsymbol{I}]^{\text{sym}} \quad \text{and} \quad \boldsymbol{W} := [\boldsymbol{J} \cdot \boldsymbol{L}]^{\text{skw}} = [\boldsymbol{L} \cdot \boldsymbol{I}]^{\text{skw}}.$$

Observe again that here no special notation is adopted to distinguish the co- and contravariant versions of $\boldsymbol{D}$ and $\boldsymbol{W}$. Then the covariant push-forward of $\boldsymbol{J} \cdot \boldsymbol{L}$ involves the covariant $\boldsymbol{c}$ as a metric and renders

$$\boldsymbol{c} \cdot \boldsymbol{l} = [\boldsymbol{c} \cdot \boldsymbol{l}]^{\text{sym}} + [\boldsymbol{c} \cdot \boldsymbol{l}]^{\text{skw}} = \boldsymbol{f}^t \cdot [\boldsymbol{D} + \boldsymbol{W}] \cdot \boldsymbol{f} = \boldsymbol{f}^t \cdot [\boldsymbol{J} \cdot \boldsymbol{L}] \cdot \boldsymbol{f}.$$

Likewise the contravariant push-forward of $\boldsymbol{L} \cdot \boldsymbol{I}$ involves the contravariant $\boldsymbol{b}$ as a metric and results in

$$\boldsymbol{l} \cdot \boldsymbol{b} = [\boldsymbol{l} \cdot \boldsymbol{b}]^{\text{sym}} + [\boldsymbol{l} \cdot \boldsymbol{b}]^{\text{skw}} = \boldsymbol{F} \cdot [\boldsymbol{D} + \boldsymbol{W}] \cdot \boldsymbol{F}^t = \boldsymbol{F} \cdot [\boldsymbol{L} \cdot \boldsymbol{I}] \cdot \boldsymbol{F}^t.$$

In summary the decomposition of the material velocity gradients into symmetric and skew symmetric contributions renders material rate-of-deformation and spin tensors, respectively, together with corresponding push-forward quantities. Thereby, the formal similarities and relations between the various tensors is intriguing.

Material Time Derivatives of Strain Measures:
On the one hand the material time derivative of the (quadratic) length of the spatial line element $\mathrm{d}\boldsymbol{x}$ involves the covariant version of the spatial rate-of-deformation tensor $\boldsymbol{d}$ or, likewise, its pull-back that proves to relate to the material time derivative of the right Cauchy-Green strain $\boldsymbol{C}$, i.e.

$$\overline{|\mathrm{d}\boldsymbol{x}|^2}^{\cdot} = 2\,\mathrm{d}\boldsymbol{x}\cdot\boldsymbol{d}\cdot\,\mathrm{d}\boldsymbol{x} = 2\,\mathrm{d}\boldsymbol{X}\cdot[\boldsymbol{C}\cdot\boldsymbol{L}]^{\mathrm{sym}}\cdot\,\mathrm{d}\boldsymbol{X} = \,\mathrm{d}\boldsymbol{X}\cdot\dot{\boldsymbol{C}}\cdot\,\mathrm{d}\boldsymbol{X}.$$

On the other hand the material time derivative of the (quadratic) length of the spatial level set normal $\mathbf{d}s$ involves the contravariant version of the spatial rate-of-deformation tensor $\boldsymbol{d}$ or, likewise, its pull-back that proves to relate to the material time derivative of the Piola strain $\boldsymbol{B}$, i.e.

$$\overline{|\mathbf{d}s|^2}^{\cdot} = -2\mathbf{d}s\cdot\boldsymbol{d}\cdot\mathbf{d}s = -2\mathbf{d}S\cdot[\boldsymbol{L}\cdot\boldsymbol{B}]^{\mathrm{sym}}\cdot\mathbf{d}S = \mathbf{d}S\cdot\dot{\boldsymbol{B}}\cdot\mathbf{d}S.$$

Indeed the co- and contravariant pull-back of the spatial rate-of-deformation tensor $\boldsymbol{d}$ compute as

$$\boldsymbol{F}^t\cdot\boldsymbol{d}\cdot\boldsymbol{F} = [\boldsymbol{C}\cdot\boldsymbol{L}]^{\mathrm{sym}} = \frac{1}{2}\dot{\boldsymbol{C}} \quad\text{and}\quad \boldsymbol{f}\cdot\boldsymbol{d}\cdot\boldsymbol{f}^t = [\boldsymbol{L}\cdot\boldsymbol{B}]^{\mathrm{sym}} = -\frac{1}{2}\dot{\boldsymbol{B}}.$$

Consequently, the material time derivatives of the co- and contravariant material Green-Lagrange and Karni-Reiner strain tensors read

$$\dot{\boldsymbol{E}} = \frac{1}{2}\dot{\boldsymbol{C}} \quad\text{and}\quad \dot{\boldsymbol{K}} = -\frac{1}{2}\dot{\boldsymbol{B}}.$$

Vice versa the co- and contravariant push-forward of the material rate-of-deformation tensor $\boldsymbol{D}$ proofs to result in

$$\boldsymbol{f}^t\cdot\boldsymbol{D}\cdot\boldsymbol{f} = [\boldsymbol{c}\cdot\boldsymbol{l}]^{\mathrm{sym}} = -\frac{1}{2}\dot{\boldsymbol{c}} \quad\text{and}\quad \boldsymbol{F}\cdot\boldsymbol{D}\cdot\boldsymbol{F}^t = [\boldsymbol{l}\cdot\boldsymbol{b}]^{\mathrm{sym}} = \frac{1}{2}\dot{\boldsymbol{b}}.$$

Thus the material time derivatives of the co- and contravariant spatial Almansi and Karni-Rainer strain tensors are expressed as

$$\dot{\boldsymbol{e}} = -\frac{1}{2}\dot{\boldsymbol{c}} \quad\text{and}\quad \dot{\boldsymbol{k}} = \frac{1}{2}\dot{\boldsymbol{b}}.$$

Again the formal similarity between the material time derivatives of the various co- and contravariant, spatial and material strain measures, respectively, shall be observed.

Left Lie Time Derivative of Spatial Tensor Fields:
The *left Lie time derivative* L_t of generic spatial tensor fields $\{\bullet\}$ is defined as

$$\mathrm{L}_t\{\bullet\} := \text{push-forward}\Big(\,\mathrm{D}_t\big(\text{pull-back of }\{\bullet\}\big)\Big).$$

As an example applying this definition to the various co- and contravariant spatial unit tensors and strain measures renders

$$\mathrm{L}_t \boldsymbol{j} = 2\,\mathrm{L}_t \boldsymbol{e} = 2\boldsymbol{d}, \quad \mathrm{L}_t \boldsymbol{c} = \boldsymbol{0} \quad \text{and} \quad \mathrm{L}_t \boldsymbol{i} = -2\,\mathrm{L}_t \boldsymbol{k} = -2\boldsymbol{d}, \quad \mathrm{L}_t \boldsymbol{b} = \boldsymbol{0}.$$

Here $\boldsymbol{j}$ and $\boldsymbol{i}$ are interpreted as either co- or contravariant spatial unit tensors, their left Lie time derivatives thus result in either $2\boldsymbol{d}$ or $-2\boldsymbol{d}$. Spatial tensor fields with vanishing left Lie time derivative (such as $\boldsymbol{c}$ and $\boldsymbol{b}$) are denoted as left *Killing* fields. Finally the relation between the left Lie and the material time derivative of generic co- and contravariant symmetric spatial tensor fields $\{\bullet\}$ results in

$$\mathrm{L}_t\{\bullet\} = \mathrm{D}_t\{\bullet\} + 2[\{\bullet\}\cdot \boldsymbol{l}]^{\mathrm{sym}} \quad \text{and} \quad \mathrm{L}_t\{\bullet\} = \mathrm{D}_t\{\bullet\} - 2[\boldsymbol{l}\cdot\{\bullet\}]^{\mathrm{sym}}.$$

A comparison of the left Lie and the material time derivative of the various co- and contravariant strain measures is clearly in agreement with the above relation. As an example consider $\mathrm{D}_t \boldsymbol{j} = \mathrm{D}_t \boldsymbol{i} = \boldsymbol{0}$, thus the covariant $\boldsymbol{j}$ results in $\mathrm{L}_t \boldsymbol{j} = 2[\boldsymbol{j}\cdot\boldsymbol{l}]^{\mathrm{sym}}$, whereas the contravariant $\boldsymbol{i}$ results in $\mathrm{L}_t \boldsymbol{i} = -2[\boldsymbol{l}\cdot\boldsymbol{i}]^{\mathrm{sym}}$.

Right Lie Time Derivative of Material Tensor Fields:
The *right Lie time derivative* R_t of generic material tensor fields $\{\bullet\}$ is defined as

$$\mathrm{R}_t\{\bullet\} := \text{pull-back}\Big(\,\mathrm{D}_t\big(\text{push-forward of } \{\bullet\}\big)\Big).$$

As an example applying this definition to the various co- and contravariant material unit tensors and strain measures renders

$$\mathrm{R}_t \boldsymbol{J} = -2\,\mathrm{R}_t \boldsymbol{E} = -2\boldsymbol{D}, \quad \mathrm{R}_t \boldsymbol{C} = \boldsymbol{0} \;\text{ and }\; \mathrm{R}_t \boldsymbol{I} = 2\,\mathrm{R}_t \boldsymbol{K} = 2\boldsymbol{D}, \quad \mathrm{R}_t \boldsymbol{B} = \boldsymbol{0}.$$

Here $\boldsymbol{J}$ and $\boldsymbol{I}$ are interpreted as either co- or contravariant material unit tensors, their right Lie time derivatives thus result in either $-2\boldsymbol{D}$ or $2\boldsymbol{D}$. Material tensor fields with vanishing right Lie time derivative (such as $\boldsymbol{C}$ and $\boldsymbol{B}$) are denoted as right *Killing* fields. Finally the relation between the right Lie and the material time derivative of generic co- and contravariant symmetric material tensor fields $\{\bullet\}$ results in

$$\mathrm{R}_t\{\bullet\} = \mathrm{D}_t\{\bullet\} - 2[\{\bullet\}\cdot \boldsymbol{L}]^{\mathrm{sym}} \quad \text{and} \quad \mathrm{R}_t\{\bullet\} = \mathrm{D}_t\{\bullet\} + 2[\boldsymbol{L}\cdot\{\bullet\}]^{\mathrm{sym}}.$$

A comparison of the right Lie and the material time derivative of the various co- and contravariant strain measures is clearly in agreement with the above relation. As an example consider $\mathrm{D}_t \boldsymbol{J} = \mathrm{D}_t \boldsymbol{I} = \boldsymbol{0}$, thus the covariant $\boldsymbol{J}$ results in $\mathrm{R}_t \boldsymbol{J} = -2[\boldsymbol{J}\cdot\boldsymbol{L}]^{\mathrm{sym}}$, whereas the contravariant $\boldsymbol{I}$ results in $\mathrm{R}_t \boldsymbol{I} = 2[\boldsymbol{L}\cdot\boldsymbol{I}]^{\mathrm{sym}}$.

Cauchy Stress:
The key to the notion of the *Cauchy stress* is the concept of the *traction vector*. The traction vector $\boldsymbol{t}_t$ denotes a distributed reaction force per unit

area da in a virtual cut-surface through a continuum body in spatial configuration. Thereby $\boldsymbol{t}_t$ is determined so as to maintain *equilibrium* for the two sub-bodies separated by the virtual cut. Then the *true* or rather Cauchy stress follows from the:

1. Cauchy Postulate:

The traction vector at a particular point $\boldsymbol{x}$ depends only on the outwards pointing normal vector $\boldsymbol{n}$ to an arbitrary virtual cut-surface through the point $\boldsymbol{x}$ in the spatial configuration

$$\boldsymbol{t}_t = \boldsymbol{t}_t(\boldsymbol{n}; \boldsymbol{x}).$$

The Cauchy postulate has axiomatic character.

2. Cauchy Lemma:

The traction vector at a particular point $\boldsymbol{x}$ is a homogeneous function of degree one in the outwards pointing normal vector $\boldsymbol{n}$ to an arbitrary virtual cut-surface through the point $\boldsymbol{x}$ in the spatial configuration

$$\boldsymbol{t}_t(-\boldsymbol{n}; \boldsymbol{x}) = -\boldsymbol{t}_t(\boldsymbol{n}; \boldsymbol{x}).$$

Remember the definition of a homogeneous function of degree n: $\alpha^n f(x) = f(\alpha x)$. The Cauchy lemma follows from equilibrium considerations across the virtual cut-surface.

3. Cauchy Theorem:

The traction vector at a particular point $\boldsymbol{x}$ follows as a linear map of the outwards pointing normal vector $\boldsymbol{n}$ to an arbitrary virtual cut-surface through the point $\boldsymbol{x}$ in the spatial configuration

$$\boldsymbol{t}_t = \boldsymbol{\sigma}(\boldsymbol{x}) \cdot \boldsymbol{n}.$$

This map is necessarily defined in terms of a spatial second-order tensor that is denoted the Cauchy stress $\boldsymbol{\sigma} = \boldsymbol{\sigma}(\boldsymbol{x})$. It describes completely the stress state at the point $\boldsymbol{x}$ in the spatial configuration. Since the Cauchy stress maps a normal vector from the cotangent space $T^*\mathcal{B}_t$ into a traction vector from the cotangent space $T^*\mathcal{B}_t$ it is a co/contra- (mixed) variant second-order tensor. In [27] Cauchy spelled out his famous theorem as

De plus, la pression ou tension exercée contre un plane quelconque se déduit très facilement, tant en grandeur qu'en direction, des pressions ou tensions exercées contre trois plans rectangulaires donnés.

The Cauchy theorem follows from the surface theorem and equilibrium considerations for an infinitesimal spatial tetraeder element. Note that the above introduction of $\boldsymbol{\sigma}$ associates its first index with the corresponding coefficient of the traction vector and its second index with the cut-surface normal. However for a first-order continuum (with zero body couples and couple tractions) the Cauchy stress turns out to be symmetric in order to satisfy the balance of angular momentum.

Piola Stress:
Instead of introducing the traction vector $\boldsymbol{t}_t$ as a distributed force per unit area $\mathrm{d}a$ in the spatial configuration it proves convenient to introduce the traction vector $\boldsymbol{t}_0$ as a distributed force per corresponding unit area $\mathrm{d}A$ in the material configuration with the equivalence

$$\boldsymbol{t}_t \,\mathrm{d}a = \boldsymbol{\sigma} \cdot \boldsymbol{n} \,\mathrm{d}a =: \boldsymbol{P} \cdot \boldsymbol{N} \,\mathrm{d}A =: \boldsymbol{t}_0 \,\mathrm{d}A.$$

Here in analogy with the Cauchy theorem a relation of $\boldsymbol{t}_0$ with the outwards pointing normal vector $\boldsymbol{N}$ to an arbitrary virtual cut-surface through point $\boldsymbol{X}$ in the material configuration has been postulated as

$$\boldsymbol{t}_0 = \boldsymbol{P}(\boldsymbol{X}) \cdot \boldsymbol{N}.$$

Thereby $\boldsymbol{P}$ is a two-point second-order tensor that is denoted the *nominal* or rather *Piola stress.* Due to the definition of the vectorial area elements $\mathrm{d}\boldsymbol{a} = \boldsymbol{n}\,\mathrm{d}a$ and $\mathrm{d}\boldsymbol{A} = \boldsymbol{N}\,\mathrm{d}A$ together with $\mathrm{d}\boldsymbol{a} = \operatorname{cof}\boldsymbol{F} \cdot \mathrm{d}\boldsymbol{A}$ the relation between the Cauchy and the Piola stress is established as

$$\boldsymbol{\sigma} \cdot \operatorname{cof} \boldsymbol{F} = \boldsymbol{P}.$$

The transformation between the Cauchy and the Piola stress is the paradigm for a *Piola transformation.* Note that the Piola stress $\boldsymbol{P}$ is not symmetric, even in the case that $\boldsymbol{\sigma} = \boldsymbol{\sigma}^t$. The transpose of the Piola stress is sometimes introduced as the *nominal stress.*

Piola Identity and Piola Transformation:
The area theorem states that the integral of the vectorial area element $\mathrm{d}\boldsymbol{a}$ over an arbitrary closed surface $\partial\mathcal{V}_t$ vanishes identically

$$\boldsymbol{0} = \int_{\partial\mathcal{V}_t} \mathrm{d}\boldsymbol{a} = \int_{\partial\mathcal{V}_0} \operatorname{cof}\boldsymbol{F} \cdot \mathrm{d}\boldsymbol{A} = \int_{\mathcal{V}_0} \operatorname{Div}(\operatorname{cof}\boldsymbol{F}) \,\mathrm{d}V.$$

Localizing the above for arbitrary $\mathcal{V}_0$ renders as point-wise statement the *Piola identity,* i.e. the material divergence of the cofactor of the deformation gradient vanishes

$$\boldsymbol{0} = \operatorname{Div}(\operatorname{cof}\boldsymbol{F}).$$

As a consequence, the material and spatial divergence of tensorial quantities that are connected by a *Piola transformation*

$$\{\bullet\} \cdot \operatorname{cof}\boldsymbol{F} \leftrightarrow \{\bullet\}$$

are related by a multiplication with the determinant of the deformation gradient

$$\mathrm{Div}(\{\bullet\} \cdot \mathrm{cof}\, \boldsymbol{F}) = J \, \mathrm{div}(\{\bullet\}).$$

The proof follows in a straightforward manner from the chain rule and the Piola identity. Thus, as the paradigmatic example, the Piola and the Cauchy stress satisfy

$$\mathrm{Div}\, \boldsymbol{P} = J \, \mathrm{div} \boldsymbol{\sigma}.$$

This is the key relation that allows in the sequel to express the balance of momentum in terms of either the two-point Piola or the spatial Cauchy stress.

Stress Measures:
Due to the relation between the Piola and the Cauchy stress it proves convenient to introduce the *Kirchhoff stress* $\boldsymbol{\tau}$ as

$$\boldsymbol{\tau} := J\boldsymbol{\sigma} = \boldsymbol{P} \cdot \boldsymbol{F}^t \quad \text{with} \quad \int_{\mathcal{V}_0} \boldsymbol{\tau} \cdots \mathrm{d}V = \int_{\mathcal{V}_t} \boldsymbol{\sigma} \cdots \mathrm{d}v.$$

The Kirchhoff stress might be considered either a contra- ($\boldsymbol{\tau}_\sharp := \boldsymbol{i} \cdot \boldsymbol{\tau}$), a co- ($\boldsymbol{\tau}_\flat := \boldsymbol{\tau} \cdot \boldsymbol{j}$), or a co/contra- (mixed) variant spatial tensor field $\boldsymbol{\tau}$. However, if possible these versions shall not be distinguished explicitly for the sake of reducing the amount of symbols used; thus the somewhat sloppy notation $\boldsymbol{\tau} \equiv \boldsymbol{\tau}^\sharp \equiv \boldsymbol{\tau}^\flat$ shall be adopted in the sequel if there is no danger of confusion. Then, the contra-, co-, and co/contra- (mixed) variant pull-backs of the Kirchhoff stress follow as

$$\boldsymbol{S} := \boldsymbol{f}_\sharp \cdot \boldsymbol{\tau} \cdot \boldsymbol{f}^t \quad \text{and} \quad \boldsymbol{T} := \boldsymbol{F}^t \cdot \boldsymbol{\tau} \cdot \boldsymbol{F}_\flat \quad \text{and} \quad \boldsymbol{M} := \boldsymbol{F}^t \cdot \boldsymbol{\tau} \cdot \boldsymbol{f}^t.$$

Here $\boldsymbol{S}$, $\boldsymbol{T}$, and $\boldsymbol{M}$ denote a contravariantly transforming material stress measure (the *Piola-Kirchhoff stress*), a covariantly transforming material stress measure (sometimes denoted the convected stress, [122]), and a co/contra- (mixed) variantly transforming material stress measure (the *Mandel stress*), respectively. Finally, a co/contra- (mixed) variantly transforming two-point (material/spatial) stress measure is introduced as $\boldsymbol{\pi} := \boldsymbol{F}^t \cdot \boldsymbol{\tau}$. It is noted that the co- and contravariant versions of the deformation gradients as used in the above have been defined as

$$\boldsymbol{F}_\flat := \boldsymbol{j} \cdot \boldsymbol{F} \quad \text{and} \quad \boldsymbol{f}_\sharp := \boldsymbol{f} \cdot \boldsymbol{i}.$$

The various relations between the introduced stress measures are assembled in the following table:

	$\boldsymbol{\tau}$	$\boldsymbol{\pi}$	$\boldsymbol{P}$	$\boldsymbol{S}$	$\boldsymbol{T}$	$\boldsymbol{M}$
$\boldsymbol{\tau}$	$\bullet$	$\boldsymbol{f}^t\cdot\boldsymbol{\pi}$	$\boldsymbol{P}\cdot\boldsymbol{F}^t$	$\boldsymbol{F}_\flat\cdot\boldsymbol{S}\cdot\boldsymbol{F}^t$	$\boldsymbol{f}^t\cdot\boldsymbol{T}\cdot\boldsymbol{f}_\sharp$	$\boldsymbol{f}^t\cdot\boldsymbol{M}\cdot\boldsymbol{F}^t$
$\boldsymbol{\pi}$	$\boldsymbol{F}^t\cdot\boldsymbol{\tau}$	$\bullet$	$\boldsymbol{F}^t\cdot\boldsymbol{P}\cdot\boldsymbol{F}^t$	$\boldsymbol{C}\cdot\boldsymbol{S}\cdot\boldsymbol{F}^t$	$\boldsymbol{T}\cdot\boldsymbol{f}_\sharp$	$\boldsymbol{M}\cdot\boldsymbol{F}^t$
$\boldsymbol{P}$	$\boldsymbol{\tau}\cdot\boldsymbol{f}^t$	$\boldsymbol{f}^t\cdot\boldsymbol{\pi}\cdot\boldsymbol{f}^t$	$\bullet$	$\boldsymbol{F}_\flat\cdot\boldsymbol{S}$	$\boldsymbol{f}^t\cdot\boldsymbol{T}\cdot\boldsymbol{B}$	$\boldsymbol{f}^t\cdot\boldsymbol{M}$
$\boldsymbol{S}$	$\boldsymbol{f}_\sharp\cdot\boldsymbol{\tau}\cdot\boldsymbol{f}^t$	$\boldsymbol{B}\cdot\boldsymbol{\pi}\cdot\boldsymbol{f}^t$	$\boldsymbol{f}_\sharp\cdot\boldsymbol{P}$	$\bullet$	$\boldsymbol{B}\cdot\boldsymbol{T}\cdot\boldsymbol{B}$	$\boldsymbol{B}\cdot\boldsymbol{M}$
$\boldsymbol{T}$	$\boldsymbol{F}^t\cdot\boldsymbol{\tau}\cdot\boldsymbol{F}_\flat$	$\boldsymbol{\pi}\cdot\boldsymbol{F}_\flat$	$\boldsymbol{F}^t\cdot\boldsymbol{P}\cdot\boldsymbol{C}$	$\boldsymbol{C}\cdot\boldsymbol{S}\cdot\boldsymbol{C}$	$\bullet$	$\boldsymbol{M}\cdot\boldsymbol{C}$
$\boldsymbol{M}$	$\boldsymbol{F}^t\cdot\boldsymbol{\tau}\cdot\boldsymbol{f}^t$	$\boldsymbol{\pi}\cdot\boldsymbol{f}^t$	$\boldsymbol{F}^t\cdot\boldsymbol{P}$	$\boldsymbol{C}\cdot\boldsymbol{S}$	$\boldsymbol{T}\cdot\boldsymbol{B}$	$\bullet$

Observe that $\boldsymbol{S}$ and $\boldsymbol{T}$ are symmetric if $\boldsymbol{\tau}_\sharp = \boldsymbol{\tau}_\sharp^t$ and $\boldsymbol{\tau}_\flat = \boldsymbol{\tau}_\flat^t$ holds, however even in this case $\boldsymbol{M}$ is only symmetric if $\boldsymbol{C}$ and $\boldsymbol{S}$ or, likewise, $\boldsymbol{B}$ and $\boldsymbol{T}$ commute.

Balance Equations:
The functions describing the state of an elastic continuous body (state functions express the dependency of selected state quantities on the remaining state quantities, the so-called state variables), i.e. the mass density ρ_0, the linear and angular momentum densities $\boldsymbol{p}_0$ and $\boldsymbol{l}_0$, and the mechanical energy density e_0 (all expressed per unit volume $\mathrm{d}V$ in the material configuration $\mathcal{B}_0$) are governed by a set of balance equations. These are the:

1. Balance of Mass:
The total *mass* $\mathsf{M}(\mathcal{V}_0)$ of an arbitrary continuous sub-body $\mathscr{V} \subseteq \mathscr{B}$ of a continuous body $\mathscr{B}$ that occupies the material sub-configuration $\mathcal{V}_0 \subseteq \mathcal{B}_0$ of the material configuration $\mathcal{B}_0$ computes in terms of either the material or spatial *mass density* ρ_0 or ρ_t, respectively, as

$$\mathsf{M}(\mathcal{V}_0) = \int_{\mathcal{V}_0} \rho_0 \,\mathrm{d}V = \int_{\mathcal{V}_t} \rho_t \,\mathrm{d}v \quad \rightsquigarrow \quad \rho_0 = J\rho_t.$$

Then the global statement for the balance of mass or rather the global statement of mass conservation reads as

$$\dot{\mathsf{M}}(\mathcal{V}_0) = 0 \quad \forall \mathcal{V}_0.$$

Upon localization for arbitrary $\mathcal{V}_0$ the local statement for the balance of mass is expressed as

$$\dot{\rho}_0 = 0 \quad \rightsquigarrow \quad \rho_0 = \rho_0(\boldsymbol{X}) \quad \text{and} \quad \dot{\rho}_t + \rho_t \,\mathrm{div}\boldsymbol{v} = 0.$$

Note that no source or fluxes are assumed in the balance of mass, however, in the case that living matter shall be modeled that allows for mass density changes such terms need to be introduced, see [99].

2. Balance of Linear Momentum:
The total *linear momentum* $\mathbf{P}(\mathcal{V}_0)$ of an arbitrary continuous sub-body $\mathscr{V} \subseteq \mathscr{B}$ of a continuous body $\mathscr{B}$ that occupies the material sub-configuration $\mathcal{V}_0 \subseteq$

$\mathcal{B}_0$ of the material configuration $\mathcal{B}_0$ computes in terms of either the material or spatial *linear momentum density* $\boldsymbol{p}_0 := \rho_0 \boldsymbol{v}$ or $\boldsymbol{p}_t := \rho_t \boldsymbol{v}$, respectively, as

$$\mathsf{P}(\mathcal{V}_0) = \int_{\mathcal{V}_0} \boldsymbol{p}_0 \, \mathrm{d}V = \int_{\mathcal{V}_t} \boldsymbol{p}_t \, \mathrm{d}v \quad \rightsquigarrow \quad \boldsymbol{p}_0 = J \boldsymbol{p}_t.$$

Then the global statement for the balance of linear momentum reads as

$$\dot{\mathsf{P}}(\mathcal{V}_0) = \boldsymbol{\mathcal{F}}(\mathcal{V}_0) \quad \forall \mathcal{V}_0 \quad \text{with} \quad \boldsymbol{\mathcal{F}}(\mathcal{V}_0) := \int_{\mathcal{V}_0} \boldsymbol{b}_0 \, \mathrm{d}V + \int_{\partial \mathcal{V}_0} \boldsymbol{t}_0 \, \mathrm{d}A.$$

Here the linear momentum source $\boldsymbol{b}_0$ and flux $\boldsymbol{t}_0 := \boldsymbol{P} \cdot \boldsymbol{N}$ denote the body force density per unit volume $\mathrm{d}V$ and the surface force density, i.e. the traction per unit area $\mathrm{d}A$, respectively. Upon localization for arbitrary $\mathcal{V}_0$ the local statement for the balance of linear momentum is eventually expressed as

$$\dot{\boldsymbol{p}}_0 = \operatorname{Div} \boldsymbol{P} + \boldsymbol{b}_0 \quad \text{with} \quad \dot{\boldsymbol{p}}_0 = \rho_0 \dot{\boldsymbol{v}} \quad \rightsquigarrow \quad \rho_t \dot{\boldsymbol{v}} = \operatorname{div} \boldsymbol{\sigma} + \boldsymbol{b}_t.$$

In the above the Cauchy theorem and the Gauss theorem have been invoked before localization in order to transform the surface integral into a volume integral. Moreover the balance of mass has been incorporated to arrive at the latter version of the balance of linear momentum in terms of spatial description quantities.

3. Balance of Angular Momentum:
The total *angular momentum* $\mathsf{L}(\mathcal{V}_0)$ of an arbitrary continuous sub-body $\mathscr{V} \subseteq \mathscr{B}$ of a continuous body $\mathscr{B}$ that occupies the material sub-configuration $\mathcal{V}_0 \subseteq \mathcal{B}_0$ of the material configuration $\mathcal{B}_0$ computes in terms of either the material or spatial *angular momentum density* $\boldsymbol{l}_0 := \boldsymbol{r} \times \boldsymbol{p}_0$ or $\boldsymbol{l}_t := \boldsymbol{r} \times \boldsymbol{p}_t$, respectively, as

$$\mathsf{L}(\mathcal{V}_0) = \int_{\mathcal{V}_0} \boldsymbol{l}_0 \, \mathrm{d}V = \int_{\mathcal{V}_t} \boldsymbol{l}_t \, \mathrm{d}v \quad \rightsquigarrow \quad \boldsymbol{l}_0 = J \boldsymbol{l}_t.$$

In the above the spatial distance vector $\boldsymbol{r}$ from a fixed spatial point $\boldsymbol{x}_0$ is defined as $\boldsymbol{r} := \boldsymbol{x} - \boldsymbol{x}_0$. Then the global statement for the balance of angular momentum reads as

$$\dot{\mathsf{L}}(\mathcal{V}_0) = \boldsymbol{\mathcal{M}}(\mathcal{V}_0) \quad \forall \mathcal{V}_0 \quad \text{with} \quad \boldsymbol{\mathcal{M}}(\mathcal{V}_0) := \int_{\mathcal{V}_0} \boldsymbol{r} \times \boldsymbol{b}_0 \, \mathrm{d}V + \int_{\partial \mathcal{V}_0} \boldsymbol{r} \times \boldsymbol{t}_0 \, \mathrm{d}A.$$

Upon localization for arbitrary $\mathcal{V}_0$ the local statement for the balance of angular momentum is eventually expressed as

$$\boldsymbol{P} \cdot \boldsymbol{F}^t = \boldsymbol{F} \cdot \boldsymbol{P}^t \quad \rightsquigarrow \quad \boldsymbol{\tau}_{\sharp,\flat} = \boldsymbol{\tau}^t_{\sharp,\flat} \quad \text{and} \quad \boldsymbol{\sigma}_{\sharp,\flat} = \boldsymbol{\sigma}^t_{\sharp,\flat}.$$

To arrive at the above result the Cauchy theorem and the Gauss theorem have been invoked before localization in order to transform the surface integral into a volume integral. Moreover the balances of mass and linear momentum have been incorporated. Observe that the balance of angular momentum as stated in the above is implicitly satisfied for symmetric Cauchy stress. Note, however, that the symmetry of the Cauchy stress is a consequence of the absence of independent body couples and couple tractions in the above statement of the balance of angular momentum. If such quantities are present the Cauchy stress results non-symmetric and the balance of angular momentum has to be considered explicitly. This is the case for generalized continuum formulations such as e.g. *micropolar*, *micromorphic*, and *higher-order continua*.

4. Balance of Mechanical Energy:
The *mechanical energy* consists of the (macroscopic) *kinetic energy* and the *elastically stored energy*. The total mechanical energy $\mathsf{E}(\mathcal{V}_0) := \mathsf{K}(\mathcal{V}_0)+\mathsf{W}(\mathcal{V}_0)$ of an arbitrary continuous sub-body $\mathscr{V} \subseteq \mathscr{B}$ of a continuous body $\mathscr{B}$ that occupies the material sub-configuration $\mathcal{V}_0 \subseteq \mathcal{B}_0$ of the material configuration $\mathcal{B}_0$ computes in terms of either the material or spatial *mechanical energy density* $e_0 := k_0 + w_0$ or $e_t := k_t + w_t$, respectively, as

$$\mathsf{E}(\mathcal{V}_0) = \int_{\mathcal{V}_0} e_0 \, \mathrm{d}V = \int_{\mathcal{V}_t} e_t \, \mathrm{d}v \quad \rightsquigarrow \quad e_0 = J e_t.$$

In the above the material and spatial kinetic energy densities are defined as $2k_0 := \rho_0 \boldsymbol{v} \cdot \boldsymbol{v}$ and $2k_t := \rho_t \boldsymbol{v} \cdot \boldsymbol{v}$, respectively, whereas w_0 and w_t denote the material and spatial elastically stored energy densities, respectively. Then the global statement for the balance of mechanical energy reads as

$$\dot{\mathsf{E}}(\mathcal{V}_0) = \mathcal{P}(\mathcal{V}_0) \quad \forall \mathcal{V}_0 \quad \text{with} \quad \mathcal{P}(\mathcal{V}_0) := \int_{\mathcal{V}_0} \boldsymbol{v} \cdot \boldsymbol{b}_0 \, \mathrm{d}V + \int_{\partial \mathcal{V}_0} \boldsymbol{v} \cdot \boldsymbol{t}_0 \, \mathrm{d}A.$$

Upon localization for arbitrary $\mathcal{V}_0$ the local statement for the balance of mechanical energy is eventually expressed as

$$\dot{e}_0 = \mathrm{Div}(\boldsymbol{v} \cdot \boldsymbol{P}) + \boldsymbol{v} \cdot \boldsymbol{b}_0 \quad \rightsquigarrow \quad \dot{w}_0 = \wp_0 \quad \text{and} \quad \dot{w}_t = \wp_t.$$

To arrive at the above result the Cauchy theorem and the Gauss theorem have been invoked before localization in order to transform the surface integral into a volume integral. Moreover the balances of mass and linear momentum resulting in $\dot{k}_0 = \boldsymbol{v} \cdot [\mathrm{Div}\, \boldsymbol{P} + \boldsymbol{b}_0]$ have been incorporated to result in the latter statements. Thereby the *stress power density* per unit volume $\mathrm{d}V$ is defined as $\wp_0 := \boldsymbol{P} : \dot{\boldsymbol{F}}$. Observe that the power supply $\mathcal{P}(\mathcal{V}_0)$ has to be supplemented by further non-mechanical contributions such as, e.g., thermal, electric, magnetic, chemical, etc. power supplies if more general situations are considered. Likewise, the purely elastically stored energy has to be exchanged by the *internal energy* for these cases.

Stress Power Density:
The stress power density $\wp_0 := \boldsymbol{P} : \dot{\boldsymbol{F}}$ as originally expressed in terms of the Piola stress and the material time derivative of the deformation gradient allows to connect alternatively various stress and strain measures so as to render *power conjugate pairings*. For contravariant stress measures and covariant strain measures power conjugate pairings read

$$\wp_0 = \boldsymbol{\tau}_\sharp : \boldsymbol{d} = \frac{1}{2}\boldsymbol{\tau}_\sharp : \mathrm{L}_t\boldsymbol{j} = \boldsymbol{\tau}_\sharp : \mathrm{L}_t\boldsymbol{e} = \frac{1}{2}\boldsymbol{S} : \dot{\boldsymbol{C}} = \boldsymbol{S} : \dot{\boldsymbol{E}}.$$

Alternatively, for covariant stress measures and contravariant strain measures power conjugate pairings read

$$\wp_0 = \boldsymbol{\tau}_\flat : \boldsymbol{d} = -\frac{1}{2}\boldsymbol{\tau}_\flat : \mathrm{L}_t\boldsymbol{i} = \boldsymbol{\tau}_\flat : \mathrm{L}_t\boldsymbol{k} = -\frac{1}{2}\boldsymbol{T} : \dot{\boldsymbol{B}} = \boldsymbol{T} : \dot{\boldsymbol{K}}.$$

Finally, for co/contra- (mixed) variant stress measures and contra/co- (mixed) variant strain measures power conjugate pairings read

$$\wp_0 = \boldsymbol{\tau} : \boldsymbol{l} = \boldsymbol{\pi} : \boldsymbol{\lambda} = \boldsymbol{M} : \boldsymbol{L} = \boldsymbol{P} : \boldsymbol{\Lambda}.$$

Likewise, the stress power density per unit volume dv in the spatial configuration $\mathcal{B}_t$ expands as $\wp_t = \boldsymbol{\sigma}_\sharp : \boldsymbol{d}$. The combination of power conjugate pairings of stress and strain measures is of importance for the determination of hyperelastic constitutive relations in the sequel.

Master Constitutive Relation:
For (hyper) elasticity the constitutive relations for the stresses in terms of the strains follow from the balance of mechanical energy, i.e. from the comparison of the material time derivative of the elastically stored energy density w_0 with the stress power density $\wp_0$. Due to the elementary definition $\wp_0 = \boldsymbol{P} : \boldsymbol{\Lambda} = \boldsymbol{P} : \dot{\boldsymbol{F}}$ the tensor function w_0 depends most generally on $\boldsymbol{F}$, i.e. the *master constitutive relation* follows as

$$w_0 = w_0(\boldsymbol{F}) \quad \rightsquigarrow \quad \boldsymbol{P} = \frac{\partial w_0}{\partial \boldsymbol{F}}.$$

Here any possible further dependence of w_0 on the material coordinates $\boldsymbol{X}$ that is needed to capture material inhomogeneities has been suppressed to ease notation. Moreover, it is noted as a remark that due to the alternative elementary definition $\wp_0 = \boldsymbol{\pi} : \boldsymbol{\lambda} = -\boldsymbol{\pi} : \dot{\boldsymbol{f}}$ the tensor function w_0 may likewise depend on $\boldsymbol{f} = \boldsymbol{F}^{-1}$, i.e.

$$w_0 = w_0(\boldsymbol{f}) \quad \rightsquigarrow \quad \boldsymbol{\pi} = -\frac{\partial w_0}{\partial \boldsymbol{f}}.$$

The dependence $w_0 = w_0(\boldsymbol{F})$ (or, likewise, $w_0 = w_0(\boldsymbol{f})$) is however not completely arbitrary but is subject to requirements of spatial objectivity, material symmetry, and covariance. These concepts are discussed in the sequel.

Spatially Superposed Rigid Body Motions:
Rigid body motions superposed to the *spatial* configuration in terms of temporarily varying rotations $\boldsymbol{Q}(t) \in \mathrm{SO}(3)$ and translations $\boldsymbol{z}(t) \in \mathrm{T}(3)$ with $\mathrm{T}(3)$ the translation group are expressed as

$$\boldsymbol{x}^* = \boldsymbol{y}^*(\boldsymbol{X}, t) = \boldsymbol{Q}(t) \cdot \boldsymbol{y}(\boldsymbol{X}, t) + \boldsymbol{z}(t) \quad \rightsquigarrow \quad \boldsymbol{F}^* = \boldsymbol{Q}(t) \cdot \boldsymbol{F}(\boldsymbol{X}, t).$$

Then, due to the orthogonality of $\boldsymbol{Q}$ the co- and contravariant material strain tensors $\boldsymbol{C}$ and $\boldsymbol{B}$ remain *invariant*, i.e. unaffected under a spatially superposed rigid body motion, i.e.

$$\boldsymbol{C}^* = \boldsymbol{F}^t \cdot \boldsymbol{Q}^t \cdot \boldsymbol{Q} \cdot \boldsymbol{F} = \boldsymbol{C} \quad \text{and} \quad \boldsymbol{B}^* = \boldsymbol{f} \cdot \boldsymbol{Q}^t \cdot \boldsymbol{Q} \cdot \boldsymbol{f}^t = \boldsymbol{B}.$$

Vice versa, the co- and contravariant spatial strain tensors $\boldsymbol{c}$ and $\boldsymbol{b}$ transform *objectively* under a spatially superposed rigid body motion, i.e.

$$\boldsymbol{c}^* = \boldsymbol{Q} \cdot \boldsymbol{f}^t \cdot \boldsymbol{f} \cdot \boldsymbol{Q}^t = \boldsymbol{Q} \cdot \boldsymbol{c} \cdot \boldsymbol{Q}^t \quad \text{and} \quad \boldsymbol{b}^* = \boldsymbol{Q} \cdot \boldsymbol{F} \cdot \boldsymbol{F}^t \cdot \boldsymbol{Q}^t = \boldsymbol{Q} \cdot \boldsymbol{b} \cdot \boldsymbol{Q}^t.$$

Recall that second-order tensors $\boldsymbol{T}$, considered as linear maps $\boldsymbol{v} = \boldsymbol{T} \cdot \boldsymbol{u}$ of vectors $\boldsymbol{u}$ into vectors $\boldsymbol{v}$, transform objectively into $\boldsymbol{T}^*$ if the transformed vectors $\boldsymbol{u}^* := \boldsymbol{Q} \cdot \boldsymbol{u}$ and $\boldsymbol{v}^* := \boldsymbol{Q} \cdot \boldsymbol{v}$ are mapped into each other as $\boldsymbol{v}^* = \boldsymbol{T}^* \cdot \boldsymbol{u}^*$. Proper transformation behavior under spatially superposed rigid body motions is alternatively denoted as *objectivity*, *frame indifference*, or *isotropy of space*. Spatial objectivity is a key requirement in the theory of constitutive modeling.

Note further that, with $\boldsymbol{\omega} := \boldsymbol{Q}^t \cdot \dot{\boldsymbol{Q}}$, the material time derivatives of $\boldsymbol{c}$ and $\boldsymbol{b}$ do not transform objectively since

$$\dot{\boldsymbol{c}}^* = \boldsymbol{Q} \cdot \left[\dot{\boldsymbol{c}} - 2[\boldsymbol{c} \cdot \boldsymbol{\omega}]^{\mathrm{sym}}\right] \cdot \boldsymbol{Q}^t \quad \textit{and} \quad \dot{\boldsymbol{b}}^* = \boldsymbol{Q} \cdot \left[\dot{\boldsymbol{b}} + 2[\boldsymbol{\omega} \cdot \boldsymbol{b}]^{\mathrm{sym}}\right] \cdot \boldsymbol{Q}^t.$$

However, since the mixed-variant spatial velocity gradient transforms as

$$\boldsymbol{l}^* = \boldsymbol{Q} \cdot [\boldsymbol{l} + \boldsymbol{\omega}] \cdot \boldsymbol{Q}^t$$

the co- and contravariant Lie time derivatives of $\boldsymbol{c}$ and $\boldsymbol{b}$, respectively, do transform objectively with

$$L_t \boldsymbol{c}^* = \boldsymbol{Q} \cdot L_t \boldsymbol{c} \cdot \boldsymbol{Q}^t \quad \textit{and} \quad L_t \boldsymbol{b}^* = \boldsymbol{Q} \cdot L_t \boldsymbol{b} \cdot \boldsymbol{Q}^t.$$

Objective time derivatives of spatial strain measures play an important role if constitutive relations shall be formulated incrementally.

Materially Superposed Rigid Body Motions:
Rigid body motions superposed to the *material* configuration in terms of temporarily varying rotations $\boldsymbol{q}(t) \in \mathrm{SO}(3)$ with $\boldsymbol{Q} = \boldsymbol{q}^{-1} = \boldsymbol{q}^t$ and translations $\boldsymbol{Z}(t) \in \mathrm{T}(3)$ are expressed as

$$\boldsymbol{X}_* = \boldsymbol{Y}_*(\boldsymbol{x},t) = \boldsymbol{q}(t)\cdot\boldsymbol{Y}(\boldsymbol{x},t) + \boldsymbol{Z}(t) \quad \rightsquigarrow \quad \boldsymbol{F}_* = \boldsymbol{F}(\boldsymbol{X},t)\cdot\boldsymbol{Q}(t).$$

Then, due to the orthogonality of $\boldsymbol{Q}$ the co- and contravariant spatial strain tensors $\boldsymbol{c}$ and $\boldsymbol{b}$ remain *invariant* under a materially superposed rigid body motion, i.e.

$$\boldsymbol{c}_* = \boldsymbol{f}^t\cdot\boldsymbol{Q}\cdot\boldsymbol{Q}^t\cdot\boldsymbol{f} = \boldsymbol{c} \quad \text{and} \quad \boldsymbol{b}_* = \boldsymbol{F}\cdot\boldsymbol{Q}\cdot\boldsymbol{Q}^t\cdot\boldsymbol{F}^t = \boldsymbol{b}.$$

Vice versa, the co- and contravariant material strain tensors $\boldsymbol{C}$ and $\boldsymbol{B}$ transform *objectively* under a materially superposed rigid body motion, i.e.

$$\boldsymbol{C}_* = \boldsymbol{Q}^t\cdot\boldsymbol{F}^t\cdot\boldsymbol{F}\cdot\boldsymbol{Q} = \boldsymbol{Q}^t\cdot\boldsymbol{C}\cdot\boldsymbol{Q} \quad \text{and} \quad \boldsymbol{B}_* = \boldsymbol{Q}^t\cdot\boldsymbol{f}\cdot\boldsymbol{f}^t\cdot\boldsymbol{Q} = \boldsymbol{Q}^t\cdot\boldsymbol{B}\cdot\boldsymbol{Q}.$$

Material strain measures transforming objectively under materially superposed rigid body motions are suited to capture anisotropic material behavior.

Covariance:
In abstract terms *covariance* requires *form invariance* of the constraints $\mathfrak{c}(\boldsymbol{T}_1,\boldsymbol{T}_2,\cdots) = 0$, i.e. the algebraic, differential, or integral equations ruling the relation between the various tensor fields $\boldsymbol{T}_1(\boldsymbol{X}),\boldsymbol{T}_2(\boldsymbol{X}),\cdots$ that describe a particular continuum (field) theory. Thereby form invariance is requested under arbitrary superposed, e.g. material, *diffeomorphisms* (differentiable mappings between manifolds)

$$\boldsymbol{x}^\star = \boldsymbol{y}^\star(\boldsymbol{X},\tau) \quad \rightsquigarrow \quad \boldsymbol{F}^\star := \mathrm{Grad}\,\boldsymbol{y}^\star.$$

Thus, with the transformation of a generic tensor field $\{\bullet\}$ under the action of $\boldsymbol{y}^\star$ denoted by $y^\star(\{\bullet\})$, covariance results in the requirement

$$y^\star(\mathfrak{c}(\boldsymbol{T}_1,\boldsymbol{T}_2,\cdots)) \doteq \mathfrak{c}(y^\star(\boldsymbol{T}_1),y^\star(\boldsymbol{T}_2),\cdots).$$

Alternatively, covariance could likewise be formulated based on arbitrary superposed spatial diffeomorphisms. As a paradigm Einstein's Theory of General Relativity/Gravitation is a covariant continuum (field) theory. Covariance embodies the case of invariance under arbitrary isometries, i.e. superposed rigid body motions as discussed in the above, but it is more general in that it allows arbitrary superposed motions or, likewise, coordinate transformations. Summarizing, the basic philosophy behind the requirement of covariance is that covariant theories shall be valid without resorting to a specific coordinate system, i.e. they shall make intrinsic tensorial sense.

Restrictions on the Elastically Stored Energy Density:
Requiring the elastically stored energy density w_0 to transform objectively under arbitrary spatially superposed rigid body motion, i.e. $w_0 \doteq w_0^*$ (scalar functions transform objectively if they remain invariant), results in the alternative restrictions

$$\begin{aligned} w_0 &\doteq w_0^* = w_0(\boldsymbol{C}; \cdots), \\ w_0 &\doteq w_0^* = w_0(\boldsymbol{B}; \cdots), \\ w_0 &\doteq w_0^* = w_0(\boldsymbol{Q} \cdot \boldsymbol{c} \cdot \boldsymbol{Q}^t; \cdots) \quad \forall \boldsymbol{Q} \in \mathrm{SO}(3), \\ w_0 &\doteq w_0^* = w_0(\boldsymbol{Q} \cdot \boldsymbol{b} \cdot \boldsymbol{Q}^t; \cdots) \quad \forall \boldsymbol{Q} \in \mathrm{SO}(3). \end{aligned}$$

Thus an arbitrary dependence on either of the material description strain measures $\boldsymbol{C}$ or $\boldsymbol{B}$ (and additive combinations with the corresponding material unit tensors $\boldsymbol{J}$ or $\boldsymbol{I}$ into, e.g., $\boldsymbol{E}$ or $\boldsymbol{K}$, respectively) automatically satisfies spatial objectivity of w_0. In contrast the dependence on either of the spatial description strain measures $\boldsymbol{c}$ or $\boldsymbol{b}$ (and additive combinations with the corresponding spatial unit tensors $\boldsymbol{j}$ or $\boldsymbol{i}$ into, e.g., $\boldsymbol{e}$ or $\boldsymbol{k}$, respectively) is subject to the above *isotropy* conditions on the tensor function w_0 in order to satisfy spatial objectivity. However, the yet unspecified arguments of w_0 indicated by the $\cdots$ allow to also capture *anisotropic* material behavior. To this end sets of co- and contravariant material *structural tensors*

$$\boldsymbol{N}_0 := \boldsymbol{J}, \quad \boldsymbol{N}_\alpha \quad \text{and} \quad \boldsymbol{M}_0 := \boldsymbol{I}, \quad \boldsymbol{M}_\alpha \quad \text{with} \quad \alpha = 1, \cdots, n_{\text{st}}$$

are introduced that capture the *material's symmetry group* $\mathrm{M}(3) \subseteq \mathrm{SO}(3)$. Here the co- and contravariant material unit tensors $\boldsymbol{J}$ and $\boldsymbol{I}$ are explicitly incorporated in order to capture the isotropic part of the response. The co- and contravariant material structural tensors satisfy the following invariance conditions related to the materials' symmetry group

$$\begin{aligned} \boldsymbol{N}_\alpha &\equiv \boldsymbol{Q}^t \cdot \boldsymbol{N}_\alpha \cdot \boldsymbol{Q} \;\; \forall \boldsymbol{Q} \in \mathrm{M}(3) \subseteq \mathrm{SO}(3), \quad \alpha = 1, \cdots, n_{\text{st}} \\ \boldsymbol{M}_\alpha &\equiv \boldsymbol{Q}^t \cdot \boldsymbol{M}_\alpha \cdot \boldsymbol{Q} \;\; \forall \boldsymbol{Q} \in \mathrm{M}(3) \subseteq \mathrm{SO}(3), \quad \alpha = 1, \cdots, n_{\text{st}}. \end{aligned}$$

In passing it is noted that the push-forward of the co- and contravariant material structural tensors into their spatial counterparts reads as

$$\boldsymbol{n}_0 = \boldsymbol{c}, \quad \boldsymbol{n}_\alpha = \boldsymbol{f}^t \cdot \boldsymbol{N}_\alpha \cdot \boldsymbol{f} \quad \text{and} \quad \boldsymbol{m}_0 = \boldsymbol{b}, \quad \boldsymbol{m}_\alpha = \boldsymbol{F} \cdot \boldsymbol{M}_\alpha \cdot \boldsymbol{F}^t.$$

Furthermore, the spatial structural tensors are Killing fields with vanishing Lie time derivative

$$\mathrm{L}_t \boldsymbol{n}_\alpha = \boldsymbol{0} \quad \text{and} \quad \mathrm{L}_t \boldsymbol{m}_\alpha = \boldsymbol{0}.$$

The material structural tensors might be thought of as dyadic products of preferred anisotropy directions. For example isotropy is characterized by the absence of any preferred direction with $n_{\text{st}} = 0$, transverse isotropy is characterized by one preferred direction that is normal to the plane of isotropy, likewise orthotropy is characterized by three preferred orthogonal directions that span the orthotropy planes (however only two of the three preferred directions are independent). Based on the material structural tensors and in accordance with the above restriction of spatial objectivity the elastically stored energy density w_0 then takes the following alternative representations

$$w_0 = w_0(\boldsymbol{C}; \boldsymbol{I}, \boldsymbol{M}_\alpha) \quad \text{and} \quad w_0 = w_0(\boldsymbol{B}; \boldsymbol{J}, \boldsymbol{N}_\alpha).$$

Finally, requiring invariance of w_0 under arbitrary materially superposed rigid body motions, i.e. the material strain measures are transformed identically to the material structural tensors, results in the following alternative restrictions

$$\begin{aligned} w_0 \doteq w_{0*} &= w_0(\boldsymbol{Q}^t \cdot \{\boldsymbol{C}; \boldsymbol{I}, \boldsymbol{M}_\alpha\} \cdot \boldsymbol{Q}) \quad \forall \boldsymbol{Q} \in \mathrm{SO}(3), \\ w_0 \doteq w_{0*} &= w_0(\boldsymbol{Q}^t \cdot \{\boldsymbol{B}; \boldsymbol{J}, \boldsymbol{N}_\alpha\} \cdot \boldsymbol{Q}) \quad \forall \boldsymbol{Q} \in \mathrm{SO}(3). \end{aligned}$$

These restrictions qualify w_0 as an *isotropic tensor function* of all of its arguments, i.e. either $\{\boldsymbol{C}; \boldsymbol{I}, \boldsymbol{M}_\alpha\}$ or $\{\boldsymbol{B}; \boldsymbol{J}, \boldsymbol{N}_\alpha\}$. Representation theorems then provide the most general dependence of isotropic tensor functions on their arguments, typically in terms of various types of invariants $I_1, I_2, I_3, \cdots$ or $i_1, i_2, i_3, \cdots$. As an example, with $n_{\mathrm{st}} = 1$ transverse isotropy is, e.g., formulated in terms of the three previously introduced invariants I_1, I_2, I_3 and the two additional invariants

$$I_4 := \boldsymbol{C} : \boldsymbol{M}_1 = \boldsymbol{j} : \boldsymbol{m}_1, I_5 := [\boldsymbol{I} \cdot \boldsymbol{C}] : [\boldsymbol{C} \cdot \boldsymbol{M}_1] = [\boldsymbol{b} \cdot \boldsymbol{j}] : [\boldsymbol{j} \cdot \boldsymbol{m}_1].$$

Requiring next covariance of the elastically stored energy density w_0 under arbitrary superposed material diffeomorphisms results in the alternative equivalent statements

$$\begin{aligned} w_0(\boldsymbol{C}; \boldsymbol{I}, \boldsymbol{M}_\alpha) &\doteq w_0(\underbrace{\boldsymbol{f}^{\star t} \cdot \boldsymbol{C} \cdot \boldsymbol{f}^\star}_{=:\boldsymbol{C}^\star}; \underbrace{\boldsymbol{F}^\star \cdot \boldsymbol{I} \cdot \boldsymbol{F}^{\star t}}_{=:\boldsymbol{I}^\star}, \underbrace{\boldsymbol{F}^\star \cdot \boldsymbol{M}_\alpha \cdot \boldsymbol{F}^{\star t}}_{=:\boldsymbol{M}^\star_\alpha}), \\ w_0(\boldsymbol{B}; \boldsymbol{J}, \boldsymbol{N}_\alpha) &\doteq w_0(\underbrace{\boldsymbol{F}^\star \cdot \boldsymbol{B} \cdot \boldsymbol{F}^{\star t}}_{=:\boldsymbol{B}^\star}; \underbrace{\boldsymbol{f}^{\star t} \cdot \boldsymbol{J} \cdot \boldsymbol{f}^\star}_{=:\boldsymbol{J}^\star}, \underbrace{\boldsymbol{f}^{\star t} \cdot \boldsymbol{N}_\alpha \cdot \boldsymbol{f}^\star}_{=:\boldsymbol{N}^\star_\alpha}). \end{aligned}$$

Here $\boldsymbol{f}^\star$ abbreviates the spatial gradient of the inverse (spatial) diffeomorphism $\boldsymbol{Y}^\star$, i.e. $\boldsymbol{f}^\star := \boldsymbol{F}^{\star -1}$. Then the particular choice $\boldsymbol{y}^\star = \boldsymbol{y}$ with $\boldsymbol{F}^\star = \boldsymbol{F}$ and $\boldsymbol{f}^\star = \boldsymbol{f}$ renders the alternative equivalent parameterizations of w_0

$$w_0(\boldsymbol{C}; \boldsymbol{I}, \boldsymbol{M}_\alpha) \doteq w_0(\boldsymbol{j}; \boldsymbol{b}, \boldsymbol{m}_\alpha) \quad \text{and} \quad w_0(\boldsymbol{B}; \boldsymbol{J}, \boldsymbol{N}_\alpha) \doteq w_0(\boldsymbol{i}; \boldsymbol{c}, \boldsymbol{n}_\alpha).$$

Finally, requiring objectivity of w_0 under arbitrary spatially superposed rigid body motions results in the following alternative restrictions

$$w_0 \doteq w_0^* = w_0(\boldsymbol{Q} \cdot \{\boldsymbol{j}; \boldsymbol{b}, \boldsymbol{m}_\alpha\} \cdot \boldsymbol{Q}^t) \quad \forall \boldsymbol{Q} \in \mathrm{SO}(3),$$
$$w_0 \doteq w_0^* = w_0(\boldsymbol{Q} \cdot \{\, \boldsymbol{i}; \boldsymbol{c},\ \boldsymbol{n}_\alpha\} \cdot \boldsymbol{Q}^t) \quad \forall \boldsymbol{Q} \in \mathrm{SO}(3).$$

These restrictions qualify w_0 again as an *isotropic tensor function* of its arguments, i.e. either $\{\boldsymbol{j}; \boldsymbol{b}, \boldsymbol{m}_\alpha\}$ or $\{\boldsymbol{i}; \boldsymbol{c}, \boldsymbol{n}_\alpha\}$, respectively. Indeed the invariants $I_1, I_2, I_3, \cdots$ and $i_1, i_2, i_3, \cdots$ parameterizing an isotropic tensor function are themselves covariant as has been partly demonstrated in the above. A comprehensive list of isotropic elastically stored energy densities is discussed in [188, 84]. Covariance requires that w_0 remains invariant under arbitrary (material) diffeomorphisms, thus its rate with respect to a time-like parameter τ has to vanish at fixed actual deformation

$$\mathrm{D}_\tau w_0|_{\boldsymbol{y}} = 0 \quad \rightsquigarrow \quad w_0 = \text{const} \quad \forall \boldsymbol{y}^\star.$$

Exploiting this condition results eventually in a relation between the derivatives of w_0 with respect to its various arguments

$$\frac{\partial w_0}{\partial \boldsymbol{C}^\star} = \boldsymbol{B}^\star \cdot \frac{\partial w_0}{\partial \boldsymbol{I}^\star} \cdot \boldsymbol{I}^\star + \boldsymbol{B}^\star \cdot \frac{\partial w_0}{\partial \boldsymbol{M}_\alpha^\star} \cdot \boldsymbol{M}_\alpha^\star = \boldsymbol{I}^\star \cdot \frac{\partial w_0}{\partial \boldsymbol{I}^\star} \cdot \boldsymbol{B}^\star + \boldsymbol{M}_\alpha^\star \cdot \frac{\partial w_0}{\partial \boldsymbol{M}_\alpha^\star} \cdot \boldsymbol{B}^\star,$$

$$\frac{\partial w_0}{\partial \boldsymbol{B}^\star} = \boldsymbol{C}^\star \cdot \frac{\partial w_0}{\partial \boldsymbol{J}^\star} \cdot \boldsymbol{J}^\star + \boldsymbol{C}^\star \cdot \frac{\partial w_0}{\partial \boldsymbol{N}_\alpha^\star} \cdot \boldsymbol{N}_\alpha^\star = \boldsymbol{J}^\star \cdot \frac{\partial w_0}{\partial \boldsymbol{J}^\star} \cdot \boldsymbol{C}^\star + \boldsymbol{N}_\alpha^\star \cdot \frac{\partial w_0}{\partial \boldsymbol{N}_\alpha^\star} \cdot \boldsymbol{C}^\star.$$

Observe the symmetry conditions for $\boldsymbol{C}^\star$ and $\boldsymbol{B}^\star$ reflected by the two alternative statements on the right-hand-sides. Then, for the particular choice $\boldsymbol{y}^\star = \boldsymbol{X}$, i.e. the identity map, the derivatives of a covariant elastically stored energy density w_0 with respect to the various material strain measures and unit tensors are related as

$$\frac{\partial w_0}{\partial \boldsymbol{C}} = \boldsymbol{B} \cdot \frac{\partial w_0}{\partial \boldsymbol{I}} \cdot \boldsymbol{I} + \boldsymbol{B} \cdot \frac{\partial w_0}{\partial \boldsymbol{M}_\alpha} \cdot \boldsymbol{M}_\alpha = \boldsymbol{I} \cdot \frac{\partial w_0}{\partial \boldsymbol{I}} \cdot \boldsymbol{B} + \boldsymbol{M}_\alpha \cdot \frac{\partial w_0}{\partial \boldsymbol{M}_\alpha} \cdot \boldsymbol{B}.$$

$$\frac{\partial w_0}{\partial \boldsymbol{B}} = \boldsymbol{C} \cdot \frac{\partial w_0}{\partial \boldsymbol{J}} \cdot \boldsymbol{J} + \boldsymbol{C} \cdot \frac{\partial w_0}{\partial \boldsymbol{N}_\alpha} \cdot \boldsymbol{N}_\alpha = \boldsymbol{J} \cdot \frac{\partial w_0}{\partial \boldsymbol{J}} \cdot \boldsymbol{C} + \boldsymbol{N}_\alpha \cdot \frac{\partial w_0}{\partial \boldsymbol{N}_\alpha} \cdot \boldsymbol{C}.$$

Likewise, for the particular choice $\boldsymbol{y}^\star = \boldsymbol{y}$, i.e. the deformation map, the derivatives of a covariant elastically stored energy density w_0 with respect to the various spatial strain measures and unit tensors are related as

$$\frac{\partial w_0}{\partial \boldsymbol{j}} = \boldsymbol{i} \cdot \frac{\partial w_0}{\partial \boldsymbol{b}} \cdot \boldsymbol{b} + \boldsymbol{i} \cdot \frac{\partial w_0}{\partial \boldsymbol{m}_\alpha} \cdot \boldsymbol{m}_\alpha = \boldsymbol{b} \cdot \frac{\partial w_0}{\partial \boldsymbol{b}} \cdot \boldsymbol{i} + \boldsymbol{m}_\alpha \cdot \frac{\partial w_0}{\partial \boldsymbol{m}_\alpha} \cdot \boldsymbol{i}.$$

$$\frac{\partial w_0}{\partial \boldsymbol{i}} = \boldsymbol{j} \cdot \frac{\partial w_0}{\partial \boldsymbol{c}} \cdot \boldsymbol{c} + \boldsymbol{j} \cdot \frac{\partial w_0}{\partial\, \boldsymbol{n}_\alpha} \cdot\ \boldsymbol{n}_\alpha = \boldsymbol{c} \cdot \frac{\partial w_0}{\partial \boldsymbol{c}} \cdot \boldsymbol{j} +\ \boldsymbol{n}_\alpha \cdot \frac{\partial w_0}{\partial\, \boldsymbol{n}_\alpha} \cdot \boldsymbol{j}.$$

As will become obvious in the sequel, these derivatives of w_0 are related to various material and spatial description stress measures.

Constitutive Relations:
Elastically stored energy densities w_0 that satisfy all requirements resulting from spatial objectivity, material symmetry, and covariance are isotropic tensor functions of their material or spatial, co- and contravariant arguments $\{\boldsymbol{C};\boldsymbol{I},\boldsymbol{M}_\alpha\} \leftrightarrow \{\boldsymbol{j};\boldsymbol{b},\boldsymbol{m}_\alpha\}$ or $\{\boldsymbol{B};\boldsymbol{J},\boldsymbol{N}_\alpha\} \leftrightarrow \{\boldsymbol{i};\boldsymbol{c},\boldsymbol{n}_\alpha\}$ (here $\leftrightarrow$ denotes covariantly equivalent sets of tensor fields). Then the comparison of the material time derivative of w_0 with the various versions of the stress power density $\wp_0$ renders appropriate constitutive relations for the material stress measures

$$\boldsymbol{S} = \frac{\partial w_0}{\partial \boldsymbol{E}} = +2\frac{\partial w_0}{\partial \boldsymbol{C}} = +2\boldsymbol{B}\cdot\frac{\partial w_0}{\partial \boldsymbol{I}}\cdot\boldsymbol{I} + 2\boldsymbol{B}\cdot\frac{\partial w_0}{\partial \boldsymbol{M}_\alpha}\cdot\boldsymbol{M}_\alpha.$$

$$\boldsymbol{T} = \frac{\partial w_0}{\partial \boldsymbol{K}} = -2\frac{\partial w_0}{\partial \boldsymbol{B}} = -2\boldsymbol{J}\cdot\frac{\partial w_0}{\partial \boldsymbol{J}}\cdot\boldsymbol{C} - 2\boldsymbol{N}_\alpha\cdot\frac{\partial w_0}{\partial \boldsymbol{N}_\alpha}\cdot\boldsymbol{C}.$$

Consequently the co/contra- (mixed) variant Mandel stress computes as

$$\boldsymbol{M} = +2\frac{\partial w_0}{\partial \boldsymbol{I}}\cdot\boldsymbol{I} + 2\frac{\partial w_0}{\partial \boldsymbol{M}_\alpha}\cdot\boldsymbol{M}_\alpha = -2\boldsymbol{J}\cdot\frac{\partial w_0}{\partial \boldsymbol{J}} - 2\boldsymbol{N}_\alpha\cdot\frac{\partial w_0}{\partial \boldsymbol{N}_\alpha}.$$

Since for a scalar function $\mathrm{L}_t w_0 \equiv \mathrm{D}_t w_0$ the comparison of the Lie time derivative of w_0 with the various versions of the stress power density $\wp_0$ renders appropriate constitutive relations for the spatial stress measures

$$\boldsymbol{\tau}_\sharp = \frac{\partial w_0}{\partial \boldsymbol{e}} = +2\frac{\partial w_0}{\partial \boldsymbol{j}} = +2\boldsymbol{i}\cdot\frac{\partial w_0}{\partial \boldsymbol{b}}\cdot\boldsymbol{b} + 2\boldsymbol{i}\cdot\frac{\partial w_0}{\partial \boldsymbol{m}_\alpha}\cdot\boldsymbol{m}_\alpha.$$

$$\boldsymbol{\tau}_\flat = \frac{\partial w_0}{\partial \boldsymbol{k}} = -2\frac{\partial w_0}{\partial \boldsymbol{i}} = -2\boldsymbol{c}\cdot\frac{\partial w_0}{\partial \boldsymbol{c}}\cdot\boldsymbol{j} - 2\boldsymbol{n}_\alpha\cdot\frac{\partial w_0}{\partial \boldsymbol{n}_\alpha}\cdot\boldsymbol{j}.$$

Consequently the co/contra- (mixed) variant Kirchhoff stress computes as

$$\boldsymbol{\tau} = +2\frac{\partial w_0}{\partial \boldsymbol{b}}\cdot\boldsymbol{b} + 2\frac{\partial w_0}{\partial \boldsymbol{m}_\alpha}\cdot\boldsymbol{m}_\alpha = -2\boldsymbol{c}\cdot\frac{\partial w_0}{\partial \boldsymbol{c}} - 2\boldsymbol{n}_\alpha\cdot\frac{\partial w_0}{\partial \boldsymbol{n}_\alpha}.$$

The relations for $\boldsymbol{\tau}$ in terms of the spatial unit tensors are due to [52], for the case of isotropy the alternative formats for $\boldsymbol{\tau}$ are due to [200, 149]. For the case of anisotropy the alternative formats for $\boldsymbol{S}$ and $\boldsymbol{\tau}$ have been proposed in [129], the corresponding alternative representations for $\boldsymbol{T}$ and $\boldsymbol{M}$ are, to the best knowledge of the author, original to this contribution. Which of the above alternative constitutive relations is chosen is essentially a matter of convenience.

Time Derivatives of Stress Measures:
The Lie time derivatives of the contra-, co-, and co/contra- (mixed) variant Kirchhoff stress are defined as

$$\mathrm{L}_t\boldsymbol{\tau}_\sharp := \boldsymbol{F}\cdot\dot{\boldsymbol{S}}\cdot\boldsymbol{F}^t, \quad \mathrm{L}_t\boldsymbol{\tau}_\flat := \boldsymbol{f}^t\cdot\dot{\boldsymbol{T}}\cdot\boldsymbol{f} \quad \text{and} \quad \mathrm{L}_t\boldsymbol{\tau} := \boldsymbol{f}^t\cdot\dot{\boldsymbol{M}}\cdot\boldsymbol{F}^t$$

and expand in the following representations

$$\mathrm{L}_t\boldsymbol{\tau}_\sharp = \dot{\boldsymbol{\tau}}_\sharp - \boldsymbol{l}\cdot\boldsymbol{\tau}_\sharp - \boldsymbol{\tau}_\sharp\cdot\boldsymbol{l}^t, \quad \mathrm{L}_t\boldsymbol{\tau}_\flat = \dot{\boldsymbol{\tau}}_\flat + \boldsymbol{l}^t\cdot\boldsymbol{\tau}_\flat + \boldsymbol{\tau}_\flat\cdot\boldsymbol{l}, \quad \mathrm{L}_t\boldsymbol{\tau} = \dot{\boldsymbol{\tau}} + \boldsymbol{l}^t\cdot\boldsymbol{\tau} - \boldsymbol{\tau}\cdot\boldsymbol{l}^t.$$

Alternatively, the Lie time derivative $\mathrm{L}_t\boldsymbol{\tau}_\sharp$ of the contravariant Kirchhoff stress is also denoted its *Oldroyd time derivative.* Likewise, with $\dot{\boldsymbol{\tau}}_\sharp = J[\dot{\boldsymbol{\sigma}}_\sharp + \boldsymbol{\sigma}_\sharp \operatorname{div}\boldsymbol{v}]$ whereby $\boldsymbol{v}$ denotes the spatial velocity field, e.g. the corresponding Lie time derivative of the contravariant Cauchy stress is expressed as

$$\mathrm{L}_t\boldsymbol{\sigma}_\sharp := j\,\mathrm{L}_t\boldsymbol{\tau}_\sharp = [\dot{\boldsymbol{\sigma}}_\sharp + \boldsymbol{\sigma}_\sharp \operatorname{div}\boldsymbol{v}] - \boldsymbol{l}\cdot\boldsymbol{\sigma}_\sharp - \boldsymbol{\sigma}_\sharp\cdot\boldsymbol{l}^t.$$

Alternatively, the Lie time derivative $\mathrm{L}_t\boldsymbol{\sigma}_\sharp$ of the contravariant Cauchy stress is also denoted its *Truesdell time derivative.* Moreover the *nominal time derivative* of the Cauchy and the Kirchhoff stress are defined as

$$\mathrm{N}_t\boldsymbol{\sigma} := j\dot{\boldsymbol{P}}\cdot\boldsymbol{F}^t = [\dot{\boldsymbol{\sigma}} + \boldsymbol{\sigma}\operatorname{div}\boldsymbol{v}] - \boldsymbol{\sigma}\cdot\boldsymbol{l}^t \quad \text{and} \quad \mathrm{N}_t\boldsymbol{\tau} := \dot{\boldsymbol{P}}\cdot\boldsymbol{F}^t = \dot{\boldsymbol{\tau}} - \boldsymbol{\tau}\cdot\boldsymbol{l}^t.$$

Note the relation $\mathrm{N}_t\boldsymbol{\sigma} = j\,\mathrm{N}_t\boldsymbol{\tau}$ between the nominal time derivative of the Kirchhoff and the Cauchy stress that again holds due to $\dot{\boldsymbol{\tau}} = J[\dot{\boldsymbol{\sigma}} + \boldsymbol{\sigma}\operatorname{div}\boldsymbol{v}]$. Finally, since the nominal time derivative of the Cauchy stress is a Piola transformation of the material time derivative of the Piola stress, and since the material time derivative commutes with the material divergence it follows that

$$\mathrm{D}_t \operatorname{Div}\boldsymbol{P} \equiv \operatorname{Div}\dot{\boldsymbol{P}} = J\operatorname{div}\mathrm{N}_t\boldsymbol{\sigma}.$$

This relations is the key to study the *incremental* balance of momentum in terms of spatial description quantities, e.g. when the wave propagation properties of an elastic material shall be analyzed.

Tangent Operators:
For incrementally linear material behavior (stress rate is a homogeneous function of degree one in the strain rate) the *tangent relation* between appropriate time derivatives of stress measures and corresponding rates of conjugated strain measures can be expressed in terms of fourth-order tensors (as counter-examples consider the cases of visco-elasticity or visco-elastoplasticity). In the sequel these fourth-order tensors shall be denoted as *tangent operators.* As the reference expression the tangent relation between the material time derivatives of the co/contra- (mixed) variant Piola stress $\boldsymbol{P}$ and the contra/co- (mixed) variant deformation gradient $\boldsymbol{F}$ reads

$$\dot{\boldsymbol{P}} = \mathbf{A} : \dot{\boldsymbol{F}}.$$

Here the tangent operator $\mathbf{A}$ is a co/contra/co/contra- (mixed) variant two-point description tensor. Likewise the tangent relation between the material time derivatives of the contravariant Piola-Kirchhoff stress $\boldsymbol{S}$ and the covariant Green-Lagrange strain $\boldsymbol{E}$ reads

$$\dot{\boldsymbol{S}} = \mathbf{C} : \dot{\boldsymbol{E}}.$$

Here the tangent operator $\mathbf{C}$ is a fully contravariant material description tensor. Then due to $\boldsymbol{P} = \boldsymbol{F}_\flat \cdot \boldsymbol{S}$ (here the covariant version of the deformation gradient $\boldsymbol{F}_\flat := \boldsymbol{j} \cdot \boldsymbol{F}$ as been included to take into account that $\boldsymbol{P}$ is co/contra- (mixed) variant) the relation between $\mathbf{A}$ and $\mathbf{C}$ follows as

$$\mathbf{A} = [\boldsymbol{F}_\flat \boxtimes \mathbf{1}_{\mathrm{m}}] : \mathbf{C} : [\boldsymbol{F}_\flat^t \boxtimes \mathbf{1}_{\mathrm{m}}^t] + \boldsymbol{j} \boxtimes \boldsymbol{S}.$$

Thereby $\mathbf{1}_{\mathrm{m}} := \boldsymbol{I} \cdot \boldsymbol{J}$ denotes the contra/co- (mixed) variant material description unit tensor. Furthermore the tangent relation between the nominal time derivative of the co/contra- (mixed) variant Kirchhoff stress $\boldsymbol{\tau}$ and the contra/co- (mixed) variant velocity gradient $\boldsymbol{l}$ reads

$$\mathrm{N}_t \boldsymbol{\tau} = \mathbf{a} : \boldsymbol{l} \quad \text{with} \quad \mathbf{a} = [\mathbf{1}_{\mathrm{s}}^t \boxtimes \boldsymbol{F}] : \mathbf{A} : [\mathbf{1}_{\mathrm{s}} \boxtimes \boldsymbol{F}^t].$$

Here the tangent operator $\mathbf{a}$ is a co/contra/co/contra- (mixed) variant spatial description tensor, whereas $\mathbf{1}_{\mathrm{s}} := \boldsymbol{i} \cdot \boldsymbol{j}$ denotes the contra/co- (mixed) variant spatial unit tensor. Likewise the tangent relation between the Lie time derivatives of the contravariant Kirchhoff stress $\boldsymbol{\tau}_\sharp$ and the covariant Almansi strain $\boldsymbol{e}$ reads

$$\mathrm{L}_t \boldsymbol{\tau}_\sharp = \mathbf{c} : \mathrm{L}_t \boldsymbol{e} \quad \text{with} \quad \mathbf{c} = [\boldsymbol{F} \boxtimes \boldsymbol{F}] : \mathbf{C} : [\boldsymbol{F}^t \boxtimes \boldsymbol{F}^t].$$

Here the tangent operator $\mathbf{c}$ is a fully contravariant spatial description tensor. Then the relation between $\mathbf{a}$ and $\mathbf{c}$ and thus the relation between $\mathbf{a}$ and $\mathbf{C}$ follows as

$$\mathbf{a} = [\boldsymbol{j} \boxtimes \mathbf{1}_{\mathrm{s}}] : \mathbf{c} : [\boldsymbol{j}^t \boxtimes \mathbf{1}_{\mathrm{s}}^t] + \boldsymbol{j} \boxtimes \boldsymbol{\tau}_\sharp = [\boldsymbol{F}_\flat \boxtimes \boldsymbol{F}] : \mathbf{C} : [\boldsymbol{F}_\flat^t \boxtimes \boldsymbol{F}^t] + \boldsymbol{j} \boxtimes \boldsymbol{\tau}_\sharp.$$

Typically the first contribution to either $\mathbf{A}$ or $\mathbf{a}$ is denoted the *material stiffness*, whereas the second contribution to either $\mathbf{A}$ or $\mathbf{a}$ is denoted the *geometric stiffness.*

Linearized Constitutive Relations:
In the case of (hyper) elasticity the co/contra/co/contra- (mixed) variant two-point tangent operator $\mathbf{A}$ of the master constitutive relation expands as the fourth-order tensor

$$\mathbf{A} = \frac{\partial^2 w_0}{\partial \boldsymbol{F} \partial \boldsymbol{F}}.$$

Accordingly, the fully contravariant material description tangent operator $\mathbf{C}$ computes as

$$\mathbf{C} = \frac{\partial^2 w_0}{\partial \boldsymbol{E} \partial \boldsymbol{E}} = 4 \frac{\partial^2 w_0}{\partial \boldsymbol{C} \partial \boldsymbol{C}}.$$

An alternative format for the material description tangent operator $\mathbf{C}$ reads

$$
\begin{aligned}
\mathsf{C} = & \; 4[\,\boldsymbol{I}\;\boxtimes\boldsymbol{B}] : \frac{\partial^2 w_0}{\partial\,\boldsymbol{I}\,\partial\,\boldsymbol{I}} : [\boldsymbol{B}\boxtimes\;\boldsymbol{I}\,] \\
& + 4[\boldsymbol{M}_\alpha\boxtimes\boldsymbol{B}] : \frac{\partial^2 w_0}{\partial\boldsymbol{M}_\alpha\partial\,\boldsymbol{I}} : [\boldsymbol{B}\boxtimes\;\boldsymbol{I}\,] \\
& + 4[\,\boldsymbol{I}\;\boxtimes\boldsymbol{B}] : \frac{\partial^2 w_0}{\partial\,\boldsymbol{I}\;\partial\boldsymbol{M}_\alpha} : [\boldsymbol{B}\boxtimes\boldsymbol{M}_\alpha] \\
& + 4[\boldsymbol{M}_\alpha\boxtimes\boldsymbol{B}] : \frac{\partial^2 w_0}{\partial\boldsymbol{M}_\alpha\partial\boldsymbol{M}_\beta} : [\boldsymbol{B}\boxtimes\boldsymbol{M}_\beta].
\end{aligned}
$$

Likewise, the corresponding fully contravariant spatial description tangent operator c is expressed as

$$
\mathsf{c} = \frac{\partial^2 w_0}{\partial\boldsymbol{e}\partial\boldsymbol{e}} = 4\frac{\partial^2 w_0}{\partial\boldsymbol{j}\partial\boldsymbol{j}}.
$$

An alternative format for the spatial description tangent operator c reads

$$
\begin{aligned}
\mathsf{c} = & \; 4[\,\boldsymbol{b}\;\boxtimes\boldsymbol{i}] : \frac{\partial^2 w_0}{\partial\,\boldsymbol{b}\,\partial\,\boldsymbol{b}} : [\boldsymbol{i}\boxtimes\;\boldsymbol{b}\,] \\
& + 4[\boldsymbol{m}_\alpha\boxtimes\boldsymbol{i}] : \frac{\partial^2 w_0}{\partial\boldsymbol{m}_\alpha\partial\,\boldsymbol{b}} : [\boldsymbol{i}\boxtimes\;\boldsymbol{b}\,] \\
& + 4[\,\boldsymbol{b}\;\boxtimes\boldsymbol{i}] : \frac{\partial^2 w_0}{\partial\,\boldsymbol{b}\;\partial\boldsymbol{m}_\alpha} : [\boldsymbol{i}\boxtimes\boldsymbol{m}_\alpha] \\
& + 4[\boldsymbol{m}_\alpha\boxtimes\boldsymbol{i}] : \frac{\partial^2 w_0}{\partial\boldsymbol{m}_\alpha\partial\boldsymbol{m}_\beta} : [\boldsymbol{i}\boxtimes\boldsymbol{m}_\beta].
\end{aligned}
$$

For the case of isotropy the alternative expression for c is due to [131]. For the case of anisotropy the alternative formats for C and c have been proposed in [129]. Which of the above alternative linearized constitutive relations is chosen is essentially a matter of convenience.

Propagating Plane Waves:
Based on an underlying homogeneous deformation state with spatially constant deformation gradient $\boldsymbol{F}$, Piola stress $\boldsymbol{P}$, and tangent operator A the incremental balance of momentum reads in the absence of body forces

$$
\operatorname{Div}\left(\mathsf{A} : \operatorname{Grad} \mathrm{d}\boldsymbol{y}\right) = \rho_0\,\ddot{\overline{\mathrm{d}\boldsymbol{y}}}.
$$

Then the conditions that allow for *propagating plane wave solutions* to the incremental balance of momentum are sought. Thereby a generic propagating plane wave solution for the incremental deformation $\mathrm{d}\boldsymbol{y}$ is expressed in terms of the (covariant, material) *wave propagation direction* $\boldsymbol{N}$, the (material) *wave propagation speed* c_0, and the (contravariant, spatial) *wave polarization vector* $\boldsymbol{m}$ as

$$
\mathrm{d}\boldsymbol{y} = \boldsymbol{m} f(\boldsymbol{N}\cdot\boldsymbol{X} \pm c_0 t)
$$

whereby f is a twice-continuously differentiable function. Thus the second spatial and temporal derivatives of $\mathrm{d}\boldsymbol{y}$ expand as

$$\operatorname{Grad}\operatorname{Grad}\mathrm{d}\boldsymbol{y} = \boldsymbol{m}\otimes\boldsymbol{N}\otimes\boldsymbol{N}f'' \quad \text{and} \quad \overline{\mathrm{d}\boldsymbol{y}}^{\,\cdot\cdot} = c_0^2\boldsymbol{m}f''.$$

Substituting these expressions in the above incremental balance of momentum results in an eigenvalue problem for the *acoustic tensor* $\boldsymbol{a}$, i.e.

$$\boldsymbol{a}(\boldsymbol{N})\cdot\boldsymbol{m} = \rho_0 c_0^2\boldsymbol{m} \quad \text{with} \quad \boldsymbol{a}(\boldsymbol{N})\cdot\boldsymbol{m} := \big[\mathsf{A} : [\boldsymbol{m}\otimes\boldsymbol{N}]\big]\cdot\boldsymbol{N}.$$

The acoustic tensor $\boldsymbol{a}$ is a second-order spatial description tensor; alternatively it allows representation in terms of the spatial description tangent operator a and the (covariant, spatial) wave propagation direction $\boldsymbol{n} := \boldsymbol{N}\cdot\boldsymbol{f}/\Lambda_N$ with $\Lambda_N^2 := \boldsymbol{N}\cdot\boldsymbol{B}\cdot\boldsymbol{N}$ the inverse stretch in the direction $\boldsymbol{N}$

$$\boldsymbol{a}(\boldsymbol{n})\cdot\boldsymbol{m} = \Lambda_N^2\big[\mathsf{a} : [\boldsymbol{m}\otimes\boldsymbol{n}]\big]\cdot\boldsymbol{n} =: J\Lambda_N^2\boldsymbol{\alpha}(\boldsymbol{n})\cdot\boldsymbol{m} \quad \rightsquigarrow \quad \boldsymbol{\alpha}(\boldsymbol{n})\cdot\boldsymbol{m} = \rho_t c_t^2\boldsymbol{m}.$$

Here $c_t := \lambda_n c_0$ with $\lambda_n^2 := \boldsymbol{n}\cdot\boldsymbol{b}\cdot\boldsymbol{n} = 1/\Lambda_N^2$ the stretch in the direction $\boldsymbol{n}$ denotes the spatial wave propagation speed. For given wave propagation direction $\boldsymbol{N}$ the wave propagation speed c_0 and the wave polarization vector $\boldsymbol{m}$ then follow as the eigenvalues and eigenvectors of this eigenvalue problem. For (hyper) elasticity the acoustic tensor is symmetric, thus for given wave propagation direction $\boldsymbol{N}$ and wave polarization vector $\boldsymbol{m}$ the (material) wave propagation speed follows as the quadratic form

$$\boldsymbol{m}\cdot\boldsymbol{a}(\boldsymbol{N})\cdot\boldsymbol{m} = \rho_0 c_0^2 \in \mathbb{R}_+.$$

The solutions $\boldsymbol{N}\cdot\boldsymbol{f} \perp \boldsymbol{m}$ and $\boldsymbol{N}\cdot\boldsymbol{f} \parallel \boldsymbol{m}$ then characterize propagating *transversal waves* and *longitudinal waves*, respectively. For elastodynamics a vanishing (material) wave propagation speed $c_0 = 0$ characterizes the loss of *strong hyperbolicity*, for elastostatics the condition for *strong ellipticity* is likewise formulated in terms of the acoustic tensor

$$\boldsymbol{m}\cdot\boldsymbol{a}(\boldsymbol{N})\cdot\boldsymbol{m} > 0 \quad \rightsquigarrow \quad \det\boldsymbol{a}(\boldsymbol{N}) > 0.$$

Thus strong ellipticity requires a positive definite acoustic tensor for all $\boldsymbol{N}$. If A satisfies the strong ellipticity condition for all $\boldsymbol{F}$ the constitutive relation $\boldsymbol{P} = \boldsymbol{P}(\boldsymbol{F})$ and the elastically stored energy density $w_0 = w_0(\boldsymbol{F})$ are said to be strongly elliptic. Accordingly the quasi-static equilibrium equations form a strongly elliptic system of partial differential equations, see [153] (for $\det\boldsymbol{a} \neq 0$ the constitutive relation, the elastically stored energy density, and the quasi-static equilibrium equations qualify as merely *elliptic*). The weaker inequality $\det\boldsymbol{a} \geq 0$, i.e. the requirement of a positive semi-definite acoustic tensor for all $\boldsymbol{N}$ is denoted the *Legendre-Hadamard condition* (semi-ellipticity) that is connected to the rank-one convexity of the elastically stored energy density, see below.

For geometrically linear isotropic elasticity the acoustic tensor $\boldsymbol{\alpha}$ *computes in terms of the Lamé constants* μ *and* λ *as*

$$\boldsymbol{\alpha} = \mu \boldsymbol{j} + [\mu + \lambda] \boldsymbol{n} \otimes \boldsymbol{n} \quad \text{with} \quad \boldsymbol{\alpha}^{-1} = \frac{1}{\mu} \boldsymbol{i} - \frac{\mu + \lambda}{\mu [2\mu + \lambda]} \boldsymbol{n}_\sharp \otimes \boldsymbol{n}_\sharp .$$

Clearly $\boldsymbol{\alpha}$ *possesses eigenvalues* μ *and* $2\mu + \lambda$ *for eigenvectors* $\boldsymbol{m} \perp \boldsymbol{n}$ *and* $\boldsymbol{m} = \boldsymbol{n}_\sharp$*, with* $\det \alpha = \mu^2 [2\mu + \lambda]$*. These eigenpairs are associated with propagating plane transversal and longitudinal waves, respectively.*

Variational Setting:
Quasi-static (hyper) elasticity can be cast into a variational framework. To this end a *total potential energy functional* $I\big(\boldsymbol{y}(\boldsymbol{X})\big)$ in terms of the deformation map $\boldsymbol{y}(\boldsymbol{X})$ with compatible deformation gradient $\boldsymbol{F} := \operatorname{Grad} \boldsymbol{y}(\boldsymbol{X})$ is introduced as

$$I\big(\boldsymbol{y}(\boldsymbol{X})\big) := \int_{\mathcal{B}_0} u_0\big(\boldsymbol{y}(\boldsymbol{X}), \boldsymbol{F}(\boldsymbol{X})\big) \, \mathrm{d}V + \int_{\partial \mathcal{B}_0} \widehat{v}_0\big(\boldsymbol{y}(\boldsymbol{X})\big) \, \mathrm{d}A.$$

Here the bulk energy density $u_0 := w_0 + v_0$ consists in an internal contribution, i.e. the elastically stored energy density $w_0 = w_0(\boldsymbol{F})$, and an external contribution $v_0 = v_0(\boldsymbol{y})$; the surface energy density is assumed for simplicity to consist only in an external contribution $\widehat{v}_0 = \widehat{v}_0(\boldsymbol{y})$ (otherwise a surface elasticity theory á la [73], see also [184, 87], needs to be considered). Then the minimum of I is sought for all *admissible variations* $\delta \boldsymbol{y}$ of $\boldsymbol{y}$

$$I\big(\boldsymbol{y}(\boldsymbol{X})\big) \to \min_{\boldsymbol{y}} .$$

Thereby admissible variations of the deformation map $\delta \boldsymbol{y}$ satisfy compatibility $\delta \boldsymbol{F} := \operatorname{Grad} \delta \boldsymbol{y}$ and homogeneous boundary conditions $\delta \boldsymbol{y} = \boldsymbol{0}$ on the Dirichlet part $\partial \mathcal{B}_0^y$ of the boundary $\partial \mathcal{B}_0 = \partial \mathcal{B}_0^y \cup \partial \mathcal{B}_0^t$ with $\emptyset = \partial \mathcal{B}_0^y \cap \partial \mathcal{B}_0^t$ (the complement to the Dirichlet part is the Neumann part $\partial \mathcal{B}_0^t$ of the boundary). According to the *direct calculus of variations*, see [48], conditions for a unique minimizer of I to exist are its *coercivity* and its *sequential weak lower semi-continuity*. In the current context the functional I is coercive, see [34], if it satisfies for $\alpha, a, b, c \in \mathbb{R}_+$ and $\beta \in \mathbb{R}$ the inequality

$$I\big(\boldsymbol{y}(\boldsymbol{X})\big) \geq \alpha \big[|\boldsymbol{F}(\boldsymbol{X})|^a + |\operatorname{cof} \boldsymbol{F}(\boldsymbol{X})|^b + |\det \boldsymbol{F}(\boldsymbol{X})|^c \big] + \beta .$$

Coercivity holds if w_0 satisfies appropriate growth conditions $I \to \infty$ for either $|\boldsymbol{F}| \to \infty$, $|\operatorname{cof} \boldsymbol{F}| \to \infty$, or $\det \boldsymbol{F} \to \infty$ together with $\det \boldsymbol{F} \to 0_+$. The functional is sequentially weakly lower semi-continuous if it satisfies for a weakly converging (minimizing) sequence $\boldsymbol{y}_\varsigma \rightharpoonup \boldsymbol{y}$ (for the precise definition of weak convergence see, e.g., [34]) the inequality

$$\liminf_{\varsigma \to \infty} I\big(\boldsymbol{y}_\varsigma(\boldsymbol{X})\big) \geq I\big(\boldsymbol{y}(\boldsymbol{X})\big) .$$

As a one-dimensional counterexample consider the (non-convex double-well) functional $I(y(X)) = \int_{\mathcal{B}_0} \left[[y'(X)]^2 - 1\right]^2 + y^2(X)\, dV$ *on* $\mathcal{B}_0 = \{X | X \in [0,1]\}$. *Then for homogeneous Dirichlet boundary conditions* $y(0) = y(1) = 0$ *the sequence of zig-zagging* y_ς *with* $y'_\varsigma = \pm 1$ *a.e. and increasingly finer micro-structure is a minimizing sequence that converges weakly to* $y = 0$. *However since* $\liminf_{\varsigma \to \infty} I(y_\varsigma) = 0 \not\geq I(0) = 1$ *the functional* I *is not sequentially weakly lower semi-continuous.*

In a one-dimensional setting sequential weak lower semi-continuity and *convexity* are synonymous, however in a multi-dimensional setting such as elasticity the necessary and sufficient condition for sequential weak lower semi-continuity is the *quasi-convexity* of the elastically stored energy w_0

$$w_0(\boldsymbol{F})|\mathcal{V}_0| \leq \int_{\mathcal{V}_0} w_0\big(\boldsymbol{F} + \operatorname{Grad} \boldsymbol{z}(\boldsymbol{X})\big)\, \mathrm{d}V \quad \forall \mathcal{V}_0 \subseteq \mathcal{B}_0, \quad \forall \boldsymbol{z}.$$

Here the inhomogeneous *fluctuations* $\boldsymbol{z}$ satisfy homogeneous Dirichlet boundary conditions on $\partial \mathcal{V}_0$, see [146]. Based on a Taylor expansion for twice differentiable w_0 quasi-convexity implies (but is not implied by) *infinitesimal quasi-convexity*

$$\int_{\mathcal{V}_0} \operatorname{Grad} \boldsymbol{z}(\boldsymbol{X}) : \mathsf{A} : \operatorname{Grad} \boldsymbol{z}(\boldsymbol{X})\, \mathrm{d}V \geq 0 \quad \forall \mathcal{V}_0 \subseteq \mathcal{B}_0.$$

Furthermore infinitesimal quasi-convexity can be shown to be equivalent to *infinitesimal rank-one-convexity*

$$[\boldsymbol{m} \otimes \boldsymbol{N}] : \mathsf{A} : [\boldsymbol{m} \otimes \boldsymbol{N}] = \boldsymbol{m} \cdot \boldsymbol{a}(\boldsymbol{N}) \cdot \boldsymbol{m} \geq 0.$$

Here $\boldsymbol{a}$ denotes the second-order *acoustic tensor* that has already been introduced earlier to characterize the *(strong) ellipticity* of elastic boundary value problems, see [153].

Due to its integral formulation quasi-convexity as a necessary and sufficient condition for sequential weak lower semi-continuity and thus for a unique minimizer of I *to exist is rather cumbersome to be evaluated. However, it has been proven in [12] that poly-convexity is a sufficient condition for quasi-convexity, see also [173]. Thereby poly-convexity requires convexity of the elastically stored energy density in each of the variables* $\boldsymbol{F}$, $\operatorname{cof} \boldsymbol{F}$, *and* $\det \boldsymbol{F}$ *separately. Poly-convexity is a weaker condition than convexity (that is a too restrictive requirement since it conflicts with i) the possibility of buckling, ii) objectivity, and iii) the growth condition for* $J \to 0$, *see [122]) and a stronger condition than quasi-convexity. It does not imply (but is implied by) convexity, vice versa it implies (but is not implied by) quasi-convexity. Poly-convexity is thus commonly accepted as the key convexity requirement for elastically stored energies. The poly-convexity of various elastically stored energy densities has extensively been studied by [169, 170].*

Alternatively, in *classical calculus of variations* the roots $\delta I = 0$ of the *first variation*

$$\delta I(\boldsymbol{y}(\boldsymbol{X})) := \frac{\mathrm{d}}{\mathrm{d}\epsilon} I(\boldsymbol{y}(\boldsymbol{X}) + \epsilon \delta \boldsymbol{y}(\boldsymbol{X}))|_{\epsilon=0} \quad \text{with} \quad \delta \boldsymbol{y}(\boldsymbol{X}) \quad \text{admissible}$$

of the functional I are determined in terms of the corresponding point-wise Euler equations. In the case of elasticity these read

$$\operatorname{Div} \frac{\partial w_0}{\partial \boldsymbol{F}} - \frac{\partial v_0}{\partial \boldsymbol{y}} \doteq \boldsymbol{0} \quad \text{and} \quad \frac{\partial w_0}{\partial \boldsymbol{F}} \cdot \boldsymbol{N} \doteq -\frac{\partial \widehat{v}_0}{\partial \boldsymbol{y}}.$$

Note that the above Dirichlet principle of the minimum of the total potential energy functional requires the strong satisfaction of (i) the kinematic field equations $\boldsymbol{F} = \operatorname{Grad} \boldsymbol{y}$, (ii) the constitutive relations, e.g. for $\boldsymbol{P} = \partial_{\boldsymbol{F}} w_0$, (iii) the potential-based relations for the body and surface force densities $\boldsymbol{b}_0 := -\partial_{\boldsymbol{y}} v_0$ and $\boldsymbol{t}_0 := -\partial_{\boldsymbol{y}} \widehat{v}_0$, and (iv) the Dirichlet boundary conditions. Then (i) the static field equations $\operatorname{Div} \boldsymbol{P} + \boldsymbol{b}_0 = \boldsymbol{0}$, and (ii) the Neumann boundary conditions $\boldsymbol{P} \cdot \boldsymbol{N} = \boldsymbol{t}_0$ are satisfied in weak (integral) sense. Finally to check whether I indeed takes a minimum at the root of $\delta I = 0$ the sign of the second variation $\delta^2 I$ at the solution dictated by the Euler equations has to be analyzed, see [100].

Four-Dimensional Formalism of First-Order Elasticity:
The pertinent relations of geometrically nonlinear first-order elasticity may be re-cast in a four-dimensional formalism in terms of appropriately defined four-vectors and -tensors. Thereby the four-dimensional formalism results in a combined representation of the balances of linear momentum and mass or mechanical energy, respectively; thus highlighting already the intimate relation between mass and energy. To this end the four-vectors of spatial and material coordinates $\underline{\boldsymbol{x}}$ and $\underline{\boldsymbol{X}}$ consisting of the ordinary space-like coordinates $\boldsymbol{x}$ and $\boldsymbol{X}$ in the first three entries and a time-like coordinate $\Xi = \xi := ct$ in the fourth entry are introduced and are related by suited four-deformation maps

$$\underline{\boldsymbol{x}} := \begin{bmatrix} \boldsymbol{x} \\ \xi \end{bmatrix} = \underline{\boldsymbol{y}}(\underline{\boldsymbol{X}}) \quad \text{and} \quad \underline{\boldsymbol{X}} := \begin{bmatrix} \boldsymbol{X} \\ \Xi \end{bmatrix} = \underline{\boldsymbol{Y}}(\underline{\boldsymbol{x}}).$$

Here c denotes a proper reference velocity that needs to be introduced for dimensional consistency. Then the corresponding four-deformation gradients follow from the application of the appropriate four-gradients to the four-deformation maps

$$\underline{\boldsymbol{F}} := \underline{\operatorname{Grad} \boldsymbol{y}} = \begin{bmatrix} \boldsymbol{F} & \boldsymbol{v}/c \\ \boldsymbol{0} & 1 \end{bmatrix} \quad \text{and} \quad \underline{\boldsymbol{f}} := \underline{\operatorname{grad} \boldsymbol{Y}} = \begin{bmatrix} \boldsymbol{f} & \boldsymbol{V}/c \\ \boldsymbol{0} & 1 \end{bmatrix}.$$

Thereby $\boldsymbol{v} := \mathrm{D}_t \boldsymbol{y}$ and $\boldsymbol{V} := \mathrm{d}_t \boldsymbol{Y}$ denote the spatial and the material velocities that are related by $\boldsymbol{v} + \boldsymbol{F} \cdot \boldsymbol{V} = \boldsymbol{0}$. Next the Piola four-stress is introduced based on the Lagrangian energy density per unit volume in the material configuration $l_0 := k_0(\boldsymbol{v}/c, 1) - w_0(\boldsymbol{F})$ with $2k_0 := \rho_0 c^2 [|\boldsymbol{v}/c|^2 + 1^2]$ the kinetic energy density and $w_0 = w_0(\boldsymbol{F})$ the elastically stored energy density as

$$\underline{\boldsymbol{P}} := -\frac{\partial l_0}{\partial \underline{\boldsymbol{F}}} = \begin{bmatrix} \boldsymbol{P} & -c\boldsymbol{p}_0 \\ \boldsymbol{0} & -\rho_0 c^2 \end{bmatrix} \quad \rightsquigarrow \quad \underline{\boldsymbol{\sigma}} := \underline{\boldsymbol{P}} \cdot \operatorname{cof} \underline{\boldsymbol{f}} = \begin{bmatrix} \boldsymbol{\sigma} - \boldsymbol{p}_t \otimes \boldsymbol{v} & -c\boldsymbol{p}_t \\ -c\boldsymbol{p}_t & -\rho_t c^2 \end{bmatrix}.$$

Note the symmetry of the Cauchy four-stress $\underline{\boldsymbol{\sigma}}$ that has been introduced as the Piola four-transformation of the Piola four-stress. Applying next the corresponding material and spatial four-divergence operators to either the Piola or the Cauchy four-stress renders collectively the appropriate versions of the balance of linear momentum and the balance of mass

$$\underline{\operatorname{Div}\boldsymbol{P}} = \begin{bmatrix} \operatorname{Div}\boldsymbol{P} - \mathrm{D}_t\boldsymbol{p}_0 \\ -c\,\mathrm{D}_t\rho_0 \end{bmatrix} \equiv J\,\underline{\operatorname{div}\boldsymbol{\sigma}} = J\begin{bmatrix} \operatorname{div}(\boldsymbol{\sigma} - \boldsymbol{p}_t \otimes \boldsymbol{v}) - \mathrm{d}_t\boldsymbol{p}_t \\ -c[\operatorname{div}\boldsymbol{p}_t + \mathrm{d}_t\rho_t] \end{bmatrix} \doteq \underline{\boldsymbol{0}}.$$

These balance equations may equivalently be expressed in terms of a pull-back by the transposed four-deformation gradient to the material manifold as

$$-\underline{\boldsymbol{F}}^t \cdot \underline{\operatorname{Div}\boldsymbol{P}} = \underline{\operatorname{Div}}\left(-l_0\underline{\boldsymbol{1}}_{\mathrm{m}} + \underline{\boldsymbol{F}}^t \cdot \frac{\partial l_0}{\partial \underline{\boldsymbol{F}}}\right) =: \underline{\operatorname{Div}\boldsymbol{\Sigma}} \doteq \underline{\boldsymbol{0}}.$$

Thereby the Eshelby four-stress $\underline{\boldsymbol{\Sigma}}$ expands in terms of the Eshelby stress $\boldsymbol{\Sigma} := -l_0 \boldsymbol{1}_{\mathrm{m}} - \boldsymbol{F}^t \cdot \boldsymbol{P}$, the material momentum density $\boldsymbol{P}_0 := -\boldsymbol{F}^t \cdot \boldsymbol{p}_0$, the material flux of mechanical energy $\boldsymbol{Q} := \boldsymbol{v} \cdot \boldsymbol{P}$ and the Hamiltonian energy density $e_0 := k_0 + w_0$ per unit volume in the material configuration as

$$\underline{\boldsymbol{\Sigma}} = \begin{bmatrix} \boldsymbol{\Sigma} & -c\boldsymbol{P}_0 \\ -\boldsymbol{Q}/c & e_0 \end{bmatrix} \quad \rightsquigarrow \quad \underline{\boldsymbol{p}} := \underline{\boldsymbol{\Sigma}} \cdot \operatorname{cof} \underline{\boldsymbol{f}} = \begin{bmatrix} \boldsymbol{p} - \boldsymbol{P}_t \otimes \boldsymbol{v} & -c\boldsymbol{P}_t \\ -[\boldsymbol{q} - e_t\boldsymbol{v}]/c & e_t \end{bmatrix}.$$

Here $\underline{\boldsymbol{p}}$ is a two-point (material/spatial) Eshelby-type four-stress, whereby $\boldsymbol{q} := \boldsymbol{Q} \cdot \operatorname{cof} \boldsymbol{f} = \boldsymbol{v} \cdot \boldsymbol{\sigma}$ denotes the spatial flux of mechanical energy. Applying finally the corresponding material and spatial four-divergence operators renders collectively the appropriate versions of the balances of linear momentum and mechanical energy

$$\underline{\operatorname{Div}\boldsymbol{\Sigma}} = \begin{bmatrix} \operatorname{Div}\boldsymbol{\Sigma} - \mathrm{D}_t\boldsymbol{P}_0 \\ -[\operatorname{Div}\boldsymbol{Q} - \mathrm{D}_t e_0]/c \end{bmatrix} \equiv J\,\underline{\operatorname{div}\boldsymbol{p}} = J\begin{bmatrix} \operatorname{div}(\boldsymbol{p} - \boldsymbol{P}_t \otimes \boldsymbol{v}) - \mathrm{d}_t\boldsymbol{P}_t \\ -[\operatorname{div}(\boldsymbol{q} - e_t\boldsymbol{v}) - \mathrm{d}_t e_t]/c \end{bmatrix} \doteq \underline{\boldsymbol{0}}.$$

The former are alternative versions of the so-called balance of material momentum that constitutes the cornerstone of configurational mechanics, see e.g. [124, 125, 126, 69, 70, 182, 183], the latter are alternative versions of the balance of mechanical energy. The relation between spatial and material time derivatives of densities is given in terms of the Reynolds transport theorem $j\,\mathrm{D}_t\{\bullet\}_0 = \mathrm{d}_t\{\bullet\}_t + \operatorname{div}(\{\bullet\}_t \otimes \boldsymbol{v})$. It is interesting to note that a pull-back of the four-balance equation by the transposed four-deformation gradient to the material manifold exchanges the balance of mass for the balance of mechanical energy. The Cauchy four-stress $\underline{\boldsymbol{\sigma}}$, the Piola four-stress $\underline{\boldsymbol{P}}$, the Eshelby four-stress $\underline{\boldsymbol{\Sigma}}$ or likewise its Piola four-transformation $\underline{\boldsymbol{p}}$ may be considered alternatives for the four-dimensional energy-momentum tensor of geometrically nonlinear elasticity. For large c (such as the speed of light) the scaled energy-momentum tensor $-\underline{\boldsymbol{\sigma}}/c^2$ degenerates in the Newtonian limit to only the entry ρ_t different from zero. Finally the density is related to Newton's gravitational potential via $\Delta\phi = 4\pi\rho_t$, thus identifying the gravitational potential ϕ as the only relevant contribution to the space-time metric. In [185] the pertinent relations for the gravito-elastically coupled problem, as is appropriate for deformable heavy masses of finite volume, are recovered from a variational setting in terms of the nonlinear deformation map and Newton's gravitational potential.

Supplement 7.2. Second-Order Elasticity in Euclidean Space
Second-order elasticity is an extension of classical first-order elasticity that essentially incorporates second gradients of the deformation map; as compared to first-order elasticity it is thus mainly characterized by the existence of double-strains and double-stresses. The motivation for second-order elasticity is manifold, e.g. (i) to introduce a length-scale into the formulation, (ii) to be able to capture boundary layer effects, and (iii) to serve as a template for the modeling of more involved constitutive behavior such as generalized plasticity. Geometrically linear second-order elasticity has been advocated in one way or the other in the early contributions [196, 197, 139, 140, 141]. The generalization to the geometrically nonlinear (and partly inelastic) case has been considered e.g. in [147, 57, 62, 88, 89, 44, 86, 15], see also further references therein. However, these contributions by-and-large concentrated on a variationally inspired setting that does not fully exploit the possibility for, e.g., various spatial, material and two-point description double-strains and double-stresses and the resulting richness of the formulation. Thus a more comprehensive account on the essential ingredients of second-order nonlinear continuum elasticity in Euclidean space shall be given in the sequel (to the best knowledge of the author for the first time). Thereby, these ingredients of second-order elasticity are traditionally grouped into i) the kinematics of finite deformations, ii) the mechanical and thermodynamical balance laws, and (iii) the constitutive relations. The intention is to come up with a formulation (and notation) that parallels the beauty of first-order elasticity as much as possible. Many aspect of the formulation are generic and are thus also valid for the continuum mechanics of second-order inelasticity. For the sake of brevity only the necessary additions to the previous exposition of first-order elasticity are outlined.

Second Deformation Gradient:
The second gradient of the nonlinear deformation map $\boldsymbol{x} = \boldsymbol{y}(\boldsymbol{X})$ with respect to the material coordinates $\boldsymbol{X}$ is denoted the *second deformation gradient*

$$\boldsymbol{G} := {}^{\Delta}\boldsymbol{F} := \operatorname{Grad} \boldsymbol{F} = \operatorname{Grad} \operatorname{Grad} \boldsymbol{y} \quad \text{with} \quad \boldsymbol{F} := \operatorname{Grad} \boldsymbol{y}.$$

Note that for notational consistency the second deformation gradient $\boldsymbol{G}$ is henceforth denoted by ${}^{\Delta}\boldsymbol{F}$ in this supplement. It is a third-order two-point (spatial/material) contra/co/co- (mixed) variant tensor that results from a Taylor expansion of the deformation map $\boldsymbol{y}(\boldsymbol{X} + \mathrm{d}\boldsymbol{X})$ at $\boldsymbol{X}$ truncated after the quadratic term, i.e.

$$\mathrm{d}\boldsymbol{y} := \boldsymbol{y}(\boldsymbol{X} + \mathrm{d}\boldsymbol{X}) - \boldsymbol{y}(\boldsymbol{X}) = \boldsymbol{F} \cdot \mathrm{d}\boldsymbol{X} + \frac{1}{2} {}^{\Delta}\boldsymbol{F} : [\mathrm{d}\boldsymbol{X} \otimes \mathrm{d}\boldsymbol{X}] + \text{h.o.t.}$$

Observe carefully that here, unlike in the case of a first-order continuum, a clear distinction is made between the spatial line element $\mathrm{d}\boldsymbol{x} = \boldsymbol{F} \cdot \mathrm{d}\boldsymbol{X}$, i.e. an element of the spatial tangent space $T\mathcal{B}_t$, and the truncated Taylor expansion $\mathrm{d}\boldsymbol{y} \neq \mathrm{d}\boldsymbol{x}$ of $\boldsymbol{y}$. The terminology *double-distortion* includes also the case of incompatible ${}^{\Delta}\boldsymbol{F}$, i.e. the generic case that ${}^{\Delta}\boldsymbol{F}$ does neither derive as the gradient of a tensor field $\boldsymbol{F}(\boldsymbol{X})$ nor as the second gradient of a vector field $\boldsymbol{y}(\boldsymbol{X})$. Likewise the spatial gradient of the inverse deformation gradient $\boldsymbol{f}$ is introduced as

$$\boldsymbol{g} := {}^{\Delta}\boldsymbol{f} := \operatorname{grad} \boldsymbol{f} = \operatorname{grad} \operatorname{grad} \boldsymbol{Y} \quad \text{with} \quad \boldsymbol{f} := \operatorname{grad} \boldsymbol{Y}.$$

Again for notational consistency the spatial gradient of the inverse deformation gradient $\boldsymbol{g}$ is henceforth denoted by ${}^{\Delta}\boldsymbol{f}$ in this supplement. Moreover, based on the inverse deformation map $\boldsymbol{X} = \boldsymbol{Y}(\boldsymbol{x})$ its second gradient $\operatorname{grad} \operatorname{grad} \boldsymbol{Y}$ with respect to the spatial coordinates $\boldsymbol{x}$ proves to coincide with ${}^{\Delta}\boldsymbol{f}$, which might therefore also be denoted as second deformation gradient. Then the relation between ${}^{\Delta}\boldsymbol{F}$ and ${}^{\Delta}\boldsymbol{f}$ computes as

$${}^{\Delta}\boldsymbol{f} = -\boldsymbol{f} \cdot {}^{\Delta}\boldsymbol{F} : [\boldsymbol{f} \boxtimes \boldsymbol{f}] \quad \text{and} \quad {}^{\Delta}\boldsymbol{F} = -\boldsymbol{F} \cdot {}^{\Delta}\boldsymbol{f} : [\boldsymbol{F} \boxtimes \boldsymbol{F}].$$

The above implies that $\boldsymbol{F} \cdot [\mathrm{d}\boldsymbol{X} - \mathrm{d}\boldsymbol{Y}] = \mathrm{d}\boldsymbol{y} - \mathrm{d}\boldsymbol{x}$ so that

$$\mathrm{d}\boldsymbol{Y} = \boldsymbol{f} \cdot \mathrm{d}\boldsymbol{x} + \frac{1}{2} {}^{\Delta}\boldsymbol{f} : [\mathrm{d}\boldsymbol{x} \otimes \mathrm{d}\boldsymbol{x}] + \text{h.o.t.}$$

Finally it proves convenient to introduce material and spatial description third-order contra/co/co- (mixed) variant *connection tensors* (see also [31, 32, 62, 44])

$${}^{\Delta}\boldsymbol{\Gamma} := \boldsymbol{f} \cdot {}^{\Delta}\boldsymbol{F} = -{}^{\Delta}\boldsymbol{f} : [\boldsymbol{F} \boxtimes \boldsymbol{F}] \quad \text{and} \quad {}^{\Delta}\boldsymbol{\gamma} := \boldsymbol{F} \cdot {}^{\Delta}\boldsymbol{f} = -{}^{\Delta}\boldsymbol{F} : [\boldsymbol{f} \boxtimes \boldsymbol{f}].$$

Note that, again for notational consistency, the third-order connection tensors $\boldsymbol{\Gamma}$ and $\boldsymbol{\gamma}$ as introduced in the main text are henceforth denoted by ${}^{\Delta}\boldsymbol{\Gamma}$ and ${}^{\Delta}\boldsymbol{\gamma}$, respectively, in this supplement. Thereby the relation between ${}^{\Delta}\boldsymbol{\Gamma}$ and ${}^{\Delta}\boldsymbol{\gamma}$ computes as

$${}^{\Delta}\boldsymbol{\gamma} = -\boldsymbol{F} \cdot {}^{\Delta}\boldsymbol{\Gamma} : [\boldsymbol{f} \boxtimes \boldsymbol{f}] \quad \text{and} \quad {}^{\Delta}\boldsymbol{\Gamma} = -\boldsymbol{f} \cdot {}^{\Delta}\boldsymbol{\gamma} : [\boldsymbol{F} \boxtimes \boldsymbol{F}].$$

Due to their definition and based on the metrics $\boldsymbol{C}$ and $\boldsymbol{c}$ the material and spatial connection tensors ${}^{\triangle}\boldsymbol{\Gamma}$ and ${}^{\triangle}\gamma$ are indeed *Riemann connection tensors* that characterize *symmetric, flat and metrically connected manifolds*, thus

$$\boldsymbol{C} \cdot {}^{\triangle}\boldsymbol{\Gamma} = \{\,\mathrm{Grad}\,\boldsymbol{C}\} \quad \text{and} \quad \boldsymbol{c} \cdot {}^{\triangle}\gamma = \{\,\mathrm{grad}\,\boldsymbol{c}\}$$

whereby the bold curly brackets $\{\{\bullet\}\}$ refer to the anti-cyclic, sign alternating index permutation $\{\bullet\}_{\{IJK\}}$ introduced in Part I. Observe that the notation for the connection tensors has been changed as compared to the rest of this treatise for notational consistency.

Covariant Double-Strain Measures:
In order to introduce covariant *double-strain* measures the (quadratic) length of the (spatial) truncated Taylor expansion

$$\mathrm{d}\boldsymbol{y} = \boldsymbol{F} \cdot \big[\mathrm{d}\boldsymbol{X} + \frac{1}{2}[{}^{\triangle}\boldsymbol{\Gamma} \cdot \mathrm{d}\boldsymbol{X}] \cdot \mathrm{d}\boldsymbol{X}\big] + \text{h.o.t.},$$

that may be expressed in terms of the material connection tensor ${}^{\triangle}\boldsymbol{\Gamma}$, is computed as

$$|\mathrm{d}\boldsymbol{y}|^2 =: \mathrm{d}\boldsymbol{X} \cdot \big[\boldsymbol{C} + [{}^{\triangle}\boldsymbol{C} \cdot \mathrm{d}\boldsymbol{X}]\big] \cdot \mathrm{d}\boldsymbol{X} + \text{h.o.t.}$$

Note that the higher-order term contains essentially the square of ${}^{\triangle}\boldsymbol{\Gamma}$, however ${}^{\triangle}\boldsymbol{\Gamma}$ is already contained in the material description third-order fully covariant double-strain tensor

$${}^{\triangle}\boldsymbol{C} := \boldsymbol{C} \cdot {}^{\triangle}\boldsymbol{\Gamma} = \boldsymbol{F}^t \cdot \boldsymbol{j} \cdot {}^{\triangle}\boldsymbol{F}.$$

Interestingly, ${}^{\triangle}\boldsymbol{C}$ is either expressed as a fully covariant connection-like tensor (in terms of the material strain metric) or in a format that resembles the definition of $\boldsymbol{C}$ if ${}^{\triangle}\boldsymbol{F}$ is only exchanged by $\boldsymbol{F}$. Similar double-strain tensors have e.g. been used in [147] for second-order continua and in [57] for the related case of micromorphic continua. Note that based on the Ricci postulate for metric connections (the covariant derivative of the metric vanishes) the material gradient of the metric $\boldsymbol{C}$ is expressed as twice the left symmetric contribution to the fully covariant connection tensor ${}^{\triangle}\boldsymbol{C}$, thus another double-strain measure in the spirit of the early proposal by [196] reads as

$$\mathrm{Grad}\,\boldsymbol{E} = \frac{1}{2}\,\mathrm{Grad}\,\boldsymbol{C} = \mathsf{I}^{\mathrm{sym}} : {}^{\triangle}\boldsymbol{C}.$$

Alternatively, by $\mathrm{d}\boldsymbol{X} = \boldsymbol{f} \cdot \mathrm{d}\boldsymbol{x}$ the (quadratic) length of $\mathrm{d}\boldsymbol{y}$ may be expressed as

$$|\mathrm{d}\boldsymbol{y}|^2 =: \mathrm{d}\boldsymbol{x} \cdot \big[\boldsymbol{j} + [{}^{\triangle}\boldsymbol{j} \cdot \mathrm{d}\boldsymbol{x}]\big] \cdot \mathrm{d}\boldsymbol{x} + \text{h.o.t.}$$

in terms of the spatial description third-order fully covariant *double-metric*

$$^{\Delta}\boldsymbol{j} := \boldsymbol{f}^t \cdot {}^{\Delta}\boldsymbol{C} : [\boldsymbol{f} \boxtimes \boldsymbol{f}] = -\boldsymbol{j} \cdot {}^{\Delta}\gamma = \boldsymbol{j} \cdot {}^{\Delta}\boldsymbol{F} : [\boldsymbol{f} \boxtimes \boldsymbol{f}].$$

Note that the spatial double-metric ${}^{\Delta}\boldsymbol{j}$ is essentially a covariant connection tensor (in terms of the Euclidean metric), whereby the negative sign follows from convention.
Likewise, the (quadratic) length of the (material) truncated Taylor expansion

$$\mathrm{d}\boldsymbol{Y} = \boldsymbol{f} \cdot \big[\, \mathrm{d}\boldsymbol{x} + \frac{1}{2}[{}^{\Delta}\gamma \cdot \mathrm{d}\boldsymbol{x}] \cdot \mathrm{d}\boldsymbol{x} \big] + \text{h.o.t.},$$

that may be expressed in terms of the spatial connection tensor ${}^{\Delta}\gamma$, follows as

$$|\mathrm{d}\boldsymbol{Y}|^2 =: \mathrm{d}\boldsymbol{x} \cdot \big[\boldsymbol{c} + [{}^{\Delta}\boldsymbol{c} \cdot \mathrm{d}\boldsymbol{x}]\big] \cdot \mathrm{d}\boldsymbol{x} + \text{h.o.t.}$$

Note again that the higher-order term contains essentially the square of ${}^{\Delta}\gamma$, however ${}^{\Delta}\gamma$ is already contained in the spatial description third-order fully covariant double-strain tensor

$${}^{\Delta}\boldsymbol{c} := \boldsymbol{c} \cdot {}^{\Delta}\gamma = \boldsymbol{f}^t \cdot \boldsymbol{J} \cdot {}^{\Delta}\boldsymbol{f}.$$

Interestingly, ${}^{\Delta}\boldsymbol{c}$ is either expressed as a fully covariant connection-like tensor (in terms of the spatial strain metric) or in a format that resembles the definition of $\boldsymbol{c}$ if ${}^{\Delta}\boldsymbol{f}$ is only exchanged by $\boldsymbol{f}$. Note again that based on the Ricci postulate for metric connections the spatial gradient of the metric $\boldsymbol{c}$ is expressed as twice the left symmetric contribution to the fully covariant connection tensor ${}^{\Delta}\boldsymbol{c}$, thus another double-strain measure in the spirit of [196] reads as

$$-\operatorname{grad} \boldsymbol{e} = \frac{1}{2} \operatorname{grad} \boldsymbol{c} = \mathbf{i}^{\mathrm{sym}} : {}^{\Delta}\boldsymbol{c}.$$

Alternatively, by $\mathrm{d}\boldsymbol{x} = \boldsymbol{F} \cdot \mathrm{d}\boldsymbol{X}$ the (quadratic) length of $\mathrm{d}\boldsymbol{Y}$ may be expressed as

$$|\mathrm{d}\boldsymbol{Y}|^2 =: \mathrm{d}\boldsymbol{X} \cdot \big[\boldsymbol{J} + [{}^{\Delta}\boldsymbol{J} \cdot \mathrm{d}\boldsymbol{X}]\big] \cdot \mathrm{d}\boldsymbol{X} + \text{h.o.t.}$$

in terms of the material description third-order fully covariant double-metric

$${}^{\Delta}\boldsymbol{J} := \boldsymbol{F}^t \cdot {}^{\Delta}\boldsymbol{c} : [\boldsymbol{F} \boxtimes \boldsymbol{F}] = -\boldsymbol{J} \cdot {}^{\Delta}\boldsymbol{\Gamma} = \boldsymbol{J} \cdot {}^{\Delta}\boldsymbol{f} : [\boldsymbol{F} \boxtimes \boldsymbol{F}].$$

Note that the material double-metric ${}^{\Delta}\boldsymbol{J}$ is basically a covariant connection tensor (in terms of the Euclidean metric), whereby the negative sign follows again from convention.
Next, on the one hand the comparison of the (quadratic) lengths of the spatial and material truncated Taylor expansions $\mathrm{d}\boldsymbol{y}$ and $\mathrm{d}\boldsymbol{Y}$ may be expressed as

$$|\mathrm{d}\boldsymbol{y}|^2 - |\mathrm{d}\boldsymbol{Y}|^2 = 2\, \mathrm{d}\boldsymbol{X} \cdot \big[\boldsymbol{E} + [{}^{\Delta}\boldsymbol{E} \cdot \mathrm{d}\boldsymbol{X}]\big] \cdot \mathrm{d}\boldsymbol{X} + \text{h.o.t.}$$

On the other hand the comparison of the (quadratic) lengths of the spatial and material truncated Taylor expansions $\mathrm{d}\boldsymbol{y}$ and $\mathrm{d}\boldsymbol{Y}$ may be stated as

$$|\mathrm{d}\boldsymbol{y}|^2 - |\mathrm{d}\boldsymbol{Y}|^2 = 2\,\mathrm{d}\boldsymbol{x} \cdot \left[\boldsymbol{e} + [{}^{\vartriangle}\boldsymbol{e} \cdot \mathrm{d}\boldsymbol{x}]\right] \cdot \mathrm{d}\boldsymbol{x} + \text{h.o.t.}$$

In the above $\boldsymbol{C}$ and $\boldsymbol{c}$ together with $\boldsymbol{E}$ and $\boldsymbol{e}$ denote the previously introduced right Cauchy-Green, Finger, Green-Lagrange, and Almansi strain tensors, respectively, moreover the third-order material and spatial double-strain tensors ${}^{\vartriangle}\boldsymbol{E}$ and ${}^{\vartriangle}\boldsymbol{e}$ have been defined for convenience as $2{}^{\vartriangle}\boldsymbol{E} := {}^{\vartriangle}\boldsymbol{C} - {}^{\vartriangle}\boldsymbol{J}$ and $2{}^{\vartriangle}\boldsymbol{e} := {}^{\vartriangle}\boldsymbol{j} - {}^{\vartriangle}\boldsymbol{c}$. The material and spatial description covariant double-strain measures and double-metrics are related by pull-back and push-forward operations

$${}^{\vartriangle}\boldsymbol{E} =: \boldsymbol{F}^t \cdot {}^{\vartriangle}\boldsymbol{e} : [\boldsymbol{F} \boxtimes \boldsymbol{F}], \quad {}^{\vartriangle}\boldsymbol{C} =: \boldsymbol{F}^t \cdot {}^{\vartriangle}\boldsymbol{j} : [\boldsymbol{F} \boxtimes \boldsymbol{F}] \quad \text{and} \quad {}^{\vartriangle}\boldsymbol{J} =: \boldsymbol{F}^t \cdot {}^{\vartriangle}\boldsymbol{c} : [\boldsymbol{F} \boxtimes \boldsymbol{F}].$$

The covariant transformation behavior qualifies the above double-strain measures and double-metrics as covariant third-order tensors.

Spectral Representation of Material Double-Strain Measures:
Due to the spectral decomposition of the (symmetric and positive definite) right Cauchy-Green strain tensor $\boldsymbol{C}$ the third-order material double-strain tensor ${}^{\vartriangle}\boldsymbol{C}$ allows the following dyadic representation

$${}^{\vartriangle}\boldsymbol{C} = \sum_{a=1}^{3} [\lambda_a]^2 \boldsymbol{N}_a \otimes \boldsymbol{G}_a \quad \text{with} \quad \boldsymbol{G}_a := \boldsymbol{N}_a \cdot {}^{\vartriangle}\boldsymbol{\Gamma}.$$

Here the three symmetric second-order tensors $\boldsymbol{G}_a$ are denoted the *material principal double-directions*. Their symmetry follows directly from the (right) symmetry of the second deformation gradient ${}^{\vartriangle}\boldsymbol{F} = \boldsymbol{F} \cdot {}^{\vartriangle}\boldsymbol{\Gamma}$, which in turn is a consequence of the symmetry of second partial derivatives. Then each symmetric principal double-direction $\boldsymbol{G}_a$ allows in turn spectral representation in terms of its real (but not necessarily positive) *principal values* γ_{ab} and the corresponding Lagrangian dyad $\boldsymbol{K}_{ab} \otimes \boldsymbol{K}_{ab}$ of its *material principal directions*

$$\boldsymbol{G}_a = \sum_{b=1}^{3} \gamma_{ab} \boldsymbol{K}_{ab} \otimes \boldsymbol{K}_{ab}.$$

The nine principle values γ_{ab} follow from the corresponding three characteristic equations in terms of the nine *material principal invariants*

$$G_{a1} = \boldsymbol{G}_a : \boldsymbol{I} \quad \text{and} \quad G_{a2} = \operatorname{cof} \boldsymbol{G}_a : \boldsymbol{J} \quad \text{and} \quad G_{a3} = \det \boldsymbol{G}_a.$$

Due to the spectral representation of the $\boldsymbol{G}_a$ the nine principal invariants may equivalently be expressed in terms of the nine *spectral invariants* γ_{ab}. Based on the spectral representation of the principal double-directions a spectral representation of the third-order double-strain tensor ${}^{\vartriangle}\boldsymbol{C}$ results eventually in

$$^{\Delta}\boldsymbol{C} = \sum_{a,b=1}^{3} [\lambda_a]^2 \gamma_{ab} \boldsymbol{N}_a \otimes \boldsymbol{K}_{ab} \otimes \boldsymbol{K}_{ab}.$$

Likewise, with $^{\Delta}\boldsymbol{J} = -\boldsymbol{J} \cdot {}^{\Delta}\boldsymbol{\Gamma}$ a spectral representation of the third-order material double-metric $^{\Delta}\boldsymbol{J}$ reads

$$^{\Delta}\boldsymbol{J} = -\sum_{a,b=1}^{3} \gamma_{ab} \boldsymbol{N}_a \otimes \boldsymbol{K}_{ab} \otimes \boldsymbol{K}_{ab}.$$

It is finally remarked that the nine combined spectral invariants $[\lambda_a]^2\gamma_{ab}$ or likewise the equivalent combined principal invariants $I_a G_{ab}$ with $a, b = 1, 2, 3$ represent the complete coordinate independent, i.e. isotropic information content of the double-strain $^{\Delta}\boldsymbol{C}$. Note, however, that the above spectral representations based on the spectral representation of $\boldsymbol{C}$ and $\boldsymbol{J}$ are not unique, but only a convenient choice.

Spectral Representation of Spatial Double-Strain Measures:
Due to the spectral decomposition of the (symmetric and positive definite) Finger strain tensor $\boldsymbol{c}$ the third-order spatial double-strain tensor $^{\Delta}\boldsymbol{c}$ allows the following dyadic representation

$$^{\Delta}\boldsymbol{c} = \sum_{a=1}^{3} [\varLambda_a]^2 \boldsymbol{n}_a \otimes \boldsymbol{g}_a \quad \text{with} \quad \boldsymbol{g}_a := \boldsymbol{n}_a \cdot {}^{\Delta}\boldsymbol{\gamma}.$$

Here the three symmetric second-order tensors $\boldsymbol{g}_a$ are denoted the *spatial principal double-directions.* Then each symmetric principal double-direction $\boldsymbol{g}_a$ allows in turn spectral representation in terms of its real (but not necessarily positive) *principal values* $\varGamma_{ab}$ and the corresponding Eulerian dyad $\boldsymbol{k}_{ab} \otimes \boldsymbol{k}_{ab}$ of its *spatial principal directions*

$$\boldsymbol{g}_a = \sum_{b=1}^{3} \varGamma_{ab} \boldsymbol{k}_{ab} \otimes \boldsymbol{k}_{ab}.$$

In passing it is noted that the spatial and material principal double-directions are related by push-forward/pull-back as

$$\boldsymbol{g}_a = -\lambda_a \boldsymbol{f}^t \cdot \boldsymbol{G}_a \cdot \boldsymbol{f} \quad \text{and} \quad \boldsymbol{G}_a = -\varLambda_a \boldsymbol{F}^t \cdot \boldsymbol{g}_a \cdot \boldsymbol{F}.$$

The nine principle values $\varGamma_{ab}$ follow from the corresponding three characteristic equations in terms of the nine *spatial principal invariants*

$$g_{a1} = \boldsymbol{g}_a : \boldsymbol{i} \quad \text{and} \quad g_{a2} = \operatorname{cof} \boldsymbol{g}_a : \boldsymbol{j} \quad \text{and} \quad g_{a3} = \det \boldsymbol{g}_a.$$

Due to the spectral representation of the $\boldsymbol{g}_a$ the nine principal invariants may equivalently be expressed in terms of the nine *spectral invariants* $\varGamma_{ab}$. Based on the spectral representation of the principal double-directions a spectral representation of the third-order double-strain tensor ${}^{\vartriangle}\boldsymbol{c}$ results eventually in

$$
{}^{\vartriangle}\boldsymbol{c} = \sum_{a,b=1}^{3} [\Lambda_a]^2 \Gamma_{ab} \boldsymbol{n}_a \otimes \boldsymbol{k}_{ab} \otimes \boldsymbol{k}_{ab}.
$$

Likewise, with ${}^{\vartriangle}\boldsymbol{j} = -\boldsymbol{j} \cdot {}^{\vartriangle}\gamma$ a spectral representation of the third-order spatial double-metric ${}^{\vartriangle}\boldsymbol{j}$ reads

$$
{}^{\vartriangle}\boldsymbol{j} = - \sum_{a,b=1}^{3} \Gamma_{ab} \boldsymbol{n}_a \otimes \boldsymbol{k}_{ab} \otimes \boldsymbol{k}_{ab}.
$$

It is remarked again that the above spectral representations based on the spectral representation of $\boldsymbol{c}$ and $\boldsymbol{j}$ are not unique, but only a convenient choice.

Contravariant Double-Strain Measures:
In addition to the covariant double-strain measures as motivated in the above a material description third-order fully *contravariant* double-strain tensor is *formally* introduced with $\mathbf{B} := [\boldsymbol{B} \boxtimes \boldsymbol{B}]$ as the fully contravariant connection-like tensor

$$
{}^{\vartriangle}\boldsymbol{B} := -{}^{\vartriangle}\boldsymbol{\Gamma} : \mathbf{B} = {}^{\vartriangle}\boldsymbol{f} : \mathbf{i} : \mathbf{f}^t,
$$

whereby the negative sign follows from convention. Note the similarity to the definition of $\boldsymbol{B}$ if ${}^{\vartriangle}\boldsymbol{f}$, $\mathbf{i} := [\boldsymbol{i} \otimes \boldsymbol{i}]$ and $\mathbf{f}^t := [\boldsymbol{f}^t \otimes \boldsymbol{f}^t]$ are only exchanged by $\boldsymbol{f}$, $\boldsymbol{i}$, and $\boldsymbol{f}^t$, respectively. The corresponding push-forward with $\mathbf{F}^t := [\boldsymbol{F}^t \otimes \boldsymbol{F}^t]$ then results in the spatial description third-order fully contravariant double-metric

$$
{}^{\vartriangle}\boldsymbol{i} := \boldsymbol{F} \cdot {}^{\vartriangle}\boldsymbol{B} : \mathbf{F}^t = {}^{\vartriangle}\gamma : \mathbf{i} = \boldsymbol{F} \cdot {}^{\vartriangle}\boldsymbol{f} : \mathbf{i}.
$$

The spatial double-metric ${}^{\vartriangle}\boldsymbol{i}$ coincides with the spatial description contravariant connection tensor (in terms of the Euclidean metric). Likewise a spatial description third-order fully *contravariant* double-strain tensor is *formally* introduced with $\mathbf{b} := [\boldsymbol{b} \boxtimes \boldsymbol{b}]$ as the fully contravariant connection-like tensor

$$
{}^{\vartriangle}\boldsymbol{b} := -{}^{\vartriangle}\gamma : \mathbf{b} = {}^{\vartriangle}\boldsymbol{F} : \mathbf{I} : \mathbf{F}^t,
$$

whereby the negative sign follows again from convention. Note the similarity to the definition of $\boldsymbol{b}$ if ${}^{\vartriangle}\boldsymbol{F}$, $\mathbf{I} := [\boldsymbol{I} \otimes \boldsymbol{I}]$ and $\mathbf{F}^t$ are only exchanged by $\boldsymbol{F}$, $\boldsymbol{I}$, and $\boldsymbol{F}^t$, respectively. The corresponding pull-back then results in the material description third-order fully contravariant double-metric

$$
{}^{\vartriangle}\boldsymbol{I} := \boldsymbol{f} \cdot {}^{\vartriangle}\boldsymbol{b} : \mathbf{f}^t = {}^{\vartriangle}\boldsymbol{\Gamma} : \mathbf{I} = \boldsymbol{f} \cdot {}^{\vartriangle}\boldsymbol{F} : \mathbf{I}.
$$

The material double-metric ${}^{\triangle}\boldsymbol{I}$ coincides with the material description contravariant connection tensor (in terms of the Euclidean metric).

Velocity Double-Gradients:
The *second spatial gradient* of the velocity field shall be denoted the contra/co/co- (mixed) variant spatial ($_t$-type) *velocity double-gradient* and expands into

$$ {}^{\triangle}_{t}\boldsymbol{l} := \operatorname{grad}\operatorname{grad}\boldsymbol{v} = \operatorname{Grad}\operatorname{Grad}\boldsymbol{v} : [\boldsymbol{f} \boxtimes \boldsymbol{f}] + \operatorname{grad}\boldsymbol{v} \cdot {}^{\triangle}\boldsymbol{\gamma}. $$

Likewise, the *second material gradient* of the velocity field shall be denoted the contra/co/co- (mixed) variant two-point ($_0$-type) velocity double-gradient and computes as

$$ {}^{\triangle}_{0}\boldsymbol{\Lambda} := \operatorname{Grad}\operatorname{Grad}\boldsymbol{v} = \operatorname{grad}\operatorname{grad}\boldsymbol{v} : [\boldsymbol{F} \boxtimes \boldsymbol{F}] + \operatorname{Grad}\boldsymbol{v} \cdot {}^{\triangle}\boldsymbol{\Gamma}. $$

Note the presence of the contra/co/co- (mixed) variant spatial and material connection tensors and the spatial and material velocity gradients in the above expressions. Then the relation between the contra/co/co- (mixed) variant spatial ($_t$-type) and two-point ($_0$-type) velocity double-gradients follows as

$$ {}^{\triangle}_{t}\boldsymbol{l} = {}^{\triangle}_{0}\boldsymbol{\Lambda} : [\boldsymbol{f} \boxtimes \boldsymbol{f}] + \boldsymbol{l} \cdot {}^{\triangle}\boldsymbol{\gamma} \quad \text{and} \quad {}^{\triangle}_{0}\boldsymbol{\Lambda} = {}^{\triangle}_{t}\boldsymbol{l} : [\boldsymbol{F} \boxtimes \boldsymbol{F}] + \boldsymbol{\Lambda} \cdot {}^{\triangle}\boldsymbol{\Gamma}. $$

Furthermore, since ${}^{\triangle}_{t}\boldsymbol{l}$ and ${}^{\triangle}_{0}\boldsymbol{\Lambda}$ are contra/co/co- (mixed) variant spatial and two-point third-order tensor fields, the following one- and two-sided pull-back/push-forward operations are possible (with τ denoting either 0 or t, respectively, i.e. $\tau = 0, t$)

$$ {}^{\triangle}_{\tau}\boldsymbol{\Lambda} = {}^{\triangle}_{\tau}\boldsymbol{l} : [\boldsymbol{F} \boxtimes \boldsymbol{F}] \quad \text{and} \quad {}^{\triangle}_{\tau}\boldsymbol{\lambda} = \boldsymbol{f} \cdot {}^{\triangle}_{\tau}\boldsymbol{l} \quad \text{and} \quad {}^{\triangle}_{\tau}\boldsymbol{L} = \boldsymbol{f} \cdot {}^{\triangle}_{\tau}\boldsymbol{l} : [\boldsymbol{F} \boxtimes \boldsymbol{F}]. $$

Observe that in contrast to ${}^{\triangle}_{t}\boldsymbol{l} = \operatorname{grad}\boldsymbol{l}$ and ${}^{\triangle}_{0}\boldsymbol{\Lambda} = \operatorname{Grad}\boldsymbol{\Lambda} = {}^{\triangle}\dot{\boldsymbol{F}}$ neither the spatial ${}^{\triangle}_{0}\boldsymbol{l}$ nor the two-point (spatial/material) ${}^{\triangle}_{t}\boldsymbol{\Lambda}$ nor the two-point (material/spatial) ${}^{\triangle}_{\tau}\boldsymbol{\lambda}$ nor the material ${}^{\triangle}_{\tau}\boldsymbol{L}$ do represent any gradient of a 'velocity gradient', nevertheless for simplicity they shall also be addressed as 'velocity double-gradients'. Note, however, that ${}^{\triangle}_{t}\boldsymbol{L}$ computes as the material time derivative of the material connection tensor, i.e. ${}^{\triangle}_{t}\boldsymbol{L} = {}^{\triangle}\dot{\boldsymbol{\Gamma}}$. The relation between the various velocity double-gradients is depicted in the following table:

	${}^{\triangle}_{\tau}\boldsymbol{l}$	${}^{\triangle}_{\tau}\boldsymbol{\Lambda}$	${}^{\triangle}_{\tau}\boldsymbol{L}$	${}^{\triangle}_{\tau}\boldsymbol{\lambda}$
${}^{\triangle}_{\tau}\boldsymbol{l}$	$\bullet$	${}^{\triangle}_{\tau}\boldsymbol{\Lambda} : [\boldsymbol{f} \boxtimes \boldsymbol{f}]$	$\boldsymbol{F} \cdot {}^{\triangle}_{\tau}\boldsymbol{L} : [\boldsymbol{f} \boxtimes \boldsymbol{f}]$	$\boldsymbol{F} \cdot {}^{\triangle}_{\tau}\boldsymbol{\lambda}$
${}^{\triangle}_{\tau}\boldsymbol{\Lambda}$	${}^{\triangle}_{\tau}\boldsymbol{l} : [\boldsymbol{F} \boxtimes \boldsymbol{F}]$	$\bullet$	$\boldsymbol{F} \cdot {}^{\triangle}_{\tau}\boldsymbol{L}$	$\boldsymbol{F} \cdot {}^{\triangle}_{\tau}\boldsymbol{\lambda} : [\boldsymbol{F} \boxtimes \boldsymbol{F}]$
${}^{\triangle}_{\tau}\boldsymbol{L}$	$\boldsymbol{f} \cdot {}^{\triangle}_{\tau}\boldsymbol{l} : [\boldsymbol{F} \boxtimes \boldsymbol{F}]$	$\boldsymbol{f} \cdot {}^{\triangle}_{\tau}\boldsymbol{\Lambda}$	$\bullet$	${}^{\triangle}_{\tau}\boldsymbol{\lambda} : [\boldsymbol{F} \boxtimes \boldsymbol{F}]$
${}^{\triangle}_{\tau}\boldsymbol{\lambda}$	$\boldsymbol{f} \cdot {}^{\triangle}_{\tau}\boldsymbol{l}$	$\boldsymbol{f} \cdot {}^{\triangle}_{\tau}\boldsymbol{\Lambda} : [\boldsymbol{f} \boxtimes \boldsymbol{f}]$	${}^{\triangle}_{\tau}\boldsymbol{L} : [\boldsymbol{f} \boxtimes \boldsymbol{f}]$	$\bullet$

Based on the above notation the relation between the contra/co/co- (mixed) variant $_t$-type and $_0$-type *velocity double-gradients* abbreviates into

$$ {}^{\triangle}_{t}\boldsymbol{l} = {}^{\triangle}_{0}\boldsymbol{l} + \boldsymbol{l}\cdot{}^{\triangle}\gamma \quad \text{and} \quad {}^{\triangle}_{0}\boldsymbol{\Lambda} = {}^{\triangle}_{t}\boldsymbol{\Lambda} + \boldsymbol{\Lambda}\cdot{}^{\triangle}\boldsymbol{\Gamma}. $$

The two-point (spatial/material) and spatial (${}_{0}$-type) velocity double-gradients contribute to map either the material or the spatial line element into the material time derivative of the (spatial) truncated Taylor expansion $\mathrm{d}\boldsymbol{y}$, i.e.

$$ \overline{\mathrm{d}\boldsymbol{y}}^{\cdot} = \boldsymbol{\Lambda}\cdot\mathrm{d}\boldsymbol{X} + \frac{1}{2}{}^{\triangle}_{0}\boldsymbol{\Lambda} : [\,\mathrm{d}\boldsymbol{X}\otimes\mathrm{d}\boldsymbol{X}] = \boldsymbol{l}\cdot\mathrm{d}\boldsymbol{x} + \frac{1}{2}{}^{\triangle}_{0}\boldsymbol{l} : [\,\mathrm{d}\boldsymbol{x}\otimes\mathrm{d}\boldsymbol{x}] + \text{h.o.t.} $$

Note that the correction term $\boldsymbol{l}\cdot{}^{\triangle}\gamma$ in the relation between the ${}^{\triangle}_{\tau}\boldsymbol{l}$ may alternatively be expressed as $\boldsymbol{l}\cdot{}^{\triangle}\gamma = -\boldsymbol{\Lambda}\cdot{}^{\triangle}\boldsymbol{\Gamma} : [\boldsymbol{f}\boxtimes\boldsymbol{f}]$ and is thus consistent with the correction term $\boldsymbol{\Lambda}\cdot{}^{\triangle}\boldsymbol{\Gamma}$ in the relation between the ${}^{\triangle}_{\tau}\boldsymbol{\Lambda}$. Finally, upon introducing the contra/contra/co/co- (mixed) variant fourth-order tensor $\mathsf{I} := \boldsymbol{l}\boxtimes\boldsymbol{1}_{\mathrm{s}} + \boldsymbol{1}_{\mathrm{s}}\boxtimes\boldsymbol{l}$ the material time derivative of ${}^{\triangle}\boldsymbol{f}$ expands into $\mathrm{D}_t{}^{\triangle}\boldsymbol{f} = -{}^{\triangle}_{t}\boldsymbol{\lambda} - {}^{\triangle}\boldsymbol{f} : \mathsf{I}$, likewise the material time derivative of the spatial connection tensor ${}^{\triangle}\gamma$ computes as $\mathrm{D}_t{}^{\triangle}\gamma = -{}^{\triangle}_{0}\boldsymbol{l} - {}^{\triangle}\gamma : \mathsf{I}$.

Spatial Rate-of-Double-Deformation and Double-Spin Tensors:
On the one hand the fully covariant version of the (${}_{0}$- or ${}_{t}$-type) *spatial velocity double-gradient* allows decomposition into *'quasi symmetric'* and *'quasi skew symmetric'* contributions

$$ \boldsymbol{j}\cdot{}^{\triangle}_{\tau}\boldsymbol{l} = [\boldsymbol{j}\cdot{}^{\triangle}_{\tau}\boldsymbol{l}]^{'\mathrm{sym}'} + [\boldsymbol{j}\cdot{}^{\triangle}_{\tau}\boldsymbol{l}]^{'\mathrm{skw}'}. $$

Thereby 'quasi symmetric' and 'quasi skew symmetric' contributions to the third-order fully covariant version of the spatial velocity double-gradient are defined as

$$ 2[\boldsymbol{j}\cdot{}^{\triangle}_{\tau}\boldsymbol{l}]^{'\mathrm{sym}'}_{'\mathrm{skw}'} := \boldsymbol{j}\cdot{}^{\triangle}_{\tau}\boldsymbol{l} \pm \boldsymbol{l}^{t}\cdot{}^{\triangle}\boldsymbol{j}. $$

Then the 'quasi symmetric' contribution is denoted the fully covariant (${}_{0}$- or ${}_{t}$-type) *spatial rate-of-double-deformation tensor* whereas the 'quasi skew symmetric' contribution is denoted the fully covariant (${}_{0}$- or ${}_{t}$-type) *spatial double-spin tensor*

$$ {}^{\triangle}_{\tau}\boldsymbol{d} := [\boldsymbol{j}\cdot{}^{\triangle}_{\tau}\boldsymbol{l}]^{'\mathrm{sym}'} \quad \text{and} \quad {}^{\triangle}_{\tau}\boldsymbol{w} := [\boldsymbol{j}\cdot{}^{\triangle}_{\tau}\boldsymbol{l}]^{'\mathrm{skw}'}. $$

Observe the relations between the (${}_{0}$- and ${}_{t}$-type) spatial rate-of-double-deformation tensors

$$ 2{}^{\triangle}_{0}\boldsymbol{d} = 2{}^{\triangle}_{t}\boldsymbol{d} - \boldsymbol{j}\cdot\boldsymbol{l}\cdot{}^{\triangle}\gamma = \boldsymbol{j}\cdot{}^{\triangle}_{t}\boldsymbol{l} - 2\boldsymbol{d}\cdot{}^{\triangle}\gamma $$

and, likewise, between the (${}_{0}$- and ${}_{t}$-type) spatial double-spin tensors

$$ 2{}^{\triangle}_{0}\boldsymbol{w} = 2{}^{\triangle}_{t}\boldsymbol{w} - \boldsymbol{j}\cdot\boldsymbol{l}\cdot{}^{\triangle}\gamma = \boldsymbol{j}\cdot{}^{\triangle}_{t}\boldsymbol{l} - 2\boldsymbol{w}\cdot{}^{\triangle}\gamma. $$

Here the expression for the spatial double-metric in terms of the fully covariant version of the spatial connection tensor ${}^{\triangle}\boldsymbol{j} = -\boldsymbol{j}\cdot{}^{\triangle}\gamma$ has been involved to obtain the results on the utmost right of the above equations. Next the covariant pull-back of the fully covariant spatial velocity double-gradient $\boldsymbol{j}\cdot{}^{\Delta}_{\tau}\boldsymbol{l}$, i.e.

$$\boldsymbol{F}^t \cdot [\boldsymbol{j} \cdot {}^{\Delta}_{\tau}\boldsymbol{l}] : [\boldsymbol{F} \boxtimes \boldsymbol{F}] = \boldsymbol{F}^t \cdot [{}^{\Delta}_{\tau}\boldsymbol{d} + {}^{\Delta}_{\tau}\boldsymbol{w}] : [\boldsymbol{F} \boxtimes \boldsymbol{F}]$$

involves the covariant right Cauchy-Green strain $\boldsymbol{C}$ as a metric and renders

$$\boldsymbol{C} \cdot {}^{\Delta}_{\tau}\boldsymbol{L} = [\boldsymbol{C} \cdot {}^{\Delta}_{\tau}\boldsymbol{L}]^{'\mathrm{sym}'} + [\boldsymbol{C} \cdot {}^{\Delta}_{\tau}\boldsymbol{L}]^{'\mathrm{skw}'}.$$

Thereby, 'quasi symmetric' and 'quasi skew symmetric' contributions to $\boldsymbol{C}\cdot{}^{\Delta}_{\tau}\boldsymbol{L}$ follow as

$$2[\boldsymbol{C} \cdot {}^{\Delta}_{\tau}\boldsymbol{L}]^{'\mathrm{sym}'}_{'\mathrm{skw}'} = \boldsymbol{C} \cdot {}^{\Delta}_{\tau}\boldsymbol{L} \pm \boldsymbol{L}^t \cdot {}^{\Delta}\boldsymbol{C}.$$

Observe finally the relations between the ($_0$- and $_t$-type) material rate-of-double-deformation tensors

$$2[\boldsymbol{C} \cdot {}^{\Delta}_{0}\boldsymbol{L}]^{'\mathrm{sym}'} = 2[\boldsymbol{C} \cdot {}^{\Delta}_{t}\boldsymbol{L}]^{'\mathrm{sym}'} + \boldsymbol{C} \cdot \boldsymbol{L} \cdot {}^{\Delta}\boldsymbol{\Gamma} = \boldsymbol{C} \cdot {}^{\Delta}_{t}\boldsymbol{L} + 2[\boldsymbol{C} \cdot \boldsymbol{L}]^{\mathrm{sym}} \cdot {}^{\Delta}\boldsymbol{\Gamma}$$

and, likewise, between the ($_0$- and $_t$-type) material double-spin tensors

$$2[\boldsymbol{C} \cdot {}^{\Delta}_{0}\boldsymbol{L}]^{'\mathrm{skw}'} = 2[\boldsymbol{C} \cdot {}^{\Delta}_{t}\boldsymbol{L}]^{'\mathrm{skw}'} + \boldsymbol{C} \cdot \boldsymbol{L} \cdot {}^{\Delta}\boldsymbol{\Gamma} = \boldsymbol{C} \cdot {}^{\Delta}_{t}\boldsymbol{L} + 2[\boldsymbol{C} \cdot \boldsymbol{L}]^{\mathrm{skw}} \cdot {}^{\Delta}\boldsymbol{\Gamma}.$$

In summary the decomposition of the spatial velocity double-gradient into 'quasi symmetric' and 'quasi skew symmetric' contributions renders spatial rate-of-double-deformation and double-spin tensors, respectively, together with corresponding pull-back quantities. Thereby, the formal similarities to the decomposition of the spatial velocity gradient are notable.

Material Rate-of-Double-Deformation and Double-Spin Tensors: On the other hand the the fully covariant version of the ($_0$- or $_t$-type) *material velocity double-gradient* allows decomposition into 'quasi symmetric' and 'quasi skew symmetric' contributions

$$\boldsymbol{J} \cdot {}^{\Delta}_{\tau}\boldsymbol{L} = [\boldsymbol{J} \cdot {}^{\Delta}_{\tau}\boldsymbol{L}]^{'\mathrm{sym}'} + [\boldsymbol{J} \cdot {}^{\Delta}_{\tau}\boldsymbol{L}]^{'\mathrm{skw}'}.$$

Thereby 'quasi symmetric' and 'quasi skew symmetric' contributions to the third-order fully covariant version of the material velocity double-gradient are defined as

$$2[\boldsymbol{J} \cdot {}^{\Delta}_{\tau}\boldsymbol{L}]^{'\mathrm{sym}'}_{'\mathrm{skw}'} := \boldsymbol{J} \cdot {}^{\Delta}_{\tau}\boldsymbol{L} \pm \boldsymbol{L}^t \cdot {}^{\Delta}\boldsymbol{J}.$$

Then the 'quasi symmetric' contribution is denoted the fully covariant ($_0$- or $_t$-type) *material rate-of-double-deformation tensor* whereas the 'quasi skew symmetric' contribution is denoted the fully covariant ($_0$- or $_t$-type) *material double-spin tensor*

$${}^{\Delta}_{\tau}\boldsymbol{D} := [\boldsymbol{J} \cdot {}^{\Delta}_{\tau}\boldsymbol{L}]^{'\mathrm{sym}'} \quad \text{and} \quad {}^{\Delta}_{\tau}\boldsymbol{W} := [\boldsymbol{J} \cdot {}^{\Delta}_{\tau}\boldsymbol{L}]^{'\mathrm{skw}'}.$$

Observe the relations between the ($_0$- and $_t$-type) material rate-of-double-deformation tensors

$$2{}^{\triangle}_{t}\boldsymbol{D} = 2{}^{\triangle}_{0}\boldsymbol{D} - \boldsymbol{J}\cdot\boldsymbol{L}\cdot{}^{\triangle}\boldsymbol{\Gamma} = \boldsymbol{J}\cdot{}^{\triangle}_{0}\boldsymbol{L} - 2\boldsymbol{D}\cdot{}^{\triangle}\boldsymbol{\Gamma}$$

and, likewise, between the ($_0$- and $_t$-type) material double-spin tensors

$$2{}^{\triangle}_{t}\boldsymbol{W} = 2{}^{\triangle}_{0}\boldsymbol{W} - \boldsymbol{J}\cdot\boldsymbol{L}\cdot{}^{\triangle}\boldsymbol{\Gamma} = \boldsymbol{J}\cdot{}^{\triangle}_{0}\boldsymbol{L} - 2\boldsymbol{W}\cdot{}^{\triangle}\boldsymbol{\Gamma}.$$

Here the expression for the material double-metric in terms of the fully covariant version of the material connection tensor ${}^{\triangle}\boldsymbol{J} = -\boldsymbol{J}\cdot{}^{\triangle}\boldsymbol{\Gamma}$ has been involved to obtain the results on the utmost right of the above equations. Next the covariant push-forward of the fully covariant material velocity double-gradient $\boldsymbol{J}\cdot{}^{\triangle}_{\tau}\boldsymbol{L}$, i.e.

$$\boldsymbol{f}^t\cdot[\boldsymbol{J}\cdot{}^{\triangle}_{\tau}\boldsymbol{L}] : [\boldsymbol{f}\boxtimes\boldsymbol{f}] = \boldsymbol{f}^t\cdot[{}^{\triangle}_{\tau}\boldsymbol{D} + {}^{\triangle}_{\tau}\boldsymbol{W}] : [\boldsymbol{f}\boxtimes\boldsymbol{f}]$$

involves the covariant Finger strain $\boldsymbol{c}$ as a metric and renders

$$\boldsymbol{c}\cdot{}^{\triangle}_{\tau}\boldsymbol{l} = [\boldsymbol{c}\cdot{}^{\triangle}_{\tau}\boldsymbol{l}]^{'\mathrm{sym}'} + [\boldsymbol{c}\cdot{}^{\triangle}_{\tau}\boldsymbol{l}]^{'\mathrm{skw}'}.$$

Thereby, 'quasi symmetric' and 'quasi skew symmetric' contributions to $\boldsymbol{c}\cdot{}^{\triangle}_{\tau}\boldsymbol{l}$ follow as

$$2[\boldsymbol{c}\cdot{}^{\triangle}_{\tau}\boldsymbol{l}]^{'\mathrm{sym}'}_{'\mathrm{skw}'} = \boldsymbol{c}\cdot{}^{\triangle}_{\tau}\boldsymbol{l} \pm \boldsymbol{l}^t\cdot{}^{\triangle}\boldsymbol{c}.$$

Observe finally the relations between the ($_0$- and $_t$-type) spatial rate-of-double-deformation tensors

$$2[\boldsymbol{c}\cdot{}^{\triangle}_{t}\boldsymbol{l}]^{'\mathrm{sym}'} = 2[\boldsymbol{c}\cdot{}^{\triangle}_{0}\boldsymbol{l}]^{'\mathrm{sym}'} + \boldsymbol{c}\cdot\boldsymbol{l}\cdot{}^{\triangle}\gamma = \boldsymbol{c}\cdot{}^{\triangle}_{0}\boldsymbol{l} + 2[\boldsymbol{c}\cdot\boldsymbol{l}]^{\mathrm{sym}}\cdot{}^{\triangle}\gamma$$

and, likewise, between the ($_0$- and $_t$-type) spatial double-spin tensors

$$2[\boldsymbol{c}\cdot{}^{\triangle}_{t}\boldsymbol{l}]^{'\mathrm{skw}'} = 2[\boldsymbol{c}\cdot{}^{\triangle}_{0}\boldsymbol{l}]^{'\mathrm{skw}'} + \boldsymbol{c}\cdot\boldsymbol{l}\cdot{}^{\triangle}\gamma = \boldsymbol{c}\cdot{}^{\triangle}_{0}\boldsymbol{l} + 2[\boldsymbol{c}\cdot\boldsymbol{l}]^{\mathrm{skw}}\cdot{}^{\triangle}\gamma.$$

In summary the decomposition of the material velocity double-gradient into 'quasi symmetric' and 'quasi skew symmetric' contributions renders material rate-of-double-deformation and double-spin tensors, respectively, together with corresponding push-forward quantities. Thereby, the formal similarities to the decomposition of the material velocity gradient are notable.

Material Time Derivative of Double-Strain Measure:
The material time derivative of the (quadratic) length of the (spatial) truncated Taylor expansion $\mathrm{d}\boldsymbol{y}$ involves the fully covariant ($_0$-type) spatial rate-of-double-deformation tensor ${}^{\triangle}_{0}\boldsymbol{d}$

$$\overline{|\,\mathrm{d}\boldsymbol{y}|^2}^{\boldsymbol{\cdot}} = 2\,\mathrm{d}\boldsymbol{x}\cdot\left[\boldsymbol{d} + [{}^{\triangle}_{0}\boldsymbol{d}^{\cdot}\,\mathrm{d}\boldsymbol{x}]\right]\cdot\mathrm{d}\boldsymbol{x} + \text{h.o.t.}$$

or, likewise, its fully covariant pull-back

$$\overline{|\,\mathrm{d}\boldsymbol{y}|^2}^{\boldsymbol{\cdot}} = 2\,\mathrm{d}\boldsymbol{X}\cdot\Big[[\boldsymbol{C}\cdot\boldsymbol{L}]^{\mathrm{sym}} + \big[[\boldsymbol{C}\cdot{}^{\triangle}_{0}\boldsymbol{L}]^{'\mathrm{sym}'}\cdot\mathrm{d}\boldsymbol{X}\big]\Big]\cdot\mathrm{d}\boldsymbol{X} + \text{h.o.t.}$$

that proves to relate to the material time derivative of the material double-strain tensor ${}^{\Delta}\boldsymbol{C}$, i.e.

$$\overline{|\,\mathrm{d}\boldsymbol{y}|^2}^{\boldsymbol{\cdot}} = \mathrm{d}\boldsymbol{X} \cdot \big[\dot{\boldsymbol{C}} + [{}^{\Delta}\dot{\boldsymbol{C}} \cdot \mathrm{d}\boldsymbol{X}]\big] \cdot \mathrm{d}\boldsymbol{X} + \text{h.o.t.}.$$

Indeed the covariant pull-back of the ($_0$-type) spatial rate-of-double-deformation tensor ${}^{\Delta}_{0}\boldsymbol{d}$ computes as

$$\boldsymbol{F}^t \cdot {}^{\Delta}_{0}\boldsymbol{d} : [\boldsymbol{F} \boxtimes \boldsymbol{F}] = [\boldsymbol{C} \cdot {}^{\Delta}_{0}\boldsymbol{L}]^{'\mathrm{sym}'} = \frac{1}{2}{}^{\Delta}\dot{\boldsymbol{C}}.$$

Consequently, the material time derivatives of the material double-strain tensor ${}^{\Delta}\boldsymbol{E}$ reads

$$2{}^{\Delta}\dot{\boldsymbol{E}} = {}^{\Delta}\dot{\boldsymbol{C}} - {}^{\Delta}\dot{\boldsymbol{J}} \quad \text{with} \quad {}^{\Delta}\dot{\boldsymbol{J}} = -\boldsymbol{J} \cdot {}^{\Delta}_{t}\boldsymbol{L}.$$

Vice versa the covariant push-forward of the material rate-of-double-deformation tensor ${}^{\Delta}_{t}\boldsymbol{D}$ proofs to result in

$$\boldsymbol{f}^t \cdot {}^{\Delta}_{t}\boldsymbol{D} : [\boldsymbol{f} \boxtimes \boldsymbol{f}] = [\boldsymbol{c} \cdot {}^{\Delta}_{t}\boldsymbol{l}]^{'\mathrm{sym}'} = -\frac{1}{2}[{}^{\Delta}\dot{\boldsymbol{c}} + {}^{\Delta}\boldsymbol{c} : \mathsf{I}].$$

Finally the material time derivative of the spatial double-strain tensor ${}^{\Delta}\boldsymbol{e}$ is expressed as

$$2{}^{\Delta}\dot{\boldsymbol{e}} = {}^{\Delta}\dot{\bar{\boldsymbol{j}}} - {}^{\Delta}\dot{\boldsymbol{c}} \quad \text{with} \quad {}^{\Delta}\dot{\bar{\boldsymbol{j}}} = 2{}^{\Delta}_{0}\boldsymbol{d} - [{}^{\Delta}\boldsymbol{j} : \mathsf{I} + \boldsymbol{l}^t \cdot {}^{\Delta}\boldsymbol{j}].$$

Again the formal similarity between the material time derivatives of the various covariant, spatial and material double-strain measures, respectively, shall be observed.

Left Lie Time Derivative of Spatial Tensor Fields:
Recall that the *left Lie time derivative* L_t of generic *spatial* tensor fields $\{\bullet\}$ was previously defined as

$$\mathrm{L}_t\{\bullet\} := \text{push-forward}\Big(\,\mathrm{D}_t\big(\text{pull-back of } \{\bullet\}\big)\Big).$$

Applying this definition as an example to the fully covariant spatial double-metric and the double-strain measures renders

$$\mathrm{L}_t{}^{\Delta}\boldsymbol{j} = 2{}^{\Delta}_{0}\boldsymbol{d} \quad \text{and} \quad \mathrm{L}_t{}^{\Delta}\boldsymbol{c} = -\boldsymbol{c} \cdot {}^{\Delta}_{t}\boldsymbol{l} \quad \rightsquigarrow \quad 2\,\mathrm{L}_t{}^{\Delta}\boldsymbol{e} = 2{}^{\Delta}_{0}\boldsymbol{d} + \boldsymbol{c} \cdot {}^{\Delta}_{t}\boldsymbol{l}.$$

In general the relation between the left Lie and the material time derivative of generic fully covariant third-order spatial tensor fields $\{\bullet\}$ results in

$$\mathrm{L}_t\{\bullet\} = \mathrm{D}_t\{\bullet\} + [\{\bullet\} : \mathsf{I} + \boldsymbol{l}^t \cdot \{\bullet\}] =: \mathrm{D}_t\{\bullet\} + 2[\{\bullet\} : \mathsf{I}]^{'\mathrm{sym}'}.$$

Here the abbreviation $2[\{\bullet\} : \mathsf{I}]^{'\mathrm{sym}'} := \{\bullet\} : \mathsf{I} + \boldsymbol{l}^t \cdot \{\bullet\}$ has been introduced. A comparison of the Lie and the material time derivative of, e.g., the fully covariant double-strain measure ${}^{\triangle}\boldsymbol{c} = \boldsymbol{f}^t \cdot \boldsymbol{J} \cdot {}^{\triangle}\boldsymbol{f}$ (thereby resorting to the previous expressions for $\mathrm{D}_t\boldsymbol{f}$ and $\mathrm{D}_t{}^{\triangle}\boldsymbol{f}$) is clearly in agreement with the above relation.

Right Lie Time Derivative of Material Tensor Fields:
Recall that the *right Lie time derivative* R_t of generic *material* tensor fields $\{\bullet\}$ was previously defined as

$$\mathrm{R}_t\{\bullet\} := \text{pull-back}\Big(\mathrm{D}_t\big(\text{push-forward of }\{\bullet\}\big)\Big).$$

Applying this definition as an example to the fully covariant material double-strain and the double-metric renders

$$\mathrm{R}_t{}^{\triangle}\boldsymbol{C} = 2[\boldsymbol{C}\cdot{}^{\triangle}_{0}\boldsymbol{L}]^{'\mathrm{sym}'} - 2[{}^{\triangle}\boldsymbol{C} : \mathsf{L}]^{'\mathrm{sym}'} \quad \text{and} \quad \mathrm{R}_t{}^{\triangle}\boldsymbol{J} = -\boldsymbol{J}\cdot{}^{\triangle}_{t}\boldsymbol{L} - 2[{}^{\triangle}\boldsymbol{J} : \mathsf{L}]^{'\mathrm{sym}'}.$$

In general the relation between the right Lie and the material time derivative of generic fully covariant third-order material tensor fields $\{\bullet\}$ results in

$$\mathrm{R}_t\{\bullet\} = \mathrm{D}_t\{\bullet\} - [\{\bullet\} : \mathsf{L} + \boldsymbol{L}^t \cdot \{\bullet\}] =: \mathrm{D}_t\{\bullet\} - 2[\{\bullet\} : \mathsf{L}]^{'\mathrm{sym}'}.$$

Here the abbreviation $2[\{\bullet\} : \mathsf{L}]^{'\mathrm{sym}'} := \{\bullet\} : \mathsf{L} + \boldsymbol{L}^t \cdot \{\bullet\}$ has been introduced.

Distortion 'Double-Cofactor' and Double-Distortion 'Cofactor':
Next fourth-order co/co/contra/contra- (mixed) variant two-point tensors are introduced that may be denoted as the *'double-cofactor' of the distortion*

$$^{\triangle}\mathrm{cof}\,\boldsymbol{F} := J\boldsymbol{f}^t \boxtimes \boldsymbol{f}^t \quad \text{and} \quad {}^{\triangle}\mathrm{cof}\,\boldsymbol{f} := j\boldsymbol{F}^t \boxtimes \boldsymbol{F}^t.$$

Obviously, ${}^{\triangle}\mathrm{cof}\,\boldsymbol{F} : {}^{\triangle}\mathrm{cof}\,\boldsymbol{f} = \mathbf{1}^t_{\mathrm{s}} \boxtimes \mathbf{1}^t_{\mathrm{s}}$ and ${}^{\triangle}\mathrm{cof}\,\boldsymbol{f} : {}^{\triangle}\mathrm{cof}\,\boldsymbol{F} = \mathbf{1}^t_{\mathrm{m}} \boxtimes \mathbf{1}^t_{\mathrm{m}}$ hold. Moreover, third-order co/co/contra- (mixed) variant two-point tensors are introduced that may be denoted as the *'cofactor' of the double-distortion*

$$\mathrm{cof}\,{}^{\triangle}\boldsymbol{F} := J{}^{\triangle}\boldsymbol{f}^T = {}^{\triangle}\boldsymbol{\gamma}^T \cdot \mathrm{cof}\,\boldsymbol{F} \quad \text{and} \quad \mathrm{cof}\,{}^{\triangle}\boldsymbol{f} := j{}^{\triangle}\boldsymbol{F}^T = {}^{\triangle}\boldsymbol{\Gamma}^T \cdot \mathrm{cof}\,\boldsymbol{f}.$$

Here ${}^{\triangle}\{\bullet\}^T$ denotes a 'transpose' of a third-order tensor that exchanges the first index with the two last indices, i.e.

$$[{}^{\triangle}\boldsymbol{F}^T]_{JK}{}^{i} = [{}^{\triangle}\boldsymbol{F}]^{i}{}_{JK}.$$

Then the relation between the 'cofactor' of the double-distortions $\mathrm{cof}\,{}^{\triangle}\boldsymbol{F}$ and $\mathrm{cof}\,{}^{\triangle}\boldsymbol{f}$ follows in terms of the 'double-cofactor' and the ordinary cofactor of the distortions as

$$\mathrm{cof}\,{}^{\triangle}\boldsymbol{F} = -{}^{\triangle}\mathrm{cof}\,\boldsymbol{F} : \mathrm{cof}\,{}^{\triangle}\boldsymbol{f} \cdot \mathrm{cof}\,\boldsymbol{F} \quad \text{and} \quad \mathrm{cof}\,{}^{\triangle}\boldsymbol{f} = -{}^{\triangle}\mathrm{cof}\,\boldsymbol{f} : \mathrm{cof}\,{}^{\triangle}\boldsymbol{F} \cdot \mathrm{cof}\,\boldsymbol{f}.$$

Finally it shall be noted that the above relations are consistent with

$$^{\Delta}\mathrm{cof}\,\boldsymbol{F} : \mathrm{cof}\,^{\Delta}\boldsymbol{f} = -^{\Delta}\boldsymbol{\gamma}^T \quad \text{and} \quad ^{\Delta}\mathrm{cof}\,\boldsymbol{f} : \mathrm{cof}\,^{\Delta}\boldsymbol{F} = -^{\Delta}\boldsymbol{\Gamma}^T.$$

The 'double-cofactor' of the distortion and the 'cofactor' of the double-distortion are merely introduced for notational convenience, e.g. to introduce Piola-type transformations of third-order tensors.

Cauchy Stress and Double-Stress:
Based on the previously established sequence of 1. the Cauchy postulate, 2. the Cauchy lemma, and 3. the Cauchy theorem a traction vector, that shall here be regarded as the *effective traction vector* $\boldsymbol{t}_t^{\text{eff}}$ (see below), at a particular point $\boldsymbol{x}$ follows as a linear map of the outwards pointing normal vector $\boldsymbol{n}$ to an arbitrary virtual cut-surface through the point $\boldsymbol{x}$ in the spatial configuration

$$\boldsymbol{t}_t^{\text{eff}} = \boldsymbol{\sigma}^{\text{eff}}(\boldsymbol{x}) \cdot \boldsymbol{n}.$$

This map is defined in terms of a spatial second-order tensor $\boldsymbol{\sigma}^{\text{eff}}$ that is denoted the *effective Cauchy stress.* As will be demonstrated below, the effective Cauchy stress expands into the *($_t$-type) Cauchy stress* minus the (spatial) divergence of the *Cauchy double-stress*

$$\boldsymbol{\sigma}^{\text{eff}} := {}_t\boldsymbol{\sigma} - \mathrm{div}^{\Delta}\boldsymbol{\sigma}.$$

In the later developments it proves convenient to also introduce the spatial *surface-type Cauchy stress* that is related to the Cauchy double-stress by

$$\boldsymbol{\sigma}' := {}^{\Delta}\boldsymbol{\sigma}(\boldsymbol{x}) \cdot \boldsymbol{n}.$$

With these preliminaries the effective traction vector per unit area $\mathrm{d}a$ is finally given in terms of the traction vector $\boldsymbol{t}_t$ and the surface Cauchy stress $\boldsymbol{\sigma}'$ as

$$\boldsymbol{t}_t^{\text{eff}} := \boldsymbol{t}_t + \varsigma_t(\boldsymbol{\sigma}') \quad \text{with} \quad \varsigma_t(\{\bullet\}) = \widehat{\mathrm{div}}\,\widehat{\{\bullet\}} = k\{\bullet\} \cdot \boldsymbol{n} + \widehat{\mathrm{div}}\{\bullet\}.$$

Here $k := -\widehat{\mathrm{div}}\,\boldsymbol{n}$ denotes the total curvature (twice the mean curvature) of the spatial cut-surface with $\widehat{\mathrm{div}}\{\bullet\}$ the spatial surface divergence operator applied to $\{\bullet\}$. Moreover, the definition $\widehat{\{\bullet\}} := \{\bullet\} - \{\bullet\} \cdot [\boldsymbol{n} \otimes \boldsymbol{n}]$ has been employed. Finally, the surface-type Cauchy stress is postulated to also satisfy a Cauchy theorem

$$\boldsymbol{t}_t' = \boldsymbol{\sigma}'(\boldsymbol{x}) \cdot \boldsymbol{n}$$

whereby $\boldsymbol{t}_t'$ denotes the spatial *double-traction vector.*

Piola Stress and Double-Stress:
Instead of introducing the effective traction vector $\boldsymbol{t}_t^{\text{eff}}$ as a distributed force per unit area $\mathrm{d}a$ in the spatial configuration it proves convenient to introduce the effective traction vector $\boldsymbol{t}_0^{\text{eff}}$ as a distributed force per corresponding unit area $\mathrm{d}A$ in the material configuration with the equivalence

$$\boldsymbol{t}_t^{\text{eff}}\,\mathrm{d}a = \boldsymbol{\sigma}^{\text{eff}} \cdot \boldsymbol{n}\,\mathrm{d}a =: \boldsymbol{P}^{\text{eff}} \cdot \boldsymbol{N}\,\mathrm{d}A =: \boldsymbol{t}_0^{\text{eff}}\,\mathrm{d}A.$$

Here in analogy with the Cauchy theorem a relation of $\boldsymbol{t}_0^{\text{eff}}$ with the outwards pointing normal vector $\boldsymbol{N}$ to an arbitrary virtual cut-surface through point $\boldsymbol{X}$ in the material configuration has been postulated as

$$\boldsymbol{t}_0^{\text{eff}} = \boldsymbol{P}^{\text{eff}}(\boldsymbol{X}) \cdot \boldsymbol{N}.$$

Thereby the two-point second-order tensor $\boldsymbol{P}^{\text{eff}}$ denotes the *effective Piola stress* that, due to $\boldsymbol{n}\,\mathrm{d}a = \operatorname{cof}\boldsymbol{F} \cdot \boldsymbol{N}\,\mathrm{d}A$, is related to the effective Cauchy stress by

$$\boldsymbol{\sigma}^{\text{eff}} \cdot \operatorname{cof}\boldsymbol{F} = \boldsymbol{P}^{\text{eff}}.$$

By the help of the chain rule it may next easily be demonstrated for generic spatial third-order tensor fields ${}^{\vartriangle}\{\bullet\}$ that

$$[\operatorname{div}{}^{\vartriangle}\{\bullet\}] \cdot \operatorname{cof}\boldsymbol{F} = \operatorname{Div}({}^{\vartriangle}\{\bullet\} : {}^{\vartriangle}\operatorname{cof}\boldsymbol{F}) - [{}^{\vartriangle}\{\bullet\} : {}^{\vartriangle}\boldsymbol{\gamma}^T] \cdot \operatorname{cof}\boldsymbol{F},$$

whereby, interestingly, the last term on the right-hand-side involving the spatial connection tensor is alternatively expressed in terms of the material connection tensor as

$$[{}^{\vartriangle}\{\bullet\} : {}^{\vartriangle}\boldsymbol{\gamma}^T] \cdot \operatorname{cof}\boldsymbol{F} = -[{}^{\vartriangle}\{\bullet\} : {}^{\vartriangle}\operatorname{cof}\boldsymbol{F}] : {}^{\vartriangle}\boldsymbol{\Gamma}^T.$$

Thus the effective Piola stress expands into the *($_0$-type) Piola stress* minus the material divergence of the *Piola double-stress*

$$\boldsymbol{P}^{\text{eff}} := {}_0\boldsymbol{P} - \operatorname{Div}{}^{\vartriangle}\boldsymbol{P}.$$

Thereby the ($_0$- and $_t$-type, respectively) Piola stress ${}_\tau\boldsymbol{P}$ (with $\tau = 0, t$) is related to the ($_t$- and $_0$-type, respectively) Cauchy stress ${}_\tau\boldsymbol{\sigma}$ (with $\tau = t, 0$) by

$$[{}_t\boldsymbol{\sigma} + {}^{\vartriangle}\boldsymbol{\sigma} : {}^{\vartriangle}\boldsymbol{\gamma}^T] \cdot \operatorname{cof}\boldsymbol{F} =: {}_t\boldsymbol{P} - {}^{\vartriangle}\boldsymbol{P} : {}^{\vartriangle}\boldsymbol{\Gamma}^T =: {}_0\boldsymbol{\sigma} \cdot \operatorname{cof}\boldsymbol{F} =: {}_0\boldsymbol{P},$$

while the Piola double-stress ${}^{\vartriangle}\boldsymbol{P}$ is related to the Cauchy double-stress ${}^{\vartriangle}\boldsymbol{\sigma}$ as

$${}^{\vartriangle}\boldsymbol{\sigma} : {}^{\vartriangle}\operatorname{cof}\boldsymbol{F} =: {}^{\vartriangle}\boldsymbol{P}.$$

In the later developments it proves convenient to also introduce the corresponding two-point *surface-type Piola stress* as

$$\boldsymbol{P}' := {}^{\vartriangle}\boldsymbol{P}(\boldsymbol{X}) \cdot \boldsymbol{N}.$$

With these preliminaries and the relations for surface quantities in [184] it may then finally be shown that

$$\varsigma_t(\boldsymbol{\sigma}')\,\mathrm{d}a = \varsigma_0(\boldsymbol{P}')\,\mathrm{d}A \quad \text{with} \quad \varsigma_0(\{\bullet\}) = \widehat{\operatorname{Div}}\,\widehat{\{\bullet\}} = K\{\bullet\} \cdot \boldsymbol{N} + \widehat{\operatorname{Div}}\{\bullet\}.$$

This is a key equivalence in order to equate $\boldsymbol{t}_t^{\text{eff}}\,\mathrm{d}a$ with $\boldsymbol{t}_0^{\text{eff}}\,\mathrm{d}A$ as postulated in the above. Here $K := -\widehat{\operatorname{Div}}\,\boldsymbol{N}$ denotes the total curvature (twice the mean curvature) of the material cut-surface with $\widehat{\operatorname{Div}}\{\bullet\}$ the material surface divergence operator applied to $\{\bullet\}$. Moreover, the definition $\widehat{\{\bullet\}} := \{\bullet\} - \{\bullet\}\cdot[\boldsymbol{N}\otimes\boldsymbol{N}]$ has been employed. Finally, the surface-type Piola stress is postulated to also satisfy a Cauchy theorem

$$\boldsymbol{t}_0' = \boldsymbol{P}'(\boldsymbol{X})\cdot\boldsymbol{N}$$

whereby $\boldsymbol{t}_0'$ denotes the material *double-traction vector.*

Piola Identity and Piola Transformation:
As a consequence from the Piola identity, the material and spatial *double-divergence* of third-order tensorial quantities that are connected by a Piola-type transformation in terms of the 'double-cofactor' of the distortion

$$^{\vartriangle}\{\bullet\} : {}^{\vartriangle}\operatorname{cof}\boldsymbol{F} \leftrightarrow {}^{\vartriangle}\{\bullet\}$$

are related by a multiplication with the determinant of the deformation gradient and a correction term involving the connection tensor

$$\operatorname{Div}\operatorname{Div}({}^{\vartriangle}\{\bullet\} : {}^{\vartriangle}\operatorname{cof}\boldsymbol{F}) = J\operatorname{div}\operatorname{div}{}^{\vartriangle}\{\bullet\} + J\operatorname{div}({}^{\vartriangle}\{\bullet\} : {}^{\vartriangle}\gamma^T).$$

The proof follows in a straightforward manner from the generic result in the previous paragraph, the chain rule and the Piola identity. By the same token the last term on the right-hand-side is alternatively expressed as

$$J\operatorname{div}({}^{\vartriangle}\{\bullet\} : {}^{\vartriangle}\gamma^T) = -\operatorname{Div}([{}^{\vartriangle}\{\bullet\} : {}^{\vartriangle}\operatorname{cof}\boldsymbol{F}] : {}^{\vartriangle}\boldsymbol{\Gamma}^T).$$

Thus, as the paradigmatic example, the ($_0$-type) Piola and the ($_t$-type) Cauchy stress together with the Piola and Cauchy double-stress satisfy

$$\operatorname{Div}{}_0\boldsymbol{P} - \operatorname{Div}\operatorname{Div}{}^{\vartriangle}\boldsymbol{P} = J\operatorname{div}_t\boldsymbol{\sigma} - J\operatorname{div}\operatorname{div}^{\vartriangle}\boldsymbol{\sigma}.$$

Alternatively, the above equates to different combinations of ($_0$- or $_t$-type) Piola and Cauchy stresses and double-stresses

$$\operatorname{Div}{}_t\boldsymbol{P} - J\operatorname{div}\operatorname{div}^{\vartriangle}\boldsymbol{\sigma} = J\operatorname{div}_0\boldsymbol{\sigma} - \operatorname{Div}\operatorname{Div}{}^{\vartriangle}\boldsymbol{P}.$$

Recall the relations between the $_0$- and $_t$-type Piola and the $_0$- and $_t$-type Cauchy stresses, respectively, as

$${}_0\boldsymbol{P} = {}_t\boldsymbol{P} - {}^{\vartriangle}\boldsymbol{P} : {}^{\vartriangle}\boldsymbol{\Gamma}^T \quad \text{and} \quad {}_t\boldsymbol{\sigma} = {}_0\boldsymbol{\sigma} - {}^{\vartriangle}\boldsymbol{\sigma} : {}^{\vartriangle}\gamma^T.$$

It is interesting to note that these relations may be re-expressed as a generalized Piola transformation in terms of the cofactor of the distortion and the 'cofactor' of the double-distortion

$$ {}_0\boldsymbol{P} = {}_t\boldsymbol{\sigma} \cdot \operatorname{cof} \boldsymbol{F} + {}^\triangle\boldsymbol{\sigma} : \operatorname{cof} {}^\triangle\boldsymbol{F} \quad \text{and} \quad {}_t\boldsymbol{\sigma} = {}_0\boldsymbol{P} \cdot \operatorname{cof} \boldsymbol{f} + {}^\triangle\boldsymbol{P} : \operatorname{cof} {}^\triangle\boldsymbol{f}. $$

The above are the key relations that allow in the sequel to express the balance of momentum in terms of either the two-point Piola or the spatial Cauchy stress and double-stress.

Stress/Double-Stress Measures:
Due to the relation between the Piola and the Cauchy double-stress it proves convenient to introduce the *Kirchhoff double-stress* ${}^\triangle\boldsymbol{\tau}$ as

$$ {}^\triangle\boldsymbol{\tau} := J\,{}^\triangle\boldsymbol{\sigma} = {}^\triangle\boldsymbol{P} : [\boldsymbol{F}^t \boxtimes \boldsymbol{F}^t]. $$

The Kirchhoff double-stress might be considered either a fully contra-, or a co/contra/contra- (mixed) variant spatial tensor field (again no special notation shall be adopted to distinguish between these options). Then the fully contra-, and the co/contra/contra- (mixed) variant pull-back of the Kirchhoff double-stress follow as

$$ {}^\triangle\boldsymbol{S} := \boldsymbol{f}_\sharp \cdot {}^\triangle\boldsymbol{\tau} : [\boldsymbol{f}^t \boxtimes \boldsymbol{f}^t] \quad \text{and} \quad {}^\triangle\boldsymbol{M} := \boldsymbol{F}^t \cdot {}^\triangle\boldsymbol{\tau} : [\boldsymbol{f}^t \boxtimes \boldsymbol{f}^t]. $$

Here ${}^\triangle\boldsymbol{S}$ and ${}^\triangle\boldsymbol{M}$ denote a contravariantly transforming material double-stress measure (that shall be denoted the *Piola-Kirchhoff double-stress*), and a co/contra/contra- (mixed) variantly transforming material double-stress measure (that shall be denoted the *Mandel double-stress*), respectively. The various relations between the introduced double-stress measures are assembled in the following table (with the abbreviations $\boldsymbol{\mathsf{F}}^t := \boldsymbol{F}^t \boxtimes \boldsymbol{F}^t$ and $\boldsymbol{\mathsf{f}}^t := \boldsymbol{f}^t \boxtimes \boldsymbol{f}^t$):

	${}^\triangle\boldsymbol{\tau}$	${}^\triangle\boldsymbol{P}$	${}^\triangle\boldsymbol{S}$	${}^\triangle\boldsymbol{M}$
${}^\triangle\boldsymbol{\tau}$	$\bullet$	${}^\triangle\boldsymbol{P} : \boldsymbol{\mathsf{F}}^t$	$\boldsymbol{F}_\flat \cdot {}^\triangle\boldsymbol{S} : \boldsymbol{\mathsf{F}}^t$	$\boldsymbol{f}^t \cdot {}^\triangle\boldsymbol{M} : \boldsymbol{\mathsf{F}}^t$
${}^\triangle\boldsymbol{P}$	${}^\triangle\boldsymbol{\tau} : \boldsymbol{\mathsf{f}}^t$	$\bullet$	$\boldsymbol{F}_\flat \cdot {}^\triangle\boldsymbol{S}$	$\boldsymbol{f}^t \cdot {}^\triangle\boldsymbol{M}$
${}^\triangle\boldsymbol{S}$	$\boldsymbol{f}_\sharp \cdot {}^\triangle\boldsymbol{\tau} : \boldsymbol{\mathsf{f}}^t$	$\boldsymbol{f}_\sharp \cdot {}^\triangle\boldsymbol{P}$	$\bullet$	$\boldsymbol{B} \cdot {}^\triangle\boldsymbol{M}$
${}^\triangle\boldsymbol{M}$	$\boldsymbol{F}^t \cdot {}^\triangle\boldsymbol{\tau} : \boldsymbol{\mathsf{f}}^t$	$\boldsymbol{F}^t \cdot {}^\triangle\boldsymbol{P}$	$\boldsymbol{C} \cdot {}^\triangle\boldsymbol{S}$	$\bullet$

Moreover, the transformations between the ${}_{\mathcal{T}}$-type Kirchhoff, Piola, Piola-Kirchhoff, and Mandel stress measures follow the pattern as outline before and are assembled in the following table:

	${}_{\mathcal{T}}\boldsymbol{\tau}$	${}_{\mathcal{T}}\boldsymbol{P}$	${}_{\mathcal{T}}\boldsymbol{S}$	${}_{\mathcal{T}}\boldsymbol{M}$
${}_{\mathcal{T}}\boldsymbol{\tau}$	$\bullet$	${}_{\mathcal{T}}\boldsymbol{P} \cdot \boldsymbol{F}^t$	$\boldsymbol{F}_\flat \cdot {}_{\mathcal{T}}\boldsymbol{S} \cdot \boldsymbol{F}^t$	$\boldsymbol{f}^t \cdot {}_{\mathcal{T}}\boldsymbol{M} \cdot \boldsymbol{F}^t$
${}_{\mathcal{T}}\boldsymbol{P}$	${}_{\mathcal{T}}\boldsymbol{\tau} \cdot \boldsymbol{f}^t$	$\bullet$	$\boldsymbol{F}_\flat \cdot {}_{\mathcal{T}}\boldsymbol{S}$	$\boldsymbol{f}^t \cdot {}_{\mathcal{T}}\boldsymbol{M}$
${}_{\mathcal{T}}\boldsymbol{S}$	$\boldsymbol{f}_\sharp \cdot {}_{\mathcal{T}}\boldsymbol{\tau} \cdot \boldsymbol{f}^t$	$\boldsymbol{f}_\sharp \cdot {}_{\mathcal{T}}\boldsymbol{P}$	$\bullet$	$\boldsymbol{B} \cdot {}_{\mathcal{T}}\boldsymbol{M}$
${}_{\mathcal{T}}\boldsymbol{M}$	$\boldsymbol{F}^t \cdot {}_{\mathcal{T}}\boldsymbol{\tau} \cdot \boldsymbol{f}^t$	$\boldsymbol{F}^t \cdot {}_{\mathcal{T}}\boldsymbol{P}$	$\boldsymbol{C} \cdot {}_{\mathcal{T}}\boldsymbol{S}$	$\bullet$

Thereby the relations between the ${}_0$- and ${}_t$-type Piola-Kirchhoff and the ${}_0$- and ${}_t$-type Mandel stresses, respectively, follow as

$$ {}_0\boldsymbol{S} = {}_t\boldsymbol{S} - {}^\Delta\boldsymbol{S} : {}^\Delta\boldsymbol{\Gamma}^T \quad \text{and} \quad {}_0\boldsymbol{M} = {}_t\boldsymbol{M} - {}^\Delta\boldsymbol{M} : {}^\Delta\boldsymbol{\Gamma}^T . $$

The ${}_0$- and ${}_t$-type stress measures are motivated on the one hand by contributing to different combinations of spatial and material divergence and double-divergence operations (as shown in the above), and on the other hand by contributing to different versions of the stress/double-stress power (as demonstrated below).

Balance Equations:
The state functions for a second-order continuum, i.e. the mass density ρ_0, the linear and angular momentum densities $\boldsymbol{p}_0$ and $\boldsymbol{l}_0$, and the mechanical energy density e_0 (all expressed per unit volume $\mathrm{d}V$ in the material configuration $\mathcal{B}_0$) are governed by a set of balance equations. These are the

1. Balance of Mass:
For a second-order continuum the global and local statements for the balance of mass are unchanged as compared to a first-order continuum.

2. Balance of Linear Momentum:
The global statement for the balance of linear momentum of an arbitrary continuous sub-body $\mathscr{V} \subseteq \mathscr{B}$ of a continuous body $\mathscr{B}$ that occupies the material sub-configuration $\mathcal{V}_0 \subseteq \mathcal{B}_0$ of the material configuration $\mathcal{B}_0$ reads as

$$ \dot{\mathsf{P}}(\mathcal{V}_0) = \boldsymbol{\mathcal{F}}(\mathcal{V}_0) \quad \forall \mathcal{V}_0 . $$

For a second-order continuum the resultant external force is defined in standard fashion

$$ \boldsymbol{\mathcal{F}}(\mathcal{V}_0) := \int_{\mathcal{V}_0} \boldsymbol{b}_0 \, \mathrm{d}V + \int_{\partial\mathcal{V}_0} \boldsymbol{t}_0 \, \mathrm{d}A . $$

Thereby, as usual, the linear momentum source $\boldsymbol{b}_0$ and flux $\boldsymbol{t}_0$ denote the body force density per unit volume $\mathrm{d}V$ and the surface force density, i.e. the (applied) traction per unit area $\mathrm{d}A$, respectively. For an arbitrary subsurface $\mathcal{S}_0 \subset \partial\mathcal{V}_0$ of the boundary $\partial\mathcal{V}_0$ to $\mathcal{V}_0$ the *surface Gauss theorem* renders

$$ \int_{\partial\mathcal{S}_0} \boldsymbol{P}' \cdot \boldsymbol{M} \, \mathrm{d}L = \int_{\mathcal{S}_0} \varsigma_0(\boldsymbol{P}') \, \mathrm{d}A \quad \text{with} \quad \varsigma_0(\{\bullet\}) := K\{\bullet\} \cdot \boldsymbol{N} + \widehat{\mathrm{Div}}\{\bullet\} . $$

The *surface-type Piola stress* $\boldsymbol{P}' := {}^\Delta\boldsymbol{P} \cdot \boldsymbol{N}$ consists of a *normal* and a *tangent (superficial)* contribution

$$ \widehat{\widehat{\boldsymbol{P}}}' := \boldsymbol{P}' \cdot [\boldsymbol{N} \otimes \boldsymbol{N}] \quad \text{and} \quad \widehat{\boldsymbol{P}}' := \boldsymbol{P}' - \widehat{\widehat{\boldsymbol{P}}}' . $$

Moreover, $\boldsymbol{M} \perp \boldsymbol{N}$ is the outwards pointing normal to $\partial\mathcal{S}_0$, thus $\boldsymbol{P}' \cdot \boldsymbol{M} = \widehat{\boldsymbol{P}}' \cdot \boldsymbol{M}$ and $\varsigma_0(\boldsymbol{P}') = \widehat{\mathrm{Div}}\, \widehat{\boldsymbol{P}}' = \widehat{J}\, \widehat{\mathrm{div}}\, \widehat{\boldsymbol{\sigma}}'$. In particular for the total boundary

$\partial\mathcal{V}_0$ to $\mathcal{V}_0$, which is a closed surface (that is assumed smooth), the left-hand-side in the above integral statement vanishes, thus the resultant external force acting on $\mathcal{V}_0$ may likewise be expressed as

$$\boldsymbol{\mathcal{F}}(\mathcal{V}_0) := \int_{\mathcal{V}_0} \boldsymbol{b}_0 \, \mathrm{d}V + \int_{\partial\mathcal{V}_0} \boldsymbol{t}_0^{\mathrm{eff}} \, \mathrm{d}A.$$

Recall that $\boldsymbol{t}_0^{\mathrm{eff}}$ denotes the *effective traction* per unit area $\mathrm{d}A$ consisting of two contributions and satisfying a Cauchy theorem (see below)

$$\boldsymbol{t}_0^{\mathrm{eff}} := \boldsymbol{t}_0 + \varsigma_0(\boldsymbol{P}') := \boldsymbol{P}^{\mathrm{eff}} \cdot \boldsymbol{N}.$$

It shall be noted that the normal part of $\boldsymbol{P}'$ denotes the (applied) double-traction ${}^{\Delta}\boldsymbol{t}_0 := \boldsymbol{t}_0' \otimes \boldsymbol{N}$ that is a characteristic feature of a second-order continuum

$${}^{\Delta}\boldsymbol{t}_0 := \widehat{\widehat{\boldsymbol{P}'}} \quad \text{with} \quad \boldsymbol{t}_0' := \boldsymbol{P}' \cdot \boldsymbol{N} = {}^{\Delta}\boldsymbol{P} : [\boldsymbol{N} \otimes \boldsymbol{N}].$$

Then upon localization for arbitrary $\mathcal{V}_0$ the local statement for the balance of linear momentum (that agrees with the classical exposition in [141]) is eventually expressed as

$$\dot{\boldsymbol{p}}_0 = \mathrm{Div}\, \boldsymbol{P}^{\mathrm{eff}} + \boldsymbol{b}_0 \quad \text{with} \quad \dot{\boldsymbol{p}}_0 = \rho_0 \dot{\boldsymbol{v}} \quad \rightsquigarrow \quad \rho_t \dot{\boldsymbol{v}} = \mathrm{div} \boldsymbol{\sigma}^{\mathrm{eff}} + \boldsymbol{b}_t.$$

Finally it shall be noted that the contribution $\varsigma_0(\boldsymbol{P}')$ in the effective traction boundary condition follows (by the help of a pill-box argument) from considering equilibrium at an arbitrary subsurface $\mathcal{S}_0 \subset \partial\mathcal{V}_0$ of the boundary $\partial\mathcal{V}_0$ to $\mathcal{V}_0$, whereby additional surface-type Piola stresses $\boldsymbol{P}'$ are allowed to act along the curve $\partial\mathcal{S}_0$, i.e.

$$\int_{\mathcal{S}_0} \boldsymbol{P}^{\mathrm{eff}} \cdot \boldsymbol{N} \, \mathrm{d}A = \int_{\mathcal{S}_0} \boldsymbol{t}_0 \, \mathrm{d}A + \int_{\partial\mathcal{S}_0} \boldsymbol{P}' \cdot \boldsymbol{M} \, \mathrm{d}L = \int_{\mathcal{S}_0} [\boldsymbol{t}_0 + \varsigma_0(\boldsymbol{P}')] \, \mathrm{d}A.$$

Localizing the above integral statement at an arbitrary point of the boundary $\partial\mathcal{V}_0$ renders the effective traction boundary condition $\boldsymbol{t}_0^{\mathrm{eff}} := \boldsymbol{P}^{\mathrm{eff}} \cdot \boldsymbol{N}$. It remains, however, to determine the effective Piola stress $\boldsymbol{P}^{\mathrm{eff}}$ in terms of the ($_0$-type) Piola stress ${}_0\boldsymbol{P}$ and and the Piola double-stress ${}^{\Delta}\boldsymbol{P}$.

3. Balance of Angular Momentum:
The global statement for the balance of angular momentum of an arbitrary continuous sub-body $\mathscr{V} \subseteq \mathscr{B}$ of a continuous body $\mathscr{B}$ that occupies the material sub-configuration $\mathcal{V}_0 \subseteq \mathcal{B}_0$ of the material configuration $\mathcal{B}_0$ reads as

$$\dot{\mathsf{L}}(\mathcal{V}_0) = \boldsymbol{\mathcal{M}}(\mathcal{V}_0) \quad \forall \mathcal{V}_0.$$

For a second-order continuum the resultant external moment also incorporates a contribution from the (applied) double-traction and is defined as

$$\boldsymbol{\mathcal{M}}(\mathcal{V}_0) := \int_{\mathcal{V}_0} \boldsymbol{r} \times \boldsymbol{b}_0 \, \mathrm{d}V + \int_{\partial\mathcal{V}_0} [\boldsymbol{r} \times \boldsymbol{t}_0 + \widehat{\widehat{\mathrm{Grad}}}\, \boldsymbol{r} \ltimes {}^{\vartriangle}\boldsymbol{t}_0] \, \mathrm{d}A.$$

Observe the additional contribution to the resultant external moment that is due to the double-traction ${}^{\vartriangle}\boldsymbol{t}_0$ crossed with the normal part of the gradient of the spatial distance vector $\boldsymbol{r}$, that is defined as

$$\widehat{\widehat{\mathrm{Grad}}}\, \boldsymbol{r} := \mathrm{Grad}\, \boldsymbol{r} \cdot [\boldsymbol{N} \otimes \boldsymbol{N}] = \mathrm{Grad}\, \boldsymbol{r} - \widehat{\mathrm{Grad}}\, \boldsymbol{r}.$$

Note that this contribution may alternatively be expressed in terms of the normal gradient of the spatial distance vector $\boldsymbol{r}$ as

$$\widehat{\widehat{\mathrm{Grad}}}\, \boldsymbol{r} \ltimes {}^{\vartriangle}\boldsymbol{t}_0 = \boldsymbol{r}' \times \boldsymbol{t}_0' \quad \text{with} \quad \boldsymbol{r}' := \nabla_{\boldsymbol{N}} \boldsymbol{r} := \mathrm{Grad}\, \boldsymbol{r} \cdot \boldsymbol{N}.$$

For an arbitrary subsurface $\mathcal{S}_0 \subset \partial\mathcal{V}_0$ of the boundary $\partial\mathcal{V}_0$ to $\mathcal{V}_0$ the surface Gauss theorem renders

$$\int_{\partial\mathcal{S}_0} \boldsymbol{r} \times [\boldsymbol{P}' \cdot \boldsymbol{M}] \, \mathrm{d}L = \int_{\mathcal{S}_0} \left[\boldsymbol{r} \times \varsigma_0(\boldsymbol{P}') + \widehat{\mathrm{Grad}}\, \boldsymbol{r} \ltimes \boldsymbol{P}'\right] \mathrm{d}A.$$

In particular for the total boundary $\partial\mathcal{V}_0$ to $\mathcal{V}_0$, which is a closed surface (that is assumed smooth), the left-hand-side in the above integral statement vanishes, thus with

$$\widehat{\widehat{\mathrm{Grad}}}\, \boldsymbol{r} \ltimes {}^{\vartriangle}\boldsymbol{t}_0 = \mathrm{Grad}\, \boldsymbol{r} \ltimes \widehat{\widehat{\boldsymbol{P}}}{}' \quad \text{and} \quad \widehat{\mathrm{Grad}}\, \boldsymbol{r} \ltimes \boldsymbol{P}' = \mathrm{Grad}\, \boldsymbol{r} \ltimes \widehat{\boldsymbol{P}}{}'$$

the resultant external moment acting on $\mathcal{V}_0$ may likewise be expressed as

$$\boldsymbol{\mathcal{M}}(\mathcal{V}_0) := \int_{\mathcal{V}_0} \boldsymbol{r} \times \boldsymbol{b}_0 \, \mathrm{d}V + \int_{\partial\mathcal{V}_0} [\boldsymbol{r} \times \boldsymbol{t}_0^{\mathrm{eff}} + \mathrm{Grad}\, \boldsymbol{r} \ltimes \boldsymbol{P}'] \, \mathrm{d}A.$$

Then upon localization for arbitrary $\mathcal{V}_0$ the local statement for the balance of angular momentum is eventually expressed as

$$\boldsymbol{0} \doteq \mathrm{Grad}\, \boldsymbol{r} \ltimes [\boldsymbol{P}^{\mathrm{eff}} + \mathrm{Div}\, {}^{\vartriangle}\boldsymbol{P}] + \mathrm{Grad}\,\mathrm{Grad}\, \boldsymbol{r} \ltimes {}^{\vartriangle}\boldsymbol{P}.$$

By *defining* next the ${}_0$-type Piola stress ${}_0\boldsymbol{P} := \boldsymbol{P}^{\mathrm{eff}} + \mathrm{Div}\, {}^{\vartriangle}\boldsymbol{P}$ the above statement transforms into the *symmetry condition* (that agrees with the classical exposition in [141])

$${}_0\boldsymbol{P} \cdot \boldsymbol{F}^t + {}^{\vartriangle}\boldsymbol{P} : {}^{\vartriangle}\boldsymbol{F}^T \doteq \boldsymbol{F} \cdot {}_0\boldsymbol{P}^t + {}^{\vartriangle}\boldsymbol{F} : {}^{\vartriangle}\boldsymbol{P}^T \quad \rightsquigarrow \quad {}_t\boldsymbol{\tau}_\sharp = {}_t\boldsymbol{\tau}_\sharp^t \;\text{ and }\; {}_t\boldsymbol{\sigma}_\sharp = {}_t\boldsymbol{\sigma}_\sharp^t.$$

Observe that for a second-order continuum the balance of angular momentum as stated in the above is implicitly satisfied for symmetric ${}_t$-type Cauchy stress, whereby the ${}_t$-type Cauchy stress is the result of an appropriate push-forward of the ${}_0$-type Piola stress and the Piola double-stress.

4. Balance of Mechanical Energy:
The global statement for the balance of mechanical energy of an arbitrary continuous sub-body $\mathscr{V} \subseteq \mathscr{B}$ of a continuous body $\mathscr{B}$ that occupies the material sub-configuration $\mathcal{V}_0 \subseteq \mathcal{B}_0$ of the material configuration $\mathcal{B}_0$ reads as

$$\dot{\mathsf{E}}(\mathcal{V}_0) = \mathcal{P}(\mathcal{V}_0) \quad \forall \mathcal{V}_0.$$

Recall that the density $e_0 = k_0 + w_0$ of the total *mechanical energy* E is composed of the *kinetic energy* density k_0 with $2k_0 = \rho_0 \boldsymbol{v} \cdot \boldsymbol{v}$ and the *elastically stored energy* density w_0. For a second-order continuum the resultant external mechanical power also incorporates a contribution from the (applied) double-traction and is defined as

$$\mathcal{P}(\mathcal{V}_0) := \int_{\mathcal{V}_0} \boldsymbol{v} \cdot \boldsymbol{b}_0 \, \mathrm{d}V + \int_{\partial \mathcal{V}_0} [\boldsymbol{v} \cdot \boldsymbol{t}_0 + \widehat{\widehat{\mathrm{Grad}}}\, \boldsymbol{v} : {}^{\Delta}\boldsymbol{t}_0] \, \mathrm{d}A.$$

Observe the additional contribution to the resultant external mechanical power that is due to the double-traction ${}^{\Delta}\boldsymbol{t}_0$ working on the normal part of the velocity gradient, that is defined as

$$\widehat{\widehat{\mathrm{Grad}}}\, \boldsymbol{v} := \mathrm{Grad}\, \boldsymbol{v} \cdot [\boldsymbol{N} \otimes \boldsymbol{N}] = \mathrm{Grad}\, \boldsymbol{v} - \widehat{\mathrm{Grad}}\, \boldsymbol{v}.$$

Note that this contribution may alternatively be expressed in terms of the normal gradient of the velocity $\boldsymbol{v}$ as

$$\widehat{\widehat{\mathrm{Grad}}}\, \boldsymbol{v} : {}^{\Delta}\boldsymbol{t}_0 = \boldsymbol{v}' \cdot \boldsymbol{t}_0' \quad \text{with} \quad \boldsymbol{v}' := \nabla_N \boldsymbol{v} := \mathrm{Grad}\, \boldsymbol{v} \cdot \boldsymbol{N}.$$

Obviously, the double-traction $\boldsymbol{t}_0'$ does only expend power on the normal gradient of the velocity $\boldsymbol{v}$. For an arbitrary subsurface $\mathcal{S}_0 \subset \partial \mathcal{V}_0$ of the boundary $\partial \mathcal{V}_0$ to $\mathcal{V}_0$ the surface Gauss theorem renders

$$\int_{\partial \mathcal{S}_0} \boldsymbol{v} \cdot [\boldsymbol{P}' \cdot \boldsymbol{M}] \, \mathrm{d}L = \int_{\mathcal{S}_0} \left[\boldsymbol{v} \cdot \varsigma_0(\boldsymbol{P}') + \widehat{\mathrm{Grad}}\, \boldsymbol{v} : \boldsymbol{P}'\right] \mathrm{d}A.$$

In particular for the total boundary $\partial \mathcal{V}_0$ to $\mathcal{V}_0$, which is a closed surface (that is assumed smooth), the left-hand-side in the above integral statement vanishes, thus with

$$\widehat{\widehat{\mathrm{Grad}}}\, \boldsymbol{v} : {}^{\Delta}\boldsymbol{t}_0 = \mathrm{Grad}\, \boldsymbol{v} : \widehat{\widehat{\boldsymbol{P}}}{}' \quad \text{and} \quad \widehat{\mathrm{Grad}}\, \boldsymbol{v} : \boldsymbol{P}' = \mathrm{Grad}\, \boldsymbol{v} : \widehat{\boldsymbol{P}}{}'$$

the resultant external mechanical power supplied to $\mathcal{V}_0$ may likewise be expressed as

$$\mathcal{P}(\mathcal{V}_0) := \int_{\mathcal{V}_0} \boldsymbol{v} \cdot \boldsymbol{b}_0 \, \mathrm{d}V + \int_{\partial \mathcal{V}_0} [\boldsymbol{v} \cdot \boldsymbol{t}_0^{\mathrm{eff}} + \mathrm{Grad}\, \boldsymbol{v} : \boldsymbol{P}'] \, \mathrm{d}A.$$

Then upon localization for arbitrary $\mathcal{V}_0$ the local statement for the balance of mechanical energy is eventually expressed as

$$\dot{w}_0 = \mathrm{Grad}\,\boldsymbol{v} : [\boldsymbol{P}^{\mathrm{eff}} + \mathrm{Div}\,{}^{\triangle}\boldsymbol{P}] + \mathrm{Grad}\,\mathrm{Grad}\,\boldsymbol{v} \,\vdots\, {}^{\triangle}\boldsymbol{P}.$$

By introducing again the $_0$-type Piola stress ${}_0\boldsymbol{P} := \boldsymbol{P}^{\mathrm{eff}} + \mathrm{Div}\,{}^{\triangle}\boldsymbol{P}$ the above statement transforms into the condition

$$\dot{w}_0 = \wp_0 \quad \text{with} \quad \wp_0 := {}_0\boldsymbol{P} : \dot{\boldsymbol{F}} + {}^{\triangle}\boldsymbol{P} \,\vdots\, {}^{\triangle}\dot{\boldsymbol{F}} \quad \text{and} \quad \dot{w}_t = \wp_t.$$

Here $\wp_0$ and $\wp_t$ denote the *stress/double-stress power densities* per unit volume $\mathrm{d}V$ or $\mathrm{d}v$, respectively.
Interestingly, the balances of linear and angular momentum follow likewise by inserting either an arbitrary time dependent rigid translation $\boldsymbol{v} = \boldsymbol{\theta}(t)$ or an arbitrary time dependent rigid rotation $\boldsymbol{v} = \boldsymbol{\theta}(t) \times \boldsymbol{r}$ into the the balance of mechanical energy

$$\boldsymbol{\theta}(t)\cdot\left[\dot{\mathbf{P}}(\mathcal{V}_0) - \boldsymbol{\mathcal{F}}(\mathcal{V}_0)\right] = 0 \quad \forall\boldsymbol{\theta}(t) \quad \text{and} \quad \boldsymbol{\theta}(t)\cdot\left[\dot{\mathbf{L}}(\mathcal{V}_0) - \boldsymbol{\mathcal{M}}(\mathcal{V}_0)\right] = 0 \quad \forall\boldsymbol{\theta}(t)$$

Here $\boldsymbol{\theta} = \boldsymbol{\theta}(t)$ is an arbitrary time dependent but spatially constant vector that may therefore be factored out from the integrals.

Stress/Double-Stress Power Density:
The stress/double-stress power density $\wp_0 = {}_0\boldsymbol{P} : \dot{\boldsymbol{F}} + {}^{\triangle}\boldsymbol{P} \,\vdots\, {}^{\triangle}\dot{\boldsymbol{F}}$ as originally expressed in terms of the $_0$-type Piola stress and the Piola double-stress in combination with the material time derivative of the first and second deformation gradients allows to connect alternatively various ($_\tau$-type) stress/double-stress and strain/double-strain measures so as to render *power conjugate pairings.* For co/contra- (mixed) variant stress/double-stress measures and and contra/co- (mixed) variant strain/double-strain measures power conjugate pairings read

$$\begin{aligned}\wp_0 &= {}_0\boldsymbol{\tau} : \boldsymbol{l} + {}^{\triangle}\boldsymbol{\tau} \,\vdots\, {}^{\triangle}_{0}\boldsymbol{l} = {}_0\boldsymbol{M} : \boldsymbol{L} + {}^{\triangle}\boldsymbol{M} \,\vdots\, {}^{\triangle}_{0}\boldsymbol{L} = {}_0\boldsymbol{P} : \boldsymbol{\Lambda} + {}^{\triangle}\boldsymbol{P} \,\vdots\, {}^{\triangle}_{0}\boldsymbol{\Lambda} \\ &= {}_t\boldsymbol{\tau} : \boldsymbol{l} + {}^{\triangle}\boldsymbol{\tau} \,\vdots\, {}^{\triangle}_{t}\boldsymbol{l} = {}_t\boldsymbol{M} : \boldsymbol{L} + {}^{\triangle}\boldsymbol{M} \,\vdots\, {}^{\triangle}_{t}\boldsymbol{L} = {}_t\boldsymbol{P} : \boldsymbol{\Lambda} + {}^{\triangle}\boldsymbol{P} \,\vdots\, {}^{\triangle}_{t}\boldsymbol{\Lambda}.\end{aligned}$$

Note that the ($_\tau$-type) contributions of the stresses and the contributions of the double-stresses to the power density in the above satisfy separately

$${}_\tau\boldsymbol{\tau} : \boldsymbol{l} = {}_\tau\boldsymbol{M} : \boldsymbol{L} = {}_\tau\boldsymbol{P} : \boldsymbol{\Lambda} \quad \text{and} \quad {}^{\triangle}\boldsymbol{\tau} \,\vdots\, {}^{\triangle}_{\tau}\boldsymbol{l} = {}^{\triangle}\boldsymbol{M} \,\vdots\, {}^{\triangle}_{\tau}\boldsymbol{L} = {}^{\triangle}\boldsymbol{P} \,\vdots\, {}^{\triangle}_{\tau}\boldsymbol{\Lambda}.$$

Thus arbitrary combinations of ($_\tau$-type) contributions of the stresses and the double-stresses to the power density are in principle possible. In particular the contribution of the symmetric ($_t$-type) stresses to the power density allows the usual alternative expressions, e.g.

$${}_t\boldsymbol{\tau}_\sharp : \boldsymbol{d} = \frac{1}{2}{}_t\boldsymbol{\tau}_\sharp : \mathrm{L}_t\boldsymbol{j} = {}_t\boldsymbol{\tau}_\sharp : \mathrm{L}_t\boldsymbol{e} = \frac{1}{2}{}_t\boldsymbol{S} : \dot{\boldsymbol{C}} = {}_t\boldsymbol{S} : \dot{\boldsymbol{E}}.$$

Note, however, that among the options for the contribution of the double-stresses to the power density only ${}^{\triangle}\boldsymbol{P} \cdot\!: {}^{\triangle}_{0}\boldsymbol{\Lambda}$ with ${}^{\triangle}_{0}\boldsymbol{\Lambda} = {}^{\triangle}\dot{\boldsymbol{F}}$ and ${}^{\triangle}\boldsymbol{M} \cdot\!: {}^{\triangle}_{t}\boldsymbol{L}$ with ${}^{\triangle}_{t}\boldsymbol{L} = {}^{\triangle}\dot{\boldsymbol{\Gamma}}$ (thus ${}^{\triangle}\boldsymbol{\tau} \cdot\!: {}^{\triangle}_{t}\boldsymbol{l} = -{}^{\triangle}\boldsymbol{\tau} \cdot\!: \mathrm{L}_t{}^{\triangle}\boldsymbol{\gamma}$) incorporate material time derivatives of double-strains. Thus, employing next the relations between $\boldsymbol{C} \cdot {}^{\triangle}_{t}\boldsymbol{L}$, ${}^{\triangle}\dot{\boldsymbol{C}}$ and $\dot{\boldsymbol{C}} \cdot {}^{\triangle}\boldsymbol{\Gamma}$ on the one hand, and the relation between $\boldsymbol{j} \cdot {}^{\triangle}_{t}\boldsymbol{l}$, $\mathrm{L}_t{}^{\triangle}\boldsymbol{j} = 2{}^{\triangle}_{0}\boldsymbol{d}$ and $\mathrm{L}_t\boldsymbol{j} \cdot {}^{\triangle}\boldsymbol{\gamma}$ on the other hand, renders alternatively the following power conjugate pairings of fully contravariant stress/double-stress measures and fully covariant strain/double-strain measures

$$
\begin{aligned}
\wp_0 &= \left[{}_t\boldsymbol{\tau}_\sharp + 2[{}^{\triangle}\boldsymbol{\tau}_\sharp : {}^{\triangle}\boldsymbol{\gamma}^T]^{\mathrm{sym}}\right] : \boldsymbol{d} + 2{}^{\triangle}\boldsymbol{\tau}_\sharp \cdot\!: {}^{\triangle}_{0}\boldsymbol{d} \\
&= \frac{1}{2}\left[{}_t\boldsymbol{\tau}_\sharp + 2[{}^{\triangle}\boldsymbol{\tau}_\sharp : {}^{\triangle}\boldsymbol{\gamma}^T]^{\mathrm{sym}}\right] : \mathrm{L}_t\boldsymbol{j} + {}^{\triangle}\boldsymbol{\tau}_\sharp \cdot\!: \mathrm{L}_t{}^{\triangle}\boldsymbol{j} \\
&= \frac{1}{2}\left[{}_t\boldsymbol{S} - 2[{}^{\triangle}\boldsymbol{S} : {}^{\triangle}\boldsymbol{\Gamma}^T]^{\mathrm{sym}}\right] : \dot{\boldsymbol{C}} + {}^{\triangle}\boldsymbol{S} \cdot\!: {}^{\triangle}\dot{\boldsymbol{C}}.
\end{aligned}
$$

Likewise, the stress/double-stress power density per unit volume $\mathrm{d}v$ in the spatial configuration $\mathcal{B}_t$ expands as

$$
\wp_t = \left[{}_t\boldsymbol{\sigma}_\sharp + 2[{}^{\triangle}\boldsymbol{\sigma}_\sharp : {}^{\triangle}\boldsymbol{\gamma}^T]^{\mathrm{sym}}\right] : \boldsymbol{d} + 2{}^{\triangle}\boldsymbol{\sigma}_\sharp \cdot\!: {}^{\triangle}_{0}\boldsymbol{d}.
$$

In the sequel it is useful to introduce the abbreviations $\boldsymbol{\tau}$ and $\boldsymbol{S}$ for the alternative (symmetric) stress measures appearing in the above power conjugate pairings as

$$
\begin{aligned}
\boldsymbol{\tau}_\sharp &:= {}_t\boldsymbol{\tau}_\sharp + 2[{}^{\triangle}\boldsymbol{\tau}_\sharp : {}^{\triangle}\boldsymbol{\gamma}^T]^{\mathrm{sym}} = {}_0\boldsymbol{\tau}_\sharp + {}^{\triangle}\boldsymbol{\gamma} : {}^{\triangle}\boldsymbol{\tau}_\sharp^T, \\
\boldsymbol{S} &:= {}_t\boldsymbol{S} - 2[{}^{\triangle}\boldsymbol{S} : {}^{\triangle}\boldsymbol{\Gamma}^T]^{\mathrm{sym}} = {}_0\boldsymbol{S} - {}^{\triangle}\boldsymbol{\Gamma} : {}^{\triangle}\boldsymbol{S}^T.
\end{aligned}
$$

Thus the stress/double stress power density takes the more intuitive format

$$
\begin{aligned}
\wp_0 &= \boldsymbol{\tau}_\sharp : \boldsymbol{d} + 2{}^{\triangle}\boldsymbol{\tau}_\sharp \cdot\!: {}^{\triangle}_{0}\boldsymbol{d} \\
&= \frac{1}{2}\boldsymbol{\tau}_\sharp : \mathrm{L}_t\boldsymbol{j} + {}^{\triangle}\boldsymbol{\tau}_\sharp \cdot\!: \mathrm{L}_t{}^{\triangle}\boldsymbol{j} \\
&= \frac{1}{2}\boldsymbol{S} : \dot{\boldsymbol{C}} + {}^{\triangle}\boldsymbol{S} \cdot\!: {}^{\triangle}\dot{\boldsymbol{C}}.
\end{aligned}
$$

It is recalled that the combination of power conjugate pairings of stress/double-stress and strain/double-strain measures is of importance for the determination of hyperelastic constitutive relations.

Master Constitutive Relation:

For second-order (hyper) elasticity the constitutive relations for the stresses and double-stresses in terms of the strains and double-strains follow from the balance of mechanical energy, i.e. from the comparison of the material time derivative of the elastically stored energy w_0 with the stress/double-stress power density $\wp_0$. Due to the elementary definition $\wp_0 = {}_0\boldsymbol{P} : \dot{\boldsymbol{F}} + {}^{\triangle}\boldsymbol{P} \cdot\!: {}^{\triangle}\dot{\boldsymbol{F}}$

the tensor function w_0 depends most generally on $\boldsymbol{F}$ and ${}^{\vartriangle}\boldsymbol{F}$, i.e. the master constitutive relation follows as

$$w_0 = w_0(\boldsymbol{F}, {}^{\vartriangle}\boldsymbol{F}) \quad \rightsquigarrow \quad {}_0\boldsymbol{P} = \frac{\partial w_0}{\partial \boldsymbol{F}} \quad \text{and} \quad {}^{\vartriangle}\boldsymbol{P} = \frac{\partial w_0}{\partial {}^{\vartriangle}\boldsymbol{F}}.$$

Here any possible further dependence of w_0 on the material coordinates $\boldsymbol{X}$ that is needed to capture material inhomogeneities has been suppressed again to ease notation. The dependence $w_0 = w_0(\boldsymbol{F}, {}^{\vartriangle}\boldsymbol{F})$ is however not completely arbitrary but is subject to requirements of spatial objectivity, material symmetry, and covariance.

Spatially Superposed Rigid Body Motions:
Rigid body motions superposed to the *spatial* configuration in terms of temporarily varying rotations $\boldsymbol{Q}(t) \in \mathrm{SO}(3)$ and translations $\boldsymbol{z}(t) \in \mathrm{T}(3)$ result in the transformed deformation gradient and second deformation gradient

$$\boldsymbol{F}^* = \boldsymbol{Q}(t) \cdot \boldsymbol{F}(\boldsymbol{X}, t) \quad \text{and} \quad {}^{\vartriangle}\boldsymbol{F}^* = \boldsymbol{Q}(t) \cdot {}^{\vartriangle}\boldsymbol{F}(\boldsymbol{X}, t).$$

Consequently, the inverse deformation gradient and its spatial gradient transform as

$$\boldsymbol{f}^* = \boldsymbol{f} \cdot \boldsymbol{Q}^t \quad \text{and} \quad {}^{\vartriangle}\boldsymbol{f}^* = {}^{\vartriangle}\boldsymbol{f} : [\boldsymbol{Q}^t \boxtimes \boldsymbol{Q}^t].$$

Then, due to the orthogonality of $\boldsymbol{Q}$ the covariant material double-strain tensors ${}^{\vartriangle}\boldsymbol{C}$ and ${}^{\vartriangle}\boldsymbol{\Gamma}$ remain *invariant*, i.e. unaffected under a spatially superposed rigid body motion, i.e.

$${}^{\vartriangle}\boldsymbol{C}^* = \boldsymbol{F}^t \cdot \boldsymbol{Q}^t \cdot \boldsymbol{Q} \cdot {}^{\vartriangle}\boldsymbol{F} = {}^{\vartriangle}\boldsymbol{C} \quad \text{and} \quad {}^{\vartriangle}\boldsymbol{\Gamma}^* = \boldsymbol{f} \cdot \boldsymbol{Q}^t \cdot \boldsymbol{Q} \cdot {}^{\vartriangle}\boldsymbol{F} = {}^{\vartriangle}\boldsymbol{\Gamma}.$$

Vice versa, the covariant spatial double-strain tensors ${}^{\vartriangle}\boldsymbol{c}$ and ${}^{\vartriangle}\boldsymbol{\gamma}$ transform *objectively* under a spatially superposed rigid body motion, i.e.

$$\begin{aligned} {}^{\vartriangle}\boldsymbol{c}^* &= \boldsymbol{Q} \cdot \boldsymbol{f}^t \cdot {}^{\vartriangle}\boldsymbol{f} : [\boldsymbol{Q}^t \boxtimes \boldsymbol{Q}^t] = \boldsymbol{Q} \cdot {}^{\vartriangle}\boldsymbol{c} : [\boldsymbol{Q}^t \boxtimes \boldsymbol{Q}^t], \\ {}^{\vartriangle}\boldsymbol{\gamma}^* &= \boldsymbol{Q} \cdot \boldsymbol{F} \ \cdot {}^{\vartriangle}\boldsymbol{f} : [\boldsymbol{Q}^t \boxtimes \boldsymbol{Q}^t] = \boldsymbol{Q} \cdot {}^{\vartriangle}\boldsymbol{\gamma} : [\boldsymbol{Q}^t \boxtimes \boldsymbol{Q}^t]. \end{aligned}$$

Finally, the material and spatial double-metrics transform under a spatially superposed rigid body motion as

$${}^{\vartriangle}\boldsymbol{J}^* = {}^{\vartriangle}\boldsymbol{J} \quad \text{and} \quad {}^{\vartriangle}\boldsymbol{j}^* = \boldsymbol{Q} \cdot {}^{\vartriangle}\boldsymbol{j} : [\boldsymbol{Q}^t \boxtimes \boldsymbol{Q}^t].$$

Third-order tensors $\boldsymbol{T}$, considered as linear maps $\boldsymbol{w} = \boldsymbol{T} : [\boldsymbol{u} \otimes \boldsymbol{v}]$ of dyads of vectors $\boldsymbol{u}$ and $\boldsymbol{v}$ into vectors $\boldsymbol{w}$, transform objectively into $\boldsymbol{T}^*$ if the transformed dyads of vectors $\boldsymbol{u}^* := \boldsymbol{Q} \cdot \boldsymbol{u}$ and $\boldsymbol{v}^* := \boldsymbol{Q} \cdot \boldsymbol{v}$ are mapped into $\boldsymbol{w}^* := \boldsymbol{Q} \cdot \boldsymbol{w}$ as $\boldsymbol{w}^* = \boldsymbol{T}^* : [\boldsymbol{u}^* \otimes \boldsymbol{v}^*]$.

Materially Superposed Rigid Body Motions:
Rigid body motions superposed to the *material* configuration in terms of temporarily varying rotations $\boldsymbol{q}(t) \in \mathrm{SO}(3)$ with $\boldsymbol{Q} = \boldsymbol{q}^{-1} = \boldsymbol{q}^t$ and translations $\boldsymbol{Z}(t) \in \mathrm{T}(3)$ result in the transformations

$$\boldsymbol{F}_* = \boldsymbol{F}(\boldsymbol{X},t) \cdot \boldsymbol{Q}(t) \quad \text{and} \quad {}^{\Delta}\boldsymbol{F}_* = {}^{\Delta}\boldsymbol{F}(\boldsymbol{X},t) : [\boldsymbol{Q}(t) \boxtimes \boldsymbol{Q}(t)].$$

Accordingly, the inverse deformation gradient and its material gradient transform as

$$\boldsymbol{f}_* = \boldsymbol{Q}^t \cdot \boldsymbol{f} \quad \text{and} \quad {}^{\Delta}\boldsymbol{f}_* = \boldsymbol{Q}^t \cdot {}^{\Delta}\boldsymbol{f}.$$

Then, due to the orthogonality of $\boldsymbol{Q}$ the covariant spatial double-strain tensors ${}^{\Delta}\boldsymbol{c}$ and ${}^{\Delta}\boldsymbol{\gamma}$ remain *invariant* under a materially superposed rigid body motion

$${}^{\Delta}\boldsymbol{c}_* = \boldsymbol{f}^t \cdot \boldsymbol{Q} \cdot \boldsymbol{Q}^t \cdot {}^{\Delta}\boldsymbol{f} = {}^{\Delta}\boldsymbol{c} \quad \text{and} \quad {}^{\Delta}\boldsymbol{\gamma}_* = \boldsymbol{F} \cdot \boldsymbol{Q} \cdot \boldsymbol{Q}^t \cdot {}^{\Delta}\boldsymbol{f} = {}^{\Delta}\boldsymbol{\gamma}.$$

Vice versa, the covariant material double-strain tensors ${}^{\Delta}\boldsymbol{C}$ and ${}^{\Delta}\boldsymbol{\Gamma}$ transform *objectively* under a materially superposed rigid body motion, i.e.

$$\begin{aligned} {}^{\Delta}\boldsymbol{C}_* &= \boldsymbol{Q}^t \cdot \boldsymbol{F}^t \cdot {}^{\Delta}\boldsymbol{F} : [\boldsymbol{Q} \boxtimes \boldsymbol{Q}] = \boldsymbol{Q}^t \cdot {}^{\Delta}\boldsymbol{C} : [\boldsymbol{Q} \boxtimes \boldsymbol{Q}], \\ {}^{\Delta}\boldsymbol{\Gamma}_* &= \boldsymbol{Q}^t \cdot \boldsymbol{f} \quad \cdot {}^{\Delta}\boldsymbol{F} : [\boldsymbol{Q} \boxtimes \boldsymbol{Q}] = \boldsymbol{Q}^t \cdot {}^{\Delta}\boldsymbol{\Gamma} : [\boldsymbol{Q} \boxtimes \boldsymbol{Q}]. \end{aligned}$$

Finally, the spatial and material double-metrics transform under a materially superposed rigid body motion as

$${}^{\Delta}\boldsymbol{j}_* = {}^{\Delta}\boldsymbol{j} \quad \text{and} \quad {}^{\Delta}\boldsymbol{J}_* = \boldsymbol{Q}^t \cdot {}^{\Delta}\boldsymbol{J} : [\boldsymbol{Q} \boxtimes \boldsymbol{Q}].$$

Material double-strain measures transforming objectively under materially superposed rigid body motions are suited to capture anisotropic material behavior.

Restrictions on the Elastically Stored Energy Density:
Requiring the elastically stored energy density w_0 to transform objectively under arbitrary spatially superposed rigid body motion (i.e. $w_0 \doteq w_0^*$) results in the alternative restrictions

$$\begin{aligned} w_0 &\doteq w_0^* = w_0(\boldsymbol{C}, {}^{\Delta}\boldsymbol{C}; \cdots), \\ w_0 &\doteq w_0^* = w_0(\boldsymbol{C}, {}^{\Delta}\boldsymbol{\Gamma}; \cdots), \\ w_0 &\doteq w_0^* = w_0(\boldsymbol{Q} \cdot \boldsymbol{c} \cdot \boldsymbol{Q}^t, \boldsymbol{Q} \cdot {}^{\Delta}\boldsymbol{c} : [\boldsymbol{Q}^t \boxtimes \boldsymbol{Q}^t]; \cdots) \quad \forall \boldsymbol{Q} \in \mathrm{SO}(3), \\ w_0 &\doteq w_0^* = w_0(\boldsymbol{Q} \cdot \boldsymbol{c} \cdot \boldsymbol{Q}^t, \boldsymbol{Q} \cdot {}^{\Delta}\boldsymbol{\gamma} : [\boldsymbol{Q}^t \boxtimes \boldsymbol{Q}^t]; \cdots) \quad \forall \boldsymbol{Q} \in \mathrm{SO}(3). \end{aligned}$$

Thus an arbitrary dependence on either of the material description double-strain measures ${}^{\Delta}\boldsymbol{C}$ or ${}^{\Delta}\boldsymbol{\Gamma}$ automatically satisfies spatial objectivity of w_0. In contrast the dependence on either of the spatial description double-strain measures ${}^{\Delta}\boldsymbol{c}$ or ${}^{\Delta}\boldsymbol{\gamma}$ is subject to the above *isotropy* conditions on the tensor function w_0 in order to satisfy spatial objectivity. However, the yet unspecified arguments of w_0 indicated by the $\cdots$ allow to also capture *anisotropic* material behavior. Based on second- and third-order material structural tensors $\boldsymbol{M}_\alpha$ and ${}^{\triangle}\boldsymbol{M}_\alpha$ and in accordance with the above restriction of spatial objectivity the elastically stored energy density w_0 then takes the following alternative representations

$$
\begin{aligned}
w_0 &= w_0(\{\boldsymbol{C};\boldsymbol{I},\boldsymbol{M}_\alpha\},\{{}^{\triangle}\boldsymbol{C};{}^{\triangle}\boldsymbol{I},{}^{\triangle}\boldsymbol{M}_\alpha\})\\
w_0 &= w_0(\{\boldsymbol{C};\boldsymbol{I},\boldsymbol{M}_\alpha\},\{{}^{\triangle}\boldsymbol{\Gamma};{}^{\triangle}\boldsymbol{I},{}^{\triangle}\boldsymbol{M}_\alpha\}).
\end{aligned}
$$

Here the contravariant material unit tensors $\boldsymbol{I}$ and ${}^{\triangle}\boldsymbol{I}$ are explicitly incorporated in order to capture the isotropic part of the response. Finally, requiring invariance of w_0 under arbitrary materially superposed rigid body motions, i.e. the material strain and double-strain measures are transformed identically to the material structural tensors, results in the following alternative restrictions

$$
\begin{aligned}
w_0 \doteq w_{0*} &= w_0(\boldsymbol{Q}^t\cdot\{\{\boldsymbol{C};\boldsymbol{I},\boldsymbol{M}_\alpha\},\{{}^{\triangle}\boldsymbol{C};{}^{\triangle}\boldsymbol{I},{}^{\triangle}\boldsymbol{M}_\alpha\}\cdot\boldsymbol{Q}\}\cdot\boldsymbol{Q}) \quad \forall\boldsymbol{Q}\in\mathrm{SO}(3),\\
w_0 \doteq w_{0*} &= w_0(\boldsymbol{Q}^t\cdot\{\{\boldsymbol{C};\boldsymbol{I},\boldsymbol{M}_\alpha\},\{{}^{\triangle}\boldsymbol{\Gamma};{}^{\triangle}\boldsymbol{I},{}^{\triangle}\boldsymbol{M}_\alpha\}\cdot\boldsymbol{Q}\}\cdot\boldsymbol{Q}) \quad \forall\boldsymbol{Q}\in\mathrm{SO}(3).
\end{aligned}
$$

These restrictions qualify w_0 as an *isotropic tensor function* of all of its arguments, i.e. either $\{\boldsymbol{C};\boldsymbol{I},\boldsymbol{M}_\alpha\}$ and $\{{}^{\triangle}\boldsymbol{C};{}^{\triangle}\boldsymbol{I},{}^{\triangle}\boldsymbol{M}_\alpha\}$ or $\{\boldsymbol{C};\boldsymbol{I},\boldsymbol{M}_\alpha\}$ and $\{{}^{\triangle}\boldsymbol{\Gamma};{}^{\triangle}\boldsymbol{I},{}^{\triangle}\boldsymbol{M}_\alpha\}$, respectively. Isotropic tensor functions typically depend on the invariants $I_1, I_2, I_3, \cdots$ and $G_{a1}, G_{a2}, G_{a3}, \cdots$ (with $a = 1,2,3$) of their arguments. Requiring next covariance of the elastically stored energy density w_0 under arbitrary superposed material diffeomorphisms results in the alternative equivalent statements

$$
\begin{aligned}
w_0 \doteq w_0(\, &\{\underbrace{\boldsymbol{f}^{\star t}\cdot\boldsymbol{C}\cdot\boldsymbol{f}^\star}_{=:\boldsymbol{C}^\star};\underbrace{\boldsymbol{F}^\star\cdot\boldsymbol{I}\cdot\boldsymbol{F}^{\star t}}_{=:\boldsymbol{I}^\star},\underbrace{\boldsymbol{F}^\star\cdot\boldsymbol{M}_\alpha\cdot\boldsymbol{F}^{\star t}}_{=:\boldsymbol{M}^\star_\alpha}\},\\
&\{\underbrace{\boldsymbol{f}^{\star t}\cdot{}^{\triangle}\boldsymbol{C}:\mathsf{f}^\star}_{=:{}^{\triangle}\boldsymbol{C}^\star};\underbrace{\boldsymbol{F}^\star\cdot{}^{\triangle}\boldsymbol{I}:\mathsf{F}^{\star t}}_{=:{}^{\triangle}\boldsymbol{I}^\star},\underbrace{\boldsymbol{F}^\star\cdot{}^{\triangle}\boldsymbol{M}_\alpha:\mathsf{F}^{\star t}}_{=:{}^{\triangle}\boldsymbol{M}^\star_\alpha}\}\,),\\
w_0 \doteq w_0(\, &\{\underbrace{\boldsymbol{f}^{\star t}\cdot\boldsymbol{C}\cdot\boldsymbol{f}^\star}_{=:\boldsymbol{C}^\star};\underbrace{\boldsymbol{F}^\star\cdot\boldsymbol{I}\cdot\boldsymbol{F}^{\star t}}_{=:\boldsymbol{I}^\star},\underbrace{\boldsymbol{F}^\star\cdot\boldsymbol{M}_\alpha\cdot\boldsymbol{F}^{\star t}}_{=:\boldsymbol{M}^\star_\alpha}\},\\
&\{\underbrace{\boldsymbol{F}^{\star t}\cdot{}^{\triangle}\boldsymbol{\Gamma}:\mathsf{f}^\star}_{=:{}^{\triangle}\boldsymbol{\Gamma}^\star};\underbrace{\boldsymbol{F}^\star\cdot{}^{\triangle}\boldsymbol{I}:\mathsf{F}^{\star t}}_{=:{}^{\triangle}\boldsymbol{I}^\star},\underbrace{\boldsymbol{F}^\star\cdot{}^{\triangle}\boldsymbol{M}_\alpha:\mathsf{F}^{\star t}}_{=:{}^{\triangle}\boldsymbol{M}^\star_\alpha}\}\,).
\end{aligned}
$$

Here $\boldsymbol{f}^\star$ abbreviates the spatial gradient of the inverse (spatial) diffeomorphism $\boldsymbol{Y}^\star$, i.e. $\boldsymbol{f}^\star := \boldsymbol{F}^{\star-1}$. Then the particular choice $\boldsymbol{y}^\star = \boldsymbol{y}$ with $\boldsymbol{F}^\star = \boldsymbol{F}$ and $\boldsymbol{f}^\star = \boldsymbol{f}$ renders the alternative equivalent parameterizations of w_0

$$
\begin{aligned}
w_0(\{\boldsymbol{C};\boldsymbol{I},\boldsymbol{M}_\alpha\},\{{}^{\triangle}\boldsymbol{C};{}^{\triangle}\boldsymbol{I},{}^{\triangle}\boldsymbol{M}_\alpha\}) &\doteq w_0(\{\boldsymbol{j};\boldsymbol{b},\boldsymbol{m}_\alpha\},\{{}^{\triangle}\boldsymbol{j};{}^{\triangle}\boldsymbol{b},{}^{\triangle}\boldsymbol{m}_\alpha\}),\\
w_0(\{\boldsymbol{C};\boldsymbol{I},\boldsymbol{M}_\alpha\},\{{}^{\triangle}\boldsymbol{\Gamma};{}^{\triangle}\boldsymbol{I},{}^{\triangle}\boldsymbol{M}_\alpha\}) &\doteq w_0(\{\boldsymbol{j};\boldsymbol{b},\boldsymbol{m}_\alpha\},\{{}^{\triangle}\boldsymbol{\gamma};{}^{\triangle}\boldsymbol{b},{}^{\triangle}\boldsymbol{m}_\alpha\}).
\end{aligned}
$$

Finally, requiring objectivity of w_0 under arbitrary spatially superposed rigid body motions results in the following alternative restrictions

$$w_0 \doteq w_0^* = w_0(\boldsymbol{Q} \cdot \{\{\boldsymbol{j}; \boldsymbol{b}, \boldsymbol{m}_\alpha\}, \{{}^\Delta\boldsymbol{j}; {}^\Delta\boldsymbol{b}, {}^\Delta\boldsymbol{m}_\alpha\} \cdot \boldsymbol{Q}^t\} \cdot \boldsymbol{Q}^t) \quad \forall \boldsymbol{Q} \in \mathrm{SO}(3),$$
$$w_0 \doteq w_0^* = w_0(\boldsymbol{Q} \cdot \{\{\boldsymbol{j}; \boldsymbol{b}, \boldsymbol{m}_\alpha\}, \{{}^\Delta\gamma; {}^\Delta\boldsymbol{b}, {}^\Delta\boldsymbol{m}_\alpha\} \cdot \boldsymbol{Q}^t\} \cdot \boldsymbol{Q}^t) \quad \forall \boldsymbol{Q} \in \mathrm{SO}(3).$$

These restrictions qualify w_0 again as an *isotropic tensor function* of its arguments, i.e. either $\{\boldsymbol{j}; \boldsymbol{b}, \boldsymbol{m}_\alpha\}$ and $\{{}^\Delta\boldsymbol{j}; {}^\Delta\boldsymbol{b}, {}^\Delta\boldsymbol{m}_\alpha\}$ or $\{\boldsymbol{j}; \boldsymbol{b}, \boldsymbol{m}_\alpha\}$ and $\{{}^\Delta\gamma; {}^\Delta\boldsymbol{b}, {}^\Delta\boldsymbol{m}_\alpha\}$, respectively.

Constitutive Relations:
Elastically stored energy densities w_0 that satisfy all requirements resulting from spatial objectivity, material symmetry, and covariance are isotropic tensor functions of either the material arguments $\{\boldsymbol{C}; \boldsymbol{I}, \boldsymbol{M}_\alpha\}$ and $\{{}^\Delta\boldsymbol{C}; {}^\Delta\boldsymbol{I}, {}^\Delta\boldsymbol{M}_\alpha\}$, i.e.

$$w_0 = w_0(\{\boldsymbol{C}; \boldsymbol{I}, \boldsymbol{M}_\alpha\}, \{{}^\Delta\boldsymbol{C}; {}^\Delta\boldsymbol{I}, {}^\Delta\boldsymbol{M}_\alpha\})$$

or the spatial arguments $\{\boldsymbol{j}; \boldsymbol{b}, \boldsymbol{m}_\alpha\}$ and $\{{}^\Delta\boldsymbol{j}; {}^\Delta\boldsymbol{b}, {}^\Delta\boldsymbol{m}_\alpha\}$, i.e.

$$w_0 = w_0(\{\boldsymbol{j}; \boldsymbol{b}, \boldsymbol{m}_\alpha\}, \{{}^\Delta\boldsymbol{j}; {}^\Delta\boldsymbol{b}, {}^\Delta\boldsymbol{m}_\alpha\}).$$

Then the comparison of the material time derivative of w_0 with the appropriate version of the stress/double-stress power density $\wp_0$ renders corresponding constitutive relations for the fully contravariant material (symmetric) stress and double-stress measures

$$\boldsymbol{S} = 2\frac{\partial w_0}{\partial \boldsymbol{C}} \quad \text{and} \quad {}^\Delta\boldsymbol{S} = \frac{\partial w_0}{\partial {}^\Delta\boldsymbol{C}}.$$

Since for a scalar function $\mathrm{L}_t w_0 \equiv \mathrm{D}_t w_0$ the comparison of the Lie time derivative of w_0 with the appropriate version of the stress/double power density $\wp_0$ renders corresponding constitutive relations for the fully contravariant spatial (symmetric) stress and double-stress measures

$$\boldsymbol{\tau}_\sharp = 2\frac{\partial w_0}{\partial \boldsymbol{j}} \quad \text{and} \quad {}^\Delta\boldsymbol{\tau}_\sharp = \frac{\partial w_0}{\partial {}^\Delta\boldsymbol{j}}.$$

Alternatively the elastically stored energy density w_0 is expressed as an isotropic tensor function of either the material arguments $\{\boldsymbol{C}; \boldsymbol{I}, \boldsymbol{M}_\alpha\}$ and $\{{}^\Delta\boldsymbol{\Gamma}; {}^\Delta\boldsymbol{I}, {}^\Delta\boldsymbol{M}_\alpha\}$, i.e.

$$w_0 = w_0(\{\boldsymbol{C}; \boldsymbol{I}, \boldsymbol{M}_\alpha\}, \{{}^\Delta\boldsymbol{\Gamma}; {}^\Delta\boldsymbol{I}, {}^\Delta\boldsymbol{M}_\alpha\})$$

or the spatial arguments $\{\boldsymbol{j}; \boldsymbol{b}, \boldsymbol{m}_\alpha\}$ and $\{{}^\Delta\gamma; {}^\Delta\boldsymbol{b}, {}^\Delta\boldsymbol{m}_\alpha\}$, i.e.

$$w_0 = w_0(\{\boldsymbol{j}; \boldsymbol{b}, \boldsymbol{m}_\alpha\}, \{{}^\Delta\gamma; {}^\Delta\boldsymbol{b}, {}^\Delta\boldsymbol{m}_\alpha\}).$$

Then the comparison of the material time derivative of w_0 with the appropriate version of the stress/double-stress power density $\wp_0$ renders corresponding constitutive relations for the material stress and double-stress measures

$$_t\boldsymbol{S} = 2\frac{\partial w_0}{\partial \boldsymbol{C}} \quad \text{and} \quad {}^{\Delta}\boldsymbol{M} = \frac{\partial w_0}{\partial {}^{\Delta}\boldsymbol{\Gamma}}.$$

Finally, with $\mathrm{L}_t{}^{\Delta}\gamma = -{}^{\Delta}_{t}\boldsymbol{l}$ the comparison of the Lie time derivative of w_0 with the appropriate version of the stress/double power density $\wp_0$ renders corresponding constitutive relations for the spatial stress and double-stress measures

$$_t\boldsymbol{\tau}_{\sharp} = 2\frac{\partial w_0}{\partial \boldsymbol{j}} \quad \text{and} \quad {}^{\Delta}\boldsymbol{\tau} = -\frac{\partial w_0}{\partial {}^{\Delta}\gamma}.$$

It is remarked that also other, more involved, constitutive relations, e.g. in terms of the double-strains ${}^{\Delta}\boldsymbol{E}$, are principally possible. Which of the above alternative constitutive relations is chosen is essentially a matter of convenience.

Variational Setting:

Quasi-static second-order (hyper) elasticity can be cast into a variational framework. To this end a total potential energy functional $I\big(\boldsymbol{y}(\boldsymbol{X})\big)$ in terms of the deformation map $\boldsymbol{y}(\boldsymbol{X})$ with compatible deformation gradient $\boldsymbol{F} := \operatorname{Grad}\boldsymbol{y}(\boldsymbol{X})$, second deformation gradient ${}^{\Delta}\boldsymbol{F} := \operatorname{Grad}\operatorname{Grad}\boldsymbol{y}(\boldsymbol{X})$ and normal gradient $\boldsymbol{y}' := \nabla_N \boldsymbol{y}(\boldsymbol{X})$ is introduced as

$$I\big(\boldsymbol{y}(\boldsymbol{X})\big) := \int_{\mathcal{B}_0} u_0\big(\boldsymbol{y}(\boldsymbol{X}), \boldsymbol{F}(\boldsymbol{X}), {}^{\Delta}\boldsymbol{F}(\boldsymbol{X})\big)\,\mathrm{d}V + \int_{\partial\mathcal{B}_0} \hat{v}_0\big(\boldsymbol{y}(\boldsymbol{X}), \boldsymbol{y}'(\boldsymbol{X})\big)\,\mathrm{d}A.$$

Here the bulk energy density $u_0 := w_0 + v_0$ consists in an internal contribution, i.e. the elastically stored energy density $w_0 = w_0(\boldsymbol{F}, {}^{\Delta}\boldsymbol{F})$, and an external contribution $v_0 = v_0(\boldsymbol{y})$; the surface energy density is assumed to consist only in an external contribution $\hat{v}_0 = \hat{v}_0(\boldsymbol{y}, \boldsymbol{y}')$. Then the minimum of I is sought for all *admissible variations* $\delta\boldsymbol{y}$ of $\boldsymbol{y}$

$$I\big(\boldsymbol{y}(\boldsymbol{X})\big) \to \min_{\boldsymbol{y}}.$$

Thereby admissible variations of the deformation map $\delta\boldsymbol{y}$ satisfy compatibility $\delta\boldsymbol{F} := \operatorname{Grad}\delta\boldsymbol{y}$ and $\delta^{\Delta}\boldsymbol{F} := \operatorname{Grad}\operatorname{Grad}\delta\boldsymbol{y}$ together with homogeneous boundary conditions $\delta\boldsymbol{y} = \boldsymbol{0}$ on the (deformation) Dirichlet part $\partial\mathcal{B}_0^y$ of the boundary

$$\partial\mathcal{B}_0 = \partial\mathcal{B}_0^y \cup \partial\mathcal{B}_0^s \quad \text{with} \quad \emptyset = \partial\mathcal{B}_0^y \cap \partial\mathcal{B}_0^s,$$

and $\delta\boldsymbol{y}' = \boldsymbol{0}$ on the (double-deformation) Dirichlet part $\partial\mathcal{B}_0^{y'}$ of the boundary

$$\partial\mathcal{B}_0 = \partial\mathcal{B}_0^{y'} \cup \partial\mathcal{B}_0^{s'} \quad \text{with} \quad \emptyset = \partial\mathcal{B}_0^{y'} \cap \partial\mathcal{B}_0^{s'}$$

(the complement to the (double-deformation) Dirichlet part $\partial\mathcal{B}_0^{y'}$ is the (double-deformation) Neumann part $\partial\mathcal{B}_0^{s'}$ of the boundary). Thus, in *classical calculus of variations* the roots $\delta I = 0$ of the *first variation*

$$\delta I(\boldsymbol{y}(\boldsymbol{X})) := \frac{\mathrm{d}}{\mathrm{d}\epsilon} I(\boldsymbol{y}(\boldsymbol{X}) + \epsilon\delta\boldsymbol{y}(\boldsymbol{X}))|_{\epsilon=0} \quad \text{with} \quad \delta\boldsymbol{y}(\boldsymbol{X}) \quad \text{admissible}$$

of the functional I are determined in terms of the corresponding point-wise Euler equations. In the case of second-order elasticity these read in the bulk (in terms of the variational derivative of w_0)

$$\operatorname{Div}\frac{\delta w_0}{\delta \boldsymbol{F}} - \frac{\partial v_0}{\partial \boldsymbol{y}} \doteq \boldsymbol{0} \quad \text{with} \quad \frac{\delta w_0}{\delta \boldsymbol{F}} := \frac{\partial w_0}{\partial \boldsymbol{F}} - \operatorname{Div}\frac{\partial w_0}{\partial^{\Delta}\boldsymbol{F}},$$

whereas the corresponding Euler equations at the boundary result in

$$\frac{\delta w_0}{\delta \boldsymbol{F}} \cdot \boldsymbol{N} \doteq -\frac{\partial \hat{v}_0}{\partial \boldsymbol{y}} + \varsigma_0\left(\frac{\partial w_0}{\partial^{\Delta}\boldsymbol{F}} \cdot \boldsymbol{N}\right) \quad \text{and} \quad \frac{\partial w_0}{\partial^{\Delta}\boldsymbol{F}} : [\boldsymbol{N} \otimes \boldsymbol{N}] \doteq -\frac{\partial \hat{v}_0}{\partial \boldsymbol{y}'}.$$

Note that the above Dirichlet principle of the minimum of the total potential energy functional requires the strong satisfaction of (i) the kinematic field equations $\boldsymbol{F} = \operatorname{Grad}\boldsymbol{y}$ and ${}^{\Delta}\boldsymbol{F} = \operatorname{Grad}\operatorname{Grad}\boldsymbol{y}$, (ii) the constitutive relations, e.g. for ${}_0\boldsymbol{P} = \partial_{\boldsymbol{F}} w_0$ and ${}^{\Delta}\boldsymbol{P} = \partial_{{}^{\Delta}\boldsymbol{F}} w_0$, (iii) the potential-based relations for the body and surface force densities $\boldsymbol{b}_0 := -\partial_{\boldsymbol{y}} v_0$, $\boldsymbol{t}_0 := -\partial_{\boldsymbol{y}}\hat{v}_0$ and $\boldsymbol{t}_0' := -\partial_{\boldsymbol{y}'}\hat{v}_0$, and (iv) the (deformation and double-deformation) Dirichlet boundary conditions. Then (i) the static field equations $\operatorname{Div}\boldsymbol{P}^{\text{eff}} + \boldsymbol{b}_0 = \boldsymbol{0}$ (with $\boldsymbol{P}^{\text{eff}} := {}_0\boldsymbol{P} - \operatorname{Div}{}^{\Delta}\boldsymbol{P}$), and (ii) the (deformation and double-deformation) Neumann boundary conditions $\boldsymbol{P}^{\text{eff}} \cdot \boldsymbol{N} = \boldsymbol{t}_0^{\text{eff}}$ (with $\boldsymbol{t}_0^{\text{eff}} := \boldsymbol{t}_0 + \varsigma_0(\boldsymbol{P}')$ where $\boldsymbol{P}' := {}^{\Delta}\boldsymbol{P} \cdot \boldsymbol{N}$ and $\varsigma_0(\boldsymbol{P}') := \widehat{\operatorname{Div}}\,\widehat{\boldsymbol{P}}'$) and $\boldsymbol{P}' \cdot \boldsymbol{N} = \boldsymbol{t}_0'$ are satisfied in weak (integral) sense.

Proof: With the above strong requirements (i)-(iv) satisfied the first variation of the functional $I(\boldsymbol{y})$ computes as

$$\delta I = \int_{\mathcal{B}_0} [\delta\boldsymbol{F} : \boldsymbol{P} + \delta^{\Delta}\boldsymbol{F} \,\vdots\, {}^{\Delta}\boldsymbol{P} - \delta\boldsymbol{y} \cdot \boldsymbol{b}_0]\,\mathrm{d}V - \int_{\partial\mathcal{B}_0} [\delta\boldsymbol{y} \cdot \boldsymbol{t}_0 + \delta\boldsymbol{y}' \cdot \boldsymbol{t}_0']\,\mathrm{d}A.$$

Applying integration by parts and the Gauss theorem to the first two terms in the above volume integral renders

$$\begin{aligned}\delta I = &-\int_{\mathcal{B}_0} [\delta\boldsymbol{y} \cdot [\operatorname{Div}\boldsymbol{P} + \boldsymbol{b}_0] + \delta\boldsymbol{F} : \operatorname{Div}{}^{\Delta}\boldsymbol{P}]\,\mathrm{d}V \\ &+ \int_{\partial\mathcal{B}_0^s} \delta\boldsymbol{y} \cdot [\boldsymbol{P} \cdot \boldsymbol{N} - \boldsymbol{t}_0^{\text{eff}}]\,\mathrm{d}A \\ &+ \int_{\partial\mathcal{B}_0^{s'}} \delta\boldsymbol{y}' \cdot [\boldsymbol{P}' \cdot \boldsymbol{N} - \boldsymbol{t}_0'\,]\,\mathrm{d}A.\end{aligned}$$

Here the term $\delta \boldsymbol{F} : \boldsymbol{P}'$ *appearing initially in the surface integral has been decomposed into* $\delta \boldsymbol{y}' \cdot \boldsymbol{P}' \cdot \boldsymbol{N} - \delta \boldsymbol{y} \cdot \varsigma_0(\boldsymbol{P}')$ *since the area integral of* $\widehat{\mathrm{Div}}(\delta \boldsymbol{y} \cdot \widehat{\boldsymbol{P}}') = \delta \widehat{\boldsymbol{F}} : \widehat{\boldsymbol{P}}' + \delta \boldsymbol{y} \cdot \widehat{\mathrm{Div}}\, \widehat{\boldsymbol{P}}'$ *over the closed boundary surface vanishes. Moreover the homogeneous (deformation and double-deformation) Dirichlet boundary conditions for the variations* $\delta \boldsymbol{y}$ *and* $\delta \boldsymbol{y}'$ *have been incorporated into the surface integrals. Applying next integration by parts and the Gauss theorem to the last term in the above volume integral renders finally*

$$\begin{aligned}
\delta I = &- \int_{\mathcal{B}_0} \delta \boldsymbol{y} \cdot [\mathrm{Div}\, \boldsymbol{P}^{\mathrm{eff}} + \boldsymbol{b}_0]\, \mathrm{d}V \\
&+ \int_{\partial \mathcal{B}_0^s} \delta \boldsymbol{y} \cdot [\boldsymbol{P}^{\mathrm{eff}} \cdot \boldsymbol{N} - \boldsymbol{t}_0^{\mathrm{eff}}]\, \mathrm{d}A \\
&+ \int_{\partial \mathcal{B}_0^{s\prime}} \delta \boldsymbol{y}' \cdot [\boldsymbol{P}' \cdot \boldsymbol{N} - \boldsymbol{t}_0']\, \mathrm{d}A.
\end{aligned}$$

Since $\delta \boldsymbol{y}$ *and* $\delta \boldsymbol{y}'$ *may be varied independently the above set of static field equations and Neumann boundary conditions are the Euler equations corresponding to* $I(\boldsymbol{y}) \to \min$.

Chapter 8
Elasto-Plasticity

TAPETENBLUME

Tapetenblume bin ich fein,
kehr wieder ohne Ende,
doch, statt im Mai'n und Mondenschein,
auf jeder der vier Wände.

Du siehst mich nimmerdar genung,
so weit du blickst im Stübchen,
und folgst du mir per Rösselsprung -
wirst du verrückt, mein Liebchen.

Christian Morgenstern

Abstract. The previous concepts of differential geometry, in particular aspects related to incompatibility, shall be applied to the kinematics of first- and second-order (nonlinear) elasto-plasticity. Thereby it shall be noted that the intermediate configuration of first- and second-order elasto-plasticity is incompatible, see Fig. 8.1. Then two cases may be considered: firstly the incompatibility of the intermediate configuration is measured based on the non-integrability of plastic tensorial quantities; secondly the situation is reversed: the incompatibility of the intermediate configuration is measured based on the non-integrability of elastic tensorial quantities. Thereby distinction can be made between straightforward measures in terms of the non-integrable plastic or elastic distortions (and double-distortions) and more involved measures in terms of the non-integrable plastic or elastic (strain) metrics (and double-metrics). The former leads to various dislocation density tensors, whereas the latter results in various incompatibility density tensors.

Comprehensive accounts on first- and second-order elasto-plasticity in Euclidean space are provided for the sake of reference at the end of the chapter in two extended supplements.

8.1 First-Order Continuum

A well-accepted assumption in the kinematics of elasto-plastic crystalline materials is the *multiplicative decomposition* of the deformation gradient into a plastic and an elastic contribution that shall be denoted the plastic and the elastic distortion, respectively

$$\boldsymbol{F} = \overline{\boldsymbol{F}} \cdot \widetilde{\boldsymbol{F}}. \tag{8.1}$$

P. Steinmann, *Geometrical Foundations of Continuum Mechanics*,
Lecture Notes in Applied Mathematics and Mechanics 2, DOI: 10.1007/978-3-662-46460-1_8

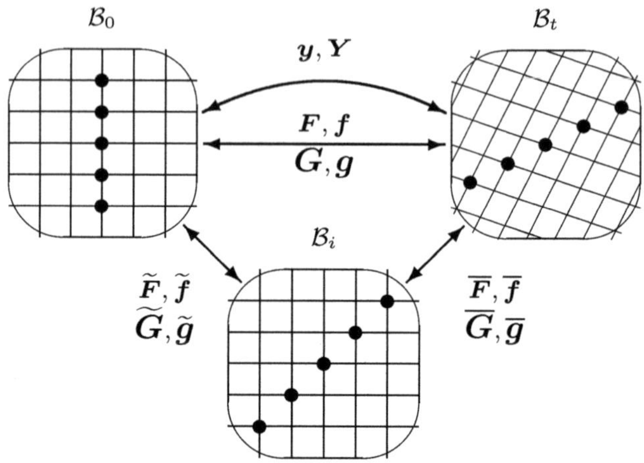

Fig. 8.1 The isoclinic intermediate configuration of first- and second-order elasto-plasticity is incompatible. Thus, based on the plastic and elastic distortions and double-distortions, measures for the incompatibility of the intermediate configuration may be established.

The multiplicative decomposition of the deformation gradient in Eq. 8.1 also allows alternative representation in terms of the inverse deformation gradient and its corresponding decomposition into an elastic and a plastic contribution

$$\boldsymbol{f} = \widetilde{\boldsymbol{f}} \cdot \overline{\boldsymbol{f}}. \tag{8.2}$$

Besides the material and the spatial configuration these decompositions also introduce the so-called *intermediate configuration*, see Fig. 8.1. The intermediate configuration is characterized by an unchanged inclination of the crystalline lattice as compared to the material configuration and is deemed accessible by locally relaxing the stress state. As a consequence the so-called *isoclinic* intermediate configuration of elasto-plasticity is generally *incompatible* and can not be patched together to a single compatible configuration in Euclidean space. The elastic distortion is then responsible for the deformation of the crystalline lattice and is thus the origin of the macroscopic stresses. In the sequel quantities referring to the plastic part of the deformation either in the material or in the intermediate configuration shall be denoted by a tilde $\{\widetilde{\bullet}\}$. Likewise quantities referring to the elastic part of the deformation either in the spatial or in the intermediate configuration shall be denoted by an overbar $\{\overline{\bullet}\}$.

8.1.1 Non-Integrable Distortion

For first-order elasto-plasticity non-integrability of either the plastic or the elastic distortion renders measures for the incompatibility of the intermediate configuration. These measures are associated with the corresponding anholonomic objects and result in dislocation density tensors.

Looking Forward: Incompatibility of Intermediate Configuration. On the one hand the incompatibility of the intermediate configuration within first-order elasto-plasticity results in an intermediate closure gap $[\![\widetilde{\boldsymbol{y}}]\!] \neq \mathbf{0}$ upon integration of the intermediate line element $\mathrm{d}\widetilde{\boldsymbol{y}} := \widetilde{\boldsymbol{F}} \cdot \mathrm{d}\boldsymbol{X}$ along arbitrary closed material circuits $\mathcal{C}_0$. This incompatibility is captured by the non-integrability of the plastic distortion

$$\widetilde{\boldsymbol{D}} := \operatorname{Curl} \widetilde{\boldsymbol{F}} \neq \mathbf{0}. \tag{8.3}$$

Recall that the Piola-type plastic dislocation density $\widetilde{\boldsymbol{D}}$ is the second-order tensor expression for the third-order anholonomic object related to the non-integrable map $\widetilde{\boldsymbol{F}} : T\mathcal{B}_0 \to T\mathcal{B}_i$ from the material to the intermediate configuration.

Looking Backward: Incompatibility of Intermediate Configuration. On the other hand the incompatibility of the intermediate configuration within first-order elasto-plasticity results in an intermediate closure gap $[\![\overline{\boldsymbol{Y}}]\!] \neq \mathbf{0}$ upon integration of the intermediate line element $\mathrm{d}\overline{\boldsymbol{Y}} := \overline{\boldsymbol{f}} \cdot \mathrm{d}\boldsymbol{x}$ along arbitrary closed spatial circuits $\mathcal{C}_t$. This incompatibility is captured by the non-integrability of the elastic distortion

$$\overline{\boldsymbol{d}} := \operatorname{curl} \overline{\boldsymbol{f}} \neq \mathbf{0}. \tag{8.4}$$

Recall that the Piola-type elastic dislocation density $\overline{\boldsymbol{d}}$ is the second-order tensor expression for the third-order anholonomic object related to the non-integrable map $\overline{\boldsymbol{f}} : T\mathcal{B}_t \to T\mathcal{B}_i$ from the spatial to the intermediate configuration.

8.1.2 Non-Integrable Metric

For a number of reasons, see the supplement on first-order elasto-plasticity below, it is often not primarily the plastic or elastic distortion but rather the plastic or elastic (strain) metric, being quadratic in the corresponding distortion, that enters the constitutive relations of first-order elasto-plasticity. Thus the failure of the plastic or elastic (strain) metric to be integrable into a compatible deformation map or rather the incompatibility of the intermediate configuration needs to be considered.

Looking Forward: Incompatibility of Intermediate Configuration. Conceptually the crystalline lattice in the intermediate configuration of elasto-plasticity possesses an unchanged inclination as compared to the material configuration. Thus distance determination and parallel transport in the intermediate configuration are of Euclidean type, see Fig. 8.2; consequently Cartesian-like coordinates may be selected with the intermediate metric coefficients coinciding with the Kronecker-symbol and a vanishing intermediate connection

$$\tilde{m}_{\alpha\beta} = \tilde{\delta}_{\alpha\beta} \quad \text{and} \quad \tilde{l}^{\alpha}{}_{\beta\gamma} = 0. \tag{8.5}$$

Note that greek indices $\alpha, \beta, \gamma, \delta$ are here reserved for the Cartesian-like coordinates and base vectors in the intermediate configuration. Thereby *Cartesian-like* refers to rectilinear but possibly anholonomic coordinates. The plastic convection of the intermediate metric and the intermediate connection by the plastic distortion then renders the *plastic Cauchy-Green tensor* and an *integrable plastic connection* in the material configuration

$$\tilde{\mathcal{M}}_{IJ} = \tilde{C}_{IJ} := \tilde{F}^{\alpha}{}_{I}\tilde{\delta}_{\alpha\beta}\tilde{F}^{\beta}{}_{J} \quad \text{and} \quad \tilde{\mathcal{L}}^{I}{}_{JK} = \tilde{f}^{I}{}_{\alpha}\tilde{F}^{\alpha}{}_{J,K}. \tag{8.6}$$

It is then easy to check that the fully covariant plastic material connection $\tilde{\mathcal{L}}_{IJK} = \tilde{C}_{IM}\tilde{\mathcal{L}}^{M}{}_{JK}$ is indeed a metric connection with respect to $\tilde{C}_{IJ}$, since

$$\tilde{C}_{IJ,K} = \tilde{F}^{\alpha}{}_{I,K}\tilde{\delta}_{\alpha\beta}\tilde{F}^{\beta}{}_{J} + \tilde{F}^{\alpha}{}_{I}\tilde{\delta}_{\alpha\beta}\tilde{F}^{\beta}{}_{J,K} = 2\tilde{\mathcal{L}}_{(IJ)K}. \tag{8.7}$$

Thus, the plastic material connection $\tilde{\mathcal{L}}_{IJK}$ decomposes into the plastic material Riemann connection based on the plastic metric $\tilde{C}_{IJ}$

$$\tilde{\mathcal{M}}_{IJK}(\tilde{C}) := \frac{1}{2}[\tilde{C}_{IJ,K} - \tilde{C}_{JK,I} + \tilde{C}_{KI,J}] \tag{8.8}$$

and the corresponding plastic material contortion based on the dislocation density

$$\tilde{\mathcal{K}}_{IJK}(\tilde{\mathcal{D}}) := \tilde{\mathcal{L}}_{IJK} - \tilde{\mathcal{M}}_{IJK}. \tag{8.9}$$

This may be verified directly by cyclic permutation of indices in Eq. 8.7 and subsequent addition of the resulting expressions

$$\begin{array}{rcl} \tilde{C}_{IJ,K} &=& + \tilde{F}^{\alpha}{}_{J}\, \tilde{\delta}_{\alpha\beta}\, \tilde{F}^{\beta}{}_{I,K} + \tilde{F}^{\alpha}{}_{I}\, \tilde{\delta}_{\alpha\beta}\, \tilde{F}^{\beta}{}_{J,K} \\ -\tilde{C}_{JK,I} &=& - \tilde{F}^{\alpha}{}_{K}\, \tilde{\delta}_{\alpha\beta}\, \tilde{F}^{\beta}{}_{J,I} - \tilde{F}^{\alpha}{}_{J}\, \tilde{\delta}_{\alpha\beta}\, \tilde{F}^{\beta}{}_{K,I} \\ \tilde{C}_{KI,J} &=& + \tilde{F}^{\alpha}{}_{I}\, \tilde{\delta}_{\alpha\beta}\, \tilde{F}^{\beta}{}_{K,J} + \tilde{F}^{\alpha}{}_{K}\, \tilde{\delta}_{\alpha\beta}\, \tilde{F}^{\beta}{}_{I,J} \\ \hline\hline \end{array}$$

$$\rightarrow \quad \tilde{\mathcal{M}}_{IJK}(\tilde{C}) = \tilde{F}^{\alpha}{}_{I}\tilde{\delta}_{\alpha\beta}\tilde{F}^{\beta}{}_{(J,K)} - \tilde{F}^{\alpha}{}_{J}\tilde{\delta}_{\alpha\beta}\tilde{F}^{\beta}{}_{[K,I]} + \tilde{F}^{\alpha}{}_{K}\tilde{\delta}_{\alpha\beta}\tilde{F}^{\beta}{}_{[I,J]} \tag{8.10}$$

and consequently the plastic material contortion reads due to Eq. 8.9

$$\rightarrow \quad \tilde{\mathcal{K}}_{IJK}(\tilde{\mathcal{D}}) = \tilde{F}^{\alpha}{}_{I}\tilde{\delta}_{\alpha\beta}\tilde{F}^{\beta}{}_{[J,K]} + \tilde{F}^{\alpha}{}_{J}\tilde{\delta}_{\alpha\beta}\tilde{F}^{\beta}{}_{[K,I]} - \tilde{F}^{\alpha}{}_{K}\tilde{\delta}_{\alpha\beta}\tilde{F}^{\beta}{}_{[I,J]}. \tag{8.11}$$

Moreover, since the plastic material connection is integrable, the associated plastic material curvature tensor vanishes identically

$$\tilde{\mathcal{R}}_{IJKL}(\tilde{\mathcal{L}}) = 2\tilde{\mathcal{L}}_{IJ[L,K]} - 2\tilde{\mathcal{L}}_{MI[K}\tilde{\mathcal{L}}^{M}{}_{JL]} \equiv 0. \tag{8.12}$$

This finding coincides of course with the postulated Euclidean character of the intermediate configuration

$$0 \equiv \tilde{r}'_{\alpha\beta\gamma\delta}(\tilde{l}) + \tilde{l}_{\alpha\beta\mu}\tilde{a}^{\mu}{}_{\delta\gamma} = \tilde{r}'_{\alpha\beta\gamma\delta}(\tilde{y}(\tilde{\mathcal{L}})) + \tilde{l}_{\alpha\beta\mu}\tilde{a}^{\mu}{}_{\delta\gamma} = \tilde{y}(\tilde{\mathcal{R}}_{IJKL}(\tilde{\mathcal{L}})). \tag{8.13}$$

Here $\tilde{r}'_{\alpha\beta\gamma\delta}(\tilde{l})$ merely abbreviates the application of the curvature format to the intermediate connection, i.e. $\tilde{r}'_{\alpha\beta\gamma\delta}(\tilde{l}) := 2\tilde{l}_{\alpha\beta[\delta.\gamma]} - 2\tilde{l}_{\mu\alpha\gamma}\tilde{l}^{\mu}{}_{\beta\delta}$. Due to the anholonomic object the curvature tensor in the intermediate configuration is then expressed as $\tilde{r}_{\alpha\beta\gamma\delta} = \tilde{r}'_{\alpha\beta\gamma\delta}(\tilde{l}) + \tilde{l}_{\alpha\beta\mu}\tilde{a}^{\mu}{}_{\delta\gamma}$. However since the intermediate connection vanishes due to the selection of Cartesian-like coordinates in the intermediate configuration, i.e. $\tilde{l}_{\alpha\beta\gamma} = 0$, the curvature tensor in the intermediate configuration also vanishes trivially, i.e. $\tilde{r}_{\alpha\beta\gamma\delta} = 0$.

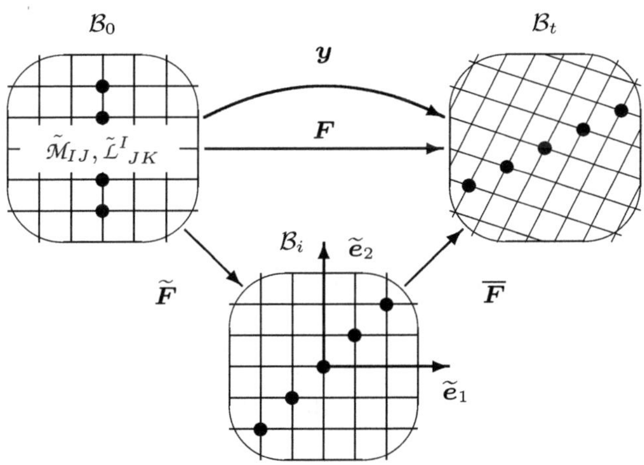

Fig. 8.2 In the intermediate configuration of elasto-plasticity distance determination and parallel transport are of Euclidean type. The plastic convection of the metric and the connection renders the plastic Cauchy-Green tensor and an integrable plastic connection in the material configuration.

Discussion 8.1. (Consequence of major symmetric part of $\tilde{\mathcal{R}}_{IJKL} = 0$)
Since the plastic material connection decomposes into the plastic material Riemann connection and the plastic material contortion, the corresponding non-vanishing material curvature tensors

$$\tilde{\mathcal{M}}_{IJKL}(\tilde{C}) := 2\tilde{\mathcal{M}}_{IJ[L,K]} - 2\tilde{\mathcal{M}}_{MI[K}\tilde{\mathcal{M}}^{M}{}_{JL]} \neq 0, \tag{8.14}$$

and

$$\tilde{\mathcal{K}}_{IJKL}(\tilde{\mathcal{D}}) := 2\tilde{\mathcal{K}}_{IJ[L,K]} - 2\tilde{\mathcal{K}}_{MI[K}\tilde{\mathcal{K}}^{M}{}_{JL]} \neq 0 \tag{8.15}$$

establish a relation between the (plastic) incompatibilities of the plastic Cauchy-Green tensor $\tilde{C}_{IJ}$ contained in the plastic Riemann curvature $\tilde{\mathcal{M}}_{IJKL}$ on the one hand and the dislocation density $\tilde{\mathcal{D}}^{\alpha L}$ contained in the plastic contortion curvature $\tilde{\mathcal{K}}_{IJKL}$ on the other hand

$$\tilde{\mathcal{M}}_{IJKL} = -\tilde{\mathcal{K}}_{IJKL} + 2\tilde{\mathcal{K}}_{MI[K}\tilde{\mathcal{M}}^{M}{}_{JL]} + 2\tilde{\mathcal{M}}_{MI[K}\tilde{\mathcal{K}}^{M}{}_{JL]}. \tag{8.16}$$

This relation is of particular relevance for cases (such as polycrystal plasticity) that are characterized by an evolution law for the plastic Cauchy-Green tensor $\tilde{C}_{IJ}$ rather than for the plastic distortion $\tilde{F}^{\alpha}{}_{J}$. Recall that the plastic Riemann curvature tensor $\tilde{\mathcal{M}}_{IJKL}$ obeys major symmetries in the index pairs IJ and KL, thus Eq. 8.16 is a result of the major symmetric part of $\tilde{\mathcal{R}}_{IJKL} = 0$. Moreover, due to the minor skew symmetries in the indices IJ and KL the fourth-order fully covariant plastic Riemann curvature tensor $\tilde{\mathcal{M}}_{IJKL}$ may likewise be expressed by its corresponding contravariant second-order double dual curvature tensor

$$\frac{1}{4}\mathcal{E}^{MIJ}\tilde{\mathcal{M}}_{IJKL}\mathcal{E}^{KLN} =: -\tilde{\mathcal{H}}^{MN}. \tag{8.17}$$

The *plastic incompatibility density tensor* $\tilde{\mathcal{H}}^{MN}$ is here defined as the negative plastic double dual Riemann curvature tensor. In ambient Euclidean space the corresponding contravariant and symmetric second-order plastic incompatibility density tensor is then expressed as

$$2\widetilde{\boldsymbol{H}}_0 = \operatorname{Inc}\widetilde{\boldsymbol{C}} := \operatorname{Curl}^t(\operatorname{Curl}^t\widetilde{\boldsymbol{C}}) + \{\operatorname{Grad}\widetilde{\boldsymbol{C}}\} \rtimes\!\ltimes \{\operatorname{Grad}\widetilde{\boldsymbol{C}}\}. \tag{8.18}$$

Here Inc denotes the *material incompatibility operator*, i.e. the geometrically nonlinear extension of the linear Kröner incompatibility operator, see [94]. Thereby the right tensor cross product of third-order tensors that results from the quadratic term $\mathcal{E}^{MIJ}\tilde{\mathcal{M}}_{AIK}\tilde{\mathcal{M}}^{A}{}_{JL}\mathcal{E}^{KLN}$ is abbreviated by $\rtimes\!\ltimes$, the bold curly brackets $\{\{\bullet\}\}$ refer to the anti-cyclic, sign alternating index permutation $2\{\bullet\}_{\{IJK\}} = \{\bullet\}_{IKJ} - \{\bullet\}_{KJI} + \{\bullet\}_{JIK}$ as introduced in Part I. The subindex 0 refers to the incompatibility density per unit volume in the material configuration, the factor two in the definition of the $\operatorname{Inc}\widetilde{\boldsymbol{C}} = 2\operatorname{Inc}\widetilde{\boldsymbol{E}}$ may be traced to the relation between the plastic right Cauchy-Green and the plastic Green-Lagrange strain $\widetilde{\boldsymbol{C}} - \widetilde{\boldsymbol{J}} =: 2\widetilde{\boldsymbol{E}}$. The relation between incompatibilities of the plastic (strain) metric and the (geometrically necessary) dislocation density as displayed in Eq. 8.16 is a key result in the geometrically linear theory of continuous dislocations that may be established comparatively straightforward, see [202]. However, for the geometrically nonlinear case it would not have been easily discovered without the tools from differential geometry. □

The consequence of the major skew symmetric part of $\tilde{\mathcal{R}}_{IJKL} = 0$ or, equivalently, of the right hand part of Eq. 8.16 is discussed below after further dislocation density tensors have been introduced.

Remark 8.1. For geometrically linear elasto-plasticity in Cartesian coordinates the plastic incompatibility density tensor reads

$$\widetilde{\boldsymbol{\eta}} = \tilde{\eta}_{mn}\boldsymbol{e}_m \otimes \boldsymbol{e}_n \doteq \boldsymbol{0} \quad \text{with} \quad \tilde{\eta}_{mn} := e_{mab}\tilde{\epsilon}_{ac,bd}e_{cdn}.$$

Due to the definition of the plastic strain as the symmetric part of the plastic distortion

$$\widetilde{\boldsymbol{\epsilon}} := \widetilde{\boldsymbol{\beta}}^{\text{sym}} \quad \rightsquigarrow \quad \widetilde{\boldsymbol{\eta}} = \operatorname{inc}\widetilde{\boldsymbol{\epsilon}} = \operatorname{inc}\widetilde{\boldsymbol{\beta}}^{\text{sym}}$$

the plastic incompatibility density tensor may alternatively be expressed in terms of the plastic dislocation density

$$\widetilde{\boldsymbol{\eta}} = \operatorname{curl}^t \operatorname{curl}^t \widetilde{\boldsymbol{\epsilon}} = \operatorname{curl}^t \operatorname{curl}^t \widetilde{\boldsymbol{\beta}}^{\text{sym}} = [\operatorname{curl}^t \operatorname{curl}^t \widetilde{\boldsymbol{\beta}}]^{\text{sym}} = [\operatorname{curl}^t \widetilde{\boldsymbol{\alpha}}^t]^{\text{sym}}.$$

Note finally that due to the compatibility of the total strain the incompatibility of the elastic strain follows as $\operatorname{inc}\overline{\boldsymbol{\epsilon}} := \overline{\boldsymbol{\eta}} = -\widetilde{\boldsymbol{\eta}}$. □

Looking Backward: Incompatibility of Intermediate Configuration. Due to its isoclinic character the intermediate configuration of elasto-plasticity is endowed with distance determination and parallel transport of Euclidean type, see Fig. 8.3; accordingly Cartesian-like coordinates may be selected again with the intermediate metric coefficients equal to the Kronecker-symbol and a zero intermediate connection

$$\bar{M}_{\alpha\beta} = \bar{\delta}_{\alpha\beta} \quad \text{and} \quad \bar{L}^{\alpha}{}_{\beta\gamma} = 0. \tag{8.19}$$

The elastic convection of the intermediate metric and the intermediate connection by the elastic distortion then renders the *elastic Finger tensor* and an *integrable elastic connection* in the spatial configuration

$$\bar{m}_{ij} = \bar{c}_{ij} := \bar{f}^{\alpha}{}_{i}\bar{\delta}_{\alpha\beta}\bar{f}^{\beta}{}_{j} \quad \text{and} \quad \bar{l}^{i}{}_{jk} = \bar{F}^{i}{}_{\alpha}\bar{f}^{\alpha}{}_{j,k}. \tag{8.20}$$

It is again easy to proof that the fully covariant elastic spatial connection $\bar{l}^{i}{}_{jk} = \bar{c}_{im}\bar{l}^{m}{}_{jk}$ is indeed a metric connection with respect to $\bar{c}_{ij}$, due to

$$\bar{c}_{ij,k} = \bar{f}^{\alpha}{}_{i,k}\bar{\delta}_{\alpha\beta}\bar{f}^{\beta}{}_{j} + \bar{f}^{\alpha}{}_{i}\bar{\delta}_{\alpha\beta}\bar{f}^{\beta}{}_{j,k} = 2\bar{l}_{(ij)k}. \tag{8.21}$$

Therefore the elastic spatial connection $\bar{l}_{ijk}$ decomposes into the elastic spatial Riemann connection based on the elastic metric $\bar{c}_{ij}$

$$\bar{m}_{ijk}(\bar{c}) := \frac{1}{2}[\bar{c}_{ij,k} - \bar{c}_{jk,i} + \bar{c}_{ki,j}] \tag{8.22}$$

and the associated elastic spatial contortion based on the dislocation density

$$\bar{k}_{ijk}(\bar{d}) := \bar{l}_{ijk} - \bar{m}_{ijk}. \tag{8.23}$$

The verification follows easily from rewriting Eq. 8.21 three times upon cyclic permutation of indices and subsequent addition to render

$$\begin{array}{rl} \bar{c}_{ij,k} &= + \bar{f}^{\alpha}{}_{j}\ \bar{\delta}_{\alpha\beta}\ \bar{f}^{\beta}{}_{i,k} + \bar{f}^{\alpha}{}_{i}\ \bar{\delta}_{\alpha\beta}\ \bar{f}^{\beta}{}_{j,k} \\ -\bar{c}_{jk,i} &= - \bar{f}^{\alpha}{}_{k}\ \bar{\delta}_{\alpha\beta}\ \bar{f}^{\beta}{}_{j,i} - \bar{f}^{\alpha}{}_{j}\ \bar{\delta}_{\alpha\beta}\ \bar{f}^{\beta}{}_{k,i} \\ \bar{c}_{ki,j} &= + \bar{f}^{\alpha}{}_{i}\ \bar{\delta}_{\alpha\beta}\ \bar{f}^{\beta}{}_{k,j} + \bar{f}^{\alpha}{}_{k}\ \bar{\delta}_{\alpha\beta}\ \bar{f}^{\beta}{}_{i,j} \\ \hline\hline \end{array} \tag{8.24}$$

$$\rightarrow \quad \bar{m}_{ijk}(\bar{c}) = \bar{f}^{\alpha}{}_{i}\bar{\delta}_{\alpha\beta}\bar{f}^{\beta}{}_{(j,k)} - \bar{f}^{\alpha}{}_{j}\bar{\delta}_{\alpha\beta}\bar{f}^{\beta}{}_{[k,i]} + \bar{f}^{\alpha}{}_{k}\bar{\delta}_{\alpha\beta}\bar{f}^{\beta}{}_{[i,j]} \tag{8.25}$$

and consequently by Eq. 8.23 the elastic spatial contortion is computed by

$$\bar{k}_{ijk}(\bar{d}) = \bar{f}^{\alpha}{}_{i}\bar{\delta}_{\alpha\beta}\bar{f}^{\beta}{}_{[j,k]} + \bar{f}^{\alpha}{}_{j}\bar{\delta}_{\alpha\beta}\bar{f}^{\beta}{}_{[k,i]} - \bar{f}^{\alpha}{}_{k}\bar{\delta}_{\alpha\beta}\bar{f}^{\beta}{}_{[i,j]}. \tag{8.26}$$

Moreover, since the elastic spatial connection is integrable, the associated elastic spatial curvature tensor vanishes identically

$$\bar{r}_{ijkl}(\bar{l}) = 2\bar{l}_{ij[l,k]} - 2\bar{l}_{mi[k}\bar{l}^{m}{}_{jl]} \equiv 0. \tag{8.27}$$

This finding coincides of course with the postulated Euclidean character of the intermediate configuration

$$0 \equiv \bar{R}'_{\alpha\beta\gamma\delta}(\bar{L}) + \bar{L}_{\alpha\beta\mu}\bar{A}^{\mu}{}_{\delta\gamma} = \bar{R}'_{\alpha\beta\gamma\delta}(\bar{Y}(\bar{l})) + \bar{L}_{\alpha\beta\mu}\bar{A}^{\mu}{}_{\delta\gamma} = \bar{Y}(\bar{r}_{ijkl}(\bar{l})). \tag{8.28}$$

Here $\bar{R}'_{\alpha\beta\gamma\delta}(\bar{L})$ merely abbreviates the application of the curvature format to the intermediate connection, i.e. $\bar{R}'_{\alpha\beta\gamma\delta}(\tilde{L}) := 2\bar{L}_{\alpha\beta[\delta.\gamma]} - 2\bar{L}_{\mu\alpha\gamma}\bar{L}^{\mu}{}_{\beta\delta}$. Due to the anholonomic object the curvature tensor in the intermediate configuration is then expressed as $\bar{R}_{\alpha\beta\gamma\delta} = \bar{R}'_{\alpha\beta\gamma\delta}(\bar{L}) + \bar{L}_{\alpha\beta\mu}\bar{A}^{\mu}{}_{\delta\gamma}$. However since the intermediate connection vanishes due to the selection of Cartesian-like coordinates in the intermediate configuration, i.e. $\bar{L}_{\alpha\beta\gamma} = 0$, the curvature tensor in the intermediate configuration also vanishes trivially, i.e. $\bar{R}_{\alpha\beta\gamma\delta} = 0$.

Discussion 8.2. (Consequence of major symmetric part of $\bar{r}_{ijkl} = 0$)
Since the elastic spatial connection decomposes into the elastic spatial Riemann connection and the elastic spatial contortion, the corresponding non-vanishing spatial curvature tensors

$$\bar{m}_{ijkl}(\bar{c}) := 2\bar{m}_{ij[l,k]} - 2\bar{m}_{mi[k}\bar{m}^{m}{}_{jl]} \neq 0, \tag{8.29}$$

and

$$\bar{k}_{ijkl}(\bar{d}) := 2\bar{k}_{ij[l,k]} - 2\bar{k}_{mi[k}\bar{k}^{m}{}_{jl]} \neq 0 \tag{8.30}$$

establish a relation between the (elastic) incompatibilities of the elastic Finger tensor $\bar{c}_{ij}$ contained in the elastic Riemann curvature $\bar{m}_{ijkl}$ on the one hand

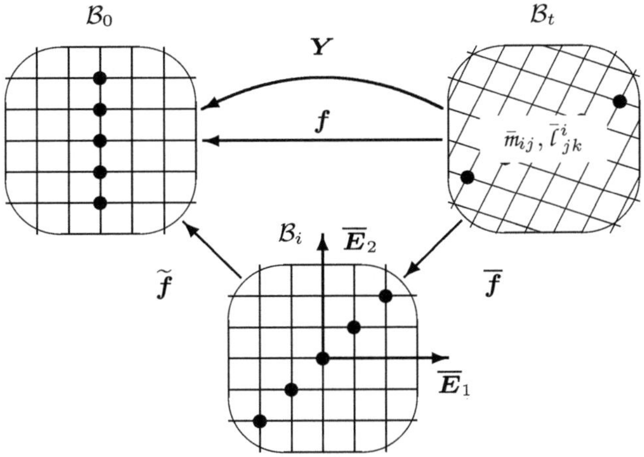

Fig. 8.3 In the intermediate configuration of elasto-plasticity distance determination and parallel transport are of Euclidean type. The elastic convection of the metric and the connection renders the elastic Finger tensor and an integrable elastic connection in the spatial configuration.

and the dislocation density $\bar{d}^{\alpha j}$ contained in the elastic contortion curvature $\bar{k}_{ijkl}$ on the other hand

$$\bar{m}_{ijkl} = -\bar{k}_{ijkl} + 2\bar{k}_{mi[k}\bar{m}^m{}_{jl]} + 2\bar{m}_{mi[k}\bar{k}^m{}_{jl]}. \qquad (8.31)$$

This relation is of particular relevance for cases (such as polycrystal plasticity) that are characterized by an evolution law for the elastic Finger tensor $\bar{c}_{ij}$ rather than for the elastic distortion $\bar{f}^\alpha{}_j$. Observe that $\bar{m}_{ijkl}$ obeys major symmetry in the index pairs ij and kl, thus Eq. 8.31 is a result of the major symmetric part of $\bar{r}_{ijkl} = 0$. Moreover, due to the minor skew symmetries in the indices ij and kl the fourth-order fully covariant elastic Riemann curvature tensor $\bar{m}_{ijkl}$ may likewise be expressed by its corresponding contravariant and symmetric second-order double dual curvature tensor

$$\frac{1}{4} e^{mij}\, \bar{m}_{ijkl}\, e^{kln} =: \bar{h}^{mn}. \qquad (8.32)$$

The *elastic incompatibility density tensor* $\bar{h}^{mn}$ is here defined as the negative elastic double dual Riemann curvature tensor. In ambient Euclidean space the corresponding contravariant second-order elastic incompatibility density tensor is then expressed as

$$-2\overline{\boldsymbol{h}}_t = \operatorname{inc}\overline{\boldsymbol{c}} := \operatorname{curl}^t(\operatorname{curl}^t \overline{\boldsymbol{c}}) + \{\operatorname{grad}\overline{\boldsymbol{c}}\} \bowtie \{\operatorname{grad}\overline{\boldsymbol{c}}\}. \qquad (8.33)$$

Here inc denotes the geometrically nonlinear *spatial incompatibility operator*, that is an extension of the linear Kröner incompatibility operator, see [94].

The subindex t refers to the incompatibility density per unit volume in the spatial configuration, the factor two in the definition of the $\operatorname{inc}\overline{\boldsymbol{c}} = -2\operatorname{Inc}\overline{\boldsymbol{e}}$ may be traced to the relation between the plastic Finger and the plastic Almansi strain $\overline{\boldsymbol{c}} - \overline{\boldsymbol{j}} =: -2\overline{\boldsymbol{e}}$. □

The consequence of the major skew symmetric part of $\bar{r}_{ijkl} = 0$ or, equivalently, of the right hand part of Eq. 8.31 is discussed below after further dislocation density tensors have been introduced.

Remark 8.2. As outlined in [181] the non-integrability of the elastic $\overline{\boldsymbol{f}}$ and the incompatibility of the elastic $\overline{\boldsymbol{c}}$ may be interpreted as the underlying cause for dislocations and macroscopic residual stresses, respectively. Moreover the (elastic) incompatibilities of $\overline{\boldsymbol{c}}$ are interrelated with the dislocation density due to Eq. 8.31. For the special case that $\overline{\boldsymbol{c}} = \boldsymbol{0}$ holds, $\overline{\boldsymbol{f}}$ degenerates to a proper orthogonal tensor and consequently the dislocation density corresponds to the *stress free* Nye lattice curvature. In the general case with $\overline{\boldsymbol{c}} \neq \boldsymbol{0}$ the non-integrability of the elastic $\overline{\boldsymbol{f}}$ corresponds to the elastic contortion while the incompatibility of the elastic $\overline{\boldsymbol{c}}$ is the source of elastic residual stresses. □

8.1.3 Dislocation Density Tensors

Plastic Anholonomic Objects versus Plastic Torsions. Associated with the incompatibility of the intermediate configuration, and referring to Eqs. 3.55.1 and 3.51.1, are the third-order plastic (Piola-Kirchhoff- and Cauchy-type) anholonomic objects

$$\tilde{\mathcal{A}}^I{}_{JK} = -2\tilde{f}^I{}_\alpha \tilde{F}^\alpha{}_{[J,K]} \quad \text{and} \quad \tilde{a}^\alpha{}_{\beta\gamma} = 2\tilde{F}^\alpha{}_{[J,K]}\tilde{f}^J{}_\beta \tilde{f}^K{}_\gamma. \tag{8.34}$$

These plastic anholonomic objects are related to the corresponding plastic material and intermediate torsions as computed from the material and intermediate plastic connections in Eqs. 8.6.2 and 8.5.2

$$\tilde{\mathcal{A}}^I{}_{JK} = -2\tilde{\mathcal{L}}^I{}_{[JK]} = -\tilde{\mathcal{T}}^I{}_{JK} \quad \text{and} \quad \tilde{a}^\alpha{}_{\beta\gamma} = \tilde{t}^\alpha{}_{\beta\gamma} - 2\tilde{l}^\alpha{}_{[\beta\gamma]} = \tilde{t}^\alpha{}_{\beta\gamma}. \tag{8.35}$$

Clearly, due to the representation of the material plastic connection in Eq. 8.6.2 its torsion coincides with the negative of the plastic (Piola-Kirchhoff-type) anholonomic object in Eq. 8.34.1; likewise due to the generic representation of the torsion in an anholonomic space in Eq. 3.52.1 the plastic torsion in the intermediate configuration coincides with the plastic (Cauchy-type) anholonomic object in Eq. 8.34.2 due to the vanishing intermediate plastic connection in Eq. 8.5.2.
As a consequence, the incompatibility of the intermediate configuration in first-order elasto-plasticity is likewise characterized in terms of the plastic torsion or in terms of the plastic anholonomic object of the underlying flat but non-symmetric geometry!

Dislocation Densities Based on Plastic Distortion. Corresponding to the plastic anholonomic objects the (contravariant) second-order Piola-Kirchhoff- and Cauchy-type dislocation density tensors are expressed in the ambient Euclidean space as

$$\widetilde{\boldsymbol{A}} = -\widetilde{\boldsymbol{f}} \cdot \widetilde{\boldsymbol{D}} \quad \text{and} \quad \widetilde{\boldsymbol{a}} = \widetilde{\boldsymbol{D}} \cdot \operatorname{cof} \widetilde{\boldsymbol{f}} \quad \text{with} \quad \widetilde{\boldsymbol{D}} = \operatorname{Curl} \widetilde{\boldsymbol{F}}. \tag{8.36}$$

Finally the continuity equation of continuum dislocation theory states that *dislocation lines* may not end within the volume, i.e. the plastic dislocation density is source free (solenoidal)

$$\operatorname{Div} \widetilde{\boldsymbol{D}} = -\widetilde{\boldsymbol{F}} \cdot [\operatorname{Div} \widetilde{\boldsymbol{A}} + \widetilde{\boldsymbol{f}} \cdot \operatorname{Grad} \widetilde{\boldsymbol{F}} : \widetilde{\boldsymbol{A}}] = \boldsymbol{0}. \tag{8.37}$$

Obviously, the plastic dislocation density being source free is a straightforward consequence from its definition as the material curl of the plastic distortion.

Discussion 8.3. (Consequence of major skew symmetric part of $\tilde{\mathcal{R}}_{IJKL} = 0$) The solenoidal character $\operatorname{Div} \widetilde{\boldsymbol{D}} = \boldsymbol{0}$ of the plastic (Piola-type) dislocation density $\widetilde{\boldsymbol{D}}$ is directly related to the algebraic Bianchy identity corresponding to Eq. 5.51: According to Eq. 5.34.2 the coefficients of the second-order dislocation density tensor follow from those of the third-order dislocation density tensor (that in turn is expressed in terms of the connection in Eq. 8.6.2) as

$$\tilde{D}^{\alpha L} = -\frac{1}{2} \tilde{D}^{\alpha}{}_{JK} \mathcal{E}^{JKL} \quad \text{with} \quad \tilde{D}^{\alpha}{}_{JK} = 2 \tilde{F}^{\alpha}{}_{I} \tilde{\mathcal{L}}^{I}{}_{[JK]}.$$

The divergence of the second-order dislocation density tensor involves the total covariant derivative in Euclidean ambient space in terms of the material Christoffel symbol $\mathcal{N}^{I}{}_{JK}$ (the Christoffel symbol related to the coordinates in the intermediate configuration $\tilde{n}^{\alpha}{}_{\beta\gamma} \equiv 0$ vanishes due to the selection of Cartesian-like coordinates, see Eq. 8.5.1). Applying the total covariant derivative to $\tilde{D}^{\alpha L}$ produces firstly

$$\tilde{D}^{\alpha L}{}_{:L} = -\tilde{F}^{\alpha}{}_{I} [\tilde{\mathcal{L}}^{I}{}_{[JK]:L} + \tilde{f}^{I}{}_{\beta} \tilde{F}^{\beta}{}_{M:L} \tilde{\mathcal{L}}^{M}{}_{[JK]}] \mathcal{E}^{JKL}.$$

Thus by resorting to Eqs. 8.6.2 and 8.34.1 and using the relation between the third and second-order dislocation density tensors the divergence statement in Eq. 8.37 is recovered. Expanding next the total covariant derivative of $\tilde{D}^{\alpha L}$ renders secondly

$$\tilde{D}^{\alpha L}{}_{:L} = -\frac{1}{2} \tilde{D}^{\alpha}{}_{JK:L} \mathcal{E}^{JKL} \equiv -\frac{1}{2} \tilde{D}^{\alpha}{}_{JK,L} \mathcal{E}^{JKL}.$$

Here, i) $\mathcal{E}^{JKL}{}_{:L} = 0$ (holds true since $\mathcal{E}^{JKL}$ is scaled by $\mathcal{M} = \det[\mathcal{M}_{IJ}]$ whereby here $\mathcal{M}_{IJ} \equiv G_{IJ}$) and ii) the right symmetry of the material Christoffel symbols have been exploited to derive at a result in terms of the partial rather than the total covariant derivative of the dislocation density. Then, application of the product rule to $\tilde{D}^{\alpha}{}_{JK}$ results in

$$\begin{aligned}\tilde{D}^{\alpha L}{}_{:L} &= -[\tilde{F}^{\alpha}{}_{I}\tilde{\mathcal{L}}^{I}{}_{[JK]}]_{,L}\mathcal{E}^{JKL} \\ &= -\tilde{F}^{\alpha}{}_{I}[\tilde{\mathcal{L}}^{I}{}_{[JK],L} + \tilde{f}^{I}{}_{\beta}\tilde{F}^{\beta}{}_{M,L}\tilde{\mathcal{L}}^{M}{}_{[JK]}]\mathcal{E}^{JKL}.\end{aligned}$$

Reassigning further the skew symmetries in the indices JKL due to the triple contraction with the permutation symbol results in an expression containing the plastic material curvature tensor

$$\tilde{D}^{\alpha L}{}_{:L} = \tilde{F}^{\alpha}{}_{I}[\tilde{\mathcal{L}}^{I}{}_{J[L,K]} + \tilde{\mathcal{L}}^{I}{}_{M[K}\tilde{\mathcal{L}}^{M}{}_{JL]}]\mathcal{E}^{JKL}.$$

Finally, the axial vector $\tilde{\mathcal{G}}^{I}$ of the plastic Einstein tensor and thus the algebraic Bianchi identity is recognized from the triple contraction of the plastic material curvature tensor with the permutation symbol

$$\tilde{D}^{\alpha L}{}_{:L} = \frac{1}{2}\tilde{F}^{\alpha}{}_{I}\tilde{\mathcal{R}}^{I}{}_{JKL}\mathcal{E}^{JKL} = 2\tilde{F}^{\alpha}{}_{I}\tilde{\mathcal{G}}^{I} = 0.$$

Summarizing, since elasto-plasticity may be considered a flat but non-symmetric geometry over the material configuration with $\tilde{\mathcal{R}}^{I}{}_{JKL} = 0$ the skew part of $\tilde{\mathcal{R}}^{I}{}_{JKL} = 0$ coincides with the corresponding algebraic Bianchi identity, compare Eq. 5.51. □

Elastic Anholonomic Objects versus Elastic Torsions. Associated with the incompatibility of the intermediate configuration, and referring to Eqs. 3.55.2 and 3.51.2, are the third-order elastic (Piola-Kirchhoff- and Cauchy-type) anholonomic objects

$$\bar{a}^{i}{}_{jk} = -2\bar{F}^{i}{}_{\alpha}\bar{f}^{\alpha}{}_{[j,k]} \quad \text{and} \quad \bar{A}^{\alpha}{}_{\beta\gamma} = 2\bar{f}^{\alpha}{}_{[j,k]}\bar{F}^{j}{}_{\beta}\bar{F}^{k}{}_{\gamma}. \tag{8.38}$$

These elastic anholonomic objects are related to the corresponding elastic intermediate and spatial torsions as computed from the spatial and intermediate connections in Eqs. 8.20.2 and 8.19.2

$$\bar{a}^{i}{}_{jk} = -2\bar{l}^{i}{}_{[jk]} = -\bar{t}^{i}{}_{jk} \quad \text{and} \quad \bar{A}^{\alpha}{}_{\beta\gamma} = \bar{T}^{\alpha}{}_{\beta\gamma} - 2\bar{L}^{\alpha}{}_{[\beta\gamma]} = \bar{T}^{\alpha}{}_{\beta\gamma}. \tag{8.39}$$

Obviously, due to the representation of the spatial elastic connection in Eq. 8.20.2 its torsion coincides with the negative of the elastic (Piola-Kirchhoff-type) anholonomic object in Eq. 8.38.1; likewise due to the generic representation of the torsion in an anholonomic space in Eq. 3.52.2 it coincides with the elastic (Cauchy-type) anholonomic object in Eq. 8.38.2 due to the vanishing intermediate elastic connection in Eq. 8.19.2.
As a consequence, the incompatibility of the intermediate configuration in first-order elasto-plasticity is likewise characterized in terms of the elastic torsion or in terms of the elastic anholonomic object of the underlying flat but non-symmetric geometry!

Dislocation Densities Based on Elastic Distortion. Corresponding to the elastic anholonomic objects the (contravariant) second-order Piola-Kirchhoff- and Cauchy-type dislocation density tensors are expressed in the ambient Euclidean space as

$$\overline{\boldsymbol{a}} = -\overline{\boldsymbol{F}} \cdot \overline{\boldsymbol{d}} \quad \text{and} \quad \overline{\boldsymbol{A}} = \overline{\boldsymbol{d}} \cdot \operatorname{cof} \overline{\boldsymbol{F}} \quad \text{with} \quad \overline{\boldsymbol{d}} = \operatorname{curl} \overline{\boldsymbol{f}}. \tag{8.40}$$

Finally the continuity equation of continuum dislocation theory states that the dislocation lines may not end within the volume, i.e. the elastic dislocation density is source free (solenoidal)

$$\operatorname{div}\overline{\boldsymbol{d}} = -\overline{\boldsymbol{f}} \cdot [\operatorname{div}\overline{\boldsymbol{a}} + \overline{\boldsymbol{F}} \cdot \operatorname{grad} \overline{\boldsymbol{f}} : \overline{\boldsymbol{a}}] = \boldsymbol{0}. \tag{8.41}$$

Obviously, the elastic dislocation density being source free is a straightforward consequence from its definition as the spatial curl of the elastic distortion.

Discussion 8.4. (Consequence of major skew symmetric part of $\bar{r}_{ijkl} = 0$) The solenoidal character $\operatorname{div}\overline{\boldsymbol{d}} = \boldsymbol{0}$ of the elastic (Piola-type) dislocation density $\overline{\boldsymbol{d}}$ is directly related to the algebraic Bianchy identity according to Eq. 5.51: According to Eq. 5.34.4 the coefficients of the second-order dislocation density tensor follow from its third-order counterpart (that in turn is expressed in terms of the connection in Eq. 8.20.2) as

$$\bar{d}^{\alpha l} = -\frac{1}{2}\bar{d}^{\alpha}{}_{jk}e^{jkl} \quad \text{with} \quad \bar{d}^{\alpha}{}_{jk} = 2\bar{f}^{\alpha}{}_{i}\bar{\ell}^{i}{}_{[jk]}.$$

The divergence of the second-order dislocation density tensor involves the total covariant derivative in Euclidean ambient space in terms of the spatial Christoffel symbol $n^{i}{}_{jk}$ (the Christoffel symbol related to the coordinates in the intermediate configuration $\bar{N}^{\alpha}{}_{\beta\gamma} \equiv 0$ vanishes due to the selection of Cartesian-like coordinates, see Eq. 8.19.1). Applying the total covariant derivative to $\bar{d}^{\alpha l}$ produces firstly

$$\bar{d}^{\alpha l}{}_{:l} = -\bar{f}^{\alpha}{}_{i}[\bar{\ell}^{i}{}_{[jk]:l} + \bar{F}^{i}{}_{\beta}\bar{f}^{\beta}{}_{m:l}\bar{\ell}^{m}{}_{[jk]}]e^{jkl}.$$

Thus by resorting to Eqs. 8.20.2 and 8.38.1 and using the relation between the third and second-order dislocation density tensors the divergence statement in Eq. 8.41 is recovered. Expanding next the total covariant derivative of $\bar{d}^{\alpha l}$ renders secondly

$$\bar{d}^{\alpha l}{}_{:l} = -\frac{1}{2}\bar{d}^{\alpha}{}_{jk:l}e^{jkl} \equiv -\frac{1}{2}\bar{d}^{\alpha}{}_{jk,l}e^{jkl}.$$

Here, i) $e^{jkl}{}_{:l} = 0$ and ii) the right symmetry of the spatial Christoffel symbol have been exploited to derive at a result in terms of the partial rather than the total covariant derivative of the dislocation density. Then, application of the product rule to $\bar{d}^{\alpha}{}_{jk}$ results in

$$\bar{d}^{\alpha l}{}_{:l} = -[\bar{f}^{\alpha}{}_{i}\bar{l}^{i}{}_{[jk]}]_{,l}e^{jkl}$$
$$= -\bar{f}^{\alpha}{}_{i}[\bar{l}^{i}{}_{[jk],l} + \bar{F}^{i}{}_{\beta}\bar{f}^{\beta}{}_{m,l}\bar{l}^{m}{}_{[jk]}]e^{jkl}.$$

Reassigning further the skew symmetries in the indices jkl due to the triple contraction with the permutation symbol results in an expression containing the elastic spatial curvature tensor in the spatial configuration

$$\bar{d}^{\alpha l}{}_{:l} = \bar{f}^{\alpha}{}_{i}[\bar{l}^{i}{}_{j[l,k]} + \bar{l}^{i}{}_{m[k}\bar{l}^{m}{}_{jl]}]e^{jkl}.$$

Finally, the axial vector $\bar{g}^{i}$ of the elastic Einstein tensor and thus the associated algebraic Bianchi identity is recognized from the triple contraction of the elastic spatial curvature tensor with the permutation symbol

$$\bar{d}^{\alpha l}{}_{:l} = \frac{1}{2}\bar{f}^{\alpha}{}_{i}\bar{r}^{i}{}_{jkl}e^{jkl} = 2\bar{f}^{\alpha}{}_{i}\bar{g}^{i} = 0.$$

Summarizing, since elasto-plasticity may be considered a flat but non-symmetric geometry over the spatial configuration with $\bar{r}^{i}{}_{jkl} = 0$ the skew part of $\bar{r}^{i}{}_{jkl} = 0$ coincides with the corresponding algebraic Bianchi identity, compare Eq. 5.51. □

Dislocation Densities in the Intermediate Configuration. Due to the multiplicative decomposition and the compatibility of the total deformation the plastic Cauchy-type and the elastic Piola-Kirchhoff-type *third-order* dislocation densities as introduced in Eqs. 8.34.2 and 8.38.2 together with the corresponding torsions in the intermediate configuration are related by the identities

$$\tilde{a}^{\alpha}{}_{\beta\gamma} \equiv \bar{A}^{\alpha}{}_{\beta\gamma} \quad \text{and} \quad \tilde{t}^{\alpha}{}_{\beta\gamma} \equiv \bar{T}^{\alpha}{}_{\beta\gamma}. \tag{8.42}$$

Proof. The multiplicative decomposition of the deformation gradient reads

$$F^{i}{}_{J} = \bar{F}^{i}{}_{\alpha}\tilde{F}^{\alpha}{}_{J}.$$

Consequently the material gradient of the deformation gradient, i.e. the double-distorsion expands into

$$F^{i}{}_{J,K} = \bar{F}^{i}{}_{\alpha,K}\tilde{F}^{\alpha}{}_{J} + \bar{F}^{i}{}_{\alpha}\tilde{F}^{\alpha}{}_{J,K} = -\bar{F}^{i}{}_{\alpha}\bar{f}^{\alpha}{}_{j,k}F^{j}{}_{J}F^{k}{}_{K} + \bar{F}^{i}{}_{\alpha}\tilde{F}^{\alpha}{}_{J,K}.$$

Here the relation $\bar{F}^{i}{}_{\alpha,K} = -\bar{F}^{i}{}_{\beta}\bar{f}^{\beta}{}_{j,K}\bar{F}^{j}{}_{\alpha}$ has been used. Multiplication with $2\tilde{f}^{J}{}_{\beta}\tilde{f}^{K}{}_{\gamma}$ and skew symmetrization in the indices JK then discovers the relation between the elastic and plastic Piola-type dislocation densities

$$\bar{d}^{\alpha}{}_{jk}\bar{F}^{j}{}_{\beta}\bar{F}^{k}{}_{\gamma} = \tilde{D}^{\alpha}{}_{JK}\tilde{f}^{J}{}_{\beta}\tilde{f}^{K}{}_{\gamma}.$$

As a consequence the relation between the elastic Piola-Kirchhoff-type and the plastic Cauchy-type third-order dislocation densities then follows as

$$\bar{A}^{\alpha}{}_{\beta\gamma} = \tilde{a}^{\alpha}{}_{\beta\gamma}$$

Clearly this relation also holds for the corresponding torsions. □

Alternatively consider the

Proof. The multiplicative decomposition of the inverse deformation gradient reads

$$f^{I}{}_{j} = \tilde{f}^{I}{}_{\alpha}\bar{f}^{\alpha}{}_{j}.$$

Consequently the spatial gradient of the deformation gradient expands into

$$f^{I}{}_{j,k} = \tilde{f}^{I}{}_{\alpha,k}\bar{f}^{\alpha}{}_{j} + \tilde{f}^{I}{}_{\alpha}\bar{f}^{\alpha}{}_{j,k} = -\tilde{f}^{I}{}_{\alpha}\tilde{F}^{\alpha}{}_{J,K}f^{J}{}_{j}f^{K}{}_{k} + \tilde{f}^{I}{}_{\alpha}\bar{f}^{\alpha}{}_{j,k}.$$

Here the relation $\tilde{f}^{I}{}_{\alpha,k} = -\tilde{f}^{I}{}_{\beta}\tilde{F}^{\beta}{}_{J,k}\tilde{f}^{J}{}_{\alpha}$ has been used. Multiplication with $2\bar{F}^{j}{}_{\beta}\bar{F}^{k}{}_{\gamma}$ and skew symmetrization in the indices jk then discovers the relation between the plastic and elastic Piola-type dislocation densities

$$\tilde{D}^{\alpha}{}_{JK}\tilde{f}^{J}{}_{\beta}\tilde{f}^{K}{}_{\gamma} = \bar{d}^{\alpha}{}_{jk}\bar{F}^{j}{}_{\beta}\bar{F}^{k}{}_{\gamma}.$$

As a consequence the relation between the plastic Cauchy-type and the elastic Piola-Kirchhoff-type dislocation densities then follows as

$$\tilde{a}^{\alpha}{}_{\beta\gamma} = \bar{A}^{\alpha}{}_{\beta\gamma}.$$

Obviously this relation also holds for the corresponding torsions. □

As a consequence the plastic Cauchy-type and the elastic Piola-Kirchhoff-type *second-order* dislocation densities in the intermediate configuration as expressed in the ambient Euclidean space are likewise related by the identity

$$\tilde{\boldsymbol{a}} \equiv \overline{\boldsymbol{A}}. \tag{8.43}$$

Proof. The second-order plastic and elastic Piola-type dislocation densities in Euclidean ambient space are computed from their third-order counterparts

$$\tilde{D}^{\alpha L} = -\frac{1}{2}\tilde{D}^{\alpha}{}_{JK}\mathcal{E}^{JKL} \quad \text{and} \quad \bar{d}^{\alpha l} = -\frac{1}{2}\bar{d}^{\alpha}{}_{jk}e^{jkl}.$$

Then the convection to the intermediate configuration according to Eqs. 8.36.2 and 8.40.2 and by considering the inverse of the relation between the third-order tensors as stated in Eqs. 8.34.2 and 8.38.2 renders

$$\tilde{a}^{\alpha\delta} = -\tilde{j}\frac{1}{2}\tilde{a}^{\alpha}{}_{\beta\gamma}\tilde{F}^{\beta}{}_{J}\tilde{F}^{\gamma}{}_{K}\mathcal{E}^{JKL}\tilde{F}^{\delta}{}_{L} \quad \text{and} \quad \bar{A}^{\alpha\delta} = -\bar{J}\frac{1}{2}\bar{A}^{\alpha}{}_{\beta\gamma}\bar{f}^{\beta}{}_{j}\bar{f}^{\gamma}{}_{k}e^{jkl}\bar{f}^{\delta}{}_{l}.$$

Here, $\tilde{j} = \det\tilde{\boldsymbol{f}}$ and $\bar{J} = \det\overline{\boldsymbol{F}}$ denote the Jacobian determinants of the (inverse) plastic and the elastic part of the deformation gradient, respectively.

Thus, due to the relation between the permutation tensors $\boldsymbol{E}$, $\widetilde{\boldsymbol{e}} \equiv \overline{\boldsymbol{E}}$ and $\boldsymbol{e}$ in the material, the intermediate and the spatial configuration, i.e.

$$\widetilde{\boldsymbol{e}} := \tilde{\jmath}[\widetilde{\boldsymbol{F}} \boxtimes \widetilde{\boldsymbol{F}}] : \boldsymbol{E} \cdot \widetilde{\boldsymbol{F}}^t \equiv \bar{J}[\overline{\boldsymbol{f}} \boxtimes \overline{\boldsymbol{f}}] : \boldsymbol{e} \cdot \overline{\boldsymbol{f}}^t =: \overline{\boldsymbol{E}} \tag{8.44}$$

together with the result as stated in Eq. 8.42.1 for the third-order dislocation densities the equality of the second-order plastic Cauchy-type and elastic Piola-Kirchhoff-type dislocation densities per unit volume in the intermediate configuration in Eq. 8.43 follows identically with

$$\tilde{a}^{\alpha\delta} \equiv \bar{A}^{\alpha\delta}.$$

Clearly this relation also holds for the corresponding second-order torsions.

□

Addendum 8.1. Anholonomic Differential Operators
Due to the anholonomicity of the intermediate configuration corresponding anholonomic differential operators and related issues have to be considered carefully.

Anholonomic Curl Operators:
It proves useful to introduce besides the curl operators as defined in Euclidean space with respect to the material and spatial coordinates also curl operators with respect to the anholonomic coordinates in the intermediate configuration, compare [130]. Thus, with the Cartesian base vectors $\overline{\boldsymbol{E}}_\gamma$ and the tangent vectors $\overline{\boldsymbol{G}}_\gamma$ in the intermediate configuration denoted by an overbar, the *elastic curl operator* as applied to the elastic part of the deformation gradient is introduced as

$$\bar{\text{Curl}}\, \overline{\boldsymbol{F}} := \bar{D}^{a\gamma} \boldsymbol{e}_a \otimes \overline{\boldsymbol{E}}_\gamma = \bar{\mathcal{D}}^{i\gamma} \boldsymbol{g}_i \otimes \overline{\boldsymbol{G}}_\gamma.$$

Here, the coefficients follow essentially the standard definition for the (material) curl operator, however, since the intermediate configuration is incompatible, the gradient with respect to the anholonomic coordinates in the intermediate configuration is expanded as the (elastic) *anholonomic gradient* $(\bullet)_{.\beta} := (\bullet)_{,b} \bar{F}^b{}_\beta$ to render

$$\bar{D}^{a\gamma} = -\bar{F}^a{}_{\alpha,b} \bar{F}^b{}_\beta \bar{E}^{\alpha\beta\gamma} \quad \text{and} \quad \bar{\mathcal{D}}^{i\gamma} = -\bar{\mathcal{F}}^i{}_{\alpha:j} \bar{\mathcal{F}}^j{}_\beta \bar{\mathcal{E}}^{\alpha\beta\gamma}.$$

Furthermore, with the Cartesian base vectors $\widetilde{\boldsymbol{e}}_\gamma$ and the tangent vectors $\widetilde{\boldsymbol{g}}_\gamma$ in the intermediate configuration denoted by a tilde, the *plastic curl operator* as applied to the plastic part of the deformation gradient is introduced as

$$\tilde{\text{curl}}\, \widetilde{\boldsymbol{f}} := \tilde{d}^{A\gamma} \boldsymbol{E}_A \otimes \widetilde{\boldsymbol{e}}_\gamma = \tilde{d}^{I\gamma} \boldsymbol{G}_I \otimes \widetilde{\boldsymbol{g}}_\gamma.$$

Note that the symbolic redundancy $\overline{\boldsymbol{E}}_\gamma \equiv \widetilde{\boldsymbol{e}}_\gamma$ and $\overline{\boldsymbol{G}}_\gamma \equiv \widetilde{\boldsymbol{g}}_\gamma$ for the base and tangent vectors in the intermediate configuration has been introduced for notational convenience. The coefficients of the plastic curl operator follow essentially the standard definition for the (spatial) curl operator, however, since the intermediate configuration is incompatible, the gradient with respect to the anholonomic coordinates in the intermediate configuration is expanded as the (plastic) anholonomic gradient $(\bullet)_{.\beta} := (\bullet)_{,B}\tilde{f}^B{}_\beta$ to render

$$\tilde{d}^{A\gamma} = -\tilde{f}^A{}_{\alpha,B}\tilde{f}^B{}_\beta\tilde{e}^{\alpha\beta\gamma} \quad \text{and} \quad \tilde{d}^{I\gamma} = -\tilde{f}^I{}_{\alpha:J}\tilde{f}^J{}_\beta\tilde{e}^{\alpha\beta\gamma}.$$

These definitions together with the relation between the permutation tensors $\boldsymbol{E}$, $\widetilde{\boldsymbol{e}} \equiv \overline{\boldsymbol{E}}$ and $\boldsymbol{e}$ in the material, the intermediate and the spatial configuration in Eq. 8.44 allow to introduce the convection of the spatial into the elastic curl operator and the material into the plastic curl operator, respectively

$$\mathrm{C\tilde{u}rl}\,\overline{\boldsymbol{F}} = -\bar{J}\overline{\boldsymbol{F}}\cdot\mathrm{curl}\,\overline{\boldsymbol{f}}\cdot\overline{\boldsymbol{f}}^t \quad \text{and} \quad \mathrm{c\tilde{u}rl}\,\widetilde{\boldsymbol{f}} = -\tilde{j}\widetilde{\boldsymbol{f}}\cdot\mathrm{Curl}\,\widetilde{\boldsymbol{F}}\cdot\widetilde{\boldsymbol{F}}^t.$$

The above relations will be exploited in the sequel to (i) introduce further versions of dislocation density tensors and to (ii) identify relations between these and the previously defined dislocation density tensors. Furthermore as a consequence of the multiplicative decomposition of the deformation gradient in Eq. 8.1, the definition of the elastic curl operator and the convection of the permutation tensors allow to decompose the material curl of the deformation gradient *additively* into elastic and plastic contributions

$$\mathrm{Curl}\,\boldsymbol{F} = \mathrm{C\tilde{u}rl}\,\overline{\boldsymbol{F}}\cdot\mathrm{cof}\,\widetilde{\boldsymbol{F}} + \overline{\boldsymbol{F}}\cdot\mathrm{Curl}\,\widetilde{\boldsymbol{F}} \doteq \boldsymbol{0}.$$

The additive decomposition of the material curl of the deformation gradient into elastic and plastic parts follows in detail from the step by step calculation

$$\begin{aligned}
\mathrm{Curl}\,\boldsymbol{F} &= -\,\mathrm{Grad}(\overline{\boldsymbol{F}}\cdot\widetilde{\boldsymbol{F}}) : \boldsymbol{E} \\
&= -\,\mathrm{grad}\,\overline{\boldsymbol{F}} : [\widetilde{\boldsymbol{F}} \boxtimes \boldsymbol{F}] : \boldsymbol{E} - \overline{\boldsymbol{F}}\cdot\mathrm{Grad}\,\widetilde{\boldsymbol{F}} : \boldsymbol{E} \\
&= -[\mathrm{grad}\,\overline{\boldsymbol{F}}\cdot\overline{\boldsymbol{F}}] : [\widetilde{\boldsymbol{F}} \boxtimes \widetilde{\boldsymbol{F}}] : \boldsymbol{E} + \overline{\boldsymbol{F}}\cdot\mathrm{Curl}\,\widetilde{\boldsymbol{F}} \\
&= -\tilde{J}[\mathrm{grad}\,\overline{\boldsymbol{F}}\cdot\overline{\boldsymbol{F}}] : [\widetilde{\boldsymbol{F}} \boxtimes \widetilde{\boldsymbol{F}}] : [\widetilde{\boldsymbol{f}} \boxtimes \widetilde{\boldsymbol{f}}] : \overline{\boldsymbol{E}}\cdot\widetilde{\boldsymbol{f}}^t + \overline{\boldsymbol{F}}\cdot\mathrm{Curl}\,\widetilde{\boldsymbol{F}} \\
&= -[\mathrm{grad}\,\overline{\boldsymbol{F}}\cdot\overline{\boldsymbol{F}}] : \overline{\boldsymbol{E}}\cdot\mathrm{cof}\,\widetilde{\boldsymbol{F}} + \overline{\boldsymbol{F}}\cdot\mathrm{Curl}\,\widetilde{\boldsymbol{F}} \\
&= \mathrm{C\tilde{u}rl}\,\overline{\boldsymbol{F}}\cdot\mathrm{cof}\,\widetilde{\boldsymbol{F}} + \overline{\boldsymbol{F}}\cdot\mathrm{Curl}\,\widetilde{\boldsymbol{F}}.
\end{aligned}$$

Likewise the multiplicative decomposition of the inverse deformation gradient in Eq. 8.2 together with the definition of the plastic curl operator and the convection of the permutation tensors allow to decompose the spatial curl of the inverse deformation gradient *additively* into plastic and elastic contributions

$$\mathrm{curl}\,\boldsymbol{f} = \mathrm{c\tilde{u}rl}\,\widetilde{\boldsymbol{f}}\cdot\mathrm{cof}\,\overline{\boldsymbol{f}} + \widetilde{\boldsymbol{f}}\cdot\mathrm{curl}\,\overline{\boldsymbol{f}} \doteq \boldsymbol{0}.$$

The additive decomposition of the material and spatial curl as applied to the total deformation gradient is in agreement with the equality of the plastic and elastic Cauchy-type dislocation densities in Eq. 8.43 as may be demonstrated in the following.

Anholonomic Piola Identity and Piola Transformation:
For non-integrable distortions $\widetilde{\boldsymbol{F}}$ and $\overline{\boldsymbol{f}}$ the corresponding *anholonomic Piola identities*, i.e. the material and spatial divergence of the cofactors $\mathrm{cof}\,\widetilde{\boldsymbol{F}}$ and

cof $\overline{\boldsymbol{f}}$, respectively, expand in terms of the Cauchy-type dislocation density tensors $\widetilde{\boldsymbol{a}}$ and $\overline{\boldsymbol{A}}$ in the intermediate configuration as

$$\text{Div}\left(\text{cof}\,\widetilde{\boldsymbol{F}}\right) = \tilde{J}\widetilde{\boldsymbol{a}} : \widetilde{\boldsymbol{e}} \quad \text{and} \quad \text{div}\left(\text{cof}\,\overline{\boldsymbol{f}}\right) = \bar{j}\overline{\boldsymbol{A}} : \overline{\boldsymbol{E}}.$$

These relations follow from straightforward calculations, see [130]; as an example consider the cofactor of the plastic distortion

$$\text{Div}\left(\text{cof}\,\widetilde{\boldsymbol{F}}\right) = \tilde{J}\big[\,\text{Div}\,\widetilde{\boldsymbol{f}}^{t} + \widetilde{\boldsymbol{f}}^{t} : \text{grad}\,\widetilde{\boldsymbol{F}}\big] = \tilde{J}\big[\text{grad}\,\widetilde{\boldsymbol{f}}^{t} : \widetilde{\boldsymbol{F}} - \widetilde{\boldsymbol{F}}^{t} : \text{grad}\,\widetilde{\boldsymbol{f}}\big] \equiv \tilde{J}\widetilde{\boldsymbol{a}} : \widetilde{\boldsymbol{e}}.$$

Thus, the material and spatial divergence of tensorial quantities that are connected by an *anholonomic Piola transformation*

$$\{\bullet\} \cdot \text{cof}\,\widetilde{\boldsymbol{F}} \leftrightarrow \{\bullet\} \quad \text{and} \quad \{\bullet\} \cdot \text{cof}\,\overline{\boldsymbol{f}} \leftrightarrow \{\bullet\},$$

respectively, are related by

$$\begin{aligned} \text{Div}\left(\{\bullet\} \cdot \text{cof}\,\widetilde{\boldsymbol{F}}\right) &= \tilde{J}\big[\,\text{d}\tilde{\text{i}}\text{v}\left(\{\bullet\}\right) + \{\bullet\} \cdot [\,\widetilde{\boldsymbol{a}} : \widetilde{\boldsymbol{e}}\,]\big], \\ \text{div}\left(\{\bullet\} \cdot \text{cof}\,\overline{\boldsymbol{f}}\,\right) &= \bar{j}\big[\,\text{D}\bar{\text{i}}\text{v}\left(\{\bullet\}\right) + \{\bullet\} \cdot [\overline{\boldsymbol{A}} : \overline{\boldsymbol{E}}]\big]. \end{aligned}$$

The proof follows directly from the chain rule and the anholonomic Piola identities, whereby $\text{d}\tilde{\text{i}}\text{v}\left(\{\bullet\}\right)$ and $\text{D}\bar{\text{i}}\text{v}\left(\{\bullet\}\right)$ denote appropriately defined *anholonomic divergence operators* as the traces of the previously introduced anholonomic (plastic and elastic) gradients $\text{gr}\tilde{\text{a}}\text{d}\left(\{\bullet\}\right) := \text{Grad}\left(\{\bullet\}\right) \cdot \widetilde{\boldsymbol{f}}$ and $\text{Gr}\bar{\text{a}}\text{d}\left(\{\bullet\}\right) := \text{grad}\left(\{\bullet\}\right) \cdot \overline{\boldsymbol{F}}$, respectively.

Dislocation Densities Based on Anholonomic Curl. Based on the anholonomic elastic intermediate curl operator an additional (contravariant) Piola-type dislocation density may be introduced in the ambient Euclidean space as

$$\overline{\boldsymbol{D}} := \text{C}\bar{\text{u}}\text{rl}\,\overline{\boldsymbol{F}}. \tag{8.45}$$

Observe that $\overline{\boldsymbol{D}}$ is a two-point tensor deriving from an anholonomic elastic intermediate curl that maps from the intermediate to the spatial configuration. As such this situation is different from the previously introduced elastic Piola-type dislocation density $\overline{\boldsymbol{d}} = \text{curl}\,\overline{\boldsymbol{f}}$ that is a two-point tensor mapping from the spatial to the intermediate configuration and that derives from the ordinary spatial curl. However, due to the convection of curl operators these two Piola-type dislocation densities are related by

$$\overline{\boldsymbol{D}} = -\overline{\boldsymbol{F}} \cdot \overline{\boldsymbol{d}} \cdot \text{cof}\,\overline{\boldsymbol{F}}. \tag{8.46}$$

Therefore, the elastic Piola-Kirchhoff- and Cauchy-type dislocation densities $\overline{\boldsymbol{A}}$ and $\overline{\boldsymbol{a}}$ as introduced in Eq. 8.40 are related to $\overline{\boldsymbol{D}}$ by

$$\overline{\boldsymbol{A}} = -\overline{\boldsymbol{f}} \cdot \overline{\boldsymbol{D}} \quad \text{and} \quad \overline{\boldsymbol{a}} = \overline{\boldsymbol{D}} \cdot \text{cof}\,\overline{\boldsymbol{f}}. \tag{8.47}$$

Likewise, based on the anholonomic plastic intermediate curl operator a further (contravariant) Piola-type dislocation density may be introduced in the ambient Euclidean space as

$$\widetilde{\boldsymbol{d}} = \text{c}\tilde{\text{u}}\text{rl}\, \widetilde{\boldsymbol{f}}. \tag{8.48}$$

Note that $\widetilde{\boldsymbol{d}}$ is a two-point tensor deriving from an anholonomic plastic intermediate curl that maps from the intermediate to the material configuration. As such this situation is different from the previously introduced plastic Piola-type dislocation density $\widetilde{\boldsymbol{D}} = \text{Curl}\, \widetilde{\boldsymbol{F}}$ that is a two-point tensor mapping from the material to the intermediate configuration and that derives from the ordinary material curl. However, due to the convection of curl operators these two Piola-type dislocation densities are related by

$$\widetilde{\boldsymbol{d}} = -\widetilde{\boldsymbol{f}} \cdot \widetilde{\boldsymbol{D}} \cdot \text{cof}\, \widetilde{\boldsymbol{f}}. \tag{8.49}$$

Thus, the plastic Piola-Kirchhoff- and Cauchy-type dislocation densities $\widetilde{\boldsymbol{A}}$ and $\widetilde{\boldsymbol{a}}$ as introduced in Eq. 8.36 are related to $\widetilde{\boldsymbol{d}}$ by

$$\widetilde{\boldsymbol{a}} = -\widetilde{\boldsymbol{F}} \cdot \widetilde{\boldsymbol{d}} \quad \text{and} \quad \widetilde{\boldsymbol{A}} = \widetilde{\boldsymbol{d}} \cdot \text{cof}\, \widetilde{\boldsymbol{F}}. \tag{8.50}$$

Summarizing, as depicted in Fig. 8.4 the multiplicative decomposition of the deformation gradient into incompatible elastic and plastic contributions in Eq. 8.1 (or likewise Eq. 8.2) results in the possible introduction of twelve (total, elastic and plastic) dislocation density tensors of Piola-, Piola-Kirchhoff- and Cauchy-type, respectively. Thereby the six Piola-type dislocation density tensors in Fig. 8.4 a) are defined as the application of the appropriate curl operator to the corresponding contribution to the deformation gradient. Accordingly, the six Piola-Kirchhoff- and Cauchy-type dislocation density tensors in Fig. 8.4 b) and c) are either a result from appropriate pull-back/push-forward operations or a result from the convection of curl operators, respectively. From the additive decomposition of the material and spatial curl of the deformation gradient and its inverse the previously derived equality of the elastic and plastic dislocation densities in the intermediate configuration finally follows in the format

$$\underbrace{\overline{\boldsymbol{f}} \cdot \overline{\boldsymbol{D}}}_{-\overline{\boldsymbol{A}}} + \underbrace{\widetilde{\boldsymbol{D}} \cdot \text{cof}\, \widetilde{\boldsymbol{f}}}_{\widetilde{\boldsymbol{a}}} = \boldsymbol{0} \quad \text{and} \quad \underbrace{\widetilde{\boldsymbol{F}} \cdot \widetilde{\boldsymbol{d}}}_{-\widetilde{\boldsymbol{a}}} + \underbrace{\overline{\boldsymbol{d}} \cdot \text{cof}\, \overline{\boldsymbol{F}}}_{\overline{\boldsymbol{A}}} = \boldsymbol{0}. \tag{8.51}$$

Obviously this is nothing but the statement of compatibility of the total deformation gradient. Finally the continuity equations of continuum dislocation theory may be restated in terms of anholonomic divergence operators

$$\text{di}\tilde{\text{v}}\, \widetilde{\boldsymbol{a}} + \widetilde{\boldsymbol{a}} \cdot [\widetilde{\boldsymbol{a}} : \widetilde{\boldsymbol{e}}] = \boldsymbol{0} \quad \text{and} \quad \text{Di}\bar{\text{v}}\, \overline{\boldsymbol{A}} + \overline{\boldsymbol{A}} \cdot [\overline{\boldsymbol{A}} : \overline{\boldsymbol{E}}] = \boldsymbol{0}. \tag{8.52}$$

These statements will be useful in the subsequent section when introducing plastic and elastic incompatibility density tensors in the intermediate configuration.

8.1.4 Incompatibility Density Tensors

Material Plastic versus Spatial Elastic Incompatibility Density. The material plastic incompatibility density computes from the material plastic Riemann curvature tensor as

$$\begin{aligned}\tilde{\mathcal{H}}^{MN} &:= -\frac{1}{4}\mathcal{E}^{MIJ}\tilde{\mathcal{M}}_{IJKL}\mathcal{E}^{KLN} \\ &= \frac{1}{2}\mathcal{E}^{MIJ}\left[\tilde{\mathcal{M}}_{IJK,L} + \tilde{\mathcal{M}}_{AIK}\tilde{\mathcal{M}}^{A}{}_{JL}\right]\mathcal{E}^{KLN} \\ &= \frac{1}{2}\mathcal{E}^{MIJ}\left[\tilde{\mathcal{M}}_{IK,JL} + \tilde{\mathcal{M}}_{AIK}\tilde{\mathcal{M}}^{A}{}_{JL}\right]\mathcal{E}^{KLN}.\end{aligned} \tag{8.53}$$

Thereby the first term translates into the $\operatorname{Curl}^t(\operatorname{Curl}^t\tilde{\boldsymbol{C}})$ contribution and the second term translates to the $\{\operatorname{Grad}\tilde{\boldsymbol{C}}\} \ggg \{\operatorname{Grad}\tilde{\boldsymbol{C}}\}$ contribution to the $\operatorname{Inc}\tilde{\boldsymbol{C}}$ as expressed in ambient Euclidean space. Likewise the spatial elastic incompatibility density computes from the spatial elastic Riemann curvature tensor as

$$\begin{aligned}\bar{h}^{mn} &:= \frac{1}{4}e^{mij}\,\bar{m}_{ijkl}\,e^{kln} \\ &= -\frac{1}{2}e^{mij}\left[\bar{m}_{ijk,l} + \bar{m}_{aik}\bar{m}^{a}{}_{jl}\right]e^{kln} \\ &= -\frac{1}{2}e^{mij}\left[\bar{m}_{ik,jl} + \bar{m}_{aik}\bar{m}^{a}{}_{jl}\right]e^{kln}.\end{aligned} \tag{8.54}$$

Thereby the first term translates into the $\operatorname{curl}^t(\operatorname{curl}^t\bar{\boldsymbol{c}})$ contribution and the second term translates to the $\{\operatorname{grad}\bar{\boldsymbol{c}}\} \ggg \{\operatorname{grad}\bar{\boldsymbol{c}}\}$ contribution to the $\operatorname{inc}\bar{\boldsymbol{c}}$ as expressed in ambient Euclidean space. Then since the total deformation is compatible, the following relations between the material plastic and the spatial elastic *Riemann connections* hold

$$\tilde{\mathcal{M}}_{IJK}(\tilde{C}) = Y(\bar{m}_{ijk}(\bar{c})) \quad\leftrightarrow\quad \bar{m}_{ijk}(\bar{c}) = y(\tilde{\mathcal{M}}_{IJK}(\tilde{C})). \tag{8.55}$$

Consequently, the material plastic and the spatial elastic *Riemann curvature tensors* are related by

$$\tilde{\mathcal{M}}_{IJKL}(\tilde{C}) = Y(\bar{m}_{ijkl}(\bar{c})) \quad\leftrightarrow\quad \bar{m}_{ijkl}(\bar{c}) = y(\tilde{\mathcal{M}}_{IJKL}(\tilde{C})). \tag{8.56}$$

Here $\tilde{C} = Y(\bar{c})$ and $\bar{c} = y(\tilde{C})$ denote the pull-back of the spatial elastic metric and the push-forward of the material plastic metric, respectively. Thus the pull-back of the spatial elastic Riemann curvature tensor by the deformation

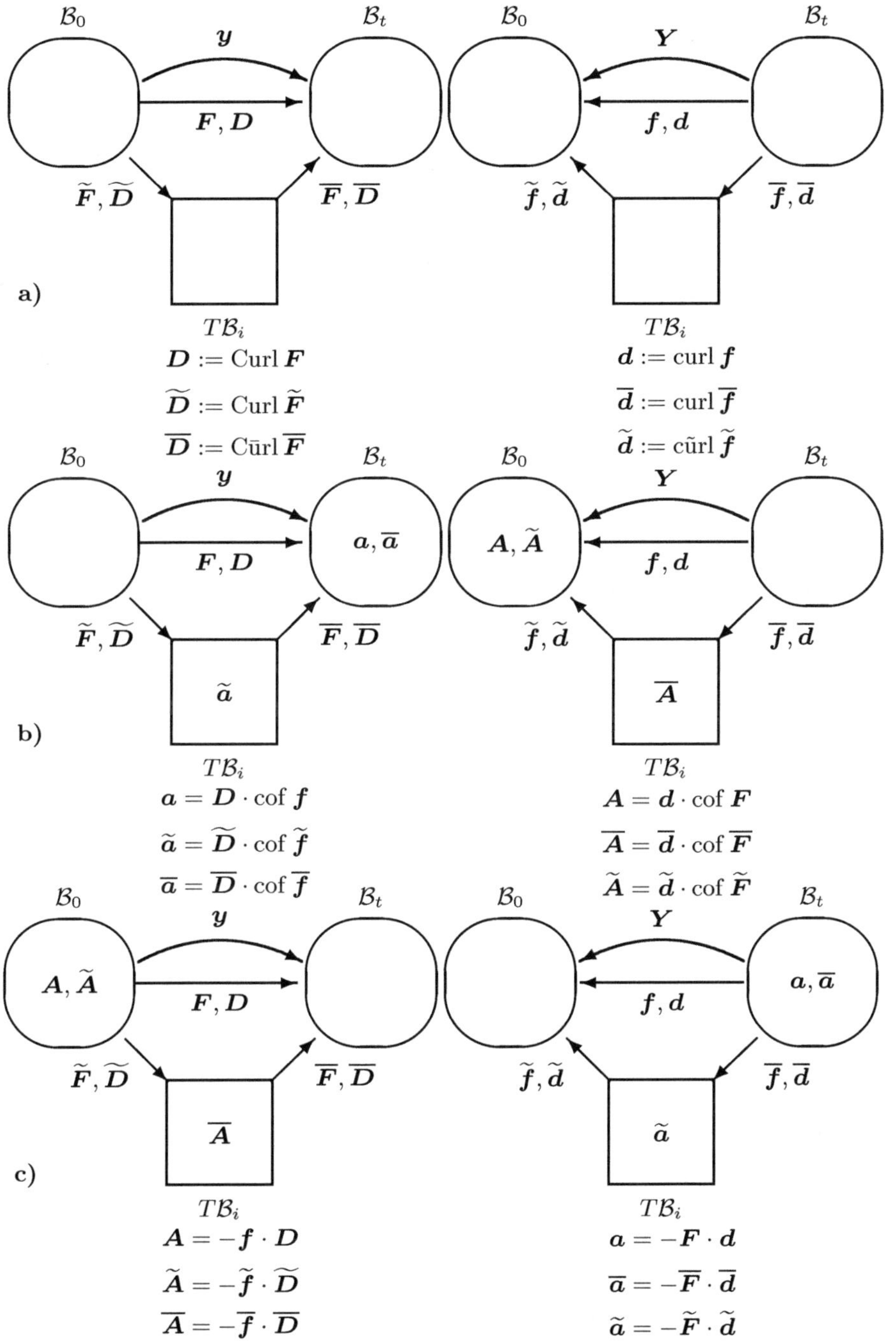

Fig. 8.4 a) Piola-type dislocation density tensors; **b)** Cauchy-type dislocation density tensors; **c)** Piola-Kirchhoff-type dislocation density tensors

gradient results in the material plastic Riemann curvature tensor, vice-versa the push-forward of the material plastic Riemann curvature tensor by the (inverse) deformation gradient results in the spatial elastic Riemann curvature tensor

$$\tilde{\mathcal{M}}_{IJKL} = F^i{}_I F^j{}_J \bar{m}_{ijkl} F^k{}_K F^l{}_L \quad \leftrightarrow \quad \bar{m}_{ijkl} = f^I{}_i f^J{}_j \tilde{\mathcal{M}}_{IJKL} f^K{}_k f^L{}_l. \tag{8.57}$$

Using these relations the contravariant push-forward of the second-order material plastic incompatibility tensor, for example, renders a multiple of the spatial elastic incompatibility tensor

$$\begin{aligned} -jF^m{}_M \tilde{\mathcal{H}}^{MN} F^n{}_N &= \frac{1}{4} j F^m{}_M \mathcal{E}^{MIJ} \tilde{\mathcal{M}}_{IJKL} \mathcal{E}^{KLN} F^n{}_N \\ &= \frac{1}{4} j \underbrace{F^m{}_M \mathcal{E}^{MIJ} F^i{}_I F^j{}_J}_{J e^{mij}} \bar{m}_{ijkl} \underbrace{F^k{}_K F^l{}_L \mathcal{E}^{KLN} F^n{}_N}_{J e^{kln}} \\ J\bar{h}^{mn} &= \frac{1}{4} J e^{mij} \bar{m}_{ijkl} e^{kln}. \end{aligned} \tag{8.58}$$

The corresponding relations for the spatial elastic and the material plastic incompatibility tensors in ambient Euclidean space

$$J\bar{\boldsymbol{h}}_t = -\boldsymbol{F} \cdot [j\widetilde{\boldsymbol{H}}_0] \cdot \boldsymbol{F}^t \quad \leftrightarrow \quad j\widetilde{\boldsymbol{H}}_0 = -\boldsymbol{f} \cdot [J\bar{\boldsymbol{h}}_t] \cdot \boldsymbol{f}^t \tag{8.59}$$

are remarkable and motivate the introduction of (Jacobian) scaled spatial elastic and material plastic incompatibility tensors, respectively

$$\bar{\boldsymbol{h}}_0 := J\bar{\boldsymbol{h}}_t \quad \text{and} \quad \widetilde{\boldsymbol{H}}_t := j\widetilde{\boldsymbol{H}}_0. \tag{8.60}$$

With these definitions the relation of the spatial elastic the material plastic incompatibility tensor in ambient Euclidean space reads as

$$\bar{\boldsymbol{h}}_0 = -\boldsymbol{F} \cdot \widetilde{\boldsymbol{H}}_t \cdot \boldsymbol{F}^t \quad \leftrightarrow \quad \widetilde{\boldsymbol{H}}_t = -\boldsymbol{f} \cdot \bar{\boldsymbol{h}}_0 \cdot \boldsymbol{f}^t. \tag{8.61}$$

These incompatibility tensors and their interrelations are depicted in Fig. 8.5. Recall that both of the material plastic and the spatial elastic incompatibility tensors measure the incompatibility of the intermediate configuration. Thereby, the difference in sign simply reflects the compatibility of the total deformation.

Intermediate Plastic Incompatibility Density. The intermediate plastic Riemann curvature tensor is defined as the (major symmetry preserving) plastic push-forward of the material plastic Riemann curvature tensor by the (inverse) plastic distortion

$$\tilde{m}_{\alpha\beta\gamma\delta} := \tilde{f}^I{}_\alpha \tilde{f}^J{}_\beta \tilde{\mathcal{M}}_{IJKL} \tilde{f}^K{}_\gamma \tilde{f}^L{}_\delta \tag{8.62}$$

Then the contravariant plastic push-forward of the second-order material plastic incompatibility tensor renders a multiple of the intermediate plastic incompatibility tensor

$$
\begin{aligned}
-\tilde{j}\tilde{F}^{\mu}{}_{M}\tilde{\mathcal{H}}^{MN}\tilde{F}^{\nu}{}_{N} &= \frac{1}{4}\tilde{j}\tilde{F}^{\mu}{}_{M}\mathcal{E}^{MIJ}\tilde{\mathcal{M}}_{IJKL}\mathcal{E}^{KLN}\tilde{F}^{\nu}{}_{N} \qquad (8.63)\\
&= \frac{1}{4}\tilde{j}\,\underbrace{\tilde{F}^{\mu}{}_{M}\mathcal{E}^{MIJ}\tilde{F}^{\alpha}{}_{I}\tilde{F}^{\beta}{}_{J}}_{\tilde{J}\tilde{e}^{\mu\alpha\beta}}\,\tilde{m}_{\alpha\beta\gamma\delta}\,\underbrace{\tilde{F}^{\gamma}{}_{K}\tilde{F}^{\delta}{}_{L}\mathcal{E}^{KLN}\tilde{F}^{\nu}{}_{N}}_{\tilde{J}\tilde{e}^{\gamma\delta\nu}}\\
\tilde{J}\tilde{h}^{\mu\nu} &:= \frac{1}{4}\tilde{J}\tilde{e}^{\mu\alpha\beta}\,\tilde{m}_{\alpha\beta\gamma\delta}\,\tilde{e}^{\gamma\delta\nu}.
\end{aligned}
$$

Consequently the corresponding relations for the (plastic Jacobian) scaled intermediate and material plastic incompatibility tensors in ambient Euclidean space

$$\widetilde{\boldsymbol{h}}_0 := \tilde{J}\widetilde{\boldsymbol{h}}_i \quad \text{and} \quad \widetilde{\boldsymbol{H}}_i := \tilde{j}\widetilde{\boldsymbol{H}}_0 \qquad (8.64)$$

take the remarkable format (whereby the change in sign shall be observed)

$$\widetilde{\boldsymbol{h}}_0 = -\widetilde{\boldsymbol{F}}\cdot\widetilde{\boldsymbol{H}}_i\cdot\widetilde{\boldsymbol{F}}^t \quad \leftrightarrow \quad \widetilde{\boldsymbol{H}}_i = -\widetilde{\boldsymbol{f}}\cdot\widetilde{\boldsymbol{h}}_0\cdot\widetilde{\boldsymbol{f}}^t. \qquad (8.65)$$

The plastic incompatibility tensors and their interrelations are depicted in Fig. 8.5. The difference in sign simply reflects the compatibility of the total deformation.

Due to the incompatibility of the intermediate configuration the intermediate plastic Riemann curvature tensor is expressed as

$$\tilde{m}_{\alpha\beta\gamma\delta} = -2\tilde{m}_{\alpha\beta[\gamma.\delta]} - 2\tilde{m}_{\sigma\alpha[\gamma}\tilde{\delta}^{\sigma\tau}\tilde{m}_{\tau\beta\delta]} - \tilde{m}_{\alpha\beta\sigma}\tilde{\delta}^{\sigma\tau}\tilde{a}_{\tau\gamma\delta}. \qquad (8.66)$$

Here the last term stems from the plastic push-forward of the material plastic Riemann curvature tensor to the incompatible (anholonomic) intermediate configuration and contributes the intermediate (Cauchy-type) plastic anholonomic object. However, since the intermediate plastic metric coefficients coincide with the Kronecker-symbol the corresponding intermediate plastic Riemann connection consists only of contributions due to the intermediate plastic anholonomic object

$$\tilde{m}_{\alpha\beta\gamma} \equiv \tilde{a}_{\{\alpha\beta\gamma\}} \quad \text{with} \quad 2\tilde{a}_{\{\alpha\beta\gamma\}} = \tilde{a}_{\alpha\gamma\beta} - \tilde{a}_{\gamma\beta\alpha} + \tilde{a}_{\beta\alpha\gamma}. \qquad (8.67)$$

Thus the intermediate plastic Riemann curvature tensor is likewise exclusively expressed in terms of the intermediate plastic anholonomic object

$$\tilde{m}_{\alpha\beta\gamma\delta} = -2\tilde{a}_{\{\alpha\beta[\gamma\}.\delta]} - 2\tilde{a}_{\{\sigma\alpha[\gamma\}}\tilde{\delta}^{\sigma\tau}\tilde{a}_{\{\tau\beta\delta]\}} - \tilde{a}_{\{\alpha\beta\sigma\}}\tilde{\delta}^{\sigma\tau}\tilde{a}_{\tau\gamma\delta}. \qquad (8.68)$$

This result clearly illuminates the interrelation between the incompatibility (of the material plastic strain metric) and the (plastic) dislocation density contained in the intermediate plastic anholonomic object. Before proceeding to compute the corresponding intermediate plastic incompatibility density tensor it proves convenient to introduce the

Lemma 8.1. *Due to the left skew symmetry of* $\tilde{a}_{\{\alpha\beta\gamma\}}$ *the following relations hold*

$$\tilde{e}^{\alpha\mu\nu}\tilde{a}_{\{\mu\nu\gamma\}} = \tilde{a}_{\mu}{}^{\mu}\tilde{\delta}_{\gamma}{}^{\alpha} - 2\tilde{a}_{\gamma}{}^{\alpha} =: 2\tilde{\kappa}^{\alpha}{}_{\gamma} \quad \textit{and} \quad \tilde{a}_{\{\alpha\beta\gamma\}} = \tilde{e}_{\alpha\beta\mu}\tilde{\kappa}^{\mu}{}_{\gamma}. \tag{8.69}$$

Since here the intermediate (dual) plastic anholonomic object $\tilde{a}_{\gamma}{}^{\mu}$ *and the intermediate (dual) plastic torsion* $\tilde{t}_{\gamma}{}^{\mu}$ *do coincide,* $\tilde{\kappa}^{\mu}{}_{\gamma}$ *is introduced as the intermediate plastic (dual) contortion. In symbolic notation the above relation between* $\tilde{\kappa}^{\alpha}{}_{\gamma}$ *and* $\tilde{a}_{\gamma}{}^{\alpha}$ *expands as*

$$[\widetilde{\boldsymbol{a}}^{t} : \widetilde{\boldsymbol{j}}]\widetilde{\boldsymbol{i}} - 2\widetilde{\boldsymbol{a}}^{t} = 2\widetilde{\boldsymbol{k}} \quad \rightsquigarrow \quad [\widetilde{\boldsymbol{k}} : \widetilde{\boldsymbol{j}}]\widetilde{\boldsymbol{i}} - \widetilde{\boldsymbol{k}} = \widetilde{\boldsymbol{a}}^{t} \tag{8.70}$$

with $2\widetilde{\boldsymbol{k}} : \widetilde{\boldsymbol{j}} = \widetilde{\boldsymbol{a}}^{t} : \widetilde{\boldsymbol{j}}$. *Here the contravariant versions of* $\widetilde{\boldsymbol{a}}$ *and* $\widetilde{\boldsymbol{k}}$ *have been invoked for the sake of demonstration.* □

Proof. Due to the right skew symmetry of the intermediate plastic anholonomic object it is easy to see that

$$\begin{aligned}\tilde{e}^{\alpha\mu\nu}\tilde{a}_{\{\mu\nu\gamma\}} &= \frac{1}{2}\tilde{e}^{\alpha\mu\nu}[-\tilde{a}_{\mu}{}^{\sigma}\tilde{e}_{\sigma\gamma\nu} + \tilde{a}_{\gamma}{}^{\sigma}\tilde{e}_{\sigma\nu\mu} - \tilde{a}_{\nu}{}^{\sigma}\tilde{e}_{\sigma\mu\gamma}] \\ &= -\tilde{a}_{\mu}{}^{\sigma}[\tilde{\delta}^{\alpha}{}_{\sigma}\tilde{\delta}^{\mu}{}_{\gamma} - \tilde{\delta}^{\alpha}{}_{\gamma}\tilde{\delta}^{\mu}{}_{\sigma}] - \tilde{a}_{\gamma}{}^{\alpha} = \tilde{a}_{\mu}{}^{\mu}\tilde{\delta}_{\gamma}{}^{\alpha} - 2\tilde{a}_{\gamma}{}^{\alpha}.\end{aligned}$$

To prove the inverse relation it is helpful to recall the following identity for the intermediate Kronecker and the intermediate permutation symbol

$$\tilde{\delta}_{\mu}{}^{\nu}\tilde{e}_{\gamma\alpha\beta} = \tilde{\delta}_{\gamma}{}^{\nu}\tilde{e}_{\mu\alpha\beta} + \tilde{\delta}_{\alpha}{}^{\nu}\tilde{e}_{\gamma\mu\beta} + \tilde{\delta}_{\beta}{}^{\nu}\tilde{e}_{\gamma\alpha\mu}.$$

Then multiplication by $\tilde{a}_{\nu}{}^{\mu}$ renders

$$\tilde{a}_{\mu}{}^{\mu}\tilde{e}_{\gamma\alpha\beta} = \tilde{a}_{\gamma}{}^{\mu}\tilde{e}_{\mu\alpha\beta} + \tilde{a}_{\alpha}{}^{\mu}\tilde{e}_{\gamma\mu\beta} + \tilde{a}_{\beta}{}^{\mu}\tilde{e}_{\gamma\alpha\mu} = -\tilde{a}_{\gamma\alpha\beta} - \tilde{a}_{\alpha\beta\gamma} - \tilde{a}_{\beta\gamma\alpha}.$$

Combining this relation with the representation of the intermediate plastic contortion renders finally

$$\tilde{e}_{\alpha\beta\mu}\tilde{\kappa}^{\mu}{}_{\gamma} = \frac{1}{2}\tilde{a}_{\mu}{}^{\mu}\tilde{e}_{\gamma\alpha\beta} + \tilde{a}_{\gamma\alpha\beta} = \frac{1}{2}[\tilde{a}_{\gamma\alpha\beta} - \tilde{a}_{\alpha\beta\gamma} - \tilde{a}_{\beta\gamma\alpha}] = \tilde{a}_{\{\alpha\beta\gamma\}}.$$

In the above derivations frequent use of the right skew symmetry of the intermediate plastic anholonomic object has been made. □

Moreover, in the sequel it will prove useful to also introduce the

Lemma 8.2. *Due to the left and right skew symmetry of* $\tilde{a}_{\{\alpha\beta[\gamma\}.\delta]}$ *the following relation holds*

$$\frac{1}{2}\tilde{e}^{\mu\alpha\beta}\tilde{a}_{\{\alpha\beta[\gamma\}.\delta]}\tilde{e}^{\gamma\delta\nu} = \tilde{\kappa}^{\mu}{}_{\gamma.\delta}\tilde{e}^{\gamma\delta\nu} = -\tilde{a}_{\gamma}{}^{(\mu}{}_{.\delta}\tilde{e}^{\gamma\delta\nu)} + \frac{1}{2}\tilde{e}^{\mu\nu\gamma}\tilde{a}_{\gamma}{}^{\delta}\tilde{e}_{\delta\alpha\beta}\tilde{a}^{\alpha\beta}. \tag{8.71}$$

With $\widetilde{\boldsymbol{a}} : \widetilde{\boldsymbol{e}} = -2\,\mathrm{axl}\,\widetilde{\boldsymbol{a}}$ *the above relation expands in symbolic notation as*

$$-\,\mathrm{curl}\,\widetilde{\boldsymbol{k}} = [\mathrm{curl}^t\,\widetilde{\boldsymbol{a}}^t]^{\mathrm{sym}} - \widetilde{\boldsymbol{e}} \cdot [\widetilde{\boldsymbol{a}} \cdot \mathrm{axl}\,\widetilde{\boldsymbol{a}}]. \tag{8.72}$$

Thus the negative of the intermediate curl applied to the intermediate plastic contortion coincides with the symmetric part of the intermediate curl applied to the transposed intermediate plastic anholonomic object plus a further correction term. □

Proof. Due to the expression for the intermediate plastic contortion in terms of the intermediate plastic anholonomic object it firstly holds that

$$\tilde{k}^{\mu}{}_{\gamma}\tilde{e}^{\gamma\delta\nu} = \frac{1}{2}[\tilde{a}_{\beta}{}^{\beta}\tilde{\delta}^{\mu}{}_{\gamma} - 2\tilde{a}_{\gamma}{}^{\mu}]\tilde{e}^{\gamma\delta\nu} = \frac{1}{2}\tilde{a}_{\beta}{}^{\beta}\tilde{e}^{\mu\delta\nu} - \tilde{a}_{\gamma}{}^{\mu}\tilde{e}^{\gamma\delta\nu}.$$

Recall next the following identity for the intermediate Kronecker and the intermediate permutation symbol

$$\tilde{\delta}_{\beta}{}^{\gamma}\tilde{e}^{\mu\delta\nu} = \tilde{\delta}_{\beta}{}^{\mu}\tilde{e}^{\gamma\delta\nu} + \tilde{\delta}_{\beta}{}^{\delta}\tilde{e}^{\mu\gamma\nu} + \tilde{\delta}_{\beta}{}^{\nu}\tilde{e}^{\mu\delta\gamma}.$$

Then multiplication by $\tilde{a}_{\gamma}{}^{\beta}$ renders

$$\tilde{a}_{\beta}{}^{\beta}\tilde{e}^{\mu\delta\nu} = \tilde{a}_{\gamma}{}^{\mu}\tilde{e}^{\gamma\delta\nu} + \tilde{a}_{\gamma}{}^{\delta}\tilde{e}^{\mu\gamma\nu} + \tilde{a}_{\gamma}{}^{\nu}\tilde{e}^{\mu\delta\gamma}.$$

Furthermore with the continuity condition $\tilde{a}_{\gamma}{}^{\delta}{}_{.\delta} = -\tilde{a}_{\gamma}{}^{\delta}\tilde{e}_{\delta\alpha\beta}\tilde{a}^{\alpha\beta}$ it follows that

$$\tilde{a}_{\beta}{}^{\beta}{}_{.\delta}\tilde{e}^{\mu\delta\nu} = \tilde{a}_{\gamma}{}^{\mu}{}_{.\delta}\tilde{e}^{\gamma\delta\nu} + \tilde{e}^{\mu\nu\gamma}\tilde{a}_{\gamma}{}^{\delta}\tilde{e}_{\delta\alpha\beta}\tilde{a}^{\alpha\beta} - \tilde{a}_{\gamma}{}^{\nu}{}_{.\delta}\tilde{e}^{\gamma\delta\mu}.$$

Combining this relation with the above representation involving the intermediate plastic contortion renders

$$\begin{aligned}\tilde{k}^{\mu}{}_{\gamma.\delta}\tilde{e}^{\gamma\delta\nu} &= \frac{1}{2}[\tilde{a}_{\gamma}{}^{\mu}{}_{.\delta}\tilde{e}^{\gamma\delta\nu} + \tilde{e}^{\mu\nu\gamma}\tilde{a}_{\gamma}{}^{\delta}\tilde{e}_{\delta\alpha\beta}\tilde{a}^{\alpha\beta} - \tilde{a}_{\gamma}{}^{\nu}{}_{.\delta}\tilde{e}^{\gamma\delta\mu}] - \tilde{a}_{\gamma}{}^{\mu}{}_{.\delta}\tilde{e}^{\gamma\delta\nu} \\ &= -\frac{1}{2}[\tilde{a}_{\gamma}{}^{\mu}{}_{.\delta}\tilde{e}^{\gamma\delta\nu} + \tilde{a}_{\gamma}{}^{\nu}{}_{.\delta}\tilde{e}^{\gamma\delta\mu}] + \frac{1}{2}\tilde{e}^{\mu\nu\gamma}\tilde{a}_{\gamma}{}^{\delta}\tilde{e}_{\delta\alpha\beta}\tilde{a}^{\alpha\beta}.\end{aligned}$$

Clearly the first term denotes the intermediate curl of the transposed intermediate plastic anholonomic object and its transpose. □

Furthermore, in the following it will finally prove helpful to introduce the

Lemma 8.3. *Due to the left skew symmetry of* $\tilde{a}_{\{\alpha\beta\gamma\}}$ *the following relation holds*

$$\tilde{e}^{\mu\alpha\beta}\tilde{a}_{\{\sigma\alpha\gamma\}}\tilde{\delta}^{\sigma\tau}\tilde{a}_{\{\tau\beta\delta\}}\tilde{e}^{\gamma\delta\nu} = \tilde{e}^{\mu\alpha\beta}\tilde{k}_{\alpha\gamma}\tilde{k}_{\beta\delta}\tilde{e}^{\gamma\delta\nu}. \tag{8.73}$$

Clearly this is the coefficient expression for the tensor cross product of the intermediate plastic contortion with itself. □

Proof. Since $\tilde{a}_{\{\sigma\alpha\gamma\}} = \tilde{e}_{\sigma\alpha o}\tilde{k}^{o}{}_{\gamma}$ and $\tilde{a}_{\{\tau\beta\delta\}} = \tilde{e}_{\tau\beta\pi}\tilde{k}^{\pi}{}_{\delta}$ it holds that

$$
\begin{aligned}
\tilde{e}^{\mu\alpha\beta}\tilde{a}_{\{\sigma\alpha\gamma\}}\tilde{\delta}^{\sigma\tau}\tilde{a}_{\{\tau\beta\delta\}}\tilde{e}^{\gamma\delta\nu} &= \tilde{e}^{\mu\alpha\beta}\tilde{e}_{\sigma\alpha o}\tilde{k}^{o}{}_{\gamma}\tilde{\delta}^{\sigma\tau}\tilde{e}_{\tau\beta\pi}\tilde{k}^{\pi}{}_{\delta}\tilde{e}^{\gamma\delta\nu} \\
&= [\tilde{\delta}^{\beta}{}_{o}\tilde{\delta}^{\mu}{}_{\sigma} - \tilde{\delta}^{\beta}{}_{\sigma}\tilde{\delta}^{\mu}{}_{o}]\tilde{k}^{o}{}_{\gamma}\tilde{\delta}^{\sigma\tau}\tilde{e}_{\tau\beta\pi}\tilde{k}^{\pi}{}_{\delta}\tilde{e}^{\gamma\delta\nu} \\
&= \tilde{k}^{\beta}{}_{\gamma}\tilde{\delta}^{\mu\tau}\tilde{e}_{\tau\beta\pi}\tilde{k}^{\pi}{}_{\delta}\tilde{e}^{\gamma\delta\nu} - \tilde{k}^{\mu}{}_{\gamma}\tilde{\delta}^{\beta\tau}\tilde{e}_{\tau\beta\pi}\tilde{k}^{\pi}{}_{\delta}\tilde{e}^{\gamma\delta\nu} \\
&= \tilde{e}^{\mu\beta\pi}\tilde{k}_{\beta\gamma}\tilde{k}_{\pi\delta}\tilde{e}^{\gamma\delta\nu}.
\end{aligned}
$$

A relabeling of the indices $\beta\pi \to \alpha\beta$ finally concludes the proof. □

With these lemmata at hand the intermediate plastic incompatibility density tensor follows as

$$
\begin{aligned}
\tilde{h}^{\mu\nu} &:= \frac{1}{4}\tilde{e}^{\mu\alpha\beta}\tilde{m}_{\alpha\beta\gamma\delta}\tilde{e}^{\gamma\delta\nu} \\
&= -\frac{1}{4}\tilde{e}^{\mu\alpha\beta}\left[2\tilde{a}_{\{\alpha\beta\gamma\}.\delta} + 2\tilde{a}_{\{\sigma\alpha\gamma\}}\tilde{\delta}^{\sigma\tau}\tilde{a}_{\{\tau\beta\delta\}} + \tilde{a}_{\{\alpha\beta\sigma\}}\tilde{\delta}^{\sigma\tau}\tilde{a}_{\tau\gamma\delta}\right]\tilde{e}^{\gamma\delta\nu} \\
&= -\tilde{k}^{\mu}{}_{\gamma.\delta}\tilde{e}^{\gamma\delta\nu} - \frac{1}{2}\tilde{e}^{\mu\alpha\beta}\tilde{k}_{\alpha\gamma}\tilde{k}_{\beta\delta}\tilde{e}^{\gamma\delta\nu} + \tilde{k}^{\mu}{}_{\sigma}\tilde{a}^{\sigma\nu} \\
&= \tilde{a}_{\gamma}{}^{(\mu}{}_{.\delta}\tilde{e}^{\gamma\delta\nu)} - \frac{1}{2}\tilde{e}^{\mu\alpha\beta}\tilde{k}_{\alpha\gamma}\tilde{k}_{\beta\delta}\tilde{e}^{\gamma\delta\nu} + \tilde{k}^{\mu}{}_{\sigma}\tilde{a}^{\sigma\nu} - \frac{1}{2}\tilde{e}^{\mu\nu\gamma}\tilde{a}_{\gamma}{}^{\delta}\tilde{e}_{\delta\alpha\beta}\tilde{a}^{\alpha\beta}.
\end{aligned}
\tag{8.74}
$$

Thus in ambient Euclidean space the corresponding contravariant second-order intermediate plastic incompatibility density tensor reads as

$$
-\widetilde{\boldsymbol{h}}_i = -\,\widetilde{\text{curl}}\,\widetilde{\boldsymbol{k}} + \frac{1}{2}\widetilde{\boldsymbol{k}} \times\!\!\times \widetilde{\boldsymbol{k}} - \widetilde{\boldsymbol{k}} \cdot \widetilde{\boldsymbol{a}}. \tag{8.75}
$$

Obviously it is the presence of the geometrically necessary dislocation density contained in the intermediate plastic anholonomic object or likewise in the intermediate plastic contortion that is the source for the corresponding incompatibility measured by a non-zero intermediate plastic Riemann curvature tensor.

Remark 8.3. The skew symmetric contributions to the various terms constituting $\widetilde{\boldsymbol{h}}_i$ have to cancel each other; indeed the skew symmetric contribution to $\widetilde{\boldsymbol{h}}_i$ vanishes since

$$
\begin{aligned}
-\left[\widetilde{\text{curl}}\,\widetilde{\boldsymbol{k}}\right] : \widetilde{\boldsymbol{e}} &= -2\widetilde{\boldsymbol{a}} \cdot \text{axl}\,\widetilde{\boldsymbol{a}}, \\
\frac{1}{2}\left[\widetilde{\boldsymbol{k}} \times\!\!\times \widetilde{\boldsymbol{k}}\right] : \widetilde{\boldsymbol{e}} &= -2\widetilde{\boldsymbol{k}} \cdot \text{axl}\,\widetilde{\boldsymbol{k}}, \\
-\left[\widetilde{\boldsymbol{k}} \cdot \widetilde{\boldsymbol{a}}\right] : \widetilde{\boldsymbol{e}} &= 2[\widetilde{\boldsymbol{k}} : \widetilde{\boldsymbol{j}}]\,\text{axl}\,\widetilde{\boldsymbol{k}}.
\end{aligned}
$$

With $[\widetilde{\boldsymbol{k}} : \widetilde{\boldsymbol{j}}]\widetilde{\boldsymbol{i}} - \widetilde{\boldsymbol{k}} = \widetilde{\boldsymbol{a}}^t$ the summation of the terms on the right-hand-sides in the above renders

$$
-2\left[\widetilde{\boldsymbol{a}} - [\widetilde{\boldsymbol{k}} : \widetilde{\boldsymbol{j}}]\widetilde{\boldsymbol{i}} + \widetilde{\boldsymbol{k}}\right] \cdot \text{axl}\,\widetilde{\boldsymbol{a}} = -4\widetilde{\boldsymbol{a}}^{\text{skw}} \cdot \text{axl}\,\widetilde{\boldsymbol{a}} = \boldsymbol{0}.
$$

Here $\text{axl}\,\widetilde{\boldsymbol{a}} \equiv \text{axl}\,\widetilde{\boldsymbol{k}}$ and the null space property $\widetilde{\boldsymbol{a}}^{\text{skw}} \cdot \text{axl}\,\widetilde{\boldsymbol{a}} = \boldsymbol{0}$ have been exploited. □

Expressed exclusively in terms of the intermediate plastic anholonomic object the second-order intermediate plastic incompatibility density tensor finally takes the expression

$$-\widetilde{\boldsymbol{h}}_i = [\widetilde{\operatorname{curl}}^t \,\widetilde{\boldsymbol{a}}^t]^{\rm sym} + \frac{1}{2}[\widetilde{\boldsymbol{a}}^t \times\!\!\times \widetilde{\boldsymbol{a}}^t]^{\rm sym} - \frac{1}{2}[\widetilde{\boldsymbol{a}}^t : \widetilde{\boldsymbol{j}}]^2 \widetilde{\boldsymbol{i}} + \widetilde{\boldsymbol{a}}^t \cdot \widetilde{\boldsymbol{a}}. \tag{8.76}$$

Here $\frac{1}{2}[\widetilde{\boldsymbol{a}}^t \times\!\!\times \widetilde{\boldsymbol{a}}^t]^{\rm skw} = \widetilde{\boldsymbol{e}} \cdot [\widetilde{\boldsymbol{a}}^t \cdot \operatorname{axl} \widetilde{\boldsymbol{a}}]$ has been exploited so that the symmetry of $\widetilde{\boldsymbol{h}}_i$ is clearly visible.

Intermediate Elastic Incompatibility Density. The intermediate elastic Riemann curvature tensor is defined as the (major symmetry preserving) elastic pull-back of the spatial elastic Riemann curvature tensor by the elastic distortion

$$\bar{\mathcal{M}}_{\alpha\beta\gamma\delta} := \bar{F}^i{}_\alpha \bar{F}^j{}_\beta \bar{m}_{ijkl} \bar{F}^k{}_\gamma \bar{F}^l{}_\delta \tag{8.77}$$

Then the contravariant elastic pull-back of the second-order spatial elastic incompatibility tensor renders a multiple of the intermediate elastic incompatibility tensor

$$\begin{aligned} \bar{J}\bar{f}^\mu{}_m \bar{h}^{mn} \bar{f}^\nu{}_n &= \frac{1}{4}\bar{J}\bar{f}^\mu{}_m e^{mij}\,\bar{m}_{ijkl}\, e^{kln}\,\bar{f}^\nu{}_n \\ &= \frac{1}{4}\underbrace{\bar{J}\,\bar{f}^\mu{}_m e^{mij}\,\bar{f}^\alpha{}_i \bar{f}^\beta{}_j}_{\bar{j}\bar{\mathcal{E}}^{\mu\alpha\beta}}\,\bar{\mathcal{M}}_{\alpha\beta\gamma\delta}\,\underbrace{\bar{f}^\gamma{}_k \bar{f}^\delta{}_l e^{kln}\,\bar{f}^\nu{}_n}_{\bar{j}\bar{\mathcal{E}}^{\gamma\delta\nu}} \\ -\bar{j}\bar{\mathcal{H}}^{\mu\nu} &:= \frac{1}{4}\bar{j}\bar{\mathcal{E}}^{\mu\alpha\beta}\bar{\mathcal{M}}_{\alpha\beta\gamma\delta}\bar{\mathcal{E}}^{\gamma\delta\nu}. \end{aligned} \tag{8.78}$$

Consequently the corresponding relations for the (elastic Jacobian) scaled intermediate and spatial elastic incompatibility tensors in ambient Euclidean space

$$\overline{\boldsymbol{H}}_t := \bar{j}\overline{\boldsymbol{H}}_i \quad \text{and} \quad \overline{\boldsymbol{h}}_i := \bar{J}\overline{\boldsymbol{h}}_t \tag{8.79}$$

take the remarkable format (whereby the change of sign shall be observed)

$$-\overline{\boldsymbol{H}}_t = \overline{\boldsymbol{f}} \cdot \overline{\boldsymbol{h}}_i \cdot \overline{\boldsymbol{f}}^t \quad \leftrightarrow \quad -\overline{\boldsymbol{h}}_i = \overline{\boldsymbol{F}} \cdot \overline{\boldsymbol{H}}_t \cdot \overline{\boldsymbol{F}}^t. \tag{8.80}$$

The elastic incompatibility tensors and their interrelations are depicted in Fig. 8.5. The difference in sign simply reflects the compatibility of the total deformation.

Due to the incompatibility of the intermediate configuration the intermediate elastic Riemann curvature tensor is expressed as

$$\bar{\mathcal{M}}_{\alpha\beta\gamma\delta} = -2\bar{\mathcal{M}}_{\alpha\beta[\gamma.\delta]} - 2\bar{\mathcal{M}}_{\sigma\alpha[\gamma}\bar{\delta}^{\sigma\tau}\bar{\mathcal{M}}_{\tau\beta\delta]} - \bar{\mathcal{M}}_{\alpha\beta\sigma}\bar{\delta}^{\sigma\tau}\bar{\mathcal{A}}_{\tau\gamma\delta}. \tag{8.81}$$

Here the last term stems from the elastic pull-back of the spatial elastic Riemann curvature tensor to the incompatible (anholonomic) intermediate configuration and contributes the intermediate (Piola-Kirchhoff-type) elastic anholonomic object. However, since the intermediate elastic metric coefficients coincide with the Kronecker-symbol the corresponding elastic Riemann connection consists only of contributions due to the intermediate elastic anholonomic object

$$\bar{\mathcal{M}}_{\alpha\beta\gamma} = \bar{\mathcal{A}}_{\{\alpha\beta\gamma\}} \quad \text{with} \quad 2\bar{\mathcal{A}}_{\{\alpha\beta\gamma\}} = \bar{\mathcal{A}}_{\alpha\gamma\beta} - \bar{\mathcal{A}}_{\gamma\beta\alpha} + \bar{\mathcal{A}}_{\beta\alpha\gamma}. \tag{8.82}$$

Thus the intermediate elastic Riemann curvature tensor is likewise exclusively expressed in terms of the intermediate elastic anholonomic object

$$\bar{\mathcal{M}}_{\alpha\beta\gamma\delta} = -2\bar{\mathcal{A}}_{\{\alpha\beta[\gamma\}.\delta]} - 2\bar{\mathcal{A}}_{\{\sigma\alpha[\gamma\}}\bar{\delta}^{\sigma\tau}\bar{\mathcal{A}}_{\{\tau\beta\delta]\}} - \bar{\mathcal{A}}_{\{\alpha\beta\sigma\}}\bar{\delta}^{\sigma\tau}\bar{\mathcal{A}}_{\tau\gamma\delta}. \tag{8.83}$$

This result illuminates again the interrelation between the incompatibility (of the spatial elastic strain metric) and the (elastic) dislocation density contained in the intermediate elastic anholonomic object. Before proceeding to compute the corresponding intermediate elastic incompatibility density tensor it proves convenient to introduce the

Lemma 8.4. *Due to the left skew symmetry of* $\bar{\mathcal{A}}_{\{\alpha\beta\gamma\}}$ *the following relations hold*

$$\bar{\mathcal{E}}^{\alpha\mu\nu}\bar{\mathcal{A}}_{\{\mu\nu\gamma\}} = \bar{\mathcal{A}}_{\mu}{}^{\mu}\bar{\delta}_{\gamma}{}^{\alpha} - 2\bar{\mathcal{A}}_{\gamma}{}^{\alpha} =: 2\bar{\mathcal{K}}^{\alpha}{}_{\gamma} \quad \text{and} \quad \bar{\mathcal{A}}_{\{\alpha\beta\gamma\}} = \bar{\mathcal{E}}_{\alpha\beta\mu}\bar{\mathcal{K}}^{\mu}{}_{\gamma}. \tag{8.84}$$

Since here the intermediate (dual) elastic anholonomic object $\bar{\mathcal{A}}_{\gamma}{}^{\mu}$ *and the intermediate (dual) elastic torsion* $\bar{\mathcal{T}}_{\gamma}{}^{\mu}$ *do coincide,* $\bar{\mathcal{K}}^{\mu}{}_{\gamma}$ *is introduced as the intermediate elastic (dual) contortion. In symbolic notation the above relation between* $\bar{\mathcal{K}}^{\alpha}{}_{\gamma}$ *and* $\bar{\mathcal{A}}_{\gamma}{}^{\alpha}$ *expands as*

$$[\overline{\boldsymbol{A}}^t : \overline{\boldsymbol{J}}]\overline{\boldsymbol{I}} - 2\overline{\boldsymbol{A}}^t = 2\overline{\boldsymbol{K}} \quad \rightsquigarrow \quad [\overline{\boldsymbol{K}} : \overline{\boldsymbol{J}}]\overline{\boldsymbol{I}} - \overline{\boldsymbol{K}} = \overline{\boldsymbol{A}}^t \tag{8.85}$$

with $2\overline{\boldsymbol{K}} : \overline{\boldsymbol{J}} = \overline{\boldsymbol{A}}^t : \overline{\boldsymbol{J}}$. *Here the contravariant versions of* $\overline{\boldsymbol{A}}$ *and* $\overline{\boldsymbol{K}}$ *have been invoked for the sake of demonstration.* □

Moreover, in the sequel it will prove useful to also introduce the

Lemma 8.5. *Due to the left and right skew symmetry of* $\bar{\mathcal{A}}_{\{\alpha\beta[\gamma\}.\delta]}$ *the following relation holds*

$$\begin{aligned} \frac{1}{2}\bar{\mathcal{E}}^{\mu\alpha\beta}\bar{\mathcal{A}}_{\{\alpha\beta[\gamma\}.\delta]}\bar{\mathcal{E}}^{\gamma\delta\nu} &= \bar{\mathcal{K}}^{\mu}{}_{\gamma.\delta}\bar{\mathcal{E}}^{\gamma\delta\nu} \\ &= -\bar{\mathcal{A}}_{\gamma}{}^{(\mu}{}_{.\delta}\bar{\mathcal{E}}^{\gamma\delta\nu)} + \frac{1}{2}\bar{\mathcal{E}}^{\mu\nu\gamma}\bar{\mathcal{A}}_{\gamma}{}^{\delta}\bar{\mathcal{E}}_{\delta\alpha\beta}\bar{\mathcal{A}}^{\alpha\beta}. \end{aligned}$$

With $\overline{\boldsymbol{A}} : \overline{\boldsymbol{E}} = -2\,\text{axl}\,\overline{\boldsymbol{A}}$ *the above relation expands in symbolic notation as*

$$-\,\text{Curl}\,\overline{\boldsymbol{K}} = [\text{Curl}^t\,\overline{\boldsymbol{A}}^t]^{\text{sym}} - \overline{\boldsymbol{E}}\cdot[\overline{\boldsymbol{A}}\cdot\text{axl}\,\overline{\boldsymbol{A}}]. \tag{8.86}$$

Thus the negative of the intermediate curl applied to the intermediate elastic contortion coincides with the symmetric part of the intermediate curl applied to the transposed intermediate elastic anholonomic object plus a further correction term. □

Furthermore, in the following it will finally prove helpful to introduce the

Lemma 8.6. *Due to the left skew symmetry of* $\bar{\mathcal{A}}_{\{\alpha\beta\gamma\}}$ *the following relation holds*

$$\bar{\mathcal{E}}^{\mu\alpha\beta}\bar{\mathcal{A}}_{\{\sigma\alpha\gamma\}}\bar{\delta}^{\sigma\tau}\bar{\mathcal{A}}_{\{\tau\beta\delta\}}\bar{\mathcal{E}}^{\gamma\delta\nu} = \bar{\mathcal{E}}^{\mu\alpha\beta}\bar{\mathcal{K}}_{\alpha\gamma}\bar{\mathcal{K}}_{\beta\delta}\bar{\mathcal{E}}^{\gamma\delta\nu}. \tag{8.87}$$

Clearly this is the coefficient expression for the tensor cross product of the intermediate elastic contortion with itself. □

The proofs of the above lemmata are analogous to the ones as developed for the representation of the intermediate plastic incompatibility tensor and are thus not repeated. Then with the above lemmata at hand the intermediate elastic incompatibility density tensor follows as

$$\begin{aligned} -\bar{\mathcal{H}}^{\mu\nu} &:= \frac{1}{4}\bar{\mathcal{E}}^{\mu\alpha\beta}\bar{\mathcal{M}}_{\alpha\beta\gamma\delta}\bar{\mathcal{E}}^{\gamma\delta\nu} \qquad (8.88)\\ &= -\frac{1}{4}\bar{\mathcal{E}}^{\mu\alpha\beta}\left[2\bar{\mathcal{A}}_{\{\alpha\beta\gamma\}.\delta} + 2\bar{\mathcal{A}}_{\{\sigma\alpha\gamma\}}\bar{\delta}^{\sigma\tau}\bar{\mathcal{A}}_{\{\tau\beta\delta\}} + \bar{\mathcal{A}}_{\{\alpha\beta\sigma\}}\bar{\delta}^{\sigma\tau}\bar{\mathcal{A}}_{\tau\gamma\delta}\right]\bar{\mathcal{E}}^{\gamma\delta\nu}\\ &= -\bar{\mathcal{K}}^{\mu}{}_{\gamma.\delta}\bar{\mathcal{E}}^{\gamma\delta\nu} - \frac{1}{2}\bar{\mathcal{E}}^{\mu\alpha\beta}\bar{\mathcal{K}}_{\alpha\gamma}\bar{\mathcal{K}}_{\beta\delta}\bar{\mathcal{E}}^{\gamma\delta\nu} + \bar{\mathcal{K}}^{\mu}{}_{\sigma}\bar{\mathcal{A}}^{\sigma\nu}\\ &= \bar{\mathcal{A}}_{\gamma}{}^{(\mu}{}_{.\delta}\bar{\mathcal{E}}^{\gamma\delta\nu)} - \frac{1}{2}\bar{\mathcal{E}}^{\mu\alpha\beta}\bar{\mathcal{K}}_{\alpha\gamma}\bar{\mathcal{K}}_{\beta\delta}\bar{\mathcal{E}}^{\gamma\delta\nu} + \bar{\mathcal{K}}^{\mu}{}_{\sigma}\bar{\mathcal{A}}^{\sigma\nu} - \frac{1}{2}\bar{\mathcal{E}}^{\mu\nu\gamma}\bar{\mathcal{A}}_{\gamma}{}^{\delta}\bar{\mathcal{E}}_{\delta\alpha\beta}\bar{\mathcal{A}}^{\alpha\beta}. \end{aligned}$$

Thus in ambient Euclidean space the corresponding contravariant second-order intermediate elastic incompatibility density tensor reads as

$$\overline{\boldsymbol{H}}_i = -\,\bar{\text{Curl}}\,\overline{\boldsymbol{K}} + \frac{1}{2}\overline{\boldsymbol{K}} \times\!\!\times \overline{\boldsymbol{K}} - \overline{\boldsymbol{K}}\cdot\overline{\boldsymbol{A}}. \tag{8.89}$$

Obviously it is the presence of the geometrically necessary dislocation density contained in the intermediate elastic anholonomic object or likewise in the intermediate elastic contortion that is the source for the corresponding incompatibility measured by a non-zero intermediate elastic Riemann curvature tensor. Expressed exclusively in terms of the intermediate elastic anholonomic object the second-order intermediate elastic incompatibility density tensor finally takes the expression

$$\overline{\boldsymbol{H}}_i = [\bar{\text{Curl}}^t\,\overline{\boldsymbol{A}}^t]^{\text{sym}} + \frac{1}{2}[\overline{\boldsymbol{A}}^t \times\!\!\times \overline{\boldsymbol{A}}^t]^{\text{sym}} - \frac{1}{2}[\overline{\boldsymbol{A}}^t : \overline{\boldsymbol{J}}]^2\overline{\boldsymbol{I}} + \overline{\boldsymbol{A}}^t\cdot\overline{\boldsymbol{A}}. \tag{8.90}$$

Here $\frac{1}{2}[\overline{\boldsymbol{A}}^t \times\!\!\times \overline{\boldsymbol{A}}^t]^{\text{skw}} = \overline{\boldsymbol{E}}\cdot[\overline{\boldsymbol{A}}^t\cdot\text{axl}\,\overline{\boldsymbol{A}}]$ has been exploited so that the symmetry of $\overline{\boldsymbol{H}}_i$ is clearly visible.

Intermediate Plastic versus Elastic Incompatibility Density. It has been demonstrated before that the plastic and elastic intermediate anholonomic objects coincide, likewise in ambient Euclidean space the corresponding second-order tensors representing the intermediate (dual) plastic and elastic anholonomic objects and therefore also the intermediate (dual) plastic and elastic contortions coincide

$$\tilde{a}_{\alpha\beta\gamma} = \bar{\mathcal{A}}_{\alpha\beta\gamma} \quad \rightarrow \quad \widetilde{\boldsymbol{a}} = \overline{\boldsymbol{A}} \quad \text{and} \quad \widetilde{\boldsymbol{k}} = \overline{\boldsymbol{K}}. \tag{8.91}$$

As a consequence also the intermediate plastic and elastic Riemann curvature tensors coincide

$$\tilde{m}_{\alpha\beta\gamma\delta} = \bar{\mathcal{M}}_{\alpha\beta\gamma\delta}. \tag{8.92}$$

However the associated intermediate plastic and elastic second-order incompatibility density tensors differ by their sign

$$\tilde{h}^{\mu\nu} + \bar{\mathcal{H}}^{\mu\nu} = 0 \quad \rightarrow \quad \widetilde{\boldsymbol{h}}_i + \overline{\boldsymbol{H}}_i = \boldsymbol{0}. \tag{8.93}$$

Clearly this identity simply reflects the compatibility of the total deformation. The plastic and elastic incompatibility tensors and their interrelations are depicted in Fig. 8.5.

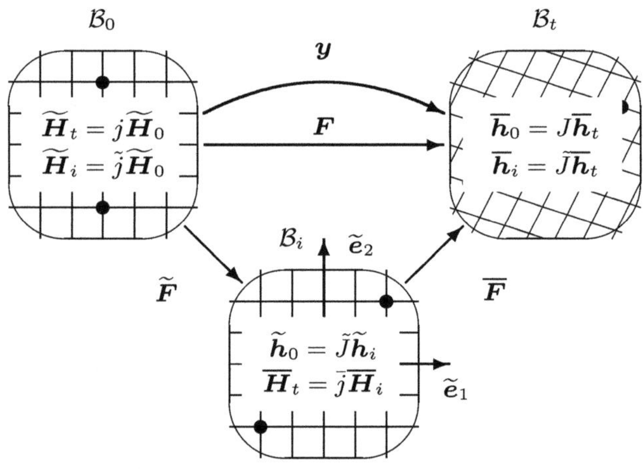

Fig. 8.5 Plastic and elastic incompatibility density tensors in the material, the spatial and the intermediate configuration. The incompatibility tensors are scaled by the Jacobians of the total, the plastic and the elastic distortions, respectively. Thereby the subindex $\tau = 0, i, t$ denotes the configuration from where or into which the corresponding scaled incompatibility tensor is mapped by contravariant push-forward or pull-back, respectively.

8.2 Second-Order Continuum

In addition to the well-accepted *multiplicative* decomposition of the (first) deformation gradient into elastic and plastic distortions, for second-order

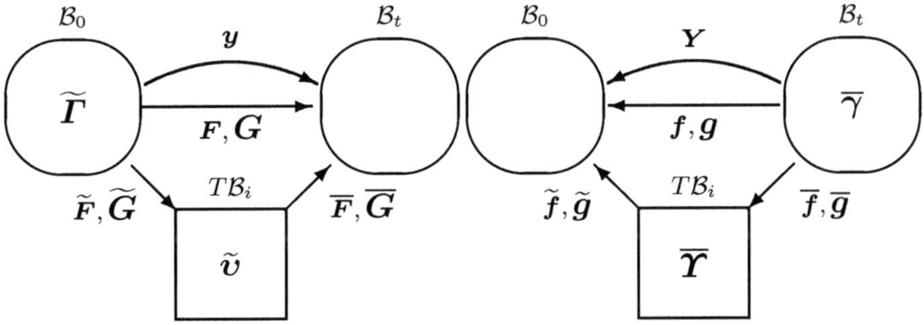

Fig. 8.6 Elastic and plastic double-distortions and double-strain measures

elasto-plasticity the second deformation gradient is decomposed *additively* into elastic and plastic double-distortions

$$\boldsymbol{F} = \overline{\boldsymbol{F}} \cdot \widetilde{\boldsymbol{F}} \quad \text{and} \quad \boldsymbol{G} = \boldsymbol{G}_{\mathrm{e}} + \boldsymbol{G}_{\mathrm{p}} \quad \text{with} \quad \boldsymbol{G}_{\mathrm{p}} =: \overline{\boldsymbol{F}} \cdot \widetilde{\boldsymbol{G}}. \tag{8.94}$$

Since the plastic double-distortion $\boldsymbol{G}_{\mathrm{p}}$ is a two-point tensor mapping from the material to the spatial configuration, the plastic double-distortion $\widetilde{\boldsymbol{G}}$ is introduced in addition as a two-point tensor mapping from the material to the intermediate configuration. Observe that here $\widetilde{\boldsymbol{G}}$ is in general assumed *incompatible* with Grad $\widetilde{\boldsymbol{F}}$. The multiplicative decomposition of the (first) deformation gradient and the additive decomposition of the second deformation gradient also allow alternative representation in terms of the inverse (first) deformation gradient and its spatial gradient

$$\boldsymbol{f} = \widetilde{\boldsymbol{f}} \cdot \overline{\boldsymbol{f}} \quad \text{and} \quad \boldsymbol{g} = \boldsymbol{g}_{\mathrm{p}} + \boldsymbol{g}_{\mathrm{e}} \quad \text{with} \quad \boldsymbol{g}_{\mathrm{e}} =: \widetilde{\boldsymbol{f}} \cdot \overline{\boldsymbol{g}}. \tag{8.95}$$

Since the elastic double-distortion $\boldsymbol{g}_{\mathrm{e}}$ is a two-point tensor mapping from the spatial to the material configuration, the elastic double-distortion $\overline{\boldsymbol{g}}$ is introduced in addition as a two-point tensor mapping form the spatial to the intermediate configuration. Observe that here $\overline{\boldsymbol{g}}$ is in general assumed incompatible with grad $\overline{\boldsymbol{f}}$.

Remark 8.4. In addition to the usual material measure of plastic strain $\widetilde{\boldsymbol{C}} = \widetilde{\boldsymbol{F}}^t \cdot \widetilde{\boldsymbol{F}}$ a material measure of plastic *double-strain* is introduced as

$$\boldsymbol{F}^t \cdot \boldsymbol{G}_{\mathrm{p}} =: \boldsymbol{C} \cdot \boldsymbol{\Gamma}_{\mathrm{p}} \quad \text{with} \quad \boldsymbol{\Gamma}_{\mathrm{p}} := \boldsymbol{f} \cdot \boldsymbol{G}_{\mathrm{p}} \equiv \widetilde{\boldsymbol{f}} \cdot \widetilde{\boldsymbol{G}} =: \widetilde{\boldsymbol{\Gamma}}.$$

Likewise, a spatial measure of elastic double-strain is introduced in addition to the usual spatial measure of elastic strain $\overline{\boldsymbol{c}} = \overline{\boldsymbol{f}}^t \cdot \overline{\boldsymbol{f}}$ as

$$\boldsymbol{f}^t \cdot \boldsymbol{g}_{\mathrm{e}} =: \boldsymbol{c} \cdot \boldsymbol{\gamma}_{\mathrm{e}} \quad \text{with} \quad \boldsymbol{\gamma}_{\mathrm{e}} := \boldsymbol{F} \cdot \boldsymbol{g}_{\mathrm{e}} \equiv \overline{\boldsymbol{F}} \cdot \overline{\boldsymbol{g}} =: \overline{\boldsymbol{\gamma}}.$$

Here $\boldsymbol{\Gamma}_{\mathrm{p}} \equiv \widetilde{\boldsymbol{\Gamma}}$ and $\boldsymbol{\gamma}_{\mathrm{e}} \equiv \overline{\boldsymbol{\gamma}}$ denote the contra/co- (mixed) variant material plastic and spatial elastic connection tensors in ambient Euclidean space. □

Remark 8.5. For the special case of a priori *constrained* second-order elasto-plasticity the elastic and the plastic distortions are compatible with the corresponding elastic and plastic double-distortions

$$\widetilde{\boldsymbol{G}} = \operatorname{Grad} \widetilde{\boldsymbol{F}} \quad \text{and} \quad \overline{\boldsymbol{g}} = \operatorname{grad} \overline{\boldsymbol{f}}.$$

For this particular case $\widetilde{\boldsymbol{G}}$ and $\overline{\boldsymbol{g}}$ are integrable, however, $\widetilde{\boldsymbol{F}}$ and $\overline{\boldsymbol{f}}$ are in general non-integrable. □

Remark 8.6. When introducing defect density tensors in the sequel, it will be useful to define also the following plastic and elastic double-distortions

$$\widetilde{\boldsymbol{g}} := -\widetilde{\boldsymbol{\Gamma}} : [\widetilde{\boldsymbol{f}} \boxtimes \widetilde{\boldsymbol{f}}] \quad \text{and} \quad \overline{\boldsymbol{G}} := -\overline{\boldsymbol{\gamma}} : [\overline{\boldsymbol{F}} \boxtimes \overline{\boldsymbol{F}}].$$

These definitions are motivated by the relation between the second deformation gradients $\boldsymbol{G}$ and $\boldsymbol{g}$. Moreover, intermediate measures of plastic and elastic double-strain are defined as

$$\widetilde{\boldsymbol{\upsilon}} := -\widetilde{\boldsymbol{G}} : [\widetilde{\boldsymbol{f}} \boxtimes \widetilde{\boldsymbol{f}}] \quad \text{and} \quad \overline{\boldsymbol{\Upsilon}} := -\overline{\boldsymbol{g}} : [\overline{\boldsymbol{F}} \boxtimes \overline{\boldsymbol{F}}].$$

Note the further useful definitions of intermediate (total) double-strains

$$\boldsymbol{\upsilon} := \overline{\boldsymbol{\upsilon}} + \widetilde{\boldsymbol{\upsilon}} \quad \text{and} \quad \boldsymbol{\Upsilon} := \overline{\boldsymbol{\Upsilon}} + \widetilde{\boldsymbol{\Upsilon}} \quad \text{with} \quad \overline{\boldsymbol{\upsilon}} := -\overline{\boldsymbol{\Upsilon}} \quad \text{and} \quad \widetilde{\boldsymbol{\Upsilon}} := -\widetilde{\boldsymbol{\upsilon}}.$$

□

The additional tensor quantities such as double-distortions and double-strain measures introduced in the above are depicted in Fig. 8.6.

8.2.1 Non-Integrable Distortion and Double-Distortion

For second-order elasto-plasticity non-integrability of either the plastic or the elastic distortion *and* double-distortion renders measures for the incompatibility of the intermediate configuration. These measures are associated with the corresponding i) anholonomic object, ii) curvature, iii.a) torsion, and iii.b) non-metricity and result in i) primary dislocation density tensors, ii) disclination density tensors, iii.a) secondary dislocation density tensors, and iii.b) point-defect density tensors.

Looking Forward: Incompatibility of Intermediate Configuration. The incompatibility of the intermediate configuration within second-order elasto-plasticity results in i) an intermediate closure gap $[\![\widetilde{\boldsymbol{y}}]\!] \neq \mathbf{0}$ upon integration of the intermediate line element $\mathrm{d}\widetilde{\boldsymbol{y}} := \widetilde{\boldsymbol{F}} \cdot \mathrm{d}\boldsymbol{X}$ along arbitrary closed material circuits $\mathcal{C}_0$, ii) a plastic distortion closure gap $[\![\widetilde{\boldsymbol{F}}]\!] \neq \boldsymbol{0}$ upon integration of the (forward) plastic differential distortion $\mathrm{d}\widetilde{\boldsymbol{F}} := \widetilde{\boldsymbol{G}} \cdot \mathrm{d}\boldsymbol{X}$ along arbitrary closed material circuits $\mathcal{C}_0$, and iii) an incompatibility between the plastic distortion $\widetilde{\boldsymbol{F}}$ and the plastic double-distortion $\widetilde{\boldsymbol{G}}$. Then the incompatibility of the intermediate configuration is captured by the non-integrability of i) the plastic distortion and ii) the plastic double-distortion

$$^{\mathrm{I}}\widetilde{\boldsymbol{D}} := \mathrm{Curl}\,\widetilde{\boldsymbol{F}} \neq \boldsymbol{0} \quad \text{and} \quad \widetilde{\boldsymbol{K}} := \mathrm{Curl}\,\widetilde{\boldsymbol{G}} \neq \boldsymbol{0}, \tag{8.96}$$

together with iii.a) the right non-symmetry of the plastic double-distortion and iii.b) the non-vanishing material plastic non-metricity

$$^{\mathrm{II}}\widetilde{\boldsymbol{D}} := -\widetilde{\boldsymbol{G}} : \boldsymbol{E} \neq \boldsymbol{0} \quad \text{and} \quad \widetilde{\boldsymbol{Q}} := 2\mathsf{I}^{\mathrm{sym}} : \left[\widetilde{\boldsymbol{F}}^t \cdot [\widetilde{\boldsymbol{G}} - \mathrm{Grad}\,\widetilde{\boldsymbol{F}}]\right] \neq \boldsymbol{0}. \tag{8.97}$$

Recall that i) the Piola-type plastic primary dislocation density $^{\mathrm{I}}\widetilde{\boldsymbol{D}}$ is the second-order tensor expression for the third-order anholonomic object related to the non-integrable map $\widetilde{\boldsymbol{F}} : T\mathcal{B}_0 \to T\mathcal{B}_i$ from the material to the intermediate configuration, ii) the Piola-type plastic disclination density $\widetilde{\boldsymbol{K}}$ is the third-order tensor expression related to the fourth-order curvature, iii.a) the Piola-type plastic secondary dislocation density $^{\mathrm{II}}\widetilde{\boldsymbol{D}}$ is the second-order tensor expression for the third-order torsion, and iii.b) the Piola-Kirchhoff-type plastic point-defect density $\widetilde{\boldsymbol{Q}}$ is the third-order tensor expression for the third-order non-metricity.

Looking Backward: Incompatibility of Intermediate Configuration. The incompatibility of the intermediate configuration within second-order elasto-plasticity results in i) an intermediate closure gap $[\![\overline{\boldsymbol{Y}}]\!] \neq \boldsymbol{0}$ upon integration of the intermediate line element $\mathrm{d}\overline{\boldsymbol{Y}} := \overline{\boldsymbol{f}} \cdot \mathrm{d}\boldsymbol{x}$ along arbitrary closed spatial circuits $\mathcal{C}_t$, ii) an elastic distortion closure gap $[\![\overline{\boldsymbol{f}}]\!] \neq \boldsymbol{0}$ upon integration of the (backward) elastic differential distortion $\mathrm{d}\overline{\boldsymbol{f}} := \overline{\boldsymbol{g}} \cdot \mathrm{d}\boldsymbol{x}$ along arbitrary closed spatial circuits $\mathcal{C}_t$, and iii) an incompatibility between the elastic distortion $\overline{\boldsymbol{f}}$ and the elastic double-distortion $\overline{\boldsymbol{g}}$. Then the incompatibility of the intermediate configuration is captured by the non-integrability of i) the elastic distortion and ii) the elastic double-distortion

$$^{\mathrm{I}}\overline{\boldsymbol{d}} := \mathrm{curl}\,\overline{\boldsymbol{f}} \neq \boldsymbol{0} \quad \text{and} \quad \overline{\boldsymbol{k}} := \mathrm{curl}\,\overline{\boldsymbol{g}} \neq \boldsymbol{0}, \tag{8.98}$$

together with iii.a) the right non-symmetry of the elastic double-distortion and iii.b) the non-vanishing spatial elastic non-metricity

$$^{\mathrm{II}}\overline{\boldsymbol{d}} := -\overline{\boldsymbol{g}} : \boldsymbol{e} \neq \boldsymbol{0} \quad \text{and} \quad \overline{\boldsymbol{q}} := 2\mathsf{i}^{\mathrm{sym}} : \left[\overline{\boldsymbol{f}}^t \cdot [\overline{\boldsymbol{g}} - \mathrm{grad}\,\overline{\boldsymbol{f}}]\right] \neq \boldsymbol{0}. \tag{8.99}$$

Recall that i) the Piola-type elastic primary dislocation density $^{\mathrm{I}}\overline{\boldsymbol{d}}$ is the second-order tensor expression for the third-order anholonomic object related to the non-integrable map $\overline{\boldsymbol{f}} : T\mathcal{B}_t \to T\mathcal{B}_i$ from the spatial to the intermediate configuration, ii) the Piola-type plastic disclination density $\overline{\boldsymbol{k}}$ is the third-order tensor expression related to the fourth-order curvature, iii.a) the Piola-type elastic secondary dislocation density ${}^{\text{II}}\overline{\boldsymbol{d}}$ is the second-order tensor expression for the third-order torsion, and iii.b) the Piola-Kirchhoff-type elastic point-defect density $\overline{\boldsymbol{q}}$ is the third-order tensor expression for the third-order non-metricity.

Remark 8.7. It is interesting to note that the plastic primary and secondary dislocation densities and the plastic disclination density jointly satisfy differential identities

$$\operatorname{Div}{}^{\text{I}}\widetilde{\boldsymbol{D}} = \boldsymbol{0} \quad \text{and} \quad \operatorname{Div}{}^{\text{II}}\widetilde{\boldsymbol{D}} = \widetilde{\boldsymbol{K}} : \boldsymbol{I} \quad \text{and} \quad \operatorname{Div}\widetilde{\boldsymbol{K}} = \boldsymbol{0}.$$

Likewise the elastic primary and secondary dislocation densities and the elastic disclination density jointly satisfy differential identities

$$\operatorname{div}{}^{\text{I}}\overline{\boldsymbol{d}} = \boldsymbol{0} \quad \text{and} \quad \operatorname{div}{}^{\text{II}}\overline{\boldsymbol{d}} = \overline{\boldsymbol{k}} : \boldsymbol{i} \quad \text{and} \quad \operatorname{div}\overline{\boldsymbol{k}} = \boldsymbol{0}.$$

Observe that the secondary dislocation densities are coupled to the disclination densities. □

Remark 8.8. For *constrained* second-order elasto-plasticity the plastic and elastic disclination densities together with the plastic and elastic point-defect densities vanish due to the compatibility of the plastic and elastic distortions with the corresponding double-distortions

$$\widetilde{\boldsymbol{K}} = \boldsymbol{0} \quad \text{and} \quad \widetilde{\boldsymbol{Q}} = \boldsymbol{0} \quad \text{and} \quad \overline{\boldsymbol{k}} = \boldsymbol{0} \quad \text{and} \quad \overline{\boldsymbol{q}} = \boldsymbol{0}.$$

Moreover the plastic or elastic primary and secondary dislocation densities, respectively, do coincide, i.e. ${}^{\text{I}}\widetilde{\boldsymbol{D}} \equiv {}^{\text{II}}\widetilde{\boldsymbol{D}}$ and ${}^{\text{I}}\overline{\boldsymbol{d}} \equiv {}^{\text{II}}\overline{\boldsymbol{d}}$. Thus for this case the discussion of non-integrability reduces to the arguments set forth for first-order elasto-plasticity. □

8.2.2 Non-Integrable Metric and Connection

For a number of reasons, see the supplement on second-order elasto-plasticity below, it is often not primarily the plastic or elastic distortion and the plastic or elastic double-distortion but rather the plastic or elastic(strain) metric, being quadratic in the corresponding distortion, and the plastic or elastic (double-strain) double-metric, viz. the plastic or elastic connection that enter the constitutive relations of second-order elasto-plasticity. Thus the failure of the plastic or elastic (strain) metric and the plastic or elastic (double-strain) double-metric, viz. the plastic or elastic connection being integrable into a compatible deformation map or rather the incompatibility of the intermediate configuration needs to be considered.

Looking Forward: Incompatibility of Intermediate Configuration. The measure i) of incompatibility of the plastic material metric (i.e. the plastic Cauchy-Green tensor) $\tilde{\mathcal{M}}_{IJ} = \tilde{C}_{IJ}$ and its relation to the primary plastic dislocation density has already been discussed in Eq. 8.16. It thus remains to state the measure ii) of non-integrability of the plastic material connection (i.e. the plastic double-strain) $\tilde{\mathcal{L}}^I{}_{JK} = \tilde{\Gamma}^I{}_{JK}$ and its relation to the plastic disclination density, iii) of right non-symmetry of the plastic material connection (i.e. the plastic double-strain) in terms of the secondary plastic dislocation density, and iv) of incompatibility of the plastic metric with the plastic material connection in terms of the plastic point-defect density.

Non-Integrability Measure i): The incompatibility of the plastic material metric $\tilde{\mathcal{M}}_{IJ} = \tilde{C}_{IJ}$ is measured in terms of the plastic Riemann curvature $\tilde{\mathcal{M}}_{IJKL}$ or rather its corresponding double dual expression

$$-{}^{\mathrm{I}}\tilde{\mathcal{H}}^{MN} := \frac{1}{4}\mathcal{E}^{MIJ}\tilde{\mathcal{M}}_{IJKL}\mathcal{E}^{KLN}.$$

In ambient Euclidean space the corresponding *primary plastic incompatibility density*, a fully contravariant second-order tensor, thus expands as

$$2\,{}^{\mathrm{I}}\widetilde{\boldsymbol{H}}_0 = \operatorname{Inc}\widetilde{\boldsymbol{C}} = \operatorname{Curl}^t(\operatorname{Curl}^t\widetilde{\boldsymbol{C}}) + \{\operatorname{Grad}\widetilde{\boldsymbol{C}}\} \mathbin{\times\!\!\!\times} \{\operatorname{Grad}\widetilde{\boldsymbol{C}}\}. \tag{8.100}$$

The subindex 0 refers to the primary plastic incompatibility density per unit volume in the material configuration, recall that the factor two in $\operatorname{Inc}\widetilde{\boldsymbol{C}} = 2\operatorname{Inc}\widetilde{\boldsymbol{E}}$ is due to the relation $\widetilde{\boldsymbol{C}} - \widetilde{\boldsymbol{J}} = 2\widetilde{\boldsymbol{E}}$.

Non-Integrability Measure ii): The non-integrability of the plastic material connection (i.e. the plastic double-strain) $\tilde{\mathcal{L}}^I{}_{JK} = \tilde{\Gamma}^I{}_{JK}$ is measured by the plastic curvature $\tilde{\mathcal{R}}^I{}_{JKL}$ or rather its corresponding dual expression

$$-{}^{\mathrm{II}}\tilde{\mathcal{H}}^I{}_J{}^N := -\frac{1}{2}\tilde{\mathcal{R}}^I{}_{JKL}\mathcal{E}^{KLN}.$$

In ambient Euclidean space the corresponding *secondary plastic incompatibility density*, a contra/co/contra- (mixed) variant third-order tensor, thus expands as

$${}^{\mathrm{II}}\widetilde{\boldsymbol{H}} = \widetilde{\boldsymbol{\Theta}} + \widetilde{\boldsymbol{\Gamma}} \rtimes \widetilde{\boldsymbol{\Gamma}}. \tag{8.101}$$

Thereby the *linear part* of the secondary plastic incompatibility density is introduced as $\widetilde{\boldsymbol{\Theta}} := \operatorname{Curl}\widetilde{\boldsymbol{\Gamma}}$. The right vector cross product of third-order tensors that results from the quadratic term $\tilde{\mathcal{L}}^I{}_{MK}\tilde{\mathcal{L}}^M{}_{JL}\mathcal{E}^{KLN}$ is abbreviated by $\rtimes$. Invoking finally the relation $\widetilde{\boldsymbol{\Theta}} = \widetilde{\boldsymbol{f}}\cdot[\widetilde{\boldsymbol{K}} - \operatorname{Grad}\widetilde{\boldsymbol{F}} \rtimes \widetilde{\boldsymbol{\Gamma}}]$ in the definition of the secondary plastic incompatibility density

$${}^{\mathrm{II}}\widetilde{\boldsymbol{H}} = \widetilde{\boldsymbol{f}} \cdot [\widetilde{\boldsymbol{K}} + [\widetilde{\boldsymbol{G}} - \operatorname{Grad}\widetilde{\boldsymbol{F}}] \rtimes \widetilde{\boldsymbol{\Gamma}}] \tag{8.102}$$

highlights its relation to the plastic disclination density $\widetilde{\boldsymbol{K}} = \operatorname{Curl} \widetilde{\boldsymbol{G}}$. Note that the term $\widetilde{\boldsymbol{G}} - \operatorname{Grad} \widetilde{\boldsymbol{F}}$ contains moreover contributions from the primary and secondary plastic dislocation densities and the plastic point defect density.

Non-Integrability Measure iii): As before the right non-symmetry of the plastic material connection (i.e. the plastic double-strain) or rather the plastic material torsion measures the (Piola-Kirchhoff-type) *secondary dislocation density*

$$^{\mathrm{II}}\widetilde{\boldsymbol{A}} := \widetilde{\boldsymbol{\Gamma}} : \boldsymbol{E} = \widetilde{\boldsymbol{f}} \cdot \widetilde{\boldsymbol{G}} : \boldsymbol{E} = -\widetilde{\boldsymbol{f}} \cdot {}^{\mathrm{II}}\widetilde{\boldsymbol{D}}. \tag{8.103}$$

It is then straightforward to show that $\operatorname{Div} {}^{\mathrm{II}}\widetilde{\boldsymbol{D}} = -\widetilde{\boldsymbol{F}} \cdot \operatorname{Div} {}^{\mathrm{II}}\widetilde{\boldsymbol{A}} - \operatorname{Grad} \widetilde{\boldsymbol{F}} : {}^{\mathrm{II}}\widetilde{\boldsymbol{A}}$ whereby the differential identity $\operatorname{Div} {}^{\mathrm{II}}\widetilde{\boldsymbol{D}} = \widetilde{\boldsymbol{K}} : \boldsymbol{I}$ holds. Finally the useful relation $\widetilde{\boldsymbol{K}} : \boldsymbol{I} = \widetilde{\boldsymbol{F}} \cdot \widetilde{\boldsymbol{\Theta}} : \boldsymbol{I} - \operatorname{Grad} \widetilde{\boldsymbol{F}} : {}^{\mathrm{II}}\widetilde{\boldsymbol{A}}$ may be obtained from the expressions in the above.

Non-Integrability Measure iv): Again as in the above the incompatibility of the plastic metric with the plastic material connection or rather the plastic non-metricity measures the *plastic point defect density*

$$\widetilde{\boldsymbol{Q}} := 2\mathsf{I}^{\mathrm{sym}} : [\widetilde{\boldsymbol{C}} \cdot \widetilde{\boldsymbol{\Gamma}}] - \operatorname{Grad} \widetilde{\boldsymbol{C}}. \tag{8.104}$$

Looking Backward: Incompatibility of Intermediate Configuration. The measure i) of incompatibility of the elastic spatial metric (i.e. the elastic Finger tensor) $\bar{m}_{ij} = \bar{c}_{ij}$ and its relation to the primary elastic dislocation density has already been discussed in Eq. 8.31. It thus remains to state the measure ii) of non-integrability of the elastic spatial connection (i.e. the elastic double-strain) $\bar{\ell}^{i}{}_{jk} = \bar{\gamma}^{i}{}_{jk}$ and its relation to the elastic disclination density, iii) of right non-symmetry of the elastic spatial connection (i.e. the elastic double-strain) in terms of the secondary elastic dislocation density, and iv) of incompatibility of the elastic metric with the elastic spatial connection in terms of the elastic point-defect density.

Non-Integrability Measure i): The incompatibility of the elastic spatial metric $\bar{m}_{ij} = \bar{c}_{ij}$ is measured in terms of the elastic Riemann curvature $\bar{m}_{ijkl}$ or rather its corresponding double dual expression

$$^{\mathrm{I}}\bar{h}^{mn} := \frac{1}{4} e^{mij}\, \bar{m}_{ijkl}\, e^{kln}.$$

In ambient Euclidean space the corresponding *primary elastic incompatibility density*, a fully contravariant second-order tensor, thus expands as

$$-2\, {}^{\mathrm{I}}\widetilde{\boldsymbol{h}}_t = \operatorname{inc} \bar{\boldsymbol{c}} = \operatorname{curl}^t(\operatorname{curl}^t \bar{\boldsymbol{c}}) + \{\operatorname{grad} \bar{\boldsymbol{c}}\} \times\!\!\times \{\operatorname{grad} \bar{\boldsymbol{c}}\}. \tag{8.105}$$

The subindex t refers to the primary elastic incompatibility density per unit volume in the spatial configuration, recall that the factor two and the minus sign in $\operatorname{inc}\overline{\boldsymbol{c}} = -2\operatorname{inc}\overline{\boldsymbol{e}}$ are due to the relation $\overline{\boldsymbol{c}} - \overline{\boldsymbol{j}} = -2\overline{\boldsymbol{e}}$.

Non-Integrability Measure ii): The non-integrability of the elastic spatial connection (i.e. the elastic double-strain) $\bar{l}^i{}_{jk} = \bar{\gamma}^i{}_{jk}$ is measured by the elastic curvature $\bar{r}^i{}_{jkl}$ or rather its corresponding dual expression

$$^{\mathrm{II}}\bar{h}^i{}_j{}^n := -\frac{1}{2}\bar{r}^i{}_{jkl}\,e^{kln}.$$

In ambient Euclidean space the corresponding *secondary elastic incompatibility density*, a contra/co/contra- (mixed) variant third-order tensor, thus expands as

$$-\,^{\mathrm{II}}\overline{\boldsymbol{h}} = \overline{\boldsymbol{\theta}} + \overline{\boldsymbol{\gamma}} \rtimes \overline{\boldsymbol{\gamma}}. \tag{8.106}$$

Thereby the *linear part* of the secondary elastic incompatibility density is introduced as $\overline{\boldsymbol{\theta}} := \operatorname{curl}\overline{\boldsymbol{\gamma}}$. The right vector cross product of third-order tensors that results from the quadratic term $\bar{l}^i{}_{mk}\bar{l}^m{}_{jl}e^{kln}$ is abbreviated by $\rtimes$. Invoking finally the relation $\overline{\boldsymbol{\theta}} = \overline{\boldsymbol{F}} \cdot [\overline{\boldsymbol{k}} - \operatorname{grad}\overline{\boldsymbol{f}} \rtimes \overline{\boldsymbol{\gamma}}]$ in the definition of the secondary elastic incompatibility density

$$-\,^{\mathrm{II}}\overline{\boldsymbol{h}} = \overline{\boldsymbol{F}} \cdot \left[\overline{\boldsymbol{k}} + [\overline{\boldsymbol{g}} - \operatorname{grad}\overline{\boldsymbol{f}}] \rtimes \overline{\boldsymbol{\gamma}}\right] \tag{8.107}$$

highlights its relation to the elastic disclination density $\overline{\boldsymbol{k}} = \operatorname{curl}\overline{\boldsymbol{g}}$. Note that the term $\overline{\boldsymbol{g}} - \operatorname{grad}\overline{\boldsymbol{f}}$ contains moreover contributions from the primary and secondary elastic dislocation densities and the elastic point defect density.

Non-Integrability Measure iii): As before the right non-symmetry of the elastic spatial connection (i.e. the elastic double-strain) or rather the elastic spatial torsion measures the (Cauchy-type) *secondary dislocation density*

$$^{\mathrm{II}}\overline{\boldsymbol{a}} := \overline{\boldsymbol{\gamma}} : \boldsymbol{e} = \overline{\boldsymbol{F}} \cdot \overline{\boldsymbol{g}} : \boldsymbol{e} = -\overline{\boldsymbol{F}} \cdot {}^{\mathrm{II}}\overline{\boldsymbol{d}}. \tag{8.108}$$

It is then straightforward to show that $\operatorname{div}{}^{\mathrm{II}}\overline{\boldsymbol{d}} = -\overline{\boldsymbol{f}} \cdot \operatorname{div}{}^{\mathrm{II}}\overline{\boldsymbol{a}} - \operatorname{grad}\overline{\boldsymbol{f}} : {}^{\mathrm{II}}\overline{\boldsymbol{a}}$ whereby the differential identity $\operatorname{div}{}^{\mathrm{II}}\overline{\boldsymbol{d}} = \overline{\boldsymbol{k}} : \boldsymbol{i}$ holds. Finally the useful relation $\overline{\boldsymbol{k}} : \boldsymbol{i} = \overline{\boldsymbol{f}} \cdot \overline{\boldsymbol{\theta}} : \boldsymbol{i} - \operatorname{grad}\overline{\boldsymbol{f}} : {}^{\mathrm{II}}\overline{\boldsymbol{a}}$ may be obtained from the expressions in the above.

Non-Integrability Measure iv): Again as in the above the incompatibility of the elastic metric with the elastic spatial connection or rather the elastic non-metricity measures the *elastic point defect density*

$$\overline{\boldsymbol{q}} := 2\mathsf{i}^{\mathrm{sym}} : [\overline{\boldsymbol{c}} \cdot \overline{\boldsymbol{\gamma}}] - \operatorname{grad}\overline{\boldsymbol{c}}. \tag{8.109}$$

Remark 8.9. It is interesting to note that the plastic primary and (linear part of the) secondary incompatibility densities and the plastic secondary dislocation density jointly satisfy differential identities

$$\operatorname{Div}\,{}^{\mathrm{I}}\widetilde{\boldsymbol{H}}_0 = \boldsymbol{0} \quad \text{and} \quad -\operatorname{Div}\,{}^{\mathrm{II}}\widetilde{\boldsymbol{A}} = \widetilde{\boldsymbol{\Theta}} : \boldsymbol{I} \quad \text{and} \quad \operatorname{Div}\widetilde{\boldsymbol{\Theta}} = \boldsymbol{0}.$$

Likewise the elastic primary and (linear part of the) secondary incompatibility densities and the elastic secondary dislocation density jointly satisfy differential identities

$$\operatorname{div}{}^{\mathrm{I}}\overline{\boldsymbol{h}}_t = \boldsymbol{0} \quad \text{and} \quad -\operatorname{div}{}^{\mathrm{II}}\overline{\boldsymbol{a}} = \overline{\boldsymbol{\theta}} : \boldsymbol{i} \quad \text{and} \quad \operatorname{div}\overline{\boldsymbol{\theta}} = \boldsymbol{0}.$$

Observe that the secondary dislocation densities are coupled to the (linear part of the) secondary incompatibility densities. □

8.2.3 Defect Density Tensors

8.2.3.1 Primary Dislocation Density Tensors

The *primary* dislocation density tensors ${}^{\mathrm{I}}\widetilde{\boldsymbol{D}}$ and ${}^{\mathrm{I}}\overline{\boldsymbol{d}}$ in second-order elasto-plasticity follow from the same definition as the dislocation density tensors $\widetilde{\boldsymbol{D}}$ and $\overline{\boldsymbol{d}}$ in first-order elasto-plasticity. Thus all its variants and the relations among them are the same as in subsection 8.1.3. The possible twelve (total, elastic and plastic) primary dislocation density tensors of Piola-, Piola-Kirchhoff- and Cauchy-type, respectively, and their interrelations are depicted in Fig. 8.7 for the sake of concreteness.

8.2.3.2 Secondary Dislocation Density Tensors

Secondary Dislocation Densities Based on Plastic Double-Distortion. The plastic (contravariant) second-order Piola-Kirchhoff- and Cauchy-type *secondary* dislocation density tensors are expressed in the ambient Euclidean space as

$${}^{\mathrm{II}}\widetilde{\boldsymbol{A}} = -\widetilde{\boldsymbol{f}} \cdot {}^{\mathrm{II}}\widetilde{\boldsymbol{D}} \quad \text{and} \quad {}^{\mathrm{II}}\widetilde{\boldsymbol{a}} = {}^{\mathrm{II}}\widetilde{\boldsymbol{D}} \cdot \operatorname{cof}\widetilde{\boldsymbol{f}} \quad \text{with} \quad {}^{\mathrm{II}}\widetilde{\boldsymbol{D}} = -\widetilde{\boldsymbol{G}} : \boldsymbol{E}. \tag{8.110}$$

Based on the definitions of the double-strain measures $\widetilde{\boldsymbol{\Gamma}}$ and $\widetilde{\boldsymbol{v}}$ together with the relation for the intermediate permutation tensor $\widetilde{\boldsymbol{e}} = [\widetilde{\boldsymbol{F}} \boxtimes \widetilde{\boldsymbol{F}}] : \boldsymbol{E} \cdot \operatorname{cof}\widetilde{\boldsymbol{f}}$ alternative expressions for ${}^{\mathrm{II}}\widetilde{\boldsymbol{A}}$ and ${}^{\mathrm{II}}\widetilde{\boldsymbol{a}}$ are eventually obtained

$${}^{\mathrm{II}}\widetilde{\boldsymbol{A}} = \widetilde{\boldsymbol{\Gamma}} : \boldsymbol{E} \quad \text{and} \quad {}^{\mathrm{II}}\widetilde{\boldsymbol{a}} = \widetilde{\boldsymbol{v}} : \widetilde{\boldsymbol{e}}. \tag{8.111}$$

Secondary Dislocation Densities Based on Elastic Double-Distortion. The elastic (contravariant) second-order Piola-Kirchhoff- and Cauchy-type *secondary* dislocation density tensors are expressed in the ambient Euclidean space as

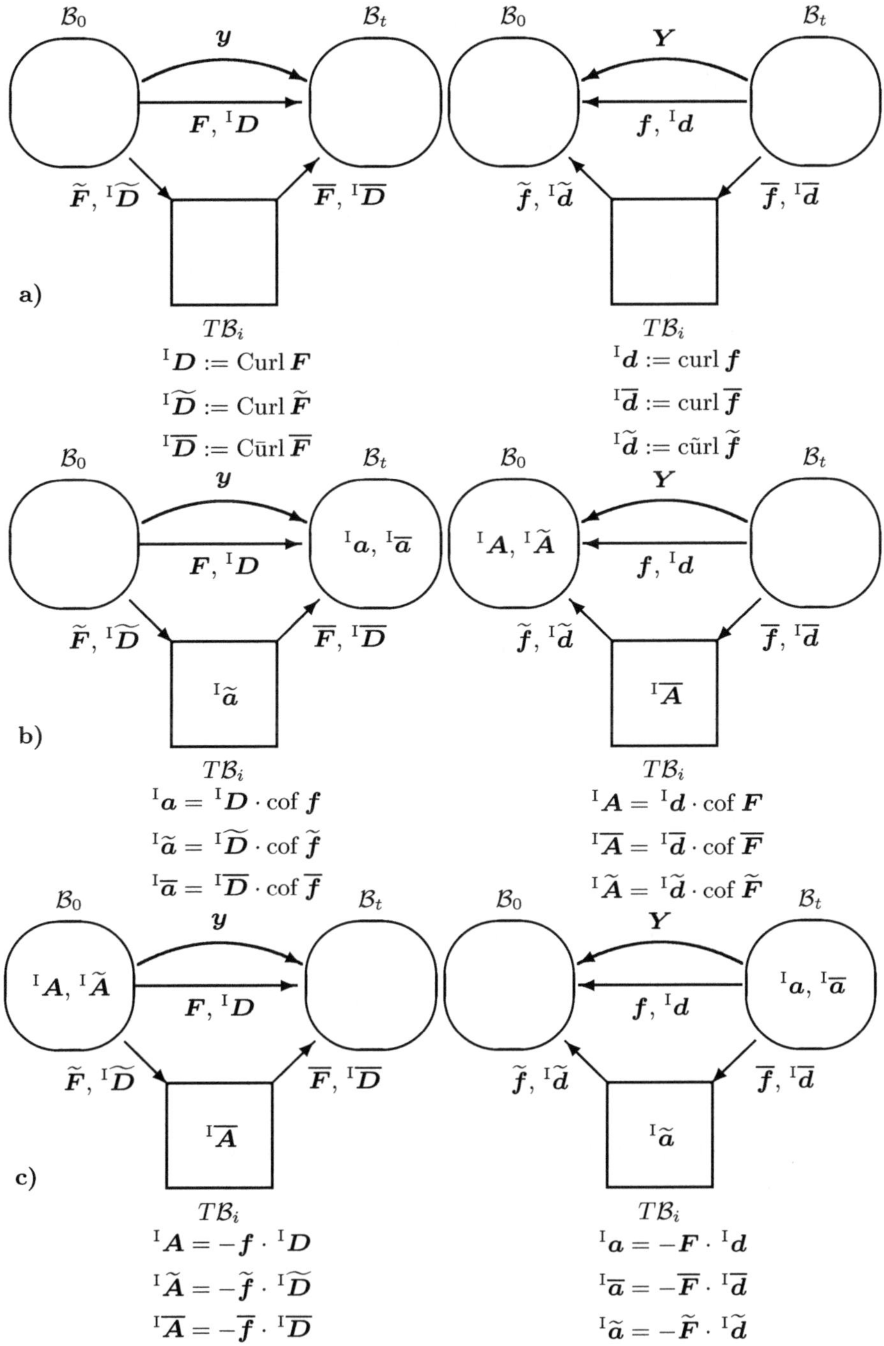

Fig. 8.7 a) Primary Piola-type dislocation density tensors; **b)** Primary Cauchy-type dislocation density tensors; **c)** Primary Piola-Kirchhoff-type dislocation density tensors

$$^{\text{II}}\overline{\boldsymbol{a}} = -\overline{\boldsymbol{F}} \cdot {}^{\text{II}}\overline{\boldsymbol{d}} \quad \text{and} \quad {}^{\text{II}}\overline{\boldsymbol{A}} = {}^{\text{II}}\overline{\boldsymbol{d}} \cdot \operatorname{cof} \overline{\boldsymbol{F}} \quad \text{with} \quad {}^{\text{II}}\overline{\boldsymbol{d}} = -\overline{\boldsymbol{g}} : \boldsymbol{e}. \tag{8.112}$$

Based on the definitions of the double-strain measures $\overline{\boldsymbol{\gamma}}$ and $\overline{\boldsymbol{\Upsilon}}$ together with the relation for the intermediate permutation tensor $\overline{\boldsymbol{E}} = [\overline{\boldsymbol{f}} \boxtimes \overline{\boldsymbol{f}}] : \boldsymbol{e} \cdot \operatorname{cof} \overline{\boldsymbol{F}}$ alternative expressions for ${}^{\text{II}}\overline{\boldsymbol{a}}$ and ${}^{\text{II}}\overline{\boldsymbol{A}}$ are eventually obtained

$$^{\text{II}}\overline{\boldsymbol{a}} = \overline{\boldsymbol{\gamma}} : \boldsymbol{e} \quad \text{and} \quad {}^{\text{II}}\overline{\boldsymbol{A}} = \overline{\boldsymbol{\Upsilon}} : \overline{\boldsymbol{E}}. \tag{8.113}$$

Secondary Dislocation Densities in the Intermediate Configuration. The plastic Cauchy-type and the elastic Piola-Kirchhoff-type *secondary* dislocation densities in the intermediate configuration as expressed in the ambient Euclidean space are related by the identity

$$^{\text{II}}\widetilde{\boldsymbol{a}} \equiv {}^{\text{II}}\overline{\boldsymbol{A}}. \tag{8.114}$$

Proof. The plastic double-distortions $\boldsymbol{g}_{\text{p}}$ and $\boldsymbol{G}_{\text{p}}$ and the elastic double-distortions $\boldsymbol{G}_{\text{e}}$ and $\boldsymbol{g}_{\text{e}}$ are related by

$$\boldsymbol{g}_{\text{p}} = -\boldsymbol{f} \cdot \boldsymbol{G}_{\text{p}} : [\boldsymbol{f} \boxtimes \boldsymbol{f}] \quad \text{and} \quad \boldsymbol{G}_{\text{e}} = -\boldsymbol{F} \cdot \boldsymbol{g}_{\text{e}} : [\boldsymbol{F} \boxtimes \boldsymbol{F}].$$

Incorporating next the definitions for $\widetilde{\boldsymbol{\upsilon}}$ and $\overline{\boldsymbol{\Upsilon}}$ together with $\boldsymbol{G}_{\text{p}} = \overline{\boldsymbol{F}} \cdot \widetilde{\boldsymbol{G}}$ and $\boldsymbol{g}_{\text{e}} = \widetilde{\boldsymbol{f}} \cdot \overline{\boldsymbol{g}}$ renders

$$[\widetilde{\boldsymbol{F}} \cdot \boldsymbol{g}_{\text{p}}] : [\overline{\boldsymbol{F}} \boxtimes \overline{\boldsymbol{F}}] = +\widetilde{\boldsymbol{\upsilon}} \quad \text{and} \quad [\overline{\boldsymbol{f}} \cdot \boldsymbol{G}_{\text{e}}] : [\widetilde{\boldsymbol{f}} \boxtimes \widetilde{\boldsymbol{f}}] = +\overline{\boldsymbol{\Upsilon}}.$$

Moreover, the definitions for $\overline{\boldsymbol{\Upsilon}}$ and $\widetilde{\boldsymbol{\upsilon}}$ take the alternative expressions

$$[\widetilde{\boldsymbol{F}} \cdot \boldsymbol{g}_{\text{e}}] : [\overline{\boldsymbol{F}} \boxtimes \overline{\boldsymbol{F}}] = -\overline{\boldsymbol{\Upsilon}} \quad \text{and} \quad [\overline{\boldsymbol{f}} \cdot \boldsymbol{G}_{\text{p}}] : [\widetilde{\boldsymbol{f}} \boxtimes \widetilde{\boldsymbol{f}}] = -\widetilde{\boldsymbol{\upsilon}}.$$

Finally, since $\boldsymbol{g} = \boldsymbol{g}_{\text{p}} + \boldsymbol{g}_{\text{e}}$ and $\boldsymbol{G} = \boldsymbol{G}_{\text{e}} + \boldsymbol{G}_{\text{p}}$ the relation

$$\widetilde{\boldsymbol{\upsilon}} : \widetilde{\boldsymbol{e}} = \overline{\boldsymbol{\Upsilon}} : \overline{\boldsymbol{E}}$$

follows immediately due to $\boldsymbol{g} : \boldsymbol{e} = \boldsymbol{0}$ and $\boldsymbol{G} : \boldsymbol{E} = \boldsymbol{0}$ and thus concludes the proof. □

Secondary Dislocation Densities Based on Intermediate Torsion. Based on the (elastic) intermediate permutation tensor an additional (contravariant) Piola-type *secondary* dislocation density may be introduced in the ambient Euclidean space as

$$^{\text{II}}\overline{\boldsymbol{D}} := -\overline{\boldsymbol{G}} : \overline{\boldsymbol{E}}. \tag{8.115}$$

Observe that ${}^{\text{II}}\overline{\boldsymbol{D}}$ is a two-point tensor mapping from the intermediate to the spatial configuration. As such this situation is different from the previously introduced elastic Piola-type dislocation density ${}^{\text{II}}\overline{\boldsymbol{d}} = -\overline{\boldsymbol{g}} : \boldsymbol{e}$ that is a two-point tensor mapping from the spatial to the intermediate configuration. However, due to the convection of permutation tensors these two Piola-type secondary dislocation densities are related by

$$ {}^{\mathrm{II}}\overline{\boldsymbol{D}} = -\overline{\boldsymbol{F}} \cdot {}^{\mathrm{II}}\overline{\boldsymbol{d}} \cdot \operatorname{cof} \overline{\boldsymbol{F}}. \tag{8.116} $$

Therefore, the elastic Piola-Kirchhoff- and Cauchy-type secondary dislocation densities ${}^{\mathrm{II}}\overline{\boldsymbol{A}}$ and ${}^{\mathrm{II}}\overline{\boldsymbol{a}}$ are related to ${}^{\mathrm{II}}\overline{\boldsymbol{D}}$ by

$$ {}^{\mathrm{II}}\overline{\boldsymbol{A}} = -\overline{\boldsymbol{f}} \cdot {}^{\mathrm{II}}\overline{\boldsymbol{D}} \quad \text{and} \quad {}^{\mathrm{II}}\overline{\boldsymbol{a}} = {}^{\mathrm{II}}\overline{\boldsymbol{D}} \cdot \operatorname{cof} \overline{\boldsymbol{f}}. \tag{8.117} $$

Likewise, based on the (plastic) intermediate permutation tensor a further (contravariant) Piola-type *secondary* dislocation density may be introduced in the ambient Euclidean space as

$$ {}^{\mathrm{II}}\widetilde{\boldsymbol{d}} = -\widetilde{\boldsymbol{g}} : \widetilde{\boldsymbol{e}}. \tag{8.118} $$

Note that ${}^{\mathrm{II}}\widetilde{\boldsymbol{d}}$ is a two-point tensor mapping from the intermediate to the material configuration. As such this situation is different from the previously introduced plastic Piola-type dislocation density ${}^{\mathrm{II}}\widetilde{\boldsymbol{D}} = -\widetilde{\boldsymbol{G}} : \boldsymbol{E}$ that is a two-point tensor mapping from the material to the intermediate configuration. However, due to the convection of permutation tensors these two Piola-type dislocation densities are related by

$$ {}^{\mathrm{II}}\widetilde{\boldsymbol{d}} = -\widetilde{\boldsymbol{f}} \cdot {}^{\mathrm{II}}\widetilde{\boldsymbol{D}} \cdot \operatorname{cof} \widetilde{\boldsymbol{f}}. \tag{8.119} $$

Thus, the plastic Piola-Kirchhoff- and Cauchy-type secondary dislocation densities ${}^{\mathrm{II}}\widetilde{\boldsymbol{A}}$ and ${}^{\mathrm{II}}\widetilde{\boldsymbol{a}}$ are related to ${}^{\mathrm{II}}\widetilde{\boldsymbol{d}}$ by

$$ {}^{\mathrm{II}}\widetilde{\boldsymbol{a}} = -\widetilde{\boldsymbol{F}} \cdot {}^{\mathrm{II}}\widetilde{\boldsymbol{d}} \quad \text{and} \quad {}^{\mathrm{II}}\widetilde{\boldsymbol{A}} = {}^{\mathrm{II}}\widetilde{\boldsymbol{d}} \cdot \operatorname{cof} \widetilde{\boldsymbol{F}}. \tag{8.120} $$

Summarizing, as depicted in Fig. 8.8 the additive decomposition of the second deformation gradients into incompatible elastic and plastic contributions results in the possible introduction of twelve (total, elastic and plastic) secondary dislocation density tensors of Piola-, Piola-Kirchhoff- and Cauchy-type, respectively. Thereby the six Piola-type secondary dislocation density tensors in Fig. 8.8 a) are defined as the negative right skew symmetric part of the corresponding contribution to the second deformation gradients. Accordingly, the six Piola-Kirchhoff- and Cauchy-type secondary dislocation density tensors in Fig. 8.8 b) and c) are either a result from appropriate pull-back/push-forward operations or a result from the convection of permutation tensors, respectively. From the additive decomposition of the intermediate double-strain that results in

$$ \overline{\boldsymbol{\Upsilon}} : \overline{\boldsymbol{E}} = \widetilde{\boldsymbol{\upsilon}} : \widetilde{\boldsymbol{e}} \tag{8.121} $$

the previously derived equality of the elastic and plastic secondary dislocation densities in the intermediate configuration finally follows in the format

$$\overline{\boldsymbol{f}} \cdot {}^{\mathrm{II}}\overline{\boldsymbol{D}} + {}^{\mathrm{II}}\widetilde{\boldsymbol{D}} \cdot \operatorname{cof} \widetilde{\boldsymbol{f}} = \boldsymbol{0} \quad \text{and} \quad \widetilde{\boldsymbol{F}} \cdot {}^{\mathrm{II}}\widetilde{\boldsymbol{d}} + {}^{\mathrm{II}}\overline{\boldsymbol{d}} \cdot \operatorname{cof} \overline{\boldsymbol{F}} = \boldsymbol{0}. \tag{8.122}$$

Obviously this is nothing but the statement of right symmetry of the total second deformation gradient.

8.2.3.3 Point-Defect Density Tensors

Point-Defect Densities Based on Plastic Distortions. Firstly, the material plastic point-defect density $\widetilde{\boldsymbol{Q}}$ as defined in Eq. 8.97 may be reformulated in terms of the material plastic metric $\widetilde{\boldsymbol{C}}$ and the material plastic double-strain $\widetilde{\boldsymbol{\Gamma}}$ as

$$\widetilde{\boldsymbol{Q}} = 2\mathsf{I}^{\mathrm{sym}} : \left[\widetilde{\boldsymbol{C}} \cdot [\widetilde{\boldsymbol{\Gamma}} - \widetilde{\boldsymbol{f}} \cdot \operatorname{Grad} \widetilde{\boldsymbol{F}}]\right]. \tag{8.123}$$

Next, by incorporating the counterparts of $\widetilde{\boldsymbol{C}}$ and $\widetilde{\boldsymbol{\Gamma}}$ in the intermediate configuration the material plastic point-defect density $\widetilde{\boldsymbol{Q}}$ may be re-expressed

$$\widetilde{\boldsymbol{Q}} = -2\mathsf{I}^{\mathrm{sym}} : [\widetilde{\boldsymbol{F}}^t \boxtimes \widetilde{\boldsymbol{F}}^t] : \left[\widetilde{\boldsymbol{j}} \cdot [\widetilde{\boldsymbol{\upsilon}} - \widetilde{\boldsymbol{F}} \cdot \widetilde{\operatorname{grad}}\, \widetilde{\boldsymbol{f}}]\right] \cdot \widetilde{\boldsymbol{F}}. \tag{8.124}$$

Thus a fully covariant plastic push-forward introduces eventually the intermediate plastic point-defect density $\widetilde{\boldsymbol{q}}$ as

$$\widetilde{\boldsymbol{q}} := -[\widetilde{\boldsymbol{f}}^t \boxtimes \widetilde{\boldsymbol{f}}^t] : \widetilde{\boldsymbol{Q}} \cdot \widetilde{\boldsymbol{f}} = 2\widetilde{\mathsf{i}}^{\mathrm{sym}} : \left[\widetilde{\boldsymbol{j}} \cdot [\widetilde{\boldsymbol{\upsilon}} - \widetilde{\boldsymbol{F}} \cdot \widetilde{\operatorname{grad}}\, \widetilde{\boldsymbol{f}}]\right]. \tag{8.125}$$

Observe that the fourth-order tensors $\mathsf{I}^{\mathrm{sym}}$ and $\widetilde{\mathsf{i}}^{\mathrm{sym}}$ are co/co/contra/contra-(mixed) variant and thus transform accordingly, i.e.

$$\widetilde{\mathsf{i}}^{\mathrm{sym}} = [\widetilde{\boldsymbol{f}}^t \boxtimes \widetilde{\boldsymbol{f}}^t] : \mathsf{I}^{\mathrm{sym}} : [\widetilde{\boldsymbol{F}}^t \boxtimes \widetilde{\boldsymbol{F}}^t]. \tag{8.126}$$

Point-Defect Densities Based on Elastic Distortions. Firstly, the spatial elastic point-defect density $\overline{\boldsymbol{q}}$ as defined in Eq. 8.99 may be reformulated in terms of the spatial elastic metric $\overline{\boldsymbol{c}}$ and the spatial elastic double-strain $\overline{\boldsymbol{\gamma}}$ as

$$\overline{\boldsymbol{q}} = 2\mathsf{i}^{\mathrm{sym}} : \left[\overline{\boldsymbol{c}} \cdot [\overline{\boldsymbol{\gamma}} - \overline{\boldsymbol{F}} \cdot \operatorname{grad} \overline{\boldsymbol{f}}]\right]. \tag{8.127}$$

Next, by incorporating the counterparts of $\overline{\boldsymbol{c}}$ and $\overline{\boldsymbol{\gamma}}$ in the intermediate configuration the spatial elastic point-defect density $\overline{\boldsymbol{q}}$ may be re-expressed

$$\overline{\boldsymbol{q}} = -2\mathsf{i}^{\mathrm{sym}} : [\overline{\boldsymbol{f}}^t \boxtimes \overline{\boldsymbol{f}}^t] : \left[\overline{\boldsymbol{J}} \cdot [\overline{\boldsymbol{\Upsilon}} - \overline{\boldsymbol{f}} \cdot \overline{\operatorname{Grad}}\, \overline{\boldsymbol{F}}]\right] \cdot \overline{\boldsymbol{f}}. \tag{8.128}$$

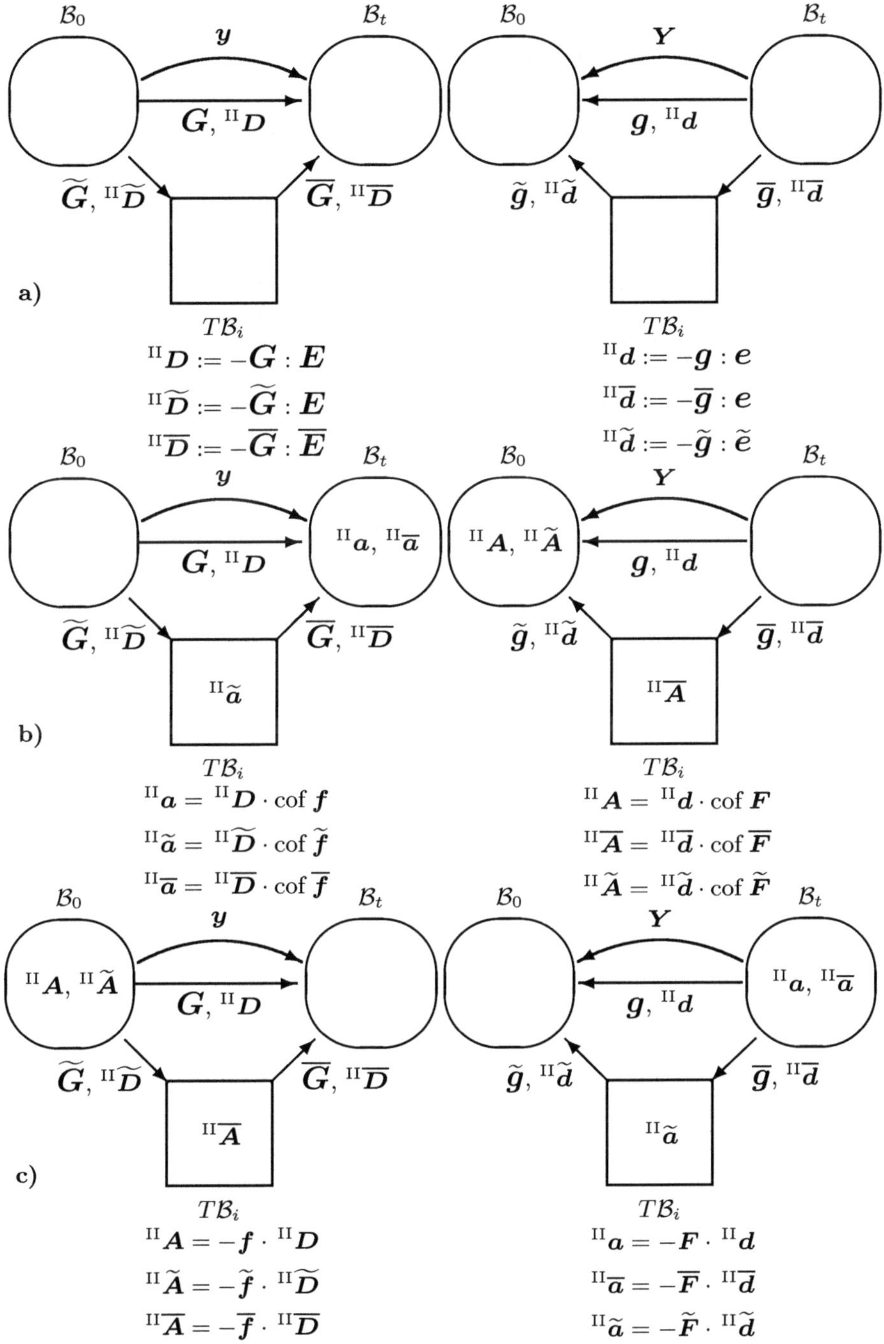

Fig. 8.8 a) Secondary Piola-type dislocation density tensors; **b)** Secondary Cauchy-type dislocation density tensors; **c)** Secondary Piola-Kirchhoff-type dislocation density tensors

Thus a fully covariant elastic pull-back introduces eventually the intermediate elastic point-defect density $\overline{\boldsymbol{Q}}$ as

$$\overline{\boldsymbol{Q}} := -[\overline{\boldsymbol{F}}^t \boxtimes \overline{\boldsymbol{F}}^t] : \overline{\boldsymbol{q}} \cdot \overline{\boldsymbol{F}} = 2\bar{\mathsf{I}}^{\text{sym}} : \left[\overline{\boldsymbol{J}} \cdot [\overline{\boldsymbol{\Upsilon}} - \overline{\boldsymbol{f}} \cdot \overline{\text{Grad}}\, \overline{\boldsymbol{F}}]\right]. \tag{8.129}$$

Observe that the fourth-order tensors i^{sym} and $\bar{\mathsf{I}}^{\text{sym}}$ are co/co/contra/contra-(mixed) variant and thus transform accordingly, i.e.

$$\bar{\mathsf{I}}^{\text{sym}} = [\overline{\boldsymbol{F}}^t \boxtimes \overline{\boldsymbol{F}}^t] : \mathsf{i}^{\text{sym}} : [\overline{\boldsymbol{f}}^t \boxtimes \overline{\boldsymbol{f}}^t]. \tag{8.130}$$

Point-Defect Densities in the Intermediate Configuration. The plastic and the elastic point-defect densities in the intermediate configuration as expressed in the ambient Euclidean space are related by the identity

$$\widetilde{\boldsymbol{q}} \equiv \overline{\boldsymbol{Q}}. \tag{8.131}$$

Proof. Based on the multiplicative decomposition of the deformation gradient the plastic push-forward of $\boldsymbol{f} \cdot \text{Grad}\, \boldsymbol{F}$ results in

$$\widetilde{\boldsymbol{F}} \cdot [\boldsymbol{f} \cdot \text{Grad}\, \boldsymbol{F}] : [\widetilde{\boldsymbol{f}} \boxtimes \widetilde{\boldsymbol{f}}] = \overline{\boldsymbol{f}} \cdot \overline{\text{Grad}}\, \overline{\boldsymbol{F}} - \widetilde{\boldsymbol{F}} \cdot \widetilde{\text{grad}}\, \widetilde{\boldsymbol{f}}.$$

Here $\text{Grad}\, \widetilde{\boldsymbol{F}} : [\widetilde{\boldsymbol{f}} \boxtimes \widetilde{\boldsymbol{f}}] = -\widetilde{\boldsymbol{F}} \cdot \widetilde{\text{grad}}\, \widetilde{\boldsymbol{f}}$ has been utilized. Next, the plastic push-forward of the material double-strain $\boldsymbol{f} \cdot \boldsymbol{G}$ reads as

$$\widetilde{\boldsymbol{F}} \cdot [\boldsymbol{f} \cdot \boldsymbol{G}] : [\widetilde{\boldsymbol{f}} \boxtimes \widetilde{\boldsymbol{f}}] = \overline{\boldsymbol{\Upsilon}} - \widetilde{\boldsymbol{v}}.$$

Therefore, upon noting that $\widetilde{\boldsymbol{j}} = \overline{\boldsymbol{J}}$ and $\widetilde{\mathsf{i}}^{\text{sym}} = \bar{\mathsf{I}}^{\text{sym}}$, the difference between the plastic and elastic point-defect densities eventually takes the expression

$$\widetilde{\boldsymbol{q}} - \overline{\boldsymbol{Q}} = -2[\widetilde{\boldsymbol{f}}^t \boxtimes \widetilde{\boldsymbol{f}}^t] : \mathsf{I}^{\text{sym}} : \left[\widetilde{\boldsymbol{C}} \cdot [\boldsymbol{f} \cdot \boldsymbol{G} - \boldsymbol{f} \cdot \text{Grad}\, \boldsymbol{F}]\right] \cdot \widetilde{\boldsymbol{f}}.$$

Noting that the total double-distortion equals the second deformation gradient so that $\boldsymbol{f} \cdot \boldsymbol{G} = \boldsymbol{f} \cdot \text{Grad}\, \boldsymbol{F}$ finally concludes the proof. □

For the sake of completeness the above derivations motivate to also introduce total material and spatial point-defect densities (that vanish however identically due to the metricity of the total deformation)

$$\boldsymbol{Q} := 2\mathsf{I}^{\text{sym}} : \left[\boldsymbol{C} \cdot [\boldsymbol{\Gamma} - \boldsymbol{f} \cdot \text{Grad}\, \boldsymbol{F}]\right] \quad \text{and} \quad \boldsymbol{q} := 2\mathsf{i}^{\text{sym}} : \left[\boldsymbol{c} \cdot [\boldsymbol{\gamma} - \boldsymbol{F} \cdot \text{grad}\, \boldsymbol{f}]\right].$$

The various point-defect density tensors are finally summarized for convenience in Fig. 8.9.

Remark 8.10. Observe finally that the material plastic metric $\widetilde{\boldsymbol{C}}$ and the spatial elastic metric $\overline{\boldsymbol{c}}$ satisfy

$$2\mathsf{I}^{\text{sym}} : [\widetilde{\boldsymbol{F}}^t \cdot \text{Grad}\, \widetilde{\boldsymbol{F}}] \equiv \text{Grad}\, \widetilde{\boldsymbol{C}} \quad \text{and} \quad 2\mathsf{i}^{\text{sym}} : [\overline{\boldsymbol{f}}^t \cdot \text{grad}\, \overline{\boldsymbol{f}}] \equiv \text{grad}\, \overline{\boldsymbol{c}}.$$

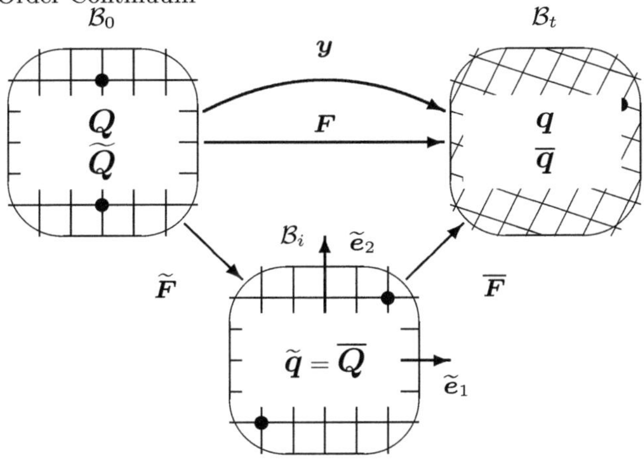

Fig. 8.9 Total, plastic and elastic point-defect density tensors in the material, the spatial and the intermediate configuration

Thus the various apparently different definitions of the point defect densities do indeed coincide. □

8.2.3.4 Disclination Density Tensors

Disclination Densities Based on Plastic Double-Distortion. Third-order plastic contra/co/contra- (mixed) variant Piola-Kirchhoff- and Cauchy-type disclination density tensors are expressed in the ambient Euclidean space as

$$\widetilde{\boldsymbol{T}} = -\widetilde{\boldsymbol{f}} \cdot \widetilde{\boldsymbol{K}} \quad \text{and} \quad \widetilde{\boldsymbol{t}} = \widetilde{\boldsymbol{K}} : [\widetilde{\boldsymbol{f}} \boxtimes \operatorname{cof} \widetilde{\boldsymbol{f}}] \quad \text{with} \quad \widetilde{\boldsymbol{K}} = \operatorname{Curl} \widetilde{\boldsymbol{G}}. \tag{8.132}$$

Disclination Densities Based on Elastic Double-Distortion. Third-order elastic contra/co/contra- (mixed) variant Piola-Kirchhoff- and Cauchy-type disclination density tensors are expressed in the ambient Euclidean space as

$$\overline{\boldsymbol{t}} = -\overline{\boldsymbol{F}} \cdot \overline{\boldsymbol{k}} \quad \text{and} \quad \overline{\boldsymbol{T}} = \overline{\boldsymbol{k}} : [\overline{\boldsymbol{F}} \otimes \operatorname{cof} \overline{\boldsymbol{F}}] \quad \text{with} \quad \overline{\boldsymbol{k}} = \operatorname{curl} \overline{\boldsymbol{g}}. \tag{8.133}$$

Addendum 8.2. Double-Anholonomic Curl Operators
The anholonomic curl operators $\widetilde{\operatorname{curl}}\{\bullet\}$ and $\overline{\operatorname{Curl}}\{\bullet\}$ with respect to the anholonomic coordinates in the intermediate configuration have been introduced already earlier. Applying their definitions to the plastic double-distortion $\widetilde{\boldsymbol{g}}$ and the elastic double-distortion $\overline{\boldsymbol{G}}$, respectively, renders

$$\widetilde{\operatorname{curl}}\,\widetilde{\boldsymbol{g}} := -\widetilde{\operatorname{grad}}\,\widetilde{\boldsymbol{g}} : \widetilde{\boldsymbol{e}} \quad \text{and} \quad \overline{\operatorname{Curl}}\,\overline{\boldsymbol{G}} := -\overline{\operatorname{Grad}}\,\overline{\boldsymbol{G}} : \overline{\boldsymbol{E}}.$$

Here the anholonomic gradient operators $\widetilde{\operatorname{grad}}\{\bullet\}$ and $\overline{\operatorname{Grad}}\{\bullet\}$ as applied to the intermediate plastic and elastic double-distortions read

$$\widetilde{\mathrm{grad}}\,\widetilde{\boldsymbol{g}} := \mathrm{Grad}\,\widetilde{\boldsymbol{g}}\cdot\widetilde{\boldsymbol{f}} \quad\text{and}\quad \overline{\mathrm{Grad}}\,\overline{\boldsymbol{G}} := \mathrm{grad}\,\overline{\boldsymbol{G}}\cdot\overline{\boldsymbol{F}}.$$

Recall next the relation between $\widetilde{\boldsymbol{G}}$ and $\widetilde{\boldsymbol{g}}$ and between $\overline{\boldsymbol{g}}$ and $\overline{\boldsymbol{G}}$ as given by

$$\widetilde{\boldsymbol{G}} = -\widetilde{\boldsymbol{F}}\cdot\widetilde{\boldsymbol{g}} : [\widetilde{\boldsymbol{F}}\boxtimes\widetilde{\boldsymbol{F}}] \quad\text{and}\quad \overline{\boldsymbol{g}} = -\overline{\boldsymbol{f}}\cdot\overline{\boldsymbol{G}} : [\overline{\boldsymbol{f}}\boxtimes\overline{\boldsymbol{f}}].$$

Then, based on the product and the chain rule, the common material and spatial curl operators as applied to the material plastic double-distortion and the spatial elastic double-distortion, respectively, are formally related to the *double-anholonomic curl operators* $\widetilde{\widetilde{\mathrm{curl}}}\{\bullet\}$ and $\overline{\overline{\mathrm{Curl}}}\{\bullet\}$ as applied to the intermediate plastic and elastic double-distortions by

$$\mathrm{Curl}\,\widetilde{\boldsymbol{G}} = -\widetilde{\boldsymbol{F}}\cdot\widetilde{\widetilde{\mathrm{curl}}}\,\widetilde{\boldsymbol{g}} : [\widetilde{\boldsymbol{F}}\boxtimes\mathrm{cof}\,\widetilde{\boldsymbol{F}}] \quad\text{and}\quad \mathrm{curl}\,\overline{\boldsymbol{g}} = -\overline{\boldsymbol{f}}\cdot\overline{\overline{\mathrm{Curl}}}\,\overline{\boldsymbol{G}} : [\overline{\boldsymbol{f}}\boxtimes\mathrm{cof}\,\overline{\boldsymbol{f}}].$$

Here the double-anholonomic curl operators denote abbreviations for

$$\widetilde{\widetilde{\mathrm{curl}}}\,\widetilde{\boldsymbol{g}} := \widetilde{\mathrm{curl}}\,\widetilde{\boldsymbol{g}} + \widetilde{\boldsymbol{a}} \quad\text{and}\quad \overline{\overline{\mathrm{Curl}}}\,\overline{\boldsymbol{G}} := \overline{\mathrm{Curl}}\,\overline{\boldsymbol{G}} + \overline{\boldsymbol{A}}.$$

The third-order tensors $\widetilde{\boldsymbol{a}}$ and $\overline{\boldsymbol{A}}$ result from the anholonomicity of $\widetilde{\boldsymbol{F}}$ and $\overline{\boldsymbol{f}}$, respectively, and follow in terms of the intermediate plastic and elastic double-strain measures as

$$\widetilde{\boldsymbol{a}} := \widetilde{\boldsymbol{f}}\cdot\Big[\widetilde{\boldsymbol{\upsilon}}\cdot{}^{\mathrm{I}}\widetilde{a} + \widetilde{\boldsymbol{\upsilon}}\rtimes[\widetilde{\boldsymbol{F}}\cdot\widetilde{\mathrm{grad}}\,\widetilde{\boldsymbol{f}}] - [\widetilde{\boldsymbol{F}}\cdot\widetilde{\mathrm{grad}}\,\widetilde{\boldsymbol{f}}]\rtimes\widetilde{\boldsymbol{\upsilon}}\Big]$$

and

$$\overline{\boldsymbol{A}} := \overline{\boldsymbol{F}}\cdot\Big[\overline{\boldsymbol{\Upsilon}}\cdot{}^{\mathrm{I}}\overline{\boldsymbol{A}} + \overline{\boldsymbol{\Upsilon}}\rtimes[\overline{\boldsymbol{f}}\cdot\overline{\mathrm{Grad}}\,\overline{\boldsymbol{F}}] - [\overline{\boldsymbol{f}}\cdot\overline{\mathrm{Grad}}\,\overline{\boldsymbol{F}}]\rtimes\overline{\boldsymbol{\Upsilon}}\Big],$$

respectively. Alternatively $\widetilde{\boldsymbol{a}}$ and $\overline{\boldsymbol{A}}$ read in terms of the material plastic and the spatial elastic double-strain measures as

$$\widetilde{\boldsymbol{a}} := \Big[\widetilde{\boldsymbol{\Gamma}}\cdot{}^{\mathrm{I}}\widetilde{\boldsymbol{A}} + \widetilde{\boldsymbol{\Gamma}}\rtimes[\widetilde{\boldsymbol{f}}\cdot\mathrm{Grad}\,\widetilde{\boldsymbol{F}}] - [\widetilde{\boldsymbol{f}}\cdot\mathrm{Grad}\,\widetilde{\boldsymbol{F}}]\rtimes\widetilde{\boldsymbol{\Gamma}}\Big] : [\widetilde{\boldsymbol{f}}\boxtimes\mathrm{cof}\,\widetilde{\boldsymbol{f}}]$$

and

$$\overline{\boldsymbol{A}} := \Big[\overline{\gamma}\cdot{}^{\mathrm{I}}\overline{\boldsymbol{a}} + \overline{\gamma}\rtimes[\overline{\boldsymbol{F}}\cdot\mathrm{grad}\,\overline{\boldsymbol{f}}] - [\overline{\boldsymbol{F}}\cdot\mathrm{Grad}\,\overline{\boldsymbol{f}}]\rtimes\overline{\gamma}\Big] : [\overline{\boldsymbol{F}}\boxtimes\mathrm{cof}\,\overline{\boldsymbol{F}}],$$

respectively. Note that $\widetilde{\boldsymbol{a}}$ and $\overline{\boldsymbol{A}}$ vanish identically in the case of compatible and integrable $\widetilde{\boldsymbol{F}}$ and $\overline{\boldsymbol{f}}$, i.e., e.g., $\mathrm{Grad}\,\widetilde{\boldsymbol{F}} = \widetilde{\boldsymbol{G}}$ and $\mathrm{grad}\,\overline{\boldsymbol{f}} = \overline{\boldsymbol{g}}$ together with ${}^{\mathrm{I}}\widetilde{\boldsymbol{A}} = \boldsymbol{0}$ and ${}^{\mathrm{I}}\overline{\boldsymbol{a}} = \boldsymbol{0}$, respectively. In this case the double-anholonomic and the anholonomic curl operators coincide and agree with the common spatial and material curl operators (with $\mathrm{Curl}\,\widetilde{\boldsymbol{G}} = \boldsymbol{0}$ and $\mathrm{curl}\,\overline{\boldsymbol{g}} = \boldsymbol{0}$ identically). Furthermore, as a consequence of the additive (and multiplicative in each term) decomposition of the second deformation gradient

$$\boldsymbol{G} = \boldsymbol{G}_{\mathrm{e}} + \boldsymbol{G}_{\mathrm{p}} = \overline{\boldsymbol{G}} : [\widetilde{\boldsymbol{F}}\boxtimes\widetilde{\boldsymbol{F}}] + \overline{\boldsymbol{F}}\cdot\widetilde{\boldsymbol{G}},$$

the definition of the elastic double-anholonomic curl operator and the convection of the permutation tensors allow to decompose the material curl of the second deformation gradient additively into elastic and plastic contributions

$$\operatorname{Curl}\boldsymbol{G}=\left[\mathrm{C\bar{\bar{u}}rl}\,\overline{\boldsymbol{G}}+\left[\overline{\boldsymbol{G}}-\mathrm{Gr\bar{a}d}\,\overline{\boldsymbol{F}}\right]\rtimes\boldsymbol{\upsilon}\right]:[\widetilde{\boldsymbol{F}}\boxtimes\operatorname{cof}\widetilde{\boldsymbol{F}}]+\overline{\boldsymbol{F}}\cdot\operatorname{Curl}\widetilde{\boldsymbol{G}}\doteq\overline{\boldsymbol{0}}.$$

The additive decomposition of the material curl of the second deformation gradient into elastic and plastic parts follows in detail from the step by step calculation

$$\begin{aligned}
\operatorname{Curl}\boldsymbol{G} &= -\operatorname{Grad}\left(\overline{\boldsymbol{G}}:[\overline{\boldsymbol{F}}\boxtimes\overline{\boldsymbol{F}}]+\overline{\boldsymbol{F}}\cdot\widetilde{\boldsymbol{G}}\right):\boldsymbol{E}\\
&=\left[\mathrm{C\bar{u}rl}\,\overline{\boldsymbol{G}}+\overline{\boldsymbol{G}}\rtimes[\widetilde{\boldsymbol{F}}\cdot\operatorname{grad}\widetilde{\boldsymbol{f}}]+\overline{\boldsymbol{G}}\cdot{}^{\mathrm{I}}\overline{\boldsymbol{A}}\right]:[\widetilde{\boldsymbol{F}}\boxtimes\operatorname{cof}\widetilde{\boldsymbol{F}}]\\
&-\left[\mathrm{Gr\bar{a}d}\,\overline{\boldsymbol{F}}\rtimes\widetilde{\boldsymbol{\upsilon}}\right]:[\widetilde{\boldsymbol{F}}\boxtimes\operatorname{cof}\widetilde{\boldsymbol{F}}]+\overline{\boldsymbol{F}}\cdot\operatorname{Curl}\widetilde{\boldsymbol{G}}\\
&=\left[\mathrm{C\bar{\bar{u}}rl}\,\overline{\boldsymbol{G}}\right]:[\widetilde{\boldsymbol{F}}\boxtimes\operatorname{cof}\widetilde{\boldsymbol{F}}]+\overline{\boldsymbol{F}}\cdot\operatorname{Curl}\widetilde{\boldsymbol{G}}\\
&+\left[\overline{\boldsymbol{G}}\rtimes\left[\widetilde{\boldsymbol{F}}\cdot\operatorname{grad}\widetilde{\boldsymbol{f}}-\overline{\boldsymbol{f}}\cdot\mathrm{Gr\bar{a}d}\,\overline{\boldsymbol{F}}\right]\right]:[\widetilde{\boldsymbol{F}}\boxtimes\operatorname{cof}\widetilde{\boldsymbol{F}}]\\
&-\left[\mathrm{Gr\bar{a}d}\,\overline{\boldsymbol{F}}\rtimes\left[\widetilde{\boldsymbol{\upsilon}}-\overline{\boldsymbol{\Upsilon}}\right]\right]:[\widetilde{\boldsymbol{F}}\boxtimes\operatorname{cof}\widetilde{\boldsymbol{F}}]
\end{aligned}$$

Noting that $\widetilde{\boldsymbol{F}}\cdot\operatorname{grad}\widetilde{\boldsymbol{f}}-\overline{\boldsymbol{f}}\cdot\mathrm{Gr\bar{a}d}\,\overline{\boldsymbol{F}}=:\boldsymbol{\upsilon}:=\widetilde{\boldsymbol{\upsilon}}-\overline{\boldsymbol{\Upsilon}}$ renders finally the result. Likewise, the additive (and multiplicative in each term) decomposition of the second deformation gradient

$$\boldsymbol{g}=\boldsymbol{g}_{\mathrm{p}}+\boldsymbol{g}_{\mathrm{e}}=\widetilde{\boldsymbol{g}}:[\overline{\boldsymbol{f}}\boxtimes\overline{\boldsymbol{f}}]+\widetilde{\boldsymbol{f}}\cdot\overline{\boldsymbol{g}},$$

the definition of the plastic double-anholonomic curl operator and the convection of the permutation tensors allow to decompose the spatial curl of the second deformation gradient additively into plastic and elastic contributions

$$\operatorname{curl}\boldsymbol{g}=\left[\mathrm{c\tilde{\tilde{u}}rl}\,\widetilde{\boldsymbol{g}}+\left[\widetilde{\boldsymbol{g}}-\mathrm{gr\tilde{a}d}\,\widetilde{\boldsymbol{f}}\right]\rtimes\boldsymbol{\Upsilon}\right]:[\overline{\boldsymbol{f}}\boxtimes\operatorname{cof}\overline{\boldsymbol{f}}]+\widetilde{\boldsymbol{f}}\cdot\operatorname{curl}\overline{\boldsymbol{g}}\doteq\overline{\boldsymbol{0}}.$$

Note that $\overline{\boldsymbol{G}}-\mathrm{Gr\bar{a}d}\,\overline{\boldsymbol{F}}$ and $\widetilde{\boldsymbol{g}}-\mathrm{gr\tilde{a}d}\,\widetilde{\boldsymbol{f}}$ vanish for the case of compatible $\overline{\boldsymbol{F}}$ and $\widetilde{\boldsymbol{f}}$, i.e. in the case of a priori constrained second-order elasto-plasticity (with also $\operatorname{Curl}\widetilde{\boldsymbol{G}}=\boldsymbol{0}$ and $\operatorname{curl}\overline{\boldsymbol{g}}=\boldsymbol{0}$ identically, however $\mathrm{C\bar{\bar{u}}rl}\,\overline{\boldsymbol{G}}=\boldsymbol{0}$ and $\mathrm{c\tilde{\tilde{u}}rl}\,\widetilde{\boldsymbol{g}}=\boldsymbol{0}$ only if $\overline{\boldsymbol{F}}$ and $\widetilde{\boldsymbol{f}}$ are in addition integrable).

Disclination Densities Based on Double-Anholonomic Curl. Based on the double-anholonomic elastic intermediate curl operator an additional contra/co/contra- (mixed) variant Piola-type disclination density may be introduced in the ambient Euclidean space as

$$\overline{\boldsymbol{K}} := \mathrm{C\bar{\bar{u}}rl}\,\overline{\boldsymbol{G}}. \tag{8.134}$$

Observe that $\overline{\boldsymbol{K}}$ is a two-point tensor mapping from the intermediate to the spatial configuration. As such this situation is different from the previously introduced elastic Piola-type disclination density $\overline{\boldsymbol{k}} = \mathrm{curl}\,\overline{\boldsymbol{g}}$ that is a two-point tensor mapping from the spatial to the intermediate configuration. However, due to the convection of curl operators these two Piola-type disclination densities are related by

$$\overline{\boldsymbol{K}} = -\overline{\boldsymbol{F}} \cdot \overline{\boldsymbol{k}} : [\overline{\boldsymbol{F}} \boxtimes \mathrm{cof}\,\overline{\boldsymbol{F}}]. \tag{8.135}$$

Therefore, the elastic Piola-Kirchhoff- and Cauchy-type disclination densities $\overline{\boldsymbol{T}}$ and $\overline{\boldsymbol{t}}$ are related to $\overline{\boldsymbol{K}}$ by

$$\overline{\boldsymbol{T}} = -\overline{\boldsymbol{f}} \cdot \overline{\boldsymbol{K}} \quad \text{and} \quad \overline{\boldsymbol{t}} = \overline{\boldsymbol{K}} : [\overline{\boldsymbol{f}} \boxtimes \mathrm{cof}\,\overline{\boldsymbol{f}}]. \tag{8.136}$$

Likewise, based on the double-anholonomic plastic intermediate curl operator a further contra/co/contra- (mixed) variant Piola-type disclination density may be introduced in the ambient Euclidean space as

$$\widetilde{\boldsymbol{k}} = \mathrm{c\tilde{\tilde{u}}rl}\,\widetilde{\boldsymbol{g}}. \tag{8.137}$$

Note that $\widetilde{\boldsymbol{k}}$ is a two-point tensor mapping from the intermediate to the material configuration. As such this situation is different from the previously introduced plastic Piola-type disclination density $\widetilde{\boldsymbol{K}} = \mathrm{Curl}\,\widetilde{\boldsymbol{G}}$ that is a two-point tensor mapping from the material to the intermediate configuration. However, due to the convection of curl operators these two Piola-type disclination densities are related by

$$\widetilde{\boldsymbol{k}} = -\widetilde{\boldsymbol{f}} \cdot \widetilde{\boldsymbol{K}} : [\widetilde{\boldsymbol{f}} \boxtimes \mathrm{cof}\,\widetilde{\boldsymbol{f}}]. \tag{8.138}$$

Thus, the plastic Piola-Kirchhoff- and Cauchy-type disclination densities $\widetilde{\boldsymbol{T}}$ and $\widetilde{\boldsymbol{t}}$ are related to $\widetilde{\boldsymbol{k}}$ by

$$\widetilde{\boldsymbol{t}} = -\widetilde{\boldsymbol{F}} \cdot \widetilde{\boldsymbol{k}} \quad \text{and} \quad \widetilde{\boldsymbol{T}} = \widetilde{\boldsymbol{k}} : [\widetilde{\boldsymbol{F}} \boxtimes \mathrm{cof}\,\widetilde{\boldsymbol{F}}]. \tag{8.139}$$

Summarizing, as depicted in Fig. 8.10 the additive decomposition of the second deformation gradients into incompatible elastic and plastic contributions results in the possible introduction of twelve (total, elastic and plastic) disclination density tensors of Piola-, Piola-Kirchhoff- and Cauchy-type, respectively. Thereby the six Piola-type disclination density tensors in Fig. 8.10 a) are defined as the application of the appropriate curl operator to the corresponding contribution to the second deformation gradients. Accordingly, the six Piola-Kirchhoff- and Cauchy-type disclination density tensors in Fig. 8.10 b) and c) are either a result from appropriate pull-back/push-forward operations or a result from the convection of curl operators, respectively.

Disclination Densities in the Intermediate Configuration. The plastic Cauchy-type and the elastic Piola-Kirchhoff-type disclination densities in the intermediate configuration as expressed in the ambient Euclidean space are related by the identities

$$\widetilde{\boldsymbol{t}} \equiv \overline{\boldsymbol{T}} + \left[\overline{\boldsymbol{\Upsilon}} - \overline{\boldsymbol{f}} \cdot \overline{\mathrm{Grad}}\, \overline{\boldsymbol{F}}\right] \rtimes \boldsymbol{\Upsilon} \quad \text{and} \quad \overline{\boldsymbol{T}} \equiv \widetilde{\boldsymbol{t}} + \left[\widetilde{\boldsymbol{\upsilon}} - \widetilde{\boldsymbol{F}} \cdot \widetilde{\mathrm{grad}}\, \widetilde{\boldsymbol{f}}\right] \rtimes \boldsymbol{\upsilon}. \tag{8.140}$$

Here the identity $\overline{\boldsymbol{\Upsilon}} - \overline{\boldsymbol{f}} \cdot \overline{\mathrm{Grad}}\, \overline{\boldsymbol{F}} = \widetilde{\boldsymbol{\upsilon}} - \widetilde{\boldsymbol{F}} \cdot \widetilde{\mathrm{grad}}\, \widetilde{\boldsymbol{f}}$ has been invoked.

Proof. From the additive decomposition of the material curl of the second deformation gradient $\boldsymbol{G}$ and the definition of the disclination densities based on the double-anholonomic curl it follows that

$$\underbrace{\overline{\boldsymbol{f}} \cdot \overline{\boldsymbol{K}}}_{-\overline{\boldsymbol{T}}} + \underbrace{\widetilde{\boldsymbol{K}} : [\widetilde{\boldsymbol{f}} \boxtimes \operatorname{cof} \widetilde{\boldsymbol{f}}]}_{\widetilde{\boldsymbol{t}}} - \left[\overline{\boldsymbol{\Upsilon}} - \overline{\boldsymbol{f}} \cdot \overline{\mathrm{Grad}}\, \overline{\boldsymbol{F}}\right] \rtimes \boldsymbol{\Upsilon} = \boldsymbol{0},$$

whereby $\boldsymbol{\Upsilon} = -\boldsymbol{\upsilon}$ has been used. Likewise from the additive decomposition of the spatial curl of the second deformation gradient $\boldsymbol{g}$ it follows that

$$\underbrace{\widetilde{\boldsymbol{F}} \cdot \widetilde{\boldsymbol{k}}}_{-\widetilde{\boldsymbol{t}}} + \underbrace{\overline{\boldsymbol{k}} : [\overline{\boldsymbol{F}} \boxtimes \operatorname{cof} \overline{\boldsymbol{F}}]}_{\overline{\boldsymbol{T}}} - \left[\widetilde{\boldsymbol{\upsilon}} - \widetilde{\boldsymbol{F}} \cdot \widetilde{\mathrm{grad}}\, \widetilde{\boldsymbol{f}}\right] \rtimes \boldsymbol{\upsilon} = \boldsymbol{0}.$$

Obviously the above relations are nothing but the statement of compatibility of the total second deformation gradient in a disguise. □

8.2.3.5 Primary Incompatibility Density Tensors

The *primary* incompatibility density tensors ${}^{\mathrm{I}}\widetilde{\boldsymbol{H}}_0$ and ${}^{\mathrm{I}}\overline{\boldsymbol{h}}_t$ in second-order elasto-plasticity follow from the same definition as the incompatibility density tensors $\widetilde{\boldsymbol{H}}_0$ and $\overline{\boldsymbol{h}}_t$ in first-order elasto-plasticity. Thus all its variants and the relations among them are the same as in subsection 8.1.4. The possible (elastic and plastic) primary incompatibility density tensors of Piola-Kirchhoff- and Cauchy-type, respectively, and their interrelations are depicted in Fig. 8.11 for the sake of concreteness.

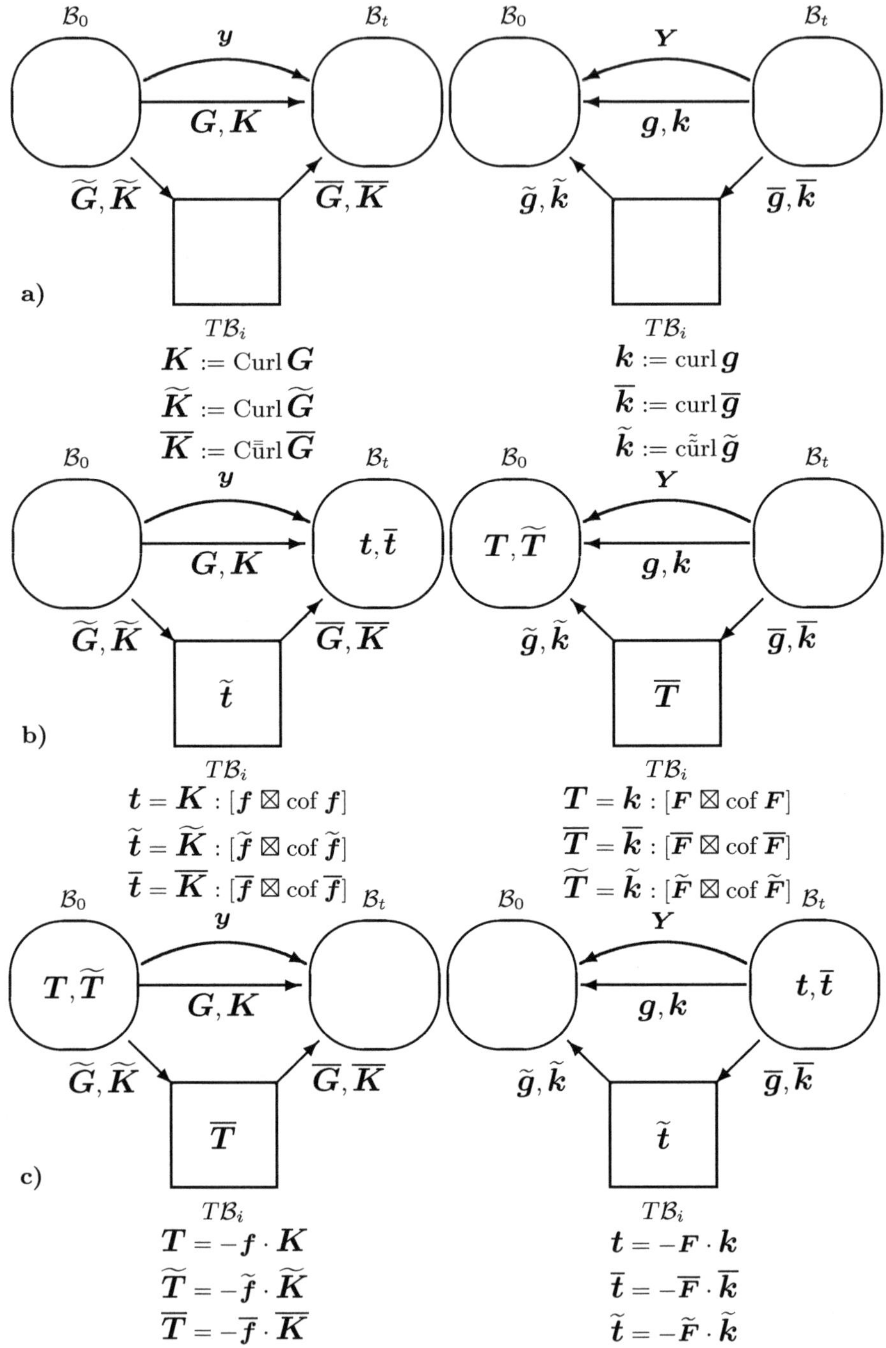

Fig. 8.10 a) Piola-type disclination density tensors; **b)** Cauchy-type disclination density tensors; **c)** Piola-Kirchhoff-type disclination density tensors

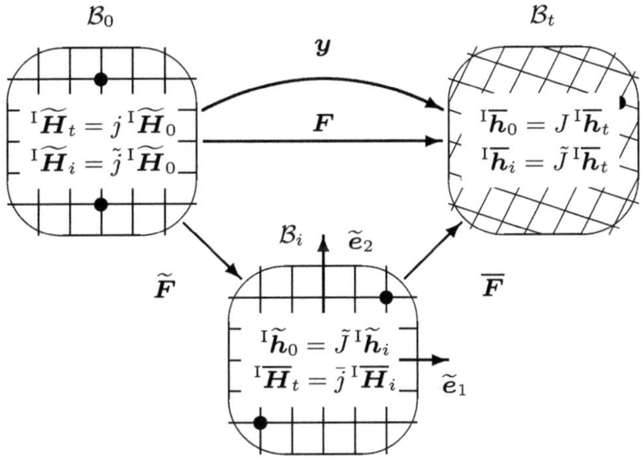

Fig. 8.11 Plastic and elastic primary incompatibility density tensors in the material, the spatial and the intermediate configuration. The primary incompatibility tensors are scaled by the Jacobians of the total, the plastic and the elastic distortions, respectively. Thereby the subindex $\tau = 0, i, t$ denotes the configuration from where or into which the corresponding scaled primary incompatibility tensor is mapped by contravariant push-forward or pull-back, respectively.

8.2.3.6 Secondary Incompatibility Density Tensors

Secondary Incompatibility Densities Based on Plastic Distortions. By incorporating the counterparts of $\widetilde{\boldsymbol{K}}$ and $\widetilde{\boldsymbol{\Gamma}}$ in the intermediate configuration the *linear part* of the material plastic secondary incompatibility density $\widetilde{\boldsymbol{\Theta}} = \widetilde{\boldsymbol{f}} \cdot [\widetilde{\boldsymbol{K}} - \operatorname{Grad} \widetilde{\boldsymbol{F}} \rtimes \widetilde{\boldsymbol{\Gamma}}]$ may be re-expressed as

$$\widetilde{\boldsymbol{\Theta}} = -\Big[\widetilde{\boldsymbol{k}} + \widetilde{\operatorname{grad}}\, \widetilde{\boldsymbol{f}} \rtimes \widetilde{\boldsymbol{v}}\Big] : [\widetilde{\boldsymbol{F}} \boxtimes \operatorname{cof} \widetilde{\boldsymbol{F}}]. \tag{8.141}$$

A contra/co/contra- (mixed) variant plastic push-forward introduces the linear part of the intermediate plastic secondary incompatibility density $\widetilde{\boldsymbol{\theta}}$ as

$$\widetilde{\boldsymbol{\theta}} := -\widetilde{\boldsymbol{F}} \cdot \widetilde{\boldsymbol{\Theta}} : [\widetilde{\boldsymbol{f}} \boxtimes \operatorname{cof} \widetilde{\boldsymbol{f}}] = \widetilde{\boldsymbol{F}} \cdot \Big[\widetilde{\boldsymbol{k}} + \widetilde{\operatorname{grad}}\, \widetilde{\boldsymbol{f}} \rtimes \widetilde{\boldsymbol{v}}\Big]. \tag{8.142}$$

In terms of the plastic intermediate Cauchy-type disclination density $\widetilde{\boldsymbol{t}}$ the linear part of the intermediate plastic secondary incompatibility density reads

$$\widetilde{\boldsymbol{\theta}} = -\widetilde{\boldsymbol{t}} + \widetilde{\boldsymbol{F}} \cdot \widetilde{\operatorname{grad}}\, \widetilde{\boldsymbol{f}} \rtimes \widetilde{\boldsymbol{v}}. \tag{8.143}$$

Correspondingly the (nonlinear) material plastic secondary incompatibility density ${}^{\mathrm{II}}\widetilde{\boldsymbol{H}} = \widetilde{\boldsymbol{\Theta}} + \widetilde{\boldsymbol{\Gamma}} \rtimes \widetilde{\boldsymbol{\Gamma}}$ may be re-expressed as

$${}^{\mathrm{II}}\widetilde{\boldsymbol{H}} = -\Big[\widetilde{\boldsymbol{k}} - [\widetilde{\boldsymbol{g}} - \widetilde{\operatorname{grad}}\, \widetilde{\boldsymbol{f}}] \rtimes \widetilde{\boldsymbol{v}}\Big] : [\widetilde{\boldsymbol{F}} \boxtimes \operatorname{cof} \widetilde{\boldsymbol{F}}]. \tag{8.144}$$

A contra/co/contra- (mixed) variant plastic push-forward introduces the (nonlinear) intermediate plastic secondary incompatibility density ${}^{\mathrm{II}}\widetilde{\boldsymbol{h}}$

$$ {}^{\mathrm{II}}\widetilde{\boldsymbol{h}} := -\widetilde{\boldsymbol{F}} \cdot {}^{\mathrm{II}}\widetilde{\boldsymbol{H}} : [\widetilde{\boldsymbol{f}} \boxtimes \operatorname{cof} \widetilde{\boldsymbol{f}}] = \widetilde{\boldsymbol{F}} \cdot \Big[\widetilde{\boldsymbol{k}} - [\widetilde{\boldsymbol{g}} - \widetilde{\operatorname{grad}}\, \widetilde{\boldsymbol{f}}] \rtimes \widetilde{\boldsymbol{v}}\Big]. \tag{8.145} $$

Finally the (nonlinear) intermediate plastic secondary incompatibility density reads in terms of $\widetilde{\boldsymbol{t}}$ as

$$ {}^{\mathrm{II}}\widetilde{\boldsymbol{h}} = -\widetilde{\boldsymbol{t}} - [\widetilde{\boldsymbol{v}} - \widetilde{\boldsymbol{F}} \cdot \widetilde{\operatorname{grad}}\, \widetilde{\boldsymbol{f}}] \rtimes \widetilde{\boldsymbol{v}}. \tag{8.146} $$

Secondary Incompatibility Densities Based on Elastic Distortions. By incorporating the counterparts of $\overline{\boldsymbol{k}}$ and $\overline{\boldsymbol{\gamma}}$ in the intermediate configuration the *linear part* of the spatial elastic secondary incompatibility density $\overline{\boldsymbol{\theta}} = \overline{\boldsymbol{F}} \cdot [\overline{\boldsymbol{k}} - \operatorname{grad} \overline{\boldsymbol{f}} \rtimes \overline{\boldsymbol{\gamma}}]$ may be re-expressed as

$$ \overline{\boldsymbol{\theta}} = -\Big[\overline{\boldsymbol{K}} + \overline{\operatorname{Grad}}\, \overline{\boldsymbol{F}} \rtimes \overline{\boldsymbol{\Upsilon}}\Big] : [\overline{\boldsymbol{f}} \boxtimes \operatorname{cof} \overline{\boldsymbol{f}}]. \tag{8.147} $$

A contra/co/contra- (mixed) variant elastic pull-back introduces the linear part of the intermediate elastic secondary incompatibility density $\overline{\boldsymbol{\Theta}}$ as

$$ \overline{\boldsymbol{\Theta}} := -\overline{\boldsymbol{f}} \cdot \overline{\boldsymbol{\theta}} : [\overline{\boldsymbol{F}} \boxtimes \operatorname{cof} \overline{\boldsymbol{F}}] = \overline{\boldsymbol{f}} \cdot \Big[\overline{\boldsymbol{K}} + \overline{\operatorname{Grad}}\, \overline{\boldsymbol{F}} \rtimes \overline{\boldsymbol{\Upsilon}}\Big]. \tag{8.148} $$

In terms of the elastic intermediate Piola-Kirchhoff-type disclination density $\overline{\boldsymbol{T}}$ the linear part of the intermediate elastic secondary incompatibility density reads

$$ \overline{\boldsymbol{\Theta}} = -\overline{\boldsymbol{T}} + \overline{\boldsymbol{f}} \cdot \overline{\operatorname{Grad}}\, \widetilde{\boldsymbol{F}} \rtimes \overline{\boldsymbol{\Upsilon}}. \tag{8.149} $$

Correspondingly the (nonlinear) spatial elastic secondary incompatibility density ${}^{\mathrm{II}}\overline{\boldsymbol{h}} = \overline{\boldsymbol{\theta}} + \overline{\boldsymbol{\gamma}} \rtimes \overline{\boldsymbol{\gamma}}$ may be re-expressed as

$$ {}^{\mathrm{II}}\overline{\boldsymbol{h}} = -\Big[\overline{\boldsymbol{K}} - [\overline{\boldsymbol{G}} - \overline{\operatorname{Grad}}\, \overline{\boldsymbol{F}}] \rtimes \overline{\boldsymbol{\Upsilon}}\Big] : [\overline{\boldsymbol{f}} \boxtimes \operatorname{cof} \overline{\boldsymbol{f}}]. \tag{8.150} $$

A contra/co/contra- (mixed) variant elastic pull-back introduces the (nonlinear) intermediate elastic secondary incompatibility density ${}^{\mathrm{II}}\overline{\boldsymbol{H}}$

$$ {}^{\mathrm{II}}\overline{\boldsymbol{H}} := -\overline{\boldsymbol{F}} \cdot {}^{\mathrm{II}}\overline{\boldsymbol{h}} : [\overline{\boldsymbol{F}} \boxtimes \operatorname{cof} \overline{\boldsymbol{F}}] = \overline{\boldsymbol{f}} \cdot \Big[\overline{\boldsymbol{K}} - [\overline{\boldsymbol{G}} - \overline{\operatorname{Grad}}\, \overline{\boldsymbol{F}}] \rtimes \overline{\boldsymbol{\Upsilon}}\Big]. \tag{8.151} $$

Finally the (nonlinear) intermediate elastic secondary incompatibility density reads in terms of $\overline{\boldsymbol{T}}$ as

$$ {}^{\mathrm{II}}\overline{\boldsymbol{H}} = -\overline{\boldsymbol{T}} - [\overline{\boldsymbol{\Upsilon}} - \overline{\boldsymbol{f}} \cdot \overline{\operatorname{Grad}}\, \overline{\boldsymbol{F}}] \rtimes \overline{\boldsymbol{\Upsilon}}. \tag{8.152} $$

Incompatibility Densities in the Intermediate Configuration. The plastic and the elastic secondary incompatibility densities in the intermediate configuration as expressed in the ambient Euclidean space are related by the identity

$$^{\text{II}}\widetilde{\boldsymbol{h}} \equiv {}^{\text{II}}\overline{\boldsymbol{H}}. \tag{8.153}$$

Proof. Equating the intermediate plastic and elastic secondary incompatibility densities (and multiplying by -1) renders

$$\widetilde{\boldsymbol{t}} + \left[\widetilde{\boldsymbol{\upsilon}} - \widetilde{\boldsymbol{F}} \cdot \widetilde{\text{grad}}\, \widetilde{\boldsymbol{f}}\right] \rtimes \widetilde{\boldsymbol{\upsilon}} = \overline{\boldsymbol{T}} + \left[\overline{\boldsymbol{\Upsilon}} - \overline{\boldsymbol{f}} \cdot \overline{\text{Grad}}\, \overline{\boldsymbol{F}}\right] \rtimes \overline{\boldsymbol{\Upsilon}}$$

Noting that $\widetilde{\boldsymbol{\upsilon}} - \widetilde{\boldsymbol{F}} \cdot \widetilde{\text{grad}}\, \widetilde{\boldsymbol{f}} = \overline{\boldsymbol{\Upsilon}} - \overline{\boldsymbol{f}} \cdot \overline{\text{Grad}}\, \overline{\boldsymbol{F}}$ together with $\overline{\boldsymbol{\Upsilon}} = -\overline{\boldsymbol{\upsilon}}$ (or $\widetilde{\boldsymbol{\upsilon}} = -\widetilde{\boldsymbol{\Upsilon}}$) and $\boldsymbol{\upsilon} = \overline{\boldsymbol{\upsilon}} + \widetilde{\boldsymbol{\upsilon}}$ (or $\boldsymbol{\Upsilon} = \overline{\boldsymbol{\Upsilon}} + \widetilde{\boldsymbol{\Upsilon}}$) and comparing with the relation between the elastic and the plastic intermediate disclination densities $\overline{\boldsymbol{T}}$ and $\widetilde{\boldsymbol{t}}$ immediately concludes the proof. □

For the sake of completeness, with $\boldsymbol{\Theta} := \text{Curl}\,\boldsymbol{\Gamma}$ and $\boldsymbol{\theta} := \text{curl}\,\boldsymbol{\gamma}$, the above derivations motivate to also introduce total material and spatial secondary incompatibility densities (that vanish however identically due to the compatibility of the total deformation)

$$^{\text{II}}\boldsymbol{H} := \boldsymbol{\Theta} + \boldsymbol{\Gamma} \rtimes \boldsymbol{\Gamma} \quad \text{and} \quad {}^{\text{II}}\boldsymbol{h} := \boldsymbol{\theta} + \boldsymbol{\gamma} \rtimes \boldsymbol{\gamma}.$$

The various secondary incompatibility density tensors are finally summarized for convenience in Fig. 8.12.

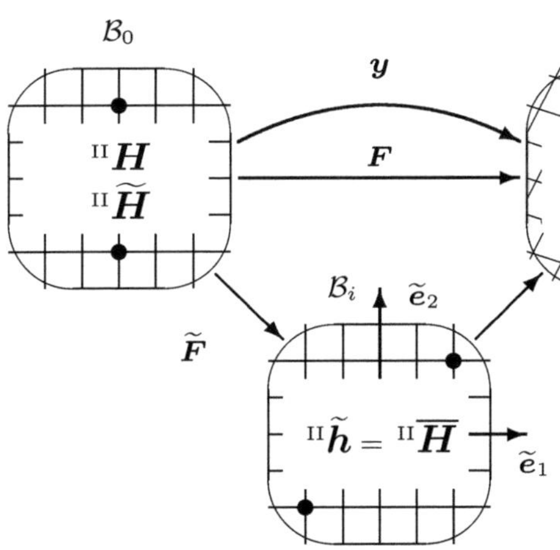

Fig. 8.12 Total, plastic and elastic secondary incompatibility density tensors in the material, the spatial and the intermediate configuration

Supplement 8.1. First-Order Elasto-Plasticity in Euclidean Space
Geometrically nonlinear first-order elasto-plasticity is typically formulated based on the multiplicative decomposition of the deformation gradient into elastic and plastic contributions. For crystalline materials the kinematics of multiplicative elasto-plasticity are motivated by the idea of a collective dislocation flow through the crystal lattice (that leaves the crystal lattice unchanged) followed by the elastic, i.e. stress producing distortion of the crystal lattice. Moreover the dislocations that get stuck in the crystal lattice do result in strong but localized stress fields that in turn hinder further plastic flow. Thus statistically stored dislocations (SSD) as well as geometrically necessary dislocations (GND) do represent obstacles for further dislocation flow and are thus responsible for the hardening of the material. The multiplicative decomposition of the deformation gradient has been advocated in the early contributions [95, 112, 121, 195] and has since then been further developed continuously. Thereby a typical focus is on suited algorithmic settings that allow for efficient and accurate computational simulations, e.g. of bulk and/or sheet forming processes. Based on the brief outline of first-order elasticity in the previous chapter the aim of this supplement is to discuss the relevant kinematic aspects of multiplicative elasto-plasticity and to develop the resulting kinetic relations that capture the evolution of irreversible inelastic deformations. Central to the derivation of the plastic evolution equations is the postulate of maximum dissipation in its various variants, e.g. within the setting of convex analysis. All kinetic relations are represented for the generic anisotropic case and may straightforwardly be detailed on a case by case basis if needed.

Multiplicative Decomposition of Deformation Gradient:
Motivated by the irreversible deformation of single crystals the deformation gradient is typically assumed to be decomposed *multiplicatively* into plastic and elastic contributions or rather *distortions*

$$\boldsymbol{F} = \overline{\boldsymbol{F}} \cdot \widetilde{\boldsymbol{F}}.$$

Thereby the *plastic distortion* $\widetilde{\boldsymbol{F}}$ defines a map from the tangent space $T\mathcal{B}_0$ of the material configuration into the tangent space $T\mathcal{B}_i$ of the *incompatible intermediate configuration*; likewise the *elastic distortion* $\overline{\boldsymbol{F}}$ defines a map from the tangent space $T\mathcal{B}_i$ of the incompatible intermediate configuration into the tangent space $T\mathcal{B}_t$ of the spatial configuration

$$\mathrm{d}\widetilde{\boldsymbol{x}} = \widetilde{\boldsymbol{F}} \cdot \mathrm{d}\boldsymbol{X} \quad \text{and} \quad \mathrm{d}\boldsymbol{x} = \overline{\boldsymbol{F}} \cdot \mathrm{d}\overline{\boldsymbol{X}} \quad \text{with} \quad \mathrm{d}\widetilde{\boldsymbol{x}} \equiv \mathrm{d}\overline{\boldsymbol{X}}.$$

Note that looking forward from the material configuration the incompatible intermediate configuration has the flavor of a spatial configuration; vice versa looking backward from the spatial configuration the incompatible intermediate configuration has the flavor of a material configuration. Thus the notation for quantities living in $\mathcal{B}_i$ is sometimes adopted to the point of view with the resulting notational redundancies such as $\mathrm{d}\widetilde{\boldsymbol{x}} \equiv \mathrm{d}\overline{\boldsymbol{X}}$. Finally, based on the plastic distortion $\widetilde{\boldsymbol{F}}$ it will later prove convenient to also introduce the *plastic dislocation density* as

$$\operatorname{Curl} \widetilde{\boldsymbol{F}}.$$

Thereby the plastic dislocation density is a second-order contra/co- (mixed) variant two-point (intermediate/material) tensor that characterizes the incompatibility of the intermediate configuration.

Elastic and Plastic Covariant Strain Measures:

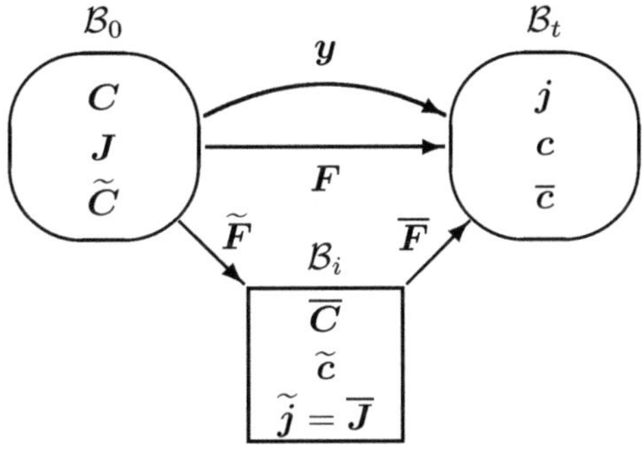

Based on the multiplicative decomposition of the deformation gradient the covariant material *plastic right Cauchy-Green strain* is defined as

$$\widetilde{\boldsymbol{C}} := \widetilde{\boldsymbol{F}}^t \cdot \widetilde{\boldsymbol{j}} \cdot \widetilde{\boldsymbol{F}} \in T^*\mathcal{B}_0 \otimes T^*\mathcal{B}_0.$$

Here $\widetilde{\boldsymbol{j}}$ denotes the covariant unit tensor of the intermediate configuration. Likewise, the covariant spatial *elastic Finger strain* is defined as

$$\overline{\boldsymbol{c}} := \overline{\boldsymbol{f}}^t \cdot \overline{\boldsymbol{J}} \cdot \overline{\boldsymbol{f}} \in T^*\mathcal{B}_t \otimes T^*\mathcal{B}_t.$$

Here $\overline{\boldsymbol{J}} \equiv \widetilde{\boldsymbol{j}}$ denotes again the covariant unit tensor of the intermediate configuration, the symbolic redundancy has been introduced for notational convenience. Finally the covariant intermediate *elastic right Cauchy-Green strain* and the covariant intermediate *plastic Finger strain* are introduced as

$$\overline{\boldsymbol{C}} := \overline{\boldsymbol{F}}^t \cdot \boldsymbol{j} \cdot \overline{\boldsymbol{F}} \in T^*\mathcal{B}_i \otimes T^*\mathcal{B}_i \quad \text{and} \quad \widetilde{\boldsymbol{c}} := \widetilde{\boldsymbol{f}}^t \cdot \boldsymbol{J} \cdot \widetilde{\boldsymbol{f}} \in T^*\mathcal{B}_i \otimes T^*\mathcal{B}_i.$$

Note the further relations between the various covariant strain measures

$$\boldsymbol{C} = \widetilde{\boldsymbol{F}}^t \cdot \overline{\boldsymbol{C}} \cdot \widetilde{\boldsymbol{F}} \quad \text{and} \quad \widetilde{\boldsymbol{C}} = \boldsymbol{F}^t \cdot \overline{\boldsymbol{c}} \cdot \boldsymbol{F} \quad \text{and} \quad \widetilde{\boldsymbol{c}} = \overline{\boldsymbol{F}}^t \cdot \boldsymbol{c} \cdot \overline{\boldsymbol{F}}.$$

The covariant transformation behavior qualifies the above strain measures and unit tensors as covariant tensors. Finally, based on the plastic right Cauchy-Green strain $\widetilde{\boldsymbol{C}}$ it will later prove convenient to also introduce the *plastic incompatibility density* as

$$\operatorname{Inc}\widetilde{\boldsymbol{C}}.$$

Thereby the plastic incompatibility density is a second-order contravariant material tensor that characterizes the incompatibility of the intermediate configuration.

Elastic and Plastic Contravariant Strain Measures:

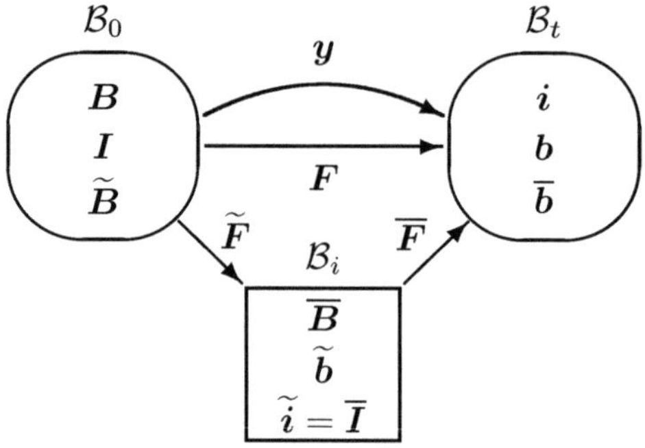

The contravariant strain measures follow as the inverses to the covariant strain measures and are thus elements of the corresponding tangent spaces. Consequently the contravariant material *plastic Piola strain* is defined as

$$\widetilde{\boldsymbol{B}} := \widetilde{\boldsymbol{f}} \cdot \widetilde{\boldsymbol{i}} \cdot \widetilde{\boldsymbol{f}}^t \in T\mathcal{B}_0 \otimes T\mathcal{B}_0.$$

Here $\widetilde{\boldsymbol{i}} = \widetilde{\boldsymbol{j}}^{-1}$ denotes the contravariant unit tensor of the intermediate configuration. Likewise, the contravariant spatial *elastic left Cauchy-Green strain* is defined as

$$\overline{\boldsymbol{b}} := \overline{\boldsymbol{F}}^t \cdot \overline{\boldsymbol{I}} \cdot \overline{\boldsymbol{F}} \in T\mathcal{B}_t \otimes T\mathcal{B}_t.$$

Here $\overline{\boldsymbol{I}} = \overline{\boldsymbol{J}}^{-1} \equiv \widetilde{\boldsymbol{i}}$ denotes again the contravariant unit tensor of the intermediate configuration. Finally the contravariant intermediate *elastic Piola strain* and the contravariant intermediate *plastic left Cauchy-Green strain* are introduced as

$$\overline{\boldsymbol{B}} := \overline{\boldsymbol{f}} \cdot \boldsymbol{i} \cdot \overline{\boldsymbol{f}}^t \in T\mathcal{B}_i \otimes T\mathcal{B}_i \quad \text{and} \quad \widetilde{\boldsymbol{b}} := \widetilde{\boldsymbol{F}} \cdot \boldsymbol{I} \cdot \widetilde{\boldsymbol{F}}^t \in T\mathcal{B}_i \otimes T\mathcal{B}_i.$$

Note the further relations between the various contravariant strain measures

$$\boldsymbol{B} = \widetilde{\boldsymbol{f}} \cdot \overline{\boldsymbol{B}} \cdot \widetilde{\boldsymbol{f}}^t \quad \text{and} \quad \widetilde{\boldsymbol{B}} = \boldsymbol{f} \cdot \overline{\boldsymbol{b}} \cdot \boldsymbol{f}^t \quad \text{and} \quad \widetilde{\boldsymbol{b}} = \overline{\boldsymbol{f}} \cdot \boldsymbol{b} \cdot \overline{\boldsymbol{f}}^t.$$

The contravariant transformation behavior qualifies the above strain measures and unit tensors as contravariant tensors.

Elastic and Plastic Distortion Rates:
Based on the material time derivatives of the elastic and plastic distortions corresponding two-point *elastic* and *plastic distortion rates* are defined as

$$\overline{\boldsymbol{\Lambda}} := \mathrm{D}_t \overline{\boldsymbol{F}} \quad \text{and} \quad \widetilde{\boldsymbol{\Lambda}} := \mathrm{D}_t \widetilde{\boldsymbol{F}}.$$

Moreover, following the notational convention as applied already to the velocity gradients, left-sided contravariant pull-back and right-sided covariant push-forward operations in terms of the appropriate elastic and plastic distortions render

$$\overline{\boldsymbol{L}} := \overline{\boldsymbol{f}} \cdot \overline{\boldsymbol{\Lambda}} \quad \text{and} \quad \overline{\boldsymbol{l}} := \overline{\boldsymbol{\Lambda}} \cdot \overline{\boldsymbol{f}} \quad \text{and} \quad \widetilde{\boldsymbol{L}} := \widetilde{\boldsymbol{f}} \cdot \widetilde{\boldsymbol{\Lambda}} \quad \text{and} \quad \widetilde{\boldsymbol{l}} := \widetilde{\boldsymbol{\Lambda}} \cdot \widetilde{\boldsymbol{f}}.$$

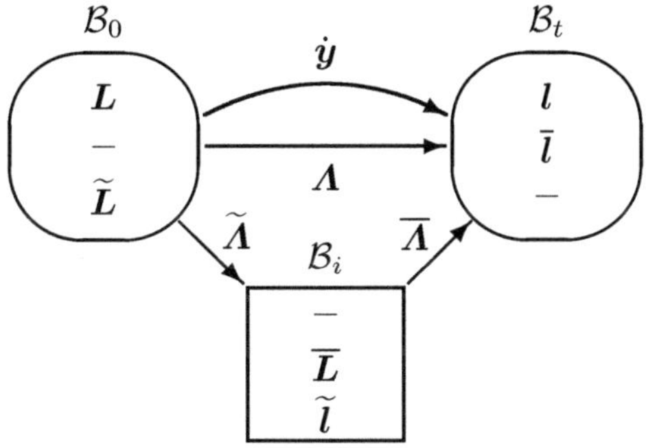

Note that the mixed variant elastic $\overline{\boldsymbol{L}}$ and plastic $\widetilde{\boldsymbol{l}}$ live in the intermediate configuration, whereas the mixed variant elastic $\overline{\boldsymbol{l}}$ and plastic $\widetilde{\boldsymbol{L}}$ live in the spatial and the material configuration, respectively. Finally, two-sided mixed variant pull-back/push-forward operations in terms of the elastic and plastic distortions introduce the two-point elastic and plastic distortion rates

$$\overline{\boldsymbol{\lambda}} := \overline{\boldsymbol{f}} \cdot \overline{\boldsymbol{\Lambda}} \cdot \overline{\boldsymbol{f}} \quad \text{and} \quad \widetilde{\boldsymbol{\lambda}} := \widetilde{\boldsymbol{f}} \cdot \widetilde{\boldsymbol{\Lambda}} \cdot \widetilde{\boldsymbol{f}}.$$

The relation between the various elastic or plastic distortion rates follows the pattern established already for the (total) velocity gradients.

Additive Decomposition of Velocity Gradients:

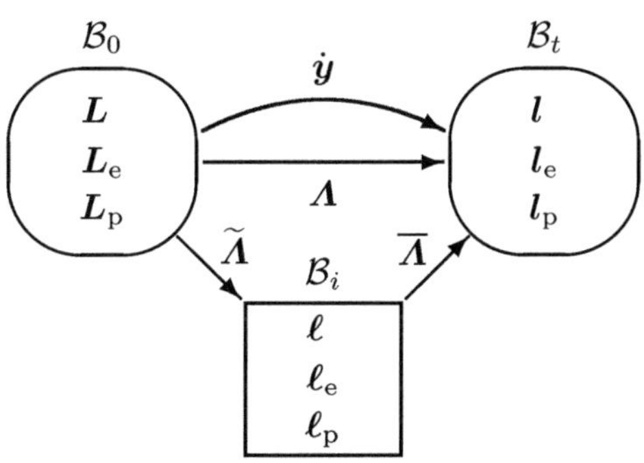

Then, with these definitions at hand and as an intermediate step, the two-point velocity gradients decompose additively into

$$\boldsymbol{\Lambda} = \overline{\boldsymbol{\Lambda}} \cdot \widetilde{\boldsymbol{F}} + \overline{\boldsymbol{F}} \cdot \widetilde{\boldsymbol{\Lambda}} =: \boldsymbol{\Lambda}_{\mathrm{e}} + \boldsymbol{\Lambda}_{\mathrm{p}}$$

and

$$\boldsymbol{\lambda} = \widetilde{\boldsymbol{f}} \cdot \overline{\boldsymbol{\lambda}} + \widetilde{\boldsymbol{\lambda}} \cdot \overline{\boldsymbol{f}} =: \boldsymbol{\lambda}_{\mathrm{e}} + \boldsymbol{\lambda}_{\mathrm{p}}.$$

As a consequence the material and the spatial velocity gradients are thus expressed as

$$\boldsymbol{L} = \boldsymbol{L}_{\mathrm{e}} + \boldsymbol{L}_{\mathrm{p}} \quad \text{with} \quad \boldsymbol{L}_{\mathrm{p}} \equiv \widetilde{\boldsymbol{L}} \quad \text{and} \quad \boldsymbol{l} = \boldsymbol{l}_{\mathrm{e}} + \boldsymbol{l}_{\mathrm{p}} \quad \text{with} \quad \boldsymbol{l}_{\mathrm{e}} \equiv \overline{\boldsymbol{l}}.$$

Here the mixed variant *plastic pull-back* of the intermediate elastic $\overline{\boldsymbol{L}}$ to the material configuration and the mixed variant *elastic push-forward* of the intermediate plastic $\widetilde{\boldsymbol{l}}$ to the spatial configuration, respectively, have been abbreviated as

$$\boldsymbol{L}_{\mathrm{e}} := \widetilde{\boldsymbol{f}} \cdot \overline{\boldsymbol{L}} \cdot \widetilde{\boldsymbol{F}} \quad \text{and} \quad \boldsymbol{l}_{\mathrm{p}} := \overline{\boldsymbol{F}} \cdot \widetilde{\boldsymbol{l}} \cdot \overline{\boldsymbol{f}}.$$

Moreover the, admittedly, symbolic redundancy for the material plastic $\widetilde{\boldsymbol{L}}$ and the spatial elastic $\overline{\boldsymbol{l}}$ has been introduced for notational convenience. Finally, the mixed variant plastic push-forward $\boldsymbol{\ell} := \widetilde{\boldsymbol{F}} \cdot \boldsymbol{L} \cdot \widetilde{\boldsymbol{f}}$ of the material velocity gradient or, equivalently, the mixed variant elastic pull-back $\boldsymbol{\ell} := \overline{\boldsymbol{f}} \cdot \boldsymbol{l} \cdot \overline{\boldsymbol{F}}$ of the spatial velocity gradient to the intermediate configuration result in

$$\boldsymbol{\ell} = \boldsymbol{\ell}_{\mathrm{e}} + \boldsymbol{\ell}_{\mathrm{p}} \quad \text{with} \quad \boldsymbol{\ell}_{\mathrm{e}} := \overline{\boldsymbol{L}} \quad \text{and} \quad \boldsymbol{\ell}_{\mathrm{p}} := \widetilde{\boldsymbol{l}}.$$

Again the symbolic redundancy for the intermediate elastic $\overline{\boldsymbol{L}}$ and the intermediate plastic $\widetilde{\boldsymbol{l}}$ has been introduced for notational convenience. Observe that only the velocity gradient $\boldsymbol{\ell}$ in the intermediate configuration is decoupled into exclusively elastic and plastic contributions. This is clearly in contrast to the (total) velocity gradients $\boldsymbol{L}$ and $\boldsymbol{l}$ in the material and the spatial configuration that include hybrid elastic-plastic terms such as $\boldsymbol{L}_{\mathrm{e}}$ or $\boldsymbol{l}_{\mathrm{p}}$, respectively.

Spatial Rate-of-Deformation Tensors:
The symmetric contributions to the fully co- and contravariant versions of the elastic and the plastic parts of the spatial velocity gradient are denoted the *spatial elastic* and *plastic rate-of-deformation tensor*, respectively, i.e.

$$\boldsymbol{d}_{\mathrm{e}} := [\boldsymbol{j} \cdot \boldsymbol{l}_{\mathrm{e}}]^{\mathrm{sym}} = [\boldsymbol{l}_{\mathrm{e}} \cdot \boldsymbol{i}]^{\mathrm{sym}} \quad \text{and} \quad \boldsymbol{d}_{\mathrm{p}} := [\boldsymbol{j} \cdot \boldsymbol{l}_{\mathrm{p}}]^{\mathrm{sym}} = [\boldsymbol{l}_{\mathrm{p}} \cdot \boldsymbol{i}]^{\mathrm{sym}}.$$

Note that the spatial elastic rate-of-deformation tensor $\boldsymbol{d}_{\mathrm{e}} \equiv \overline{\boldsymbol{d}}$ computes from exclusively elastic contributions

$$\boldsymbol{d}_{\mathrm{e}} \equiv \overline{\boldsymbol{d}} := [\boldsymbol{j} \cdot \overline{\boldsymbol{l}}]^{\mathrm{sym}} = [\overline{\boldsymbol{l}} \cdot \boldsymbol{i}]^{\mathrm{sym}}$$

which is in contrast to $\boldsymbol{d}_{\mathrm{p}}$ that computes from the hybrid elastic-plastic $\boldsymbol{l}_{\mathrm{p}}$.

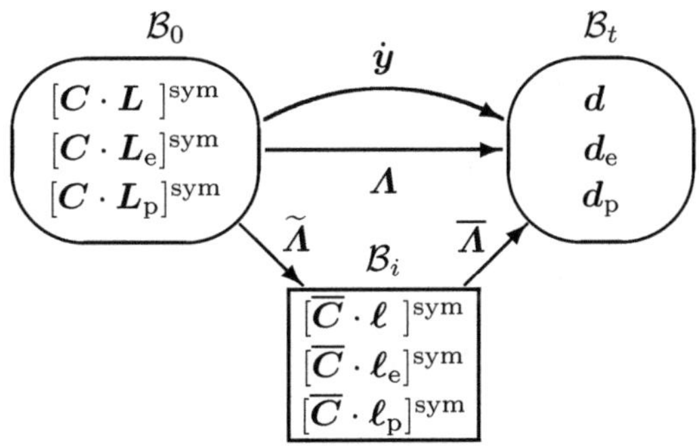

Then the covariant elastic and total pull-back of the spatial *elastic* rate-of-deformation tensor involve as metric the covariant $\overline{\boldsymbol{C}}$ and $\boldsymbol{C}$, respectively, and render

$$[\overline{\boldsymbol{C}}\cdot\boldsymbol{\ell}_{\rm e}]^{\rm sym}=\overline{\boldsymbol{F}}^t\cdot\boldsymbol{d}_{\rm e}\cdot\overline{\boldsymbol{F}}\quad\text{and}\quad[\boldsymbol{C}\cdot\boldsymbol{L}_{\rm e}]^{\rm sym}=\boldsymbol{F}^t\cdot\boldsymbol{d}_{\rm e}\cdot\boldsymbol{F}.$$

Accordingly the covariant elastic and total pull-back of the spatial *plastic* rate-of-deformation tensor result in

$$[\overline{\boldsymbol{C}}\cdot\boldsymbol{\ell}_{\rm p}]^{\rm sym}=\overline{\boldsymbol{F}}^t\cdot\boldsymbol{d}_{\rm p}\cdot\overline{\boldsymbol{F}}\quad\text{and}\quad[\boldsymbol{C}\cdot\boldsymbol{L}_{\rm p}]^{\rm sym}=\boldsymbol{F}^t\cdot\boldsymbol{d}_{\rm p}\cdot\boldsymbol{F}.$$

Observe finally that the covariant elastic and total pull-back of the spatial (total) rate-of-deformation tensor $\boldsymbol{d}:=[\boldsymbol{j}\cdot\boldsymbol{l}]^{\rm sym}$ lead to

$$[\overline{\boldsymbol{C}}\cdot\boldsymbol{\ell}]^{\rm sym}=\overline{\boldsymbol{F}}^t\cdot\boldsymbol{d}\cdot\overline{\boldsymbol{F}}\quad\text{and}\quad[\boldsymbol{C}\cdot\boldsymbol{L}]^{\rm sym}=\boldsymbol{F}^t\cdot\boldsymbol{d}\cdot\boldsymbol{F}.$$

In summary the covariant spatial (total) rate-of-deformation tensor (and its elastic and total pull-back) decomposes additively into elastic and plastic contributions.

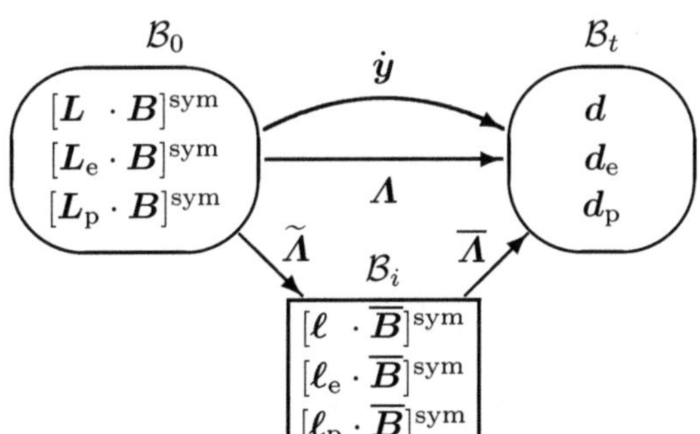

Likewise the contravariant elastic and total pull-back of the spatial *elastic* rate-of-deformation tensor involve as metric the contravariant $\overline{\boldsymbol{B}}$ and $\boldsymbol{B}$, respectively, and render

$$[\boldsymbol{\ell}_{\rm e}\cdot\overline{\boldsymbol{B}}]^{\rm sym}=\overline{\boldsymbol{f}}\cdot\boldsymbol{d}_{\rm e}\cdot\overline{\boldsymbol{f}}^t\quad\text{and}\quad[\boldsymbol{L}_{\rm e}\cdot\boldsymbol{B}]^{\rm sym}=\boldsymbol{f}\cdot\boldsymbol{d}_{\rm e}\cdot\boldsymbol{f}^t.$$

Accordingly the contravariant elastic and total pull-back of the spatial *plastic* rate-of-deformation tensor result in

$$[\boldsymbol{\ell}_\mathrm{p} \cdot \overline{\boldsymbol{B}}]^\mathrm{sym} = \overline{\boldsymbol{f}} \cdot \boldsymbol{d}_\mathrm{p} \cdot \overline{\boldsymbol{f}}^t \quad \text{and} \quad [\boldsymbol{L}_\mathrm{p} \cdot \boldsymbol{B}]^\mathrm{sym} = \boldsymbol{f} \cdot \boldsymbol{d}_\mathrm{p} \cdot \boldsymbol{f}^t.$$

Observe finally that the contravariant elastic and total pull-back of the spatial (total) rate-of-deformation tensor $\boldsymbol{d} := [\boldsymbol{l} \cdot \boldsymbol{i}]^\mathrm{sym}$ lead to

$$[\boldsymbol{\ell} \cdot \overline{\boldsymbol{B}}]^\mathrm{sym} = \overline{\boldsymbol{f}} \cdot \boldsymbol{d} \cdot \overline{\boldsymbol{f}}^t \quad \text{and} \quad [\boldsymbol{L} \cdot \boldsymbol{B}]^\mathrm{sym} = \boldsymbol{f} \cdot \boldsymbol{d} \cdot \boldsymbol{f}^t.$$

In summary the contravariant spatial (total) rate-of-deformation tensor (and its elastic and total pull-back) decomposes additively into elastic and plastic contributions.

Material Rate-of-Deformation Tensors:
The symmetric contributions to the fully co- and contravariant versions of the elastic and the plastic parts of the material velocity gradient are denoted the *material elastic* and *plastic rate-of-deformation tensor*, respectively, i.e.

$$\boldsymbol{D}_\mathrm{e} := [\boldsymbol{J} \cdot \boldsymbol{L}_\mathrm{e}]^\mathrm{sym} = [\boldsymbol{L}_\mathrm{e} \cdot \boldsymbol{I}]^\mathrm{sym} \quad \text{and} \quad \boldsymbol{D}_\mathrm{p} := [\boldsymbol{J} \cdot \boldsymbol{L}_\mathrm{p}]^\mathrm{sym} = [\boldsymbol{L}_\mathrm{p} \cdot \boldsymbol{I}]^\mathrm{sym}.$$

Note that the material plastic rate-of-deformation tensor $\boldsymbol{D}_\mathrm{p} \equiv \widetilde{\boldsymbol{D}}$ computes from exclusively plastic contributions

$$\boldsymbol{D}_\mathrm{p} \equiv \widetilde{\boldsymbol{D}} := [\boldsymbol{J} \cdot \widetilde{\boldsymbol{L}}]^\mathrm{sym} = [\widetilde{\boldsymbol{L}} \cdot \boldsymbol{I}]^\mathrm{sym}$$

which is in contrast to $\boldsymbol{D}_\mathrm{e}$ that computes from the hybrid elastic-plastic $\boldsymbol{L}_\mathrm{e}$.

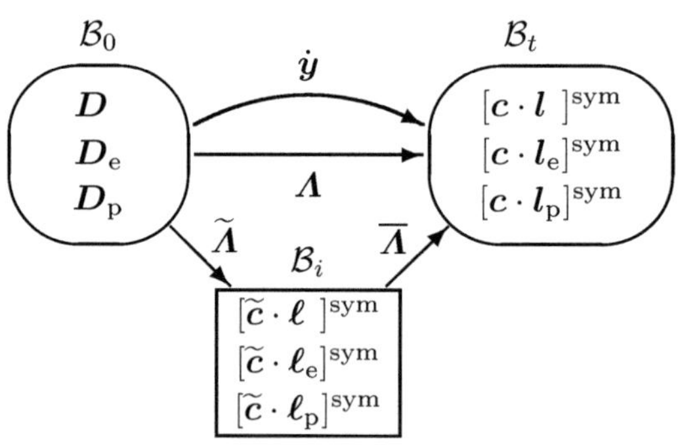

Then the covariant plastic and total push-forward of the material *elastic* rate-of-deformation tensor involve as metric the covariant $\widetilde{\boldsymbol{c}}$ and $\boldsymbol{c}$, respectively, and render

$$[\widetilde{\boldsymbol{c}} \cdot \boldsymbol{\ell}_\mathrm{e}]^\mathrm{sym} = \widetilde{\boldsymbol{f}}^t \cdot \boldsymbol{D}_\mathrm{e} \cdot \widetilde{\boldsymbol{f}} \quad \text{and} \quad [\boldsymbol{c} \cdot \boldsymbol{l}_\mathrm{e}]^\mathrm{sym} = \boldsymbol{f}^t \cdot \boldsymbol{D}_\mathrm{e} \cdot \boldsymbol{f}.$$

Accordingly the covariant plastic and total push-forward of the material *plastic* rate-of-deformation tensor result in

$$[\widetilde{\boldsymbol{c}} \cdot \boldsymbol{\ell}_\mathrm{p}]^\mathrm{sym} = \widetilde{\boldsymbol{f}}^t \cdot \boldsymbol{D}_\mathrm{p} \cdot \widetilde{\boldsymbol{f}} \quad \text{and} \quad [\boldsymbol{c} \cdot \boldsymbol{l}_\mathrm{p}]^\mathrm{sym} = \boldsymbol{f}^t \cdot \boldsymbol{D}_\mathrm{p} \cdot \boldsymbol{f}.$$

Observe finally that the covariant plastic and total push-forward of the material (total) rate-of-deformation tensor $\boldsymbol{D} := [\boldsymbol{J} \cdot \boldsymbol{L}]^\mathrm{sym}$ lead to

$$[\widetilde{\boldsymbol{c}} \cdot \boldsymbol{\ell}]^\mathrm{sym} = \widetilde{\boldsymbol{f}}^t \cdot \boldsymbol{D} \cdot \widetilde{\boldsymbol{f}} \quad \text{and} \quad [\boldsymbol{c} \cdot \boldsymbol{l}]^\mathrm{sym} = \boldsymbol{f}^t \cdot \boldsymbol{D} \cdot \boldsymbol{f}.$$

In summary the covariant material (total) rate-of-deformation tensor (and its plastic and total push-forward) decomposes additively into elastic and plastic contributions.

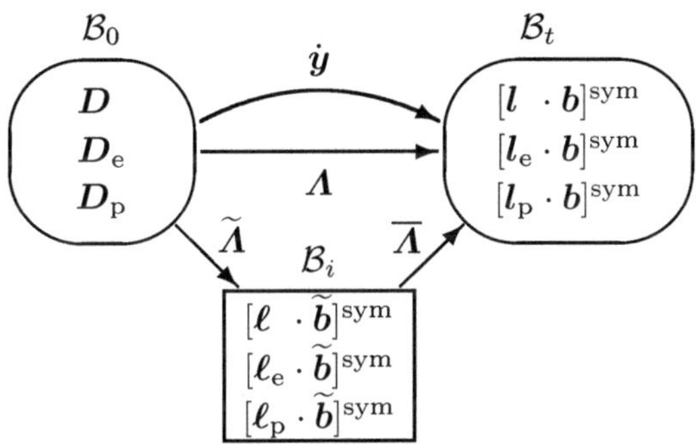

Likewise the contravariant plastic and total push-forward of the material *elastic* rate-of-deformation tensor involve as metric the contravariant $\widetilde{\boldsymbol{b}}$ and $\boldsymbol{b}$, respectively, and render

$$[\boldsymbol{\ell}_\mathrm{e} \cdot \widetilde{\boldsymbol{b}}]^\mathrm{sym} = \widetilde{\boldsymbol{F}} \cdot \boldsymbol{D}_\mathrm{e} \cdot \widetilde{\boldsymbol{F}}^t \quad \text{and} \quad [\boldsymbol{l}_\mathrm{e} \cdot \boldsymbol{b}]^\mathrm{sym} = \boldsymbol{F} \cdot \boldsymbol{D}_\mathrm{e} \cdot \boldsymbol{F}^t.$$

Accordingly the contravariant plastic and total push-forward of the material *plastic* rate-of-deformation tensor result in

$$[\boldsymbol{\ell}_\mathrm{p} \cdot \widetilde{\boldsymbol{b}}]^\mathrm{sym} = \widetilde{\boldsymbol{F}} \cdot \boldsymbol{D}_\mathrm{p} \cdot \widetilde{\boldsymbol{F}}^t \quad \text{and} \quad [\boldsymbol{l}_\mathrm{p} \cdot \boldsymbol{b}]^\mathrm{sym} = \boldsymbol{F} \cdot \boldsymbol{D}_\mathrm{p} \cdot \boldsymbol{F}^t.$$

Observe finally that the contravariant plastic and total push-forward of the material (total) rate-of-deformation tensor $\boldsymbol{D} := [\boldsymbol{L} \cdot \boldsymbol{I}]^\mathrm{sym}$ lead to

$$[\boldsymbol{\ell} \cdot \widetilde{\boldsymbol{b}}]^\mathrm{sym} = \widetilde{\boldsymbol{F}} \cdot \boldsymbol{D} \cdot \widetilde{\boldsymbol{F}}^t \quad \text{and} \quad [\boldsymbol{l} \cdot \boldsymbol{b}]^\mathrm{sym} = \boldsymbol{F} \cdot \boldsymbol{D} \cdot \boldsymbol{F}^t.$$

In summary the contravariant material (total) rate-of-deformation tensor (and its plastic and total push-forward) decomposes additively into elastic and plastic contributions.

Intermediate Rate-of-Deformation Tensors:
The symmetric contributions to the fully co- and contravariant versions of the elastic and the plastic parts of the intermediate velocity gradient are denoted the *intermediate elastic* and *plastic rate-of-deformation tensor*, respectively, i.e.

$$\overline{\boldsymbol{D}} := [\overline{\boldsymbol{J}} \cdot \overline{\boldsymbol{L}}]^{\text{sym}} = [\overline{\boldsymbol{L}} \cdot \overline{\boldsymbol{I}}]^{\text{sym}} \quad \text{and} \quad \widetilde{\boldsymbol{d}} := [\widetilde{\boldsymbol{j}} \cdot \widetilde{\boldsymbol{l}}]^{\text{sym}} = [\widetilde{\boldsymbol{l}} \cdot \widetilde{\boldsymbol{i}}]^{\text{sym}}.$$

Note that the intermediate elastic and plastic rate-of-deformation tensors compute from either exclusively elastic or exclusively plastic contributions, respectively.

Then the covariant plastic pull-back and elastic push-forward of the intermediate *elastic* rate-of-deformation tensor involve as metric the covariant $\widetilde{\boldsymbol{C}}$ and $\overline{\boldsymbol{c}}$, respectively, and render

$$[\widetilde{\boldsymbol{C}} \cdot \boldsymbol{L}_{\text{e}}]^{\text{sym}} = \widetilde{\boldsymbol{F}}^t \cdot \overline{\boldsymbol{D}} \cdot \widetilde{\boldsymbol{F}} \quad \text{and} \quad [\overline{\boldsymbol{c}} \cdot \boldsymbol{l}_{\text{e}}]^{\text{sym}} = \overline{\boldsymbol{f}}^t \cdot \overline{\boldsymbol{D}} \cdot \overline{\boldsymbol{f}}.$$

Accordingly the covariant plastic pull-back and elastic push-forward of the intermediate *plastic* rate-of-deformation tensor result in

$$[\widetilde{\boldsymbol{C}} \cdot \boldsymbol{L}_{\text{p}}]^{\text{sym}} = \widetilde{\boldsymbol{F}}^t \cdot \widetilde{\boldsymbol{d}} \cdot \widetilde{\boldsymbol{F}} \quad \text{and} \quad [\overline{\boldsymbol{c}} \cdot \boldsymbol{l}_{\text{p}}]^{\text{sym}} = \overline{\boldsymbol{f}}^t \cdot \widetilde{\boldsymbol{d}} \cdot \overline{\boldsymbol{f}}.$$

Observe finally that the covariant plastic pull-back and elastic push-forward of the intermediate (total) rate-of-deformation tensor $\overline{\boldsymbol{D}} + \widetilde{\boldsymbol{d}}$ lead to

$$[\widetilde{\boldsymbol{C}} \cdot \boldsymbol{L}]^{\text{sym}} = \widetilde{\boldsymbol{F}}^t \cdot [\overline{\boldsymbol{D}} + \widetilde{\boldsymbol{d}}] \cdot \widetilde{\boldsymbol{F}} \quad \text{and} \quad [\overline{\boldsymbol{c}} \cdot \boldsymbol{l}]^{\text{sym}} = \overline{\boldsymbol{f}}^t \cdot [\overline{\boldsymbol{D}} + \widetilde{\boldsymbol{d}}] \cdot \overline{\boldsymbol{f}}.$$

In summary the covariant intermediate (total) rate-of-deformation tensor (and its plastic pull-back and elastic push-forward) decomposes additively into elastic and plastic contributions.

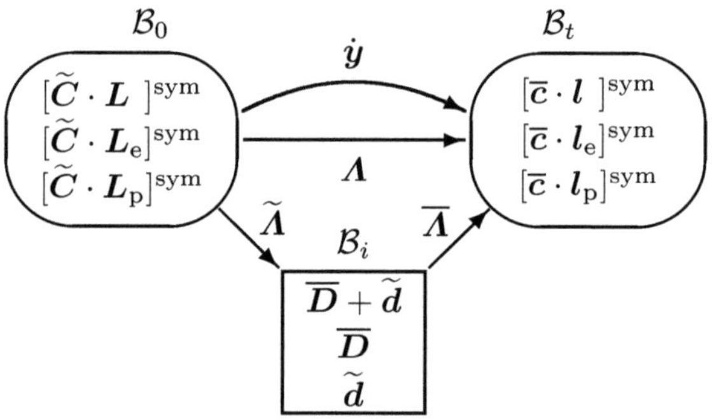

Likewise the contravariant plastic pull-back and elastic push-forward of the intermediate *elastic* rate-of-deformation tensor involve as metric the contravariant $\widetilde{\boldsymbol{B}}$ and $\overline{\boldsymbol{b}}$, respectively, and render

$$[\boldsymbol{L}_{\text{e}} \cdot \widetilde{\boldsymbol{B}}]^{\text{sym}} = \widetilde{\boldsymbol{f}} \cdot \overline{\boldsymbol{D}} \cdot \widetilde{\boldsymbol{f}}^t \quad \text{and} \quad [\boldsymbol{l}_{\text{e}} \cdot \overline{\boldsymbol{b}}]^{\text{sym}} = \overline{\boldsymbol{F}} \cdot \overline{\boldsymbol{D}} \cdot \overline{\boldsymbol{F}}^t.$$

Accordingly the contravariant plastic pull-black and elastic push-forward of the intermediate *plastic* rate-of-deformation tensor result in

$$[\boldsymbol{L}_{\text{p}} \cdot \widetilde{\boldsymbol{B}}]^{\text{sym}} = \widetilde{\boldsymbol{f}} \cdot \widetilde{\boldsymbol{d}} \cdot \widetilde{\boldsymbol{f}}^t \quad \text{and} \quad [\boldsymbol{l}_{\text{p}} \cdot \overline{\boldsymbol{b}}]^{\text{sym}} = \overline{\boldsymbol{F}} \cdot \widetilde{\boldsymbol{d}} \cdot \overline{\boldsymbol{F}}^t.$$

Observe finally that the contravariant plastic pull-back and elastic push-forward of the intermediate (total) rate-of-deformation tensor $\overline{\boldsymbol{D}} + \widetilde{\boldsymbol{d}}$ lead to

$$[\boldsymbol{L} \cdot \widetilde{\boldsymbol{B}}]^{\mathrm{sym}} = \widetilde{\boldsymbol{f}} \cdot [\overline{\boldsymbol{D}} + \widetilde{\boldsymbol{d}}] \cdot \widetilde{\boldsymbol{f}}^t \quad \text{and} \quad [\boldsymbol{l} \cdot \overline{\boldsymbol{b}}]^{\mathrm{sym}} = \overline{\boldsymbol{F}} \cdot [\overline{\boldsymbol{D}} + \widetilde{\boldsymbol{d}}] \cdot \overline{\boldsymbol{F}}^t.$$

In summary the contravariant intermediate (total) rate-of-deformation tensor (and its plastic pull-back and elastic push-forward) decomposes additively into elastic and plastic contributions.

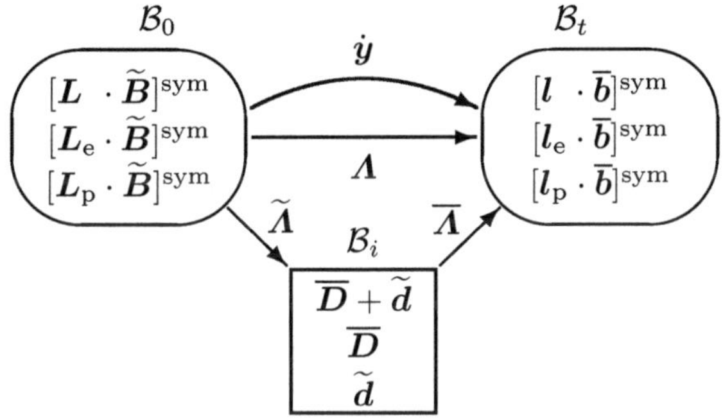

Material Time Derivative: Elastic and Plastic Strain Measures: The material time derivatives of the *material* plastic right Cauchy-Green strain $\widetilde{\boldsymbol{C}}$ and the material plastic Piola strain $\widetilde{\boldsymbol{B}}$ compute as the co- and contravariant *plastic pull-back* of the intermediate *plastic* rate-of-deformation tensor $\widetilde{\boldsymbol{d}}$ as

$$\widetilde{\boldsymbol{F}}^t \cdot \widetilde{\boldsymbol{d}} \cdot \widetilde{\boldsymbol{F}} = [\widetilde{\boldsymbol{C}} \cdot \widetilde{\boldsymbol{L}}]^{\mathrm{sym}} = \frac{1}{2}\,\mathrm{D}_t \widetilde{\boldsymbol{C}} \quad \text{and} \quad \widetilde{\boldsymbol{f}} \cdot \widetilde{\boldsymbol{d}} \cdot \widetilde{\boldsymbol{f}}^t = [\widetilde{\boldsymbol{L}} \cdot \widetilde{\boldsymbol{B}}]^{\mathrm{sym}} = -\frac{1}{2}\,\mathrm{D}_t \widetilde{\boldsymbol{B}}.$$

Likewise the material time derivatives of the *intermediate* elastic right Cauchy-Green strain $\overline{\boldsymbol{C}}$ and the intermediate elastic Piola strain $\overline{\boldsymbol{B}}$ compute as the co- and contravariant *elastic pull-back* of the spatial *elastic* rate-of-deformation tensor $\overline{\boldsymbol{d}} \equiv \boldsymbol{d}_{\mathrm{e}}$ as

$$\overline{\boldsymbol{F}}^t \cdot \overline{\boldsymbol{d}} \cdot \overline{\boldsymbol{F}} = [\overline{\boldsymbol{C}} \cdot \overline{\boldsymbol{L}}]^{\mathrm{sym}} = \frac{1}{2}\,\mathrm{D}_t \overline{\boldsymbol{C}} \quad \text{and} \quad \overline{\boldsymbol{f}} \cdot \overline{\boldsymbol{d}} \cdot \overline{\boldsymbol{f}}^t = [\overline{\boldsymbol{L}} \cdot \overline{\boldsymbol{B}}]^{\mathrm{sym}} = -\frac{1}{2}\,\mathrm{D}_t \overline{\boldsymbol{B}}.$$

Moreover the material time derivatives of the *intermediate* plastic Finger strain $\widetilde{\boldsymbol{c}}$ and the intermediate plastic left Cauchy-Green strain $\widetilde{\boldsymbol{b}}$ compute as the co- and contravariant *plastic push-forward* of the material *plastic* rate-of-deformation tensor $\widetilde{\boldsymbol{D}} \equiv \boldsymbol{D}_{\mathrm{p}}$ as

$$\widetilde{\boldsymbol{f}}^t \cdot \widetilde{\boldsymbol{D}} \cdot \widetilde{\boldsymbol{f}} = [\widetilde{\boldsymbol{c}} \cdot \widetilde{\boldsymbol{l}}]^{\mathrm{sym}} = -\frac{1}{2}\,\mathrm{D}_t \widetilde{\boldsymbol{c}} \quad \text{and} \quad \widetilde{\boldsymbol{F}} \cdot \widetilde{\boldsymbol{D}} \cdot \widetilde{\boldsymbol{F}}^t = [\widetilde{\boldsymbol{l}} \cdot \widetilde{\boldsymbol{b}}]^{\mathrm{sym}} = \frac{1}{2}\,\mathrm{D}_t \widetilde{\boldsymbol{b}}.$$

Finally the material time derivatives of the *spatial* elastic Finger strain $\overline{\boldsymbol{c}}$ and the spatial elastic left Cauchy-Green strain $\overline{\boldsymbol{b}}$ compute as the co- and contravariant *elastic push-forward* of the intermediate *elastic* rate-of-deformation tensor $\overline{\boldsymbol{D}}$ as

$$\overline{\boldsymbol{f}}^t \cdot \overline{\boldsymbol{D}} \cdot \overline{\boldsymbol{f}} = [\overline{\boldsymbol{c}} \cdot \overline{\boldsymbol{l}}]^{\text{sym}} = -\frac{1}{2}\,\mathrm{D}_t \overline{\boldsymbol{c}} \quad \text{and} \quad \overline{\boldsymbol{F}} \cdot \overline{\boldsymbol{D}} \cdot \overline{\boldsymbol{F}}^t = [\overline{\boldsymbol{l}} \cdot \overline{\boldsymbol{b}}]^{\text{sym}} = \frac{1}{2}\,\mathrm{D}_t \overline{\boldsymbol{b}}.$$

Observe the intriguing formal similarity between the material time derivatives of the various co- and contravariant, spatial, intermediate and material, as well as total, elastic and plastic strain measures, respectively.

Left Lie Time Derivatives: Spatial & Intermediate Tensor Fields: Recall the (total) left Lie time derivative of the various co- and contravariant spatial unit tensors and (total and elastic) strain measures

$$\begin{aligned} \mathrm{L}_t \boldsymbol{j} &= \ \ 2\boldsymbol{d}, \quad \mathrm{L}_t \boldsymbol{c} = \boldsymbol{0} \quad \text{and} \quad \mathrm{L}_t \overline{\boldsymbol{c}} = \ \ 2[\overline{\boldsymbol{c}} \cdot \boldsymbol{l}_{\mathrm{p}}]^{\text{sym}} \\ \mathrm{L}_t \boldsymbol{i} &= -2\boldsymbol{d}, \quad \mathrm{L}_t \boldsymbol{b} = \boldsymbol{0} \quad \text{and} \quad \mathrm{L}_t \overline{\boldsymbol{b}} = -2[\boldsymbol{l}_{\mathrm{p}} \cdot \overline{\boldsymbol{b}}]^{\text{sym}}. \end{aligned}$$

Then the *elastic left Lie time derivative* $\overline{\mathrm{L}}_t$ of generic *spatial* tensor fields $\{\bullet\}$ is defined likewise as

$$\overline{\mathrm{L}}_t\{\bullet\} := \text{elastic push-forward}\Big(\,\mathrm{D}_t\big(\text{elastic pull-back of } \{\bullet\}\big)\Big).$$

As an example applying this definition to the various co- and contravariant spatial unit tensors and (total and elastic) strain measures renders

$$\begin{aligned} \overline{\mathrm{L}}_t \boldsymbol{j} &= \ \ 2\overline{\boldsymbol{d}}, \quad \overline{\mathrm{L}}_t \boldsymbol{c} = -2[\boldsymbol{c} \cdot \boldsymbol{l}_{\mathrm{p}}]^{\text{sym}} \quad \text{and} \quad \overline{\mathrm{L}}_t \overline{\boldsymbol{c}} = \boldsymbol{0} \\ \overline{\mathrm{L}}_t \boldsymbol{i} &= -2\overline{\boldsymbol{d}}, \quad \overline{\mathrm{L}}_t \boldsymbol{b} = \ \ 2[\boldsymbol{l}_{\mathrm{p}} \cdot \boldsymbol{b}]^{\text{sym}} \quad \text{and} \quad \overline{\mathrm{L}}_t \overline{\boldsymbol{b}} = \boldsymbol{0}. \end{aligned}$$

Thereby the relation between the elastic left Lie and the material time derivative of generic co- and contravariant symmetric spatial tensor fields $\{\bullet\}$ holds as

$$\overline{\mathrm{L}}_t\{\bullet\} = \mathrm{D}_t\{\bullet\} + 2[\{\bullet\} \cdot \boldsymbol{l}_{\mathrm{e}}]^{\text{sym}} \quad \text{and} \quad \overline{\mathrm{L}}_t\{\bullet\} = \mathrm{D}_t\{\bullet\} - 2[\boldsymbol{l}_{\mathrm{e}} \cdot \{\bullet\}]^{\text{sym}}.$$

Accordingly, the relation between the (total) left Lie and the elastic left Lie time derivative follows as

$$\mathrm{L}_t\{\bullet\} = \overline{\mathrm{L}}_t\{\bullet\} + 2[\{\bullet\} \cdot \boldsymbol{l}_{\mathrm{p}}]^{\text{sym}} \quad \text{and} \quad \mathrm{L}_t\{\bullet\} = \overline{\mathrm{L}}_t\{\bullet\} - 2[\boldsymbol{l}_{\mathrm{p}} \cdot \{\bullet\}]^{\text{sym}}.$$

Moreover, the *plastic left Lie time derivative* $\widetilde{\mathrm{L}}_t$ of generic *intermediate* tensor fields $\{\bullet\}$ is defined in analogy as

$$\widetilde{\mathrm{L}}_t\{\bullet\} := \text{plastic push-forward}\Big(\,\mathrm{D}_t\big(\text{plastic pull-back of } \{\bullet\}\big)\Big).$$

As an example applying this definition to the various co- and contravariant intermediate unit tensors and (total and elastic) strain measures renders

$$\begin{aligned}
\widetilde{\mathrm{L}}_t\overline{\boldsymbol{C}} &= 2[\overline{\boldsymbol{C}}\cdot\boldsymbol{\ell}]^{\mathrm{sym}}, \quad \widetilde{\mathrm{L}}_t\widetilde{\boldsymbol{c}} = \boldsymbol{0} \quad \text{and} \quad \widetilde{\mathrm{L}}_t\widetilde{\boldsymbol{j}} = 2\widetilde{\boldsymbol{d}}\\
\widetilde{\mathrm{L}}_t\overline{\boldsymbol{B}} &= -2[\boldsymbol{\ell}\cdot\overline{\boldsymbol{B}}]^{\mathrm{sym}}, \quad \widetilde{\mathrm{L}}_t\widetilde{\boldsymbol{b}} = \boldsymbol{0} \quad \text{and} \quad \widetilde{\mathrm{L}}_t\widetilde{\boldsymbol{i}} = -2\widetilde{\boldsymbol{d}}.
\end{aligned}$$

Thus the relation between the plastic left Lie and the material time derivative of generic co- and contravariant symmetric intermediate tensor fields $\{\bullet\}$ holds as

$$\widetilde{\mathrm{L}}_t\{\bullet\} = \mathrm{D}_t\{\bullet\} + 2[\{\bullet\}\cdot\boldsymbol{\ell}_{\mathrm{p}}]^{\mathrm{sym}} \quad \text{and} \quad \widetilde{\mathrm{L}}_t\{\bullet\} = \mathrm{D}_t\{\bullet\} - 2[\boldsymbol{\ell}_{\mathrm{p}}\cdot\{\bullet\}]^{\mathrm{sym}}.$$

A comparison of the elastic or plastic, respectively, left Lie and the material time derivative of the various co- and contravariant strain measures is obviously in agreement with the relations in the above. As an example consider $\mathrm{D}_t\boldsymbol{j} = \mathrm{D}_t\boldsymbol{i} = \boldsymbol{0}$, thus the covariant $\boldsymbol{j}$ results in $\overline{\mathrm{L}}_t\boldsymbol{j} = 2[\boldsymbol{j}\cdot\boldsymbol{l}_{\mathrm{e}}]^{\mathrm{sym}}$, whereas the contravariant $\boldsymbol{i}$ results in $\overline{\mathrm{L}}_t\boldsymbol{i} = -2[\boldsymbol{l}_{\mathrm{e}}\cdot\boldsymbol{i}]^{\mathrm{sym}}$. Likewise $\mathrm{D}_t\widetilde{\boldsymbol{j}} = \mathrm{D}_t\widetilde{\boldsymbol{i}} = \boldsymbol{0}$, thus the covariant $\widetilde{\boldsymbol{j}}$ results in $\widetilde{\mathrm{L}}_t\widetilde{\boldsymbol{j}} = 2[\widetilde{\boldsymbol{j}}\cdot\boldsymbol{\ell}_{\mathrm{p}}]^{\mathrm{sym}}$, whereas the contravariant $\widetilde{\boldsymbol{i}}$ results in $\widetilde{\mathrm{L}}_t\widetilde{\boldsymbol{i}} = -2[\boldsymbol{\ell}_{\mathrm{p}}\cdot\widetilde{\boldsymbol{i}}]^{\mathrm{sym}}$.

Right Lie Time Derivatives: Material & Intermediate Tensor Fields:
Recall the (total) right Lie time derivative of the various co- and contravariant material unit tensors and (total and elastic) strain measures

$$\begin{aligned}
\mathrm{R}_t\boldsymbol{J} &= -2\boldsymbol{D}, \quad \mathrm{R}_t\boldsymbol{C} = \boldsymbol{0} \quad \text{and} \quad \mathrm{R}_t\widetilde{\boldsymbol{C}} = -2[\widetilde{\boldsymbol{C}}\cdot\boldsymbol{L}_{\mathrm{e}}]^{\mathrm{sym}}\\
\mathrm{R}_t\boldsymbol{I} &= 2\boldsymbol{D}, \quad \mathrm{R}_t\boldsymbol{B} = \boldsymbol{0} \quad \text{and} \quad \mathrm{R}_t\widetilde{\boldsymbol{B}} = 2[\boldsymbol{L}_{\mathrm{e}}\cdot\widetilde{\boldsymbol{B}}]^{\mathrm{sym}}.
\end{aligned}$$

Then the *plastic right Lie time derivative* $\widetilde{\mathrm{R}}_t$ of generic *material* tensor fields $\{\bullet\}$ is defined likewise as

$$\widetilde{\mathrm{R}}_t\{\bullet\} := \text{plastic pull-back}\Big(\,\mathrm{D}_t\big(\text{plastic push-forward of }\{\bullet\}\big)\Big).$$

As an example applying this definition to the various co- and contravariant material unit tensors and (total and elastic) strain measures renders

$$\begin{aligned}
\widetilde{\mathrm{R}}_t\boldsymbol{J} &= -2\widetilde{\boldsymbol{D}}, \quad \widetilde{\mathrm{R}}_t\boldsymbol{C} = 2[\boldsymbol{C}\cdot\boldsymbol{L}_{\mathrm{e}}]^{\mathrm{sym}} \quad \text{and} \quad \widetilde{\mathrm{R}}_t\widetilde{\boldsymbol{C}} = \boldsymbol{0}\\
\widetilde{\mathrm{R}}_t\boldsymbol{I} &= 2\widetilde{\boldsymbol{D}}, \quad \widetilde{\mathrm{R}}_t\boldsymbol{B} = -2[\boldsymbol{L}_{\mathrm{e}}\cdot\boldsymbol{B}]^{\mathrm{sym}} \quad \text{and} \quad \widetilde{\mathrm{R}}_t\widetilde{\boldsymbol{B}} = \boldsymbol{0}.
\end{aligned}$$

Thereby the relation between the plastic right Lie and the material time derivative of generic co- and contravariant symmetric material tensor fields $\{\bullet\}$ holds as

$$\widetilde{\mathrm{R}}_t\{\bullet\} = \mathrm{D}_t\{\bullet\} - 2[\{\bullet\}\cdot\boldsymbol{L}_{\mathrm{p}}]^{\mathrm{sym}} \quad \text{and} \quad \widetilde{\mathrm{R}}_t\{\bullet\} = \mathrm{D}_t\{\bullet\} + 2[\boldsymbol{L}_{\mathrm{p}}\cdot\{\bullet\}]^{\mathrm{sym}}.$$

Accordingly, the relation between the (total) right Lie and the plastic right Lie time derivative follows as

$$\mathrm{R}_t\{\bullet\} = \widetilde{\mathrm{R}}_t\{\bullet\} - 2[\{\bullet\}\cdot \boldsymbol{L}_\mathrm{e}]^{\mathrm{sym}} \quad \text{and} \quad \mathrm{R}_t\{\bullet\} = \widetilde{\mathrm{R}}_t\{\bullet\} + 2[\boldsymbol{L}_\mathrm{e}\cdot \{\bullet\}]^{\mathrm{sym}}.$$

Moreover, the *elastic right Lie time derivative* $\overline{\mathrm{R}}_t$ of generic *intermediate* tensor fields $\{\bullet\}$ is defined in analogy as

$$\overline{\mathrm{R}}_t\{\bullet\} := \text{elastic pull-back}\Big(\,\mathrm{D}_t\big(\text{elastic push-forward of } \{\bullet\}\big)\Big).$$

As an example applying this definition to the various co- and contravariant intermediate unit tensors and (total and elastic) strain measures renders

$$\begin{array}{llll} \overline{\mathrm{R}}_t\widetilde{\boldsymbol{c}} = -2[\widetilde{\boldsymbol{c}}\cdot\boldsymbol{\ell}]^{\mathrm{sym}}, & \overline{\mathrm{R}}_t\overline{\boldsymbol{C}} = \boldsymbol{0} & \text{and} & \overline{\mathrm{R}}_t\overline{\boldsymbol{J}} = -2\overline{\boldsymbol{D}} \\ \overline{\mathrm{R}}_t\widetilde{\boldsymbol{b}} = \;\;\, 2[\boldsymbol{\ell}\cdot\widetilde{\boldsymbol{b}}]^{\mathrm{sym}}, & \overline{\mathrm{R}}_t\overline{\boldsymbol{B}} = \boldsymbol{0} & \text{and} & \overline{\mathrm{R}}_t\overline{\boldsymbol{I}} = \;\;\, 2\overline{\boldsymbol{D}}. \end{array}$$

Thus the relation between the elastic right Lie and the material time derivative of generic co- and contravariant symmetric intermediate tensor fields $\{\bullet\}$ holds as

$$\overline{\mathrm{R}}_t\{\bullet\} = \mathrm{D}_t\{\bullet\} - 2[\{\bullet\}\cdot \boldsymbol{\ell}_\mathrm{e}]^{\mathrm{sym}} \quad \text{and} \quad \overline{\mathrm{R}}_t\{\bullet\} = \mathrm{D}_t\{\bullet\} + 2[\boldsymbol{\ell}_\mathrm{e}\cdot \{\bullet\}]^{\mathrm{sym}}.$$

A comparison of the plastic or elastic, respectively, right Lie and the material time derivative of the various co- and contravariant strain measures is obviously in agreement with the relations in the above. As an example consider $\mathrm{D}_t\boldsymbol{J} = \mathrm{D}_t\boldsymbol{I} = \boldsymbol{0}$, thus the covariant $\boldsymbol{J}$ results in $\widetilde{\mathrm{R}}_t\boldsymbol{J} = -2[\boldsymbol{J}\cdot\boldsymbol{L}_\mathrm{p}]^{\mathrm{sym}}$, whereas the contravariant $\boldsymbol{I}$ results in $\widetilde{\mathrm{R}}_t\boldsymbol{I} = 2[\boldsymbol{L}_\mathrm{p}\cdot\boldsymbol{I}]^{\mathrm{sym}}$. Likewise $\mathrm{D}_t\overline{\boldsymbol{J}} = \mathrm{D}_t\overline{\boldsymbol{I}} = \boldsymbol{0}$, thus the covariant $\overline{\boldsymbol{J}}$ results in $\overline{\mathrm{R}}_t\overline{\boldsymbol{J}} = -2[\overline{\boldsymbol{J}}\cdot\boldsymbol{\ell}_\mathrm{e}]^{\mathrm{sym}}$, whereas the contravariant $\overline{\boldsymbol{I}}$ results in $\overline{\mathrm{R}}_t\overline{\boldsymbol{I}} = 2[\boldsymbol{\ell}_\mathrm{e}\cdot\overline{\boldsymbol{I}}]^{\mathrm{sym}}$.

Material↔Lie Time Derivatives of Strain Measures:

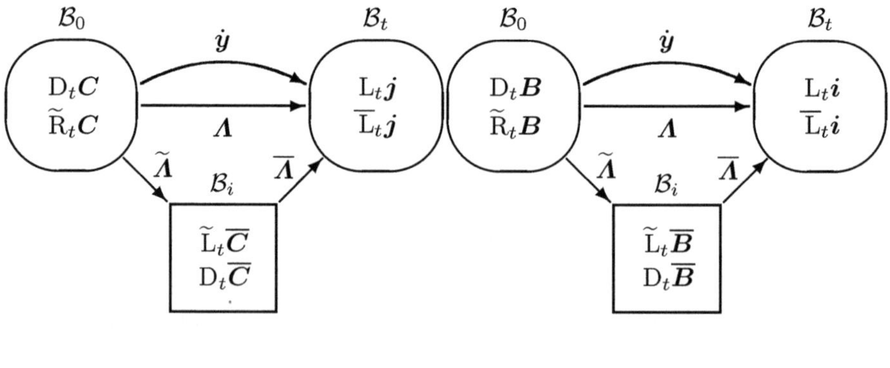

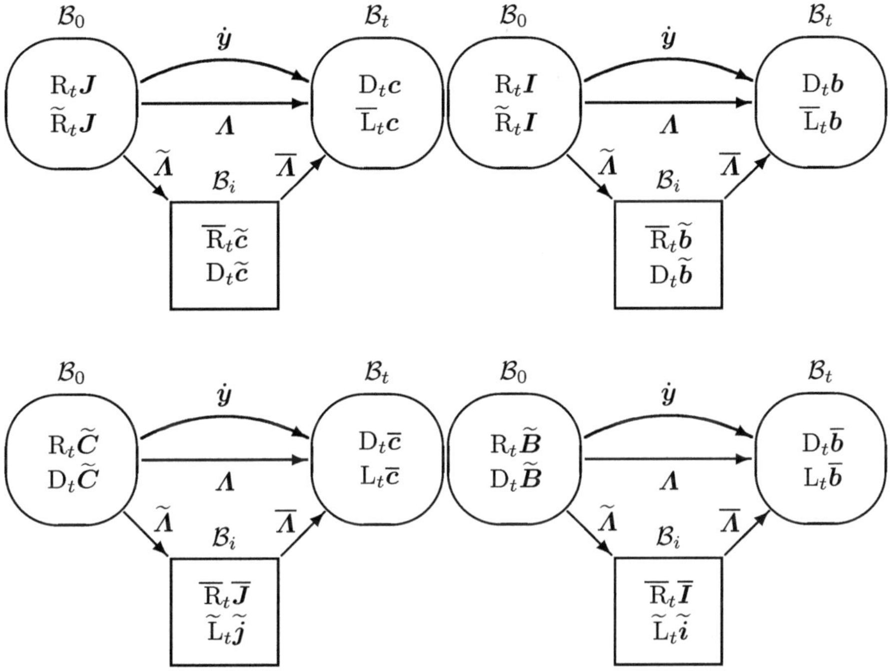

The relations between the various types of time derivatives for the total, elastic and plastic strain measures are depicted in the above cartoons. Thereby it is interesting to note that the (left and right) Lie time derivatives of the material, intermediate and spatial unit tensors coincide with the corresponding rate-of-deformation tensors.

Stress Measures:

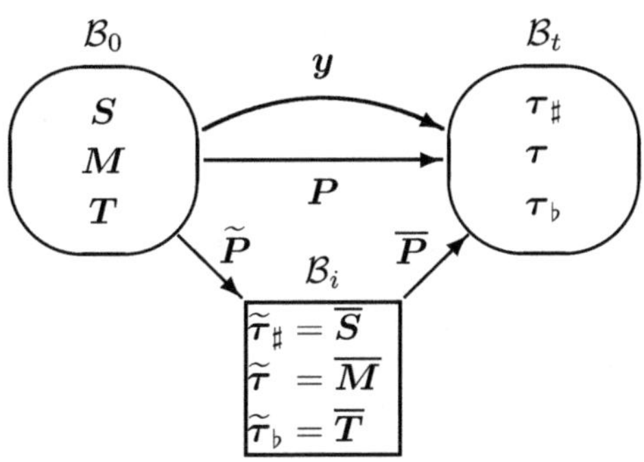

Right-sided contravariant plastic push-forward and left-sided covariant elastic pullback of the (total) Piola stress results in the two-point *elastic* and *plastic Piola stress*, respectively, as

$$\overline{\boldsymbol{P}} := \boldsymbol{P} \cdot \widetilde{\boldsymbol{F}}^t \quad \text{and} \quad \widetilde{\boldsymbol{P}} := \overline{\boldsymbol{F}}^t \cdot \boldsymbol{P}.$$

Then the co/contra- (mixed) variant two-point *elastic* Piola stress may be related to the contravariantly transforming intermediate Piola-Kirchhoff stress $\overline{\boldsymbol{S}}$, the covariantly transforming intermediate stress $\overline{\boldsymbol{T}}$, and the co/contra- (mixed) variantly transforming intermediate Mandel stress $\overline{\boldsymbol{M}}$, respectively, as

$$\overline{\boldsymbol{S}} := \overline{\boldsymbol{f}}_\sharp \cdot \overline{\boldsymbol{P}} \quad \text{and} \quad \overline{\boldsymbol{T}} := \overline{\boldsymbol{F}}^t \cdot \overline{\boldsymbol{P}} \cdot \overline{\boldsymbol{C}} \quad \text{and} \quad \overline{\boldsymbol{M}} := \overline{\boldsymbol{F}}^t \cdot \overline{\boldsymbol{P}}.$$

Accordingly the spatial co/contra- (mixed) variant Kirchhoff stress $\boldsymbol{\tau}$ is related to the two-point elastic Piola stress and the stress measures in the intermediate configuration by

$$\boldsymbol{\tau} = \overline{\boldsymbol{P}} \cdot \overline{\boldsymbol{F}}^t = \overline{\boldsymbol{F}}_\flat \cdot \overline{\boldsymbol{S}} \cdot \overline{\boldsymbol{F}}^t = \overline{\boldsymbol{f}}^t \cdot \overline{\boldsymbol{T}} \cdot \overline{\boldsymbol{f}}_\sharp = \overline{\boldsymbol{f}}^t \cdot \overline{\boldsymbol{M}} \cdot \overline{\boldsymbol{F}}^t.$$

Likewise the co/contra- (mixed) variant two-point *plastic* Piola stress may be related to the contravariantly transforming material Piola-Kirchhoff stress $\boldsymbol{S}$, the covariantly transforming material stress $\boldsymbol{T}$, and the co/contra- (mixed) variantly transforming material Mandel stress $\boldsymbol{M}$, respectively

$$\boldsymbol{S} := \widetilde{\boldsymbol{f}}_\sharp \cdot \widetilde{\boldsymbol{P}} \quad \text{and} \quad \boldsymbol{T} := \widetilde{\boldsymbol{F}}^t \cdot \widetilde{\boldsymbol{P}} \cdot \widetilde{\boldsymbol{C}} \quad \text{and} \quad \boldsymbol{M} := \widetilde{\boldsymbol{F}}^t \cdot \widetilde{\boldsymbol{P}}.$$

Accordingly the intermediate co/contra- (mixed) variant Kirchhoff stress $\widetilde{\boldsymbol{\tau}}$ is related to the two-point plastic Piola stress and the stress measures in the material configuration by

$$\widetilde{\boldsymbol{\tau}} := \widetilde{\boldsymbol{P}} \cdot \widetilde{\boldsymbol{F}}^t = \widetilde{\boldsymbol{F}}_\flat \cdot \boldsymbol{S} \cdot \widetilde{\boldsymbol{F}}^t = \widetilde{\boldsymbol{f}}^t \cdot \boldsymbol{T} \cdot \widetilde{\boldsymbol{f}}_\sharp = \widetilde{\boldsymbol{f}}^t \cdot \boldsymbol{M} \cdot \widetilde{\boldsymbol{F}}^t.$$

Finally, it shall be noted that the intermediate co/contra- (mixed) variant Kirchhoff and Mandel stresses do coincide

$$\widetilde{\boldsymbol{\tau}} \equiv \overline{\boldsymbol{M}} \quad \rightsquigarrow \quad \widetilde{\boldsymbol{\tau}}_\sharp := \overline{\boldsymbol{B}} \cdot \widetilde{\boldsymbol{\tau}} \equiv \overline{\boldsymbol{S}} \quad \text{and} \quad \widetilde{\boldsymbol{\tau}}_\flat := \widetilde{\boldsymbol{\tau}} \cdot \overline{\boldsymbol{C}} \equiv \overline{\boldsymbol{T}}.$$

The symbolic redundancy for the intermediate stress measures has been introduced again for notational convenience. It is noted that the co- and contravariant versions of the elastic and plastic distortions as used in the above have been defined as

$$\overline{\boldsymbol{F}}_\flat := \boldsymbol{j} \cdot \overline{\boldsymbol{F}} \quad \text{and} \quad \overline{\boldsymbol{f}}_\sharp := \overline{\boldsymbol{f}} \cdot \boldsymbol{i} \quad \text{and} \quad \widetilde{\boldsymbol{F}}_\flat := \widetilde{\boldsymbol{j}} \cdot \widetilde{\boldsymbol{F}} \quad \text{and} \quad \widetilde{\boldsymbol{f}}_\sharp := \widetilde{\boldsymbol{f}} \cdot \widetilde{\boldsymbol{i}}.$$

The relations between the various stress measures in the material and the intermediate configuration or in the intermediate and the spatial configuration follows the pattern established already for the relations between the various stress measures in the material and the spatial configuration.

Stress Power Density:
Due to the additive decomposition of the contra/co- (mixed) variant velocity gradients into elastic and plastic parts the stress power density in terms of co/contra- (mixed) variant stress measures decomposes accordingly

$$\wp_0 = \overline{\wp}_0 + \widetilde{\wp}_0 := \boldsymbol{\tau} : \overline{\boldsymbol{l}} + \widetilde{\boldsymbol{\tau}} : \widetilde{\boldsymbol{l}} = \overline{\boldsymbol{M}} : \overline{\boldsymbol{L}} + \boldsymbol{M} : \widetilde{\boldsymbol{L}} = \overline{\boldsymbol{P}} : \overline{\boldsymbol{\Lambda}} + \widetilde{\boldsymbol{P}} : \widetilde{\boldsymbol{\Lambda}}.$$

Then the elastic part of the stress power density expresses in terms of contravariant stress measures and covariant strain measures as

$$\overline{\wp}_0 = \boldsymbol{\tau}_\sharp : \overline{\boldsymbol{d}} = \frac{1}{2}\boldsymbol{\tau}_\sharp : \overline{\mathrm{L}_t \boldsymbol{j}} = \frac{1}{2}\overline{\boldsymbol{S}} : \mathrm{D}_t\overline{\boldsymbol{C}} = \boldsymbol{S} : [\boldsymbol{C} \cdot \boldsymbol{L}_\mathrm{e}]^{\mathrm{sym}},$$

whereas the plastic part of the stress power density expresses in terms of contravariant stress measures and covariant strain measures as

$$\widetilde{\wp}_0 = \boldsymbol{\tau}_\sharp : \boldsymbol{d}_\mathrm{p} = \boldsymbol{\tau}_\sharp : [\boldsymbol{j} \cdot \boldsymbol{l}_\mathrm{p}]^{\mathrm{sym}} = \overline{\boldsymbol{S}} : [\overline{\boldsymbol{C}} \cdot \boldsymbol{\ell}_\mathrm{p}]^{\mathrm{sym}} = \boldsymbol{S} : [\boldsymbol{C} \cdot \boldsymbol{L}_\mathrm{p}]^{\mathrm{sym}}.$$

It is noted that for the *special case* that $\overline{\boldsymbol{S}}$ commutes with $\overline{\boldsymbol{C}}$, i.e. $\overline{\boldsymbol{S}}$ and $\overline{\boldsymbol{C}}$ share the same principal directions (elastic isotropy), so that $\widetilde{\boldsymbol{i}} \cdot \widetilde{\boldsymbol{\tau}} = [\widetilde{\boldsymbol{i}} \cdot \widetilde{\boldsymbol{\tau}}]^t$ holds, the plastic part of the stress power density takes the alternative expression

$$\widetilde{\wp}_0 \overset{\mathrm{iso}}{=} \widetilde{\boldsymbol{\tau}} : [\widetilde{\boldsymbol{i}} \cdot \widetilde{\boldsymbol{d}}] = \frac{1}{2}\widetilde{\boldsymbol{\tau}} : [\widetilde{\boldsymbol{i}} \cdot \widetilde{\mathrm{L}_t \boldsymbol{j}}] = \frac{1}{2}\boldsymbol{\tau} : [\overline{\boldsymbol{b}} \cdot \mathrm{L}_t\overline{\boldsymbol{c}}] = \frac{1}{2}\boldsymbol{M} : [\widetilde{\boldsymbol{B}} \cdot \mathrm{D}_t\widetilde{\boldsymbol{C}}].$$

Observe the equivalence $\widetilde{\boldsymbol{i}} \equiv \overline{\boldsymbol{I}}$ and $\widetilde{\boldsymbol{j}} \equiv \overline{\boldsymbol{J}}$ due to the symbolic redundancy that has been introduced for notational convenience. Likewise the elastic part of the stress power density expresses in terms of covariant stress measures and contravariant strain measures as

$$\overline{\wp}_0 = \boldsymbol{\tau}_\flat : \overline{\boldsymbol{d}} = -\frac{1}{2}\boldsymbol{\tau}_\flat : \overline{\mathrm{L}_t \boldsymbol{i}} = -\frac{1}{2}\overline{\boldsymbol{T}} : \mathrm{D}_t\overline{\boldsymbol{B}} = \boldsymbol{T} : [\boldsymbol{L}_\mathrm{e} \cdot \boldsymbol{B}]^{\mathrm{sym}},$$

whereas the plastic part of the stress power density expresses in terms of covariant stress measures and contravariant strain measures as

$$\widetilde{\wp}_0 = \boldsymbol{\tau}_\flat : \boldsymbol{d}_\mathrm{p} = \boldsymbol{\tau}_\flat : [\boldsymbol{l}_\mathrm{p} \cdot \boldsymbol{i}]^{\mathrm{sym}} = \overline{\boldsymbol{T}} : [\boldsymbol{\ell}_\mathrm{p} \cdot \overline{\boldsymbol{B}}]^{\mathrm{sym}} = \boldsymbol{T} : [\boldsymbol{L}_\mathrm{p} \cdot \boldsymbol{B}]^{\mathrm{sym}}.$$

It is noted again that for the *special case* that $\overline{\boldsymbol{T}}$ commutes with $\overline{\boldsymbol{B}}$, i.e. $\overline{\boldsymbol{T}}$ and $\overline{\boldsymbol{B}}$ share the same principal directions (elastic isotropy), so that $\widetilde{\boldsymbol{\tau}} \cdot \widetilde{\boldsymbol{j}} = [\widetilde{\boldsymbol{\tau}} \cdot \widetilde{\boldsymbol{j}}]^t$ holds, the plastic part of the stress power density takes the alternative expression

$$\widetilde{\wp}_0 \overset{\mathrm{iso}}{=} \widetilde{\boldsymbol{\tau}} : [\widetilde{\boldsymbol{d}} \cdot \widetilde{\boldsymbol{j}}] = -\frac{1}{2}\widetilde{\boldsymbol{\tau}} : [\widetilde{\mathrm{L}_t \boldsymbol{i}} \cdot \widetilde{\boldsymbol{j}}] = -\frac{1}{2}\boldsymbol{\tau} : [\mathrm{L}_t\overline{\boldsymbol{b}} \cdot \overline{\boldsymbol{c}}] = -\frac{1}{2}\boldsymbol{M} : [\mathrm{D}_t\widetilde{\boldsymbol{B}} \cdot \widetilde{\boldsymbol{C}}].$$

Observe again the equivalence $\widetilde{\boldsymbol{i}} \equiv \overline{\boldsymbol{I}}$ and $\widetilde{\boldsymbol{j}} \equiv \overline{\boldsymbol{J}}$ due to the symbolic redundancy that has been introduced for notational convenience. The combination of power conjugate pairings of stress and elastic and plastic strain measures is of importance for the determination of constitutive relations in the sequel.

Dissipation Inequality:
The second law of thermodynamics requires positive *entropy production* for irreversible processes. In an *isothermal* continuum setting this requirement boils down to the statement that the stress power (which coincides with the external working for quasi-static cases) is never smaller than the change of *free energy* Ψ stored in the body. Thus for an arbitrary continuous sub-body $\mathscr{V} \subseteq \mathscr{B}$ of a continuous body $\mathscr{B}$ that occupies the material sub-configuration $\mathcal{V}_0 \subseteq \mathcal{B}_0$ of the material configuration $\mathcal{B}_0$ the global statement for the *dissipation inequality* reads

$$\int_{\mathcal{V}_0} \wp_0 \, \mathrm{d}V \geq \dot{\Psi}(\mathcal{V}_0) := \int_{\mathcal{V}_0} \mathrm{D}_t \psi_0 \, \mathrm{d}V.$$

Here ψ_0 denotes the free energy density per unit volume $\mathrm{d}V$ in the material configuration $\mathcal{B}_0$. The free energy density is the result of a Legendre transformation that exchanges the *entropy density* as an argument of the *internal energy density* for the *absolute temperature* θ, thus in general the free energy density is parameterized as $\psi_0 = \psi_0(\boldsymbol{F}, \theta, \cdots)$, whereby the $\cdots$ denote the dependence on further *internal variables*, see below. For the here considered isothermal case, however, the absolute temperature is a given parameter $\theta = \theta_{\text{ref}}$ and thus drops out from the list of arguments of the free energy density, i.e. here $\psi_0 = \psi_0(\boldsymbol{F}, \cdots)$. Here and in the following any possible further dependence of ψ_0 on the material coordinates $\boldsymbol{X}$ that is needed to capture material inhomogeneities has been suppressed to ease notation. The dissipation inequality follows from balancing the change of entropy contained in $\mathcal{V}_0$ by the entropy flux across $\partial\mathcal{V}_0$ and the entropy source in $\mathcal{V}_0$ together with the (positive) entropy production in $\mathcal{V}_0$. Then based on the classical Clausius-Duhem assumption that relates the entropy flux and source to the heat flux and source, respectively, divided by the absolute temperature in the sense of an integrating denominator, the incorporation of the balance of energy and the introduction of the free energy renders eventually the above statement for the dissipation inequality.

Master Constitutive Relations:
For (hyper) elasto-plasticity the constitutive relations for the stresses in terms of the strains follow from a Coleman-Noll-type exploitation of the dissipation inequality, i.e. from the comparison of the material time derivative of the free energy density ψ_0 with the stress power density $\wp_0$ and the observation that $\boldsymbol{F}$ and $\mathrm{D}_t\boldsymbol{F}$ may conceptually be controlled independently. Thereby, due to the elementary definition $\wp_0 = \boldsymbol{P} : \mathrm{D}_t\boldsymbol{F}$, the tensor function ψ_0, i.e. the free energy density, depends most generally on $\boldsymbol{F}$ and, in order to capture the contribution of the microstructure intrinsic to the material (due to, for example, irreversible rearrangements, statistically stored dislocations, geometrically necessary dislocations), on $\widetilde{\boldsymbol{F}}$ and a set of *internal variables* $\mathbb{F}$. The master constitutive relations for the Piola stress and the *thermodynamic force* $\mathbb{P}$ that drives the evolution of the set of internal variables $\mathbb{F}$ then follow as

$$\psi_0 = \psi_0(\boldsymbol{F}, \widetilde{\boldsymbol{F}}, \mathbb{F}) \quad \rightsquigarrow \quad \boldsymbol{P} = \frac{\partial \psi_0}{\partial \boldsymbol{F}} \quad \text{and} \quad \mathbb{P} := -\frac{\partial \psi_0}{\partial \mathbb{F}}.$$

Moreover the remaining *reduced dissipation inequality* reads

$$-\frac{\partial \psi_0}{\partial \widetilde{\boldsymbol{F}}} : \mathrm{D}_t \widetilde{\boldsymbol{F}} + \mathbb{P} \circ \mathrm{D}_t \mathbb{F} \geq 0$$

with $\circ$ denoting henceforth an appropriate scalar product. Then, to be more specific, the collective dependence of ψ_0 on the deformation gradient $\boldsymbol{F}$ and the plastic distortion $\widetilde{\boldsymbol{F}}$ is typically assumed to be expressed via the elastic distortion $\overline{\boldsymbol{F}} = \boldsymbol{F} \cdot \widetilde{\boldsymbol{f}}$. Thus, based on the expression for the stress power density $\wp_0 = \overline{\boldsymbol{P}} : \mathrm{D}_t \overline{\boldsymbol{F}} + \widetilde{\boldsymbol{P}} : \mathrm{D}_t \widetilde{\boldsymbol{F}}$, the master constitutive relations specify to

$$\psi_0 = \psi_0(\overline{\boldsymbol{F}}, \mathbb{F}) \quad \rightsquigarrow \quad \overline{\boldsymbol{P}} = \frac{\partial \psi_0}{\partial \overline{\boldsymbol{F}}} \quad \text{and} \quad \mathbb{P} := -\frac{\partial \psi_0}{\partial \mathbb{F}}$$

whereby the remaining reduced dissipation inequality now appears as

$$\widetilde{\boldsymbol{P}} : \mathrm{D}_t \widetilde{\boldsymbol{F}} + \mathbb{P} \circ \mathrm{D}_t \mathbb{F} \geq 0.$$

In the above $\widetilde{\boldsymbol{P}}$ and $\mathbb{P}$ collectively denote the thermodynamic forces that drive the evolution of $\widetilde{\boldsymbol{F}}$ and the set of internal variables $\mathbb{F}$. In passing observe the interesting relation for $\widetilde{\boldsymbol{P}}$ that follows directly from the chain rule

$$\widetilde{\boldsymbol{P}} := -\frac{\partial \psi_0}{\partial \widetilde{\boldsymbol{F}}} = -\frac{\partial \psi_0}{\partial \overline{\boldsymbol{F}}} : \frac{\partial \overline{\boldsymbol{F}}}{\partial \widetilde{\boldsymbol{F}}} = \overline{\boldsymbol{F}}^t \cdot \overline{\boldsymbol{P}} \cdot \widetilde{\boldsymbol{f}}^t.$$

Then for elasto-plasticity the *admissible domain* $\mathbb{A}$ for the *generalized thermodynamic force* $\boldsymbol{\mathcal{P}} := \{\widetilde{\boldsymbol{P}}, \mathbb{P}\}$ is characterized in the spirit of a *standard dissipative material formulation* [75] by a convex set in terms of the *yield function* $\phi = \phi(\boldsymbol{\mathcal{P}}; \boldsymbol{\mathcal{F}})$ as

$$\mathbb{A} := \{\boldsymbol{\mathcal{P}} |\ \phi(\boldsymbol{\mathcal{P}}; \boldsymbol{\mathcal{F}}) \leq 0\ \}.$$

The dependence $\phi = \phi(\boldsymbol{\mathcal{P}}; \boldsymbol{\mathcal{F}})$ is however not completely arbitrary but is subject to requirements of spatial objectivity and material symmetry, see the discussion below. Moreover the yield function is convex in $\boldsymbol{\mathcal{P}}$, i.e. $\phi(\theta \boldsymbol{\mathcal{P}}_1 + [1-\theta]\boldsymbol{\mathcal{P}}_2) \leq \theta\phi(\boldsymbol{\mathcal{P}}_1) + [1-\theta]\phi(\boldsymbol{\mathcal{P}}_2)$ for $\{\boldsymbol{\mathcal{P}}_1, \boldsymbol{\mathcal{P}}_2\} \in \mathbb{A}$ and $\theta \in (0,1)$; and is negative for $\boldsymbol{\mathcal{P}}$=zero. The evolution of the *generalized internal variable* $\boldsymbol{\mathcal{F}} := \{\widetilde{\boldsymbol{F}}, \mathbb{F}\}$ may then be derived from the *postulate of maximum dissipation* as the constrained optimization problem

$$\boldsymbol{\mathcal{P}} = \arg\max \left(\boldsymbol{\mathcal{P}}^* \circ \dot{\boldsymbol{\mathcal{F}}} \,\middle|\, \boldsymbol{\mathcal{P}}^* \in \mathbb{A} \right) \quad \text{and} \quad \boldsymbol{\mathcal{P}} \circ \dot{\boldsymbol{\mathcal{F}}} \geq 0.$$

The corresponding optimality conditions render the *associated evolution law* for the generalized internal variable $\boldsymbol{\mathcal{F}}$ in terms of the positive Lagrange or rather *plastic multiplier* γ

$$\dot{\mathcal{F}} = \gamma \mathcal{N} \quad \text{with} \quad \mathcal{N} = \mathcal{N}(\mathcal{P}; \mathcal{F}) := \frac{\partial \phi}{\partial \mathcal{P}}$$

together with the *Karush-Kuhn-Tucker* complementary conditions that distinguish *plastic loading* from *elastic unloading*

$$\gamma \geq 0 \quad \text{and} \quad \phi(\mathcal{P}; \mathcal{F}) \leq 0 \quad \text{and} \quad \gamma \phi(\mathcal{P}; \mathcal{F}) = 0.$$

The formulation of elasto-plasticity is termed *associated* if an associated evolution law for $\mathcal{F}$ is used. It is remarked, however, that in the more general case a *non-associated evolution law* for $\mathcal{F}$ is sometimes simply *postulated* as

$$\dot{\mathcal{F}} = \gamma \mathcal{M} \quad \text{with} \quad \mathcal{M} = \mathcal{M}(\mathcal{P}; \mathcal{F}).$$

In the case of plastic loading with $\gamma > 0$ the plastic multiplier γ may be computed from the so-called *consistency condition*

$$\gamma \dot{\phi}(\mathcal{P}; \mathcal{F}) = 0 \quad \rightsquigarrow \quad \gamma = -\mathcal{D}^{-1} \mathcal{N} \circ \frac{\partial \mathcal{P}}{\partial \boldsymbol{F}} : \dot{\boldsymbol{F}}.$$

Here the denominator $\mathcal{D}$ takes the explicit dependence of the yield condition on the set of generalized internal variables $\mathcal{F}$ into account and is given in terms of the *total derivative* of the yield function

$$\mathcal{D} := \mathcal{N} \circ \mathcal{A} \circ \mathcal{M} + \frac{\partial \phi}{\partial \mathcal{F}} \circ \mathcal{M} =: \frac{\mathrm{d}\phi}{\mathrm{d}\mathcal{F}} \circ \mathcal{M}.$$

Finally the elasto-plastic tangent relation between the material time derivatives of the Piola stress and the deformation gradient follows as

$$\dot{\boldsymbol{P}} = \mathbf{A}_{\text{ep}} : \dot{\boldsymbol{F}} \quad \text{with} \quad \mathbf{A}_{\text{ep}} := \mathbf{A} - \mathcal{D}^{-1} \frac{\partial \boldsymbol{P}}{\partial \mathcal{F}} \circ \mathcal{M} \otimes \mathcal{N} \circ \frac{\partial \mathcal{P}}{\partial \boldsymbol{F}}.$$

In the above the Hessian-type tensors of second partial derivatives of the free energy density with respect to $\boldsymbol{F}$ and $\mathcal{F}$ have been introduced

$$\mathbf{A} := \frac{\partial^2 \psi_0}{\partial \boldsymbol{F} \partial \boldsymbol{F}} \quad \text{and} \quad \mathcal{A} := -\frac{\partial^2 \psi_0}{\partial \mathcal{F} \partial \mathcal{F}}.$$

Observe the major symmetry properties of $\mathbf{A}_{\text{ep}}$ in the case of associated elasto-plasticity.

Convex Analysis Setting:

The constitutive relations of elasto-plasticity may be reformulated within the framework of convex analysis, for a detailed account refer to [145, 123, 76]. Recall that the admissible domain $\mathbb{A}^{\mathcal{F}}$ is a convex set in the space of generalized thermodynamic forces $\mathcal{P}$ that contains $\mathcal{P} =$ zero and is given as

$$\mathbb{A}^{\mathcal{F}}(\mathcal{P}) := \{\mathcal{P} | \, \phi^{\mathcal{F}}(\mathcal{P}) \leq 0 \,\} \quad \text{with} \quad \phi^{\mathcal{F}}(\mathcal{P}) := \phi(\mathcal{P}; \mathcal{F}).$$

Then the *support function* of $\mathbb{A}^{\mathcal{F}}$ in the space of fluxes (material time derivatives) of the generalized internal variables $\dot{\boldsymbol{\mathcal{F}}}$ is defined as

$$d_0^{\mathcal{F}}(\dot{\boldsymbol{\mathcal{F}}}) := \max\left(\boldsymbol{\mathcal{P}} \circ \dot{\boldsymbol{\mathcal{F}}} \middle| \; \boldsymbol{\mathcal{P}} \in \mathbb{A}^{\mathcal{F}}\right) \quad \text{with} \quad d_0^{\mathcal{F}}(\dot{\boldsymbol{\mathcal{F}}}) := d_0(\dot{\boldsymbol{\mathcal{F}}}; \boldsymbol{\mathcal{F}}).$$

In the context of elasto-plasticity the support function d_0 is denoted the (density of the) *dissipation potential*, whereby d_0 is homogeneous of degree one in $\dot{\boldsymbol{\mathcal{F}}}$ for elasto-plasticity. Moreover the *plastic potential* p_0 is introduced as the *characteristic function* (or indicator function, [162]) of the convex set $\mathbb{A}^{\mathcal{F}}$, i.e.

$$p_0^{\mathcal{F}}(\boldsymbol{\mathcal{P}}) := \chi_{\mathbb{A}}^{\mathcal{F}}(\boldsymbol{\mathcal{P}}) \quad \text{with} \quad \chi_{\mathbb{A}}^{\mathcal{F}}(\boldsymbol{\mathcal{P}}) := \chi_{\mathbb{A}}(\boldsymbol{\mathcal{P}}; \boldsymbol{\mathcal{F}}) := \begin{cases} 0 & \boldsymbol{\mathcal{P}} \in \mathbb{A}^{\mathcal{F}} \\ +\infty & \boldsymbol{\mathcal{P}} \notin \mathbb{A}^{\mathcal{F}} \end{cases}.$$

Then the dissipation potential and the plastic potential are conjugate potentials in the sense of Legendre-Fenchel transformations

$$\begin{aligned} d_0^{\mathcal{F}}(\dot{\boldsymbol{\mathcal{F}}}) &:= \max\left(\boldsymbol{\mathcal{P}} \circ \dot{\boldsymbol{\mathcal{F}}} - p_0^{\mathcal{F}}(\boldsymbol{\mathcal{P}}) \middle| \; \boldsymbol{\mathcal{P}}\right), \\ p_0^{\mathcal{F}}(\boldsymbol{\mathcal{P}}) &:= \max\left(\boldsymbol{\mathcal{P}} \circ \dot{\boldsymbol{\mathcal{F}}} - d_0^{\mathcal{F}}(\dot{\boldsymbol{\mathcal{F}}}) \middle| \; \dot{\boldsymbol{\mathcal{F}}}\right). \end{aligned}$$

With these preliminaries at hand, the generalized thermodynamic forces $\boldsymbol{\mathcal{P}}$ and the fluxes $\dot{\boldsymbol{\mathcal{F}}}$ of the generalized internal variables $\boldsymbol{\mathcal{F}}$ follow as the *subdifferential* ∂ of the dissipation potential and the plastic potential, respectively, [143, 144]

$$\boldsymbol{\mathcal{P}} \in \partial d_0(\dot{\boldsymbol{\mathcal{F}}}; \boldsymbol{\mathcal{F}}) \quad \text{and} \quad \dot{\boldsymbol{\mathcal{F}}} \in \partial p_0(\boldsymbol{\mathcal{P}}; \boldsymbol{\mathcal{F}}).$$

Note that the subdifferential of the characteristic function, i.e. of the plastic potential p_0 coincides with the *normal cone* of the convex set $\mathbb{A}$. Here the subdifferentials of d_0 and p_0 are defined as

$$\begin{aligned} \partial d_0^{\mathcal{F}}(\dot{\boldsymbol{\mathcal{F}}}) &:= \{\boldsymbol{\mathcal{P}} |\; d_0^{\mathcal{F}}(\dot{\boldsymbol{\mathcal{F}}}^*) \geq d_0^{\mathcal{F}}(\dot{\boldsymbol{\mathcal{F}}}) + \boldsymbol{\mathcal{P}} \circ [\dot{\boldsymbol{\mathcal{F}}}^* - \dot{\boldsymbol{\mathcal{F}}}] \quad \forall \dot{\boldsymbol{\mathcal{F}}}^* \}, \\ \partial p_0^{\mathcal{F}}(\boldsymbol{\mathcal{P}}) &:= \{\dot{\boldsymbol{\mathcal{F}}} |\; p_0^{\mathcal{F}}(\boldsymbol{\mathcal{P}}^*) \geq p_0^{\mathcal{F}}(\boldsymbol{\mathcal{P}}) + [\boldsymbol{\mathcal{P}}^* - \boldsymbol{\mathcal{P}}] \circ \dot{\boldsymbol{\mathcal{F}}} \quad \forall \boldsymbol{\mathcal{P}}^* \}. \end{aligned}$$

For the sake of concreteness, consider the simplest possible geometrically linear one-dimensional setting of ideal plasticity. Then, in the so-called dual formulation [76], plasticity is described by a yield function $\phi(\sigma) = |\sigma| - \sigma_{\rm y}$, *a plastic evolution law* $\dot{\epsilon}_{\rm p} = \gamma \partial \phi = \gamma \sigma / |\sigma|$ *and a set of Karush-Kuhn-Tucker conditions* $\gamma \geq 0, \phi \leq 0, \gamma \phi = 0$. *In the case of loading the plastic multiplier follows here as* $\gamma = |\dot{\epsilon}_{\rm p}|$. *The corresponding reduced dissipation inequality is given by* $\sigma \dot{\epsilon}_{\rm p} = \gamma \sigma_{\rm y} = |\dot{\epsilon}_{\rm p}| \sigma_{\rm y} \geq 0$. *The plastic evolution law may equivalently be expressed in the so-called primal formulation [76] based on the plastic potential* $\dot{\epsilon}_{\rm p} \in \partial p$, *which in turn is here expressed in terms of the characteristic or rather indicator function* $p(\sigma) = \chi_{\{\phi(\sigma)\}}$ *with* $\chi_{\{\phi \leq 0\}} = 0$ *and* $\chi_{\{\phi > 0\}} = +\infty$ *with subdifferential* $\partial \chi_{\{\phi < 0\}} = 0$ *and* $\partial \chi_{\{\phi = 0\}} \in [0, \infty[$.

Then we may introduce the dissipation potential $d(\dot{\epsilon}_{\rm p})$ as the Legendre-Fenchel transformation of the plastic potential $d(\dot{\epsilon}_{\rm p}) = \sup_\sigma\{\sigma\dot{\epsilon}_{\rm p} - p(\sigma)\}$, which results here in $d(\dot{\epsilon}_{\rm p}) = |\dot{\epsilon}_{\rm p}|\sigma_{\rm y}$, acting as a potential for the stress (thermodynamical force) $\sigma \in \partial d$. As a result we obtain here $\sigma = \sigma_{\rm y}\dot{\epsilon}_{\rm p}/|\dot{\epsilon}_{\rm p}|$ for $|\dot{\epsilon}_{\rm p}| > 0$ (i.e. the plastic evolution law) and $\sigma \in [-\sigma_{\rm y}, +\sigma_{\rm y}]$ (i.e. the yield condition) for $\dot{\epsilon}_{\rm p} = 0$.

Local Plasticity: Choice of Internal Variables:
As a possibility the set of internal variables $\mathbb{F}$ may for example consist of the length of the *plastic path* and the modulus of the actual plastic distortion

$$\boldsymbol{\mathcal{F}} = \{\widetilde{\boldsymbol{F}}, \mathbb{F}\} \quad \text{with} \quad \mathbb{F} := \left\{\alpha_{\rm iso} := \int_0^t |\,\mathrm{D}_\tau \widetilde{\boldsymbol{F}}|\,\mathrm{d}\tau,\ \alpha_{\rm kin} := |\widetilde{\boldsymbol{F}}|\right\}.$$

These variables are suited to capture isotropic and kinematic (Prager-type) hardening (or softening). The square of the modulus of the plastic distortion expands into $|\widetilde{\boldsymbol{F}}|^2 = \widetilde{\boldsymbol{F}}_\flat : \widetilde{\boldsymbol{F}}_\sharp$ (with $\widetilde{\boldsymbol{F}}_\flat := \widetilde{\boldsymbol{j}} \cdot \widetilde{\boldsymbol{F}}$ and $\widetilde{\boldsymbol{F}}_\sharp := \widetilde{\boldsymbol{F}} \cdot \boldsymbol{I}$), thus the material time derivative of of the internal variables in $\mathbb{F}$ computes as

$$\mathrm{D}_t\mathbb{F} = \left\{\mathrm{D}_t\alpha_{\rm iso} = |\,\mathrm{D}_t\widetilde{\boldsymbol{F}}|,\ \mathrm{D}_t\alpha_{\rm kin} = \mathrm{D}_t\widetilde{\boldsymbol{F}} : \frac{\widetilde{\boldsymbol{F}}_\flat \cdot \boldsymbol{I}}{|\widetilde{\boldsymbol{F}}|}\right\}.$$

Then the thermodynamic force $\mathbb{P}$ that drives the evolution of $\mathbb{F}$ derives as

$$\boldsymbol{\mathcal{P}} = \{\widetilde{\boldsymbol{P}}, \mathbb{P}\} \quad \text{with} \quad -\,\mathbb{P} = \left\{\beta_{\rm iso} := \frac{\partial\psi_0}{\partial\alpha_{\rm iso}},\ \beta_{\rm kin} := \frac{\partial\psi_0}{\partial\alpha_{\rm kin}}\right\}.$$

Consequently, the reduced dissipation inequality expands as

$$\boldsymbol{\mathcal{P}} \circ \mathrm{D}_t\boldsymbol{\mathcal{F}} = \left[\widetilde{\boldsymbol{P}} - \widetilde{\boldsymbol{P}}_{\rm kin}\right] : \mathrm{D}_t\widetilde{\boldsymbol{F}} - \beta_{\rm iso}|\,\mathrm{D}_t\widetilde{\boldsymbol{F}}| \geq 0.$$

Obviously the contribution due to kinematic hardening acts as a *back-stress*

$$\widetilde{\boldsymbol{P}}_{\rm kin} := \beta_{\rm kin}\frac{\widetilde{\boldsymbol{F}}_\flat \cdot \boldsymbol{I}}{|\widetilde{\boldsymbol{F}}|}$$

(shifting the origin of the yield surface in the space of plastic Piola stresses), whereas the contribution due to isotropic hardening acts as a *drag-stress* $\beta_{\rm iso}$ (changing the diameter of the admissible domain). Thus the yield function may be stated as

$$\phi(\widetilde{\boldsymbol{P}}, \beta_{\rm iso}, \beta_{\rm kin}) = \varphi\left(\widetilde{\boldsymbol{P}} - \widetilde{\boldsymbol{P}}_{\rm kin}\right) - [\sigma_{\rm y} + \beta_{\rm iso}] \leq 0.$$

The parametric dependence of φ on $\boldsymbol{\mathcal{F}}$ has been suppressed for the ease of notation. In the sequel the abbreviation $\widetilde{\boldsymbol{P}}_{\rm rel} := \widetilde{\boldsymbol{P}} - \widetilde{\boldsymbol{P}}_{\rm kin}$ shall be adopted. The yield function $\varphi : \widetilde{\boldsymbol{P}}_{\rm rel} \mapsto [0, \infty]$ is a *gauge*, i.e. a convex

($\varphi(\theta\widetilde{\boldsymbol{P}}_{\text{rel}^1} + [1-\theta]\widetilde{\boldsymbol{P}}_{\text{rel}^2}) \leq \theta\varphi(\widetilde{\boldsymbol{P}}_{\text{rel}^1}) + [1-\theta]\phi(\widetilde{\boldsymbol{P}}_{\text{rel}^2})$ for $\theta \in (0,1)$) and positively homogeneous function ($\varphi(a\widetilde{\boldsymbol{P}}_{\text{rel}}) = a\varphi(\widetilde{\boldsymbol{P}}_{\text{rel}})$ for $a > 0$) that is positive ($\varphi(\widetilde{\boldsymbol{P}}_{\text{rel}}) \geq 0$) and zero at the origin ($\varphi(\mathbf{0}) = 0$), whereas σ_{y} is a positive constant (the *initial yield stress*). Accordingly, the associated plastic evolution equations for the internal variables in $\boldsymbol{\mathcal{F}}$ read as

$$\mathrm{D}_t\widetilde{\boldsymbol{F}} = \gamma\widetilde{\boldsymbol{N}} := \gamma\frac{\partial\varphi}{\partial\widetilde{\boldsymbol{P}}_{\text{rel}}} \quad \text{and} \quad \mathrm{D}_t\alpha_{\text{iso}} = \gamma \quad \text{and} \quad \mathrm{D}_t\alpha_{\text{kin}} = \gamma\widetilde{\boldsymbol{N}} : \frac{\widetilde{\boldsymbol{F}}_\flat \cdot \boldsymbol{I}}{|\widetilde{\boldsymbol{F}}|}.$$

Observe that due to φ being positive homogeneous the evolution equations for $\widetilde{\boldsymbol{F}}$, α_{iso} and α_{kin} are mutually consistent. It is mentioned that more sophisticated kinematic hardening models, e.g. of Frederick-Armstrong-type, are possible.

Generalized Plasticity: Dislocation Density=Internal Variable: Motivated by the insight that *kinematically necessary dislocations* hinder further plastic flow, the dislocation density tensor $\operatorname{Curl}\widetilde{\boldsymbol{F}}$ that captures the incompatibility of the intermediate configuration may be added to an *extended* set of internal variables $\mathbb{F}_{\text{ext}}$ for single crystal plasticity, correspondingly the thermodynamic force $\mathbb{P}_{\text{ext}}$ is also extended

$$\mathbb{F}_{\text{ext}} := \left\{\alpha_{\text{iso}},\ \alpha_{\text{kin}},\ \widetilde{\boldsymbol{\alpha}}_{\text{dis}} := \operatorname{Curl}\widetilde{\boldsymbol{F}}\right\},\ -\mathbb{P}_{\text{ext}} = \left\{\beta_{\text{iso}},\ \beta_{\text{kin}},\ \widetilde{\boldsymbol{\beta}}_{\text{dis}} := \frac{\partial\psi_0}{\partial\widetilde{\boldsymbol{\alpha}}_{\text{dis}}}\right\}.$$

Consequently, since $\mathrm{D}_t\widetilde{\boldsymbol{\alpha}}_{\text{dis}} = \operatorname{Curl}\mathrm{D}_t\widetilde{\boldsymbol{F}}$ and based on integration by parts the reduced dissipation inequality may be formulated in terms of a *dislocation-induced back-stress* $\operatorname{Curl}\widetilde{\boldsymbol{\beta}}_{\text{dis}}$ contained in $\widetilde{\boldsymbol{P}}^\star$

$$\boldsymbol{\mathcal{P}} \circ \mathrm{D}_t\boldsymbol{\mathcal{F}} = \left[\widetilde{\boldsymbol{P}}^\star - \widetilde{\boldsymbol{P}}_{\text{kin}}\right] : \mathrm{D}_t\widetilde{\boldsymbol{F}} - \beta_{\text{iso}}|\,\mathrm{D}_t\widetilde{\boldsymbol{F}}| \geq 0,$$

where $\boldsymbol{\mathcal{P}} := \{\widetilde{\boldsymbol{P}}^\star, \mathbb{P}\}$ with $\widetilde{\boldsymbol{P}}^\star := \widetilde{\boldsymbol{P}} - \operatorname{Curl}\widetilde{\boldsymbol{\beta}}_{\text{dis}}$ and $\boldsymbol{\mathcal{F}} := \{\widetilde{\boldsymbol{F}}, \mathbb{F}\}$, together with a dissipation inequality at the elasto-plastic boundary

$$\mathrm{D}_t\widetilde{\boldsymbol{F}} : [\widetilde{\boldsymbol{\beta}}_{\text{dis}} \cdot \operatorname{spn}\boldsymbol{N}] \leq 0.$$

Note that $\widetilde{\boldsymbol{P}}^\star$ may alternatively be considered the *variational derivative* of the free energy density with respect to the plastic distortion. As a valid choice the dissipation inequality at the boundary may be satisfied trivially by appropriate boundary conditions, e.g. homogeneous Neumann-type boundary conditions $\widetilde{\boldsymbol{\beta}}_{\text{dis}} \cdot \operatorname{spn}\boldsymbol{N} = \mathbf{0}$. Finally the yield function may be stated as

$$\phi(\widetilde{\boldsymbol{P}}^\star, \beta_{\text{iso}}, \beta_{\text{kin}}) = \varphi\left(\widetilde{\boldsymbol{P}}^\star - \widetilde{\boldsymbol{P}}_{\text{kin}}\right) - [\sigma_{\text{y}} + \beta_{\text{iso}}] \leq 0.$$

Interestingly, the corresponding associated plastic evolution equations for the internal variables in $\boldsymbol{\mathcal{F}}$ are unchanged as compared to the purely local case.

Generalized plasticity involving the plastic dislocation density tensor as an internal variable has been proposed e.g. in [181, 128, 41, 42].
Example: For geometrically linear generalized single crystal plasticity the generalized Schmid stress acting on slip system ς *may be defined as* $\varphi_\varsigma := \boldsymbol{s}_\varsigma \cdot \widetilde{\boldsymbol{\sigma}}^\star \cdot \boldsymbol{m}_\varsigma$ *with* $\boldsymbol{s}_\varsigma$ *and* $\boldsymbol{m}_\varsigma$ *the slip direction and the slip plane normal, respectively. Then the contribution of the free energy density due to the dislocation density may be formulated as* $2\psi_{\text{dis}} = G\ell^2 \widetilde{\boldsymbol{\alpha}}_{\text{dis}} : \widetilde{\boldsymbol{\alpha}}_{\text{dis}}$ *with* G *the shear modulus and* ℓ *an internal length, thus the thermodynamic force corresponding to* $\widetilde{\boldsymbol{\alpha}}_{\text{dis}}$ *reads as* $\widetilde{\boldsymbol{\beta}}_{\text{dis}} = G\ell^2 \widetilde{\boldsymbol{\alpha}}_{\text{dis}}$*. As a result, the yield function for slip system* ς *reads as* $\phi_\varsigma = \boldsymbol{s}_\varsigma \cdot \widetilde{\boldsymbol{\sigma}} \cdot \boldsymbol{m}_\varsigma - [\sigma_{\text{y}}^\varsigma + \beta_{\text{iso}}^\varsigma + G\ell^2 \boldsymbol{s}_\varsigma \cdot \operatorname{curl} \widetilde{\boldsymbol{\alpha}}_{\text{dis}} \cdot \boldsymbol{m}_\varsigma]$*. It is emphasized once again that the corresponding associated flow rule for the plastic distortion is completely unaffected by the incorporation of higher (second) gradients* $\operatorname{curl} \widetilde{\boldsymbol{\alpha}}_{\text{dis}}$ *of the plastic distortion into the yield function.*

Generalized Plasticity: Incompatibility Density=Internal Variable: Motivated by the insight that *incompatibilities* hinder further plastic flow, the incompatibility density tensor $\operatorname{Inc} \widetilde{\boldsymbol{C}}$ that captures the incompatibility of the intermediate configuration may be added to an *extended* set of internal variables $\mathbb{F}_{\text{ext}}$ for polycrystal plasticity, correspondingly the thermodynamic force $\mathbb{P}_{\text{ext}}$ is also extended

$$\mathbb{F}_{\text{ext}} := \left\{\alpha_{\text{iso}},\, \alpha_{\text{kin}},\, \widetilde{\boldsymbol{\alpha}}_{\text{inc}} := \operatorname{Inc} \widetilde{\boldsymbol{C}}\right\}, \quad -\mathbb{P}_{\text{ext}} = \left\{\beta_{\text{iso}},\, \beta_{\text{kin}},\, \widetilde{\boldsymbol{\beta}}_{\text{inc}} := \frac{\partial \psi_0}{\partial \widetilde{\boldsymbol{\alpha}}_{\text{inc}}}\right\}.$$

For the sake of simplicity, only the linear contribution $\operatorname{Curl}^t \operatorname{Curl}^t \widetilde{\boldsymbol{C}}$ to the incompatibility density tensor $\operatorname{Inc} \widetilde{\boldsymbol{C}}$ shall be considered in the sequel. Consequently, since $\mathrm{D}_t \widetilde{\boldsymbol{\alpha}}_{\text{inc}} = \operatorname{Inc} \mathrm{D}_t \widetilde{\boldsymbol{C}}$ and based on integration by parts the reduced dissipation inequality may be formulated in terms of a *incompatibility-induced back-stress* $\operatorname{Inc} \widetilde{\boldsymbol{\beta}}_{\text{inc}}$ contained in $\widetilde{\boldsymbol{P}}^\star$

$$\boldsymbol{\mathcal{P}} \circ \mathrm{D}_t \boldsymbol{\mathcal{F}} = \left[\widetilde{\boldsymbol{P}}^\star - \widetilde{\boldsymbol{P}}_{\text{kin}}\right] : \mathrm{D}_t \widetilde{\boldsymbol{F}} - \beta_{\text{iso}} |\, \mathrm{D}_t \widetilde{\boldsymbol{F}}| \geq 0,$$

where $\boldsymbol{\mathcal{P}} := \{\widetilde{\boldsymbol{P}}^\star, \mathbb{P}\}$ with $\widetilde{\boldsymbol{P}}^\star := \widetilde{\boldsymbol{P}} - 2\widetilde{\boldsymbol{F}} \cdot \operatorname{Inc} \widetilde{\boldsymbol{\beta}}_{\text{inc}}$ and $\boldsymbol{\mathcal{F}} := \{\widetilde{\boldsymbol{F}}, \mathbb{F}\}$, together with a dissipation inequality at the elasto-plastic boundary

$$\mathrm{D}_t \operatorname{Curl}^t \widetilde{\boldsymbol{C}} : [\widetilde{\boldsymbol{\beta}}_{\text{inc}} \cdot \operatorname{spn} \boldsymbol{N}] + \mathrm{D}_t \widetilde{\boldsymbol{C}} : [\operatorname{Curl}^t \widetilde{\boldsymbol{\beta}}_{\text{inc}} \cdot \operatorname{spn} \boldsymbol{N}] \leq 0.$$

Note that $\widetilde{\boldsymbol{P}}^\star$ may alternatively be considered the *variational derivative* of the free energy density with respect to the plastic distortion. As a valid choice the dissipation inequality at the boundary may be satisfied trivially by appropriate boundary conditions, e.g. homogeneous Neumann-type boundary conditions $\widetilde{\boldsymbol{\beta}}_{\text{inc}} \cdot \operatorname{spn} \boldsymbol{N} = \boldsymbol{0}$ and $\operatorname{Curl}^t \widetilde{\boldsymbol{\beta}}_{\text{inc}} \cdot \operatorname{spn} \boldsymbol{N} = \boldsymbol{0}$. Finally the yield function may be stated as

$$\phi(\widetilde{\boldsymbol{P}}^\star, \beta_{\text{iso}}, \beta_{\text{kin}}) = \varphi\left(\widetilde{\boldsymbol{P}}^\star - \widetilde{\boldsymbol{P}}_{\text{kin}}\right) - [\sigma_{\text{y}} + \beta_{\text{iso}}] \leq 0.$$

Observe again, that the corresponding associated plastic evolution equations for the internal variables in $\boldsymbol{\mathcal{F}}$ are unchanged as compared to the purely local case. The geometrically linear version of generalized plasticity involving the plastic incompatibility density tensor as an internal variable has been proposed in [128].

Example: For geometrically linear generalized v. Mises plasticity the generalized equivalent stress may be defined as $\varphi := |\operatorname{dev}\widetilde{\boldsymbol{\sigma}}^{\star}|$ *with* $\operatorname{dev}\{\bullet\}$ *the deviatoric part of* $\{\bullet\}$*. Then the contribution of the free energy density due to the incompatibility density may be formulated as* $2\psi_{\text{inc}} = G\ell^4\widetilde{\boldsymbol{\alpha}}_{\text{inc}} : \widetilde{\boldsymbol{\alpha}}_{\text{inc}}$ *with* G *the shear modulus and* ℓ *an internal length, thus the thermodynamic force corresponding to* $\widetilde{\boldsymbol{\alpha}}_{\text{inc}}$ *reads as* $\widetilde{\boldsymbol{\beta}}_{\text{inc}} = G\ell^4\widetilde{\boldsymbol{\alpha}}_{\text{inc}}$*. As a result, the generalized v. Mises yield function expands as* $\phi = |\operatorname{dev}\widetilde{\boldsymbol{\sigma}} - G\ell^4 \operatorname{inc}\widetilde{\boldsymbol{\alpha}}_{\text{inc}}| - [\sigma_{\text{y}} + \beta_{\text{iso}}]$*. It is emphasized once again that the corresponding associated flow rule for the plastic strain is completely unaffected by the incorporation of higher (fourth) gradients* $\operatorname{inc}\widetilde{\boldsymbol{\alpha}}_{\text{inc}}$ *of the plastic strain into the yield function.*

Spatially Superposed Rigid Body Motions:
Rigid body motions superposed to the *spatial* configuration in terms of temporarily varying rotations $\boldsymbol{Q}(t) \in \text{SO}(3)$ and translations $\boldsymbol{z}(t) \in \text{T}(3)$ result in the transformed deformation gradient and the transformed elastic distortion

$$\boldsymbol{F}^* = \boldsymbol{Q}(t) \cdot \boldsymbol{F}(\boldsymbol{X}, t) \quad \text{and} \quad \overline{\boldsymbol{F}}^* = \boldsymbol{Q}(t) \cdot \overline{\boldsymbol{F}}(\boldsymbol{X}, t).$$

Vice versa the plastic distortion remains unaffected by a spatially superposed rigid body motion (a conceptually possible rigid body rotation superposed to the intermediate configuration remains in general undetermined and is here, as a rule, set to the same rotation as suffered by the material configuration, i.e. to zero)

$$\widetilde{\boldsymbol{F}}^* = \widetilde{\boldsymbol{F}}(\boldsymbol{X}, t).$$

Then, the contravariant material Piola-Kirchhoff stress $\boldsymbol{S}$, the covariant material stress $\boldsymbol{T}$ and the co/contra- (mixed) variant material Mandel stress $\boldsymbol{M}$ remain *invariant*, i.e. unaffected under a spatially superposed rigid body motion

$$\boldsymbol{S}^* = \boldsymbol{S} \quad \text{and} \quad \boldsymbol{T}^* = \boldsymbol{T} \quad \text{and} \quad \boldsymbol{M}^* = \boldsymbol{M}.$$

Likewise also the corresponding plastic push-forwards to the intermediate configuration in terms of the plastic distortion $\widetilde{\boldsymbol{F}}$ trivially remain invariant, i.e.

$$\overline{\boldsymbol{S}}^* = \overline{\boldsymbol{S}} \quad \text{and} \quad \overline{\boldsymbol{T}}^* = \overline{\boldsymbol{T}} \quad \text{and} \quad \overline{\boldsymbol{M}}^* = \overline{\boldsymbol{M}}.$$

Contrarily, due to the orthogonality of $\boldsymbol{Q}$ the contra-, co- and co/contra- (mixed) variant versions of the spatial Kirchhoff stress transform *objectively* under a spatially superposed rigid body motion, for example

$$\boldsymbol{\tau}^* = \boldsymbol{Q} \cdot \boldsymbol{\tau} \cdot \boldsymbol{Q}^t.$$

Moreover, the co- and contravariant spatial strain tensors $\overline{\boldsymbol{c}}$ and $\overline{\boldsymbol{b}}$ transform *objectively* under a spatially superposed rigid body motion, i.e.

$$\overline{\boldsymbol{c}}^* = \boldsymbol{Q} \cdot \overline{\boldsymbol{f}}^t \cdot \overline{\boldsymbol{f}} \cdot \boldsymbol{Q}^t = \boldsymbol{Q} \cdot \overline{\boldsymbol{c}} \cdot \boldsymbol{Q}^t \quad \text{and} \quad \overline{\boldsymbol{b}}^* = \boldsymbol{Q} \cdot \overline{\boldsymbol{F}} \cdot \overline{\boldsymbol{F}}^t \cdot \boldsymbol{Q}^t = \boldsymbol{Q} \cdot \overline{\boldsymbol{b}} \cdot \boldsymbol{Q}^t.$$

Vice versa, the co- and contravariant intermediate strain tensors $\overline{\boldsymbol{C}}$ and $\overline{\boldsymbol{B}}$ remain *invariant* under a spatially superposed rigid body motion, i.e.

$$\overline{\boldsymbol{C}}^* = \overline{\boldsymbol{F}}^t \cdot \boldsymbol{Q}^t \cdot \boldsymbol{Q} \cdot \overline{\boldsymbol{F}} = \overline{\boldsymbol{C}} \quad \text{and} \quad \overline{\boldsymbol{B}}^* = \overline{\boldsymbol{f}} \cdot \boldsymbol{Q}^t \cdot \boldsymbol{Q} \cdot \overline{\boldsymbol{f}}^t = \overline{\boldsymbol{B}}.$$

Likewise, the co- and contravariant intermediate strain tensors $\widetilde{\boldsymbol{c}}$ and $\widetilde{\boldsymbol{b}}$ trivially remain *invariant* under a spatially superposed rigid body motion, i.e.

$$\widetilde{\boldsymbol{c}}^* = \widetilde{\boldsymbol{f}}^t \cdot \widetilde{\boldsymbol{f}} = \widetilde{\boldsymbol{c}} \quad \text{and} \quad \widetilde{\boldsymbol{b}}^* = \widetilde{\boldsymbol{F}} \cdot \widetilde{\boldsymbol{F}}^t = \widetilde{\boldsymbol{b}}.$$

Finally, also the co- and contravariant material strain tensors $\widetilde{\boldsymbol{C}}$ and $\widetilde{\boldsymbol{B}}$ trivially remain *invariant* under a spatially superposed rigid body motion, i.e.

$$\widetilde{\boldsymbol{C}}^* = \widetilde{\boldsymbol{F}}^t \cdot \widetilde{\boldsymbol{F}} = \widetilde{\boldsymbol{C}} \quad \text{and} \quad \widetilde{\boldsymbol{B}}^* = \widetilde{\boldsymbol{f}} \cdot \widetilde{\boldsymbol{f}}^t = \widetilde{\boldsymbol{B}}.$$

Recall that spatial objectivity is a key requirement in the theory of constitutive modeling.

Materially Superposed Rigid Body Motions:
Rigid body motions superposed to the *material* configuration in terms of temporarily varying rotations $\boldsymbol{q}(t) \in \mathrm{SO}(3)$ with $\boldsymbol{Q} = \boldsymbol{q}^{-1} = \boldsymbol{q}^t$ and translations $\boldsymbol{Z}(t) \in \mathrm{T}(3)$ result in the transformed deformation gradient and transformed plastic distortion

$$\boldsymbol{F}_* = \boldsymbol{F}(\boldsymbol{X}, t) \cdot \boldsymbol{Q}(t) \quad \text{and} \quad \widetilde{\boldsymbol{F}}_* = \boldsymbol{q}(t) \cdot \widetilde{\boldsymbol{F}}(\boldsymbol{X}, t) \cdot \boldsymbol{Q}(t).$$

Here the viewpoint is that the intermediate configuration rotates under materially superposed rigid body motion in complete slavery with the material configuration (recall that a conceptually possible rigid body rotation superposed to the intermediate configuration remains in general undetermined and may thus be selected arbitrarily). As a consequence the elastic distortion also transforms under a materially superposed rigid body motion as

$$\overline{\boldsymbol{F}}_* = \overline{\boldsymbol{F}}(\boldsymbol{X}, t) \cdot \boldsymbol{Q}(t).$$

Then, the contra-, co- and co/contra- (mixed) variant versions of the spatial Kirchhoff stress remain *invariant* under a materially superposed rigid body motion, for example

$$\boldsymbol{\tau}_* = \boldsymbol{\tau}.$$

Contrarily, due to the orthogonality of $\boldsymbol{Q}$ the contravariant intermediate Piola-Kirchhoff stress $\overline{\boldsymbol{S}}$, the covariant intermediate stress $\overline{\boldsymbol{T}}$ and the co/contra- (mixed) variant intermediate Mandel stress $\overline{\boldsymbol{M}}$ transform *objectively* under a materially superposed rigid body motion, i.e.

$$\overline{\boldsymbol{S}}_* = \boldsymbol{Q}^t \cdot \overline{\boldsymbol{S}} \cdot \boldsymbol{Q} \quad \text{and} \quad \overline{\boldsymbol{T}}_* = \boldsymbol{Q}^t \cdot \overline{\boldsymbol{T}} \cdot \boldsymbol{Q} \quad \text{and} \quad \overline{\boldsymbol{M}}_* = \boldsymbol{Q}^t \cdot \overline{\boldsymbol{M}} \cdot \boldsymbol{Q}.$$

Likewise also the corresponding plastic pull-backs to the material configuration in terms of the plastic distortion $\widetilde{\boldsymbol{F}}$ transform *objectively* under a materially superposed rigid body motion, i.e.

$$\boldsymbol{S}_* = \boldsymbol{Q}^t \cdot \boldsymbol{S} \cdot \boldsymbol{Q} \quad \text{and} \quad \boldsymbol{T}_* = \boldsymbol{Q}^t \cdot \boldsymbol{T} \cdot \boldsymbol{Q} \quad \text{and} \quad \boldsymbol{M}_* = \boldsymbol{Q}^t \cdot \boldsymbol{M} \cdot \boldsymbol{Q}.$$

Moreover, the co- and contravariant spatial strain tensors $\overline{\boldsymbol{c}}$ and $\overline{\boldsymbol{b}}$ remain *invariant* under a materially superposed rigid body motion, i.e.

$$\overline{\boldsymbol{c}}_* = \overline{\boldsymbol{f}}^t \cdot \boldsymbol{Q} \cdot \boldsymbol{Q}^t \cdot \overline{\boldsymbol{f}} = \overline{\boldsymbol{c}} \quad \text{and} \quad \overline{\boldsymbol{b}}_* = \overline{\boldsymbol{F}} \cdot \boldsymbol{Q} \cdot \boldsymbol{Q}^t \cdot \overline{\boldsymbol{F}}^t = \overline{\boldsymbol{b}}.$$

Vice versa, the co- and contravariant intermediate strain tensors $\overline{\boldsymbol{C}}$ and $\overline{\boldsymbol{B}}$ transform *objectively* under a materially superposed rigid body motion, i.e.

$$\overline{\boldsymbol{C}}_* = \boldsymbol{Q}^t \cdot \overline{\boldsymbol{F}}^t \cdot \overline{\boldsymbol{F}} \cdot \boldsymbol{Q} = \boldsymbol{Q}^t \cdot \overline{\boldsymbol{C}} \cdot \boldsymbol{Q} \quad \text{and} \quad \overline{\boldsymbol{B}}_* = \boldsymbol{Q}^t \cdot \overline{\boldsymbol{f}} \cdot \overline{\boldsymbol{f}}^t \cdot \boldsymbol{Q} = \boldsymbol{Q}^t \cdot \overline{\boldsymbol{B}} \cdot \boldsymbol{Q}.$$

Likewise, the co- and contravariant intermediate strain tensors $\widetilde{\boldsymbol{c}}$ and $\widetilde{\boldsymbol{b}}$ transform *objectively* under a materially superposed rigid body motion, i.e.

$$\widetilde{\boldsymbol{c}}_* = \boldsymbol{q} \cdot \widetilde{\boldsymbol{f}}^t \cdot \boldsymbol{Q} \cdot \boldsymbol{Q}^t \cdot \widetilde{\boldsymbol{f}} \cdot \boldsymbol{q}^t = \boldsymbol{q} \cdot \widetilde{\boldsymbol{c}} \cdot \boldsymbol{q}^t \quad \text{and} \quad \widetilde{\boldsymbol{b}}_* = \boldsymbol{q} \cdot \widetilde{\boldsymbol{F}} \cdot \boldsymbol{Q} \cdot \boldsymbol{Q}^t \cdot \widetilde{\boldsymbol{F}}^t \cdot \boldsymbol{q}^t = \boldsymbol{q} \cdot \widetilde{\boldsymbol{b}} \cdot \boldsymbol{q}^t.$$

Finally, also the co- and contravariant material strain tensors $\widetilde{\boldsymbol{C}}$ and $\widetilde{\boldsymbol{B}}$ transform *objectively* under a materially superposed rigid body motion, i.e.

$$\widetilde{\boldsymbol{C}}_* = \boldsymbol{Q}^t \cdot \widetilde{\boldsymbol{F}}^t \cdot \boldsymbol{q}^t \cdot \boldsymbol{q} \cdot \widetilde{\boldsymbol{F}} \cdot \boldsymbol{Q} = \boldsymbol{Q}^t \cdot \widetilde{\boldsymbol{C}} \cdot \boldsymbol{Q} \quad \text{and} \quad \widetilde{\boldsymbol{B}}_* = \boldsymbol{Q}^t \cdot \widetilde{\boldsymbol{f}} \cdot \boldsymbol{q}^t \cdot \boldsymbol{q} \cdot \widetilde{\boldsymbol{f}}^t \cdot \boldsymbol{Q} = \boldsymbol{Q}^t \cdot \widetilde{\boldsymbol{B}} \cdot \boldsymbol{Q}.$$

Intermediate and material stress and strain measures transforming objectively under materially superposed rigid body motions are suited to capture anisotropic material behavior.

Restrictions on the Free Energy Density:
Requiring the free energy density ψ_0 to transform objectively under arbitrary spatially superposed rigid body motion, i.e. $\psi_0 \doteq \psi_0^*$, results in the alternative restrictions

$$\begin{aligned}
\psi_0 \doteq \psi_0^* &= \psi_0(\overline{\boldsymbol{C}}; \cdots), \\
\psi_0 \doteq \psi_0^* &= \psi_0(\overline{\boldsymbol{B}}; \cdots), \\
\psi_0 \doteq \psi_0^* &= \psi_0(\boldsymbol{Q} \cdot \overline{\boldsymbol{c}} \cdot \boldsymbol{Q}^t; \cdots) \quad \forall \boldsymbol{Q} \in \mathrm{SO}(3), \\
\psi_0 \doteq \psi_0^* &= \psi_0(\boldsymbol{Q} \cdot \overline{\boldsymbol{b}} \cdot \boldsymbol{Q}^t; \cdots) \quad \forall \boldsymbol{Q} \in \mathrm{SO}(3).
\end{aligned}$$

Thus an arbitrary dependence on either of the intermediate description strain measures $\overline{\boldsymbol{C}}$ or $\overline{\boldsymbol{B}}$ automatically satisfies spatial objectivity of ψ_0. In contrast the dependence on either of the spatial description strain measures $\overline{\boldsymbol{c}}$ or $\overline{\boldsymbol{b}}$ is subject to the above *isotropy* conditions on the tensor function ψ_0 in order to satisfy spatial objectivity. However, the yet unspecified arguments of ψ_0 indicated by the $\cdots$ allow to also capture *anisotropic* material behavior. To this end sets of co- and contravariant intermediate *structural tensors*

$$\overline{\boldsymbol{N}}_0 := \overline{\boldsymbol{J}}, \quad \overline{\boldsymbol{N}}_\alpha \quad \text{and} \quad \overline{\boldsymbol{M}}_0 := \overline{\boldsymbol{I}}, \quad \overline{\boldsymbol{M}}_\alpha \quad \text{with} \quad \alpha = 1, \cdots, n_{\text{st}}$$

are introduced that capture the *material's symmetry group* $\mathrm{M}(3) \subseteq \mathrm{SO}(3)$. Here the co- and contravariant intermediate unit tensors $\overline{\boldsymbol{J}}$ and $\overline{\boldsymbol{I}}$ are explicitly incorporated in order to capture the isotropic part of the response. The co- and contravariant intermediate structural tensors satisfy the following invariance conditions related to the materials' symmetry group

$$\begin{aligned} \overline{\boldsymbol{N}}_\alpha &\equiv \boldsymbol{Q}^t \cdot \overline{\boldsymbol{N}}_\alpha \cdot \boldsymbol{Q} \ \ \forall \boldsymbol{Q} \in \mathrm{M}(3) \subseteq \mathrm{SO}(3), \quad \alpha = 1, \cdots, n_{\text{st}} \\ \overline{\boldsymbol{M}}_\alpha &\equiv \boldsymbol{Q}^t \cdot \overline{\boldsymbol{M}}_\alpha \cdot \boldsymbol{Q} \ \ \forall \boldsymbol{Q} \in \mathrm{M}(3) \subseteq \mathrm{SO}(3), \quad \alpha = 1, \cdots, n_{\text{st}}. \end{aligned}$$

In passing it is noted that the pull-back and push-forward of the co- and contravariant intermediate structural tensors into their material and spatial counterparts reads as

$$\begin{aligned} \boldsymbol{N}_0 &= \widetilde{\boldsymbol{C}}, \quad \boldsymbol{N}_\alpha = \widetilde{\boldsymbol{F}}^t \cdot \overline{\boldsymbol{N}}_\alpha \cdot \widetilde{\boldsymbol{F}} \quad \text{and} \quad \boldsymbol{n}_0 = \overline{\boldsymbol{c}}, \quad \boldsymbol{n}_\alpha = \overline{\boldsymbol{f}}^t \cdot \overline{\boldsymbol{N}}_\alpha \cdot \overline{\boldsymbol{f}} \ , \\ \boldsymbol{M}_0 &= \widetilde{\boldsymbol{B}}, \quad \boldsymbol{M}_\alpha = \widetilde{\boldsymbol{f}} \ \cdot \overline{\boldsymbol{M}}_\alpha \cdot \widetilde{\boldsymbol{f}}^t \quad \text{and} \quad \boldsymbol{m}_0 = \overline{\boldsymbol{b}}, \quad \boldsymbol{m}_\alpha = \overline{\boldsymbol{F}} \ \cdot \overline{\boldsymbol{M}}_\alpha \cdot \overline{\boldsymbol{F}}^t. \end{aligned}$$

Furthermore, the spatial structural tensors are (elastic) Killing fields with vanishing elastic Lie time derivative (i.e. the intermediate structural tensors are assumed constant, however, constant material structural tensors would be a valid alternative modeling assumption that shall not be pursued in the sequel)

$$\widetilde{\mathrm{R}}_t \boldsymbol{N}_\alpha = \boldsymbol{0} \quad \text{and} \quad \widetilde{\mathrm{R}}_t \boldsymbol{M}_\alpha = \boldsymbol{0} \quad \text{and} \quad \overline{\mathrm{L}}_t \boldsymbol{n}_\alpha = \boldsymbol{0} \quad \text{and} \quad \overline{\mathrm{L}}_t \boldsymbol{m}_\alpha = \boldsymbol{0}.$$

Based on the intermediate structural tensors and in accordance with the above restriction of spatial objectivity the free energy density ψ_0 then takes the following alternative representations

$$\psi_0 = \psi_0(\overline{\boldsymbol{C}}; \overline{\boldsymbol{I}}, \overline{\boldsymbol{M}}_\alpha) \quad \text{and} \quad \psi_0 = \psi_0(\overline{\boldsymbol{B}}; \overline{\boldsymbol{J}}, \overline{\boldsymbol{N}}_\alpha).$$

Finally, requiring invariance of ψ_0 under arbitrary materially superposed rigid body motions, i.e. the intermediate strain measures are transformed identically to the intermediate structural tensors, results in the following alternative restrictions

$$\psi_0 \doteq \psi_{0*} = \psi_0(\boldsymbol{Q}^t \cdot \{\overline{\boldsymbol{C}}; \overline{\boldsymbol{I}}, \overline{\boldsymbol{M}}_\alpha\} \cdot \boldsymbol{Q}) \quad \forall \boldsymbol{Q} \in \mathrm{SO}(3),$$
$$\psi_0 \doteq \psi_{0*} = \psi_0(\boldsymbol{Q}^t \cdot \{\overline{\boldsymbol{B}}; \overline{\boldsymbol{J}}, \overline{\boldsymbol{N}}_\alpha\} \cdot \boldsymbol{Q}) \quad \forall \boldsymbol{Q} \in \mathrm{SO}(3).$$

These restrictions qualify ψ_0 as an *isotropic tensor function* of all of its arguments, i.e. either $\{\overline{\boldsymbol{C}}; \overline{\boldsymbol{I}}, \overline{\boldsymbol{M}}_\alpha\}$ or $\{\overline{\boldsymbol{B}}; \overline{\boldsymbol{J}}, \overline{\boldsymbol{N}}_\alpha\}$. Requiring next covariance of the free energy density ψ_0 under arbitrary superposed *homeomorphisms* (recall that the elastic and plastic distortions do not necessarily derive from a compatible vector field) results in the alternative equivalent statements

$$\psi_0(\overline{\boldsymbol{C}}; \overline{\boldsymbol{I}}, \overline{\boldsymbol{M}}_\alpha) \doteq \psi_0(\underbrace{\boldsymbol{f}^{\star t} \cdot \overline{\boldsymbol{C}} \cdot \boldsymbol{f}^\star}_{=:\overline{\boldsymbol{C}}^\star}; \underbrace{\boldsymbol{F}^\star \cdot \overline{\boldsymbol{I}} \cdot \boldsymbol{F}^{\star t}}_{=:\overline{\boldsymbol{I}}^\star}, \underbrace{\boldsymbol{F}^\star \cdot \overline{\boldsymbol{M}}_\alpha \cdot \boldsymbol{F}^{\star t}}_{=:\overline{\boldsymbol{M}}^\star_\alpha}),$$
$$\psi_0(\overline{\boldsymbol{B}}; \overline{\boldsymbol{J}}, \overline{\boldsymbol{N}}_\alpha) \doteq \psi_0(\underbrace{\boldsymbol{F}^\star \cdot \overline{\boldsymbol{B}} \cdot \boldsymbol{F}^{\star t}}_{=:\overline{\boldsymbol{B}}^\star}; \underbrace{\boldsymbol{f}^{\star t} \cdot \overline{\boldsymbol{J}} \cdot \boldsymbol{f}^\star}_{=:\overline{\boldsymbol{J}}^\star}, \underbrace{\boldsymbol{f}^{\star t} \cdot \overline{\boldsymbol{N}}_\alpha \cdot \boldsymbol{f}^\star}_{=:\overline{\boldsymbol{N}}^\star_\alpha}).$$

Here $\boldsymbol{f}^\star$ abbreviates the inverse $\boldsymbol{f}^\star := \boldsymbol{F}^{\star -1}$ of the homeomorphism $\boldsymbol{F}^\star$. Then the particular choices $\boldsymbol{F}^\star = \widetilde{\boldsymbol{f}}$ with $\boldsymbol{f}^\star = \widetilde{\boldsymbol{F}}$ and $\boldsymbol{F}^\star = \overline{\boldsymbol{F}}$ with $\boldsymbol{f}^\star = \overline{\boldsymbol{f}}$ render the alternative equivalent parameterizations of ψ_0

$$\psi_0(\overline{\boldsymbol{C}}; \overline{\boldsymbol{I}}, \overline{\boldsymbol{M}}_\alpha) \doteq \psi_0(\boldsymbol{C}; \widetilde{\boldsymbol{B}}, \boldsymbol{M}_\alpha) \quad \text{and} \quad \psi_0(\overline{\boldsymbol{B}}; \overline{\boldsymbol{J}}, \overline{\boldsymbol{N}}_\alpha) \doteq \psi_0(\boldsymbol{B}; \widetilde{\boldsymbol{C}}, \boldsymbol{N}_\alpha),$$
$$\psi_0(\overline{\boldsymbol{C}}; \overline{\boldsymbol{I}}, \overline{\boldsymbol{M}}_\alpha) \doteq \psi_0(\,\boldsymbol{j}\,;\, \overline{\boldsymbol{b}},\, \boldsymbol{m}_\alpha) \quad \text{and} \quad \psi_0(\overline{\boldsymbol{B}}; \overline{\boldsymbol{J}}, \overline{\boldsymbol{N}}_\alpha) \doteq \psi_0(\,\boldsymbol{i}\,;\, \overline{\boldsymbol{c}},\, \boldsymbol{n}_\alpha).$$

Finally, requiring objectivity of ψ_0 under arbitrary spatially superposed rigid body motions results in the following alternative restrictions

$$\psi_0 \doteq \psi_0^* = \psi_0(\boldsymbol{Q} \cdot \{\boldsymbol{j}; \overline{\boldsymbol{b}}, \boldsymbol{m}_\alpha\} \cdot \boldsymbol{Q}^t) \quad \forall \boldsymbol{Q} \in \mathrm{SO}(3),$$
$$\psi_0 \doteq \psi_0^* = \psi_0(\boldsymbol{Q} \cdot \{\,\boldsymbol{i}; \overline{\boldsymbol{c}},\, \boldsymbol{n}_\alpha\} \cdot \boldsymbol{Q}^t) \quad \forall \boldsymbol{Q} \in \mathrm{SO}(3).$$

These restrictions qualify ψ_0 again as an *isotropic tensor function* of its arguments, i.e. either $\{\boldsymbol{j}; \overline{\boldsymbol{b}}, \boldsymbol{m}_\alpha\}$ or $\{\boldsymbol{i}; \overline{\boldsymbol{c}}, \boldsymbol{n}_\alpha\}$, respectively. Covariance requires that ψ_0 remains invariant under arbitrary homeomorphisms, thus its rate with respect to a time-like parameter τ has to vanish at fixed actual distortions

$$\mathrm{D}_\tau \psi_0|_{\overline{\boldsymbol{F}}} = 0 \quad \rightsquigarrow \quad \psi_0 = \text{const} \quad \forall \boldsymbol{F}^\star.$$

Exploiting this condition results eventually in a relation between the derivatives of ψ_0 with respect to its various arguments

$$\frac{\partial \psi_0}{\partial \overline{\boldsymbol{C}}^\star} = \overline{\boldsymbol{B}}^\star \cdot \frac{\partial \psi_0}{\partial \overline{\boldsymbol{I}}^\star} \cdot \overline{\boldsymbol{I}}^\star + \overline{\boldsymbol{B}}^\star \cdot \frac{\partial \psi_0}{\partial \overline{\boldsymbol{M}}^\star_\alpha} \cdot \overline{\boldsymbol{M}}^\star_\alpha = \overline{\boldsymbol{I}}^\star \cdot \frac{\partial \psi_0}{\partial \overline{\boldsymbol{I}}^\star} \cdot \overline{\boldsymbol{B}}^\star + \overline{\boldsymbol{M}}^\star_\alpha \cdot \frac{\partial \psi_0}{\partial \overline{\boldsymbol{M}}^\star_\alpha} \cdot \overline{\boldsymbol{B}}^\star,$$

$$\frac{\partial \psi_0}{\partial \overline{\boldsymbol{B}}^\star} = \overline{\boldsymbol{C}}^\star \cdot \frac{\partial \psi_0}{\partial \overline{\boldsymbol{J}}^\star} \cdot \overline{\boldsymbol{J}}^\star + \overline{\boldsymbol{C}}^\star \cdot \frac{\partial \psi_0}{\partial \overline{\boldsymbol{N}}^\star_\alpha} \cdot \overline{\boldsymbol{N}}^\star_\alpha = \overline{\boldsymbol{J}}^\star \cdot \frac{\partial \psi_0}{\partial \overline{\boldsymbol{J}}^\star} \cdot \overline{\boldsymbol{C}}^\star + \overline{\boldsymbol{N}}^\star_\alpha \cdot \frac{\partial \psi_0}{\partial \overline{\boldsymbol{N}}^\star_\alpha} \cdot \overline{\boldsymbol{C}}^\star.$$

Observe the symmetry conditions for $\overline{\boldsymbol{C}}^\star$ and $\overline{\boldsymbol{B}}^\star$ reflected by the two alternative statements on the right-hand-sides. Then, for the particular choice

$\boldsymbol{F}^\star = \boldsymbol{1}_{\rm i}$, i.e. the identity map in the intermediate configuration, the derivatives of a covariant free energy density ψ_0 with respect to the various intermediate strain measures and unit tensors are related as

$$\frac{\partial\psi_0}{\partial\overline{\boldsymbol{C}}} = \overline{\boldsymbol{B}}\cdot\frac{\partial\psi_0}{\partial\overline{\boldsymbol{I}}}\cdot\overline{\boldsymbol{I}} + \overline{\boldsymbol{B}}\cdot\frac{\partial\psi_0}{\partial\overline{\boldsymbol{M}}_\alpha}\cdot\overline{\boldsymbol{M}}_\alpha = \overline{\boldsymbol{I}}\cdot\frac{\partial\psi_0}{\partial\overline{\boldsymbol{I}}}\cdot\overline{\boldsymbol{B}} + \overline{\boldsymbol{M}}_\alpha\cdot\frac{\partial\psi_0}{\partial\overline{\boldsymbol{M}}_\alpha}\cdot\overline{\boldsymbol{B}}.$$

$$\frac{\partial\psi_0}{\partial\overline{\boldsymbol{B}}} = \overline{\boldsymbol{C}}\cdot\frac{\partial\psi_0}{\partial\overline{\boldsymbol{J}}}\cdot\overline{\boldsymbol{J}} + \overline{\boldsymbol{C}}\cdot\frac{\partial\psi_0}{\partial\overline{\boldsymbol{N}}_\alpha}\cdot\overline{\boldsymbol{N}}_\alpha = \overline{\boldsymbol{J}}\cdot\frac{\partial\psi_0}{\partial\overline{\boldsymbol{J}}}\cdot\overline{\boldsymbol{C}} + \overline{\boldsymbol{N}}_\alpha\cdot\frac{\partial\psi_0}{\partial\overline{\boldsymbol{N}}_\alpha}\cdot\overline{\boldsymbol{C}}.$$

Likewise, for the particular choice $\boldsymbol{F}^\star = \overline{\boldsymbol{F}}$, i.e. the elastic distortion, the derivatives of a covariant free energy density ψ_0 with respect to the various spatial strain measures and unit tensors are related as

$$\frac{\partial\psi_0}{\partial\boldsymbol{j}} = \boldsymbol{i}\cdot\frac{\partial\psi_0}{\partial\overline{\boldsymbol{b}}}\cdot\overline{\boldsymbol{b}} + \boldsymbol{i}\cdot\frac{\partial\psi_0}{\partial\boldsymbol{m}_\alpha}\cdot\boldsymbol{m}_\alpha = \overline{\boldsymbol{b}}\cdot\frac{\partial\psi_0}{\partial\overline{\boldsymbol{b}}}\cdot\boldsymbol{i} + \boldsymbol{m}_\alpha\cdot\frac{\partial\psi_0}{\partial\boldsymbol{m}_\alpha}\cdot\boldsymbol{i}.$$

$$\frac{\partial\psi_0}{\partial\boldsymbol{i}} = \boldsymbol{j}\cdot\frac{\partial\psi_0}{\partial\overline{\boldsymbol{c}}}\cdot\overline{\boldsymbol{c}} + \boldsymbol{j}\cdot\frac{\partial\psi_0}{\partial\,\boldsymbol{n}_\alpha}\cdot\,\boldsymbol{n}_\alpha = \overline{\boldsymbol{c}}\cdot\frac{\partial\psi_0}{\partial\overline{\boldsymbol{c}}}\cdot\boldsymbol{j} + \,\boldsymbol{n}_\alpha\cdot\frac{\partial\psi_0}{\partial\,\boldsymbol{n}_\alpha}\cdot\boldsymbol{j}.$$

Finally, for the particular choice $\boldsymbol{F}^\star = \widetilde{\boldsymbol{f}}$, i.e. the (inverse) plastic distortion, the derivatives of a covariant free energy density ψ_0 with respect to the various material strain measures and unit tensors are related as

$$\frac{\partial\psi_0}{\partial\boldsymbol{C}} = \boldsymbol{B}\cdot\frac{\partial\psi_0}{\partial\widetilde{\boldsymbol{B}}}\cdot\widetilde{\boldsymbol{B}} + \boldsymbol{B}\cdot\frac{\partial\psi_0}{\partial\boldsymbol{M}_\alpha}\cdot\boldsymbol{M}_\alpha = \widetilde{\boldsymbol{B}}\cdot\frac{\partial\psi_0}{\partial\widetilde{\boldsymbol{B}}}\cdot\boldsymbol{B} + \boldsymbol{M}_\alpha\cdot\frac{\partial\psi_0}{\partial\boldsymbol{M}_\alpha}\cdot\boldsymbol{B}.$$

$$\frac{\partial\psi_0}{\partial\boldsymbol{B}} = \boldsymbol{C}\cdot\frac{\partial\psi_0}{\partial\widetilde{\boldsymbol{C}}}\cdot\widetilde{\boldsymbol{C}} + \boldsymbol{C}\cdot\frac{\partial\psi_0}{\partial\,\boldsymbol{N}_\alpha}\cdot\,\boldsymbol{N}_\alpha = \widetilde{\boldsymbol{C}}\cdot\frac{\partial\psi_0}{\partial\widetilde{\boldsymbol{C}}}\cdot\boldsymbol{C} + \,\boldsymbol{N}_\alpha\cdot\frac{\partial\psi_0}{\partial\,\boldsymbol{N}_\alpha}\cdot\boldsymbol{C}.$$

As will become obvious in the sequel, these derivatives of ψ_0 are related to various material, intermediate and spatial description stress measures.

Stress Response:
Free energy densities ψ_0 that satisfy all requirements resulting from spatial objectivity, material symmetry, and covariance are isotropic tensor functions of their material, intermediate or spatial, co- and contravariant arguments

$$\{\boldsymbol{C};\widetilde{\boldsymbol{B}},\boldsymbol{M}_\alpha\} \leftrightarrow \{\overline{\boldsymbol{C}};\overline{\boldsymbol{I}},\overline{\boldsymbol{M}}_\alpha\} \leftrightarrow \{\boldsymbol{j};\overline{\boldsymbol{b}},\boldsymbol{m}_\alpha\},$$

$$\{\boldsymbol{B};\widetilde{\boldsymbol{C}},\ \boldsymbol{N}_\alpha\} \leftrightarrow \{\overline{\boldsymbol{B}};\overline{\boldsymbol{J}},\ \overline{\boldsymbol{N}}_\alpha\} \leftrightarrow \{\boldsymbol{i};\overline{\boldsymbol{c}},\ \boldsymbol{n}_\alpha\}.$$

Then the comparison of the material time derivative of ψ_0 with the various versions of the elastic part of the stress power density $\overline{\wp}_0$ renders appropriate constitutive relations for the intermediate stress measures

$$\overline{\boldsymbol{S}} = +2\frac{\partial\psi_0}{\partial\overline{\boldsymbol{C}}} = +2\overline{\boldsymbol{B}}\cdot\frac{\partial\psi_0}{\partial\overline{\boldsymbol{I}}}\cdot\overline{\boldsymbol{I}} + 2\overline{\boldsymbol{B}}\cdot\frac{\partial\psi_0}{\partial\overline{\boldsymbol{M}}_\alpha}\cdot\overline{\boldsymbol{M}}_\alpha = \widetilde{\boldsymbol{\tau}}_\sharp.$$

$$\overline{\boldsymbol{T}} = -2\frac{\partial \psi_0}{\partial \overline{\boldsymbol{B}}} = -2\overline{\boldsymbol{J}} \cdot \frac{\partial \psi_0}{\partial \overline{\boldsymbol{J}}} \cdot \overline{\boldsymbol{C}} - 2\overline{\boldsymbol{N}}_\alpha \cdot \frac{\partial \psi_0}{\partial \overline{\boldsymbol{N}}_\alpha} \cdot \overline{\boldsymbol{C}} = \widetilde{\boldsymbol{\tau}}_\flat.$$

Consequently the co/contra- (mixed) variant intermediate Mandel stress computes as

$$\overline{\boldsymbol{M}} = +2\frac{\partial \psi_0}{\partial \overline{\boldsymbol{I}}} \cdot \overline{\boldsymbol{I}} + 2\frac{\partial \psi_0}{\partial \overline{\boldsymbol{M}}_\alpha} \cdot \overline{\boldsymbol{M}}_\alpha = -2\overline{\boldsymbol{J}} \cdot \frac{\partial \psi_0}{\partial \overline{\boldsymbol{J}}} - 2\overline{\boldsymbol{N}}_\alpha \cdot \frac{\partial \psi_0}{\partial \overline{\boldsymbol{N}}_\alpha} = \widetilde{\boldsymbol{\tau}}.$$

Then since for a scalar function $\widetilde{\mathrm{R}}_t\psi_0 \equiv \mathrm{D}_t\psi_0$ the comparison of the plastic (right) Lie time derivative of ψ_0 with the various versions of the elastic part of the stress power density $\overline{\wp}_0$ renders appropriate constitutive relations for the material stress measures

$$\boldsymbol{S} = +2\frac{\partial \psi_0}{\partial \boldsymbol{C}} = +2\boldsymbol{B} \cdot \frac{\partial \psi_0}{\partial \widetilde{\boldsymbol{B}}} \cdot \widetilde{\boldsymbol{B}} + 2\boldsymbol{B} \cdot \frac{\partial \psi_0}{\partial \boldsymbol{M}_\alpha} \cdot \boldsymbol{M}_\alpha.$$

$$\boldsymbol{T} = -2\frac{\partial \psi_0}{\partial \boldsymbol{B}} = -2\widetilde{\boldsymbol{C}} \cdot \frac{\partial \psi_0}{\partial \widetilde{\boldsymbol{C}}} \cdot \boldsymbol{C} - 2\overline{\boldsymbol{N}}_\alpha \cdot \frac{\partial \psi_0}{\partial \overline{\boldsymbol{N}}_\alpha} \cdot \boldsymbol{C}.$$

Consequently the co/contra- (mixed) variant material Mandel stress computes as

$$\boldsymbol{M} = +2\frac{\partial \psi_0}{\partial \widetilde{\boldsymbol{B}}} \cdot \widetilde{\boldsymbol{B}} + 2\frac{\partial \psi_0}{\partial \overline{\boldsymbol{M}}_\alpha} \cdot \overline{\boldsymbol{M}}_\alpha = -2\widetilde{\boldsymbol{C}} \cdot \frac{\partial \psi_0}{\partial \widetilde{\boldsymbol{C}}} - 2\overline{\boldsymbol{N}}_\alpha \cdot \frac{\partial \psi_0}{\partial \overline{\boldsymbol{N}}_\alpha}.$$

Finally since for a scalar function $\overline{\mathrm{L}}_t\psi_0 \equiv \mathrm{D}_t\psi_0$ the comparison of the elastic (left) Lie time derivative of ψ_0 with the various versions of the elastic part of the stress power density $\overline{\wp}_0$ renders appropriate constitutive relations for the spatial stress measures

$$\boldsymbol{\tau}_\sharp = +2\frac{\partial \psi_0}{\partial \boldsymbol{j}} = +2\boldsymbol{i} \cdot \frac{\partial \psi_0}{\partial \overline{\boldsymbol{b}}} \cdot \overline{\boldsymbol{b}} + 2\boldsymbol{i} \cdot \frac{\partial \psi_0}{\partial \boldsymbol{m}_\alpha} \cdot \boldsymbol{m}_\alpha.$$

$$\boldsymbol{\tau}_\flat = -2\frac{\partial \psi_0}{\partial \boldsymbol{i}} = -2\overline{\boldsymbol{c}} \cdot \frac{\partial \psi_0}{\partial \overline{\boldsymbol{c}}} \cdot \boldsymbol{j} - 2\boldsymbol{n}_\alpha \cdot \frac{\partial \psi_0}{\partial \boldsymbol{n}_\alpha} \cdot \boldsymbol{j}.$$

Consequently the co/contra- (mixed) variant Kirchhoff stress computes as

$$\boldsymbol{\tau} = +2\frac{\partial \psi_0}{\partial \overline{\boldsymbol{b}}} \cdot \overline{\boldsymbol{b}} + 2\frac{\partial \psi_0}{\partial \boldsymbol{m}_\alpha} \cdot \boldsymbol{m}_\alpha = -2\overline{\boldsymbol{c}} \cdot \frac{\partial \psi_0}{\partial \overline{\boldsymbol{c}}} - 2\boldsymbol{n}_\alpha \cdot \frac{\partial \psi_0}{\partial \boldsymbol{n}_\alpha}.$$

For the case of isotropy the alternative formats for $\boldsymbol{\tau}$ are due to [177, 175], for the case of anisotropy the alternative formats for the various material, intermediate and spatial stress measures are, to the best knowledge of the author, original to this contribution. Which of the above alternative constitutive relations is chosen is essentially a matter of convenience.

Restrictions on the Yield Function:
Requiring the yield function ϕ to transform objectively under arbitrary spatially superposed rigid body motions (i.e. $\phi \doteq \phi^*$) results in the alternative restrictions

$$\begin{aligned} \phi \doteq \phi^* &= \phi(\boldsymbol{M}; \cdots), \\ \phi \doteq \phi^* &= \phi(\overline{\boldsymbol{M}}; \cdots), \\ \phi \doteq \phi^* &= \phi(\boldsymbol{Q} \cdot \boldsymbol{\tau} \cdot \boldsymbol{Q}^t; \cdots) \quad \forall \boldsymbol{Q} \in \mathrm{SO}(3). \end{aligned}$$

Thus an arbitrary dependence on the material description Mandel stress $\boldsymbol{M}$ or the intermediate description Mandel stress $\overline{\boldsymbol{M}}$ automatically satisfies spatial objectivity of ϕ. In contrast the dependence on the spatial description Kirchhoff stress $\boldsymbol{\tau}$ is subject to the above *isotropy* condition on the tensor function ϕ in order to satisfy spatial objectivity. However, the yet unspecified arguments of ϕ indicated by the $\cdots$ allow to also capture *anisotropic* yielding behavior. To this end a set of contra/co- (mixed) variant intermediate *anisotropy tensors*

$$\overline{\boldsymbol{A}}_0 := \mathbf{1}_{\mathrm{i}}, \quad \overline{\boldsymbol{A}}_\alpha \quad \text{with} \quad \alpha = 1, \cdots, n_{\mathrm{at}}$$

is introduced that captures the *yielding anisotropy group* $\mathrm{A}(3) \subseteq \mathrm{SO}(3)$. Here the contra/co- (mixed) variant intermediate unit tensor $\mathbf{1}_{\mathrm{i}}$ is explicitly incorporated in order to capture the isotropic part of the yielding behavior. The contra/co- (mixed) variant intermediate anisotropy tensors satisfy the following invariance conditions related to the yielding anisotropy group

$$\overline{\boldsymbol{A}}_\alpha \equiv \boldsymbol{Q}^t \cdot \overline{\boldsymbol{A}}_\alpha \cdot \boldsymbol{Q} \quad \forall \boldsymbol{Q} \in \mathrm{A}(3) \subseteq \mathrm{SO}(3), \quad \alpha = 1, \cdots, n_{\mathrm{at}}.$$

In passing it is noted that the pull-back and push-forward of the contra/co- (mixed) variant intermediate anisotropy tensors into their material and spatial counterparts read as

$$\boldsymbol{A}_0 = \mathbf{1}_{\mathrm{m}}, \quad \boldsymbol{A}_\alpha = \widetilde{\boldsymbol{f}} \cdot \overline{\boldsymbol{A}}_\alpha \cdot \widetilde{\boldsymbol{F}} \quad \text{and} \quad \boldsymbol{a}_0 = \mathbf{1}_{\mathrm{s}}, \quad \boldsymbol{a}_\alpha = \overline{\boldsymbol{F}} \cdot \overline{\boldsymbol{A}}_\alpha \cdot \overline{\boldsymbol{f}}.$$

Furthermore, the spatial anisotropy tensors are (elastic) Killing fields with vanishing elastic Lie time derivative (i.e. the intermediate anisotropy tensors are assumed constant; as an example the *isoclinic* intermediate configuration of single crystal plasticity is characterized by constant Schmid tensors)

$$\widetilde{\mathrm{R}}_t \boldsymbol{A}_\alpha = \mathbf{0} \quad \text{and} \quad \overline{\mathrm{L}}_t \boldsymbol{a}_\alpha = \mathbf{0}.$$

The anisotropy tensors might be thought of as dyadic products of preferred anisotropy directions. Thereby the anisotropy tensors may in general be selected independently from the choice of the structural tensors that enter the free energy density, however it is also well conceivable that the material's symmetry group $\mathrm{M}(3)$ and the yielding anisotropy group $\mathrm{A}(3)$ do indeed coincide. Based on the material and intermediate contra/co- (mixed) variant anisotropy tensors $\boldsymbol{A}_\alpha$ and $\overline{\boldsymbol{A}}_\alpha$ and in accordance with the above restriction of spatial objectivity the yield function ϕ then takes the following alternative representations

$$\phi = \phi(\boldsymbol{M}; \boldsymbol{A}_\alpha) \quad \text{and} \quad \phi = \phi(\overline{\boldsymbol{M}}; \overline{\boldsymbol{A}}_\alpha).$$

Here the explicit dependence on the contra/co- (mixed) variant unit tensors $\boldsymbol{1}_\mathrm{m}$ and $\boldsymbol{1}_\mathrm{i}$ has been suppressed for the ease of representation. Finally, requiring invariance of ϕ under arbitrary materially superposed rigid body motions, i.e. the material and the intermediate Mandel stress are transformed identically to the material and intermediate anisotropy tensors, results in the following restrictions

$$\begin{aligned} \phi \doteq \phi_* &= \phi(\boldsymbol{Q}^t \cdot \{\boldsymbol{M}; \boldsymbol{A}_\alpha\} \cdot \boldsymbol{Q}) \quad \forall \boldsymbol{Q} \in \mathrm{SO}(3), \\ \phi \doteq \phi_* &= \phi(\boldsymbol{Q}^t \cdot \{\overline{\boldsymbol{M}}; \overline{\boldsymbol{A}}_\alpha\} \cdot \boldsymbol{Q}) \quad \forall \boldsymbol{Q} \in \mathrm{SO}(3). \end{aligned}$$

These restrictions qualify ϕ as an *isotropic tensor function* of its arguments, i.e. either $\{\boldsymbol{M}; \boldsymbol{A}_\alpha\}$ or $\{\overline{\boldsymbol{M}}; \overline{\boldsymbol{A}}_\alpha\}$. Representation theorems provide the most general dependence of isotropic tensor functions on their arguments, for example in terms of various types of *basic invariants* $M_1, M_2, M_3, \cdots$. As an example, with $n_\mathrm{at} = 1$ transverse isotropic yielding behavior is captured in terms of

$$\begin{aligned} M_n &:= \boldsymbol{M}^n : \boldsymbol{1}_\mathrm{m} = \overline{\boldsymbol{M}}^n : \boldsymbol{1}_\mathrm{i} = \boldsymbol{\tau}^n : \boldsymbol{1}_\mathrm{s}\,, \\ M_{3+m} &:= \boldsymbol{M}^m : \boldsymbol{A}_1 = \overline{\boldsymbol{M}}^m : \overline{\boldsymbol{A}}_1 = \boldsymbol{\tau}^m : \boldsymbol{a}_1 \end{aligned}$$

with $n = 1, 2, 3$ and $m = 1, 2$. Interestingly these invariants may likewise be expressed in terms of either the material or intermediate contravariant $\boldsymbol{S}$ or $\overline{\boldsymbol{S}}$ and the corresponding covariant $\boldsymbol{C}$ or $\overline{\boldsymbol{C}}$ (or their spatial pendants $\boldsymbol{\tau}_\sharp$ and $\boldsymbol{j}$) or the material or intermediate covariant $\boldsymbol{T}$ or $\overline{\boldsymbol{T}}$ and the corresponding contravariant $\boldsymbol{B}$ or $\overline{\boldsymbol{B}}$ (or their spatial pendants $\boldsymbol{\tau}_\flat$ and $\boldsymbol{i}$) as

$$\begin{aligned} M_n &= \big[\boldsymbol{S} \cdot [\boldsymbol{C} \cdot \boldsymbol{S}]^{n-1}\big] : \boldsymbol{C}, \\ &= \big[\overline{\boldsymbol{S}} \cdot [\overline{\boldsymbol{C}} \cdot \overline{\boldsymbol{S}}]^{n-1}\big] : \overline{\boldsymbol{C}} = \big[\boldsymbol{\tau}_\sharp \cdot [\boldsymbol{j} \cdot \boldsymbol{\tau}_\sharp]^{n-1}\big] : \boldsymbol{j}, \\ &= \big[[\boldsymbol{T} \cdot \boldsymbol{B}]^{n-1} \cdot \boldsymbol{T}\big] : \boldsymbol{B}, \\ &= \big[[\overline{\boldsymbol{T}} \cdot \overline{\boldsymbol{B}}]^{n-1} \cdot \overline{\boldsymbol{T}}\big] : \overline{\boldsymbol{B}} = \big[\,[\boldsymbol{\tau}_\flat \cdot \boldsymbol{i}]^{n-1} \cdot \boldsymbol{\tau}_\flat\big] : \boldsymbol{i}\,, \end{aligned}$$

$$\begin{aligned} M_{3+m} &= \big[\boldsymbol{S} \cdot [\boldsymbol{C} \cdot \boldsymbol{S}]^{m-1}\big] : [\boldsymbol{C} \cdot \boldsymbol{A}_1], \\ &= \big[\overline{\boldsymbol{S}} \cdot [\overline{\boldsymbol{C}} \cdot \overline{\boldsymbol{S}}]^{m-1}\big] : [\overline{\boldsymbol{C}} \cdot \overline{\boldsymbol{A}}_1] = \big[\boldsymbol{\tau}_\sharp \cdot [\boldsymbol{j} \cdot \boldsymbol{\tau}_\sharp]^{m-1}\big] : [\boldsymbol{j} \cdot \boldsymbol{a}_1], \\ &= \big[[\boldsymbol{T} \cdot \boldsymbol{B}]^{m-1} \cdot \boldsymbol{T}\big] : [\boldsymbol{A}_1 \cdot \boldsymbol{B}], \\ &= \big[[\overline{\boldsymbol{T}} \cdot \overline{\boldsymbol{B}}]^{m-1} \cdot \overline{\boldsymbol{T}}\big] : [\overline{\boldsymbol{A}}_1 \cdot \overline{\boldsymbol{B}}] = \big[\,[\boldsymbol{\tau}_\flat \cdot \boldsymbol{i}]^{m-1} \cdot \boldsymbol{\tau}_\flat\big] : [\boldsymbol{a}_1 \cdot \boldsymbol{i}\,]. \end{aligned}$$

Thus valid alternative versions of the yield function ϕ read

$$\phi(\boldsymbol{S}\,;\boldsymbol{C},[\boldsymbol{C}\cdot\boldsymbol{A}_\alpha]) \equiv \phi(\overline{\boldsymbol{S}}\,;\overline{\boldsymbol{C}},[\overline{\boldsymbol{C}}\cdot\overline{\boldsymbol{A}}_\alpha]) \equiv \phi(\boldsymbol{\tau}_\sharp;\boldsymbol{j},[\boldsymbol{j}\cdot\boldsymbol{a}_\alpha]),$$
$$\phi(\boldsymbol{T};\boldsymbol{B},[\boldsymbol{A}_\alpha\cdot\boldsymbol{B}]) \equiv \phi(\overline{\boldsymbol{T}};\overline{\boldsymbol{B}},[\overline{\boldsymbol{A}}_\alpha\cdot\overline{\boldsymbol{B}}]) \equiv \phi(\boldsymbol{\tau}_\flat;\boldsymbol{i}\,,[\boldsymbol{a}_\alpha\cdot\boldsymbol{i}\,]).$$

Requiring finally objectivity of ϕ under arbitrary spatially superposed rigid body motions results in the following alternative restrictions

$$\phi \doteq \phi^* = \phi(\boldsymbol{Q}\cdot\{\boldsymbol{\tau}\;; \qquad \boldsymbol{a}_\alpha\,\}\cdot\boldsymbol{Q}^t) \quad \forall\boldsymbol{Q}\in\mathrm{SO}(3),$$
$$\phi \doteq \phi^* = \phi(\boldsymbol{Q}\cdot\{\boldsymbol{\tau}_\sharp;\boldsymbol{j},[\boldsymbol{j}\cdot\boldsymbol{a}_\alpha]\}\cdot\boldsymbol{Q}^t) \quad \forall\boldsymbol{Q}\in\mathrm{SO}(3),$$
$$\phi \doteq \phi^* = \phi(\boldsymbol{Q}\cdot\{\boldsymbol{\tau}_\flat;\boldsymbol{i}\,,[\boldsymbol{a}_\alpha\cdot\boldsymbol{i}]\}\cdot\boldsymbol{Q}^t) \quad \forall\boldsymbol{Q}\in\mathrm{SO}(3).$$

These restrictions qualify ϕ again as an *isotropic tensor function* of its arguments, i.e. either $\{\boldsymbol{\tau};\boldsymbol{a}_\alpha\}$, $\{\boldsymbol{\tau}_\sharp;\boldsymbol{j},[\boldsymbol{j}\cdot\boldsymbol{a}_\alpha]\}$ or $\{\boldsymbol{\tau}_\flat;\boldsymbol{i},[\boldsymbol{a}_\alpha\cdot\boldsymbol{i}]\}$, respectively.

Plastic Evolution Laws:
Upon considering for simplicity and without loss of generality the case of *ideal plasticity* (the consideration of hardening/softening does not pose any substantial difficulties and is thus omitted for the sake of conciseness in the sequel) the reduced dissipation inequality is simply expressed in terms of the plastic part of the stress power density as

$$\widetilde{\wp}_0 \geq 0.$$

Then, based on the introduction of a yield function ϕ in terms of either of the co/contra- (mixed) variant spatial, intermediate or material stress measures, respectively, i.e.

$$\phi = \phi(\boldsymbol{\tau};\boldsymbol{a}_\alpha) \quad \text{or} \quad \phi = \phi(\overline{\boldsymbol{M}};\overline{\boldsymbol{A}}_\alpha) \quad \text{or} \quad \phi = \phi(\boldsymbol{M};\boldsymbol{A}_\alpha)$$

the exploitation of the postulate of maximum dissipation as a constrained optimization problem

$$\widetilde{\wp}_0 \geq \widetilde{\wp}_0^* \geq 0 \quad \forall\{\boldsymbol{\tau}^*/\overline{\boldsymbol{M}}^*/\boldsymbol{M}^*\}\in\mathbb{A}.$$

results in alternative but equivalent versions of associated plastic evolution laws

$$\boldsymbol{l}_\mathrm{p} = \gamma\frac{\partial\phi}{\partial\boldsymbol{\tau}} \quad \text{or} \quad \boldsymbol{\ell}_\mathrm{p} = \gamma\frac{\partial\phi}{\partial\overline{\boldsymbol{M}}} \quad \text{or} \quad \boldsymbol{L}_\mathrm{p} = \gamma\frac{\partial\phi}{\partial\boldsymbol{M}}.$$

Next, based on the chain rule and the symmetry of $\boldsymbol{\tau}_\sharp$, $\overline{\boldsymbol{S}}$ and $\boldsymbol{S}$, the following relations hold

$$\frac{\partial\phi}{\partial\boldsymbol{\tau}_\sharp} = \left[\boldsymbol{j}\cdot\frac{\partial\phi}{\partial\boldsymbol{\tau}}\right]^{\mathrm{sym}}, \quad \frac{\partial\phi}{\partial\overline{\boldsymbol{S}}} = \left[\overline{\boldsymbol{C}}\cdot\frac{\partial\phi}{\partial\overline{\boldsymbol{M}}}\right]^{\mathrm{sym}} \quad \text{and} \quad \frac{\partial\phi}{\partial\boldsymbol{S}} = \left[\boldsymbol{C}\cdot\frac{\partial\phi}{\partial\boldsymbol{M}}\right]^{\mathrm{sym}}.$$

Consequently, and in accordance with the various representations of the plastic part of the stress power, further alternative but equivalent versions of associated plastic evolution laws emerge

$$[\boldsymbol{j}\cdot\boldsymbol{l}_{\rm p}]^{\rm sym}=\gamma\frac{\partial\phi}{\partial\boldsymbol{\tau}_\sharp}\quad\text{or}\quad[\overline{\boldsymbol{C}}\cdot\boldsymbol{\ell}_{\rm p}]^{\rm sym}=\gamma\frac{\partial\phi}{\partial\overline{\boldsymbol{S}}}\quad\text{or}\quad[\boldsymbol{C}\cdot\boldsymbol{L}_{\rm p}]^{\rm sym}=\gamma\frac{\partial\phi}{\partial\boldsymbol{S}}.$$

For the *special case* of elastic isotropy, i.e. $\overline{\boldsymbol{S}}$ and $\overline{\boldsymbol{C}}$ commute, the plastic evolution laws specialize to the useful format

$$\frac{1}{2}\overline{\boldsymbol{b}}\cdot\mathrm{L}_t\overline{\boldsymbol{c}}=\gamma\frac{\partial\phi}{\partial\boldsymbol{\tau}}\quad\text{or}\quad\frac{1}{2}\overline{\boldsymbol{I}}\cdot\widetilde{\mathrm{L}}_t\overline{\boldsymbol{J}}=\gamma\frac{\partial\phi}{\partial\overline{\boldsymbol{M}}}\quad\text{or}\quad\frac{1}{2}\widetilde{\boldsymbol{B}}\cdot\mathrm{D}_t\widetilde{\boldsymbol{C}}=\gamma\frac{\partial\phi}{\partial\boldsymbol{M}}.$$

Likewise, based on the chain rule and the symmetry of $\boldsymbol{\tau}_\flat$, $\overline{\boldsymbol{T}}$ and $\boldsymbol{T}$, the following relations hold

$$\frac{\partial\phi}{\partial\boldsymbol{\tau}_\flat}=\left[\frac{\partial\phi}{\partial\boldsymbol{\tau}}\cdot\boldsymbol{i}\right]^{\rm sym},\quad\frac{\partial\phi}{\partial\overline{\boldsymbol{T}}}=\left[\frac{\partial\phi}{\partial\overline{\boldsymbol{M}}}\cdot\overline{\boldsymbol{B}}\right]^{\rm sym}\quad\text{and}\quad\frac{\partial\phi}{\partial\boldsymbol{T}}=\left[\frac{\partial\phi}{\partial\boldsymbol{M}}\cdot\boldsymbol{B}\right]^{\rm sym}.$$

Consequently, and again in accordance with the various representations of the plastic part of the stress power, further alternative but equivalent versions of associated plastic evolution laws emerge

$$[\boldsymbol{l}_{\rm p}\cdot\boldsymbol{i}]^{\rm sym}=\gamma\frac{\partial\phi}{\partial\boldsymbol{\tau}_\flat}\quad\text{or}\quad[\boldsymbol{\ell}_{\rm p}\cdot\overline{\boldsymbol{B}}]^{\rm sym}=\gamma\frac{\partial\phi}{\partial\overline{\boldsymbol{T}}}\quad\text{or}\quad[\boldsymbol{L}_{\rm p}\cdot\boldsymbol{B}]^{\rm sym}=\gamma\frac{\partial\phi}{\partial\boldsymbol{T}}.$$

For the *special case* of elastic isotropy, i.e. $\overline{\boldsymbol{T}}$ and $\overline{\boldsymbol{B}}$ commute, the plastic evolution laws specialize to the useful format

$$-\frac{1}{2}\mathrm{L}_t\overline{\boldsymbol{b}}\cdot\overline{\boldsymbol{c}}=\gamma\frac{\partial\phi}{\partial\boldsymbol{\tau}}\quad\text{or}\quad-\frac{1}{2}\widetilde{\mathrm{L}}_t\overline{\boldsymbol{I}}\cdot\overline{\boldsymbol{J}}=\gamma\frac{\partial\phi}{\partial\overline{\boldsymbol{M}}}\quad\text{or}\quad-\frac{1}{2}\mathrm{D}_t\widetilde{\boldsymbol{B}}\cdot\overline{\boldsymbol{C}}=\gamma\frac{\partial\phi}{\partial\boldsymbol{M}}.$$

Which of the above alternative but equivalent plastic evolution laws is chosen is essentially a matter of convenience. Since $\boldsymbol{\ell}_{\rm p}$ and $\boldsymbol{L}_{\rm p}$ contain the material time derivative of the plastic distortion in an elementary fashion, their evolutions laws are well-suited for time integration, consider the example of single crystal plasticity. In the special case of isotropy the particular format of the plastic evolution laws lends itself to an exponential-type time integration algorithm that leads to a very convenient computational implementation, see [175, 177].

A penalty regularization of the postulate of maximum dissipation that results in an unconstrained optimization problem reads

$$\min\left(-\wp_0^*+\frac{1}{2\eta}\langle\phi^*\rangle^2|\ \{\boldsymbol{\tau}^*/\overline{\boldsymbol{M}}^*/\boldsymbol{M}^*\}\right)$$

and renders the regularized version of the plastic multiplier in terms of the viscosity $\eta\in(0,\infty)$ *and the overstress function* $\langle\phi\rangle$ *as*

$$\gamma:=\frac{1}{\eta}\langle\phi\rangle.$$

Here $\langle\{\bullet\}\rangle$ denote the Macaulay brackets that extract only the positive part of $\{\bullet\}$. A penalty regularization of the postulate of maximum dissipation thus leads to visco-plastic overstress Perzyna-type models. In particular the solution to the regularized problem converges to the solution of the non-regularized problem as $\eta \to 0$. More details concerning this type of penalty regularization may be found in [176].

Time Derivatives of Stress Measures:
The elastic (left) Lie time derivative of the spatial contravariant Kirchhoff stress, the material time derivative of the intermediate contravariant Piola-Kirchhoff stress and the (right) plastic Lie time derivative of the material contravariant Piola-Kirchhoff stress are defined as

$$\overline{\mathrm{L}}_t \boldsymbol{\tau}_\sharp := \overline{\boldsymbol{F}} \cdot \dot{\overline{\boldsymbol{S}}} \cdot \overline{\boldsymbol{F}}^t \quad \text{and} \quad \mathrm{D}_t \overline{\boldsymbol{S}} := \dot{\overline{\boldsymbol{S}}} \quad \text{and} \quad \widetilde{\mathrm{R}}_t \boldsymbol{S} := \widetilde{\boldsymbol{f}} \cdot \dot{\overline{\boldsymbol{S}}} \cdot \widetilde{\boldsymbol{f}}^t.$$

Then the Lie time derivative of the spatial contravariant Kirchhoff stress, the plastic Lie time derivative of the intermediate contravariant Piola-Kirchhoff stress and the material time derivative of the material contravariant Piola-Kirchhoff stress take the useful expressions

$$\mathrm{L}_t \boldsymbol{\tau}_\sharp = \overline{\mathrm{L}}_t \boldsymbol{\tau}_\sharp - [\boldsymbol{l}_\mathrm{p} \cdot \boldsymbol{\tau}_\sharp]^{\mathrm{sym}}, \; \widetilde{\mathrm{L}}_t \overline{\boldsymbol{S}} = \mathrm{D}_t \overline{\boldsymbol{S}} - [\boldsymbol{\ell}_\mathrm{p} \cdot \overline{\boldsymbol{S}}]^{\mathrm{sym}}, \; \mathrm{D}_t \boldsymbol{S} = \widetilde{\mathrm{R}}_t \boldsymbol{S} - [\boldsymbol{L}_\mathrm{p} \cdot \boldsymbol{S}]^{\mathrm{sym}}.$$

These relations serve as the starting point to develop the elasto-plastic linearized constitutive relations, see below.

Tangent Operators:
The *elasto-plastic* tangent relation between the material time derivatives of the material contravariant Piola-Kirchhoff stress $\boldsymbol{S}$ and the material covariant Cauchy-Green strain $\boldsymbol{C}$ reads

$$2\,\mathrm{D}_t \boldsymbol{S} = \mathbf{C}_\mathrm{ep} : \mathrm{D}_t \boldsymbol{C}.$$

As an auxiliary definition the *elastic* tangent relation between the plastic (right) Lie time derivatives of $\boldsymbol{S}$ and $\boldsymbol{C}$ is introduced

$$2\,\widetilde{\mathrm{R}}_t \boldsymbol{S} = \mathbf{C}_\mathrm{e} \; : \widetilde{\mathrm{R}}_t \boldsymbol{C}.$$

The elasto-plastic and elastic tangent operators $\mathbf{C}_\mathrm{ep}$ and $\mathbf{C}_\mathrm{e}$ are fully contravariant material description tensors. Note that the relation of the elasto-plastic tangent operator $\mathbf{C}_\mathrm{ep}$ to the corresponding $\mathbf{A}_\mathrm{ep}$ in terms of material and geometric stiffnesses is as before. Next the *elasto-plastic* tangent relation between the plastic (left) Lie time derivatives of the intermediate contravariant Piola-Kirchhoff stress $\overline{\boldsymbol{S}}$ and the intermediate covariant elastic Cauchy-Green strain $\overline{\boldsymbol{C}}$ reads

$$2\,\widetilde{\mathrm{L}}_t \overline{\boldsymbol{S}} = \overline{\mathbf{C}}_\mathrm{ep} : \widetilde{\mathrm{L}}_t \overline{\boldsymbol{C}} \quad \text{with} \quad \overline{\mathbf{C}}_\mathrm{ep} = [\widetilde{\boldsymbol{F}} \boxtimes \widetilde{\boldsymbol{F}}] : \mathbf{C}_\mathrm{ep} : [\widetilde{\boldsymbol{F}}^t \boxtimes \widetilde{\boldsymbol{F}}^t].$$

As an auxiliary definition the *elastic* tangent relation between the material time derivatives of $\overline{\boldsymbol{S}}$ and $\overline{\boldsymbol{C}}$ is introduced

$$2\,\mathrm{D}_t\overline{\boldsymbol{S}} = \overline{\mathbf{C}}_{\mathrm{e}} : \mathrm{D}_t\overline{\boldsymbol{C}} \quad \text{with} \quad \overline{\mathbf{C}}_{\mathrm{e}} = [\widetilde{\boldsymbol{F}} \boxtimes \widetilde{\boldsymbol{F}}] : \mathbf{C}_{\mathrm{e}} : [\widetilde{\boldsymbol{F}}^t \boxtimes \widetilde{\boldsymbol{F}}^t].$$

The elasto-plastic and elastic tangent operators $\overline{\mathbf{C}}_{\mathrm{ep}}$ and $\overline{\mathbf{C}}_{\mathrm{e}}$ are fully contravariant intermediate description tensors. Finally the *elasto-plastic* tangent relation between the (left) Lie time derivatives of the spatial contravariant Kirchhoff stress $\boldsymbol{\tau}_\sharp$ and the spatial covariant metric $\boldsymbol{j}$ reads

$$2\,\mathrm{L}_t\boldsymbol{\tau}_\sharp = \mathbf{c}_{\mathrm{ep}} : \mathrm{L}_t\boldsymbol{j} \quad \text{with} \quad \mathbf{c}_{\mathrm{ep}} = [\overline{\boldsymbol{F}} \boxtimes \overline{\boldsymbol{F}}] : \overline{\mathbf{C}}_{\mathrm{ep}} : [\overline{\boldsymbol{F}}^t \boxtimes \overline{\boldsymbol{F}}^t].$$

As an auxiliary definition the *elastic* tangent relation between the elastic (left) Lie time derivatives of $\boldsymbol{\tau}_\sharp$ and $\boldsymbol{j}$ is introduced

$$2\,\overline{\mathrm{L}}_t\boldsymbol{\tau}_\sharp = \mathbf{c}_{\mathrm{e}} : \overline{\mathrm{L}}_t\boldsymbol{j} \quad \text{with} \quad \mathbf{c}_{\mathrm{e}} = [\overline{\boldsymbol{F}} \boxtimes \overline{\boldsymbol{F}}] : \overline{\mathbf{C}}_{\mathrm{e}} : [\overline{\boldsymbol{F}}^t \boxtimes \overline{\boldsymbol{F}}^t].$$

The elasto-plastic and elastic tangent operators $\mathbf{c}_{\mathrm{ep}}$ and $\mathbf{c}_{\mathrm{e}}$ are fully contravariant spatial description tensors. Note that the relation of the elasto-plastic tangent operator $\mathbf{c}_{\mathrm{ep}}$ to the corresponding $\mathbf{a}_{\mathrm{ep}}$ in terms of material and geometric stiffnesses is as before.

Elasto-Plastic Linearized Constitutive Relations:

As an auxiliary result the fully contravariant material, intermediate and spatial description *elastic* tangent operators $\mathbf{C}_{\mathrm{e}}$, $\overline{\mathbf{C}}_{\mathrm{e}}$ and $\mathbf{c}_{\mathrm{e}}$ compute as

$$\mathbf{C}_{\mathrm{e}} = 4\frac{\partial^2\psi_0}{\partial\boldsymbol{C}\partial\boldsymbol{C}} \quad \text{and} \quad \overline{\mathbf{C}}_{\mathrm{e}} = 4\frac{\partial^2\psi_0}{\partial\overline{\boldsymbol{C}}\partial\overline{\boldsymbol{C}}} \quad \text{and} \quad \mathbf{c}_{\mathrm{e}} = 4\frac{\partial^2\psi_0}{\partial\boldsymbol{j}\partial\boldsymbol{j}}.$$

An alternative format for the intermediate description elastic tangent operator $\overline{\mathbf{C}}_{\mathrm{e}}$ reads

$$\begin{aligned}\overline{\mathbf{C}}_{\mathrm{e}} &= 4[\,\overline{\boldsymbol{I}} \boxtimes \overline{\boldsymbol{B}}] : \frac{\partial^2\psi_0}{\partial\,\overline{\boldsymbol{I}}\,\partial\,\overline{\boldsymbol{I}}} : [\overline{\boldsymbol{B}} \boxtimes \overline{\boldsymbol{I}}\,] \\ &+ 4[\overline{\boldsymbol{M}}_\alpha \boxtimes \overline{\boldsymbol{B}}] : \frac{\partial^2\psi_0}{\partial\overline{\boldsymbol{M}}_\alpha\partial\,\overline{\boldsymbol{I}}} : [\overline{\boldsymbol{B}} \boxtimes \overline{\boldsymbol{I}}\,] \\ &+ 4[\,\overline{\boldsymbol{I}} \boxtimes \overline{\boldsymbol{B}}] : \frac{\partial^2\psi_0}{\partial\,\overline{\boldsymbol{I}}\,\partial\overline{\boldsymbol{M}}_\alpha} : [\overline{\boldsymbol{B}} \boxtimes \overline{\boldsymbol{M}}_\alpha] \\ &+ 4[\overline{\boldsymbol{M}}_\alpha \boxtimes \overline{\boldsymbol{B}}] : \frac{\partial^2\psi_0}{\partial\overline{\boldsymbol{M}}_\alpha\partial\overline{\boldsymbol{M}}_\beta} : [\overline{\boldsymbol{B}} \boxtimes \overline{\boldsymbol{M}}_\beta].\end{aligned}$$

Likewise, an alternative format for the spatial description tangent operator $\mathbf{c}_{\mathrm{e}}$ reads

$$
\begin{aligned}
\mathsf{c}_{\mathrm{e}} = & 4[\,\bar{\boldsymbol{b}}\boxtimes\boldsymbol{i}] : \frac{\partial^2\psi_0}{\partial\,\bar{\boldsymbol{b}}\,\partial\,\bar{\boldsymbol{b}}} : [\boldsymbol{i}\boxtimes\bar{\boldsymbol{b}}\,] \\
& + 4[\boldsymbol{m}_\alpha\boxtimes\boldsymbol{i}] : \frac{\partial^2\psi_0}{\partial\boldsymbol{m}_\alpha\partial\,\bar{\boldsymbol{b}}} : [\boldsymbol{i}\boxtimes\bar{\boldsymbol{b}}\,] \\
& + 4[\,\bar{\boldsymbol{b}}\boxtimes\boldsymbol{i}] : \frac{\partial^2\psi_0}{\partial\,\bar{\boldsymbol{b}}\,\partial\boldsymbol{m}_\alpha} : [\boldsymbol{i}\boxtimes\boldsymbol{m}_\alpha] \\
& + 4[\boldsymbol{m}_\alpha\boxtimes\boldsymbol{i}] : \frac{\partial^2\psi_0}{\partial\boldsymbol{m}_\alpha\partial\boldsymbol{m}_\beta} : [\boldsymbol{i}\boxtimes\boldsymbol{m}_\beta].
\end{aligned}
$$

With these preliminary results at hand the fully contravariant material, intermediate and spatial description *elasto-plastic* tangent operators C_{ep}, $\overline{\mathsf{C}}_{\mathrm{ep}}$ and c_{ep} compute after straightforward intermediate steps (see below) eventually as

$$
\begin{aligned}
\mathsf{C}_{\mathrm{ep}} &= \mathsf{C}_{\mathrm{e}} - 4D^{-1}\frac{\mathrm{d}\phi}{\mathrm{d}\boldsymbol{C}}\otimes\frac{\mathrm{d}\phi}{\mathrm{d}\boldsymbol{C}}, \\
\overline{\mathsf{C}}_{\mathrm{ep}} &= \overline{\mathsf{C}}_{\mathrm{e}} - 4D^{-1}\frac{\mathrm{d}\phi}{\mathrm{d}\overline{\boldsymbol{C}}}\otimes\frac{\mathrm{d}\phi}{\mathrm{d}\overline{\boldsymbol{C}}}, \\
\mathsf{c}_{\mathrm{ep}} &= \ \mathsf{c}_{\mathrm{e}} - 4D^{-1}\frac{\mathrm{d}\phi}{\mathrm{d}\boldsymbol{j}}\otimes\frac{\mathrm{d}\phi}{\mathrm{d}\boldsymbol{j}}.
\end{aligned}
$$

Here the denominator D is given for the simplifying case of ideal plasticity by

$$
D := 2\frac{\mathrm{d}\phi}{\mathrm{d}\boldsymbol{C}} : \frac{\partial\phi}{\partial\boldsymbol{S}} \equiv 2\frac{\mathrm{d}\phi}{\mathrm{d}\overline{\boldsymbol{C}}} : \frac{\partial\phi}{\partial\overline{\boldsymbol{S}}} \equiv 2\frac{\mathrm{d}\phi}{\mathrm{d}\boldsymbol{j}} : \frac{\partial\phi}{\partial\boldsymbol{\tau}_\sharp},
$$

whereby the *total derivatives* of the yield function ϕ have been defined as

$$
\begin{aligned}
\frac{\mathrm{d}\phi}{\mathrm{d}\boldsymbol{C}} &:= \frac{\partial\phi}{\partial\boldsymbol{C}} + \frac{1}{2}\mathsf{C}_{\mathrm{e}} : \frac{\partial\phi}{\partial\boldsymbol{S}}, \\
\frac{\mathrm{d}\phi}{\mathrm{d}\overline{\boldsymbol{C}}} &:= \frac{\partial\phi}{\partial\overline{\boldsymbol{C}}} + \frac{1}{2}\overline{\mathsf{C}}_{\mathrm{e}} : \frac{\partial\phi}{\partial\overline{\boldsymbol{S}}}, \\
\frac{\mathrm{d}\phi}{\mathrm{d}\boldsymbol{j}} &:= \frac{\partial\phi}{\partial\boldsymbol{j}} + \frac{1}{2}\,\mathsf{c}_{\mathrm{e}} : \frac{\partial\phi}{\partial\boldsymbol{\tau}_\sharp}.
\end{aligned}
$$

Note the fully contravariant character of the elasto-plastic tangent operators that is implicit to the formulation in terms of the total derivatives of the yield function with respect to covariant strain measures. Moreover, observe the formal similarity of the elasto-plastic tangent operators of the geometrically nonlinear case with the familiar format of the elasto-plastic tangent operator of the geometrically linear case. Equivalent expression for the intermediate and the spatial elasto-plastic tangent operators have been formulated as generalization of the geometrically linear Prandtl-Reuss tensors in [132, 137].

The derivation of the elasto-plastic tangent operators follows from the subsequent five intermediate steps:

i) recall the representation for the Lie time derivative of $\boldsymbol{\tau}_\sharp$*, the plastic Lie time derivative of* $\overline{\boldsymbol{S}}$ *and the material time derivative of* $\boldsymbol{S}$

$$\begin{aligned} 2\,\mathrm{L}_t\boldsymbol{\tau}_\sharp &= 2\overline{\mathrm{L}}_t\boldsymbol{\tau}_\sharp - 4[\boldsymbol{l}_\mathrm{p}\cdot\boldsymbol{\tau}_\sharp]^{\mathrm{sym}},\\ 2\,\widetilde{\mathrm{L}}_t\overline{\boldsymbol{S}} &= 2\,\mathrm{D}_t\overline{\boldsymbol{S}} - 4[\boldsymbol{\ell}_\mathrm{p}\cdot\overline{\boldsymbol{S}}\,]^{\mathrm{sym}},\\ 2\,\mathrm{D}_t\boldsymbol{S} &= 2\,\widetilde{\mathrm{R}}_t\boldsymbol{S} - 4[\boldsymbol{L}_\mathrm{p}\cdot\boldsymbol{S}\,]^{\mathrm{sym}}; \end{aligned}$$

ii) insert the elastic tangent relations for $\overline{\mathrm{L}}_t\boldsymbol{\tau}_\sharp$*,* $\mathrm{D}_t\overline{\boldsymbol{S}}$ *and* $\widetilde{\mathrm{R}}_t\boldsymbol{S}$ *in terms of the elastic tangent operators* $\mathbf{c}_\mathrm{e}$*,* $\overline{\mathbf{C}}_\mathrm{e}$ *and* $\mathbf{c}_\mathrm{e}$ *and the associated plastic evolution laws for* $\boldsymbol{l}_\mathrm{p}$*,* $\boldsymbol{\ell}_\mathrm{p}$ *and* $\boldsymbol{L}_\mathrm{p}$

$$\begin{aligned} 2\,\mathrm{L}_t\boldsymbol{\tau}_\sharp &= \mathbf{c}_\mathrm{e} : \overline{\mathrm{L}}_t\boldsymbol{j} - 4[\frac{\partial\phi}{\partial\boldsymbol{\tau}}\cdot\boldsymbol{\tau}_\sharp]^{\mathrm{sym}}\gamma,\\ 2\,\widetilde{\mathrm{L}}_t\overline{\boldsymbol{S}} &= \overline{\mathbf{C}}_\mathrm{e} : \mathrm{D}_t\overline{\boldsymbol{C}} - 4[\frac{\partial\phi}{\partial\overline{\boldsymbol{M}}}\cdot\overline{\boldsymbol{S}}\,]^{\mathrm{sym}}\gamma,\\ 2\,\mathrm{D}_t\boldsymbol{S} &= \mathbf{C}_\mathrm{e} : \widetilde{\mathrm{R}}_t\boldsymbol{C} - 4[\frac{\partial\phi}{\partial\boldsymbol{M}}\cdot\boldsymbol{S}\,]^{\mathrm{sym}}\gamma; \end{aligned}$$

iii) incorporate the additive elasto-plastic kinematic decompositions

$$\begin{aligned} \overline{\mathrm{L}}_t\,\boldsymbol{j} &= \mathrm{L}_t\boldsymbol{j} - 2[\boldsymbol{j}\cdot\boldsymbol{l}_\mathrm{p}]^{\mathrm{sym}},\\ \mathrm{D}_t\overline{\boldsymbol{C}} &= \widetilde{\mathrm{L}}_t\overline{\boldsymbol{C}} - 2[\overline{\boldsymbol{C}}\cdot\boldsymbol{\ell}_\mathrm{p}]^{\mathrm{sym}},\\ \widetilde{\mathrm{R}}_t\boldsymbol{C} &= \mathrm{D}_t\boldsymbol{C} - 2[\boldsymbol{C}\cdot\boldsymbol{L}_\mathrm{p}]^{\mathrm{sym}} \end{aligned}$$

and relations following from the chain rule and the symmetry of $\boldsymbol{j}$*,* $\overline{\boldsymbol{C}}$ *and* $\boldsymbol{C}$

$$\frac{\partial\phi}{\partial\boldsymbol{j}} = \left[\frac{\partial\phi}{\partial\boldsymbol{\tau}}\cdot\boldsymbol{\tau}_\sharp\right]^{\mathrm{sym}}, \quad \frac{\partial\phi}{\partial\overline{\boldsymbol{C}}} = \left[\frac{\partial\phi}{\partial\overline{\boldsymbol{M}}}\cdot\overline{\boldsymbol{S}}\right]^{\mathrm{sym}} \quad \text{and} \quad \frac{\partial\phi}{\partial\boldsymbol{C}} = \left[\frac{\partial\phi}{\partial\boldsymbol{M}}\cdot\boldsymbol{S}\right]^{\mathrm{sym}}$$

to render

$$\begin{aligned} 2\,\mathrm{L}_t\boldsymbol{\tau}_\sharp &= \mathbf{c}_\mathrm{e} : \mathrm{L}_t\boldsymbol{j} - 2\,\mathbf{c}_\mathrm{e} : [\boldsymbol{j}\cdot\boldsymbol{l}_\mathrm{p}]^{\mathrm{sym}} - 4\frac{\partial\phi}{\partial\boldsymbol{j}}\gamma,\\ 2\,\widetilde{\mathrm{L}}_t\overline{\boldsymbol{S}} &= \overline{\mathbf{C}}_\mathrm{e} : \widetilde{\mathrm{L}}_t\overline{\boldsymbol{C}} - 2\overline{\mathbf{C}}_\mathrm{e} : [\overline{\boldsymbol{C}}\cdot\boldsymbol{\ell}_\mathrm{p}]^{\mathrm{sym}} - 4\frac{\partial\phi}{\partial\overline{\boldsymbol{C}}}\gamma,\\ 2\,\mathrm{D}_t\boldsymbol{S} &= \mathbf{C}_\mathrm{e} : \mathrm{D}_t\boldsymbol{C} - 2\mathbf{C}_\mathrm{e} : [\boldsymbol{C}\cdot\boldsymbol{L}_\mathrm{p}]^{\mathrm{sym}} - 4\frac{\partial\phi}{\partial\boldsymbol{C}}\gamma; \end{aligned}$$

iv) take the associated plastic evolution laws for $[\boldsymbol{j}\cdot\boldsymbol{l}_\mathrm{p}]^{\mathrm{sym}}$*,* $[\overline{\boldsymbol{C}}\cdot\boldsymbol{\ell}_\mathrm{p}]^{\mathrm{sym}}$ *and* $[\boldsymbol{C}\cdot\boldsymbol{L}_\mathrm{p}]^{\mathrm{sym}}$ *into account and introduce the abbreviations defined in the above for the total derivative of the yield surface*

$$2\,\mathrm{L}_t\boldsymbol{\tau}_\sharp = \mathbf{c}_\mathrm{e} : \mathrm{L}_t\boldsymbol{j} \;-4\frac{\mathrm{d}\phi}{\mathrm{d}\boldsymbol{j}}\gamma,$$

$$2\,\widetilde{\mathrm{L}}_t\overline{\boldsymbol{S}} = \overline{\mathbf{C}}_\mathrm{e} : \widetilde{\mathrm{L}}_t\overline{\boldsymbol{C}} - 4\frac{\mathrm{d}\phi}{\mathrm{d}\overline{\boldsymbol{C}}}\gamma,$$

$$2\,\mathrm{D}_t\boldsymbol{S} = \mathbf{C}_\mathrm{e} : \mathrm{D}_t\boldsymbol{C} - 4\frac{\mathrm{d}\phi}{\mathrm{d}\boldsymbol{C}}\gamma;$$

v) compute the plastic multiplier γ from the consistency condition and introduce the abbreviation defined in the above for the denominator D

$$\overline{\mathrm{L}}_t\phi = \frac{\mathrm{d}\phi}{\mathrm{d}\boldsymbol{j}} : \mathrm{L}_t\boldsymbol{j} \;- D\gamma \doteq 0,$$

$$\mathrm{D}_t\phi = \frac{\mathrm{d}\phi}{\mathrm{d}\overline{\boldsymbol{C}}} : \widetilde{\mathrm{L}}_t\overline{\boldsymbol{C}} - D\gamma \doteq 0,$$

$$\widetilde{\mathrm{R}}_t\phi = \frac{\mathrm{d}\phi}{\mathrm{d}\boldsymbol{C}} : \mathrm{D}_t\boldsymbol{C} - D\gamma \doteq 0.$$

Combination of the last two results renders finally the elasto-plastic tangent operators as given in the above.

Elasto-Plastic Acoustic Tensor:
The particular format of the elasto-plastic tangent operator as a rank-one update of the elastic tangent operator carries over to the corresponding elasto-plastic acoustic tensor. With σ_n the normal contribution of the Cauchy stress in a cut-surface with orientation $\boldsymbol{n}$ (a covariant vector) the Cauchy-type elastic acoustic tensor follows as

$$\boldsymbol{\alpha}_\mathrm{e}(\boldsymbol{n}) \cdot \boldsymbol{m} = j\left[\boldsymbol{j} \cdot \mathbf{c}_\mathrm{e} : [\boldsymbol{m}_\flat \otimes \boldsymbol{n}]\right] \cdot \boldsymbol{n} + \sigma_n \boldsymbol{m}_\flat.$$

Likewise the elasto-plastic Cauchy-type acoustic tensor decomposes into a material and a geometric contribution

$$\boldsymbol{\alpha}_\mathrm{ep}(\boldsymbol{n}) \cdot \boldsymbol{m} = j\left[\boldsymbol{j} \cdot \mathbf{c}_\mathrm{ep} : [\boldsymbol{m}_\flat \otimes \boldsymbol{n}]\right] \cdot \boldsymbol{n} + \sigma_n \boldsymbol{m}_\flat.$$

Thus the elasto-plastic acoustic tensor may be expressed in terms of the elastic acoustic tensor

$$\boldsymbol{\alpha}_\mathrm{ep}(\boldsymbol{n}) \cdot \boldsymbol{m} = \left[\boldsymbol{\alpha}_\mathrm{e}(\boldsymbol{n}) - 4jD^{-1}\boldsymbol{j} \cdot \frac{\mathrm{d}\phi}{\mathrm{d}\boldsymbol{j}} \cdot \boldsymbol{n} \otimes \boldsymbol{n} \cdot \frac{\mathrm{d}\phi}{\mathrm{d}\boldsymbol{j}} \cdot \boldsymbol{j}\right] \cdot \boldsymbol{m}.$$

Due to the rank-one update format the determinant of the elasto-plastic acoustic tensor is easily computable

$$\det \boldsymbol{\alpha}_\mathrm{ep}(\boldsymbol{n}) = \det \boldsymbol{\alpha}_\mathrm{e}(\boldsymbol{n}) \left[1 - 4jD^{-1}\boldsymbol{n} \cdot \frac{\mathrm{d}\phi}{\mathrm{d}\boldsymbol{j}} \cdot \boldsymbol{j} \cdot [\boldsymbol{\alpha}_\mathrm{e}(\boldsymbol{n})]^{-1} \cdot \boldsymbol{j} \cdot \frac{\mathrm{d}\phi}{\mathrm{d}\boldsymbol{j}} \cdot \boldsymbol{n}\right]$$

As a result the loss of strong ellipticity for quasi-static problems is indicated at a given state if the following inequality is satisfied for an orientation $\boldsymbol{n}$

$$0 \leq 4 \max_{|\boldsymbol{n}|=1} \left(\boldsymbol{n} \cdot \frac{\mathrm{d}\phi}{\mathrm{d}\boldsymbol{j}} \cdot \boldsymbol{j} \cdot [\boldsymbol{\alpha}_{\mathrm{e}}(\boldsymbol{n})]^{-1} \cdot \boldsymbol{j} \cdot \frac{\mathrm{d}\phi}{\mathrm{d}\boldsymbol{j}} \cdot \boldsymbol{n} \right) - JD.$$

A more detailed analysis of the geometrically nonlinear elasto-plastic acoustic tensor including hardening is provided in [189], a general account on bifurcation and instability in inelastic solids is found in [16].

For geometrically linear isotropic ideal v. Mises elasto-plasticity the loss of strong ellipticity condition reads in terms of the Lamé constants μ and λ together with the deviatoric stress $\boldsymbol{\sigma}_{\mathrm{dev}}$ as

$$0 \leq \max_{|\boldsymbol{n}|=1} \left(\boldsymbol{n} \cdot \boldsymbol{\sigma}^2_{\mathrm{dev}} \cdot \boldsymbol{n} - \frac{\mu + \lambda}{2\mu + \lambda} [\boldsymbol{n} \cdot \boldsymbol{\sigma}_{\mathrm{dev}} \cdot \boldsymbol{n}]^2 \right) - \frac{|\boldsymbol{\sigma}_{\mathrm{dev}}|^2}{2}$$

As an example for the planar case of pure shear it holds that $\boldsymbol{\sigma}_{\mathrm{dev}} \sim \lceil +1, -1 \rfloor$ and $\boldsymbol{\sigma}^2_{\mathrm{dev}} \sim \lceil +1, +1 \rfloor$ with $|\boldsymbol{\sigma}_{\mathrm{dev}}|^2 \sim 2$. Thus with $\boldsymbol{n} \hat{=} [\cos\theta, \sin\theta]$ the loss of strong ellipticity condition reduces to

$$0 \geq \max_{\theta} \left([\cos^2\theta - \sin^2\theta]^2 \right)$$

Clearly, this conditions is only satisfied for $\theta = \pm\pi/4$. Likewise for the planar case of simple shear it holds that $\boldsymbol{\sigma}_{\mathrm{dev}} \sim \lfloor +1, +1 \rceil$ and $\boldsymbol{\sigma}^2_{\mathrm{dev}} \sim \lceil +1, +1 \rfloor$ with $|\boldsymbol{\sigma}_{\mathrm{dev}}|^2 \sim 2$. Thus the loss of strong ellipticity condition reduces to

$$0 \geq \max_{\theta} \left([\cos\theta \sin\theta]^2 \right)$$

Obviously, this condition is only satisfied for $\theta = \pm 0, \pm\pi/2$ which fits to the case of simple shear being rotated by $\pi/4$ with respect to the case of pure shear. Similar analyses can be carried out along the same lines for more general situations including hardening and geometric nonlinearities.

Incremental Variational Setting:
Quasi-static associated (hyper) elasto-plasticity can be cast into an *incremental variational setting.* To this end a *total power functional*

$$\dot{I} = \dot{I}(\dot{\boldsymbol{y}}(\boldsymbol{X}), \dot{\boldsymbol{\mathcal{F}}}(\boldsymbol{X}); \boldsymbol{y}(\boldsymbol{X}), \boldsymbol{\mathcal{F}}(\boldsymbol{X}))$$

in terms of the velocity field $\dot{\boldsymbol{y}}(\boldsymbol{X})$ with compatible velocity gradient $\dot{\boldsymbol{F}} := \operatorname{Grad} \dot{\boldsymbol{y}}(\boldsymbol{X})$ and the fluxes (material time derivatives) of the generalized internal variables $\dot{\boldsymbol{\mathcal{F}}}(\boldsymbol{X})$ is introduced as

$$\dot{I} := \int_{\mathcal{B}_0} \dot{u}_0(\dot{\boldsymbol{y}}(\boldsymbol{X}), \dot{\boldsymbol{F}}(\boldsymbol{X}), \dot{\boldsymbol{\mathcal{F}}}(\boldsymbol{X}))\, \mathrm{d}V + \int_{\partial\mathcal{B}_0} \hat{\dot{v}}_0(\dot{\boldsymbol{y}}(\boldsymbol{X}))\, \mathrm{d}A.$$

Here the bulk power density $\dot{u}_0 := \dot{\psi}_0 + \dot{v}_0 + d_0$ consists firstly in the material time derivative of the free energy density in terms of the Piola stress and the generalized thermodynamic forces

$$\dot{\psi}_0 = \dot{\psi}_0(\dot{\boldsymbol{F}}, \dot{\boldsymbol{\mathcal{F}}}; \boldsymbol{F}, \boldsymbol{\mathcal{F}}) = \frac{\partial \psi_0}{\partial \boldsymbol{F}} : \dot{\boldsymbol{F}} + \frac{\partial \psi_0}{\partial \boldsymbol{\mathcal{F}}} : \dot{\boldsymbol{\mathcal{F}}} =: \boldsymbol{P}(\boldsymbol{F}, \boldsymbol{\mathcal{F}}) : \dot{\boldsymbol{F}} - \boldsymbol{\mathcal{P}}(\boldsymbol{F}, \boldsymbol{\mathcal{F}}) : \dot{\boldsymbol{\mathcal{F}}},$$

secondly an external contribution in terms of the body force density

$$\dot{v}_0 = \dot{v}_0(\dot{\boldsymbol{y}}; \boldsymbol{y}) = \frac{\partial v_0}{\partial \boldsymbol{y}} : \dot{\boldsymbol{y}} =: -\boldsymbol{b}_0(\boldsymbol{y}) \cdot \dot{\boldsymbol{y}},$$

and finally the (density of the) *dissipation potential* $d_0 = d_0(\dot{\boldsymbol{\mathcal{F}}}; \boldsymbol{\mathcal{F}})$; the surface power density consists in an external contribution in terms of the surface force density

$$\dot{\widehat{v}}_0 = \dot{\widehat{v}}_0(\dot{\boldsymbol{y}}; \boldsymbol{y}) = \frac{\partial \widehat{v}_0}{\partial \boldsymbol{y}} : \dot{\boldsymbol{y}} =: -\boldsymbol{s}_0(\boldsymbol{y}) \cdot \dot{\boldsymbol{y}}.$$

Thus the total power functional takes the more concrete format

$$\dot{I} := \int_{\mathcal{B}_0} [\boldsymbol{P} : \dot{\boldsymbol{F}} - \boldsymbol{b}_0 \cdot \dot{\boldsymbol{y}}] \,\mathrm{d}V - \int_{\partial\mathcal{B}_0} \boldsymbol{s}_0 \cdot \dot{\boldsymbol{y}} \,\mathrm{d}A - \int_{\mathcal{B}_0} \left[\boldsymbol{\mathcal{P}} : \dot{\boldsymbol{\mathcal{F}}} - d_0\right] \mathrm{d}V.$$

Then the minimum of $\dot{I}$ is sought for all *admissible variations* $\delta\dot{\boldsymbol{y}}$ of the velocities $\dot{\boldsymbol{y}}$ and $\delta\dot{\boldsymbol{\mathcal{F}}}$ of the fluxes $\dot{\boldsymbol{\mathcal{F}}}$

$$\dot{I}\big(\dot{\boldsymbol{y}}(\boldsymbol{X}), \dot{\boldsymbol{\mathcal{F}}}(\boldsymbol{X}); \boldsymbol{y}(\boldsymbol{X}), \boldsymbol{\mathcal{F}}(\boldsymbol{X})\big) \to \min_{\dot{\boldsymbol{y}}, \dot{\boldsymbol{\mathcal{F}}}}.$$

Thereby admissible variations of the velocity field $\delta\dot{\boldsymbol{y}}$ satisfy compatibility $\delta\dot{\boldsymbol{F}} := \operatorname{Grad} \delta\dot{\boldsymbol{y}}$ and homogeneous boundary conditions $\delta\dot{\boldsymbol{y}} = \boldsymbol{0}$ on the Dirichlet part $\partial\mathcal{B}_0^y$ of the boundary $\partial\mathcal{B}_0$. In contrast, for the case of *local plasticity* admissible variations of the fluxes $\delta\dot{\boldsymbol{\mathcal{F}}}$ are not constrained by particular compatibility requirements or boundary conditions. Then minimization with respect to the velocities $\dot{\boldsymbol{y}}$ renders as point-wise Euler equations the well-known equilibrium equations

$$\operatorname{Div} \boldsymbol{P} + \boldsymbol{b}_0 \doteq \boldsymbol{0} \quad \text{and} \quad \boldsymbol{P} \cdot \boldsymbol{N} \doteq \boldsymbol{s}_0,$$

whereas minimization with respect to the fluxes $\delta\dot{\boldsymbol{\mathcal{F}}}$ results in

$$\boldsymbol{\mathcal{P}} \in \frac{\partial d_0}{\partial \dot{\boldsymbol{\mathcal{F}}}} \quad \leftrightarrow \quad \dot{\boldsymbol{\mathcal{F}}} \in \frac{\partial p_0}{\partial \boldsymbol{\mathcal{P}}}.$$

The equivalence of the two differential inclusions in the above is a fundamental result for convex, proper, lower semi-continuous potentials in convex analysis. Observe that the minimum of $\dot{I}$ with respect to the fluxes $\dot{\mathcal{F}}$ may be re-expressed in terms of the *plastic potential* p_0

$$-\min\Big(\mathcal{P} : \dot{\mathcal{F}} - d_0^{\mathcal{F}}(\dot{\mathcal{F}})\Big|\ \dot{\mathcal{F}}\Big) = \max\Big(\mathcal{P} : \dot{\mathcal{F}} - d_0^{\mathcal{F}}(\dot{\mathcal{F}})\Big|\ \dot{\mathcal{F}}\Big) = p_0^{\mathcal{F}}(\mathcal{P}).$$

Thus a *condensed total power functional* $\dot{I}' = \dot{I}'(\dot{\boldsymbol{y}}(\boldsymbol{X}); \boldsymbol{y}(\boldsymbol{X}), \mathcal{F}(\boldsymbol{X}))$ in which the fluxes of the generalized internal variables have been eliminated by pre-minimization with respect to $\dot{\mathcal{F}}$ may be defined as

$$\dot{I}' := \int_{\mathcal{B}_0} [\boldsymbol{P} : \dot{\boldsymbol{F}} - \boldsymbol{b}_0 \cdot \dot{\boldsymbol{y}}]\, \mathrm{d}V - \int_{\partial\mathcal{B}_0} \boldsymbol{s}_0 \cdot \dot{\boldsymbol{y}}\, \mathrm{d}A + \int_{\mathcal{B}_0} p_0^{\mathcal{F}}(\mathcal{P})\, \mathrm{d}V.$$

Then the equilibrium equations obviously follow from minimization of $\dot{I}'$ with respect to the velocities $\dot{\boldsymbol{y}}$

$$\dot{I}'(\dot{\boldsymbol{y}}(\boldsymbol{X}); \boldsymbol{y}(\boldsymbol{X}), \mathcal{F}(\boldsymbol{X})) \to \min_{\dot{\boldsymbol{y}}}.$$

It shall be noted that due to $p_0^{\mathcal{F}} = \chi_{\mathcal{A}}^{\mathcal{F}}$ the volume term including the plastic potential p_0 assures that the generalized thermodynamic forces $\mathcal{P}$ remain inside the admissible domain $\mathbb{A}^{\mathcal{F}}$. Incremental variational settings of elasto-plasticity have been studied intensively e.g. by [157, 138, 74] and constitute the basis for effective algorithmic treatments of elasto-plastic boundary value problems, see e.g. [150, 156, 136]. Moreover discrete incremental variational settings of elasto-plasticity allow the analysis of emerging micro-structures in the spirit of [13] due to the loss of quasi-convexity, see e.g. [155, 21, 133, 134, 135].

Supplement 8.2. Second-Order Elasto-Plasticity in Euclidean Space
Geometrically nonlinear second-order elasto-plasticity shall here be formulated based on an *unconstrained* additive decomposition of the second deformation gradient into elastic and plastic contributions. These are denoted as the elastic and plastic double-distortions. Kinematically related approaches have been taken, e.g., by [180, 31, 62, 32, 43, 15]. Alternatively, based on the multiplicative decomposition of the deformation gradient into elastic and plastic distortions and the corresponding elastic and plastic distortion gradients, a *constrained* additive decomposition of the second deformation gradient is conceptually also possible. However, thereby the resulting elastic and plastic double-distortions are constrained since they ought to be compatible with the multiplicative decomposition of the deformation gradient. Approaches that fall into this category are, e.g., by [193, 118, 44]. Second-order elasto-plasticity formulated in terms of a constrained additive decomposition of the second deformation gradient requires satisfaction of corresponding compatibility conditions between the evolution equations for the plastic distortion and the plastic distortion gradient. In contrast the here advocated unconstrained additive decomposition of the second deformation gradient into elastic and plastic double-distortions is in general independent from the multiplicative decomposition of the deformation gradient. Thus the unconstrained additive decomposition of the second deformation gradient allows for more modeling flexibility as compared to the alternatively possible constrained additive decomposition. Various modeling options for the geometrically linear kinematics of generalized plasticity are discussed at length in [82].

Additive Decomposition of Second Deformation Gradient:
The second deformation gradient ${}^{\vartriangle}\boldsymbol{F}$ (a third-order two-point (spatial/material) contra/co/co- (mixed) variant tensor) is decomposed *additively* into *elastic* and *plastic double-distortions*

$$ {}^{\vartriangle}\boldsymbol{F} = \operatorname{Grad} \boldsymbol{F} = {}^{\vartriangle}\boldsymbol{F}_{\mathrm{e}} + {}^{\vartriangle}\boldsymbol{F}_{\mathrm{p}}. $$

Thereby, due to the additive decomposition the elastic and plastic double-distortions ${}^{\vartriangle}\boldsymbol{F}_{\mathrm{e}}$ and ${}^{\vartriangle}\boldsymbol{F}_{\mathrm{p}}$ are, quite like the second deformation gradient ${}^{\vartriangle}\boldsymbol{F}$, both third-order two-point (spatial/material) contra/co/co- (mixed) variant tensors. Likewise the second deformation gradient ${}^{\vartriangle}\boldsymbol{f}$ (a third-order two-point (material/spatial) contra/co/co- (mixed) variant tensor) is decomposed additively into elastic and plastic double-distortions

$$ {}^{\vartriangle}\boldsymbol{f} = \operatorname{grad} \boldsymbol{f} = {}^{\vartriangle}\boldsymbol{f}_{\mathrm{e}} + {}^{\vartriangle}\boldsymbol{f}_{\mathrm{p}}. $$

Thereby, due to the additive decomposition the elastic and plastic double-distortions ${}^{\vartriangle}\boldsymbol{f}_{\mathrm{e}}$ and ${}^{\vartriangle}\boldsymbol{f}_{\mathrm{p}}$ are, again quite like the second deformation gradient ${}^{\vartriangle}\boldsymbol{f}$, both third-order two-point (material/spatial) contra/co/co- (mixed) variant tensors. Then the relation between ${}^{\vartriangle}\boldsymbol{F}_{\bullet}$ and ${}^{\vartriangle}\boldsymbol{f}_{\bullet}$ (with the bullet denoting here and henceforth $\bullet = \mathrm{e}, \mathrm{p}$) follows the already established relation between ${}^{\Delta}\boldsymbol{F}$ and ${}^{\Delta}\boldsymbol{f}$ and thus computes as

$$ {}^{\Delta}\boldsymbol{f}_\bullet = -\boldsymbol{f} \cdot {}^{\Delta}\boldsymbol{F}_\bullet : [\boldsymbol{f} \boxtimes \boldsymbol{f}] \quad \text{and} \quad {}^{\Delta}\boldsymbol{F}_\bullet = -\boldsymbol{F} \cdot {}^{\Delta}\boldsymbol{f}_\bullet : [\boldsymbol{F} \boxtimes \boldsymbol{F}]. $$

Note that the additive decomposition of the second deformation gradient into elastic and plastic double-distortions is in general independent from the multiplicative decomposition of the deformation gradient $\boldsymbol{F} = \overline{\boldsymbol{F}} \cdot \widetilde{\boldsymbol{F}}$ into elastic and plastic distortions. Thus the unconstrained additive decomposition of ${}^{\Delta}\boldsymbol{F}$ allows for more modeling flexibility as compared to the alternatively possible constrained additive decomposition.

Elastic and Plastic Connection Tensors:

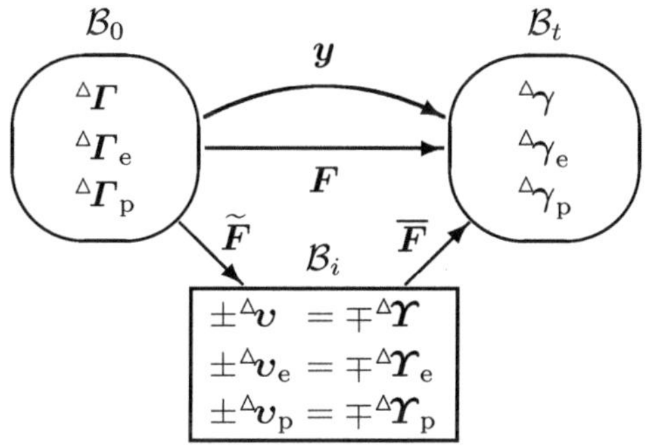

Moreover it proves convenient to introduce material and spatial description third-order contra/co/co- (mixed) variant *elastic* and *plastic connection tensors*

$$ {}^{\Delta}\boldsymbol{\Gamma}_\bullet := \boldsymbol{f} \cdot {}^{\Delta}\boldsymbol{F}_\bullet = -{}^{\Delta}\boldsymbol{f}_\bullet : [\boldsymbol{F} \boxtimes \boldsymbol{F}] \quad \text{and} \quad {}^{\Delta}\gamma_\bullet := \boldsymbol{F} \cdot {}^{\Delta}\boldsymbol{f}_\bullet = -{}^{\Delta}\boldsymbol{F}_\bullet : [\boldsymbol{f} \boxtimes \boldsymbol{f}], $$

whereby the relation between the ${}^{\Delta}\boldsymbol{\Gamma}_\bullet$ and the ${}^{\Delta}\gamma_\bullet$ computes as

$$ {}^{\Delta}\gamma_\bullet = -\boldsymbol{F} \cdot {}^{\Delta}\boldsymbol{\Gamma}_\bullet : [\boldsymbol{f} \boxtimes \boldsymbol{f}] \quad \text{and} \quad {}^{\Delta}\boldsymbol{\Gamma}_\bullet = -\boldsymbol{f} \cdot {}^{\Delta}\gamma_\bullet : [\boldsymbol{F} \boxtimes \boldsymbol{F}]. $$

Consequently the additive decomposition of the total connection tensors reads

$$ {}^{\Delta}\boldsymbol{\Gamma} = {}^{\Delta}\boldsymbol{\Gamma}_\mathrm{e} + {}^{\Delta}\boldsymbol{\Gamma}_\mathrm{p} \quad \text{and} \quad {}^{\Delta}\gamma = {}^{\Delta}\gamma_\mathrm{p} + {}^{\Delta}\gamma_\mathrm{e}. $$

Moreover, corresponding intermediate description elastic and plastic connection tensors ${}^{\Delta}\boldsymbol{\upsilon}_\bullet$ and ${}^{\Delta}\boldsymbol{\Upsilon}_\bullet$ may be defined in analogy as

$$ {}^{\Delta}\boldsymbol{\upsilon}_\bullet := -\widetilde{\boldsymbol{F}} \cdot {}^{\Delta}\boldsymbol{\Gamma}_\bullet : [\widetilde{\boldsymbol{f}} \boxtimes \widetilde{\boldsymbol{f}}] \quad \text{and} \quad {}^{\Delta}\boldsymbol{\Upsilon}_\bullet := -\overline{\boldsymbol{f}} \cdot {}^{\Delta}\gamma_\bullet : [\overline{\boldsymbol{F}} \boxtimes \overline{\boldsymbol{F}}]. $$

Correspondingly, with the plastic push-forward of the total ${}^{\Delta}\boldsymbol{\Gamma}$ and the elastic pull-back of the total ${}^{\Delta}\gamma$ into

$$ {}^{\Delta}\boldsymbol{\upsilon} := -\widetilde{\boldsymbol{F}} \cdot {}^{\Delta}\boldsymbol{\Gamma} : [\widetilde{\boldsymbol{f}} \boxtimes \widetilde{\boldsymbol{f}}] \quad \text{and} \quad {}^{\Delta}\boldsymbol{\Upsilon} := -\overline{\boldsymbol{f}} \cdot {}^{\Delta}\gamma : [\overline{\boldsymbol{F}} \boxtimes \overline{\boldsymbol{F}}], $$

the additive decomposition of the total intermediate connection tensors reads

$$ {}^{\triangle}\boldsymbol{v} =: {}^{\triangle}\boldsymbol{v}_{\mathrm{e}} + {}^{\triangle}\boldsymbol{v}_{\mathrm{p}} \quad \text{and} \quad {}^{\triangle}\boldsymbol{\Upsilon} =: {}^{\triangle}\boldsymbol{\Upsilon}_{\mathrm{e}} + {}^{\triangle}\boldsymbol{\Upsilon}_{\mathrm{p}}. $$

Here the symbolic redundancy for the total, elastic and plastic intermediate connection tensors has been introduced as

$$ \pm{}^{\triangle}\boldsymbol{v} \equiv \mp{}^{\triangle}\boldsymbol{\Upsilon} \quad \text{and} \quad \pm{}^{\triangle}\boldsymbol{v}_{\bullet} \equiv \mp{}^{\triangle}\boldsymbol{\Upsilon}_{\bullet} $$

for notational convenience, the benefit of which may be seen later. A similar additive decomposition of the intermediate description connection tensor into elastic and plastic parts has been advocated by [15].

Elastic and Plastic Covariant Double-Strain Measures:

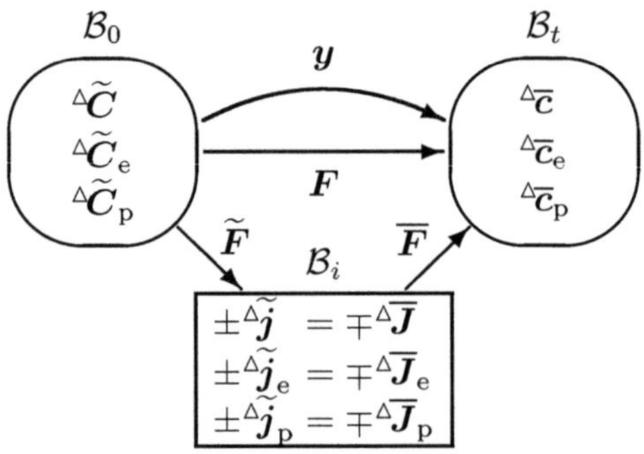

In most of the literature (see, e.g., [31, 62, 32, 44, 15]) on second-order elasto-plasticity one of the connection tensors is chosen as the relevant measure of higher-order strain. However, based on the additive decomposition of the second deformation gradients into elastic and plastic double-distortions and the resulting additive decomposition of the connection tensors into elastic and plastic contributions, corresponding covariant elastic and plastic double-strains and double-metrics may be defined.First the covariant material double-strains ${}^{\triangle}\widetilde{\boldsymbol{C}}_{\bullet}$, that are based on the material plastic strain metric $\widetilde{\boldsymbol{C}}$, together with the covariant intermediate double-metrics $\pm{}^{\triangle}\widetilde{\boldsymbol{j}}_{\bullet} = \mp{}^{\triangle}\overline{\boldsymbol{J}}_{\bullet}$, that are based on the intermediate Euclidean metric $\widetilde{\boldsymbol{j}} = \overline{\boldsymbol{J}}$, and the covariant spatial double-strains ${}^{\triangle}\overline{\boldsymbol{c}}_{\bullet}$, that are based on the spatial elastic strain metric $\overline{\boldsymbol{c}}$, are defined as

$$ \begin{array}{llll} {}^{\triangle}\widetilde{\boldsymbol{C}} := \widetilde{\boldsymbol{C}} \cdot {}^{\triangle}\boldsymbol{\Gamma}\,, & {}^{\triangle}\widetilde{\boldsymbol{j}} := -\widetilde{\boldsymbol{j}} \cdot {}^{\triangle}\boldsymbol{v}\,, & {}^{\triangle}\overline{\boldsymbol{J}} := -\overline{\boldsymbol{J}} \cdot {}^{\triangle}\boldsymbol{\Upsilon}\,, & {}^{\triangle}\overline{\boldsymbol{c}} := \overline{\boldsymbol{c}} \cdot {}^{\triangle}\boldsymbol{\gamma}\,, \\ {}^{\triangle}\widetilde{\boldsymbol{C}}_{\mathrm{e}} := \widetilde{\boldsymbol{C}} \cdot {}^{\triangle}\boldsymbol{\Gamma}_{\mathrm{e}}, & {}^{\triangle}\widetilde{\boldsymbol{j}}_{\mathrm{e}} := -\widetilde{\boldsymbol{j}} \cdot {}^{\triangle}\boldsymbol{v}_{\mathrm{e}}, & {}^{\triangle}\overline{\boldsymbol{J}}_{\mathrm{e}} := -\overline{\boldsymbol{J}} \cdot {}^{\triangle}\boldsymbol{\Upsilon}_{\mathrm{e}}, & {}^{\triangle}\overline{\boldsymbol{c}}_{\mathrm{e}} := \overline{\boldsymbol{c}} \cdot {}^{\triangle}\boldsymbol{\gamma}_{\mathrm{e}}, \\ {}^{\triangle}\widetilde{\boldsymbol{C}}_{\mathrm{p}} := \widetilde{\boldsymbol{C}} \cdot {}^{\triangle}\boldsymbol{\Gamma}_{\mathrm{p}}, & {}^{\triangle}\widetilde{\boldsymbol{j}}_{\mathrm{p}} := -\widetilde{\boldsymbol{j}} \cdot {}^{\triangle}\boldsymbol{v}_{\mathrm{p}}, & {}^{\triangle}\overline{\boldsymbol{J}}_{\mathrm{p}} := -\overline{\boldsymbol{J}} \cdot {}^{\triangle}\boldsymbol{\Upsilon}_{\mathrm{p}}, & {}^{\triangle}\overline{\boldsymbol{c}}_{\mathrm{p}} := \overline{\boldsymbol{c}} \cdot {}^{\triangle}\boldsymbol{\gamma}_{\mathrm{p}}. \end{array} $$

Accordingly the so defined material double-strain and its covariant push-forwards into the intermediate double-metric and the spatial double-strain also decompose additively into elastic and plastic contributions

$$ {}^{\Delta}\widetilde{\boldsymbol{C}} = {}^{\Delta}\widetilde{\boldsymbol{C}}_{\rm e} + {}^{\Delta}\widetilde{\boldsymbol{C}}_{\rm p} \quad \text{and} \quad \begin{array}{c} {}^{\Delta}\widetilde{\boldsymbol{j}} = {}^{\Delta}\widetilde{\boldsymbol{j}}_{\rm e} + {}^{\Delta}\widetilde{\boldsymbol{j}}_{\rm p} \\ {}^{\Delta}\overline{\boldsymbol{J}} = {}^{\Delta}\overline{\boldsymbol{J}}_{\rm e} + {}^{\Delta}\overline{\boldsymbol{J}}_{\rm p} \end{array} \quad \text{and} \quad {}^{\Delta}\overline{\boldsymbol{c}} = {}^{\Delta}\overline{\boldsymbol{c}}_{\rm e} + {}^{\Delta}\overline{\boldsymbol{c}}_{\rm p}. $$

Obviously, the double-strains and double-metrics introduced in the above take the interpretation as fully covariant versions of the previously introduced material, intermediate and spatial connection tensors. Corresponding pull-back/push-forward relations may be established for these fully covariant versions of the total, the elastic and the plastic connection tensors; summarizing these read as

$$ \begin{aligned} -{}^{\Delta}\widetilde{\boldsymbol{C}} &= -\widetilde{\boldsymbol{F}}^{t} \cdot {}^{\Delta}\widetilde{\boldsymbol{j}} : [\widetilde{\boldsymbol{F}} \boxtimes \widetilde{\boldsymbol{F}}] = \widetilde{\boldsymbol{F}}^{t} \cdot {}^{\Delta}\overline{\boldsymbol{J}} : [\widetilde{\boldsymbol{F}} \boxtimes \widetilde{\boldsymbol{F}}] = \boldsymbol{F}^{t} \cdot {}^{\Delta}\overline{\boldsymbol{c}} : [\boldsymbol{F} \boxtimes \boldsymbol{F}], \\ -{}^{\Delta}\widetilde{\boldsymbol{C}}_{\rm e} &= -\widetilde{\boldsymbol{F}}^{t} \cdot {}^{\Delta}\widetilde{\boldsymbol{j}}_{\rm e} : [\widetilde{\boldsymbol{F}} \boxtimes \widetilde{\boldsymbol{F}}] = \widetilde{\boldsymbol{F}}^{t} \cdot {}^{\Delta}\overline{\boldsymbol{J}}_{\rm e} : [\widetilde{\boldsymbol{F}} \boxtimes \widetilde{\boldsymbol{F}}] = \boldsymbol{F}^{t} \cdot {}^{\Delta}\overline{\boldsymbol{c}}_{\rm e} : [\boldsymbol{F} \boxtimes \boldsymbol{F}], \\ -{}^{\Delta}\widetilde{\boldsymbol{C}}_{\rm p} &= -\widetilde{\boldsymbol{F}}^{t} \cdot {}^{\Delta}\widetilde{\boldsymbol{j}}_{\rm p} : [\widetilde{\boldsymbol{F}} \boxtimes \widetilde{\boldsymbol{F}}] = \widetilde{\boldsymbol{F}}^{t} \cdot {}^{\Delta}\overline{\boldsymbol{J}}_{\rm p} : [\widetilde{\boldsymbol{F}} \boxtimes \widetilde{\boldsymbol{F}}] = \boldsymbol{F}^{t} \cdot {}^{\Delta}\overline{\boldsymbol{c}}_{\rm p} : [\boldsymbol{F} \boxtimes \boldsymbol{F}]. \end{aligned} $$

Second the role of the plastic or the elastic strain metric and the corresponding Euclidean metric may be exchanged.

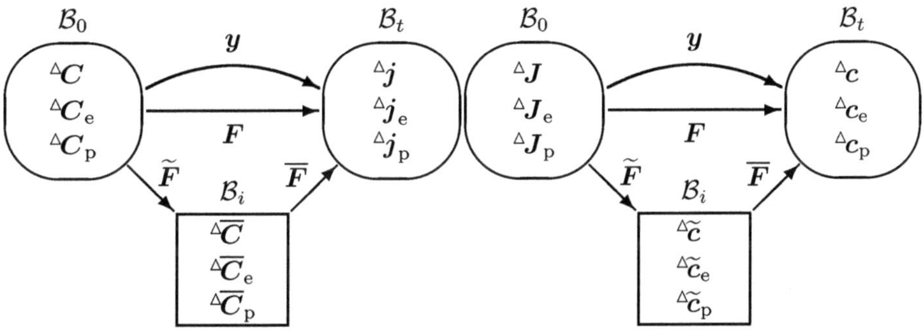

Then the covariant material double-strains ${}^{\Delta}\boldsymbol{C}_{\bullet}$, that are based on the material strain metric $\boldsymbol{C}$, together with the covariant intermediate double-strains ${}^{\Delta}\overline{\boldsymbol{C}}_{\bullet}$, that are based on the intermediate elastic strain metric $\overline{\boldsymbol{C}}$, and the covariant spatial double-metrics ${}^{\Delta}\boldsymbol{j}_{\bullet}$, that are based on the spatial Euclidean metric $\boldsymbol{j}$, are defined as

$$ \begin{aligned} {}^{\Delta}\boldsymbol{C} &:= \boldsymbol{C} \cdot {}^{\Delta}\boldsymbol{\Gamma}\ , & {}^{\Delta}\overline{\boldsymbol{C}} &:= \overline{\boldsymbol{C}} \cdot {}^{\Delta}\boldsymbol{\Upsilon}\ , & {}^{\Delta}\boldsymbol{j} &:= -\boldsymbol{j} \cdot {}^{\Delta}\boldsymbol{\gamma}\ , \\ {}^{\Delta}\boldsymbol{C}_{\rm e} &:= \boldsymbol{C} \cdot {}^{\Delta}\boldsymbol{\Gamma}_{\rm e}, & {}^{\Delta}\overline{\boldsymbol{C}}_{\rm e} &:= \overline{\boldsymbol{C}} \cdot {}^{\Delta}\boldsymbol{\Upsilon}_{\rm e}, & {}^{\Delta}\boldsymbol{j}_{\rm e} &:= -\boldsymbol{j} \cdot {}^{\Delta}\boldsymbol{\gamma}_{\rm e}, \\ {}^{\Delta}\boldsymbol{C}_{\rm p} &:= \boldsymbol{C} \cdot {}^{\Delta}\boldsymbol{\Gamma}_{\rm p}, & {}^{\Delta}\overline{\boldsymbol{C}}_{\rm p} &:= \overline{\boldsymbol{C}} \cdot {}^{\Delta}\boldsymbol{\Upsilon}_{\rm p}, & {}^{\Delta}\boldsymbol{j}_{\rm p} &:= -\boldsymbol{j} \cdot {}^{\Delta}\boldsymbol{\gamma}_{\rm p}. \end{aligned} $$

Accordingly the so defined covariant material double-strain and its covariant push-forwards into the intermediate double-strain and the spatial double-metric also decompose additively into elastic and plastic contributions

$$ {}^{\Delta}\boldsymbol{C} = {}^{\Delta}\boldsymbol{C}_{\rm e} + {}^{\Delta}\boldsymbol{C}_{\rm p} \quad \text{and} \quad {}^{\Delta}\overline{\boldsymbol{C}} = {}^{\Delta}\overline{\boldsymbol{C}}_{\rm e} + {}^{\Delta}\overline{\boldsymbol{C}}_{\rm p} \quad \text{and} \quad {}^{\Delta}\boldsymbol{j} = {}^{\Delta}\boldsymbol{j}_{\rm e} + {}^{\Delta}\boldsymbol{j}_{\rm p}. $$

Alternatively the covariant material elastic and plastic double-strains ${}^{\vartriangle}\boldsymbol{C}_{\bullet}$ read

$$ {}^{\vartriangle}\boldsymbol{C}_{\mathrm{e}} := \boldsymbol{F}^t \cdot \boldsymbol{j} \cdot {}^{\vartriangle}\boldsymbol{F}_{\mathrm{e}} \quad \text{and} \quad {}^{\vartriangle}\boldsymbol{C}_{\mathrm{p}} := \boldsymbol{F}^t \cdot \boldsymbol{j} \cdot {}^{\vartriangle}\boldsymbol{F}_{\mathrm{p}}. $$

Obviously, the double-strains and double-metrics introduced in the above take the interpretation as fully covariant versions of the previously introduced material, intermediate and spatial connection tensors. Corresponding pull-back/push-forward relations may be established for these fully covariant versions of the total, the elastic and the plastic connection tensors; summarizing these read as

$$ \begin{aligned} {}^{\vartriangle}\boldsymbol{C} &= \widetilde{\boldsymbol{F}}^t \cdot {}^{\vartriangle}\overline{\boldsymbol{C}} : [\widetilde{\boldsymbol{F}} \boxtimes \widetilde{\boldsymbol{F}}] = \boldsymbol{F}^t \cdot {}^{\vartriangle}\boldsymbol{j} : [\boldsymbol{F} \boxtimes \boldsymbol{F}], \\ {}^{\vartriangle}\boldsymbol{C}_{\mathrm{e}} &= \widetilde{\boldsymbol{F}}^t \cdot {}^{\vartriangle}\overline{\boldsymbol{C}}_{\mathrm{e}} : [\widetilde{\boldsymbol{F}} \boxtimes \widetilde{\boldsymbol{F}}] = \boldsymbol{F}^t \cdot {}^{\vartriangle}\boldsymbol{j}_{\mathrm{e}} : [\boldsymbol{F} \boxtimes \boldsymbol{F}], \\ {}^{\vartriangle}\boldsymbol{C}_{\mathrm{p}} &= \widetilde{\boldsymbol{F}}^t \cdot {}^{\vartriangle}\overline{\boldsymbol{C}}_{\mathrm{p}} : [\widetilde{\boldsymbol{F}} \boxtimes \widetilde{\boldsymbol{F}}] = \boldsymbol{F}^t \cdot {}^{\vartriangle}\boldsymbol{j}_{\mathrm{p}} : [\boldsymbol{F} \boxtimes \boldsymbol{F}]. \end{aligned} $$

Likewise the covariant material double-metrics ${}^{\vartriangle}\boldsymbol{J}_{\bullet}$, that are based on the material Euclidean metric $\boldsymbol{J}$, together with the covariant intermediate double-strains ${}^{\vartriangle}\widetilde{\boldsymbol{c}}_{\bullet}$, that are based on the intermediate plastic strain metric $\widetilde{\boldsymbol{c}}$, and the covariant spatial double-strains ${}^{\vartriangle}\boldsymbol{c}_{\bullet}$, that are based on the spatial strain metric $\boldsymbol{c}$, are defined as

$$ \begin{aligned} &{}^{\vartriangle}\boldsymbol{J} := -\boldsymbol{J} \cdot {}^{\vartriangle}\boldsymbol{\Gamma}\ , && {}^{\vartriangle}\widetilde{\boldsymbol{c}} := \widetilde{\boldsymbol{c}} \cdot {}^{\vartriangle}\boldsymbol{\upsilon}\ , && {}^{\vartriangle}\boldsymbol{c} := \boldsymbol{c} \cdot {}^{\vartriangle}\boldsymbol{\gamma}\ , \\ &{}^{\vartriangle}\boldsymbol{J}_{\mathrm{e}} := -\boldsymbol{J} \cdot {}^{\vartriangle}\boldsymbol{\Gamma}_{\mathrm{e}}, && {}^{\vartriangle}\widetilde{\boldsymbol{c}}_{\mathrm{e}} := \widetilde{\boldsymbol{c}} \cdot {}^{\vartriangle}\boldsymbol{\upsilon}_{\mathrm{e}}, && {}^{\vartriangle}\boldsymbol{c}_{\mathrm{e}} := \boldsymbol{c} \cdot {}^{\vartriangle}\boldsymbol{\gamma}_{\mathrm{e}}, \\ &{}^{\vartriangle}\boldsymbol{J}_{\mathrm{p}} := -\boldsymbol{J} \cdot {}^{\vartriangle}\boldsymbol{\Gamma}_{\mathrm{p}}, && {}^{\vartriangle}\widetilde{\boldsymbol{c}}_{\mathrm{p}} := \widetilde{\boldsymbol{c}} \cdot {}^{\vartriangle}\boldsymbol{\upsilon}_{\mathrm{p}}, && {}^{\vartriangle}\boldsymbol{c}_{\mathrm{p}} := \boldsymbol{c} \cdot {}^{\vartriangle}\boldsymbol{\gamma}_{\mathrm{p}}. \end{aligned} $$

Accordingly the so defined covariant material double-metric and its covariant push-forwards into the intermediate double-strain and the material double-strain also decompose additively into elastic and plastic contributions

$$ {}^{\vartriangle}\boldsymbol{J} = {}^{\vartriangle}\boldsymbol{J}_{\mathrm{e}} + {}^{\vartriangle}\boldsymbol{J}_{\mathrm{p}} \quad \text{and} \quad {}^{\vartriangle}\widetilde{\boldsymbol{c}} = {}^{\vartriangle}\widetilde{\boldsymbol{c}}_{\mathrm{e}} + {}^{\vartriangle}\widetilde{\boldsymbol{c}}_{\mathrm{p}} \quad \text{and} \quad {}^{\vartriangle}\boldsymbol{c} = {}^{\vartriangle}\boldsymbol{c}_{\mathrm{e}} + {}^{\vartriangle}\boldsymbol{c}_{\mathrm{p}}. $$

Alternatively the covariant spatial elastic and plastic double-strains ${}^{\vartriangle}\boldsymbol{c}_{\bullet}$ read

$$ {}^{\vartriangle}\boldsymbol{c}_{\mathrm{e}} := \boldsymbol{f}^t \cdot \boldsymbol{J} \cdot {}^{\vartriangle}\boldsymbol{f}_{\mathrm{e}} \quad \text{and} \quad {}^{\vartriangle}\boldsymbol{c}_{\mathrm{p}} := \boldsymbol{f}^t \cdot \boldsymbol{J} \cdot {}^{\vartriangle}\boldsymbol{f}_{\mathrm{p}}. $$

Obviously, the double-strains and double-metrics introduced in the above take the interpretation as fully covariant versions of the previously introduced material, intermediate and spatial connection tensors. Corresponding pull-back/push-forward relations may be established for these fully covariant versions of the total, the elastic and the plastic connection tensors; summarizing these read as

$$ \begin{aligned} {}^{\vartriangle}\boldsymbol{J} &= \widetilde{\boldsymbol{F}}^t \cdot {}^{\vartriangle}\widetilde{\boldsymbol{c}} : [\widetilde{\boldsymbol{F}} \boxtimes \widetilde{\boldsymbol{F}}] = \boldsymbol{F}^t \cdot {}^{\vartriangle}\boldsymbol{c} : [\boldsymbol{F} \boxtimes \boldsymbol{F}], \\ {}^{\vartriangle}\boldsymbol{J}_{\mathrm{e}} &= \widetilde{\boldsymbol{F}}^t \cdot {}^{\vartriangle}\widetilde{\boldsymbol{c}}_{\mathrm{e}} : [\widetilde{\boldsymbol{F}} \boxtimes \widetilde{\boldsymbol{F}}] = \boldsymbol{F}^t \cdot {}^{\vartriangle}\boldsymbol{c}_{\mathrm{e}} : [\boldsymbol{F} \boxtimes \boldsymbol{F}], \\ {}^{\vartriangle}\boldsymbol{J}_{\mathrm{p}} &= \widetilde{\boldsymbol{F}}^t \cdot {}^{\vartriangle}\widetilde{\boldsymbol{c}}_{\mathrm{p}} : [\widetilde{\boldsymbol{F}} \boxtimes \widetilde{\boldsymbol{F}}] = \boldsymbol{F}^t \cdot {}^{\vartriangle}\boldsymbol{c}_{\mathrm{p}} : [\boldsymbol{F} \boxtimes \boldsymbol{F}]. \end{aligned} $$

Note that the above nine sets of three double-strains and double-metrics follow from the combination of the nine total, elastic, and plastic covariant strain measures as introduced earlier with the nine total, elastic, and plastic as well as material, intermediate, and spatial contra/co/co- (mixed) variant connection tensors.

Additive Decomposition of Velocity Double-Gradients:
Based on ${}^{\triangle}\boldsymbol{F} = {}^{\triangle}\boldsymbol{F}_{\rm e} + {}^{\triangle}\boldsymbol{F}_{\rm p}$ the two-point ${}_0$-type velocity double-gradient may be decomposed additively into elastic and plastic ${}_0$-type *double-distortion rates*

$$\overset{\triangle}{\underset{0}{}}\!\boldsymbol{\Lambda} = \overset{\triangle}{\underset{0}{}}\!\boldsymbol{\Lambda}_{\rm e} + \overset{\triangle}{\underset{0}{}}\!\boldsymbol{\Lambda}_{\rm p} \quad \text{with} \quad \overset{\triangle}{\underset{0}{}}\!\boldsymbol{\Lambda}_{\rm e} := {\rm D}_t {}^{\triangle}\boldsymbol{F}_{\rm e} \quad \text{and} \quad \overset{\triangle}{\underset{0}{}}\!\boldsymbol{\Lambda}_{\rm p} := {\rm D}_t {}^{\triangle}\boldsymbol{F}_{\rm p}.$$

Besides the ${}_0$-type double-distortion rates, elastic and plastic two-point ${}_t$-type double-distortion rates are introduced (with $\bullet = {\rm e}, {\rm p}$) as

$$\overset{\triangle}{\underset{t}{}}\!\boldsymbol{\Lambda}_\bullet = \overset{\triangle}{\underset{0}{}}\!\boldsymbol{\Lambda}_\bullet - \boldsymbol{\Lambda}_\bullet \cdot {}^{\triangle}\boldsymbol{\Gamma}.$$

Then the relation between the spatial, the intermediate, the two-point and the material versions of the elastic and plastic ${}_\tau$-type double-distortion rates (with $\tau = 0, t$) follows the pattern established already for the velocity double-gradients and is summarized in the following table

	$\overset{\triangle}{\underset{\tau}{}}\!\boldsymbol{l}_\bullet$	$\overset{\triangle}{\underset{\tau}{}}\!\boldsymbol{\ell}_\bullet$	$\overset{\triangle}{\underset{\tau}{}}\!\boldsymbol{\Lambda}_\bullet$	$\overset{\triangle}{\underset{\tau}{}}\!\boldsymbol{L}_\bullet$
$\overset{\triangle}{\underset{\tau}{}}\!\boldsymbol{l}_\bullet$	$\bullet$	$\overline{\boldsymbol{F}} \cdot \overset{\triangle}{\underset{\tau}{}}\!\boldsymbol{\ell}_\bullet : [\overline{\boldsymbol{f}} \boxtimes \overline{\boldsymbol{f}}]$	$\overset{\triangle}{\underset{\tau}{}}\!\boldsymbol{\Lambda}_\bullet : [\boldsymbol{f} \boxtimes \boldsymbol{f}]$	$\boldsymbol{F} \cdot \overset{\triangle}{\underset{\tau}{}}\!\boldsymbol{L}_\bullet : [\boldsymbol{f} \boxtimes \boldsymbol{f}]$
$\overset{\triangle}{\underset{\tau}{}}\!\boldsymbol{\ell}_\bullet$	$\overline{\boldsymbol{f}} \cdot \overset{\triangle}{\underset{\tau}{}}\!\boldsymbol{l}_\bullet : [\overline{\boldsymbol{F}} \boxtimes \overline{\boldsymbol{F}}]$	$\bullet$	$\overline{\boldsymbol{f}} \cdot \overset{\triangle}{\underset{\tau}{}}\!\boldsymbol{\Lambda}_\bullet : [\widetilde{\boldsymbol{f}} \boxtimes \widetilde{\boldsymbol{f}}]$	$\widetilde{\boldsymbol{F}} \cdot \overset{\triangle}{\underset{\tau}{}}\!\boldsymbol{L}_\bullet : [\widetilde{\boldsymbol{f}} \boxtimes \widetilde{\boldsymbol{f}}]$
$\overset{\triangle}{\underset{\tau}{}}\!\boldsymbol{\Lambda}_\bullet$	$\overset{\triangle}{\underset{\tau}{}}\!\boldsymbol{l}_\bullet : [\boldsymbol{F} \boxtimes \boldsymbol{F}]$	$\overline{\boldsymbol{F}} \cdot \overset{\triangle}{\underset{\tau}{}}\!\boldsymbol{\ell}_\bullet : [\widetilde{\boldsymbol{F}} \boxtimes \widetilde{\boldsymbol{F}}]$	$\bullet$	$\boldsymbol{F} \cdot \overset{\triangle}{\underset{\tau}{}}\!\boldsymbol{L}_\bullet$
$\overset{\triangle}{\underset{\tau}{}}\!\boldsymbol{L}_\bullet$	$\boldsymbol{f} \cdot \overset{\triangle}{\underset{\tau}{}}\!\boldsymbol{l}_\bullet : [\boldsymbol{F} \boxtimes \boldsymbol{F}]$	$\widetilde{\boldsymbol{f}} \cdot \overset{\triangle}{\underset{\tau}{}}\!\boldsymbol{\ell}_\bullet : [\widetilde{\boldsymbol{F}} \boxtimes \widetilde{\boldsymbol{F}}]$	$\boldsymbol{f} \cdot \overset{\triangle}{\underset{\tau}{}}\!\boldsymbol{\Lambda}_\bullet$	$\bullet$

Here the elastic and plastic material ${}_t$-type double-distortion rates follow as

$$\overset{\triangle}{\underset{t}{}}\!\boldsymbol{L}_\bullet = \overset{\triangle}{\underset{0}{}}\!\boldsymbol{L}_\bullet - \boldsymbol{L}_\bullet \cdot {}^{\triangle}\boldsymbol{\Gamma}.$$

Moreover the elastic and plastic intermediate ${}_t$-type double-distortion rates are expressed as

$$\overset{\triangle}{\underset{t}{}}\!\boldsymbol{\ell}_\bullet = \overset{\triangle}{\underset{0}{}}\!\boldsymbol{\ell}_\bullet + \boldsymbol{\ell}_\bullet \cdot {}^{\triangle}\boldsymbol{v}.$$

Finally the elastic and plastic spatial ${}_t$-type double-distortion rates read

$$\overset{\triangle}{\underset{t}{}}\!\boldsymbol{l}_\bullet = \overset{\triangle}{\underset{0}{}}\!\boldsymbol{l}_\bullet + \boldsymbol{l}_\bullet \cdot {}^{\triangle}\boldsymbol{\gamma}.$$

Note that the elastic and plastic ${}_\tau$-type double-distortion rates are in general completely independent from the elastic and plastic distortion rates and distortion gradient rates that emerge as a consequence of the multiplicative decomposition of the deformation gradient into elastic and plastic distortions.

Spatial Rate-of-Double-Deformation Tensors:

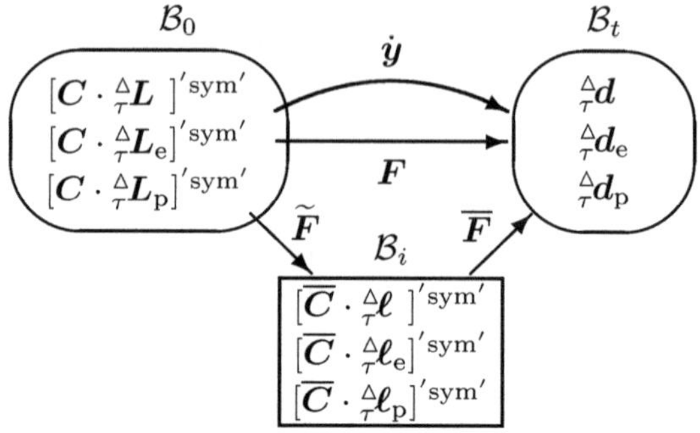

The 'quasi symmetric' contributions to the fully covariant versions of the elastic and plastic parts of the spatial velocity double-gradient are denoted the spatial *elastic* and *plastic rate-of-double-deformation tensor*, respectively, i.e.

$${}^{\triangle}_{\tau}\boldsymbol{d}_{\bullet} := [\boldsymbol{j}\cdot{}^{\triangle}_{\tau}\boldsymbol{l}_{\bullet}]^{'\mathrm{sym}'} \quad \text{with} \quad 2[\boldsymbol{j}\cdot{}^{\triangle}_{\tau}\boldsymbol{l}_{\bullet}]^{'\mathrm{sym}'} := \boldsymbol{j}\cdot{}^{\triangle}_{\tau}\boldsymbol{l}_{\bullet} + \boldsymbol{l}^{t}\cdot{}^{\triangle}\boldsymbol{j}_{\bullet}.$$

Accordingly the spatial (total) rate-of-double-deformation tensor ${}^{\triangle}_{\tau}\boldsymbol{d}$ decomposes additively into elastic and plastic contributions

$${}^{\triangle}_{\tau}\boldsymbol{d} = {}^{\triangle}_{\tau}\boldsymbol{d}_{\mathrm{e}} + {}^{\triangle}_{\tau}\boldsymbol{d}_{\mathrm{p}}.$$

Then the covariant elastic and total pull-backs of the spatial elastic and plastic rate-of-double-deformation tensors involve as metric the covariant $\overline{\boldsymbol{C}}$ and $\boldsymbol{C}$, respectively, and render

$$[\overline{\boldsymbol{C}}\cdot{}^{\triangle}_{\tau}\boldsymbol{\ell}_{\bullet}]^{'\mathrm{sym}'} = \overline{\boldsymbol{F}}^{t}\cdot{}^{\triangle}_{\tau}\boldsymbol{d}_{\bullet} : [\overline{\boldsymbol{F}}\boxtimes\overline{\boldsymbol{F}}] \quad \text{and} \quad [\boldsymbol{C}\cdot{}^{\triangle}_{\tau}\boldsymbol{L}_{\bullet}]^{'\mathrm{sym}'} = \boldsymbol{F}^{t}\cdot{}^{\triangle}_{\tau}\boldsymbol{d}_{\bullet} : [\boldsymbol{F}\boxtimes\boldsymbol{F}].$$

Thereby, 'quasi symmetric' contributions to $\overline{\boldsymbol{C}}\cdot{}^{\triangle}_{\tau}\boldsymbol{\ell}_{\bullet}$ and $\boldsymbol{C}\cdot{}^{\triangle}_{\tau}\boldsymbol{L}_{\bullet}$ follow as

$$2[\overline{\boldsymbol{C}}\cdot{}^{\triangle}_{\tau}\boldsymbol{\ell}_{\bullet}]^{'\mathrm{sym}'} = \overline{\boldsymbol{C}}\cdot{}^{\triangle}_{\tau}\boldsymbol{\ell}_{\bullet} + \boldsymbol{\ell}^{t}\cdot{}^{\triangle}\overline{\boldsymbol{C}}_{\bullet} \quad \text{and} \quad 2[\boldsymbol{C}\cdot{}^{\triangle}_{\tau}\boldsymbol{L}_{\bullet}]^{'\mathrm{sym}'} = \boldsymbol{C}\cdot{}^{\triangle}_{\tau}\boldsymbol{L}_{\bullet} + \boldsymbol{L}^{t}\cdot{}^{\triangle}\boldsymbol{C}_{\bullet}.$$

Note the relations between the ${}_0$- and ${}_t$-type material rate-of-double-deformation tensors

$$2[\boldsymbol{C}\cdot{}^{\triangle}_{0}\boldsymbol{L}_{\bullet}]^{'\mathrm{sym}'} = 2[\boldsymbol{C}\cdot{}^{\triangle}_{t}\boldsymbol{L}_{\bullet}]^{'\mathrm{sym}'} + \boldsymbol{C}\cdot\boldsymbol{L}_{\bullet}\cdot{}^{\triangle}\boldsymbol{\Gamma}$$

Observe finally that the covariant elastic and total pull-backs of the spatial (total) rate-of-double-deformation tensor ${}^{\triangle}_{\tau}\boldsymbol{d} := [\boldsymbol{j}\cdot{}^{\triangle}_{\tau}\boldsymbol{l}]^{'\mathrm{sym}'}$ lead to

$$[\overline{\boldsymbol{C}}\cdot{}^{\triangle}_{\tau}\boldsymbol{\ell}]^{'\mathrm{sym}'} = \overline{\boldsymbol{F}}^{t}\cdot{}^{\triangle}_{\tau}\boldsymbol{d} : [\overline{\boldsymbol{F}}\boxtimes\overline{\boldsymbol{F}}] \quad \text{and} \quad [\boldsymbol{C}\cdot{}^{\triangle}_{\tau}\boldsymbol{L}]^{'\mathrm{sym}'} = \boldsymbol{F}^{t}\cdot{}^{\triangle}_{\tau}\boldsymbol{d} : [\boldsymbol{F}\boxtimes\boldsymbol{F}].$$

In summary the elastic and total pull-backs of the covariant spatial (total) rate-of-double-deformation tensor also decompose additively into elastic and plastic contributions.

Material Rate-of-Double-Deformation Tensors:

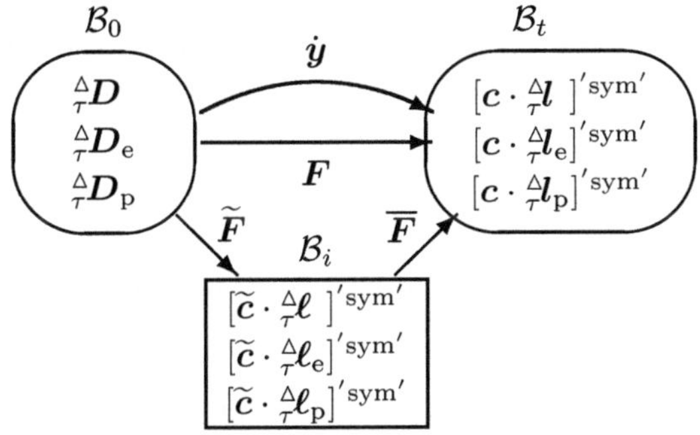

The 'quasi symmetric' contributions to the fully covariant versions of the elastic and plastic parts of the material velocity double-gradient are denoted the material *elastic* and *plastic rate-of-double-deformation tensor*, respectively, i.e.

$$ {}^{\triangle}_{\tau}\boldsymbol{D}_{\bullet} := [\boldsymbol{J}\cdot{}^{\triangle}_{\tau}\boldsymbol{L}_{\bullet}]^{'\mathrm{sym}'} \quad \text{with} \quad 2[\boldsymbol{J}\cdot{}^{\triangle}_{\tau}\boldsymbol{L}_{\bullet}]^{'\mathrm{sym}'} := \boldsymbol{J}\cdot{}^{\triangle}_{\tau}\boldsymbol{L}_{\bullet} + \boldsymbol{L}^{t}\cdot{}^{\triangle}\boldsymbol{J}_{\bullet}. $$

Accordingly the material (total) rate-of-double-deformation tensor ${}^{\triangle}_{\tau}\boldsymbol{D}$ decomposes additively into elastic and plastic contributions

$$ {}^{\triangle}_{\tau}\boldsymbol{D} = {}^{\triangle}_{\tau}\boldsymbol{D}_{\mathrm{e}} + {}^{\triangle}_{\tau}\boldsymbol{D}_{\mathrm{p}}. $$

Then the covariant plastic and total push-forwards of the material elastic and plastic rate-of-double-deformation tensors involve as metric the covariant $\widetilde{\boldsymbol{c}}$ and $\boldsymbol{c}$, respectively, and render

$$ [\widetilde{\boldsymbol{c}}\cdot{}^{\triangle}_{\tau}\boldsymbol{\ell}_{\bullet}]^{'\mathrm{sym}'} = \widetilde{\boldsymbol{f}}^{t}\cdot{}^{\triangle}_{\tau}\boldsymbol{D}_{\bullet} : [\widetilde{\boldsymbol{f}}\boxtimes\widetilde{\boldsymbol{f}}] \quad \text{and} \quad [\boldsymbol{c}\cdot{}^{\triangle}_{\tau}\boldsymbol{l}_{\bullet}]^{'\mathrm{sym}'} = \boldsymbol{f}^{t}\cdot{}^{\triangle}_{\tau}\boldsymbol{D}_{\bullet} : [\boldsymbol{f}\boxtimes\boldsymbol{f}]. $$

Thereby, 'quasi symmetric' contributions to $\widetilde{\boldsymbol{c}}\cdot{}^{\triangle}_{\tau}\boldsymbol{\ell}_{\bullet}$ and $\boldsymbol{c}\cdot{}^{\triangle}_{\tau}\boldsymbol{l}_{\bullet}$ follow as

$$ 2[\widetilde{\boldsymbol{c}}\cdot{}^{\triangle}_{\tau}\boldsymbol{\ell}_{\bullet}]^{'\mathrm{sym}'} = \widetilde{\boldsymbol{c}}\cdot{}^{\triangle}_{\tau}\boldsymbol{\ell}_{\bullet} + \boldsymbol{\ell}^{t}\cdot{}^{\triangle}\widetilde{\boldsymbol{c}}_{\bullet} \quad \text{and} \quad 2[\boldsymbol{c}\cdot{}^{\triangle}_{\tau}\boldsymbol{l}_{\bullet}]^{'\mathrm{sym}'} = \boldsymbol{c}\cdot{}^{\triangle}_{\tau}\boldsymbol{l}_{\bullet} + \boldsymbol{l}^{t}\cdot{}^{\triangle}\boldsymbol{c}_{\bullet}. $$

Note the relations between the ${}_{t}$- and ${}_{0}$-type spatial rate-of-double-deformation tensors

$$ 2[\boldsymbol{c}\cdot{}^{\triangle}_{t}\boldsymbol{l}_{\bullet}]^{'\mathrm{sym}'} = 2[\boldsymbol{c}\cdot{}^{\triangle}_{0}\boldsymbol{l}_{\bullet}]^{'\mathrm{sym}'} + \boldsymbol{c}\cdot\boldsymbol{l}_{\bullet}\cdot{}^{\triangle}\boldsymbol{\gamma}. $$

Observe finally that the covariant plastic and total push-forwards of the material (total) rate-of-double-deformation tensor ${}^{\triangle}_{\tau}\boldsymbol{D} := [\boldsymbol{J}\cdot{}^{\triangle}_{\tau}\boldsymbol{L}]^{'\mathrm{sym}'}$ lead to

$$ [\widetilde{\boldsymbol{c}}\cdot{}^{\triangle}_{\tau}\boldsymbol{\ell}]^{'\mathrm{sym}'} = \widetilde{\boldsymbol{f}}^{t}\cdot{}^{\triangle}_{\tau}\boldsymbol{D} : [\widetilde{\boldsymbol{f}}\boxtimes\widetilde{\boldsymbol{f}}] \quad \text{and} \quad [\boldsymbol{c}\cdot{}^{\triangle}_{\tau}\boldsymbol{l}]^{'\mathrm{sym}'} = \boldsymbol{f}^{t}\cdot{}^{\triangle}_{\tau}\boldsymbol{D} : [\boldsymbol{f}\boxtimes\boldsymbol{f}]. $$

In summary the plastic and total push-forwards of the covariant material (total) rate-of-double-deformation tensor also decompose additively into elastic and plastic contributions.

Intermediate Rate-of-Double-Deformation Tensors:
The 'quasi symmetric' and 'quasi skew symmetric' contributions to the fully covariant versions of the elastic and plastic parts of the intermediate velocity double-gradient are denoted the intermediate *elastic* and *plastic rate-of-double-deformation tensors*, respectively

$${}^{\triangle}_{\tau}\overline{\boldsymbol{D}}_{\bullet} := [\overline{\boldsymbol{J}}\cdot{}^{\triangle}_{\tau}\boldsymbol{\ell}_{\bullet}]^{'\text{sym}'} = [\widetilde{\boldsymbol{j}}\cdot{}^{\triangle}_{\tau}\boldsymbol{\ell}_{\bullet}]^{'\text{skw}'} \quad \text{and} \quad {}^{\triangle}_{\tau}\widetilde{\boldsymbol{d}}_{\bullet} := [\widetilde{\boldsymbol{j}}\cdot{}^{\triangle}_{\tau}\boldsymbol{\ell}_{\bullet}]^{'\text{sym}'} = [\overline{\boldsymbol{J}}\cdot{}^{\triangle}_{\tau}\boldsymbol{\ell}_{\bullet}]^{'\text{skw}'}.$$

Thereby, 'quasi symmetric' and 'quasi skew symmetric' contributions to either $\overline{\boldsymbol{J}}\cdot{}^{\triangle}_{\tau}\boldsymbol{\ell}_{\bullet}$ or $\widetilde{\boldsymbol{j}}\cdot{}^{\triangle}_{\tau}\boldsymbol{\ell}_{\bullet}$ follow as

$$2[\overline{\boldsymbol{J}}\cdot{}^{\triangle}_{\tau}\boldsymbol{\ell}_{\bullet}]^{'\text{sym}'}_{'\text{skw}'} = \overline{\boldsymbol{J}}\cdot{}^{\triangle}_{\tau}\boldsymbol{\ell}_{\bullet} \pm \boldsymbol{\ell}^{t}\cdot{}^{\triangle}\overline{\boldsymbol{J}}_{\text{e}} \quad \text{and} \quad 2[\widetilde{\boldsymbol{j}}\cdot{}^{\triangle}_{\tau}\boldsymbol{\ell}_{\bullet}]^{'\text{sym}'}_{'\text{skw}'} = \widetilde{\boldsymbol{j}}\cdot{}^{\triangle}_{\tau}\boldsymbol{\ell}_{\bullet} \pm \boldsymbol{\ell}^{t}\cdot{}^{\triangle}\widetilde{\boldsymbol{j}}_{\text{p}}.$$

The introduction of two versions of intermediate rate-of-double-deformation tensors that are based on either the 'quasi symmetric' or the 'quasi skew symmetric' contributions to the fully covariant versions of the elastic and plastic parts of the intermediate velocity double-gradient is possible since in the intermediate configuration $\widetilde{\boldsymbol{j}} = \overline{\boldsymbol{J}}$ but $\pm{}^{\triangle}\widetilde{\boldsymbol{j}}_{\bullet} = \mp{}^{\triangle}\overline{\boldsymbol{J}}_{\bullet}$!

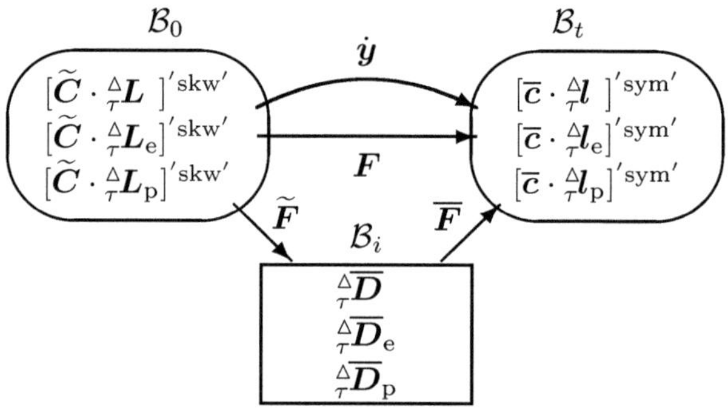

Furthermore the intermediate (total) rate-of-double-deformation tensors are introduced as the sum of the elastic and plastic contributions, i.e.

$${}^{\triangle}_{\tau}\overline{\boldsymbol{D}} := {}^{\triangle}_{\tau}\overline{\boldsymbol{D}}_{\text{e}} + {}^{\triangle}_{\tau}\overline{\boldsymbol{D}}_{\text{p}} \quad \text{and} \quad {}^{\triangle}_{\tau}\widetilde{\boldsymbol{d}} := {}^{\triangle}_{\tau}\widetilde{\boldsymbol{d}}_{\text{e}} + {}^{\triangle}_{\tau}\widetilde{\boldsymbol{d}}_{\text{p}}.$$

Then the covariant plastic pull-back and the covariant elastic push-forward of the intermediate elastic and plastic rate-of-double-deformation tensors involve as metric the covariant $\widetilde{\boldsymbol{C}}$ and $\overline{\boldsymbol{c}}$, respectively, and render

$$[\widetilde{\boldsymbol{C}}\cdot{}^{\triangle}_{\tau}\boldsymbol{L}_{\bullet}]^{'\text{skw}'} = \widetilde{\boldsymbol{F}}^{t}\cdot{}^{\triangle}_{\tau}\overline{\boldsymbol{D}}_{\bullet} : [\widetilde{\boldsymbol{F}}\boxtimes\widetilde{\boldsymbol{F}}] \quad \text{and} \quad [\overline{\boldsymbol{c}}\cdot{}^{\triangle}_{\tau}\boldsymbol{l}_{\bullet}]^{'\text{sym}'} = \overline{\boldsymbol{f}}^{t}\cdot{}^{\triangle}_{\tau}\overline{\boldsymbol{D}}_{\bullet} : [\overline{\boldsymbol{f}}\boxtimes\overline{\boldsymbol{f}}].$$

Likewise the covariant plastic pull-back and the covariant elastic push-forward of the intermediate elastic and plastic rate-of-double-deformation tensors result in

$$[\widetilde{\boldsymbol{C}}\cdot{}^{\triangle}_{\tau}\boldsymbol{L}_{\bullet}]^{'\text{sym}'} = \widetilde{\boldsymbol{F}}^{t}\cdot{}^{\triangle}_{\tau}\widetilde{\boldsymbol{d}}_{\bullet} : [\widetilde{\boldsymbol{F}}\boxtimes\widetilde{\boldsymbol{F}}] \quad \text{and} \quad [\overline{\boldsymbol{c}}\cdot{}^{\triangle}_{\tau}\boldsymbol{l}_{\bullet}]^{'\text{skw}'} = \overline{\boldsymbol{f}}^{t}\cdot{}^{\triangle}_{\tau}\widetilde{\boldsymbol{d}}_{\bullet} : [\overline{\boldsymbol{f}}\boxtimes\overline{\boldsymbol{f}}].$$

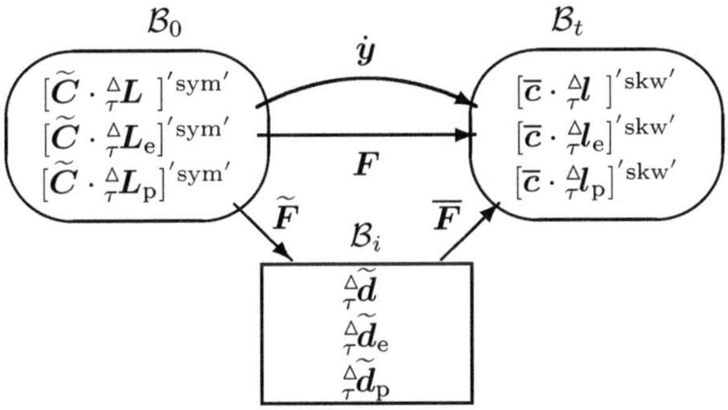

Thereby, 'quasi symmetric' and 'quasi skew symmetric' contributions to either $\widetilde{\boldsymbol{C}} \cdot {}_{T}^{\Delta}\boldsymbol{L}_{\bullet}$ or $\overline{\boldsymbol{c}} \cdot {}_{T}^{\Delta}\boldsymbol{l}_{\bullet}$ follow as

$$2[\widetilde{\boldsymbol{C}} \cdot {}_{T}^{\Delta}\boldsymbol{L}_{\bullet}]^{'\mathrm{sym}'}_{'\mathrm{skw}'} = \widetilde{\boldsymbol{C}} \cdot {}_{T}^{\Delta}\boldsymbol{L}_{\bullet} \pm \boldsymbol{L}^{t} \cdot {}^{\Delta}\widetilde{\boldsymbol{C}}_{\bullet} \quad \text{and} \quad 2[\overline{\boldsymbol{c}} \cdot {}_{T}^{\Delta}\boldsymbol{l}_{\bullet}]^{'\mathrm{sym}'}_{'\mathrm{skw}'} = \overline{\boldsymbol{c}} \cdot {}_{T}^{\Delta}\boldsymbol{l}_{\bullet} \pm \boldsymbol{l}^{t} \cdot {}^{\Delta}\overline{\boldsymbol{c}}_{\bullet}.$$

Observe finally that the covariant plastic pull-back and the covariant elastic push-forward of the intermediate (total) rate-of-double-deformation tensors ${}_{T}^{\Delta}\overline{\boldsymbol{D}} = [\overline{\boldsymbol{J}} \cdot {}_{T}^{\Delta}\boldsymbol{\ell}]^{'\mathrm{sym}'} = [\widetilde{\boldsymbol{j}} \cdot {}_{T}^{\Delta}\boldsymbol{\ell}]^{'\mathrm{skw}'}$ and ${}_{T}^{\Delta}\widetilde{\boldsymbol{d}} = [\widetilde{\boldsymbol{j}} \cdot {}_{T}^{\Delta}\boldsymbol{\ell}]^{'\mathrm{sym}'} = [\overline{\boldsymbol{J}} \cdot {}_{T}^{\Delta}\boldsymbol{\ell}]^{'\mathrm{skw}'}$ lead to

$$[\widetilde{\boldsymbol{C}} \cdot {}_{T}^{\Delta}\boldsymbol{L}]^{'\mathrm{skw}'} = \widetilde{\boldsymbol{F}}^{t} \cdot {}_{T}^{\Delta}\overline{\boldsymbol{D}} : [\widetilde{\boldsymbol{F}} \boxtimes \widetilde{\boldsymbol{F}}] \quad \text{and} \quad [\overline{\boldsymbol{c}} \cdot {}_{T}^{\Delta}\boldsymbol{l}]^{'\mathrm{sym}'} = \overline{\boldsymbol{f}}^{t} \cdot {}_{T}^{\Delta}\overline{\boldsymbol{D}} : [\overline{\boldsymbol{f}} \boxtimes \overline{\boldsymbol{f}}].$$

$$[\widetilde{\boldsymbol{C}} \cdot {}_{T}^{\Delta}\boldsymbol{L}]^{'\mathrm{sym}'} = \widetilde{\boldsymbol{F}}^{t} \cdot {}_{T}^{\Delta}\widetilde{\boldsymbol{d}} \;: [\widetilde{\boldsymbol{F}} \boxtimes \widetilde{\boldsymbol{F}}] \quad \text{and} \quad [\overline{\boldsymbol{c}} \cdot {}_{T}^{\Delta}\boldsymbol{l}]^{'\mathrm{skw}'} = \overline{\boldsymbol{f}}^{t} \cdot {}_{T}^{\Delta}\widetilde{\boldsymbol{d}} \;: [\overline{\boldsymbol{f}} \boxtimes \overline{\boldsymbol{f}}].$$

In summary the covariant intermediate (total) rate-of-double-deformation tensors (and their plastic pull-back and elastic push-forward) decompose additively into elastic and plastic contributions.

Left Lie Time Derivatives: Spatial & Intermediate Connections: Recall that the (total) *left Lie time derivative* L_t of generic *spatial* tensor fields $\{\bullet\}$ was previously defined as

$$\mathrm{L}_t\{\bullet\} := \text{push-forward}\Big(\mathrm{D}_t\big(\text{pull-back of } \{\bullet\}\big)\Big).$$

Applying this definition to the spatial description third-order contra/co/co-(mixed) variant (total, elastic and plastic) connection tensors ${}^{\Delta}\boldsymbol{\gamma}_{\bullet}$ renders

$$\mathrm{L}_t{}^{\Delta}\boldsymbol{\gamma}_{\bullet} = -\boldsymbol{F} \cdot \mathrm{D}_t{}^{\Delta}\boldsymbol{\Gamma}_{\bullet} : [\boldsymbol{f} \boxtimes \boldsymbol{f}].$$

Thereby the relation between the left Lie and the material time derivative of the (total, elastic and plastic) ${}^{\Delta}\boldsymbol{\gamma}_{\bullet}$ holds as

$$\mathrm{L}_t{}^{\Delta}\boldsymbol{\gamma}_{\bullet} = \mathrm{D}_t{}^{\Delta}\boldsymbol{\gamma}_{\bullet} + {}^{\Delta}\boldsymbol{\gamma}_{\bullet} : \mathsf{I} - \boldsymbol{l} \cdot {}^{\Delta}\boldsymbol{\gamma}_{\bullet}.$$

Detailing these relations to the spatial total, elastic and plastic connection tensors (with $\boldsymbol{\xi} := \boldsymbol{l}_{\mathrm{e}} \cdot {}^{\Delta}\boldsymbol{\gamma}_{\mathrm{p}} - \boldsymbol{l}_{\mathrm{p}} \cdot {}^{\Delta}\boldsymbol{\gamma}_{\mathrm{e}}$) results in

$$\mathrm{L}_t{}^{\Delta}\boldsymbol{\gamma} = -{}^{\Delta}_{t}\boldsymbol{l} \quad \text{and} \quad \mathrm{L}_t{}^{\Delta}\boldsymbol{\gamma}_{\mathrm{e}} = -{}^{\Delta}_{t}\boldsymbol{l}_{\mathrm{e}} + \boldsymbol{\xi} \quad \text{and} \quad \mathrm{L}_t{}^{\Delta}\boldsymbol{\gamma}_{\mathrm{p}} = -{}^{\Delta}_{t}\boldsymbol{l}_{\mathrm{p}} - \boldsymbol{\xi}.$$

Furthermore the *elastic left Lie time derivative* $\overline{\mathrm{L}}_t$ of generic *spatial* tensor fields $\{\bullet\}$ was previously defined as

$$\overline{\mathrm{L}}_t\{\bullet\} := \text{elastic push-forward}\Big(\mathrm{D}_t(\text{elastic pull-back of }\{\bullet\})\Big).$$

Applying this definition to the spatial description third-order contra/co/co-(mixed) variant (total, elastic and plastic) connection tensors ${}^{\Delta}\boldsymbol{\gamma}_{\bullet}$ renders

$$\overline{\mathrm{L}}_t{}^{\Delta}\boldsymbol{\gamma}_{\bullet} = -\overline{\boldsymbol{F}} \cdot \mathrm{D}_t{}^{\Delta}\boldsymbol{\Upsilon}_{\bullet} : [\overline{\boldsymbol{f}} \boxtimes \overline{\boldsymbol{f}}].$$

Thereby the relation between the elastic left Lie and the material time derivative of the (total, elastic and plastic) ${}^{\Delta}\boldsymbol{\gamma}_{\bullet}$ holds as

$$\overline{\mathrm{L}}_t{}^{\Delta}\boldsymbol{\gamma}_{\bullet} = \mathrm{D}_t{}^{\Delta}\boldsymbol{\gamma}_{\bullet} + {}^{\Delta}\boldsymbol{\gamma}_{\bullet} : \mathsf{l}_{\mathrm{e}} - \boldsymbol{l}_{\mathrm{e}} \cdot {}^{\Delta}\boldsymbol{\gamma}_{\bullet}.$$

Accordingly, the relation between the (total) left Lie and the elastic left Lie time derivative of the (total, elastic and plastic) ${}^{\Delta}\boldsymbol{\gamma}_{\bullet}$ follows as

$$\mathrm{L}_t{}^{\Delta}\boldsymbol{\gamma}_{\bullet} = \overline{\mathrm{L}}_t{}^{\Delta}\boldsymbol{\gamma}_{\bullet} + {}^{\Delta}\boldsymbol{\gamma}_{\bullet} : \mathsf{l}_{\mathrm{p}} - \boldsymbol{l}_{\mathrm{p}} \cdot {}^{\Delta}\boldsymbol{\gamma}_{\bullet}.$$

Finally the *plastic left Lie time derivative* $\widetilde{\mathrm{L}}_t$ of generic *intermediate* tensor fields $\{\bullet\}$ was previously defined as

$$\widetilde{\mathrm{L}}_t\{\bullet\} := \text{plastic push-forward}\Big(\mathrm{D}_t(\text{plastic pull-back of }\{\bullet\})\Big).$$

Applying this definition to the intermediate description third-order contra/co/co-(mixed) variant (total, elastic and plastic) connection tensors ${}^{\Delta}\boldsymbol{v}_{\bullet}$ renders

$$\widetilde{\mathrm{L}}_t{}^{\Delta}\boldsymbol{v}_{\bullet} = -\widetilde{\boldsymbol{F}} \cdot \mathrm{D}_t{}^{\Delta}\boldsymbol{\Gamma}_{\bullet} : [\widetilde{\boldsymbol{f}} \boxtimes \widetilde{\boldsymbol{f}}].$$

Thereby the relation between the plastic left Lie and the material time derivative of the (total, elastic and plastic) ${}^{\Delta}\boldsymbol{v}_{\bullet}$ holds as

$$\widetilde{\mathrm{L}}_t{}^{\Delta}\boldsymbol{v}_{\bullet} = \mathrm{D}_t{}^{\Delta}\boldsymbol{v}_{\bullet} + {}^{\Delta}\boldsymbol{v}_{\bullet} : \widetilde{\mathsf{l}} - \widetilde{\boldsymbol{l}} \cdot {}^{\Delta}\boldsymbol{v}_{\bullet}.$$

Detailing these relations to the intermediate total, elastic and plastic connection tensors (with $\boldsymbol{\psi} := \boldsymbol{\ell}_{\mathrm{e}} \cdot {}^{\Delta}\boldsymbol{v}_{\mathrm{p}} - \boldsymbol{\ell}_{\mathrm{p}} \cdot {}^{\Delta}\boldsymbol{v}_{\mathrm{e}}$) results in

$$\widetilde{\mathrm{L}}_t{}^{\Delta}\boldsymbol{v} = -{}^{\Delta}_{t}\boldsymbol{\ell} \quad \text{and} \quad \widetilde{\mathrm{L}}_t{}^{\Delta}\boldsymbol{v}_{\mathrm{e}} = -{}^{\Delta}_{t}\boldsymbol{\ell}_{\mathrm{e}} + \boldsymbol{\psi} \quad \text{and} \quad \widetilde{\mathrm{L}}_t{}^{\Delta}\boldsymbol{v}_{\mathrm{p}} = -{}^{\Delta}_{t}\boldsymbol{\ell}_{\mathrm{p}} - \boldsymbol{\psi}.$$

It is noted that the *material time derivative* D_t of the *material* total, elastic and plastic connection tensors (with $\boldsymbol{\Xi} := \boldsymbol{L}_{\mathrm{e}} \cdot {}^{\Delta}\boldsymbol{\Gamma}_{\mathrm{p}} - \boldsymbol{L}_{\mathrm{p}} \cdot {}^{\Delta}\boldsymbol{\Gamma}_{\mathrm{e}}$) expands as

$$\mathrm{D}_t{}^{\Delta}\boldsymbol{\Gamma} = {}^{\Delta}_{t}\boldsymbol{L} \quad \text{and} \quad \mathrm{D}_t{}^{\Delta}\boldsymbol{\Gamma}_{\mathrm{e}} = {}^{\Delta}_{t}\boldsymbol{L}_{\mathrm{e}} + \boldsymbol{\Xi} \quad \text{and} \quad \mathrm{D}_t{}^{\Delta}\boldsymbol{\Gamma}_{\mathrm{p}} = {}^{\Delta}_{t}\boldsymbol{L}_{\mathrm{p}} - \boldsymbol{\Xi}.$$

Remarkably, the concise representations for the various time derivatives of the connection tensors in the above rely on the ${}_t$-type double-distortion rates, however the corresponding expressions in terms of the ${}_0$-type double-distortion rates are straightforward.

Right Lie Time Derivatives: Material & Intermediate Connections: Recall that the (total) *right Lie time derivative* R_t of generic *material* tensor fields $\{\bullet\}$ was previously defined as

$$\mathrm{R}_t\{\bullet\} := \text{pull-back}\Big(\,\mathrm{D}_t(\text{push-forward of }\{\bullet\})\Big).$$

Applying this definition to the material description third-order contra/co/co-(mixed) variant (total, elastic and plastic) connection tensors ${}^{\Delta}\boldsymbol{\Gamma}_{\bullet}$ renders

$$\mathrm{R}_t{}^{\Delta}\boldsymbol{\Gamma}_{\bullet} = -\boldsymbol{f}\cdot \mathrm{D}_t{}^{\Delta}\boldsymbol{\gamma}_{\bullet} : [\boldsymbol{F}\boxtimes\boldsymbol{F}].$$

Thereby the relation between the right Lie and the material time derivative of the (total, elastic and plastic) ${}^{\Delta}\boldsymbol{\Gamma}_{\bullet}$ holds as

$$\mathrm{R}_t{}^{\Delta}\boldsymbol{\Gamma}_{\bullet} = \mathrm{D}_t{}^{\Delta}\boldsymbol{\Gamma}_{\bullet} - {}^{\Delta}\boldsymbol{\Gamma}_{\bullet} : \mathsf{L} + \boldsymbol{L}\cdot{}^{\Delta}\boldsymbol{\Gamma}_{\bullet}.$$

Furthermore the *plastic right Lie time derivative* $\widetilde{\mathrm{R}}_t$ of generic *material* tensor fields $\{\bullet\}$ was previously defined as

$$\widetilde{\mathrm{R}}_t\{\bullet\} := \text{plastic pull-back}\Big(\,\mathrm{D}_t(\text{plastic push-forward of }\{\bullet\})\Big).$$

Applying this definition to the material description third-order contra/co/co-(mixed) variant (total, elastic and plastic) connection tensors ${}^{\Delta}\boldsymbol{\Gamma}_{\bullet}$ renders

$$\widetilde{\mathrm{R}}_t{}^{\Delta}\boldsymbol{\Gamma}_{\bullet} = -\widetilde{\boldsymbol{f}}\cdot \mathrm{D}_t{}^{\Delta}\boldsymbol{v}_{\bullet} : [\widetilde{\boldsymbol{F}}\boxtimes\widetilde{\boldsymbol{F}}].$$

Thereby the relation between the plastic right Lie and the material time derivative of the (total, elastic and plastic) ${}^{\Delta}\boldsymbol{\Gamma}_{\bullet}$ holds as

$$\widetilde{\mathrm{R}}_t{}^{\Delta}\boldsymbol{\Gamma}_{\bullet} = \mathrm{D}_t{}^{\Delta}\boldsymbol{\Gamma}_{\bullet} - {}^{\Delta}\boldsymbol{\Gamma}_{\bullet} : \mathsf{L}_{\mathrm{p}} + \boldsymbol{L}_{\mathrm{p}}\cdot{}^{\Delta}\boldsymbol{\Gamma}_{\bullet}.$$

Accordingly, the relation between the (total) right Lie and the plastic left Lie time derivative of the (total, elastic and plastic) ${}^{\Delta}\boldsymbol{\Gamma}_{\bullet}$ follows as

$$\mathrm{R}_t{}^{\Delta}\boldsymbol{\Gamma}_{\bullet} = \widetilde{\mathrm{R}}_t{}^{\Delta}\boldsymbol{\Gamma}_{\bullet} - {}^{\Delta}\boldsymbol{\Gamma}_{\bullet} : \mathsf{L}_{\mathrm{e}} + \boldsymbol{L}_{\mathrm{e}}\cdot{}^{\Delta}\boldsymbol{\Gamma}_{\bullet}.$$

Finally the *elastic right Lie time derivative* $\overline{\mathrm{R}}_t$ of generic *intermediate* tensor fields $\{\bullet\}$ was previously defined as

$$\overline{\mathrm{R}}_t\{\bullet\} := \text{elastic pull-back}\Big(\,\mathrm{D}_t(\text{elastic push-forward of }\{\bullet\})\Big).$$

Applying this definition to the intermediate description third-order contra/co/co- (mixed) variant (total, elastic and plastic) connection tensors ${}^{\Delta}\boldsymbol{\Upsilon}_{\bullet}$ renders

$$\overline{\mathrm{R}}_t{}^{\Delta}\boldsymbol{\Upsilon}_{\bullet} = -\overline{\boldsymbol{f}} \cdot \mathrm{D}_t{}^{\Delta}\boldsymbol{\gamma}_{\bullet} : [\overline{\boldsymbol{F}} \boxtimes \overline{\boldsymbol{F}}].$$

Thereby the relation between the elastic right Lie and the material time derivative of the (total, elastic and plastic) ${}^{\Delta}\boldsymbol{\Upsilon}_{\bullet}$ holds as

$$\overline{\mathrm{R}}_t{}^{\Delta}\boldsymbol{\Upsilon}_{\bullet} = \mathrm{D}_t{}^{\Delta}\boldsymbol{\Upsilon}_{\bullet} - {}^{\Delta}\boldsymbol{\Upsilon}_{\bullet} : \overline{\mathsf{L}} + \overline{\boldsymbol{L}} \cdot {}^{\Delta}\boldsymbol{\Upsilon}_{\bullet}.$$

Moreover the relation between the plastic left and the elastic right Lie time derivative of the (total, elastic and plastic) ${}^{\Delta}\boldsymbol{\Upsilon}_{\bullet} \equiv -{}^{\Delta}\boldsymbol{\upsilon}_{\bullet}$ results in

$$\widetilde{\mathrm{L}}_t{}^{\Delta}\boldsymbol{\Upsilon}_{\bullet} - {}^{\Delta}\boldsymbol{\Upsilon}_{\bullet} : \tilde{\mathsf{l}} + \tilde{\boldsymbol{l}} \cdot {}^{\Delta}\boldsymbol{\Upsilon}_{\bullet} = \overline{\mathrm{R}}_t{}^{\Delta}\boldsymbol{\Upsilon}_{\bullet} + {}^{\Delta}\boldsymbol{\Upsilon}_{\bullet} : \overline{\mathsf{L}} - \overline{\boldsymbol{L}} \cdot {}^{\Delta}\boldsymbol{\Upsilon}_{\bullet}.$$

Note that all right Lie time derivatives in the above may easily be further detailed based on either the left Lie or the material time derivatives of the (total, elastic and plastic) connection tensors that are in turn expressible in terms of either the $_t$- or the $_0$-type double-distortion rates.

Time Derivatives: Total, Elastic, Plastic Double-Strain Measures: Based on the material and the (left and right) Lie time derivatives of the (total, elastic and plastic) strain measures and connection tensors the corresponding time derivatives of the (total, elastic and plastic) double-strain measures may be computed:
The material time derivatives of the material double-strains ${}^{\Delta}\boldsymbol{C}_{\bullet} = \boldsymbol{C} \cdot {}^{\Delta}\boldsymbol{\Gamma}_{\bullet}$ follow as

$$\begin{aligned}
\mathrm{D}_t{}^{\Delta}\boldsymbol{C} &= 2[\boldsymbol{C} \cdot \boldsymbol{L}]^{\mathrm{sym}} \cdot {}^{\Delta}\boldsymbol{\Gamma} + \boldsymbol{C} \cdot {}^{\Delta}_{t}\boldsymbol{L} \\
\mathrm{D}_t{}^{\Delta}\boldsymbol{C}_{\mathrm{e}} &= 2[\boldsymbol{C} \cdot \boldsymbol{L}]^{\mathrm{sym}} \cdot {}^{\Delta}\boldsymbol{\Gamma}_{\mathrm{e}} + \boldsymbol{C} \cdot [{}^{\Delta}_{t}\boldsymbol{L}_{\mathrm{e}} + \boldsymbol{\Xi}] \\
\mathrm{D}_t{}^{\Delta}\boldsymbol{C}_{\mathrm{p}} &= 2[\boldsymbol{C} \cdot \boldsymbol{L}]^{\mathrm{sym}} \cdot {}^{\Delta}\boldsymbol{\Gamma}_{\mathrm{p}} + \boldsymbol{C} \cdot [{}^{\Delta}_{t}\boldsymbol{L}_{\mathrm{p}} - \boldsymbol{\Xi}].
\end{aligned}$$

Interestingly the expressions for the material time derivatives of the material double-strains ${}^{\Delta}\boldsymbol{C}_{\bullet}$ may be rearranged into a format that resembles the material time derivative $\mathrm{D}_t\boldsymbol{C} = 2[\boldsymbol{C} \cdot \boldsymbol{L}]^{\mathrm{sym}}$ of the right Cauchy-Green strain

$$\begin{aligned}
\mathrm{D}_t{}^{\Delta}\boldsymbol{C} &= 2[\boldsymbol{C} \cdot {}^{\Delta}_{0}\boldsymbol{L}\ ]^{'\mathrm{sym}'} = 2\boldsymbol{F}^t \cdot {}^{\Delta}_{0}\boldsymbol{d}\ : [\boldsymbol{F} \boxtimes \boldsymbol{F}] \\
\mathrm{D}_t{}^{\Delta}\boldsymbol{C}_{\mathrm{e}} &= 2[\boldsymbol{C} \cdot {}^{\Delta}_{0}\boldsymbol{L}_{\mathrm{e}}]^{'\mathrm{sym}'} = 2\boldsymbol{F}^t \cdot {}^{\Delta}_{0}\boldsymbol{d}_{\mathrm{e}} : [\boldsymbol{F} \boxtimes \boldsymbol{F}] \\
\mathrm{D}_t{}^{\Delta}\boldsymbol{C}_{\mathrm{p}} &= 2[\boldsymbol{C} \cdot {}^{\Delta}_{0}\boldsymbol{L}_{\mathrm{p}}]^{'\mathrm{sym}'} = 2\boldsymbol{F}^t \cdot {}^{\Delta}_{0}\boldsymbol{d}_{\mathrm{p}} : [\boldsymbol{F} \boxtimes \boldsymbol{F}].
\end{aligned}$$

Moreover the $\mathrm{D}_t{}^{\Delta}\boldsymbol{C}_{\bullet}$ are the pull-backs of twice the corresponding spatial rate-of-double-deformation tensors ${}^{\Delta}_{0}\boldsymbol{d}_{\bullet}$. The material time derivatives of the material double-strains ${}^{\Delta}\widetilde{\boldsymbol{C}}_{\bullet} = \widetilde{\boldsymbol{C}} \cdot {}^{\Delta}\boldsymbol{\Gamma}_{\bullet}$ read

$$\begin{aligned}
\mathrm{D}_t{}^{\Delta}\widetilde{\boldsymbol{C}} &= 2[\widetilde{\boldsymbol{C}}\cdot\boldsymbol{L}_{\mathrm{p}}]^{\mathrm{sym}}\cdot{}^{\Delta}\boldsymbol{\Gamma} + \widetilde{\boldsymbol{C}}\cdot{}^{\Delta}_{t}\boldsymbol{L}\\
\mathrm{D}_t{}^{\Delta}\widetilde{\boldsymbol{C}}_{\mathrm{e}} &= 2[\widetilde{\boldsymbol{C}}\cdot\boldsymbol{L}_{\mathrm{p}}]^{\mathrm{sym}}\cdot{}^{\Delta}\boldsymbol{\Gamma}_{\mathrm{e}} + \widetilde{\boldsymbol{C}}\cdot[{}^{\Delta}_{t}\boldsymbol{L}_{\mathrm{e}}+\boldsymbol{\Xi}]\\
\mathrm{D}_t{}^{\Delta}\widetilde{\boldsymbol{C}}_{\mathrm{p}} &= 2[\widetilde{\boldsymbol{C}}\cdot\boldsymbol{L}_{\mathrm{p}}]^{\mathrm{sym}}\cdot{}^{\Delta}\boldsymbol{\Gamma}_{\mathrm{p}} + \widetilde{\boldsymbol{C}}\cdot[{}^{\Delta}_{t}\boldsymbol{L}_{\mathrm{p}}-\boldsymbol{\Xi}].
\end{aligned}$$

The material time derivatives of the material double-metrics ${}^{\Delta}\boldsymbol{J}_{\bullet} = -\boldsymbol{J}\cdot{}^{\Delta}\boldsymbol{\Gamma}_{\bullet}$ expand into

$$\begin{aligned}
\mathrm{D}_t{}^{\Delta}\boldsymbol{J} &= -\boldsymbol{J}\cdot{}^{\Delta}_{t}\boldsymbol{L}\\
\mathrm{D}_t{}^{\Delta}\boldsymbol{J}_{\mathrm{e}} &= -\boldsymbol{J}\cdot[{}^{\Delta}_{t}\boldsymbol{L}_{\mathrm{e}}+\boldsymbol{\Xi}]\\
\mathrm{D}_t{}^{\Delta}\boldsymbol{J}_{\mathrm{p}} &= -\boldsymbol{J}\cdot[{}^{\Delta}_{t}\boldsymbol{L}_{\mathrm{p}}-\boldsymbol{\Xi}].
\end{aligned}$$

The corresponding total and plastic right Lie time derivatives R_t and $\widetilde{\mathrm{R}}_t$ of the material double-strain measures may finally be computed from the generic expressions for R_t and $\widetilde{\mathrm{R}}_t$ applied to fully covariant third-order material tensor fields $\{\bullet\}$

$$\mathrm{R}_t\{\bullet\} = \widetilde{\mathrm{R}}_t\{\bullet\} - 2[\{\bullet\}:\mathsf{L}_{\mathrm{e}}]^{'\mathrm{sym}'} \quad \text{with} \quad \widetilde{\mathrm{R}}_t\{\bullet\} = \mathrm{D}_t\{\bullet\} - 2[\{\bullet\}:\mathsf{L}_{\mathrm{p}}]^{'\mathrm{sym}'}.$$

Here the following abbreviations have been introduced

$$2[\{\bullet\}:\mathsf{L}_{\bullet}]^{'\mathrm{sym}'} := \{\bullet\}:\mathsf{L}_{\bullet} + \boldsymbol{L}^{t}_{\bullet}\cdot\{\bullet\}.$$

The left Lie time derivatives of the spatial double-metrics ${}^{\Delta}\boldsymbol{j}_{\bullet} = -\boldsymbol{j}\cdot{}^{\Delta}\boldsymbol{\gamma}_{\bullet}$ render

$$\begin{aligned}
\mathrm{L}_t{}^{\Delta}\boldsymbol{j} &= -2[\boldsymbol{j}\cdot\boldsymbol{l}]^{\mathrm{sym}}\cdot{}^{\Delta}\boldsymbol{\gamma} + \boldsymbol{j}\cdot{}^{\Delta}_{t}\boldsymbol{l}\\
\mathrm{L}_t{}^{\Delta}\boldsymbol{j}_{\mathrm{e}} &= -2[\boldsymbol{j}\cdot\boldsymbol{l}]^{\mathrm{sym}}\cdot{}^{\Delta}\boldsymbol{\gamma}_{\mathrm{e}} + \boldsymbol{j}\cdot[{}^{\Delta}_{t}\boldsymbol{l}_{\mathrm{e}}-\boldsymbol{\xi}]\\
\mathrm{L}_t{}^{\Delta}\boldsymbol{j}_{\mathrm{p}} &= -2[\boldsymbol{j}\cdot\boldsymbol{l}]^{\mathrm{sym}}\cdot{}^{\Delta}\boldsymbol{\gamma}_{\mathrm{p}} + \boldsymbol{j}\cdot[{}^{\Delta}_{t}\boldsymbol{l}_{\mathrm{p}}+\boldsymbol{\xi}].
\end{aligned}$$

Interestingly the expressions for the left Lie time derivatives of the spatial double-metrics ${}^{\Delta}\boldsymbol{j}_{\bullet}$ may be rearranged into a format that resembles the left Lie time derivative $\mathrm{L}_t\boldsymbol{j} = 2[\boldsymbol{j}\cdot\boldsymbol{l}]^{\mathrm{sym}}$ of the covariant spatial metric

$$\begin{aligned}
\mathrm{L}_t{}^{\Delta}\boldsymbol{j} &= 2[\boldsymbol{j}\cdot{}^{\Delta}_{0}\boldsymbol{l}\;]^{'\mathrm{sym}'} = 2\,{}^{\Delta}_{0}\boldsymbol{d}\\
\mathrm{L}_t{}^{\Delta}\boldsymbol{j}_{\mathrm{e}} &= 2[\boldsymbol{j}\cdot{}^{\Delta}_{0}\boldsymbol{l}_{\mathrm{e}}]^{'\mathrm{sym}'} = 2\,{}^{\Delta}_{0}\boldsymbol{d}_{\mathrm{e}}\\
\mathrm{L}_t{}^{\Delta}\boldsymbol{j}_{\mathrm{p}} &= 2[\boldsymbol{j}\cdot{}^{\Delta}_{0}\boldsymbol{l}_{\mathrm{p}}]^{'\mathrm{sym}'} = 2\,{}^{\Delta}_{0}\boldsymbol{d}_{\mathrm{p}}.
\end{aligned}$$

Moreover the $\mathrm{L}_t{}^{\Delta}\boldsymbol{j}_{\bullet}$ coincide with twice the corresponding spatial rate-of-double-deformation tensors ${}^{\Delta}_{0}\boldsymbol{d}_{\bullet}$. The left Lie time derivatives of the spatial double-strains ${}^{\Delta}\overline{\boldsymbol{c}}_{\bullet} = \overline{\boldsymbol{c}}\cdot{}^{\Delta}\boldsymbol{\gamma}_{\bullet}$ read

$$\begin{aligned}
\mathrm{L}_t{}^{\Delta}\overline{\boldsymbol{c}} &= 2[\overline{\boldsymbol{c}}\cdot\boldsymbol{l}_{\mathrm{p}}]^{\mathrm{sym}}\cdot{}^{\Delta}\boldsymbol{\gamma} - \overline{\boldsymbol{c}}\cdot{}^{\Delta}_{t}\boldsymbol{l}\\
\mathrm{L}_t{}^{\Delta}\overline{\boldsymbol{c}}_{\mathrm{e}} &= 2[\overline{\boldsymbol{c}}\cdot\boldsymbol{l}_{\mathrm{p}}]^{\mathrm{sym}}\cdot{}^{\Delta}\boldsymbol{\gamma}_{\mathrm{e}} - \overline{\boldsymbol{c}}\cdot[{}^{\Delta}_{t}\boldsymbol{l}_{\mathrm{e}}-\boldsymbol{\xi}]\\
\mathrm{L}_t{}^{\Delta}\overline{\boldsymbol{c}}_{\mathrm{p}} &= 2[\overline{\boldsymbol{c}}\cdot\boldsymbol{l}_{\mathrm{p}}]^{\mathrm{sym}}\cdot{}^{\Delta}\boldsymbol{\gamma}_{\mathrm{p}} - \overline{\boldsymbol{c}}\cdot[{}^{\Delta}_{t}\boldsymbol{l}_{\mathrm{p}}+\boldsymbol{\xi}].
\end{aligned}$$

The left Lie time derivatives of the spatial double-strains ${}^{\Delta}\boldsymbol{c}_{\bullet} = \boldsymbol{c}\cdot{}^{\Delta}\boldsymbol{\gamma}_{\bullet}$ expand into

$$\begin{aligned} \mathrm{L}_t{}^{\Delta}\boldsymbol{c} &= -\boldsymbol{c}\cdot {}^{\Delta}_{t}\boldsymbol{l} \\ \mathrm{L}_t{}^{\Delta}\boldsymbol{c}_{\mathrm{e}} &= -\boldsymbol{c}\cdot [{}^{\Delta}_{t}\boldsymbol{l}_{\mathrm{e}} - \boldsymbol{\xi}] \\ \mathrm{L}_t{}^{\Delta}\boldsymbol{c}_{\mathrm{p}} &= -\boldsymbol{c}\cdot [{}^{\Delta}_{t}\boldsymbol{l}_{\mathrm{p}} + \boldsymbol{\xi}]. \end{aligned}$$

The corresponding material and elastic left Lie time derivatives D_t and $\overline{\mathrm{L}}_t$ of the spatial double-strain measures may finally be computed from the generic expressions for D_t and $\overline{\mathrm{L}}_t$ applied to fully covariant third-order spatial tensor fields $\{\bullet\}$

$$\mathrm{D}_t\{\bullet\} = \overline{\mathrm{L}}_t\{\bullet\} - 2[\{\bullet\} : \mathsf{l}_{\mathrm{e}}]^{'\mathrm{sym}'} \quad \text{with} \quad \overline{\mathrm{L}}_t\{\bullet\} = \mathrm{L}_t\{\bullet\} - 2[\{\bullet\} : \mathsf{l}_{\mathrm{p}}]^{'\mathrm{sym}'}.$$

Here the following abbreviations have been introduced

$$2[\{\bullet\} : \mathsf{l}_{\bullet}]^{'\mathrm{sym}'} := \{\bullet\} : \mathsf{l}_{\bullet} + \boldsymbol{l}_{\bullet}^{t}\cdot\{\bullet\}.$$

The plastic left Lie time derivatives of the intermediate double-strains ${}^{\Delta}\overline{\boldsymbol{C}}_{\bullet} = \overline{\boldsymbol{C}}\cdot{}^{\Delta}\boldsymbol{\Upsilon}_{\bullet}$ render

$$\begin{aligned} \widetilde{\mathrm{L}}_t{}^{\Delta}\overline{\boldsymbol{C}} &= 2[\overline{\boldsymbol{C}}\cdot\boldsymbol{\ell}]^{\mathrm{sym}}\cdot{}^{\Delta}\boldsymbol{\Upsilon} + \overline{\boldsymbol{C}}\cdot{}^{\Delta}_{t}\boldsymbol{\ell} \\ \widetilde{\mathrm{L}}_t{}^{\Delta}\overline{\boldsymbol{C}}_{\mathrm{e}} &= 2[\overline{\boldsymbol{C}}\cdot\boldsymbol{\ell}]^{\mathrm{sym}}\cdot{}^{\Delta}\boldsymbol{\Upsilon}_{\mathrm{e}} + \overline{\boldsymbol{C}}\cdot[{}^{\Delta}_{t}\boldsymbol{\ell}_{\mathrm{e}} + \boldsymbol{\Psi}] \\ \widetilde{\mathrm{L}}_t{}^{\Delta}\overline{\boldsymbol{C}}_{\mathrm{p}} &= 2[\overline{\boldsymbol{C}}\cdot\boldsymbol{\ell}]^{\mathrm{sym}}\cdot{}^{\Delta}\boldsymbol{\Upsilon}_{\mathrm{p}} + \overline{\boldsymbol{C}}\cdot[{}^{\Delta}_{t}\boldsymbol{\ell}_{\mathrm{p}} - \boldsymbol{\Psi}]. \end{aligned}$$

Interestingly the expressions for the plastic left Lie time derivatives of the intermediate double-strains ${}^{\Delta}\overline{\boldsymbol{C}}_{\bullet}$ may be rearranged into a format that resembles the plastic left Lie time derivative $\widetilde{\mathrm{L}}_t\overline{\boldsymbol{C}} = 2[\overline{\boldsymbol{C}}\cdot\boldsymbol{\ell}]^{\mathrm{sym}}$ of the elastic right Cauchy-Green strain

$$\begin{aligned} \widetilde{\mathrm{L}}_t{}^{\Delta}\overline{\boldsymbol{C}} &= 2[\overline{\boldsymbol{C}}\cdot{}^{\Delta}_{0}\boldsymbol{\ell}\,]^{'\mathrm{sym}'} = 2\overline{\boldsymbol{F}}^{t}\cdot{}^{\Delta}_{0}\boldsymbol{d} : [\overline{\boldsymbol{F}}\boxtimes\overline{\boldsymbol{F}}] \\ \widetilde{\mathrm{L}}_t{}^{\Delta}\overline{\boldsymbol{C}}_{\mathrm{e}} &= 2[\overline{\boldsymbol{C}}\cdot{}^{\Delta}_{0}\boldsymbol{\ell}_{\mathrm{e}}]^{'\mathrm{sym}'} = 2\overline{\boldsymbol{F}}^{t}\cdot{}^{\Delta}_{0}\boldsymbol{d}_{\mathrm{e}} : [\overline{\boldsymbol{F}}\boxtimes\overline{\boldsymbol{F}}] \\ \widetilde{\mathrm{L}}_t{}^{\Delta}\overline{\boldsymbol{C}}_{\mathrm{p}} &= 2[\overline{\boldsymbol{C}}\cdot{}^{\Delta}_{0}\boldsymbol{\ell}_{\mathrm{p}}]^{'\mathrm{sym}'} = 2\overline{\boldsymbol{F}}^{t}\cdot{}^{\Delta}_{0}\boldsymbol{d}_{\mathrm{p}} : [\overline{\boldsymbol{F}}\boxtimes\overline{\boldsymbol{F}}]. \end{aligned}$$

Moreover the $\widetilde{\mathrm{L}}_t{}^{\Delta}\overline{\boldsymbol{C}}_{\bullet}$ are the elastic pull-backs of twice the corresponding spatial rate-of-double-deformation tensors ${}^{\Delta}_{0}\boldsymbol{d}_{\bullet}$. The plastic left Lie time derivatives of the intermediate double-metrics ${}^{\Delta}\overline{\boldsymbol{J}}_{\bullet} = -\overline{\boldsymbol{J}}\cdot{}^{\Delta}\boldsymbol{\Upsilon}_{\bullet}$ read

$$\begin{aligned} \widetilde{\mathrm{L}}_t{}^{\Delta}\overline{\boldsymbol{J}} &= -2[\overline{\boldsymbol{J}}\cdot\boldsymbol{\ell}_{\mathrm{p}}]^{\mathrm{sym}}\cdot{}^{\Delta}\boldsymbol{\Upsilon} - \overline{\boldsymbol{J}}\cdot{}^{\Delta}_{t}\boldsymbol{\ell} \\ \widetilde{\mathrm{L}}_t{}^{\Delta}\overline{\boldsymbol{J}}_{\mathrm{e}} &= -2[\overline{\boldsymbol{J}}\cdot\boldsymbol{\ell}_{\mathrm{p}}]^{\mathrm{sym}}\cdot{}^{\Delta}\boldsymbol{\Upsilon}_{\mathrm{e}} - \overline{\boldsymbol{J}}\cdot[{}^{\Delta}_{t}\boldsymbol{\ell}_{\mathrm{e}} + \boldsymbol{\Psi}] \\ \widetilde{\mathrm{L}}_t{}^{\Delta}\overline{\boldsymbol{J}}_{\mathrm{p}} &= -2[\overline{\boldsymbol{J}}\cdot\boldsymbol{\ell}_{\mathrm{p}}]^{\mathrm{sym}}\cdot{}^{\Delta}\boldsymbol{\Upsilon}_{\mathrm{p}} - \overline{\boldsymbol{J}}\cdot[{}^{\Delta}_{t}\boldsymbol{\ell}_{\mathrm{p}} - \boldsymbol{\Psi}]. \end{aligned}$$

Alternatively, since $\pm{}^{\Delta}\widetilde{\boldsymbol{j}}_{\bullet} = \mp{}^{\Delta}\widetilde{\boldsymbol{J}}_{\bullet}$ with ${}^{\Delta}\widetilde{\boldsymbol{j}}_{\bullet} = -\widetilde{\boldsymbol{j}}\cdot{}^{\Delta}\boldsymbol{v}_{\bullet}$ (together with $\pm{}^{\Delta}\boldsymbol{v}_{\bullet} = \mp{}^{\Delta}\boldsymbol{\Upsilon}_{\bullet}$ and $\pm\boldsymbol{\psi} = \mp\boldsymbol{\Psi}$) the plastic left Lie time derivatives of the intermediate double-metrics take the expressions

$$
\begin{aligned}
\widetilde{\mathrm{L}}_t{}^{\Delta}\widetilde{\boldsymbol{j}} &= -2[\widetilde{\boldsymbol{j}}\cdot\boldsymbol{\ell}_{\mathrm{p}}]^{\mathrm{sym}}\cdot{}^{\Delta}\boldsymbol{v} + \widetilde{\boldsymbol{j}}\cdot{}^{\Delta}_{t}\boldsymbol{\ell}\\
\widetilde{\mathrm{L}}_t{}^{\Delta}\widetilde{\boldsymbol{j}}_{\mathrm{e}} &= -2[\widetilde{\boldsymbol{j}}\cdot\boldsymbol{\ell}_{\mathrm{p}}]^{\mathrm{sym}}\cdot{}^{\Delta}\boldsymbol{v}_{\mathrm{e}} + \widetilde{\boldsymbol{j}}\cdot[{}^{\Delta}_{t}\boldsymbol{\ell}_{\mathrm{e}} - \boldsymbol{\psi}]\\
\widetilde{\mathrm{L}}_t{}^{\Delta}\widetilde{\boldsymbol{j}}_{\mathrm{p}} &= -2[\widetilde{\boldsymbol{j}}\cdot\boldsymbol{\ell}_{\mathrm{p}}]^{\mathrm{sym}}\cdot{}^{\Delta}\boldsymbol{v}_{\mathrm{p}} + \widetilde{\boldsymbol{j}}\cdot[{}^{\Delta}_{t}\boldsymbol{\ell}_{\mathrm{p}} + \boldsymbol{\psi}].
\end{aligned}
$$

The plastic left Lie time derivatives of the intermediate double-strains ${}^{\Delta}\widetilde{\boldsymbol{c}}_{\bullet} = \widetilde{\boldsymbol{c}}\cdot{}^{\Delta}\boldsymbol{v}_{\bullet}$ render

$$
\begin{aligned}
\widetilde{\mathrm{L}}_t{}^{\Delta}\widetilde{\boldsymbol{c}} &= -\widetilde{\boldsymbol{c}}\cdot{}^{\Delta}_{t}\boldsymbol{\ell}\\
\widetilde{\mathrm{L}}_t{}^{\Delta}\widetilde{\boldsymbol{c}}_{\mathrm{e}} &= -\widetilde{\boldsymbol{c}}\cdot[{}^{\Delta}_{t}\boldsymbol{\ell}_{\mathrm{e}} - \boldsymbol{\psi}]\\
\widetilde{\mathrm{L}}_t{}^{\Delta}\widetilde{\boldsymbol{c}}_{\mathrm{p}} &= -\widetilde{\boldsymbol{c}}\cdot[{}^{\Delta}_{t}\boldsymbol{\ell}_{\mathrm{p}} + \boldsymbol{\psi}].
\end{aligned}
$$

The corresponding elastic right Lie and material time derivatives $\overline{\mathrm{R}}_t$ and D_t of the intermediate double-strain measures may finally be computed from the generic expressions for $\overline{\mathrm{R}}_t$ and D_t applied to fully covariant third-order intermediate tensor fields $\{\bullet\}$

$$
\overline{\mathrm{R}}_t\{\bullet\} = \mathrm{D}_t\{\bullet\} - 2[\{\bullet\} : \overline{\mathsf{L}}]^{'\mathrm{sym}'} \quad \text{with} \quad \mathrm{D}_t\{\bullet\} = \widetilde{\mathrm{L}}_t\{\bullet\} - 2[\{\bullet\} : \widetilde{\mathsf{l}}]^{'\mathrm{sym}'}.
$$

Here the following abbreviations have been introduced

$$
2[\{\bullet\} : \overline{\mathsf{L}}]^{'\mathrm{sym}'} := \{\bullet\} : \overline{\mathsf{L}} + \overline{\boldsymbol{L}}^{t}\cdot\{\bullet\} \quad \text{and} \quad 2[\{\bullet\} : \widetilde{\mathsf{l}}]^{'\mathrm{sym}'} := \{\bullet\} : \widetilde{\mathsf{l}} + \widetilde{\boldsymbol{l}}^{t}\cdot\{\bullet\}.
$$

In summary, to each of the three sets of either material, intermediate or spatial double-strain measures (each of these sets itself consisting of three members) three different types of time derivatives may be applied. Thus in total $3\times3\times3\times3 = 81(!)$ different double-strain rates may be listed, however this seems too much of book keeping for this supplement. Note, however, that all of these double-strain rates allow representation incorporating, in most of the cases among other correction terms, appropriate rate-of-deformation tensors.

Material↔Lie Time Derivatives of Double-Strain Measures:

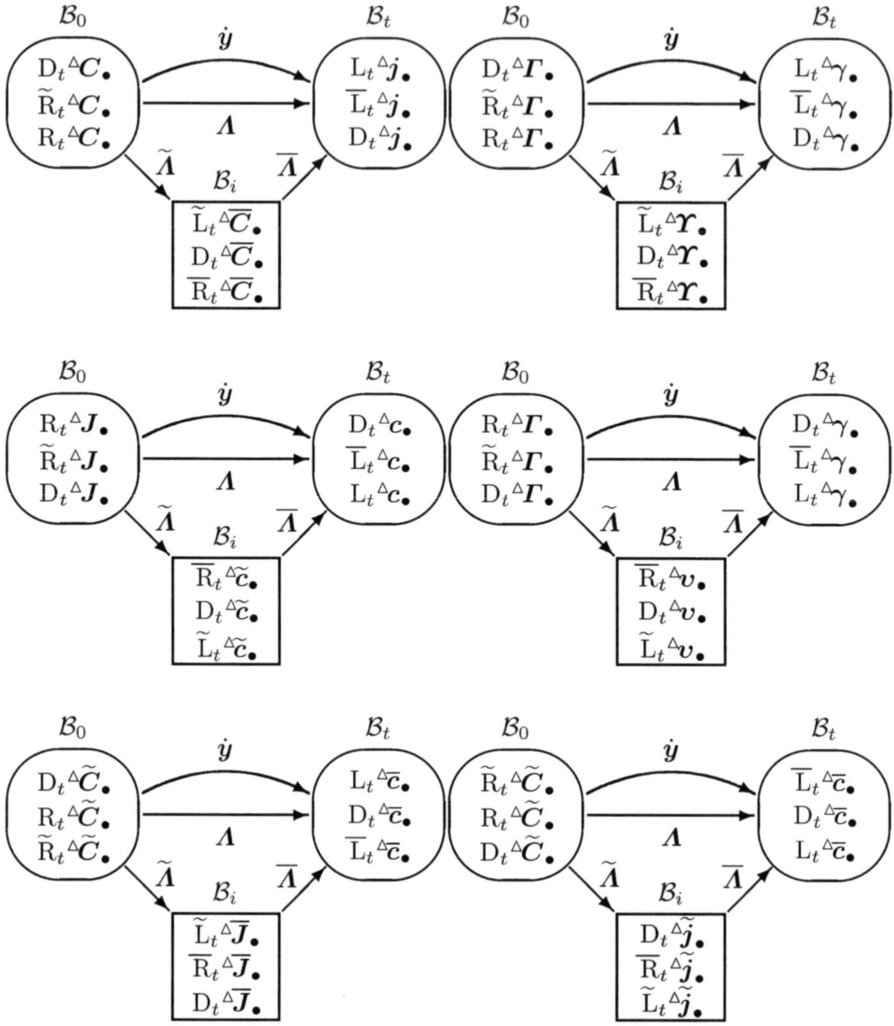

The relations between the various types of time derivatives for the total, elastic and plastic double-strain measures are depicted in the above cartoons. Thereby it is interesting to note that only the left Lie time derivative of the spatial double-metrics may be related to the corresponding rate-of-double-deformation tensors without further correction terms.

Double-Stress Measures:

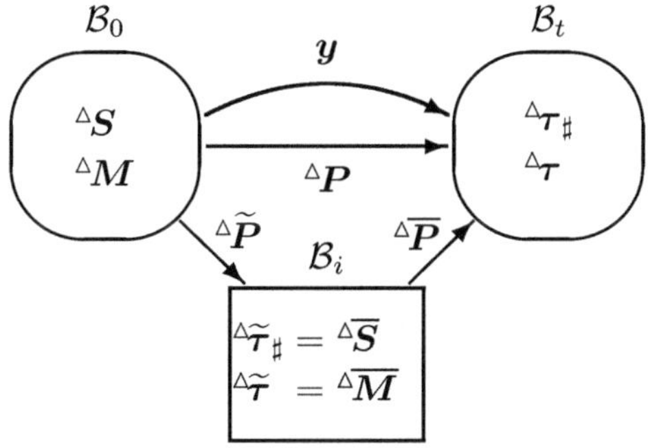

Right-sided contravariant plastic push-forward and left-sided covariant elastic pullback of the (total) Piola double-stress results in the two-point *elastic* and *plastic Piola double-stress, respectively, as*

$$
{}^{\vartriangle}\overline{\boldsymbol{P}} := {}^{\vartriangle}\boldsymbol{P} : \widetilde{\mathsf{F}}^t \quad \textit{and} \quad {}^{\vartriangle}\widetilde{\boldsymbol{P}} := \overline{\boldsymbol{F}}^t \cdot {}^{\vartriangle}\boldsymbol{P}.
$$

Here the abbreviation $\widetilde{\mathsf{F}} := \widetilde{\boldsymbol{F}} \boxtimes \widetilde{\boldsymbol{F}}$ *with* $\widetilde{\mathsf{F}}^t := \widetilde{\boldsymbol{F}}^t \boxtimes \widetilde{\boldsymbol{F}}^t$ *was introduced.* Then the co/contra/contra- (mixed) variant two-point *elastic* Piola double-stress may be related to the contravariantly transforming intermediate Piola-Kirchhoff double-stress ${}^{\vartriangle}\overline{\boldsymbol{S}}$ and the co/contra/contra- (mixed) variantly transforming intermediate Mandel double-stress ${}^{\vartriangle}\overline{\boldsymbol{M}}$, respectively, as

$$
{}^{\vartriangle}\overline{\boldsymbol{S}} := \overline{\boldsymbol{f}}_\sharp \cdot {}^{\vartriangle}\overline{\boldsymbol{P}} \quad \text{and} \quad {}^{\vartriangle}\overline{\boldsymbol{M}} := \overline{\boldsymbol{F}}^t \cdot {}^{\vartriangle}\overline{\boldsymbol{P}}.
$$

Accordingly the spatial co/contra/contra- (mixed) variant Kirchhoff double-stress ${}^{\vartriangle}\boldsymbol{\tau}$ is related to the two-point elastic Piola double-stress and the double-stress measures in the intermediate configuration by

$$
{}^{\vartriangle}\boldsymbol{\tau} = {}^{\vartriangle}\overline{\boldsymbol{P}} : \overline{\mathsf{F}}^t = \overline{\boldsymbol{F}}_\flat \cdot {}^{\vartriangle}\overline{\boldsymbol{S}} : \overline{\mathsf{F}}^t = \overline{\boldsymbol{f}}^t \cdot {}^{\vartriangle}\overline{\boldsymbol{M}} \cdot \overline{\mathsf{F}}^t.
$$

Here the abbreviation $\overline{\mathsf{F}} := \overline{\boldsymbol{F}} \boxtimes \overline{\boldsymbol{F}}$ with $\overline{\mathsf{F}}^t := \overline{\boldsymbol{F}}^t \boxtimes \overline{\boldsymbol{F}}^t$ was introduced. Likewise the co/contra/contra- (mixed) variant two-point *plastic* Piola double-stress may be related to the contravariantly transforming material Piola-Kirchhoff double-stress ${}^{\vartriangle}\boldsymbol{S}$ and the co/contra/contra- (mixed) variantly transforming material Mandel double-stress $\boldsymbol{M}$, respectively

$$
{}^{\vartriangle}\boldsymbol{S} := \widetilde{\boldsymbol{f}}_\sharp \cdot {}^{\vartriangle}\widetilde{\boldsymbol{P}} \quad \text{and} \quad {}^{\vartriangle}\boldsymbol{M} := \widetilde{\boldsymbol{F}}^t \cdot {}^{\vartriangle}\widetilde{\boldsymbol{P}}.
$$

Accordingly the intermediate co/contra/contra- (mixed) variant Kirchhoff double-stress ${}^{\vartriangle}\widetilde{\boldsymbol{\tau}}$ is related to the two-point plastic Piola double-stress and the double-stress measures in the material configuration by

$$
{}^{\vartriangle}\widetilde{\boldsymbol{\tau}} := {}^{\vartriangle}\widetilde{\boldsymbol{P}} : \widetilde{\mathsf{F}}^t = \widetilde{\boldsymbol{F}}_\flat \cdot {}^{\vartriangle}\boldsymbol{S} : \widetilde{\mathsf{F}}^t = \widetilde{\boldsymbol{f}}^t \cdot {}^{\vartriangle}\boldsymbol{M} : \widetilde{\mathsf{F}}^t.
$$

Finally, it shall be noted that the intermediate co/contra/contra- (mixed) variant Kirchhoff and Mandel double-stresses do coincide

$$ {}^{\Delta}\widetilde{\boldsymbol{\tau}} \equiv {}^{\Delta}\overline{\boldsymbol{M}} \quad \rightsquigarrow \quad {}^{\Delta}\widetilde{\boldsymbol{\tau}}_{\sharp} := \overline{\boldsymbol{B}} \cdot {}^{\Delta}\widetilde{\boldsymbol{\tau}} \equiv {}^{\Delta}\overline{\boldsymbol{S}}. $$

The symbolic redundancy for the intermediate double-stress measures has been introduced again for notational convenience. The relations between the various double-stress measures in the material and the intermediate configuration or in the intermediate and the spatial configuration follows the pattern established already for the relations between the various double-stress measures in the material and the spatial configuration.

Stress Measures:

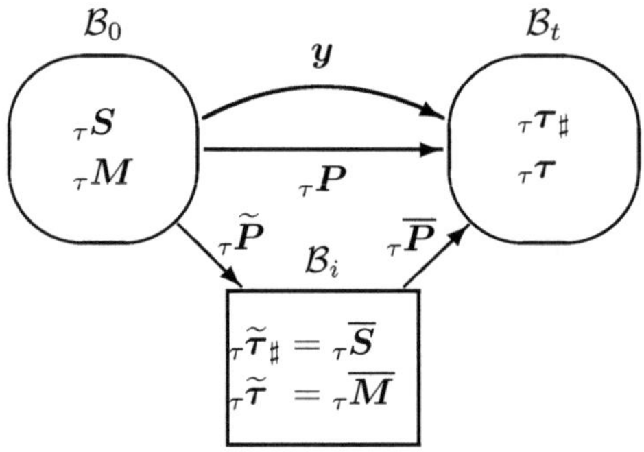

In addition to the ${}_0$-type and ${}_t$-type Kirchhoff, Piola, Piola-Kirchhoff, and Mandel stress measures, that are related via the spatial and material (total) connection tensors, respectively,

$$ \begin{aligned} {}_t\boldsymbol{\tau} &= {}_0\boldsymbol{\tau} - {}^{\Delta}\boldsymbol{\tau} : {}^{\Delta}\boldsymbol{\gamma}^T, \\ {}_0\boldsymbol{P} &= {}_t\boldsymbol{P} - {}^{\Delta}\boldsymbol{P} : {}^{\Delta}\boldsymbol{\Gamma}^T, \\ {}_0\boldsymbol{S} &= {}_t\boldsymbol{S} - {}^{\Delta}\boldsymbol{S} : {}^{\Delta}\boldsymbol{\Gamma}^T, \\ {}_0\boldsymbol{M} &= {}_t\boldsymbol{M} - {}^{\Delta}\boldsymbol{M} : {}^{\Delta}\boldsymbol{\Gamma}^T, \end{aligned} $$

corresponding plastic and elastic counterparts of ${}_0$-type and ${}_t$-type are introduced. Then, on the one hand, based on a left-sided elastic pull-back of ${}_0\boldsymbol{P}$, the corresponding relation between the two-point (intermediate/material) ${}_{\tau}$-type *plastic Piola stresses* reads

$$ {}_0\widetilde{\boldsymbol{P}} = {}_t\widetilde{\boldsymbol{P}} - {}^{\Delta}\widetilde{\boldsymbol{P}} : {}^{\Delta}\boldsymbol{\Gamma}^T. $$

An additional right-sided plastic push-forward involves the *intermediate connection tensor* and results in the relation between (the contravariant and) the co/contra- (mixed) variant intermediate ${}_{\tau}$-type *plastic Kirchhoff stresses*

$$ {}_t\widetilde{\boldsymbol{\tau}} = {}_0\widetilde{\boldsymbol{\tau}} - {}^{\Delta}\widetilde{\boldsymbol{\tau}} : {}^{\Delta}\boldsymbol{\upsilon}^T. $$

Note next that the co/contra- (mixed) variant intermediate plastic Kirchhoff and the elastic Mandel stresses do coincide

$$_\tau\widetilde{\boldsymbol{\tau}} \equiv {}_\tau\overline{\boldsymbol{M}} \quad \rightsquigarrow \quad {}_\tau\widetilde{\boldsymbol{\tau}}_\sharp := \overline{\boldsymbol{B}} \cdot {}_\tau\widetilde{\boldsymbol{\tau}} \equiv {}_\tau\overline{\boldsymbol{S}}.$$

Then, on the other hand, based on a right-sided plastic push-forward of ${}_\tau\boldsymbol{P}$ or, likewise, a left-sided elastic push-forward of ${}_\tau\overline{\boldsymbol{M}}$, the corresponding relation between the two-point (spatial/intermediate) $_\tau$-type *elastic Piola stresses* reads

$${}_0\overline{\boldsymbol{P}} = {}_t\overline{\boldsymbol{P}} - {}^\vartriangle\overline{\boldsymbol{P}} : {}^\vartriangle\boldsymbol{\Upsilon}^T.$$

An additional left-sided elastic pull-back involves the *intermediate connection tensors* and results in the relation between the intermediate $_\tau$-type *elastic Piola-Kirchhoff stresses*

$${}_0\overline{\boldsymbol{S}} = {}_t\overline{\boldsymbol{S}} - {}^\vartriangle\overline{\boldsymbol{S}} : {}^\vartriangle\boldsymbol{\Upsilon}^T$$

and the intermediate $_\tau$-type *elastic Mandel stresses*

$${}_0\overline{\boldsymbol{M}} = {}_t\overline{\boldsymbol{M}} - {}^\vartriangle\overline{\boldsymbol{M}} : {}^\vartriangle\boldsymbol{\Upsilon}^T.$$

The various $_\tau$-type stress measures are motivated by contributing to different versions of the stress/double-stress power (as demonstrated below). Thereby the relations between the various $_\tau$-type stress measures in the material and the intermediate configuration or in the intermediate and the spatial configuration follows the pattern established already for the relations between the various stress measures in the material and the spatial configuration.

Stress/Double-Stress Power Density:
Due to the additive decomposition of the contra/co- (mixed) variant velocity gradients and the contra/co/co- (mixed) variant velocity double-gradients into elastic and plastic parts

$$\boldsymbol{\Lambda} = \boldsymbol{\Lambda}_\mathrm{e} + \boldsymbol{\Lambda}_\mathrm{p} \quad \text{and} \quad \overset{\vartriangle}{_\tau}\boldsymbol{\Lambda} = \overset{\vartriangle}{_\tau}\boldsymbol{\Lambda}_\mathrm{e} + \overset{\vartriangle}{_\tau}\boldsymbol{\Lambda}_\mathrm{p}$$

(whereby $\tau = 0, t$) the stress/double-stress power density as expressed in terms of co/contra- (mixed) variant stress and co/contra/contra- (mixed) variant double-stress measures expands into the elementary representation

$$\wp_0 = {}_\tau\boldsymbol{P} : \boldsymbol{\Lambda} + {}^\vartriangle\boldsymbol{P} \mathbin{\therefore} \overset{\vartriangle}{_\tau}\boldsymbol{\Lambda} = [{}_\tau\boldsymbol{P} : \boldsymbol{\Lambda}_\mathrm{e} + {}^\vartriangle\boldsymbol{P} \mathbin{\therefore} \overset{\vartriangle}{_\tau}\boldsymbol{\Lambda}_\mathrm{e}] + [{}_\tau\boldsymbol{P} : \boldsymbol{\Lambda}_\mathrm{p} + {}^\vartriangle\boldsymbol{P} \mathbin{\therefore} \overset{\vartriangle}{_\tau}\boldsymbol{\Lambda}_\mathrm{p}].$$

Alternatively, based on the expansions of the elastic and the plastic parts of the velocity gradient

$$\boldsymbol{\Lambda}_\mathrm{e} = \overline{\boldsymbol{\Lambda}} \cdot \widetilde{\boldsymbol{F}} \quad \text{and} \quad \boldsymbol{\Lambda}_\mathrm{p} = \overline{\boldsymbol{F}} \cdot \widetilde{\boldsymbol{\Lambda}}$$

the elementary representation of the stress/double-stress power density may be rewritten in terms of the elastic and plastic Piola stresses

$$\wp_0 = [{}_{\tau}\overline{\boldsymbol{P}} : \overline{\boldsymbol{\Lambda}} + {}^{\vartriangle}\boldsymbol{P} \mathbin{\therefore} {}^{\vartriangle}_{\tau}\boldsymbol{\Lambda}_{\mathrm{e}}] + [{}_{\tau}\widetilde{\boldsymbol{P}} : \widetilde{\boldsymbol{\Lambda}} + {}^{\vartriangle}\boldsymbol{P} \mathbin{\therefore} {}^{\vartriangle}_{\tau}\boldsymbol{\Lambda}_{\mathrm{p}}].$$

Note that the contributions of the double-stresses (corresponding to ${}^{\vartriangle}\boldsymbol{P} \mathbin{\therefore} {}^{\vartriangle}_{\tau}\boldsymbol{\Lambda}$) to the elastic and plastic parts of the stress/double-stress power density are unaffected by this rewriting. By suited one- and two-sided push-forward and pull-back operations the elementary representation of the stress/double-stress power density then takes the alternative expressions

$$\begin{aligned}\wp_0 &= [{}_{\tau}\boldsymbol{\tau} : \boldsymbol{l}_{\mathrm{e}} + {}^{\vartriangle}\boldsymbol{\tau} \mathbin{\therefore} {}^{\vartriangle}_{\tau}\boldsymbol{l}_{\mathrm{e}}] \mid = [{}_{\tau}\widetilde{\boldsymbol{\tau}} : \boldsymbol{\ell}_{\mathrm{e}} + {}^{\vartriangle}\widetilde{\boldsymbol{\tau}} \mathbin{\therefore} {}^{\vartriangle}_{\tau}\boldsymbol{\ell}_{\mathrm{e}}] \mid = [{}_{\tau}\boldsymbol{M} : \boldsymbol{L}_{\mathrm{e}} + {}^{\vartriangle}\boldsymbol{M} \mathbin{\therefore} {}^{\vartriangle}_{\tau}\boldsymbol{L}_{\mathrm{e}}] \\ &+ [{}_{\tau}\boldsymbol{\tau} : \boldsymbol{l}_{\mathrm{p}} + {}^{\vartriangle}\boldsymbol{\tau} \mathbin{\therefore} {}^{\vartriangle}_{\tau}\boldsymbol{l}_{\mathrm{p}}] \mid + [{}_{\tau}\widetilde{\boldsymbol{\tau}} : \boldsymbol{\ell}_{\mathrm{p}} + {}^{\vartriangle}\widetilde{\boldsymbol{\tau}} \mathbin{\therefore} {}^{\vartriangle}_{\tau}\boldsymbol{\ell}_{\mathrm{p}}] \mid + [{}_{\tau}\boldsymbol{M} : \boldsymbol{L}_{\mathrm{p}} + {}^{\vartriangle}\boldsymbol{M} \mathbin{\therefore} {}^{\vartriangle}_{\tau}\boldsymbol{L}_{\mathrm{p}}].\end{aligned}$$

Observe that in the above the ${}_{\tau}$-type contributions of the stresses (corresponding to ${}_{\tau}\boldsymbol{P} : \boldsymbol{\Lambda}$) and the double-stresses (corresponding to ${}^{\vartriangle}\boldsymbol{P} \mathbin{\therefore} {}^{\vartriangle}_{\tau}\boldsymbol{\Lambda}$) to the elementary representation of the elastic and plastic parts of the stress/double-stress power density satisfy separately

$$\begin{aligned}{}_{\tau}\boldsymbol{\tau} : \boldsymbol{l}_{\mathrm{e}} = {}_{\tau}\widetilde{\boldsymbol{\tau}} : \boldsymbol{\ell}_{\mathrm{e}} = {}_{\tau}\boldsymbol{M} : \boldsymbol{L}_{\mathrm{e}} \quad &\text{and} \quad {}^{\vartriangle}\boldsymbol{\tau} \mathbin{\therefore} {}^{\vartriangle}_{\tau}\boldsymbol{l}_{\mathrm{e}} = {}^{\vartriangle}\widetilde{\boldsymbol{\tau}} \mathbin{\therefore} {}^{\vartriangle}_{\tau}\boldsymbol{\ell}_{\mathrm{e}} = {}^{\vartriangle}\boldsymbol{M} \mathbin{\therefore} {}^{\vartriangle}_{\tau}\boldsymbol{L}_{\mathrm{e}} \\ {}_{\tau}\boldsymbol{\tau} : \boldsymbol{l}_{\mathrm{p}} = {}_{\tau}\widetilde{\boldsymbol{\tau}} : \boldsymbol{\ell}_{\mathrm{p}} = {}_{\tau}\boldsymbol{M} : \boldsymbol{L}_{\mathrm{p}} \quad &\text{and} \quad {}^{\vartriangle}\boldsymbol{\tau} \mathbin{\therefore} {}^{\vartriangle}_{\tau}\boldsymbol{l}_{\mathrm{p}} = {}^{\vartriangle}\widetilde{\boldsymbol{\tau}} \mathbin{\therefore} {}^{\vartriangle}_{\tau}\boldsymbol{\ell}_{\mathrm{p}} = {}^{\vartriangle}\boldsymbol{M} \mathbin{\therefore} {}^{\vartriangle}_{\tau}\boldsymbol{L}_{\mathrm{p}}.\end{aligned}$$

Thus arbitrary combinations of ${}_{\tau}$-type contributions of the stresses and the double-stresses to the stress/double-stress power density are in principle possible. In particular the contributions ${}_{t}\boldsymbol{\tau} : \boldsymbol{l}_{\mathrm{e}} = {}_{t}\boldsymbol{\tau}_{\sharp} : \boldsymbol{d}_{\mathrm{e}}$ and ${}_{t}\boldsymbol{\tau} : \boldsymbol{l}_{\mathrm{p}} = {}_{t}\boldsymbol{\tau}_{\sharp} : \boldsymbol{d}_{\mathrm{p}}$ of the symmetric ${}_{t}$-type stresses to the stress/double-stress power density allow the usual alternative expressions

$$\begin{aligned}{}_{t}\boldsymbol{\tau}_{\sharp} : \boldsymbol{d}_{\mathrm{e}} &= {}_{t}\boldsymbol{\tau}_{\sharp} : [\boldsymbol{j} \cdot \boldsymbol{l}_{\mathrm{e}}]^{\mathrm{sym}} = {}_{t}\overline{\boldsymbol{S}} : [\overline{\boldsymbol{C}} \cdot \boldsymbol{\ell}_{\mathrm{e}}]^{\mathrm{sym}} = {}_{t}\boldsymbol{S} : [\boldsymbol{C} \cdot \boldsymbol{L}_{\mathrm{e}}]^{\mathrm{sym}}, \\ {}_{t}\boldsymbol{\tau}_{\sharp} : \boldsymbol{d}_{\mathrm{p}} &= {}_{t}\boldsymbol{\tau}_{\sharp} : [\boldsymbol{j} \cdot \boldsymbol{l}_{\mathrm{p}}]^{\mathrm{sym}} = {}_{t}\overline{\boldsymbol{S}} : [\overline{\boldsymbol{C}} \cdot \boldsymbol{\ell}_{\mathrm{p}}]^{\mathrm{sym}} = {}_{t}\boldsymbol{S} : [\boldsymbol{C} \cdot \boldsymbol{L}_{\mathrm{p}}]^{\mathrm{sym}}.\end{aligned}$$

Note, however, that among the options for the contribution of the double-stresses to the power density only ${}^{\vartriangle}\boldsymbol{P} \mathbin{\therefore} {}^{\vartriangle}_{0}\boldsymbol{\Lambda}_{\bullet}$ with ${}^{\vartriangle}_{0}\boldsymbol{\Lambda}_{\bullet} = {}^{\vartriangle}\dot{\boldsymbol{F}}_{\bullet}$ and ${}^{\vartriangle}\boldsymbol{M} \mathbin{\therefore} {}^{\vartriangle}_{t}\boldsymbol{L}_{\bullet}$ with ${}^{\vartriangle}_{t}\boldsymbol{L}_{\mathrm{e}} = {}^{\vartriangle}\dot{\boldsymbol{\Gamma}}_{\mathrm{e}} - \boldsymbol{\Xi}$ and ${}^{\vartriangle}_{t}\boldsymbol{L}_{\mathrm{p}} = {}^{\vartriangle}\dot{\boldsymbol{\Gamma}}_{\mathrm{p}} + \boldsymbol{\Xi}$ (where $\boldsymbol{\Xi} := \boldsymbol{L}_{\mathrm{e}} \cdot {}^{\vartriangle}\boldsymbol{\Gamma}_{\mathrm{p}} - \boldsymbol{L}_{\mathrm{p}} \cdot {}^{\vartriangle}\boldsymbol{\Gamma}_{\mathrm{e}}$) incorporate material time derivatives of double-strains. Thus, employing next the relations between $\boldsymbol{C} \cdot {}^{\vartriangle}_{t}\boldsymbol{L}_{\bullet}$, ${}^{\vartriangle}\dot{\boldsymbol{C}}_{\bullet}$ and $\dot{\boldsymbol{C}} \cdot {}^{\vartriangle}\boldsymbol{\Gamma}_{\bullet}$ on the one hand, and the relation between $\boldsymbol{j} \cdot {}^{\vartriangle}_{t}\boldsymbol{l}_{\bullet}$, $\mathrm{L}_t{}^{\vartriangle}\boldsymbol{j}_{\bullet} = 2{}^{\vartriangle}_{0}\boldsymbol{d}_{\bullet}$ and $\mathrm{L}_t\boldsymbol{j} \cdot {}^{\vartriangle}\boldsymbol{\gamma}_{\bullet}$ on the other hand, renders alternatively the following power conjugate pairings of fully contravariant stress/double-stress measures and fully covariant strain/double-strain measures

$$
\begin{aligned}
\wp_0 &= \big[{}_t\boldsymbol{\tau}_\sharp + 2[{}^\Delta\boldsymbol{\tau}_\sharp : {}^\Delta\boldsymbol{\gamma}^T]^{\rm sym}\big] : \boldsymbol{d}_{\rm e} + 2{}^\Delta\boldsymbol{\tau}_\sharp \,\vdots\, {}^\Delta_0\boldsymbol{d}_{\rm e} \\
&+ \big[{}_t\boldsymbol{\tau}_\sharp + 2[{}^\Delta\boldsymbol{\tau}_\sharp : {}^\Delta\boldsymbol{\gamma}^T]^{\rm sym}\big] : \boldsymbol{d}_{\rm p} + 2{}^\Delta\boldsymbol{\tau}_\sharp \,\vdots\, {}^\Delta_0\boldsymbol{d}_{\rm p} \\
&= \big[{}_t\boldsymbol{\tau}_\sharp + 2[{}^\Delta\boldsymbol{\tau}_\sharp : {}^\Delta\boldsymbol{\gamma}^T]^{\rm sym}\big] : [\,\boldsymbol{j} \cdot \boldsymbol{l}_{\rm e}]^{\rm sym} + 2{}^\Delta\boldsymbol{\tau}_\sharp \,\vdots\, [\,\boldsymbol{j} \cdot {}^\Delta_0\boldsymbol{l}_{\rm e}]^{'\rm sym'} \\
&+ \big[{}_t\boldsymbol{\tau}_\sharp + 2[{}^\Delta\boldsymbol{\tau}_\sharp : {}^\Delta\boldsymbol{\gamma}^T]^{\rm sym}\big] : [\,\boldsymbol{j} \cdot \boldsymbol{l}_{\rm p}]^{\rm sym} + 2{}^\Delta\boldsymbol{\tau}_\sharp \,\vdots\, [\,\boldsymbol{j} \cdot {}^\Delta_0\boldsymbol{l}_{\rm p}]^{'\rm sym'} \\
&= \big[{}_t\widetilde{\boldsymbol{\tau}}_\sharp + 2[{}^\Delta\widetilde{\boldsymbol{\tau}}_\sharp : {}^\Delta\boldsymbol{\upsilon}^T]^{\rm sym}\big] : [\overline{\boldsymbol{C}} \cdot \boldsymbol{\ell}_{\rm e}]^{\rm sym} + 2{}^\Delta\widetilde{\boldsymbol{\tau}}_\sharp \,\vdots\, [\overline{\boldsymbol{C}} \cdot {}^\Delta_0\boldsymbol{\ell}_{\rm e}]^{'\rm sym'} \\
&+ \big[{}_t\widetilde{\boldsymbol{\tau}}_\sharp + 2[{}^\Delta\widetilde{\boldsymbol{\tau}}_\sharp : {}^\Delta\boldsymbol{\upsilon}^T]^{\rm sym}\big] : [\overline{\boldsymbol{C}} \cdot \boldsymbol{\ell}_{\rm p}]^{\rm sym} + 2{}^\Delta\widetilde{\boldsymbol{\tau}}_\sharp \,\vdots\, [\overline{\boldsymbol{C}} \cdot {}^\Delta_0\boldsymbol{\ell}_{\rm p}]^{'\rm sym'} \\
&\equiv \big[{}_t\overline{\boldsymbol{S}} - 2[{}^\Delta\overline{\boldsymbol{S}} : {}^\Delta\boldsymbol{\Upsilon}^T]^{\rm sym}\big] : [\overline{\boldsymbol{C}} \cdot \boldsymbol{\ell}_{\rm e}]^{\rm sym} + 2{}^\Delta\overline{\boldsymbol{S}} \,\vdots\, [\overline{\boldsymbol{C}} \cdot {}^\Delta_0\boldsymbol{\ell}_{\rm e}]^{'\rm sym'} \\
&+ \big[{}_t\overline{\boldsymbol{S}} - 2[{}^\Delta\overline{\boldsymbol{S}} : {}^\Delta\boldsymbol{\Upsilon}^T]^{\rm sym}\big] : [\overline{\boldsymbol{C}} \cdot \boldsymbol{\ell}_{\rm p}]^{\rm sym} + 2{}^\Delta\overline{\boldsymbol{S}} \,\vdots\, [\overline{\boldsymbol{C}} \cdot {}^\Delta_0\boldsymbol{\ell}_{\rm p}]^{'\rm sym'} \\
&= \big[{}_t\boldsymbol{S} - 2[{}^\Delta\boldsymbol{S} : {}^\Delta\boldsymbol{\Gamma}^T]^{\rm sym}\big] : [\boldsymbol{C} \cdot \boldsymbol{L}_{\rm e}]^{\rm sym} + 2{}^\Delta\boldsymbol{S} \,\vdots\, [\boldsymbol{C} \cdot {}^\Delta_0\boldsymbol{L}_{\rm e}]^{'\rm sym'} \\
&+ \big[{}_t\boldsymbol{S} - 2[{}^\Delta\boldsymbol{S} : {}^\Delta\boldsymbol{\Gamma}^T]^{\rm sym}\big] : [\boldsymbol{C} \cdot \boldsymbol{L}_{\rm p}]^{\rm sym} + 2{}^\Delta\boldsymbol{S} \,\vdots\, [\boldsymbol{C} \cdot {}^\Delta_0\boldsymbol{L}_{\rm p}]^{'\rm sym'}.
\end{aligned}
$$

In the sequel it is useful to introduce the abbreviations $\boldsymbol{\tau}_\sharp$, $\widetilde{\boldsymbol{\tau}}_\sharp \equiv \overline{\boldsymbol{S}}$ and $\boldsymbol{S}$ for the alternative (symmetric) stress measures appearing in the above power conjugate pairings as

$$
\begin{aligned}
\boldsymbol{\tau}_\sharp &:= {}_t\boldsymbol{\tau}_\sharp + 2[{}^\Delta\boldsymbol{\tau}_\sharp : {}^\Delta\boldsymbol{\gamma}^T]^{\rm sym} = {}_0\boldsymbol{\tau}_\sharp + {}^\Delta\boldsymbol{\gamma} : {}^\Delta\boldsymbol{\tau}_\sharp^T, \\
\widetilde{\boldsymbol{\tau}}_\sharp &:= {}_t\widetilde{\boldsymbol{\tau}}_\sharp + 2[{}^\Delta\widetilde{\boldsymbol{\tau}}_\sharp : {}^\Delta\boldsymbol{\upsilon}^T]^{\rm sym} = {}_0\widetilde{\boldsymbol{\tau}}_\sharp + {}^\Delta\boldsymbol{\upsilon} : {}^\Delta\widetilde{\boldsymbol{\tau}}_\sharp^T, \\
\equiv \overline{\boldsymbol{S}} &:= {}_t\overline{\boldsymbol{S}} - 2[{}^\Delta\overline{\boldsymbol{S}} : {}^\Delta\boldsymbol{\Upsilon}^T]^{\rm sym} = {}_0\overline{\boldsymbol{S}} - {}^\Delta\boldsymbol{\Upsilon} : {}^\Delta\overline{\boldsymbol{S}}^T, \\
\boldsymbol{S} &:= {}_t\boldsymbol{S} - 2[{}^\Delta\boldsymbol{S} : {}^\Delta\boldsymbol{\Gamma}^T]^{\rm sym} = {}_0\boldsymbol{S} - {}^\Delta\boldsymbol{\Gamma} : {}^\Delta\boldsymbol{S}^T.
\end{aligned}
$$

Thus finally the stress/double stress power density takes the more intuitive format

$$
\begin{aligned}
\wp_0 &= \boldsymbol{\tau}_\sharp : \boldsymbol{d}_{\rm e} + 2{}^\Delta\boldsymbol{\tau}_\sharp \,\vdots\, {}^\Delta_0\boldsymbol{d}_{\rm e} \\
&+ \boldsymbol{\tau}_\sharp : \boldsymbol{d}_{\rm p} + 2{}^\Delta\boldsymbol{\tau}_\sharp \,\vdots\, {}^\Delta_0\boldsymbol{d}_{\rm p} \\
&= \boldsymbol{\tau}_\sharp : [\,\boldsymbol{j} \cdot \boldsymbol{l}_{\rm e}]^{\rm sym} + 2{}^\Delta\boldsymbol{\tau}_\sharp \,\vdots\, [\,\boldsymbol{j} \cdot {}^\Delta_0\boldsymbol{l}_{\rm e}]^{'\rm sym'} \\
&+ \boldsymbol{\tau}_\sharp : [\,\boldsymbol{j} \cdot \boldsymbol{l}_{\rm p}]^{\rm sym} + 2{}^\Delta\boldsymbol{\tau}_\sharp \,\vdots\, [\,\boldsymbol{j} \cdot {}^\Delta_0\boldsymbol{l}_{\rm p}]^{'\rm sym'} \\
&= \widetilde{\boldsymbol{\tau}}_\sharp : [\overline{\boldsymbol{C}} \cdot \boldsymbol{\ell}_{\rm e}]^{\rm sym} + 2{}^\Delta\widetilde{\boldsymbol{\tau}}_\sharp \,\vdots\, [\overline{\boldsymbol{C}} \cdot {}^\Delta_0\boldsymbol{\ell}_{\rm e}]^{'\rm sym'} \\
&+ \widetilde{\boldsymbol{\tau}}_\sharp : [\overline{\boldsymbol{C}} \cdot \boldsymbol{\ell}_{\rm p}]^{\rm sym} + 2{}^\Delta\widetilde{\boldsymbol{\tau}}_\sharp \,\vdots\, [\overline{\boldsymbol{C}} \cdot {}^\Delta_0\boldsymbol{\ell}_{\rm p}]^{'\rm sym'} \\
&\equiv \overline{\boldsymbol{S}} : [\overline{\boldsymbol{C}} \cdot \boldsymbol{\ell}_{\rm e}]^{\rm sym} + 2{}^\Delta\overline{\boldsymbol{S}} \,\vdots\, [\overline{\boldsymbol{C}} \cdot {}^\Delta_0\boldsymbol{\ell}_{\rm e}]^{'\rm sym'} \\
&+ \overline{\boldsymbol{S}} : [\overline{\boldsymbol{C}} \cdot \boldsymbol{\ell}_{\rm p}]^{\rm sym} + 2{}^\Delta\overline{\boldsymbol{S}} \,\vdots\, [\overline{\boldsymbol{C}} \cdot {}^\Delta_0\boldsymbol{\ell}_{\rm p}]^{'\rm sym'} \\
&= \boldsymbol{S} : [\boldsymbol{C} \cdot \boldsymbol{L}_{\rm e}]^{\rm sym} + 2{}^\Delta\boldsymbol{S} \,\vdots\, [\boldsymbol{C} \cdot {}^\Delta_0\boldsymbol{L}_{\rm e}]^{'\rm sym'} \\
&+ \boldsymbol{S} : [\boldsymbol{C} \cdot \boldsymbol{L}_{\rm p}]^{\rm sym} + 2{}^\Delta\boldsymbol{S} \,\vdots\, [\boldsymbol{C} \cdot {}^\Delta_0\boldsymbol{L}_{\rm p}]^{'\rm sym'}.
\end{aligned}
$$

It is recalled that the combination of power conjugate pairings of stress/double-stress and strain/double-strain measures is of importance for the determination of elasto-plastic constitutive relations.

Dissipation Inequality:
Recall that in an *isothermal* continuum setting the requirement of positive *entropy production* boils down to the statement that the stress/double-stress power (which coincides with the external working for quasi-static cases) is never smaller than the change of *free energy* Ψ stored in the body. Thus for an arbitrary continuous sub-body $\mathscr{V} \subseteq \mathscr{B}$ of a continuous body $\mathscr{B}$ that occupies the material sub-configuration $\mathcal{V}_0 \subseteq \mathcal{B}_0$ of the material configuration $\mathcal{B}_0$ the global statement for the *dissipation inequality* reads

$$\int_{\mathcal{V}_0} \wp_0 \, \mathrm{d}V \geq \dot{\Psi}(\mathcal{V}_0) := \int_{\mathcal{V}_0} \mathrm{D}_t \psi_0 \, \mathrm{d}V.$$

Here ψ_0 denotes again the free energy density per unit volume $\mathrm{d}V$ in the material configuration $\mathcal{B}_0$; in general the free energy density is parameterized as $\psi_0 = \psi_0(\boldsymbol{F}, {}^{\Delta}\boldsymbol{F}, \theta, \cdots)$, whereby the $\cdots$ denote the dependence on further *internal variables*. For the here considered isothermal case, however, the absolute temperature is a given parameter $\theta = \theta_{\mathrm{ref}}$ and thus drops out from the list of arguments of the free energy density, i.e. here $\psi_0 = \psi_0(\boldsymbol{F}, {}^{\Delta}\boldsymbol{F}, \cdots)$. Here and in the following any possible further dependence of ψ_0 on the material coordinates $\boldsymbol{X}$ that is needed to capture material inhomogeneities has been suppressed to ease notation.

Master Constitutive Relations:
For second-order (hyper) elasto-plasticity the constitutive relations for the stresses and the double-stresses in terms of the strains and the double-strains follow from a Coleman-Noll-type exploitation of the dissipation inequality, i.e. from the comparison of the material time derivative of the free energy density ψ_0 with the stress/double-stress power density $\wp_0$ and the observation that $\boldsymbol{F}$ and $\mathrm{D}_t\boldsymbol{F}$ as well as ${}^{\Delta}\boldsymbol{F}$ and $\mathrm{D}_t{}^{\Delta}\boldsymbol{F}$ may conceptually be controlled independently. Thereby, due to the elementary definition $\wp_0 = {}_0\boldsymbol{P} : \mathrm{D}_t\boldsymbol{F} + {}^{\Delta}\boldsymbol{P} \,\cdot\!: \mathrm{D}_t{}^{\Delta}\boldsymbol{F}$, the tensor function ψ_0, i.e. the free energy density, depends most generally on $\boldsymbol{F}$, ${}^{\Delta}\boldsymbol{F}$ and, in order to capture the contribution of the microstructure intrinsic to the material, on $\widetilde{\boldsymbol{F}}$, ${}^{\Delta}\boldsymbol{F}_{\mathrm{p}}$ and a set of *internal variables* $\mathbb{F}$. The master constitutive relations for the ${}_0$-type Piola stress and the Piola double-stress then follow as

$$\psi_0 = \psi_0(\boldsymbol{F}, \widetilde{\boldsymbol{F}}, {}^{\Delta}\boldsymbol{F}, {}^{\Delta}\boldsymbol{F}_{\mathrm{p}}, \mathbb{F}) \quad \rightsquigarrow \quad {}_0\boldsymbol{P} = \frac{\partial \psi_0}{\partial \boldsymbol{F}} \quad \text{and} \quad {}^{\Delta}\boldsymbol{P} = \frac{\partial \psi_0}{\partial {}^{\Delta}\boldsymbol{F}}.$$

The *thermodynamic force* $\mathbb{P}$ that drives the evolution of the set of internal variables $\mathbb{F}$ takes the same format as in first-order elasto-plasticity

$$\mathbb{P} := -\frac{\partial \psi_0}{\partial \mathbb{F}}.$$

Moreover the remaining *reduced dissipation inequality* reads

$$-\frac{\partial\psi_0}{\partial\widetilde{\boldsymbol{F}}} : \mathrm{D}_t\widetilde{\boldsymbol{F}} - \frac{\partial\psi_0}{\partial{}^\vartriangle\boldsymbol{F}_\mathrm{p}} \because \mathrm{D}_t{}^\vartriangle\boldsymbol{F}_\mathrm{p} + \mathbb{P}\circ\mathrm{D}_t\mathbb{F} \geq 0$$

with $\circ$ denoting an appropriate scalar product. Then, to be more specific, the collective dependence of ψ_0 on the deformation gradient $\boldsymbol{F}$, the second deformation gradient ${}^\vartriangle\boldsymbol{F}$, the plastic distortion $\widetilde{\boldsymbol{F}}$ and the plastic double-distortion ${}^\vartriangle\boldsymbol{F}_\mathrm{p}$ is assumed to be expressed via the (multiplicative) elastic distortion $\overline{\boldsymbol{F}} = \boldsymbol{F}\cdot\widetilde{\boldsymbol{f}}$ and the (additive) elastic double-distortion ${}^\vartriangle\boldsymbol{F}_\mathrm{e} = {}^\vartriangle\boldsymbol{F} - {}^\vartriangle\boldsymbol{F}_\mathrm{p}$. Thus, based on the appropriate expression for the stress/double-stress power density $\wp_0 = {}_0\overline{\boldsymbol{P}} : \mathrm{D}_t\overline{\boldsymbol{F}} + {}^\vartriangle\boldsymbol{P} \because \mathrm{D}_t{}^\vartriangle\boldsymbol{F}_\mathrm{e} + {}_0\widetilde{\boldsymbol{P}} : \mathrm{D}_t\widetilde{\boldsymbol{F}} + {}^\vartriangle\boldsymbol{P} \because \mathrm{D}_t{}^\vartriangle\boldsymbol{F}_\mathrm{p}$, the master constitutive relations specify to

$$\psi_0 = \psi_0(\overline{\boldsymbol{F}}, {}^\vartriangle\boldsymbol{F}_\mathrm{e}, \mathbb{F}) \quad\rightsquigarrow\quad {}_0\overline{\boldsymbol{P}} = \frac{\partial\psi_0}{\partial\overline{\boldsymbol{F}}} \quad\text{and}\quad {}^\vartriangle\boldsymbol{P} = \frac{\partial\psi_0}{\partial{}^\vartriangle\boldsymbol{F}_\mathrm{e}}$$

whereby the remaining reduced dissipation inequality now appears as

$${}_0\widetilde{\boldsymbol{P}} : \mathrm{D}_t\widetilde{\boldsymbol{F}} + {}^\vartriangle\boldsymbol{P} \because \mathrm{D}_t{}^\vartriangle\boldsymbol{F}_\mathrm{p} + \mathbb{P}\circ\mathrm{D}_t\mathbb{F} \geq 0.$$

In the above ${}_0\widetilde{\boldsymbol{P}}$, ${}^\vartriangle\boldsymbol{P}$ and $\mathbb{P}$ collectively denote the thermodynamic forces that drive the evolution of $\widetilde{\boldsymbol{F}}$, ${}^\vartriangle\boldsymbol{F}_\mathrm{p}$ and the set of internal variables $\mathbb{F}$. In passing observe the interesting relation for ${}_0\widetilde{\boldsymbol{P}}$ that follows directly from the chain rule

$${}_0\widetilde{\boldsymbol{P}} := -\frac{\partial\psi_0}{\partial\widetilde{\boldsymbol{F}}} = -\frac{\partial\psi_0}{\partial\overline{\boldsymbol{F}}} : \frac{\partial\overline{\boldsymbol{F}}}{\partial\widetilde{\boldsymbol{F}}} = \overline{\boldsymbol{F}}^t \cdot {}_0\overline{\boldsymbol{P}} \cdot \widetilde{\boldsymbol{f}}^t.$$

Likewise for ${}^\vartriangle\boldsymbol{P}$ it holds trivially

$${}^\vartriangle\boldsymbol{P} := -\frac{\partial\psi_0}{\partial{}^\vartriangle\boldsymbol{F}_\mathrm{p}} = -\frac{\partial\psi_0}{\partial{}^\vartriangle\boldsymbol{F}_\mathrm{e}} \because \frac{\partial{}^\vartriangle\boldsymbol{F}_\mathrm{e}}{\partial{}^\vartriangle\boldsymbol{F}_\mathrm{p}} = {}^\vartriangle\boldsymbol{P}.$$

Then for second-order elasto-plasticity the *admissible domain* $\mathbb{A}$ for the *generalized thermodynamic force* $\boldsymbol{\mathcal{P}} := \{\widetilde{\boldsymbol{P}}, {}^\vartriangle\boldsymbol{P}, \mathbb{P}\}$ is characterized in the spirit of a *standard dissipative material formulation* [75] by a convex set in terms of the *yield function* $\phi = \phi(\boldsymbol{\mathcal{P}}; \boldsymbol{\mathcal{F}})$ as

$$\mathbb{A} := \{\boldsymbol{\mathcal{P}} |\ \phi(\boldsymbol{\mathcal{P}}; \boldsymbol{\mathcal{F}}) \leq 0\,\}.$$

The evolution of the *generalized internal variable* $\boldsymbol{\mathcal{F}} := \{\widetilde{\boldsymbol{F}}, {}^\vartriangle\boldsymbol{F}_\mathrm{p}, \mathbb{F}\}$ may then be derived from the *postulate of maximum dissipation* as the constrained optimization problem

$$\boldsymbol{\mathcal{P}} = \arg\max\left(\boldsymbol{\mathcal{P}}^*\circ\dot{\boldsymbol{\mathcal{F}}}\,\middle|\,\boldsymbol{\mathcal{P}}^*\in\mathbb{A}\right) \quad\text{and}\quad \boldsymbol{\mathcal{P}}\circ\dot{\boldsymbol{\mathcal{F}}} \geq 0.$$

The corresponding optimality conditions render the *associated evolution law* for the generalized internal variable $\boldsymbol{\mathcal{F}}$ in terms of the positive Lagrange or rather *plastic multiplier* γ

$$\dot{\boldsymbol{\mathcal{F}}} = \gamma \boldsymbol{\mathcal{N}} \quad \text{with} \quad \boldsymbol{\mathcal{N}} = \boldsymbol{\mathcal{N}}(\boldsymbol{\mathcal{P}}; \boldsymbol{\mathcal{F}}) := \frac{\partial \phi}{\partial \boldsymbol{\mathcal{P}}}$$

together with the *Karush-Kuhn-Tucker* complementary conditions that distinguish *plastic loading* from *elastic unloading*

$$\gamma \geq 0 \quad \text{and} \quad \phi(\boldsymbol{\mathcal{P}}; \boldsymbol{\mathcal{F}}) \leq 0 \quad \text{and} \quad \gamma \phi(\boldsymbol{\mathcal{P}}; \boldsymbol{\mathcal{F}}) = 0.$$

It is remarked that in the more general case a *non-associated evolution law* for $\boldsymbol{\mathcal{F}}$ may simply be *postulated* as

$$\dot{\boldsymbol{\mathcal{F}}} = \gamma \boldsymbol{\mathcal{M}} \quad \text{with} \quad \boldsymbol{\mathcal{M}} = \boldsymbol{\mathcal{M}}(\boldsymbol{\mathcal{P}}; \boldsymbol{\mathcal{F}}).$$

In the case of plastic loading with $\gamma > 0$ the plastic multiplier γ may be computed from the so-called *consistency condition*

$$\gamma \dot{\phi}(\boldsymbol{\mathcal{P}}; \boldsymbol{\mathcal{F}}) = 0 \quad \rightsquigarrow \quad \gamma = -\mathcal{D}^{-1} \boldsymbol{\mathcal{N}} \circ \left[\frac{\partial \boldsymbol{\mathcal{P}}}{\partial \boldsymbol{F}} : \dot{\boldsymbol{F}} + \frac{\partial \boldsymbol{\mathcal{P}}}{\partial {}^{\Delta}\boldsymbol{F}} \because {}^{\Delta}\dot{\boldsymbol{F}} \right].$$

Here the denominator $\mathcal{D}$ takes the explicit dependence of the yield condition on the set of generalized internal variables $\boldsymbol{\mathcal{F}}$ into account and is given in terms of the *total derivative* of the yield function

$$\mathcal{D} := \boldsymbol{\mathcal{N}} \circ \boldsymbol{\mathcal{A}} \circ \boldsymbol{\mathcal{M}} + \frac{\partial \phi}{\partial \boldsymbol{\mathcal{F}}} \circ \boldsymbol{\mathcal{M}} =: \frac{\mathrm{d}\phi}{\mathrm{d}\boldsymbol{\mathcal{F}}} \circ \boldsymbol{\mathcal{M}}.$$

Finally the elasto-plastic tangent relation between the material time derivatives of the ${}_0$-type Piola stress and the Piola double-stress and the first and second deformation gradients follows as

$$\begin{bmatrix} {}_0\dot{\boldsymbol{P}} \\ {}^{\Delta}\dot{\boldsymbol{P}} \end{bmatrix} = \begin{bmatrix} \mathbf{A}_{\mathrm{ep}} & \mathbf{A}^{\Delta}_{\mathrm{ep}} \\ {}^{\Delta}\mathbf{A}_{\mathrm{ep}} & {}^{\Delta}\mathbf{A}^{\Delta}_{\mathrm{ep}} \end{bmatrix} \begin{matrix} : \\ \because \end{matrix} \begin{bmatrix} \dot{\boldsymbol{F}} \\ {}^{\Delta}\dot{\boldsymbol{F}} \end{bmatrix}.$$

Here the elasto-plastic tangent operators are given as

$$\begin{bmatrix} \mathbf{A}_{\mathrm{ep}} & \mathbf{A}^{\Delta}_{\mathrm{ep}} \\ {}^{\Delta}\mathbf{A}_{\mathrm{ep}} & {}^{\Delta}\mathbf{A}^{\Delta}_{\mathrm{ep}} \end{bmatrix} = \begin{bmatrix} \mathbf{A} & \mathbf{A}^{\Delta} \\ {}^{\Delta}\mathbf{A} & {}^{\Delta}\mathbf{A}^{\Delta} \end{bmatrix} - \mathcal{D}^{-1} \begin{bmatrix} \dfrac{\partial_0 \boldsymbol{P}}{\partial \boldsymbol{\mathcal{F}}} \circ \boldsymbol{\mathcal{M}} \\ \dfrac{\partial^{\Delta} \boldsymbol{P}}{\partial \boldsymbol{\mathcal{F}}} \circ \boldsymbol{\mathcal{M}} \end{bmatrix} \otimes \begin{bmatrix} \boldsymbol{\mathcal{N}} \circ \dfrac{\partial \boldsymbol{\mathcal{P}}}{\partial \boldsymbol{F}} \\ \boldsymbol{\mathcal{N}} \circ \dfrac{\partial \boldsymbol{\mathcal{P}}}{\partial {}^{\Delta}\boldsymbol{F}} \end{bmatrix}.$$

In the above the Hessian-type tensors of second partial derivatives of the free energy density with respect to $\boldsymbol{F}$, ${}^{\Delta}\boldsymbol{F}$ and $\boldsymbol{\mathcal{F}}$ have been introduced as

$$\begin{bmatrix} \mathbf{A} & \mathbf{A}^{\Delta} \\ {}^{\Delta}\mathbf{A} & {}^{\Delta}\mathbf{A}^{\Delta} \end{bmatrix} := \begin{bmatrix} \dfrac{\partial^2 \psi_0}{\partial \boldsymbol{F} \partial \boldsymbol{F}} & \dfrac{\partial^2 \psi_0}{\partial \boldsymbol{F} \partial^{\Delta} \boldsymbol{F}} \\ \dfrac{\partial^2 \psi_0}{\partial^{\Delta} \boldsymbol{F} \partial \boldsymbol{F}} & \dfrac{\partial^2 \psi_0}{\partial^{\Delta} \boldsymbol{F} \partial^{\Delta} \boldsymbol{F}} \end{bmatrix} \quad \text{and} \quad \boldsymbol{\mathcal{A}} := -\frac{\partial^2 \psi_0}{\partial \boldsymbol{\mathcal{F}} \partial \boldsymbol{\mathcal{F}}}.$$

Observe the major symmetry properties of the extended elasto-plastic tangent operator consisting of A_{ep}, $\mathsf{A}^{\Delta}_{\text{ep}}$, ${}^{\Delta}\mathsf{A}_{\text{ep}}$ and ${}^{\Delta}\mathsf{A}^{\Delta}_{\text{ep}}$ in the case of associated second-order elasto-plasticity.

Spatially Superposed Rigid Body Motions:
Rigid body motions superposed to the *spatial* configuration in terms of temporarily varying rotations $\boldsymbol{Q}(t) \in \text{SO}(3)$ and translations $\boldsymbol{z}(t) \in \text{T}(3)$ result in the transformed deformation gradient and elastic distortion together with the transformed second deformation gradient and elastic (and plastic) double-distortion

$$\boldsymbol{F}^* = \boldsymbol{Q}(t) \cdot \boldsymbol{F}(\boldsymbol{X}, t), \quad \overline{\boldsymbol{F}}^* = \boldsymbol{Q}(t) \cdot \overline{\boldsymbol{F}}(\boldsymbol{X}, t) \quad \text{and} \quad {}^{\Delta}\boldsymbol{F}^*_{\bullet} = \boldsymbol{Q}(t) \cdot {}^{\Delta}\boldsymbol{F}_{\bullet}(\boldsymbol{X}, t).$$

Recall that the plastic distortion remains unaffected by a spatially superposed rigid body motion

$$\widetilde{\boldsymbol{F}}^* = \widetilde{\boldsymbol{F}}(\boldsymbol{X}, t).$$

Furthermore the inverse deformation gradient, its spatial gradient together with the corresponding elastic distortion and elastic (and plastic) double-distortion transform as

$$\boldsymbol{f}^* = \boldsymbol{f} \cdot \boldsymbol{Q}^t \quad \text{and} \quad \overline{\boldsymbol{f}}^* = \overline{\boldsymbol{f}} \cdot \boldsymbol{Q}^t \quad \text{and} \quad {}^{\Delta}\boldsymbol{f}^*_{\bullet} = {}^{\Delta}\boldsymbol{f}_{\bullet} : [\boldsymbol{Q}^t \boxtimes \boldsymbol{Q}^t].$$

Consequently, the material and intermediate description (total, elastic and plastic) connection tensors ${}^{\Delta}\boldsymbol{\Gamma}_{\bullet}$ and $\pm{}^{\Delta}\boldsymbol{\upsilon}_{\bullet} = \mp{}^{\Delta}\boldsymbol{\Upsilon}_{\bullet}$ remain invariant, i.e.

$${}^{\Delta}\boldsymbol{\Gamma}^*_{\bullet} = \boldsymbol{f} \cdot \boldsymbol{Q}^t \cdot \boldsymbol{Q} \cdot {}^{\Delta}\boldsymbol{F}_{\bullet} = {}^{\Delta}\boldsymbol{\Gamma}_{\bullet} \quad \text{and} \quad {}^{\Delta}\boldsymbol{\upsilon}^*_{\bullet} = {}^{\Delta}\boldsymbol{\upsilon}_{\bullet},$$

whereas the spatial description (total, elastic and plastic) connection tensors ${}^{\Delta}\boldsymbol{\gamma}_{\bullet}$ transform objectively under a spatially superposed rigid body motion

$${}^{\Delta}\boldsymbol{\gamma}^*_{\bullet} = \boldsymbol{Q} \cdot \boldsymbol{F} \cdot {}^{\Delta}\boldsymbol{f}_{\bullet} : [\boldsymbol{Q}^t \boxtimes \boldsymbol{Q}^t] = \boldsymbol{Q} \cdot {}^{\Delta}\boldsymbol{\gamma}_{\bullet} : [\boldsymbol{Q}^t \boxtimes \boldsymbol{Q}^t].$$

Then, the contravariant material Piola-Kirchhoff double-stress ${}^{\Delta}\boldsymbol{S}$ and the co/contra/contra- (mixed) variant material Mandel double-stress ${}^{\Delta}\boldsymbol{M}$ remain *invariant*, i.e. unaffected under a spatially superposed rigid body motion

$${}^{\Delta}\boldsymbol{S}^* = {}^{\Delta}\boldsymbol{S} \quad \text{and} \quad {}^{\Delta}\boldsymbol{M}^* = {}^{\Delta}\boldsymbol{M}.$$

Likewise also the corresponding plastic push-forwards to the intermediate configuration in terms of the plastic distortion $\widetilde{\boldsymbol{F}}$ trivially remain invariant, i.e.

$${}^{\Delta}\overline{\boldsymbol{S}}^* = {}^{\Delta}\overline{\boldsymbol{S}} \quad \text{and} \quad {}^{\Delta}\overline{\boldsymbol{M}}^* = {}^{\Delta}\overline{\boldsymbol{M}}.$$

Contrarily, due to the orthogonality of $\boldsymbol{Q}$ the contra- and co/contra/contra- (mixed) variant versions of the spatial Kirchhoff double-stress transform *objectively* under a spatially superposed rigid body motion, for example

$$^{\Delta}\boldsymbol{\tau}^{*} = \boldsymbol{Q} \cdot {}^{\Delta}\boldsymbol{\tau} : [\boldsymbol{Q}^{t} \boxtimes \boldsymbol{Q}^{t}].$$

Moreover, the covariant material (total, elastic and plastic) double-strains and double-metrics ${}^{\Delta}\boldsymbol{C}_{\bullet}$, ${}^{\Delta}\widetilde{\boldsymbol{C}}_{\bullet}$ and ${}^{\Delta}\boldsymbol{J}_{\bullet}$ remain *invariant* under a spatially superposed rigid body motion, i.e.

$${}^{\Delta}\boldsymbol{C}^{*}_{\bullet} = {}^{\Delta}\boldsymbol{C}_{\bullet} \quad \text{and} \quad {}^{\Delta}\widetilde{\boldsymbol{C}}^{*}_{\bullet} = {}^{\Delta}\widetilde{\boldsymbol{C}}_{\bullet} \quad \text{and} \quad {}^{\Delta}\boldsymbol{J}^{*}_{\bullet} = {}^{\Delta}\boldsymbol{J}_{\bullet}.$$

Likewise, the covariant intermediate (total, elastic and plastic) double-strains and double-metrics ${}^{\Delta}\overline{\boldsymbol{C}}_{\bullet}$, $\pm{}^{\Delta}\widetilde{\boldsymbol{j}}_{\bullet} = \mp{}^{\Delta}\overline{\boldsymbol{J}}_{\bullet}$ and ${}^{\Delta}\widetilde{\boldsymbol{c}}_{\bullet}$ trivially remain *invariant* under a spatially superposed rigid body motion, i.e.

$${}^{\Delta}\overline{\boldsymbol{C}}^{*}_{\bullet} = {}^{\Delta}\overline{\boldsymbol{C}}_{\bullet} \quad \text{and} \quad {}^{\Delta}\widetilde{\boldsymbol{j}}^{*}_{\bullet} = {}^{\Delta}\widetilde{\boldsymbol{j}}_{\bullet} \quad \text{and} \quad {}^{\Delta}\widetilde{\boldsymbol{c}}^{*}_{\bullet} = {}^{\Delta}\widetilde{\boldsymbol{c}}_{\bullet}.$$

Vice versa, the covariant spatial (total, elastic and plastic) double-strains and double-metrics ${}^{\Delta}\boldsymbol{j}_{\bullet}$, ${}^{\Delta}\overline{\boldsymbol{c}}_{\bullet}$ and ${}^{\Delta}\boldsymbol{c}_{\bullet}$ transform *objectively* under a spatially superposed rigid body motion, i.e.

$$\begin{aligned}
{}^{\Delta}\boldsymbol{j}^{*}_{\bullet} &= -\boldsymbol{Q} \cdot \boldsymbol{j} \cdot \boldsymbol{Q}^{t} \cdot \boldsymbol{Q} \cdot {}^{\Delta}\boldsymbol{\gamma}_{\bullet} : [\boldsymbol{Q}^{t} \boxtimes \boldsymbol{Q}^{t}] = \boldsymbol{Q} \cdot \boldsymbol{j}_{\bullet} : [\boldsymbol{Q}^{t} \boxtimes \boldsymbol{Q}^{t}],\\
{}^{\Delta}\overline{\boldsymbol{c}}^{*}_{\bullet} &= \quad \boldsymbol{Q} \cdot \overline{\boldsymbol{c}} \cdot \boldsymbol{Q}^{t} \cdot \boldsymbol{Q} \cdot {}^{\Delta}\boldsymbol{\gamma}_{\bullet} : [\boldsymbol{Q}^{t} \boxtimes \boldsymbol{Q}^{t}] = \boldsymbol{Q} \cdot \overline{\boldsymbol{c}}_{\bullet} : [\boldsymbol{Q}^{t} \boxtimes \boldsymbol{Q}^{t}],\\
{}^{\Delta}\boldsymbol{c}^{*}_{\bullet} &= \quad \boldsymbol{Q} \cdot \boldsymbol{c} \cdot \boldsymbol{Q}^{t} \cdot \boldsymbol{Q} \cdot {}^{\Delta}\boldsymbol{\gamma}_{\bullet} : [\boldsymbol{Q}^{t} \boxtimes \boldsymbol{Q}^{t}] = \boldsymbol{Q} \cdot \boldsymbol{c}_{\bullet} : [\boldsymbol{Q}^{t} \boxtimes \boldsymbol{Q}^{t}].
\end{aligned}$$

Recall that spatial objectivity is a key requirement in the theory of constitutive modeling.

Materially Superposed Rigid Body Motions:
Rigid body motions superposed to the *material* configuration in terms of temporarily varying rotations $\boldsymbol{q}(t) \in \mathrm{SO}(3)$ with $\boldsymbol{Q} = \boldsymbol{q}^{-1} = \boldsymbol{q}^{t}$ and translations $\boldsymbol{Z}(t) \in \mathrm{T}(3)$ result in the transformed deformation gradient and plastic distortion together with the transformed second deformation gradient and plastic (and elastic) double-distortion

$$\boldsymbol{F}_{*} = \boldsymbol{F}(\boldsymbol{X},t) \cdot \boldsymbol{Q}(t), \quad \widetilde{\boldsymbol{F}}_{*} = \boldsymbol{q}(t) \cdot \widetilde{\boldsymbol{F}}(\boldsymbol{X},t) \cdot \boldsymbol{Q}(t), \quad {}^{\Delta}\boldsymbol{F}_{\bullet *} = {}^{\Delta}\boldsymbol{F}_{\bullet}(\boldsymbol{X},t) : \mathbf{Q}(t).$$

Here the abbreviation $\mathbf{Q} := \boldsymbol{Q} \boxtimes \boldsymbol{Q}$ has been introduced for a more concise presentation. Recall that due to viewpoint of the intermediate configuration rotating in complete slavery with the material configuration the elastic distortion also transforms under a materially superposed rigid body motion as

$$\overline{\boldsymbol{F}}_{*} = \overline{\boldsymbol{F}}(\boldsymbol{X},t) \cdot \boldsymbol{Q}(t).$$

Furthermore the inverse deformation gradient, its spatial gradient together with the corresponding plastic distortion and plastic (and elastic) double-distortion transform as

$$\boldsymbol{f}_{*} = \boldsymbol{Q}^{t} \cdot \boldsymbol{f} \quad \text{and} \quad \widetilde{\boldsymbol{f}}_{*} = \boldsymbol{Q}^{t} \cdot \widetilde{\boldsymbol{f}} \cdot \boldsymbol{q}^{t} \quad \text{and} \quad {}^{\Delta}\boldsymbol{f}_{\bullet *} = \boldsymbol{Q}^{t} \cdot {}^{\Delta}\boldsymbol{f}_{\bullet}.$$

Consequently, the spatial description (total, elastic and plastic) connection tensors ${}^{\vartriangle}\gamma_{\bullet}$ remain invariant, i.e.

$$ {}^{\vartriangle}\gamma_{\bullet *} = \boldsymbol{F} \cdot \boldsymbol{Q} \cdot \boldsymbol{Q}^t \cdot {}^{\vartriangle}\boldsymbol{f}_{\bullet} = {}^{\vartriangle}\gamma_{\bullet}, $$

whereas the material and intermediate description (total, elastic and plastic) connection tensors ${}^{\vartriangle}\boldsymbol{\Gamma}_{\bullet}$ and $\pm{}^{\vartriangle}\boldsymbol{v}_{\bullet} = \mp{}^{\vartriangle}\boldsymbol{\Upsilon}_{\bullet}$ transform objectively under a materially superposed rigid body motion

$$ {}^{\vartriangle}\boldsymbol{\Gamma}_{\bullet *} = \boldsymbol{Q}^t \cdot \boldsymbol{f} \cdot {}^{\vartriangle}\boldsymbol{F}_{\bullet} : \mathbf{Q} = \boldsymbol{Q}^t \cdot {}^{\vartriangle}\boldsymbol{\Gamma}_{\bullet} : \mathbf{Q} \quad \text{and} \quad {}^{\vartriangle}\boldsymbol{v}_{\bullet *} = \boldsymbol{Q}^t \cdot {}^{\vartriangle}\boldsymbol{v}_{\bullet} : \mathbf{Q}. $$

Then, the contra- and co/contra- (mixed) variant versions of the spatial Kirchhoff double-stress remain *invariant* under a materially superposed rigid body motion, for example

$$ {}^{\vartriangle}\boldsymbol{\tau}_{*} = {}^{\vartriangle}\boldsymbol{\tau}. $$

Contrarily, due to the orthogonality of $\boldsymbol{Q}$ the contravariant intermediate Piola-Kirchhoff double-stress ${}^{\vartriangle}\overline{\boldsymbol{S}}$ and the co/contra- (mixed) variant intermediate Mandel double- stress ${}^{\vartriangle}\overline{\boldsymbol{M}}$ transform *objectively* under a materially superposed rigid body motion, i.e.

$$ {}^{\vartriangle}\overline{\boldsymbol{S}}_{*} = \boldsymbol{Q}^t \cdot {}^{\vartriangle}\overline{\boldsymbol{S}} : [\boldsymbol{Q} \boxtimes \boldsymbol{Q}] \quad \text{and} \quad {}^{\vartriangle}\overline{\boldsymbol{M}}_{*} = \boldsymbol{Q}^t \cdot {}^{\vartriangle}\overline{\boldsymbol{M}} : [\boldsymbol{Q} \boxtimes \boldsymbol{Q}]. $$

Likewise also the corresponding plastic pull-backs to the material configuration in terms of the plastic distortion $\widetilde{\boldsymbol{F}}$ transform *objectively* under a materially superposed rigid body motion, i.e.

$$ {}^{\vartriangle}\boldsymbol{S}_{*} = \boldsymbol{Q}^t \cdot {}^{\vartriangle}\boldsymbol{S} : [\boldsymbol{Q} \boxtimes \boldsymbol{Q}] \quad \text{and} \quad {}^{\vartriangle}\boldsymbol{M}_{*} = \boldsymbol{Q}^t \cdot {}^{\vartriangle}\boldsymbol{M} : [\boldsymbol{Q} \boxtimes \boldsymbol{Q}]. $$

Moreover, the covariant material (total, elastic and plastic) double-strains and double-metrics ${}^{\vartriangle}\boldsymbol{C}_{\bullet}$, ${}^{\vartriangle}\widetilde{\boldsymbol{C}}_{\bullet}$ and ${}^{\vartriangle}\boldsymbol{J}_{\bullet}$ transform *objectively* under a materially superposed rigid body motion, i.e.

$$ \begin{aligned} {}^{\vartriangle}\boldsymbol{C}_{\bullet *} &= \boldsymbol{Q}^t \cdot \boldsymbol{C} \cdot \boldsymbol{Q} \cdot \boldsymbol{Q}^t \cdot {}^{\vartriangle}\boldsymbol{\Gamma}_{\bullet} : [\boldsymbol{Q} \boxtimes \boldsymbol{Q}] = \boldsymbol{Q}^t \cdot {}^{\vartriangle}\boldsymbol{C}_{\bullet} : [\boldsymbol{Q} \boxtimes \boldsymbol{Q}], \\ {}^{\vartriangle}\widetilde{\boldsymbol{C}}_{\bullet *} &= \boldsymbol{Q}^t \cdot \widetilde{\boldsymbol{C}} \cdot \boldsymbol{Q} \cdot \boldsymbol{Q}^t \cdot {}^{\vartriangle}\boldsymbol{\Gamma}_{\bullet} : [\boldsymbol{Q} \boxtimes \boldsymbol{Q}] = \boldsymbol{Q}^t \cdot {}^{\vartriangle}\widetilde{\boldsymbol{C}}_{\bullet} : [\boldsymbol{Q} \boxtimes \boldsymbol{Q}], \\ {}^{\vartriangle}\boldsymbol{J}_{\bullet *} &= -\boldsymbol{Q}^t \cdot \boldsymbol{J} \cdot \boldsymbol{Q} \cdot \boldsymbol{Q}^t \cdot {}^{\vartriangle}\boldsymbol{\Gamma}_{\bullet} : [\boldsymbol{Q} \boxtimes \boldsymbol{Q}] = \boldsymbol{Q}^t \cdot {}^{\vartriangle}\boldsymbol{J}_{\bullet} : [\boldsymbol{Q} \boxtimes \boldsymbol{Q}]. \end{aligned} $$

Likewise, the covariant intermediate (total, elastic and plastic) double-strains and double-metrics ${}^{\vartriangle}\overline{\boldsymbol{C}}_{\bullet}$, $\pm{}^{\vartriangle}\widetilde{\boldsymbol{\jmath}}_{\bullet} = \mp{}^{\vartriangle}\overline{\boldsymbol{J}}_{\bullet}$ and ${}^{\vartriangle}\widetilde{\boldsymbol{c}}_{\bullet}$ trivially transform *objectively* under a materially superposed rigid body motion, i.e.

$$ \begin{aligned} {}^{\vartriangle}\overline{\boldsymbol{C}}_{\bullet *} &= \boldsymbol{Q}^t \cdot \overline{\boldsymbol{C}} \cdot \boldsymbol{Q} \cdot \boldsymbol{Q}^t \cdot {}^{\vartriangle}\boldsymbol{\Upsilon}_{\bullet} : [\boldsymbol{Q} \boxtimes \boldsymbol{Q}] = \boldsymbol{Q}^t \cdot {}^{\vartriangle}\overline{\boldsymbol{C}}_{\bullet} : [\boldsymbol{Q} \boxtimes \boldsymbol{Q}], \\ {}^{\vartriangle}\widetilde{\boldsymbol{\jmath}}_{\bullet *} &= -\boldsymbol{Q}^t \cdot \widetilde{\boldsymbol{\jmath}} \cdot \boldsymbol{Q} \cdot \boldsymbol{Q}^t \cdot {}^{\vartriangle}\boldsymbol{v}_{\bullet} : [\boldsymbol{Q} \boxtimes \boldsymbol{Q}] = \boldsymbol{Q}^t \cdot {}^{\vartriangle}\widetilde{\boldsymbol{\jmath}}_{\bullet} : [\boldsymbol{Q} \boxtimes \boldsymbol{Q}], \\ {}^{\vartriangle}\widetilde{\boldsymbol{c}}_{\bullet *} &= \boldsymbol{Q}^t \cdot \widetilde{\boldsymbol{c}} \cdot \boldsymbol{Q} \cdot \boldsymbol{Q}^t \cdot {}^{\vartriangle}\boldsymbol{v}_{\bullet} : [\boldsymbol{Q} \boxtimes \boldsymbol{Q}] = \boldsymbol{Q}^t \cdot {}^{\vartriangle}\widetilde{\boldsymbol{c}}_{\bullet} : [\boldsymbol{Q} \boxtimes \boldsymbol{Q}]. \end{aligned} $$

Vice versa, the covariant spatial (total, elastic and plastic) double-strains and double-metrics ${}^{\vartriangle}\boldsymbol{j}_\bullet$, ${}^{\vartriangle}\overline{\boldsymbol{c}}_\bullet$ and ${}^{\vartriangle}\boldsymbol{c}_\bullet$ remain *invariant* under a materially superposed rigid body motion, i.e.

$$ {}^{\vartriangle}\boldsymbol{j}_{\bullet *} = \boldsymbol{j}_\bullet \quad \text{and} \quad {}^{\vartriangle}\overline{\boldsymbol{c}}_{\bullet *} = \overline{\boldsymbol{c}}_\bullet \quad \text{and} \quad {}^{\vartriangle}\boldsymbol{c}_{\bullet *} = \boldsymbol{c}_\bullet . $$

Intermediate and material double-stress and double-strain measures transforming objectively under materially superposed rigid body motions are suited to capture anisotropic material behavior.

Restrictions on the Free Energy Density:
Requiring the free energy density ψ_0 to transform objectively under arbitrary spatially superposed rigid body motion, i.e. $\psi_0 \doteq \psi_0^*$, results in the alternative restrictions

$$
\begin{aligned}
\psi_0 &\doteq \psi_0^* = \psi_0(\overline{\boldsymbol{C}}, {}^{\vartriangle}\boldsymbol{C}_\mathrm{e}; \cdots), \\
\psi_0 &\doteq \psi_0^* = \psi_0(\overline{\boldsymbol{C}}, {}^{\vartriangle}\boldsymbol{\Gamma}_\mathrm{e}; \cdots), \\
\psi_0 &\doteq \psi_0^* = \psi_0(\boldsymbol{Q}\cdot\overline{\boldsymbol{c}}\cdot\boldsymbol{Q}^t, \boldsymbol{Q}\cdot{}^{\vartriangle}\boldsymbol{c}_\mathrm{e} : [\boldsymbol{Q}^t \boxtimes \boldsymbol{Q}^t]; \cdots) \quad \forall \boldsymbol{Q} \in \mathrm{SO}(3), \\
\psi_0 &\doteq \psi_0^* = \psi_0(\boldsymbol{Q}\cdot\overline{\boldsymbol{c}}\cdot\boldsymbol{Q}^t, \boldsymbol{Q}\cdot{}^{\vartriangle}\boldsymbol{\gamma}_\mathrm{e} : [\boldsymbol{Q}^t \boxtimes \boldsymbol{Q}^t]; \cdots) \quad \forall \boldsymbol{Q} \in \mathrm{SO}(3).
\end{aligned}
$$

Thus an arbitrary dependence on the intermediate description strain measure $\overline{\boldsymbol{C}}$ and on either of the material description elastic double-strain measures ${}^{\vartriangle}\boldsymbol{C}_\mathrm{e}$ or ${}^{\vartriangle}\boldsymbol{\Gamma}_\mathrm{e}$ automatically satisfies spatial objectivity of ψ_0. In contrast the dependence on the spatial description strain measure $\overline{\boldsymbol{c}}$ and on either of the spatial description elastic double-strain measures ${}^{\vartriangle}\boldsymbol{c}_\mathrm{e}$ or ${}^{\vartriangle}\boldsymbol{\gamma}_\mathrm{e}$ is subject to the above *isotropy* conditions on the tensor function ψ_0 in order to satisfy spatial objectivity. However, the yet unspecified arguments of ψ_0 indicated by the $\cdots$ allow to also capture *anisotropic* material behavior. Based on (constant) intermediate and material structural tensors $\overline{\boldsymbol{M}}_\alpha$ and ${}^{\vartriangle}\boldsymbol{M}_\alpha$ and in accordance with the above restriction of spatial objectivity the free energy density ψ_0 then takes the following alternative representations

$$
\begin{aligned}
\psi_0 &= \psi_0(\{\overline{\boldsymbol{C}}; \overline{\boldsymbol{I}}, \overline{\boldsymbol{M}}_\alpha\}, \{{}^{\vartriangle}\boldsymbol{C}_\mathrm{e}; {}^{\vartriangle}\boldsymbol{I}, {}^{\vartriangle}\boldsymbol{M}_\alpha\}) \\
\psi_0 &= \psi_0(\{\overline{\boldsymbol{C}}; \overline{\boldsymbol{I}}, \overline{\boldsymbol{M}}_\alpha\}, \{{}^{\vartriangle}\boldsymbol{\Gamma}_\mathrm{e}; {}^{\vartriangle}\boldsymbol{I}, {}^{\vartriangle}\boldsymbol{M}_\alpha\}).
\end{aligned}
$$

The above groupings indicated by the curly brackets $\{\bullet\}$ emphasize the understanding that the $\overline{\boldsymbol{M}}_\alpha$ and the ${}^{\vartriangle}\boldsymbol{M}_\alpha$ capture the independent material's symmetry groups separately for the strains in the intermediate configuration and for the double-strains in the material configuration. Here the contravariant intermediate and material unit tensors $\overline{\boldsymbol{I}}$ and ${}^{\vartriangle}\boldsymbol{I}$ are explicitly incorporated in order to capture the isotropic part of the response. Finally, requiring invariance of ψ_0 under arbitrary materially superposed rigid body motions, i.e. the intermediate strain and the material double-strain measures are transformed identically to the corresponding intermediate and material structural tensors, results in the following alternative restrictions

$$\psi_0 \doteq \psi_{0*} = \psi_0(\boldsymbol{Q}^t \cdot \{\{\overline{\boldsymbol{C}}; \overline{\boldsymbol{I}}, \overline{\boldsymbol{M}}_\alpha\}, \{{}^\vartriangle\boldsymbol{C}_\mathrm{e}; {}^\vartriangle\boldsymbol{I}, {}^\vartriangle\boldsymbol{M}_\alpha\} \cdot \boldsymbol{Q}\} \cdot \boldsymbol{Q}) \quad \forall \boldsymbol{Q} \in \mathrm{SO}(3),$$
$$\psi_0 \doteq \psi_{0*} = \psi_0(\boldsymbol{Q}^t \cdot \{\{\overline{\boldsymbol{C}}; \overline{\boldsymbol{I}}, \overline{\boldsymbol{M}}_\alpha\}, \{{}^\vartriangle\boldsymbol{\Gamma}_\mathrm{e}; {}^\vartriangle\boldsymbol{I}, {}^\vartriangle\boldsymbol{M}_\alpha\} \cdot \boldsymbol{Q}\} \cdot \boldsymbol{Q}) \quad \forall \boldsymbol{Q} \in \mathrm{SO}(3).$$

These restrictions qualify ψ_0 as an *isotropic tensor function* of all of its arguments, i.e. either $\{\overline{\boldsymbol{C}}; \overline{\boldsymbol{I}}, \overline{\boldsymbol{M}}_\alpha\}$ and $\{{}^\vartriangle\boldsymbol{C}_\mathrm{e}; {}^\vartriangle\boldsymbol{I}, {}^\vartriangle\boldsymbol{M}_\alpha\}$ or $\{\overline{\boldsymbol{C}}; \overline{\boldsymbol{I}}, \overline{\boldsymbol{M}}_\alpha\}$ and ${}^\vartriangle\boldsymbol{\Gamma}_\mathrm{e}; {}^\vartriangle\boldsymbol{I}, {}^\vartriangle\boldsymbol{M}_\alpha\}$. Requiring next covariance of the free energy density ψ_0 under arbitrary superposed material homeomorphisms results in the alternative equivalent statements

$$\begin{aligned}\psi_0 \doteq \psi_0(\ &\{\underbrace{\overline{\boldsymbol{f}}^{\star t} \cdot \overline{\boldsymbol{C}} \cdot \overline{\boldsymbol{f}}^{\star}}_{=:\overline{\boldsymbol{C}}^\star}; \underbrace{\overline{\boldsymbol{F}}^{\star} \cdot \overline{\boldsymbol{I}} \cdot \overline{\boldsymbol{F}}^{\star t}}_{=:\overline{\boldsymbol{I}}^\star}, \underbrace{\overline{\boldsymbol{F}}^{\star} \cdot \overline{\boldsymbol{M}}_\alpha \cdot \overline{\boldsymbol{F}}^{\star t}}_{=:\overline{\boldsymbol{M}}^\star_\alpha}\},\\ &\{\underbrace{\boldsymbol{f}^{\star t} \cdot {}^\vartriangle\boldsymbol{C}_\mathrm{e} : \mathsf{f}^\star}_{=:{}^\vartriangle\boldsymbol{C}^\star_\mathrm{e}}; \underbrace{\boldsymbol{F}^\star \cdot {}^\vartriangle\boldsymbol{I} : \mathsf{F}^{\star t}}_{=:{}^\vartriangle\boldsymbol{I}^\star}, \underbrace{\boldsymbol{F}^\star \cdot {}^\vartriangle\boldsymbol{M}_\alpha : \mathsf{F}^{\star t}}_{=:{}^\vartriangle\boldsymbol{M}^\star_\alpha}\}\ ),\\ \psi_0 \doteq \psi_0(\ &\{\underbrace{\overline{\boldsymbol{f}}^{\star t} \cdot \overline{\boldsymbol{C}} \cdot \overline{\boldsymbol{f}}^{\star}}_{=:\overline{\boldsymbol{C}}^\star}; \underbrace{\overline{\boldsymbol{F}}^{\star} \cdot \overline{\boldsymbol{I}} \cdot \overline{\boldsymbol{F}}^{\star t}}_{=:\overline{\boldsymbol{I}}^\star}, \underbrace{\overline{\boldsymbol{F}}^{\star} \cdot \overline{\boldsymbol{M}}_\alpha \cdot \overline{\boldsymbol{F}}^{\star t}}_{=:\overline{\boldsymbol{M}}^\star_\alpha}\},\\ &\{\underbrace{\boldsymbol{F}^{\star t} \cdot {}^\vartriangle\boldsymbol{\Gamma}_\mathrm{e} : \mathsf{f}^\star}_{=:{}^\vartriangle\boldsymbol{\Gamma}^\star_\mathrm{e}}; \underbrace{\boldsymbol{F}^\star \cdot {}^\vartriangle\boldsymbol{I} : \mathsf{F}^{\star t}}_{=:{}^\vartriangle\boldsymbol{I}^\star}, \underbrace{\boldsymbol{F}^\star \cdot {}^\vartriangle\boldsymbol{M}_\alpha : \mathsf{F}^{\star t}}_{=:{}^\vartriangle\boldsymbol{M}^\star_\alpha}\}\ ).\end{aligned}$$

Here $\boldsymbol{f}^\star$ and $\overline{\boldsymbol{f}}^\star$ abbreviate the inverses $\boldsymbol{f}^\star := \boldsymbol{F}^{\star-1}$ and $\overline{\boldsymbol{f}}^\star := \overline{\boldsymbol{F}}^{\star-1}$ of the homeomorphisms $\boldsymbol{F}^\star$ and $\overline{\boldsymbol{F}}^\star := \boldsymbol{F}^\star \cdot \widetilde{\boldsymbol{f}}$. Then the particular choices i) $\boldsymbol{F}^\star = \boldsymbol{id}$ with $\overline{\boldsymbol{F}}^\star = \widetilde{\boldsymbol{f}}$ (and $\overline{\boldsymbol{f}}^\star = \widetilde{\boldsymbol{F}}$), ii) $\boldsymbol{F}^\star = \widetilde{\boldsymbol{F}}$ with $\overline{\boldsymbol{F}}^\star = \boldsymbol{id}$ (and $\overline{\boldsymbol{f}}^\star = \boldsymbol{id}$), and iii) $\boldsymbol{F}^\star = \boldsymbol{F}$ with $\overline{\boldsymbol{F}}^\star = \overline{\boldsymbol{F}}$ (and $\overline{\boldsymbol{f}}^\star = \overline{\boldsymbol{f}}$) render the alternative equivalent parameterizations of ψ_0

$$\begin{aligned}\psi_0(\{\overline{\boldsymbol{C}}; \overline{\boldsymbol{I}}, \overline{\boldsymbol{M}}_\alpha\}, \{{}^\vartriangle\boldsymbol{C}_\mathrm{e}; {}^\vartriangle\boldsymbol{I}, {}^\vartriangle\boldsymbol{M}_\alpha\}) &\doteq \psi_0(\{\boldsymbol{C}; \widetilde{\boldsymbol{B}}, \boldsymbol{M}_\alpha\}, \{{}^\vartriangle\boldsymbol{C}_\mathrm{e}; {}^\vartriangle\boldsymbol{I}, {}^\vartriangle\boldsymbol{M}_\alpha\}),\\ &\doteq \psi_0(\{\overline{\boldsymbol{C}};\ \overline{\boldsymbol{I}}, \overline{\boldsymbol{M}}_\alpha\}, \{{}^\vartriangle\overline{\boldsymbol{C}}_\mathrm{e}; {}^\vartriangle\widetilde{\boldsymbol{b}}, {}^\vartriangle\overline{\boldsymbol{M}}_\alpha\}),\\ &\doteq \psi_0(\{\ \boldsymbol{j}\ ;\ \overline{\boldsymbol{b}},\ \boldsymbol{m}_\alpha\}, \{\ {}^\vartriangle\boldsymbol{j}_\mathrm{e}\ ; {}^\vartriangle\boldsymbol{b},\ {}^\vartriangle\boldsymbol{m}_\alpha\}),\\ \psi_0(\{\overline{\boldsymbol{C}}; \overline{\boldsymbol{I}}, \overline{\boldsymbol{M}}_\alpha\}, \{{}^\vartriangle\boldsymbol{\Gamma}_\mathrm{e}; {}^\vartriangle\boldsymbol{I}, {}^\vartriangle\boldsymbol{M}_\alpha\}) &\doteq \psi_0(\{\boldsymbol{C}; \widetilde{\boldsymbol{B}}, \boldsymbol{M}_\alpha\}, \{{}^\vartriangle\boldsymbol{\Gamma}_\mathrm{e}\ ; {}^\vartriangle\boldsymbol{I}, {}^\vartriangle\boldsymbol{M}_\alpha\}),\\ &\doteq \psi_0(\{\overline{\boldsymbol{C}};\ \overline{\boldsymbol{I}}, \overline{\boldsymbol{M}}_\alpha\}, \{{}^\vartriangle\boldsymbol{v}_\mathrm{e}\ ; {}^\vartriangle\widetilde{\boldsymbol{b}}, {}^\vartriangle\overline{\boldsymbol{M}}_\alpha\}),\\ &\doteq \psi_0(\{\ \boldsymbol{j}\ ;\ \overline{\boldsymbol{b}},\ \boldsymbol{m}_\alpha\}, \{{}^\vartriangle\boldsymbol{\gamma}_\mathrm{e}\ ; {}^\vartriangle\boldsymbol{b},\ {}^\vartriangle\boldsymbol{m}_\alpha\}).\end{aligned}$$

Finally, requiring objectivity of ψ_0 under arbitrary spatially superposed rigid body motions results in the following alternative restrictions

$$\psi_0 \doteq \psi_0^* = \psi_0(\boldsymbol{Q} \cdot \{\{\boldsymbol{j}; \overline{\boldsymbol{b}}, \boldsymbol{m}_\alpha\}, \{{}^\vartriangle\boldsymbol{j}_\mathrm{e}; {}^\vartriangle\boldsymbol{b}, {}^\vartriangle\boldsymbol{m}_\alpha\} \cdot \boldsymbol{Q}^t\} \cdot \boldsymbol{Q}^t) \quad \forall \boldsymbol{Q} \in \mathrm{SO}(3),$$
$$\psi_0 \doteq \psi_0^* = \psi_0(\boldsymbol{Q} \cdot \{\{\boldsymbol{j}; \overline{\boldsymbol{b}}, \boldsymbol{m}_\alpha\}, \{{}^\vartriangle\boldsymbol{\gamma}_\mathrm{e}; {}^\vartriangle\boldsymbol{b}, {}^\vartriangle\boldsymbol{m}_\alpha\} \cdot \boldsymbol{Q}^t\} \cdot \boldsymbol{Q}^t) \quad \forall \boldsymbol{Q} \in \mathrm{SO}(3).$$

These restrictions qualify ψ_0 again as an *isotropic tensor function* of its arguments, i.e. either $\{\boldsymbol{j}; \overline{\boldsymbol{b}}, \boldsymbol{m}_\alpha\}$ and $\{{}^\vartriangle\boldsymbol{j}_\mathrm{e}; {}^\vartriangle\boldsymbol{b}, {}^\vartriangle\boldsymbol{m}_\alpha\}$ or $\{\boldsymbol{j}; \overline{\boldsymbol{b}}, \boldsymbol{m}_\alpha\}$ and $\{{}^\vartriangle\boldsymbol{\gamma}_\mathrm{e}; {}^\vartriangle\boldsymbol{b}, {}^\vartriangle\boldsymbol{m}_\alpha\}$, respectively.

Stress Response:
Free energy densities ψ_0 that satisfy all requirements resulting from spatial objectivity, material symmetry, and covariance are isotropic tensor functions of the material, intermediate or spatial, co- and contravariant arguments

$$\{\boldsymbol{C}; \widetilde{\boldsymbol{B}}, \boldsymbol{M}_\alpha\}, \{{}^\Delta\boldsymbol{C}_{\rm e}; {}^\Delta\boldsymbol{I}, {}^\Delta\boldsymbol{M}_\alpha\} \leftrightarrow \{\overline{\boldsymbol{C}}; \overline{\boldsymbol{I}}, \overline{\boldsymbol{M}}_\alpha\}, \{{}^\Delta\overline{\boldsymbol{C}}_{\rm e}; {}^\Delta\widetilde{\boldsymbol{b}}, {}^\Delta\overline{\boldsymbol{M}}_\alpha\}$$
$$\leftrightarrow \{\boldsymbol{j}\,;\, \overline{\boldsymbol{b}},\, \boldsymbol{m}_\alpha\}, \{\,{}^\Delta\boldsymbol{j}_{\rm e}\,; {}^\Delta\boldsymbol{b},\, {}^\Delta\boldsymbol{m}_\alpha\}.$$

Then since for a scalar function $\mathrm{D}_t\psi_0 \equiv \widetilde{\mathrm{L}}_t\psi_0$ the comparison of the suited time derivative of ψ_0 with the appropriate version of the elastic part of the stress/double-stress power density $\wp_0$ renders corresponding constitutive relations for the fully contravariant intermediate (symmetric) stress and double-stress measures

$$\overline{\boldsymbol{S}} = 2\frac{\partial\psi_0}{\partial\overline{\boldsymbol{C}}} = \widetilde{\boldsymbol{\tau}}_\sharp \quad \text{and} \quad {}^\Delta\overline{\boldsymbol{S}} = \frac{\partial\psi_0}{\partial{}^\Delta\overline{\boldsymbol{C}}_{\rm e}} = {}^\Delta\widetilde{\boldsymbol{\tau}}_\sharp.$$

Moreover since for a scalar function $\widetilde{\mathrm{R}}_t\psi_0 \equiv \mathrm{D}_t\psi_0$ the comparison of the suited time derivative of ψ_0 with the appropriate version of the elastic part of the stress/double-stress power density $\wp_0$ renders corresponding constitutive relations for the fully contravariant material (symmetric) stress and double-stress measures

$$\boldsymbol{S} = 2\frac{\partial\psi_0}{\partial\boldsymbol{C}} \quad \text{and} \quad {}^\Delta\boldsymbol{S} = \frac{\partial\psi_0}{\partial{}^\Delta\boldsymbol{C}_{\rm e}}.$$

Finally since for a scalar function $\overline{\mathrm{L}}_t\psi_0 \equiv \mathrm{L}_t\psi_0$ the comparison of the suited time derivative of ψ_0 with the appropriate version of the elastic part of the stress/double-stress power density $\wp_0$ renders corresponding constitutive relations for the fully contravariant spatial (symmetric) stress and double-stress measures

$$\boldsymbol{\tau}_\sharp = 2\frac{\partial\psi_0}{\partial\boldsymbol{j}} \quad \text{and} \quad {}^\Delta\boldsymbol{\tau}_\sharp = \frac{\partial\psi_0}{\partial{}^\Delta\boldsymbol{j}_{\rm e}}.$$

Alternatively the free energy density ψ_0 is expressed as an isotropic tensor functions of the material, intermediate or spatial, co- and contravariant arguments

$$\{\boldsymbol{C}; \widetilde{\boldsymbol{B}}, \boldsymbol{M}_\alpha\}, \{{}^\Delta\boldsymbol{\Gamma}_{\rm e}\,; {}^\Delta\boldsymbol{I}, {}^\Delta\boldsymbol{M}_\alpha\} \leftrightarrow \{\overline{\boldsymbol{C}}; \overline{\boldsymbol{I}}, \overline{\boldsymbol{M}}_\alpha\}, \{{}^\Delta\boldsymbol{\upsilon}_{\rm e}\,; {}^\Delta\widetilde{\boldsymbol{b}}, {}^\Delta\overline{\boldsymbol{M}}_\alpha\}$$
$$\leftrightarrow \{\boldsymbol{j}\,;\, \overline{\boldsymbol{b}},\, \boldsymbol{m}_\alpha\}, \{{}^\Delta\boldsymbol{\gamma}_{\rm e}\,; {}^\Delta\boldsymbol{b},\, {}^\Delta\boldsymbol{m}_\alpha\}.$$

Then, with ${}^\Delta_t\boldsymbol{\ell}_{\rm e} = \widetilde{\mathrm{L}}_t{}^\Delta\boldsymbol{\Upsilon}_{\rm e} - \boldsymbol{\Psi} = -\widetilde{\mathrm{L}}_t{}^\Delta\boldsymbol{\upsilon}_{\rm e} + \boldsymbol{\psi}$ (and ${}^\Delta_t\boldsymbol{\ell}_{\rm p} = \widetilde{\mathrm{L}}_t{}^\Delta\boldsymbol{\Upsilon}_{\rm p} + \boldsymbol{\Psi} = -\widetilde{\mathrm{L}}_t{}^\Delta\boldsymbol{\upsilon}_{\rm p} - \boldsymbol{\psi}$), the comparison of the suited time derivative of ψ_0 with the appropriate version of the elastic part of the stress/double-stress power density $\wp_0$ renders corresponding constitutive relations for the intermediate stress and double-stress measures

$$ {}_t\overline{\boldsymbol{S}} = 2\frac{\partial \psi_0}{\partial \overline{\boldsymbol{C}}} = {}_t\widetilde{\boldsymbol{\tau}}_\sharp \quad \text{and} \quad {}^\Delta\overline{\boldsymbol{M}} = \frac{\partial \psi_0}{\partial {}^\Delta\boldsymbol{\Upsilon}_\text{e}} = {}^\Delta\widetilde{\boldsymbol{\tau}} = -\frac{\partial \psi_0}{\partial {}^\Delta\boldsymbol{\upsilon}_\text{e}}. $$

Moreover, with ${}^\Delta_t\boldsymbol{L}_\text{e} = \text{D}_t{}^\Delta\boldsymbol{\Gamma}_\text{e} - \boldsymbol{\Xi}$ (and ${}^\Delta_t\boldsymbol{L}_\text{p} = \text{D}_t{}^\Delta\boldsymbol{\Gamma}_\text{p} + \boldsymbol{\Xi}$), the comparison of the suited time derivative of ψ_0 with the appropriate version of the elastic part of the stress/double-stress power density $\wp_0$ renders corresponding constitutive relations for the material stress and double-stress measures

$$ {}_t\boldsymbol{S} = 2\frac{\partial \psi_0}{\partial \boldsymbol{C}} \quad \text{and} \quad {}^\Delta\boldsymbol{M} = \frac{\partial \psi_0}{\partial {}^\Delta\boldsymbol{\Gamma}_\text{e}}. $$

Finally, with ${}^\Delta_t\boldsymbol{l}_\text{e} = -\,\text{L}_t{}^\Delta\gamma_\text{e} + \boldsymbol{\xi}$ (and ${}^\Delta_t\boldsymbol{l}_\text{p} = -\,\text{L}_t{}^\Delta\gamma_\text{p} - \boldsymbol{\xi}$), the comparison of the suited time derivative of ψ_0 with the appropriate version of the elastic part of the stress/double-stress power density $\wp_0$ renders corresponding constitutive relations for the spatial stress and double-stress measures

$$ {}_t\boldsymbol{\tau}_\sharp = 2\frac{\partial \psi_0}{\partial \boldsymbol{j}} \quad \text{and} \quad {}^\Delta\boldsymbol{\tau} = -\frac{\partial \psi_0}{\partial {}^\Delta\gamma_\text{e}}. $$

Which of the above alternative constitutive relations is chosen is essentially a matter of convenience.

Restrictions on the Yield Function:
Requiring the yield function ϕ to transform objectively under arbitrary spatially superposed rigid body motions (i.e. $\phi \doteq \phi^*$) results in the alternative restrictions

$$ \begin{aligned} \phi &\doteq \phi^* = \phi({}_\tau\boldsymbol{M}, {}^\Delta\boldsymbol{M}; \cdots), \\ \phi &\doteq \phi^* = \phi({}_\tau\overline{\boldsymbol{M}}, {}^\Delta\overline{\boldsymbol{M}}; \cdots), \\ \phi &\doteq \phi^* = \phi(\boldsymbol{Q} \cdot {}_\tau\boldsymbol{\tau} \cdot \boldsymbol{Q}^t, \boldsymbol{Q} \cdot {}^\Delta\boldsymbol{\tau} : [\boldsymbol{Q}^t \boxtimes \boldsymbol{Q}^t]; \cdots) \quad \forall \boldsymbol{Q} \in \text{SO}(3). \end{aligned} $$

Thus an arbitrary dependence on the material description ${}_\tau$-type Mandel stress ${}_\tau\boldsymbol{M}$ and double-stress ${}^\Delta\boldsymbol{M}$ or the intermediate description ${}_\tau$-type Mandel stress ${}_\tau\overline{\boldsymbol{M}}$ and double-stress ${}^\Delta\overline{\boldsymbol{M}}$ automatically satisfies spatial objectivity of ϕ. In contrast the dependence on the spatial description ${}_\tau$-type Kirchhoff stress ${}_\tau\boldsymbol{\tau}$ and double-stress ${}^\Delta\boldsymbol{\tau}$ is subject to the above *isotropy* condition on the tensor function ϕ in order to satisfy spatial objectivity. However, the yet unspecified arguments of ϕ indicated by the $\cdots$ allow to also capture *anisotropic* yielding behavior. Based on material and intermediate contra/co- and contra/co/co- (mixed) variant anisotropy tensors $\boldsymbol{A}_\alpha$, $\overline{\boldsymbol{A}}_\alpha$ and ${}^\Delta\boldsymbol{A}_\alpha$, ${}^\Delta\overline{\boldsymbol{A}}_\alpha$ and in accordance with the above restriction of spatial objectivity the yield function ϕ then takes the following alternative representations

$$ \phi = \phi(\{{}_\tau\boldsymbol{M}; \boldsymbol{A}_\alpha\}, \{{}^\Delta\boldsymbol{M}; {}^\Delta\boldsymbol{A}_\alpha\}) \quad \text{and} \quad \phi = \phi(\{{}_\tau\overline{\boldsymbol{M}}; \overline{\boldsymbol{A}}_\alpha\}, \{{}^\Delta\overline{\boldsymbol{M}}; {}^\Delta\overline{\boldsymbol{A}}_\alpha\}). $$

Here the explicit dependence on the contra/co- and contra/co/co- (mixed) variant unit tensors $\boldsymbol{A}_0 = \boldsymbol{1}_\text{m}$, $\overline{\boldsymbol{A}}_0 = \boldsymbol{1}_\text{i}$ and ${}^\Delta\boldsymbol{A}_0 = {}^\Delta\boldsymbol{1}_\text{m} := {}^\Delta\boldsymbol{\Gamma}$, ${}^\Delta\overline{\boldsymbol{A}}_0 =$

${}^{\vartriangle}\mathbf{1}_{\mathrm{i}} := {}^{\vartriangle}\boldsymbol{\Upsilon}$ has been suppressed for the ease of representation. Next, requiring invariance of ϕ under arbitrary materially superposed rigid body motions, i.e. the material and the intermediate ${}_{\tau}$-type Mandel stress and Mandel double-stress are transformed identically to the material and intermediate anisotropy tensors, results in the following restrictions

$$
\begin{aligned}
\phi \doteq \phi_{*} &= \phi(\boldsymbol{Q}^{t} \cdot \{\{{}_{\tau}\boldsymbol{M}; \boldsymbol{A}_{\alpha}\}, \{{}^{\vartriangle}\boldsymbol{M} \cdot \boldsymbol{Q}; {}^{\vartriangle}\boldsymbol{A}_{\alpha} \cdot \boldsymbol{Q}\}\} \cdot \boldsymbol{Q}) \quad \forall \boldsymbol{Q} \in \mathrm{SO}(3), \\
\phi \doteq \phi_{*} &= \phi(\boldsymbol{Q}^{t} \cdot \{\{{}_{\tau}\overline{\boldsymbol{M}}; \overline{\boldsymbol{A}}_{\alpha}\}, \{{}^{\vartriangle}\overline{\boldsymbol{M}} \cdot \boldsymbol{Q}; {}^{\vartriangle}\overline{\boldsymbol{A}}_{\alpha} \cdot \boldsymbol{Q}\}\} \cdot \boldsymbol{Q}) \quad \forall \boldsymbol{Q} \in \mathrm{SO}(3).
\end{aligned}
$$

These restrictions qualify ϕ as an *isotropic tensor function* of its arguments, i.e. either $\{{}_{\tau}\boldsymbol{M}; \boldsymbol{A}_{\alpha}\}$ and $\{{}^{\vartriangle}\boldsymbol{M}; {}^{\vartriangle}\boldsymbol{A}_{\alpha}\}$ or likewise $\{{}_{\tau}\overline{\boldsymbol{M}}; \overline{\boldsymbol{A}}_{\alpha}\}$ and $\{{}^{\vartriangle}\overline{\boldsymbol{M}}; {}^{\vartriangle}\overline{\boldsymbol{A}}_{\alpha}\}$. Representation theorems provide the most general dependence of isotropic tensor functions on their arguments, as a result valid alternative versions of the yield function ϕ read

$$
\begin{aligned}
\phi &= \phi(\{\boldsymbol{S}\,; \boldsymbol{C}, [\boldsymbol{C} \cdot \boldsymbol{A}_{\alpha}]\}, \{{}^{\vartriangle}\boldsymbol{S}\,; {}^{\vartriangle}\boldsymbol{C}, [\boldsymbol{C} \cdot {}^{\vartriangle}\boldsymbol{A}_{\alpha}]\}) \\
&\equiv \phi(\{\overline{\boldsymbol{S}}\,; \overline{\boldsymbol{C}}, [\overline{\boldsymbol{C}} \cdot \overline{\boldsymbol{A}}_{\alpha}]\}, \{{}^{\vartriangle}\overline{\boldsymbol{S}}\,; {}^{\vartriangle}\overline{\boldsymbol{C}}, [\overline{\boldsymbol{C}} \cdot {}^{\vartriangle}\overline{\boldsymbol{A}}_{\alpha}]\}) \\
&\equiv \phi(\{\boldsymbol{\tau}_{\sharp}; \boldsymbol{j}\,, [\,\boldsymbol{j} \cdot\, \boldsymbol{a}_{\alpha}]\}, \{{}^{\vartriangle}\boldsymbol{\tau}_{\sharp}; {}^{\vartriangle}\boldsymbol{j}\,, [\,\boldsymbol{j} \cdot\, {}^{\vartriangle}\boldsymbol{a}_{\alpha}]\}).
\end{aligned}
$$

Requiring finally objectivity of ϕ under arbitrary spatially superposed rigid body motions with $\boldsymbol{Q} \in \mathrm{SO}(3)$ results in the following alternative restrictions

$$
\begin{aligned}
\phi \doteq \phi^{*} &= \phi(\boldsymbol{Q} \cdot \{\{{}_{\tau}\boldsymbol{\tau}\,; \quad \boldsymbol{a}_{\alpha}\}, \{{}^{\vartriangle}\boldsymbol{\tau} \cdot \boldsymbol{Q}^{t}; \quad {}^{\vartriangle}\boldsymbol{a}_{\alpha} \cdot \boldsymbol{Q}^{t}\}\} \cdot \boldsymbol{Q}^{t}), \\
\phi \doteq \phi^{*} &= \phi(\boldsymbol{Q} \cdot \{\{\boldsymbol{\tau}_{\sharp}; \boldsymbol{j}, [\boldsymbol{j} \cdot \boldsymbol{a}_{\alpha}]\}, \{{}^{\vartriangle}\boldsymbol{\tau}_{\sharp} \cdot \boldsymbol{Q}^{t}; {}^{\vartriangle}\boldsymbol{j} \cdot \boldsymbol{Q}^{t}, [\boldsymbol{j} \cdot {}^{\vartriangle}\boldsymbol{a}_{\alpha} \cdot \boldsymbol{Q}^{t}]\}\} \cdot \boldsymbol{Q}^{t}).
\end{aligned}
$$

These restrictions qualify ϕ again as an *isotropic tensor function* of its arguments, i.e. either $\{{}_{\tau}\boldsymbol{\tau}; \boldsymbol{a}_{\alpha}\}$ and $\{{}^{\vartriangle}\boldsymbol{\tau}; {}^{\vartriangle}\boldsymbol{a}_{\alpha}\}$ or likewise $\{\boldsymbol{\tau}_{\sharp}; \boldsymbol{j}, [\boldsymbol{j} \cdot \boldsymbol{a}_{\alpha}]\}$ and $\{{}^{\vartriangle}\boldsymbol{\tau}_{\sharp}; {}^{\vartriangle}\boldsymbol{j}, [\boldsymbol{j} \cdot {}^{\vartriangle}\boldsymbol{a}_{\alpha}]\}$, respectively.

Plastic Evolution Laws:
Upon considering for simplicity and without loss of generality the case of *ideal plasticity* the reduced dissipation inequality is simply expressed in terms of the plastic part of the stress/double stress power density. Then, based on the introduction of a yield function ϕ in terms of either of the co/contra- and co/contra/contra- (mixed) variant spatial, intermediate or material ${}_{\tau}$-type stress measures and double-stress measures, respectively, i.e.

$$
\begin{aligned}
\phi = \phi(\{{}_{\tau}\boldsymbol{\tau}; \boldsymbol{a}_{\alpha}\}, \{{}^{\vartriangle}\boldsymbol{\tau}; {}^{\vartriangle}\boldsymbol{a}_{\alpha}\}) \quad \text{or} \quad & \phi = \phi(\{{}_{\tau}\overline{\boldsymbol{M}}; \overline{\boldsymbol{A}}_{\alpha}\}, \{{}^{\vartriangle}\overline{\boldsymbol{M}}; {}^{\vartriangle}\overline{\boldsymbol{A}}_{\alpha}\}) \\
\text{or} \quad & \phi = \phi(\{{}_{\tau}\boldsymbol{M}; \boldsymbol{A}_{\alpha}\}, \{{}^{\vartriangle}\boldsymbol{M}; {}^{\vartriangle}\boldsymbol{A}_{\alpha}\})
\end{aligned}
$$

the exploitation of the postulate of maximum dissipation as a constrained optimization problem results in alternative but equivalent versions of associated plastic evolution laws for the plastic part of the velocity gradient

$$\boldsymbol{l}_{\mathrm{p}} = \gamma \frac{\partial \phi}{\partial_\tau \boldsymbol{\tau}} \quad \text{or} \quad \boldsymbol{\ell}_{\mathrm{p}} = \gamma \frac{\partial \phi}{\partial_\tau \overline{\boldsymbol{M}}} \quad \text{or} \quad \boldsymbol{L}_{\mathrm{p}} = \gamma \frac{\partial \phi}{\partial_\tau \boldsymbol{M}}$$

and the plastic part of the $_\tau$-type velocity double-gradient

$$\overset{\Delta}{_\tau}\boldsymbol{l}_{\mathrm{p}} = \gamma \frac{\partial \phi}{\partial^\Delta \boldsymbol{\tau}} \quad \text{or} \quad \overset{\Delta}{_\tau}\boldsymbol{\ell}_{\mathrm{p}} = \gamma \frac{\partial \phi}{\partial^\Delta \overline{\boldsymbol{M}}} \quad \text{or} \quad \overset{\Delta}{_\tau}\boldsymbol{L}_{\mathrm{p}} = \gamma \frac{\partial \phi}{\partial^\Delta \boldsymbol{M}}.$$

Likewise, based on the introduction of a yield function ϕ in terms of either of the fully contravariant spatial, intermediate or material stress measures and double-stress measures, respectively, i.e.

$$\begin{aligned} \phi = \phi(\{\boldsymbol{\tau}_\sharp; \cdots\}, \{^\Delta\boldsymbol{\tau}_\sharp; \cdots\}) \quad &\text{or} \quad \phi = \phi(\{\overline{\boldsymbol{S}}; \cdots\}, \{^\Delta\overline{\boldsymbol{S}}; \cdots\}) \\ &\text{or} \quad \phi = \phi(\{\boldsymbol{S}; \cdots\}, \{^\Delta\boldsymbol{S}; \cdots\}) \end{aligned}$$

the exploitation of the postulate of maximum dissipation as a constrained optimization problem renders, in accordance with the various representations of the plastic part of the stress power, further alternative but equivalent versions of associated plastic evolution laws

$$[\boldsymbol{j} \cdot \boldsymbol{l}_{\mathrm{p}}]^{\mathrm{sym}} = \gamma \frac{\partial \phi}{\partial \boldsymbol{\tau}_\sharp} \quad \text{or} \quad [\overline{\boldsymbol{C}} \cdot \boldsymbol{\ell}_{\mathrm{p}}]^{\mathrm{sym}} = \gamma \frac{\partial \phi}{\partial \overline{\boldsymbol{S}}} \quad \text{or} \quad [\boldsymbol{C} \cdot \boldsymbol{L}_{\mathrm{p}}]^{\mathrm{sym}} = \gamma \frac{\partial \phi}{\partial \boldsymbol{S}}.$$

and

$$\begin{aligned} 2[\boldsymbol{j} \cdot \overset{\Delta}{_0}\boldsymbol{l}_{\mathrm{p}}]^{'\mathrm{sym}'} = \gamma \frac{\partial \phi}{\partial^\Delta \boldsymbol{\tau}_\sharp} \quad &\text{or} \quad 2[\overline{\boldsymbol{C}} \cdot \overset{\Delta}{_0}\boldsymbol{\ell}_{\mathrm{p}}\,]^{'\mathrm{sym}'} = \gamma \frac{\partial \phi}{\partial^\Delta \overline{\boldsymbol{S}}} \\ &\text{or} \quad 2[\boldsymbol{C} \cdot \overset{\Delta}{_0}\boldsymbol{L}_{\mathrm{p}}]^{'\mathrm{sym}'} = \gamma \frac{\partial \phi}{\partial^\Delta \boldsymbol{S}}. \end{aligned}$$

Which of the above alternative but equivalent plastic evolution laws is chosen is essentially a matter of convenience.

Part IV
Epilogue

...summarizes the previously derived integrability conditions for first- and second-order elasticity and non-integrability measures for first- and second-order elasto-plasticity. Thereby the aim is to distinguish formulations in terms of the distortions (and double-distortions) from formulations in terms of the metric (and the connection).

Chapter 9
Integrability and Non-Integrability in a Nutshell

DENKMALSWUNSCH

Setze mir ein Denkmal, cher,
ganz aus Zucker, tief im Meer.

Ein Süßwassersee, zwar kurz,
werd ich dann nach meinem Sturz:

doch so lang, daß Fische, hundert,
nehmen einen Schluck verwundert.-

Diese ißt in Hamburg und
Bremen dann des Menschen Mund.-

Wiederum in eure Kreise
komm ich so auf gute Weise,

während, werd ich Stein und Erz,
nur ein Vogel seinen Sterz

oder gar ein Mensch von Wert
seinen Witz auf mich entleert.

Christian Morgenstern

Abstract. The integrability conditions for the spatial and material configurations in first- and second-order elasticity may be stated either in terms of the distortions and double-distortions or, likewise, in terms of the (strain) metrics and connections (corresponding to the double-metrics/strains). In analogy, the lack of integrability for the intermediate configuration in first- and second-order elasto-plasticity is captured by non-integrability measures either in terms of the plastic or elastic distortions and double-distortions or, likewise, in terms of the plastic or elastic (strain) metrics and connections. It is emphasized that the integrability conditions and non-integrability measures have clear counterparts in differential geometry. This chapter aims in a concise summary of the previously derived relations and in a juxtaposition of the various cases considered.

9.1 First-Order Continuum

9.1.1 Integrability Conditions

Elasticity in a first-order continuum may be considered a flat, symmetric and metrically connected geometry, i.e. an Euclidean geometry.

P. Steinmann, *Geometrical Foundations of Continuum Mechanics*,
Lecture Notes in Applied Mathematics and Mechanics 2, DOI: 10.1007/978-3-662-46460-1_9

i) On the one hand the *integrability conditions* for the (forward and backward) *distortions* $\boldsymbol{F}$ and $\boldsymbol{f}$ (see Eqs. 7.1 and 7.2) correspond to vanishing *anholonomic objects* and are expressed in ambient Euclidean space as

$$\operatorname{Curl}\boldsymbol{F} := -\operatorname{Grad}\boldsymbol{F} : \boldsymbol{E} \doteq \boldsymbol{0} \quad \text{and} \quad \operatorname{curl}\boldsymbol{f} := -\operatorname{grad}\boldsymbol{f} : \boldsymbol{e} \doteq \boldsymbol{0}. \tag{9.1}$$

ii) On the other hand the *integrability conditions* for the (Cauchy-Green and Finger) *strains* $\boldsymbol{C}$ and $\boldsymbol{c}$ correspond to vanishing *Riemann curvatures* (see Eqs. 7.8 and 7.15) and are expressed in ambient Euclidean space as

$$\operatorname{Inc}\boldsymbol{C} := \operatorname{Curl}^t(\operatorname{Curl}^t\boldsymbol{C}) + \{\operatorname{Grad}\boldsymbol{C}\} \rtimes\!\!\ltimes \{\operatorname{Grad}\boldsymbol{C}\} \doteq \boldsymbol{0} \tag{9.2}$$
$$\text{and}$$
$$\operatorname{inc}\boldsymbol{c} := \operatorname{curl}^t(\operatorname{curl}^t\boldsymbol{c}) + \{\operatorname{grad}\boldsymbol{c}\} \rtimes\!\!\ltimes \{\operatorname{grad}\boldsymbol{c}\} \doteq \boldsymbol{0}.$$

Here Inc and inc denote the material and spatial *incompatibility operators*, i.e. the geometrically nonlinear extension of the linear Kröner incompatibility operator, see [94]. Thereby the right tensor cross product of third-order tensors that results from the quadratic terms $\mathcal{E}^{MIJ}\mathcal{M}_{OIK}\mathcal{M}^{O}{}_{JL}\mathcal{E}^{KLN}$ and $e^{mij}m_{oik}m^{o}{}_{jl}e^{kln}$ is abbreviated by $\rtimes\!\!\ltimes$, the bold curly brackets $\{\{\bullet\}\}$ refer to the anti-cyclic, sign alternating index permutation $2\{\bullet\}_{\{IJK\}} = \{\bullet\}_{IKJ} - \{\bullet\}_{KJI} + \{\bullet\}_{JIK}$ or $2\{\bullet\}_{\{ijk\}} = \{\bullet\}_{ikj} - \{\bullet\}_{kji} + \{\bullet\}_{jik}$, respectively.
In summary the integrability conditions for elasticity in a first-order continuum are stated in terms of tensorial quantities that have clear counterparts in differential geometry.

9.1.2 Non-Integrability Measures

The modeling of *mono-crystalline* elasto-plasticity within a first-order continuum formulation basically rests on a nine-dimensional temporal evolution law for the (non-symmetric) plastic distortion. Thus, after integration in time, the plastic distortion (or, for given total distortion, likewise the elastic distortion) may directly be considered for the analysis of non-integrability. In contrast the modeling of *poly-crystalline* (or rather phenomenological) elasto-plasticity within a first-order continuum formulation rests on a six-dimensional temporal evolution law for the (symmetric) plastic strain metric. Thus, after integration in time, the plastic strain metric (or, for given total distortion, likewise the elastic strain metric) needs to be considered when analyzing non-integrability.

Mono-Crystalline Elasto-Plasticity. On the one hand, based on the non-integrable plastic and elastic distortions (and the corresponding integrable but (right) non-symmetric plastic and elastic connections), mono-crystalline first-order elasto-plasticity may be considered a flat, non-symmetric and metrically connected geometry, i.e. a Cartan geometry.

Thereby the *non-integrability measures* for the (forward) plastic and (backward) elastic *distortions* $\widetilde{\boldsymbol{F}}$ and $\overline{\boldsymbol{f}}$ (see Eqs. 8.3 and 8.4) correspond to non-vanishing plastic and elastic *anholonomic objects* (that are here related to the plastic and elastic torsions) and are expressed in ambient Euclidean space as

$$\widetilde{\boldsymbol{D}} := \operatorname{Curl} \widetilde{\boldsymbol{F}} \neq \mathbf{0} \quad \text{and} \quad \overline{\boldsymbol{d}} := \operatorname{curl} \overline{\boldsymbol{f}} \neq \mathbf{0}. \tag{9.3}$$

Here plastic and elastic anholonomic objects (and correspondingly plastic and elastic torsions) are due to the plastic and elastic (geometrically necessary) dislocation densities.

Poly-Crystalline Elasto-Plasticity. On the other hand, based on the incompatible plastic and elastic (strain) metrics (and the corresponding non-integrable Riemann connections), poly-crystalline first-order elasto-plasticity may alternatively be considered a non-flat (curved), symmetric and metrically connected geometry, i.e. a Riemann geometry.

Thereby the *non-integrability measures* for the plastic (Cauchy-Green) and elastic (Finger) *strains* $\widetilde{\boldsymbol{C}}$ and $\overline{\boldsymbol{c}}$ (see Eqs. 8.18 and 8.33) correspond to non-vanishing plastic and elastic *Riemann curvatures* and are expressed in ambient Euclidean space as

$$2\widetilde{\boldsymbol{H}}_0 := \operatorname{Inc} \widetilde{\boldsymbol{C}} = \operatorname{Curl}^t(\operatorname{Curl}^t \widetilde{\boldsymbol{C}}) + \{\operatorname{Grad} \widetilde{\boldsymbol{C}}\} \rtimes\!\!\!\ltimes \{\operatorname{Grad} \widetilde{\boldsymbol{C}}\} \neq \mathbf{0} \tag{9.4}$$

and

$$-2\overline{\boldsymbol{h}}_t := \operatorname{inc} \overline{\boldsymbol{c}} = \operatorname{curl}^t(\operatorname{curl}^t \overline{\boldsymbol{c}}) + \{\operatorname{grad} \overline{\boldsymbol{c}}\} \rtimes\!\!\!\ltimes \{\operatorname{grad} \overline{\boldsymbol{c}}\} \neq \mathbf{0}.$$

Here the plastic and elastic Riemann curvatures are due to the incompatibilities of the plastic and elastic (strain) metrics (that are in turn a consequence of the plastic and elastic, geometrically necessary, dislocation densities).

In summary the non-integrability measures for elasto-plasticity in a first-order continuum are stated in terms of tensorial quantities that have clear counterparts in differential geometry.

9.2 Second-Order Continuum

9.2.1 Integrability Conditions

Elasticity in a second-order continuum may be considered a flat, symmetric and metrically connected geometry, i.e. an Euclidean geometry.

i) On the one hand the *integrability conditions* for the (forward and backward) *distortions* $\boldsymbol{F}$ and $\boldsymbol{f}$ correspond to vanishing *anholonomic objects* and are expressed together with the *integrability conditions* for the (forward and backward) *double-distortions* $\boldsymbol{G}$ and $\boldsymbol{g}$ in ambient Euclidean space (see Eqs. 7.17 and 7.19) as

$$\operatorname{Curl} \boldsymbol{F} := -\operatorname{Grad} \boldsymbol{F} : \boldsymbol{E} \doteq \boldsymbol{0} \quad \text{and} \quad \operatorname{Curl} \boldsymbol{G} := -\operatorname{Grad} \boldsymbol{G} : \boldsymbol{E} \doteq \boldsymbol{0} \tag{9.5}$$

or

$$\operatorname{curl} \boldsymbol{f} := -\operatorname{grad} \boldsymbol{f} : \boldsymbol{e} \doteq \boldsymbol{0} \quad \text{and} \quad \operatorname{curl} \boldsymbol{g} := -\operatorname{grad} \boldsymbol{g} : \boldsymbol{e} \doteq \boldsymbol{0}.$$

Moreover, the *compatibility* of the (forward and backward) *distortions* with the corresponding *double-distortions* corresponds to vanishing *torsions* and *non-metricities* and is expressed in ambient Euclidean space (see Eqs. 7.18 and 7.20) as

$$-\boldsymbol{G} : \boldsymbol{E} \doteq \boldsymbol{0} \quad \text{and} \quad 2\mathsf{I}^{\mathrm{sym}} : \left[\boldsymbol{F}^t \cdot [\boldsymbol{G} - \operatorname{Grad} \boldsymbol{F}]\right] \doteq \boldsymbol{0} \tag{9.6}$$

or

$$-\boldsymbol{g} : \boldsymbol{e} \doteq \boldsymbol{0} \quad \text{and} \quad 2\mathsf{i}^{\mathrm{sym}} : \left[\boldsymbol{f}^t \cdot [\boldsymbol{g} - \operatorname{grad} \boldsymbol{f}]\right] \doteq \boldsymbol{0}.$$

ii) On the other hand the *integrability conditions* for the (Cauchy-Green and Finger) *strains* $\boldsymbol{C}$ and $\boldsymbol{c}$ and the (material and spatial) *connections* $\boldsymbol{\Gamma}$ (with $\boldsymbol{G} =: \boldsymbol{F} \cdot \boldsymbol{\Gamma}$) and $\boldsymbol{\gamma}$ (with $\boldsymbol{g} =: \boldsymbol{f} \cdot \boldsymbol{\gamma}$) correspond to vanishing *Riemann curvatures* (see Eqs. 7.21 and 7.25) and *curvatures* (see Eqs. 7.22 and 7.26) and are expressed in ambient Euclidean space as

$$\operatorname{Inc} \boldsymbol{C} := \operatorname{Curl}^t(\operatorname{Curl}^t \boldsymbol{C}) + \{\operatorname{Grad} \boldsymbol{C}\} \Join \{\operatorname{Grad} \boldsymbol{C}\} \doteq \boldsymbol{0} \tag{9.7}$$

and

$$\operatorname{Curl} \boldsymbol{\Gamma} + \boldsymbol{\Gamma} \rtimes \boldsymbol{\Gamma} \doteq \boldsymbol{0}$$

or

$$\operatorname{inc} \boldsymbol{c} := \operatorname{curl}^t(\operatorname{curl}^t \boldsymbol{c}) + \{\operatorname{grad} \boldsymbol{c}\} \Join \{\operatorname{grad} \boldsymbol{c}\} \doteq \boldsymbol{0}$$

and

$$\operatorname{curl} \boldsymbol{\gamma} + \boldsymbol{\gamma} \rtimes \boldsymbol{\gamma} \doteq \boldsymbol{0}.$$

Thereby the right vector cross product of third-order tensors that results from the quadratic terms $\mathcal{L}^I{}_{MK}\mathcal{L}^M{}_{JL}\mathcal{E}^{KLN}$ and $\ell^i{}_{mk}\ell^m{}_{jl}e^{kln}$, respectively, is abbreviated by $\rtimes$.

Moreover, the *integrability conditions* for the (forward and backward) *distortions compatible to the connections* correspond to vanishing *torsions* (see Eqs. 7.23 and 7.27), whereas the *compatibility* of the (material and spatial) *connections* with the corresponding *strains* correspond to vanishing *non-metricities* (see Eqs. 7.24 and 7.28); these conditions are expressed in ambient Euclidean space as

$$-\boldsymbol{\Gamma} : \boldsymbol{E} \doteq \mathbf{0} \quad \text{and} \quad 2\mathsf{I}^{\text{sym}} : [\boldsymbol{C} \cdot \boldsymbol{\Gamma}] - \operatorname{Grad} \boldsymbol{C} \doteq \mathbf{0} \tag{9.8}$$

or

$$-\boldsymbol{\gamma} : \boldsymbol{e} \doteq \mathbf{0} \quad \text{and} \quad 2\mathsf{i}^{\text{sym}} : [\boldsymbol{c} \cdot \boldsymbol{\gamma}] - \operatorname{grad} \boldsymbol{c} \doteq \mathbf{0}.$$

In summary the integrability conditions for elasticity in a second-order continuum are stated in terms of four tensorial quantities that have clear counterparts in differential geometry.

9.2.2 Non-Integrability Measures

The modeling of *mono-crystalline* elasto-plasticity within a second-order continuum formulation basically rests on a nine-dimensional temporal evolution law for the (non-symmetric) plastic distortion and a 27-dimensional temporal evolution law for the plastic double-distortion. Thus, after integration in time, the plastic distortion and the plastic double-distortion (or likewise the elastic distortion and the elastic double-distortion) may directly be considered for the analysis of non-integrability. In contrast the modeling of *poly-crystalline* (or rather phenomenological) elasto-plasticity within a second-order continuum formulation rests on a six-dimensional temporal evolution law for the (symmetric) plastic strain metric and a 27-dimensional temporal evolution law for the plastic double-strain (or rather the plastic connection). Thus, after integration in time, the plastic strain metric and the plastic double-strain metric (or likewise the elastic strain metric and the elastic double-strain metric) need to be considered when analyzing non-integrability.

Mono-Crystalline Elasto-Plasticity. On the one hand, based on the non-integrable plastic and elastic distortions and double-distortions, mono-crystalline second-order elasto-plasticity may be considered a non-flat (curved), non-symmetric and non-metrically connected geometry, i.e. a general metric geometry.

Thereby the *non-integrability measures* for the (forward) plastic and (backward) elastic *distortions* $\widetilde{\boldsymbol{F}}$ and $\overline{\boldsymbol{f}}$, corresponding to non-vanishing plastic and elastic *anholonomic objects*, are expressed together with the *non-integrability measures* for the (forward) plastic and (backward) elastic *double-distortions* $\widetilde{\boldsymbol{G}}$ and $\widetilde{\boldsymbol{g}}$, corresponding to non-vanishing plastic and elastic *'curvatures'*, in ambient Euclidean space (see Eqs. 8.96 and 8.98) as

$$^{\text{I}}\widetilde{\boldsymbol{D}} := \operatorname{Curl} \widetilde{\boldsymbol{F}} \neq \mathbf{0} \quad \text{and} \quad \widetilde{\boldsymbol{K}} := \operatorname{Curl} \widetilde{\boldsymbol{G}} \neq \mathbf{0} \tag{9.9}$$

and

$$^{\text{I}}\overline{\boldsymbol{d}} := \operatorname{curl} \overline{\boldsymbol{f}} \neq \mathbf{0} \quad \text{and} \quad \overline{\boldsymbol{k}} := \operatorname{curl} \overline{\boldsymbol{g}} \neq \mathbf{0}.$$

Moreover, the *incompatibilities* of the (forward) plastic and (backward) elastic *distortions* with the corresponding (forward) plastic and (backward) elastic *double-distortions* correspond to non-vanishing plastic and elastic *torsions* and non-vanishing plastic and elastic *non-metricities* and are expressed in ambient Euclidean space (see Eqs. 8.97 and 8.99) as

$$^{\mathrm{II}}\widetilde{\boldsymbol{D}} := -\widetilde{\boldsymbol{G}} : \boldsymbol{E} \neq \mathbf{0} \quad \text{and} \quad \widetilde{\boldsymbol{Q}} := 2\mathsf{I}^{\mathrm{sym}} : [\widetilde{\boldsymbol{F}}^t \cdot [\widetilde{\boldsymbol{G}} - \operatorname{Grad} \widetilde{\boldsymbol{F}}]] \neq \mathbf{0} \quad (9.10)$$

and

$$^{\mathrm{II}}\overline{\boldsymbol{d}} := -\overline{\boldsymbol{g}} : \boldsymbol{e} \neq \mathbf{0} \quad \text{and} \quad \overline{\boldsymbol{q}} := 2\mathsf{i}^{\mathrm{sym}} : [\overline{\boldsymbol{f}}^t \cdot [\overline{\boldsymbol{g}} - \operatorname{grad} \overline{\boldsymbol{f}}]] \neq \mathbf{0}.$$

Here plastic and elastic anholonomic objects are due to the plastic and elastic primary dislocation densities; plastic and elastic torsions are due to plastic and elastic secondary dislocation densities; plastic and elastic 'curvatures' are due to plastic and elastic disclination densities; and plastic and elastic non-metricities are due to plastic and elastic point-defect densities.

Poly-Crystalline Elasto-Plasticity. On the other hand, based on the non-integrable plastic and elastic (strain) metrics and non-integrable plastic and elastic connections (double-strains), poly-crystalline second-order elasto-plasticity may be considered a non-flat (curved), non-symmetric and non-metrically connected geometry, i.e. a general metric geometry. Therein the incompatible plastic and elastic (strain) metrics (and the corresponding non-integrable Riemann connections) themselves represent a non-flat (curved), symmetric and metrically connected geometry, i.e. a Riemann geometry.

Thereby the *non-integrability measures* for the plastic (Cauchy-Green) and elastic (Finger) *strains* $\widetilde{\boldsymbol{C}}$ and $\overline{\boldsymbol{c}}$ (see Eqs. 8.100 and 8.105), corresponding to non-vanishing plastic and elastic *Riemann curvatures*, are expressed together with the *non-integrability measures* for the (material) plastic and (spatial) elastic *connections* $\widetilde{\boldsymbol{\Gamma}}$ (with $\widetilde{\boldsymbol{G}} =: \widetilde{\boldsymbol{F}} \cdot \widetilde{\boldsymbol{\Gamma}}$) and $\overline{\boldsymbol{\gamma}}$ (with $\overline{\boldsymbol{g}} =: \overline{\boldsymbol{f}} \cdot \overline{\boldsymbol{\gamma}}$), corresponding to non-vanishing plastic and elastic *curvatures* (see Eqs. 8.101 and 8.106), in ambient Euclidean space as

$$2\,^{\mathrm{I}}\widetilde{\boldsymbol{H}}_0 := \operatorname{Inc} \widetilde{\boldsymbol{C}} = \operatorname{Curl}^t(\operatorname{Curl}^t \widetilde{\boldsymbol{C}}) + \{\operatorname{Grad} \widetilde{\boldsymbol{C}}\} \times\!\!\times \{\operatorname{Grad} \widetilde{\boldsymbol{C}}\} \neq \mathbf{0} \quad (9.11)$$

and

$$^{\mathrm{II}}\widetilde{\boldsymbol{H}} := \operatorname{Curl} \widetilde{\boldsymbol{\Gamma}} + \widetilde{\boldsymbol{\Gamma}} \rtimes \widetilde{\boldsymbol{\Gamma}} \neq \mathbf{0}$$

and

$$-2\,^{\mathrm{I}}\overline{\boldsymbol{h}}_t := \operatorname{inc} \overline{\boldsymbol{c}} = \operatorname{curl}^t(\operatorname{curl}^t \overline{\boldsymbol{c}}) + \{\operatorname{grad} \overline{\boldsymbol{c}}\} \times\!\!\times \{\operatorname{grad} \overline{\boldsymbol{c}}\} \neq \mathbf{0}$$

and

$$-\,^{\mathrm{II}}\overline{\boldsymbol{h}} := \operatorname{curl} \overline{\boldsymbol{\gamma}} + \overline{\boldsymbol{\gamma}} \rtimes \overline{\boldsymbol{\gamma}} \neq \mathbf{0}.$$

Moreover, the *non-integrability measures* for the (forward) plastic and (backward) elastic *distortions compatible to the connections* (see Eqs. 8.103 and 8.108) correspond to non-vanishing plastic and elastic *torsions*, whereas the *incompatibilities* of the (material) plastic and (spatial) elastic *connections* with the corresponding plastic (Cauchy-Green) and elastic (Finger) *strains* (see Eqs. 8.104 and 8.109) are captured by non-vanishing plastic and elastic *non-metricities*; these measures are expressed in ambient Euclidean space as

$$ {}^{\mathrm{II}}\widetilde{\boldsymbol{A}} := -\widetilde{\boldsymbol{\Gamma}} : \boldsymbol{E} \neq \boldsymbol{0} \quad \text{and} \quad \widetilde{\boldsymbol{Q}} := 2\mathsf{I}^{\mathrm{sym}} : [\widetilde{\boldsymbol{C}} \cdot \widetilde{\boldsymbol{\Gamma}}] - \mathrm{Grad}\,\widetilde{\boldsymbol{C}} \neq \boldsymbol{0} \qquad (9.12) $$

and

$$ {}^{\mathrm{II}}\overline{\boldsymbol{a}} := -\overline{\boldsymbol{\gamma}} : \boldsymbol{e} \neq \boldsymbol{0} \quad \text{and} \quad \overline{\boldsymbol{q}} := 2\mathsf{i}^{\mathrm{sym}} : [\overline{\boldsymbol{c}} \cdot \overline{\boldsymbol{\gamma}}] - \mathrm{grad}\,\overline{\boldsymbol{c}} \neq \boldsymbol{0}. $$

Here plastic and elastic Riemann curvatures are due to the incompatibilities of the plastic and elastic (strain) metrics (that are in turn a consequence of the plastic and elastic primary dislocation densities); plastic and elastic curvatures are due to plastic and elastic disclination densities; plastic and elastic torsions are due to plastic and elastic secondary dislocation densities; and plastic and elastic non-metricities are due to plastic and elastic point-defect densities.

In summary the non-integrability measures for elasto-plasticity in a second-order continuum are stated in terms of four tensorial quantities that have clear counterparts in differential geometry.

References

1. Abraham, R., Marsden, J., Ratiu, T.S.: Manifolds, tensor analysis, and applications, vol. 75. Springer, New York (1988)
2. Abraham, R., Marsden, J.E.: Foundations of Mechanics. Addison-Wesley, Reading (1978)
3. Acharya, A.: A model of crystal plasticity based on the theory of continuously distributed dislocations. J. Mech. Phys. Solids 49, 761–784 (2001)
4. Acharya, A., Bassani, J.L.: Lattice incompatibility and a gradient theory of crystal plasticity. J. Mech. Phys. Solids 48, 1565–1595 (2000)
5. Aifantis, E.: On the role of gradients in the localization of deformation and fracture. Int. J. Engng. Sci. 30, 1279–1299 (1992)
6. Anthony, K.H.: Die Reduktion von nichteuklidischen geometrischen Objekten in eine euklidische Form und physikalische Deutung der Reduktion durch Eigenspannungszustände in Kristallen. Arch. Rat. Mech. Anal. 37, 161–180 (1970)
7. Anthony, K.H.: Die Theorie der Disklinationen. Arch. Rat. Mech. Anal. 39, 43–88 (1970)
8. Anthony, K.H.: Die Theorie der nichtmetrischen Spannungen in Kristallen. Arch. Rat. Mech. Anal. 40, 50–78 (1971)
9. Antman, S.S.: Non-linear problems of elasticity. Springer, New York (1995)
10. Arnold, V.I.: Mathematical Methods of Classical Mechanics. Springer, New York (1989)
11. Arramon, Y.P., Mehrabadi, M.M., Martin, D.W., Cowin, S.C.: A multidimensional anisotropic strength criterion based on kelvin modes. Int. J. Solids Struct. 37, 2915–2935 (2000)
12. Ball, J.M.: Convexity conditions and existence theorems in nonlinear elasticity. Arch. Rat. Mech. Anal. 63, 337–403 (1977)
13. Ball, J.M., James, R.D.: Fine phase mixtures as minimizers of energy. Arch. Rat. Mech. Anal. 100, 13–52 (1987)
14. Becker, M.: Incompatibility and Instability Based Size Effects in Crystals and Composites at Finite Elastoplastic Strains, *Bericht*, vol. I-18. Institut für Mechanik, Lehrstuhl I (2006)
15. Bertram, A.: Finite gradient elasticity and plasticity: A constitutive mechanical framework. Cont. Mech (2014), doi:10.1007/s00161-014-0387-0
16. Bigoni, D.: Nonlinear Solid Mechanics: Bifurcation Theory and Material Instability. Cambridge University Press (2012)

17. Bilby, B.A., Bullough, R., Smith, E.: Continuous distributions of dislocations: A new application of the methods of non-riemannian geometry. Proc. Roy. Soc. A231, 263–237 (1955)
18. Bilby, B.A., Smith, E.: Continuous distributions of dislocations iii. Proc. Roy, 481–505 (1956)
19. Bloom, F.: Modern differential geometric techniques in the theory of continuous distributions of dislocations. Springer-Verlag (1979)
20. do Carmo, M.P.: Differential forms and applications. Springer, Berlin (1994)
21. Carstensen, C., Hackl, K., Mielke, A.: Non–convex potentials and microstructures in finite–strain plasticity. Proc. R. Soc. A 458, 299–317 (2002)
22. Cartan, E.: Sur une généralisation de la notion de courbure de Riemann et les espace à torsion. Comptes Rend. 174, 593–575 (1922)
23. Cartan, E.: La Théorie des Groupes Finis et Continus et la Géométrie Différentielle Traitées par la Méthode du Repère Mobile. Gauthier-Villars, Paris (1937)
24. Cartan, E.: Les systèmes différentiels extérieurs et leurs applications géométriques. Actualités Scientific 994 (1945)
25. Cartan, E.: On manifolds with an affine connection and the theory of general relativity. In: Monographs and Textbooks in Physical Science, Bibliopolis, Napoli (1986)
26. Cartan, H.: Formes Différentielles. Herrmann, Paris (1967)
27. Cauchy, A.L.: Recherches sur l'équilibre et le mouvement intérieur des corps solides ou fluides, élastiques ou non élastiques. In: Bulletin de la Société Philomatique, pp. 9–13 (1823)
28. Cermelli, P., Gurtin, M.: On the characterization of geometrically necessary dislocations in finite plasticity. J. Mech. Phys. Solids 49, 1539–1568 (2001)
29. Chadwick, P.: Continuum Mechanics. Consise theory and problems. Dover, Minola (1999)
30. Chadwick, P., Vianello, M., Cowin, S.C.: A new proof that the number of linear elastic symmetries is eight. J. Mech. Phys. Solids 49, 2471–2492 (2001)
31. Chambon, R., Caillerie, D., Tamagnini, C.: A finite deformation second gradient theory of plasticity. Comptes Rendus de l'Académie des Sciences-Series IIb-Mechanics 329, 797–802 (2001)
32. Chambon, R., Caillerie, D., Tamagnini, C.: A strain space gradient plasticity theory for finite strain. Comp. Meth. Appl. Mech. Engng. 193, 2797–2826 (2004)
33. Christoffel, E.B.: Ueber die Transformation der homogenen Differentialausdrücke zweiten Grades. J. Reine Angew. Mathe. 70, 46–70 (1869)
34. Ciarlet, P.G.: Mathematical Elasticity. Three dimensional elasticity, vol. I. Elsevier, Amsterdam (1988)
35. Ciarlet, P.G.: An Introduction to Differential Geometry with Applications to Elasticity. Springer (2006)
36. Ciufolini, I., Wheeler, J.A.: Gravitation and Inertia. Princeton University Press, Princeton (1995)
37. Clayton, J.D.: Nonlinear Mechanics of Crystals. Springer, Dordrecht (2011)
38. Clayton, J.D.: On anholonomic deformation, geometry, and differentiation. Math. Mech. Solids 17, 702–735 (2012)
39. Clayton, J.D., Bammann, D.J., McDowell, D.L.: Anholonomic configuration spaces and metric tensors in finite elastoplasticity. Int. J. Non-Linear Mech. 39, 1039–1049 (2004)

40. Clayton, J.D., McDowell, D.L., Bammann, D.J.: Modeling dislocations and disclinations with finite micropolar elastoplasticity. Int. J. Plast. 22, 210–256 (2006)
41. Cleja-Ţigoiu, S.: Couple stresses and non-Riemannian plastic connection in finite elasto-plasticity. Zeitschrift für angewandte Mathematik und Physik ZAMP 53, 996–1013 (2002)
42. Cleja-Ţigoiu, S.: Small elastic strains in finite elasto-plastic materials with continuously distributed dislocations. Theor. Appl. Mech. (28-29), 93–112 (2002)
43. Cleja-Ţigoiu, S.: Elasto-plastic materials with lattice defects modeled by second order deformations with non-zero curvature. Int. J. Frac. 166, 61–75 (2010)
44. Cleja-Ţigoiu, S.: Non-local elasto-viscoplastic models with dislocations in finite elasto-plasticity. Math. Mech. Solids 18, 349–372 (2011)
45. Cleja-Ţigoiu, S., Fortune, D., Vallee, C.: Torsion equation in anisotropic elastoplastic materials with continuously distributed dislocations. Math. Mech. Solids 13, 667–689 (2008)
46. Cowin, S.C., Mehrabadi, M.M.: On the identification of material symmetry for anisotropic elastic materials. Quart. J. Mech. Appl. Math. 40, 451–476 (1987)
47. Cowin, S.C., Mehrabadi, M.M.: The structure of the linear anisotropic elastic symmetries. J. Mech. Phys. Solids 40, 1459–1471 (1992)
48. Dacorogna, B.: Direct Methods in the Calculus of Variations. Springer, Berlin (1989)
49. Dassios, G., Lindell, I.V.: On the Helmholtz decomposition for polyadics. Quart. Appl. Math. 59, 787–796 (2001)
50. Derezin, S.V., Zubov, L.M.: Disclinations in nonlinear elasticity. Z. Angew. Math. Mech. 91, 433–442 (2011)
51. Dłużewski, P.: On geometry and continuum thermodynamics of movement of structural defects. Mech. Mat. 22, 23–41 (1996)
52. Doyle, T.C., Ericksen, J.L.: Nonlinear elasticity. In: Advances in Applied Mechanics IV, pp. 53–116. Academic Press, Inc., New York (1956)
53. Edelen, D., Laws, N.: On the thermodynamics of systems with nonlocality. Arch. Rat. Mech. Anal. 43, 24–35 (1971)
54. Epstein, M.: The geometrical language of continuum mechanics. Cambridge University Press (2010)
55. Epstein, M., Maugin, G.A.: The energy-momentum tensor and material uniformity in finite elasticity. Acta Mech. 83, 127–133 (1990)
56. Ericksen, J.L.: Tensor Fields, Handbuch der Physik, vol. III/1. Springer, Berlin (1960)
57. Eringen, A.C.: Microcontinuum Field Theories. Foundations and Solids, vol. I. Springer, New York (1999)
58. Euclid: The elements, vol. I-XIII. Alexandria (-300?)
59. Feynman, R.P., Leighton, R.B., Sands, M.: The Feynman lectures on physics, vol. I-III. Pearson Education (2006)
60. Flanders, H.: Differential Forms with Applications to the Physical Sciences. Dover, New York (1989)
61. Forest, S.: The micromorphic approach for gradient elasticity, viscoplasticity and damage. ASCE J. Engng. Mech. 135, 117–131 (2009)
62. Forest, S., Sievert, R.: Elastoviscoplastic constitutive frameworks for generalized continua. Acta Mech. 160, 71–111 (2003)
63. Frank, F.C.: On the theory of liquid crystals. Discuss. Faraday Soc. 25, 19–28 (1958)

64. Frankel, T.: The geometry of physics: an introduction. Cambridge University Press, Cambridge (2004)
65. Grammenoudis, P., Tsakmakis, C.: Incompatible deformations - plastic intermediate configuration. Z. Angew. Math. Mech. 88, 403–432 (2008)
66. Grammenoudis, P., Tsakmakis, C.: Plastic intermediate configuration and related spatial differential operators in micromorphic plasticity. Math. Mech. Solids 15, 515–538 (2010)
67. Green, A.E., Zerna, W.: Theoretical Elasticity. Dover, New York (1968)
68. Gurtin, M.: An Introduction to continuum mechanics. Academic Press, New York (1981)
69. Gurtin, M.: On the nature of configurational forces. Arch. Rat. Mech. Anal. 131, 67–100 (1995)
70. Gurtin, M.: Configurational forces as basic concepts of continuum physics. Springer, New York (2000)
71. Gurtin, M.: A gradient theory of single-crystal viscoplasticity that accounts for geometrically necessary dislocations. J. Mech. Phys. Solids 50, 5–32 (2002)
72. Gurtin, M., Fried, E., Anand, L.: The Mechanics and thermodynamics of continua. Cambridge University Press, Cambridge (2010)
73. Gurtin, M., Murdoch, A.I.: A continuum theory of elastic material surfaces. Arch. Rat. Mech. Anal. 57, 291–323 (1975)
74. Hackl, K., Fischer, F.D.: On the relation between the principle of maximum dissipation and inelastic evolution given by dissipation potentials. Proc. R. Soc. A 464(2008), 117–132 (2089)
75. Halphen, B., Nguyen, Q.S.: Sur les matériaux standard généralisés. Journal Méc 14, 39–63 (1975)
76. Han, W., Reddy, B.D.: Plasticity: mathematical theory and numerical analysis. Springer, New York (2013)
77. Haupt, P.: Continuum Mechanics and Theory of Materials. Springer, Berlin (2000)
78. Hehl, F.W., Obukhov, Y.N.: Foundations of classical electrodynamics: Charge, flux, and metric, vol. 33. Springer (2003)
79. Hehl, F.W., Obukhov, Y.N.: Élie Cartan's torsion in geometry and in field theory, an essay. Annales de la Fondation Louis de Broglie 32, 157–194 (2007)
80. Helmholtz, H.: Über Integrale der hydrodynamischen Gleichungen, welche den Wirbelbewegungen entsprechen. J. Reine Angew. Math. 55, 25–55 (1858)
81. Hilbert, D.: Die Grundlagen der Physik. Königl. Gesell. Wiss. Göttingen. Nachr. Math.-Phys. Kl., 395–407 (1915)
82. Hirschberger, B., Steinmann, P.: Classification of concepts in thermodynamically consistent generalized plasticity. ASCE J. Engng. Mech. 135, 156–170 (2009)
83. Holzapfel, G.A.: Nonlinear Solid Mechanics. Wiley, Chichester (2000)
84. Hossain, M., Steinmann, P.: More hyperelastic models for rubber-like materials: consistent tangent operators and comparative study. J. Mech. Behav. Mater. 22, 27–50 (2013)
85. Itskov, M.: Tensor algebra and tensor analysis for engineers. Springer, Berlin (2007)
86. Javili, A., Dell'Isola, F., Steinmann, P.: Geometrically nonlinear higher-gradient elasticity with energetic boundaries. J. Mech. Phys. Solids 61, 2381–2401 (2013)

87. Javili, A., McBride, A., Steinmann, P.: Thermomechanics of solids with lower-dimensional energetics: on the importance of surface, interface and curve structures at the nanoscale. Appl. Mech. Rev. 65 (2013), doi:10.1115/1.4023
88. Kirchner, N., Steinmann, P.: A unifying treatise on variational principles for gradient and micromorphic continua. Philos. Mag. 85, 3875–3895 (2005)
89. Kirchner, N., Steinmann, P.: On the material setting of gradient hyperelasticity. Math. Mech. Solids 12, 559–580 (2007)
90. Klingbeil, E.: Tensorrechnung für Ingenieure, vol. 197. Bibliographisches Institut, Hochschultaschenbücher-Verlag (1966)
91. Kondo, K.: On the geometrical and physical foundation of the theory of yielding, pp. 2:41–2:47. Publisher (1952)
92. Kondo, K.: On the analytical and physical foundations of the theory of dislocations and yielding by the differential geometry of continua. Int. J. Engng. Sci. 2, 219–251 (1964)
93. Kreyszig, E.: Differential Geometry. Dover, New York (1991)
94. Kröner, E.: Kontinuumstheorie der Versetzungen und Eigenspannungen. Erg. Angew. Math. 5, 1–179 (1958)
95. Kröner, E.: Allgemeine Kontinuumstheorie der Versetzungen und Eigenspannungen. Arch. Rat. Mech. Anal. 4, 237–334 (1960)
96. Kröner, E.: Differential geometry of defects in condensed systems of particles with only translational mobility. Int. J. Engng. Sci. 19, 1507–1515 (1981)
97. Kröner, E., Anthony, K.H.: Dislocations and disclinations in material structures: The basic topological concepts. Annu. Rev. Mater. Sci. 5, 43–72 (1975)
98. Kröner, E., Seeger, A.: Nichtlineare Elastizitätstheorie der Versetzungen und Eigenspannungen. Arch. Rat. Mech. Anal. 3, 97–119 (1959)
99. Kuhl, E., Steinmann, P.: Mass and volume specific views on thermodynamics for open systems. Proc. Royal Soc. Lond. A. 459, 2547–2568 (2003)
100. Lanczos, C.: The Variational Principles of Mechanics. Dover, New York (1986)
101. Lazar, M.: Dislocation theory as a 3-dimensional translational gauge theory. Ann. Phys (Leipzig) 9, 461–473 (2000)
102. Lazar, M.: An elastoplastic theory of dislocations as a physical field theory with torsion. J. Phys. A: Math. Gen. 35, 1983–2004 (2002)
103. Lazar, M.: Twist disclination in the field theory of elastoplasticity. J. Phys. Condens. Matter 15, 6781–6800 (2003)
104. Lazar, M.: Wedge disclination in the field theory of elastoplasticity. Phys. Lett. A 311, 416–425 (2003)
105. Lazar, M., Hehl, F.W.: Cartan's spiral staircase in physics and, in particular, in the gauge theory of dislocations. Found. Phys. 40, 1298–1325 (2010)
106. Lazar, M., Maugin, G.A.: Defects in gradient micropolar elasticity II: edge dislocation and wedge disclination. J. Mech. Phys. Solids 52, 2285–2307 (2004)
107. Lazar, M., Maugin, G.A.: Nonsingular stress and strain fields of dislocations and disclinations in first strain gradient elasticity. Int. J. Engng. Sci. 43, 1157–1184 (2005)
108. Lazar, M., Maugin, G.A.: On microcontinuum field theories: the Eshelby stress tensor and incompatibility conditions. Phil. Mag. 87, 3853–3870 (2007)
109. Le, K.C., Stumpf, H.: A model of elastoplastic bodies with continuously distributed dislocations. Int. J. Plast. 12, 611–627 (1996)
110. Le, K.C., Stumpf, H.: Nonlinear continuum theory of dislocations. Int. J. Engng. Sci. 34, 339–358 (1996)

111. Le, K.C., Stumpf, H.: On the determination of the crystal reference in nonlinear continuum theory of dislocations. Proc. Royal Soc. London A 452, 359–371 (1996)
112. Lee, E.H.: Elastic-plastic deformations at finite strains. J. Appl. Mech., ASME 36, 1–6 (1969)
113. Lee, J.M.: Riemannian manifolds: an introduction to curvature, vol. 176. Springer (1997)
114. Li, J.C., Gilman, J.J.: Disclination loops in polymers. J. Appl. Phys. 41, 4248–4256 (1970)
115. Liebe, T., Menzel, A., Steinmann, P.: Theory and numerics of a thermodynamically consistent framework for geometrically non-linear gradient plasticity. Int. J. Engng. Sci. 41, 1603–1629 (2003)
116. Liebe, T., Steinmann, P.: Theory and numerics of a thermodynamically consistent framework for geometrically linear gradient plasticity. Int. J. Num. Meth. Engng. 51, 1437–1467 (2001)
117. Lovelock, D., Rund, H.: Tensors, Differential Forms and Variational Principles. Dover, New York (1989)
118. Luscher, D., McDowell, D., Bronkhorst, C.: A second gradient theoretical framework for hierarchical multiscale modeling of materials. Int. J. Plasticity 26, 1248–1275 (2010)
119. Mahnken, R.: Anisotropy in geometrically non-linear elasticity with generalized Seth–Hill strain tensors projected to invariant subspaces. Commun. Num. Meth. Engng. 21(8), 405–418 (2005)
120. Malvern, L.E.: Introduction to the Mechanics of a Continuous Medium. Prentice-Hall, New Jersey (1969)
121. Mandel, J.: Plasticité classique et viscoplasticité.: CISM Courses and Lectures, 97, Udine 1771. Springer, Berlin (1972)
122. Marsden, J.E., Hughes, T.J.R.: Mathematical Foundations of Elasticity. Dover, New York (1994)
123. Maugin, G.A.: The thermodynamics of plasticity and fracture. Cambridge University Press, Cambridge (1992)
124. Maugin, G.A.: Material inhomogeneities in elasticity. Chapman and Hall, London (1993)
125. Maugin, G.A.: Material forces: concepts and applications. ASME Appl. Mech. Rev. 48, 213–245 (1995)
126. Maugin, G.A.: Configurational forces: thermomechanics, physics, mathematics, and numerics. CRC Press, Boca Raton (2011)
127. Mehrabadi, M.M., Cowin, S.C.: Eigentensors of linear anisotropic elastic materials. Quart. J. Mech. Appl. Math. 43, 15–41 (1990)
128. Menzel, A., Steinmann, P.: On the continuum formulation of higher gradient plasticity for single and polycrystals. J. Mech. Phys. Solids 48, 1777–1796 (2000)
129. Menzel, A., Steinmann, P.: A view on anisotropic finite hyper-elasticity. European J. Mech. A. 22, 71–87 (2003)
130. Menzel, A., Steinmann, P.: On configurational forces in multiplicative elastoplasticity. Int. J. Solids Struct. 44, 4442–4471 (2007)
131. Miehe, C.: Aspects of the formulation and finite element implementation of large strain isotropic elasticity. Int. J. Num. Meth. Engng. 54, 1981–2004 (1994)

132. Miehe, C.: On the representation of Prandtl-Reuss tensors within the framework of multiplicative elastoplasticity. Int. J. Plast. 10, 609–621 (1994)
133. Miehe, C., Lambrecht, M.: Analysis of microstructure development in shearbands by energy relaxation of incremental stress potentials: Large-strain theory for standard dissipative solids. Int. J. Num. Meth. Engng. 58, 1–41 (2003)
134. Miehe, C., Lambrecht, M.: A two-scale finite element relaxation analysis of shear bands in non-convex inelastic solids: small-strain theory for standard dissipative materials. Comp. Meth. Appl. Mech. Engng. 192(5), 473–508 (2003)
135. Miehe, C., Lambrecht, M., Gürses, E.: Analysis of material instabilities in inelastic solids by incremental energy minimization and relaxation methods: evolving deformation microstructures in finite plasticity. J. Mech. Phys. Solids 52, 2725–2769 (2004)
136. Miehe, C., Schotte, J., Lambrecht, M.: Homogenization of inelastic solid materials at finite strains based on incremental minimization principles. J. Mech. Phys. Solids 50(10), 2123–2167 (2002)
137. Miehe, C., Stein, E.: A canonical model of multiplicative elasto-plasticity formulation and aspects of the numerical implementation. Eur. J. Mech. A/Solids 11, 25–43 (1992)
138. Mielke, A.: Energetic formulation of multiplicative elasto-plasticity using dissipation distances. Cont. Mech. Thermodyn. 15, 351–382 (2003)
139. Mindlin, R.D.: Micro-structure in linear elasticity. Arch. Rat. Mech. Anal. 16, 51–78 (1964)
140. Mindlin, R.D.: Second gradient of strain and surface-tension in linear elasticity. Int. J. Solids Structures 1, 417–438 (1965)
141. Mindlin, R.D., Eshel, N.N.: On first strain-gradient theories in linear elasticity. Int. J. Solids Structures 4, 109–124 (1968)
142. Misner, C.W., Thorne, K.S., Wheeler, J.A.: Gravitation. Freeman and Company, New York (1998)
143. Moreau, J.J.: Fonctionelles sous-différentiables. CR Acad. Sci. Paris 257, 4117–4119 (1963)
144. Moreau, J.J.: La notion de sur-potentiel et les liaisons unilatérales en élastostatique. CR Acad. Sc. Paris 267, 954–957 (1968)
145. Moreau, J.J.: Sur les lois de frottement, de plasticité et de viscosité. CR Acad. Sc. Paris 271, 608–611 (1970)
146. Morrey, C.B.: Quasiconvexity and the semicontinuity of multiple integrands. Pacific J. Math. 2, 25–53 (1952)
147. Murdoch, A.I.: Symmetry considerations for materials of second grade. J. Elasticity 9, 43–50 (1979)
148. Murdoch, A.I.: Physical foundations of continuum mechanics. Cambridge University Press, Cambridge (2012)
149. Murnaghan, F.D.: Finite deformations of an elastic body. Am. J. Math. 59, 235–260 (1937)
150. Nguyen, Q.S.: On the elastic plastic initial-boundary value problem and its numerical integration. Int. J. Num. Meth. Engng. 11, 817–832 (1977)
151. Noll, W.: Materially uniform simple bodies with inhomogeneities. Arch. Rat. Mech. Anal. 27, 1–32 (1967)
152. Nye, J.F.: Some geometrical relations in dislocated crystals. Acta Metallurgica 1, 153–162 (1953)
153. Ogden, R.W.: Non-Linear Elastic Deformations. Dover, Mineola (1997)

154. O'Neill, B.: Elementary differential geometry. Academic Press, Burlington, etc. (2006)
155. Ortiz, M., Repetto, E.A.: Nonconvex energy minimization and dislocation structures in ductile single crystals. J. Mech. Phys. Solids 47, 397–462 (1999)
156. Ortiz, M., Stainier, L.: The variational formulation of viscoplastic constitutive updates. Comp. Meth. Appl. Mech. Engng. 171(3), 419–444 (1999)
157. Petryk, H.: Incremental energy minimization in dissipative solids. Comptes Rendus Mecanique 331(7), 469–474 (2003)
158. Polizotto, C., Borino, G.: A thermodynamics-based formulation of gradient-dependent plasticity. Eur. J. Mech. A/Solids 17, 741–761 (1998)
159. Polizotto, C., Borino, G., Fuschi, P.: A thermodynamically consistent formulation of nonlocal and gradient plasticity. Mech. Res. Comm. 25, 75–82 (1998)
160. Reddy, B.D., Ebobisse, F., McBride, A.: Well-posedness of a model of strain gradient plasticity for plastically irrotational materials. Int. J. Plast. 24, 55–73 (2008)
161. Ricci, M.M.G., Levi-Civita, T.: Méthodes de calcul différentiel absolu et leurs applications. Math. Annal. 54, 125–201 (1900)
162. Rockafellar, R.T.: Convex analysis. Princeton University Press, Princeton, N.J (1997)
163. Romanov, A.E., Vladimirov, V.I.: Disclinations in solids. Phys. Stat. Sol. 78, 11–34 (1983)
164. Schade, H., Neemann, K.: Tensoranalysis. Walter de Gruyter (2009)
165. Schmid, M., Biedermann, A., Stadler, H., Slama, C., Varga, P.: Mismatch dislocations caused by preferential sputtering of a platinum-nickel alloy surface. Appl. Phys. A 55, 468–475 (1992)
166. Schmid, M., Biedermann, A., Stadler, H., Varga, P.: Lattice mismatch dislocations in a preferentially sputtered alloy studied by scanning tunneling microscopy. Physical Rev. Lett. 69, 925–928 (1992)
167. Schouten, J.A.: Ricci-Calculus. Springer, Berlin (1954)
168. Schouten, J.A.: Tensor Analysis for Physicists. Dover, New York (1989)
169. Schröder, J., Neff, P.: Invariant formulation of hyperelastic transverse isotropy based on polyconvex free energy functions. Int. J. Solids Struct. 40
170. Schröder, J., Neff, P., Ebbing, V.: Anisotropic polyconvex energies on the basis of crystallographic motivated structural tensors. J. Mech. Phys. Solids 56
171. Schutz, B.F.: Geometrical methods of mathematical physics. Cambridge University Press, Cambridge (1984)
172. Seth, B.R.: Generalized strain measures with application to physical problems. In: Reiner, M., Abir, D. (eds.) Second-Order Effects in Elasticity, Plasticity and Fluid Dynamic, pp. 162–172 (1964)
173. Silhavy, M.: The mechanics and thermomechanics of continuous media. Springer, Berlin (1997)
174. Simmonds, J.G.: A brief on tensor analysis. Springer, New York (1994)
175. Simo, J.C.: Algorithms for static and dynamic multiplicative plasticity that preserve the classical return mapping schemes of the infinitesimal theory. Comp. Meth. Appl. Mech. Engng. 99, 61–112 (1992)
176. Simo, J.C., Hughes, T.J.R.: Computational Inelasticity. Springer, New York (1998)
177. Simo, J.C., Miehe, C.: Associative coupled thermoplasticity at finite strains: formulation, numerical analysis and implementation. Comp. Meth. Appl. Mech. Engng. 98, 41–104 (1992)

178. Spencer, A.J.M.: Continuum Mechanics. Dover, Minola, New York (2004)
179. Spivak, M.: A Comprehensive Introduction to Differential Geometry, vol. I, II. Perish, Inc., Houston (1979)
180. Steinmann, P.: A micropolar theory of finite deformation and finite rotation multiplicative elastoplasticity. Int. J. Solids Struct. 31, 1063–1084 (1994)
181. Steinmann, P.: Views on multiplicative elastoplasticity and the continuum theory of dislocations. Int. J. Engng. Sci. 34, 1717–1735 (1996)
182. Steinmann, P.: On spatial and material settings of hyperelastodynamics. Acta Mech 156, 193–218 (2002)
183. Steinmann, P.: On spatial and material settings of thermo-hyperelastodynamics. J. Elast. 66, 109–157 (2002)
184. Steinmann, P.: On boundary potential energies in deformational and configurational mechanics. J. Mech. Phys. Solids 56, 772–800 (2008)
185. Steinmann, P.: Geometrically nonlinear gravito-elasticity: Hyperelastostatics coupled to newtonian gravitation. Int. J. Engng. Sci. 49, 1452–1460 (2011)
186. Steinmann, P.: On the roots of continuum mechanics in differential geometry – a review –. In: Altenbach, H., Eremeyev, V.A. (eds.) Generalized Continua – from the Theory to Engineering Applications, CISM Courses and Lectures, vol. 541, pp. 1–64. Springer, Wien (2012)
187. Steinmann, P.: On generalized crystal-plasticity based on defect-densities in a second-order continuum. GAMM-Mitteilungen 36, 203–218 (2013)
188. Steinmann, P., Hossain, M., Possart, G.: Hyperelastic models for rubber-like materials: consistent tangent operators and suitability for Treloar's data. Arch. Appl. Mech. 82, 1183–1217 (2012)
189. Steinmann, P., Larsson, R., Runesson, K.: On the localization properties of multiplicative hyperelasto-plastic continua with strong discontinuities. Int. J. Solids Struct. 34, 969–990 (1997)
190. Sternberg, S.Z.: Lectures on differential geometry. Englewood Cliffs Prentice Hall (1964)
191. Sutcliffe, S.: Spectral decomposition of the elasticity tensor. J. Appl. Mech. 59, 762–773 (1992)
192. Svendsen, B.: Continuum thermodynamic models for crystal plasticity including the effects of geometrically-necessary dislocations. J. Mech. Phys. Solids 50, 1297–1329 (2002)
193. Svendsen, B., Neff, P., Menzel, A.: On constitutive and configurational aspects of models for gradient continua with microstructure. ZAMM-J. Appl. Math. Mech. 89, 687–697 (2009)
194. Tadmor, E.B., Miller, R.E., Elliott, R.S.: Continuum mechanics and thermodynamics. Cambridge University Press, Cambridge (2012)
195. Teodosiu, C., Sidoroff, F.: A finite theory of elastoviscoplasticity of single crystals. Int. J. Engng. Sci. 14, 713–723 (1976)
196. Toupin, R.A.: Elastic materials with couple-stresses. Arch. Rat. Mech. Anal. 11, 385–414 (1962)
197. Toupin, R.A.: Theories of elasticity with couple-stress. Arch. Rat. Mech. Anal. 17, 85–112 (1964)
198. Träuble, H., Essmann, U.: Fehler im Flußliniengitter von Supraleitern zweiter Art. Phys. Stat. Sol. 25, 373–393 (1968)
199. Trautman, A.: Einstein-Cartan theory. In: Francoise, J.P., Naber, G.L., Tsou, S.T. (eds.) Encyclopedia of Mathematical Physics, pp. 189–195. Elsevier, Oxford (2006)

200. Truesdell, C., Noll, W.: The non-linear field theories of mechanics. Springer, Berlin (2004)
201. Wang, C.C.: On the geometric structures of simple bodies, a mathematical foundation for the theory of continuous distributions of dislocations. Arch. Rat. Mech. Anal. 27, 33–94 (1967)
202. de Wit, R.: A view of the relation between the continuum theory of lattice defects and non-euclidean geometry in the linear approximation. Int. J. Engng. Sci. 19, 1475–1506 (1981)
203. Yavari, A., Goriely, A.: Riemann–Cartan geometry of nonlinear dislocation mechanics. Arch. Rat. Mech. Anal. 205, 59–118 (2012)
204. Yavari, A., Goriely, A.: Weyl geometry and the nonlinear mechanics of distributed point defects. Proc. Royal Soc. London A 468, 3902–3922 (2012)
205. Yavari, A., Goriely, A.: Riemann–Cartan geometry of nonlinear disclination mechanics. Math. Mech. Solids 18, 91–102 (2013)
206. Zubov, L.M.: Nonlinear theory of dislocations and disclinations in elastic bodies, vol. 47. Springer (1997)

Index

MIX
Papier aus verantwortungsvollen Quellen
Paper from responsible sources
FSC® C105338

If you have any concerns about our products, you can contact us on
ProductSafety@springernature.com

In case Publisher is established outside the EU, the EU authorized representative is:
Springer Nature Customer Service Center GmbH
Europaplatz 3, 69115 Heidelberg, Germany

Printed by Libri Plureos GmbH
in Hamburg, Germany